T0180753

Lecture Notes in Computer Science 14225

The series Lecture Notes in Computer Science (LNCS), including its subseries Lecture Notes in Artificial Intelligence (LNAI) and Lecture Notes in Bioinformatics (LNBI), has established itself as a medium for the publication of new developments in computer science and information technology research, teaching, and education.

LNCS enjoys close cooperation with the computer science R & D community, the series counts many renowned academics among its volume editors and paper authors, and collaborates with prestigious societies. Its mission is to serve this international community by providing an invaluable service, mainly focused on the publication of conference and workshop proceedings and postproceedings. LNCS commenced publication in 1973.

Hayit Greenspan · Anant Madabhushi ·
Parvin Mousavi · Septimiu Salcudean ·
James Duncan · Tanveer Syeda-Mahmood ·
Russell Taylor
Editors

Medical Image Computing and Computer Assisted Intervention – MICCAI 2023

26th International Conference
Vancouver, BC, Canada, October 8–12, 2023
Proceedings, Part VI

Editors
Hayit Greenspan
Icahn School of Medicine, Mount Sinai,
NYC, NY, USA

Tel Aviv University
Tel Aviv, Israel

Parvin Mousavi
Queen's University
Kingston, ON, Canada

James Duncan
Yale University
New Haven, CT, USA

Russell Taylor
Johns Hopkins University
Baltimore, MD, USA

Anant Madabhushi
Emory University
Atlanta, GA, USA

Septimiu Salcudean
The University of British Columbia
Vancouver, BC, Canada

Tanveer Syeda-Mahmood
IBM Research
San Jose, CA, USA

ISSN 0302-9743 ISSN 1611-3349 (electronic)
Lecture Notes in Computer Science
ISBN 978-3-031-43986-5 ISBN 978-3-031-43987-2 (eBook)
https://doi.org/10.1007/978-3-031-43987-2

This Springer imprint is published by the registered company Springer Nature Switzerland AG
The registered company address is: Gewerbestrasse 11, 6330 Cham, Switzerland

Preface

We are pleased to present the proceedings for the 26th International Conference on Medical Image Computing and Computer-Assisted Intervention (MICCAI). After several difficult years of virtual conferences, this edition was held in a mainly in-person format with a hybrid component at the Vancouver Convention Centre, in Vancouver, BC, Canada October 8–12, 2023. The conference featured 33 physical workshops, 15 online workshops, 15 tutorials, and 29 challenges held on October 8 and October 12. Co-located with the conference was also the 3rd Conference on Clinical Translation on Medical Image Computing and Computer-Assisted Intervention (CLINICCAI) on October 10.

MICCAI 2023 received the largest number of submissions so far, with an approximately 30% increase compared to 2022. We received 2365 full submissions of which 2250 were subjected to full review. To keep the acceptance ratios around 32% as in previous years, there was a corresponding increase in accepted papers leading to 730 papers accepted, with 68 orals and the remaining presented in poster form. These papers comprise ten volumes of Lecture Notes in Computer Science (LNCS) proceedings as follows:

- Part I, LNCS Volume 14220: Machine Learning with Limited Supervision and Machine Learning – Transfer Learning
- Part II, LNCS Volume 14221: Machine Learning – Learning Strategies and Machine Learning – Explainability, Bias, and Uncertainty I
- Part III, LNCS Volume 14222: Machine Learning – Explainability, Bias, and Uncertainty II and Image Segmentation I
- Part IV, LNCS Volume 14223: Image Segmentation II
- Part V, LNCS Volume 14224: Computer-Aided Diagnosis I
- Part VI, LNCS Volume 14225: Computer-Aided Diagnosis II and Computational Pathology
- Part VII, LNCS Volume 14226: Clinical Applications – Abdomen, Clinical Applications – Breast, Clinical Applications – Cardiac, Clinical Applications – Dermatology, Clinical Applications – Fetal Imaging, Clinical Applications – Lung, Clinical Applications – Musculoskeletal, Clinical Applications – Oncology, Clinical Applications – Ophthalmology, and Clinical Applications – Vascular
- Part VIII, LNCS Volume 14227: Clinical Applications – Neuroimaging and Microscopy
- Part IX, LNCS Volume 14228: Image-Guided Intervention, Surgical Planning, and Data Science
- Part X, LNCS Volume 14229: Image Reconstruction and Image Registration

The papers for the proceedings were selected after a rigorous double-blind peer-review process. The MICCAI 2023 Program Committee consisted of 133 area chairs and over 1600 reviewers, with representation from several countries across all major continents. It also maintained a gender balance with 31% of scientists who self-identified as women. With an increase in the number of area chairs and reviewers, the reviewer load on the experts was reduced this year, keeping to 16–18 papers per area chair and about 4–6 papers per reviewer. Based on the double-blinded reviews, area chairs' recommendations, and program chairs' global adjustments, 308 papers (14%) were provisionally accepted, 1196 papers (53%) were provisionally rejected, and 746 papers (33%) proceeded to the rebuttal stage. As in previous years, Microsoft's Conference Management Toolkit (CMT) was used for paper management and organizing the overall review process. Similarly, the Toronto paper matching system (TPMS) was employed to ensure knowledgeable experts were assigned to review appropriate papers. Area chairs and reviewers were selected following public calls to the community, and were vetted by the program chairs.

Among the new features this year was the emphasis on clinical translation, moving Medical Image Computing (MIC) and Computer-Assisted Interventions (CAI) research from theory to practice by featuring two clinical translational sessions reflecting the real-world impact of the field in the clinical workflows and clinical evaluations. For the first time, clinicians were appointed as Clinical Chairs to select papers for the clinical translational sessions. The philosophy behind the dedicated clinical translational sessions was to maintain the high scientific and technical standard of MICCAI papers in terms of methodology development, while at the same time showcasing the strong focus on clinical applications. This was an opportunity to expose the MICCAI community to the clinical challenges and for ideation of novel solutions to address these unmet needs. Consequently, during paper submission, in addition to MIC and CAI a new category of "Clinical Applications" was introduced for authors to self-declare.

MICCAI 2023 for the first time in its history also featured dual parallel tracks that allowed the conference to keep the same proportion of oral presentations as in previous years, despite the 30% increase in submitted and accepted papers.

We also introduced two new sessions this year focusing on young and emerging scientists through their Ph.D. thesis presentations, and another with experienced researchers commenting on the state of the field through a fireside chat format.

The organization of the final program by grouping the papers into topics and sessions was aided by the latest advancements in generative AI models. Specifically, Open AI's GPT-4 large language model was used to group the papers into initial topics which were then manually curated and organized. This resulted in fresh titles for sessions that are more reflective of the technical advancements of our field.

Although not reflected in the proceedings, the conference also benefited from keynote talks from experts in their respective fields including Turing Award winner Yann LeCun and leading experts Jocelyne Troccaz and Mihaela van der Schaar.

We extend our sincere gratitude to everyone who contributed to the success of MICCAI 2023 and the quality of its proceedings. In particular, we would like to express our profound thanks to the MICCAI Submission System Manager Kitty Wong whose meticulous support throughout the paper submission, review, program planning, and proceeding preparation process was invaluable. We are especially appreciative of the effort and dedication of our Satellite Events Chair, Bennett Landman, who tirelessly coordinated the organization of over 90 satellite events consisting of workshops, challenges and tutorials. Our workshop chairs Hongzhi Wang, Alistair Young, tutorial chairs Islem

Rekik, Guoyan Zheng, and challenge chairs, Lena Maier-Hein, Jayashree Kalpathy-Kramer, Alexander Seitel, worked hard to assemble a strong program for the satellite events. Special mention this year also goes to our first-time Clinical Chairs, Drs. Curtis Langlotz, Charles Kahn, and Masaru Ishii who helped us select papers for the clinical sessions and organized the clinical sessions.

We acknowledge the contributions of our Keynote Chairs, William Wells and Alejandro Frangi, who secured our keynote speakers. Our publication chairs, Kevin Zhou and Ron Summers, helped in our efforts to get the MICCAI papers indexed in PubMed. It was a challenging year for fundraising for the conference due to the recovery of the economy after the COVID pandemic. Despite this situation, our industrial sponsorship chairs, Mohammad Yaqub, Le Lu and Yanwu Xu, along with Dekon's Mehmet Eldegez, worked tirelessly to secure sponsors in innovative ways, for which we are grateful.

An active body of the MICCAI Student Board led by Camila Gonzalez and our 2023 student representatives Nathaniel Braman and Vaishnavi Subramanian helped put together student-run networking and social events including a novel Ph.D. thesis 3-minute madness event to spotlight new graduates for their careers. Similarly, Women in MICCAI chairs Xiaoxiao Li and Jayanthi Sivaswamy and RISE chairs, Islem Rekik, Pingkun Yan, and Andrea Lara further strengthened the quality of our technical program through their organized events. Local arrangements logistics including the recruiting of University of British Columbia students and invitation letters to attendees, was ably looked after by our local arrangement chairs Purang Abolmaesumi and Mehdi Moradi. They also helped coordinate the visits to the local sites in Vancouver both during the selection of the site and organization of our local activities during the conference. Our Young Investigator chairs Marius Linguraru, Archana Venkataraman, Antonio Porras Perez put forward the startup village and helped secure funding from NIH for early career scientist participation in the conference. Our communications chair, Ehsan Adeli, and Diana Cunningham were active in making the conference visible on social media platforms and circulating the newsletters. Niharika D'Souza was our cross-committee liaison providing note-taking support for all our meetings. We are grateful to all these organization committee members for their active contributions that made the conference successful.

We would like to thank the MICCAI society chair, Caroline Essert, and the MICCAI board for their approvals, support and feedback, which provided clarity on various aspects of running the conference. Behind the scenes, we acknowledge the contributions of the MICCAI secretariat personnel, Janette Wallace, and Johanne Langford, who kept a close eye on logistics and budgets, and Diana Cunningham and Anna Van Vliet for including our conference announcements in a timely manner in the MICCAI society newsletters. This year, when the existing virtual platform provider indicated that they would discontinue their service, a new virtual platform provider Conference Catalysts was chosen after due diligence by John Baxter. John also handled the setup and coordination with CMT and consultation with program chairs on features, for which we are very grateful. The physical organization of the conference at the site, budget financials, fund-raising, and the smooth running of events would not have been possible without our Professional Conference Organization team from Dekon Congress & Tourism led by Mehmet Eldegez. The model of having a PCO run the conference, which we used at

MICCAI, significantly reduces the work of general chairs for which we are particularly grateful.

Finally, we are especially grateful to all members of the Program Committee for their diligent work in the reviewer assignments and final paper selection, as well as the reviewers for their support during the entire process. Lastly, and most importantly, we thank all authors, co-authors, students/postdocs, and supervisors for submitting and presenting their high-quality work, which played a pivotal role in making MICCAI 2023 a resounding success.

With a successful MICCAI 2023, we now look forward to seeing you next year in Marrakesh, Morocco when MICCAI 2024 goes to the African continent for the first time.

October 2023

Tanveer Syeda-Mahmood
James Duncan
Russ Taylor
General Chairs

Hayit Greenspan
Anant Madabhushi
Parvin Mousavi
Septimiu Salcudean
Program Chairs

Organization

General Chairs

Tanveer Syeda-Mahmood	IBM Research, USA
James Duncan	Yale University, USA
Russ Taylor	Johns Hopkins University, USA

Program Committee Chairs

Hayit Greenspan	Tel-Aviv University, Israel and Icahn School of Medicine at Mount Sinai, USA
Anant Madabhushi	Emory University, USA
Parvin Mousavi	Queen's University, Canada
Septimiu Salcudean	University of British Columbia, Canada

Satellite Events Chair

Bennett Landman	Vanderbilt University, USA

Workshop Chairs

Hongzhi Wang	IBM Research, USA
Alistair Young	King's College, London, UK

Challenges Chairs

Jayashree Kalpathy-Kramer	Harvard University, USA
Alexander Seitel	German Cancer Research Center, Germany
Lena Maier-Hein	German Cancer Research Center, Germany

Tutorial Chairs

Islem Rekik	Imperial College London, UK
Guoyan Zheng	Shanghai Jiao Tong University, China

Clinical Chairs

Curtis Langlotz	Stanford University, USA
Charles Kahn	University of Pennsylvania, USA
Masaru Ishii	Johns Hopkins University, USA

Local Arrangements Chairs

Purang Abolmaesumi	University of British Columbia, Canada
Mehdi Moradi	McMaster University, Canada

Keynote Chairs

William Wells	Harvard University, USA
Alejandro Frangi	University of Manchester, UK

Industrial Sponsorship Chairs

Mohammad Yaqub	MBZ University of Artificial Intelligence, Abu Dhabi
Le Lu	DAMO Academy, Alibaba Group, USA
Yanwu Xu	Baidu, China

Communication Chair

Ehsan Adeli	Stanford University, USA

Publication Chairs

Ron Summers	National Institutes of Health, USA
Kevin Zhou	University of Science and Technology of China, China

Young Investigator Chairs

Marius Linguraru	Children's National Institute, USA
Archana Venkataraman	Boston University, USA
Antonio Porras	University of Colorado Anschutz Medical Campus, USA

Student Activities Chairs

Nathaniel Braman	Picture Health, USA
Vaishnavi Subramanian	EPFL, France

Women in MICCAI Chairs

Jayanthi Sivaswamy	IIIT, Hyderabad, India
Xiaoxiao Li	University of British Columbia, Canada

RISE Committee Chairs

Islem Rekik	Imperial College London, UK
Pingkun Yan	Rensselaer Polytechnic Institute, USA
Andrea Lara	Universidad Galileo, Guatemala

Submission Platform Manager

Kitty Wong	The MICCAI Society, Canada

Virtual Platform Manager

John Baxter	INSERM, Université de Rennes 1, France

Cross-Committee Liaison

Niharika D'Souza	IBM Research, USA

Program Committee

Sahar Ahmad	University of North Carolina at Chapel Hill, USA
Shadi Albarqouni	University of Bonn and Helmholtz Munich, Germany
Angelica Aviles-Rivero	University of Cambridge, UK
Shekoofeh Azizi	Google, Google Brain, USA
Ulas Bagci	Northwestern University, USA
Wenjia Bai	Imperial College London, UK
Sophia Bano	University College London, UK
Kayhan Batmanghelich	University of Pittsburgh and Boston University, USA
Ismail Ben Ayed	ETS Montreal, Canada
Katharina Breininger	Friedrich-Alexander-Universität Erlangen-Nürnberg, Germany
Weidong Cai	University of Sydney, Australia
Geng Chen	Northwestern Polytechnical University, China
Hao Chen	Hong Kong University of Science and Technology, China
Jun Cheng	Institute for Infocomm Research, A*STAR, Singapore
Li Cheng	University of Alberta, Canada
Albert C. S. Chung	University of Exeter, UK
Toby Collins	Ircad, France
Adrian Dalca	Massachusetts Institute of Technology and Harvard Medical School, USA
Jose Dolz	ETS Montreal, Canada
Qi Dou	Chinese University of Hong Kong, China
Nicha Dvornek	Yale University, USA
Shireen Elhabian	University of Utah, USA
Sandy Engelhardt	Heidelberg University Hospital, Germany
Ruogu Fang	University of Florida, USA

Aasa Feragen	Technical University of Denmark, Denmark
Moti Freiman	Technion - Israel Institute of Technology, Israel
Huazhu Fu	IHPC, A*STAR, Singapore
Adrian Galdran	Universitat Pompeu Fabra, Barcelona, Spain
Zhifan Gao	Sun Yat-sen University, China
Zongyuan Ge	Monash University, Australia
Stamatia Giannarou	Imperial College London, UK
Yun Gu	Shanghai Jiao Tong University, China
Hu Han	Institute of Computing Technology, Chinese Academy of Sciences, China
Daniel Hashimoto	University of Pennsylvania, USA
Mattias Heinrich	University of Lübeck, Germany
Heng Huang	University of Pittsburgh, USA
Yuankai Huo	Vanderbilt University, USA
Mobarakol Islam	University College London, UK
Jayender Jagadeesan	Harvard Medical School, USA
Won-Ki Jeong	Korea University, South Korea
Xi Jiang	University of Electronic Science and Technology of China, China
Yueming Jin	National University of Singapore, Singapore
Anand Joshi	University of Southern California, USA
Shantanu Joshi	UCLA, USA
Leo Joskowicz	Hebrew University of Jerusalem, Israel
Samuel Kadoury	Polytechnique Montreal, Canada
Bernhard Kainz	Friedrich-Alexander-Universität Erlangen-Nürnberg, Germany and Imperial College London, UK
Davood Karimi	Harvard University, USA
Anees Kazi	Massachusetts General Hospital, USA
Marta Kersten-Oertel	Concordia University, Canada
Fahmi Khalifa	Mansoura University, Egypt
Minjeong Kim	University of North Carolina, Greensboro, USA
Seong Tae Kim	Kyung Hee University, South Korea
Pavitra Krishnaswamy	Institute for Infocomm Research, Agency for Science Technology and Research (A*STAR), Singapore
Jin Tae Kwak	Korea University, South Korea
Baiying Lei	Shenzhen University, China
Xiang Li	Massachusetts General Hospital, USA
Xiaoxiao Li	University of British Columbia, Canada
Yuexiang Li	Tencent Jarvis Lab, China
Chunfeng Lian	Xi'an Jiaotong University, China

Reviewers

Alaa Eldin Abdelaal
John Abel
Kumar Abhishek
Shahira Abousamra
Mazdak Abulnaga
Burak Acar
Abdoljalil Addeh
Ehsan Adeli
Sukesh Adiga Vasudeva
Seyed-Ahmad Ahmadi
Euijoon Ahn
Faranak Akbarifar
Alireza Akhondi-asl
Saad Ullah Akram
Daniel Alexander
Hanan Alghamdi
Hassan Alhajj
Omar Al-Kadi
Max Allan
Andre Altmann
Pablo Alvarez
Charlems Alvarez-Jimenez
Jennifer Alvén
Lidia Al-Zogbi
Kimberly Amador
Tamaz Amiranashvili
Amine Amyar
Wangpeng An
Vincent Andrearczyk
Manon Ansart
Sameer Antani
Jacob Antunes
Michel Antunes
Guilherme Aresta
Mohammad Ali Armin
Kasra Arnavaz
Corey Arnold
Janan Arslan
Marius Arvinte
Muhammad Asad
John Ashburner
Md Ashikuzzaman
Shahab Aslani
Mehdi Astaraki
Angélica Atehortúa
Benjamin Aubert
Marc Aubreville
Paolo Avesani
Sana Ayromlou
Reza Azad
Mohammad Farid Azampour
Qinle Ba
Meritxell Bach Cuadra
Hyeon-Min Bae
Matheus Baffa
Cagla Bahadir
Fan Bai
Jun Bai
Long Bai
Pradeep Bajracharya
Shafa Balaram
Yaël Balbastre
Yutong Ban
Abhirup Banerjee
Soumyanil Banerjee
Sreya Banerjee
Shunxing Bao
Omri Bar
Adrian Barbu
Joao Barreto
Adrian Basarab
Berke Basaran
Michael Baumgartner
Siming Bayer
Roza Bayrak
Aicha BenTaieb
Guy Ben-Yosef
Sutanu Bera
Cosmin Bercea
Jorge Bernal
Jose Bernal
Gabriel Bernardino
Riddhish Bhalodia
Jignesh Bhatt
Indrani Bhattacharya
Binod Bhattarai
Lei Bi
Qi Bi
Cheng Bian
Gui-Bin Bian
Carlo Biffi
Alexander Bigalke
Benjamin Billot
Manuel Birlo
Ryoma Bise
Daniel Blezek
Stefano Blumberg
Sebastian Bodenstedt
Federico Bolelli
Bhushan Borotikar
Ilaria Boscolo Galazzo
Alexandre Bousse
Nicolas Boutry
Joseph Boyd
Behzad Bozorgtabar
Nadia Brancati
Clara Brémond Martin
Stéphanie Bricq
Christopher Bridge
Coleman Broaddus
Rupert Brooks
Tom Brosch
Mikael Brudfors
Ninon Burgos
Nikolay Burlutskiy
Michal Byra
Ryan Cabeen
Mariano Cabezas
Hongmin Cai
Tongan Cai
Zongyou Cai
Liane Canas
Bing Cao
Guogang Cao
Weiguo Cao
Xu Cao
Yankun Cao
Zhenjie Cao

Jaime Cardoso
M. Jorge Cardoso
Owen Carmichael
Jacob Carse
Adrià Casamitjana
Alessandro Casella
Angela Castillo
Kate Cevora
Krishna Chaitanya
Satrajit Chakrabarty
Yi Hao Chan
Shekhar Chandra
Ming-Ching Chang
Peng Chang
Qi Chang
Yuchou Chang
Hanqing Chao
Simon Chatelin
Soumick Chatterjee
Sudhanya Chatterjee
Muhammad Faizyab Ali Chaudhary
Antong Chen
Bingzhi Chen
Chen Chen
Cheng Chen
Chengkuan Chen
Eric Chen
Fang Chen
Haomin Chen
Jianan Chen
Jianxu Chen
Jiazhou Chen
Jie Chen
Jintai Chen
Jun Chen
Junxiang Chen
Junyu Chen
Li Chen
Liyun Chen
Nenglun Chen
Pingjun Chen
Pingyi Chen
Qi Chen
Qiang Chen
Runnan Chen
Shengcong Chen
Sihao Chen
Tingting Chen
Wenting Chen
Xi Chen
Xiang Chen
Xiaoran Chen
Xin Chen
Xiongchao Chen
Yanxi Chen
Yixiong Chen
Yixuan Chen
Yuanyuan Chen
Yuqian Chen
Zhaolin Chen
Zhen Chen
Zhenghao Chen
Zhennong Chen
Zhihao Chen
Zhineng Chen
Zhixiang Chen
Chang-Chieh Cheng
Jiale Cheng
Jianhong Cheng
Jun Cheng
Xuelian Cheng
Yupeng Cheng
Mark Chiew
Philip Chikontwe
Eleni Chiou
Jungchan Cho
Jang-Hwan Choi
Min-Kook Choi
Wookjin Choi
Jaegul Choo
Yu-Cheng Chou
Daan Christiaens
Argyrios Christodoulidis
Stergios Christodoulidis
Kai-Cheng Chuang
Hyungjin Chung
Matthew Clarkson
Michaël Clément
Dana Cobzas
Jaume Coll-Font
Olivier Colliot
Runmin Cong
Yulai Cong
Laura Connolly
William Consagra
Pierre-Henri Conze
Tim Cootes
Teresa Correia
Baris Coskunuzer
Alex Crimi
Can Cui
Hejie Cui
Hui Cui
Lei Cui
Wenhui Cui
Tolga Cukur
Tobias Czempiel
Javid Dadashkarimi
Haixing Dai
Tingting Dan
Kang Dang
Salman Ul Hassan Dar
Eleonora D'Arnese
Dhritiman Das
Neda Davoudi
Tareen Dawood
Sandro De Zanet
Farah Deeba
Charles Delahunt
Herve Delingette
Ugur Demir
Liang-Jian Deng
Ruining Deng
Wenlong Deng
Felix Denzinger
Adrien Depeursinge
Mohammad Mahdi Derakhshani
Hrishikesh Deshpande
Adrien Desjardins
Christian Desrosiers
Blake Dewey
Neel Dey
Rohan Dhamdhere

Maxime Di Folco
Songhui Diao
Alina Dima
Hao Ding
Li Ding
Ying Ding
Zhipeng Ding
Nicola Dinsdale
Konstantin Dmitriev
Ines Domingues
Bo Dong
Liang Dong
Nanqing Dong
Siyuan Dong
Reuben Dorent
Gianfranco Doretto
Sven Dorkenwald
Haoran Dou
Mitchell Doughty
Jason Dowling
Niharika D'Souza
Guodong Du
Jie Du
Shiyi Du
Hongyi Duanmu
Benoit Dufumier
James Duncan
Joshua Durso-Finley
Dmitry V. Dylov
Oleh Dzyubachyk
Mahdi (Elias) Ebnali
Philip Edwards
Jan Egger
Gudmundur Einarsson
Mostafa El Habib Daho
Ahmed Elazab
Idris El-Feghi
David Ellis
Mohammed Elmogy
Amr Elsawy
Okyaz Eminaga
Ertunc Erdil
Lauren Erdman
Marius Erdt
Maria Escobar
Hooman Esfandiari
Nazila Esmaeili
Ivan Ezhov
Alessio Fagioli
Deng-Ping Fan
Lei Fan
Xin Fan
Yubo Fan
Huihui Fang
Jiansheng Fang
Xi Fang
Zhenghan Fang
Mohammad Farazi
Azade Farshad
Mohsen Farzi
Hamid Fehri
Lina Felsner
Chaolu Feng
Chun-Mei Feng
Jianjiang Feng
Mengling Feng
Ruibin Feng
Zishun Feng
Alvaro Fernandez-Quilez
Ricardo Ferrari
Lucas Fidon
Lukas Fischer
Madalina Fiterau
Antonio Foncubierta-Rodríguez
Fahimeh Fooladgar
Germain Forestier
Nils Daniel Forkert
Jean-Rassaire Fouefack
Kevin François-Bouaou
Wolfgang Freysinger
Bianca Freytag
Guanghui Fu
Kexue Fu
Lan Fu
Yunguan Fu
Pedro Furtado
Ryo Furukawa
Jin Kyu Gahm
Mélanie Gaillochet
Francesca Galassi
Jiangzhang Gan
Yu Gan
Yulu Gan
Alireza Ganjdanesh
Chang Gao
Cong Gao
Linlin Gao
Zeyu Gao
Zhongpai Gao
Sara Garbarino
Alain Garcia
Beatriz Garcia Santa Cruz
Rongjun Ge
Shiv Gehlot
Manuela Geiss
Salah Ghamizi
Negin Ghamsarian
Ramtin Gharleghi
Ghazal Ghazaei
Florin Ghesu
Sayan Ghosal
Syed Zulqarnain Gilani
Mahdi Gilany
Yannik Glaser
Ben Glocker
Bharti Goel
Jacob Goldberger
Polina Golland
Alberto Gomez
Catalina Gomez
Estibaliz Gómez-de-Mariscal
Haifan Gong
Kuang Gong
Xun Gong
Ricardo Gonzales
Camila Gonzalez
German Gonzalez
Vanessa Gonzalez Duque
Sharath Gopal
Karthik Gopinath
Pietro Gori
Michael Götz
Shuiping Gou

Maged Goubran
Sobhan Goudarzi
Mark Graham
Alejandro Granados
Mara Graziani
Thomas Grenier
Radu Grosu
Michal Grzeszczyk
Feng Gu
Pengfei Gu
Qiangqiang Gu
Ran Gu
Shi Gu
Wenhao Gu
Xianfeng Gu
Yiwen Gu
Zaiwang Gu
Hao Guan
Jayavardhana Gubbi
Houssem-Eddine Gueziri
Dazhou Guo
Hengtao Guo
Jixiang Guo
Jun Guo
Pengfei Guo
Wenzhangzhi Guo
Xiaoqing Guo
Xueqi Guo
Yi Guo
Vikash Gupta
Praveen Gurunath Bharathi
Prashnna Gyawali
Sung Min Ha
Mohamad Habes
Ilker Hacihaliloglu
Stathis Hadjidemetriou
Fatemeh Haghighi
Justin Haldar
Noura Hamze
Liang Han
Luyi Han
Seungjae Han
Tianyu Han
Zhongyi Han
Jonny Hancox
Lasse Hansen
Degan Hao
Huaying Hao
Jinkui Hao
Nazim Haouchine
Michael Hardisty
Stefan Harrer
Jeffry Hartanto
Charles Hatt
Huiguang He
Kelei He
Qi He
Shenghua He
Xinwei He
Stefan Heldmann
Nicholas Heller
Edward Henderson
Alessa Hering
Monica Hernandez
Kilian Hett
Amogh Hiremath
David Ho
Malte Hoffmann
Matthew Holden
Qingqi Hong
Yoonmi Hong
Mohammad Reza Hosseinzadeh Taher
William Hsu
Chuanfei Hu
Dan Hu
Kai Hu
Rongyao Hu
Shishuai Hu
Xiaoling Hu
Xinrong Hu
Yan Hu
Yang Hu
Chaoqin Huang
Junzhou Huang
Ling Huang
Luojie Huang
Qinwen Huang
Sharon Xiaolei Huang
Weijian Huang
Xiaoyang Huang
Yi-Jie Huang
Yongsong Huang
Yongxiang Huang
Yuhao Huang
Zhe Huang
Zhi-An Huang
Ziyi Huang
Arnaud Huaulmé
Henkjan Huisman
Alex Hung
Jiayu Huo
Andreas Husch
Mohammad Arafat Hussain
Sarfaraz Hussein
Jana Hutter
Khoi Huynh
Ilknur Icke
Kay Igwe
Abdullah Al Zubaer Imran
Muhammad Imran
Samra Irshad
Nahid Ul Islam
Koichi Ito
Hayato Itoh
Yuji Iwahori
Krithika Iyer
Mohammad Jafari
Srikrishna Jaganathan
Hassan Jahanandish
Andras Jakab
Amir Jamaludin
Amoon Jamzad
Ananya Jana
Se-In Jang
Pierre Jannin
Vincent Jaouen
Uditha Jarayathne
Ronnachai Jaroensri
Guillaume Jaume
Syed Ashar Javed
Rachid Jennane
Debesh Jha
Ge-Peng Ji

Luping Ji
Zexuan Ji
Zhanghexuan Ji
Haozhe Jia
Hongchao Jiang
Jue Jiang
Meirui Jiang
Tingting Jiang
Xiajun Jiang
Zekun Jiang
Zhifan Jiang
Ziyu Jiang
Jianbo Jiao
Zhicheng Jiao
Chen Jin
Dakai Jin
Qiangguo Jin
Qiuye Jin
Weina Jin
Baoyu Jing
Bin Jing
Yaqub Jonmohamadi
Lie Ju
Yohan Jun
Dinkar Juyal
Manjunath K N
Ali Kafaei Zad Tehrani
John Kalafut
Niveditha Kalavakonda
Megha Kalia
Anil Kamat
Qingbo Kang
Po-Yu Kao
Anuradha Kar
Neerav Karani
Turkay Kart
Satyananda Kashyap
Alexander Katzmann
Lisa Kausch
Maxime Kayser
Salome Kazeminia
Wenchi Ke
Youngwook Kee
Matthias Keicher
Erwan Kerrien
Afifa Khaled
Nadieh Khalili
Farzad Khalvati
Bidur Khanal
Bishesh Khanal
Pulkit Khandelwal
Maksim Kholiavchenko
Ron Kikinis
Benjamin Killeen
Daeseung Kim
Heejong Kim
Jaeil Kim
Jinhee Kim
Jinman Kim
Junsik Kim
Minkyung Kim
Namkug Kim
Sangwook Kim
Tae Soo Kim
Younghoon Kim
Young-Min Kim
Andrew King
Miranda Kirby
Gabriel Kiss
Andreas Kist
Yoshiro Kitamura
Stefan Klein
Tobias Klinder
Kazuma Kobayashi
Lisa Koch
Satoshi Kondo
Fanwei Kong
Tomasz Konopczynski
Ender Konukoglu
Aishik Konwer
Thijs Kooi
Ivica Kopriva
Avinash Kori
Kivanc Kose
Suraj Kothawade
Anna Kreshuk
AnithaPriya Krishnan
Florian Kromp
Frithjof Kruggel
Thomas Kuestner
Levin Kuhlmann
Abhay Kumar
Kuldeep Kumar
Sayantan Kumar
Manuela Kunz
Holger Kunze
Tahsin Kurc
Anvar Kurmukov
Yoshihiro Kuroda
Yusuke Kurose
Hyuksool Kwon
Aymen Laadhari
Jorma Laaksonen
Dmitrii Lachinov
Alain Lalande
Rodney LaLonde
Bennett Landman
Daniel Lang
Carole Lartizien
Shlomi Laufer
Max-Heinrich Laves
William Le
Loic Le Folgoc
Christian Ledig
Eung-Joo Lee
Ho Hin Lee
Hyekyoung Lee
John Lee
Kisuk Lee
Kyungsu Lee
Soochahn Lee
Woonghee Lee
Étienne Léger
Wen Hui Lei
Yiming Lei
George Leifman
Rogers Jeffrey Leo John
Juan Leon
Bo Li
Caizi Li
Chao Li
Chen Li
Cheng Li
Chenxin Li
Chnegyin Li

Dawei Li
Fuhai Li
Gang Li
Guang Li
Hao Li
Haofeng Li
Haojia Li
Heng Li
Hongming Li
Hongwei Li
Huiqi Li
Jian Li
Jieyu Li
Kang Li
Lin Li
Mengzhang Li
Ming Li
Qing Li
Quanzheng Li
Shaohua Li
Shulong Li
Tengfei Li
Weijian Li
Wen Li
Xiaomeng Li
Xingyu Li
Xinhui Li
Xuelu Li
Xueshen Li
Yamin Li
Yang Li
Yi Li
Yuemeng Li
Yunxiang Li
Zeju Li
Zhaoshuo Li
Zhe Li
Zhen Li
Zhenqiang Li
Zhiyuan Li
Zhjin Li
Zi Li
Hao Liang
Libin Liang
Peixian Liang
Yuan Liang
Yudong Liang
Haofu Liao
Hongen Liao
Wei Liao
Zehui Liao
Gilbert Lim
Hongxiang Lin
Li Lin
Manxi Lin
Mingquan Lin
Tiancheng Lin
Yi Lin
Zudi Lin
Claudia Lindner
Simone Lionetti
Chi Liu
Chuanbin Liu
Daochang Liu
Dongnan Liu
Feihong Liu
Fenglin Liu
Han Liu
Huiye Liu
Jiang Liu
Jie Liu
Jinduo Liu
Jing Liu
Jingya Liu
Jundong Liu
Lihao Liu
Mengting Liu
Mingyuan Liu
Peirong Liu
Peng Liu
Qin Liu
Quan Liu
Rui Liu
Shengfeng Liu
Shuangjun Liu
Sidong Liu
Siyuan Liu
Weide Liu
Xiao Liu
Xiaoyu Liu
Xingtong Liu
Xinwen Liu
Xinyang Liu
Xinyu Liu
Yan Liu
Yi Liu
Yihao Liu
Yikang Liu
Yilin Liu
Yilong Liu
Yiqiao Liu
Yong Liu
Yuhang Liu
Zelong Liu
Zhe Liu
Zhiyuan Liu
Zuozhu Liu
Lisette Lockhart
Andrea Loddo
Nicolas Loménie
Yonghao Long
Daniel Lopes
Ange Lou
Brian Lovell
Nicolas Loy Rodas
Charles Lu
Chun-Shien Lu
Donghuan Lu
Guangming Lu
Huanxiang Lu
Jingpei Lu
Yao Lu
Oeslle Lucena
Jie Luo
Luyang Luo
Ma Luo
Mingyuan Luo
Wenhan Luo
Xiangde Luo
Xinzhe Luo
Jinxin Lv
Tianxu Lv
Fei Lyu
Ilwoo Lyu
Mengye Lyu

Jin Pan
Yongsheng Pan
Egor Panfilov
Jiaxuan Pang
Joao Papa
Constantin Pape
Bartlomiej Papiez
Nripesh Parajuli
Hyunjin Park
Akash Parvatikar
Tiziano Passerini
Diego Patiño Cortés
Mayank Patwari
Angshuman Paul
Rasmus Paulsen
Yuchen Pei
Yuru Pei
Tao Peng
Wei Peng
Yige Peng
Yunsong Peng
Matteo Pennisi
Antonio Pepe
Oscar Perdomo
Sérgio Pereira
Jose-Antonio Pérez-Carrasco
Mehran Pesteie
Terry Peters
Eike Petersen
Jens Petersen
Micha Pfeiffer
Dzung Pham
Hieu Pham
Ashish Phophalia
Tomasz Pieciak
Antonio Pinheiro
Pramod Pisharady
Theodoros Pissas
Szymon Płotka
Kilian Pohl
Sebastian Pölsterl
Alison Pouch
Tim Prangemeier
Prateek Prasanna
Raphael Prevost
Juan Prieto
Federica Proietto Salanitri
Sergi Pujades
Elodie Puybareau
Talha Qaiser
Buyue Qian
Mengyun Qiao
Yuchuan Qiao
Zhi Qiao
Chenchen Qin
Fangbo Qin
Wenjian Qin
Yulei Qin
Jie Qiu
Jielin Qiu
Peijie Qiu
Shi Qiu
Wu Qiu
Liangqiong Qu
Linhao Qu
Quan Quan
Tran Minh Quan
Sandro Queirós
Prashanth R
Febrian Rachmadi
Daniel Racoceanu
Mehdi Rahim
Jagath Rajapakse
Kashif Rajpoot
Keerthi Ram
Dhanesh Ramachandram
João Ramalhinho
Xuming Ran
Aneesh Rangnekar
Hatem Rashwan
Keerthi Sravan Ravi
Daniele Ravì
Sadhana Ravikumar
Harish Raviprakash
Surreerat Reaungamornrat
Samuel Remedios
Mengwei Ren
Sucheng Ren
Elton Rexhepaj
Mauricio Reyes
Constantino Reyes-Aldasoro
Abel Reyes-Angulo
Hadrien Reynaud
Razieh Rezaei
Anne-Marie Rickmann
Laurent Risser
Dominik Rivoir
Emma Robinson
Robert Robinson
Jessica Rodgers
Ranga Rodrigo
Rafael Rodrigues
Robert Rohling
Margherita Rosnati
Łukasz Roszkowiak
Holger Roth
José Rouco
Dan Ruan
Jiacheng Ruan
Daniel Rueckert
Danny Ruijters
Kanghyun Ryu
Ario Sadafi
Numan Saeed
Monjoy Saha
Pramit Saha
Farhang Sahba
Pranjal Sahu
Simone Saitta
Md Sirajus Salekin
Abbas Samani
Pedro Sanchez
Luis Sanchez Giraldo
Yudi Sang
Gerard Sanroma-Guell
Rodrigo Santa Cruz
Alice Santilli
Rachana Sathish
Olivier Saut
Mattia Savardi
Nico Scherf
Alexander Schlaefer
Jerome Schmid

Adam Schmidt
Julia Schnabel
Lawrence Schobs
Julian Schön
Peter Schueffler
Andreas Schuh
Christina Schwarz-Gsaxner
Michaël Sdika
Suman Sedai
Lalithkumar Seenivasan
Matthias Seibold
Sourya Sengupta
Lama Seoud
Ana Sequeira
Sharmishtaa Seshamani
Ahmed Shaffie
Jay Shah
Keyur Shah
Ahmed Shahin
Mohammad Abuzar Shaikh
S. Shailja
Hongming Shan
Wei Shao
Mostafa Sharifzadeh
Anuja Sharma
Gregory Sharp
Hailan Shen
Li Shen
Linlin Shen
Mali Shen
Mingren Shen
Yiqing Shen
Zhengyang Shen
Jun Shi
Xiaoshuang Shi
Yiyu Shi
Yonggang Shi
Hoo-Chang Shin
Jitae Shin
Keewon Shin
Boris Shirokikh
Suzanne Shontz
Yucheng Shu
Hanna Siebert
Alberto Signoroni
Wilson Silva
Julio Silva-Rodríguez
Margarida Silveira
Walter Simson
Praveer Singh
Vivek Singh
Nitin Singhal
Elena Sizikova
Gregory Slabaugh
Dane Smith
Kevin Smith
Tiffany So
Rajath Soans
Roger Soberanis-Mukul
Hessam Sokooti
Jingwei Song
Weinan Song
Xinhang Song
Xinrui Song
Mazen Soufi
Georgia Sovatzidi
Bella Specktor Fadida
William Speier
Ziga Spiclin
Dominik Spinczyk
Jon Sporring
Pradeeba Sridar
Chetan L. Srinidhi
Abhishek Srivastava
Lawrence Staib
Marc Stamminger
Justin Strait
Hai Su
Ruisheng Su
Zhe Su
Vaishnavi Subramanian
Gérard Subsol
Carole Sudre
Dong Sui
Heung-Il Suk
Shipra Suman
He Sun
Hongfu Sun
Jian Sun
Li Sun
Liyan Sun
Shanlin Sun
Kyung Sung
Yannick Suter
Swapna T. R.
Amir Tahmasebi
Pablo Tahoces
Sirine Taleb
Bingyao Tan
Chaowei Tan
Wenjun Tan
Hao Tang
Siyi Tang
Xiaoying Tang
Yucheng Tang
Zihao Tang
Michael Tanzer
Austin Tapp
Elias Tappeiner
Mickael Tardy
Giacomo Tarroni
Athena Taymourtash
Kaveri Thakoor
Elina Thibeau-Sutre
Paul Thienphrapa
Sarina Thomas
Stephen Thompson
Karl Thurnhofer-Hemsi
Cristiana Tiago
Lin Tian
Lixia Tian
Yapeng Tian
Yu Tian
Yun Tian
Aleksei Tiulpin
Hamid Tizhoosh
Minh Nguyen Nhat To
Matthew Toews
Maryam Toloubidokhti
Minh Tran
Quoc-Huy Trinh
Jocelyne Troccaz
Roger Trullo

Chialing Tsai
Apostolia Tsirikoglou
Puxun Tu
Samyakh Tukra
Sudhakar Tummala
Georgios Tziritas
Vladimír Ulman
Tamas Ungi
Régis Vaillant
Jeya Maria Jose Valanarasu
Vanya Valindria
Juan Miguel Valverde
Fons van der Sommen
Maureen van Eijnatten
Tom van Sonsbeek
Gijs van Tulder
Yogatheesan Varatharajah
Madhurima Vardhan
Thomas Varsavsky
Hooman Vaseli
Serge Vasylechko
S. Swaroop Vedula
Sanketh Vedula
Gonzalo Vegas Sanchez-Ferrero
Matthew Velazquez
Archana Venkataraman
Sulaiman Vesal
Mitko Veta
Barbara Villarini
Athanasios Vlontzos
Wolf-Dieter Vogl
Ingmar Voigt
Sandrine Voros
Vibashan VS
Trinh Thi Le Vuong
An Wang
Bo Wang
Ce Wang
Changmiao Wang
Ching-Wei Wang
Dadong Wang
Dong Wang
Fakai Wang
Guotai Wang
Haifeng Wang
Haoran Wang
Hong Wang
Hongxiao Wang
Hongyu Wang
Jiacheng Wang
Jing Wang
Jue Wang
Kang Wang
Ke Wang
Lei Wang
Li Wang
Liansheng Wang
Lin Wang
Ling Wang
Linwei Wang
Manning Wang
Mingliang Wang
Puyang Wang
Qiuli Wang
Renzhen Wang
Ruixuan Wang
Shaoyu Wang
Sheng Wang
Shujun Wang
Shuo Wang
Shuqiang Wang
Tao Wang
Tianchen Wang
Tianyu Wang
Wenzhe Wang
Xi Wang
Xiangdong Wang
Xiaoqing Wang
Xiaosong Wang
Yan Wang
Yangang Wang
Yaping Wang
Yi Wang
Yirui Wang
Yixin Wang
Zeyi Wang
Zhao Wang
Zichen Wang
Ziqin Wang
Ziyi Wang
Zuhui Wang
Dong Wei
Donglai Wei
Hao Wei
Jia Wei
Leihao Wei
Ruofeng Wei
Shuwen Wei
Martin Weigert
Wolfgang Wein
Michael Wels
Cédric Wemmert
Thomas Wendler
Markus Wenzel
Rhydian Windsor
Adam Wittek
Marek Wodzinski
Ivo Wolf
Julia Wolleb
Ka-Chun Wong
Jonghye Woo
Chongruo Wu
Chunpeng Wu
Fuping Wu
Huaqian Wu
Ji Wu
Jiangjie Wu
Jiong Wu
Junde Wu
Linshan Wu
Qing Wu
Weiwen Wu
Wenjun Wu
Xiyin Wu
Yawen Wu
Ye Wu
Yicheng Wu
Yongfei Wu
Zhengwang Wu
Pengcheng Xi
Chao Xia
Siyu Xia
Wenjun Xia
Lei Xiang

Tiange Xiang
Deqiang Xiao
Li Xiao
Xiaojiao Xiao
Yiming Xiao
Zeyu Xiao
Hongtao Xie
Huidong Xie
Jianyang Xie
Long Xie
Weidi Xie
Fangxu Xing
Shuwei Xing
Xiaodan Xing
Xiaohan Xing
Haoyi Xiong
Yujian Xiong
Di Xu
Feng Xu
Haozheng Xu
Hongming Xu
Jiangchang Xu
Jiaqi Xu
Junshen Xu
Kele Xu
Lijian Xu
Min Xu
Moucheng Xu
Rui Xu
Xiaowei Xu
Xuanang Xu
Yanwu Xu
Yanyu Xu
Yongchao Xu
Yunqiu Xu
Zhe Xu
Zhoubing Xu
Ziyue Xu
Kai Xuan
Cheng Xue
Jie Xue
Tengfei Xue
Wufeng Xue
Yuan Xue
Zhong Xue
Ts Faridah Yahya
Chaochao Yan
Jiangpeng Yan
Ming Yan
Qingsen Yan
Xiangyi Yan
Yuguang Yan
Zengqiang Yan
Baoyao Yang
Carl Yang
Changchun Yang
Chen Yang
Feng Yang
Fengting Yang
Ge Yang
Guanyu Yang
Heran Yang
Huijuan Yang
Jiancheng Yang
Jiewen Yang
Peng Yang
Qi Yang
Qiushi Yang
Wei Yang
Xin Yang
Xuan Yang
Yan Yang
Yanwu Yang
Yifan Yang
Yingyu Yang
Zhicheng Yang
Zhijian Yang
Jiangchao Yao
Jiawen Yao
Lanhong Yao
Linlin Yao
Qingsong Yao
Tianyuan Yao
Xiaohui Yao
Zhao Yao
Dong Hye Ye
Menglong Ye
Yousef Yeganeh
Jirong Yi
Xin Yi
Chong Yin
Pengshuai Yin
Yi Yin
Zhaozheng Yin
Chunwei Ying
Youngjin Yoo
Jihun Yoon
Chenyu You
Hanchao Yu
Heng Yu
Jinhua Yu
Jinze Yu
Ke Yu
Qi Yu
Qian Yu
Thomas Yu
Weimin Yu
Yang Yu
Chenxi Yuan
Kun Yuan
Wu Yuan
Yixuan Yuan
Paul Yushkevich
Fatemeh Zabihollahy
Samira Zare
Ramy Zeineldin
Dong Zeng
Qi Zeng
Tianyi Zeng
Wei Zeng
Kilian Zepf
Kun Zhan
Bokai Zhang
Daoqiang Zhang
Dong Zhang
Fa Zhang
Hang Zhang
Hanxiao Zhang
Hao Zhang
Haopeng Zhang
Haoyue Zhang
Hongrun Zhang
Jiadong Zhang
Jiajin Zhang
Jianpeng Zhang

Jiawei Zhang
Jingqing Zhang
Jingyang Zhang
Jinwei Zhang
Jiong Zhang
Jiping Zhang
Ke Zhang
Lefei Zhang
Lei Zhang
Li Zhang
Lichi Zhang
Lu Zhang
Minghui Zhang
Molin Zhang
Ning Zhang
Rongzhao Zhang
Ruipeng Zhang
Ruisi Zhang
Shichuan Zhang
Shihao Zhang
Shuai Zhang
Tuo Zhang
Wei Zhang
Weihang Zhang
Wen Zhang
Wenhua Zhang
Wenqiang Zhang
Xiaodan Zhang
Xiaoran Zhang
Xin Zhang
Xukun Zhang
Xuzhe Zhang
Ya Zhang
Yanbo Zhang
Yanfu Zhang
Yao Zhang
Yi Zhang
Yifan Zhang
Yixiao Zhang
Yongqin Zhang
You Zhang
Youshan Zhang
Yu Zhang
Yubo Zhang
Yue Zhang
Yuhan Zhang
Yulun Zhang
Yundong Zhang
Yunlong Zhang
Yuyao Zhang
Zheng Zhang
Zhenxi Zhang
Ziqi Zhang
Can Zhao
Chongyue Zhao
Fenqiang Zhao
Gangming Zhao
He Zhao
Jianfeng Zhao
Jun Zhao
Li Zhao
Liang Zhao
Lin Zhao
Mengliu Zhao
Mingbo Zhao
Qingyu Zhao
Shang Zhao
Shijie Zhao
Tengda Zhao
Tianyi Zhao
Wei Zhao
Yidong Zhao
Yiyuan Zhao
Yu Zhao
Zhihe Zhao
Ziyuan Zhao
Haiyong Zheng
Hao Zheng
Jiannan Zheng
Kang Zheng
Meng Zheng
Sisi Zheng
Tianshu Zheng
Yalin Zheng
Yefeng Zheng
Yinqiang Zheng
Yushan Zheng
Aoxiao Zhong
Jia-Xing Zhong
Tao Zhong
Zichun Zhong
Hong-Yu Zhou
Houliang Zhou
Huiyu Zhou
Kang Zhou
Qin Zhou
Ran Zhou
S. Kevin Zhou
Tianfei Zhou
Wei Zhou
Xiao-Hu Zhou
Xiao-Yun Zhou
Yi Zhou
Youjia Zhou
Yukun Zhou
Zongwei Zhou
Chenglu Zhu
Dongxiao Zhu
Heqin Zhu
Jiayi Zhu
Meilu Zhu
Wei Zhu
Wenhui Zhu
Xiaofeng Zhu
Xin Zhu
Yonghua Zhu
Yongpei Zhu
Yuemin Zhu
Yan Zhuang
David Zimmerer
Yongshuo Zong
Ke Zou
Yukai Zou
Lianrui Zuo
Gerald Zwettler

Outstanding Area Chairs

Mingxia Liu	University of North Carolina at Chapel Hill, USA
Matthias Wilms	University of Calgary, Canada
Veronika Zimmer	Technical University Munich, Germany

Outstanding Reviewers

Kimberly Amador	University of Calgary, Canada
Angela Castillo	Universidad de los Andes, Colombia
Chen Chen	Imperial College London, UK
Laura Connolly	Queen's University, Canada
Pierre-Henri Conze	IMT Atlantique, France
Niharika D'Souza	IBM Research, USA
Michael Götz	University Hospital Ulm, Germany
Meirui Jiang	Chinese University of Hong Kong, China
Manuela Kunz	National Research Council Canada, Canada
Zdravko Marinov	Karlsruhe Institute of Technology, Germany
Sérgio Pereira	Lunit, South Korea
Lalithkumar Seenivasan	National University of Singapore, Singapore

Honorable Mentions (Reviewers)

Kumar Abhishek	Simon Fraser University, Canada
Guilherme Aresta	Medical University of Vienna, Austria
Shahab Aslani	University College London, UK
Marc Aubreville	Technische Hochschule Ingolstadt, Germany
Yaël Balbastre	Massachusetts General Hospital, USA
Omri Bar	Theator, Israel
Aicha Ben Taieb	Simon Fraser University, Canada
Cosmin Bercea	Technical University Munich and Helmholtz AI and Helmholtz Center Munich, Germany
Benjamin Billot	Massachusetts Institute of Technology, USA
Michal Byra	RIKEN Center for Brain Science, Japan
Mariano Cabezas	University of Sydney, Australia
Alessandro Casella	Italian Institute of Technology and Politecnico di Milano, Italy
Junyu Chen	Johns Hopkins University, USA
Argyrios Christodoulidis	Pfizer, Greece
Olivier Colliot	CNRS, France

Lei Cui	Northwest University, China
Neel Dey	Massachusetts Institute of Technology, USA
Alessio Fagioli	Sapienza University, Italy
Yannik Glaser	University of Hawaii at Manoa, USA
Haifan Gong	Chinese University of Hong Kong, Shenzhen, China
Ricardo Gonzales	University of Oxford, UK
Sobhan Goudarzi	Sunnybrook Research Institute, Canada
Michal Grzeszczyk	Sano Centre for Computational Medicine, Poland
Fatemeh Haghighi	Arizona State University, USA
Edward Henderson	University of Manchester, UK
Qingqi Hong	Xiamen University, China
Mohammad R. H. Taher	Arizona State University, USA
Henkjan Huisman	Radboud University Medical Center, the Netherlands
Ronnachai Jaroensri	Google, USA
Qiangguo Jin	Northwestern Polytechnical University, China
Neerav Karani	Massachusetts Institute of Technology, USA
Benjamin Killeen	Johns Hopkins University, USA
Daniel Lang	Helmholtz Center Munich, Germany
Max-Heinrich Laves	Philips Research and ImFusion GmbH, Germany
Gilbert Lim	SingHealth, Singapore
Mingquan Lin	Weill Cornell Medicine, USA
Charles Lu	Massachusetts Institute of Technology, USA
Yuhui Ma	Chinese Academy of Sciences, China
Tejas Sudharshan Mathai	National Institutes of Health, USA
Felix Meissen	Technische Universität München, Germany
Mingyuan Meng	University of Sydney, Australia
Leo Milecki	CentraleSupelec, France
Marc Modat	King's College London, UK
Tiziano Passerini	Siemens Healthineers, USA
Tomasz Pieciak	Universidad de Valladolid, Spain
Daniel Rueckert	Imperial College London, UK
Julio Silva-Rodríguez	ETS Montreal, Canada
Bingyao Tan	Nanyang Technological University, Singapore
Elias Tappeiner	UMIT - Private University for Health Sciences, Medical Informatics and Technology, Austria
Jocelyne Troccaz	TIMC Lab, Grenoble Alpes University-CNRS, France
Chialing Tsai	Queens College, City University New York, USA
Juan Miguel Valverde	University of Eastern Finland, Finland
Sulaiman Vesal	Stanford University, USA

Contents – Part VI

Computer-Aided Diagnosis II

Computational Pathology

Computer-Aided Diagnosis II

Mining Negative Temporal Contexts for False Positive Suppression in Real-Time Ultrasound Lesion Detection

Haojun Yu[1], Youcheng Li[1], QuanLin Wu[2,5], Ziwei Zhao[2], Dengbo Chen[4], Dong Wang[1], and Liwei Wang[1,3](✉)

[1] National Key Laboratory of General Artificial Intelligence, School of Intelligence Science and Technology, Peking University, Beijing, China
{haojunyu,wanglw}@pku.edu.cn
[2] Center of Data Science, Peking University, Beijing, China
[3] Center for Machine Learning Research, Peking University, Beijing, China
[4] Yizhun Medical AI Co., Ltd., Beijing, China
[5] Pazhou Laboratory, Huangpu, Guangdong, China

Abstract. During ultrasonic scanning processes, real-time lesion detection can assist radiologists in accurate cancer diagnosis. However, this essential task remains challenging and underexplored. General-purpose real-time object detection models can mistakenly report obvious false positives (FPs) when applied to ultrasound videos, potentially misleading junior radiologists. One key issue is their failure to utilize negative symptoms in previous frames, denoted as *negative temporal contexts* (NTC) [15]. To address this issue, we propose to extract contexts from previous frames, including NTC, with the guidance of inverse optical flow. By aggregating extracted contexts, we endow the model with the ability to suppress FPs by leveraging NTC. We call the resulting model *UltraDet*. The proposed UltraDet demonstrates significant improvement over previous state-of-the-arts and achieves real-time inference speed. We release the code, checkpoints, and high-quality labels of the CVA-BUS dataset [9] in https://github.com/HaojunYu1998/UltraDet.

Keywords: Ultrasound Video · Real-time Lesion Detection · Negative Temporal Context · False Positive Suppression

1 Introduction

Ultrasound is a widely-used imaging modality for clinical cancer screening. Deep Learning has recently emerged as a promising approach for ultrasound lesion detection. While previous works focused on lesion detection in still images [25] and offline videos [9,11,22], this paper explores real-time ultrasound video lesion detection. Real-time lesion prompts can assist radiologists during scanning, thus

Supplementary Information The online version contains supplementary material available at https://doi.org/10.1007/978-3-031-43987-2_1.

H. Greenspan et al. (Eds.): MICCAI 2023, LNCS 14225, pp. 3–13, 2023.
https://doi.org/10.1007/978-3-031-43987-2_1

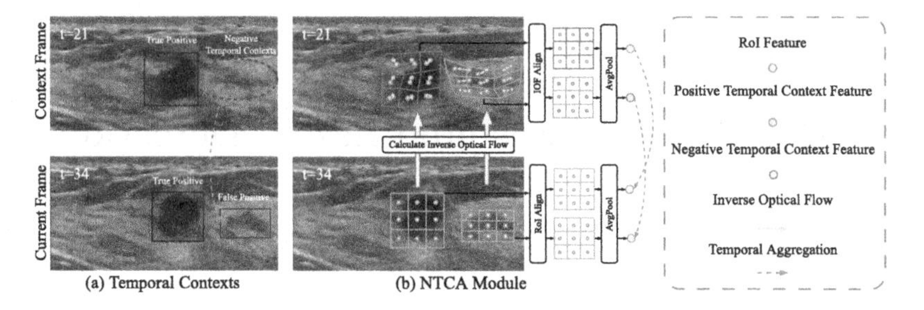

Fig. 1. Illustration of Negative Temporal Context Aggregation (NTCA) module. (a) Our motivation: mining negative temporal contexts for FP suppression. (b) The NTCA module leverages temporal contexts to suppress the FP. (Color figure online)

being more helpful to improve the accuracy of diagnosis. This task requires the model to infer faster than 30 frames per second (FPS) [19] and only previous frames are available for current frame processing.

Previous general-purpose detectors [1,2] report simple and obvious FPs when applied to ultrasound videos, *e.g.* the red box in Fig. 1(a). These FPs, attributable to non-lesion anatomies, can mislead junior readers. These anatomies appear like lesions in certain frames, but typically show negative symptoms in adjacent frames when scanned from different positions. So experienced radiologists will refer to corresponding regions in previous frames, denoted as *temporal contexts* (TC), to help restrain FPs. If TC of a lesion-like region exhibit negative symptoms, denoted as *negative temporal contexts* (NTC), radiologists are less likely to report it as a lesion [15]. Although important, the utilization of NTC remains unexplored. In natural videos, as transitions from non-objects to objects are implausible, previous works [1,2,20] only consider inter-object relationships. As shown in Sect. 4.4, the inability to utilize NTC is a key issue leading to the FPs reported by general-purpose detectors.

To address this issue, we propose a novel *UltraDet* model to leverage NTC. For each Region of Interest (RoI) $\mathcal{R}$ proposed by a basic detector, we extract temporal contexts from previous frames. To compensate for inter-frame motion, we generate deformed grids by applying inverse optical flow to the original regular RoI grids, illustrated in Fig. 1. Then we extract the RoI features from the deformed grids in previous frames and aggregate them into $\mathcal{R}$. We call the overall process *Negative Temporal Context Aggregation* (NTCA). The NTCA module leverages RoI-level NTC which are crucial for radiologists but ignored in previous works, thereby effectively improving the detection performance in a reliable and interpretable way. We plug the NTCA module into a basic real-time detector to form *UltraDet*. Experiments on CVA-BUS dataset [9] demonstrate that UltraDet, with real-time inference speed, significantly outperforms previous works, reducing about 50% FPs at a recall rate of 0.90.

Our contributions are four-fold. (1) We identify that the failure of general-purpose detectors on ultrasound videos derives from their incapability of utilizing negative temporal contexts. (2) We propose a novel UltraDet model, incorporating an NTCA module that effectively leverages NTC for FP suppression. (3) We conduct extensive experiments to demonstrate the proposed UltraDet significantly outperforms the previous state-of-the-arts. (4) We release high-quality labels of the CVA-BUS dataset [9] to facilitate future research.

2 Related Works

Real-Time Video Object Detection is typically achieved by single-frame detectors, often with temporal information aggregation modules. One-stage detectors [5,8,16,21] use only intra-frame information, DETR-based detectors [20,26] and Faster R-CNN-based detectors [1,2,7,14,23,28] are also widely utilized in video object detection. They aggregate temporal information by mining inter-object relationships without considering NTC.

Ultrasound Lesion Detection [10] can assist radiologists in clinical practice. Previous works have explored lesion detection in still images [25] and offline videos [9,11,22]. Real-time video lesion detection is underexplored. In previous works, YOLO series [17,24] and knowledge distillation [19] are used to speed up inference. However, these works use single-frame detectors or post-process methods while learnable inter-frame aggregation modules are not adopted. Thus their performances are far from satisfactory.

Optical Flow [3] is used to guide ultrasound segmentation [12], motion estimation [4] and elastography [13]. For the first time, we use inverse optical flow to guide temporal context information extraction.

3 Method

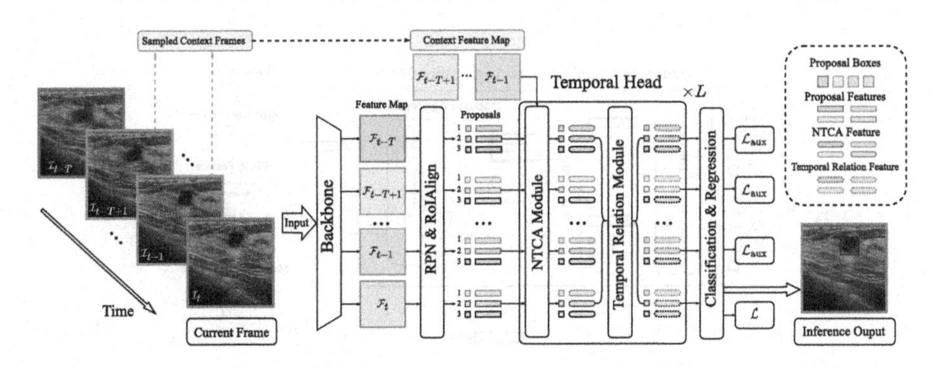

Fig. 2. Illustration of UltraDet model. The yellow and green frames are sampled as context frames, and their feature maps are inputs of the NTCA module. (Color figure online)

In real-time video lesion detection, given the current frame $\mathcal{I}_t$ and a sequence of T previous frames as $\{\mathcal{I}_\tau\}_{\tau=t-T}^{t-1}$, the goal is to detect lesions in $\mathcal{I}_t$ by exploiting the temporal information in previous frames as illustrated in Fig. 2.

3.1 Basic Real-Time Detector

The basic real-time detector comprises three main components: a lightweight backbone (*e.g.* ResNet34 [6]), a Region Proposal Network (RPN) [14], and a Temporal Relation head [2]. The backbone is responsible for extracting feature map $\mathcal{F}_\tau$ of frame $\mathcal{I}_\tau$. The RPN generates proposals consisting of boxes $\mathcal{B}_\tau$ and proposal features $\mathcal{Q}_\tau$ using RoI Align and average pooling:

$$\mathcal{Q}_\tau = \text{AvgPool}\left(\text{RoIAlign}(\mathcal{F}_\tau, \mathcal{B}_\tau)\right) \tag{1}$$

where $\tau = t-T, \cdots, t-1, t$. To aggregate temporal information, proposals from all $T+1$ frames are fed into the Temporal Relation head and updated with inter-lesion information extracted via a relation operation [7]:

$$\mathcal{Q}^l = \mathcal{Q}^{l-1} + \text{Relation}(\mathcal{Q}^{l-1}, \mathcal{B}) \tag{2}$$

where $l = 1, \cdots, L$ represent layer indices, $\mathcal{B}$ and $\mathcal{Q}$ are the concatenation of all $\mathcal{B}_\tau$ and $\mathcal{Q}_\tau$, and $\mathcal{Q}^0 = \mathcal{Q}$. We call this basic real-time detector *BasicDet*. The BasicDet is conceptually similar to RDN [2] but does not incorporate relation distillation since the number of lesions and proposals in this study is much smaller than in natural videos.

3.2 Negative Temporal Context Aggregation

In this section, we present the Negative Temporal Context Aggregation (NTCA) module. We sample T_{ctxt} context frames from T previous frames, then extract temporal contexts (TC) from context frames and aggregate them into proposals. We illustrate the NTCA module in Fig. 3 and elaborate on details as follows.

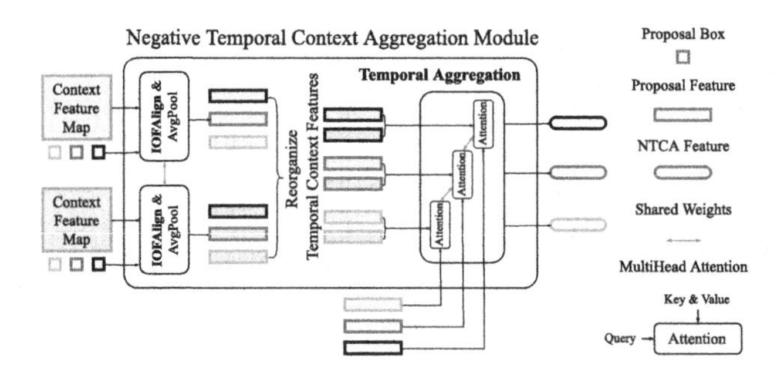

Fig. 3. Illustration of the Negative Temporal Context Aggregation module.

Inverse Optical Flow Align. We propose the Inverse Optical Flow Align (IOF Align) to extract TC features. For the current frame $\mathcal{I}_t$ and a sampled context frame $\mathcal{I}_\tau$ with $\tau < t$, we extract TC features from the context feature map $\mathcal{F}_\tau$ with the corresponding regions. We use inverse optical flow $\mathcal{O}_{t\to\tau} \in \mathbb{R}^{H\times W\times 2}$

to transform the RoIs from frame t to τ: $\mathcal{O}_{t\to\tau} = \text{FlowNet}(\mathcal{I}_t, \mathcal{I}_\tau)$ where H, W represent height and width of feature maps. The $\text{FlowNet}(\mathcal{I}_t, \mathcal{I}_\tau)$ is a fixed network [3] to predict optical flow from $\mathcal{I}_t$ to $\mathcal{I}_\tau$. We refer to $\mathcal{O}_{t\to\tau}$ as *inverse optical flow* because it represents the optical flow in inverse chronological order from t to τ. We conduct IOF Align and average pooling to extract $\mathcal{C}_{t,\tau}$:

$$\mathcal{C}_{t,\tau} = \text{AvgPool}\left(\text{IOFAlign}(\mathcal{F}_\tau, \mathcal{B}_t, \mathcal{O}_{t\to\tau})\right) \tag{3}$$

where $\text{IOFAlign}(\mathcal{F}_\tau, \mathcal{B}_t, \mathcal{O}_{t\to\tau})$ extracts context features in $\mathcal{F}_\tau$ from deformed grids generated by applying offsets $\mathcal{O}_{t\to\tau}$ to the original regular grids in $\mathcal{B}_t$, which is illustrated in the Fig. 1(b).

Temporal Aggregation. We concatenate $\mathcal{C}_{t,\tau}$ in all T_{ctxt} context frames to form $\mathcal{C}_t$ and enhance proposal features by fusing $\mathcal{C}_t$ into $\mathcal{Q}_t$:

$$\mathcal{Q}^l_{\text{ctxt},t} = \mathcal{Q}^{l-1}_{\text{ctxt},t} + \text{Attention}(\mathcal{Q}^{l-1}_{\text{ctxt},t}, \mathcal{C}_t, \mathcal{C}_t) \tag{4}$$

where $l = 1, \cdots, L$ represent layer indices, $\mathcal{Q}^0_{\text{ctxt},t} = \mathcal{Q}_t$, and $\text{Attention}(Q, K, V)$ is Multi-head Attention [18]. We refer to the concatenation of all TC-enhanced proposal features in $T+1$ frames as $\mathcal{Q}_{\text{ctxt}}$. To extract consistent TC, the context frames of T previous frames are shared with the current frame.

3.3 UltraDet for Real-Time Lesion Detection

We integrate the NTCA module into the BasicDet introduced in Sect. 3.1 to form the UltraDet model, which is illustrated in Fig. 2. The head of UltraDet consists of stacked NTCA and relation modules:

$$\mathcal{Q}^l = \mathcal{Q}^l_{\text{ctxt}} + \text{Relation}(\mathcal{Q}^l_{\text{ctxt}}, \mathcal{B}). \tag{5}$$

During training, we apply regression and classification losses $\mathcal{L} = \mathcal{L}_{\text{reg}} + \mathcal{L}_{\text{cls}}$ to the current frame. To improve training efficiency, we apply auxiliary losses $\mathcal{L}_{\text{aux}} = \mathcal{L}$ to all previous T frames. During inference, the UltraDet model uses the current frame and T previous frames as inputs and generates predictions only for the current frame. This design endows the UltraDet with the ability to perform real-time lesion detection.

4 Experiments

4.1 Dateset

CVA-BUS Dateset. We use the open source CVA-BUS dataset that consists of 186 valid videos, which is proposed in CVA-Net [9]. We split the dataset into train-val (154 videos) and test (32 videos) sets. In the train-val split, there are 21423 frames with 170 lesions. In the test split, there are 3849 frames with 32 lesions. We focus on the lesion detection task and do not utilize the benign/malignant classification labels provided in the original dataset.

High-Quality Labels. The bounding box labels provided in the original CVA-BUS dataset are unsteady and sometimes inaccurate, leading to jiggling and inaccurate model predictions. We provide a new version of high-quality labels that are re-annotated by experienced radiologists. We reproduce all baselines using our high-quality labels to ensure a fair comparison. Visual comparisons of two versions of labels are available in supplementary materials. To facilitate future research, we will release these high-quality labels.

Table 1. Quantitative results of real-time lesion detection on CVA-BUS [9].

Model	Type	Pr80	Pr90	FP80	FP90	AP50	R@16	FPS
		One-Stage Detectors						
YOLOX [5]	Image	$69.7_{3.7}$	$43.4_{7.7}$	$23.8_{4.8}$	$87.6_{24.5}$	$80.4_{1.6}$	$97.5_{0.5}$	**59.8**
RetinaNet [8]	Image	$75.7_{2.5}$	$57.2_{2.9}$	$9.3_{2.0}$	$32.8_{6.5}$	$84.5_{1.0}$	$95.1_{0.6}$	53.6
FCOS [16]	Image	$87.2_{2.2}$	$72.2_{5.1}$	$11.0_{2.4}$	$23.0_{3.7}$	$89.5_{1.4}$	$98.8_{0.3}$	56.1
DeFCN [21]	Image	$81.5_{1.8}$	$67.5_{2.3}$	$21.1_{3.2}$	$33.4_{4.3}$	$86.4_{1.3}$	$\mathbf{99.3}_{0.3}$	51.2
Track-YOLO [24]	Video	$75.1_{2.7}$	$47.0_{3.1}$	$18.1_{1.9}$	$74.2_{14.7}$	$80.1_{1.0}$	$94.7_{0.9}$	46.0
		DETR-Based Detectors						
DeformDETR [27]	Image	$90.1_{3.2}$	$72.7_{10.6}$	$5.6_{2.2}$	$37.8_{20.9}$	$90.5_{2.0}$	$98.7_{0.3}$	33.8
TransVOD [26]	Video	$92.5_{2.2}$	$77.5_{7.2}$	$3.1_{1.3}$	$23.7_{11.5}$	$90.1_{1.8}$	$98.4_{0.4}$	24.2
CVA-Net [9]	Video	$92.3_{2.6}$	$80.2_{6.1}$	$4.7_{2.6}$	$19.6_{5.6}$	$\mathbf{91.6}_{1.9}$	$98.6_{0.8}$	23.1
PTSEFormer [20]	Video	$93.3_{1.9}$	$85.4_{6.0}$	$2.8_{1.1}$	$12.5_{9.8}$	$91.5_{1.6}$	$97.9_{1.2}$	9.1
		FasterRCNN-Based Detectors						
FasterRCNN [14]	Image	$91.3_{0.9}$	$75.2_{3.6}$	$6.9_{1.4}$	$34.4_{6.7}$	$88.0_{1.4}$	$92.4_{1.0}$	49.2
RelationNet [7]	Image	$91.4_{1.3}$	$79.2_{2.9}$	$6.2_{2.0}$	$24.4_{5.6}$	$87.6_{1.7}$	$92.4_{0.9}$	42.7
FGFA [28]	Video	$92.9_{1.5}$	$82.2_{4.1}$	$4.4_{1.6}$	$13.3_{3.7}$	$90.5_{1.1}$	$93.6_{0.9}$	33.8
SELSA [23]	Video	$91.6_{1.7}$	$80.2_{2.5}$	$7.5_{1.5}$	$23.3_{5.5}$	$89.2_{1.1}$	$92.6_{0.8}$	43.8
MEGA [1]	Video	$93.9_{1.5}$	$86.9_{2.3}$	$3.1_{1.7}$	$11.7_{3.0}$	$90.9_{1.0}$	$93.6_{0.7}$	40.2
BasicDet (RDN) [2]	Video	$92.4_{1.0}$	$83.6_{2.2}$	$3.8_{1.2}$	$13.4_{3.2}$	$88.7_{1.4}$	$92.7_{0.6}$	42.2
UltraDet (Ours)	Video	$\mathbf{95.7}_{1.2}$	$\mathbf{90.8}_{1.4}$	$\mathbf{1.9}_{0.4}$	$\mathbf{5.7}_{1.6}$	$\mathbf{91.6}_{1.6}$	$93.8_{1.3}$	30.4

4.2 Evaluation Metrics

Pr_{80}, Pr_{90}. In clinical applications, it is important for detection models to be sensitive. So we provide frame-level precision values with high recall rates of 0.80 and 0.90, which we denote as Pr_{80} and Pr_{90}, respectively.

FP_{80}, FP_{90}. We further report lesion-level FP rates as critical metrics. Frame-level FPs are linked by IoU scores to form FP sequences [24]. The number of FP sequences per minute at recall rates of 0.80 and 0.90 are reported as FP_{80} and FP_{90}, respectively. The unit of lesion-level FP rates is seq/min.

AP_{50}. We provide AP_{50} instead of mAP or AP_{75} because the IoU threshold of 0.50 is sufficient for lesion localization in clinical practice. Higher thresholds like 0.75 or 0.90 are impractical due to the presence of blurred lesion edges.

R@16. To evaluate the highest achievable sensitivity, we report the frame-level average recall rates of Top-16 proposals, denoted as R@16.

4.3 Implementation Details

UltraDet Settings. We use FlowNetS [3] as the fixed FlowNet in IOF Align and share the same finding with previous works [4,12,13] that the FlowNet trained on natural datasets generalizes well on ultrasound datasets. We set the pooling stride in the FlowNet to 4, the number of UltraDet head layers $L = 2$, the number of previous frames $T = 15$ and $T_{\text{ctxt}} = 2$, and the number of proposals is 16. We cached intermediate results of previous frames and reuse them to speed up inference. Other hyper-parameters are listed in supplementary materials.

Shared Settings. All models are built in PyTorch framework and trained using eight NVIDIA GeForce RTX 3090 GPUs. We use ResNet34 [6] as backbones and set the number of training iterations to 10,000. We set the feature dimensions of detection heads to 256 and baselines are re-implemented to utilize only previous frames. We refer to our code for more details.

4.4 Main Results

Quantitative Results. We compare performances of real-time detectors with the UltraDet in Table 1. We perform 4-fold cross-validation and report the mean values and standard errors on the test set to mitigate fluctuations. The UltraDet outperforms all previous state-of-the-art in terms of precision and FP rates. Especially, the Pr90 of UltraDet achieves 90.8%, representing a 5.4% absolute improvement over the best competitor, PTSEFormer [20]. Moreover, the FP90 of UltraDet is 5.7 seq/min, reducing about 50% FPs of the best competitor, PTSEFormer. Although CVA-Net [9] achieve comparable AP50 with our method, we significantly improve precision and FP rates over the CVA-Net [9].

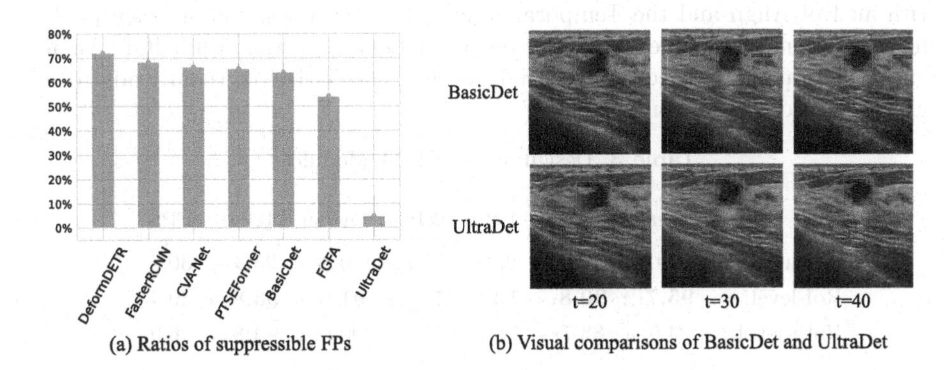

Fig. 4. (a) Ratios of FPs that are suppressible by leveraging NTC. (b) Visual comparisons of BasicDet and UltraDet prediction results at recall 0.90. Blue boxes are true positives and red boxes are FPs. (Color figure online)

Importance of NTC. In Fig. 4(a), we illustrate the FP ratios that can be suppressed by using NTC. The determination of whether FPs can be inhibited by NTC is based on manual judgments of experienced radiologists. We find that about 50%–70% FPs of previous methods are suppressible. However, by utilizing NTC in our UltraDet, we are able to effectively prevent this type of FPs.

Inference Speed. We run inference using one NVIDIA GeForce RTX 3090 GPU and report the inference speed in Table 1. The UltraDet achieves an inference speed of 30.4 FPS and already meets the 30 FPS requirement. Using TensorRT, we further optimize the speed to 35.2 FPS, which is sufficient for clinical applications [19].

Qualitative Results. Figure 4(b) visually compares BasicDet and UltraDet. The BasicDet reports FPs at $t = 30$ and 40 as it fails to leverage NTC when $t = 20$, while the UltraDet successfully suppresses FPs with the NTCA module.

4.5 Ablation Study

Table 2. Ablation study of each NTCA sub-module.

IOFAlign	TempAgg	Pr80	Pr90	FP80	FP90	AP50	R@16	FPS
-	-	$92.4_{1.0}$	$83.6_{2.2}$	$3.8_{1.2}$	$13.4_{3.2}$	$88.7_{1.4}$	$92.7_{0.6}$	42.2
-	✓	$93.7_{1.8}$	$84.3_{1.4}$	$3.4_{1.0}$	$12.5_{0.8}$	$90.0_{1.9}$	$93.0_{1.3}$	37.2
✓	-	$94.5_{2.3}$	$88.7_{2.2}$	$2.6_{0.6}$	$9.0_{1.5}$	$90.5_{1.9}$	$92.9_{1.4}$	32.3
✓	✓	$\mathbf{95.7}_{1.2}$	$\mathbf{90.8}_{1.4}$	$\mathbf{1.9}_{0.4}$	$\mathbf{5.7}_{1.6}$	$\mathbf{91.6}_{1.6}$	$\mathbf{93.8}_{1.3}$	30.4

Effectiveness of Each Sub-module. We ablate the effectiveness of each sub-module of the NTCA module in Table 2. Specifically, we replace the IOF Align with an RoI Align and the Temporal Aggregation with a simple average pooling in the temporal dimension. The results demonstrate that both IOF Align and Temporal Aggregation are crucial, as removing either of them leads to a noticeable drop in performance.

Table 3. Design of the NTCA Module.

Num	Pr80	Pr90	FP80	FP90	AP50	R@16	FPS
Feature-level	$94.0_{0.9}$	$84.6_{2.9}$	$2.9_{0.8}$	$11.7_{3.0}$	$90.8_{0.7}$	$93.3_{0.6}$	30.6
RoI-level	$\mathbf{95.7}_{1.2}$	$\mathbf{90.8}_{1.4}$	$\mathbf{1.9}_{0.4}$	$\mathbf{5.7}_{1.6}$	$\mathbf{91.6}_{1.6}$	$\mathbf{93.8}_{1.3}$	30.4
Both-level	$94.6_{1.0}$	$88.7_{1.8}$	$2.5_{0.9}$	$7.9_{2.4}$	$90.8_{1.5}$	$\mathbf{93.8}_{0.9}$	26.9

Design of the NTCA Module. Besides RoI-level TC aggregation in UltraDet, feature-level aggregation is also feasible. We plug the optical flow feature warping proposed in FGFA [28] into the BasicDet and report the results in Table 3. We find RoI-level aggregation is more effective than feature-level, and both-level aggregation provides no performance gains. This conclusion agrees with radiologists' skills to focus more on local regions instead of global information.

5 Conclusion

In this paper, we address the clinical challenge of real-time ultrasound lesion detection. We propose a novel Negative Temporal Context Aggregation (NTCA) module, imitating radiologists' diagnosis processes to suppress FPs. The NTCA module leverages negative temporal contexts that are essential for FP suppression but ignored in previous works, thereby being more effective in suppressing FPs. We plug the NTCA module into a BasicDet to form the UltraDet model, which significantly improves the precision and FP rates over previous state-of-the-arts while achieving real-time inference speed. The UltraDet has the potential to become a real-time lesion detection application and assist radiologists in more accurate cancer diagnosis in clinical practice.

Acknowledgements. This work is supported by National Key R&D Program of China (2022ZD0114900) and National Science Foundation of China (NSFC62276005).

References

1. Chen, Y., Cao, Y., Hu, H., Wang, L.: Memory enhanced global-local aggregation for video object detection. In: Proceedings of the IEEE/CVF Conference on Computer Vision and Pattern Recognition, pp. 10337–10346 (2020)
2. Deng, J., Pan, Y., Yao, T., Zhou, W., Li, H., Mei, T.: Relation distillation networks for video object detection. In: Proceedings of the IEEE/CVF International Conference on Computer Vision, pp. 7023–7032 (2019)
3. Dosovitskiy, A., et al.: FlowNet: learning optical flow with convolutional networks. In: Proceedings of the IEEE International Conference on Computer Vision, pp. 2758–2766 (2015)
4. Evain, E., Faraz, K., Grenier, T., Garcia, D., De Craene, M., Bernard, O.: A pilot study on convolutional neural networks for motion estimation from ultrasound images. IEEE Trans. Ultrason. Ferroelectr. Freq. Control **67**(12), 2565–2573 (2020)
5. Ge, Z., Liu, S., Wang, F., Li, Z., Sun, J.: YOLOX: exceeding yolo series in 2021. arXiv preprint arXiv:2107.08430 (2021)
6. He, K., Zhang, X., Ren, S., Sun, J.: Deep residual learning for image recognition. In: Proceedings of the IEEE Conference on Computer Vision and Pattern Recognition, pp. 770–778 (2016)
7. Hu, H., Gu, J., Zhang, Z., Dai, J., Wei, Y.: Relation networks for object detection. In: Proceedings of the IEEE Conference on Computer Vision and Pattern Recognition, pp. 3588–3597 (2018)
8. Lin, T.Y., Goyal, P., Girshick, R., He, K., Dollár, P.: Focal loss for dense object detection. In: Proceedings of the IEEE International Conference on Computer Vision, pp. 2980–2988 (2017)
9. Lin, Z., Lin, J., Zhu, L., Fu, H., Qin, J., Wang, L.: A new dataset and a baseline model for breast lesion detection in ultrasound videos. In: Wang, L., Dou, Q., Fletcher, P.T., Speidel, S., Li, S. (eds.) MICCAI 2022, Part III. LNCS, vol. 13433, pp. 614–623. Springer, Cham (2022). https://doi.org/10.1007/978-3-031-16437-8_59

10. Liu, S., et al.: Deep learning in medical ultrasound analysis: a review. Engineering **5**(2), 261–275 (2019)
11. Movahedi, M.M., Zamani, A., Parsaei, H., Tavakoli Golpaygani, A., Haghighi Poya, M.R.: Automated analysis of ultrasound videos for detection of breast lesions. Middle East J. Cancer **11**(1), 80–90 (2020)
12. Nguyen, A., et al.: End-to-end real-time catheter segmentation with optical flow-guided warping during endovascular intervention. In: 2020 IEEE International Conference on Robotics and Automation (ICRA), pp. 9967–9973. IEEE (2020)
13. Peng, B., Xian, Y., Jiang, J.: A convolution neural network-based speckle tracking method for ultrasound elastography. In: 2018 IEEE International Ultrasonics Symposium (IUS), pp. 206–212. IEEE (2018)
14. Ren, S., He, K., Girshick, R., Sun, J.: Faster R-CNN: towards real-time object detection with region proposal networks. In: Advances in Neural Information Processing Systems, vol. 28 (2015)
15. Spak, D.A., Plaxco, J., Santiago, L., Dryden, M., Dogan, B.: BI-RADS® fifth edition: a summary of changes. Diagn. Interv. Imaging **98**(3), 179–190 (2017)
16. Tian, Z., Shen, C., Chen, H., He, T.: FCOS: fully convolutional one-stage object detection. In: Proceedings of the IEEE/CVF International Conference on Computer Vision, pp. 9627–9636 (2019)
17. Tiyarattanachai, T., et al.: The feasibility to use artificial intelligence to aid detecting focal liver lesions in real-time ultrasound: a preliminary study based on videos. Sci. Rep. **12**(1), 7749 (2022)
18. Vaswani, A., et al.: Attention is all you need. In: Advances in Neural Information Processing Systems, vol. 30 (2017)
19. Vaze, S., Xie, W., Namburete, A.I.: Low-memory CNNs enabling real-time ultrasound segmentation towards mobile deployment. IEEE J. Biomed. Health Inform. **24**(4), 1059–1069 (2020)
20. Wang, H., Tang, J., Liu, X., Guan, S., Xie, R., Song, L.: PTSEFormer: progressive temporal-spatial enhanced transformer towards video object detection. In: Avidan, S., Brostow, G., Cissé, M., Farinella, G.M., Hassner, T. (eds.) ECCV 2022, Part VIII. LNCS, vol. 13668, pp. 732–747. Springer, Cham (2022). https://doi.org/10.1007/978-3-031-20074-8_42
21. Wang, J., Song, L., Li, Z., Sun, H., Sun, J., Zheng, N.: End-to-end object detection with fully convolutional network. In: Proceedings of the IEEE/CVF Conference on Computer Vision and Pattern Recognition, pp. 15849–15858 (2021)
22. Wang, Y., et al.: Key-frame guided network for thyroid nodule recognition using ultrasound videos. In: Wang, L., Dou, Q., Fletcher, P.T., Speidel, S., Li, S. (eds.) MICCAI 2022, Part IV. LNCS, vol. 13434, pp. 238–247. Springer, Cham (2022). https://doi.org/10.1007/978-3-031-16440-8_23
23. Wu, H., Chen, Y., Wang, N., Zhang, Z.: Sequence level semantics aggregation for video object detection. In: Proceedings of the IEEE/CVF International Conference on Computer Vision, pp. 9217–9225 (2019)
24. Wu, X., et al.: CacheTrack-YOLO: real-time detection and tracking for thyroid nodules and surrounding tissues in ultrasound videos. IEEE J. Biomed. Health Inform. **25**(10), 3812–3823 (2021)
25. Yap, M.H., et al.: Automated breast ultrasound lesions detection using convolutional neural networks. IEEE J. Biomed. Health Inform. **22**(4), 1218–1226 (2017)
26. Zhou, Q., et al.: TransVOD: end-to-end video object detection with spatial-temporal transformers. IEEE Trans. Pattern Anal. Mach. Intell. **45**(6), 7853–7869 (2023)

27. Zhu, X., Su, W., Lu, L., Li, B., Wang, X., Dai, J.: Deformable DETR: deformable transformers for end-to-end object detection. arXiv preprint arXiv:2010.04159 (2020)
28. Zhu, X., Wang, Y., Dai, J., Yuan, L., Wei, Y.: Flow-guided feature aggregation for video object detection. In: Proceedings of the IEEE International Conference on Computer Vision, pp. 408–417 (2017)

Combat Long-Tails in Medical Classification with Relation-Aware Consistency and Virtual Features Compensation

Li Pan[1], Yupei Zhang[2], Qiushi Yang[3], Tan Li[4], and Zhen Chen[5](✉)

[1] The Chinese University of Hong Kong, Sha Tin, Hong Kong
[2] The Centre for Intelligent Multidimensional Data Analysis (CIMDA), Pak Shek Kok, Hong Kong
[3] City University of Hong Kong, Kowloon, Hong Kong
[4] Department of Computer Science, The Hang Seng University of Hong Kong, Sha Tin, Hong Kong
[5] Centre for Artificial Intelligence and Robotics (CAIR), HKISI-CAS, Pak Shek Kok, Hong Kong
zhen.chen@cair-cas.org.hk

Abstract. Deep learning techniques have achieved promising performance for computer-aided diagnosis, which is beneficial to alleviate the workload of clinicians. However, due to the scarcity of diseased samples, medical image datasets suffer from an inherent imbalance, and lead diagnostic algorithms biased to majority categories. This degrades the diagnostic performance, especially in recognizing rare categories. Existing works formulate this challenge as long-tails and adopt decoupling strategies to mitigate the effect of the biased classifier. But these works only use the imbalanced dataset to train the encoder and resample data to re-train the classifier by discarding the samples of head categories, thereby restricting the diagnostic performance. To address these problems, we propose a Multi-view Relation-aware Consistency and Virtual Features Compensation (MRC-VFC) framework for long-tailed medical image classification in two stages. In the first stage, we devise a Multi-view Relation-aware Consistency (MRC) for representation learning, which provides the training of encoders with unbiased guidance in addition to the imbalanced supervision. In the second stage, to produce an impartial classifier, we propose the Virtual Features Compensation (VFC) to recalibrate the classifier by generating massive balanced virtual features. Compared with the resampling, VFC compensates the minority classes to optimize an unbiased classifier with preserving complete knowledge of the majority ones. Extensive experiments on two long-tailed public benchmarks confirm that our MRC-VFC framework remarkably outperforms state-of-the-art algorithms.

Keywords: Class Imbalance · Dermoscopy · Representation Learning

L. Pan and Y. Zhang—Equal contribution.

H. Greenspan et al. (Eds.): MICCAI 2023, LNCS 14225, pp. 14–23, 2023.
https://doi.org/10.1007/978-3-031-43987-2_2

1 Introduction

Recent years have witnessed the great success of deep learning techniques in various applications on computer-aided diagnosis [5,6,9,23]. However, the challenge of class imbalance inherently exists in medical datasets due to the scarcity of target diseases [7], where normal samples are significantly more than diseased samples. This challenge leads the model training biased to the majority categories [13] and severely impairs the performance of diagnostic models in real-world scenarios [8,18]. Therefore, it is urgent to improve the performance of diagnostic models in clinical applications, especially to achieve balanced recognition of minority categories.

Technically, the issue of class imbalance is formulated as a long-tailed problem in existing works [10,12,15], where a few head classes contain numerous samples while the tail classes comprise only a few instances [28]. To address this issue, most of the previous methods have typically attempted to rebalance the data distribution through under-sampling the head classes [2], over-sampling the tail classes [21], or reweighting the contribution of different classes during the optimization process [4,8]. Nevertheless, these resampling methods can encounter a decrease in performance on certain datasets since the total information volume of the dataset is either unchanged or even reduced [29]. Recent advantages in long-tailed medical image classification have been achieved by two-stage methods, which first train the model on the entire dataset and then fine-tune the classifier in the second stage using rebalancing techniques to counteract the class imbalance [12,14,15,19]. By decoupling the training of encoders and classifiers, the two-stage methods can recalibrate the biased classifiers and utilize all of the training samples to enhance representation learning for the encoder.

Although the aforementioned decoupling methods [14,15] have somewhat alleviated the long-tails, the classification performance degradation in the minority classes remains unsolved, which can be attributed to two challenges. First, in the first stage, the decoupling methods train the model on the imbalanced dataset, which is insufficient for representation learning in the rare classes due to the scarcity of samples [17]. To this end, improving the first-stage training strategy to render effective supervision on representation learning is in great demand. The second problem lies in the second stage, where decoupling methods freeze the pre-trained encoder and fine-tune the classifier [14,15]. Traditional rebalancing techniques, such as resampling and reweighting, are used by the decoupling methods to eliminate the bias in the classifier. However, these rebalancing strategies have intrinsic drawbacks, e.g., resampling-based methods discard the samples of head classes, and reweighting cannot eliminate the imbalance with simple coefficients [26]. Thus, a novel approach that can perform balanced classifier training by generating abundant features is desired to recalibrate the classifier and preserve the representation quality of the encoder.

To address the above two challenges, we propose the MRC-VFC framework that adopts the decoupling strategy to enhance the first-stage representation learning with Multi-view Relation-aware Consistency (MRC) and recalibrate the classifier using Virtual Features Compensation (VFC). Specifically, in the first

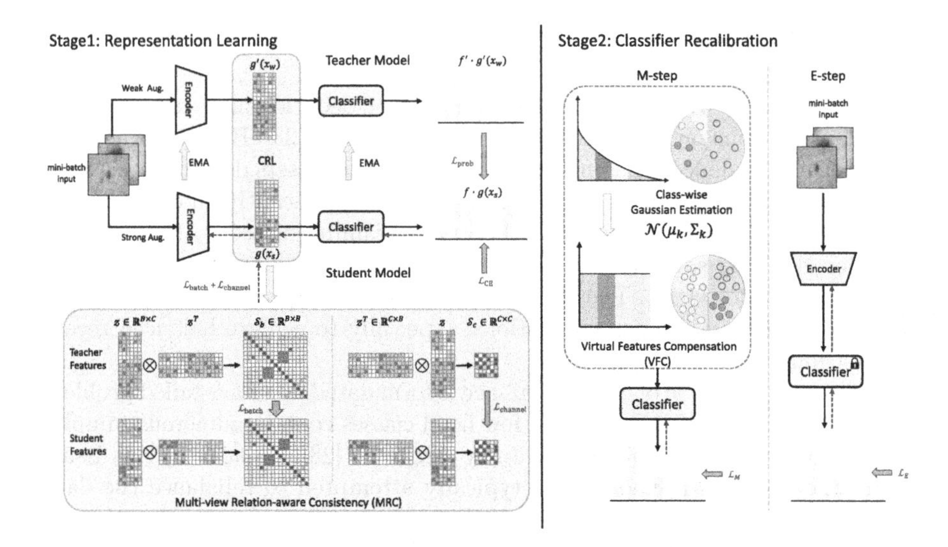

Fig. 1. The MRC-VFC framework. In stage 1, we perform the representation learning with the MRC module for the encoder on the imbalanced dataset. In stage 2, we recalibrate the classifier with VFC in two-step of the expectation and maximization.

stage, to boost the representation learning under limited samples, we build a two-stream architecture to perform representation learning with the MRC module, which encourages the model to capture semantic information from images under different data perturbations. In the second stage, to recalibrate the classifier, we propose to generate virtual features from multivariate Gaussian distribution with the expectation-maximization algorithm, which can compensate for tail classes and preserves the correlations among features. In this way, the proposed MRC-VFC framework can rectify the biases in the encoder and classifier, and construct a balanced and representative feature space to improve the performance for rare diseases. Experiments on two public dermoscopic datasets prove that our MRC-VFC framework outperforms state-of-the-art methods for long-tailed diagnosis.

2 Methodology

As illustrated in Fig. 1, our MRC-VFC framework follows the decoupling strategy [14,31] to combat the long-tailed challenges in two stages. In the first stage, we introduce the Multi-view Relation-aware Consistency (MRC) to boost representation learning for the encoder g. In the second stage, the proposed Virtual Features Compensation (VFC) recalibrates the classifier f by generating massive balanced virtual features, which compensates the tails classes without dropping the samples of the head classes. By enhancing the encoder with MRC and recalibrating the classifier with VFC, our MRC-VFC framework can perform effective and balanced training on long-tailed medical datasets.

2.1 Multi-view Relation-Aware Consistency

The representation learning towards the decoupling models is insufficient [28, 29]. To boost the representation learning, we propose the Multi-view Relation-aware Consistency to encourage the encoder to apprehend the inherent semantic features of the input images under different data augmentations. Specifically, we build a student neural network $f \cdot g$ for the strong augmented input $\boldsymbol{x}_s$ and duplicate a teacher model $f' \cdot g'$ for the weak augmented input $\boldsymbol{x}_w$. The two models are constrained by the MRC module to promote the consistency for different perturbations of the same input. The parameters of the teacher model are updated via an exponential moving average of the student parameters [24].

To motivate the student model to learn from the data representations but the ill distributions, we propose multi-view constraints on the consistency of two models at various phases. A straightforward solution is to encourage identical predictions for different augmentations of the same input image, as follows:

$$\mathcal{L}_{\text{prob}} = \frac{1}{B}\text{KL}(f \cdot g(\boldsymbol{x}_s),\ f' \cdot g'(\boldsymbol{x}_w)), \tag{1}$$

where $\text{KL}(\cdot,\cdot)$ refers to the Kullback-Leibler divergence to measure the difference between two outputs. As this loss function calculates the variance of classifier output, the supervision for the encoders is less effective. To this end, the proposed MRC measures the sample-wise and channel-wise similarity between the feature maps of two encoders to regularize the consistency of the encoders. We first define the correlations of individuals and feature channels as $\mathcal{S}_b(\boldsymbol{z}) = \boldsymbol{z} \cdot \boldsymbol{z}^\intercal$ and $\mathcal{S}_c(\boldsymbol{z}) = \boldsymbol{z}^\intercal \cdot \boldsymbol{z}$, where $\boldsymbol{z} = g(\boldsymbol{x}_s) \in \mathbb{R}^{B \times C}$ is the output features of the encoder, and B and C are the batch size and channel number. $\mathcal{S}_b(\boldsymbol{z})$ denotes the Gram matrix of feature $\boldsymbol{z}$, representing the correlations among individuals, and $\mathcal{S}_c(\boldsymbol{z})$ indicates the similarities across feature channels. Thus, the consistency between the feature maps of two models can be defined as:

$$\mathcal{L}_{\text{batch}} = \frac{1}{B}||\mathcal{S}_b(g(\boldsymbol{x}_s)) - \mathcal{S}_b(g'(\boldsymbol{x}_w))||_2, \tag{2}$$

$$\mathcal{L}_{\text{channel}} = \frac{1}{C}||\mathcal{S}_c(g(\boldsymbol{x}_s)) - \mathcal{S}_c(g'(\boldsymbol{x}_w))||_2. \tag{3}$$

Furthermore, we also adopt the cross-entropy loss $\mathcal{L}_{\text{CE}} = \frac{1}{B}L(f \cdot g(\boldsymbol{x}_w),\ y)$, where y denotes the ground truth, between the predictions and ground truth to ensure that the optimization will not be misled to a trivial solution. The overall loss function is summarized as $\mathcal{L}_{\text{stage1}} = \mathcal{L}_{\text{CE}} + \lambda_1\mathcal{L}_{\text{batch}} + \lambda_2\mathcal{L}_{\text{channel}} + \lambda_3\mathcal{L}_{\text{prob}}$, where λ_1, λ_2 and λ_3 are coefficients to control the trade-off of each loss term. By introducing extra semantic constraints, the MRC can enhance the representation capacity of encoders. The feature space generated by the encoders is more balanced with abundant semantics, thereby facilitating the MRC-VFC framework to combat long-tails in medical diagnosis.

2.2 Virtual Features Compensation

Recalling the introduction of decoupling methods, the two-stage methods [14] decouple the training of the encoder and classifier to eliminate the bias in the classifier while retaining the representation learning of the encoder. However, most existing decoupling approaches [12,15] employ the resampling strategy in the second stage to rebalance the data class distribution, causing the intrinsic drawbacks of the resampling of discarding the head class samples. To handle this issue, we propose Virtual Features Compensation, which generates virtual features $z_k \in \mathbb{R}^{N_k \times C}$ for each class k under multivariate Gaussian distribution [1] to combat the long-tailed problem. Different from existing resampling methods [2], the feature vectors produced by the VFC module preserve the correlations among classes and the semantic information from the encoder. Given the k-th class, we first calculate the class-wise Gaussian distribution with mean $\boldsymbol{\mu}_k$ and covariance $\boldsymbol{\Sigma}_k$, as follows:

$$\boldsymbol{\mu}_k = \frac{1}{N_k} \sum_{\boldsymbol{x} \in X_k} g^I(\boldsymbol{x}), \ \boldsymbol{\Sigma}_k = \frac{1}{N_k - 1} \sum_{\boldsymbol{x} \in X_k} (\boldsymbol{x} - \boldsymbol{\mu}_k)^\intercal (\boldsymbol{x} - \boldsymbol{\mu}_k), \tag{4}$$

where X_k denotes the set of all samples in the k-th class, and $g^I(\cdot)$ denotes the encoder trained in the first stage on the imbalanced dataset and N_k is the sample number of the k-th class. We then randomly sample R feature vectors for each category from the corresponding Gaussian distribution $\mathcal{N}(\boldsymbol{\mu}_k, \boldsymbol{\Sigma}_k)$ to build the unbiased feature space, as $\{V_k \in \mathbb{R}^{R \times C}\}_{k=1}^{K}$. We re-initialize the classifier and then calibrated it under cross-entropy loss, as follows:

$$\mathcal{L}_{\text{stage2}}^{M} = \frac{1}{RK} \sum_{k=1}^{K} \sum_{\boldsymbol{v}_i \in V_k} L_{\text{CE}}(f(\boldsymbol{v}_i), y), \tag{5}$$

where K is the number of categories in the dataset. As the Gaussian distribution is calculated according to the statistics from the first-stage feature space, to further alleviate the potential bias, we employ the expectation-maximization algorithm [20] to iteratively fine-tune the classifier and encoder. At the expectation step, we freeze the classifier and supervise the encoder with extra balancing constraints to avoid being re-contaminated by the long-tailed label space. Thus, we adopt the generalized cross-entropy (GCE) loss [30] for the expectation step as follows:

$$\mathcal{L}_{\text{stage2}}^{E} = \frac{1}{N} \sum_{\boldsymbol{x} \in X} \frac{(1 - (f \cdot g^I(\boldsymbol{x}) y)^q)}{q}, \tag{6}$$

where q is a hyper-parameter to control the trade-off between the imbalance calibration and the classification task. At the maximization step, we freeze the encoder and train the classifier on the impartial feature space. By enriching the semantic features with balanced virtual features, our MRC-VFC framework can improve the classification performance in long-tailed datasets, especially the performance of minority categories.

Table 1. Comparison with state-of-the-art algorithms on the *ISIC-2019-LT* dataset.

ISIC-2019-LT			
Methods	Acc (%) @ Factor = 100	Acc (%) @ Factor = 200	Acc (%) @ Factor = 500
CE	56.91	53.77	43.89
RS	61.41	55.12	47.76
MixUp [27]	59.85	54.23	43.11
GCE+SR [32]	64.57	58.28	54.36
Seesaw loss [26]	68.82	65.84	62.92
Focal loss [16]	67.54	65.93	61.66
CB loss [8]	67.54	66.70	61.89
FCD [15]	70.15	68.82	63.59
FS [12]	71.97	69.30	65.22
Ours *w/o* MRC	75.04	73.13	70.13
Ours *w/o* VFC	72.91	71.07	67.48
Ours	**77.41**	**75.98**	**74.62**

3 Experiments

3.1 Datasets

To evaluate the performance on long-tailed medical image classification, we construct two dermatology datasets from ISIC[1] [25] following [12]. In particular, we construct the *ISIC-2019-LT* dataset as the long-tailed version of ISIC 2019 challenge[2], which includes 8 diagnostic categories of dermoscopic images. We sample the subset from Pareto distribution [8] as $N_c = N_0(r^{-(k-1)})^c$, where the imbalance factor $r = N_0/N_{k-1}$ is defined as the sample number of the head class N_0 divided by the tail one N_{k-1}. We adopt three imbalance factors for *ISIC-2019-LT*, as $r = \{100, 200, 500\}$. Furthermore, the *ISIC-Archive-LT* dataset [12] is sampled from ISIC Archive with a larger imbalance factor $r \approx 1000$ and contains dermoscopic images of 14 classes. We randomly split these two datasets into train, validation and test sets as 7:1:2.

3.2 Implementation Details

We implement the proposed MRC-VFC framework with the PyTorch library [22], and employ ResNet-18 [11] as the encoder for both long-tailed datasets. All the experiments are done on four NVIDIA GeForce GTX 1080 Ti GPUs with a batch size of 128. All images are resized to 224×224 pixels. In the first stage of MRC-VFC, we train the model using Stochastic Gradient Descent (SGD) with a learning rate of 0.01. For the strong augmentation [3], we utilize the random flip, blur, rotate, distortion, color jitter, grid dropout, and normalization, and adopt the random flip and the same normalization for the weak augmentation. In the second stage, we use SGD with a learning rate of 1×10^{-5} for optimizing the classifier and 1×10^{-6} for optimizing the encoder, respectively. The loss weights

[1] https://www.isic-archive.com/.
[2] https://challenge.isic-archive.com/landing/2019/.

Table 2. Comparison with state-of-the-art algorithms on the *ISIC-Archive-LT* dataset.

ISIC-Archive-LT				
Methods	Head (Acc%)	Medium (Acc%)	Tail (Acc%)	All (Acc%)
CE	**71.31**	49.22	38.17	52.90
RS	70.17	55.29	34.29	53.25
GCE+SR [32]	64.93	57.26	38.22	53.47
Seesaw loss [26]	70.26	55.98	42.14	59.46
Focal loss [16]	69.57	56.21	39.65	57.81
CB loss [8]	64.98	57.01	61.61	61.20
FCD [15]	66.39	61.17	60.54	62.70
FS [12]	68.69	58.74	64.48	63.97
Ours *w/o* MRC	69.06	62.14	65.12	65.44
Ours *w/o* VFC	65.11	62.35	67.30	64.92
Ours	69.71	**63.47**	**70.34**	**67.84**

λ_1, λ_2 and λ_3 in the first stage are set as 10, 10 and 5, and the q in the second stage is set as 0.8. We set training epochs as 100 for the first stage and 500 for the second stage. The source code is available at https://github.com/jhonP-Li/MRC_VFC.

3.3 Comparison on ISIC-2019-LT Dataset

We evaluate the performance of our MRC-VFC framework with state-of-the-art methods for long-tailed medical image classification, including (i) baselines: fine-tuning classification models with cross-entropy loss (CE), random data resampling methods (RS), and MixUp [27]; (ii) recent loss reweighting methods: Generalized Cross-Entropy with Sparse Regularization (GCE+SR) [32], Seesaw loss [26], focal loss [16], and Class-Balancing (CB) loss [8]; (iii) recent works for long-tailed medical image classification: Flat-aware Cross-stage Distillation (FCD) [15], and Flexible Sampling (FS) [12].

As illustrated in Table 1, we compare our MRC-VFC framework with the aforementioned methods on the *ISIC-2019-LT* dataset under different imbalance factors. Among these methods, our MRC-VFC framework achieves the best performance with an accuracy of 77.41%, 75.98%, and 74.62% under the imbalance factor of 100, 200, and 500, respectively. Noticeably, compared with the state-of-the-art decoupling method FCD [15] on long-tailed medical image classification, our MRC-VFC framework surpasses it by a large margin of 11.03% accuracy when the imbalance factor is 500, demonstrating the effectiveness of representation learning and virtual features compensation in our framework. Furthermore, our MRC-VFC framework outperforms FS [12], which improves the resampling strategy and achieves the best performance on the *ISIC-2019-LT* dataset, with an accuracy increase of 9.4% under imbalance factor = 500. These experimental results demonstrate the superiority of our MRC-VFC framework over existing approaches in long-tailed medical image classification tasks.

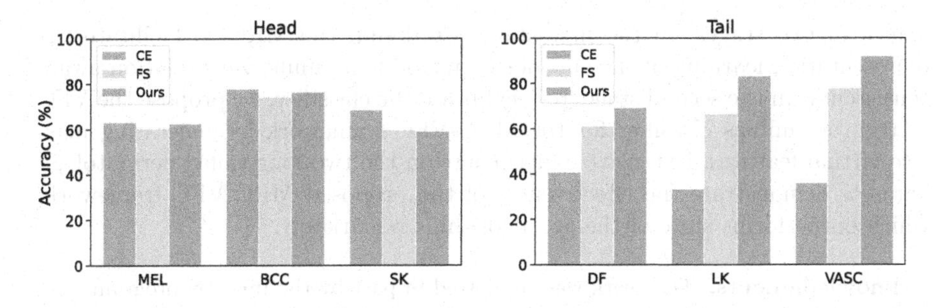

Fig. 2. The performance comparison of head/tail classes in *ISIC-Archive-LT* dataset.

Ablation Study. We perform the ablation study to validate the effectiveness of our proposed MRC and VFC modules on two long-tailed datasets. As shown in Table 1 and 2, both MRC and VFC modules remarkably improve the performance over the baselines. In particular, we apply two ablative baselines of the proposed MRC-VFC framework by disabling the MRC (denoted as Ours *w/o* MRC) and the VFC (denoted as Ours *w/o* VFC) individually. In detail, as shown in Table 1, when the imbalance factor is 500, the accuracy increases by 4.49% and 7.14% for MRC and VFC, respectively. In addition, as illustrated in Table 2, the mean accuracy of all classes in the *ISIC-Archive-LT* shows an improvement of 2.40% and 2.92% for MRC and VFC correspondingly. The ablation study verifies the effectiveness of our MRC and VFC modules.

3.4 Comparison on ISIC-Archive-LT Dataset

To comprehensively evaluate our MRC-VFC framework, we further perform the comparison with state-of-the-art algorithms on a more challenging *ISIC-Archive-LT* dataset for long-tailed diagnosis. As illustrated in Table 2, our MRC-VFC framework achieves the best overall performance with an accuracy of 67.84% among state-of-the-art algorithms, and results in a balanced performance over different classes, *i.e.*, 69.71% for head classes and 70.34% for tail classes. Compared with the advanced decoupling method [15] for medical image diagnosis, our MRC-VFC framework significantly improves the accuracy with 4.73% in medium classes and 8.87% in tail classes, respectively.

Performance Analysis on Head/Tail Classes. We further present the performance of several head and tail classes in Fig. 2. Our MRC-VFC framework outperforms FS [12] on both tail and head classes, and significantly promotes the performance of tail classes, thereby effectively alleviating the affect of long-tailed problems on medical image diagnosis. These comparisons confirm the advantage of our MRC-VFC framework in more challenging long-tailed scenarios.

4 Conclusion

To address the long-tails in computer-aided diagnosis, we propose the MRC-VFC framework to improve medical image classification with balanced performance in two stages. In the first stage, we design the MRC to facilitate the representation learning of the encoder by introducing multi-view relation-aware consistency. In the second stage, to recalibrate the classifier, we propose the VFC to train an unbias classifier for the MRC-VFC framework by generating massive virtual features. Extensive experiments on the two long-tailed dermatology datasets demonstrate the effectiveness of the proposed MRC-VFC framework, which outperforms state-of-the-art algorithms remarkably.

Acknowledgments. This work was supported in part by the InnoHK program.

References

1. Ahrendt, P.: The multivariate gaussian probability distribution. Technical report, Technical University of Denmark, p. 203 (2005)
2. Buda, M., Maki, A., Mazurowski, M.A.: A systematic study of the class imbalance problem in convolutional neural networks. Neural Netw. **106**, 249–259 (2018)
3. Buslaev, A., Iglovikov, V.I., Khvedchenya, E., Parinov, A., Druzhinin, M., Kalinin, A.A.: Albumentations: fast and flexible image augmentations. Information **11**(2), 125 (2020)
4. Cao, K., Wei, C., Gaidon, A., Arechiga, N., Ma, T.: Learning imbalanced datasets with label-distribution-aware margin loss. In: NeurIPS, vol. 32 (2019)
5. Chen, Z., Guo, X., Woo, P.Y., Yuan, Y.: Super-resolution enhanced medical image diagnosis with sample affinity interaction. IEEE Trans. Med. Imaging **40**(5), 1377–1389 (2021)
6. Chen, Z., Guo, X., Yang, C., Ibragimov, B., Yuan, Y.: Joint spatial-wavelet dual-stream network for super-resolution. In: Martel, A.L., et al. (eds.) MICCAI 2020. LNCS, vol. 12265, pp. 184–193. Springer, Cham (2020). https://doi.org/10.1007/978-3-030-59722-1_18
7. Chen, Z., Yang, C., Zhu, M., Peng, Z., Yuan, Y.: Personalized retrogress-resilient federated learning toward imbalanced medical data. IEEE Trans. Med. Imaging **41**(12), 3663–3674 (2022)
8. Cui, Y., Jia, M., Lin, T.Y., Song, Y., Belongie, S.: Class-balanced loss based on effective number of samples. In: CVPR, pp. 9268–9277 (2019)
9. De Fauw, J., et al.: Clinically applicable deep learning for diagnosis and referral in retinal disease. Nat. Med. **24**(9), 1342–1350 (2018)
10. Esteva, A., et al.: Dermatologist-level classification of skin cancer with deep neural networks. Nature **542**(7639), 115–118 (2017)
11. He, K., Zhang, X., Ren, S., Sun, J.: Deep residual learning for image recognition. In: CVPR, pp. 770–778 (2016)
12. Ju, L., et al.: Flexible sampling for long-tailed skin lesion classification. In: Wang, L., Dou, Q., Fletcher, P.T., Speidel, S., Li, S. (eds.) MICCAI 2022. LNCS, vol. 13433, pp. 462–471. Springer, Cham (2022). https://doi.org/10.1007/978-3-031-16437-8_44
13. Kang, B., Li, Y., Xie, S., Yuan, Z., Feng, J.: Exploring balanced feature spaces for representation learning. In: ICLR (2021)
14. Kang, B., et al.: Decoupling representation and classifier for long-tailed recognition. In: ICLR (2020)

15. Li, J., et al.: Flat-aware cross-stage distilled framework for imbalanced medical image classification. In: Wang, L., Dou, Q., Fletcher, P.T., Speidel, S., Li, S. (eds.) MICCAI 2022. LNCS, vol. 13433, pp. 217–226. Springer, Cham (2022). https://doi.org/10.1007/978-3-031-16437-8_21
16. Lin, T.Y., Goyal, P., Girshick, R., He, K., Dollár, P.: Focal loss for dense object detection. In: ICCV, pp. 2980–2988 (2017)
17. Liu, J., Sun, Y., Han, C., Dou, Z., Li, W.: Deep representation learning on long-tailed data: a learnable embedding augmentation perspective. In: CVPR, pp. 2970–2979 (2020)
18. Liu, Z., Miao, Z., Zhan, X., Wang, J., Gong, B., Yu, S.X.: Large-scale long-tailed recognition in an open world. In: CVPR, pp. 2537–2546 (2019)
19. Luo, M., Chen, F., Hu, D., Zhang, Y., Liang, J., Feng, J.: No fear of heterogeneity: classifier calibration for federated learning with non-IID data. In: NeurIPS, vol. 34, pp. 5972–5984 (2021)
20. Moon, T.K.: The expectation-maximization algorithm. IEEE Sig. Process. Mag. **13**(6), 47–60 (1996)
21. More, A.: Survey of resampling techniques for improving classification performance in unbalanced datasets. arXiv preprint arXiv:1608.06048 (2016)
22. Paszke, A., et al.: PyTorch: an imperative style, high-performance deep learning library. In: NeurIPS, vol. 32 (2019)
23. Srinidhi, C.L., Ciga, O., Martel, A.L.: Deep neural network models for computational histopathology: a survey. Med. Image Anal. **67**, 101813 (2021)
24. Tarvainen, A., Valpola, H.: Mean teachers are better role models: weight-averaged consistency targets improve semi-supervised deep learning results. In: Advances in Neural Information Processing Systems, vol. 30 (2017)
25. Tschandl, P., Rosendahl, C., Kittler, H.: The HAM10000 dataset, a large collection of multi-source dermatoscopic images of common pigmented skin lesions. Sci. Data **5**(1), 1–9 (2018)
26. Wang, J., et al.: Seesaw loss for long-tailed instance segmentation. In: CVPR, pp. 9695–9704 (2021)
27. Zhang, H., Cisse, M., Dauphin, Y.N., Lopez-Paz, D.: mixup: beyond empirical risk minimization. arXiv preprint arXiv:1710.09412 (2017)
28. Zhang, Y., Kang, B., Hooi, B., Yan, S., Feng, J.: Deep long-tailed learning: a survey. arXiv preprint arXiv:2110.04596 (2021)
29. Zhang, Y., Wei, X.S., Zhou, B., Wu, J.: Bag of tricks for long-tailed visual recognition with deep convolutional neural networks. In: AAAI, vol. 35, pp. 3447–3455 (2021)
30. Zhang, Z., Sabuncu, M.: Generalized cross entropy loss for training deep neural networks with noisy labels. In: NeurIPS, vol. 31 (2018)
31. Zhou, B., Cui, Q., Wei, X.S., Chen, Z.M.: BBN: bilateral-branch network with cumulative learning for long-tailed visual recognition. In: CVPR, pp. 9719–9728 (2020)
32. Zhou, X., Liu, X., Wang, C., Zhai, D., Jiang, J., Ji, X.: Learning with noisy labels via sparse regularization. In: ICCV, pp. 72–81 (2021)

Towards Novel Class Discovery: A Study in Novel Skin Lesions Clustering

Wei Feng[1,2,4,6], Lie Ju[1,2,4,6], Lin Wang[1,2,4,6], Kaimin Song[6], and Zongyuan Ge[1,2,3,4,5](✉)

[1] Faculty of Engineering, Monash University, Melbourne, Australia
zongyuan.ge@monash.edu
[2] Monash Medical AI Group, Monash University, Melbourne, Australia
[3] AIM for Health Lab, Monash University, Melbourne, VIC, Australia
[4] Airdoc-Monash Research Lab, Monash University, Suzhou, China
[5] Faculty of IT, Monash University, Melbourne, VIC, Australia
[6] Airdoc LLC, Beijing, China
https://www.monash.edu/mmai-group

Abstract. Existing deep learning models have achieved promising performance in recognizing skin diseases from dermoscopic images. However, these models can only recognize samples from predefined categories, when they are deployed in the clinic, data from new unknown categories are constantly emerging. Therefore, it is crucial to automatically discover and identify new semantic categories from new data. In this paper, we propose a new novel class discovery framework for automatically discovering new semantic classes from dermoscopy image datasets based on the knowledge of known classes. Specifically, we first use contrastive learning to learn a robust and unbiased feature representation based on all data from known and unknown categories. We then propose an uncertainty-aware multi-view cross pseudo-supervision strategy, which is trained jointly on all categories of data using pseudo labels generated by a self-labeling strategy. Finally, we further refine the pseudo label by aggregating neighborhood information through local sample similarity to improve the clustering performance of the model for unknown categories. We conducted extensive experiments on the dermatology dataset ISIC 2019, and the experimental results show that our approach can effectively leverage knowledge from known categories to discover new semantic categories. We also further validated the effectiveness of the different modules through extensive ablation experiments. Our code will be released soon.

Keywords: Novel Class Discovery · Skin Lesion Recognition · Deep Learning

1 Introduction

Automatic identification of lesions from dermoscopic images is of great importance for the diagnosis of skin cancer [16,22]. Currently, deep learning mod-

Supplementary Information The online version contains supplementary material available at https://doi.org/10.1007/978-3-031-43987-2_3.

H. Greenspan et al. (Eds.): MICCAI 2023, LNCS 14225, pp. 24–33, 2023.
https://doi.org/10.1007/978-3-031-43987-2_3

els, especially those based on deep convolution neural networks, have achieved remarkable success in this task [17,18,22]. However, this comes at the cost of a large amount of labeled data that needs to be collected for each class. To alleviate the labeling burden, semi-supervised learning has been proposed to exploit a large amount of unlabeled data to improve performance in the case of limited labeled data [10,15,19]. However, it still requires a small amount of labeled data for each class, which is often impossible in real practice. For example, there are roughly more than 2000 named dermatological diseases today, of which more than 200 are common, and new dermatological diseases are still emerging, making it impractical to annotate data from scratch for each new disease category [20]. However, since there is a correlation between new and known diseases, a priori knowledge from known diseases is expected to help automatically identify new diseases [9].

One approach to address the above problem is novel class discovery (NCD) [7,9,24], which aims to transfer knowledge from known classes to discover new semantic classes. Most NCD methods follow a two-stage scheme: 1) a stage of fully supervised training on known category data and 2) a stage of clustering on unknown categories [7,9,24]. For example, Han et al. [9] further introduced self-supervised learning in the first stage to learn general feature representations. They also used ranking statistics to compute pairwise similarity for clustering. Zhong et al. [24] proposed OpenMix based on the mixup strategy [21] to further exploit the information from known classes to improve the performance of unsupervised clustering. Fini et al. [7] proposed UNO, which unifies multiple objective functions into a holistic framework to achieve better interaction of information between known and unknown classes. Zhong et al. [23] used neighborhood information in the embedding space to learn more discriminative representations. However, most of these methods require the construction of a pairwise similarity prediction task to perform clustering based on pairwise similarity pseudo labels between samples. In this process, the generated pseudo labels are usually noisy, which may affect the clustering process and cause error accumulation. In addition, they only consider the global alignment of samples to the category center, ignoring the local inter-sample alignment thus leading to poor clustering performance.

In this paper, we propose a new novel class discovery framework to automatically discover novel disease categories. Specifically, we first use contrastive learning to pretrain the model based on all data from known and unknown categories to learn a robust and general semantic feature representation. Then, we propose an uncertainty-aware multi-view cross-pseudo-supervision strategy to perform clustering. It first uses a self-labeling strategy to generate pseudo-labels for unknown categories, which can be treated homogeneously with ground truth labels. The cross-pseudo-supervision strategy is then used to force the model to maintain consistent prediction outputs for different views of unlabeled images. In addition, we propose to use prediction uncertainty to adaptively adjust the contribution of the pseudo labels to mitigate the effects of noisy pseudo labels. Finally, to encourage local neighborhood alignment and further refine the pseudo

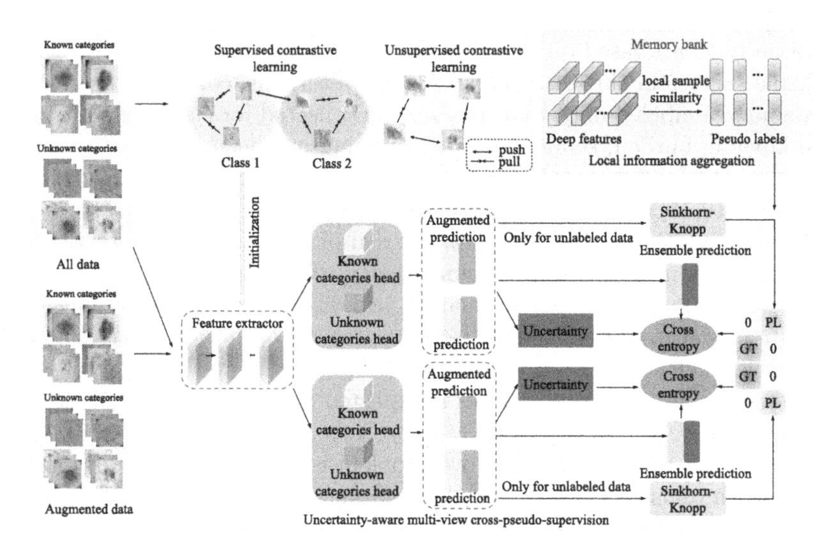

Fig. 1. The overall framework of our proposed novel class discovery algorithm.

labels, we propose a local information aggregation module to aggregate the information of the neighborhood samples to boost the clustering performance. We conducted extensive experiments on the dermoscopy dataset ISIC 2019, and the experimental results show that our method outperforms other state-of-the-art comparison algorithms by a large margin. In addition, we also validated the effectiveness of different components through extensive ablation experiments.

2 Methodology

Given an unlabeled dataset $\{x_i^u\}_{i=1}^{N^u}$ with N^u images, where x_i^u is the ith unlabeled image. Our goal is to automatically cluster the unlabeled data into C^u clusters. In addition, we also have access to a labeled dataset $\{x_i^l, y_i^l\}_{i=1}^{N^l}$ with N^l images, where x_i^l is the ith labeled image and $y_i^l \in \mathcal{Y} = \{1, \ldots, C^l\}$ is its corresponding label. In the novel class discovery task, the known and unknown classes are disjoint, i.e., $C^l \cap C^u = \varnothing$. However, the known and unknown classes are similar, and we aim to use the knowledge of the known classes to help the clustering of the unknown classes. The overall framework of our proposed novel class discovery algorithm is shown in Fig. 1. Specifically, we first learn general and robust feature representations through contrastive learning. Then, the uncertainty-aware multi-view cross-pseudo-supervision strategy is used for joint training on all category data. Finally, the local information aggregation module benefits the NCD by aggregating the useful information of the neighborhood samples.

Contrastive Learning. To achieve a robust feature representation for the NCD task, we first use noise contrastive learning [8] to pretrain the feature extractor network, which effectively avoids model over-fitting to known categories. Specifically, we use x_i and x_i' to represent different augmented versions of the same image in a mini-batch. The unsupervised contrastive loss can be formulated as:

$$L_i^{ucl} = -\log \frac{\exp\left(z_i \cdot z_i'/\tau\right)}{\sum_n \mathbb{1}_{[n \neq i]} \exp\left(z_i \cdot z_n/\tau\right)} \tag{1}$$

where $z_i = E(x_i)$ is the deep feature representation of the image x_i, E is the feature extractor network, and τ is the temperature value. $\mathbb{1}$ is the indicator function.

In addition, to help the feature extractor learn semantically meaningful feature representations, we introduce supervised contrastive learning [12] for labeled known category data, which can be denoted as:

$$L_i^{scl} = -\frac{1}{|N(i)|} \sum_{q \in N(i)} \log \frac{\exp\left(z_i \cdot z_q/\tau\right)}{\sum_n \mathbb{1}_{[n \neq i]} \exp\left(z_i \cdot z_n/\tau\right)} \tag{2}$$

where $N(i)$ represents the sample set with the same label as x_i in a mini-batch data. $|N(i)|$ represents the number of samples.

The overall contrastive loss can be expressed as: $L_{cl} = (1-\mu)\sum_{i \in B} L_i^{ucl} + \mu \sum_{i \in B_l} L_i^{scl}$, where μ denotes the balance coefficient. B_l is the labeled subset of mini-batch data.

Uncertainty-Aware Multi-view Cross-Pseudo-Supervision. We now describe how to train uniformly on known and unknown categories using the uncertainty-aware multi-view cross-pseudo-supervision strategy. Specifically, we construct two parallel classification models M_1 and M_2, both of them composed of a feature extractor and two category classification heads, using different initialization parameters. For an original image x_i, we generate two augmented versions of x_i, x_i^{v1} and x_i^{v2}. We then feed these two augmented images into M_1 and M_2 to obtain the predictions for x_i^{v1} and x_i^{v2}:

$$p_{i,1}^{v1} = M_1(x_i^{v1}), p_{i,1}^{v2} = M_1(x_i^{v2}), p_{i,2}^{v1} = M_2(x_i^{v1}), p_{i,2}^{v2} = M_2(x_i^{v2}). \tag{3}$$

The prediction outputs are obtained by concatenating the outputs of the two classification heads and then passing a softmax layer [7]. Then, we can compute the ensemble predicted output of M_1 and M_2: $p_i^{M_1} = \left(p_{i,1}^{v1} + p_{i,1}^{v2}\right)/2$, $p_i^{M_2} = \left(p_{i,2}^{v1} + p_{i,2}^{v2}\right)/2$.

Next, we need to obtain training targets for all data. For an input image x_i, if x_i is from the known category, we construct the training target as one hot vector, where the first C^l elements are ground truth labels and the last C^u elements are 0. If x_i is from the unknown category, we set the first C^l elements to 0 and use pseudo labels for the remaining C^u elements.

We follow the self-labeling method in [1,3] to generate pseudo labels. Specifically, the parameters in the unknown category classification head can be viewed as prototypes of each category, and our training goal is to distribute a set of samples uniformly to each prototype while maximizing the similarity between samples and prototypes [1]. Let $\mathbf{P} = \left[p_1^u; \ldots; p_{B_u}^u\right] \in \mathbb{R}^{B_u \times C^u}$ denotes the ensemble prediction of data of unknown categories in a mini-batch, where B_u represents the number of samples. Here we only consider the output of the unknown categories head due to the samples coming from unknown categories [7]. We obtain the pseudo label by optimizing the following objective:

$$\max_{\mathbf{Y} \in \mathcal{S}} \operatorname{tr}\left(\mathbf{Y}\mathbf{P}^{\top}\right) + \delta H(\mathbf{Y}) \tag{4}$$

where $\mathbf{Y} = \left[y_1^u; \ldots; y_{B_u}^u\right] \in \mathbb{R}^{B_u \times C^u}$ will assign B_u unknown category samples to C^u category prototypes uniformly, i.e., each category prototype will be selected B_u/C^u times on average. $\mathcal{S}$ is the search space. H is the entropy function used to control the smoothness of $\mathbf{Y}$. δ is the hyperparameter. The solution to this objective can be calculated by the Sinkhorn-Knopp algorithm [6]. After generating the pseudo-labels, we can combine them with the ground truth labels of known categories as training targets for uniform training.

To mitigate the effect of noisy pseudo labels, we propose to use prediction uncertainty [14] to adaptively adjust the weights of pseudo labels. Specifically, we first compute the variance of the predicted outputs of the models for the different augmented images via KL-divergence:

$$V_1 = E\left[p_{i,1}^{v1} \log\left(\frac{p_{i,1}^{v1}}{p_{i,1}^{v2}}\right)\right], V_2 = E\left[p_{i,2}^{v1} \log\left(\frac{p_{i,2}^{v1}}{p_{i,2}^{v2}}\right)\right], \tag{5}$$

where E represents the expected value. If the variance of the model's predictions for different augmented images is large, the pseudo label may be of low quality, and vice versa. Then, based on the prediction variance of the two models, the multi-view cross-pseudo supervision loss can be formulated as:

$$L_{cps} = E\left[e^{-V_1} L_{ce}\left(p^{M_2}, y^{v1}\right) + V_1\right] + E\left[e^{-V_2} L_{ce}\left(p^{M_1}, y^{v2}\right) + V_2\right] \tag{6}$$

where L_{ce} denotes the cross-entropy loss. y^{v1} and y^{v2} are the training targets.

Local Information Aggregation. After the cross-pseudo-supervision training described above, we are able to assign the instances to their corresponding clustering centers. However, it ignores the alignment between local neighborhood samples, i.e., the samples are susceptible to interference from some irrelevant semantic factors such as background and color. Here, we propose a local information aggregation to enhance the alignment of local samples. Specifically, as shown in Fig. 1, we maintain a first-in-first-out memory bank $\mathcal{M} = \left\{z_k^m, y_k^m\right\}_{k=1}^{N^m}$ during the training process, which contains the features of N^m most recent samples and their pseudo labels. For each sample in the current batch, we compute the similarity between its features and the features of each sample in the memory bank:

$$d_k = \frac{\exp\left(z \cdot z_k^m\right)}{\sum_{k=1}^{N^m} \exp\left(z \cdot z_k^m\right)}. \tag{7}$$

Then based on this feature similarity, we obtain the final pseudo labels as: $y^u = \rho y^u + (1-\rho) \sum_{k=1}^{N^m} d_k y_k^m$, where ρ is the balance coefficient. By aggregating the information of the neighborhood samples, we are able to ensure consistency between local samples, which further improves the clustering performance.

3 Experiments

Dataset. To validate the effectiveness of the proposed algorithm, we conduct experiments on the widely used public dermoscopy challenge dataset ISIC 2019 [4,5]. The dataset contains a total of 25,331 dermoscopic images from eight categories: Melanoma (MEL), Melanocytic Nevus (NV), Basal Cell Carcinoma (BCC), Actinic Keratosis (AK), Benign Keratosis (BKL), Dermatofibroma (DF), Vascular Lesion (VASC), and Squamous Cell Carcinoma (SCC). Since the dataset suffers from severe category imbalance, we randomly sampled 500 samples from those major categories (MEL, NV, BCC, BKL) to maintain category balance. Then, we construct the NCD task where we treat 50% of the categories (AK, MEL, NV, BCC) as known categories and the remaining 50% of the categories (BKL, SCC, DF, VASC) as unknown categories. We also swap the known and unknown categories to form a second NCD task. For task 1 and task 2, we report the average performance of 5 runs.

Implementation Details. We used ResNet-18 [11] as the backbone of the classification model. The known category classification head is an *l2*-normalized linear classifier with C^l output units. The unknown category classification head consists of a projection layer with 128 output units, followed by an *l2*-normalized linear classifier with C^u output units. In the first contrastive learning pre-training step, we used SGD optimizer to train the model for 200 epochs and gradually decay the learning rate starting from 0.1 and dividing it by 5 at the epochs 60, 120, and 180. μ is set to 0.5, τ is set to 0.5. In the joint training phase, we fix the parameters of the previous feature extractor and only fine-tune the parameters of the classification head. We use the SGD optimizer to train the model for 200 epochs with linear warm-up and cosine annealing ($lr_{\text{base}} = 0.1$, $lr_{\text{min}} = 0.001$), and the weight decay is set to 1.5×10^{-4}. For data augmentation, we use random horizontal/vertical flipping, color jitter, and Gaussian blurring following [7]. For pseudo label, we use the Sinkhorn-Knopp algorithm with hyperparameters inherited from [7]: $\delta = 0.05$ and the number of iterations is 3. We use a memory bank $\mathcal{M}$ of size 100 and the hyperparameter ρ is set to 0.6. The batch size in all experiments is 512. In the inference phase, we only use the output of the unknown category classification head of M_1 [9]. Following [9,23,24], we report the clustering performance on the unlabeled unknown category dataset. We assume that the number of unknown categories is known and it can also be obtained by the category number estimation method proposed in [9].

Table 1. Clustering performance of different comparison algorithms on different tasks.

Method	Task1			Task2		
	ACC	NMI	ARI	ACC	NMI	ARI
Baseline	0.4685	0.2107	0.1457	0.3899	0.0851	0.0522
RankStats [9]	0.5652	0.2571	0.2203	0.4284	0.1164	0.1023
RankStats+ [9]	0.5845	0.2633	0.2374	0.4362	0.1382	0.1184
OpenMix [24]	0.6083	0.2863	0.2512	0.4684	0.1519	0.1488
NCL [23]	0.5941	0.2802	0.2475	0.4762	0.1635	0.1573
UNO [7]	0.6131	0.3016	0.2763	0.4947	0.1692	0.1796
Ours	**0.6654**	**0.3372**	**0.3018**	**0.5271**	**0.1826**	**0.2033**

Following [2,9], we use the average clustering accuracy (ACC), normalized mutual information (NMI) and adjusted rand index (ARI) to evaluate the clustering performance of different algorithms. Specifically, we first match the clustering assignment and ground truth labels by the Hungarian algorithm [13]. After the optimal assignment is determined, we then compute each metric. We implement all algorithms based on the PyTorch framework and conduct experiments on 8 RTX 3090 GPUs.

Comparison with State-of-the-Art Methods. We compare our algorithms with some state-of-the-art NCD methods, including RankStats [9], RankStats+ (RankStats with incremental learning) [9], OpenMix [24], NCL [23], UNO [7]. we also compare with the benchmark method (Baseline), which first trains a model using known category data and then performs clustering on unknown category data. Table 1 shows the clustering performance of each comparison algorithm on different NCD tasks. It can be seen that the clustering performance of the benchmark method is poor, which indicates that the model pre-trained using only the known category data does not provide a good clustering of the unknown category. Moreover, the state-of-the-art NCD methods can improve the clustering performance, which demonstrates the effectiveness of the currently popular two-stage solution. However, our method outperforms them, mainly due to the fact that they need to generate pairwise similarity pseudo labels through features obtained based on self-supervised learning, while ignoring the effect of noisy pseudo labels. Compared with the best comparison algorithm UNO, our method yields 5.23% ACC improvement, 3.56% NMI improvement, and 2.55% ARI improvement on Task1, and 3.24% ACC improvement, 1.34% NMI improvement, and 2.37% ARI improvement on Task2, which shows that our method is able to provide more reliable pseudo labels for NCD.

Ablation Study of Each Key Component. We performed ablation experiments to verify the effectiveness of each component. As shown in Table 2, CL is contrastive learning, UMCPS is uncertainty-aware multi-view cross-pseudo-supervision, and LIA is the local information aggregation module. It can be observed that CL brings a significant performance gain, which indicates that

Table 2. Ablation study of each key component.

Method			Task1			Task2		
CL	UMCPS	LIA	ACC	NMI	ARI	ACC	NMI	ARI
✗	✗	✗	0.4685	0.2107	0.1457	0.3899	0.0851	0.0522
✓			0.5898	0.2701	0.2375	0.4402	0.1465	0.1322
✓	✓		0.6471	0.3183	0.2821	0.5012	0.1732	0.1851
✓		✓	0.6255	0.3122	0.2799	0.4893	0.1688	0.1781
✓	✓	✓	**0.6654**	**0.3372**	**0.3018**	**0.5271**	**0.1826**	**0.2033**

Table 3. Ablation study of contrastive learning and uncertainty-aware multi-view cross-pseudo-supervision.

Method	Task1			Task2		
	ACC	NMI	ARI	ACC	NMI	ARI
Baseline	0.4685	0.2107	0.1457	0.3899	0.0851	0.0522
SCL	0.5381	0.2362	0.1988	0.4092	0.1121	0.1003
UCL	0.5492	0.2482	0.2151	0.4291	0.1173	0.1174
SCL+UCL	**0.5898**	**0.2701**	**0.2375**	**0.4402**	**0.1465**	**0.1322**
w/o CPS	0.6021	0.2877	0.2688	0.4828	0.1672	0.1629
CPS	0.6426	0.3201	0.2917	0.5082	0.1703	0.1902
UMCPS	**0.6654**	**0.3372**	**0.3018**	**0.5271**	**0.1826**	**0.2033**

contrastive learning helps to learn a general and robust feature representation for NCD. In addition, UMCPS also improves the clustering performance of the model, which indicates that unified training helps to the category information interaction. LIA further improves the clustering performance, which indicates that local information aggregation helps to provide better pseudo labels. Finally, our algorithm incorporates each component to achieve the best performance.

Ablation Study of Contrastive Learning. We further examined the effectiveness of each component in contrastive learning. Recall that the contrastive learning strategy includes supervised contrastive learning for the labeled known category data and unsupervised contrastive learning for all data. As shown in Table 3, it can be observed that both components improve the clustering performance of the model, which indicates that SCL helps the model to learn semantically meaningful feature representations, while UCL makes the model learn robust unbiased feature representations and avoid its overfitting to known categories.

Uncertainty-Aware Multi-view Cross-Pseudo-Supervision. We also examine the effectiveness of uncertainty-aware multi-view cross-pseudo-supervision. We compare it with 1) w/o CPS, which does not use cross-pseudo-supervision, and 2) CPS, which uses cross-pseudo-supervision but not the uncertainty to control the contribution of the pseudo label. As shown in Table 3, it can be seen that CPS outperforms w/o CPS, which indicates that CPS encourages the model to maintain consistent predictions for different augmented versions of the input images, and enhances the generalization performance of the model. UMCPS achieves the best clustering performance, which shows its ability to use uncertainty to alleviate the effect of noisy pseudo labels and avoid causing error accumulation.

4 Conclusion

In this paper, we propose a novel class discovery framework for discovering new dermatological classes. Our approach consists of three key designs. First, contrastive learning is used to learn a robust feature representation. Second, uncertainty-aware multi-view cross-pseudo-supervision strategy is trained uniformly on data from all categories, while prediction uncertainty is used to alleviate the effect of noisy pseudo labels. Finally, the local information aggregation module further refines the pseudo label by aggregating the neighborhood information to improve the clustering performance. Extensive experimental results validate the effectiveness of our approach. Future work will be to apply this framework to other medical image analysis tasks.

References

1. Asano, Y.M., Rupprecht, C., Vedaldi, A.: Self-labelling via simultaneous clustering and representation learning. arXiv preprint arXiv:1911.05371 (2019)
2. Cao, K., Brbic, M., Leskovec, J.: Open-world semi-supervised learning. arXiv preprint arXiv:2102.03526 (2021)
3. Caron, M., Bojanowski, P., Joulin, A., Douze, M.: Deep clustering for unsupervised learning of visual features. In: Ferrari, V., Hebert, M., Sminchisescu, C., Weiss, Y. (eds.) Computer Vision – ECCV 2018. LNCS, vol. 11218, pp. 139–156. Springer, Cham (2018). https://doi.org/10.1007/978-3-030-01264-9_9
4. Codella, N.C., et al.: Skin lesion analysis toward melanoma detection: a challenge at the 2017 international symposium on biomedical imaging (ISBI), hosted by the international skin imaging collaboration (ISIC). In: 2018 IEEE 15th International Symposium on Biomedical Imaging (ISBI 2018), pp. 168–172. IEEE (2018)
5. Combalia, M., et al.: BCN20000: dermoscopic lesions in the wild. arXiv preprint arXiv:1908.02288 (2019)
6. Cuturi, M.: Sinkhorn distances: lightspeed computation of optimal transport. In: Advances in Neural Information Processing Systems, vol. 26 (2013)
7. Fini, E., Sangineto, E., Lathuilière, S., Zhong, Z., Nabi, M., Ricci, E.: A unified objective for novel class discovery. In: Proceedings of the IEEE/CVF International Conference on Computer Vision, pp. 9284–9292 (2021)
8. Gutmann, M., Hyvärinen, A.: Noise-contrastive estimation: a new estimation principle for unnormalized statistical models. In: Proceedings of the Thirteenth International Conference on Artificial Intelligence and Statistics. JMLR Workshop and Conference Proceedings, pp. 297–304 (2010)
9. Han, K., Rebuffi, S.A., Ehrhardt, S., Vedaldi, A., Zisserman, A.: AutoNovel: automatically discovering and learning novel visual categories. IEEE Trans. Pattern Anal. Mach. Intell. **44**(10), 6767–6781 (2021)

10. Hang, W., et al.: Local and global structure-aware entropy regularized mean teacher model for 3D left atrium segmentation. In: Martel, A.L., et al. (eds.) MICCAI 2020, Part I. LNCS, vol. 12261, pp. 562–571. Springer, Cham (2020). https://doi.org/10.1007/978-3-030-59710-8_55
11. He, K., Zhang, X., Ren, S., Sun, J.: Deep residual learning for image recognition. In: Proceedings of the IEEE Conference on Computer Vision and Pattern Recognition, pp. 770–778 (2016)
12. Khosla, P., et al.: Supervised contrastive learning. In: Advances in Neural Information Processing Systems, vol. 33, pp. 18661–18673 (2020)
13. Kuhn, H.W.: The Hungarian method for the assignment problem. Naval Res. Logist. Q. **2**(1–2), 83–97 (1955)
14. Li, Z., Togo, R., Ogawa, T., Haseyama, M.: Learning intra-domain style-invariant representation for unsupervised domain adaptation of semantic segmentation. Pattern Recogn. **132**, 108911 (2022)
15. Liu, F., Tian, Y., Chen, Y., Liu, Y., Belagiannis, V., Carneiro, G.: ACPL: anti-curriculum pseudo-labelling for semi-supervised medical image classification. In: Proceedings of the IEEE/CVF Conference on Computer Vision and Pattern Recognition, pp. 20697–20706 (2022)
16. Mahbod, A., Schaefer, G., Ellinger, I., Ecker, R., Pitiot, A., Wang, C.: Fusing fine-tuned deep features for skin lesion classification. Comput. Med. Imaging Graph. **71**, 19–29 (2019)
17. Tang, P., Liang, Q., Yan, X., Xiang, S., Zhang, D.: GP-CNN-DTEL: global-part CNN model with data-transformed ensemble learning for skin lesion classification. IEEE J. Biomed. Health Inform. **24**(10), 2870–2882 (2020)
18. Yao, P., et al.: Single model deep learning on imbalanced small datasets for skin lesion classification. IEEE Trans. Med. Imaging **41**(5), 1242–1254 (2021)
19. You, C., Zhou, Y., Zhao, R., Staib, L., Duncan, J.S.: SimCVD: simple contrastive voxel-wise representation distillation for semi-supervised medical image segmentation. IEEE Trans. Med. Imaging **41**(9), 2228–2237 (2022)
20. Zhang, B., et al.: Opportunities and challenges: classification of skin disease based on deep learning. Chin. J. Mech. Eng. **34**(1), 1–14 (2021)
21. Zhang, H., Cisse, M., Dauphin, Y.N., Lopez-Paz, D.: mixup: beyond empirical risk minimization. arXiv preprint arXiv:1710.09412 (2017)
22. Zhang, J., Xie, Y., Xia, Y., Shen, C.: Attention residual learning for skin lesion classification. IEEE Trans. Med. Imaging **38**(9), 2092–2103 (2019)
23. Zhong, Z., Fini, E., Roy, S., Luo, Z., Ricci, E., Sebe, N.: Neighborhood contrastive learning for novel class discovery. In: Proceedings of the IEEE/CVF Conference on Computer Vision and Pattern Recognition, pp. 10867–10875 (2021)
24. Zhong, Z., Zhu, L., Luo, Z., Li, S., Yang, Y., Sebe, N.: OpenMix: reviving known knowledge for discovering novel visual categories in an open world. In: Proceedings of the IEEE/CVF Conference on Computer Vision and Pattern Recognition, pp. 9462–9470 (2021)

Joint Segmentation and Sub-pixel Localization in Structured Light Laryngoscopy

Jann-Ole Henningson[1(✉)], Marion Semmler[2], Michael Döllinger[2], and Marc Stamminger[1]

[1] Friedrich-Alexander-Universität Erlangen-Nürnberg, Erlangen, Germany
jann-ole.henningson@fau.de

[2] Division of Phoniatrics and Pediatric Audiology at the Department of Otorhinolaryngology, Head and Neck Surgery, University Hospital Erlangen, Friedrich-Alexander-Universität Erlangen-Nürnberg, 91054 Erlangen, Germany

Abstract. In recent years, phoniatric diagnostics has seen a surge of interest in structured light-based high-speed video endoscopy, as it enables the observation of oscillating human vocal folds in vertical direction. However, structured light laryngoscopy suffers from practical problems: specular reflections interfere with the projected pattern, mucosal tissue dilates the pattern, and lastly the algorithms need to deal with huge amounts of data generated by a high-speed video camera. To address these issues, we propose a neural approach for the joint semantic segmentation and keypoint detection in structured light high-speed video endoscopy that improves the robustness, accuracy, and performance of current human vocal fold reconstruction pipelines. Major contributions are the reformulation of one channel of a semantic segmentation approach as a single-channel heatmap regression problem, and the prediction of sub-pixel accurate 2D point locations through weighted least squares in a fully-differentiable manner with negligible computational cost. Lastly, we expand the publicly available Human Laser Endoscopic dataset to also include segmentations of the human vocal folds itself. The source code and dataset are available at: github.com/Henningson/SSSLsquared

Keywords: Human Vocal Folds · Laryngoscopy · Keypoint Detection · Semantic Segmentation

1 Introduction

The voice is an essential aspect of human communication and plays a critical role in expressing emotions, conveying information, and establishing personal connections. An impaired function of the voice can have significant negative

Supplementary Information The online version contains supplementary material available at https://doi.org/10.1007/978-3-031-43987-2_4.

H. Greenspan et al. (Eds.): MICCAI 2023, LNCS 14225, pp. 34–43, 2023.
https://doi.org/10.1007/978-3-031-43987-2_4

impacts on an individual. Malign changes of human vocal folds are conventionally observed by the use of (high-speed) video endoscopy that measures their 2D deformation in image space. However, it was shown that their dynamics contain a significant vertical deformation. This led to the development of varying methods for the 3D reconstruction of human vocal folds during phonation. In these works, especially active reconstruction systems have been researched that project a pattern onto the surface of vocal folds [10,16,19,20]. All these systems need to deal with issues introduced by the structured light system, in particular a) specular reflections from the endoscopic light source which occlude the structured light pattern (cp. Fig. 1), b) a dilation of the structured light pattern through subsurface scattering effects in mucosal tissue, and c) vasts amount of data generated by high-speed cameras recording with up to 4000 frames per second. Furthermore, the introduction of a structured light source, e.g. a laser projection unit (LPU), increases the form-factor of the endoscope, which makes recording uncomfortable for the patient. In current systems, the video processing happens offline, which means that often unnecessarily long footage is recorded to be sure that an appropriate sequence is contained, or—even worse—that a patient has to show up again, because the recorded sequence is not of sufficient quality. Ideally, recording should thus happen in a very short time (seconds) and provide immediate feedback to the operator, which is only possible if the segmentation and pattern detection happen close to real time.

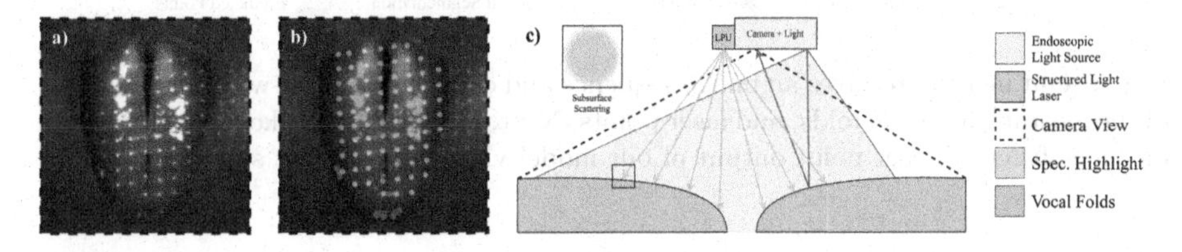

Fig. 1. Recording setup. From left to right: a) single frame of a recorded video containing laser dots and specular reflections, b) same frame with highlighted laser dots and specular reflections, c) illustration of recording setup with endoscope and laser projection unit (LPU) looking at the vocal folds.

To address all these practical issues, we present a novel method for the highly efficient and accurate segmentation, localization, and tracking of human vocal folds and projection patterns in laser-supported high-speed video endoscopy. An overview of our pipeline is shown in Fig. 2. It is based on two stages: First, a convolutional neural network predicts a segmentation of the vocal folds, the glottal area, and projected laser dots. Secondly, we compute sub-pixel accurate 2D point locations based on the pixel-level laser dot class probabilities in a weighted least-squares manner that further increase prediction accuracy. This approach can provide immediate feedback about the success of the recording to the physician, e.g. in form of the number of successfully tracked laser dots. Furthermore, this method can not only be used in vocal fold 3D reconstruction pipelines but also allows for the analysis of clinically relevant 2D features.

2 Related Work

Our work can properly be assumed to be simultaneously a heatmap regression as well as a semantic segmentation approach. Deep learning based medical semantic segmentation has been extensively studied in recent years with novel architectures, loss-functions, regularizations, data augmentations, holistic training and optimization approaches [23]. Here, we will focus on the specific application of semantic segmentation in laryngoscopy.

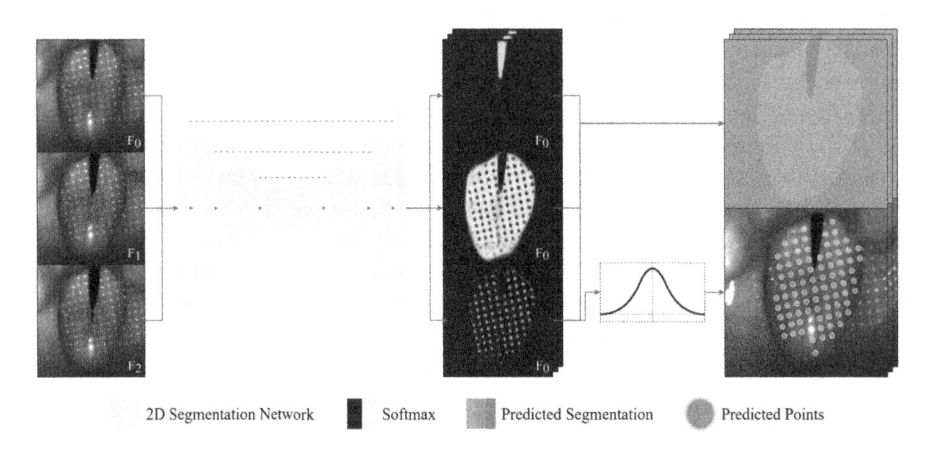

Fig. 2. Our method receives an image sequence and computes a pixel-wise classification of the glottal gap, vocal folds, and laser points. Next, we estimate 2D keypoint locations on the softmaxed laser point output of our model via weighted least squares.

Deep Learning in Laryngoscopy. In the realm of laryngoscopy works involving deep learning are few and far between. Most of these works focus on the segmentation of the glottal gap over time. The glottal dynamics give information about the patients underlying conditions, all the while being an easily detectable feature. Fehling et al. were the first to propose a CNN-based method that also infers a segmentation of the human vocal folds itself [8]. Their method uses a general U-Net architecture extended with Long Short-Term Memory cells to also take temporal information into account. Pedersen et al. [17] use off-the-shelf U-Nets to estimate glottal gap and vocal folds itself. Cho et al. [3] compare different segmentation architectures including CNN6, VGG16, Inception V3 and XCeption. Döllinger et al. [4] have shown that pretraining CNNs for human vocal fold segmentation can boost the respective CNNs performance. To the best of our knowledge, no publication has specifically targeted segmentation and detection in structured light laryngoscopy via deep learning.

Keypoint Detection can generally be separated into regression-based approaches that infer keypoint positions directly from the images [5,22,26] and heatmap-based approaches that model the likelihood of the existence of a keypoint, i.e. landmark, via channel-wise 2D Gaussians and determining channel-wise global maxima via $argmax()$. This leads to obvious quantization errors, that are addressed in recent works [1,7,25]. Most related to our approach is the work by Sharan et al. that proposes a method for determining the position of sutures by reformulating a single-channel binary segmentation to find local maxima and calculating their positions through general center of gravity estimation [21]. In case of human vocal folds, there have been works regarding laser dot detection in structured light laryngoscopy. However, these suffer from either a manual labeling step [14,16,19,20] or work only on a per-image basis [10], thus being susceptible to artifacts introduced through specular highlights. In a similar vein, none of the mentioned methods apply promising deep learning techniques.

Fig. 3. We extract windows around local maxima through dilation filtering and fit a Gaussian function into the window to estimate sub-pixel accurate keypoints.

3 Method

Given an isotropic light source and a material having subsurface scattering properties, the energy density of the penetrating light follows an exponential falloff [12]. In case of laser-based structured light endoscopy, this means that we can observe a bleeding effect, i.e. diffusion, of the respective collimated laser beams in mucosal tissue. Note that the energy density of laser beams is Gaussian distributed as well, amplifying this effect. Our method is now based on the assumption, that a pixel-level (binary- or multi-class) classifier will follow this exponential falloff in its predictions, and we can estimate sub-pixel accurate point positions through Gaussian fitting (cp. Fig. 3). This allows us to model the keypoint detection as a semantic segmentation task, such that we can jointly estimate the human vocal folds, the glottal gap, as well as the laserpoints' positions using only a single inference step. The method can be properly divided into two parts. At first, an arbitrary hourglass-style CNN (in our case we use the U-Net architecture) estimates semantic segmentations of the glottal gap, vocal folds and the 2D laserdots position for a fixed videosequence in a supervised manner. Next, we extract these local maxima lying above a certain threshold and use weighted least squares to fit a Gaussian function into the windowed regions. In this way, we can estimate semantic segmentations as well as the position of keypoints in a single inference pass, which significantly speeds up computation.

Feature Extraction. Current heatmap regression approaches use the *argmax*(.) or *topk*(.) functions to estimate channel-wise global maxima, i.e. a single point per channel. In case of an 31 by 31 LPU this would necessitate such an approach to estimate 961 channels; easily exceeding maintainable memory usage. Thus our method needs to allow multiple points to be on a single channel. However, this makes taking the *argmax*(.) or *topk*(.) functions to extract local maxima infeasible, or outright impossible. Thus, to extract an arbitrary amount of local maxima adhering to a certain quality, we use dilation filtering on a thresholded and Gaussian blurred image. More precisely we calculate $\mathbf{I}_T = \mathbf{I} > [T(\mathbf{I}, \theta) * \mathbf{G} \oplus \mathbf{B}]$, where $T(x, y)$ depicts a basic thresholding operation, $\mathbf{G}$ a Gaussian kernel, $\mathbf{B}$ a general box kernel with a 0 at the center, and $\oplus$ the dilation operator. Finally we can easily retrieve local maxima by just extracting every non-zero element of $\mathbf{I}_T$. We then span a window I_{ij} of size $k \in N$ around the non-zero discrete points $\mathbf{p}_i$. For improved comprehensibility, we are dropping the subscripts in further explanations, and explain the algorithm for a single point $\mathbf{p}$. Next, we need to estimate a Gaussian function. Note that, a Gaussian function is of the form $f(x) = Ae^{-(x-\mu)^2/2\sigma^2}$, where $x = \mu$ is the peak, A the peaks height, and σ defines the functions width. Gaussian fitting approaches can generally be separated into two types: the first ones employ non-linear least squares optimization techniques [24], while others use the result of Caruana et al. in that the logarithm of a Gaussian is a polynomial equation [2].

Caruanas Algorithm. As stated previously, Caruanas algorithm is based on the observation that by taking the logarithm of the Gaussian function, we generate a polynomial equation (Eq. 1).

$$\ln(f(x)) = \ln(A) + \frac{-(x-\mu)^2}{2\sigma^2} = \ln(A) - \frac{-\mu^2}{2\sigma^2} + \frac{2\mu x}{2\sigma^2} - \frac{x^2}{2\sigma^2} \tag{1}$$

Note that the last equation is in polynomial form $\ln(y) = ax^2 + bx + c$, with $c = \ln(A) - \frac{-\mu^2}{2\sigma^2}$, $b = \frac{\mu}{\sigma^2}$ and $a = \frac{-1}{2\sigma^2}$. By defining the error function $\delta = \ln(f(x)) - (ax^2 + bx + c)$ and differentiating the sum of residuals gives a linear system of equations (Eq. 2).

$$\begin{bmatrix} N & \sum x & \sum x^2 \\ \sum x & \sum x^2 & \sum x^3 \\ \sum x^2 & \sum x^3 & \sum x^4 \end{bmatrix} \begin{bmatrix} a \\ b \\ c \end{bmatrix} = \begin{bmatrix} \sum \ln(\hat{y}) \\ \sum x \ln(\hat{y}) \\ \sum x^2 \ln(\hat{y}) \end{bmatrix} \tag{2}$$

After solving the linear system, we can finally retrieve μ, σ and A with $\mu = -b/2c$, $\sigma = \sqrt{-1/2c}$, and $A = e^{a-b^2/4c}$.

Guos Algorithm. Due to the logarithmic nature of Caruanas algorithm, it is very susceptible towards outliers. Guo [9] addresses these problems through a weighted least-squares regimen, by introducing an additive noise term η and reformulating the cost function to (Eq. 3).

$$\epsilon = y[\ln(y + \eta) - (a + bx + cx^2)] \approx y[\ln(y) - (a + bx + cx^2)] + \eta. \tag{3}$$

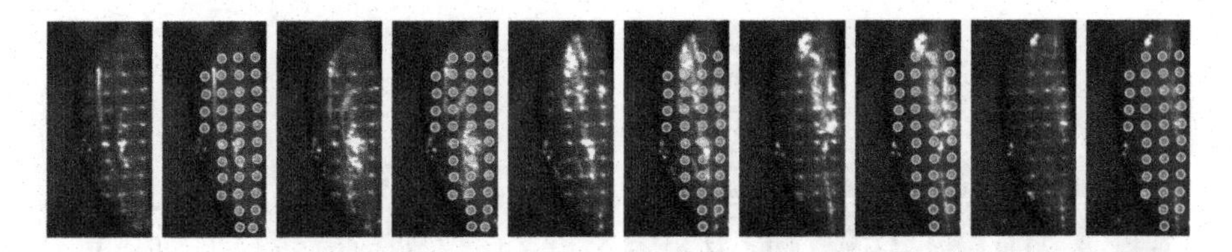

Fig. 4. In-vivo dynamics of a human vocal fold in laser-based structured light endoscopy. Due to the endoscopic light source and moist mucosal tissue, we see an abundance of specular reflections occluding the laserdots. In green: keypoint detection of our approach. It successfully detects laser dots in occluded regions. (Color figure online)

Similarly, by differentiating the sum of ϵ^2, we retrieve a linear system of the form given in Eq. 4.

$$\begin{bmatrix} \sum \hat{y}^2 & \sum x\hat{y}^2 & \sum x^2\hat{y}^2 \\ \sum x\hat{y}^2 & \sum x^2\hat{y}^2 & \sum x^3\hat{y}^2 \\ \sum x^2\hat{y}^2 & \sum x^3\hat{y}^2 & \sum x^4\hat{y}^2 \end{bmatrix} \begin{bmatrix} a \\ b \\ c \end{bmatrix} = \begin{bmatrix} \sum \hat{y}^2 \ln(\hat{y}) \\ \sum x\hat{y}^2 \ln(\hat{y}) \\ \sum x^2\hat{y}^2 \ln(\hat{y}) \end{bmatrix} \tag{4}$$

The parameters of the Gaussian function μ, σ and A can be calculated similar to Caruanas algorithm. Recall that polynomial regression has an analytical solution, where the vector of estimated polynomial regression coefficients is $\beta = (\mathbf{X}^\mathbf{T}\mathbf{X})^{-1}\mathbf{X}^\mathbf{T}\hat{y}$. This formulation is easily parallelizable on a GPU and fully differentiable. Hence, we can efficiently compute the Gaussian coefficients necessary for determining the subpixel position of local maxima. Finally, we can calculate the points position $\hat{\mathbf{p}}$ by simple addition using Eq. 5.

$$\hat{\mathbf{p}} = \boldsymbol{\mu} + \mathbf{p} \tag{5}$$

4 Evaluation

We implemented our code in Python (3.10.6), using the PyTorch (1.12.1) [15] and kornia (0.6.8) [6] libraries. We evaluate our code as well as the comparison methods on an Nvidia RTX 3080 GPU. We follow the respective methods closely for training. For data augmentation, we use vertical and horizontal flipping, affine and perspective transformations, as well as gamma correction and brightness modulation. We opted to use data augmentation strategies that mimic data that is likely to occur in laryngoscopy. We evaluate all approaches using a k-fold cross validation scheme with $k = 5$, on an expanded version of the publicly available Human-Laser Endoscopic (HLE) dataset [10] that also includes segmentations of the human vocal fold itself. We evaluate on subjects that were not contained in the training sets. The data generation scheme for the human vocal fold segmentation is described further below. For our baseline, we use dilation filtering and the moment method on images segmented using the ground-truth labels similar to [10,19,20]. We opt to use the ground-truth labels to show how

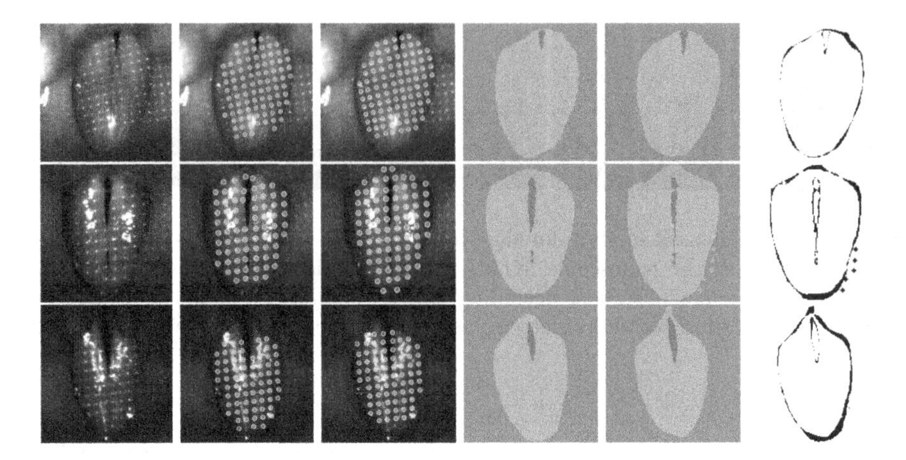

Fig. 5. Qualitative Assessment of the inferred point positions and segmentations. Left to right: Input, Predicted Keypoints, GT Keypoints, Predicted Segmentation, GT Segmentation, Pixelwise Error.

good these methods may become given perfect information. We train our approach via Stochastic Gradient Descent, with a learning rate of 10^{-1} for 100 epochs and update the learning rate via a polynomial learning rate scheduler similar to nnU-Net [11]. For estimating the keypoints we use a window size of 7, a Gaussian blur of size 5 and set the keypoint threshold to 0.7.

HLE++. The HLE dataset is a publicly available dataset consisting of 10 labeled in-vivo recordings of human vocal folds during phonation [10], where each recording contains one healthy subject. The labels include segmentations of the glottal gap, the glottal mid- and outline as well as the 2D image space positions of the laser dots projected onto the superior surface of the vocal folds. We expand the dataset to also include segmentation labels for the vocal folds itself. Due to the high framerate of the recordings, motion stemming from the manually recording physician is minimal and can be assumed to be linear. Thus, to generate the vocal fold segmentation masks, we generate a segmentation mask of the vocal fold region manually and calculate its centroid in the first and last frame of each video. Then, we linearly interpolate between the measured centroids. To account for the glottal gap inside each frame, we set $F_i = F_i \setminus G_i$, where F_i is the vocal fold segmentation at frame i and G_i the glottal segmentation, respectively.

Quantitative and Qualitative Evaluation. Table 1 shows a quantitative evaluation of different neural network architectures that have been used in laryngoscopy [8,18] or medical keypoint detection tasks [21]. In general, we could reformulate this segmentation task as a 3D segmentation task, in which we model the width and height of the images as first and second dimension, and lastly time as our third dimension. However, due to the poor optimization of 3D Convolutions on GPUs [13] and the large amounts of data generated in high-speed video

Table 1. Quantitative evaluation of the precision and F1-score of predicted keypoints, IoU and DICE score of inferred segmentations as well as the inference time for a single image and the frames per second on an Nvidia RTX 3080 GPU.

	Precision↑	F1-Score↑	IoU↑	DICE↑	Inf. Speed(ms)↓	FPS↑
Baseline	0.64	0.6923	✗	✗	✗	✗
U-LSTM [8]	0.70 ± 0.41	0.58 ± 0.32	0.52 ± 0.18	0.77 ± 0.08	65.57 ± 0.31	15
U-Net [18]	**0.92** ± 0.08	**0.88** ± 0.04	**0.68** ± 0.08	**0.88** ± 0.02	4.54 ± 0.03	220
Sharan [21]	0.17 ± 0.19	0.16 ± 0.17	✗	✗	5.97 ± 0.25	168
2.5D U-Net	0.90 ± 0.08	0.81 ± 0.05	0.65 ± 0.06	0.87 ± 0.02	**1.08** ± 0.01	**926**

recordings, this would create a serious bottleneck in real-time 3D pipelines. Thus, we also evaluate a 2.5D U-Net architecture, in which we employ channel-wise 3D convolutions inside the bottleneck as well as the output layers. This allows us to predict segmentations sequence-wise; drastically lowering inference times on a per frame basis while keeping the number of floating point operations minimal. Interestingly, this architecture achieves similar results to a standard U-Net architecture (see Table 1). However, frame jumps can be seen inbetween sequences. Since HLE was generated with a single recording unit, we assume that a properly trained 2D-CNN can infer occluded point positions based on the surrounding laser points as well as the observed topology of the vocal folds itself. However, we believe that it generalizes less well to arbitrary point patterns than a network architecture including temporal information. For the segmentation tasks, we measure the foreground IoU and DICE scores. For the keypoints, we evaluate the precision and the F1-score. We count a keypoint prediction as true positive, when its distance to its closest ground-truth point does not exceed 2 pixels. In Fig. 4 an example of a prediction over 5 frames is given, showing that neural networks can infer proper point positions even in case of occlusions. A further qualitative assessment of point predictions, vocal fold and glottal gap segmentations is given in Fig. 5.

5 Conclusion

We presented a method for the simultaneous segmentation and laser dot localization in structured light high-speed video laryngoscopy. The general idea is that we can dedicate one channel of the output of a general multiclass segmentation model to learn Gaussian heatmaps depicting the locations of multiple unlabeled keypoints. To robustly handle noise, we propose to use Gaussian regression on a per local-maxima basis that estimates sub-pixel accurate keypoints with negligible computational overhead. Our pipeline is very accurate and robust, and can give feedback about the success of a recording within a fraction of a second. Additionally, we extended the publicly available HLE Dataset to include segmentations of the human vocal fold itself. For future work, it would be beneficial to investigate how this method generalizes to arbitrary projection patterns and non-healthy subjects.

Acknowledgements. We thank **Dominik Penk** and **Bernhard Egger** for their valuable feedback.This work was supported by Deutsche Forschungsgemeinschaft (DFG, German Research Foundation) under grant STA662/6-1, Project-ID 448240908 and (partly) funded by the DFG - SFB 1483 - Project-ID 442419336, EmpkinS. The authors gratefully acknowledge the scientific support and HPC resources provided by the Erlangen National High Performance Computing Center of the Friedrich-Alexander-Universität Erlangen-Nürnberg.

References

1. Bulat, A., Sanchez, E., Tzimiropoulos, G.: Subpixel heatmap regression for facial landmark localization. In: 32nd British Machine Vision Conference 2021, BMVC 2021, vol. 2021(32), pp. 22–25 (2021). https://arxiv.org/abs/2111.02360
2. Caruana, R.A., Searle, R.B., Shupack, S.I.: Additional capabilities of a fast algorithm for the resolution of spectra. Anal. Chem. **60**(18), 1896–1900 (1988). https://doi.org/10.1021/ac00169a011
3. Cho, W.K., Choi, S.H.: Comparison of convolutional neural network models for determination of vocal fold normality in laryngoscopic images. J. Voice **36**(5), 590–598 (2022). https://doi.org/10.1016/j.jvoice.2020.08.003, https://www.sciencedirect.com/science/article/pii/S0892199720302927
4. Döllinger, M., et al.: Re-training of convolutional neural networks for glottis segmentation in endoscopic high-speed videos. Appl. Sci. **12**(19), 9791 (2022). https://doi.org/10.3390/app12199791, https://www.mdpi.com/2076-3417/12/19/9791
5. Duffner, S., Garcia, C.: A connexionist approach for robust and precise facial feature detection in complex scenes, pp. 316–321 (2005). https://doi.org/10.1109/ISPA.2005.195430
6. Riba, E., Mishkin, D., Ponsa, D., Rublee, E., Bradski, G.: Kornia: an open source differentiable computer vision library for pytorch. In: Winter Conference on Applications of Computer Vision (2020). https://arxiv.org/pdf/1910.02190.pdf
7. Earp, S.W.F., Samacoïts, A., Jain, S., Noinongyao, P., Boonpunmongkol, S.: Subpixel face landmarks using heatmaps and a bag of tricks. CoRR abs/2103.03059 (2021). https://arxiv.org/abs/2103.03059
8. Fehling, M.K., Grosch, F., Schuster, M.E., Schick, B., Lohscheller, J.: Fully automatic segmentation of glottis and vocal folds in endoscopic laryngeal high-speed videos using a deep convolutional LSTM network. PLOS ONE **15**, 1–29 (2020). https://doi.org/10.1371/journal.pone.0227791
9. Guo, H.: A simple algorithm for fitting a gaussian function [DSP tips and tricks]. IEEE Signal Process. Mag. **28**(5), 134–137 (2011). https://doi.org/10.1109/MSP.2011.941846
10. Henningson, J.O., Stamminger, M., Döllinger, M., Semmler, M.: Real-time 3D reconstruction of human vocal folds via high-speed laser-endoscopy. In: Wang, L., Dou, Q., Fletcher, P.T., Speidel, S., Li, S. (eds.) MICCAI 2022. Lecture Notes in Computer Science, vol. 13437, pp. 3–12. Springer, Cham (2022). https://doi.org/10.1007/978-3-031-16449-1_1
11. Isensee, F., Jaeger, P.F., Kohl, S.A.A., Petersen, J., Maier-Hein, K.H.: nnU-Net: a self-configuring method for deep learning-based biomedical image segmentation. Nat. Meth. **18**(2), 203–211 (2021)

12. Jensen, H.W., Marschner, S.R., Levoy, M., Hanrahan, P.: A practical model for subsurface light transport. In: Proceedings of the 28th Annual Conference on Computer Graphics and Interactive Techniques, pp. 511–518. SIGGRAPH 2001, Association for Computing Machinery, New York, NY, USA (2001). https://doi.org/10.1145/383259.383319
13. Jiang, J., Huang, D., Du, J., Lu, Y., Liao, X.: Optimizing small channel 3D convolution on GPU with tensor core. Parallel Comput. **113**(C), 102954 (2022). https://doi.org/10.1016/j.parco.2022.102954
14. Luegmair, G., Mehta, D., Kobler, J., Döllinger, M.: Three-dimensional optical reconstruction of vocal fold kinematics using high-speed videomicroscopy with a laser projection system. IEEE Trans. Med. Imaging **34**, 2572–2582 (2015). https://doi.org/10.1109/TMI.2015.2445921
15. Paszke, A., et al.: Pytorch: an imperative style, high-performance deep learning library. In: Advances in Neural Information Processing Systems, vol. 32, pp. 8024–8035. Curran Associates, Inc. (2019). http://papers.neurips.cc/paper/9015-pytorch-an-imperative-style-high-performance-deep-learning-library.pdf
16. Patel, R., Donohue, K., Lau, D., Unnikrishnan, H.: In vivo measurement of pediatric vocal fold motion using structured light laser projection. J. Voice: Off. J. Voice Found. **27**, 463–472 (2013). https://doi.org/10.1016/j.jvoice.2013.03.004
17. Pedersen, M., Larsen, C., Madsen, B., Eeg, M.: Localization and quantification of glottal gaps on deep learning segmentation of vocal folds. Sci. Rep. **13**, 878 (2023). https://doi.org/10.1038/s41598-023-27980-y
18. Ronneberger, O., Fischer, P., Brox, T.: U-Net: convolutional networks for biomedical image segmentation. In: Navab, N., Hornegger, J., Wells, W.M., Frangi, A.F. (eds.) MICCAI 2015. LNCS, vol. 9351, pp. 234–241. Springer, Cham (2015). https://doi.org/10.1007/978-3-319-24574-4_28, http://lmb.informatik.uni-freiburg.de/Publications/2015/RFB15a, (arXiv:1505.04597 [cs.CV])
19. Semmler, M., Kniesburges, S., Birk, V., Ziethe, A., Patel, R., Döllinger, M.: 3D reconstruction of human laryngeal dynamics based on endoscopic high-speed recordings. IEEE Trans. Med. Imaging **35**(7), 1615–1624 (2016). https://doi.org/10.1109/TMI.2016.2521419
20. Semmler, M., et al.: Endoscopic laser-based 3D imaging for functional voice diagnostics. Appl. Sci. **7**, 600 (2017). https://doi.org/10.3390/app7060600
21. Sharan, L., et al.: Point detection through multi-instance deep heatmap regression for sutures in endoscopy. Int. J. Comput. Assist. Radiol. Surg. **16**, 2107–2117 (2021). https://doi.org/10.1007/s11548-021-02523-w
22. Sun, P., Min, J.K., Xiong, G.: Globally tuned cascade pose regression via back propagation with application in 2D face pose estimation and heart segmentation in 3D CT images. ArXiv abs/1503.08843 (2015)
23. Ulku, I., Akagündüz, E.: A survey on deep learning-based architectures for semantic segmentation on 2D images. Appl. Artif. Intell. **36**(1), 2032924 (2022). https://doi.org/10.1080/08839514.2022.2032924
24. Ypma, T.J.: Historical development of the Newton-Raphson method. SIAM Rev. **37**(4), 531–551 (1995). http://www.jstor.org/stable/2132904
25. Yu, B., Tao, D.: Heatmap regression via randomized rounding. IEEE Trans. Pattern Anal. Mach. Intell. **44**(11), 8276–8289 (2021)
26. Zhang, J., Liu, M., Shen, D.: Detecting anatomical landmarks from limited medical imaging data using two-stage task-oriented deep neural networks. IEEE Trans. Image Process. **26**(10), 4753–4764 (2017). https://doi.org/10.1109/TIP.2017.2721106

A Style Transfer-Based Augmentation Framework for Improving Segmentation and Classification Performance Across Different Sources in Ultrasound Images

Bin Huang[1,7], Ziyue Xu[2], Shing-Chow Chan[3], Zhong Liu[1], Huiying Wen[1], Chao Hou[1], Qicai Huang[1], Meiqin Jiang[1], Changfeng Dong[4], Jie Zeng[5], Ruhai Zou[6], Bingsheng Huang[7(✉)], Xin Chen[1(✉)], and Shuo Li[8]

[1] School of Biomedical Engineering, Shenzhen University Medical School, Shenzhen University, Shenzhen 518060, China
chenxin@szu.edu.cn
[2] Nvidia Corporation, Bethesda, MD 20814, USA
[3] Department of Electrical and Electronic Engineering, The University of Hong Kong, Hong Kong, SAR, China
[4] Institute of Hepatology, Shenzhen Third People's Hospital, Shenzhen 518000, China
[5] Department of Medical Ultrasonics, The Third Affiliated Hospital, Sun Yat-sen University, Guangzhou 510000, China
[6] State Key Laboratory of Oncology in South China, Collaborative Innovation Center of Cancer Medicine, Department of Ultrasound, Sun Yat-sen University Cancer Center, Guangzhou 510000, China
[7] Medical AI Lab, School of Biomedical Engineering, Shenzhen University Medical School, Shenzhen University, Shenzhen 518000, China
huangb@szu.edu.cn
[8] Department of Biomedical Engineering, Case Western Reserve University, Cleveland, OH, USA

Abstract. Ultrasound imaging can vary in style/appearance due to differences in scanning equipment and other factors, resulting in degraded segmentation and classification performance of deep learning models for ultrasound image analysis. Previous studies have attempted to solve this problem by using style transfer and augmentation techniques, but these methods usually require a large amount of data from multiple sources and source-specific discriminators, which are not feasible for medical datasets with limited samples. Moreover, finding suitable augmentation methods for ultrasound data can be difficult. To address these challenges, we propose a novel style transfer-based augmentation framework that consists of three components: mixed style augmentation (MixStyleAug), feature augmentation (FeatAug), and mask-based style augmentation (MaskAug). MixStyleAug uses a style transfer network to transform the

Supplementary Information The online version contains supplementary material available at https://doi.org/10.1007/978-3-031-43987-2_5.

H. Greenspan et al. (Eds.): MICCAI 2023, LNCS 14225, pp. 44–53, 2023.
https://doi.org/10.1007/978-3-031-43987-2_5

style of a training image into various reference styles, which enriches the information from different sources for the network. FeatAug augments the styles at the feature level to compensate for possible style variations, especially for small-size datasets with limited styles. MaskAug leverages segmentation masks to highlight the key regions in the images, which enhances the model's generalizability. We evaluate our framework on five ultrasound datasets collected from different scanners and centers. Our framework outperforms previous methods on both segmentation and classification tasks, especially on small-size datasets. Our results suggest that our framework can effectively improve the performance of deep learning models across different ultrasound sources with limited data.

Keywords: Ultrasound · Segmentation · Classification · Style transfer · Data augmentation

1 Introduction

Classification and segmentation are two common tasks that use deep learning techniques to solve clinical problems [1,2]. However, training deep learning models reliably usually requires a large amount of data samples. Models trained with limited data are susceptible to overfitting and possible variations due to small

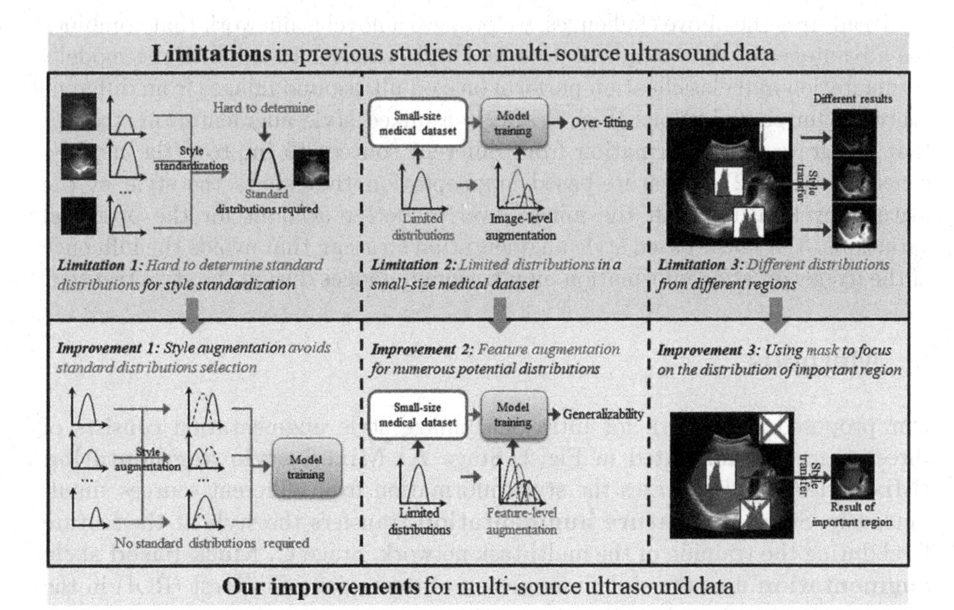

Fig. 1. Limitations of previous studies and our improvements for multi-source ultrasound data.

sample size, which can lead to poor performance across different sources. Different sources refer to the same modality collected from different scanners. In medical imaging, one of the main reasons for poor performance is the variation in the imaging process, such as the type of scanner, the settings, the protocol, *etc.* [3]. This can cause changes in the intensity distributions of the images [4,5]. While training deep learning models with a large number of high-quality data could potentially address this problem, this approach is often challenging due to limited resources and difficulties in collecting medical images, as well as the manual annotation required by experienced radiologists or experts with professional domain knowledge. Thus, limited labeled data is commonly used to model the classification and segmentation network.

To prevent overfitting and improve generalization, data augmentation [3,6–10] has been proposed to generate more similar but different samples for the training dataset. Very often, this can be done by applying various transformations to the training images to create new images that reflect natural variations within each class. However, the model's performance across different sources heavily depends on the augmentation strategies. Another popular technique is style transfer [11], which adapts the style of test images to match the selected reference images (standard distributions) [4,5,12,13]. However, these methods have a limitation that their performance depends on the quality of the reference images. Moreover, these methods tend to transfer the style of the whole images, which may introduce irrelevant distribution information for medical imaging applications, as shown in Fig. 1. This problem is more severe in ultrasound images due to the presence of acoustic shadow.

To address the above challenges, we propose a novel framework that combines the advantages of data augmentation and style transfer to enhance the model's segmentation and classification performance on ultrasound images from different sources. Our contributions (Fig. 1) are: 1) a mixed style augmentation strategy that integrates the information from different sources to improve the model's generalizability. 2) A feature-based augmentation that shifts the style at the feature level rather than the image level to better account for the potential variations. 3) a mask-based style augmentation strategy that avoids the influence of the irrelevant style information on ultrasound images during the style transfer.

2 Methods

Our proposed framework for ultrasonic image style augmentation consists of three stages, as illustrated in Fig. 2. Stage **A. Mixed style augmentation (MixStyleAug)** integrates the style information from different sources simultaneously. Stage **B. Feature augmentation** transfers the style at the feature level during the training of the multi-task network. Stage **C. Mask-based style augmentation** uses the style information of the region of interest (ROI) in the ultrasound image based on the segmentation results.

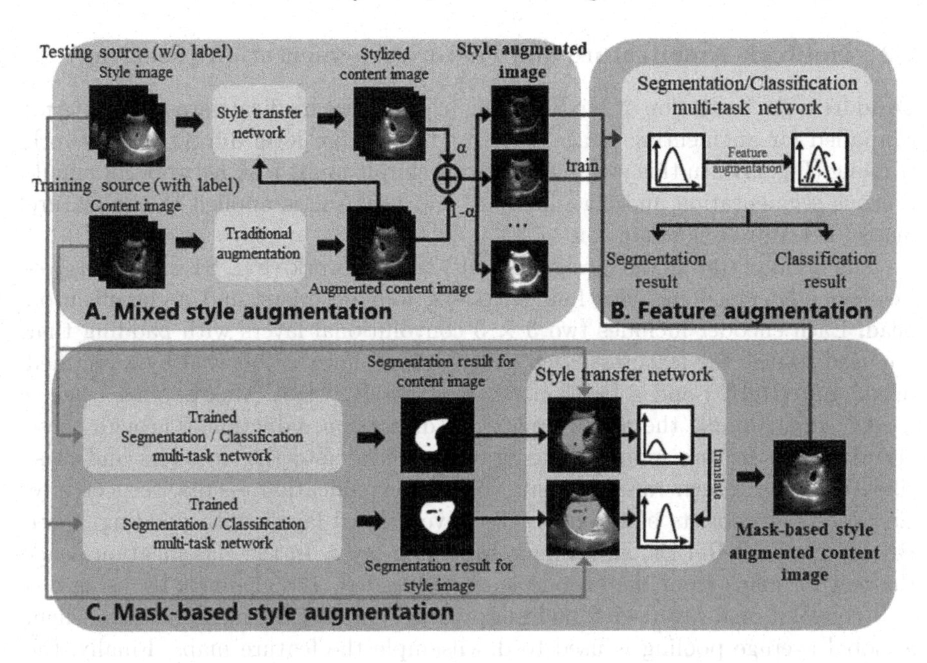

Fig. 2. Overview of our proposed style transfer-based augmentation framework. The whole framework consists of mixed style augmentation, feature augmentation, and mask-based style augmentation.

2.1 Mixed Style Augmentation (MixStyleAug)

To improve the performance of the multi-task network, we design MixStyleAug, combining traditional transformations and style transfer to incorporate image information from target sources during training (Fig. 2A). In this method, the content and the style images are sampled from training and target sources, respectively. Firstly, the traditional augmentation is applied to transform the content image, which can prevent overfitting. The traditional augmentation includes rotation, translation, scaling, and deformation transformations. Next, we translate the style of the augmented content image to that of the style image using the WCT^2 [14] style transfer network, generating a stylized content image. Finally, inspired by AugMix [15], we mix the stylized and augmented content images using random weights to create a style-augmented image that includes information from the training source. MixStyleAug allows the augmented training dataset to implicitly contain information from multiple sources, improving the model's performance across different sources. However, this method requires a large number of available images as reference styles for style augmentation, making it impractical for small-sized datasets.

2.2 Network Architecture and Feature Augmentation (FeatAug)

To address the limitation of MixStyleAug in small-size medical datasets, FeatAug is applied for augmenting image styles at the feature level during the network training (Fig. 2B). In this work, we design a simple multi-task network for simultaneous segmentation and classification, and FeatAug is applied to the feature maps for feature augmentation.

The architecture of our designed multi-task network (Fig. S1 in the *Supplementary Materials*) includes four encoders, four decoders, and a classification head. Each encoder includes two 3 × 3 convolutional layers with padding that are used to fuse the features. Each convolutional layer is followed by a rectified linear unit (ReLU) and a batch normalization (BN) [16]. Max-pooling layer is used to downsample the feature maps for dimension reduction. Through these encoders, the feature maps are generated and fed into the decoders and classification head to generate segmentation and classification results, respectively. Each decoder consists of three 3 × 3 convolutional layers with padding, three BN layers, three ReLUs, and a max-unpooling layer. In the classification head, the feature maps from the encoders are reduced to 128 channels by using a 3 × 3 convolutional layer with padding followed by ReLU and BN layer. Then, a global average pooling is used to downsample the feature maps. Finally, the features are fed into a fully connected layer followed by a sigmoid layer to output the classification result.

Previous studies reported that changing the mean and standard deviation of the feature maps could lead to different image styles [17,18]. Thus, we design a module to randomly alter these values to augment the styles at the feature level. To avoid over-augmentation at the feature level, this module is randomly applied with a 50% probability after the residual connection in each encoder. The module is defined as follows:

$$A' = \frac{A - \mu_A}{\sigma_A} \cdot \left(\sigma_A + \mathcal{N}(\mu, \sigma)\right) + \left(\mu_A + \mathcal{N}(\mu, \sigma)\right) \tag{1}$$

where A indicates the feature map, A' indicates the augmented feature map, μ_A indicates the mean of feature map A, σ_A indicates the standard deviation of feature map A, and $\mathcal{N}(\mu, \sigma)$ indicates a value randomly generated from a normal distribution with mean μ and standard deviation σ. In this study, the μ and σ of the normal distribution were empirically set to 0 and 0.1 according to preliminary experimental results, respectively.

2.3 Mask-Based Style Augmentation (MaskAug)

In general, the style transfer uses the style information of the entire image, but this approach may not be ideal when the regions outside of the ROIs contain conflicting style information as compared to the regions within the ROIs, as illustrated in Fig. 1. To mitigate the impact of irrelevant or even adverse style information, we propose a mask-based augmentation technique (MaskAug) that emphasize the ROIs in the ultrasound image during style transfer network training.

Figure 2C shows the pipeline of MaskAug and the steps are: 1) Content and style images are randomly chosen from training and target sources, respectively. 2) A trained multi-task network, which has been trained for several epochs and will be updated in the later epochs, is used to automatically generate ROIs of these images. 3) The content image, style image and their ROIs are input to the style transfer network. 4) During the style transfer, the intensity distribution of the ROI in the content image is changed to that of the style image. 5) Finally, mask-based style augmented images are produced and these images are then input to the multi-task network for further training.

2.4 Loss Function and Implementation Details

We utilized cross-entropy (CE) as the primary loss function for segmentation and classification during the training stage. Additionally, Dice loss [19] was computed as an auxiliary loss for segmentation. These loss functions are defined as:

$$\mathcal{L}_m = \mathcal{L}_{CE}^{Seg} + \mathcal{L}_{Dice}^{Seg} + \mathcal{L}_{CE}^{Cls} \tag{2}$$

where $\mathcal{L}_{CE}$ denotes CE loss, $\mathcal{L}_{Dice}$ denotes Dice loss, $\mathcal{L}_m$ denotes the loss for the multi-task network optimization, $\mathcal{L}^{Seg}$ denotes the loss computed from the segmentation result, and $\mathcal{L}^{Cls}$ denotes the loss computed from the classification result.

We adopted Pytorch to implement the proposed framework, and the multi-task network was trained on Nvidia RTX 3070 with 8 GB memory. During training, the batch size was set to 16, the maximum epoch number was 300, and the initial learning rate was set to 0.0005. We decayed the learning rate with cosine annealing [20] for each epoch, and the minimum learning rate was set to 0.000001. The restart epoch of cosine annealing was set to 300, ensuring that the learning rate monotonically decreased during the training process. For optimization, we used the AdamW optimizer [21] in our experiments. The whole training takes about 6 h and the inference time for a sample is about 0.2 s.

3 Experimental Results and Discussion

Datasets and Evaluation Metrics. We evaluated our framework on five ultrasound datasets (each representing a source) collected from multiple centers using different ultrasound scanners, including three liver datasets and two thyroid nodules datasets. A detailed description of the collected datasets is provided in Table S1 of the *Supplementary Materials*. We used the dataset with the largest sample size as the training source to prevent overfitting, while the other datasets were the target sources. For each datasets, we randomly split 20% of the samples for test, and used the remaining 80% for training the network. All the results in this study are based on the test set. In the training set, 20% data was randomly selected as validation set. In the data preprocessing, the input images were resized to 224×224 and were normalized by dividing 255.

AUROC is used to evaluate the classification performance. DSC is used to assess the performance of the segmentation. The DSC is defined as:

$$DSC = \frac{2TP}{FP + 2TP + FN} \tag{3}$$

where TP refers to the pixels where both the predicted results and the gold standard are positive, FP refers to the pixels where the predicted results are positive and the gold standard are negative, and FN refers to the pixels where the predicted results are negative and the gold standard are positive.

Table 1. Comparison of segmentation and classification performance of different augmentation methods in five ultrasound datasets in terms of DSC (%) and AUROC (×100%). Training/Target: Training/Target source datasets. MixStyleAug: mixed style augmentation. FeatAug: feature augmentation. MaskAug: mask-based style augmentation. LD: liver dataset. TD: thyroid nodule dataset.

Method	Metric	LD1	LD2	LD3	TD1	TD2
		Training	Target	Target	Training	Target
Traditional Augmentation	DSC	94.5	88.3	89.6	64.0	63.1
	AUROC	86.6	61.3	65.6	72.6	62.3
MixStyleAug	DSC	94.0	87.3	91.1	62.8	65.7
	AUROC	87.9	64.0	68.9	78.1	62.3
MixStyleAug+FeatAug	DSC	94.0	86.9	90.2	63.9	65.2
	AUROC	89.7	66.3	68.8	**85.5**	**64.2**
MixStyleAug+FeatAug+MaskAug	DSC	**94.8**	**89.7**	**91.2**	**77.9**	**65.9**
	AUROC	**92.3**	**67.3**	**69.3**	83.0	62.4

Ablation Study. We evaluated the effects of MixStyleAug, FeatAug, and MaskAug by training a multi-task network with different combinations of these augmentation strategies. Table 1 shows that MixStyleAug improves the segmentation and classification performance on the target sources compared to traditional augmentation. Furthermore, The combination of FeatAug and MixStyleAug improves the classification performance slightly in the liver datasets and significantly in the thyroid nodule datasets. This improvement is due to the style transfer at the feature level, which make the augmented features more similar to the target sources.

Using MaskAug improved both segmentation and classification performance on both training and target sources, compared to the combination of FeatAug and MixStyleAug. This resulted in excellent performance. Figure 3 shows that the mask-based stylized content image has a more similar distribution to the style image than the other images, which helps the model perform better on both training and target sources.

Comparison with Previous Studies. We compared our proposed method with BigAug [3], the style augmentation method by Hesse *et al.* [8], AutoAug [10], and UDA [22] on our collected datasets. Table 2 shows that our method performs excellently on both training and target sources. Unlike BigAug [3], our method uses style augmentation instead of intensity transformations, which avoids a drop in classification performance. Hesse *et al.* [8] only uses training sources for style

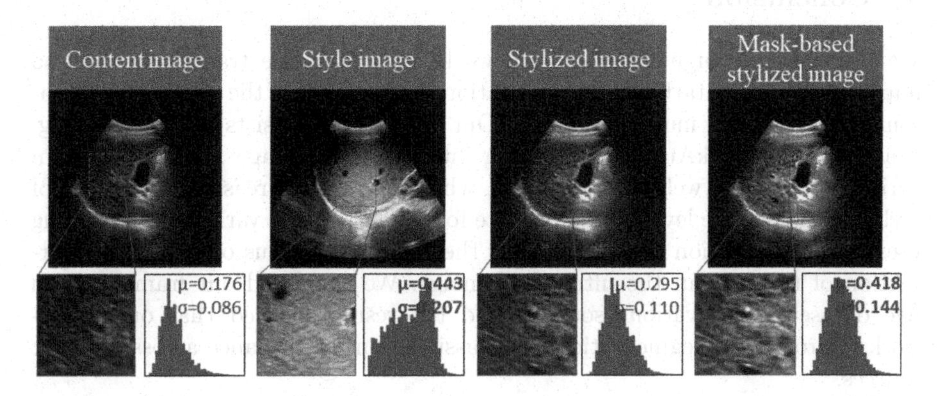

Fig. 3. Illustrations of the conventional style transfer and mask-based style transfer in an ultrasound image. A neural style transfer network is used to translate the content image to the style image, resulting in a stylized image with reference to the style of the entire style image. In contrast, mask-based stylized images are generated with reference to the style of the liver substance in the stylized image. The histogram shows the intensity distribution of the liver region, with μ and σ representing the mean and standard deviation of the liver parenchyma in the ultrasound image, respectively.

Table 2. Segmentation and classification performance of our proposed framework and previous studies in five ultrasound datasets in terms of DSC (%) and AUROC (×100%). Training/Target: Training/Target source datasets. LD: liver dataset. TD: thyroid nodules dataset. UDA: unsupervised domain adaptation.

Method	Metric	LD1	LD2	LD3	TD1	TD2
		Training	Target	Target	Training	Target
BigAug [3]	DSC	93.8	88.0	90.8	54.5	56.4
	AUROC	82.1	59.2	51.9	62.7	**63.9**
Hesse *et al.* [8]	DSC	92.6	86.9	91.3	61.4	62.7
	AUROC	79.6	64.9	68.3	71.7	58.8
AutoAug [10]	DSC	93.9	87.6	**91.4**	68.7	53.5
	AUROC	87.1	65.3	67.8	76.9	39.2
UDA [22]	DSC	94.2	57.7	64.9	65.1	35.2
	AUROC	84.6	61.9	67.9	67.2	55.3
Proposed method	DSC	**94.8**	**89.7**	91.2	**77.9**	**65.9**
	AUROC	**92.3**	**67.3**	**69.3**	**83.0**	62.4

augmentation, which fail to improve performance on target sources, especially in classification tasks, when using a small-sized, single-source training dataset. Our method outperforms AutoAug [10], which relies on large samples to obtain the optimal augmentation strategy. UDA [22] is hard to train with a small-sized dataset due to overfitting and the complex adversarial training.

4 Conclusion

We proposed an augmentation framework based on style transfer method to improve the segmentation and classification performance of the network on ultrasound images from multiple sources. Our framework consists of MixStyleAug, FeatAug, and MaskAug. MixStyleAug integrates the image information from various sources for well generalization, while FeatAug increases the number of styles at the feature level to compensate for potential style variations. MaskAug uses the segmentation results to guide the network to focus on the style information of the ROI in the ultrasound image. We evaluated our framework on five datasets from various sources, and the results showed that our framework improved the segmentation and classification performance across different sources.

References

1. Litjens, G., et al.: A survey on deep learning in medical image analysis. Med. Image Anal. **42**, 60–88 (2017)
2. Shen, D., Wu, G., Suk, H.-I.: Deep learning in medical image analysis. Annu. Rev. Biomed. Eng. **19**(1), 221–248 (2017)
3. Zhang, L., et al.: Generalizing deep learning for medical image segmentation to unseen domains via deep stacked transformation. IEEE Trans. Med. Imaging **39**(7), 2531–2540 (2020)
4. Liu, Z., et al.: Remove appearance shift for ultrasound image segmentation via fast and universal style transfer. In: 2020 IEEE 17th International Symposium on Biomedical Imaging (ISBI) (2020)
5. Gao, Y., et al.: A universal intensity standardization method based on a many-to-one weak-paired cycle generative adversarial network for magnetic resonance images. IEEE Trans. Med. Imaging **38**(9), 2059–2069 (2019)
6. Isensee, F., et al.: nnU-Net: a self-configuring method for deep learning-based biomedical image segmentation. Nat. Methods **18**(2), 203–211 (2021)
7. Jackson, P.T., et al.: Style augmentation: data augmentation via style randomization. In: CVPR Workshops (2019)
8. Hesse, L.S., et al.: Intensity augmentation to improve generalizability of breast segmentation across different MRI scan protocols. IEEE Trans. Biomed. Eng. **68**(3), 759–770 (2021)
9. Yamashita, R., et al.: Learning domain-agnostic visual representation for computational pathology using medically-irrelevant style transfer augmentation. IEEE Trans. Med. Imaging **40**(12), 3945–3954 (2021)
10. Cubuk, E. D., et al.: Autoaugment: learning augmentation policies from data. arXiv preprint arXiv:1805.09501 (2018)

11. Jing, Y., et al.: Neural style transfer: a review. IEEE Trans. Visual. Comput. Graph. **26**(11), 3365–3385 (2020)
12. Andreini, P., et al.: Image generation by GAN and style transfer for agar plate image segmentation. Comput. Methods Prog. Biomed. **184**(105268) (2020)
13. Yang, X., et al.: Generalizing deep models for ultrasound image segmentation. In: Frangi, A.F., Schnabel, J.A., Davatzikos, C., Alberola-López, C., Fichtinger, G. (eds.) MICCAI 2018. LNCS, vol. 11073, pp. 497–505. Springer, Cham (2018). https://doi.org/10.1007/978-3-030-00937-3_57
14. Yoo, J., et al.: Photorealistic style transfer via wavelet transforms. In: Proceedings of the IEEE/CVF International Conference on Computer Vision (2019)
15. Hendrycks, D., et al.: Augmix: asimple data processing method to improve robustness and uncertainty. arXiv preprint arXiv:1912.02781 (2019)
16. Ioffe, S., Szegedy, C.: Batch normalization: accelerating deep network training by reducing internal covariate shift. arXiv preprint arXiv:1502.03167 (2015)
17. Huang, X., Belongie, S.: Arbitrary style transfer in real-time with adaptive instance normalization. In: Proceedings of the IEEE International Conference on Computer Vision (2017)
18. Tang, Z., et al.: Crossnorm and selfnorm for generalization under distribution shifts. In: Proceedings of the IEEE/CVF International Conference on Computer Vision (2021)
19. Milletari, F., Navab, N., Ahmadi, S.: V-net: fully convolutional neural networks for volumetric medical image segmentation. In: 2016 Fourth International Conference on 3D Vision (3DV) (2016)
20. Loshchilov, I., Hutter, F.: SGDR: stochastic gradient descent with warm restarts. arXiv preprint arXiv:1608.03983 (2016)
21. Loshchilov, I., Hutter, F.: Decoupled weight decay regularization. In: International Conference on Learning Representations, New Orleans (2019)
22. Ganin, Y., Lempitsky, V.: Unsupervised domain adaptation by backpropagation. In: Proceedings of the 32nd International Conference on Machine Learning, pp. 1180–1189 (2015)

HC-Net: Hybrid Classification Network for Automatic Periodontal Disease Diagnosis

Lanzhuju Mei[1,2,4], Yu Fang[1,2,4], Zhiming Cui[1], Ke Deng[3], Nizhuan Wang[1], Xuming He[2], Yiqiang Zhan[4], Xiang Zhou[4], Maurizio Tonetti[3](✉), and Dinggang Shen[1,4,5](✉)

[1] School of Biomedical Engineering, ShanghaiTech University, Shanghai, China
dgshen@shanghaitech.edu.cn
[2] School of Information Science and Technology, ShanghaiTech University, Shanghai, China
[3] Shanghai Ninth People's Hospital, Shanghai Jiao Tong University, Shanghai, China
maurizio.tonetti@ergoperio.eu
[4] Shanghai United Imaging Intelligence Co. Ltd., Shanghai, China
[5] Shanghai Clinical Research and Trial Center, Shanghai, China

Abstract. Accurate periodontal disease classification from panoramic X-ray images is of great significance for efficient clinical diagnosis and treatment. It has been a challenging task due to the subtle evidence in radiography. Recent methods attempt to estimate bone loss on these images to classify periodontal diseases, relying on the radiographic manual annotations to supervise segmentation or keypoint detection. However, these radiographic annotations are inconsistent with the clinical golden standard of probing measurements and thus can lead to measurement errors and unstable classifications. In this paper, we propose a novel hybrid classification framework, HC-Net, for accurate periodontal disease classification from X-ray images, which consists of three components, i.e., tooth-level classification, patient-level classification, and a learnable adaptive noisy-OR gate. Specifically, in the tooth-level classification, we first introduce instance segmentation to capture each tooth, and then classify the periodontal disease in the tooth level. As for the patient level, we exploit a multi-task strategy to jointly learn patient-level classification and classification activation map (CAM) that reflects the confidence of local lesion areas upon the panoramic X-ray image. Eventually, the adaptive noisy-OR gate obtains a hybrid classification by integrating predictions from both levels. Extensive experiments on the dataset collected from real-world clinics demonstrate that our proposed HC-Net achieves state-of-the-art performance in periodontal disease classification and shows great application potential. Our code is available at https://github.com/ShanghaiTech-IMPACT/Periodental_Disease.

1 Introduction

Periodontal disease is a set of inflammatory gum infections damaging the soft tissues in the oral cavity, and one of the most common issues for oral health [4].

H. Greenspan et al. (Eds.): MICCAI 2023, LNCS 14225, pp. 54–63, 2023.
https://doi.org/10.1007/978-3-031-43987-2_6

If not diagnosed and treated promptly, it can develop into irreversible loss of the bone and tissue that support the teeth, eventually causing tooth loosening or even falling out. Thus, it is of great significance to accurately classify periodontal disease in an early stage. However, in clinics, dentists have to measure the clinical attachment loss (CAL) of each tooth by manual probing, and eventually determine the severity and progression of periodontal disease mainly based on the most severe area [14]. This is excessively time-consuming, laborious, and over-dependent on the clinical experience of experts. Therefore, it is essential to develop an efficient automatic method for accurate periodontal disease diagnosis from radiography, i.e., panoramic X-ray images.

With the development of computer techniques, computer-aided diagnosis has been widely applied for lesion detection [8,9] and pathological classification [10] in medical image analysis. However, periodontal disease diagnosis from panoramic X-ray images is a very challenging task. While reliable diagnosis can only be provided from 3D probing measurements of each tooth (i.e. clinical golden standard), evidence is highly subtle to be recognized from radiographic images. Panoramic X-ray images make it even more difficult with only 2D information, along with severe tooth occlusion and distortion. Moreover, due to this reason, it is extremely hard to provide confident and consistent radiographic annotations on these images, even for the most experienced experts. Many researchers have already attempted to directly measure radiographic bone loss from panoramic X-ray images for periodontal disease diagnosis. Chang et al. [2] employ a multi-task framework to simulate clinical probing, by detecting bone level, cementoenamel junction (CEJ) level, and tooth long axis. Jiang et al. [7] propose a two-stage network to calculate radiographic bone loss with tooth segmentation and keypoint object detection. Although these methods provide feasible strategies, they still rely heavily on radiographic annotations that are actually not convincing. These manually-labeled landmarks are hard to accurately delineate and usually inconsistent with clinical diagnosis by probing measurements. For this reason, the post-estimated radiographic bone loss is easily affected by prediction errors and noises, which can lead to incorrect and unstable diagnosis.

To address the aforementioned challenges and limitations of previous methods, we propose HC-Net, a novel hybrid classification framework for automatic periodontal disease diagnosis from panoramic X-ray images, which significantly learns from clinical probing measurements instead of any radiographic manual annotations. The framework learns upon both tooth-level and patient-level with three major components, including tooth-level classification, patient-level classification, and an adaptive noisy-OR gate. Specifically, tooth-level classification first applies tooth instance segmentation, then extracts features from each tooth and predicts a tooth-wise score. Meanwhile, patient-level classification provides patient-wise prediction with a multi-task strategy, simultaneously learning a classification activation map (CAM) to show the confidence of local lesion areas upon the panoramic X-ray image. Most importantly, a learnable adaptive noisy-OR gate is designed to integrate information from both levels, with the tooth-level

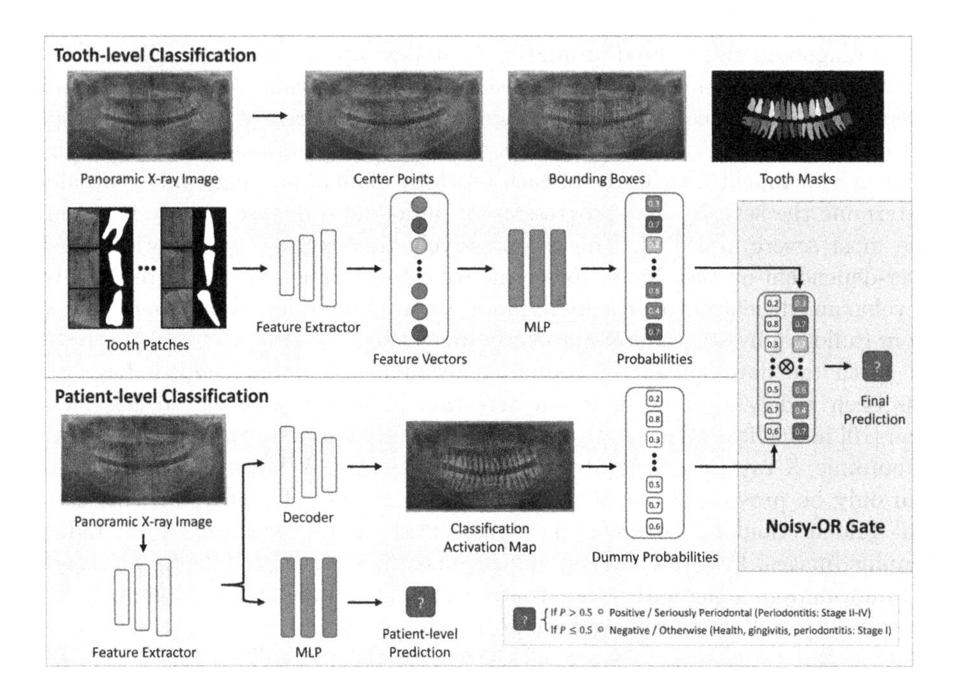

Fig. 1. Illustration of our HC-Net.

scores and patient-level CAM. Note that our classification is only supervised by the clinical golden standard, i.e., probing measurements. We provide comprehensive learning and integration on both tooth-level and patient-level classification, eventually contributing to confident and stable diagnosis. Our proposed HC-Net is validated on the dataset from real-world clinics. Experiments have demonstrated the outstanding performance of our hybrid structure for periodontal disease diagnosis compared to state-of-the-art methods.

2 Method

An overview of our proposed framework, HC-Net, is shown in Fig. 1. We first formulate our task (Sect. 2.1), and then elaborate the details of tooth-level classification (Sect. 2.2), patient-level classification (Sect. 2.3), and adaptive noisy-OR gate (Sect. 2.4), respectively.

2.1 Task Formulation and Method Overview

In this paper, we aim to classify each patient into seriously periodontal (including periodontitis stage II-IV) or not (including health, gingivitis, and periodontitis stage I), abbreviated below as 'positive' or 'negative'. We collect a set of panoramic X-ray images $\mathcal{X} = \{\mathcal{X}_1, \mathcal{X}_2, \dots, \mathcal{X}_N\}$ with their patient-level labels $\mathcal{Y} = \{\mathcal{Y}_1, \mathcal{Y}_2, \ldots, \mathcal{Y}_N\}$ from clinical diagnosis, where $\mathcal{Y}_i \in \{0, 1\}$ indicates whether the i-th patient is negative (0) or positive (1). For the i-th patient, we acquire corresponding tooth-level labels $\mathcal{T}_i = \{\mathcal{T}_i^1, \mathcal{T}_i^2, \ldots, \mathcal{T}_i^{K_i}\}$ from clinical golden standard, where K_i denotes the number of teeth, and $\mathcal{T}_i^j \in \{0, 1\}$ indicates whether the j-th tooth of the i-th patient is positive or negative.

Briefly, our goal is to build a learning-based framework to predict the probability $\mathcal{P}_i \in [0, 1]$ of the i-th patient from panoramic X-ray image. An intuitive solution is to directly perform patient-level classification upon panoramic X-ray images. However, it fails to achieve stable and satisfying results (See Sect. 3.2), mainly for the following two reasons. Firstly, evidence is subtle to be recognized in the large-scale panoramic X-ray image (notice that clinical diagnosis relies on tedious probing around each tooth). Secondly, as we supervise the classification with clinical golden standard (i.e., probing measurements), a mapping should be well designed and established from radiography to this standard, since the extracted discriminative features based on radiography may not be well consistent with the golden standard. Therefore, as shown in Fig. 1, we propose a novel hybrid classification framework to learn upon both tooth-level and patient-level, and a learnable adaptive noisy-OR gate that integrates the predictions from both labels and returns the final classification (i.e., positive or negative).

2.2 Tooth-Level Classification

Given the panoramic X-ray image of the i-th patient, we propose a two-stage structure for tooth-level classification, which first captures each tooth with tooth instance segmentation, and then predicts the classification of each tooth. Tooth instance segmentation aims to efficiently detect each tooth with its centroid, bounding box, and mask, which are later used to enhance tooth-level learning. It introduces a detection network with Hourglass [12] as the backbone, followed by three branches, including tooth center regression, bounding box regression, and tooth semantic segmentation. Specifically, the first branch generates tooth center heatmap $\mathcal{H}$. We obtain the filtered heatmap $\tilde{\mathcal{H}}$ to get center points for each tooth, by a kernel that retains the peak value for every 8-adjacent, described as

$$\mathcal{H}_{p_c} = \begin{cases} \mathcal{H}_{p_c}, \text{ if } \mathcal{H}_{p_c} \geq \mathcal{H}_{\mathbf{p}_j}, \forall \mathbf{p}_j \in \mathbf{p} \\ 0, \text{ otherwise} \end{cases}, \quad (1)$$

where we denote $\mathbf{p} = \{p_c + e_i\}_{i=1}^8$ as the set of 8-adjacent, where p_c is the center point and $\{e_i\}_{i=1}^8$ is the set of direction vectors. The second branch then uses the center points and image features generated by the backbone to regress the bounding box offsets. The third branch utilizes each bounding box to crop the original panoramic X-ray image and segment each tooth. Eventually, with the image patch and corresponding mask $\mathcal{A}_i^j$ for the j-th tooth of the i-th patient, we employ a classification network (i.e., feature extractor and MLP) to predict the probability $\tilde{\mathcal{T}}_i^j$, if the tooth being positive. To train the tooth-level framework, we design a multi-term objective function to supervise the learning process. Specifically, for tooth center regression, we employ the focal loss of [16] to calculate the heatmap error, denoted as $\mathcal{L}_{ctr}$. For bounding box regression, we utilize L1 loss to calculate the regression error, denoted as $\mathcal{L}_{bbx}$. For tooth semantic segmentation, we jointly compute the cross-entropy loss and dice loss, denoted as $\mathcal{L}_{seg} = 0.5 \times (\mathcal{L}_{seg_{CE}} + \mathcal{L}_{seg_{Dice}})$. We finally supervise the tooth-level classification with a cross-entropy loss, denoted as $\mathcal{L}_{cls_t}$. Therefore, the total loss of the tooth-level classification is formulated as $\mathcal{L}_{tooth} = \mathcal{L}_{ctr} + 0.1 \times \mathcal{L}_{bbx} + \mathcal{L}_{seg} + \mathcal{L}_{cls_t}$.

2.3 Patient-Level Classification

As described in Sect. 2.1, although patient-level diagnosis is our final goal, direct classification is not a satisfying solution, and thus we propose a hybrid classification network on both tooth-level and patient-level. Additionally, to enhance patient-level classification, we introduce a multi-task strategy that simultaneously predicts the patient-level classification and a classification activation map (CAM). The patient-level framework first utilizes a backbone network to extract image features for its following two branches. One branch directly determines whether the patient is positive or negative through an MLP, which makes the extracted image features more discriminative. We mainly rely on the other branch, which transforms the image features into CAM to provide local confidence upon the panoramic X-ray image.

Specifically, for the i-th patient, with the predicted area $\{\mathcal{A}_i^j\}_{j=1}^{K_i}$ of each tooth and the CAM $\mathcal{M}_i$, the intensity $\mathcal{I}$ of the j-th tooth can be obtained, described as

$$\mathcal{I}_i^j = \mathcal{C}(\mathcal{M}_i, \mathcal{A}_i^j), \tag{2}$$

where $\mathcal{C}(\cdot, *)$ denotes the operation that crops $\cdot$ with the area of $*$. To supervise the CAM, we generate a distance map upon the panoramic X-ray image, based on Euclidean Distance Transform with areas of positive tooth masks. In this way, we train patient-level classification in a multi-task scheme, jointly with direct classification and CAM regression, which increases the focus on possible local areas of lesions and contributes to accurate classification. We introduce two terms to train the patient-level framework, including a cross-entropy loss $\mathcal{L}_{cls_p}$ to supervise the classification, and a mean squared loss $\mathcal{L}_{CAM}$ to supervise the regression for CAM. Eventually, the total loss of the patient-level classification is $\mathcal{L}_{patient} = \mathcal{L}_{cls_p} + \mathcal{L}_{CAM}$.

2.4 Learnable Adaptive Noisy-OR Gate

We finally present a learnable adaptive noisy-OR gate [13] to integrate tooth-level classification and patient-level classification. To further specify the confidence of local lesion areas on CAM, we propose to learn dummy probabilities $\mathcal{D}_i^j$ for each tooth with its intensity $\mathcal{I}_i^j$

$$\mathcal{D}_i^j = \Phi(\mathcal{I}_i^j), \tag{3}$$

where Φ denotes the pooling operation.

In this way, as shown in Fig. 1, we have obtained tooth-wise probabilities predicted from both tooth-level (i.e., probabilities $\{\tilde{\mathcal{T}}_i^j\}_{j=1}^{K_i}$) and patient-level (i.e., dummy probabilities $\{\mathcal{D}_i^j\}_{j=1}^{K_i}$). We then formulate the final diagnosis as hybrid classification, by designing a novel learnable adaptive noisy-OR gate to aggregate these probabilities, described as

$$\tilde{\mathcal{Y}}_i = 1 - \prod_{j \in \mathcal{G}_i} \mathcal{D}_i^j (1 - \tilde{\mathcal{T}}_i^j), \tag{4}$$

where $\tilde{\mathcal{Y}}_i$ is the final prediction of the i-th patient, $\mathcal{G}_i$ is the subset of tooth numbers. We employ the binary cross entropy loss $\mathcal{L}_{gate}$ to supervise the learning of adaptive noisy-OR Gate. Eventually, the total loss $\mathcal{L}$ of our complete hybrid classification framework is formulated as $\mathcal{L} = \mathcal{L}_{tooth} + \mathcal{L}_{patient} + \mathcal{L}_{gate}$.

3 Experiments

3.1 Dataset and Evaluation Metrics

To evaluate our framework, we collect 426 panoramic X-ray images of different patients from real-world clinics, with the same size of 2903 × 1536. Each patient has corresponding clinical records of golden standard, measured and diagnosed by experienced experts. We randomly split these 426 scans into three sets, including 300 for training, 45 for validation, and 81 for testing. To quantitatively evaluate the classification performance of our method, we report the following metrics, including accuracy, F1 score, and AUROC. Accuracy directly reflects the performance of classification. F1 score further supports the accuracy with the harmonic mean of precision and recall. AUROC additionally summarizes the performance over all possible classification thresholds.

3.2 Comparison with Other Methods

We mainly compare our proposed HC-Net with several state-of-the-art classification networks, which can be adapted for periodontal disease diagnosis. ResNet [5], DenseNet [6], and vision transformer [3] are three of the most representative classification methods, which are used to perform patient-level classification as competing methods. We implement TC-Net as an approach for tooth-level classification, which extracts features respectively from all tooth patches, and all features are concatenated together to directly predict the diagnosis. Moreover, we notice the impressive performance of the multi-task strategy in medical imaging classification tasks [15], and thus adopt MTL [1] to perform multi-task learning scheme. Note that we do not include [2,7] in our comparisons, as they do not consider the supervision by golden standard and heavily rely on unconvincing radiographic manual annotations, which actually cannot be applied in clinics. We employed the well-studied CenterNet [16] for tooth instance segmentation, achieving promising detection (mAP50 of 93%) and segmentation (DICE of 91%) accuracy.

Table 1. Quantitative comparison with representative classification methods for periodontal disease classification.

Method	Accuracy (%)	F1 Score (%)	AUROC (%)
ResNet [5]	80.25	83.33	91.24
DenseNet [6]	86.42	87.36	93.30
TC-Net	85.19	87.50	93.49
xViTCOS [11]	86.42	88.17	94.24
MTL [1]	87.65	89.80	95.12
HC-Net	**92.59**	**93.61**	**95.81**

As shown in Table 1, our HC-Net outperforms all other methods by a large margin. Compared to the patient-level classification methods (such as, ResNet [5], DenseNet [6] and transformer-based xViTCOS [11]) and the tooth-level classification method (TC-Net), MTL [1] achieves better performance and robustness in terms of all metrics, showing the significance of learning from both levels with multi-task strategy. Compared to MTL, we exploit the multi-task strategy with CAM in the patient-level, and design an effective adaptive noisy-OR gate to integrate both levels. Although the DeLong test doesn't show a significant difference, the boosting of all metrics (e.g., accuracy increase from 87.65% to 92.59%) demonstrates the contributions of our better designs that can aggregate both levels more effectively.

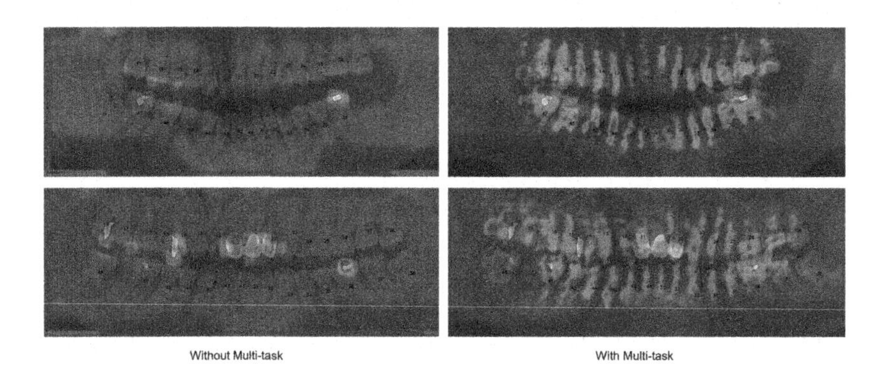

Fig. 2. Illustration of classification activation maps to validate the effectiveness of multi-task strategy with CAM. The first and second columns are visualized respectively from B-Net and M-Net.

3.3 Ablation Studies

We conduct ablative experiments to validate the effectiveness of each module in HC-Net, including patient-level multi-task strategy with classification activation map (CAM) and hybrid classification with adaptive noisy-OR gate. We first define the baseline network, called B-Net, with only patient-level classification. Then, we enhance B-Net with the multi-task strategy, denoted as M-Net, which involves CAM for joint learning on the patient level. Eventually, we extend B-Net to our full framework HC-Net, introducing the tooth-level classification and the adaptive noisy-OR gate.

Effectiveness of Multi-task Strategy with CAM. We mainly compare M-Net to B-Net to validate the multi-task strategy with CAM. We show the classification activation area of both methods as the qualitative results in Fig. 2. Obviously, the activation area of B-Net is almost evenly distributed, while M-Net concentrates more on the tooth area. It shows great potential in locating evidence on local areas of the large-scale panoramic X-ray image, which discriminates the features to support classification. Eventually, it contributes to more accurate qualitative results, as shown in Table 2 and Fig. 3.

Table 2. Quantitative comparison for ablation study.

Method	Accuracy (%)	F1 Score (%)	AUROC (%)
B-Net	86.42	87.36	93.30
M-Net	90.12	91.67	95.37
HC-Net	92.59	93.61	95.81

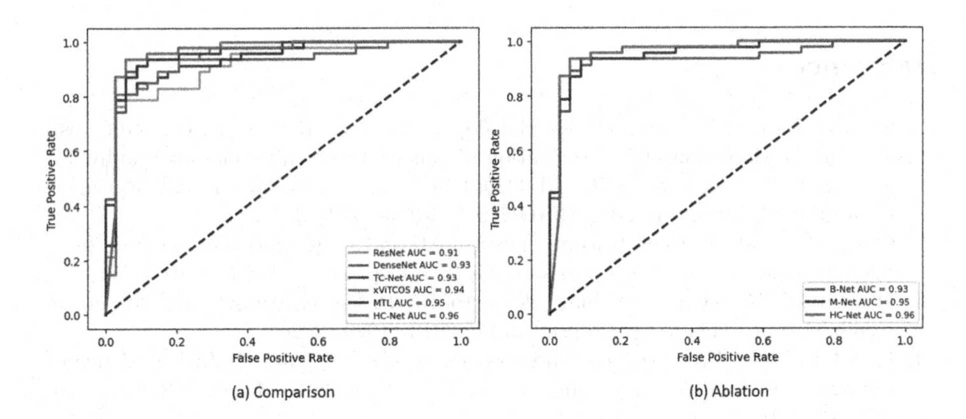

Fig. 3. Illustration of the ROC curves for the comparison and ablation.

Effectiveness of Hybrid Classification with Noisy-OR Gate. We eventually utilize hybrid classification with adaptive noisy-OR Gate, comparing our full framework HC-Net to M-Net. In Table 2 and Fig. 3, we observe that all metrics are dramatically improved. Specifically, the accuracy and F1 score are boosted from 90.12% and 91.67%, to 92.59% and 93.61%, respectively. Note that the AUROC is also significantly increased to 95.81%, which verifies that hybrid classification with noisy-OR gate can improve both the accuracy and robustness of our framework.

3.4 Implementation Details

Our framework is implemented based on the PyTorch platform and is trained with a total of 200 epochs on the NVIDIA A100 GPU with 80GB memory. The feature extractors are based on DenseNet [6]. We use the Adam optimizer with the initial learning rate of 0.001, which is divided by 10 every 50 epochs. Note that in the learnable noisy-or gate, we utilize the probabilities of the top 3 teeth to make predictions for the final outcome.

4 Conclusion

We propose a hybrid classification network, HC-Net, for automatic periodontal disease diagnosis from panoramic X-ray images. In tooth-level, we introduce instance segmentation to help extract features for tooth-level classification. In patient-level, we adopt the multi-task strategy that jointly learns the patient-level classification and CAM. Eventually, a novel learnable adaptable noisy-OR gate integrates both levels to return the final diagnosis. Notice that we significantly utilize the clinical golden standard instead of unconvincing radiographic annotations. Extensive experiments have demonstrated the effectiveness of our proposed HC-Net, indicating the potential to be applied in real-world clinics.

References

1. Sainz de Cea, M.V., Diedrich, K., Bakalo, R., Ness, L., Richmond, D.: Multi-task learning for detection and classification of cancer in screening mammography. In: Martel, A.L., et al. (eds.) MICCAI 2020. LNCS, vol. 12266, pp. 241–250. Springer, Cham (2020). https://doi.org/10.1007/978-3-030-59725-2_24
2. Chang, H.J., et al.: Deep learning hybrid method to automatically diagnose periodontal bone loss and stage periodontitis. Sci. Rep. **10**(1), 1–8 (2020)
3. Dosovitskiy, A., et al.: An image is worth 16×16 words: transformers for image recognition at scale. arXiv preprint arXiv:2010.11929 (2020)
4. Eke, P.I., Dye, B.A., Wei, L., Thornton-Evans, G.O., Genco, R.J.: Prevalence of periodontitis in adults in the united states: 2009 and 2010. J. Dent. Res. **91**(10), 914–920 (2012)
5. He, K., Zhang, X., Ren, S., Sun, J.: Deep residual learning for image recognition. In: Proceedings of the IEEE Conference on Computer Vision and Pattern Recognition, pp. 770–778 (2016)

6. Huang, G., Liu, Z., Van Der Maaten, L., Weinberger, K.Q.: Densely connected convolutional networks. In: Proceedings of the IEEE Conference on Computer Vision and Pattern Recognition, pp. 4700–4708 (2017)
7. Jiang, L., Chen, D., Cao, Z., Wu, F., Zhu, H., Zhu, F.: A two-stage deep learning architecture for radiographic assessment of periodontal bone loss (2021)
8. Kim, J., Lee, H.S., Song, I.S., Jung, K.H.: DeNTNet: deep neural transfer network for the detection of periodontal bone loss using panoramic dental radiographs. Sci. Rep. **9**(1), 1–9 (2019)
9. Krois, J., et al.: Deep learning for the radiographic detection of periodontal bone loss. Sci. Rep. **9**(1), 1–6 (2019)
10. Madabhushi, A., Lee, G.: Image analysis and machine learning in digital pathology: challenges and opportunities. Med. Image Anal. **33**, 170–175 (2016)
11. Mondal, A.K., Bhattacharjee, A., Singla, P., Prathosh, A.: xViTCOS: explainable vision transformer based COVID-19 screening using radiography. IEEE J. Trans. Eng. Health Med. **10**, 1–10 (2021)
12. Newell, A., Yang, K., Deng, J.: Stacked hourglass networks for human pose estimation. In: Leibe, B., Matas, J., Sebe, N., Welling, M. (eds.) ECCV 2016. LNCS, vol. 9912, pp. 483–499. Springer, Cham (2016). https://doi.org/10.1007/978-3-319-46484-8_29
13. Srinivas, S.: A generalization of the noisy-or model. In: Uncertainty in Artificial Intelligence, pp. 208–215. Elsevier (1993)
14. Tonetti, M.S., Greenwell, H., Kornman, K.S.: Staging and grading of periodontitis: framework and proposal of a new classification and case definition. J. Periodontol. **89**, S159–S172 (2018)
15. Zhao, Y., Wang, X., Che, T., Bao, G., Li, S.: Multi-task deep learning for medical image computing and analysis: a review. Comput. Biol. Med. **153**, 106496 (2022)
16. Zhou, X., Wang, D., Krähenbühl, P.: Objects as points. arXiv preprint arXiv:1904.07850 (2019)

SegNetr: Rethinking the Local-Global Interactions and Skip Connections in U-Shaped Networks

Junlong Cheng, Chengrui Gao, Fengjie Wang, and Min Zhu(✉)

College of Computer Science, Sichuan University, Chengdu 610065, China
zhumin@scu.edu.cn

Abstract. Recently, U-shaped networks have dominated the field of medical image segmentation due to their simple and easily tuned structure. However, existing U-shaped segmentation networks: 1) mostly focus on designing complex self-attention modules to compensate for the lack of long-term dependence based on convolution operation, which increases the overall number of parameters and computational complexity of the network; 2) simply fuse the features of encoder and decoder, ignoring the connection between their spatial locations. In this paper, we rethink the above problem and build a lightweight medical image segmentation network, called SegNetr. Specifically, we introduce a novel SegNetr block that can perform local-global interactions dynamically at any stage and with only linear complexity. At the same time, we design a general information retention skip connection (IRSC) to preserve the spatial location information of encoder features and achieve accurate fusion with the decoder features. We validate the effectiveness of SegNetr on four mainstream medical image segmentation datasets, with 59% and 76% fewer parameters and GFLOPs than vanilla U-Net, while achieving segmentation performance comparable to state-of-the-art methods. Notably, the components proposed in this paper can be applied to other U-shaped networks to improve their segmentation performance.

Keywords: Local-global interactions · Information retention skip connection · Medical image segmentation · U-shaped networks

1 Introduction

Medical image segmentation has been one of the key aspects in developing automated assisted diagnosis systems, which aims to separate objects or structures in medical images for independent analysis and processing. Normally, segmentation needs to be performed manually by professional physicians, which is time-consuming and error-prone. In contrast, developing computer-aided segmentation algorithms can be faster and more accurate for batch processing. The approach represented by U-Net [1] is a general architecture for medical image segmentation, which generates a hierarchical feature representation of the

H. Greenspan et al. (Eds.): MICCAI 2023, LNCS 14225, pp. 64–74, 2023.
https://doi.org/10.1007/978-3-031-43987-2_7

image through a top-down encoder path and uses a bottom-up decoder path to map the learned feature representation to the original resolution to achieve pixel-by-pixel classification. After U-Net, U-shaped methods based on **Convolutional Neural Networks** (CNN) have been extended for various medical image segmentation tasks [2–9]. They either enhance the feature representation capabilities of the encoder-decoder or carefully design the attention module to focus on specific content in the image. Although these extensions can improve the benchmark approach, the local nature of the convolution limits them to capturing long-term dependencies, which is critical for medical image segmentation. Recently, segmentation methods based on U-shaped networks have undergone significant changes driven by **Transformer** [10,11]. Chen et al [12] proposed the first Transformer-based U-shaped segmentation network. Cao et al [13] extended the Swin Transformer [14] directly to the U-shaped structure. The above methods suffer from high computational and memory cost explosion when the feature map size becomes large. In addition, some researchers have tried to build **Hybrid Networks** by combining the advantages of CNN and Transformer, such as UNeXt [15], TransFuse [16], MedT [17], and FAT-Net [18]. Similar to these works, we redesign the window-based local-global interaction and insert it into a pure convolutional framework to compensate for the deficiency of convolution in capturing global features and to reduce the high computational cost arising from self-attention operations.

Skip connection is the most basic operation for fusing shallow and deep features in U-shaped networks. Considering that this simple fusion does not fully exploit the information, researchers have proposed some novel ways of skip connection [19–22]. UNet++ [19] design a series of dense skip connections to reduce the semantic gap between the encoder and decoder sub-network feature maps. SegNet [20] used the maximum pooling index to determine the location information to avoid the ambiguity problem during up-sampling using deconvolution. BiO-Net [21] proposed bi-directional skip connections to reuse building blocks in a cyclic manner. UCTransNet [22] designed a Transformer-based channel feature fusion method to bridge the semantic gap between shallow and deep features. Our approach focuses on the connection between the spatial locations of the encoder and decoder, preserving more of the original features to help recover the resolution of the feature map in the upsampling phase, and thus obtaining a more accurate segmentation map.

By reviewing the above multiple successful cases based on U-shaped structure, we believe that the efficiency and performance of U-shaped networks can be improved by improving the following two aspects: **(i) local-global interactions.** Often networks need to deal with objects of different sizes in medical images, and local-global interactions can help the network understand the content of the images more accurately. **(ii) Spatial connection between encoder-decoder.** Semantically stronger and positionally more accurate features can be obtained using the spatial information between encoder-decoders. Based on the above analysis, this paper rethinks the design of the U-shaped network. Specifically, we construct lightweight SegNetr (**Seg**mentation **Net**work with Transformer) blocks to dynamically learn local-global information over non-overlapping windows and maintain linear complexity. We propose information retention skip connection (IRSC), which focuses on the connection between encoder and decoder spatial locations, retaining more original features to help recover the resolution of the feature map in the up-sampling phase. In summary, the contributions of this paper can be summarized as follows: 1) We propose a lightweight U-shape SegNetr segmentation network with less computational cost and better segmentation performance. 2) We investigate the potential deficiency of the traditional U-shaped framework for skip connection and improve a skip connection with information retention. 3) When we apply the components proposed in this paper to other U-shaped methods, the segmentation performance obtains a consistent improvement.

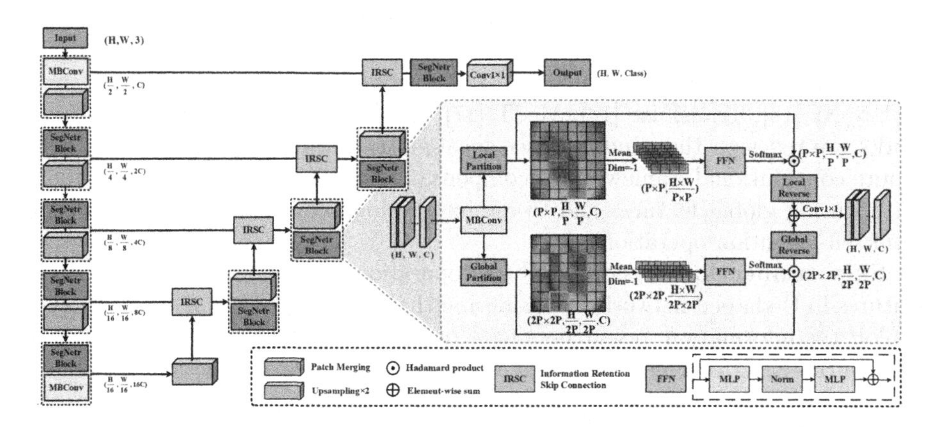

Fig. 1. Overview of the SegNetr approach. SegNetr blocks interact through parallel local and global branches. IRSC preserves the positional information of encoder features and achieves accurate fusion with decoder features.

2 Method

As shown in Fig. 1, SegNetr is a hierarchical U-shaped network with important components including SegNetr blocks and IRSC. To make the network more lightweight, we use MBConv [24] as the base convolutional building block. SegNetr blocks implement dynamic local-global interaction in the encoder and decoder stages. Patch merging [14] is used to reduce the resolution by a factor of two without losing the original image information. IRSC is used to fuse encoder and decoder features, reducing the detailed information lost by the network as the depth deepens. Note that by changing the number of channels, we can get the smaller version of SegNetr-S (C = 32) and the standard version of SegNetr (C = 64). Next, we will explain in detail the important components in SegNetr.

2.1 SegNetr Block

The self-attention mechanism with global interactions is one of the keys to Transformer's success, but computing the attention matrix over the entire space requires a quadratic complexity. Inspired by the window attention method [14, 23], we construct SegNetr blocks that require only linear complexity to implement local-global interactions. Let the input feature map be $X \in R^{H \times W \times C}$. We first extract the feature $X_{MBConv} \in R^{H \times W \times C}$ using MBConv [24], which provides non-explicit position encoding compared to the usual convolutional layer.

Local interaction can be achieved by calculating the attention matrix of non-overlapping small patches (P for patch size). First, we divide X_{MBConv} into a series of patches $(\frac{H \times W}{P \times P}, P, P, C)$ that are spatially continuous (Fig. 1 shows the patch size for $P = 2$) using a computationally costless local partition (LP) operation. Then, we average the information of the channel dimensions and flatten the spatial dimensions to obtain $(\frac{H \times W}{P \times P}, P \times P)$, which is fed into the FFN [11] for linear computation. Since the importance of the channel aspect is weighed in MBConv [24], we focus on the computation of spatial attention when performing local interactions. Finally, we use Softamx to obtain the spatial dimensional probability distribution and weight the input features X_{MBConv}. This approach is not only beneficial for parallel computation, but also focuses more purely on the importance of the local space.

Considering that local interactions are not sufficient and may have underfitting problems, we also design parallel **global interaction** branches. First, we use the global partition (GP) operation to aggregate non-contiguous patches on the space. GP adds the operation of window displacement to LP with the aim of changing the overall distribution of features in space (The global branch in Fig. 1 shows the change in patch space location after displacement). The displacement rules are one window to the left for odd patches in the horizontal direction (and vice versa for even patches to the right), and one window up for odd patches in the vertical direction (and vice versa for even patches down). Note that the displacement of patches does not have any computational cost and only memory changes occur. Compared to the sliding window operation of [14], our approach is more global in nature. Then, we decompose the spatially shifted feature map into $2P$ $(\frac{H \times W}{2P \times 2P}, 2P, 2P, C)$ patches and perform global attention computation (similar to the local interaction branch). Even though the global interaction computes the attention matrix over a larger window relative to the local interaction operation, the amount of computation required is much smaller than that of the standard self-attention model.

The local and global branches are finally fused by weighted summation, before which the feature map shape needs to be recovered by LP and GP reversal operations (i.e., local reverse (LR) and global reverse (GR)). In addition, our approach also employs efficient designs of Transformer, such as Norm, feed-forward networks (FFN) and residual connections. Most Transformer models use fixed-size patches [11–14, 24], but this approach limits them to focus on a wider range of regions in the early stages. This paper alleviates this problem by applying dynamically sized patches. In the encoder stage, we compute local

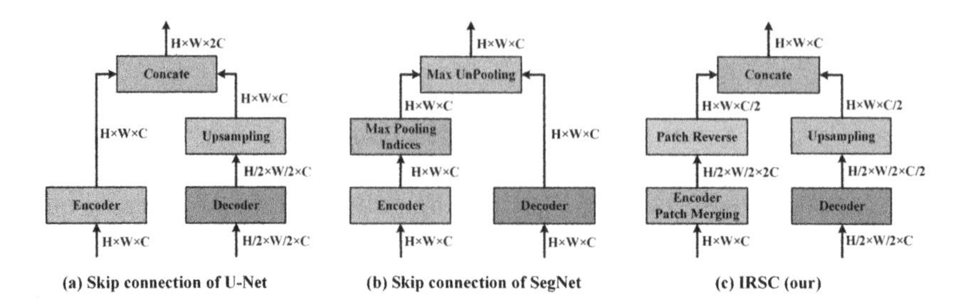

Fig. 2. Comparison of skip connections of U-Net, SegNet and our SegNetr. Our method does not incorporate redundant computable modules, but the patch reverse (PR) provides spatial location information.

attention using patches of $(8, 4, 2, 1)$ in turn, and the global branch expands patches to the size of $(16, 8, 4, 2)$. To reduce the hyper-parameter setting, the patches of the decoder stage are of the same size as the encoder patches of the corresponding stage.

2.2 Information Retention Skip Connection

Figure 2 shows three different types of skip connections. U-Net splices the channel dimensions at the corresponding stages of the encoder and decoder, allowing the decoder to retain more high-resolution detail information when performing up-sampling. SegNet assists the decoder to recover the feature map resolution by retaining the position information of the down-sampling process in the encoder. We design the IRSC to consider both of these features, i.e., to preserve the location information of encoder features while achieving the fusion of shallow and deep features. Specifically, the patch merging (PM) operation in the encoder reduces the resolution of the input feature map $X_{in} \in R^{H \times W \times C}$ to twice the original one, while the channel dimension is expanded to four times the original one to obtain $X_{PM} \in R^{\frac{H}{2} \times \frac{W}{2} \times 4C}$. The essence of the PM operation is to convert the information in the spatial dimension into a channel representation without any computational cost and retaining all the information of the input features. The patch reverse (PR) in IRSC is used to recover the spatial resolution of the encoder, and it is a reciprocal operation with PM. We alternately select half the number of channels of X_{PM} (i.e., $\frac{H}{2} \times \frac{W}{2} \times 2C$) as the input of PR, which can reduce the redundant features in the encoder on the one hand and align the number of feature channels in the decoder on the other hand. PR reduces the problem of information loss to a large extent compared to traditional up-sampling methods, while providing accurate location information. Finally, the output features $X_{PR} \in R^{H \times W \times \frac{C}{2}}$ of PR are fused with the up-sampled features of the decoder for the next stage of learning.

3 Experiments and Discussion

Datasets. To verify the validity of SegNetr, we selected four datasets, ISIC2017 [25], PH2 [26], TNSCUI [27] and ACDC [28], for benchmarking. ISIC2017 consists of 2000 training images, 200 validation images, and 600 test images. The PH2 and ISIC2017 tasks are the same, but this dataset contains only 200 images without any specific test set, so we use a five-fold cross-validation approach to validate the different models. The TNSCUI dataset has 3644 ultrasound images of thyroid nodules, which we randomly divided into a 6:2:2 ratio for training, validation, and testing. The ACDC contains Cardiac MRI images from 150 patients, and we obtained a total of 1489 slice images from 150 3D images, of which 951 were used for training and 538 for testing. Unlike the three datasets mentioned above, the ACDC dataset contains three categories: left ventricle (LV), right ventricle (RV), and myocardium (Myo). We use this dataset to explore the performance of different models for multi-category segmentation.

Implementation Details. We implement the SegNetr method based on the PyTorch framework by training on an NVIDIA 3090 GPU with 24 GB of memory. Use the Adam optimizer with a fixed learning rate of 1e−4. All networks use a cross-entropy loss function and an input image resolution of 224 × 224, and training is stopped when 200 epochs are iteratively optimized. We use the source code provided by the authors to conduct experiments with the same dataset, and data enhancement strategy. In addition, we use the IoU and Dice metrics to evaluate the segmentation performance, while giving the number of parameters and GFLOPs for the comparison models.

Table 1. Quantitative results on ISIC2017 and PH2 datasets.

Network	ISIC2017		PH2		Params	GFLOPs
	IoU	Dice	IoU	Dice		
U-Net [1]	0.736	0.825	0.878 ± 0.025	0.919 ± 0.045	29.59 M	41.83
SegNet [20]	0.696	0.821	0.880 ± 0.020	0.934 ± 0.012	17.94 M	22.35
UNet++ [19]	0.753	0.840	0.883 ± 0.013	0.936 ± 0.008	25.66 M	28.77
FAT-Net [18]	0.765	0.850	0.895 ± 0.019	0.943 ± 0.011	28.23 M	42.83
ResGANet [5]	0.764	0.862	–	–	39.21 M	65.10
nnU-Net [29]	0.760	0.843	–	–	–	–
Swin-UNet [13]	0.767	0.850	0.872 ± 0.022	0.927 ± 0.014	25.86 M	5.86
TransUNet [12]	0.775	0.847	0.887 ± 0.020	0.937 ± 0.012	88.87 M	24.63
UNeXt-L [15]	0.754	0.840	0.884 ± 0.021	0.936 ± 0.013	3.80 M	1.08
SegNetr-S	0.752	0.838	0.889 ± 0.018	0.939 ± 0.011	3.60 M	2.71
SegNetr	0.775	0.856	0.905 ± 0.023	0.948 ± 0.014	12.26 M	10.18

3.1 Comparison with State-of-the-Arts

ISIC2017 and PH2 Results. As shown in Table 1, we compared SegNetr with the baseline U-Net and eight other state-of-the-art methods [5,12,13,15,18–20,29]. On the ISIC2017 dataset, SegNetr and TransUNet obtained the highest IoU (0.775), which is 3.9% higher than the baseline U-Net. Even SegNetr-S with a smaller number of parameters can obtain a segmentation performance similar to that of its UNeXt-L counterpart. By observing the experimental results of PH2, we found that the Transformer-based method Swin-UNet segmentation has the worst performance, which is directly related to the data volume of the target dataset. Our method obtains the best segmentation performance on this dataset and keeps the overhead low. Although we use an attention method based on window displacement, the convolutional neural network has a better inductive bias, so the dependence on the amount of data is smaller compared to Transformer-based methods such as Swin-UNet or TransUNet.

Table 2. Quantitative results on TNSCUI and ACDC datasets.

Network	TNSCUI	ACDC/IoU			Average IoU (Dice)	Params	GFLOPs
	IoU (Dice)	RV	Myo	LV			
U-Net [1]	0.718 (0.806)	0.743	0.717	0.861	0.774 (0.834)	29.59 M	41.83
SegNet [20]	0.726 (0.819)	0.738	0.720	0.864	0.774 (0.836)	17.94 M	22.35
FAT-Net [18]	0.751 (0.842)	0.743	0.702	0.859	0.768 (0.834)	28.23 M	42.83
Swin-UNet [13]	0.744 (0.835)	0.754	0.722	0.865	0.780 (0.843)	25.86 M	5.86
TransUNet [12]	0.746 (0.837)	0.750	0.715	0.866	0.777 (0.838)	88.87 M	24.63
EANet [30]	0.751 (0.839)	0.742	0.732	0.864	0.779 (0.839)	47.07 M	98.63
UNeXt [15]	0.655 (0.749)	0.697	0.646	0.814	0.719 (0.796)	1.40 M	0.44
UNeXt-L [15]	0.693 (0.794)	0.719	0.675	0.840	0.744 (0.815)	3.80 M	1.08
SegNetr-S	0.707 (0.804)	0.723	0.692	0.845	0.753 (0.821)	3.60 M	2.71
SegNetr	0.767 (0.850)	0.761	0.738	0.872	0.791 (0.847)	12.26 M	10.18

TNSCUI and ACDC Results. As shown in Table 2, SegNetr's IoU and Dice are 1.6% and 0.8 higher than those of the dual encoder FATNet, respectively, while the GFLOPs are 32.65 less. In the ACDC dataset, the left ventricle is easier to segment, with an IoU of 0.861 for U-Net, but 1.1% worse than SegNetr. The myocardium is in the middle of the left and right ventricles in an annular pattern, and our method is 0.6% higher IoU than the EANet that focuses on the boundary segmentation mass. In addition, we observe the segmentation performance of the four networks UNeXt, UNeXt-L, SegNetr-S and SegNetr to find that the smaller parameters may limit the learning ability of the network. The proposed method in this paper shows competitive segmentation performance on all four datasets, indicating that our method has good generalization performance and robustness. Additional qualitative results are in the supplementary.

In addition, Fig. 3 provides qualitative examples that demonstrate the effectiveness and robustness of our proposed method. The results show that SegNetr is capable of accurately describing skin lesions with less data, and achieves multi-class segmentation with minimized under-segmentation and over-segmentation.

3.2 Ablation Study

Effect of Local-Global Interactions. The role of local-global interactions in SegNetr can be understood from Table 3. The overall parameters of the network are less when there is no local or global interaction, but the segmentation performance is also greatly affected. With the addition of local or global interactions, the segmentation performance of the network for different categories

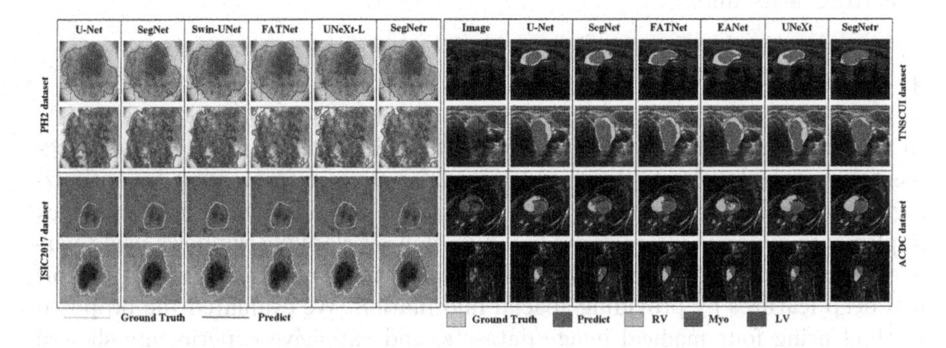

Fig. 3. Qualitative experimental results of different methods on four datasets.

Table 3. Ablation study of local-global interactions on the ACDC dataset.

Settings	ACDC			Average IoU (Dice)	Params	GFLOPs
	RV	Myo	LV			
Without	0.750(0.799)	0.720(0.816)	0.861(0.897)	0.777(0.837)	10.93 M	9.75
Only local	0.753(0.800)	0.733(0.825)	0.868(0.904)	0.785(0.843)	12.22 M	10.18
Only global	0.756(0.803)	0.734(0.827)	0.875(0.909)	0.788(0.846)	11.64 M	10.17
Series	0.761(0.809)	0.732(0.824)	0.871(0.907)	0.788(0.846)	12.26 M	10.18
Parallel	0.761(0.807)	0.738(0.828)	0.872(0.907)	0.791(0.847)	12.26 M	10.18

Table 4. Ablation study of patch size (left) and IRSC (right) on TNSCUI and ISIC2017 datasets.

Patch size	TNSCUI		Params	GFLOPs	Network+IRSC	ISIC2017	
	IoU	Dice				IoU	Dice
(2,2,2,2)	0.751	0.835	54.34 M	10.38	UNeXt-L	0.760(+0.6%)	0.843(+0.3%)
(4,4,4,2)	0.762	0.841	14.32 M	10.22	U-Net	0.744(+0.8%)	0.839(+1.4%)
(8,4,4,2)	0.762	0.843	11.96 M	10.18	UNet++	0.763(+1.0%)	0.845(+0.5%)
(8,4,2,1)	0.767	0.850	12.26 M	10.18	SegNet	0.712(+1.6%)	0.829(+0.8%)

is improved. In addition, similar performance can be obtained by running the local-global interaction modules in series and parallel, but the series connection leads to lower computational efficiency and affects the running speed.

Effect of Patch Size. As shown in Table 4 (left), different patch size significantly affects the efficiency and parameters of the model. The number of parameters reaches 54.34 M when patches of size 2 are used in each phase, which is an increase of 42.08 M compared to using dynamic patches of size (8, 4, 2, 1). Based on this ablation study, we recommend the use of $[\frac{Resolution}{14}]$ patches size at different stages.

Effect of IRSC. Table 4 (right) shows the experimental results of replacing the skip connections of UNeXt, U-Net, U-Net++, and SegNet with IRSC. These methods get consistent improvement with the help of IRSC, which clearly shows that IRSC is useful.

4 Conclusion

In this study, we introduce a novel framework SegNetr for medical image segmentation, which achieves segmentation performance improvement by optimizing local-global interactions and skip connections. Specifically, the SegNetr block implements dynamic interactions based on non-overlapping windows using parallel local and global branches, and IRSC enables more accurate fusion of shallow and deep features by providing spacial information. We evaluated the proposed method using four medical image datasets, and extensive experiments showed that SegNetr is able to obtain challenging experimental results while maintaining a small number of parameters and GFLOPs. The proposed framework is general and flexible that we believe it can be easily extended to other U-shaped networks.

References

1. Ronneberger, O., Fischer, P., Brox, T.: U-Net: convolutional networks for biomedical image segmentation. In: Navab, N., Hornegger, J., Wells, W.M., Frangi, A.F. (eds.) MICCAI 2015. LNCS, vol. 9351, pp. 234–241. Springer, Cham (2015). https://doi.org/10.1007/978-3-319-24574-4_28
2. Ma, Q., Zu, C., Wu, X., Zhou, J., Wang, Y.: Coarse-to-fine segmentation of organs at risk in nasopharyngeal carcinoma radiotherapy. In: de Bruijne, M., et al. (eds.) MICCAI 2021. LNCS, vol. 12901, pp. 358–368. Springer, Cham (2021). https://doi.org/10.1007/978-3-030-87193-2_34
3. Han, Z., Jian, M., Wang, G.G.: ConvUNeXt: an efficient convolution neural network for medical image segmentation. KBS **253**, 109512 (2022)
4. Oktay, O., Schlemper, J., Folgoc, L.L., et al.: Attention U-Net: learning where to look for the pancreas. arXiv preprint arXiv:1804.03999 (2018)
5. Cheng, J., Tian, S., Yu, L., et al.: ResGANet: residual group attention network for medical image classification and segmentation. Med. Image Anal. **76**, 102313 (2022)

6. Wang, K., Zhan, B., Zu, C., et al.: Semi-supervised medical image segmentation via a tripled-uncertainty guided mean teacher model with contrastive learning. Med. Image Anal. **79**, 102447 (2022)
7. Gu, Z., Cheng, J., Fu, H., et al.: Ce-net: context encoder network for 2D medical image segmentation. IEEE TMI **38**(10), 2281–2292 (2019)
8. Wu, Y., et al.: D-former: a U-shaped dilated transformer for 3D medical image segmentation. Neural Comput. Appl. **35**, 1–14 (2022). https://doi.org/10.1007/s00521-022-07859-1
9. Cheng, J., Tian, S., Yu, L., et al.: A deep learning algorithm using contrast-enhanced computed tomography (CT) images for segmentation and rapid automatic detection of aortic dissection. BSPC **62**, 102145 (2020)
10. Dosovitskiy, A., et al.: An image is worth 16×16 words: transformers for image recognition at scale. In: ICLR, pp. 3–7 (2021)
11. Vaswani, A., et al.: Attention is all you need. In: NIPS, vol. 30 (2017)
12. Chen, J., et al.: TransUNet: transformers make strong encoders for medical image segmentation. arXiv preprint arXiv:2102.04306 (2021)
13. Cao, H., et al.: Swin-Unet: Unet-like pure transformer for medical image segmentation. In: Karlinsky, L., Michaeli, T., Nishino, K. (eds.) ECCV 2022. Lecture Notes in Computer Science, vol. 13803, pp. 205–218. Springer, Cham (2023). https://doi.org/10.1007/978-3-031-25066-8_9
14. Liu, Z., et al.: Swin transformer: hierarchical vision transformer using shifted windows. In: IEEE ICCV, pp. 10012–10022 (2021)
15. Valanarasu, J.M.J., Patel, V.M.: UNeXt: MLP-based rapid medical image segmentation network. In: Wang, L., Dou, Q., Fletcher, P.T., Speidel, S., Li, S. (eds.) MICCAI 2022. Lecture Notes in Computer Science, vol. 13435, pp. 23–33. Springer, Cham (2022). https://doi.org/10.1007/978-3-031-16443-9_3
16. Zhang, Y., Liu, H., Hu, Q.: TransFuse: fusing transformers and CNNs for medical image segmentation. In: de Bruijne, M., et al. (eds.) MICCAI 2021. LNCS, vol. 12901, pp. 14–24. Springer, Cham (2021). https://doi.org/10.1007/978-3-030-87193-2_2
17. Valanarasu, J.M.J., Oza, P., Hacihaliloglu, I., Patel, V.M.: Medical transformer: gated axial-attention for medical image segmentation. In: de Bruijne, M., Zheng, Y., Essert, C. (eds.) MICCAI 2021. LNCS, vol. 12901, pp. 36–46. Springer, Cham (2021). https://doi.org/10.1007/978-3-030-87193-2_4
18. Wu, H., Chen, S., Chen, G., et al.: FAT-Net: feature adaptive transformers for automated skin lesion segmentation. Med. Image Anal. **76**, 102327 (2022)
19. Zhou, Z., Rahman Siddiquee, M.M., Tajbakhsh, N., Liang, J.: UNet++: a nested U-Net architecture for medical image segmentation. In: Stoyanov, D., et al. (eds.) DLMIA/ML-CDS -2018. LNCS, vol. 11045, pp. 3–11. Springer, Cham (2018). https://doi.org/10.1007/978-3-030-00889-5_1
20. Badrinarayanan, V., Kendall, A., Cipolla, R.: SegNet: a deep convolutional encoder-decoder architecture for image segmentation. IEEE TPAMI **39**(12), 2481–2495 (2017)
21. Xiang, T., Zhang, C., Liu, D., Song, Y., Huang, H., Cai, W.: BiO-Net: learning recurrent bi-directional connections for encoder-decoder architecture. In: Martel, A.L., et al. (eds.) MICCAI 2020. LNCS, vol. 12261, pp. 74–84. Springer, Cham (2020). https://doi.org/10.1007/978-3-030-59710-8_8
22. Wang, H., et, al.: UCTransNet: rethinking the skip connections in U-Net from a channel-wise perspective with transformer. In: AAAI, vol. 36(3), pp. 2441–2449 (2022)

23. Tu, Z., et al.: MaxViT: multi-axis vision transformer. In: Avidan, S., Brostow, G., Cissé, M., Farinella, G.M., Hassner, T. (eds.) ECCV 2022. Lecture Notes in Computer Science, vol. 13684, pp. 459–479. Springer, Cham (2022)
24. Tan, M., Le, Q.: EfficientNet: rethinking model scaling for convolutional neural networks. In: ICML, PP. 6105–6114 (2019)
25. Quang, N.H.: Automatic skin lesion analysis towards melanoma detection. In: IES, pp. 106–111. IEEE (2017)
26. Mendonça, T., et al.: PH 2-A dermoscopic image database for research and benchmarking. In: EMBC, pp. 5437–5440. IEEE (2013)
27. Pedraza, L., et al.: An open access thyroid ultrasound image database. In: SPIE, vol. 9287, pp. 188–193 (2015)
28. Bernard, O., Lalande, A., Zotti, C., et al.: Deep learning techniques for automatic MRI cardiac multi-structures segmentation and diagnosis: is the problem solved? IEEE TMI **37**(11), 2514–2525 (2018)
29. Isensee, F., Jaeger, P.F., Kohl, S.A.A., et al.: nnU-Net: a self-configuring method for deep learning-based biomedical image segmentation. Nat. Methods **18**(2), 203–211 (2021)
30. Wang, K., Zhang, X., Zhang, X., et al.: EANet: iterative edge attention network for medical image segmentation. Pattern Recogn. **127**, 108636 (2022)

Gradient and Feature Conformity-Steered Medical Image Classification with Noisy Labels

Xiaohan Xing[1], Zhen Chen[2], Zhifan Gao[3], and Yixuan Yuan[4](✉)

[1] Department of Radiation Oncology, Stanford University, Stanford, CA, USA
[2] Centre for Artificial Intelligence and Robotics (CAIR), Hong Kong Institute of Science & Innovation, Chinese Academy of Sciences, NT, Hong Kong SAR, China
[3] School of Biomedical Engineering, Sun Yat-sen University, Guangdong, China
[4] Department of Electronic Engineering, Chinese University of Hong Kong, NT, Hong Kong SAR, China
yxyuan@ee.cuhk.edu.hk

Abstract. Noisy annotations are inevitable in clinical practice due to the requirement of labeling efforts and expert domain knowledge. Therefore, medical image classification with noisy labels is an important topic. A recently advanced paradigm in learning with noisy labels (LNL) first selects clean data with small-loss criterion, then formulates the LNL problem as semi-supervised learning (SSL) task and employs Mixup to augment the dataset. However, the small-loss criterion is vulnerable to noisy labels and the Mixup operation is prone to accumulate errors in pseudo labels. To tackle these issues, we present a two-stage framework with novel criteria for clean data selection and a more advanced Mixup method for SSL. In the clean data selection stage, based on the observation that gradient space reflects optimization dynamics and feature space is more robust to noisy labels, we propose two novel criteria, i.e., *Gradient Conformity-based Selection* (GCS) and *Feature Conformity-based Selection* (FCS), to select clean samples. Specifically, the GCS and FCS criteria identify clean data that better aligns with the class-wise optimization dynamics in the gradient space and principal eigenvector in the feature space. In the SSL stage, to effectively augment the dataset while mitigating disturbance of unreliable pseudo-labels, we propose a *Sample Reliability-based Mixup* (SRMix) method which selects mixup partners based on their spatial reliability, temporal stability, and prediction confidence. Extensive experiments demonstrate that the proposed framework outperforms state-of-the-art methods on two medical datasets with synthetic and real-world label noise. The code is available at https://github.com/hathawayxxh/FGCS-LNL.

Keywords: Label noise · Gradient conformity · Feature eigenvector conformity · Mixup

---

Supplementary Information The online version contains supplementary material available at https://doi.org/10.1007/978-3-031-43987-2_8.

H. Greenspan et al. (Eds.): MICCAI 2023, LNCS 14225, pp. 75–84, 2023.
https://doi.org/10.1007/978-3-031-43987-2_8

1 Introduction

Deep neural networks (DNNs) have achieved remarkable success in medical image classification. However, the great success of DNNs relies on a large amount of training data with high-quality annotations, which is practically infeasible. The annotation of medical images requires expert domain knowledge, and suffers from large intra- and inter-observer variability even among experts, thus noisy annotations are inevitable in clinical practice. Due to the strong memorization ability, DNNs can easily over-fit the corrupted labels and degrade performance [1,2], thus it is crucial to train DNNs that are robust to noisy labels.

An effective paradigm in learning with noisy labels (LNL) first selects clean samples, then formulates the LNL problem as semi-supervised learning (SSL) task by regarding the clean samples as a labeled set and noisy samples as an unlabeled set [3]. However, both the clean data selection and SSL stages in existing methods have some drawbacks. In the clean data selection stage, most existing studies rely on the small-loss [3,4] or high-confidence criteria [5] of individual samples, but neglect the global contextual information and high-order topological correlations among samples, thus unavoidably resulting in confirmation bias [6]. Besides, the above criteria in the output space are directly supervised and easily affected by corrupted labels [7]. Previous studies indicate that optimization dynamics (characterized by sample gradients) can reflect the true class information [8] and feature space is more robust to noisy labels, thus can provide more robust criteria for clean data selection [6,7,9]. Therefore, we aim to achieve more accurate clean data selection by exploring the topological correlation and contextual information in the robust gradient and feature spaces.

In the SSL stage, most existing studies [3,10] estimate pseudo labels for all samples and employ Mixup [11,12] to linearly interpolate the input samples and their pseudo labels for model training. Compared with previous methods that train DNNs by reweighting samples [13] or utilizing clean data only [4,14], the Mixup [12] operation can effectively augment the dataset and regularize the model from over-fitting. However, as the pseudo labels of noisy datasets cannot be always reliable, the traditional Mixup method which randomly chooses the mixup partner for each sample may accumulate errors in pseudo labels. Therefore, it is highly desirable to design a novel Mixup method that can select reliable mixup partners and mitigate the interference of unreliable pseudo labels.

In this paper, we present a novel two-stage framework to combat noisy labels in medical image classification. In the clean data selection stage, we propose a gradient and feature conformity-based method to identify the samples with clean labels. Specifically, the *Gradient Conformity-based Selection* (GCS) criterion selects clean samples that show higher conformity with the principal gradient of its labeled class. The *Feature Conformity-based Selection* (FCS) criterion identifies clean samples that show better alignment with the feature eigenvector of its labeled class. In the SSL stage, we propose a *Sample Reliability-based Mixup* (SRMix) to augment the training data without aggravating the error accumulation of pseudo labels. Specifically, SRMix interpolates each sample with reliable mixup partners which are selected based on their spatial reliability, temporal stability, and prediction confidence. Our main contributions are as follows:

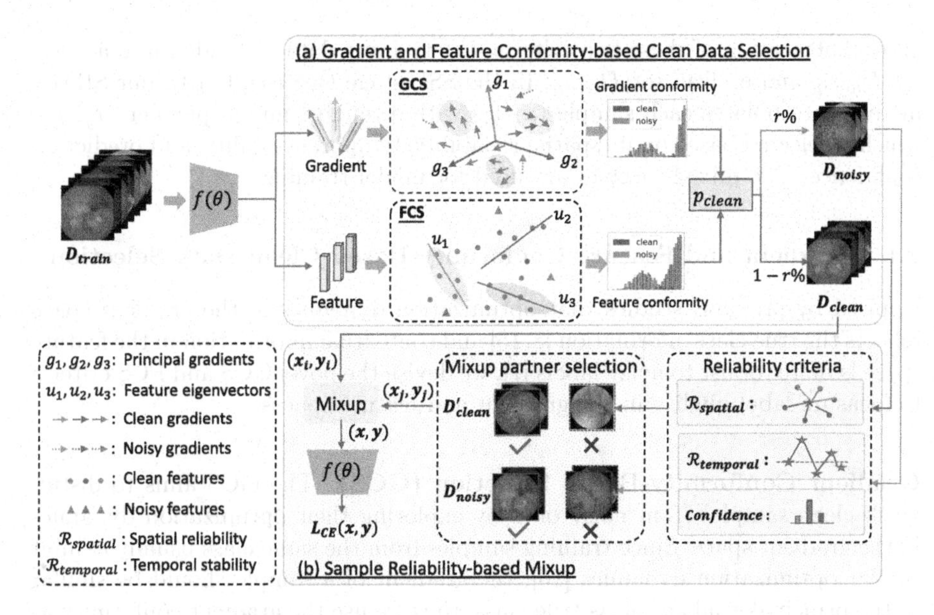

Fig. 1. The framework of our method. (a) Gradient and feature conformity-based clean data selection module, including GCS and FCS criteria, divides training samples into a clean set D_{clean} and a noisy set D_{noisy}. (b) Sample Reliability-based Mixup (SRMix) module interpolates each sample (x_i, y_i) with a reliable mixup partner (x_j, y_j).

- We devise two novel criteria (i.e., GCS and FCS) to improve clean data selection by exploring the topological correlation and contextual information in the gradient and feature spaces.
- We propose a novel SRMix method that selects reliable mixup partners to mitigate the error accumulation of pseudo labels and improve model training.
- Extensive experiments show that our proposed framework is effective in combating label noise and outperforms state-of-the-art methods on two medical datasets with both synthetic and real-world label noise.

2 Method

An overview of our proposed two-stage framework is shown in Fig. 1. The training dataset is denoted as $D_{train} \in \{(x_i, y_i)\}_{i=1}^{N}$, where the given label y_i could be noisy or clean. In the clean data selection stage, we propose a gradient and feature conformity-based method to distinguish clean samples from the noisy dataset. As shown in Fig. 1 (a), the GCS computes the principal gradient of each class (i.e., g_1, g_2, g_3) to represent its optimization dynamics, and measures the label quality of each sample by its gradient conformity with the class-wise principal gradient. The FCS computes the principal feature eigenvector of each class to reflect its contextual information, and measures the label quality of each sample by its feature conformity with the class-wise feature eigenvector. Based on the integration of these two criteria, the training data is divided into a noisy set D_{noisy} and a clean set D_{clean}. In the SSL stage (see Fig. 1 (b)), our SRMix module interpolates each sample (x_i, y_i) with a reliable mixup partner (x_j, y_j), which is selected based on its spatial reliability, temporal stability, and prediction confidence. The mixed samples are used for model training.

2.1 Gradient and Feature Conformity-Based Clean Data Selection

Inspired by previous studies that optimization dynamics in the gradient space reflects the true class information [8, 15] and contextual information in the feature space is more robust to noisy labels [7], we devise the novel GCS and FCS criteria to measure label quality in the gradient and feature spaces.

Gradient Conformity-Based Selection (GCS). The GCS aims to distinguish clean samples from noisy ones by exploring their optimization dynamics in the gradient space. Since training samples from the same class usually exhibit similar optimization dynamics [15], the gradient of a sample should be similar to the principal gradient of its true class, thus we use the gradient conformity as a criterion to evaluate the quality of its given label. Specifically, for each sample x_i, its gradient $g(x_i)$ is computed as:

$$g(x_i) = \frac{\partial(-logp^{'}(x_i))}{\partial f(x_i)}, \quad p^{'}(x_i) = max\sum\nolimits_{x_j} p(x_j), \forall x_j \in KNN\{x_i\}, \tag{1}$$

where $f(x_i)$ is the feature vector of the sample x_i, and x_j denotes the K-Nearest Neighbors (KNN) of x_i. $p^{'}(x_i)$ is the probability of the most likely true class predicted by its KNN neighbors. Therefore, the gradient $g(x_i)$ is very likely to reflect the true class information and optimization dynamics of x_i. For each class, we select $\alpha\%$ samples with the smallest loss as an anchor set $\mathcal{A}_c$, which is depicted in the shaded areas of the GCS in Fig. 1 (a). Then, the principal gradient of the c-th class is computed as:

$$g_c = \frac{1}{N_c \cdot \alpha\%}\sum\nolimits_{x_i} g(x_i), \quad x_i \in \mathcal{A}_c, \tag{2}$$

which is the average gradient of all samples in the anchor set $\mathcal{A}_c$ of the c-th class. Then, we can measure the similarity between the gradient of the sample x_i and the principal gradient of class y_i with the cosine similarity $s_g(x_i) = cos < g(x_i), g_{y_i} >$. For the sample x_i, if y_i is a noisy label, $g(x_i)$ should be consistent with the principal gradient of its true class and diverge from g_{y_i}, thus yielding small $s_g(x_i)$. By fitting Gaussian mixture models (GMM) on the similarity score $s_g(x_i)$, we can get $c_g(x_i) = GMM(s_g(x_i))$, which represents the clean probability of the sample x_i decided by the GCS criterion. To the best of our knowledge, this is the first work that explores gradient conformity for clean data selection.

Feature Conformity-Based Selection (FCS). Since feature space is more robust to noisy labels than the output space [7], our FCS criterion explores high-order topological information in the feature space and utilizes the feature conformity with class-wise principal eigenvectors as a criterion to select clean samples. Specifically, for each class, we compute the gram matrix as:

$$M_c = \sum_{x_i} f(x_i) \cdot f(x_i)^T, \quad x_i \in \mathcal{A}_c, \tag{3}$$

where $f(x_i)$ denotes the feature vector of the sample x_i in the anchor set $\mathcal{A}_c$ of the c-th class. Then, we perform eigen-decomposition on the gram matrix: $M_c = U_c \cdot \Sigma_c \cdot U_c^T$, where U_c is the eigenvector matrix and Σ_c is a diagonal matrix composed of eigenvalues. The principal eigenvector u_c of U_c is utilized to represent the distribution and contextual information of the c-th class. Then, for each sample x_i, we measure its label quality based on the conformity of its feature $f(x_i)$ with the principal eigenvector u_{y_i} of its given label: $s_f(x_i) = cos < f(x_i), u_{y_i} >$. Samples that better align with the principal eigenvectors of their labeled class are more likely to be clean. According to the FCS criterion, the clean probability of the sample x_i is obtained by $c_f(x_i) = GMM(s_f(x_i))$. Compared with existing methods that utilize class-wise average features to represent contextual information [7], the eigenvectors in our method can better explore the high-order topological information among samples and are less affected by noisy features.

Integration of GCS and FCS. Finally, we average the clean probabilities estimated by the GCS and FCS criteria to identify clean data. As shown in Fig. 1 (a), for a dataset with the noise rate of $r\%$, we divide all samples into a clean set D_{clean} (i.e., $(1 - r\%)$ samples with higher clean probabilities) and a noisy set D_{noisy} (i.e., $r\%$ samples with lower clean probabilities).

2.2 Sample Reliability-Based Mixup (SRMix)

By regarding D_{clean} as a labeled set and D_{noisy} as an unlabeled set, we can formulate the LNL task into an SSL problem and employ Mixup [12] to generate mixed samples for model training [3]. In the traditional Mixup [12], each sample (x_i, y_i) is linearly interpolated with another sample (x_j, y_j) randomly chosen from the mini-batch. However, the pseudo labels of noisy datasets cannot be always reliable and the Mixup operation will aggravate the error accumulation of pseudo labels. To mitigate the error accumulation of pseudo labels, we propose a *Sample Reliability-based Mixup* (SRMix) method, which selects mixup partners based on their spatial reliability, temporal stability, and prediction confidence.

Intuitively, samples with reliable pseudo labels should have consistent predictions with their neighboring samples, stable predictions along sequential training epochs, and high prediction confidence. As shown in Fig. 1 (b), we select reliable mixup partners for each sample based on the triple criteria. First, for each sample x_j, we define the spatial reliability as:

$$\mathcal{R}_{spatial}(x_j) = 1 - Normalize(||p(x_j) - \frac{1}{K}\sum_{x_k} p(x_k)||_2^2), \forall x_k \in KNN\{x_j\} \quad (4)$$

where $p(x_j)$ and $p(x_k)$ are the pseudo labels of sample x_j and its neighbor x_k. $Normalize()$ denotes the min-max normalization over all samples in each batch. If the pseudo label of a sample is more consistent with its neighbors, a higher $\mathcal{R}_{spatial}(x_j)$ will be assigned, and vice versa. Second, for each sample x_j, we keep the historical sequence of its predictions in the past T epochs, e.g., the prediction sequence at the t-th epoch is defined as $P_t(x_j) = [p_{t-T+1}(x_j), ..., p_{t-1}(x_j), p_t(x_j)]$. The temporal stability of x_j can be defined as:

$$\mathcal{R}_{temporal}(x_j) = 1 - Normalize\left(\sqrt{\frac{1}{T}\sum_{n=0}^{T-1}(p_{t-n}(x_j) - \bar{p}(x_j))^2}\right), \quad (5)$$

where $\bar{p}(x_j)$ is the average prediction of the historical sequence. According to Eq. (5), a sample with smaller variance or fluctuation over time will be assigned with a larger $\mathcal{R}_{temporal}(x_j)$, and vice versa. Finally, the overall sample reliability is defined as: $\mathcal{R}(x_j) = \mathcal{R}_{spatial}(x_j) \cdot \mathcal{R}_{temporal}(x_j) \cdot max(p(x_j))$, where $max(p(x_j))$ denotes the prediction confidence of the pseudo label of the sample x_j. The possibility of x_j being chosen as a mixup partner is set as

$$p_m(x_j) = \begin{cases} \mathcal{R}(x_j), & \mathcal{R}(x_j) \geq \tau_R \\ 0, & \text{else}, \end{cases} \quad (6)$$

where τ_R is a predefined threshold to filter out unreliable mixup partners. For each sample x_i, we select a mixup partner x_j with the probability $p_m(x_j)$ defined in Eq. (6), and linearly interpolate their inputs and pseudo labels to generate a mixed sample (x, y). The mixed sample is fed into the network and trained with cross-entropy loss $\mathcal{L}_{CE}(x, y)$. Considering the estimation of sample reliability might be inaccurate in the initial training stage, we employ the traditional Mixup in the first 5 epochs and utilize our proposed SRMix for the rest training epochs. Compared with the traditional Mixup, the SRMix can effectively mitigate error accumulation of the pseudo labels and promote model training.

3 Experiments

3.1 Datasets and Implementation Details

WCE Dataset with Synthetic Label Noise. The Wireless Capsule Endoscopy (WCE) dataset [16] contains 1,812 images, including 600 normal images, 605 vascular lesions, and 607 inflammatory frames. We perform 5-fold cross-validation to evaluate our method. Following the common practice in the LNL community [3,4,10], we employ symmetric and pairflip label noise with diverse settings on the training set to simulate errors in the annotation process. The symmetric noise rate is set as 20%, 40%, 50%, and the pairflip noise rate is set as 40%. The model performance is measured by the average Accuracy (ACC) and Area Under the Curve (AUC) on the 5-fold test data.

Histopathology Dataset with Real-World Label Noise. The histopathology image dataset is collected from Chaoyang Hospital [10] and is annotated by 3 professional pathologists. There are 1,816 normal, 1,163 serrated, 2,244 adenocarcinoma, and 937 adenoma samples of colon slides in total. The samples with the consensus of 3 pathologists are selected as the test set, including 705 normal, 321 serrated, 840 adenocarcinoma, and 273 adenoma samples. The rest samples are utilized to construct the training set, with randomly selected opinions from one of the three doctors used as the noisy labels. The model performance is measured by the average Accuracy (ACC), F1 Score (F1), Precision, and Recall on 3 independent runs.

Table 1. Comparison with state-of-the-art LNL methods on the WCE dataset under diverse noise settings. Best and second-best results are **highlighted** and <u>underlined</u>.

Method	20% Sym.		40% Sym.	
	Accuracy (%)	AUC (%)	Accuracy (%)	AUC (%)
CE (Standard)	86.98 ± 1.32	96.36 ± 0.77	65.84 ± 1.31	83.03 ± 0.98
Co-teaching (NeurIPS 2018) [4]	91.97 ± 1.48	98.45 ± 0.48	84.27 ± 2.51	94.72 ± 0.93
Coteaching+ (ICML2019) [14]	91.86 ± 0.47	98.16 ± 0.26	73.92 ± 1.37	88.00 ± 0.89
DivideMix (ICLR 2020) [3]	<u>93.98</u> ± 2.27	98.52 ± 0.73	<u>89.79</u> ± 1.57	<u>96.79</u> ± 0.92
EHN-NSHE (TMI 2022) [10]	92.71 ± 0.87	<u>98.56</u> ± 0.35	82.40 ± 2.12	94.22 ± 1.17
SFT (ECCV 2022) [19]	93.32 ± 1.09	**98.66** ± 0.50	84.33 ± 4.64	94.59 ± 1.61
TSCSI (ECCV 2022) [7]	90.07 ± 2.21	97.63 ± 1.02	85.65 ± 4.94	95.61 ± 2.18
Our method	**94.65** ± 2.08	<u>98.56</u> ± 0.69	**92.44** ± 2.38	**97.70** ± 0.76
Method	50% Sym.		40% Pairflip	
	Accuracy (%)	AUC (%)	Accuracy (%)	AUC (%)
CE (Standard)	52.59 ± 0.76	70.36 ± 0.44	62.36 ± 2.84	81.05 ± 1.49
Co-teaching (NeurIPS 2018) [4]	77.51 ± 2.23	92.14 ± 1.25	82.01 ± 1.40	93.99 ± 0.71
Coteaching+ (ICML2019) [14]	58.00 ± 1.33	75.23 ± 1.24	61.26 ± 1.54	79.12 ± 0.99
DivideMix (ICLR 2020) [3]	<u>82.40</u> ± 1.93	<u>92.67</u> ± 1.05	<u>88.30</u> ± 3.33	96.28 ± 1.60
EHN-NSHE (TMI 2022) [10]	70.14 ± 3.41	85.01 ± 2.70	80.74 ± 4.84	92.15 ± 2.63
SFT (ECCV 2022) [19]	66.94 ± 2.54	83.55 ± 1.94	80.52 ± 6.55	90.65 ± 4.32
TSCSI (ECCV 2022) [7]	76.50 ± 5.04	79.75 ± 6.49	75.28 ± 2.29	<u>96.37</u> ± 3.03
Our method	**86.59** ± 2.26	**95.25** ± 1.38	**92.38** ± 2.74	**97.80** ± 1.03

Implementation Details. Our method follows the baseline framework of DivideMix [3] and adopts the pre-trained ResNet-50 [17] for feature extraction. We implement our method and all comparison methods on NVIDIA RTX 2080ti GPU using PyTorch [18]. For the WCE dataset, our method is trained for 40 epochs with an initial learning rate set to 0.0001 and divided by 10 after 20 epochs. For the histopathology dataset, the network is trained for 20 epochs with the learning rate set to 0.0001. For both datasets, the network is trained by Adam optimizer with $\beta_1 = 0.9$ and $\beta_2 = 0.999$, and batch size of 16. The number of neighbors K is set as 10 in Eq. (1) and Eq. (4). Length T of the historical sequence is set as 3. The reliability threshold τ_R is set as 0.2 for the WCE dataset and 0.05 for the histopathology dataset.

Table 2. Comparison with state-of-the-art methods on the histopathology dataset with real-world label noise. Best and second-best results are **highlighted** and underlined.

Method	ACC (%)	F1 (%)	Precision (%)	Recall (%)
CE (Standard)	80.36 ± 1.29	73.00 ± 0.84	76.47 ± 3.32	72.13 ± 0.29
Co-teaching (NeurIPS 2018) [4]	80.57 ± 0.55	72.39 ± 1.05	76.58 ± 1.40	71.33 ± 1.76
Coteaching+ (ICML2019) [14]	82.15 ± 0.34	74.63 ± 0.30	77.04 ± 0.56	73.94 ± 0.37
DivideMix (ICLR 2020) [3]	82.89 ± 0.80	77.36 ± 1.08	78.31 ± 1.74	76.77 ± 0.76
EHN-NSHE (TMI 2022) [10]	83.06 ± 0.28	76.68 ± 0.39	78.53 ± 0.41	75.00 ± 0.42
SFT (ECCV 2022) [19]	82.68 ± 0.97	76.65 ± 0.89	78.53 ± 1.45	75.64 ± 0.73
TSCSI (ECCV 2022) [7]	82.03 ± 2.12	75.54 ± 2.05	78.51 ± 3.60	74.57 ± 1.78
Our method	**84.29** ± 0.70	**78.98** ± 0.81	**80.34** ± 0.85	**78.19** ± 0.99

Table 3. Ablation study on the WCE dataset under 40% symmetric and pairflip noises.

	GCS	FCS	SRMix	40% Sym.		40% Pairflip	
				ACC (%)	AUC (%)	ACC (%)	AUC (%)
1				89.79 ± 1.57	96.79 ± 0.92	88.30 ± 3.33	96.28 ± 1.60
2	✓			91.17 ± 1.36	97.43 ± 0.78	91.12 ± 2.99	97.36 ± 1.13
3		✓		90.56 ± 2.03	96.89 ± 1.03	91.01 ± 2.23	97.22 ± 1.06
4			✓	90.84 ± 1.66	96.74 ± 1.25	89.40 ± 2.70	96.27 ± 1.63
5	✓	✓		92.11 ± 2.31	97.44 ± 0.92	91.72 ± 3.01	97.60 ± 1.06
6	✓	✓	✓	92.44 ± 2.38	97.70 ± 0.76	92.38 ± 2.74	97.80 ± 1.03

3.2 Experimental Results

Comparison with State-of-the-Art Methods. We first evaluate our method on the WCE dataset under diverse synthetic noise settings and show the results in Table 1. We compare with three well-known LNL methods (i.e., Co-teaching [4], Coteaching+ [14], and DivideMix [3]) and three state-of-the-art LNL methods (i.e., EHN-NSHE [10], SFT [19], and TSCSI [7]). As shown in Table 1, our method outperforms existing methods under all noise settings, and the performance gain is more significant under severe noise settings (e.g., noise rate $\geq$ 40%). Under the four settings, our method outperforms the second-best model by 0.67%, 2.65%, 4.19%, and 4.08% in accuracy. These results indicate the effectiveness of our method.

We then evaluate our method on the histopathology dataset with real-world label noise. As shown in Table 2, our method outperforms existing state-of-the-art methods, indicating the capability of our method in dealing with complex real-world label noise.

Ablation Study. To quantitatively analyze the contribution of the proposed components (i.e., GCS, FCS, and SRMix) in combating label noise, we perform an ablation study on the WCE dataset under 40% symmetric and pairflip noise. As shown in Table 3, compared with the DivideMix baseline (line 1) [3], replacing the small-loss criterion by GCS or FCS both improve the model performance significantly (lines 2–3), and their combination leads to further performance gains (line 5). Furthermore, better performance can be achieved by replacing the traditional Mixup with our proposed SRMix method (line 1 *vs.* line 4, line 5 *vs.* line 6). These results indicate that filtering out unreliable mixup partners can effectively improve the model's capacity in combating label noise.

More comprehensive analysis of the GCS and FCS criteria is provided in the supplementary material. Figure S1 demonstrates that compared with the normalized loss [3], the GCS and FCS criteria are more distinguishable between the clean and noisy data. This is consistent with the improvement of clean data selection accuracy in Fig. S2. As shown in Fig. S3, both the feature and gradient of each sample are aligned with the center of its true class, further validating the rationality of using gradient and feature conformity for clean data selection.

4 Conclusion

In this paper, we present a two-stage framework to combat label noise in medical image classification tasks. In the first stage, we propose two novel criteria (i.e., GCS and FCS) that select clean data based on their conformity with the class-wise principal gradients and feature eigenvectors. By exploring contextual information and high-order topological correlations in the gradient space and feature space, our GCS and FCS criteria enable more accurate clean data selection and benefit LNL tasks. In the second stage, to mitigate the error accumulation of pseudo labels, we propose an SRMix method that interpolates input samples with reliable mixup partners which are selected based on their spatial reliability, temporal stability, and prediction confidence. Extensive experiments on two datasets with both diverse synthetic and real-world label noise indicate the effectiveness of our method.

Acknowledgements. This work was supported by National Natural Science Foundation of China 62001410 and Innovation and Technology Commission-Innovation and Technology Fund ITS/100/20.

References

1. Zhang, C., Bengio, S., Hardt, M., Recht, B., Vinyals, O.: Understanding deep learning (still) requires rethinking generalization. Commun. ACM **64**(3), 107–115 (2021)
2. Liu, S., Niles-Weed, J., Razavian, N., Fernandez-Granda, C.: Early-learning regularization prevents memorization of noisy labels. NeurIPS **33**, 20331–20342 (2020)
3. Li, J., Socher, R., Hoi, S.C.H.: DivideMix: learning with noisy labels as semi-supervised learning. arXiv preprint arXiv:2002.07394 (2020)
4. Han, B., et al.: Co-teaching: robust training of deep neural networks with extremely noisy labels. In: NeurIPS, vol. 31 (2018)
5. Bai, Y., Liu, T.: Me-momentum: extracting hard confident examples from noisily labeled data. In: ICCV, pp. 9312–9321 (2021)
6. Li, J., Li, G., Liu, F., Yu, Y.: Neighborhood collective estimation for noisy label identification and correction. arXiv preprint arXiv:2208.03207 (2022)
7. Zhao, G., Li, G., Qin, Y., Liu, F., Yu, Y.: Centrality and consistency: two-stage clean samples identification for learning with instance-dependent noisy labels. In: Avidan, S., Brostow, G., Cissé, M., Farinella, G.M., Hassner, T. (eds.) Computer Vision - ECCV 2022. ECCV 2022. LNCS, vol. 13685, pp 21–37. Springer, Cham (2022). https://doi.org/10.1007/978-3-031-19806-9_2
8. Tang, H., Jia, K.: Towards discovering the effectiveness of moderately confident samples for semi-supervised learning. In: CVPR, pp. 14658–14667 (2022)
9. Iscen, A., Valmadre, J., Arnab, A., Schmid, C.: Learning with neighbor consistency for noisy labels. In: CVPR, pp. 4672–4681 (2022)
10. Zhu, C., Chen, W., Peng, T., Wang, Y., Jin, M.: Hard sample aware noise robust learning for histopathology image classification. IEEE Trans. Med. Imaging **41**(4), 881–894 (2021)
11. Berthelot, D., Carlini, N., Goodfellow, I., Papernot, N., Oliver, A., Raffel, C.A.:. MixMatch: a holistic approach to semi-supervised learning. In: NeurIPS, vol. 32 (2019)
12. Zhang, H., Cisse, M., Dauphin, Y.N., Lopez-Paz, D.: mixup: beyond empirical risk minimization. arXiv preprint arXiv:1710.09412 (2017)
13. Jiang, L., Zhou, Z., Leung, T., Li, L.J., Fei-Fei, L.: MentorNet: learning data-driven curriculum for very deep neural networks on corrupted labels. In: ICML, pp. 2304–2313. PMLR (2018)
14. Yu, X., Han, B., Yao, J., Niu, G., Tsang, I., Sugiyama, M.: How does disagreement help generalization against label corruption? In: ICML, pp. 7164–7173. PMLR (2019)
15. Arpit, D., et al.: A closer look at memorization in deep networks. In: ICML, pp. 233–242. PMLR (2017)
16. Dray, X., et al.: Cad-cap: UNE base de données française à vocation internationale, pour le développement et la validation d'outils de diagnostic assisté par ordinateur en vidéocapsule endoscopique du grêle. Endoscopy **50**(03), 000441 (2018)
17. He, K., Zhang, X., Ren, S., Sun, J.: Deep residual learning for image recognition. In: CVPR, pp. 770–778 (2016)
18. Paszke, A., et al.: PyTorch: an imperative style, high-performance deep learning library. In: NeurIPS, vol. 32 (2019)
19. Wei, Q., Sun, H., Lu, X., Yin, Y.: Self-filtering: a noise-aware sample selection for label noise with confidence penalization. arXiv preprint arXiv:2208.11351 (2022)

Multi-modality Contrastive Learning for Sarcopenia Screening from Hip X-rays and Clinical Information

Qiangguo Jin[1], Changjiang Zou[1], Hui Cui[2], Changming Sun[3], Shu-Wei Huang[4], Yi-Jie Kuo[4,5], Ping Xuan[6], Leilei Cao[1], Ran Su[7], Leyi Wei[8], Henry B. L. Duh[2], and Yu-Pin Chen[4,5](✉)

[1] School of Software, Northwestern Polytechnical University, Shaanxi, China
[2] Department of Computer Science and Information Technology, La Trobe University, Melbourne, Australia
[3] CSIRO Data61, Sydney, Australia
[4] Department of Orthopedics, Wan Fang Hospital, Taipei Medical University, Taipei, Taiwan
99231@w.tmu.edu.tw
[5] Department of Orthopedics, School of Medicine, College of Medicine, Taipei Medical University, Taipei, Taiwan
[6] Department of Computer Science, School of Engineering, Shantou University, Guangdong, China
[7] School of Computer Software, College of Intelligence and Computing, Tianjin University, Tianjin, China
[8] School of Software, Shandong University, Shandong, China

Abstract. Sarcopenia is a condition of age-associated muscle degeneration that shortens the life expectancy in those it affects, compared to individuals with normal muscle strength. Accurate screening for sarcopenia is a key process of clinical diagnosis and therapy. In this work, we propose a novel multi-modality contrastive learning (MM-CL) based method that combines hip X-ray images and clinical parameters for sarcopenia screening. Our method captures the long-range information with Non-local CAM Enhancement, explores the correlations in visual-text features via Visual-text Feature Fusion, and improves the model's feature representation ability through Auxiliary contrastive representation. Furthermore, we establish a large in-house dataset with 1,176 patients to validate the effectiveness of multi-modality based methods. Significant performances with an AUC of 84.64%, ACC of 79.93%, F1 of 74.88%, SEN of 72.06%, SPC of 86.06%, and PRE of 78.44%, show that our method outperforms other single-modality and multi-modality based methods.

Keywords: Sarcopenia screening · Contrastive learning · Multi-modality feature fusion

1 Introduction

Sarcopenia is a progressive and skeletal muscle disorder associated with loss of muscle mass, strength, and function [1,5,8]. The presence of sarcopenia increases the risk of

H. Greenspan et al. (Eds.): MICCAI 2023, LNCS 14225, pp. 85–94, 2023.
https://doi.org/10.1007/978-3-031-43987-2_9

hospitalization and the cost of care during hospitalization. A systematic analysis of the world's population showed that the prevalence of sarcopenia is approximately 10% in healthy adults over the age of 60 [5,13]. However, the development of sarcopenia is insidious, without overt symptoms in the early stages, which means that the potential number of patients at risk for adverse outcomes is very high. Thus, early identification, screening, and diagnosis are of great necessity to improve treatment outcomes, especially for elderly people.

The development of effective, reproducible, and cost-effective algorithms for reliable quantification of muscle mass is critical for diagnosing sarcopenia. However, automatically identifying sarcopenia is a challenging task due to several reasons. First, the subtle contrast between muscle and fat mass in the leg region makes it difficult to recognize sarcopenia from X-ray images. Second, although previous clinical studies [7,11] show that patient information, such as age, gender, education level, smoking and drinking status, physical activity (PA), and body mass index (BMI), is crucial for correct sarcopenia diagnosis, there is no generalizable standard. It is of great importance to develop a computerized predictive model that can fuse and mine diagnostic features from heterogeneous hip X-rays and tabular data containing patient information. Third, the number of previous works on sarcopenia diagnosis is limited, resulting in limited usable data.

Deep learning attracted intensive research interests in various medical diagnosis domains [17,18]. For instance, Zhang et al. [19] proposed an attention residual learning CNN model (ARLNet) for skin lesion classification to leverage multiple ARL blocks to tackle the challenge of data insufficiency, inter-class similarities, and intra-class variations. For multi-modality based deep learning, PathomicFusion (PF) [3] fused multimodal histology images and genomic (mutations, CNV, and RNA-Seq) features for survival outcome prediction in an end-to-end manner. Based on PF [3], Braman et al. [2] proposed a deep orthogonal fusion model to combine information from multiparametric MRI exams, biopsy-based modalities, and clinical variables into a comprehensive multimodal risk score. Despite the recent success in various medical imaging analysis tasks [2,3,19], sarcopenia diagnosis by deep learning based algorithms is still under study. To the best of our knowledge, recent work by Ryu et al. [12] is the most relevant to our proposed method. Ryu et al. [12] first used three ensembled deep learning models to test appendicular lean mass (ALM), handgrip strength (HGS), and chair rise test (CRT) performance using chest X-ray images. Then they built machine learning models to aggregate predicted ALM, HGS, and CRT performance values along with basic tabular features to diagnose sarcopenia. However, the major drawback of their work lies in the complex two-stage workflow and the tedious ensemble training. Besides, since sarcopenia is defined by low appendicular muscle mass, measuring muscle wasting through hip X-ray images, which have the greatest proportion of muscle mass, is much more appropriate for screening sarcopenia.

In this work, we propose a multi-modality contrastive learning (MM-CL)[1] model for sarcopenia diagnosis from hip X-rays and clinical information. Different from Ryu et al.'s model [12], our MM-CL can process multi-modality images and clinical data and screen sarcopenia in an end-to-end fashion. The overall framework is given in Fig. 1.

[1] Source code will be released at https://github.com/qgking/MM-CL.git.

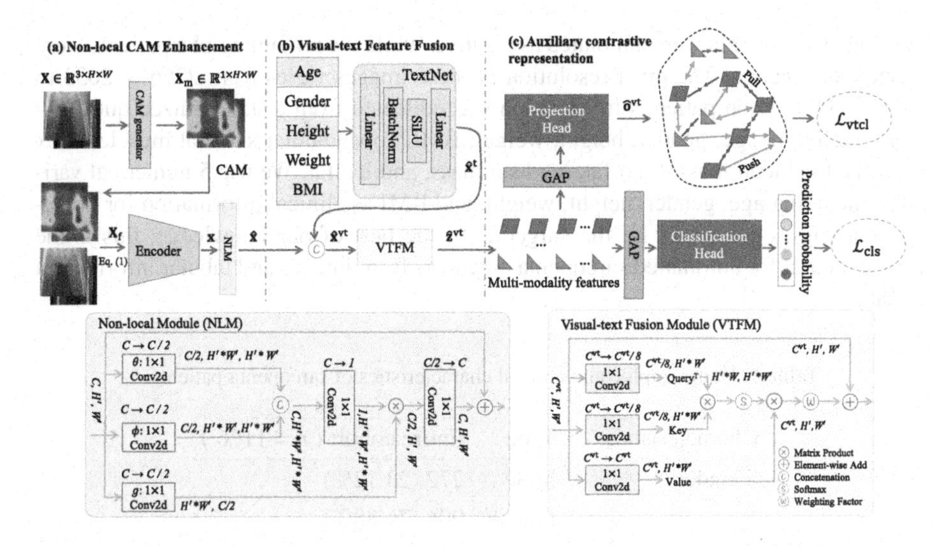

Fig. 1. The framework of our proposed MM-CL. MM-CL is composed of (a) Non-local CAM Enhancement, (b) Visual-text Feature Fusion, and (c) Auxiliary contrastive representation. The output feature size of each block is given in the channel size × height × width ($C \times H \times W$) format. GAP denotes the global average pooling, and CAM denotes the class activation map.

The major components include Non-local CAM Enhancement (NLC), Visual-text Feature Fusion (VFF), and Auxiliary contrastive representation (ACR) modules. Non-local CAM Enhancement enables the network to capture global long-range information and assists the network to concentrate on semantically important regions generated by class activation maps (CAM). Visual-text Feature Fusion encourages the network to improve the multi-modality feature representation ability. Auxiliary contrastive representation utilizes unsupervised learning and thus improves its ability for discriminative representation in the high-level latent space. The main contributions of this paper are summarized as follows. First, we propose a multi-modality contrastive learning model, which enhances the feature representation ability via integrating extra global knowledge, fusing multi-modality information, and joint unsupervised and supervised learning. Second, to address the absence of multi-modality datasets for sarcopenia screening, we select 1,176 patients from the Taipei Municipal Wanfang Hospital. To the best of our knowledge, our dataset is the largest for automated sarcopenia diagnosis from images and tabular information to date. Third, we experimentally show the superiority of the proposed method for predicting sarcopenia from hip X-rays and clinical information.

2 Data Collection

In this retrospective study, we collected anonymized data from patients who underwent sarcopenia examinations at the Taipei Municipal Wanfang Hospital. The data collection was approved by an institutional review board. The demographic and clinical characteristics of this dataset are shown in Table 1. 490 of 1,176 eligible patients who had developed sarcopenia were annotated as positive, while the remaining 686 patients were labeled as negative. The pixel resolution of these images varies from 2266 × 2033 to 3408 × 3408. Each patient's information was collected from a standardized questionnaire, including age, gender, height, weight, BMI, appendicular skeletal muscle index (ASMI), total lean mass, total fat, leg lean mass, and leg fat. We use 5 numerical variables including age, gender, height, weight, and BMI as clinical information for boosting learning as suggested by the surgeon. To the best of our knowledge, this is the largest dataset for automated sarcopenia diagnosis from images and tabular information to date.

Table 1. Demographic and clinical characteristics of sarcopenia patients.

Characteristics	Type	Entire cohort ($n = 1176$)
Gender	Male	272 (23.12%)
	Female	904 (76.88%)
Age at diagnosis⋆		71 [63–81]
BMI⋆		22.8 [20.5–25.2]
Height (cm)⋆		155.9 [150.2–162]
Weight (kg)⋆		55 [49.5–63]

Note: ⋆ indicates the median values [interquartile range, 25th–75th percentile].

3 Methodology

As shown in Fig. 1, MM-CL consists of three major components. The Non-local CAM Enhancement module is proposed to force the network to learn from attentional spatial regions learned from class activation map (CAM) [20] to enhance the global feature representation ability. Then, we fuse the heterogeneous images and tabular data by integrating clinical variables through a Visual-text Feature Fusion module. Finally, we present an unsupervised contrastive representation learning strategy to assist the supervised screening by Auxiliary contrastive representation.

3.1 Non-local CAM Enhancement

Considering the large proportion of muscle regions in hip X-ray images, capturing long-range dependencies is of great importance for sarcopenia screening. In this work, we adopt the non-local module [16] (NLM) and propose using coarse CAM localization maps as extra information to accelerate learning. We have two hypotheses. First, the long-range dependency of the left and right legs should be well captured; Second, the CAM may highlight part of muscle regions, providing weak supervision to accelerate the convergence of the network. Figure 1(a) shows the overall structure of the Non-local CAM Enhancement.

CAM Enhancement: First, each training image $\mathbf{X} \in \mathbb{R}^{3\times H\times W}$ is sent to the CAM generator as shown in Fig. 1(a) to generate coarse localization map $\mathbf{X}_\mathrm{m} \in \mathbb{R}^{1\times H\times W}$. We use the Smooth Grad-CAM++ [10] technique to generate CAM via the ResNet18 [9] architecture. After the corresponding CAM is generated, the training image $\mathbf{X}$ is enhanced by its coarse localization map $\mathbf{X}_\mathrm{m}$ via smooth attention to the downstream precise prediction network. The output image $\mathbf{X}_\mathrm{f}$ is obtained as:

$$\mathbf{X}_\mathrm{f} = \mathbf{X} \cdot (1 + \mathrm{sigmoid}(\mathbf{X}_\mathrm{m})), \tag{1}$$

where sigmoid denotes the Sigmoid function. The downstream main encoder is identical to ResNet18.

Non-local Module: Given a hip X-ray image $\mathbf{X}$ and the corresponding CAM map $\mathbf{X}_\mathrm{m}$, we apply the backbone of ResNet18 to extract the high-level feature maps $\mathbf{x} \in \mathbb{R}^{C\times H'\times W'}$. The feature maps are then treated as inputs for the non-local module. For output $\hat{\mathbf{x}}_i$ from position index i, we have

$$\hat{\mathbf{x}}_i = \sum_{j=1}^{H'W'} a_{ij} g(\mathbf{x}_j) + \mathbf{x}_i, \quad a_{ij} = \mathrm{ReLU}\left(\mathbf{w}_f^T \operatorname{concat}(\theta(\mathbf{x}_i), \phi(\mathbf{x}_j))\right), \tag{2}$$

where concat denotes concatenation, $\mathbf{w}_f$ is a weight vector that projects the concatenated vector to a scalar, ReLU is the ReLU function, a_{ij} denotes the non-local feature attention that represents correlations between the features at two locations (i.e., $\mathbf{x}_i$ and $\mathbf{x}_j$), θ, ϕ, and g are mapping functions as shown in Fig. 1(a).

3.2 Visual-Text Feature Fusion

After capturing the global information, we aim to fuse the visual and text features in the high-level latent space. We hypothesize that the clinical data may have a positive effect to boost the visual prediction performance. The overall structure of this strategy is given in Fig. 1(b). We extract the clinical features using a simple network, termed as TextNet. Finally, we propose a visual-text fusion module inspired by self-attention [15] to fuse the concatenated visual-text features.

Visual-Text Fusion Module: In order to learn from clinical data, we first encode 5 numerical variables as a vector and send it to TextNet. As shown in Fig. 1(b), TextNet consists of two linear layers, a batch normalization layer, and a sigmoid linear unit (SiLU) layer. We then expand and reshape the output feature $\hat{\mathbf{x}}^\mathrm{t} \in \mathbb{R}^{C^\mathrm{t}}$ of TextNet to fit the size of $C \times H' \times W'$. The text and visual representations are then concatenated as $\hat{\mathbf{x}}^\mathrm{vt} = \operatorname{concat}(\hat{\mathbf{x}}, \operatorname{reshape}(\hat{\mathbf{x}}^\mathrm{t}))$ before sending it to the visual-text fusion module, where $\hat{\mathbf{x}} \in \mathbb{R}^{C\times H'\times W'}$ denotes the output features from Non-local CAM Enhancement. Feature vector $\hat{\mathbf{x}}_i^\mathrm{vt} \in \mathbb{R}^{C^\mathrm{vt}}$ encodes information about the combination of a specific location i in image and text features with $C^\mathrm{vt} = C + C^\mathrm{t}$. The visual-text self-attention module first produces a set of query, key, and value by 1×1 convolutional transformations as $\mathbf{q}_i = \mathbf{W}_\mathrm{q}\hat{\mathbf{x}}_i^\mathrm{vt}$, $\mathbf{k}_i = \mathbf{W}_\mathrm{k}\hat{\mathbf{x}}_i^\mathrm{vt}$, and $\mathbf{v}_i = \mathbf{W}_\mathrm{v}\hat{\mathbf{x}}_i^\mathrm{vt}$ at each spatial location i, where $\mathbf{W}_\mathrm{q}$, $\mathbf{W}_\mathrm{k}$, and $\mathbf{W}_{\mathrm{v}}$ are part of the model parameters to be learned. We compute the visual-text self-attentive feature $\hat{\mathbf{z}}_i^{\mathrm{vt}}$ at position i as

$$\hat{\mathbf{z}}_i^{\mathrm{vt}} = \sum_{j=1}^{H'W'} s_{ij}\mathbf{v}_j + \mathbf{v}_i, \quad s_{ij} = \mathrm{Softmax}\left(\mathbf{q}_j^{\mathrm{T}} \cdot \mathbf{k}_i\right). \tag{3}$$

The softmax operation indicates the attention across each visual and text pair in the multi-modality feature.

3.3 Auxiliary Contrastive Representation

Inspired by unsupervised representation learning [4], we present a contrastive representation learning strategy that encourages the supervised model to pull similar data samples close to each other and push the different data samples away in the high-level embedding space. By such means, the feature representation ability in the embedding space could be further improved.

During the training stage, given N samples in a mini-batch, we obtain $2N$ samples by applying different augmentations (AutoAugment [6]) on each sample. Two augmented samples from the same sample are regarded as positive pairs, and others are treated as negative pairs. Thus, we have a positive sample and $2N-2$ negative samples for each patch. We apply global average pooling and linear transformations (Projection Head in Fig. 1(c)) to the visual-text embeddings $\hat{\mathbf{z}}^{\mathrm{vt}}$ in sequence, and obtain transformed features $\hat{\mathbf{o}}^{\mathrm{vt}}$. Let $\hat{\mathbf{o}}^{\mathrm{vt+}}$ and $\hat{\mathbf{o}}^{\mathrm{vt-}}$ denote the positive and negative embeddings of $\hat{\mathbf{o}}^{\mathrm{vt}}$, the formula of contrastive loss is defined as

$$\mathcal{L}_{\mathrm{vtcl}} = -\log \frac{\exp\left(sim\left(\hat{\mathbf{o}}^{\mathrm{vt}}, \hat{\mathbf{o}}^{\mathrm{vt+}}\right)/\tau\right)}{\exp\left(sim\left(\hat{\mathbf{o}}^{\mathrm{vt}}, \hat{\mathbf{o}}^{\mathrm{vt+}}\right)/\tau\right) + \sum_{\hat{\mathbf{o}}^{\mathrm{vt-}} \in \mathcal{N}} \exp\left(sim\left(\hat{\mathbf{o}}^{\mathrm{vt}}, \hat{\mathbf{o}}^{\mathrm{vt-}}\right)/\tau\right)}, \tag{4}$$

where $\mathcal{N}$ is the set of negative counterparts of $\hat{\mathbf{o}}^{\mathrm{vt}}$, the $sim(\cdot,\cdot)$ is the cosine similarity between two representations, and τ is the temperature scaling parameter. Note that all the visual-text embeddings in the loss function are ℓ_2-normalized.

Finally, we integrate the auxiliary contrastive learning branch into the main Classification Head as shown in Fig. 1(c), which is a set of linear layers. We use weighted cross-entropy loss $\mathcal{L}_{\mathrm{cls}}$ as our classification loss. The overall loss function is calculated as $\mathcal{L}_{\mathrm{total}} = \mathcal{L}_{\mathrm{cls}} + \beta\mathcal{L}_{\mathrm{vtcl}}$, where β is a weight factor.

4 Experiments and Results

4.1 Implementation Details and Evaluation Measures

Our method is implemented in PyTorch using an NVIDIA RTX 3090 graphic card. We set the batch size to 32. Adam optimizer is used with a polynomial learning rate policy, where the initial learning rate 2.5×10^{-4} is multiplied by $\left(1 - \frac{epoch}{total\_epoch}\right)^{power}$ with $power$ as 0.9. The total number of training epochs is set to 100, and early stopping is adopted to avoid overfitting. Weight factor β is set to 0.01. The temperature constant

τ is set to 0.5. Visual images are cropped to 4/5 of the original height and resized to 224×224 after different online augmentation. The backbone is initialized with the weights pretrained on ImageNet.

Extensive 5-fold cross-validation is conducted for sarcopenia diagnosis. We report the diagnosis performance using comprehensive quantitative metrics including area under the receiver operating characteristic curve (AUC), F1 score (F1), accuracy (ACC), sensitivity (SEN), specificity (SPC), and precision (PRE).

4.2 Quantitative and Qualitative Comparison

We implement several state-of-the-art single-modality (ResNet, ARLNet, MaxNet [2], Support vector machine (SVM), and K-nearest neighbors(KNN)) and multi-modality methods (PF [3]) to demonstrate the effectiveness of our MM-CL. For a fair comparison, we use the same training settings.

Table 2. Sarcopenia diagnosis performance of recently proposed methods.

Method	Modality	AUC (%)	ACC (%)	F1 (%)	SEN (%)	SPC (%)	PRE (%)
ResNet18 [9]	Image	76.86	72.53	64.58	60.56	79.35	69.46
ARLNet [19]	Image	76.73	72.87	65.33	61.72	80.78	69.08
MaxNet [2]	Text	80.05	72.44	58.30	47.05	90.68	78.10
SVM	Text	82.72	73.63	65.47	60.43	83.26	72.08
KNN	Text	80.69	73.21	66.94	65.59	78.80	68.80
PF [3]	Multi	77.88	73.98	67.33	65.11	80.30	70.65
MM-CL	Multi	**84.64**	**79.93**	**74.88**	**72.06**	**86.06**	**78.44**

Table 2 outlines the performance of all methods. As shown, our model achieves the best AUC of 84.64% and ACC of 79.93% among all the methods in comparison.

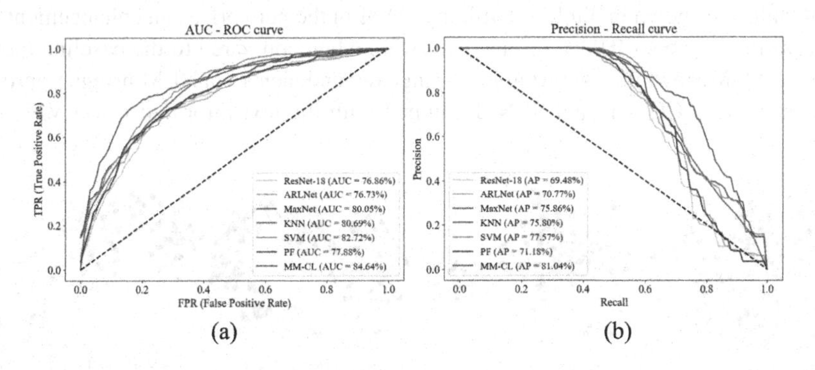

Fig. 2. AUC-ROC (a) and Precision-Recall (b) curves for comparison with state-of-the-art methods.

When compared to the single-modality models, MM-CL outperforms state-of-the-art approaches by at least 6% on ACC. Among all the single-modality models, MaxNet [2], SVM, and KNN gain better results than image-only models. When compared to the multi-modality models, MM-CL also performs better than these methods by a large margin, which proves the effectiveness of our proposed modules.

We further visualize the AUC-ROC and Precision-Recall curves to intuitively show the improved performance. As shown in Fig. 2, the MM-CL achieves the best AUC and average precision (AP), which demonstrates the effectiveness of the proposed MM-CL.

We have three observations: (1) Multi-modality based models outperform single-modality based methods, and we explain this finding that multiple modalities complement each other with useful information. (2) MaxNet [2] gains worse results than traditional machine learning methods. One primary reason is that MaxNet contains a large number of parameters to be learned, the tabular information only includes 5 factors, which could result in overfitting. (3) With the help of NLC, VFF, and ACR, our MM-CL achieves substantial improvement over all the other methods.

4.3 Ablation Study of the Proposed Method

Table 3. Sarcopenia diagnosis performance with ablation studies.

Modality		NLC		VFF	ACR	AUC (%)	ACC (%)	F1 (%)	SEN (%)	SPC (%)	PRE (%)
Image	Text	CAM	NLM								
√						76.86	72.53	64.58	60.56	79.35	69.46
√		√				77.09	73.46	65.55	60.83	82.55	71.53
√		√	√			77.86	73.80	66.23	62.85	81.22	70.93
√	√	√	√	√		84.21	79.16	75.13	76.63	80.69	74.03
√	√	√	√	√	√	84.64	79.93	74.88	72.06	86.06	78.44

We also conduct ablation studies to validate each proposed component i.e., NLC, VFF, and ACR. CAM/NLM of NLC denotes the CAM enhancement/non-local module. Results are shown in Table 3. Utilizing CAM in the network as an enhancement for optimization improves 0.93% for average ACC, when compared to the baseline model (ResNet18). Meanwhile, capturing long-range dependencies via NLM brings improvement on AUC, ACC, F1, and SEN. Equipped with the text information via VFF, our

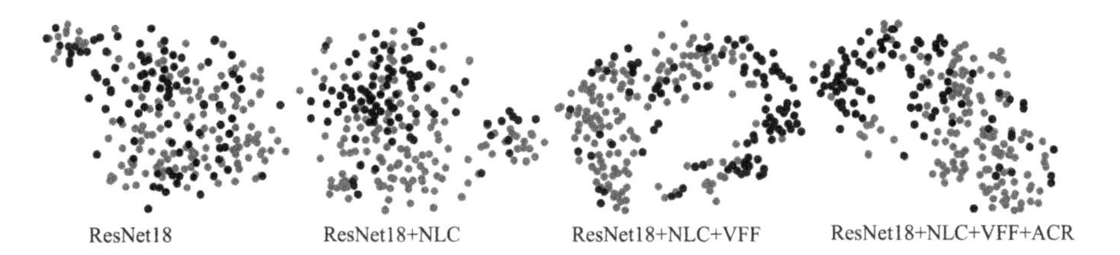

Fig. 3. Visual interpretation of high-level features using t-SNE. The red and blue circles are sarcopenia and non-sarcopenia instances respectively. (Color figure online)

method can lead to significant performance gains on ACC compared with image-only experiments, e.g., 79.16% vs. 73.80%. Lastly, applying ACR to the network improves the average ACC score from 79.16% to 79.93%.

We also visualize the ability of feature representation in the high-level semantic latent feature space before the final classification via t-SNE [14]. As can be seen in Fig. 3, by gradually adding the proposed modules, the feature representation ability of our model becomes more and more powerful, and the high-level features are better clustered.

Our first finding is that fusing visual and text knowledge brings significant improvement, which demonstrates that the extra tabular information could help substantially in learning. Second, incorporating unsupervised contrastive learning in the supervised learning framework could also improve the feature representation ability of the model.

5 Conclusions

In conclusion, we propose a multi-modality contrastive learning model for sarcopenia screening using hip X-ray images and clinical information. The proposed model consists of a Non-local CAM Enhancement module, a Visual-text Feature Fusion module, and an Auxiliary contrastive representation for improving the feature representation ability of the network. Moreover, we collect a large dataset for screening sarcopenia from heterogeneous data. Comprehensive experiments and explanations demonstrate the superiority of the proposed method. Our future work includes the extension of our approach to other multi-modality diagnosis tasks in the medical imaging domain.

Acknowledgment. This work was supported by the Fundamental Research Funds for the Central Universities, the National Natural Science Foundation of China [Grant No. 62201460 and No. 62072329], and the National Key Technology R&D Program of China [Grant No. 2018YFB1701700].

References

1. Ackermans, L.L., et al.: Screening, diagnosis and monitoring of sarcopenia: when to use which tool? Clinical Nutrition ESPEN (2022)
2. Braman, N., Gordon, J.W.H., Goossens, E.T., Willis, C., Stumpe, M.C., Venkataraman, J.: Deep orthogonal fusion: multimodal prognostic biomarker discovery integrating radiology, pathology, genomic, and clinical data. In: de Bruijne, M., et al. (eds.) MICCAI 2021. LNCS, vol. 12905, pp. 667–677. Springer, Cham (2021). https://doi.org/10.1007/978-3-030-87240-3_64
3. Chen, R.J., et al.: Pathomic fusion: an integrated framework for fusing histopathology and genomic features for cancer diagnosis and prognosis. IEEE Trans. Med. Imaging **41**(4), 757–770 (2020)
4. Chen, T., Kornblith, S., Norouzi, M., Hinton, G.: A simple framework for contrastive learning of visual representations. In: International Conference on Machine Learning, pp. 1597–1607. PMLR (2020)
5. Cruz-Jentoft, A.J., et al.: Sarcopenia: revised European consensus on definition and diagnosis. Age Ageing **48**(1), 16–31 (2019)

6. Cubuk, E.D., Zoph, B., Mane, D., Vasudevan, V., Le, Q.V.: AutoAugment: learning augmentation strategies from data. In: Proceedings of the IEEE/CVF Conference on Computer Vision and Pattern Recognition, pp. 113–123 (2019)
7. Dodds, R.M., Granic, A., Davies, K., Kirkwood, T.B., Jagger, C., Sayer, A.A.: Prevalence and incidence of sarcopenia in the very old: findings from the Newcastle 85+ study. J. Cachexia, Sarcopenia Muscle **8**(2), 229–237 (2017)
8. Giovannini, S., et al.: Sarcopenia: diagnosis and management, state of the art and contribution of ultrasound. J. Clin. Med. **10**(23), 5552 (2021)
9. He, K., Zhang, X., Ren, S., Sun, J.: Deep residual learning for image recognition. In: Proceedings of the IEEE Conference on Computer Vision and Pattern Recognition, pp. 770–778 (2016)
10. Omeiza, D., Speakman, S., Cintas, C., Weldermariam, K.: Smooth grad-CAM++: an enhanced inference level visualization technique for deep convolutional neural network models. arXiv preprint arXiv:1908.01224 (2019)
11. Pang, B.W.J., et al.: Prevalence and associated factors of Sarcopenia in Singaporean adults-the Yishun Study. J. Am. Med. Direct. Assoc. **22**(4), e1-885 (2021)
12. Ryu, J., Eom, S., Kim, H.C., Kim, C.O., Rhee, Y., You, S.C., Hong, N.: Chest X-ray-based opportunistic screening of sarcopenia using deep learning. J. Cachexia, Sarcopenia Muscle **14**(1), 418–428 (2022)
13. Shafiee, G., Keshtkar, A., Soltani, A., Ahadi, Z., Larijani, B., Heshmat, R.: Prevalence of sarcopenia in the world: a systematic review and meta-analysis of general population studies. J. Diab. Metab. Disord. **16**(1), 1–10 (2017)
14. Van Der Maaten, L.: Accelerating t-SNE using tree-based algorithms. J. Mach. Learn. Res. **15**(1), 3221–3245 (2014)
15. Vaswani, A., et al.: Attention is all you need. In: Advances in Neural Information Processing Systems, vol. 30 (2017)
16. Wang, X., Girshick, R., Gupta, A., He, K.: Non-local neural networks. In: Proceedings of the IEEE Conference on Computer Vision and Pattern Recognition, pp. 7794–7803 (2018)
17. Yan, K., Guo, Y., Liu, B.: Pretp-2l: identification of therapeutic peptides and their types using two-layer ensemble learning framework. Bioinformatics **39**(4), btad125 (2023)
18. Yan, K., Lv, H., Guo, Y., Peng, W., Liu, B.: samppred-gat: prediction of antimicrobial peptide by graph attention network and predicted peptide structure. Bioinformatics **39**(1), btac715 (2023)
19. Zhang, J., Xie, Y., Xia, Y., Shen, C.: Attention residual learning for skin lesion classification. IEEE Trans. Med. Imaging **38**(9), 2092–2103 (2019)
20. Zhou, B., Khosla, A., Lapedriza, A., Oliva, A., Torralba, A.: Learning deep features for discriminative localization. In: Proceedings of the IEEE Conference on Computer Vision and Pattern Recognition, pp. 2921–2929 (2016)

DiffMIC: Dual-Guidance Diffusion Network for Medical Image Classification

Yijun Yang[1,2], Huazhu Fu[3], Angelica I. Aviles-Rivero[4], Carola-Bibiane Schönlieb[4], and Lei Zhu[1,2](✉)

[1] The Hong Kong University of Science and Technology (Guangzhou), Nansha, Guangzhou, Guangdong, China
leizhu@ust.hk

[2] The Hong Kong University of Science and Technology, Hong Kong, China

[3] Institute of High Performance Computing, Agency for Science, Technology and Research, Singapore, Singapore

[4] University of Cambridge, Cambridge, UK

Abstract. Diffusion Probabilistic Models have recently shown remarkable performance in generative image modeling, attracting significant attention in the computer vision community. However, while a substantial amount of diffusion-based research has focused on generative tasks, few studies have applied diffusion models to general medical image classification. In this paper, we propose the first diffusion-based model (named DiffMIC) to address general medical image classification by eliminating unexpected noise and perturbations in medical images and robustly capturing semantic representation. To achieve this goal, we devise a dual conditional guidance strategy that conditions each diffusion step with multiple granularities to improve step-wise regional attention. Furthermore, we propose learning the mutual information in each granularity by enforcing Maximum-Mean Discrepancy regularization during the diffusion forward process. We evaluate the effectiveness of our DiffMIC on three medical classification tasks with different image modalities, including placental maturity grading on ultrasound images, skin lesion classification using dermatoscopic images, and diabetic retinopathy grading using fundus images. Our experimental results demonstrate that DiffMIC outperforms state-of-the-art methods by a significant margin, indicating the universality and effectiveness of the proposed model. Our code is publicly available at https://github.com/scott-yjyang/DiffMIC.

Keywords: diffusion probabilistic model · medical image classification · placental maturity · skin lesion · diabetic retinopathy

Supplementary Information The online version contains supplementary material available at https://doi.org/10.1007/978-3-031-43987-2_10.

H. Greenspan et al. (Eds.): MICCAI 2023, LNCS 14225, pp. 95–105, 2023.
https://doi.org/10.1007/978-3-031-43987-2_10

1 Introduction

Medical image analysis plays an indispensable role in clinical therapy because of the implications of digital medical imaging in modern healthcare [5]. Medical image classification, a fundamental step in the analysis of medical images, strives to distinguish medical images from different modalities based on certain criteria. An automatic and reliable classification system can help doctors interpret medical images quickly and accurately. Massive solutions for medical image classification have been developed over the past decades in the literature, most of which adopt deep learning ranging from popular CNNs to vision transformers [8,9,22,23]. These methods have the potential to reduce the time and effort required for manual classification and improve the accuracy and consistency of results. However, medical images with diverse modalities still challenge existing methods due to the presence of various ambiguous lesions and fine-grained tissues, such as ultrasound (US), dermatoscopic, and fundus images. Moreover, generating medical images under hardware limitations can cause noisy and blurry effects, which can degrade image quality and thus demand a more effective feature representation modeling for robust classifications.

Recently, Denoising Diffusion Probabilistic Models (DDPM) [14] have achieved excellent results in image generation and synthesis tasks [2,6,21,26] by iteratively improving the quality of a given image. Specifically, DDPM is a generative model based on a Markov chain, which models the data distribution by simulating a diffusion process that evolves the input data towards a target distribution. Although a few pioneer works tried to adopt the diffusion model for image segmentation and object detection tasks [1,4,12,29], their potential for high-level vision has yet to be fully explored.

Motivated by the achievements of diffusion probabilistic models in generative image modeling, **1) we present a novel Denoising Diffusion-based model named DiffMIC** for accurate classification of diverse medical image modalities. As far as we know, we are the first to propose a Diffusion-based model for general medical image classification. Our method can appropriately eliminate undesirable noise in medical images as the diffusion process is stochastic in nature for each sampling step. **2) In particular, we introduce a Dual-granularity Conditional Guidance (DCG) strategy** to guide the denoising procedure, conditioning each step with both global and local priors in the diffusion process. By conducting the diffusion process on smaller patches, our method can distinguish critical tissues with fine-grained capability. **3) Moreover, we introduce Condition-specific Maximum-Mean Discrepancy (MMD) regularization** to learn the mutual information in the latent space for each granularity, enabling the network to model a robust feature representation shared by the whole image and patches. **4) We evaluate the effectiveness of DiffMIC on three 2D medical image classification tasks** including placental maturity grading, skin lesion classification, and diabetic retinopathy grading. The experimental results demonstrate that our diffusion-based classification method consistently and significantly surpasses state-of-the-art methods for all three tasks.

2 Method

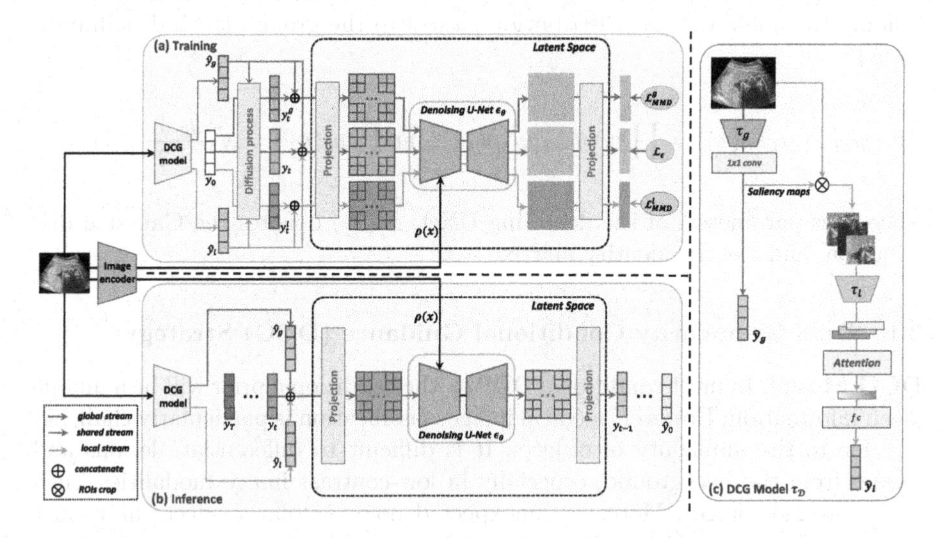

Fig. 1. Overview of our DiffMIC framework. (a) The training phase (forward process) and (b) The inference phase (reverse process) are constructed, respectively. (The noise of feature embedding is greater with the darker color.) (c) The DCG Model $\tau_\mathcal{D}$ guides the diffusion process by the dual priors from the raw image and ROIs.

Figure 1 shows the schematic illustration of our network for medical image classification. Given an input medical image x, we pass it to an image encoder to obtain the image feature embedding $\rho(x)$, and a dual-granularity conditional guidance (DCG) model to produce the global prior $\hat{y}_g$ and local prior $\hat{y}_l$. At the training stage, we apply the diffusion process on ground truth y_0 and different priors to generate three noisy variables y_t^g, y_t^l, and y_t (the global prior for y_t^g, the local prior for y_t^l, and dual priors for y_t). Then, we combine the three noisy variables y_t^g, y_t^l, and y_t and their respective priors and project them into a latent space, respectively. We further integrate three projected embeddings with the image feature embedding $\rho(x)$ in the denoising U-Net, respectively, and predict the noise distribution sampled for y_t^g, y_t^l, and y_t. We devise condition-specific maximum-mean discrepancy (MMD) regularization loss on the predicted noise of y_t^g and y_t^l, and employ the noise estimation loss by mean squared error (MSE) on the predicted noise of y_t to collaboratively train our DiffMIC network.

Diffusion Model. Following DDPM [14], our diffusion model also has two stages: a forward diffusion stage (training) and a reverse diffusion stage (inference). In the forward process, the ground truth response variable y_0 is added Gaussian noise through the diffusion process conditioned by time step t sampled from a uniform distribution of $[1, T]$, and such noisy variables are denoted as $\{y_1, ..., y_t, .., y_T\}$. As suggested by the standard implementation of DDPM, we adopt a UNet as the denoising network to parameterize the reverse diffusion process and learn the noise distribution in the forward process. In the reverse diffusion process, the trained UNet ϵ_θ generates the final prediction $\hat{y}_0$ by transforming the noisy variable distribution $p_\theta(y_T)$ to the ground truth distribution $p_\theta(y_0)$:

$$p_\theta(y_{0:T-1}|y_T, \rho(x)) = \prod_{t=1}^{T} p_\theta(y_{t-1}|y_t, \rho(x)), \quad \text{and} \quad p_\theta(y_T) = \mathcal{N}(\frac{\hat{y}_g + \hat{y}_l}{2}, \mathbb{I}), \tag{1}$$

where θ is parameters of the denoising UNet, $\mathcal{N}(\cdot, \cdot)$ denotes the Gaussian distribution, and $\mathbb{I}$ is the identity matrix.

2.1 Dual-Granularity Conditional Guidance (DCG) Strategy

DCG Model. In most conditional DDPM, the conditional prior will be a unique given information. However, medical image classification is particularly challenging due to the ambiguity of objects. It is difficult to differentiate lesions and tissues from the background, especially in low-contrast image modalities, such as ultrasound images. Moreover, unexpected noise or blurry effects may exist in regions of interest (ROIs), thereby hindering the understanding of high-level semantics. Taking only a raw image x as the condition in each diffusion step will be insufficient to robustly learn the fine-grained information, resulting in classification performance degradation.

To alleviate this issue, we design a Dual-granularity Conditional Guidance (DCG) for encoding each diffusion step. Specifically, we introduce a DCG model $\tau_\mathcal{D}$ to compute the global and local conditional priors for the diffusion process. Similar to the diagnostic process of a radiologist, we can obtain a holistic understanding from the global prior and also concentrate on areas corresponding to lesions from the local prior when removing the negative noise effects. As shown in Fig. 1 (c), for the global stream, the raw image data x is fed into the global encoder τ_g and then a 1×1 convolutional layer to generate a saliency map of the whole image. The global prior $\hat{y}_g$ is then predicted from the whole saliency map by averaging the responses. For the local stream, we further crop the ROIs whose responses are significant in the saliency map of the whole image. Each ROI is fed into the local encoder τ_l to obtain a feature vector. We then leverage the gated attention mechanism [15] to fuse all feature vectors from ROIs to obtain a weighted vector, which is then utilized for computing the local prior $\hat{y}_l$ by one linear layer.

Denoising Model. The noisy variable y_t is sampled in the diffusion process based on the global and local priors computed by the DCG model following:

$$y_t = \sqrt{\bar{\alpha}_t} y_0 + \sqrt{1 - \bar{\alpha}_t} \epsilon + (1 - \sqrt{\bar{\alpha}_t})(\hat{y}_g + \hat{y}_l), \tag{2}$$

where $\epsilon \sim \mathcal{N}(0, I)$, $\bar{\alpha}_t = \prod_t \alpha_t, \alpha_t = 1 - \beta_t$ with a linear noise schedule $\{\beta_t\}_{t=1:T} \in (0,1)^T$. After that, we feed the concatenated vector of the noisy variable y_t and dual priors into our denoising model UNet ϵ_θ to estimate the noise distribution, which is formulated as:

$$\epsilon_\theta(\rho(x), y_t, \hat{y}_g, \hat{y}_l, t) = D(E(f([y_t, \hat{y}_g, \hat{y}_l]), \rho(x), t), t), \tag{3}$$

where $f(\cdot)$ denotes the projection layer to the latent space. $[\cdot]$ is the concatenation operation. $E(\cdot)$ and $D(\cdot)$ are the encoder and decoder of UNet. Note that the image feature embedding $\rho(x)$ is further integrated with the projected noisy embedding in the UNet to make the model focus on high-level semantics and thus obtain more robust feature representations. In the forward process, we seek to minimize the noise estimation loss $\mathcal{L}_\epsilon$:

$$\mathcal{L}_\epsilon = ||\epsilon - \epsilon_\theta(\rho(x), y_t, \hat{y}_g, \hat{y}_l, t)||^2. \tag{4}$$

Our method improves the vanilla diffusion model by conditioning each step estimation function on priors that combine information derived from the raw image and ROIs.

2.2 Condition-Specific MMD Regularization

Maximum-Mean Discrepancy (MMD) is to measure the similarity between two distributions by comparing all of their moments [11,17], which can be efficiently achieved by a kernel function. Inspired by InfoVAE [31], we introduce an additional pair of condition-specific MMD regularization loss to learn mutual information between the sampled noise distribution and the Gaussian distribution. To be specific, we sample the noisy variable y_t^g from the diffusion process at time step t conditioned only by the global prior and then compute an MMD-regularization loss as:

$$\begin{aligned} &\mathcal{L}^g_{MMD}(n||m) = \mathbb{K}(n, n^{'}) - 2\mathbb{K}(m, n) + \mathbb{K}(m, m^{'}), \\ &\qquad \text{with } n = \epsilon, \;\; m = \epsilon_\theta(\rho(x), \sqrt{\bar{\alpha}_t} y_0 + \sqrt{1-\bar{\alpha}_t}\epsilon + (1-\sqrt{\bar{\alpha}_t})\hat{y}_g, \hat{y}_g, t), \end{aligned} \tag{5}$$

where $\mathbb{K}(\cdot,\cdot)$ is a positive definite kernel to reproduce distributions in the Hilbert space. The condition-specific MMD regularization is also applied on the local prior, as shown in Fig. 1 (a). While the general noise estimation loss $\mathcal{L}_\epsilon$ captures the complementary information from both priors, the condition-specific MMD regularization maintains the mutual information between each prior and target distribution. This also helps the network better model the robust feature representation shared by dual priors and converge faster in a stable way.

2.3 Training and Inference Scheme

Total Loss. By adding the noise estimation loss and the MMD-regularization loss, we compute the total loss $\mathcal{L}_{diff}$ of our denoising network as follows:

$$\mathcal{L}_{diff} = \mathcal{L}_\epsilon + \lambda(\mathcal{L}^g_{MMD} + \mathcal{L}^l_{MMD}), \tag{6}$$

where λ is a balancing hyper-parameter, and it is empirically set as λ=0.5.

Training Details. The diffusion model in this study leverages a standard DDPM training process, where the diffusion time step t is selected from a uniform distribution of $[1, T]$, and the noise is linearly scheduled with $\beta_1 = 1 \times 10^{-4}$ and $\beta_T = 0.02$. We adopt ResNet18 as the image encoder $\rho(\cdot)$. Following [12], we concatenate y_t,$\hat{y}_g$,$\hat{y}_l$, and apply a linear layer with an output dimension of 6144 to obtain the fused vector in the latent space. To condition the response embedding on the timestep, we perform a Hadamard product between the fused vector and a timestep embedding. We then integrate the image feature embedding and response embedding by performing another Hadamard product between them. The output vector is sent through two consecutive fully-connected layers, each followed by a Hadamard product with a timestep embedding. Finally, we use a fully-connected layer to predict the noise with an output dimension of classes. It is worth noting that all fully-connected layers are accompanied by a batch normalization layer and a Softplus non-linearity, with the exception of the output layer. For the DCG model τ_D, the backbone of its global and local stream is ResNet. We adopt the standard cross-entropy loss as the objective of the DCG model. We jointly train the denoising diffusion model and DCG model after pretraining the DCG model 10 epochs for warm-up, thereby resulting in an end-to-end DiffMIC for medical image classification.

Inference Stage. As displayed in Fig. 1 (b), given an input image x, we first feed it into the DCG model to obtain dual priors $\hat{y}_g, \hat{y}_l$. Then, following the pipeline of DDPM, the final prediction $\hat{y}_0$ is iteratively denoised from the random prediction y_T using the trained UNet conditioned by dual priors $\hat{y}_g, \hat{y}_l$ and the image feature embedding $\rho(x)$.

3 Experiments

Datasets and Evaluation: We evaluate the effectiveness of our network on an in-home dataset and two public datasets, e.g., PMG2000, HAM10000 [27], and APTOS2019 [16]. **(a) PMG2000.** We collect and annotate a benchmark dataset (denoted as PMG2000) for placental maturity grading (PMG) with four categories[1]. PMG2000 is composed of 2,098 ultrasound images, and we randomly divide the entire dataset into a training part and a testing part at an 8:2 ratio. **(b) HAM10000.** HAM10000 [27] is from the Skin Lesion Analysis Toward Melanoma Detection 2018 challenge, and it contains 10,015 skin lesion images with predefined 7 categories. **(c) APTOS2019.** In APTOS2019 [16], A total of 3,662 fundus images have been labeled to classify diabetic retinopathy into five different categories. Following the same protocol in [10], we split HAM10000 and APTOS2019 into a train part and a test part at a 7:3 ratio. These three datasets are with different medical image modalities. PMG2000 is gray-scale and class-balanced ultrasound images; HAM10000 is colorful but class-imbalanced dermatoscopic images; and APTOS2019 is another class-imbalanced

[1] Our data collection is approved by the Institutional Review Board (IRB).

Table 1. Quantitative comparison to SOTA methods on three classification tasks. The best results are marked in bold font.

(a) PMG2000

Methods		ResNet [13]	ViT [7]	Swin [19]	PVT [28]	GMIC [24]	**Our DiffMIC**
PMG2000	Accuracy	0.879	0.886	0.893	0.907	0.900	**0.931**
	F1-score	0.881	0.890	0.892	0.902	0.901	**0.926**

(b) HAM10000 and APTOS2019

Methods		LDAM [3]	OHEM [25]	MTL [18]	DANIL [10]	CL [20]	ProCo [30]	**Our DiffMIC**
HAM10000	Accuracy	0.857	0.818	0.811	0.825	0.865	0.887	**0.906**
	F1-score	0.734	0.660	0.667	0.674	0.739	0.763	**0.816**
APTOS2019	Accuracy	0.813	0.813	0.813	0.825	0.825	0.837	**0.858**
	F1-score	0.620	0.631	0.632	0.660	0.652	0.674	**0.716**

dataset with colorful Fundus images. Moreover, we introduce two widely-used metrics Accuracy and F1-score to quantitatively compare our framework against existing SOTA methods.

Implementation Details: Our framework is implemented with the PyTorch on one NVIDIA RTX 3090 GPU. We center-crop the image and then resize the spatial resolution of the cropped image to 224×224. Random flipping and rotation for data augmentation are implemented during the training processing. In all experiments, we extract six 32×32 ROI patches from each image. We trained our network end-to-end using the batch size of 32 and the Adam optimizer. The initial learning rate for the denoising model U-Net is set as 1×10^{-3}, while for the DCG model (see Sect. 2.1) it is set to 2×10^{-4} when training the entire network. Following [20], the number of training epochs is set as 1,000 for all three datasets. In inference, we empirically set the total diffusion time step T as 100 for PMG2000, 250 for HAM10000, and 60 for APTOS2019, which is much smaller than most of the existing works [12,14]. The average running time of our DiffMIC is about 0.056 s for classifying an image with a spatial resolution of 224×224.

Comparison with State-of-the-Art Methods: In Table 1(a), we compare our DiffMIC against many state-of-the-art CNNs and transformer-based networks, including ResNet, Vision Transformer (ViT), Swin Transformer (Swin), Pyramid Transformer (PVT), and a medical image classification method (i.e., GMIC) on PMG2000. Apparently, PVT has the largest Accuracy of 0.907, and the largest F1-score of 0.902 among these methods. More importantly, our method further outperforms PVT. It improves the Accuracy from 0.907 to 0.931, and the F1-score from 0.902 to 0.926.

Note that both HAM10000 and APTOS2019 have a class imbalance issue. Hence, we compare our DiffMIC against state-of-the-art long-tailed medical image classification methods, and report the comparison results in Table 1(b). For HAM10000, our method produces a promising improvement over the second-best method ProCo of 0.019 and 0.053 in terms of Accuracy and F1-score, respectively. For APTOS2019, our method obtains a considerable improvement over ProCo of 0.021 and 0.042 in Accuracy and F1-score respectively.

Table 2. Effectiveness of each module in our DiffMIC on the PMG2000 dataset.

	Diffusion	DCG	MMD-reg	Accuracy	F1-score
basic	–	–	–	0.879	0.881
C1	✓	–	–	0.906	0.899
C2	✓	✓	–	0.920	0.914
Our method	✓	✓	✓	**0.931**	**0.926**

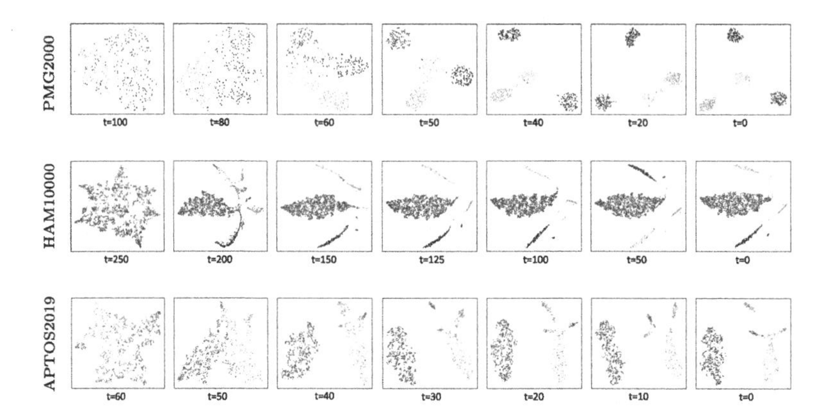

Fig. 2. t-SNE obtained from the denoised feature embedding by the diffusion reverse process during inference on three datasets. t is the current diffusion time step for inference. As the time step encoding progresses, the noise is gradually removed, thereby obtaining a clear distribution of classes; see the last column (please zoom in).

Ablation Study: Extensive experiments are conducted to evaluate the effectiveness of major modules of our network. To do so, we build three baseline networks from our method. The first baseline (denoted as "basic") is to remove all diffusion operations and the MMD regularization loss from our network. It means that "basic" is equal to the classical ResNet18. Then, we apply the vanilla diffusion process onto "basic" to construct another baseline network (denoted as "C1"), and further add our dual-granularity conditional guidance into the diffusion process to build a baseline network, which is denoted as "C2". Hence, "C2" is equal to removing the MMD regularization loss from our network for image classification. Table 2 reports the Accuracy and F1-score results of our method and three baseline networks on our PMG2000 dataset. Apparently, compared to "basic", "C1" has an Accuracy improvement of 0.027 and an F1-score improvement of 0.018, which indicates that the diffusion mechanism can learn more discriminate features for medical image classification, thereby improving the PMG performance. Moreover, the better Accuracy and F1-score results of "C2" over "C1" demonstrates that introducing our dual-granularity conditional guidance into the vanilla diffusion process can benefit the PMG performance. Furthermore, our method outperforms "C2" in terms of Accuracy and F1-score, which indicates that exploring the MMD regularization loss in the diffusion process can further help to enhance the PMG results.

Visualization of Our Diffusion Procedure: To illustrate the diffusion reverse process guided by our dual-granularity conditional encoding, we used the t-SNE tool to visualize the denoised feature embeddings at consecutive time steps. Figure 2 presents the results of this process on all three datasets. As the time step encoding progresses, the denoise diffusion model gradually removes noise from the feature representation, resulting in a clearer distribution of classes from the Gaussian distribution. The total number of time steps required for inference depends on the complexity of the dataset.

4 Conclusion

This work presents a diffusion-based network (DiffMIC) to boost medical image classification. The main idea of our DiffMIC is to introduce dual-granularity conditional guidance over vanilla DDPM, and enforce condition-specific MMD regularization to improve classification performance. Experimental results on three medical image classification datasets with diverse image modalities show the superior performance of our network over state-of-the-art methods. As the first diffusion-based model for general medical image classification, our DiffMIC has the potential to serve as an essential baseline for future research in this area.

Acknowledgments. This research is supported by Guangzhou Municipal Science and Technology Project (Grant No. 2023A03J0671), the National Research Foundation, Singapore under its AI Singapore Programme (AISG Award No: AISG2-TC-2021-003), A*STAR AME Programmatic Funding Scheme Under Project A20H4b0141, and A*STAR Central Research Fund.

References

1. Amit, T., Nachmani, E., Shaharbany, T., Wolf, L.: SegDiff: image segmentation with diffusion probabilistic models. arXiv preprint arXiv:2112.00390 (2021)
2. Batzolis, G., Stanczuk, J., Schönlieb, C.B., Etmann, C.: Conditional image generation with score-based diffusion models. arXiv preprint arXiv:2111.13606 (2021)
3. Cao, K., Wei, C., Gaidon, A., Arechiga, N., Ma, T.: Learning imbalanced datasets with label-distribution-aware margin loss. In: Advances in Neural Information Processing Systems, vol. 32 (2019)
4. Chen, S., Sun, P., Song, Y., Luo, P.: DiffusionDet: diffusion model for object detection. arXiv preprint arXiv:2211.09788 (2022)
5. De Bruijne, M.: Machine learning approaches in medical image analysis: from detection to diagnosis (2016)
6. Dhariwal, P., Nichol, A.: Diffusion models beat GANs on image synthesis. Adv. Neural Inf. Process. Syst. **34**, 8780–8794 (2021)
7. Dosovitskiy, A., et al.: An image is worth 16x16 words: transformers for image recognition at scale. arXiv preprint arXiv:2010.11929 (2020)
8. Esteva, A., et al.: Dermatologist-level classification of skin cancer with deep neural networks. Nature **542**(7639), 115–118 (2017)

9. Esteva, A., et al.: A guide to deep learning in healthcare. Nat. Med. **25**(1), 24–29 (2019)
10. Gong, L., Ma, K., Zheng, Y.: Distractor-aware neuron intrinsic learning for generic 2d medical image classifications. In: Martel, A.L., et al. (eds.) MICCAI 2020. LNCS, vol. 12262, pp. 591–601. Springer, Cham (2020). https://doi.org/10.1007/978-3-030-59713-9_57
11. Gretton, A., Borgwardt, K., Rasch, M., Schölkopf, B., Smola, A.: A kernel method for the two-sample-problem. In: Advances in Neural Information Processing Systems, vol. 19 (2006)
12. Han, X., Zheng, H., Zhou, M.: Card: classification and regression diffusion models. arXiv preprint arXiv:2206.07275 (2022)
13. He, K., Zhang, X., Ren, S., Sun, J.: Deep residual learning for image recognition. In: Proceedings of the IEEE Conference on Computer Vision and Pattern Recognition, pp. 770–778 (2016)
14. Ho, J., Jain, A., Abbeel, P.: Denoising diffusion probabilistic models. Adv. Neural Inf. Process. Syst. **33**, 6840–6851 (2020)
15. Ilse, M., Tomczak, J., Welling, M.: Attention-based deep multiple instance learning. In: International Conference on Machine Learning, pp. 2127–2136. PMLR (2018)
16. Karthik, Maggie, S.D.: Aptos 2019 blindness detection (2019). https://kaggle.com/competitions/aptos2019-blindness-detection
17. Li, Y., Swersky, K., Zemel, R.: Generative moment matching networks. In: International Conference on Machine Learning, pp. 1718–1727. PMLR (2015)
18. Liao, H., Luo, J.: A deep multi-task learning approach to skin lesion classification. arXiv preprint arXiv:1812.03527 (2018)
19. Liu, Z., et al.: Swin transformer: hierarchical vision transformer using shifted windows. In: Proceedings of the IEEE/CVF International Conference on Computer Vision, pp. 10012–10022 (2021)
20. Marrakchi, Y., Makansi, O., Brox, T.: Fighting class imbalance with contrastive learning. In: de Bruijne, M., et al. (eds.) MICCAI 2021. LNCS, vol. 12903, pp. 466–476. Springer, Cham (2021). https://doi.org/10.1007/978-3-030-87199-4_44
21. Nichol, A.Q., Dhariwal, P.: Improved denoising diffusion probabilistic models. In: Meila, M., Zhang, T. (eds.) Proceedings of the 38th International Conference on Machine Learning. Proceedings of Machine Learning Research, vol. 139, pp. 8162–8171. PMLR, 18–24 July 2021
22. Rajpurkar, P., Chen, E., Banerjee, O., Topol, E.J.: AI in health and medicine. Nat. Med. **28**(1), 31–38 (2022)
23. Shamshad, F., et al.: Transformers in medical imaging: a survey. arXiv, January 2022
24. Shen, Y., et al.: An interpretable classifier for high-resolution breast cancer screening images utilizing weakly supervised localization. Med. Image Anal. **68**, 101908 (2021)
25. Shrivastava, A., Gupta, A., Girshick, R.: Training region-based object detectors with online hard example mining. In: Proceedings of the IEEE Conference on Computer Vision and Pattern Recognition, pp. 761–769 (2016)
26. Singh, J., Gould, S., Zheng, L.: High-fidelity guided image synthesis with latent diffusion models. arXiv preprint arXiv:2211.17084 (2022)
27. Tschandl, P., Rosendahl, C., Kittler, H.: The ham10000 dataset, a large collection of multi-source dermatoscopic images of common pigmented skin lesions. Sci. Data **5**(1), 1–9 (2018)

28. Wang, W., et al.: Pyramid vision transformer: a versatile backbone for dense prediction without convolutions. In: Proceedings of the IEEE/CVF International Conference on Computer Vision, pp. 568–578 (2021)
29. Wolleb, J., Sandkühler, R., Bieder, F., Valmaggia, P., Cattin, P.C.: Diffusion models for implicit image segmentation ensembles. In: International Conference on Medical Imaging with Deep Learning, pp. 1336–1348. PMLR (2022)
30. Yang, Z., et al.: ProCo: prototype-aware contrastive learning for long-tailed medical image classification. In: Wang, L., Dou, Q., Fletcher, P.T., Speidel, S., Li, S. (eds.) Medical Image Computing and Computer Assisted Intervention - MICCAI 2022, MICCAI 2022. LNCS, vol. 13438. pp. 173–182. Springer, Cham (2022). https://doi.org/10.1007/978-3-031-16452-1_17
31. Zhao, S., Song, J., Ermon, S.: Infovae: information maximizing variational autoencoders. arXiv preprint arXiv:1706.02262 (2017)

Whole-Heart Reconstruction with Explicit Topology Integrated Learning

Huilin Yang[1,2], Roger Tam[1], and Xiaoying Tang[2](✉)

[1] The University of British Columbia, Vancouver, BC V6T 1Z4, Canada
huiliny1@student.ubc.ca

[2] Southern University of Science and Technology, Shenzhen 518055, China
tangxy@sustech.edu.cn

Abstract. Reconstruction and visualization of cardiac structures play significant roles in computer-aided clinical practice as well as scientific research. With the advancement of medical imaging techniques, computing facilities, and deep learning models, automatically generating whole-heart meshes directly from medical imaging data becomes feasible and shows great potential. Existing works usually employ a point cloud metric, namely the Chamfer distance, as the optimization objective when reconstructing the whole-heart meshes, which nevertheless does not take the cardiac topology into consideration. Here, we propose a novel *currents*-represented surface loss to optimize the reconstructed mesh topology. Due to *currents*'s favorable property of encoding the topology of a whole surface, our proposed pipeline delivers whole-heart reconstruction results with correct topology and comparable or even higher accuracy.

Keywords: Whole Heart Reconstruction · Topology · Deep Learning

1 Introduction

With the advent of advanced medical imaging technologies, such as computed tomography (CT) and magnetic resonance (MR), non-invasive visualizations of various human organs and tissues become feasible and are widely utilized in clinical practice [8,22]. Cardiac CT imaging and MR imaging play important roles in the understanding of cardiac anatomy, diagnosis of cardiac diseases [20] and

Supported by the National Natural Science Foundation of China (62071210); the Shenzhen Science and Technology Program (RCYX20210609103056042); the Shenzhen Science and Technology Innovation Committee (KCXFZ2020122117340001); the Shenzhen Basic Research Program (JCYJ20200925153847004, JCYJ20190809 120205578).

Supplementary Information The online version contains supplementary material available at https://doi.org/10.1007/978-3-031-43987-2_11.

H. Greenspan et al. (Eds.): MICCAI 2023, LNCS 14225, pp. 106–115, 2023.
https://doi.org/10.1007/978-3-031-43987-2_11

multimodal visualizations [9]. Potential applications range from patient-specific treatment planning, virtual surgery, morphology assessment to biomedical simulations [2,17]. However, in traditional procedures, visualizing human organs usually requires significant expert efforts and could take up to dozens of hours depending on the specific organs of interest [7,24], which makes large-cohort studies prohibitive and limits clinical applications [16].

Empowered by the great feature extraction ability of deep neural networks (DNNs) and the strong parallel computing power of graph processing units (GPUs), automated visualizations of cardiac organs have been extensively explored in recent years [18,24]. These methods typically follow a common processing flow that requires a series of post-processing steps to produce acceptable reconstruction results. Specifically, the organs of interest are first segmented from medical imaging data. After that, an isosurface generation algorithm, such as marching cubes [15], is utilized to create 3D visualizations typically with staircase appearance, followed by smoothing filters to create smooth meshes. Finally, manual corrections or connected component analyses [11] are applied to remove artifacts and improve topological correctness. The entire flow is not optimized in an end-to-end fashion, which might introduce and accumulate multi-step errors or still demand non-trivial manual efforts.

In such context, automated approaches that can directly and efficiently generate cardiac shapes from medical imaging data are highly desired. Recently, various DNN works [1,3,4,12–14,19] delve into this topic and achieve promising outcomes. In particular, the method depicted in [1] performs predictions of cardiac ventricles using both cine MR and patient metadata based on statistical shape modeling (SSM). Similarly, built on SSM, [4] uses 2D cine MR slices to generate five cardiac meshes. Another approach proposed in [19] employs distortion energy to produce meshes of the aortic valves. Inspiringly, graph neural network (GNN) based methods [12–14] are shown to be capable of simultaneously reconstructing seven cardiac organs in a single pass, producing whole-heart meshes that are suitable for computational simulations of cardiac functioning. The training processes for these aforementioned methods are usually optimized via the Chamfer distance (CD) loss, a point cloud based evaluation metric. Such type of point cloud based losses is first calculated for each individual vertex, followed by an average across all vertices, which nonetheless does not take the overall mesh topology into consideration. This could result in suboptimal or even incorrect topology in the reconstructed mesh, which is undesirable.

To solve this issue, we introduce a novel surface loss that inherently considers the topology of the two to-be-compared meshes in the loss function, with a goal of optimizing the anatomical topology of the reconstructed mesh. The surface loss is defined by a computable norm on *currents* [6] and is originally introduced in [23] for diffeomorphic surface registration, which has extensive applicability in shape analysis and disease diagnosis [5,21]. Motivated by its inherent ability to characterize and quantify a mesh's topology, we make use of it to minimize the topology-considered overall difference between a reconstructed mesh and its corresponding ground truth mesh. Such *currents* guided supervision ensures effective and efficient whole-heart mesh reconstructions of seven cardiac organs, with high reconstruction accuracy and correct anatomical topology being attained.

2 Methodology

Figure 1 illustrates the proposed end-to-end pipeline, consisting of a voxel feature extraction module (top panel) and a deformation module (middle panel). The inputs contain a CT or MR volume accompanied by seven initial spherical meshes. To be noted, the seven initial spherical meshes are the same for all training and testing cases. A volume encoder followed by a decoder is employed as the voxel feature extraction module, which is supervised by a segmentation loss comprising binary cross entropy (BCE) and Dice. This ensures that the extracted features explicitly encode the characteristics of the regions of interest (ROIs). For the deformation module, a GNN is utilized to map coordinates of the mesh vertices, combine and map trilinearly-interpolated voxel features indexed at each mesh vertex, extract mesh features, and deform the initial meshes to reconstruct the whole-heart meshes. There are three deformation blocks that progressively deform the initial meshes. Each deformation block is optimized on three types of losses: a surface loss for both accuracy and topology correctness purposes,

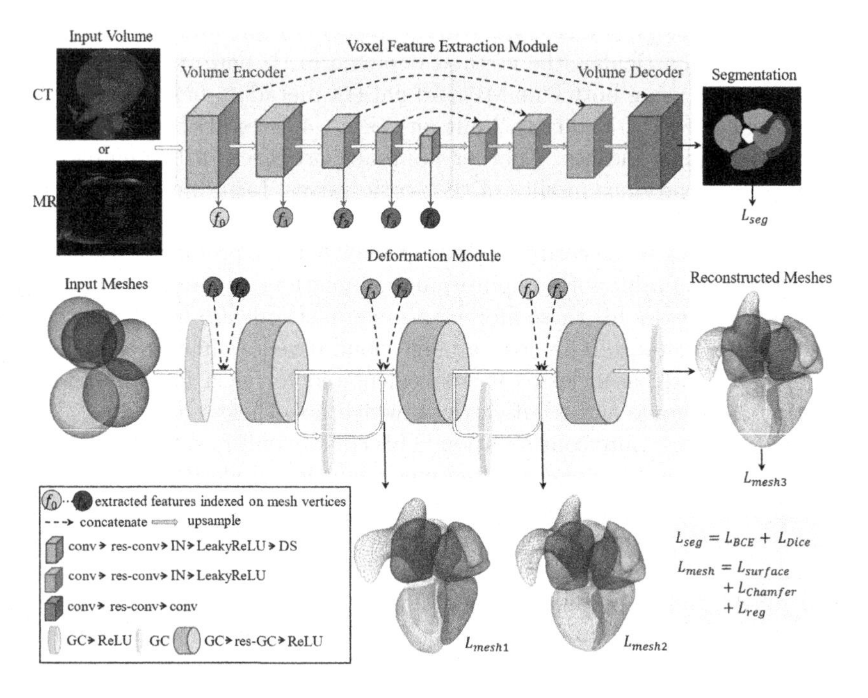

Fig. 1. Illustration of the proposed pipeline. Abbreviations are as follows: instance normalization (IN), downsampling (DS), convolution (conv), graph convolution (GC), residual convolution (res-conv), and residual graph convolution (res-GC).

a point cloud loss for an accuracy purpose, and three regularization losses for a smoothness purpose. The network structure details of the two modules are detailed in the supplementary material.

For an input CT or MR volume, it passes into the voxel feature extraction module to predict binary segmentation for the to-be-reconstructed ROIs. Meanwhile, the initial spherical meshes enter into the first deformation block along with the trilinearly-interpolated voxel features to predict the vertex-wise displacements of the initial meshes. Then, the updated meshes go through the following blocks for subsequent deformations. The third deformation block finally outputs the reconstructed whole-heart meshes. The three deformation blocks follow the same process, except for the meshes they deform and the trilinearly-interpolated voxel features they operate on. In the first deformation block, we use high-level voxel features, f_3 and f_4, obtained from the deepest layers of the volume encoder. In the second deformation block, the middle-level voxel features, f_1 and f_2, are employed. As for the last deformation block, its input meshes are usually quite accurate and only need to be locally refined. Thus, low-level voxel features are employed to supervise this refining process.

Surface Representation as *Currents*. Keeping in line with [23], we employ a generalized distribution from geometric measure theory, namely *currents* [6], to represent surfaces. Specifically, surfaces are represented as objects in a linear space equipped with a computable norm. Given a triangular mesh S embedded in $\mathbb{R}^3$, it can be associated with a linear functional on the space of 2-form via the following equation

$$S(\omega) = \int_S \omega(x)(u_x^1, u_x^2) d\sigma(x), \tag{1}$$

where for each $x \in S$ u_x^1 and u_x^2 form an orthonormal basis of the tangent plane at x. $\omega(x)$ is a skew-symmetric bilinear function on $\mathbb{R}^3$. $d\sigma(x)$ represents the basic element of surface area. Subsequently, a surface can be represented as *currents* in the following expression

$$S(\omega) = \sum_f \int_f \bar{\omega}(x) \cdot (u_x^1 \times u_x^2) d\sigma_f(x), \tag{2}$$

where $S(\omega)$ denotes the *currents* representation of the surface. f denotes each face of S and σ_f is the surface measure on f. $\bar{\omega}(x)$ is the vectorial representation of $\omega(x)$, with $\cdot$ and $\times$ respectively representing dot product and cross product. After the *currents* representation is established, an approximation of ω over each face can be obtained by using its value at the face center.

Let f_v^1, f_v^2, f_v^3 denote the three vertices of a face f, $e^1 = f_v^2 - f_v^3$, $e^2 = f_v^3 - f_v^1$, $e^3 = f_v^1 - f_v^2$ are the edges, $c(f) = \frac{1}{3}(f_v^1 + f_v^2 + f_v^3)$ is the center of the face and $N(f) = \frac{1}{2}(e^2 \times e^3)$ is the normal vector of the face with its length being equal to the face area. Then, ω can be approximated over the face by its value at the face center, resulting in $S(\omega) \approx \sum_f \bar{\omega}(c(f)) \cdot N(f)$. In fact, the approximation is a sum of linear evaluation functionals $C(S) = \sum_f \delta_{c(f)}^{N(f)}$ associated with a Reproducing Kernel Hilbert Space (RKHS) under the constraints presented elsewhere [23].

Thus, S_ε, the discrepancy between two surfaces S and T, can be approximately calculated via the RKHS as below

$$S_\varepsilon = ||C(S) - C(T)||^2_{W^*} = \sum_{f,g} N(f)^T k_W(c(g), c(f)) N(g) - 2\sum_{f,q} N(f)^T k_W(c(q), c(f)) N(q) + \sum_{q,r} N(q)^T k_W(c(q), c(r)) N(r), \quad (3)$$

where W^* is the dual space of a Hilbert space $(W, \langle \cdot, \cdot \rangle_W)$ of differential 2-forms and $|| \, ||^2$ is l_2-norm. $()^T$ denotes the transpose operator. f, g index the faces of S and q, r index the faces of T. k_W is an isometry between W^* and W, and we have $\langle \delta_x^\xi, \delta_y^\eta \rangle_{W^*} = k_W(x, y)\xi \cdot \eta$ [23]. The first and third terms enforce the structural integrity of the two surfaces, while the middle term penalizes the geometric and spatial discrepancies between them. With this preferable property, Eq. 3 fulfills the topology correctness purpose, the key of this proposed pipeline.

Surface Loss. As in [23], we choose a Gaussian kernel as the instance of k_W. Namely, $k_W(x, y) = exp(-\frac{||x-y||^2}{\sigma_W^2})$, where x and y are the centers of two faces and σ_W is a scale controlling parameter that controls the affecting scale between the two faces. Therefore, the surface loss can be expressed as

$$L_{surface} = \sum_{t_1,t_2} exp(-\frac{||c(t_1) - c(t_2)||^2}{\sigma_W^2}) N(t_1)^T \cdot N(t_2) - 2\sum_{t,p} exp(-\frac{||c(t) - c(p)||^2}{\sigma_W^2}) N(t)^T \cdot N(p) + \sum_{p_1,p_2} exp(-\frac{||c(p_1) - c(p_2)||^2}{\sigma_W^2}) N(p_1)^T \cdot N(p_2), \quad (4)$$

where t_1, t_2, t and p_1, p_2, p respectively index faces on the reconstructed surfaces S_R and those on the corresponding ground truth surfaces S_T. $L_{surface}$ not only considers each face on the surfaces but also its corresponding direction. When the reconstructed surfaces are exactly the same as the ground truth, the surface loss $L_{surface}$ should be 0. Otherwise, $L_{surface}$ is a bounded positive value [23]. Minimizing $L_{surface}$ enforces the reconstructed surfaces to be progressively close to the ground truth as the training procedure develops.

Figure 2 illustrates how σ_W controls the affecting scale of a face on a surface. The three surfaces are identical meshes of a left atrium structure except for the affecting scale (shown in different colors) on them. There are three colored circles (red, blue, and green) respectively representing the centers of three faces on the surfaces, and the arrowed vectors on these circles denote the corresponding face normals. The color bar ranges from 0 to 1, with 0 representing no effect and 1 representing the most significant effect. From Fig. 2, the distance between the blue circle and the red one is closer than that between the blue circle and the green one, and the effect between the red circle and the blue one is accordingly larger than that between the red circle and the green one. With σ_W varying from a large value to a small one, the effects between the red face and other

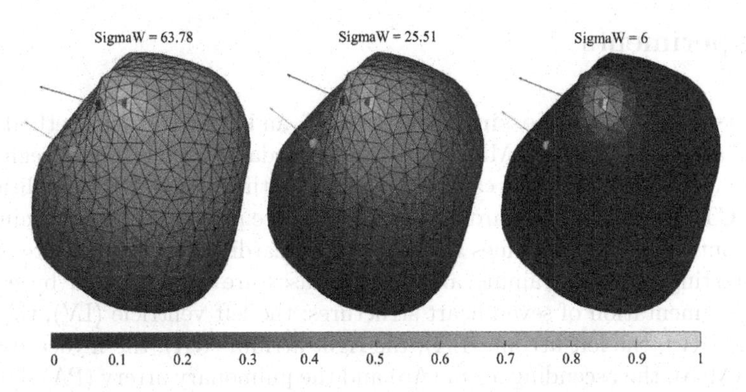

Fig. 2. Illustration of how σ_W controls the affecting scale in the surface loss. The three surfaces represent an identical left atrium structure with different σ_W controls. Red, blue, and green circles denote three centers of the same surface faces. The varying colors represent the magnitudes of effects between the red circle and all other face centers on the same surface.

remaining faces become increasingly small. In this way, we are able to control the acting scale of the surface loss via changing the value of σ_W. Assigning σ_W a value that covers the entire surface results in a global topology encoding of the surface, while assigning a small value that only covers neighbors shall result in a topology encoding that focuses on local geometries.

Loss Function. In addition to the surface loss we introduce above, we also involve two segmentation losses L_{BCE} and L_{Dice}, one point cloud loss L_{CD}, and three regularization losses $L_{laplace}$, L_{edge}, and L_{normal} that comply with [13]. The total loss function can be expressed as:

$$L_{total} = L_{seg} + L_{mesh_1} + L_{mesh_2} + L_{mesh_3}, \tag{5}$$

$$L_{seg} = w_s(L_{BCE} + L_{Dice}), \tag{6}$$

$$L_{mesh} = L_{surface}^{w_1} \cdot L_{CD}^{w_2} \cdot L_{laplace}^{w_3} \cdot L_{edge}^{w_4} \cdot L_{normal}^{w_5}. \tag{7}$$

where w_s is the weight for the segmentation loss, and w_1, w_2, w_3 and w_4 are respectively the weights for the surface loss, the Chamfer distance, the Laplace loss, and the edge loss. The geometric mean is adopted to combine the five individual mesh losses to accommodate their different magnitudes.

L_{seg} ensures useful feature learning of the ROIs. $L_{surface}$ enforces the integrity of the reconstructed meshes and makes them topologically similar to the ground truth. L_{CD} makes the point cloud representation of the reconstructed meshes to be close to that of the ground truth. Additionally, $L_{laplace}$, L_{edge}, and L_{normal} are employed for the smoothness consideration of the reconstructed meshes.

3 Experiments

Datasets and Preprocessing. We evaluate and validate our method on a publicly-accessible dataset MM-WHS (multi-modality whole heart segmentation) [24], which contains 3D cardiac images of both CT and MR modalities. 20 cardiac CT volumes and 20 cardiac MR volumes are provided in the training set. 40 held-out cardiac CT volumes and 40 held-out cardiac MR volumes are offered in the testing set. All training and testing cases are accompanied by expert-labeled segmentation of seven heart structures: the left ventricle (LV), the right ventricle (RV), the left atrium (LA), the right atrium (RA), the myocardium of the LV (Myo), the ascending aorta (Ao) and the pulmonary artery (PA). For preprocessing, we follow [13] to perform resizing, intensity normalization, and data augmentation (random rotation, scaling, shearing, and elastic warping) for each training case. Data characteristics and preprocessing details are summarised in the supplementary material.

Evaluation Metrics. In order to compare with existing state-of-the-art (SOTA) methods, four metrics as in [13] are employed for evaluation, including Dice, Jaccard, average symmetric surface distance (ASSD), and Hausdorff distance (HD). Furthermore, intersected mesh facets are detected by TetGen [10] and used for quantifying self-intersection (SI).

Table 1. Comparisons with two SOTA methods on the MM-WHS CT test data. The MeshDeform [13] results are obtained from our self-reimplementation, while the Voxel2Mesh results are directly copied from [13] since its code has not been open sourced yet. WH denotes whole heart.

		Myo	LA	LV	RA	RV	Ao	PA	WH
Dice (↑)	**Ours**	**0.888**	**0.870**	0.928	**0.928**	**0.904**	**0.948**	**0.841**	**0.908**
	MeshDeform	0.883	0.864	0.928	0.921	0.885	0.925	0.814	0.899
	Voxel2Mesh	0.501	0.748	0.669	0.717	0.698	0.555	0.491	0.656
Jaccard (↑)	**Ours**	0.803	**0.829**	0.869	**0.867**	**0.829**	**0.902**	**0.737**	**0.832**
	MeshDeform	**0.806**	0.766	**0.870**	0.855	0.801	0.861	0.704	0.819
	Voxel2Mesh	0.337	0.600	0.510	0.570	0.543	0.397	0.337	0.491
ASSD (mm) (↓)	**Ours**	1.528	**1.106**	**0.962**	**1.727**	**1.211**	**0.593**	**1.344**	**1.308**
	MeshDeform	**1.474**	1.137	0.966	1.750	1.320	0.729	2.020	1.333
	Voxel2Mesh	3.412	3.147	4.973	3.638	4.300	4.326	5.857	4.287
HD (mm) (↓)	**Ours**	**12.588**	11.019	**13.616**	14.279	11.136	**5.369**	**8.789**	**16.934**
	MeshDeform	13.143	**9.177**	13.823	**14.140**	**7.66**	5.408	9.664	17.681
	Voxel2Mesh	15.526	13.683	22.146	16.834	18.390	19.419	35.322	37.065
SI (%) (↓)	**Ours**	**0.009**	0.007	0.011	0.004	0.019	0.003	0.038	**0.013**
	MeshDeform	0.014	0.006	0.017	0.007	0.024	0.005	0.049	0.017
	Voxel2Mesh	0.269	**0.000**	**0.000**	**0.003**	**0.000**	**0.000**	**0.020**	0.042

Results. We compare our method with two SOTA methods on the five evaluation metrics. Ours and MeshDeform [13] are trained on the same dataset consisting of 16 CT and 16 MR data that are randomly selected from the MM-WHS training set with 60 augmentations for each, and the remaining 4 CT and 4 MR are used for validation. Evaluations are performed on the encrypted testing set with the officially provided executables. We reimplement MeshDeform [13] with Pytorch according to the publicly available Tensorflow version. Please note the Voxel2Mesh results are directly obtained from [13] since its code has not been open sourced yet. Training settings are detailed in the supplementary material.

Table 1 shows evaluation results on the seven heart structures and the whole heart of the MM-WHS CT testing set. Our method achieves the best results in most entries. For SI, Voxel2Mesh holds the best results in most entries because of its unpooling operations in each deformation procedure, in which topological information is additionally used. However, as described in [13], Voxel2Mesh may easily encounter out-of-memory errors for its increasing vertices along the reconstruction process. More results for the MR data can be found in the supplementary material.

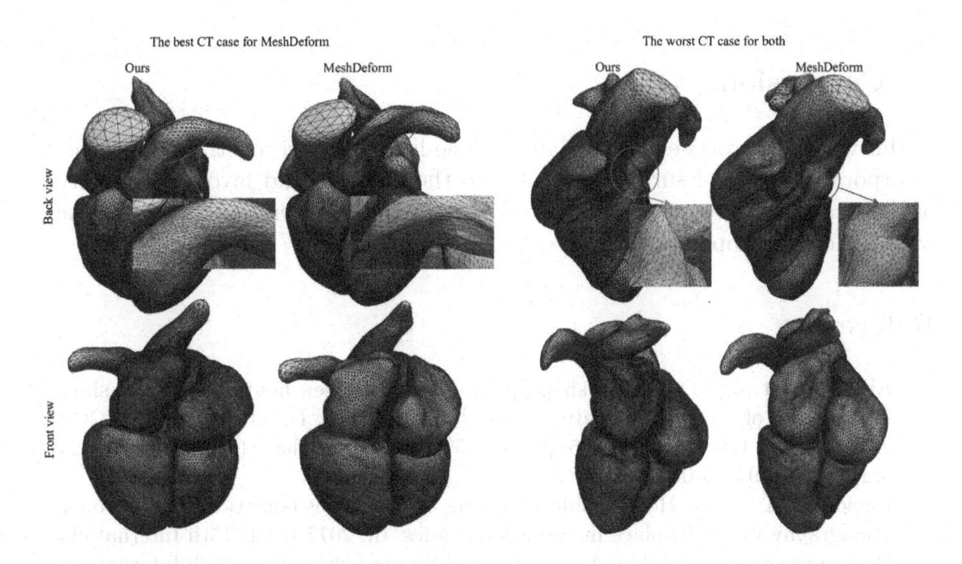

Fig. 3. Demonstration of the best as well as the worst cases for MeshDeform with respect to Dice and our results on the same cases.

Figure 3 shows the best and the worst CT results for MeshDeform with respect to Dice and our results on the same cases. Noticeably, the best case for MeshDeform is not the best for our method. For that best case of MeshDeform, we can see obvious folded areas on the mesh of PA, while our method yields more satisfactory visualization results. As for the worst case, both methods obtain unsatisfactory visualizations. However, the two structures (PA and RV) obtained from MeshDeform intersect with each other, leading to significant topological errors. Our method does not have such topology issues.

Ablation Study. For the ablation study, we train a model without the surface loss while keeping the rest the same. Table 2 shows the ablation analysis results on the CT data, which apparently validates the effectiveness of the surface loss.

Table 2. A comparison of reconstruction accuracy on the MM-WHS CT test data for the proposed method with and without (w.o.) the surface loss.

		Myo	LA	LV	RA	RV	Ao	PA	WH
Dice (↑)	Ours	**0.888**	**0.870**	0.928	**0.928**	**0.904**	**0.948**	**0.841**	**0.908**
	Ours (w.o. L_s)	0.881	0.861	**0.932**	0.927	0.902	0.947	0.829	0.905
Jaccard (↑)	Ours	**0.803**	**0.829**	0.869	**0.867**	**0.775**	**0.902**	**0.737**	**0.832**
	Ours (w.o. L_s)	0.793	0.824	**0.875**	0.866	0.762	0.899	0.719	0.827
HD (mm) (↓)	Ours	**12.588**	11.019	13.616	**14.279**	**11.136**	**5.369**	**8.789**	**16.934**
	Ours (w.o. L_s)	14.347	**10.954**	**10.100**	14.901	13.225	6.781	10.484	17.602

4 Conclusion

In this work, we propose and validate a whole-heart mesh reconstruction method incorporating a novel surface loss. Due to the intrinsic and favorable property of the *currents* representation, our method is able to generate accurate meshes with the correct topology.

References

1. Attar, R., et al.: 3D cardiac shape prediction with deep neural networks: simultaneous use of images and patient metadata. In: Shen, D., et al. (eds.) MICCAI 2019, Part II. LNCS, vol. 11765, pp. 586–594. Springer, Cham (2019). https://doi.org/10.1007/978-3-030-32245-8_65
2. Bucioli, A.A., et al. Holographic real time 3D heart visualization from coronary tomography for multi-place medical diagnostics. In: 2017 IEEE 15th International Conference on Dependable, Autonomic and Secure Computing, 15th International Conference on Pervasive Intelligence and Computing, 3rd International Conference on Big Data Intelligence and Computing and Cyber Science and Technology Congress (DASC/PiCom/DataCom/CyberSciTech), pp. 239–244. IEEE, 6 November 2017
3. Beetz, M., Banerjee, A., Grau, V.: Biventricular surface reconstruction from cine MRI contours using point completion networks. In: 2021 IEEE 18th International Symposium on Biomedical Imaging (ISBI), pp. 105–109. IEEE, April 2021
4. Banerjee, A., Zacur, E., Choudhury, R. P., Grau, V.: Automated 3D whole-heart mesh reconstruction from 2D cine MR slices using statistical shape model. In: 2022 44th Annual International Conference of the IEEE Engineering in Medicine and Biology Society (EMBC), pp. 1702–1706. IEEE, July 2022
5. Charon, N., Younes, L.: Shape spaces: From geometry to biological plausibility. In: Handbook of Mathematical Models and Algorithms in Computer Vision and Imaging: Mathematical Imaging and Vision, pp. 1929–1958 (2023)

6. De Rham, G.: Variétés différentiables: formes, courants, formes harmoniques, vol. 3. Editions Hermann (1973)
7. Fischl, B.: FreeSurfer. Neuroimage **62**(2), 774–781 (2012)
8. Garvey, C.J., Hanlon, R.: Computed tomography in clinical practice. BMJ **324**(7345), 1077–1080 (2002)
9. González Izard, S., Sánchez Torres, R., Alonso Plaza, O., Juanes Mendez, J.A., García-Peñalvo, F.J.: NextMed: automatic imaging segmentation, 3D reconstruction, and 3D model visualization platform using augmented and virtual reality. Sensors **20**(10), 2962 (2020)
10. Hang, S.: TetGen, a delaunay-based quality tetrahedral mesh generator. ACM Trans. Math. Softw **41**(2), 11 (2015)
11. He, L., Ren, X., Gao, Q., Zhao, X., Yao, B., Chao, Y.: The connected-component labeling problem: a review of state-of-the-art algorithms. Pattern Recogn. **70**, 25–43 (2017)
12. Kong, F., Shadden, S.C.: Whole heart mesh generation for image-based computational simulations by learning free-from deformations. In: de Bruijne, M., et al. (eds.) MICCAI 2021. LNCS, vol. 12904, pp. 550–559. Springer, Cham (2021). https://doi.org/10.1007/978-3-030-87202-1_53
13. Kong, F., Wilson, N., Shadden, S.: A deep-learning approach for direct whole-heart mesh reconstruction. Med. Image Anal. **74**, 102222 (2021)
14. Kong, F., Shadden, S.C.: Learning whole heart mesh generation from patient images for computational simulations. IEEE Trans. Med. Imaging **42**(2), 533–545 (2022)
15. Lorensen, W.E., Cline, H.E.: Marching cubes: a high resolution 3d surface construction algorithm. ACM siggraph Comput. Graph. **21**(4), 163–169 (1987)
16. Mittal, R., et al.: Computational modeling of cardiac hemodynamics: current status and future outlook. J. Comput. Phys. **305**, 1065–1082 (2016)
17. Prakosa, A., et al.: Personalized virtual-heart technology for guiding the ablation of infarct-related ventricular tachycardia. Nat. Biomed. Eng. **2**(10), 732–740 (2018)
18. Painchaud, N., Skandarani, Y., Judge, T., Bernard, O., Lalande, A., Jodoin, P.M.: Cardiac segmentation with strong anatomical guarantees. IEEE Trans. Med. Imaging **39**(11), 3703–3713 (2020)
19. Pak, D.H., et al.: Distortion energy for deep learning-based volumetric finite element mesh generation for aortic valves. In: de Bruijne, M., et al. (eds.) MICCAI 2021, Part VI. LNCS, vol. 12906, pp. 485–494. Springer, Cham (2021). https://doi.org/10.1007/978-3-030-87231-1_47
20. Stokes, M.B., Roberts-Thomson, R.: The role of cardiac imaging in clinical practice. Aust. Prescriber **40**(4), 151 (2017)
21. Tang, X., Holland, D., Dale, A.M., Younes, L., Miller, M.I., Initiative, A.D.N.: Shape abnormalities of subcortical and ventricular structures in mild cognitive impairment and Alzheimer's disease: detecting, quantifying, and predicting. Hum. Brain Mapping **35**(8), 3701–3725 (2014)
22. Tsougos, I.: Advanced MR Neuroimaging: from Theory to Clinical Practice. CRC Press, Boca Raton (2017)
23. Vaillant, M., Glaunès, J.: Surface matching via currents. In: Christensen, G.E., Sonka, M. (eds.) IPMI 2005. LNCS, vol. 3565, pp. 381–392. Springer, Heidelberg (2005). https://doi.org/10.1007/11505730_32
24. Zhuang, X., Shen, J.: Multi-scale patch and multi-modality atlases for whole heart segmentation of MRI. Med. Image Anal. **31**, 77–87 (2016)

Enhancing Automatic Placenta Analysis Through Distributional Feature Recomposition in Vision-Language Contrastive Learning

Yimu Pan[1(✉)], Tongan Cai[1], Manas Mehta[1], Alison D. Gernand[1], Jeffery A. Goldstein[2], Leena Mithal[3], Delia Mwinyelle[4], Kelly Gallagher[1], and James Z. Wang[1]

[1] The Pennsylvania State University, University Park, PA, USA
ymp5078@psu.edu
[2] Northwestern University, Chicago, IL, USA
[3] Lurie Children's Hospital, Chicago, IL, USA
[4] The University of Chicago, Chicago, IL, USA

Abstract. The placenta is a valuable organ that can aid in understanding adverse events during pregnancy and predicting issues post-birth. Manual pathological examination and report generation, however, are laborious and resource-intensive. Limitations in diagnostic accuracy and model efficiency have impeded previous attempts to automate placenta analysis. This study presents a novel framework for the automatic analysis of placenta images that aims to improve accuracy and efficiency. Building on previous vision-language contrastive learning (VLC) methods, we propose two enhancements, namely Pathology Report Feature Recomposition and Distributional Feature Recomposition, which increase representation robustness and mitigate feature suppression. In addition, we employ efficient neural networks as image encoders to achieve model compression and inference acceleration. Experiments validate that the proposed approach outperforms prior work in both performance and efficiency by significant margins. The benefits of our method,

Research reported in this publication was supported by the National Institute of Biomedical Imaging and Bioengineering of the National Institutes of Health (NIH) under award number R01EB030130 and the College of Information Sciences and Technology of The Pennsylvania State University. The content is solely the responsibility of the authors and does not necessarily represent the official views of the NIH. This work used computing resources at the Pittsburgh Supercomputer Center through allocation IRI180002 from the Advanced Cyberinfrastructure Coordination Ecosystem: Services & Support (ACCESS) program, which is supported by National Science Foundation grants Nos. 2138259, 2138286, 2138307, 2137603, and 2138296.

Supplementary Information The online version contains supplementary material available at https://doi.org/10.1007/978-3-031-43987-2_12.

H. Greenspan et al. (Eds.): MICCAI 2023, LNCS 14225, pp. 116–126, 2023.
https://doi.org/10.1007/978-3-031-43987-2_12

including enhanced efficacy and deployability, may have significant implications for reproductive healthcare, particularly in rural areas or low- and middle-income countries.

Keywords: Placenta Analysis · Representation · Vision-Language

1 Introduction

World Bank data from 2020 suggests that while the infant mortality rate in high-income countries is as low as 0.4%, the number is over ten times higher in low-income countries (approximately 4.7%). This stark contrast underlines the necessity for accessible healthcare. The placenta, as a vital organ connecting the fetus to the mother, has discernable features such as meconium staining, infections, and inflammation. These can serve as indicators of adverse pregnancy outcomes, including preterm delivery, growth restriction, respiratory or neurodevelopmental conditions, and even neonatal deaths [9].

In a clinical context, these adverse outcomes are often signaled by morphological changes in the placenta, identifiable through pathological analysis [19]. Timely conducted placental pathology can reduce the risks of serious consequences of pregnancy-related infections and distress, ultimately improving the well-being of newborns and their families. Unfortunately, traditional placenta pathology examination is resource-intensive, requiring specialized equipment and expertise. It is also a time-consuming task, where a full exam can easily take several days, limiting its widespread applications even in developed countries. To overcome these challenges, researchers have been exploring the use of automatic placenta analysis tools that rely on photographic images. By enabling broader and more timely placental analysis, these tools could help reduce infant fatalities and improve the quality of life for families with newborns.

Related Work. Considerable progress has been made in segmenting [17,20,23] and classifying [1,8,10,13,15,21,26] placenta images using histopathological, ultrasound, or MRI data. However, these methods are dependent on expensive and bulky equipment, restricting the accessibility of reproductive healthcare. Only limited research has been conducted on the gross analysis of post-birth placenta photographs, which have a lower equipment barrier. AI-PLAX [4] combines handcrafted features and deep learning, and a more recent study [29] relies on deep learning and domain adaptation. Unfortunately, both are constrained by issues such as data scarcity and single modality, which hinder their robustness and generalizability. To address these, Pan et al. [16] incorporated vision-and-language contrastive learning (VLC) using pathology reports. However, their method struggles with variable-length reports and is computationally demanding, making it impractical for low-resource communities.

With growing research in vision-and-language and contrastive learning [18, 28], recent research has focused on improving the performance and efficiency of VLC approaches. They propose new model architectures [2,24], better visual representation [7,27], loss function design [14,16], or sampling strategies [5,12].

However, these methods are still not suitable for variable-length reports and are inefficient in low-resource settings.

Our Contributions. We propose a novel framework for more accurate and efficient computer-aided placenta analysis. Our framework introduces two key enhancements: Pathology Report Feature Recomposition, a first in the medical VLC domain that captures features from pathology reports of variable lengths, and Distributional Feature Recomposition, which provides a more robust, distribution-aware representation. We demonstrate that our approach improves representational power and surpasses previous methods by a significant performance margin, without additional data. Furthermore, we boost training and testing efficiency by eliminating the large language model (LLM) from the training process and incorporating more efficient encoders. To the best of our knowledge, this is the first study to improve both the efficiency and performance of VLC training techniques for placenta analysis.

2 Dataset

We use the exact dataset from Pan et al. [16] collected using a professional photography instrument in the pathology department of the Northwestern Memorial Hospital (Chicago) from 2014 to 2018 and an iPad in 2021. There are three parts of the dataset: 1) the pre-training dataset, containing 10,193 image-and-text pairs; 2) the primary fine-tuning dataset, comprising 2,811 images labeled for five tasks: *meconium*, *fetal inflammatory response* (FIR), *maternal inflammatory response* (MIR), and *histologic chorioamnionitis*, and *neonatal sepsis*; and 3) the iPad evaluation dataset, consisting of 52 images from an iPad labeled for MIR and *clinical chorioamnionitis*. As with the original study, we assess the effectiveness of our method on the primary dataset, while utilizing iPad images to evaluate the robustness against distribution shifts. All images contain the fetal side of a placenta, the cord, and a ruler for scale. The pre-training data is also accompanied by a corresponding text sequence for the image containing a part of the corresponding pathology report as shown in Fig. 1. A detailed breakdown of the images is provided in the supplementary materials.

3 Method

This section aims to provide an introduction to the background, intuition, and specifics of the proposed methods. An overview is given in Fig. 1.

3.1 Problem Formulation

Our tasks are to train an encoder to produce placenta features and a classifier to classify them. Formally, we aim to learn a function f^v using a learned function f^u, such that for any pair of input $(\mathbf{x}_i, \mathbf{t}_i)$ and a similarity function $\mathtt{sim}$, we have

$$\mathtt{sim}(\mathbf{u}_i, \mathbf{v}_i) > \mathtt{sim}(\mathbf{u}_i, \mathbf{v}_j),\ i \neq j\ , \tag{1}$$

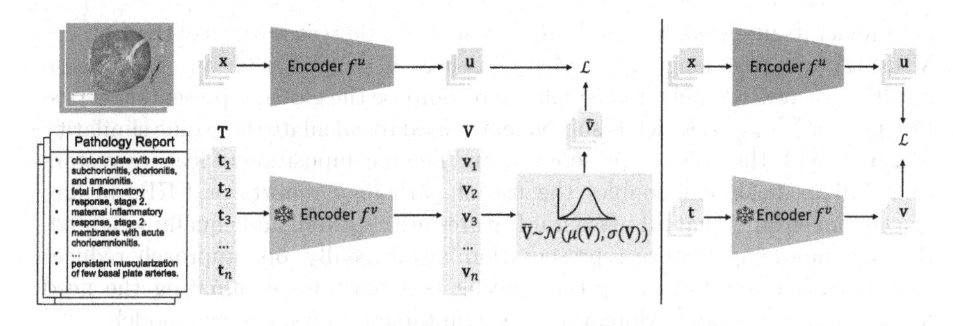

Fig. 1. A diagram illustrating the difference between the proposed approach (left) and the traditional VLC approach (right). $\mathbf{x}$ and $\mathbf{t}$ are images and text inputs, respectively. One sample input image and text are shown on the left. The loss function is defined as $\mathcal{L} = \frac{1}{N}\sum_{i=1}^{N}\left(\lambda\tilde{\ell}_i^{(u\to v)} + (1-\lambda)\tilde{\ell}_i^{(v\to u)}\right)$, following the notations in Sec. 3.

where $\mathtt{sim}(\mathbf{u}, \mathbf{v})$ represents the cosine similarity between the two feature vectors $\mathbf{u} = f^u(\mathbf{x})$, $\mathbf{v} = f^v(\mathbf{t})$. The objective function for achieving inequality (1) is:

$$\ell_i^{(v\to u)} = -\log \frac{\exp(\mathtt{sim}(\mathbf{u}_i, \mathbf{v}_i)/\tau)}{\sum_{k=1}^{N}\exp(\mathtt{sim}(\mathbf{u}_i, \mathbf{v}_k)/\tau)}, \tag{2}$$

where τ is the temperature hyper-parameter and N is the mini-batch size.

To train a classifier, we aim to learn a function f_t^c using the learned function f^v for each task $t \in [1:T]$, such that for a pair of input $(\mathbf{x}_i, l_i^t)$, $f_t^c(f^v(\mathbf{x}_i)) = l_i^t$.

3.2 Pathology Report Feature Recomposition

Traditional VLC approaches for medical image and text analysis, such as ConVIRT [28], encode the entire natural language medical report or electronic health record (EHR) associated with each patient into a single vector representation using a language model. However, solely relying on a pre-trained language model presents two significant challenges. First, the encoding process can result in suppression of important features in the report as the encoder is allowed to ignore certain placental features to minimize loss, leading to a single dominant feature influencing the objective (1), rather than the consideration of all relevant features in the report. Second, the length of the pathology report may exceed the capacity of the text encoder, causing truncation (e.g., a BERT [6] usually allows 512 sub-word tokens during training). Moreover, recent LLMs may handle text length but not feature suppression. Our method seeks to address both challenges simultaneously.

Our approach addresses the limitations of traditional VLC methods in the medical domain by first decomposing the placenta pathology report into set $\mathbf{T}$ of arbitrary size, where each $\mathbf{t}_i \in \mathbf{T}$ represents a distinct placental feature; the individual items depicted in the pathology report in Fig. 1 correspond to distinct placental features. Since the order of items in a pathology report does not impact its integrity, we obtain the set of vector representations of the features $\mathbf{V}$ using an expert language model f^v, where $\mathbf{v}_i = f^v(\mathbf{t}_i)$ for $\mathbf{v}_i \in \mathbf{V}$. These resulting vectors are weighted equally to recompose the global representation (see Fig. 1), $\bar{\mathbf{v}} = \sum_{\mathbf{v} \in \mathbf{V}} \mathbf{v}$, which is subsequently used to calculate the cosine similarity $\mathtt{sim}(\mathbf{u}, \bar{\mathbf{v}})$ with the image representation $\mathbf{u}$. The recomposition of feature vectors from full medical text enables the use of pathology reports or EHRs of any length and ensures that all placental features are captured and equally weighted, thereby improving feature representation. Additionally, our approach reduces computational resources by precomputing text features, eliminating the need for an LLM in training. Moreover, it is adaptable to any language model.

3.3 Distributional Feature Recomposition

Since our pathology reports are decomposed and encoded as a set of feature vectors, to ensure an accurate representation, it is necessary to consider potential limitations associated with vector operations. In the context of vector summation, we anticipate similar representations when two sets differ only slightly. However, even minor changes in individual features within the set can significantly alter the overall representation. This is evident in the substantial difference between $\bar{\mathbf{v}}_1$ and $\bar{\mathbf{v}}_2$ in Fig. 2, despite $\mathbf{V}_1$ and $\mathbf{V}_2$ differing by only one vector magnitude. On the other hand, two distinct sets may result in the same representation, as shown by $\bar{\mathbf{v}}_1$ and $\bar{\mathbf{v}}_3$ in Fig. 2, even when the individual feature vectors have drastically different meanings. Consequently, it is crucial to develop a method that ensures $\mathtt{sim}(\mathbf{V}_1, \mathbf{V}_2) > \mathtt{sim}(\mathbf{V}_1, \mathbf{V}_3)$.

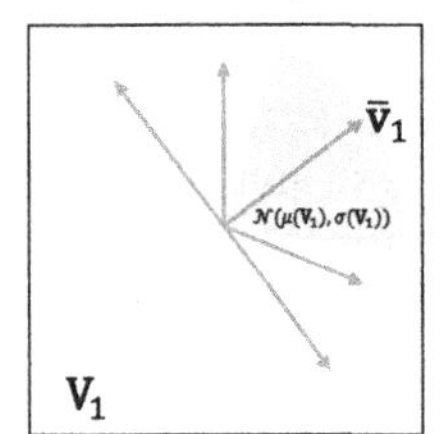

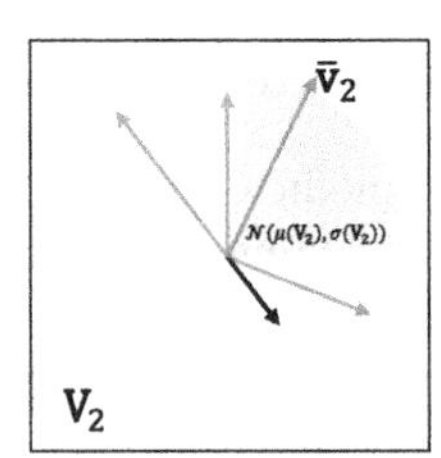

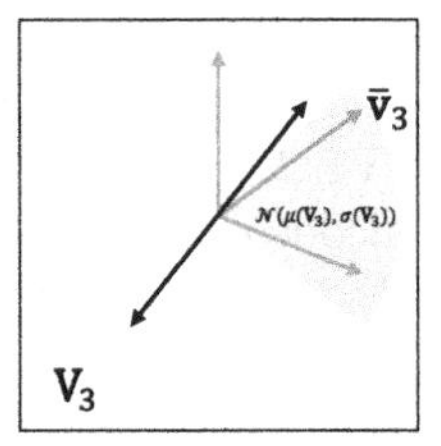

Fig. 2. A diagram illustrating the idea of the proposed distributional feature recomposition. $\bar{\mathbf{v}}_i$ denotes the point estimate sum of the placenta pathological text vectors set $\mathbf{V}_i$. $\mathcal{N}(\mu(\mathbf{V}_i), \sigma(\mathbf{V}_i))$ represents the distribution of the mean placental feature estimated from each $\mathbf{V}_i$. The dark vectors represent the changing vectors from $\mathbf{V}_1$.

To address these limitations, we extend the feature recomposition in Sect. 3.2 to *Distributional Feature Recomposition* that estimates a stable high-dimensional vector space defined by each set of features. We suggest utilizing the distribution $\mathcal{N}(\mu(\mathbf{V}), \sigma(\mathbf{V}))$ of the feature vectors $\mathbf{V}$, instead of point estimates (single vector sum) as a more comprehensive representation, where $\mu(\mathbf{V})$ and $\sigma(\mathbf{V})$ denote the mean and standard deviation, respectively. As shown by the shaded area in Fig. 2, the proposed distributional feature recomposition is more stable and representative than the point estimate sum of vector: $\mathcal{N}(\mu(\mathbf{V}_1), \sigma(\mathbf{V}_1))$ is similar to $\mathcal{N}(\mu(\mathbf{V}_2), \sigma(\mathbf{V}_2))$, but significantly different from $\mathcal{N}(\mu(\mathbf{V}_3), \sigma(\mathbf{V}_3))$.

Implementation-wise, we employ bootstrapping to estimate the distribution of the mean vector. We assume that the vectors adhere to a normal distribution with zero covariance between dimensions. During each training iteration, we randomly generate a new bootstrapped sample set $\tilde{\mathbf{V}}$ from the estimated normal distribution $\mathcal{N}(\mu(\mathbf{V}), \sigma(\mathbf{V}))$. Note that a slightly different sample set is generated in each training epoch to cover the variations in the feature distribution. We can therefore represent this distribution by the vector $\tilde{\mathbf{v}} = \sum_{\mathbf{v} \in \tilde{\mathbf{V}}} \mathbf{v}$, the sum of the sampled vectors, which captures the mean feature distribution in its values and carries the feature variation through epochs. By leveraging a sufficient amount of training data and running multiple epochs, we anticipate achieving a reliable estimation. The distributional feature recomposition not only inherits the scalability and efficiency of the traditional sum of vector approach but also provides a more robust estimate of the distribution of the mean vector, resulting in improved representational power and better generalizability.

3.4 Efficient Neural Networks

Efficient models, which are smaller and faster neural networks, facilitate easy deployment across a variety of devices, making them beneficial for low-resource communities. EfficientNet [22] and MobileNetV3 [11] are two notable examples of such networks. These models achieve comparable or better performance than state-of-the-art ResNet on ImageNet. However, efficient models generally have shallower network layers and can underperform when the features are more difficult to learn, particularly in medical applications [25]. To further demonstrate the representation power of our proposed method and expedite the diagnosis process, we experimentally substitute our image backbone with two efficient models, EfficientNet-B0 and MobileNetV3-Large-1.0, both of which exhibit highly competitive performance on ImageNet when compared to the original ResNet50. This evaluation serves two purposes: First, to test the applicability of our proposed method across different models, and second, to provide a more efficient and accessible placenta analysis model.

4 Experiments

4.1 Implementation

We implemented the proposed methods and baselines using the Python/PyTorch framework and deployed the system on a computing server. For input images, we used PlacentaNet [3] for segmentation and applied random augmentations such as random rotation and color jittering. We used a pre-trained BERT[1] [6] as our text encoder. EfficientNet-B0 and MobileNetV3-Large-1.0 followed official PyTorch implementations. All models and baselines were trained for 400 epochs.

[1] https://tfhub.dev/google/experts/bert/pubmed/2.

The encoder in the last epoch was saved and evaluated on their task-specific performance on the test set, measured by the AUC-ROC scores (area under the ROC curve). To ensure the reliability of the results, each evaluation experiment was repeated five times using different fine-tuning dataset random splits. The same testing procedure was adopted for all our methods. We masked all iPad images using the provided manual segmentation masks. For more information, please refer to the supplementary material.

4.2 Results

We compare our proposed methods (**Ours**) with three strong baselines: a ResNet-50 classification network, the ConVIRT [28] Medical VLC framework, and Pan et al. The mean results and confidence intervals (CIs) reported for each of the experiments on the two datasets are shown in Table 1. Some qualitative examples are in the supplementary material.

Table 1. AUC-ROC scores (in %) for placenta analysis tasks. The mean and 95% CI of five random splits. The highest means are in bold and the second-highest means are underlined. Primary stands for the main placenta dataset, and iPad stands for the iPad dataset. (*Mecon.*: meconium; *H.Chorio.*: histologic chorioamnionitis; *C.Chorio.*: clinical chorioamnionitis)

Method	Primary Task					iPad Task	
	Mecon.	FIR	MIR	H.Chorio.	Sepsis	MIR	C.Chorio
Supervised (ResNet-50)	77.0±2.9	74.2±3.3	68.5±3.4	67.4±2.7	88.4±2.0	50.8±21.6	47.0±16.7
ConVIRT (ResNet-50)	77.5±2.7	76.5±2.6	69.2±2.8	68.0±2.5	89.2±3.6	52.5±25.7	50.7±6.6
Pan et al. (ResNet-50)	79.4±1.3	77.4±3.4	70.3±4.0	68.9±5.0	<u>89.8</u>±2.8	<u>61.9</u>±14.4	53.6±4.2
Ours (ResNet-50)	<u>81.3</u>±2.3	**81.3**±3.0	**75.0**±1.6	**72.3**±2.6	**92.0**±0.9	**74.9**±5.0	<u>59.9</u>±4.5
Ours (EfficientNet)	79.7±1.5	78.5±3.9	71.5±2.6	67.8±2.8	87.7±4.1	58.7±13.3	**61.2**±4.6
Ours (MobileNet)	**81.4**±1.6	<u>80.5</u>±4.0	<u>73.3</u>±1.1	<u>70.9</u>±3.6	88.4±3.6	58.3±10.1	52.3±11.2

Our performance-optimized method with the ResNet backbone consistently outperforms all other methods in all placental analysis tasks. These results confirm the effectiveness of our approach in reducing feature suppression and enhancing representational power. Moreover, compared to Pan et al., our method generally has lower variation across different random splits, indicating that our training method can improve the stability of learned representations. Furthermore, the qualitative examples provided in the supplementary material show that incorrect predictions are often associated with incorrect salient locations.

Table 2 shows the speed improvements of our method. Since the efficiency of Pan et al. and ConVIRT is the same, we only present one of them for brevity. By removing the LLM during training, our method reduces the training time by a factor of 2.0. Moreover, the efficient version (e.g., MobileNet encoder) of our method has 2.4 to 4.1 times the throughput of the original model while still outperforming the traditional baseline approaches in most of the tasks, as shown

Table 2. Training and inference efficiency metrics. All these measurements are performed on a Tesla V100 GPU with a batch size of 32 at full precision (fp32). ResNet-50s have the same inference efficiency and the number of parameters. (*#params*: number of parameters; *Time*: total training time in hours; *throughput*: examples/second; *TFLOPS*: Tera FLoating-point Operations/second). Improvements are in green.

Method	#params↓	Training	Inference	
		Time↓	Throughput↑	TFLOPS↓
Pan et al. (ResNet-50)	27.7M	38 hrs	–	–
Ours (ResNet-50)	27.7M	20 hrs ÷1.9	334	4.12
Ours (EfficientNet)	6.9M÷4.01	19 hrs÷2.0	822×2.46	0.40÷10.3
Ours (MobileNet)	7.1M÷3.90	18 hrs÷2.1	1368×4.10	0.22÷18.7

in Table 1. These results further support the superiority of the proposed representation and training method in terms of both training and testing efficiency.

4.3 Ablation

To better understand the improvements, we conduct a component-wise ablation study. We use the ConVIRT method (instead of Pan et al.) as the starting point to keep the loss function the same. We report the mean AUC-ROC across all tasks to minimize the effects of randomness.

Table 3. Mean AUC-ROC scores over placenta analysis tasks on the primary dataset. The mean and 95% CI of five random splits. *+Recomposition* means the use of Pathology Report Feature Recomposition over the baseline, ∼*+Distributional* stands for the further adoption of the Distributional Feature Recomposition. Improvements are in green. The abbreviations follow Table 1.

	Mecon.	FIR	MIR	H. Chorio.	Sepsis	Mean
Baseline (ConVIRT)	77.5±2.7	76.5±2.6	69.2±2.8	68.0±2.5	89.2±3.6	76.1
+ Recomposition	80.8±1.9	80.2±3.1	74.6±1.8	71.8±3.2	92.0±1.4	79.9+3.8
∼ + Distributional	81.3±2.3	81.3±3.0	75.0±1.6	72.3±2.6	92.0±0.9	80.4+4.3

As shown in Table 3, the text feature recomposition resulted in a significant improvement in performance since it treats all placental features equally to reduce the feature suppression problem. Moreover, applying distributional feature recomposition further improved performance, indicating that using a distribution to represent a set produces a more robust representation than a simple sum. Additionally, even the efficient version of our approach outperformed the performance version that was trained using the traditional VLC method. These improvements demonstrate the effectiveness of the proposed methods across different model architectures. However, we observed that the additional improvement from the distributional method was relatively small compared to that from the recomposition method. This may be due to the fact that the feature suppression problem is more prevalent than the misleading representation problem, or that the improvements may not be linearly proportional to the effectiveness–it may be more challenging to improve a better-performing model.

5 Conclusions and Future Work

We presented a novel automatic placenta analysis framework that achieves improved performance and efficiency. Additionally, our framework can accommodate architectures of different sizes, resulting in better-performing models that are faster and smaller, thereby enabling a wider range of applications. The framework demonstrated clear performance advantages over previous work without requiring additional data, while significantly reducing the model size and computational cost. These improvements have the potential to promote the clinical deployment of automated placenta analysis, which is particularly beneficial for resource-constrained communities.

Nonetheless, we acknowledge the large variance and performance drop when evaluating the iPad images. Hence, further research is required to enhance the model's robustness, and a larger external validation dataset is essential. Moreover, the performance of the image encoder is heavily reliant on the pre-trained language model, and our framework does not support online training of the language model. We aim to address these limitations in our future work.

References

1. Asadpour, V., Puttock, E.J., Getahun, D., Fassett, M.J., Xie, F.: Automated placental abruption identification using semantic segmentation, quantitative features, SVM, ensemble and multi-path CNN. Heliyon **9**(2), e13577:1–13 (2023)
2. Bakkali, S., Ming, Z., Coustaty, M., Rusiñol, M., Terrades, O.R.: VLCDoC: vision-language contrastive pre-training model for cross-modal document classification. Pattern Recogn. **139**(109419), 1–11 (2023)
3. Chen, Y., Wu, C., Zhang, Z., Goldstein, J.A., Gernand, A.D., Wang, J.Z.: PlacentaNet: automatic morphological characterization of placenta photos with deep learning. In: Shen, D., et al. (eds.) MICCAI 2019. LNCS, vol. 11764, pp. 487–495. Springer, Cham (2019). https://doi.org/10.1007/978-3-030-32239-7_54
4. Chen, Y., et al.: AI-PLAX: AI-based placental assessment and examination using photos. Comput. Med. Imaging Graph. **84**(101744), 1–15 (2020)
5. Cui, Q., et al.: Contrastive vision-language pre-training with limited resources. In: Avidan, S., Brostow, G., Cissé, M., Farinella, G.M., Hassner, T. (eds) Computer Vision - ECCV 2022. ECCV 2022. LNCS, vol. 13696, pp. 236–253. Springer, Cham (2022). https://doi.org/10.1007/978-3-031-20059-5_14
6. Devlin, J., Chang, M.W., Lee, K., Toutanova, K.: BERT: pre-training of deep bidirectional transformers for language understanding. arXiv preprint arXiv:1810.04805 (2018)

7. Dong, X., et al.: MaskCLIP: masked self-distillation advances contrastive language-image pretraining. arXiv preprint arXiv:2208.12262 (2022)
8. Dormer, J.D., et al.: CascadeNet for hysterectomy prediction in pregnant women due to placenta accreta spectrum. In: Proceedings of SPIE-the International Society for Optical Engineering, vol. 12032, pp. 156–164. SPIE (2022)
9. Goldstein, J.A., Gallagher, K., Beck, C., Kumar, R., Gernand, A.D.: Maternal-fetal inflammation in the placenta and the developmental origins of health and disease. Front. Immunol. **11**(531543), 1–14 (2020)
10. Gupta, K., Balyan, K., Lamba, B., Puri, M., Sengupta, D., Kumar, M.: Ultrasound placental image texture analysis using artificial intelligence to predict hypertension in pregnancy. J. Matern.-Fetal Neonatal. Med. **35**(25), 5587–5594 (2022)
11. Howard, A., et al.: Searching for MobileNetV3. In: Proceedings of the IEEE/CVF International Conference on Computer Vision, pp. 1314–1324 (2019)
12. Jia, C., et al.: Scaling up visual and vision-language representation learning with noisy text supervision. In: Proceedings of the International Conference on Machine Learning, pp. 4904–4916. PMLR (2021)
13. Khodaee, A., Grynspan, D., Bainbridge, S., Ukwatta, E., Chan, A.D.: Automatic placental distal villous hypoplasia scoring using a deep convolutional neural network regression model. In: Proceedings of the IEEE International Instrumentation and Measurement Technology Conference (I2MTC), pp. 1–5. IEEE (2022)
14. Li, T., et al.: Addressing feature suppression in unsupervised visual representations. In: Proceedings of the IEEE/CVF Winter Conference on Applications of Computer Vision, pp. 1411–1420 (2023)
15. Mobadersany, P., Cooper, L.A., Goldstein, J.A.: GestAltNet: aggregation and attention to improve deep learning of gestational age from placental whole-slide images. Lab. Invest. **101**(7), 942–951 (2021)
16. Pan, Y., Gernand, A.D., Goldstein, J.A., Mithal, L., Mwinyelle, D., Wang, J.Z.: Vision-language contrastive learning approach to robust automatic placenta analysis using photographic images. In: Wang, L., Dou, Q., Fletcher, P.T., Speidel, S., Li, S. (eds.) Medical Image Computing and Computer Assisted Intervention - MICCAI 2022. MICCAI 2022. Lecture Notes in Computer Science, vol. 13433, pp 707–716. Springer, Cham (2022). https://doi.org/10.1007/978-3-031-16437-8_68
17. Pietsch, M., et al.: APPLAUSE: automatic prediction of PLAcental health via U-net segmentation and statistical evaluation. Med. Image Anal. **72**(102145), 1–11 (2021)
18. Radford, A., et al.: Learning transferable visual models from natural language supervision. In: Proceedings of the International Conference on Machine Learning, pp. 8748–8763. PMLR (2021)
19. Roberts, D.J.: Placental pathology, a survival guide. Arch. Pathol. Labor. Med. **132**(4), 641–651 (2008)
20. Specktor-Fadida, B., et al.: A bootstrap self-training method for sequence transfer: state-of-the-art placenta segmentation in fetal MRI. In: Sudre, C.H., et al. (eds.) UNSURE/PIPPI -2021. LNCS, vol. 12959, pp. 189–199. Springer, Cham (2021). https://doi.org/10.1007/978-3-030-87735-4_18
21. Sun, H., Jiao, J., Ren, Y., Guo, Y., Wang, Y.: Multimodal fusion model for classifying placenta ultrasound imaging in pregnancies with hypertension disorders. Pregnancy Hypertension **31**, 46–53 (2023)
22. Tan, M., Le, Q.: EfficientNet: rethinking model scaling for convolutional neural networks. In: Proceedings of the International Conference on Machine Learning, pp. 6105–6114. PMLR (2019)

23. Wang, Y., Li, Y.Z., Lai, Q.Q., Li, S.T., Huang, J.: RU-net: an improved U-Net placenta segmentation network based on ResNet. Comput. Methods Program. Biomed. **227**(107206), 1–7 (2022)
24. Wen, K., Xia, J., Huang, Y., Li, L., Xu, J., Shao, J.: COOKIE: contrastive cross-modal knowledge sharing pre-training for vision-language representation. In: Proceedings of the IEEE/CVF International Conference on Computer Vision, pp. 2208–2217 (2021)
25. Yang, Y., et al.: A comparative analysis of eleven neural networks architectures for small datasets of lung images of COVID-19 patients toward improved clinical decisions. Comput. Biol. Med. **139**(104887), 1–26 (2021)
26. Ye, Z., Xuan, R., Ouyang, M., Wang, Y., Xu, J., Jin, W.: Prediction of placenta accreta spectrum by combining deep learning and radiomics using T2WI: A multicenter study. Abdom. Radiol. **47**(12), 4205–4218 (2022)
27. Zhang, P., et al.: Vinvl: revisiting visual representations in vision-language models. In: Proceedings of the IEEE/CVF Conference on Computer Vision and Pattern Recognition, pp. 5579–5588 (2021)
28. Zhang, Y., Jiang, H., Miura, Y., Manning, C.D., Langlotz, C.P.: Contrastive learning of medical visual representations from paired images and text. In: Proceedings of the Machine Learning for Healthcare Conference, pp. 2–25. PMLR (2022)
29. Zhang, Z., Davaasuren, D., Wu, C., Goldstein, J.A., Gernand, A.D., Wang, J.Z.: Multi-region saliency-aware learning for cross-domain placenta image segmentation. Pattern Recogn. Lett. **140**, 165–171 (2020)

Frequency-Mixed Single-Source Domain Generalization for Medical Image Segmentation

Heng Li[1,3], Haojin Li[1,2], Wei Zhao[3,4](✉), Huazhu Fu[5], Xiuyun Su[3,4], Yan Hu[2], and Jiang Liu[1,2,3,6](✉)

[1] Research Institute of Trustworthy Autonomous Systems, Southern University of Science and Technology, Shenzhen, China

[2] Department of Computer Science and Engineering, Southern University of Science and Technology, Shenzhen, China

[3] Medical Intelligence and Innovation Academy, Southern University of Science and Technology, Shenzhen, China

[4] Southern University of Science and Technology Hospital, Shenzhen, China

zhaow3@sustech.edu.cn

[5] Institute of High Performance Computing, Agency for Science, Technology and Research, Singapore, Singapore

[6] Guangdong Provincial Key Laboratory of Brain-inspired Intelligent Computation, Southern University of Science and Technology, Shenzhen, China

liuj@sustech.edu.cn

Abstract. The annotation scarcity of medical image segmentation poses challenges in collecting sufficient training data for deep learning models. Specifically, models trained on limited data may not generalize well to other unseen data domains, resulting in a domain shift issue. Consequently, domain generalization (DG) is developed to boost the performance of segmentation models on unseen domains. However, the DG setup requires multiple source domains, which impedes the efficient deployment of segmentation algorithms in clinical scenarios. To address this challenge and improve the segmentation model's generalizability, we propose a novel approach called the Frequency-mixed Single-source Domain Generalization method (FreeSDG). By analyzing the frequency's effect on domain discrepancy, FreeSDG leverages a mixed frequency spectrum to augment the single-source domain. Additionally, self-supervision is constructed in the domain augmentation to learn robust context-aware representations for the segmentation task. Experimental results on five datasets of three modalities demonstrate the effectiveness of the proposed algorithm. FreeSDG outperforms state-of-the-art methods and significantly improves the segmentation model's generalizability. Therefore, FreeSDG provides a promising solution for enhancing the generalization of medical image segmentation models, especially when annotated data is scarce. The code is available at https://github.com/liamheng/Non-IID_Medical_Image_Segmentation.

Keywords: Medical image segmentation · single-source domain generalization · domain augmentation · frequency spectrum

H. Greenspan et al. (Eds.): MICCAI 2023, LNCS 14225, pp. 127–136, 2023.
https://doi.org/10.1007/978-3-031-43987-2_13

1 Introduction

Due to the superiority in image representation, tremendous success has been achieved in medical image segmentation through recent advancements of deep learning [10]. Nevertheless, sufficient labeled training data is necessary for deep learning to learn state-of-the-art segmentation networks, resulting in the burden of costly and labor-intensive pixel-accurate annotations [2]. Consequently, annotation scarcity has become a pervasive bottleneck for clinically deploying deep networks, and existing similar datasets have been resorted to alleviate the annotation burden. However, networks trained on a single-source dataset may suffer performance dropping when applied to clinical datasets, since neural networks are sensitive to domain shifts.

Consequently, domain adaptation (DA) and DG [14] have been leveraged to mitigate the impact of domain shifts between source and target domains/datasets. Unfortunately, DA relies on a strong assumption that source and target data are simultaneously accessible [4], which does not always hold in practice. Thereby, DG has been introduced to overcome the absence of target data, which learns a robust model from distinct source domains to generalize to any target domain. To efficiently transfer domain knowledge across various source domains, FACT [12] has been designed to adapt the domains by swapping the low-frequency spectrum of one with the other. Considering privacy protection in medical scenarios, federated learning and continuous frequency space interpolation were combined to achieve DG on medical image segmentation [5]. More recently, single-source domain generalization (SDG) [9] has been proposed to implement DG without accessing multi-source domains. Based on global intensity non-linear augmentation (GIN) and interventional pseudocorrelation augmentation (IPA), a causality-inspired SDG was designed in [7]. Although DG has boosted the clinical practice of deep neural networks, troublesome challenges still remain in clinical deployment. 1) Data from multi-source domains are commonly required to implement DG, which is costly and even impractical to collect in clinics. 2) Medical data sharing is highly concerned, accessing multi-source domains exacerbates the risk of data breaching. 3) Additional generative networks may constrain algorithms' efficiency and versatility, negatively impacting clinical deployment.

To circumvent the above challenges, a frequency-mixed single-source domain generalization strategy, called FreeSDG, is proposed in this paper to learn generalizable segmentation models from a single-source domain. Specifically, the impact of frequency on domain discrepancy is first explored to test our hypotheses on domain augmentation. Then based on the hypotheses, diverse frequency views are extracted from medical images and mixed to augment the single-source domain. Simultaneously, a self-supervised task is posed from frequency views to learn robust context-aware representations. Such that the representations are injected into the vanilla segmentation task to train segmentation networks for out-of-domain inference. Our main contributions are summarised as follows:

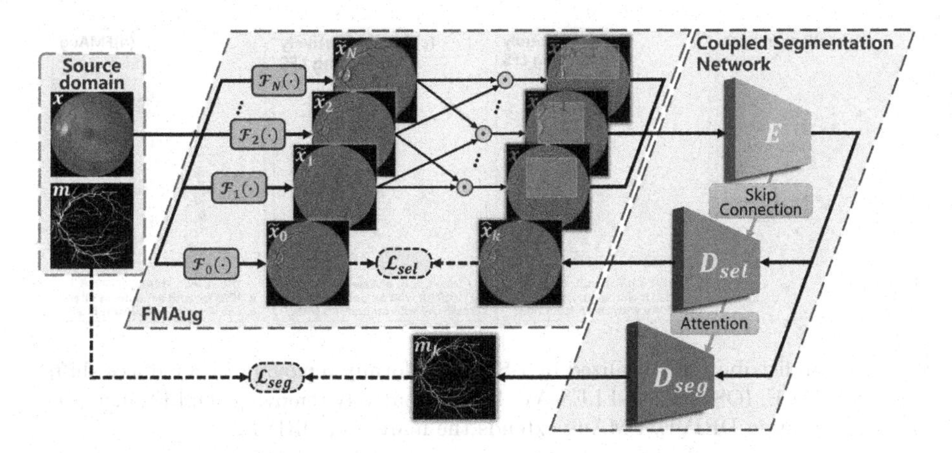

Fig. 1. Overview of FreeSDG, which learns a generalizable segmentation network from a single-source domain. FMAug extends the domain margin by mixing patches (orange boxes) from diverse frequency views, and poses a self-supervised task to learn context-aware representations. The representations are injected into segmentation using attention mechanisms in the coupled network to achieve a generalizable model. (Color figure online)

- We design an efficient SDG algorithm named FreeSDG for medical image segmentation by exploring the impact of frequency on domain discrepancy and mixing frequency views for domain augmentation.
- Through identifying the frequency factor for domain discrepancy, a frequency-mixed domain augmentation (FMAug) is proposed to extend the margin of the single-source domain.
- A self-supervised task is tailored with FMAug to learn robust context-aware representations, which are injected into the segmentation task.
- Experiments on various medical image modalities demonstrate the effectiveness of the proposed approach, by which data dependency is alleviated and superior performance is presented when compared with state-of-the-art DG algorithms in medical image segmentation.

2 Methodology

Aiming to robustly counter clinical data from unknown domains, an SDG algorithm for medical image segmentation is proposed, as shown in Fig. 1. A generalizable segmentation network is attempted to be produced from a single-source domain $(x, m) \sim \mathbb{D}(x, m)$, where $m \in \mathbb{R}^{H \times W}$ is the segmentation mask for the image $x \in \mathbb{R}^{H \times W \times 3}$. By mixing frequency spectrums, FMAug is executed to augment the single-source domain, and self-supervision is simultaneously acquired to learn context-aware representations. Thus a medical image segmentation network capable of out-of-domain generalization is implemented from a single-source domain.

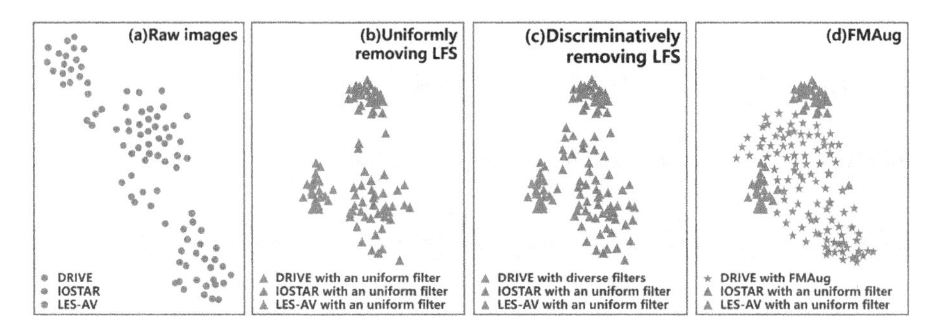

Fig. 2. Data distribution visualized by t-SNE. Uniformly removing LFS reduces shifts between DRIVE, IOSTAR, and LES-AV. Discriminatively removing the LFS increases the discrepancy in DRIVE. FMAug extends the margin of DRIVE.

2.1 Frequency-controlled Domain Discrepancy

Generalizable algorithms have been developed using out-of-domain knowledge to circumvent the clinical performance dropping caused by domain shifts. Nevertheless, extra data dependency is often inevitable in developing the generalizable algorithms, limiting their clinical deployment. To alleviate the data dependency, a single source generalization strategy is designed inspired by the Fourier domain adaption [13] and generalization [12].

According to [12,13], the domain shifts between the source and target could be reduced by swapping/integrating the low-frequency spectrum (LFS) of one with the other. Thus we post two hypotheses:

1) uniformly removing the LFS reduces inter- and inner-domain shifts;
2) discriminatively removing the LFS from a single domain increases inner-domain discrepancy.

Various frequency views are thus extracted from medical images with changing parameters to verify the above hypotheses. Denote the frequency filter with parameters θ_n as $\mathcal{F}_n(\cdot)$, where $n \in \mathbb{R}^{N+1}$ refers to the index of parameters. Following [3,4], a frequency view acquired with θ_n from an image x is given by $\tilde{x}_n = \mathcal{F}_n(x) = x - x * g(r_n, \sigma_n)$, where $g(r_n, \sigma_n)$ denotes a Gaussian filter with radius $r_n \in [5, 50]$ and spatial constant $\sigma_n \in [2, 22]$. Then the frequency views are converted to vectors by a pre-trained ResNet-18 and t-SNE is employed to demonstrate the domain discrepancy controlled by the low-frequency spectrum.

As shown in Fig. 2, compared to the raw images, the distribution of various datasets is more clustered after the uniform LFS removement, which indicates domain shift reduction. While the domain discrepancy in DRIVE is increased by discriminatively removing the LFS. Accordingly, these hypotheses can be leveraged to implement SDG.

2.2 Frequency-mixed Domain Augmentation

Motivated by the hypotheses, domain augmentation is implemented by $\mathcal{F}_n(\cdot)$ with perturbed parameters. Moreover, the local-frequency-mix is executed to further extend the domain margin, as shown in Fig. 2 (d). As exhibited in the blue block of Fig. 1, random patches are cut from a frequency view and mixed with diverse ones to conduct FMAug, which is given by

$$\bar{x}_k = \mathcal{M}(\tilde{x}_i, \tilde{x}_j) = M \odot \tilde{x}_i + (1 - M) \odot \tilde{x}_j, \tag{1}$$

where $M \in 0, 1^{W \times H}$ is a binary mask controlling where to drop out and fill in from two images, and $\odot$ is element-wise multiplication. $k = (i-1) \times N + (j-1)$ denotes the index of the augmentation outcomes, where $i, j \in \mathbb{R}^N, i \neq j$.

Notably, self-supervision is simultaneously acquired from FMAug, where only patches from N frequency views $\tilde{x}_n, n \in \mathbb{R}^N$ are mixed, and the rest one $\tilde{x}_0$ is cast as a specific view to be reconstructed from the mixed ones, where $(r_n, \sigma_n) = (27, 9)$. Under the self-supervision, an objective function for learning context-aware representations from view reconstruction is defined as

$$\mathcal{L}_{sel} = \mathbb{E}\left[\sum_{k=1}^{K} \|\tilde{x}_0 - \hat{x}_k\|_1\right]. \tag{2}$$

where $\hat{x}_k$ refers to the view reconstructed from $\bar{x}_k$, $K = N \times (N-1)$.

Consequently, FMAug not only extends the domain discrepancy and margin, but also poses a self-supervised pretext task to learn generalizable context-aware representations from view reconstruction.

2.3 Coupled Segmentation Network

As the FMAug promises domain-augmented training data and generalizable context-aware representations, a segmentation model capable of out-of-domain inference is waiting to be learned. To inject the context-aware representations into the segmentation model seamlessly, a coupled network is designed with attention mechanisms (shown in the purple block of Fig. 1), which utilize the most relevant parts of representation in a flexible manner.

Concretely, the network comprises an encoder E and two decoders D_{sel}, D_{seg}, where skip connection bridges E and D_{sel} while D_{sel} marries D_{seg} using attention mechanisms. For the above pretext task, E and D_{sel} compose a U-Net architecture to reconstruct $\tilde{x}_0$ from $\bar{x}_k$ with the objective function given in Eq. 2. On the other hand, the segmentation task shares E with the pretext task, and introduces representations from D_{sel} to D_{seg}. The features outcomes from the l-th layer of D_{seg} are given by

$$f_{seg}^{l} = D_{seg}^{l}([f_{seg}^{l-1}, f_{sel}^{l-1}]),\ l = 1, 2, ..., L, \tag{3}$$

where f_{sel}^{l} refers to the features from the l-th layer of D_{sel}. Additionally, attention modules are implemented to properly couple the features from D_{sel} and D_{seg}. D_{seg}^{l} imports and concatenates f_{seg}^{l-1} and f_{sel}^{l-1} as a tensor. Subsequently, the efficient channel and spatial attention modules proposed by [11] are executed to couple the representations learned from the pretext and segmentation task. Then convolutional layers are used to generate the final outcome f^l_{seg}. Accordingly, denote the segmentation result from $\bar{x}_k$ as $\hat{m}_k$, the objective function for segmentation task is given by

$$\mathcal{L}_{seg} = \mathbb{E}\left[\sum_{k=1}^{K}[-m \log \hat{m}_k - (1-m)\log(1-\hat{m}_k)]\right]. \tag{4}$$

where m denotes the ground-truth segmentation mask corresponding to the original source sample x. Therefore, the overall objective function for the network is defined as

$$\mathcal{L}_{total} = \mathcal{L}_{sel}(E, D_{sel}) + \alpha \mathcal{L}_{seg}(E, D_{sel}, D_{seg}), \tag{5}$$

where α is the hyper-parameter to balance $\mathcal{L}_{sel}$ and $\mathcal{L}_{seg}$.

3 Experiments

Implementation: Five image datasets of three modalities were collected to conduct segmentation experiments on fundus vessels and articular cartilage. For fundus vessels, training was based on 1) DRIVE[1] and 2) EyePACS[2], where DRIVE is a vessel segmentation dataset on fundus photography used as the single source domain to learn a generalizable segmentation model, EyePACS is a tremendous fundus photography dataset employed as extra multiple source domains to implement DG-based algorithms. 3) LES-AV[3] and 4) IOSTAR[4] are vessel segmentation datasets respectively on fundus photography and Scanning Laser Ophthalmoscopy (SLO), which were used to verify the generalizability of models learned from DRIVE. For articular cartilage, 5) ultrasound images of joints with cartilage masks were collected by Southern University of Science and Technology Hospital, under disparate settings to validate the algorithm's effectiveness in multiple medical scenarios, where the training, generalization, and test splits respectively contain 517, 7530, 1828 images.

The image data were resized to 512 × 512, the training batch size was 2, and Adam optimizer was used. The model was trained according to an early-stop mechanism, which means the optimal parameter on the validation set was selected in the total 200 epochs, where the learning rate is 0.001 in the first 80 epochs and decreases linearly to 0 in the last 120 epochs. The encoder and two decoders are constructed based on the U-net architecture with 8 layers. The comparisons were conducted with the same setting and were quantified by DICE and Matthews's correlation coefficient (Mcc).

Comparison and Ablation Study: The effectiveness of the proposed algorithm is demonstrated in comparison with state-of-the-art methods and an ablation study. The Fourior-based DG methods FACT [12], FedDG [5], and the

---
[1] http://www.isi.uu.nl/Research/Databases/DRIVE/.
[2] https://www.kaggle.com/c/diabetic-retinopathy-detection.
[3] https://figshare.com/articles/dataset/LES-AV_dataset/11857698/1.
[4] http://www.retinacheck.org/datasets.

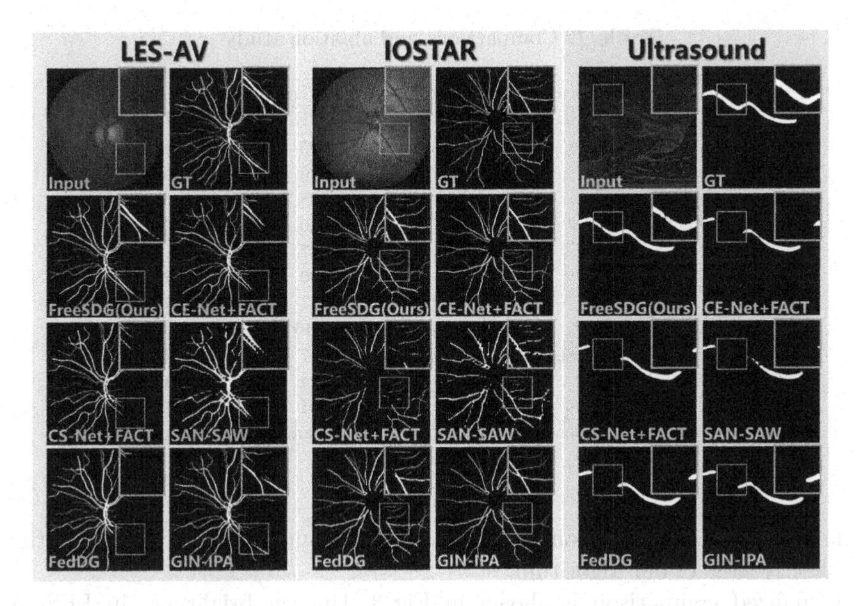

Fig. 3. Segmentation comparison in three medical image modalities.

whitening-based DG method SAN-SAW [8], as well as the SGD method GIN-IPA [7] were compared, where CE-Net [1] and CS-Net [6] were served as the base models cooperated with FACT [12]. Then in the ablation study, FMAug, self-supervised learning (SSL), and attention mechanisms (ATT) were respectively removed from the proposed algorithm.

(1) Comparison. Quantified comparison of our algorithm with the competing methods is summarized in Table 1, where segmentation results in three modalities and data dependency are exhibited. Due to the domain shifts between DRIVE and LES-AV as well as IOSTAR, interior performance are presented by CE-Net [1] and CS-Net [6], which are only learned from DRIVE without DG. Due to the substantial domain discrepancy, EyePACS were treated as multiple source domains to implement DG. FACT [12] boosts the generalization by transferring LFS across the multi-source domains, and efficiently promotes the performance of CE-Net [1] and CS-Net [6]. FedDG [5] were then respectively trained using DRIVE perturbed by EyePACS. As SAN-SAW [8] was designed for region structure segmentation, it appears redundant in the vessel structure task. Thanks to coupling federated learning and contrastive learning, reasonable performance are provided by FedDG [5]. GIN-IPA [7] and our FreeSDG were learned based on the single source domain of DRIVE. Through augmenting the source domain with intensity variance and consistency constraint, GIN-IPA [7] performs decently on out-of-domain inference. The proposed FreeSDG allows for learning efficient segmentation models only from DRIVE. Therefore, our FreeSDG outperforms the state-of-the-art methods without extra data dependency. Additionally, an iden-

Table 1. Comparisons and ablation study

Algorithms	*Dependency**		LES-AV		IOSTAR		Ultrasound	
	IID	*MSD*	DICE	Mcc	DICE	Mcc	DICE	Mcc
CE-Net	⋆		0.636	0.618	0.505	0.514	0.788	0.796
CS-Net	⋆		0.593	0.559	0.520	0.521	0.699	0.721
CE-Net+FACT		⋆	0.730	0.711	0.728	0.705	0.846	0.846
CS-Net+FACT		⋆	0.725	0.705	0.580	0.572	0.829	0.827
SAN-SAW		⋆	0.629	0.599	0.617	0.585	0.819	0.822
Feddg		⋆	0.745	0.725	0.720	0.697	0.872	0.871
GIN-IPA			0.683	0.665	0.641	0.650	0.827	0.824
FreeSDG(ours)			**0.795**	**0.778**	**0.736**	**0.716**	**0.913**	**0.912**
FreeSDG w/o FMAug, SSL, ATT			0.720	0.705	0.687	0.665	0.875	0.873
FreeSDG w/o SSL, ATT			0.751	0.734	0.724	0.701	0.881	0.881
FreeSDG w/o ATT			0.777	0.760	0.731	0.709	0.898	0.897

* Independent and identically distributed data (*IID*) and multi-source domains (*MSD*).

tical situation is observed from the results of ultrasound data, further validating the effectiveness of our algorithm.

Visualized comparison is shown in Fig. 3. Uneven brightness in LES-AV impacts the segmentation performance, vessels in the highlight box are ignored by most algorithms. Cooperating with FACT [12], CE-Net [1] achieves impressive performance. The remarkable performance of GIN-IPA [7] indicates that SDG is a promising paradigm for generalizable segmentation. In the cross-modality segmentation in IOSTAR, CE-Net [1] married with FACT [12] and GIN-IPA [7] still performs outstandingly. In addition, decent segmentation is also observed from FedDG [5] via DG with multi-source domains. FreeSDG efficiently recognizes the variational vessels in LES-AV and IOSTAR, indicating its robustness and generalizability in the quantitative comparison. Furthermore, FreeSDG outperforms the competing methods in accurately segmenting low-contrast cartilage of ultrasound images. In nutshell, our SDG strategy promises FreeSDG prominent performance without extra data dependency.

(2) Ablation Study. According to Table 1, the ablation study also validates the effectiveness of the three designed modules. Through FMAug, an augmented source domain with adequate discrepancy is constructed for training generalizable models. Robust context-aware representations are extracted from self-supervised learning, boosting the downstream segmentation task. Attention mechanisms seamlessly inject the context-aware representations into segmentation, further improving the proposed algorithm. Therefore, a promising segmentation model for medical images is learned from a single-source domain.

4 Conclusion

Pixel-accurate annotations have long been a common bottleneck for developing medical image segmentation networks. Segmentation models learned from a single-source dataset always suffer performance dropping on out-of-domain data. Leveraging DG solutions bring extra data dependency, limiting the deployment of segmentation models. In this paper, we proposed a novel SDG strategy called FreeSDG that leverages a frequency-based domain augmentation technique to extend the single-source domain discrepancy and injects robust representations learned from self-supervision into the network to boost segmentation performance. Our experimental results demonstrated that the proposed algorithm outperforms state-of-the-art methods without requiring extra data dependencies, providing a promising solution for developing accurate and generalizable medical image segmentation models. Overall, our approach enables the development of accurate and generalizable segmentation models from a single-source dataset, presenting the potential to be deployed in real-world clinical scenarios.

Acknowledgment. This work was supported in part by Guangdong Basic and Applied Basic Research Foundation (2020A1515110286), the National Natural Science Foundation of China (82102189, 82272086), Guangdong Provincial Department of Education (2020ZDZX3043), Guangdong Provincial Key Laboratory (2020B121201001), Shenzhen Natural Science Fund (JCYJ20200109140820699, 20200925174052004), Shenzhen Science and Technology Program (SGDX202111 23114204007), Agency for Science, Technology and Research (A*STAR) Advanced Manufacturing and Engineering (AME) Programmatic Fund (A20H4b0141) and Central Research Fund (CRF).

References

1. Gu, Z., et al.: Ce-net: context encoder network for 2d medical image segmentation. IEEE Trans. Med. Imaging **38**(10), 2281–2292 (2019)
2. Jiang, H., Gao, M., Li, H., Jin, R., Miao, H., Liu, J.: Multi-learner based deep meta-learning for few-shot medical image classification. IEEE J. Biomed. Health Inf. **27**(1), 17–28 (2022)
3. Li, H., et al.: Structure-consistent restoration network for cataract fundus image enhancement. In: Wang, L., Dou, Q., Fletcher, P.T., Speidel, S., Li, S. (eds.) Medical Image Computing and Computer Assisted Intervention - MICCAI 2022. MICCAI 2022. LNCS, vol. 13432, pp. 487–496. Springer, Cham (2022). https://doi.org/10.1007/978-3-031-16434-7_47
4. Li, H., et al.: An annotation-free restoration network for cataractous fundus images. IEEE Trans. Med. Imaging **41**(7), 1699–1710 (2022)
5. Liu, Q., Chen, C., Qin, J., Dou, Q., Heng, P.A.: FedDG: federated domain generalization on medical image segmentation via episodic learning in continuous frequency space. In: Proceedings of the IEEE/CVF Conference on Computer Vision and Pattern Recognition, pp. 1013–1023 (2021)
6. Mou, L., et al.: CS-net: channel and spatial attention network for curvilinear structure segmentation. In: Shen, D., et al. (eds.) MICCAI 2019. LNCS, vol. 11764, pp. 721–730. Springer, Cham (2019). https://doi.org/10.1007/978-3-030-32239-7_80
7. Ouyang, C., et al.: Causality-inspired single-source domain generalization for medical image segmentation. IEEE Trans. Med. Imaging **42**(4), 1095–1106 (2022)
8. Peng, D., Lei, Y., Hayat, M., Guo, Y., Li, W.: Semantic-aware domain generalized segmentation. In: Proceedings of the IEEE/CVF Conference on Computer Vision and Pattern Recognition, pp. 2594–2605 (2022)
9. Peng, X., Qiao, F., Zhao, L.: Out-of-domain generalization from a single source: an uncertainty quantification approach. IEEE Trans. Pattern Anal. Mach. Intell. (2022)

10. Qiu, Z., Hu, Y., Zhang, J., Chen, X., Liu, J.: FGAM: a pluggable light-weight attention module for medical image segmentation. Comput. Biol. Med. **146**, 105628 (2022)
11. Woo, S., Park, J., Lee, J.Y., Kweon, I.S.: CBAM: convolutional block attention module. In: Proceedings of the European Conference on Computer vision (ECCV), pp. 3–19 (2018)
12. Xu, Q., Zhang, R., Zhang, Y., Wang, Y., Tian, Q.: A fourier-based framework for domain generalization. In: Proceedings of the IEEE/CVF Conference on Computer Vision and Pattern Recognition, pp. 14383–14392 (2021)
13. Yang, Y., Soatto, S.: FDA: fourier domain adaptation for semantic segmentation. In: Proceedings of the IEEE/CVF Conference on Computer Vision and Pattern Recognition, pp. 4085–4095 (2020)
14. Zhou, K., Liu, Z., Qiao, Y., Xiang, T., Loy, C.C.: Domain generalization: a survey. IEEE Trans. Pattern Anal. Mach. Intell. **45**, 4396–4415 (2022)

A Semantic-Guided and Knowledge-Based Generative Framework for Orthodontic Visual Outcome Preview

Yizhou Chen[1] and Xiaojun Chen[1,2](✉)

[1] Institute of Biomedical Manufacturing and Life Quality Engineering, School of Mechanical Engineering, Shanghai Jiao Tong University, Shanghai, China
[2] Institute of Medical Robotics, Shanghai Jiao Tong University, Shanghai, China
xiaojunchen@sjtu.edu.cn

Abstract. Orthodontic treatment typically lasts for two years, and its outcome cannot be predicted intuitively in advance. In this paper, we propose a semantic-guided and knowledge-based generative framework to predict the visual outcome of orthodontic treatment from a single frontal photo. The framework involves four steps. Firstly, we perform tooth semantic segmentation and mouth cavity segmentation and extract category-specific teeth contours from frontal images. Secondly, we deform the established tooth-row templates to match the projected contours with the detected ones to reconstruct 3D teeth models. Thirdly, we apply a teeth alignment algorithm to simulate the orthodontic treatment. Finally, we train a semantic-guided generative adversarial network to predict the visual outcome of teeth alignment. Quantitative tests are conducted to evaluate the proposed framework, and the results are as follows: the tooth semantic segmentation model achieves a mean intersection of union of 0.834 for the anterior teeth, the average symmetric surface distance error of our 3D teeth reconstruction method is 0.626 mm on the test cases, and the image generation model has an average Fréchet inception distance of 6.847 over all the test images. These evaluation results demonstrate the practicality of our framework in orthodontics.

Keywords: Semantic segmentation · 3D teeth reconstruction · Generative adversarial network (GAN)

1 Introduction

Orthodontic treatment aims to correct misaligned teeth and restore normal occlusion. Patients are required to wear dental braces or clear aligners for a duration of one to three years, reported by [21], with only a vague expectation of the treatment result. Therefore, a generative framework is needed to enable patients to preview treatment outcomes and assist those considering orthodontic

Supplementary Information The online version contains supplementary material available at https://doi.org/10.1007/978-3-031-43987-2_14.

H. Greenspan et al. (Eds.): MICCAI 2023, LNCS 14225, pp. 137–147, 2023.
https://doi.org/10.1007/978-3-031-43987-2_14

treatment in making decisions. Such framework may involve multiple research fields, such as semantic segmentation, 3D reconstruction, and image generation.

Deep learning methods have achieved great success in image-related tasks. In the field of tooth semantic segmentation, there exist plenty of studies targeting on different data modalities, such as dental mesh scanning [31], point cloud [26,30], cone beam CT image [5,6], panoramic dental X-ray image [28], and 2D natural image [32]. Regarding image generation, the family of generative adversarial networks (GAN) [8,13–15,23] and the emerging diffusion models [11,22,25] can generate diverse high-fidelity images. Although the diffusion models can overcome the mode collapse problem and excel in image diversity compared to GAN [7], their repeated reverse process at inference stage prolongs the execution time excessively, limiting their application in real-time situations.

When it comes to 3D teeth reconstruction, both template-based and deep-learning-based frameworks offer unique benefits. Wu et al. employed a template-based approach by adapting their pre-designed teeth template to match teeth contours extracted from a set of images [29]. Similarly, Wirtz et al. proposed an optimization-based pipeline that uses five intra-oral photos to restore the 3D arrangement of teeth [27]. Liang et al. restored 3D teeth using convolution neural networks (CNN) from a single panoramic radiograph [18]. However, while deep CNNs have a strong generalization ability compared to template-based methods, it often struggles to precisely and reasonably restore occluded objects.

Predicting the smiling portrait after orthodontic treatment has recently gained much attention. Yang et al. developed three deep neural networks to extract teeth contours from smiling images, arrange 3D teeth models, and generate images of post-treatment teeth arrangement, respectively [19]. However, their framework requires a single frontal smiling image and the corresponding unaligned 3D teeth model from dental scanning, which may be difficult for general users to obtain. In contrast, Chen et al. proposed a StyleGAN generator with a latent space editing method that utilizes GAN inversion to discover the optimal aligned teeth appearance from a single image [3]. Although their method takes only a frontal image as input and manipulates the teeth structure and appearance implicitly in image space, it may overestimate the treatment effect and result in inaccurate visual outcomes.

In this study, we propose an explainable generative framework, which is semantic-guided and knowledge-based, to predict teeth alignment after orthodontic treatment. Previous works have either required 3D teeth model as additional input and predicted its alignment using neural networks [19], or directly utilized an end-to-end StyleGAN to predict the final orthodontic outcome [3]. In contrast, our approach requires only a single frontal image as input, restores the 3D teeth model through a template-based algorithm, and explicitly incorporates orthodontists' experience, resulting in a more observable and explainable process. Our contributions are therefore three-fold: 1) we introduce a region-boundary feature fusion module to enhance the 2D tooth semantic segmentation results; 2) we employ statistical priors and reconstruct 3D teeth models from teeth semantic boundaries extracted in a single frontal image; 3) by incorporating an orthodontic simulation algorithm and a pSpGAN style encoder [23], we can yield more realistic and explainable visualization of the post-treatment teeth appearance.

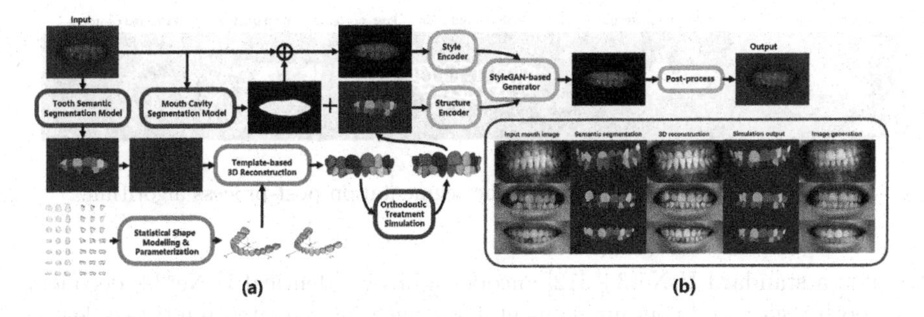

Fig. 1. An overview of (a) the proposed generative framework for orthodontic treatment outcome prediction, and (b) the framework's outputs at each stage with images from left to right: input mouth image, tooth semantic segmentation map, input image overlaid with semi-transparent 3D reconstruction teeth mesh, projection of orthodontic treatment simulation output, and orthodontic visual outcome prediction.

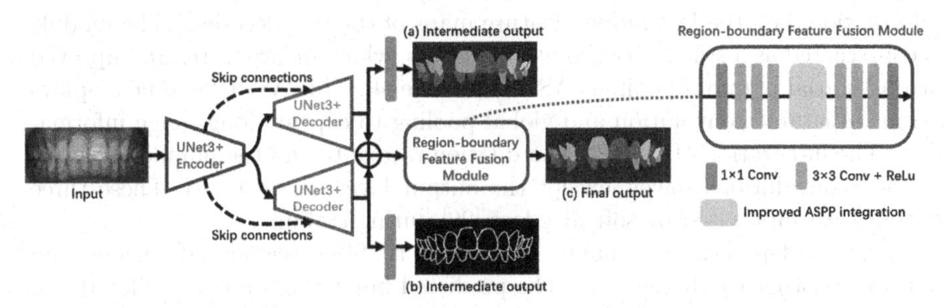

Fig. 2. The proposed tooth semantic segmentation model with (a) coarse region segmentation, (b) auxiliary boundary segmentation, and (c) final region segmentation generated by the region-boundary feature fusion module.

2 Method

The proposed generative framework (Fig. 1) consists of four parts: semantic segmentation in frontal images, template-based 3D teeth reconstruction, orthodontic treatment simulation, and semantic-guided image generation of mouth cavity.

2.1 Semantic Segmentation in Frontal Images

The tooth areas and mouth cavity are our region of interest for semantic segmentation in each frontal image. As a preprocessing step, the rectangle mouth area is firstly extracted from frontal images by the minimal bounding box that encloses the facial key points around the mouth, which are detected by dlib toolbox [17]. We then use two separate segmentation models to handle these mouth images, considering one-hot encoding and the integrity of mouth cavity mask. A standard U-Net [24] is trained with soft dice loss [20] to predict mouth cavity.

The tooth semantic segmentation model (Fig. 2) is a dual-branch U-Net3+ based network that predicts tooth regions and contours simultaneously. We

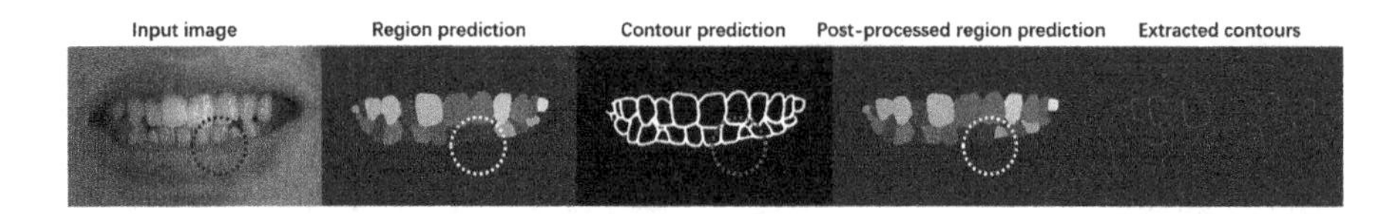

Fig. 3. Illustration of the effect of the segmentation post-process algorithm.

employ a standard U-Net3+ [12] encoder and two identical U-Net3+ decoders for tooth region and contour segmentation. Such inter-related multi-task learning can enhance the performance of each task and mitigate overfitting. The teeth are manually labeled using FDI World Dental Federation notation, resulting in a total of 33 classes, including background.

To generate a more precise tooth segmentation map, we introduce the region-boundary feature fusion module, which merges the tooth region and boundary information, i.e., the last hidden feature maps of the two decoders. The module is constructed as a stack of convolutional layers, which incorporates an improved atrous spatial pyramid pooling (ASPP) module [4]. This ASPP module employs atrous separable convolution and global pooling to capture long-range information. The integration of ASPP has a dilation rate of 6, and the number of filters in each convolutional layer, except the output layer, is set to 64. These three outputs are supervised by soft dice loss [20] during training.

Some post-process techniques are added to filter segmented regions and obtain smoother tooth contours. The predicted binary contours are dilated and used to divide the semantic segmentation map into multiple connected regions. The pixels in each region are classified by its dominant tooth label. Background components are ignored and small ones are removed. Once two regions have duplicate labels, a drop-or-relabel strategy is performed on the region away from the center of central incisors. The connected regions are processed sequentially from central incisors to the third molars. Figure 3 shows the changes of tooth region prediction before and after the post process.

2.2 Template-Based 3D Teeth Reconstruction

To achieve 3D tooth reconstruction from a single frontal image, we deform the parametric templates of the upper and lower tooth rows to match the projected contours with the extracted teeth contours in semantic segmentation.

The parametric tooth-row template is a statistical model that characterizes the shape, scale, and pose of each tooth in a tooth row. To describe the shape of each tooth, we construct a group of morphable shape models [1]. We model the pose (orientation and position) of each tooth as a multivariate normal distribution as Wu et al. [29] did. Additionally, we suppose that the scales of all teeth follows a multivariate normal distribution. The mean shape and average of the tooth scales and poses are used to generate a standard tooth-row template.

The optimization-based 3D teeth reconstruction following [29] is an iterative process alternating between searching for point correspondences between the projected and segmented teeth contours and updating the parameters of the tooth-row templates. The parameters that require estimation are the camera parameters, the relative pose between the upper and lower tooth rows, and the scales, poses, and shape parameters of each tooth.

The point correspondences are established by Eq. (1) considering the semantic information in teeth contours, where c_i^τ and n_i^τ are the position and normal of the detected contour point i of tooth τ in image space, $\hat{c_j}^\tau$ and $\hat{n_j}^\tau$ are those of the projected contour point j, $< \cdot, \cdot >$ denotes inner product, and $\sigma_{angle} = 0.3$ is a fine-tuned hyper parameter in [29].

$$\hat{c_i}^\tau = \underset{\hat{c_j}^\tau}{\arg\min} ||c_i^\tau - \hat{c_j}^\tau||_2^2 \cdot \exp\left[-\left(\frac{< n_i^\tau, \hat{n_j}^\tau >}{\sigma_{angle}}\right)^2\right] \tag{1}$$

We use $\mathcal{L}$ as the objective function to minimize, expressed in Eq. (2), which comprises an image-space contour loss [29] and a regularization term $\mathcal{L}_{prior}$ described by Mahalanobis distance in probability space, where N is the number of detected contour points, $\lambda_n = 50$ and $\lambda_p = 25$ are the fine-tuned weights.

$$\mathcal{L} = \frac{1}{N}\sum_\tau\sum_i \left(||c_i^\tau - \hat{c_i}^\tau||_2^2 + \lambda_n < c_i^\tau - \hat{c_i}^\tau, \hat{n_i}^\tau >^2\right) + \lambda_p\, \mathcal{L}_{prior} \tag{2}$$

The regularization term $\mathcal{L}_{prior}$ in Eq. (3) is the negative log likelihood of the distributions of the vector of tooth scales, denoted by $\boldsymbol{s}$, the pose vector of tooth τ, denoted by $\boldsymbol{p}^\tau$, and the shape vector of tooth τ, denoted by $\boldsymbol{b}^\tau$. The covariance matrices Σ_s and Σ_p^τ are obtained in building tooth-row templates.

$$\mathcal{L}_{prior} = (\boldsymbol{s} - \overline{\boldsymbol{s}})^T \Sigma_s^{-1} (\boldsymbol{s} - \overline{\boldsymbol{s}}) + \sum_\tau \left[(\boldsymbol{p}^\tau - \overline{\boldsymbol{p}}^\tau)^T {\Sigma_p^\tau}^{-1} (\boldsymbol{p}^\tau - \overline{\boldsymbol{p}}^\tau) + ||\boldsymbol{b}^\tau||_2^2\right] \tag{3}$$

During optimization, we first optimize the camera parameters and the relative pose of tooth rows for 10 iterations and optimize all parameters for 20 iterations. Afterward, we use Poisson surface reconstruction [16] to transform the surface point clouds into 3D meshes.

2.3 Orthodontic Treatment Simulation

We implement a naive teeth alignment algorithm to mimic orthodontic treatment. The symmetrical beta function (Eq. (4)) is used to approximate the dental arch curve of a tooth row [2]. Its parameters W and D can be estimated through linear regression by fitting the positions of tooth landmarks [2].

$$\beta(x; D, W) = 3.0314 * D * \left[\frac{1}{2} + \frac{x}{W}\right]^{0.8} \left[\frac{1}{2} - \frac{x}{W}\right]^{0.8} \tag{4}$$

We assume that the established dental arch curves are parallel to the occlusal plane. Each reconstructed tooth is translated towards its expected position in

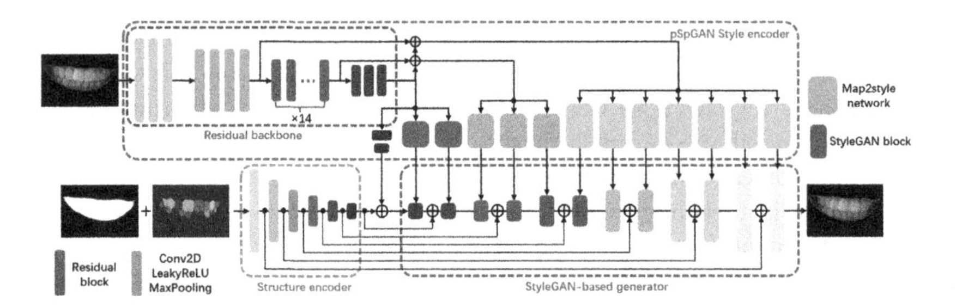

Fig. 4. Architecture of the semantic-guided image generation model. The multi-level style feature maps are extracted from the residual backbone and encoded into twelve 512-dim style vectors through the map2style networks [23], structural information are compressed and skip connected through the structure encoder, and structure and style features are entangled in the StyleGAN-based generator with weight modulation [15].

the dental arch and rotated to its standard orientation while preserving its shape. The teeth gaps are then reduced along the dental arch curve with collision detection performed. The relative pose between tooth rows is re-calculated to achieve a normal occlusion. Finally, the aligned 3D teeth models are projected with the same camera parameters to generate the semantic image output of the simulation.

2.4 Semantic-Guided Image Generation

The idea behind the semantic-guided image generation model is to decompose an image into an orthogonal representation of its style and structure. By manipulating the structural or style information, we can control the characteristics of the generated image. Improving upon [19], we replace the naive style encoder in [19] with the PixelStylePixel style encoder [23] to capture the multi-level style features and use semantic teeth image instead of teeth contours as input to better guide the generation process. The architecture is illustrated in Fig. 4. At training stage, the model learns the style and structural encoding of teeth images and attempts to restore the original image. Gradient penalty [9] and path length regularization [15] are applied to stabilize the training process. We use the same loss function as [19] did and take a standard UNet encoder connected with dense layers as the discriminator. At inference stage, the semantic teeth image output from orthodontic simulation is used to control the generated teeth structure. To remove the boundary artifacts, we dilate the input mouth cavity map and use Gaussian filtering to smooth the sharp edges after image patching.

3 Experiments and Results

3.1 Dataset and Implementation Details

We collected 225 digital dental scans with labelled teeth and their intra-oral photos, as well as 5610 frontal intra-oral images, of which 3300 were labelled,

Table 1. Segmentation accuracy on the test data measured by mean intersection of union for different groups of tooth labels and for different network architectures where DB denotes dual-branch architecture and RBFF denotes region-boundary feature fusion. Group A has 32 tooth classes, group B has 28 classes with the third molars excluded, and group C has 24 classes with the second and third molars excluded.

Settings	UNet			UNet3+		
	Baseline	DB	RBFF+DB	Baseline	DB	RBFF+DB
Group A	0.679	0.697	0.708	0.686	0.699	0.730
Group B	0.764	0.774	0.789	0.766	0.780	0.803
Group C	0.800	0.809	0.820	0.800	0.816	0.834

Table 2. Teeth reconstruction error (avg.± std.) on all the teeth of the 95 test cases (ASSD: average symmetric surface distance, HD: Hausdorff distance, CD: Chamfer distance, DSC: Dice similarity coefficient).

Methods	ASSD(mm)↓	HD(mm)↓	CD(mm^2)↓	DSC↑
Wirtz et al. [27]	$0.848_{\pm 0.379}$ [27]	$2.627_{\pm 0.915}$ [27]	—	$0.659_{\pm 0.140}$ [27]
Nearest retrieval	$0.802_{\pm 0.355}$	$2.213_{\pm 0.891}$	$2.140_{\pm 2.219}$	$0.653_{\pm 0.158}$
Ours	$0.626_{\pm 0.265}$	$1.776_{\pm 0.723}$	$1.272_{\pm 1.364}$	$0.732_{\pm 0.125}$

and 4330 smiling images, of which 2000 were labelled, from our partner hospitals. The digital dental scans were divided into two groups, 130 scans for building morphable shape models and tooth-row templates and the remaining 95 scans for 3D teeth reconstruction evaluation. The labelled 3300 intra-oral images and 2000 smiling images were randomly split into training (90%) and labelled test (10%) datasets. The segmentation accuracy was computed on the labelled test data, and the synthetic image quality was evaluated on the unlabelled test data.

All the models were trained and evaluated on an NVIDIA GeForce RTX 3090 GPU. We trained the segmentation models for 100 epochs with a batch size of 4, and trained the image generation models for 300 epochs with a batch size of 8. The input and output size of the image segmentation and generation models are 256 × 256. The training was started from scratch and the learning rate was set to 10^{-4}. We saved the models with the minimal loss on the labelled test data. At the inference stage, our method takes approximately 15 s to run a single case on average, with the 3D reconstruction stage accounting for the majority of the execution time, tests performed solely on an Intel 12700H CPU.

3.2 Evaluation

Ablation Study of Tooth Segmentation Model. We conduct an ablation study to explore the improvement of segmentation accuracy brought by the dual-branch network architecture and the region-boundary feature fusion module. Segmentation accuracy is measured by the mean intersection over union (mIoU)

Table 3. Average Fréchet inception distance of different generators on the test data.

Model	TSynNet	Contour-guided pSpGAN	Semantic-guided pSpGAN
Test smiling images	11.343	7.292	6.501
All test images	20.133	7.832	6.847

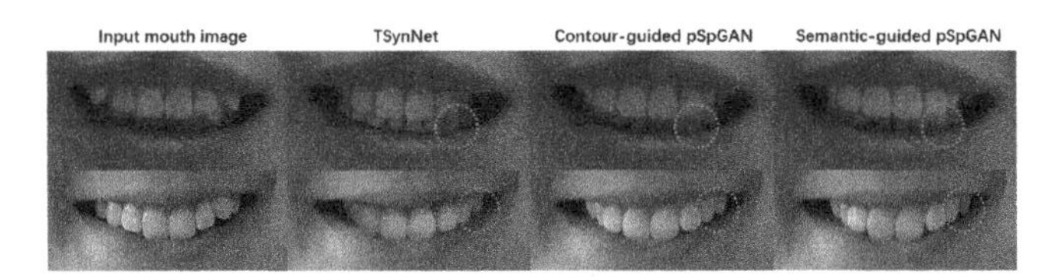

Fig. 5. Comparison of the orthodontic treatment outcome predictions generated by different models: TSynNet [19], contour-guided pSpGAN, and semantic-guided pSpGAN.

Fig. 6. Teeth alignment predictions of the proposed generative framework on frontal images in the Flickr-Faces-HQ data set.

metric for different groups of tooth labels. The results listed in Table 1 show that the proposed region-boundary feature fusion module assisted with dual-branch architecture can further enhance the segmentation accuracy for UNet and its variant. Our tooth segmentation model can predict quite accurately the region of the frontal teeth with a mIoU of 0.834.

Accuracy of 3D Teeth Reconstruction. We reconstruct the 3D teeth models of the 95 test cases from their intra-oral photos. The restored teeth models are aligned with their ground truth by global similarity registration. We compare the reconstruction error using different metrics, shown in Table 2, with the method of [27] that reconstructs teeth models from five intra-oral photos and the nearest retrieval that selects the most similar teeth mesh in the 135 teeth meshes for building tooth-row templates. The results show that our teeth reconstruction method significantly outperforms the method of [27] and nearest retrieval.

Image Generation Quality. We use Fréchet inception distance (FID) [10] to evaluate the quality of images generated different generators on the unlabelled test data, results listed in Table 3. The multi-level style features captured by pSpGAN improve greatly the image quality from the quantitative comparison (Table 3) and the visual perception (Fig. 5). Our semantic-guided pSpGAN that takes semantic teeth image as input can further increase the constrast of different teeth and yield sharper boundaries. We test our framework on some images in Flickr-Faces-HQ dataset [14] to visualize virtual teeth alignment, shown in Fig. 6.

4 Conclusion

In conclusion, we develop a semantic-guided generative framework to predict the orthodontic treatment visual outcome. It comprises tooth semantic segmentation, template-based 3D teeth reconstruction, orthodontic treatment simulation, and semantic-guided mouth cavity generation. The results of quantitative tests show that the proposed framework has a potential for orthodontic application.

Acknowledgement. This work was supported by grants from the National Natural Science Foundation of China (81971709; M-0019; 82011530141), the Foundation of Science and Technology Commission of Shanghai Municipality (20490740700; 22Y11911700), Shanghai Pudong Science and Technology Development Fund (PKX2021-R04), Shanghai Jiao Tong University Foundation on Medical and Technological Joint Science Research (YG2021ZD21; YG2021QN72; YG2022QN056; YG2023ZD19; YG2023ZD15), and the Funding of Xiamen Science and Technology Bureau (No. 3502Z20221012).

References

1. Blanz, V., Vetter, T.: A morphable model for the synthesis of 3D faces. In: Proceedings of the 26th Annual Conference on Computer Graphics and Interactive Techniques, pp. 187–194 (1999)
2. Braun, S., Hnat, W.P., Fender, D.E., Legan, H.L.: The form of the human dental arch. Angle Orthod. **68**(1), 29–36 (1998)
3. Chen, B., Fu, H., Zhou, K., Zheng, Y.: Orthoaligner: image-based teeth alignment prediction via latent style manipulation. IEEE Trans. Visual Comput. Graphics **29**, 3617–3629 (2022)
4. Chen, L.C., Zhu, Y., Papandreou, G., Schroff, F., Adam, H.: Encoder-decoder with atrous separable convolution for semantic image segmentation. In: Proceedings of the European Conference on Computer Vision (ECCV), pp. 801–818 (2018)
5. Chen, Y., et al.: Automatic segmentation of individual tooth in dental CBCT images from tooth surface map by a multi-task FCN. IEEE Access **8**, 97296–97309 (2020)
6. Chung, M., et al.: Pose-aware instance segmentation framework from cone beam CT images for tooth segmentation. Comput. Biol. Med. **120**, 103720 (2020)
7. Dhariwal, P., Nichol, A.: Diffusion models beat GANs on image synthesis. Adv. Neural. Inf. Process. Syst. **34**, 8780–8794 (2021)
8. Goodfellow, I., et al.: Generative adversarial networks. Commun. ACM **63**(11), 139–144 (2020)

9. Gulrajani, I., Ahmed, F., Arjovsky, M., Dumoulin, V., Courville, A.C.: Improved training of Wasserstein GANs. In: Advances in Neural Information Processing Systems, vol. 30 (2017)
10. Heusel, M., Ramsauer, H., Unterthiner, T., Nessler, B., Hochreiter, S.: GANs trained by a two time-scale update rule converge to a local NASH equilibrium. In: Advances in Neural Information Processing Systems, vol. 30 (2017)
11. Ho, J., Jain, A., Abbeel, P.: Denoising diffusion probabilistic models. Adv. Neural. Inf. Process. Syst. **33**, 6840–6851 (2020)
12. Huang, H., et al.: Unet 3+: a full-scale connected UNet for medical image segmentation. In: ICASSP 2020–2020 IEEE International Conference on Acoustics, Speech and Signal Processing (ICASSP), pp. 1055–1059 (2020)
13. Karras, T., et al.: Alias-free generative adversarial networks. Adv. Neural. Inf. Process. Syst. **34**, 852–863 (2021)
14. Karras, T., Laine, S., Aila, T.: A style-based generator architecture for generative adversarial networks. In: Proceedings of the IEEE/CVF Conference on Computer Vision and Pattern Recognition, pp. 4401–4410 (2019)
15. Karras, T., Laine, S., Aittala, M., Hellsten, J., Lehtinen, J., Aila, T.: Analyzing and improving the image quality of stylegan. In: Proceedings of the IEEE/CVF Conference on Computer Vision and Pattern Recognition, pp. 8110–8119 (2020)
16. Kazhdan, M., Bolitho, M., Hoppe, H.: Poisson surface reconstruction. In: Proceedings of the Fourth Eurographics Symposium on Geometry Processing, vol. 7 (2006)
17. King, D.E.: Dlib-ml: a machine learning toolkit. J. Mach. Learn. Res. **10**, 1755–1758 (2009)
18. Liang, Y., Song, W., Yang, J., Qiu, L., Wang, K., He, L.: X2Teeth: 3D teeth reconstruction from a single panoramic radiograph. In: Martel, A.L., et al. (eds.) MICCAI 2020, Part II. LNCS, vol. 12262, pp. 400–409. Springer, Cham (2020). https://doi.org/10.1007/978-3-030-59713-9_39
19. Lingchen, Y., et al.: iorthopredictor: model-guided deep prediction of teeth alignment. ACM Trans. Graphics **39**(6), 216 (2020)
20. Ma, J., et al.: Loss odyssey in medical image segmentation. Med. Image Anal. **71**, 102035 (2021)
21. Mavreas, D., Athanasiou, A.E.: Factors affecting the duration of orthodontic treatment: a systematic review. Eur. J. Orthod. **30**(4), 386–395 (2008)
22. Nichol, A., et al.: Glide: towards photorealistic image generation and editing with text-guided diffusion models. arXiv preprint arXiv:2112.10741 (2021)
23. Richardson, E., et al.: Encoding in style: a stylegan encoder for image-to-image translation. In: Proceedings of the IEEE/CVF Conference on Computer Vision and Pattern Recognition, pp. 2287–2296 (2021)
24. Ronneberger, O., Fischer, P., Brox, T.: U-Net: convolutional networks for biomedical image segmentation. In: Navab, N., Hornegger, J., Wells, W.M., Frangi, A.F. (eds.) MICCAI 2015, Part III. LNCS, vol. 9351, pp. 234–241. Springer, Cham (2015). https://doi.org/10.1007/978-3-319-24574-4_28
25. Saharia, C., et al.: Palette: image-to-image diffusion models. In: ACM SIGGRAPH 2022 Conference Proceedings, pp. 1–10 (2022)
26. Tian, Y., et al.: 3D tooth instance segmentation learning objectness and affinity in point cloud. ACM Trans. Multimedia Comput. Commun. Appl. (TOMM) **18**(4), 1–16 (2022)
27. Wirtz, A., Jung, F., Noll, M., Wang, A., Wesarg, S.: Automatic model-based 3-D reconstruction of the teeth from five photographs with predefined viewing directions. In: Medical Imaging 2021: Image Processing, vol. 11596, pp. 198–212 (2021)

28. Wirtz, A., Mirashi, S.G., Wesarg, S.: Automatic teeth segmentation in panoramic x-ray images using a coupled shape model in combination with a neural network. In: Frangi, A.F., Schnabel, J.A., Davatzikos, C., Alberola-López, C., Fichtinger, G. (eds.) MICCAI 2018, Part IV. LNCS, vol. 11073, pp. 712–719. Springer, Cham (2018). https://doi.org/10.1007/978-3-030-00937-3_81
29. Wu, C., et al.: Model-based teeth reconstruction. ACM Trans. Graph. **35**(6), 220–1 (2016)
30. Zanjani, F.G., et al.: Deep learning approach to semantic segmentation in 3D point cloud intra-oral scans of teeth. In: International Conference on Medical Imaging with Deep Learning, pp. 557–571 (2019)
31. Zhao, Q., et al.: Automatic 3D teeth semantic segmentation with mesh augmentation network. In: 2022 3rd International Conference on Pattern Recognition and Machine Learning, pp. 136–142 (2022)
32. Zhu, G., Piao, Z., Kim, S.C.: Tooth detection and segmentation with mask R-CNN. In: 2020 International Conference on Artificial Intelligence in Information and Communication (ICAIIC), pp. 070–072 (2020)

A Multi-task Method for Immunofixation Electrophoresis Image Classification

Yi Shi[1], Rui-Xiang Li[1], Wen-Qi Shao[2], Xin-Cen Duan[2], Han-Jia Ye[1(✉)], De-Chuan Zhan[1], Bai-Shen Pan[2], Bei-Li Wang[2], Wei Guo[2], and Yuan Jiang[1]

[1] State Key Laboratory for Novel Software Technology, Nanjing University, Nanjing, China
yehj@lamda.nju.edu.cn

[2] Department of Laboratory Medicine, Zhongshan Hospital, Fudan University, Shanghai, China

Abstract. In the field of plasma cell disorders diagnosis, the detection of abnormal monoclonal (M) proteins through Immunofixation Electrophoresis (IFE) is a widely accepted practice. However, the classification of IFE images into nine distinct categories is a complex task due to the significant class imbalance problem. To address this challenge, a two-sub-task classification approach is proposed, which divides the classification task into the determination of severe and mild cases, followed by their combination to produce the final result. This strategy is based on the expert understanding that the nine classes are different combinations of severe and mild cases. Additionally, the examination of the dense band co-location on the electrophoresis lane and other lanes is crucial in the expert evaluation of the image class. To incorporate this expert knowledge into the model training, inner-task and inter-task regularization is introduced. The effectiveness of the proposed method is demonstrated through experiments conducted on approximately 15,000 IFE images, resulting in interpretable visualization outcomes that are in alignment with expert expectations. Codes are available at https://github.com/shiy19/IFE-classification.

Keywords: IFE image classification · Multi-task learning · Task-related regularization · Class imbalance · Expert knowledge

1 Introduction

The utilization of immunofixation electrophoresis (IFE) as a laboratory technique has been widely adopted for the identification and characterization of abnormal proteins in blood or urine specimens [11,15,25]. This technique is commonly employed to detect monoclonal gammopathy, a condition characterized

---

Supplementary Information The online version contains supplementary material available at https://doi.org/10.1007/978-3-031-43987-2_15.

H. Greenspan et al. (Eds.): MICCAI 2023, LNCS 14225, pp. 148–158, 2023.
https://doi.org/10.1007/978-3-031-43987-2_15

by the presence of deviant monoclonal (M) proteins, which can aid in the diagnosis of multiple myeloma, Waldenstrom's macroglobulinemia, and other plasma cell disorders [2,16,18,26]. The electrophoresis process separates serum proteins into distinct lanes, which are then treated with specific antisera against 'IgG', 'IgA', 'IgM', 'κ', and 'λ'. Precipitin bands indicative of abnormal M-proteins exhibit greater density and darkness than their normal counterparts. The electrophoresis (ELP) lane, which represents a mixture of multiple proteins, serves as a reference for the recognition of various M-proteins. For human experts, identification of an M-protein is achieved by observing the co-location of bands between the ELP lane, heavy chain lanes ('G', 'A', and 'M'), and light chain lanes ('κ' and 'λ'). For example, as shown in the top of Fig. 1(a), if the dense (dark) band of the ELP lane can be aligned with that of the heavy chain lane 'G' and the light chain lane 'κ', this sample can be identified as 'IgG-κ positive'. In the same way, we can identify the other two samples in Fig. 1(a) as 'λ positive' and 'Negative'.

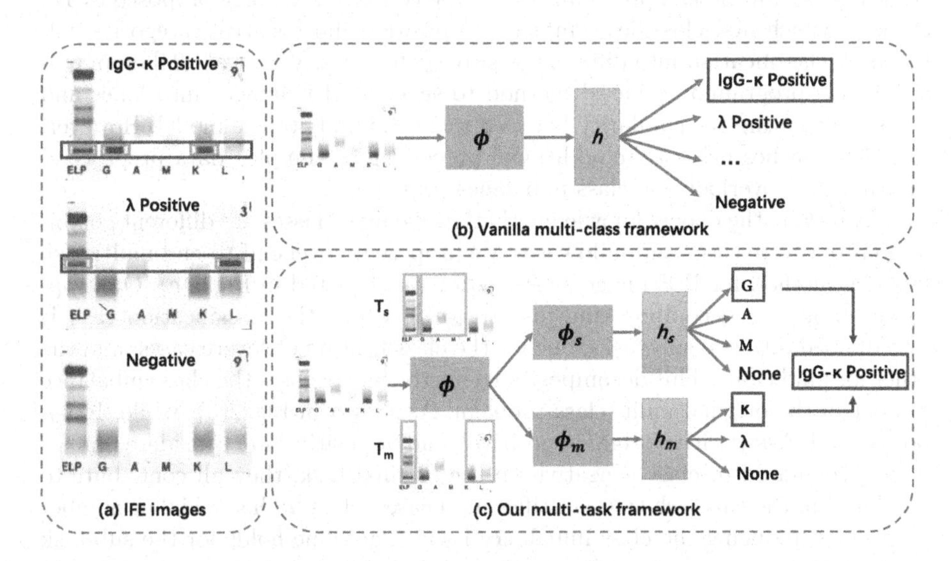

Fig. 1. (a) Three kinds of IFE images: 'IgG-κ positive' (severe case and mild case simultaneously), 'λ positive' (only mild case), and 'Negative'. The area in the red box is the judgment area marked by our experts, and the area in the green box is the dense band. (b) An illustration of vanilla multi-class framework, which directly performs a 9-class classification for IFE images. (c) An illustration of our proposed multi-task framework, which splits the task into two sub-tasks: T_s and T_m to deal with the classification of severe cases and mild cases, respectively. T_s and T_m can focus on corresponding areas, as shown in the orange boxes. ϕ and h represent feature extractors and classifiers, respectively. (Color figure online)

Recently, the application of machine learning for image recognition has gained attention as a means of handling large-scale samples and providing objective and quality results [19,20,23,28]. While machine learning has been successfully applied to a wide range of medical images, such as Computed Tomography (CT) [4,13], histopathological images [8,31], magnetic resonance images [3,12], ultrasound images [17,29], and mammograms [21,30], its application to IFE presents unique challenges. IFE images encompass nine distinct yet interrelated classes, including 'Negative', two types of positive with mild cases ('κ', 'λ'), and six types of positive with both severe and mild cases ('IgG-κ', 'IgG-λ', 'IgA-κ', 'IgA-λ', 'IgM-κ', 'IgM-λ')[1], which can be identified as different combinations of three severe cases ('IgG', 'IgA', and 'IgM') and two mild cases ('κ' and 'λ'). The class imbalance phenomenon [1,9] is common in IFE images, as demonstrated by our dataset collected from a large hospital, where the predominant class 'Negative' comprises over 10,000 samples, while the minor class 'κ positive' has only 63 samples. Conventional multi-class classification approaches struggle to accurately learn the features of minor classes, leading to subpar performance, particularly for minor classes. Previous studies have approached IFE image classification as a multi-class problem, such as the two-stage strategy proposed by Hu et al. [6] which first classifies samples into positive and negative categories, followed by classification into different positive categories. Wei et al. [24] employed a dynamic programming based method to segment IFE images into lanes and strips, calculating the similarity between them as inputs to the model. However, these approaches necessitate additional processing or complex data preprocessing, and they overlook the class imbalance problem.

Inspired by the expert knowledge that all the nine classes are different combinations of severe cases and mild cases, we propose a novel end-to-end multi-task learning method for IFE image classification, as depicted in Fig. 1(c). Our approach employs a decompose-and-fuse strategy, where the classification task is first divided into two sub-tasks: one for the classification of severe cases, and the other for mild cases. This decomposition effectively mitigates the class imbalance present in the original multi-class problem. As shown in Fig. 2(a), while 'IgG-κ positive', 'IgA-κ positive', 'IgM-κ positive', and 'κ positive' are considered minor classes compared to class 'Negative' in the original task, they all contribute to 'κ'. Thus, in the sub-task that classifies mild cases, class 'κ' has a higher number of samples, reducing the class imbalance issue. The same holds for the sub-task that categorizes severe cases. Due to the reduced class imbalance in the two subtasks, improved performance can be achieved. The results from the two sub-tasks are then combined to produce good performance in the original 9-class classification problem. Additionally, we have integrated an attention module into the convolution layers at the height level of the image to enhance the model's focus on relevant image features and improve visualization and explanation. Drawing upon expert knowledge, we have implemented two forms of regularization to more accurately model the relationships within and between tasks, as depicted in Fig. 2(b, c). The inner-task regularization smoothens the attention map of negative samples and sharpens that of positive samples, which aligns with the expert

[1] Samples with multiple positives are rare, *e.g.*, with 'IgG-κ' and 'IgM-λ' simultaneously. These samples are not considered in this study.

understanding that only positive samples exhibit co-located dense bands. For a sample with both severe and mild cases simultaneously (namely S&M-sample), the inter-task regularization reduces the gap between its attention maps from the two sub-tasks, which corresponds to the expert knowledge that dense bands co-exist on both the severe lane and the mild lane for a S&M-sample.

In summary, our contributions are three-fold: (1) we propose an end-to-end multi-task method for IFE image classification that effectively addresses the class imbalance issue, (2) we implement two types of regularization mechanisms to more accurately model relationships within and between tasks, and (3) our extensive experiments and visualization demonstrate the effectiveness and explainability of our proposed method.

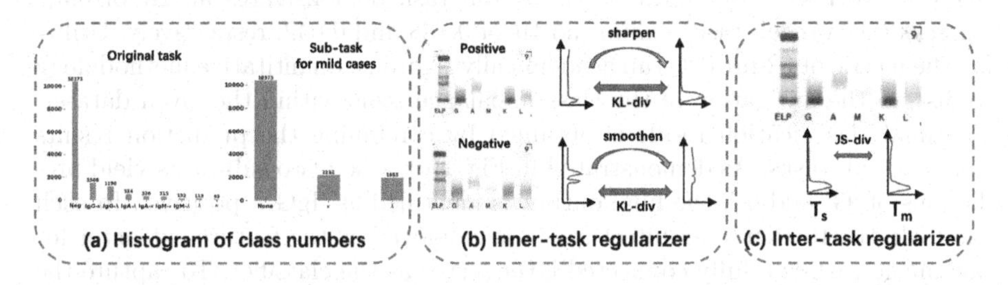

Fig. 2. (a) Histogram of class numbers for the original multi-class task and our sub-task that classifies mild cases. (b) An illustration of our proposed inner-task regularizer. We sharpen/smoothen the attention map of positive/negative samples first, and minimize its KL-divergence with the original attention map. (c) An illustration of our proposed inter-task regularizer. For positive samples with severe and mild case simultaneously, we minimize the JS-divergence between the attention maps of sub-task T_s and sub-task T_m.

2 Method

In this section, we describe the methodology for the development of a novel multi-task learning approach for the classification of IFE images. Based on the two-sub-task framework, our approach employs both within-task and between-task regularization on the attention module to address several critical challenges, such as alleviating class imbalance and modeling different patterns of severe and mild cases, which can incorporate expert knowledge into the model training.

2.1 Multi-task Framework

Directly addressing the classification of IFE images as a multi-class (9-class) problem may face a severe class imbalance phenomenon, which makes the classification of IFE images a challenging task. Inspired by expert knowledge, we decompose the multi-class problem into two sub-tasks to mitigate this issue.

The first sub-task is to classify heavy lanes into four classes ('None', 'G', 'A', and 'M'), while the second is to classify light lanes into three classes ('None', 'κ', and 'λ'). If a sample does not possess the corresponding dense bands in its heavy or light lane that match with the ELP lane, it is classified as 'None'. This decomposition approach ensures that different minor classes of the original 9-class task that shares the same severe or mild cases are combined into larger minor classes for the two sub-tasks, which can be proven by the histogram in Fig. 2(a). Define imbalance ratio (IR) as follows:

$$IR = N_{\min}/N_{\max} \tag{1}$$

Here, $N_{\min}$ and $N_{\max}$ represent minority class size and majority class size, respectively. In reviewing our dataset, the initial task demonstrates an IR of 0.06, whereas the two sub-tasks exhibit an IR of 0.188 and 0.042, respectively. Utilizing the metric of IR provides an academically rigorous, quantitative methodology to discern the mitigation of the class imbalance issue within the given dataset. The final classification result is obtained by combining the prediction results from two sub-tasks. As demonstrated in Fig. 1(c), when two sub-tasks yield predictions of 'G' and 'κ', the final output is identified as 'IgG-κ positive'. In each sub-task, the ResNet18 architecture [5] is employed as the feature extractor for raw images, while a fully connected layer serves as the classifier. To capture the relationship between the two sub-tasks, and to ensure that they both learn the basic features of the raw image, such as the dense band in the lane, the first three blocks of the ResNet18 model are shared between the two sub-tasks. This sharing of blocks was found to provide the best performance empirically, and further details can be found in the supplementary material. Formally speaking, for an instance $\mathbf{x}_i \in \mathbb{R}^D$, we split its ground-truth label $\mathbf{y}_i \in \mathbb{S}(9)$ into $\mathbf{y}_{is} \in \mathbb{S}(4)$ and $\mathbf{y}_{im} \in \mathbb{S}(3)$. Here, $\mathbb{S}(k)$ means the set of $\{1, 2, \cdots, k\}$, and subscripts s and m represent the task that classifies severe cases and mild cases, respectively. Let $\boldsymbol{\phi}(\cdot)$ and $\boldsymbol{h}(\cdot)$ represent the feature extractor and the classifier, respectively. The logit vectors of a sample $\mathbf{x}_i$ in two sub-tasks are obtained with $\hat{\mathbf{f}}_{is} = \boldsymbol{h}_s(\boldsymbol{\phi}_s(\mathbf{x}_i))$ and $\hat{\mathbf{f}}_{im} = \boldsymbol{h}_m(\boldsymbol{\phi}_m(\mathbf{x}_i))$. The predicted label for each sub-task is then obtained by taking the maximum value from the corresponding logit vector. $\boldsymbol{\phi}_s$ and $\boldsymbol{h}_s$ ($\boldsymbol{\phi}_m$ and $\boldsymbol{h}_m$) of the sub-task are trained with cross-entropy loss $\mathbf{L}^{\mathbf{s}}_{\mathbf{CE}}$ ($\mathbf{L}^{\mathbf{m}}_{\mathbf{CE}}$), which minimizes the discrepancy between the posterior class probability with the ground-truth.

2.2 Task-Related Regularization

Horizontal Attention. As depicted in Fig. 1(a), human experts identify the valid part of an IFE image as a horizontal rectangular region that displays the alignment of dense bands between the ELP lane and the other lanes. To encourage the model to focus on this area, an attention mechanism has been integrated into the convolutional layer of the proposed method. Attention mechanisms [7,22,27] are widely used in Convolutional Neural Network (CNN) models to focus on specific channels or spatial regions. However, for IFE image classification, only attention to the horizontal direction is required. The horizontal attention mechanism used in this study differs from conventional spatial attention in that it only focuses on a 1D direction (the horizontal direction). To calculate the horizontal attention, average-pooling AvgPool($\mathbf{F}$) and max-pooling MaxPool($\mathbf{F}$) are applied along the channel axis and the width axis, followed by a standard convolution layer and a sigmoid function σ to generate a horizontal attention map $\mathbf{M}(\mathbf{F}) \in \mathbb{R}^H$:$\mathbf{M}(\mathbf{F}) = \sigma\left(\text{Conv}([\text{AvgPool}(\mathbf{F}); \text{MaxPool}(\mathbf{F})])\right)$. Here, $\mathbf{F}$ represents the feature vector, and H is the height of the image. Compared to spatial attention, this attention map encodes the horizontal region to be emphasized, enabling the model to better capture the relationship between the dense bands in different lanes. Furthermore, the horizontal attention mechanism provides a convenient way to visualize the model results, making it more interpretable, as illustrated in Fig. 3.

Inner-Task Regularization. To effectively model the patterns in different samples, it is crucial to understand the desired appearance of the attention map for positive and negative samples. As previously discussed, for positive samples, the model should emphasize the horizontal region that displays the alignment of dense bands. However, obtaining human-labeled attention areas for the training phase is a time-consuming and labor-intensive task. To address this issue, the attention map is differentiated between positive and negative samples. The goal is to sharpen the attention map for positive samples and smoothen it for negative samples. The attention map $\mathbf{M}(\mathbf{F})$ can be viewed as the distribution of attention intensity along the height dimension of the image, as depicted in Fig. 2(b). The distribution is adjusted to be sharper or flatter by applying a Softmax operator with a temperature parameter τ: $\overline{\mathbf{M}(\mathbf{F})} = \mathbf{Softmax}\left(\mathbf{M}(\mathbf{F})/\tau\right)$. The inner-task regularization $\mathbf{Reg}^{\mathbf{s}}_{\mathbf{inner}}$ and $\mathbf{Reg}^{\mathbf{m}}_{\mathbf{inner}}$ can be designed using the Kullback-Leibler divergence (KLdiv) [10] as follows:

$$\begin{aligned} \mathbf{Reg}^{\mathbf{s}}_{\mathbf{inner}} &= \mathbf{KLdiv}(\overline{\mathbf{M_s}(\mathbf{F_s})} \cdot \mathbf{M_s}(\mathbf{F_s})) \\ \mathbf{Reg}^{\mathbf{m}}_{\mathbf{inner}} &= \mathbf{KLdiv}(\overline{\mathbf{M_m}(\mathbf{F_m})} \cdot \mathbf{M_m}(\mathbf{F_m})) \end{aligned} \quad . \tag{2}$$

Here, $\mathbf{KLdiv}(A \cdot B) = A \cdot (\log A - \log B)$. The inner-task attention mechanism allows the model to differentiate the attention map between positive and negative samples.

Inter-Task Regularization. It is important to note that the alignment of dense bands between heavy lanes, light lanes, and the ELP lane is a crucial factor in the identification of S&M-samples by human experts. To model this relationship in the proposed method, the two sub-tasks are encouraged to emphasize similar horizontal areas. Specifically, for all S&M-samples, the inter-task regularization can be designed with the Jensen-Shannon Divergence (JSdiv) [14] as follows :

$$\mathbf{Reg_{inter}} = \mathbf{JSdiv}(\mathbf{M_s}(\mathbf{F_s}) \cdot \mathbf{M_m}(\mathbf{F_m})) \tag{3}$$

Here, $\mathbf{JSdiv}(A \cdot B) = \left(\mathbf{KLdiv}(A \cdot C) + \mathbf{KLdiv}(B \cdot C)\right)/2$, and $C = (A + B)/2$. The inter-task regularizer enables the model to better capture the patterns of S&M-samples by modeling the relationship between the two sub-tasks.

2.3 Overall Objective

The overall optimization objective for the proposed method can be expressed as:

$$\mathbf{L}_{\mathbf{CE}}^{\mathbf{s}} + \mathbf{L}_{\mathbf{CE}}^{\mathbf{m}} + \lambda_{inner} \cdot (\mathbf{Reg}_{\mathbf{inner}}^{\mathbf{s}} + \mathbf{Reg}_{\mathbf{inner}}^{\mathbf{m}}) + \lambda_{inter} \cdot \mathbf{Reg}_{\mathbf{inter}} \, . \tag{4}$$

Here, λ_{inner} and λ_{inter} are hyper-parameters.

In conclusion, the proposed multi-task method mitigates the class imbalance problem in the classification of IFE images by decomposing the task into two separate sub-tasks. The use of inner-task and inter-task regularization on the horizontal attention mechanism allows the model to capture the relationships between sub-tasks and learn discriminative patterns for both positive and negative samples.

3 Experiment

Dataset. We introduce a new IFE image dataset, which comprises 14540 images along with their respective protein presence class labels. As defined in Eq. 1, the dataset exhibits an imbalance ratio of 0.006, which underscores a pronounced instance of class imbalance within the dataset. Our dataset is collected from a large hospital between January 2020 and December 2022, and it reflects the real-world environment, as the images usually contain certain types of noise, such as left and right offset of lanes and non-pure white background color. Unlike previous works [6,24] that perform preprocessing operations, such as cutting the images into different horizontal or vertical pieces, the raw images are utilized in the proposed method, resulting in an end-to-end approach. The training and test sets are stratified based on a 4:1 ratio, and the dataset characteristics can be found in the supplementary material. The images are resized to 224×224 as the input for the model.

Table 1. Performance comparison on IFE dataset. The best performances are in bold.

Methods	Acc (%)	F1-score(%)	Fnr (%)	M1-Acc (%)	M3-Acc (%)
Multi-class	95.12 ± 0.04	85.16 ± 0.60	10.10 ± 0.41	88.46 ± 3.84	73.83 ± 2.04
Collocative Net [24]	93.07 ± 0.05	81.80 ± 0.08	14.22 ± 0.48	80.77 ± 3.84	59.97 ± 0.65
Two stage DNN [6]	92.54 ± 0.63	75.68 ± 0.96	9.31 ± 0.60	42.30 ± 3.85	42.03 ± 0.64
Our Method	$\mathbf{95.43 \pm 0.10}$	$\mathbf{86.68 \pm 0.24}$	$\mathbf{8.54 \pm 0.18}$	$\mathbf{92.31 \pm 0.00}$	$\mathbf{76.10 \pm 1.28}$

Table 2. Ablation study on the multi-task framework and two regularization.

Multi-task	$\text{Reg}_{\text{inner}}$	$\text{Reg}_{\text{inter}}$	Acc (%)	F1-score (%)	Fnr (%)	M1-Acc (%)	M3-Acc (%)
✗	✗	✗	95.12 ± 0.04	85.16 ± 0.60	10.10 ± 0.41	88.46 ± 3.84	73.83 ± 2.04
✓	✗	✗	95.23 ± 0.01	85.79 ± 0.35	9.80 ± 0.12	92.31 ± 0.00	73.60 ± 0.46
✓	✓	✗	95.37 ± 0.05	86.16 ± 0.15	9.14 ± 0.42	92.31 ± 0.00	75.99 ± 1.39
✓	✗	✓	95.24 ± 0.08	85.70 ± 0.85	9.26 ± 0.06	88.46 ± 3.84	75.00 ± 0.01
✓	✓	✓	$\mathbf{95.43 \pm 0.10}$	$\mathbf{86.68 \pm 0.24}$	$\mathbf{8.54 \pm 0.18}$	$\mathbf{92.31 \pm 0.00}$	$\mathbf{76.10 \pm 1.28}$

Implementation Details. All models are implemented using the PyTorch framework and trained on an NVIDIA GeforceRTX 2080ti GPU with 11 GB memory. The network is trained for 100 epochs with a batch size of 64, and the optimization is performed using Stochastic Gradient Descent (SGD) with momentum as the optimizer. The initial learning rate is set to 0.03 for the first 5 epochs and then cosine-decayed for the remaining epochs. The training stage takes 1.2 h on one GPU and occupies 4G display memory. The hyper-parameters of regularization are tuned based on cross-validation on the training set, with the best hyper-parameters set to $\lambda_{\text{inner}} = 0.3$, $\tau = 5, \lambda_{\text{inter}} = 0.3$.

Evaluation Metric and Comparison Methods. We assess the efficacy via a suite of evaluation metrics, including accuracy (acc), F1-score, false negative rate (fnr), top-1 minor class accuracy (M1-Acc), and top-3 minor class accuracy (M3-Acc). These metrics are widely adopted for classification problems, with the false negative rate being of particular significance in the medical domain as it quantifies the likelihood of misidentifying positive samples as negative. The top-1 and top-3 minor class accuracy metrics evaluate the model's performance on underrepresented classes. A ResNet18-based multi-class classification approach is selected as the baseline and tuned using an identical range of hyper-parameters as the proposed method. The baseline is trained for 100 epochs with a batch size of 64, an initial learning rate of 0.1, and a decay scheduler consistent with the proposed method. Additionally, the proposed method is compared against Two Stage DNN [6] and Collocative Net [24], the state-of-the-art methods in the IFE recognition domain, using our dataset. The results are obtained by adhering to the code released by the authors. The dataset is preprocessed manually prior to network training in accordance with the methodology described in the respective papers.

Performance Evaluation. The methodology proposed in this study is thoroughly evaluated and compared to both the baseline method and other state-of-the-art techniques on the IFE dataset, as presented in Table 1. The results reveal that our method outperforms all other methods in terms of all evaluation metrics, particularly with respect to top1 and top3 minor class accuracy, thus highlighting its efficacy in addressing the class imbalance problem and achieving improved performance for underrepresented classes. Our empirical analysis demonstrates that Collocative Net [24] and Two Stage DNN [6] exhibit inferior performance compared to the conventional multi-class approach. This can be attributed to the excessive manual intervention employed during preprocessing and model development, which is not suitable for handling real-world IFE datasets containing various forms of noise, such as lane offsets and impurity pixels in the background.

Visualization. The proposed horizontal attention mechanism provides a useful tool for verifying the consistency of model-learned knowledge with expert experience. As illustrated in Fig. 3, the areas with the highest attention intensity are highlighted in orange, while red boxes indicate the judgment areas of our experts. The results show that the attention regions of positive samples largely

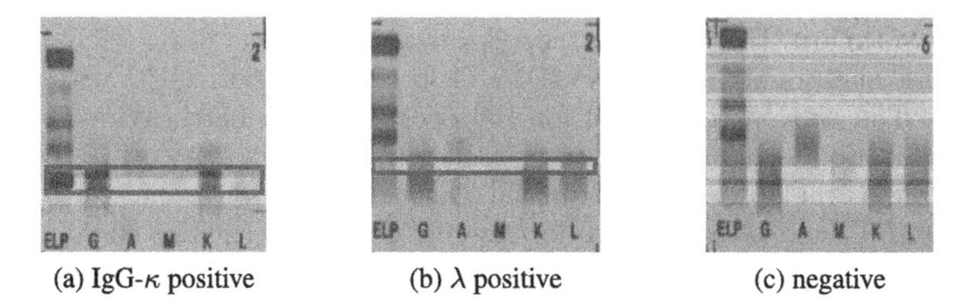

(a) IgG-κ positive (b) λ positive (c) negative

Fig. 3. Visualization of horizontal attention maps on different types of IFE images. The orange areas are the most attention part learned by our model, and the areas in red boxes are the judgment areas marked by our experts. (Color figure online)

overlap with the expert judgment areas, while the attention regions of negative samples are more dispersed, indicating a lack of specific areas of interest. These findings demonstrate that the knowledge learned by our model is consistent with expert experience.

Ablation Study. An ablation study is conducted to examine the contribution of key components of the proposed model, including the multi-task framework and two regularization methods. The results, presented in Table 2, indicate that the use of the multi-task framework results in significant performance gains, particularly for minority classes, and that the introduction of the two regularization methods further improves the overall performance of the model. Further exploration of the effects of different hyper-parameters can be found in the supplementary materials.

4 Conclusion

In this study, we investigate the issue of classifying immunofixation electrophoresis (IFE) images and contend that conventional multi-class classification strategies may not be adequate due to the considerable class imbalance issue. To address this issue, we introduce a multi-task framework that leverages expert knowledge. Our method effectively captures different patterns of both severe and mild cases through the use of inner-task and inter-task regularization. The empirical evaluation of our method, conducted on a dataset of approximately 15,000 IFE images, demonstrates its superiority over other approaches. The results also provide interpretable visualizations that align with expert annotations and highlight the potential for improved accuracy and reliability in the classification of IFE images. This development has the potential to significantly advance the diagnostic capabilities of monoclonal protein disorders in the medical field.

Acknowledgments. This research was supported by National Key R&D Program of China (2020AAA0109401), NSFC (62176117, 61921006, 62006112), Collaborative Innovation Center of Novel Software Technology and Industrialization.

References

1. Abd Elrahman, S.M., Abraham, A.: A review of class imbalance problem. J. Network Innovative Comput. **1**(2013), 332–340 (2013)
2. Cowan, A.J., et al.: Diagnosis and management of multiple myeloma: a review. JAMA **327**(5), 464–477 (2022)
3. Dong, S., et al.: Multi-scale super-resolution magnetic resonance spectroscopic imaging with adjustable sharpness. In: Wang, L., Dou, Q., Fletcher, P.T., Speidel, S., Li, S. (eds.) MICCAI 2022. LNCS, vol. 13436, pp. 410–420. Springer, Cham (2022). https://doi.org/10.1007/978-3-031-16446-0_39
4. Gu, Y., et al.: BMD-GAN: bone mineral density estimation using x-ray image decomposition into projections of bone-segmented quantitative computed tomography using hierarchical learning. In: Wang, L., Dou, Q., Fletcher, P.T., Speidel, S., Li, S. (eds.) MICCAI 2022. LNCS, vol. 13436, pp. 644–654. Springer, Cham (2022). https://doi.org/10.1007/978-3-031-16446-0_61
5. He, K., Zhang, X., Ren, S., Sun, J.: Deep residual learning for image recognition. In: CVPR, pp. 770–778 (2016)
6. Hu, H., et al.: Expert-level immunofixation electrophoresis image recognition based on explainable and generalizable deep learning. Clin. Chem. **69**(2), 130–139 (2023)
7. Hu, J., Shen, L., Sun, G.: Squeeze-and-excitation networks. In: CVPR, pp. 7132–7141 (2018)
8. Jimenez, G., et al.: Visual deep learning-based explanation for Neuritic plaques segmentation in Alzheimer's disease using weakly annotated whole slide histopathological images. In: Wang, L., Dou, Q., Fletcher, P.T., Speidel, S., Li, S. (eds.) MICCAI 2022. LNCS, vol. 13432, pp. 336–344. Springer, Cham (2022). https://doi.org/10.1007/978-3-031-16434-7_33
9. Johnson, J.M., Khoshgoftaar, T.M.: Survey on deep learning with class imbalance. J. Big Data **6**(1), 1–54 (2019). https://doi.org/10.1186/s40537-019-0192-5
10. Joyce, J.M.: Kullback-leibler divergence. In: International Encyclopedia of Statistical Science, pp. 720–722 (2011)
11. Keren, D.F.: High-Resolution Electrophoresis and Immunofixation: Techniques and Interpretation (2017)
12. Kuzmina, E., Razumov, A., Rogov, O.Y., Adalsteinsson, E., White, J., Dylov, D.V.: Autofocusing+: noise-resilient motion correction in magnetic resonance imaging. In: Wang, L., Dou, Q., Fletcher, P.T., Speidel, S., Li, S. (eds.) MICCAI 2022. LNCS, vol. 13436, pp. 365–375. Springer, Cham (2022). https://doi.org/10.1007/978-3-031-16446-0_35
13. López Diez, P., et al.: Deep reinforcement learning for detection of inner ear abnormal anatomy in computed tomography. In: Wang, L., Dou, Q., Fletcher, P.T., Speidel, S., Li, S. (eds.) MICCAI 2022. LNCS, vol. 13433, pp. 697–706. Springer, Cham (2022). https://doi.org/10.1007/978-3-031-16437-8_67
14. Menéndez, M., Pardo, J., Pardo, L., Pardo, M.: The jensen-shannon divergence. J. Franklin Inst. **334**(2), 307–318 (1997)
15. Moore, A.R., Avery, P.R.: Protein characterization using electrophoresis and immunofixation; a case-based review of dogs and cats. Veterinary Clin. Pathol. **48**, 29–44 (2019)
16. Moreau, P., et al.: Treatment of relapsed and refractory multiple myeloma: recommendations from the international myeloma working group. Lancet Oncol. **22**(3), e105–e118 (2021)

17. Oh, S., Kim, M.G., Kim, Y., Jung, G., Kwon, H., Bae, H.M.: Sensor geometry generalization to untrained conditions in quantitative ultrasound imaging. In: Wang, L., Dou, Q., Fletcher, P.T., Speidel, S., Li, S. (eds.) MICCAI 2022. LNCS, vol. 13436, pp. 780–789. Springer, Cham (2022). https://doi.org/10.1007/978-3-031-16446-0_74
18. Rajkumar, S.V., Kumar, S.: Multiple myeloma current treatment algorithms. Blood Cancer J. **10**(9), 94 (2020)
19. Shin, Y., et al.: Digestive organ recognition in video capsule endoscopy based on temporal segmentation network. In: Wang, L., Dou, Q., Fletcher, P.T., Speidel, S., Li, S. (eds.) MICCAI 2022. LNCS, vol. 13437, pp. 136–146. Springer, Cham (2022). https://doi.org/10.1007/978-3-031-16449-1_14
20. Shui, Z., et al.: End-to-end cell recognition by point annotation. In: Wang, L., Dou, Q., Fletcher, P.T., Speidel, S., Li, S. (eds.) MICCAI 2022. LNCS, vol. 13434, pp. 109–118. Springer, Cham (2022). https://doi.org/10.1007/978-3-031-16440-8_11
21. Tardy, M., Mateus, D.: Leveraging multi-task learning to cope with poor and missing labels of mammograms. Front. Radiol. **1**, 796078 (2022)
22. Wang, X., Girshick, R., Gupta, A., He, K.: Non-local neural networks. In: CVPR, pp. 7794–7803 (2018)
23. Wang, Y., et al.: Key-frame guided network for thyroid nodule recognition using ultrasound videos. In: Wang, L., Dou, Q., Fletcher, P.T., Speidel, S., Li, S. (eds.) MICCAI 2022. LNCS, vol. 13434, pp. 238–247. Springer, Cham (2022). https://doi.org/10.1007/978-3-031-16440-8_23
24. Wei, X.Y., et al.: Deep collocative learning for immunofixation electrophoresis image analysis. IEEE Trans. Med. Imaging **40**(7), 1898–1910 (2021)
25. Wilhite, D., Arfa, A., Cotter, T., Savage, N.M., Bollag, R.J., Singh, G.: Multiple myeloma: detection of free monoclonal light chains by modified immunofixation electrophoresis with antisera against free light chains. Pract. Lab. Med. **27**, e00256 (2021)
26. Willrich, M.A., Murray, D.L., Kyle, R.A.: Laboratory testing for monoclonal gammopathies: focus on monoclonal gammopathy of undetermined significance and smoldering multiple myeloma. Clin. Biochem. **51**, 38–47 (2018)
27. Woo, S., Park, J., Lee, J.Y., Kweon, I.S.: CBAM: convolutional block attention module. In: ECCV, pp. 3–19 (2018)
28. Xia, C., Wang, J., Qin, Y., Gu, Y., Chen, B., Yang, J.: An end-to-end combinatorial optimization method for r-band chromosome recognition with grouping guided attention. In: Wang, L., Dou, Q., Fletcher, P.T., Speidel, S., Li, S. (eds.) MICCAI 2023. LNCS, vol. 13434, pp. 3–13. Springer, Cham (2022). https://doi.org/10.1007/978-3-031-16440-8_1
29. Xie, Y., Liao, H., Zhang, D., Chen, F.: Uncertainty-aware cascade network for ultrasound image segmentation with ambiguous boundary. In: Wang, L., Dou, Q., Fletcher, P.T., Speidel, S., Li, S. (eds.) MICCAI 2022. LNCS, vol. 13434, pp. 268–278. Springer, Cham (2022). https://doi.org/10.1007/978-3-031-16440-8_26
30. Yu, X., Pang, W., Xu, Q., Liang, M.: Mammographic image classification with deep fusion learning. Sci. Rep. **10**(1), 14361 (2020)
31. Zeng, B., et al.: Semi-supervised PR virtual staining for breast histopathological images. In: Wang, L., Dou, Q., Fletcher, P.T., Speidel, S., Li, S. (eds.) MICCAI 2022. LNCS, vol. 13432, pp. 232–241. Springer, Cham (2022)

Thinking Like Sonographers: A Deep CNN Model for Diagnosing Gout from Musculoskeletal Ultrasound

Zhi Cao[1], Weijing Zhang[2], Keke Chen[2], Di Zhao[2], Daoqiang Zhang[1], Hongen Liao[3], and Fang Chen[1](✉)

[1] Key Laboratory of Brain-Machine Intelligence Technology, Ministry of Education, Nanjing University of Aeronautics and Astronautics, Nanjing 210016, China
chenfang@nuaa.edu.cn

[2] Department of Ultrasound, Nanjing Drum Tower Hosptial, The Affiliated Hospital of Nanjing University Medical School, Nanjing 210008, China

[3] Department of Biomedical Engineering, School of Medicine, Tsinghua University, Beijing 100084, China

Abstract. We explore the potential of deep convolutional neural network (CNN) models for differential diagnosis of gout from musculoskeletal ultrasound (MSKUS), as no prior study on this topic is known. Our exhaustive study of state-of-the-art (SOTA) CNN image classification models for this problem reveals that they often fail to learn the gouty MSKUS features, including the double contour sign, tophus, and snowstorm, which are essential for sonographers' decisions. To address this issue, we establish a framework to adjust CNNs to "think like sonographers" for gout diagnosis, which consists of three novel components: (1) Where to adjust: Modeling sonographers' gaze map to emphasize the region that needs adjust; (2) What to adjust: Classifying instances to systematically detect predictions made based on unreasonable/biased reasoning and adjust; (3) How to adjust: Developing a training mechanism to balance gout prediction accuracy and attention reasonability for improved CNNs. The experimental results on clinical MSKUS datasets demonstrate the superiority of our method over several SOTA CNNs.

Keywords: Musculoskeletal ultrasound · Gout diagnosis · Gaze tracking · Reasonability

1 Introduction

Gout is the most common inflammatory arthritis and musculoskeletal ultrasound (MSKUS) scanning is recommended to diagnose gout due to the non-ionizing radiation, fast imaging speed, and non-invasive characteristics of MSKUS [7]. However, misdiagnosis of gout can occur frequently when a patient's clinical characteristics are atypical. Traditional MSKUS diagnosis relies on the experience of the radiologist which is time-consuming and labor-intensive. Although

Supplementary Information The online version contains supplementary material available at https://doi.org/10.1007/978-3-031-43987-2_16.

H. Greenspan et al. (Eds.): MICCAI 2023, LNCS 14225, pp. 159–168, 2023.
https://doi.org/10.1007/978-3-031-43987-2_16

convolutional neural networks (CNNs) based ultrasound classification models have been successfully used for diseases such as thyroid nodules and breast cancer, conspicuously absent from these successful applications is the use of CNNs for gout diagnosis from MSKUS images.

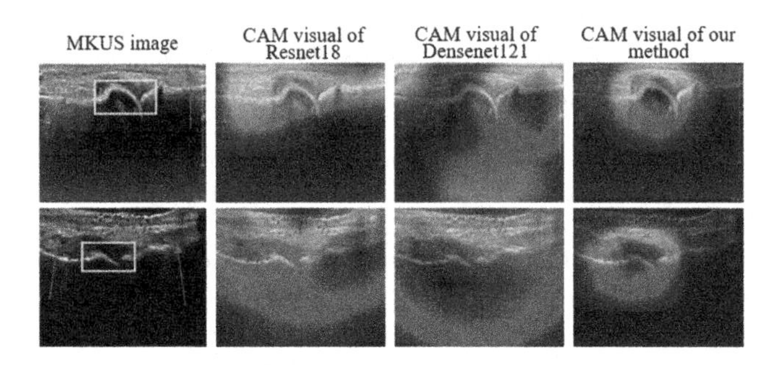

Fig. 1. (a) MSKUS images. Yellow boxes denote the gaze areas of the sonographers and red arrows denote the surrounding fascial tissues; (b) Grad-Cam visual of ResNet18; (c) Grad-Cam visual of Densenet121; (d) Grad-Cam visual of our method. (Color figure online)

There are significant challenges in CNN based gout diagnosis. Firstly, the gout-characteristics contain various types including double contour sign, synovial hypertrophy, synovial effusion, synovial dislocation and bone erosion, and these gout-characteristics are small and difficult to localize in MSKUS. Secondly, the surrounding fascial tissues such as the muscle, sarcolemma and articular capsule have similar visual traits with gout-characteristics, and we found the existing CNN models can't accurately pay attention to the gout-characteristics that radiologist doctors pay attention to during the diagnosis process (as shown in Fig. 1). Due to these issues, SOTA CNN models often fail to learn the gouty MSKUS features which are key factors for sonographers' decision.

In medical image analysis, recent works have attempted to inject the recorded gaze information of clinicians into deep CNN models for helping the models to predict correctly based on lesion area. Mall et al. [9,10] modeled the visual search behavior of radiologists for breast cancer using CNN and injected human visual attention into CNN to detect missing cancer in mammography. Wang et al. [15] demonstrated that the eye movement of radiologists can be a new supervision form to train the CNN model. Cai et al. [3,4] developed the SonoNet [1] model, which integrates eye-gaze data of sonographers and used Generative Adversarial Networks to address the lack of eye-gaze data. Patra et al. [11] proposed the use of a teacher-student knowledge transfer framework for US image analysis, which combines doctor's eye-gaze data with US images as input to a large teacher model, whose outputs and intermediate feature maps are used to condition a student model. Although these methods have led to promising results, they can be difficult to implement due to the need to collect doctors' eye movement data for each image, along with certain restrictions on the network structure.

Different from the existing studies, we propose a novel framework to adjust the general CNNs to "think like sonographers" from three different levels. (1) Where to adjust: Modeling sonographers' gaze map to emphasize the region that needs adjust; (2) What to adjust: Classify the instances to systemically detect predictions made based on unreasonable/biased reasoning and adjust; (3) How to adjust: Developing a training mechanism to strike the balance between gout prediction accuracy and attention reasonability.

2 Method

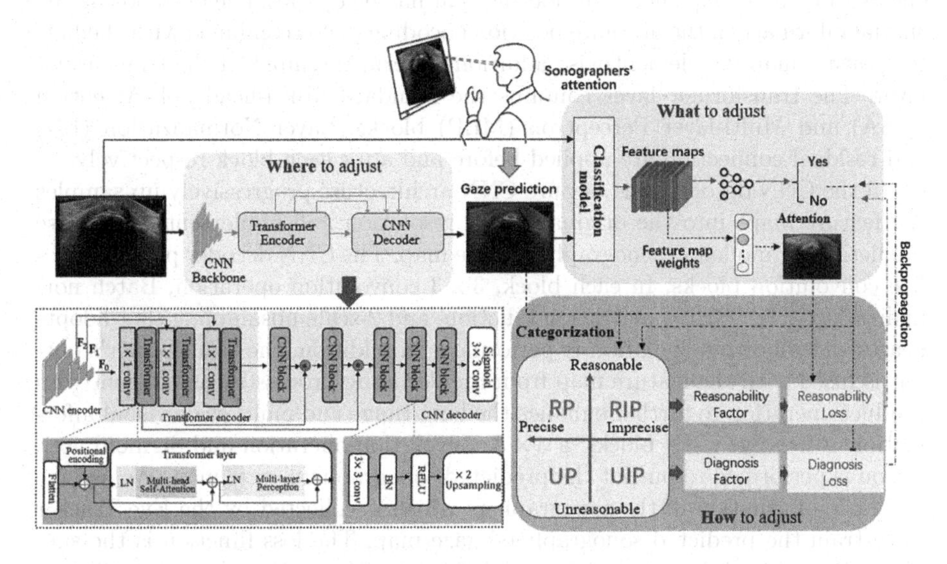

Fig. 2. The overall framework of the proposed method.

Figure 2 presents the overall framework, which controls CNNs to "think like sonographers" for gout diagnosis from three levels. 1) Where to adjust: we model the sonographers' gaze map to emphasize the region that needs control. This part learns the eye gaze information of the sonographers which is collected by the Eye-Tracker. 2) What to adjust: we divide instances into four categories to reflect whether the model prediction given to the instance is reasonable and precise. 3) How to adjust: a training mechanism is developed to strike the balance between gout diagnosis and attention accuracy for improving CNN.

2.1 Where to Adjust

It is essential to obtain the gaze map corresponding to each MSKUS to emphasize the region where gouty features are obvious. Inspired by studies of saliency model [8], we integrate transformer into CNNs to capture multi-scale and long-range contextual visual information for modeling sonographers' gaze map. This gaze map learns the eye gaze information, collected by the Eye-Tracker, of the sonographers when they perform diagnosis. As shown in Fig. 2, this part consists of a CNN encoder for extracting multi-scale feature, a transformer-encoder for capturing long-range dependency, and a CNN decoder for predicting gaze map.

The MSKUS image $I_0 \in \mathbf{R}^{H \times W \times 3}$ is first input into CNN encoder that contains five convolution blocks. The output feature maps from the deeper last three convolution blocks are denoted as $\mathbf{F}_0, \mathbf{F}_1, \mathbf{F}_2$ and are respectively fed into transformer encoders to enhance the long-range and contextual information. During the transformer-encoder, we first flatten the feature maps produced by the CNN encoder into a 1D sequence. Considering that flatten operation leads to losing the spatial information, the absolute position encoding [14] is combined with the flatten feature map via element-wise addition to form the input of the transformer layer. The transformer layer contains the standard Multi-head Self-Attention (MSA) and Multi-layer Perceptron (MLP) blocks. Layer Normalization (LN) and residual connection are applied before and after each block respectively.

In the CNN decoder part, a pure CNN architecture progressively up-samples the feature maps into the original image resolution and implements pixel-wise prediction for modeling sonographers' gaze map. The CNN decoder part includes five convolution blocks. In each block, 3×3 convolution operation, Batch normalization (BN), RELU activation function, and 2-scale upsampling that adopts nearest-neighbor interpolation is performed. In addition, the transformer's output is fused with the feature map from the decoding process by an element-wise product operation to further enhance the long-range and multi-scale visual information. After five CNN blocks, a 3×3 convolution operation and Sigmoid activation is performed to output the predicted sonographers' gaze map. We use the eye gaze information of the sonographers which is collected by the Eye-Tracker to restrain the predicted sonographers' gaze map. The loss function is the sum of the Normalized Scanpath Saliency (NSS), the Linear Correlation Coefficient (CC), Kullback-Leibler divergence (KLD) and Similarity (SIM) [2].

2.2 What to Adjust

Common CNN classification models for gout diagnosis often fail to learn the gouty MSKUS features including the double contour sign, tophus, and snowstorm which are key factors for sonographers' decision. A CAM for a particular category indicates the discriminative regions used by the CNN to identify that category. Inspired by CAM technique, it is needed to decide whether the attention region given to an CNN model is reasonable for diagnosis of gout. We firstly use the Grad-CAM technique [12] to acquire the salient attention region S_{CAM} that CNN model perceives for differential diagnosis of gout. To ensure the scale of the attention region S_{CAM} is the same as the sonographers' gaze map S_{sono} which is modeled by saliency model, we normalize S_{CAM} to the values between 0 and 1, get $\widetilde{S}_{CAM}$. Then we make bit-wise intersection over union(IoU) operations with the S_{sono} and $\widetilde{S}_{CAM}$ to measure how well the two maps overlap. Note that we only calculate the part of $\widetilde{S}_{CAM}$ that is greater than 0.5. For instances whose IoU is less than 50%, we consider that the model's prediction for that instance is unreasonable. As shown in Fig. 3, when CNN do prediction, we can divide the instances into four categories:

RP: Reasonable Precise: The attention region focusses on the gouty features which are important for sonographers' decision, and the diagnosis is precise.
RIP: Reasonable Imprecise: Although attention region focusses on the gouty features, while the diagnosis result is imprecise.
UP: Unreasonable Precise: Although the gout diagnosis is precise, amount of attention is given to irrelevant feature of MSKUS image.
UIP: Unreasonable Imprecise: The attention region focusses on irrelevant features, and the diagnosis is imprecise.

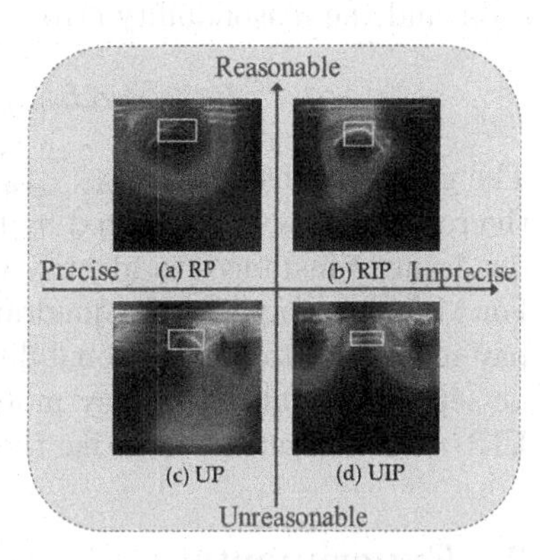

Fig. 3. Four categories: (a) RP (b) RIP (c) UP (d) UIP. Yellow boxes denote the gaze areas of the sonographers.

Our target of adjustment is to reduce imprecise and unreasonable predictions. In this way, CNNs not only finish correct gout diagnosis, but also acquire the attention region that agreements with the sonographers' gaze map.

2.3 How to Adjust

We proposed a training mechanism (Algorithm 1) which can strike the balance between the gout diagnosis error and the reasonability error of attention region to promote the CNNs to "think like sonographers". In addition to reducing the diagnosis error, we also want to minimize the difference between sonographers' gaze map S_{sono} and normalized salient attention region $\widetilde{S}_{CAM}$, which directly leads to our target:

Algorithm 1: Proposed training mechanism

Input:D_{train}, O_{target}, M_{attn} generated from saliency model,
W_{base} parameters of base model needed to be adjusted
Output:Optimized model parameters W^*
$O_{pred}, M_{cam} = model(D_{train} | W_{base})$;
split D_{train} into 4 categories based on (O_{pred}, O_{target}) and (M_{cam}, M_{attn});
set the α base on the ratio of 4 categories;
$W^* = W_{base}$;
for $epoch = \{1 \ldots, N\}$ **do**
  $W^* = train(W^*, D_{train}, M_{attn}, \alpha)$;
end

$$\boldsymbol{min}(\boldsymbol{L}_{diagnosis} + \boldsymbol{L}_{reasonability})$$

$$\boldsymbol{L}_{reasonability} = L_1(\widetilde{S}_{CAM}, S_{sono})$$

The total loss function can be expressed as the weighted sum the gout diagnosis error and the reasonability error, as follows:

$$\boldsymbol{L}_{total} = \alpha \boldsymbol{L}_{diagnosis} + (1 - \alpha)\boldsymbol{L}_{reasonability}$$

The gout diagnosis error $\boldsymbol{L}_{diagnosis}$ is calculated by the Cross-entropy loss, and the reasonability is calculated by the L1-loss. This training mechanism uses the quadrant of instances to identify whether samples' attention needs to adjusted. For MSKUS sample in the quadrant of UP, α can be set 0.2 to control the CNN pay more attention to reasonability. Correspondingly, for sample in RIP, α can be set 0.8 to make CNN pay more attention to precise. For sample in RP and UIP, α can be set 0.5 to strike the balance between accuracy and reasonability.

3 Experiments

MSKUS Dataset Collection. The MSKUS data were collected for patients suspected of metatarsal gout in Nanjing Drum Tower Hosptial. Informed written consent was obtained at the time of recruitment. Dataset totally contains 1127 US images from different patients including 509 gout images and 618 healthy images. The resolution of the MSKUS images were resized to 224×224. During experiments, we randomly divided 10% of the dataset into testing sets, then the remaining data was divided equally into two parts for the different phases of the training. We used 5-fold cross validation to divide the training sets and validation sets.

Gaze Data Collection. We collected the eye movement data with the Tobii 4C eye-tracker operating at 90 Hz. The MSKUS images were displayed on a 1920×1080 27-inch LCD screen. The eye tracker was attached beneath the screen with a magnetic mounting bracket. Sonographers were seated in front of the screen and free to adjust the chair's height and the display's inclination. Binary maps of the same size as the corresponding MSKUS images were generated using the gaze data, with the pixel corresponding to the point of gaze marked with a'1' and the other pixels marked with a'0'. A sonographer gaze map S was generated for each binary map by convolving it with a truncated Gaussian Kernel $G(\sigma_{x,y})$, where G has 299 pixels along x dimension, and 119 pixels along y dimension.

Evaluation Metrics. Five metrics were used to evaluate model performance: Accuracy (ACC), Area Under Curve (AUC), Correlation Coefficient (CC), Similarity (SIM) and Kullback-Leibler divergence (KLD) [2]. ACC and AUC were implemented to assess the gout classification performance of each model, while CC, SIM, and KLD were used to evaluate the similarity of the areas that the model and sonographers focus on during diagnoses.

Evaluation of "Thinking like Sonographers" Mechanism. To evaluate the effectiveness of our proposed mechanism of "Thinking like Sonographers" (TLS) that combines "where to adjust", "what to adjust" and "how to adjust", we compared the gout diagnosis results of several classic CNN classification [5, 6, 13] models without/with our TLS mechanism. The results, shown in Table 1, revealed that using our TLS mechanism led to a significant improvement in all metrics. Specifically, for ACC and AUC, the model with our TLS mechanism achieved better results than the model without it. Resnet34 with TLS acquired the highest improvement in ACC with a 4.41% increase, and Resnet18 with TLS had a 0.027 boost in AUC. Our TLS mechanism consistently performed well in improving the gout classification performance of the CNN models. More comparison results were shown in Appendix Fig. A1 and Fig. A2.

Table 1. The performances of models training wi/wo our mechanism in MSKUS.

Method	ACC↑ (%)	AUC↑	CC↑	SIM↑	KLD↓
Resnet18 wo TLS	86.46 ± 3.90	0.941 ± 0.023	0.161 ± 0.024	0.145 ± 0.011	3.856 ± 0.235
Resnet18 wi TLS	89.13 ± 3.32	$\mathbf{0.968 \pm 0.005}$	0.404 ± 0.004	0.281 ± 0.003	$\mathbf{1.787 \pm 0.017}$
Resnet34 wo TLS	83.15 ± 3.78	0.922 ± 0.024	0.190 ± 0.054	0.151 ± 0.020	3.500 ± 0.530
Resnet34 wi TLS	87.56 ± 1.89	0.947 ± 0.018	0.376 ± 0.006	0.252 ± 0.004	1.951 ± 0.043
Resnet50 wo TLS	88.82 ± 1.16	0.956 ± 0.008	0.189 ± 0.024	0.157 ± 0.013	4.019 ± 0.522
Resnet50 wi TLS	89.61 ± 3.05	0.967 ± 0.011	0.402 ± 0.028	0.298 ± 0.020	2.133 ± 0.232
Vgg16 wo TLS	89.13 ± 3.20	0.958 ± 0.021	0.221 ± 0.089	0.182 ± 0.044	3.461 ± 0.776
Vgg16 wi TLS	$\mathbf{91.50 \pm 2.88}$	0.966 ± 0.020	$\mathbf{0.416 \pm 0.020}$	$\mathbf{0.305 \pm 0.013}$	1.932 ± 0.084
DenseNet121 wo TLS	88.82 ± 2.08	0.956 ± 0.015	0.175 ± 0.030	0.152 ± 0.011	3.822 ± 0.599
DenseNet121 wi TLS	89.45 ± 1.62	0.965 ± 0.010	0.368 ± 0.011	0.239 ± 0.007	1.991 ± 0.062

The CC, SIM, and KLD metrics were utilized to assess the similarity between the CAMs of classification models and the collected gaze maps, providing an indication of whether the model was able to "think" like a sonographer. Table 1 showed that the models with our TLS mechanism achieved significantly better results in terms of CC and SIM (i.e., higher is better), as well as a decline of more than 1.50 in KLD (lower is better), when compared to the original models. This indicated that the models with TLS focused on the areas shown to be similar to the actual sonographers. Furthermore, Fig. 4 illustrated the qualitative results of CAMs of models with and without TLS mechanism. The original models without TLS paid more attention to noise, textures, and artifacts, resulting in unreasonable gout diagnosis. With TLS, however, models could focus on the crucial areas in lesions, allowing them to think like sonographers.

Stability Under Different Gaze Maps via t-Test. To evaluate the prediction's stability under the predicted gaze map from the generation model in "Where to adjust", we conducted three t-test studies. Specifically, we trained two classification models (M_C and M_P), using the actual collected gaze maps, and the predicted maps from the generation model, respectively. During the testing, we used the collected maps as input for M_C and M_P to get classification results R_{CC} and R_{PC}. Similarly, we used the predicted maps as input for M_C andM_P

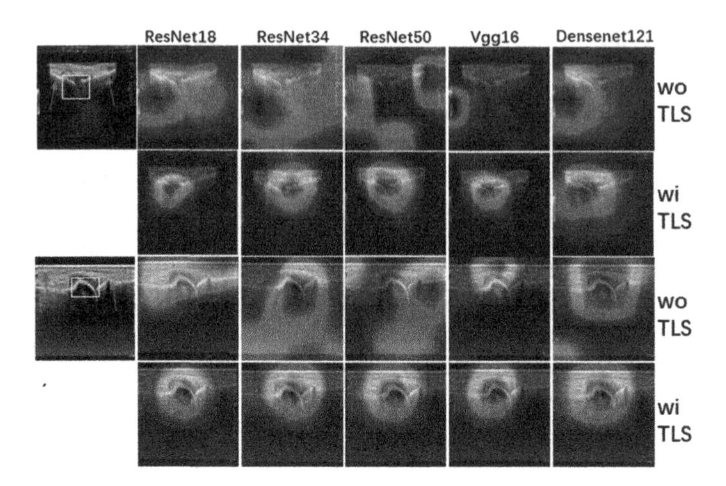

Fig. 4. Grad-CAM for ResNet18, ResNet34, ResNet50, vgg16 and Densenet121. Yellow boxes denote sonographers' gaze areas and red arrows denote the fascial tissues. (Color figure online)

to get the results R_{CP} and R_{PP}. Then, we conducted three T-tests: (1) between R_{CC} and R_{PC}; (2) between R_{CP} and R_{PP}; and (3) between R_{CC} and R_{PP}.

As shown in Table 2, the p-values of t-test (1)(2) and (3) are all greater than 0.005, suggesting that no significant difference was observed between the classification results obtained from different generative strategies. This implied that our training mechanism was model-insensitive. Consequently, it was possible to use predicted gaze maps for both the training and testing phases of the classification models without any notable performance decrease. This removed the need to collect eye movement maps during the training and testing phases, significantly lightening the workload of data collection. Therefore, our TLS mechanism, which involved predicting the gaze maps, could potentially be used in clinical environments. This would allow us to bypass the need to collect the real gaze maps of the doctors while classifying newly acquired US images, and thus improved the clinical implications of our mechanism, "Thinking like Sonographers".

Table 2. Statistical test results

Method/p-value	ResNet18	ResNet34	ResNet50	Vgg16	DenseNet121
t-test(1)	0.4219	0.8719	0.8701	0.6281	0.4428
t-test(2)	0.4223	0.8700	0.8725	0.6272	0.4434
t-test(3)	0.4192	0.8714	0.8727	0.6191	0.4446

4 Conclusion

In this study, we propose a framework to adjust CNNs to "think like sonographers", and diagnose gout from MSKUS images. The mechanism of "thinking like sonographers" contains three levels: where to adjust, what to adjust, and how to adjust. The proposed design not only steers CNN models as we intended, but also helps the CNN classifier focus on the crucial gout features. Extensive experiments show that our framework, combined with the mechanism of "thinking like sonographers" improves performance over the baseline deep classification architectures. Additionally, we can bypass the need to collect the real gaze maps of the doctors during the classification of newly acquired MSKUS images, thus our method has good clinical application values.

Acknowledgment. This work was supported in part by National Nature Science Foundation of China grants (62271246, U20A20389, 82027807, U22A2051), Key Research and Development Plan of Jiangsu Province (No. BE2022842), National Key Research and Development Program of China (2022YFC2405200).

References

1. Baumgartner, C.F., et al.: Sononet: real-time detection and localisation of fetal standard scan planes in freehand ultrasound. IEEE Trans. Med. Imaging **36**(11), 2204–2215 (2017)
2. Bylinskii, Z., Judd, T., Oliva, A., Torralba, A., Durand, F.: What do different evaluation metrics tell us about saliency models? IEEE Trans. Pattern Anal. Mach. Intell. **41**(3), 740–757 (2018)
3. Cai, Y., Sharma, H., Chatelain, P., Noble, J.A.: Multi-task SonoEyeNet: detection of fetal standardized planes assisted by generated sonographer attention maps. In: Frangi, A.F., Schnabel, J.A., Davatzikos, C., Alberola-López, C., Fichtinger, G. (eds.) MICCAI 2018. LNCS, vol. 11070, pp. 871–879. Springer, Cham (2018). https://doi.org/10.1007/978-3-030-00928-1_98
4. Cai, Y., Sharma, H., Chatelain, P., Noble, J.A.: Sonoeyenet: standardized fetal ultrasound plane detection informed by eye tracking. In: 2018 IEEE 15th International Symposium on Biomedical Imaging (ISBI 2018), pp. 1475–1478. IEEE (2018)
5. He, K., Zhang, X., Ren, S., Sun, J.: Deep residual learning for image recognition. In: Proceedings of the IEEE Conference on Computer Vision and Pattern Recognition, pp. 770–778 (2016)
6. Huang, G., Liu, Z., Van Der Maaten, L., Weinberger, K.Q.: Densely connected convolutional networks. In: Proceedings of the IEEE Conference on Computer Vision and Pattern Recognition, pp. 4700–4708 (2017)
7. Liu, S., et al.: Deep learning in medical ultrasound analysis: a review. Engineering **5**(2), 261–275 (2019)
8. Lou, J., Lin, H., Marshall, D., Saupe, D., Liu, H.: Transalnet: towards perceptually relevant visual saliency prediction. Neurocomputing **494**, 455–467 (2022)
9. Mall, S., Brennan, P.C., Mello-Thoms, C.: Modeling visual search behavior of breast radiologists using a deep convolution neural network. J. Med. Imaging **5**(3), 035502–035502 (2018)

10. Mall, S., Krupinski, E., Mello-Thoms, C.: Missed cancer and visual search of mammograms: what feature-based machine-learning can tell us that deep-convolution learning cannot. In: Medical Imaging 2019: Image Perception, Observer Performance, and Technology Assessment, vol. 10952, pp. 281–287. SPIE (2019)
11. Patra, A., et al.: Efficient ultrasound image analysis models with sonographer gaze assisted distillation. In: Shen, D., et al. (eds.) MICCAI 2019. LNCS, vol. 11767, pp. 394–402. Springer, Cham (2019). https://doi.org/10.1007/978-3-030-32251-9_43
12. Selvaraju, R.R., Cogswell, M., Das, A., Vedantam, R., Parikh, D., Batra, D.: Grad-cam: visual explanations from deep networks via gradient-based localization. In: Proceedings of the IEEE International Conference on Computer Vision, pp. 618–626 (2017)
13. Simonyan, K., Zisserman, A.: Very deep convolutional networks for large-scale image recognition. arXiv preprint arXiv:1409.1556 (2014)
14. Vaswani, A., et al.: Attention is all you need. In: Advances in Neural Information Processing Systems, vol. 30 (2017)
15. Wang, S., Ouyang, X., Liu, T., Wang, Q., Shen, D.: Follow my eye: using gaze to supervise computer-aided diagnosis. IEEE Trans. Med. Imaging **41**(7), 1688–1698 (2022)

Thyroid Nodule Diagnosis in Dynamic Contrast-Enhanced Ultrasound via Microvessel Infiltration Awareness

Haojie Han[1], Hongen Liao[2], Daoqiang Zhang[1], Wentao Kong[3], and Fang Chen[1](✉)

[1] Key Laboratory of Brain-Machine Intelligence Technology, Ministry of Education, Nanjing University of Aeronautics and Astronautics, Nanjing 211106, China
chenfang@nuaa.edu.cn

[2] Department of Biomedical Engineering, School of Medicine, Tsinghua University, Beijing 10084, China

[3] Department of Ultrasound, Affiliated Drum Tower Hospital, Nanjing University Medical School, Nanjing 21008, China

Abstract. Dynamic contrast-enhanced ultrasound (CEUS) video with microbubble contrast agents reflects the microvessel distribution and dynamic microvessel perfusion, and may provide more discriminative information than conventional gray ultrasound (US). Thus, CEUS video has vital clinical value in differentiating between malignant and benign thyroid nodules. In particular, the CEUS video can show numerous neovascularisations around the nodule, which constantly infiltrate the surrounding tissues. Although the infiltrative of microvessel is ambiguous on CEUS video, it causes the tumor size and margin to be larger on CEUS video than on conventional gray US and may promote the diagnosis of thyroid nodules. In this paper, we propose a novel framework to diagnose thyroid nodules based on dynamic CEUS video by considering microvessel infiltration and via segmented confidence mapping assists diagnosis. Specifically, the Temporal Projection Attention (TPA) is proposed to complement and interact with the semantic information of microvessel perfusion from the time dimension of dynamic CEUS. In addition, we employ a group of confidence maps with a series of flexible Sigmoid Alpha Functions (SAF) to aware and describe the infiltrative area of microvessel for enhancing diagnosis. The experimental results on clinical CEUS video data indicate that our approach can attain an diagnostic accuracy of 88.79% for thyroid nodule and perform better than conventional methods. In addition, we also achieve an optimal dice of 85.54% compared to other classical segmentation methods. Therefore, consideration of dynamic microvessel perfusion and infiltrative expansion is helpful for CEUS-based diagnosis and segmentation of thyroid nodules. The datasets and codes will be available.

Supplementary Information The online version contains supplementary material available at https://doi.org/10.1007/978-3-031-43987-2_17.

H. Greenspan et al. (Eds.): MICCAI 2023, LNCS 14225, pp. 169–179, 2023.
https://doi.org/10.1007/978-3-031-43987-2_17

Keywords: Contrast-enhanced ultrasound · Thyroid nodule · Dynamic perfusion · Infiltration Awareness

1 Introduction

Contrast-enhanced ultrasound (CEUS) as a modality of functional imaging has the ability to assess the intensity of vascular perfusion and haemodynamics in the thyroid nodule, thus considered a valuable new approach in the determination of benign vs. malignant nodules [1]. In practice, CEUS video allows the dynamic observation of microvascular perfusion through intravenous injection of contrast agents. According to clinical experience, for thyroid nodules diagnosis, there are two characteristic that are important when analyzing CEUS video. 1) Dynamic microvessel perfusion. As shown in Fig. 1(A), clinically acquired CEUS records the dynamic relative intensity changes (microvessel perfusion pattern) throughout the whole examination [2]. 2) Infiltrative expansion of microvessel. Many microvessels around nodules are constantly infiltrating and growing into the surrounding tissue. As shown in Fig. 1(B), based on the difference in lesion size displayed by the two modalities, clinical practice shows that gray US underestimates the size of lesions, and CEUS video overestimates the size of some lesions [3]. Although the radiologist's cognition of microvascular invasive expansion is fuzzy, they think it may promote diagnosing thyroid nodules [1].

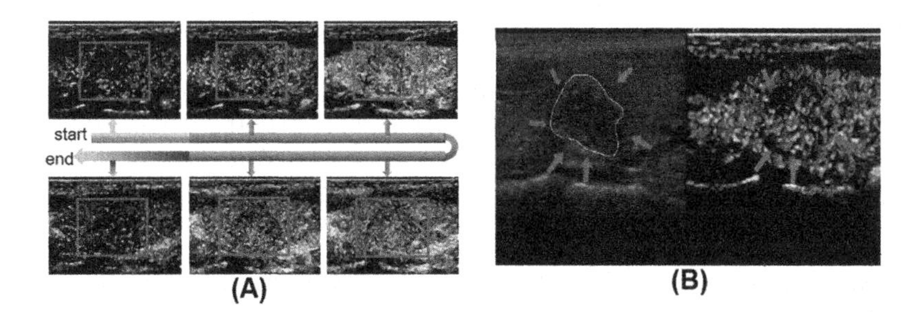

Fig. 1. (A) The dynamic cropped frames in CEUS video. Radiologists identify the area of lesions by comparing various frames, but each individual frame does not effectively describe the lesions area with precision. From the start to the end of the timeline, the three colours represent the change in intensity of the CEUS video over three different time periods. The red border represents the area of the lesion. (B) The size of thyroid nodules described on CEUS is significantly larger than detected through gray US. The yellow line indicates the lesion area labeled by radiologists on gray US, while the red arrow corresponds to the infiltration area or continuous lesion expansion on CEUS. (Color figure online)

Currently, CEUS-based lesion diagnosis methods mainly use the convolution neural network (CNN) to extract spatial-temporal features from dynamic CEUS. Wan et al. [4] proposed a hierarchical temporal attention network which

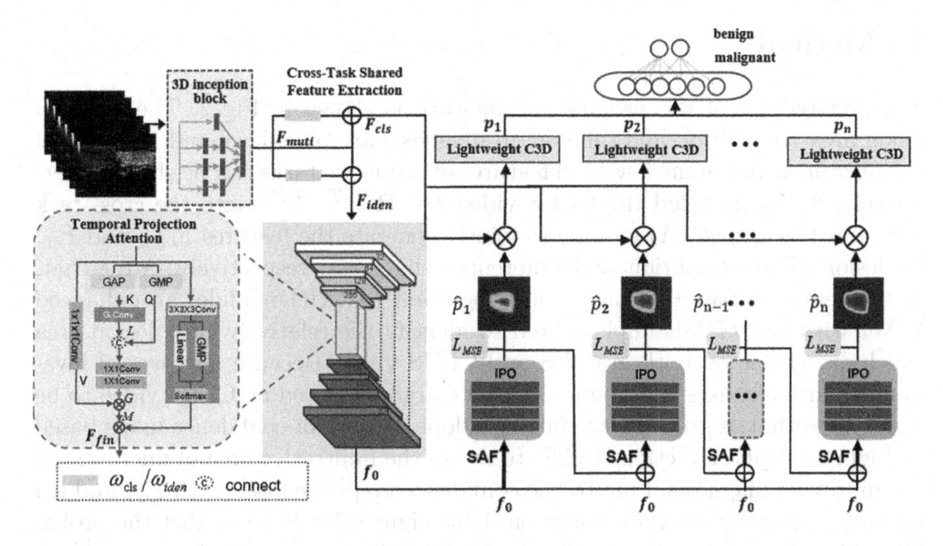

Fig. 2. The overall architecture which contains four parts: cross-task shared feature extraction, temporal-based lesions area recognition, microvessel infiltration awareness and thyroid nodules diagnosis. Here,$\hat{P} = \{\hat{p}_1, \hat{p}_2 \ldots \hat{p}_n\}$ represents the infiltration process of microvessels from gray US to CEUS, and $\hat{p}_n$ is the final segmentation result.

used the spatial feature enhancement for disease diagnosis based on dynamic CEUS. Furthermore, by combing the US modality, Chen et al. [5] proposed a domain-knowledge-guided temporal attention module for breast cancer diagnosis. However, due to artifacts in CEUS, SOTA classification methods often fail to learn regions where thyroid nodules are prominent (As in Appendix Fig. A1) [6]. Even the SOTA segmentation methods cannot accurately identify the lesion area for blurred lesion boundaries, thus, the existing automatic diagnosis network using CEUS still requires manual labeling of pixel-level labels which will lose key information around the tissues [7]. In particular, few studies have developed the CEUS video based diagnostic model inspired by the dynamic microvessel perfusion, or these existing methods generally ignore the influence of microvessel infiltrative expansion. Whether the awareness of infiltrative area information can be helpful in the improvement of diagnostic accuracy is still unexplored.

Here, we propose an explanatory framework for the diagnosis of thyroid nodules based on dynamic CEUS video, which considers the dynamic perfusion characteristics and the amplification of the lesion region caused by microvessel infiltration. Our contributions are twofolds. First, the Temporal Projection Attention (TPA) is proposed to complement and interact with the semantic information of microvessel perfusion from the time dimension. Second, we adopt a group of confidence maps instead of binary masks to perceive the infiltrative expansion area from gray US to CEUS of microvessels for improving diagnosis.

2 Method

The architecture of the proposed framework is shown in Fig. 2. The tasks of lesion area recognition and differential diagnosis are pixel-level and image-level classifications, and some low-level features of these two tasks can be shared interactively [8]. We first fed the CEUS video $I \in \mathbf{R}^{C\times T\times H\times W}$ into the cross-task feature extraction (CFA) module to jointly generate the features F_{iden} and F_{cls} for lesion area recognition and differential diagnosis, respectively. After that, in the temporal-based lesions area recognition (TLAR) module, an enhanced V-Net with the TPA is implemented to identify the relatively clear lesion area which are visible on both gray US and CEUS video. Because microvessel invasion expansion causes the tumor size and margin depicted by CEUS video to be larger than that of gray US, we further adopt a group of confidence maps based on Sigmoid Alpha Functions (SAF) to aware the infiltrative area of microvessels for improving diagnosis. Finally, the confidence maps are fused with F_{cls} and fed into a diagnosis subnetwork based on lightweight C3D [9] to predict the probability of benign and malignant. In the CFA, we first use the 3D inception block to extract multi-scale features F_{muti}. The 3D inception block has 4 branches with cascaded 3D convolutions. Multiple receptive fields are obtained through different branches, and then group normalization and ReLU activation are performed to obtain multi-scale features F_{muti}. Then, we use the cross-task feature adaptive unit to generate the features F_{iden} and F_{cls} required for lesions area recognition and thyroid nodules diagnosis via the following formula [10]:

$$[F_{iden}, F_{cls}] = [\omega_{iden}, \omega_{cls}] * F_{muti} + [F_{muti}, F_{muti}] \tag{1}$$

where $\omega_{iden}, \omega_{cls}$ are the learnable weights.

2.1 Temporal-Based Lesions Area Recognition (TLAR)

The great challenge of automatic recognition of lesion area from CEUS video is that the semantic information of the lesion area is different in the CEUS video of the different microvessel perfusion periods. Especially in the perfusion period and the regression period, the semantic information of lesions cannot be fully depicted in an isolated CEUS frame. Thus, the interactive fusion of semantic information of the whole microvessel perfusion period will promote the identification of the lesion area, and we design the Temporal Projection Attention (TPA) to realize this idea. We use V-Net as the backbone, which consists of four encoder/decoder blocks for TLAR, and the TPA is used in the bottleneck of the V-Net.

Temporal Projection Attention (TPA). Given a feature $F_{4th} \in \mathbf{R}^{C\times T\times \frac{H}{16}\times \frac{W}{16}}$ after four down-sampling operations in encoder, its original 3D feature map is projected [11] to 2D plane to get keys and queries: $K, Q \in \mathbf{R}^{C\times \frac{H}{16}\times \frac{W}{16}}$, and we use global average pooling (GAP) and global maximum pooling (GMP) as temporal projection operations. Here, $V \in \mathbf{R}^{C\times T\times \frac{H}{16}\times \frac{W}{16}}$ is obtained by a single convolution. This operation can also filter out the irrelevant background and display the key information of the lesions. After the temporal projection, a group convolution with a group size of 4 is employed on K to extract the local temporal attention $L \in \mathbf{R}^{C\times\frac{H}{16}\times\frac{W}{16}}$. Then, we concatenate L with Q to further obtain the global attention $G \in \mathbf{R}^{C\times 1\times\frac{H}{16}\times\frac{W}{16}}$ by two consecutive 1×1 2D convolutions and dimension expend. Those operations are described as follows:

$$K = Q = GAP(F_{4th}) + GMP(F_{4th}) \tag{2}$$

$$G = Expend(Conv(\sigma(Conv(\sigma(Gonv(K)) \oplus Q)))) \tag{3}$$

where $Gonv(\cdot)$ is the group convolution, σ denotes the normalization, "$\oplus$" is the concatenation operation. The global attention G encodes not only the contextual information within isolated query-key pairs but also the attention inside the keys [12]. After that, based on the 2D global attention G, we multiply V and G to calculate the global temporal fusion attention map $M \in \mathbf{R}^{C\times T\times\frac{H}{16}\times\frac{W}{16}}$ to enhance the feature representation.

Meanwhile, to make better use of the channel information, we use $3 \times 3 \times 3$ $Conv$ to get enhanced channel feature $F''_{4th} \in \mathbf{R}^{C_1\times T\times\frac{H}{16}\times\frac{W}{16}}$. Then, we use parallel average pooling and full connection operation to reweight the channel information of F''_{4th} to obtain the reweighted feature $F'_{4th} \in \mathbf{R}^{C\times T\times\frac{H}{16}\times\frac{W}{16}}$. The obtained global temporal fusion attention maps M are fused with the reweighted feature F'_{4th} to get the output features F_{fin}. Finally, we input F_{fin} into the decoder of the TLAR to acquire the feature map of lesion.

2.2 Microvessel Infiltration Awareness (MIA)

We design a MIA module to learn the infiltrative areas of microvessel. The tumors and margin depicted by CEUS may be larger than those depicted by gray US because of continuous infiltrative expansion. Inspired by the continuous infiltrative expansion, a series of flexible Sigmoid Alpha Functions (SAF) simulate the infiltrative expansion of microvessels by establishing the distance maps from the pixel to lesion boundary. Here, the distance maps [13] are denoted as the initial probability distribution P_D. Then, we utilize Iterative Probabilistic Optimization (IPO) unit to produce a set of optimized probability maps $\hat{P} = \{\hat{p}_1, \hat{p}_2 \ldots \hat{p}_n\}$ to aware the microvessel infiltration for thyroid nodules diagnosis. Based on SAF and IPO, CEUS-based diagnosis of thyroid nodules can make full use of the ambiguous information caused by microvessel infiltration.

Sigmoid Alpha Function (SAF). It is generally believed that the differentiation between benign and malignant thyroid nodules is related to the pixels around the boundaries of the lesion [14], especially in the infiltrative areas of microvessel [3]. Therefore, we firstly build the initial probability distribution P_D based on the distance between the pixels and the annotation boundaries by using SAF in order to aware the infiltrative areas. Here, SAF is defined as follows:

$$SAF_{(i,j)} = C * \left(\frac{2}{1 + e^{\frac{-\alpha D(i,j)}{max(D(i,j))}}} - 1 \right) \tag{4}$$

$$C = \left(1 + e^{-\alpha}\right) / \left(1 - e^{-\alpha}\right); \quad \alpha \in (0, +\infty) \tag{5}$$

where α is the conversion factor for generating initial probability distribution P_D (when $\alpha \to \infty$, the generated P_D is binary mask); C is used to control the function value within the range of $[0, 1]$; (i, j) is the coordinate point in feature map; $D(i, j)$ indicates the shortest distance from (i, j) to lesion's boundaries.

Iterative Probabilistic Optimization (IPO) Unit. Based on the fact that IncepText [15] has experimentally demonstrated that asymmetric convolution can effectively solve the problem of highly variable size and aspect ratio, we use asymmetric convolution in the IPO unit. Asymmetric convolution-based IPO unit can optimize the initial distribution P_D to generate optimized probability maps $\hat{P}$ that can reflect the confidence of benign and malignant diagnosis. Specifically, with the IPO, our network can make full use of the prediction information in low-level iteration layer, which may improve the prediction accuracy of high-level iteration layer. In addition, the parameters in the high-level iteration layer can be optimized through the back-propagation gradient from the high-level iteration layer. IPO unit can be shown as the following formula:

$$\hat{p}_1 = ConvBlock(SAF(f_0, \alpha_1)) \tag{6}$$

$$\hat{p}_i = ConvBlock(SAF((f_0 \oplus \hat{p}_{i-1}), \alpha_i)) \quad i \in (1, n] \tag{7}$$

where "$\oplus$" represents the concatenation operation; *ConvBlock* consists of a group of asymmetric convolutions (e.g., Conv1 $\times$ 5, Conv5 $\times$ 1 and Conv1 $\times$ 1); n denotes the number of the layers of IPO unit. With the lesion's feature map f_0 from the TLAR module, the initial distribution P_D obtained by SAF is fed into the first optimize layer of IPO unit to produce the first optimized probability map $\hat{p}_1$. Then, $\hat{p}_1$ is contacted with f_0, and used to generate optimized probability map $\hat{p}_2$ through the continuous operation based on SAF and the second optimize layer of IPO unit. The optimized probability map $\hat{p}_{i-1}$ provides prior information for producing the next probability map $\hat{p}_i$. In this way, we can get a group of probability map $\hat{P}$ to aware the microvascular infiltration.

2.3 Loss Function

With continuous probability map $\hat{P}$ obtained from MIA, $\hat{P}$ are multiplied with the feature F_{cls}. Then, these maps are fed into a lightweight C3D to predict the probability of benign and malignant, as shown in Fig. 2. We use the mean square error L_{MSE} to constrain the generation of $\hat{P}$. Assuming that the generated $\hat{P}$ is ready to supervise the classification network, we want to ensure that the probability maps can accurately reflect the classification confidence. Thus, we design a task focus loss L_{ta} to generate confidence maps P, as follows:

$$L_{MSE} = \sum_{i=1}^{n} \frac{1}{\Omega} \sum_{p \in \Omega} \|\boldsymbol{g}_i(pi), \hat{\boldsymbol{p}}_i(pi)\|_2 \tag{8}$$

$$L_{ta} = \frac{1}{2\sigma^2} \sum_{i=1}^{n} \|\boldsymbol{p}_i - \widehat{\boldsymbol{p}}_i\|_2^2 + \log \sigma \tag{9}$$

where $\boldsymbol{g}_i$ is the label of $\hat{\boldsymbol{p}}_i$, which is generated by the operation of $SAF(D_{(i,j)}, \alpha_i)$; pi denotes pixel in the image domain Ω, σ is a learnable parameter to eliminate the hidden uncertainty information.

For differentiating malignant and benign, we employ a hybrid loss L_{total} that consists of the cross-entropy loss L_{cls}, the loss of L_{MSE} computing optimized probability maps $\hat{P}$, and task focus loss L_{ta}. The L_{total} is denoted as follows:

$$L_{total} = \lambda_1 \cdot L_{cls} + \lambda_2 \cdot L_{MSE} + \lambda_3 \cdot L_{ta} \tag{10}$$

where $\lambda_1, \lambda_2, \lambda_3$ are the hyper-parameters to balance the corresponding loss. As the weight parameter, we set $\lambda_1, \lambda_2, \lambda_3$ are 0.5,0.2,0.3 in the experiments.

3 Experiments

Dataset. Our dataset contained 282 consecutive patients who underwent thyroid nodule examination at Nanjing Drum Tower Hospital. All patients performed dynamic CEUS examination by an experienced sonographer using an iU22 scanner (Philips Healthcare, Bothell, WA) equipped with a linear transducer L9-3 probe. These 282 cases included 147 malignant nodules and 135 benign nodules. On the one hand, the percutaneous biopsy based pathological examination was implemented to determine the ground-truth of malignant and benign. On the other hand, a sonographer with more than 10 years of experience manually annotated the nodule lesion mask to obtain the pixel-level ground-truth of thyroid nodules segmentation. All data were approved by the Institutional Review Board of Nanjing Drum Tower Hospital, and all patients signed the informed consent before enrollment into the study.

Implementation Details. Our network was implemented using Pytorch framework with the single 12 GB GPU of NVIDIA RTX 3060. During training, we first pre-trained the TALR backbone via dice loss for 30 epochs and used Adam optimizer with learning rate of 0.0001. Then, we loaded the pre-trained weights to train the whole model for 100 epochs and used Adam optimizer with learning rate of 0.0001. Here, we set batch-size to 4 during the entire training process The CEUS consisted the full wash-in and wash-out phases, and the resolution of each frame was (600×800). In addition, we carried out data augmentation, including random rotation and cropping, and we resize the resolution of input

Table 1. Quantitative lesion recognition results are compared with SOTA methods and ablation experiments.

Network	UNet3D [18]	V-Net [16]	TransUNet [17]	V-Net+ TPA	V-Net+ TPA+SAF	V-Net+ TPA+IPO	Ours
DICE(%)↑	73.63 ± 5.54	77.94 ± 4.77	72.84 ± 6.88	81.22 ± 4.18	82.96 ± 4.11	83.32 ± 4.03	$\mathbf{85.54 \pm 4.93}$
Recall(%)↑	79.96 ± 4.39	83.17 ± 5.05	77.69 ± 5.18	83.96 ± 3.66	88.71 ± 3.92	89.45 ± 3.77	$\mathbf{90.40 \pm 5.93}$
IOU(%)↑	60.56 ± 5.83	65.20 ± 5.09	59.83 ± 6.19	69.96 ± 3.58	72.63 ± 3.15	73.26 ± 4.03	$\mathbf{74.99 \pm 4.72}$

frames to (224 × 224). We adopted 5-fold cross-validation to achieve quantitative evaluation. Three indexes including Dice, Recall, and IOU, were used to evaluate the lesion recognition task, while five indexes, namely average accuracy (ACC), sensitivity (Se), specificity (Sp), F1-score (F1), and AUC, were used to evaluate the diagnosis task.

Experimental Results. As in Table 1, we compared our method with SOTA method including V-Net, Unet3D, TransUnet. For the task of identifying lesions, the index of Recall is important, because information in irrelevant regions can be discarded, but it will be disastrous to lose any lesion information. V-Net achieved the highest Recall scores compared to others; thus, it was chosen as the backbone of TLAR. Table 1 revealed that the modules (TPA, SAF, and IPO) used in the network greatly improved the segmentation performance compared to baseline, increasing Dice and Recall scores by 7.60% and 7.23%, respectively. For the lesion area recognition task, our method achieved the highest Dice of 85.54% and Recall of 90.40%, and the visualized results were shown in Fig. 3.

Table 2. Quantitative diagnostic results are compared with SOTA methods and ablation experiments.

Network	ACC (%) ↑	Se (%) ↑	Sp (%) ↑	F1 (%) ↑	AUC (%) ↑
C3D+Mask [19]	76.61 ± 4.13	88.05 ± 3.73	76.92 ± 3.17	77.52 ± 3.32	88.05 ± 3.01
R3D+Mask [20]	77.01 ± 3.25	87.17 ± 4.08	79.49 ± 3.88	87.05 ± 3.93	83.10 ± 2.35
R2plus1D+Mask [21]	78.88 ± 2.81	85.02 ± 2.49	77.45 ± 3.43	87.75 ± 2.79	81.10 ± 3.82
ConvLSTM+Mask [22]	78.59 ± 4.44	84.37 ± 3.66	76.26 ± 4.26	80.22 ± 3.92	85.95 ± 3.73
Baseline+Mask	79.29 ± 2.58	89.83 ± 1.80	80.84 ± 3.04	86.01 ± 2.00	88.25 ± 2.84
Baseline+TLAR	81.10 ± 2.24	84.97 ± 1.28	81.58 ± 2.74	82.49 ± 1.81	88.67 ± 1.96
Baseline+TLAR+SAF	84.15 ± 1.78	89.90 ± 0.94	79.08 ± 1.85	83.95 ± 1.79	89.90 ± 1.97
Baseline+TLAR+IPO	86.56 ± 2.45	92.58 ± 2.38	79.93 ± 2.53	86.41 ± 1.36	93.33 ± 2.74
Ours	$\mathbf{88.79 \pm 1.40}$	$\mathbf{94.26 \pm 1.68}$	$\mathbf{88.37 \pm 1.80}$	$\mathbf{90.41 \pm 1.85}$	$\mathbf{94.54 \pm 1.54}$

To evaluate the effectiveness of the baseline of lightweight C3D, we compared the results with SOTA video classification methods including C3D, R3D, R2plus1D and ConvLSTM. For fair comparison, all methods used the manually annotated lesion mask to assist the diagnosis. Experimental results in Table 2 revealed that our baseline network could be useful for the diagnosis. With the effective baseline, the introduced modules including TLAR, SAF and IPO further improved the diagnosis accuracy, increasing the accuracy by 9.5%. The awareness of microvascular infiltration using SAF and IPO unit was helpful for CEUS-based diagnosis, as it could improve the diagnosis accuracy by 7.69% (As in Table 2). As in Appendix Fig. A1, although SOTA method fails to focus on lesion areas, our method can pinpoint discriminating lesion areas.

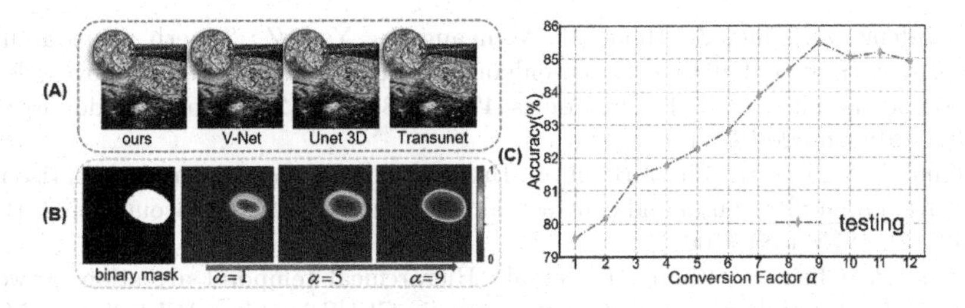

Fig. 3. (A)Comparison of the visualised results with the SOTA method, green and red contours is the automatically recognized area and ground-truth. By enlarging the local details, we show that our model can obtain the optimal result (More visuals provided in Appendix Fig. A2 and Fig. A3.). (B) Microvascular infiltration was simulated from gray US to CEUS via a set of confidence maps. (C) Influence of α values. (Color figure online)

Influence of α Values. The value of α in *SAF* is associated with simulating microvessel infiltration. Figure 3 (C) showed that the diagnosis accuracy increased along with the increment of α and then tended to become stable when α was close to 9. Therefore, for balancing the efficiency and performance, the number of IPO was set as n = 3 and α was set as $\alpha = \{1, 5, 9\}$ to generate a group of confidence maps that can simulate the process of microvessel infiltration. (More details about the setting of n is in Appendix Fig. A4 of the supplementary material.)

4 Conclusion

The microvessel infiltration leads to the observation that the lesions detected on CEUS tend to be larger than those on gray US. Considering the microvessel infiltration, we propose an method for thyroid nodule diagnosis based on CEUS videos. Our model utilizes a set of confidence maps to recreate the lesion expansion process; it effectively captures the ambiguous information caused by microvessel infiltration, thereby improving the accuracy of diagnosis. This method is an attempt to eliminate the inaccuracy of diagnostic task due to the fact that gray US underestimates lesion size and CEUS generally overestimates lesion size. To the best of our knowledge, this is the first attempt to develop an automated diagnostic tool for thyroid nodules that takes into account the effects of microvessel infiltration. The way in which we fully exploit the information in time dimension through TPA also makes the model more clinically explanatory.

References

1. Radzina, M., Ratniece, M., Putrins, D.S., et al.: Performance of contrast-enhanced ultrasound in thyroid nodules: review of current state and future perspectives. Cancers **13**(21), 5469 (2021)

2. Yongfeng, Z., Ping, Z., Hong, P., Wengang, L., Yan, Z.: Superb microvascular imaging compared with contrast-enhanced ultrasound to assess microvessels in thyroid nodules. J. Med. Ultrasonics **47**(2), 287–297 (2020). https://doi.org/10.1007/s10396-020-01011-z
3. Jiang, Y.X., Liu, H., Liu, J.B., et al.: Breast tumor size assessment: comparison of conventional ultrasound and contrast-enhanced ultrasound. Ultrasound Med. Biol. **33**(12), 1873–1881 (2007)
4. Wan, P., Chen, F., Zhang, D., et al.: Hierarchical temporal attention network for thyroid nodule recognition using dynamic CEUS imaging. IEEE Trans. Med. Imaging **40**(6), 1646–1660 (2021)
5. Chen, C., Wang, Y., et al.: Domain knowledge powered deep learning for breast cancer diagnosis based on contrast-enhanced ultrasound videos. IEEE Trans. Med. Imaging **40**(9), 2439–2451 (2021)
6. Manh, V. T., Zhou, J., Jia, X., Lin, Z., et al.: Multi-attribute attention network for interpretable diagnosis of thyroid nodules in ultrasound images. IEEE Trans. Ultrason. Ferroelect. Frequency Control 69(9), 2611–2620 (2022)
7. Moon, W.K., Lee, Y.W., et al.: Computer-aided prediction of axillary lymph node status in breast cancer using tumor surrounding tissue features in ultrasound images. Comput. Meth. Programs Biomed. **146**, 143–150 (2017)
8. Golts, A., Livneh, I., Zohar, Y., et al.: Simultaneous detection and classification of partially and weakly supervised cells. In: Karlinsky, L., et al. (eds.) ECCV 2022. LNCS, vol. 13803, pp. 313–329. Springer, Cham (2023). https://doi.org/10.1007/978-3-031-25066-8_16
9. Wang, Y., Li, Z., Cui, X., Zhang, L., et al.: Key-frame guided network for thyroid nodule recognition using ultrasound videos. In: Wang, L., et al. (eds.) MICCAI 2022. LNCS, vol. 13434, pp. 238–247. Springer, Cham (2022). https://doi.org/10.1007/978-3-031-16440-8_23
10. Wang, X., Jiang, L., Li, L., et al.: Joint learning of 3D lesion segmentation and classification for explainable COVID-19 diagnosis. IEEE Trans. Med. Imaging **40**(9), 2463–2476 (2021)
11. Wang, H., Zhu, Y., Green, B., Adam, H., Yuille, A., Chen, L.-C.: Axial-DeepLab: stand-alone axial-attention for panoptic segmentation. In: Vedaldi, A., Bischof, H., Brox, T., Frahm, J.-M. (eds.) ECCV 2020. LNCS, vol. 12349, pp. 108–126. Springer, Cham (2020). https://doi.org/10.1007/978-3-030-58548-8_7
12. Jiang, Y., Zhang, Z., Qin, S., et al.: APAUNet: axis projection attention UNet for small target in 3D medical segmentation. In: Proceedings of the Asian Conference on Computer Vision, pp. 283–298 (2022)
13. Zhang, S., Zhu, X., Chen, L., Hou, J., et al.: Arbitrary shape text detection via segmentation with probability maps. IEEE Trans. Pattern Anal. Mach. Intell. **14** (2022). https://doi.org/10.1109/TPAMI.2022.3176122
14. Gómez-Flores, W., et al.: Assessment of the invariance and discriminant power of morphological features under geometric transformations for breast tumor classification. Comput. Methods Programs Biomed. **185**, 105173 (2020)
15. Yang, Q., et al.: Inceptext: a new inception-text module with deformable psroipooling for multi-oriented scene text detection. In: IJCAI, pp. 1071–1077(2018)
16. Milletari, F., Navab, N., Ahmadi, S.A.: V-Net: fully convolutional neural networks for volumetric medical image segmentation. In: IEEE Fourth International Conference on 3D Vision (3DV), pp. 565–571 (2016)
17. Chen, J., et al.: TransuNet: transformers make strong encoders for medical image segmentation. arXiv preprint arXiv:2102.04306 (2021)

18. Ronneberger, O., Fischer, P., Brox, T.: U-Net: convolutional networks for biomedical image segmentation. In: Navab, N., Hornegger, J., Wells, W.M., Frangi, A.F. (eds.) MICCAI 2015. LNCS, vol. 9351, pp. 234–241. Springer, Cham (2015). https://doi.org/10.1007/978-3-319-24574-4_28
19. Tran, D., et al.: Learning spatiotemporal features with 3D convolutional networks. In: Proceedings of the IEEE International Conference on Computer Vision, pp. 4489–4497 (2015)
20. Tran, D., Wang, H., et al.: A closer look at spatiotemporal convolutions for action recognition. In: Proceedings of the IEEE Conference on Computer Vision and Pattern Recognition, pp. 6450–6459 (2018)
21. Tran, D., Wang, H., et al.: A closer look at spatiotemporal convolutions for action recognition . In: Proceedings of the IEEE conference on Computer Vision and Pattern Recognition, pp. 6450–6459 (2018)
22. Mutegeki, R., Han, D. S., et al.: A CNN-LSTM approach to human activity recognition. In: 2020 International Conference on Artificial Intelligence in Information and Communication (ICAIIC), pp. 362–366 (2022)

Polar Eyeball Shape Net for 3D Posterior Ocular Shape Representation

Jiaqi Zhang[1,2], Yan Hu[2(✉)], Xiaojuan Qi[1], Ting Meng[3], Lihui Wang[4], Huazhu Fu[5], Mingming Yang[3(✉)], and Jiang Liu[2(✉)]

[1] The University of Hong Kong, Pokfulam, Hong Kong
[2] Research Institute of Trustworthy Autonomous Systems and Department of Computer Science and Engineering, Southern University of Science and Technology, Shenzhen, China
{huy3,liuj}@sustech.edu.cn
[3] Department of Ophthalmology, Shenzhen People's Hospital, Shenzhen, China
ming4622@163.com
[4] Institute of Semiconductors, Guangdong Academy of Sciences, Guangzhou, China
[5] Institute of High Performance Computing, Agency for Science, Technology and Research, Singapore, Singapore

Abstract. The shape of the posterior eyeball is a crucial factor in many clinical applications, such as myopia prevention, surgical planning, and disease screening. However, current shape representations are limited by their low resolution or small field of view, providing insufficient information for surgeons to make accurate decisions. This paper proposes a novel task of reconstructing complete 3D posterior shapes based on small-FOV OCT images and introduces a novel Posterior Eyeball Shape Network (PESNet) to accomplish this task. The proposed PESNet is designed with dual branches that incorporate anatomical information of the eyeball as guidance. To capture more detailed information, we introduce a Polar Voxelization Block (PVB) that transfers sparse input point clouds to a dense representation. Furthermore, we propose a Radius-wise Fusion Block (RFB) that fuses correlative hierarchical features from the two branches. Our qualitative results indicate that PESNet provides a well-represented complete posterior eyeball shape with a chamfer distance of 9.52, SSIM of 0.78, and Density of 0.013 on the self-made posterior ocular shape dataset. We also demonstrate the effectiveness of our model by testing it on patients' data. Overall, our proposed PESNet offers a significant improvement over existing methods in accurately reconstructing the complete 3D posterior eyeball shape. This achievement has important implications for clinical applications.

Keywords: Posterior eyeball shape · 3D reconstruction · Polar transformation · Anatomical prior guided

J. Zhang and Y. Hu—Co-first authors.

Supplementary Information The online version contains supplementary material available at https://doi.org/10.1007/978-3-031-43987-2_18.

H. Greenspan et al. (Eds.): MICCAI 2023, LNCS 14225, pp. 180–190, 2023.
https://doi.org/10.1007/978-3-031-43987-2_18

1 Introduction

Posterior eyeball shape (PES) is related to various ophthalmic diseases, such as glaucoma, high myopia, and retinoblastoma [6,8,25]. Ophthalmologists often define the retinal pigment epithelium (RPE) layer represented as the shape of the posterior eyeball [3,5,19]. The changes in PES are helpful for surgeons to further specifically optimize the treatments of myopia, surgical planning, and diagnosis result [2,12,20]. For example, the PES can assist ophthalmologists to determine the expansion types of myopia, 6 types inferred to [18,22], in order to prevent further increase of myopic degree at an early stage [5]. Besides, the precise PES facilitates surgeons to estimate the cut length of the rectus during surgical planning, optimizing operative outcomes and avoiding refractive error after strabismus surgery [14]. Moreover, posterior eyeball shape is a sensitive indicator to facilitate fundus disease screening, such as glaucoma and tilted disc syndrome (TDS) [3,12].

In ophthalmic clinics, existing representations of posterior eyeball shape are mainly based on two medical imaging devices, including Magnetic Resonance Imaging (MRI) and Optical Coherence Tomography (OCT). MRI cannot become the preferred screening device in ophthalmic clinics due to its being expensive and time-consuming. Even most ophthalmic hospitals do not equip MR devices. Moreover, the resolution of ocular MRI is $0.416 \times 0.416 \times 0.399\,\text{mm}^2$, while the depth of the retina is around $250\,\mu\text{m}$ [1]. Thus, limited by the resolution of MRI, surgeons only infer the approximate posterior shape from the outer edge of the sclera or inner edge of the retina, as the retinal layers cannot be distinguished from MR images [2,7,10,18,23,26]. Most common OCT devices applied in eye hospitals can provide clear retinal layer imaging, but their field of view (FOV) is very limited, ranging from 4.5*4.5 to 13*13 mm, which nearly equals 2%–19% area of the entire posterior eyeball. Unlike OCT volume correlation tasks that are impacted by flatten distortion of A-scans [13], some researchers proposed to roughly infer shape changes of the posterior eyeball based on the simplified geometric shapes or non-quantitative topography in 2-dimensional (2D) obtained from retinal landmarks or RPE segmented lines [17,20,27]. However, such 2D expressions can not assist surgeons in making decisions. Therefore, we propose a novel task constructing a 3-dimensional (3D) posterior eyeball shape based on small-FOV OCT images.

As far as we know, there is no existing work on 3D eyeball reconstruction just using local OCT images. We refer several 3D reconstruction methods using natural images. For example, NeRF and Multi-View Stereo [15,16] need multi-view images, which are difficult to obtain in the ophthalmic field. The point cloud completion task [28] recovers local details based on global information, which is different from our target. As the limited FOV of common retinal imaging devices makes capturing global eyeballs difficult, we reconstruct global shapes from local OCT information. Moreover, the curved shape of the eyeball determines that its imaging information is often concentrated in a local area of the image where most of the pixels are empty, resulting in sparse information. Directly downsampling as in most existing point-based algorithms often leads to surface details lost limited by the number of input points. Therefore, to address the above problems, we first propose to adopt a standard posterior eyeball template to provide medical prior knowledge for the construction; then we propose coordinate transformation to handle the sparse representations of retinal in OCT images in this paper. Our main contributions are summarized as follows:

1) We define a novel task to reconstruct the complete posterior eyeball shape within large FOV only based on local OCT images. We then propose a novel Posterior Eyeball Shape Network (PESNet) leveraging the anatomical prior of the eyeball structure to accomplish the task.
2) We propose a Polar Voxelization Block (PVB) to provide a dense representation of the posterior eyeball in the polar coordinate system to address the problems of high computation cost and detail loss caused by sparse distributions of input point clouds. Besides, we propose a Radius-wise fusion block (RFB) to fuse rich hierarchical Radius-channel features from the dual branch of PESNet.
3) Our proposed PESNet is qualitatively and quantitatively validated on our self-made posterior ocular shape dataset. It is also tested on patients' data.

2 Method

The overview of our proposed Posterior Eyeball Shape Network (PESNet) is shown in Fig. 1(a). The subarea point cloud (size $N_1 * 3$) from local OCT images and the template point cloud (size $N_2 * 3$) from the template OCT images are taken as inputs. N_1 and N_2 are points number. The inputs are transformed to the polar grid with size (B, C, R, U, V) by our proposed Polar Voxelization Block (PVB), respectively. Then two polar grids are input into the dual-branch architecture of PESNet, whose detail is shown in Fig. 1(b). After a 2D ocular surface map (OSM) is predicted from the network, we convert it to a 3D point cloud as our predicted posterior eyeball shape using inverse polar transformation.

2.1 Dual-Branch Architecture of PESNet

The input point cloud is derived from limited FOV OCT images, and the structures between the local area and the complete posterior eyeball are totally different. Thus, considering the structural regularity of ocular shape in OCT images, we propose to aggregate structural prior knowledge for the reconstruction. As shown in Fig. 1(b), the proposed PESNet is constructed by the Shape regression branch (S-branch) and Anatomical prior branch (A-branch) with the same RBs, whose details are shown in Fig. 1(c). Inspired by ResNet [9], we propose the RBs to gradually extract hierarchical features from the R-channel of input grids with the reduction of the R-dimension. The last layer of RB1 to RB4 uses anisotropic 3D convolution layer with $(2 \times 3 \times 3)$ kernel, $(2, 3, 3)$ stride, and $(0, 1, 1)$ padding. Features from every RB in S-branch are fused with corresponding features from A-branch by our RFB. For RB5, the last layer adopts

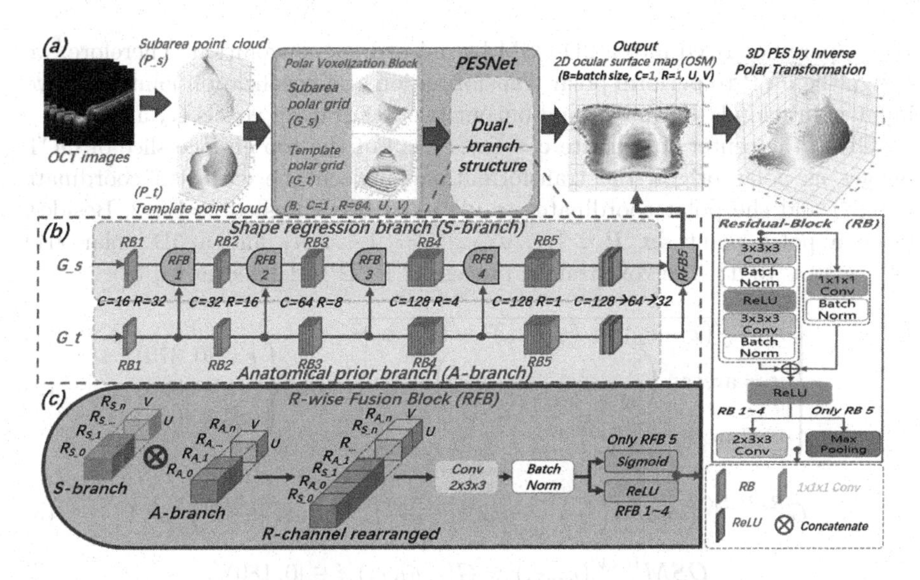

Fig. 1. The proposed PESNet for posterior eyeball reconstruction based on the local point cloud derived from OCT images. (a) The whole pipeline of the reconstruction process. (b) Dual-branch structure of PESNet, including an anatomical prior branch to extract structure features from the ocular template. (c) The key components of PESNet include R-wise Fusion Block (RFB), Residual-Block (RB), and used layers.

a maxpooling layer by setting kernel as $(4, 1, 1)$ to reduce the R-dimension to $\mathbf{1}$ for reducing the computational cost. Then $(1 \times 1 \times 1)$ convolution layers are adopted to further reduce the feature channels from 128 to 32. Finally, a 2D OSM with size $(B, 1, 1, N, N)$ is predicted from the dual-branch architecture ($B = batch\ size$, $N = 180$ in this paper) by the last RFB5. For the loss of our PESNet, we adopt the combination of smooth L1 loss $\mathcal{L}1_{smooth}$ and perceptual loss $\mathcal{L}_{percep}$ [11] as $Loss = \alpha\mathcal{L}1_s + \mathcal{L}_{perc}$, and smooth L1 loss is defined as:

$$\mathcal{L}1_s = \frac{1}{UV}\sum_{u=0}^{N-1}\sum_{v=0}^{N-1}\begin{cases}(\mathcal{O}(u,v) - \hat{\mathcal{O}}(u,v))^2 * 0.5, if|\mathcal{O}(u,v) - \hat{\mathcal{O}}(u,v)| < 1 \\ |\mathcal{O}(u,v) - \hat{\mathcal{O}}(u,v)| - 0.5, otherwise\end{cases} \quad (1)$$

where $\mathcal{O}(u, v)$ and $\hat{\mathcal{O}}(u, v)$ denote the predicted and ground truth (GT) OSM, respectively. The (u, v) and U, V represent the positional index and size of OSM. The α denotes a manual set weight value.

2.2 Polar Voxelization Block and R-Wise Fusion Block

Polar Voxelization Block (PVB). The curve shape of the eyeball determines the sparse layer labels of OCT in cartesian coordinates. Sparse convolution significantly increases memory. Inspired by the shape of the eyeball, we propose to use polar transformation to obtain the dense representation of the input. Due to the hierarchical and regular distribution of RPE points in polar grid, the 3D grid is parameterized into a 2D OSM by anisotropic convolution. Therefore, we design the PVB to perform polar transformation and voxelization jointly for two input point clouds P_s and P_t and output two 3D grids of voxels G_s and G_t.

Given the center coordinate $\mathbf{c} = (c_x, c_y, c_z)$ of the top en-face slice of OCT images as polar origin, the transformation between the cartesian coordinate (x, y, z) and the polar coordinates (r, u, v) can be expressed as Eq. 2. Besides, given a polar point set $\boldsymbol{P} : \{(r_i, u_i, v_i)|i = 1, \ldots, n\}$ and a 3D polar grid $\boldsymbol{G}(r, u, v) \in \{0, 1\}$, the voxelization process of PVB is defined in Eq. 3:

$$\begin{cases} r = \sqrt{(x - c_x)^2 + (y - c_y)^2 + (z - c_z)^2} \\ u = \arctan\left(\frac{z - c_z}{\sqrt{(x-c_x)^2 + (y-c_y)^2}}\right) + \frac{\pi}{2} \\ v = \arctan\left(\frac{y - c_y}{x - c_x}\right) + \frac{\pi}{2} \end{cases} \quad and \begin{cases} r \in [0, 370) \\ u \in [0, \pi) \\ v \in [0, \pi) \end{cases} \tag{2}$$

$$\boldsymbol{G}^{R \times U \times V}\left(round\left(\frac{r_i}{\phi}\right), round\left(\frac{u_i \times 180}{\pi}\right), round\left(\frac{v_i \times 180}{\pi}\right)\right) = 1 \tag{3}$$

$$\boldsymbol{OSM}^{U \times V}(u_i, v_i) = \boldsymbol{G}(r, u_i, v_i), i \in [0, 180) \tag{4}$$

where u,v and r are the elevation angle, azimuth angle, and radius in polar coordinates. The extra $\frac{\pi}{2}$ term is added to make the angle range correspond to the integer index of height and width of OSM. Besides, R,U,V denote the depth, height, and width of the 3D grid ($R = 64, U$ and $V = 180$ in our paper), which make grid $\boldsymbol{G}$ hold enough resolution with acceptable memory overhand. In addition, ϕ is a manual set number to control the depth-dimension of $\boldsymbol{G}$. Since we choose the center of the top en-face slice as the origin, the RPE voxels distribute in layers along the R-axis of $\boldsymbol{G}$. Therefore, we can compress the 3D polar grid $\boldsymbol{G}(r, u, v)$ into a 2D OSM along the R-axis of grid using Eq. 4, where $\boldsymbol{OSM}(u, v)$ denotes the pixel value at position (u, v), which represents the radius from the polar origin to an ocular surface point along the elevation and azimuth angles specified by index (u, v). The entire posterior eyeball shape can be reconstructed once given $\boldsymbol{OSM}$ and the center coordinate C. This representation of ocular shape is not only much denser than a conventional voxel-based grid (370∗300∗256 in our paper) in cartesian coordinates but also maintains more points and spatial structure than common point-based methods.

R-wise Fusion Block (RFB)

The conventional feature fusion methods (denoted as F in Table 1) often concatenate two feature volumes along a certain axis and simultaneously compute the correlation among R-, U-, and V-dimensions with convolution. They do not consider the feature distribution in the dimension. However, for our task, the hierarchical features distributed in the same channel along R-axis have a higher relevance. Separately computing relevance between features inside the R channel is able to enhance the structure guidance of prior information. Thus, we propose RFBs to fuse the feature volumes from S-branch and A-branch. As shown in Fig. 1(c), the first four blocks (RFB1 to RFB4) first concatenate two feature volumes along the R-dimension and then rearrange the channel order in the R-dimension of combined volume to place the correlated features together. Finally, the relevant hierarchical features from S- and A-branches are fused by a convolution layer with $(2 \times 3 \times 3)$ kernel and $(2, 1, 1)$ stride followed by BatchNorm and ReLU layers. The output volume of RFB has the same size as each input one. It's worth noting that the last RFB rearranges the order in feature- instead of R-dimension, since R-dimension is already equal to 1, and the last activation layer is Sigmoid to ensure the intensity values of output OSM ranging in $[0, 1]$.

3 Experiments

Dataset: We build a Posterior Ocular Shape (POS) dataset based on Ultra-widefield ($24 \times 20\,\text{mm}^2$) swept-source OCT device (BM-400K BMizar; TowardPi Medical Technology). The POS dataset contains 55 eyes of thirty-three healthy participants. Simulating local regions captured by common OCT devices, we sample five subareas from every ultra-widefield sample. Both subarea point clouds and ground truth of OSM are derived from segmented RPE label images from all samples (more details are in supplementary). The POS dataset is randomly split into training, validation, and test sets with 70%, 20% and 10%, respectively. Referring to [4,24], the template is obtained by averaging 7 high-quality GT OSMs without scanning defects and gaps from the training set and then perform Gaussian smoothing to reduce textual details until it does not change further. Finally, the averaged OSM is converted into the template point cloud by inverse polar transformation.

Implementation Details: The intensity values of all GT OSMs are normalized from $[0, 63]$ to $[0, 1]$ for loss calculation. We use unified center coordinates as the polar origin for polar transformation since all OCT samples are aligned to the macular fovea during acquisition. Since the background pixels in predicted OSM are hardly equal to 0, which cause many noise points in reconstructed point cloud, only intensity values that are not equal to $OSM(5, 5)$ are converted to points. During training, we set α in the loss function as a linear decay weight after the first 50 epochs, adam optimizer with a base learning rate of 0.001, and a batch size of 10 for 300 epochs. The model achieving the lowest loss on the validation set is saved for evaluation on the testing set. We implemented our PESNet using the PyTorch framework and trained on 2 A100 GPUs.

Evaluation Metrics: Since our proposal employs 2D OSM regression in polar coordinates to achieve 3D reconstruction in cartesian coordinates, we evaluate its performance using both 2D and 3D metrics. Specifically, we use Chamfer Distance (**CD**) to measure the mean distance between predicted point clouds and its GT point cloud, and Structural Similarity Index Measure (**SSIM**) as 2D metrics to evaluate the OSM regression performance. Additionally, we introduce point density (**Density**) as an extra metric to evaluate the number ratio of total points between predicted and GT point clouds, defined as $Density = |1 - N_{pred}/N_{GT}|$, where N_{pred} is the number of predicted point clouds, N_{GT} is the number of GT point clouds. The lower Density, the fewer defects in the valid region of OSMs.

Method Comparison and Ablation Study: The effectiveness of the key components of PESNet is investigated through a comparison and ablation study, as presented in Fig. 2 (other samples are shown in supplementary) and Table 1. To provide a benchmark, we re-implemented PointNet, a pioneering point-based method (details in the supplementary material). Compared to PointNet (Fig. 2(a)), the visual results in OSM and a reconstructed point cloud of 1-branch or 2-branch networks show more details in the valid region and reasonable shape, with better evaluation metrics, as hierarchical structure information is preserved in the dense representation by PVB. Compared to that by 1-branch (Fig. 2(b)), the visual results by 2-branch networks are much closer to the posterior eyeball shape, as the adopted prior knowledge provides structure assistance for reconstruction. As shown in Fig. 2, after incorporating RFB and unshare-weight, the 2D and 3D figures express the curve structure of the eyeball with more information, especially the macular and disc (in the bottom of the eyeball) are back to

Table 1. Method comparison and ablation studies on Posterior Ocular Shape Dataset.

Method	CD ↓	SSIM ↑	Density ↓
PointNet [21]	59.81	0.433	0.669
1-branch	13.57	0.310	0.714
2-branch F^5	10.14	0.504	0.140
2-branch F^{1-5}	11.68	0.744	0.048
$*$2-branch F^{1-5} with unshare-weight	11.12	0.764	0.059
2-branch RFB^5	9.65	0.287	0.339
2-branch RFB^{1-5}	9.84	0.710	0.173
$*$2-branch RFB^{1-5} with unshare-weight	**9.52**	**0.781**	**0.013**

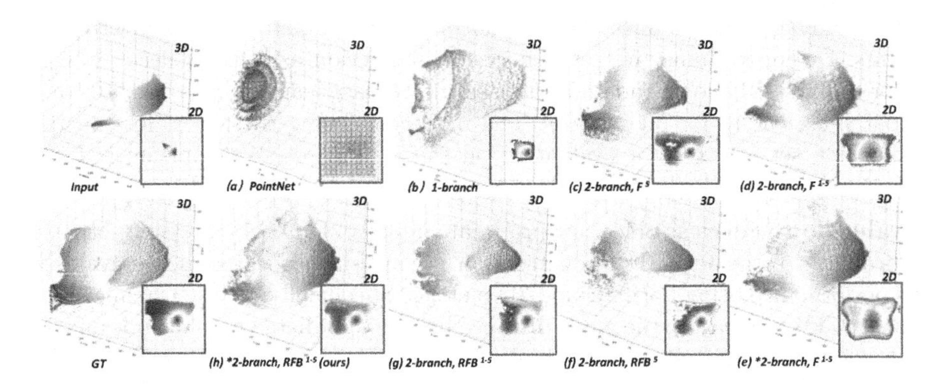

Fig. 2. The visualization results of the comparison. The main- and sub-images show the reconstructed point clouds in cartesian coordinate and its corresponding output OSMs, respectively, where F^5 and F^{1-5} denote common late- and multi-state fusion, RFB^5 and RFB^{1-5} denotes late- and multi-state RFB, and $*$ means unshare-weight.

Table 2. Ablation study about loss functions.

Loss function	CD ↓	SSIM ↑	Density ↓
$\mathcal{L}1_{smooth}$	65.49	0.287	0.446
$\mathcal{L}_{perc}$	9.95	0.729	0.085
$\mathcal{L}1_{smooth} + L_{ssim}$	49.14	0.259	1.615
$\mathcal{L}1_{smooth} + \mathcal{L}_{grad}$	56.04	0.305	0.544
$\mathcal{L}1_{grad} + \mathcal{L}_{perc}$	38.47	0.239	0.099
$\mathcal{L}1_{smooth} + \mathcal{L}_{ssim} + \mathcal{L}_{grad} + \mathcal{L}_{perc}$	74.65	0.161	0.824
$\mathcal{L}1_{smooth} + \mathcal{L}_{perc}$	**9.52**	**0.781**	**0.013**

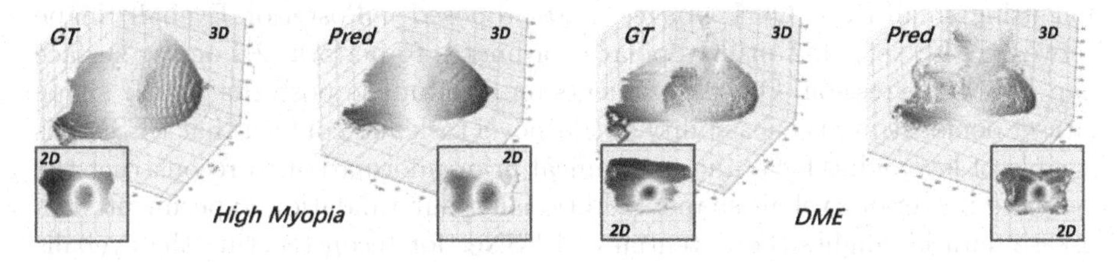

Fig. 3. The test results with the data from a high myopia case and a diabetic macular edema (DME) case, the shape point cloud and its OSM of GT are shown on the left side, and those of predicted results are shown on the right side.

the right position. The visualization shows that multi-state fusion can extract more structural details in R-dimension from multi-scale feature volumes of two branches, which can improve the scope and textual details of the valid regions in OSMs. As shown in Table 1, our 2-branch variant exceeds 1-branch ones by $3.43, 0.19, 0.57$ at CD, SSIM, and Density after adding A-branch, which further proves that prior knowledge guides to reconstruct a more complete valid region. If incorporating the RFB, the 3D shape (expressed by CD) can be closer to GT. In our PESNet, we adopt unshare-weight for two branches, which produces the best metrics in the last row. Therefore, our proposed algorithm reconstructs the eyeball close to GT considering the specificity of the eyeball.

Ablation Study About Loss Function: We compare different loss functions applied in our task, and the results as shown in Table 2. The results reveal two trends: 1): $\mathcal{L}1_{smooth}$ can ensure the correct optimization direction, impelling the OSMs area other than the background close 1, but we find it cannot effectively learn the accurate scope of OSMs from experiments. 2): $\mathcal{L}_{perc}$ can achieve sub-optimal results, but is too unstable to keep the right direction of optimization, which highly increase training time and decrease reproducibility. For other combinations, their OSMs show either wrong optimization direction or limited scope of valid regions which are also revealed by metrics. Therefore, our loss function combine the advantages of $\mathcal{L}1_{smooth}$ and $\mathcal{L}_{perc}$, and the extra α makes $\mathcal{L}_{perc}$ further descent stably. Our combination obtains the best metrics and visualization results, proving that it is the most suitable choice for our task.

Validation on Disease Sample: Although our POS dataset only contains the data of healthy eyeballs, we still test our PESNet on two disease cases: one is high myopia and the other is Diabetic Macular Edema. As shown in Fig. 3, although the template lack of priors from the disease eyeball, the range of pixel value and scope of region are close to GT. The result shows the potential of our PESNet to handle the posterior shape reconstruction for the diseased eyeballs. We are collecting more patients' data.

4 Conclusion

In this study, we introduced the task of posterior eyeball shape 3D reconstruction using small-FOV OCT images, and proposed a Posterior Eyeball Shape Network (PESNet) that utilizes polar coordinates to perform 2D ocular surface map (OSM) regression. Our experiments on a self-made posterior ocular shape dataset demonstrate the feasibility of this novel task and confirm that PESNet is capable of leveraging local and anatomical prior information to reconstruct the complete posterior eyeball shape. Additionally, our validation experiment with disease data highlights the potential of PESNet for reconstructing the eyeballs of patients with ophthalmic diseases.

Acknowledgment. This work was supported in part by General Program of National Natural Science Foundation of China (82272086 and 82102189), Guangdong Basic and Applied Basic Research Foundation (2021A1515012195), Shenzhen Stable Support Plan Program (20220815111736001 and 20200925174052004), and Agency for Science, Technology and Research (A*STAR) Advanced Manufacturing and Engineering (AME) Programmatic Fund (A20H4b0141) and Central Research Fund (CRF).

References

1. Alamouti, B., Funk, J.: Retinal thickness decreases with age: an oct study. Br. J. Ophthalmol. **87**(7), 899–901 (2003)
2. Atchison, D.A., et al.: Eye shape in emmetropia and myopia. Investig. Ophthalmol. Vis. Sci. **45**(10), 3380–3386 (2004)
3. Belghith, A., et al.: Structural change can be detected in advanced-glaucoma eyes. Investig. Ophthalmol. Vis. Sci. **57**(9), OCT511–OCT518 (2016)
4. Bongratz, F., Rickmann, A.M., Pölsterl, S., Wachinger, C.: Vox2cortex: fast explicit reconstruction of cortical surfaces from 3D MRI scans with geometric deep neural networks. In: Proceedings of the IEEE/CVF Conference on Computer Vision and Pattern Recognition, pp. 20773–20783 (2022)
5. Brennan, N.A., Toubouti, Y.M., Cheng, X., Bullimore, M.A.: Efficacy in myopia control. Prog. Retin. Eye Res. **83**, 100923 (2021)
6. Ciller, C., et al.: Multi-channel MRI segmentation of eye structures and tumors using patient-specific features. PLoS ONE **12**(3), e0173900 (2017)
7. Ciller, C., et al.: Automatic segmentation of the eye in 3D magnetic resonance imaging: a novel statistical shape model for treatment planning of retinoblastoma. Int. J. Radiat. Oncol. Biol. Phys. **92**(4), 794–802 (2015)

8. Guo, X., et al.: Three-dimensional eye shape, myopic maculopathy, and visual acuity: the Zhongshan ophthalmic center-brien holden vision institute high myopia cohort study. Ophthalmology **124**(5), 679–687 (2017)
9. He, K., Zhang, X., Ren, S., Sun, J.: Deep residual learning for image recognition. In: Proceedings of the IEEE Conference on Computer Vision and Pattern Recognition, pp. 770–778 (2016)
10. Ishii, K., Iwata, H., Oshika, T.: Quantitative evaluation of changes in eyeball shape in emmetropization and myopic changes based on elliptic fourier descriptors. Investig. Ophthalmol. Vis. Sci. **52**(12), 8585–8591 (2011)
11. Johnson, J., Alahi, A., Fei-Fei, L.: Perceptual losses for real-time style transfer and super-resolution. In: Leibe, B., Matas, J., Sebe, N., Welling, M. (eds.) ECCV 2016. LNCS, vol. 9906, pp. 694–711. Springer, Cham (2016). https://doi.org/10.1007/978-3-319-46475-6_43
12. Kim, Y.C., Moon, J.S., Park, H.Y.L., Park, C.K.: Three dimensional evaluation of posterior pole and optic nerve head in tilted disc. Sci. Rep. **8**(1), 1–11 (2018)
13. Kuo, A.N., et al.: Correction of ocular shape in retinal optical coherence tomography and effect on current clinical measures. Am. J. Ophthalmol. **156**(2), 304–311 (2013)
14. Leshno, A., Mezad-Koursh, D., Ziv-Baran, T., Stolovitch, C.: A paired comparison study on refractive changes after strabismus surgery. J. Am. Assoc. Pediatr. Ophthalmol. Strabismus **21**(6), 460–462 (2017)
15. Liu, J., et al.: Planemvs: 3D plane reconstruction from multi-view stereo. In: Proceedings of the IEEE/CVF Conference on Computer Vision and Pattern Recognition (CVPR), pp. 8665–8675 (2022)
16. Mildenhall, B., Srinivasan, P.P., Tancik, M., Barron, J.T., Ramamoorthi, R., Ng, R.: NERF: representing scenes as neural radiance fields for view synthesis. Commun. ACM **65**(1), 99–106 (2021)
17. Miyake, M., et al.: Analysis of fundus shape in highly myopic eyes by using curvature maps constructed from optical coherence tomography. PLoS ONE **9**(9), e107923 (2014)
18. Moriyama, M., et al.: Topographic analyses of shape of eyes with pathologic myopia by high-resolution three-dimensional magnetic resonance imaging. Ophthalmology **118**(8), 1626–1637 (2011)
19. Palchunova, K., et al.: Precise retinal shape measurement by alignment error and eye model calibration. Opt. Rev. **29**(3), 188–196 (2022)
20. Park, Y., Kim, Y.C., Ahn, Y.J., Park, S.H., Shin, S.Y.: Morphological change of the posterior pole following the horizontal strabismus surgery with swept source optical coherence tomography. Sci. Rep. **11**(1), 1–11 (2021)
21. Qi, C.R., Su, H., Mo, K., Guibas, L.J.: Pointnet: deep learning on point sets for 3D classification and segmentation. In: Proceedings of the IEEE Conference on Computer Vision and Pattern Recognition, pp. 652–660 (2017)
22. Rozema, J., Dankert, S., Iribarren, R., Lanca, C., Saw, S.M.: Axial growth and lens power loss at myopia onset in Singaporean children. Investig. Ophthalmol. Vis. Sci. **60**(8), 3091–3099 (2019)
23. Singh, K.D., Logan, N.S., Gilmartin, B.: Three-dimensional modeling of the human eye based on magnetic resonance imaging. Investig. Ophthalmol. Vis. Sci. **47**(6), 2272–2279 (2006)
24. Sun, L., Shao, W., Zhang, D., Liu, M.: Anatomical attention guided deep networks for ROI segmentation of brain MR images. IEEE Trans. Med. Imaging **39**(6), 2000–2012 (2019)

25. Tatewaki, Y., et al.: Morphological prediction of glaucoma by quantitative analyses of ocular shape and volume using 3-dimensional T2-weighted MR images. Sci. Rep. **9**(1), 15148 (2019)
26. Verkicharla, P.K., Mathur, A., Mallen, E.A., Pope, J.M., Atchison, D.A.: Eye shape and retinal shape, and their relation to peripheral refraction. Ophthalmic Physiol. Opt. **32**(3), 184–199 (2012)
27. Wang, Y.X., Panda-Jonas, S., Jonas, J.B.: Optic nerve head anatomy in myopia and glaucoma, including parapapillary zones alpha, beta, gamma and delta: histology and clinical features. Prog. Retin. Eye Res. **83**, 100933 (2021)
28. Xiang, P., et al.: Snowflake point deconvolution for point cloud completion and generation with skip-transformer. IEEE Trans. Pattern Anal. Mach. Intell. **45**(5), 6320–6338 (2022)

DiffDP: Radiotherapy Dose Prediction via a Diffusion Model

Zhenghao Feng[1], Lu Wen[1], Peng Wang[1], Binyu Yan[1], Xi Wu[2], Jiliu Zhou[1,2], and Yan Wang[1](✉)

[1] School of Computer Science, Sichuan University, Chengdu, China
wangyanscu@hotmail.com

[2] School of Computer Science, Chengdu University of Information Technology, Chengdu, China

Abstract. Currently, deep learning (DL) has achieved the automatic prediction of dose distribution in radiotherapy planning, enhancing its efficiency and quality. However, existing methods suffer from the over-smoothing problem for their commonly used L_1 or L_2 loss with posterior average calculations. To alleviate this limitation, we innovatively introduce a diffusion-based dose prediction (DiffDP) model for predicting the radiotherapy dose distribution of cancer patients. Specifically, the DiffDP model contains a forward process and a reverse process. In the forward process, DiffDP gradually transforms dose distribution maps into Gaussian noise by adding small noise and trains a noise predictor to predict the noise added in each timestep. In the reverse process, it removes the noise from the original Gaussian noise in multiple steps with the well-trained noise predictor and finally outputs the predicted dose distribution map. To ensure the accuracy of the prediction, we further design a structure encoder to extract anatomical information from patient anatomy images and enable the noise predictor to be aware of the dose constraints within several essential organs, i.e., the planning target volume and organs at risk. Extensive experiments on an in-house dataset with 130 rectum cancer patients demonstrate the superiority of our method.

Keywords: Radiotherapy Treatment · Dose Prediction · Diffusion Model · Deep Learning

1 Introduction

Radiotherapy, one of the mainstream treatments for cancer patients, has gained notable advancements in past decades. For promising curative effect, a high-quality radiotherapy plan is demanded to distribute sufficient dose of radiation to the planning target volume (PTV) while minimizing the radiation hazard to organs at risk (OARs). To achieve this, radiotherapy plans need to be manually adjusted by the dosimetrists in a trial-and-error manner, which is extremely labor-intensive and time-consuming [1, 2]. Additionally, the quality of treatment plans might be variable among radiologists due to their different expertise and experience [3]. Consequently, it is essential to develop a robust methodology to automatically predict the dose distribution for cancer patients, relieving the burden on dosimetrists and accelerating the radiotherapy procedure.

Z. Feng and L. Wen—Contribute equally to this work.

H. Greenspan et al. (Eds.): MICCAI 2023, LNCS 14225, pp. 191–201, 2023.
https://doi.org/10.1007/978-3-031-43987-2_19

Recently, the blossom of deep learning (DL) has promoted the automatic medical image processing tasks [4–6], especially for dose prediction [7–14]. For example, Nguyen *et al.* [7] modified the traditional 2D UNet [15] to predict the dose of prostate cancer patients. Wang *et al.* [10] utilized a progressive refinement UNet (PRUNet) to refine the predictions from low resolution to high resolution. Besides the above UNet-based frameworks, Song *et al.* [11] employed the deepLabV3+ [16] to excavate contextual information from different scales, thus obtaining accuracy improvements in the dose prediction of rectum cancer. Mahmood *et al.* [12] utilized a generative adversarial network (GAN)-based method to predict the dose maps of oropharyngeal cancer. Furthermore, Zhan *et al.* [13] designed a multi-organ constraint loss to enforce the deep model to better consider the dose requirements of different organs. Following the idea of multi-task learning, Tan *et al.* [8] utilized isodose line and gradient information to promote the performance of dose prediction of rectum cancer. To ease the burden on the delineation of PTV and OARs, Li *et al.* [17] constructed an additional segmentation task to provide the dose prediction task with essential anatomical knowledge.

Although the above methods have achieved good performance in predicting dose distribution, they suffer from the over-smoothing problem. These DL-based dose prediction methods always apply the L_1 or L_2 loss to guide the model optimization which calculates a posterior mean of the joint distribution between the predictions and the ground truth [17, 18], leading to the over-smoothed predicted images without important high-frequency details [19]. We display predicted dose maps from multiple deep models in Fig. 1. As shown, compared with the ground truth, i.e., (5) in Fig. 1, the predictions from (1) to (3) are blurred with fewer high-frequency details, such as ray shapes. These high-frequency features formed by ray penetration reveal the ray directions and dose attenuation with the aim of killing the cancer cells while protecting the OARs as much as possible, which are critical for radiotherapy. Consequently, exploring an automatic method to generate high-quality predictions with rich high-frequency information is important to improve the performance of dose prediction.

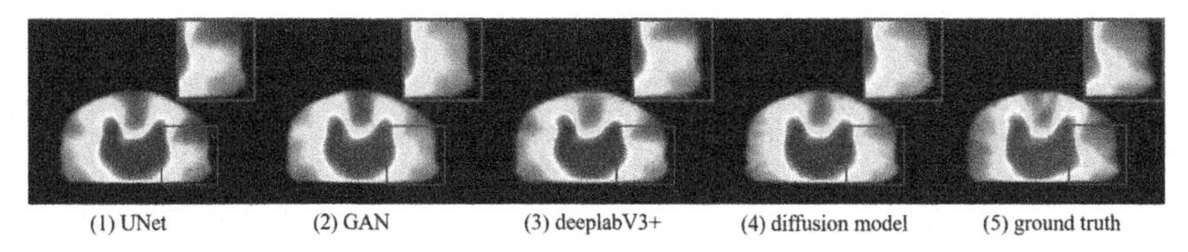

Fig. 1. Instances from a rectum cancer patient. (1) –(4): Dose maps predicted by UNet, GAN, deepLabV3+, and diffusion model.

Currently, diffusion model [20] has verified its remarkable potential in modeling complex image distributions in some vision tasks [21–23]. Unlike other DL models, the diffusion model is trained without any extra assumption about target data distribution, thus evading the average effect and alleviating the over-smoothing problem [24]. Figure 1 (4) provides an example in which the diffusion-based model predicts a dose map with shaper and clearer boundaries of ray-penetrated areas. Therefore, introducing a diffusion model to the dose prediction task is a worthwhile endeavor.

In this paper, we investigate the feasibility of applying a diffusion model to the dose prediction task and propose a diffusion-based model, called DiffDP, to automatically predict the clinically acceptable dose distribution for rectum cancer patients. Specifically, the DiffDP consists of a forward process and a reverse process. In the forward process, the model employs a Markov chain to gradually transform dose distribution maps with complex distribution into Gaussian distribution by progressively adding pre-defined noise. Then, in the reverse process, given a pure Gaussian noise, the model gradually removes the noise in multiple steps and finally outputs the predicted dose map. In this procedure, a noise predictor is trained to predict the noise added in the corresponding step of the forward process. To further ensure the accuracy of the predicted dose distribution for both the PTV and OARs, we design a DL-based structure encoder to extract the anatomical information from the CT image and the segmentation masks of the PTV and OARs. Such anatomical information can indicate the structure and relative position of organs. By incorporating the anatomical information, the noise predictor can be aware of the dose constraints among PTV and OARs, thus distributing more appropriate dose to them and generating more accurate dose distribution maps.

Overall, the contributions of this paper can be concluded as follows: (1) We propose a novel diffusion-based model for dose prediction in cancer radiotherapy to address the over-smoothing issue commonly encountered in existing DL-based dose prediction methods. *To the best of our knowledge, we are the first to introduce the diffusion model for this task.* (2) We introduce a structure encoder to extract the anatomical information available in the CT images and organ segmentation masks, and exploit the anatomical information to guide the noise predictor in the diffusion model towards generating more precise predictions. (3) The proposed DiffDP is extensively evaluated on a clinical dataset consisting of 130 rectum cancer patients, and the results demonstrate that our approach outperforms other state-of-the-art methods.

2 Methodology

An overview of the proposed diffDP model is illustrated in Fig. 2, containing two Markov chain processes: a forward process and a reverse process. An image set of cancer patient is defined as $\{x, y\}$, where $x \in R^{H\times W\times(2+o)}$ represents the structure images, "2" signifies the CT image and the segmentation mask of the PTV, and o denotes the total number of segmentation mask of OARs. Meanwhile, $y \in R^{H\times W\times 1}$ is the corresponding dose distribution map for x. Concretely, the forward process produces a sequence of noisy images $\{y_0, y_1, \ldots, y_T\}, y_0 = y$ by gradually adding a small amount of noise to y in T steps with the noise increased at each step and a noise predictor f is constructed to predict the noise added to y_{t-1} by treating y_t, anatomic information from x and embedding of step t as input. To obtain the anatomic information, a structure encoder g is designed to extract the crucial feature representations from the structure images. Then, in the reverse process, the model progressively deduces the dose distribution map by iteratively denoising from y_T using the well-trained noise predictor.

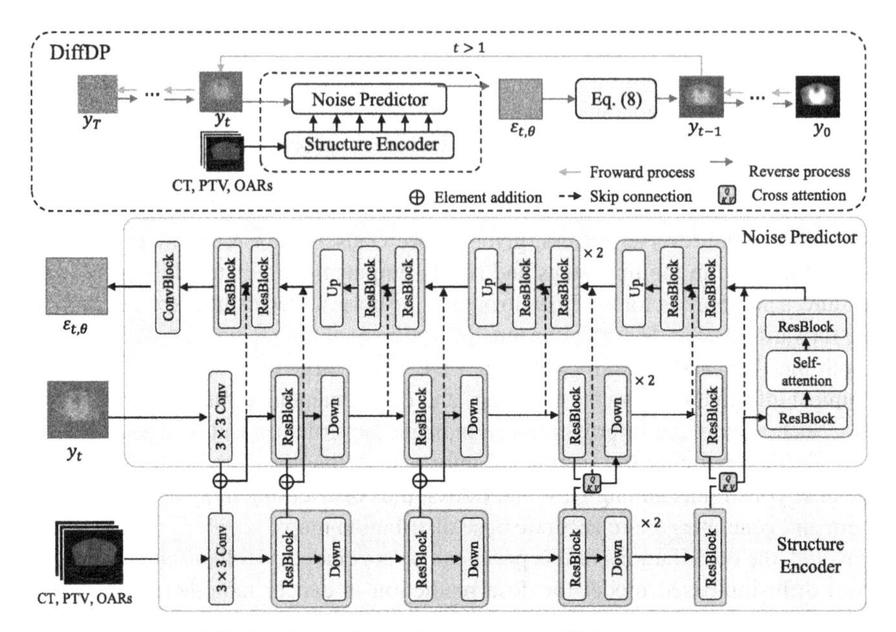

Fig. 2. Illustration of the proposed DiffDP network.

2.1 Diffusion Model

The framework of DiffDP is designed following the Denoising Diffusion Probabilistic Models (DDPM) [25] which contains a forward process and a reverse process. By utilizing both processes, the DiffDP model can progressively transform the Gaussian noise into complex data distribution.

Forward Process. In the forward process, the DiffDP model employs the Markov chain to progressively add noise to the initial dose distribution map $y_0 \sim q(y_0)$ until the final disturbed image y_T becomes completely Gaussian noise which is represented as $y_T \sim \mathcal{N}(y_T \mid 0, I)$. This forward process can be formulated as:

$$q(y_{1:T} \mid y_0) := \Pi_{t=1}^{T} q(y_t \mid y_{t-1}), \tag{1}$$

$$q(y_t \mid y_{t-1}) := N\left(y_t; \sqrt{\alpha_t} y_{t-1}, (1 - \alpha_t)I\right), \tag{2}$$

where α_t is the unlearnable standard deviation of the noise added to y_{t-1}.

Herein, the $\alpha_t (t = 1, \ldots, T)$ could accumulate during the forward process, which can be treated as the noise intensity $\gamma_t = \prod_{i=1}^{t} \alpha_i$. Based on this, we can directly obtain the distribution of y_t at any step t from y_0 through the following formula:

$$q(y_t \mid y_0) = \mathcal{N}\left(y_t; \sqrt{\gamma_t} y_0, (1 - \gamma_t)I\right), \tag{3}$$

where the disturbed image y_t is sampled using:

$$y_t = \sqrt{\gamma_t} y_0 + \sqrt{1 - \gamma_t} \varepsilon_t, \tag{4}$$

in which $\varepsilon_t \sim \mathcal{N}(0, I)$ is random noise sampled from normal Gaussian distribution.

Reverse Process. The reverse process also harnesses the Markov chain to progressively convert the latent variable distribution $p_\theta(y_T)$ into distribution $p_\theta(y_0)$ parameterized by θ. Corresponding to the forward process, the reverse one is a denoising transformation under the guidance of structure images x that begins with a standard Gaussian distribution $y_T \sim \mathcal{N}(y_T \mid 0, I)$. This reverse inference process can be formulated as:

$$p_\theta(y_{0:T} \mid y_t, x) = p(y_T) \prod\nolimits_{t=1}^{T} p_\theta(y_{t-1} \mid y_t, x), \tag{5}$$

$$p_\theta(y_{t-1} \mid y_t, x) = \mathcal{N}\left(y_{t-1}; \mu_\theta(x, y_t, \gamma_t), \sigma_t^2 I\right). \tag{6}$$

where $\mu_\theta(x, y_t, t)$ is a learned mean, and σ_t is a unlearnable standard deviation. Following the idea of [16], we parameterize the mean of μ_θ as:

$$\mu_\theta(x, y_t, \gamma_t) = \frac{1}{\sqrt{\alpha_t}}\left(y_t - \frac{1-\alpha_t}{\sqrt{1-\gamma_t}}\varepsilon_{t,\theta}\right), \tag{7}$$

where $\varepsilon_{t,\theta}$ is a function approximator intended to predict ε_t from the input x, y_t and γ_t. Consequently, the reverse inference at two adjacent steps can be expressed as:

$$y_{t-1} \leftarrow \frac{1}{\sqrt{\alpha_t}}\left(y_t - \frac{1-\alpha_t}{\sqrt{1-\gamma_t}}\varepsilon_{t,\theta}\right) + \sqrt{1-\alpha_t} z_t, \tag{8}$$

where $z_t \sim \mathcal{N}(0, I)$ is a random noise sampled from normal Gaussian distribution. More derivation processes can be found in the original paper of diffusion model [25].

2.2 Structure Encoder

Vanilla diffusion model has difficulty preserving essential structural information and produce unstable results when predicting dose distribution maps directly from noise with a simple condition mechanism. To address this, we design a structure encoder g that effectively extracts the anatomical information from the structure images guiding the noise predictor to generate more accurate dose maps by incorporating extracted structural knowledge. Concretely, the structure encoder includes five operation steps, each with a residual block (ResBlock) and a Down block, except for the last one. The ResBlock consists of two convolutional blocks (ConvBlock), each containing a 3×3 convolutional (Conv) layer, a GroupNorm (GN) layer, and a Swish activation function. The residual connections are reserved for preventing gradient vanishment in the training. The Down block includes a 3×3 Conv layer with a stride of 2. It takes structure image x as input, which includes the CT image and segmentation masks of PTV and OARs, and evacuates the compact feature representation in different levels to improve the accuracy of dose prediction. The structure encoder is pre-trained by L_1 loss and the corresponding feature representation $x_e = g(x)$ is then fed into the noise predictor.

2.3 Noise Predictor

The purpose of the noise predictor $f(x_e, y_t, \gamma_t)$ is to predict the noise added on the distribution map y_t with the guidance of the feature representation x_e extracted from the structure images x and current noise intensity γ_t in each step t. Inspired by the great achievements of UNet [15], we employ a six-level UNet to construct the noise predictor. Specifically, the encoder holds the similar architecture with the structure encoder while the decoder comprises five deconvolution blocks to fulfill the up-sampling operation, and each contains an Up block and two ResBlock, except for the last one which discards the UP block. In each Up block, the Nearest neighbor up-sampling and a Conv layer with a kernel size of 1 are used. A bottleneck with two Resblocks and a self-attention module is embedded between the encoder and decoder.

In the encoding procedure, to guide the noise predictor with essential anatomical structure, the feature representations respectively extracted from the structure images x and noisy image y_t are simultaneously fed into the noise predictor. Firstly, y_t is encoded into feature maps through a convolutional layer. Then, these two feature maps are fused by element-wise addition, allowing the structure information in x to be transferred to the noise predictor. The following two down-sampling operations retain the addition operation to complete information fusion, while the last three use a cross-attention mechanism to gain similarity-based structure guidance at deeper levels.

In the decoding procedure, the noise predictor restores the feature representations captured by the encoder to the final output, i.e., the noise $\varepsilon_{t,\theta} = f(x_e, y_t, \gamma_t)$ in step t. The skip connections between the encoder and decoder are reserved for multi-level feature reuse and aggregation.

2.4 Objective Function

The main purpose of the DiffDP model is to train the noise predictor f and structure encoder g, so that the predicted noise $\varepsilon_{t,\theta} = f(g(x), y_t, \gamma_t)$ in the reverse process can approximate the added noise ε_t in the forward process. To achieve this, we define the objective function as:

$$\min_{\theta} \mathbb{E}_{(x,y)}\mathbb{E}_{\varepsilon,\gamma}\left\| f\left(g(x), \underbrace{\sqrt{\gamma}y_0 + \sqrt{1-\gamma}\varepsilon_t}_{y_t}, \gamma_t \right) - \varepsilon_t \right\|, \varepsilon_t \sim \mathcal{N}(0, I) \quad (9)$$

For a clearer understanding, the training procedure is summarized in Algorithm 1.

Algorithm 1: Training procedure

1: **Input:** Input image pairs $P = \{(x_i, y_i)\}_{i=1}^{I}$ where x is the structure image and y is the corresponding dose distribution map, the total number of diffusion steps T.
2: **Initialize:** Randomly initialize the noise predictor f and pre-trained structure encoder g.
3: **Repeat**
4: Sample $(x, y) \sim P$
5: Sample $\varepsilon_t \sim \mathcal{N}(0, I)$, and $t \sim Uniform(\{1, \ldots, T\})$
6: Perform the gradient step on Equation (9)
7: **until** converged

2.5 Training Details

We accomplish the proposed network in the PyTorch framework. All of our experiments are conducted through one NVIDIA RTX 3090 GPU with 24 GB memory and a batch size of 16 with an Adaptive moment estimation (Adam) optimizer. We train the whole model for 1500 epochs (about 1.5M training steps) where the learning rate is initialized to 1e−4 and reset to 5e−5 after 1200 epochs. The parameter T is set to 1000. Additionally, the noise intensity is initialized to 1e−2 and decayed to 1e−4 linearly along with the increase of steps.

3 Experiments and Results

Dataset and Evaluations. We measure the performance of our model on an in-house rectum cancer dataset which contains 130 patients who underwent volumetric modulated arc therapy (VMAT) treatment at West China Hospital. Concretely, for every patient, the CT images, PTV segmentation, OARs segmentations, and the clinically planned dose distribution are included. Additionally, there are four OARs of rectum cancer containing the bladder, femoral head R, femoral head L, and small intestine. We randomly select 98 patients for model training, 10 patients for validation, and the remaining 22 patients for test. The thickness of the CTs is 3 mm and all the images are resized to the resolution of 256×256 before the training procedure.

We measure the performance of our proposed model with multiple metrics. Considering Dm represents the minimal absorbed dose covering m% percentage volume of PTV, we involve D_{98}, D_2, maximum dose (D_{max}), and mean dose (D_{mean}) as metrics. Besides, the heterogeneity index (HI) is used to quantify dose heterogeneity [26]. To quantify performance more directly, we calculate the difference (Δ) of these metrics between the ground truth and the predicted results. More intuitively, we involve the dose volume histogram (DVH) [27] as another essential metric of dose prediction performance. When the DVH curves of the predictions are closer to the ground truth, we can infer higher prediction accuracy.

Comparison with State-of-the-Art Methods. To verify the superior accuracy of our proposed model, we select multiple state-of-the-art (SOTA) models in dose prediction, containing UNet (2017) [7], GAN (2018) [12], deepLabV3+ (2020) [11], C3D (2021) [9], and PRUNet (2022) [10], for comparison. The quantitative comparison results are listed in Table. 1 where our method outperforms the existing SOTAs in terms of all metrics. Specifically, compared with deepLabV3+ with the second-best accuracy in ΔHI (0.0448) and ΔD_{98} (0.0416), the results generated by the proposed are 0.0035 and 0.0014 lower, respectively. As for ΔD_2 and ΔD_{max}, our method gains overwhelming performance with 0.0008 and 0.0005, respectively. Moreover, the paired t-test is conducted to investigate the significance of the results. The p-values between the proposed and other SOTAs are almost all less than 0.05, indicating that the enhancement of performance is statistically meaningful.

Besides the quantitative results, we also present the DVH curves derived by compared methods in Fig. 3. The results are compared on PTV as well as two OARs: bladder and small intestine. Compared with other methods, the disparity between the DVH curves of

Table 1. Quantitative comparison results with state-of-the-art methods in terms of ΔHI, ΔD_{98}, ΔD_2, and ΔD_{max}. * means our method is significantly better than compared method with p < 0.05 via paired t-test.

Methods	ΔHI	ΔD_{98}	ΔD_2	ΔD_{max}
UNet [3]	0.0494(5.8E−3)*	0.0428(5.2E−3)*	0.0048(1.1E−5)*	0.0186(2.6E−5)*
GAN [8]	0.0545(5.2E−3)*	0.0431(4.3E−3)*	0.0253(2.0E−5)*	0.0435(3.5E−6)*
deepLabV3 + [7]	0.0448(4.8E−3)*	0.0416(4.2E−3)	0.0036(7.8E−6)*	0.0139(8.2E−6)*
C3D [5]	0.0460(5.6E−3)*	0.0400(4.9E−3)	0.0077(1.8E−5)*	0.0206(3.0E−5)*
PRUNet [6]	0.0452(5.2E−3)*	0.0407(4.5E−3)	0.0088(6.2E−5)*	0.0221(4.3E−5)*
Proposed	**0.0413(4.5E−3)**	**0.0392(4.1E−3)**	**0.0008(1.1E−5)**	**0.0005(4.4E−6)**

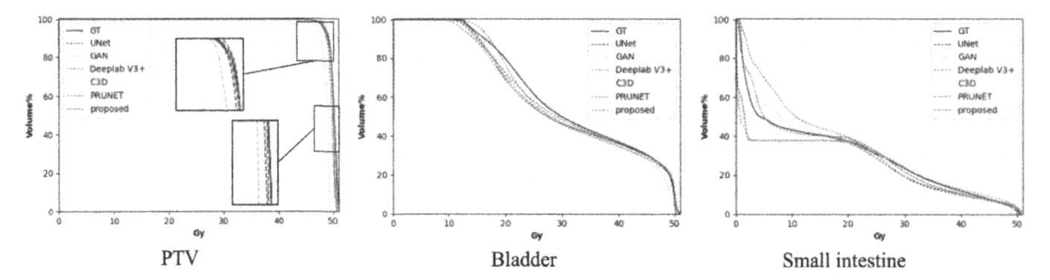

Fig. 3. Visual comparison of DVH curves by our method and SOTA methods, including DVH curves of PTV, bladder, and small intestine.

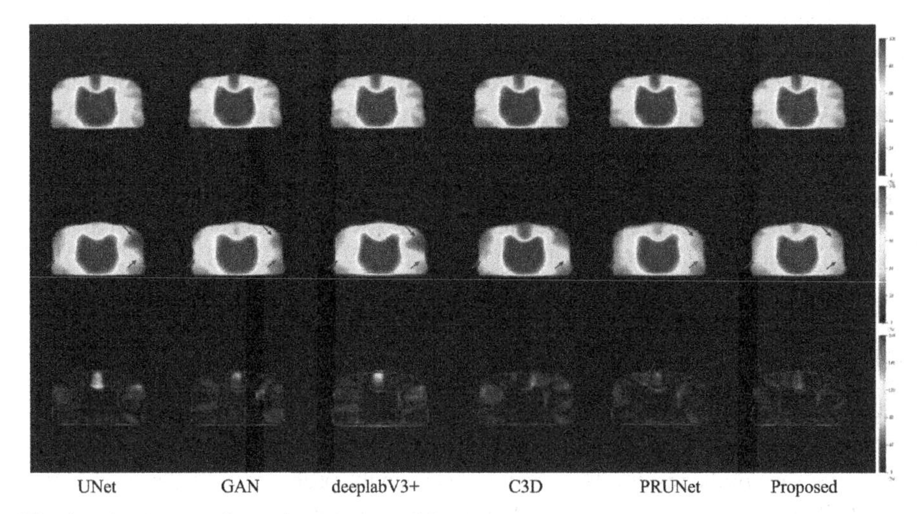

Fig. 4. Visual comparison with SOTA models. From top to bottom: ground truth, predicted dose map and corresponding error maps.

our method and the ground truth is the smallest, demonstrating the superior performance of the proposed.

Furthermore, we display the visualization comparison in Fig. 4. As we can see, the proposed model achieves the best visual quality with clearer and sharper high-frequency details (as indicated by red arrows). Furthermore, the error map of the proposed is the darkest, suggesting the least disparity compared with the ground truth.

Ablation Study. To study the contributions of key components of the proposed method, we conduct the ablation experiments by 1) removing the structure encoder from the proposed method and concatenating the anatomical images x and noisy image y_t together as the original input for diffusion model (denoted as Baseline); 2) the proposed DiffDP model. The quantitative results are given in Table 2. We can clearly see the performance for all metrics is enhanced with the structure encoder, demonstrating its effectiveness in the proposed model.

Table 2. Ablation study of our method in terms of ΔHI, ΔD_{98}, ΔD_2, and ΔD_{mean}. * means our method is significantly better than other variants with $p < 0.05$ via paired t-test.

Methods	ΔHI	ΔD_{98}	ΔD_2	ΔD_{mean}
Baseline	0.0444(4.7E−3)*	0.0426(4.2E−3)	0.0021(1.1E−5)*	0.0246(7.5E−4)*
Proposed	**0.0413(4.5E−3)**	**0.0392(4.1E−3)**	**0.0008(1.1E−5)**	**0.0154(6.5E−4)**

4 Conclusion

In this paper, we introduce a novel diffusion-based dose prediction (DiffDP) model for predicting the radiotherapy dose distribution of cancer patients. The proposed method involves a forward and a reverse process to generate accurate prediction by progressively transferring the Gaussian noise into a dose distribution map. Moreover, we propose a structure encoder to extract anatomical information from patient anatomy images and enable the model to concentrate on the dose constraints within several essential organs. Extensive experiments on an in-house dataset with 130 rectum cancer patients demonstrate the superiority of our method.

Acknowledgement. This work is supported by the National Natural Science Foundation of China (NSFC 62371325, 62071314), Sichuan Science and Technology Program 2023YFG0263, 2023YFG0025, 2023NSFSC0497.

References

1. Murakami, Y., et al.: Possibility of chest wall dose reduction using volumetric-modulated arc therapy (VMAT) in radiation-induced rib fracture cases: comparison with stereotactic body radiation therapy (SBRT). J. Radiat. Res. **59**(3), 327–332 (2018)

2. Wang, K., et al.: Semi-supervised medical image segmentation via a tripled-uncertainty guided mean teacher model with contrastive learning. Med. Image Anal. **79**, 102447 (2022)
3. Nelms, B.E., et al.: Variation in external beam treatment plan quality: an inter-institutional study of planners and planning systems. Pract. Radiat. Oncol. **2**(4), 296–305 (2012)
4. Shi, Y., et al.: ASMFS: adaptive-similarity-based multi-modality feature selection for classification of Alzheimer's disease. Pattern Recogn. **126**, 108566 (2022)
5. Wang, Y., et al.: 3D auto-context-based locality adaptive multi-modality GANs for PET synthesis. IEEE Trans. Med. Imaging **38**(6), 1328–1339 (2018)
6. Zhang, J., Wang, L., Zhou, L., Li, W.: Beyond covariance: SICE and kernel based visual feature representation. Int. J. Comput. Vision **129**, 300–320 (2021)
7. Nguyen, D., et al.: Dose prediction with U-Net: a feasibility study for predicting dose distributions from contours using deep learning on prostate IMRT patients. arXiv preprint arXiv: 1709.09233, 17 (2017)
8. Tan, S., et al.: Incorporating isodose lines and gradient information via multi-task learning for dose prediction in radiotherapy. In: de Bruijne, M., et al. (eds.) Medical Image Computing and Computer Assisted Intervention–MICCAI 2021: 24th International Conference, Proceedings, Part VII, vol. 24, pp. 753–763. Springer, Cham (2021). https://doi.org/10.1007/978-3-030-87234-2_71
9. Liu, S., Zhang, J., Li, T., Yan, H., Liu, J.: A cascade 3D U-Net for dose prediction in radiotherapy. Med. Phys. **48**(9), 5574–5582 (2021)
10. Wang, J., et al.: VMAT dose prediction in radiotherapy by using progressive refinement UNet. Neurocomputing **488**, 528–539 (2022)
11. Song, Y., et al.: Dose prediction using a deep neural network for accelerated planning of rectal cancer radiotherapy. Radiother. Oncol. **149**, 111–116 (2020)
12. Mahmood, R., Babier, A., McNiven, A., Diamant, A., Chan, T.C.: Automated treatment planning in radiation therapy using generative adversarial networks. In: Machine Learning for Healthcare Conference, pp. 484–499 (2018)
13. Zhan, B., et al.: Multi-constraint generative adversarial network for dose prediction in radiotherapy. Med. Image Anal. **77**, 102339 (2022)
14. Wen, L., et al.: A transformer-embedded multi-task model for dose distribution prediction. Int. J. Neural Syst., 2350043 (2023)
15. Ronneberger, O., Fischer, P., Brox, T.: U-Net: convolutional networks for biomedical image segmentation. In: Navab, N., Hornegger, J., Wells, W., Frangi, A. (eds.) Medical Image Computing and Computer-Assisted Intervention–MICCAI 2015: 18th International Conference, Munich, Germany, 5–9 October 2015, Proceedings, Part III, vol. 18, pp. 234–241. Springer, Cham (2015). https://doi.org/10.1007/978-3-319-24574-4_28
16. Chen, L.C., Zhu, Y., Papandreou, G., Schroff, F., Adam, H.: Encoder-decoder with atrous separable convolution for semantic image segmentation. In: Ferrari, V., Hebert, M., Sminchisescu, C., Weiss, Y. (eds.) Proceedings of the European Conference on Computer Vision (ECCV), pp. 801–818. Springer, Cham (2018). https://doi.org/10.1007/978-3-030-01234-2_49
17. Li, H., et al.: Explainable attention guided adversarial deep network for 3D radiotherapy dose distribution prediction. Knowl.-Based Syst. **241**, 108324 (2022)
18. Wen, L., et al.: Multi-level progressive transfer learning for cervical cancer dose prediction. Pattern Recogn. **141**, 109606 (2023)
19. Xie, Y., Yuan, M., Dong, B., Li, Q.: Diffusion model for generative image denoising. arXiv preprint arXiv:2302.02398 (2023)
20. Sohl-Dickstein, J., Weiss, E., Maheswaranathan, N., Ganguli, S.: Deep unsupervised learning using nonequilibrium thermodynamics. In: International Conference on Machine Learning, pp. 2256–2265 (2015)

21. Wolleb, J., Bieder, F., Sandkühler, R., Cattin, P.C.: Diffusion models for medical anomaly detection. In: Wang, L., Dou, Q., Fletcher, P.T., Speidel, S., Li, S. (eds.) Medical Image Computing and Computer Assisted Intervention–MICCAI 2022: 25th International Conference, Proceedings, Part VIII, pp. 35–45. Springer, Cham (2022). https://doi.org/10.1007/978-3-031-16452-1_4
22. Kim, B., Ye, J.C.: Diffusion deformable model for 4D temporal medical image generation. In: Wang, L., Dou, Q., Fletcher, P.T., Speidel, S., Li, S. (eds.) Medical Image Computing and Computer Assisted Intervention–MICCAI 2022: 25th International Conference, Proceedings, Part I, pp. 539–548. Springer, Cham (2022). https://doi.org/10.1007/978-3-031-16431-6_51
23. Zhang, J., Zhou, L., Wang, L., Liu, M., Shen, D.: Diffusion kernel attention network for brain disorder classification. IEEE Trans. Med. Imaging **41**(10), 2814–2827 (2022)
24. Li, H., et al.: SRDiff: single image super-resolution with diffusion probabilistic models. Neurocomputing **479**, 47–59 (2022)
25. Ho, J., Jain, A., Abbeel, P.: Denoising diffusion probabilistic models. Adv. Neural. Inf. Process. Syst. **33**, 6840–6851 (2020)
26. Helal, A., Omar, A.: Homogeneity index: effective tool. In: Advances in Neural Information Processing Systems, vol. 33, pp. 6840–6851 (2020). Ho, J., Jain, A., Abbeel, P.: Denoising diffusion probabilistic models
27. Graham, M.V., et al.: Clinical dose–volume histogram analysis for pneumonitis after 3D treatment for non-small cell lung cancer (NSCLC). Int. J. Radiat. Oncol. Biol. Phys. **45**(2), 323–329 (1999)

A Novel Multi-task Model Imitating Dermatologists for Accurate Differential Diagnosis of Skin Diseases in Clinical Images

Yan-Jie Zhou[1,2](✉), Wei Liu[1,2], Yuan Gao[1,2], Jing Xu[1,2], Le Lu[1], Yuping Duan[3], Hao Cheng[4], Na Jin[4], Xiaoyong Man[5], Shuang Zhao[6], and Yu Wang[1](✉)

[1] DAMO Academy, Alibaba Group, Hangzhou, China
zhouyanjie.zyj@alibaba-inc.com, Flimanadam@gmail.com
[2] Hupan Lab, Hangzhou, China
[3] School of Mathematical Sciences, Beijing Normal University, Beijing, China
[4] Sir Run Run Shaw Hospital, Hangzhou, China
[5] The Second Affiliated Hospital Zhejiang University School of Medicine, Hangzhou, China
[6] Xiangya Hospital Central South University, Changsha, China

Abstract. Skin diseases are among the most prevalent health issues, and accurate computer-aided diagnosis methods are of importance for both dermatologists and patients. However, most of the existing methods overlook the essential domain knowledge required for skin disease diagnosis. A novel multi-task model, namely **DermImitFormer**, is proposed to fill this gap by imitating dermatologists' diagnostic procedures and strategies. Through multi-task learning, the model simultaneously predicts body parts and lesion attributes in addition to the disease itself, enhancing diagnosis accuracy and improving diagnosis interpretability. The designed lesion selection module mimics dermatologists' zoom-in action, effectively highlighting the local lesion features from noisy backgrounds. Additionally, the presented cross-interaction module explicitly models the complicated diagnostic reasoning between body parts, lesion attributes, and diseases. To provide a more robust evaluation of the proposed method, a large-scale clinical image dataset of skin diseases with significantly more cases than existing datasets has been established. Extensive experiments on three different datasets consistently demonstrate the state-of-the-art recognition performance of the proposed approach.

Keywords: Skin disease · Multi-task learning · Vision transformer

1 Introduction

As the largest organ in the human body, the skin is an important barrier protecting the internal organs and tissues from harmful external substances, such as

H. Greenspan et al. (Eds.): MICCAI 2023, LNCS 14225, pp. 202–212, 2023.
https://doi.org/10.1007/978-3-031-43987-2_20

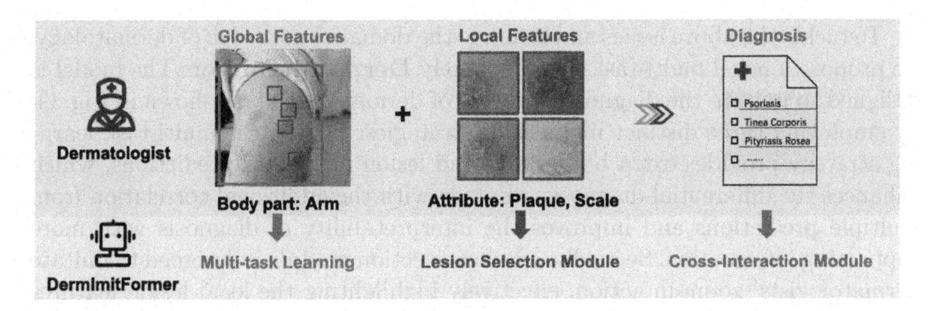

Fig. 1. The relationship between dermatologists' diagnostic procedures and our proposed model (best viewed in color).

sun exposure, pollution, and microorganisms [8,10]. In recent years, the increasing number of deaths by skin diseases has aroused widespread public concern [16,17]. Due to the complexity of skin diseases and the shortage of dermatological expertise resources, developing an automatic and accurate skin disease diagnosis framework is of great necessity.

Among non-invasive skin imaging techniques, dermoscopy is currently widely used in the diagnosis of many skin diseases [1,7], but it is technically demanding and not necessary for many common skin diseases. Clinical images, on the contrary, can be easily acquired through consumer-grade cameras, increasingly utilized in teledermatology, but their diagnostic value is underestimated. Recently, deep learning-based methods have received great attention in clinical skin disease image recognition and achieved promising results [3,5,11,18,20,23,25,26]. Sun *et al.* [18] released a clinical image dataset of skin diseases, namely SD-198, containing 6,584 images from 198 different categories. The results demonstrate that deep features from convolutional neural networks (CNNs) outperform handcrafted features in exploiting structural and semantic information. Gupta *et al.* [5] proposed a dual stream network that employs class activation maps to localize discriminative regions of the skin disease and exploit local features from detected regions to improve classification performance.

Although these approaches have achieved impressive results, most of them neglect the domain knowledge of dermatology and lack interpretability in diagnosis basis and results. In a typical inspection, dermatologists give an initial evaluation with the consideration of both global information, e.g. body part, and local information, e.g. the attributes of skin lesions, and further information including the patient's medical history or additional examination is required to draw a diagnostic conclusion from several possible skin diseases. Recognizing skin diseases from clinical images presents various challenges that can be summarized as follows: (1) Clinical images taken by portable electronic devices (e.g. mobile phones) often have cluttered backgrounds, posing difficulty in accurately locating lesions. (2) Skin diseases exhibit high intra-class variability in lesion appearance, but low inter-class variability, thereby making discrimination challenging. (3) The diagnostic reasoning of dermatologists is empirical and complicated, which makes it hard to simulate and model.

To tackle the above issues and leverage the domain knowledge of dermatology, we propose a novel multi-task model, namely **DermImitFormer**. The model is designed to imitate the diagnostic process of dermatologists (as shown in Fig. 1), by employing three distinct modules or strategies. Firstly, the multi-task learning strategy provides extra body parts and lesion attributes predictions, which enhances the differential diagnosis accuracy with the additional correlation from multiple predictions and improves the interpretability of diagnosis with more supporting information. Secondly, a lesion selection module is designed to imitate dermatologists' zoom-in action, effectively highlighting the local lesion features from noisy backgrounds. Thirdly, a cross-interaction module explicitly models the complicated diagnostic reasoning between body parts, lesion attributes, and diseases, increasing the feature alignments and decreasing gradient conflicts from different tasks. Last but not least, we build a new dataset containing 57,246 clinical images. The dataset includes 49 most common skin diseases, covering 80% of the consultation scenarios, 15 body parts, and 27 lesion attributes, following the International League of Dermatological Societies (ILDS) guideline [13].

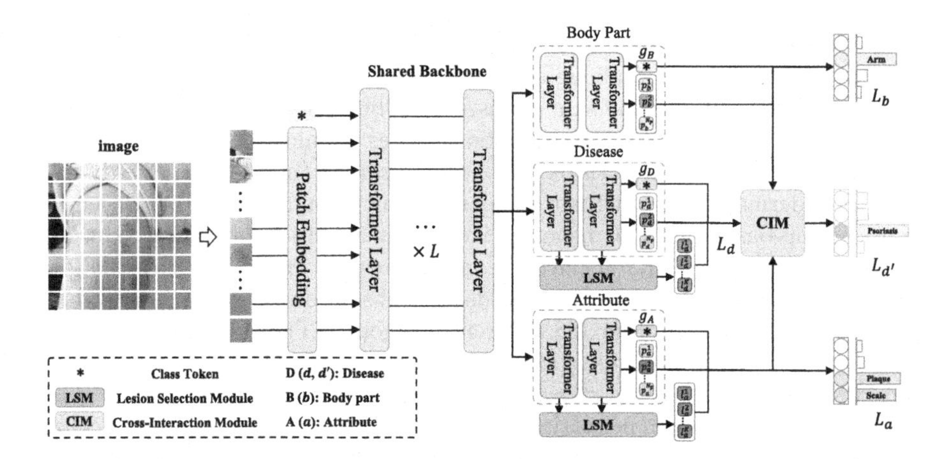

Fig. 2. The overall architecture of the multi-task imitation model (DermImitFormer) with shared backbone and task-specific heads.

The main contributions can be summarized as follows: (1) A novel multi-task model DermImitFormer is proposed to imitate dermatologists' diagnostic processes, providing outputs of diseases, body parts, and lesion attributes for improved clinical interpretability and accuracy. (2) A lesion selection module is presented to encourage the model to learn more distinctive lesion features. A cross-interaction module is designed to effectively fuse three different feature representations. (3) A large-scale clinical image dataset of skin diseases is established, containing significantly more cases than existing datasets, and closer to the real data distribution of clinical routine. More importantly, our proposed approach achieves the leading recognition performance on three different datasets.

2 Method

The architecture of the proposed multi-task model DermImitFormer is shown in Fig. 2. It takes the clinical image as input and outputs the classification results of skin diseases, body parts, and attributes in an end-to-end manner. During diagnostic processes, dermatologists consider local and global contextual features of the entire clinical image, including shape, size, distribution, texture, location, etc. To effectively capture these visual features, we use the vision transformer (ViT) [4] as the shared backbone. Three separate task-specific heads are then utilized to predict diseases, body parts, and attributes, respectively, with each head containing two independent ViT layers. In particular, in the task-specific heads of diseases and attributes, the extracted features of each layer are separated into the image features and the patch features. These two groups of features are fed into the lesion selection module (LSM), to select the most informative lesion tokens. Finally, the feature representations of diseases, body parts, and attributes are delivered to the cross-interaction module (CIM) to generate a more comprehensive representation for the final differential diagnosis.

Shared Backbone. Following the ViT model, an input image X is divided to N_p squared patches $\{x_n, n \in \{1, 2, ..., N_p\}\}$, where $N_p = (H \times W)/P^2$, P is the side length of a squared patch, H and W are the height and width of the image, respectively. Then, the patches are flattened and linearly projected into patch tokens with a learnable position embedding, denoted as $t_n, n \in \{1, 2, ..., N_p\}$. Together with an extra class token t_0, the network inputs are represented as $t_n \in \mathbb{R}^D, n \in \{0, 1, ..., N_p\}$ with a dimension of D. Finally, the tokens are fed to L consecutive transformer layers to obtain the preliminary image features.

Lesion Selection Module. As introduced above, skin diseases have high variability in lesion appearance and distribution. Thus, it requires the model to concentrate on lesion patches so as to describe the attributes and associated diseases precisely. The multi-head self-attention (MHSA) block in ViT generates global attention, weighing the informativeness of each token. Inspired by [19], we introduce a lesion selection module (LSM), which guides the transformer encoder to select the tokens that are most relevant to lesions at different levels. Specifically, for each attention head in MHSA blocks, we compute the attention matrix $\boldsymbol{A}^m = \text{Softmax}(\mathcal{Q}\mathcal{K}^T/\sqrt{D}) \in \mathbb{R}^{(N_p+1)\times(N_p+1)}$, where $m \in \{1, 2, ..., N_h\}$, N_h denoting the number of heads, $\mathcal{Q}$ and $\mathcal{K}$ the *Query* and *Key* representations of the block inputs, respectively. The first row calculates the similarities between the class token and each patch token. As the class token is utilized for classification, the higher the value, the more informative each token is. We apply softmax to the first row and the first column of $\boldsymbol{A}^m$, denoted as $a_{0,n}^m$ and $a_{n,0}^m, n \in \{1, 2, ..., N_p\}$, representing the attention scores between the class token and other tokens:

$$a_{0,n}^{m} = \frac{e^{A_{0,n}^{m}}}{\sum_{i=1}^{N_p} e^{A_{0,i}^{m}}}, \quad a_{n,0}^{m} = \frac{e^{A_{n,0}^{m}}}{\sum_{i=1}^{N_p} e^{A_{i,0}^{m}}}, \quad s_n = \frac{1}{N_h} \sum_{m=1}^{N_h} a_{0,n}^{m} \cdot a_{n,0}^{m} \tag{1}$$

The mutual attention score s_n is calculated across all attention heads. Thereafter, we select the top K tokens according to s_n for two task heads as $\boldsymbol{l}_d^k$ and $\boldsymbol{l}_a^k, k \in \{1, 2, ..., K\}$.

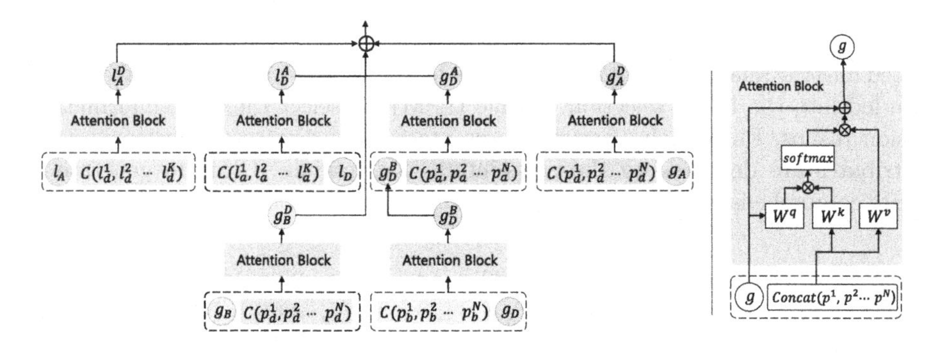

Fig. 3. Schematic of cross-interaction module.

Cross-Interaction Module. A diagnostic process of skin diseases takes multiple visual information into account, which is relatively complicated and difficult to model in an analytical way. Simple fusion operations such as concatenation are insufficient to simulate the diagnostic logic. Thus, partially inspired by [21], the CIM is designed to learn complicated correlations between disease, body part, and attribute. The detailed module schematic is shown in Fig. 3. Firstly, the features of body-part and disease are integrated to enhance global representations by a cross-attention block. For example, the fusion between the class token of disease and patch tokens of body-part is:

$$z_b = LN(GAP(\boldsymbol{p}_b^1, \boldsymbol{p}_b^2, ..., \boldsymbol{p}_b^{N_p})) \tag{2}$$

$$\mathcal{Q} = LN\left(\boldsymbol{g}_D\right) \boldsymbol{W}_{BD}^{\mathcal{Q}}, \quad \mathcal{K} = \boldsymbol{z}_b \boldsymbol{W}_{BD}^{\mathcal{K}}, \quad \mathcal{V} = \boldsymbol{z}_b \boldsymbol{W}_{BD}^{\mathcal{V}} \tag{3}$$

$$\boldsymbol{g}_D^B = LN\left(\boldsymbol{g}_D\right) + \text{linear}(\text{softmax}(\frac{\mathcal{Q}\mathcal{K}^T}{\sqrt{F/N_h}})\mathcal{V}) \tag{4}$$

where $\boldsymbol{g}_B$, $\boldsymbol{g}_D$ are the class token, $\boldsymbol{p}_b^i$, $\boldsymbol{p}_d^i, i \in \{1, 2, ..., N_p\}$ the corresponding patch tokens. GAP and LN denote the global average pooling and layer normalization, respectively. $\boldsymbol{W}_{BD}^{\mathcal{Q}}, \boldsymbol{W}_{BD}^{\mathcal{K}}, \boldsymbol{W}_{BD}^{\mathcal{V}} \in \mathcal{R}^{F \times F}$ denote learnable parameters. F denotes the dimension of features. $\boldsymbol{g}_B^D$ is computed from the patch tokens of disease and the class token of body-part in the same fashion. Similarly, we can obtain the fused class tokens ($\boldsymbol{g}_A^D$ and $\boldsymbol{g}_D^A$) and the fused local class tokens ($\boldsymbol{l}_A^D$ and $\boldsymbol{l}_D^A$) between attribute and disease. Note that the disease class token $\boldsymbol{g}_D$ is replaced by $\boldsymbol{g}_D^B$ in the later computations, and local class tokens $\boldsymbol{l}_A$ and $\boldsymbol{l}_D$ in Fig. 3 are generated by GAP on selected local patch tokens from LSM. Finally, these mutually enhanced features from CIM are concatenated together to generate more accurate predictions of diseases, body parts, and attributes.

Learning and Optimization. We argue that joint training can enhance the feature representation for each task. Thus, we define a multi-task loss as follows:

Table 1. Ablation study for DermImitFormer on Derm-49 dataset. D, B, and A denote the task-specific head of diseases, body parts, and attributes, respectively.

Dimension	LSM	Fusion		F1-score (%)			Accuracy(%)
		Concat	CIM	Disease	Body part	Attribute	Disease
D				76.2	-	-	80.4
D	✓			77.8	-	-	82.0
D + B	✓	✓		78.1	85.0	-	82.4
D + A	✓	✓		78.4	-	68.7	82.6
D + B + A	✓	✓		79.1	85.1	69.0	82.9
D + B + A	✓		✓	**79.5**	**85.9**	**70.4**	**83.3**

$$\mathcal{L}_x = -\frac{1}{N_s}\sum_{i=1}^{N_s}\sum_{j=1}^{n_x} y_{ij}\log(p_{ij}), \quad x \in \{d, d'\} \tag{5}$$

$$\mathcal{L}_h = -\frac{1}{N_s}\sum_{i=1}^{N_s}\sum_{j=1}^{n_h} y_{ij}\log(p_{ij}) + (1-y_{ij})\log(1-p_{ij}), \quad h \in \{a, b\} \tag{6}$$

$$\mathcal{L} = \mathcal{L}_d + \mathcal{L}_{d'} + \mathcal{L}_b + \mathcal{L}_a \tag{7}$$

where N_s denotes the number of samples, n_x, n_h the number of classes for each task, and p_{ij}, y_{ij} the prediction and label, respectively. Notably, body parts and attributes are defined as multi-label classification tasks, optimized with the binary cross-entropy loss, as shown in Eq. 6. The correspondence of x and h is shown in Fig. 2.

3 Experiment

Datasets. The proposed DermImitFormer is evaluated on three different clinical skin image datasets including an in-house dataset and two public benchmarks. (1) **Derm-49:** We establish a large-scale clinical image dataset of skin diseases, collected from three cooperative hospitals and a teledermatology platform. The 57,246 images in the dataset were annotated with the diagnostic ground truth of skin disease, body parts, and lesions attributes from the patient records. We clean up the ground truth into 49 skin diseases, 15 body parts, and 27 lesion attributes following the ILDS guidelines [13]. (2) **SD-198** [18]**:** It is one of the largest publicly available datasets in this field containing 198 skin diseases and 6,584 clinical images collected through digital cameras or mobile phones. (3) **PAD-UFES-20** [15]**:** The dataset contains 2,298 samples of 6 skin diseases. Each sample contains a clinical image and a set of metadata with labels such as diseases and body parts.

Implementation Details. The DermImitFormer is initialized with the pre-trained ViT-B/16 backbone and optimized with SGD method (initial learning rate 0.003, momentum 0.95, and weight decay 10^{-5}) for 100 epochs on 4 NVIDIA Tesla V100 GPUs with a batch size of 96. We define the input size i.e. $H = W = 384$ that produces a total of 576 spatial tokens i.e. $N_p = 576$ for a ViT-B backbone. $K = 24$ in the LSM module. For data augmentation, we employed the Cutmix [24] with a probability of 0.5 and Beta(0.3, 0.3) during optimization. We adopt precision, recall, F1-score, and accuracy as the evaluation metrics.

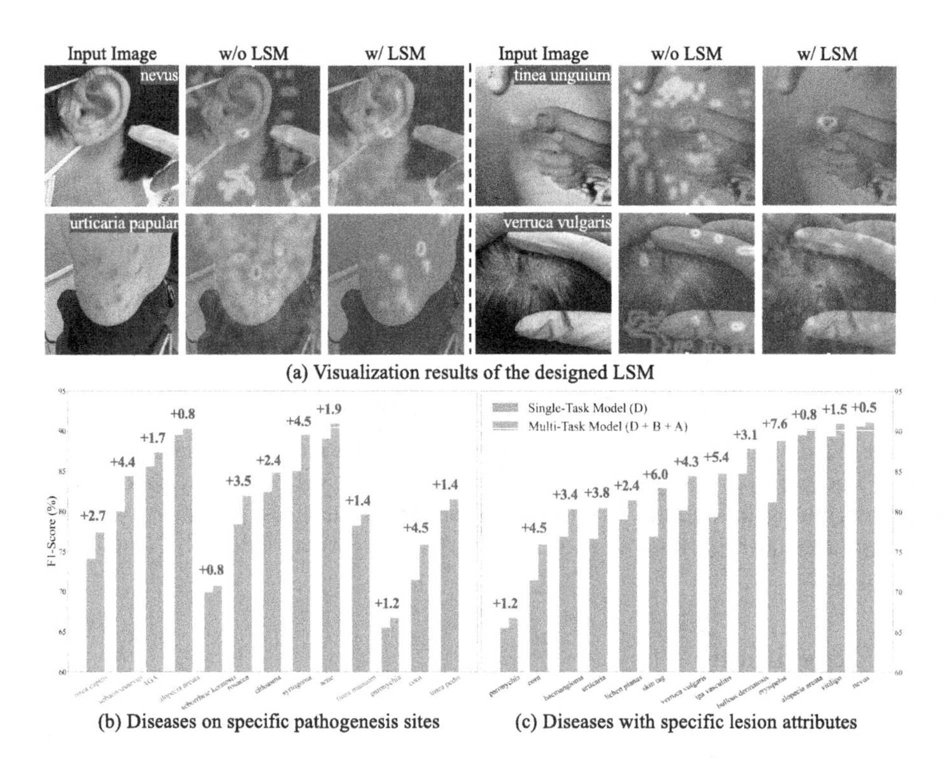

Fig. 4. Comparative results on Derm-49 dataset. Red, blue, green, and purple fonts denote diseases on heads, faces, hands, and feet (best viewed in color).

Ablation Study. The experiment is conducted based on ViT-B/16 and the results are reported in Table 1. (1) **LSM:** Quantitative results demonstrate that the designed LSM yields 1.6% improvement in accuracy. Qualitative results are shown in Fig. 4(a), which depicts the attention maps obtained from the last transformer layer. Without LSM, vision transformers would struggle of localizing lesions and produce noisy attention maps. With LSM, the attention maps are more discriminative and lesions are localized precisely, regardless of variations in terms of scale and distribution. (2) **Multi-task learning:** The models are trained with a shared backbone and different combinations of task-specific heads. The results show that multi-task learning (D+B+A) increases the F1-score from 77.8 to 79.1. (3) **CIM:** Quantitative results show that the presented CIM can further improve the F1-score of diseases to 79.5. Notably, the p-value of 1.03e-05 (<0.01) is calculated by comparing the results of 5-fold cross-validation with the baseline, illustrating the significance of our model. In particular, the representation with fused features of body parts and attributes can improve the recognition performance of diseases. As shown in Fig. 4(b) and (c), statistics show that the classification performance of these diseases is improved by the multi-task learning strategy and CIM. For instance, rosacea and tinea versicolor share the same attributes of macule and papular, but rosacea typically affects the face. By fusing the representation of body parts, the F1-score of rosacea is increased by 4.5%. Similarly, our model improves the recognition accuracy of diseases with distinctive lesion attributes such as skin tags, urticaria, etc. Meanwhile, the extra information about body parts and attributes improves the interpretability of diagnoses.

Table 2. Comparison to state-of-the-art methods on Derm-49 dataset (top) and two public benchmarks: SD-198 dataset (mid), PAD-UFES-20 dataset (bottom).

Datasets	Methods	F1-score (%)	Precision (%)	Recall (%)	Accuracy (%)
Derm-49	DX [6]	72.6 ± 2.3	73.7 ± 0.6	72.2 ± 3.1	73.4 ± 0.7
	ViT-Base [4]	75.9 ± 0.8	80.6 ± 0.7	72.9 ± 0.9	80.4 ± 0.4
	Swin-Base [12]	76.6 ± 0.6	83.5 ± 1.1	71.0 ± 0.9	80.6 ± 0.5
	DermImitFomer	**78.8 ± 0.5**	**83.5 ± 0.6**	**74.6 ± 1.1**	**82.6 ± 0.5**
SD-198	SPBL [23]	66.2 ± 1.6	71.4 ± 1.7	65.7 ± 1.6	67.8 ± 1.8
	Aux-D [22]	68.0 ± 1.0	67.9 ± 1.0	69.2 ± 0.9	-
	Dual Stream [5]	70.9 ± 1.2	73.1 ± 1.4	69.2 ± 1.1	71.4 ± 1.1
	TPC [9]	63.2 ± 1.6	65.6 ± 1.7	64.7 ± 1.6	-
	IASN [3]	68.6 ± 0.7	71.9 ± 0.8	70.0 ± 0.9	70.7 ± 0.8
	PCCT [2]	65.2 ± 1.6	68.4 ± 1.4	66.0 ± 1.5	-
	DermImitFomer-ST	**73.6 ± 2.6**	**76.1 ± 2.6**	**75.1 ± 2.2**	**74.5 ± 2.6**
PAD-UFES-20	PAD [15]	71.0 ± 2.9	73.4 ± 2.9	70.8 ± 2.8	70.7 ± 2.8
	T-Enc [14]	-	-	-	61.6 ± 5.1
	ResNet-50 [15]	67.8 ± 3.7	72.0 ± 4.1	67.0 ± 4.1	67.1 ± 4.1
	ViT-Base [4]	69.9 ± 1.4	69.4 ± 1.5	70.4 ± 2.2	70.6 ± 1.8
	Swin-Base [12]	72.1 ± 2.5	72.0 ± 2.9	72.7 ± 2.6	72.7 ± 2.5
	DermImitFomer-ST	73.6 ± 2.8	72.8 ± 3.2	74.4 ± 2.4	74.4 ± 2.4
	DermImitFomer	**74.5 ± 2.5**	**73.9 ± 2.9**	**75.0 ± 2.1**	**75.0 ± 2.1**

Results. To evaluate the effectiveness of our proposed DermImitFormer, we conduct a comparison with various state-of-the-art methods on three different datasets. The results are reported in Table 2. (1) **Derm-49:** Compared with other state-of-the-art approaches, our proposed DermImitFormer achieves the leading classification performance in our established dataset with the 5-fold cross-validation splits. (2) **SD-198:** Since the dataset does not contain labels of lesion attributes and body parts, the proposed DermImitFormer in Single-Task mode (w/o CIM) is implemented in the experiment. The result is based on the provided 5-fold cross-validation splits. Quantitative results in Table 2(mid) demonstrate that our proposed DermImitFormer-ST achieves state-of-the-art classification performance. In contrast to other approaches, our model can precisely localize more discriminative lesion regions and thus has superior classification accuracy. (3) **PAD-UFES-20:** The dataset contains labels of diseases and body parts. Thus, the proposed DermImitFormer with different modes is evaluated in the experiment by the 5-fold cross-validation splits. Quantitative results in Table 2 (bottom) demonstrate that our proposed model outperforms the CNN-based [14,15], and transformer-based methods [4,12], achieving the state-of-the-art classification performance. In particular, the performance of DermImitFormer is better than that of DermImitFormer-ST in Single-Task mode (w/o CIM), which further indicates the effectiveness of the multi-task learning strategy and CIM.

4 Conclusion

In this work, DermImitFormer, a multi-task model, has been proposed to better utilize dermatologists' domain knowledge by mimicking their subjective diagnostic procedures. Extensive experiments demonstrate that our approach achieves state-of-the-art recognition performance in two public benchmarks and a large-scale in-house dataset, which highlights the potential of our approach to be employed in real clinical environments and showcases the value of leveraging domain knowledge in the development of machine learning models.

Acknowledgement. This work was supported by National Key R&D Program of China (2020YFC2008703) and the Project of Intelligent Management Software for Multimodal Medical Big Data for New Generation Information Technology, the Ministry of Industry and Information Technology of the People's Republic of China (TC210804V).

References

1. Binder, M., et al.: Epiluminescence microscopy: a useful tool for the diagnosis of pigmented skin lesions for formally trained dermatologists. Arch. Dermatol. **131**(3), 286–291 (1995)
2. Chen, K., Lei, W., Zhang, R., Zhao, S., Zheng, W.S., Wang, R.: PCCT: progressive class-center triplet loss for imbalanced medical image classification. arXiv preprint arXiv:2207.04793 (2022)

3. Chen, X., Li, D., Zhang, Y., Jian, M.: Interactive attention sampling network for clinical skin disease image classification. In: Ma, H., et al. (eds.) PRCV 2021. LNCS, vol. 13021, pp. 398–410. Springer, Cham (2021). https://doi.org/10.1007/978-3-030-88010-1_33
4. Dosovitskiy, A., et al.: An image is worth 16x16 words: transformers for image recognition at scale. arXiv preprint arXiv:2010.11929 (2020)
5. Gupta, K., Krishnan, M., Narayanan, A., Narayan, N.S., et al.: Dual stream network with selective optimization for skin disease recognition in consumer grade images. In: Proceedings of the International Conference on Pattern Recognition (ICPR), pp. 5262–5269. IEEE (2021)
6. Jalaboi, R., Faye, F., Orbes-Arteaga, M., Jørgensen, D., Winther, O., Galimzianova, A.: Dermx: an end-to-end framework for explainable automated dermatological diagnosis. Med. Image Anal. **83**, 102647 (2023)
7. Kittler, H., Pehamberger, H., Wolff, K., Binder, M.: Diagnostic accuracy of dermoscopy. Lancet Oncol. **3**(3), 159–165 (2002)
8. Kshirsagar, P.R., Manoharan, H., Shitharth, S., Alshareef, A.M., Albishry, N., Balachandran, P.K.: Deep learning approaches for prognosis of automated skin disease. Life **12**(3), 426 (2022)
9. Lei, W., Zhang, R., Yang, Y., Wang, R., Zheng, W.S.: Class-center involved triplet loss for skin disease classification on imbalanced data. In: Proceedings of the International Symposium on Biomedical Imaging (ISBI), pp. 1–5. IEEE (2020)
10. Li, L.F., Wang, X., Hu, W.J., Xiong, N.N., Du, Y.X., Li, B.S.: Deep learning in skin disease image recognition: a review. IEEE Access **8**, 208264–208280 (2020)
11. Liu, Y., et al.: A deep learning system for differential diagnosis of skin diseases. Nat. Med. **26**(6), 900–908 (2020)
12. Liu, Z., et al.: Swin transformer: hierarchical vision transformer using shifted windows. In: Proceedings of the IEEE/CVF International Conference on Computer Vision (ICCV), pp. 10012–10022 (2021)
13. Nast, A., Griffiths, C.E., Hay, R., Sterry, W., Bolognia, J.L.: The 2016 international league of dermatological societies' revised glossary for the description of cutaneous lesions. Br. J. Dermatol. **174**(6), 1351–1358 (2016)
14. Ou, C., et al.: A deep learning based multimodal fusion model for skin lesion diagnosis using smartphone collected clinical images and metadata. Front. Surg. **9**, 1029991 (2022)
15. Pacheco, A.G., Krohling, R.A.: The impact of patient clinical information on automated skin cancer detection. Comput. Biol. Med. **116**, 103545 (2020)
16. Rogers, H.W., Weinstock, M.A., Feldman, S.R., Coldiron, B.M.: Incidence estimate of nonmelanoma skin cancer (keratinocyte carcinomas) in the us population, 2012. JAMA Dermatol. **151**(10), 1081–1086 (2015)
17. Siegel, R.L., Miller, K.D., Fuchs, H.E., Jemal, A.: Cancer statistics, 2022. CA Cancer J. Clin. **72**(1), 7–33 (2022)
18. Sun, X., Yang, J., Sun, M., Wang, K.: A benchmark for automatic visual classification of clinical skin disease images. In: Leibe, B., Matas, J., Sebe, N., Welling, M. (eds.) ECCV 2016. LNCS, vol. 9910, pp. 206–222. Springer, Cham (2016). https://doi.org/10.1007/978-3-319-46466-4_13
19. Wang, J., Yu, X., Gao, Y.: Feature fusion vision transformer for fine-grained visual categorization. arXiv preprint arXiv:2107.02341 (2021)
20. Wu, J., et al.: Learning differential diagnosis of skin conditions with co-occurrence supervision using graph convolutional networks. In: Martel, A.L., et al. (eds.) MICCAI 2020. LNCS, vol. 12262, pp. 335–344. Springer, Cham (2020). https://doi.org/10.1007/978-3-030-59713-9_33

21. Xu, J., et al.: Remixformer: a transformer model for precision skin tumor differential diagnosis via multi-modal imaging and non-imaging data. In: Wang, L., Dou, Q., Fletcher, P.T., Speidel, S., Li, S. (eds.) MICCAI 2022. LNCS, vol. 13433, pp. 624–633. Springer, Cham (2022). https://doi.org/10.1007/978-3-031-16437-8_60
22. Xu, Z., Zhuang, J., Zhang, R., Wang, R., Guo, X., Zheng, W.S.: Auxiliary decoder and classifier for imbalanced skin disease diagnosis. J. Phys. Conf. Ser. **1631**(1), 012046 (2020)
23. Yang, J., et al.: Self-paced balance learning for clinical skin disease recognition. IEEE Trans. Neural Netw. Learn. Syst. **31**(8), 2832–2846 (2019)
24. Yun, S., Han, D., Oh, S.J., Chun, S., Choe, J., Yoo, Y.: Cutmix: regularization strategy to train strong classifiers with localizable features. In: Proceedings of the IEEE/CVF International Conference on Computer Vision (ICCV), pp. 6023–6032 (2019)
25. Zhang, J., Xie, Y., Wu, Q., Xia, Y.: Medical image classification using synergic deep learning. Med. Image Anal. **54**, 10–19 (2019)
26. Zhang, J., Xie, Y., Xia, Y., Shen, C.: Attention residual learning for skin lesion classification. IEEE Trans. Med. Imaging **38**(9), 2092–2103 (2019)

Wall Thickness Estimation from Short Axis Ultrasound Images via Temporal Compatible Deformation Learning

Ang Zhang[1,2,3], Guijuan Peng[4], Jialan Zheng[1,2,3], Jun Cheng[1,2,3], Xiaohua Liu[4], Qian Liu[4], Yuanyuan Sheng[4], Yingqi Zheng[4], Yumei Yang[4], Jie Deng[4], Yingying Liu[4(✉)], Wufeng Xue[1,2,3(✉)], and Dong Ni[1,2,3]

[1] National-Regional Key Technology Engineering Laboratory for Medical Ultrasound, School of Biomedical Engineering, Shenzhen University Medical School, Shenzhen University, Shenzhen, China
xuewf@szu.edu.cn
[2] Medical Ultrasound Image Computing (MUSIC) Laboratory, Shenzhen University, Shenzhen, China
[3] Marshall Laboratory of Biomedical Engineering, Shenzhen University, Shenzhen, China
[4] Department of Ultrasound, Shenzhen People's Hospital (The Second Clinical Medical College, Jinan University; The First Affiliated Hospital, Southern University of Science and Technology), Shenzhen, China
yingyingliu@ext.jnu.edu.cn

Abstract. Structural parameters of the heart, such as left ventricular wall thickness (LVWT), have important clinical significance for cardiac disease. In clinical practice, it requires tedious labor work to be obtained manually from ultrasound images and results in large variations between experts. Great challenges exist to automatize this procedure: the myocardium boundary is sensitive to heavy noise and can lead to irregular boundaries; the temporal dynamics in the ultrasound video are not well retained. In this paper, we propose a Temporally Compatible Deformation learning network, named *TC-Deformer*, to detect the myocardium boundaries and estimate LVWT automatically. Specifically, we first propose a two-stage deformation learning network to estimate the myocardium boundaries by deforming a prior myocardium template. A global affine transformation is first learned to shift and scale the template. Then a dense deformation field is learned to adjust locally the template to match the myocardium boundaries. Second, to make the deformation learning of different frames become compatible in the temporal dynamics, we adopt the mean parameters of affine transformation for all frames and propose a bi-direction deformation learning to guarantee that the deformation fields across the whole sequences can be applied to both the myocardium boundaries and the ultrasound images. Experimental results on an ultrasound dataset of 201 participants show that the proposed method can achieve good boundary detection of basal, middle, and apical myocardium, and lead to accurate estimation of the LVWT, with a mean absolute error of less than 1.00 mm. When compared with

H. Greenspan et al. (Eds.): MICCAI 2023, LNCS 14225, pp. 213–222, 2023.
https://doi.org/10.1007/978-3-031-43987-2_21

human methods, our TC-Deformer performs better than the junior cardiologists and is on par with the middle-level cardiologists.

Keywords: Wall thickness · Segmentation · Deformation learning

1 Introduction

Cardiovascular disease is a leading cause of death in the world. Accurate quantification of left ventricular wall thicknesses (LVWT) from ultrasound images is among the most clinically important and significant indices for evaluating cardiac function and diagnosis of cardiac diseases [2,6]. Figure 1 illustrates short axis (SAX) ultrasound images of basal, middle, and apical myocardium, with the corresponding LVWTs according to the 16-segment myocardium model. In clinical practice, obtaining reliable clinical information mainly depends on radiologists to manually draw the contours of the endocardium and epicardium of the left ventricle (LV). It is time-consuming and laborious. Efforts have been devoted to the automatic estimation of LVWTs, where great challenge exists. First, the myocardium boundary is sensitive to heavy noise, especially for the apical and basal SAX images, and can lead to irregular boundaries and undermine the estimation of LVWTs. Second, the temporal dynamics in the ultrasound video are difficult to be modeled, leading to prediction results that are not well compatible with the temporal dynamics of the whole video.

Existing work can be divided into two categories: segmentation-based and direct-regression methods. The direct-regression methods to learn the regress LVWTs from cardiac images directly without identifying the contours first. [5,15] proposed end-to-end cardiac index quantification frameworks based on cascaded convolutional autoencoders and regression networks, using only the values of cardiac indices for supervision. [4] proposed a two-stage network that learns the LV contours first and then estimates the LV indices with a new network. [16] proposed a residual recurrent neural network further improves the estimation by modeling the temporal and spatial of the LV myocardium to achieve accurate frame-by-frame LVWT estimation. However, these models lack explicit temporal dynamic modeling of the whole sequence.

Segmentation-based methods segment the myocardium first and then calculate cardiac parameters. To the best of our knowledge, existing segmentation work mainly focus on apical views to evaluate the ejection fraction, and rare work exists for short-axis views. [14] utilizes the underlying motion information to assist in improving segmentation results by accurately predicting optical flow fields. [12,13] proposed appearance-level and shape-level co-learning (CLAS) to enhance the temporal consistency of the predicted masks across the whole sequence and accuracy. This method effectively improves the accuracy and consistency of myocardial segmentation. [1] proposed to introduce residual structure into U-net and [3] proposed a hybrid framework combining a convolutional encoder-decoder structure and a transformer. [7,17] proposed a multi-attention mechanism to guide the network to capture features effectively while suppressing

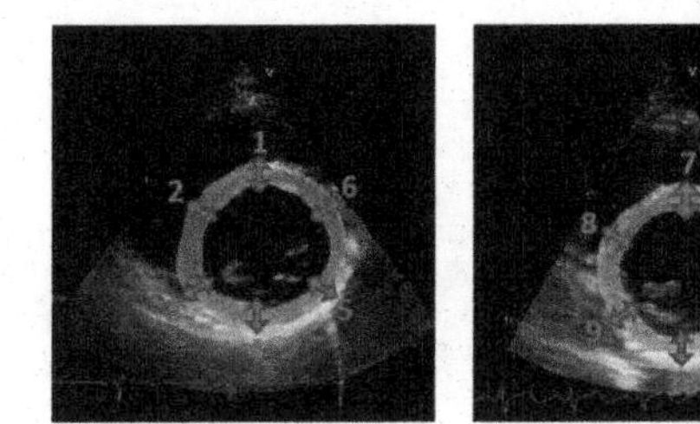

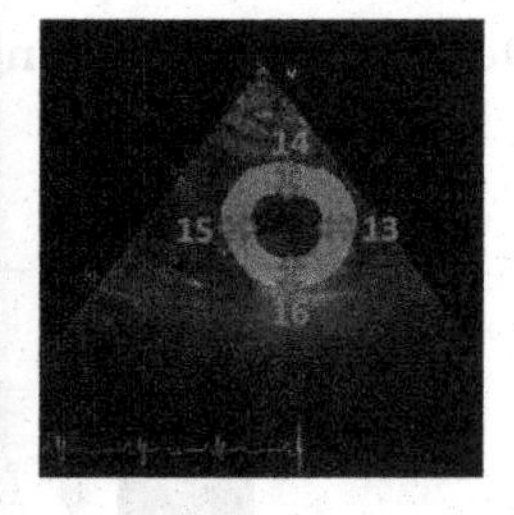

Fig. 1. Illustration of LVWT to be quantified for short-axis view cardiac image and 16 segments.Basal (left), middle (middle), and apical (right). Regional wall thicknesses (black arrows). 1–16: 16 segments.

noise, and integrated deep supervision mechanism and spatial pyramid feature fusion to enhance feature extraction. However, these models are not robust to the heavy noise in the SAX images, which may lead to irregular boundaries and undermines the estimation of LVWT.

To overcome the above mention challenges and inspired [8] where template transformer was employed for image segmentation, we propose a novel Temporal-Compatible Deformation learning network for myocardium boundary detection from ultrasound SAX images. The primary contributions of this paper are as follows. 1) To overcome the irregular boundaries caused by the heavy noise, we propose a two-stage deformation learning network for myocardium boundary detection. A global affine transformation and a local deformation are used to deform the prior myocardium template to match the myocardium boundary. 2) To make the template-deformed myocardium boundaries compatible across the whole sequence, we propose a bi-direction deformation learning to guarantee that the deformation fields across the whole sequences can be applied to both the myocardium boundaries and the ultrasound images. 3) The proposed TC-deformer achieves excellent performance for LVWT estimation, with an error of less than 1.00 mm, and is comparable with middle-level cardiologists.

2 Methods

The structures of the myocardium in ultrasound SAX images generally follow a circular ring shape, which is a vital characteristic of prior knowledge in short-axis myocardial segmentation, especially for ultrasound images with heavy noisy myocardium boundaries. In this paper, we propose a novel method, Temporal Compatibility Deformation Learning, named *TC-Deformer*, to achieve accurate and plausible myocardium contours and LVWTs estimation. The details are described as follows.

2.1 Deformation Learning

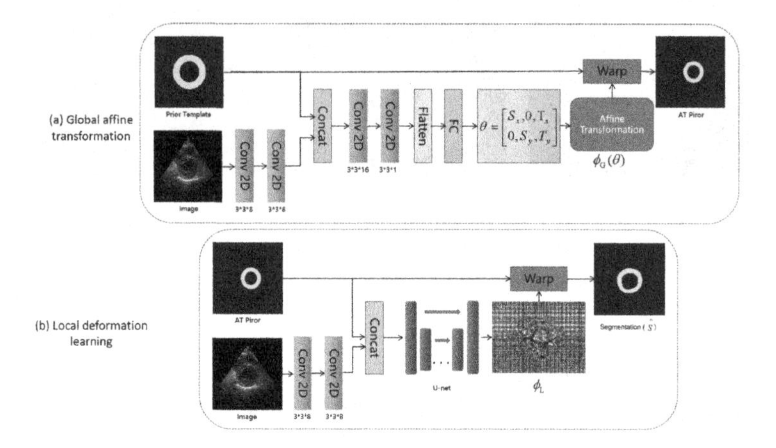

Fig. 2. (a)Overview of the proposed global affine transformation. (b)Overview of the proposed local deformation learning.

Global Affine Transformation. Our deformation learning consists of two stages: global affine transformation and local deformation learning. In this subsection, we will describe our global affine transformation. Considering the diversity of the cardiac structure and data gaps from different machines, we compute the mean value of the thickness measurements and the center position of the circle according to the annotated information to generate the prior template. As shown in Fig. 2(a), the prior template (P) is first concatenated with the convolution features extracted from the SAX image to learn the global affine parameters $\theta = \{(\Delta x, \Delta y), s\}$, which represent the shift and scale from the prior template to the myocardium boundaries in the SAX image. Then we get the affine prior (AT) $S_{AT} = \phi_G(P)$ by warping the prior template (P) with the global affine parameters θ. The loss function is as follows:

$$Loss_{AT} = -S \cdot logS_{AT} + (1 - \frac{|S \cdot S_{AT}|}{|S| + |S_{AT}|}) \tag{1}$$

where S is the ground truth of the myocardium segmentation. However, the shapes warping from the prior template with global affine parameters are far from precise due to the individual variation. So, we introduce the local deformation learning to get a precise myocardium shape with the learned dense deformation field.

Local Deformation Learning. In this part, we aim to learn a dense deformation field to adjust the AT prior locally to match the myocardium boundaries. As shown in Fig. 2(b), the AT prior is first concatenated with the image convolution feature to learn a dense deformation field $\phi_L \in R^{256\times256\times2}$, which represents the pixel-level displacement along both horizontal and vertical directions. Our local deformation learning considers the prior shape, the prior position, and the image feature simultaneously, which can help the network learn a more precise deformation field and get a local adjustment of the template.

Finally, we take the segmentation $\hat{S} = \phi_L(S_{AT})$ warping from the AT prior with the dense deformation filed as the final myocardium segmentation result. The loss function is as follows:

$$Loss_{seg} = -S \cdot log\hat{S} + (1 - \frac{|S \cdot \hat{S}|}{|S| + |\hat{S}|}) \quad (2)$$

where S is the ground truth of the myocardium segmentation.

2.2 TC-Deformer

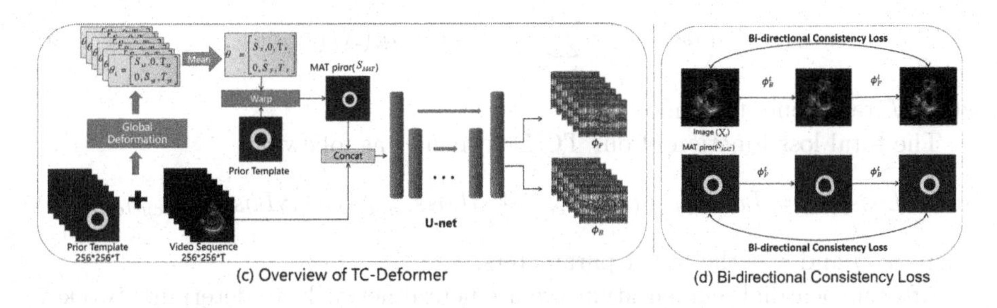

Fig. 3. (c) Overview of the proposed TC-Deformer network framework. (d) Schematic diagram of bi-directional consistency loss for images and segmentation.

In this work, only frames at the end-systolic (ED) and end-diastolic (ES) phases are annotated by cardiologists for each ultrasound SAX video. To make the template-deformed myocardium boundaries compatible across the whole sequence, we propose a bi-direction deformation learning to guarantee that the deformation fields across the whole sequence can be applied to both the myocardium boundaries and the ultrasound images in the sequence.

As shown in Fig. 3, first of all, we use the mean $\bar{\theta}$ of the affine transformation parameters $\{\theta_1, \theta_2, \theta_3, ..., \theta_{T-2}, \theta_{T-1}, \theta_T\}$ of all frames in the sequence as a video-level parameter and obtain the mean affine transformation prior template (MAT prior). Next, the MAT prior is combined with the sequential images to learn a series of dense bi-direction deformation fields. Let ϕ_F^t be the forward deformation and ϕ_B^t the backward deformation for the frame t. Our bi-direction deformation learning (as shown in Fig. 3(b)) aims to constrain that for each frame X_t, after forward deformation and backward deformation, we can still obtain the original images. We adopt the structural similarity metric (SSIM) [11] in image quality assessment to quantify the deformation error. For the image cycle, the loss function is as follows:

$$Loss_{imgcycle} = \sum_{t=1}^{T} (1 - SSIM(X_t, \phi_F^t (\phi_B^t (X_t))) \tag{3}$$

Similarly, for the MAT prior S_{MAT}, we can have the shape consistency constraint when applied to the bi-direction deformation procedure:

$$Loss_{shapecycle} = \sum_{t=1}^{T} (1 - \frac{|S_{MAT} \cdot \phi_B^t (\phi_F^t (S_{MAT}))|}{|S_{MAT}| + |\phi_B^t (\phi_F^t (S_{MAT}))|}) \tag{4}$$

As the temporally compatible deformation started with a common template, we assume that when warped backward, all the frames will have a similar appearance. So, we introduce a centralization loss, to minimize the deviation between those backward deformed frames:

$$Loss_{cent} = \sum_{t=1}^{T} |\phi_B^t(X_t) - \bar{\phi}_B(X_t)|^2 \tag{5}$$

where T represents the time.

The total loss function of our TC-Deformer is as follows:

$$Loss_{total} = Loss_{seg} + \alpha Loss_{cent} + \beta Loss_{imgcycle} + \gamma Loss_{shapecycle} \tag{6}$$

where α, β and γ is the hyper-parameters.

After myocardial segmentation, we use neural networks to determine two key points, combined with the centroid of the segmentation, to divided it into 16-segments according to the 16-segment model of the American Heart Association and calculated the corresponding LVWTs.

3 Experiments and Results

3.1 Dataset Description and Experimental Setup

Datasets. In this experiment, we trained our method on ultrasound SAX videos of 141 participants and tested with 60 participants. All the data was collected with the GE Vivid E95 from the Shenzhen People's Hospital and this study was approved by local institutional review boards. For each participant, videos of the basal, middle, and apical SAX views were acquired with multiple cardiac cycles. The mask of the myocardium at the ED and ES phases from one cardiac cycle was annotated by experts (including two senior, three middle-level, and three junior cardiologists). We compare the LVWT of each group with the average wall thickness of the senior doctors to get the average of the errors for different doctors as result. For the training dataset, the senior cardiologists conducted quality control for junior and middle-level cardiologists. All images were annotated by experienced doctors using the Pair annotation software package [9](https://www.aipair.com.cn/en/, Version 2.7, RayShape, Shenzhen, China).

Experimental Setup. We resized the images to the same size 256 × 256 and used the Adam optimization strategy during model training. The training is in two stages and the initial learning rate was 0.0001, the total epoch number is 100, and the batch size is 4. The hyperparameters α, β, and γ were set to be 0.1, 0.5, and 0.5, respectively, according to a small validation set. The models are implemented with PyTorch on the NVIDIA A100 Tensor Core GPU.

3.2 Results

Table 1. Quantitative comparison results of the average Dice, Hausdorff distance(HD), FLOPs.

	Dice	HD (mm)	FLOPs (G)
Unet [10]	0.827	3.73	**40.46**
CLAS [12]	0.842	2.98	208.47
Deformer	0.852	**2.64**	41.46
TC-Deformer	**0.854**	2.92	41.46

Table 1 shows the segmentation performance on the test set in terms of Dice, Hausdorff distance (HD), and floating-point operations per second (FLOPs). We can conclude that the proposed method achieves excellent segmentation performance and outperforms U-net and the state-of-the-art CLAS, while costing much less computation. Figure 4 shows the segmentation results of myocardium compared with four methods. It indicates that our method has a more reasonable myocardium shape than others, which is important to LVWT.

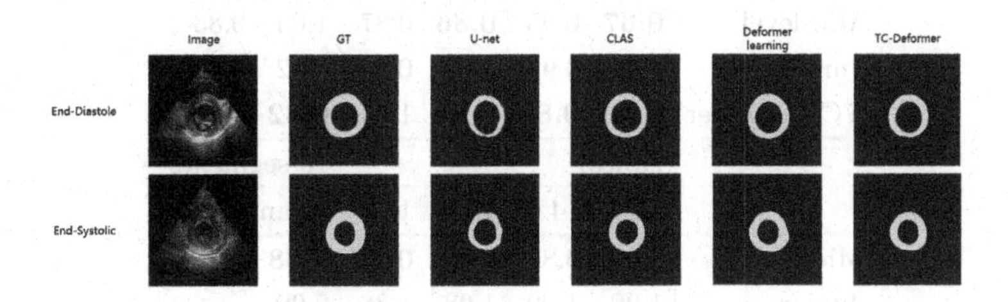

Fig. 4. Visualization results of four segmentation methods.

Figure 5 shows the segmentation results of the myocardium in one cardiac cycle. It indicates that our method can obtain smoother contours for the middle frames than CLAS, implying that the temporally compatible deformation learning has a better temporal consistency.

Table 2 shows the MAE of the measurements in LVWT for middle-level and junior cardiologists, as while as for the proposed TD-Deformer. It indicates our results of the measurements are better than the junior groups and comparable

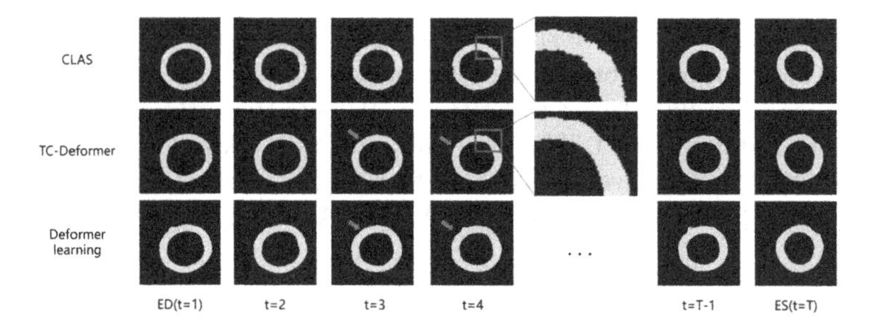

Fig. 5. Visualization of the segmentation results for one cardiac cycle.

with the middle-level group. The error of LVWT estimation is less than 1.00 mm. Figure 6 shows the absolute error of the predicted results for 16 segments. We can conclude that for all 16 segments, the prediction results of TC-Deformer are stable and accurate.

Table 2. The MAE of LVWT (mm)

	Basal					
	1	2	3	4	5	6
Mid-level	0.86	**0.67**	0.93	0.89	**0.87**	0.85
Junior	1.02	0.87	1.03	**0.87**	0.91	0.90
TC-Deformer	**0.80**	0.80	**0.86**	0.91	0.92	**0.85**
	Middle					
	7	8	9	10	11	12
Mid-level	**0.67**	0.95	**0.86**	0.87	1.04	0.83
Junior	0.73	0.95	1.02	**0.81**	1.02	0.93
TC-Deformer	1.02	**0.82**	0.90	1.04	**0.82**	**0.83**
	Apical				16-segments	
	13	14	15	16	mean	
Mid-level	0.90	0.89	0.96	**0.97**	**0.88**	
Junior	1.20	1.22	1.08	1.36	0.99	
TC-Deformer	**0.87**	**0.97**	**0.94**	1.18	0.91	

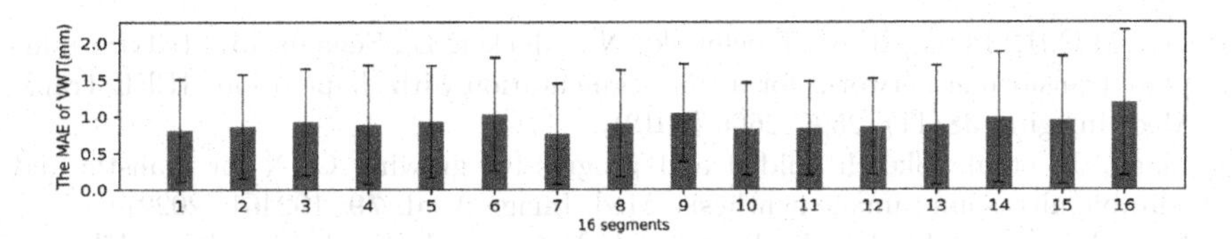

Fig. 6. The MAE of prediction LVWT indices as well as their corresponding standard deviation.

4 Conclusion

In this paper, We propose a Temporally Compatible Deformation learning network, named TC-Deformer, to detect the myocardium boundaries and estimate regional left ventricle wall thickness. Our method is designed to avoid the irregular contours that can happen in ultrasound images with heavy noise and can incorporate the temporal dynamics of the myocardium in one cardiac cycle into the deformation field learning. When validated with a dataset of 201 patients, our method achieves less than 1.00 mm estimation error for all 16 myocardium segments and outperforms existing state-of-the-art methods.

Acknowledgement. The work is partially supported by the Natural Science Foundation of China (62171290), the Shenzhen Science and Technology Program (20220810145705001, JCYJ20190808115419619, SGDX20201103095613036), Medical Scientific Research Foundation of Guangdong Province (No. A2021370).

References

1. Amer, A., Ye, X., Janan, F.: ResDUnet: a deep learning-based left ventricle segmentation method for echocardiography. IEEE Access **9**, 159755–159763 (2021)
2. Chen, L., Su, Y., Yang, X., Li, C., Yu, J.: Clinical study on LVO-based evaluation of left ventricular wall thickness and volume of AHCM patients. J. Radiat. Res. Appl. Sci. **16**(2), 100545 (2023)
3. Deng, K., et al.: TransBridge: a lightweight transformer for left ventricle segmentation in echocardiography. In: Noble, J.A., Aylward, S., Grimwood, A., Min, Z., Lee, S.-L., Hu, Y. (eds.) ASMUS 2021. LNCS, vol. 12967, pp. 63–72. Springer, Cham (2021). https://doi.org/10.1007/978-3-030-87583-1_7
4. Du, X., Tang, R., Yin, S., Zhang, Y., Li, S.: Direct segmentation-based full quantification for left ventricle via deep multi-task regression learning network. IEEE J. Biomed. Health Inf. **23**(3), 942–948 (2018)
5. Ge, R., et al.: PV-LVNet: direct left ventricle multitype indices estimation from 2d echocardiograms of paired apical views with deep neural networks. Med. Image Anal. **58**, 101554 (2019)
6. Karamitsos, T.D., Francis, J.M., Myerson, S., Selvanayagam, J.B., Neubauer, S.: The role of cardiovascular magnetic resonance imaging in heart failure. J. Am. Coll. Cardiol. **54**(15), 1407–1424 (2009)
7. Leclerc, S., et al.: LU-Net: a multistage attention network to improve the robustness of segmentation of left ventricular structures in 2-D echocardiography. IEEE Trans. Ultrason. Ferroelectr. Freq. Control **67**(12), 2519–2530 (2020)

8. Lee, M.C.H., Petersen, K., Pawlowski, N., Glocker, B., Schaap, M.: TeTrIS: template transformer networks for image segmentation with shape priors. IEEE Trans. Med. Imaging **38**(11), 2596–2606 (2019)
9. Liang, J., et al.: Sketch guided and progressive growing GAN for realistic and editable ultrasound image synthesis. Med. Image Anal. **79**, 102461 (2022)
10. Ronneberger, O., Fischer, P., Brox, T.: U-Net: convolutional networks for biomedical image segmentation. In: Navab, N., Hornegger, J., Wells, W.M., Frangi, A.F. (eds.) MICCAI 2015. LNCS, vol. 9351, pp. 234–241. Springer, Cham (2015). https://doi.org/10.1007/978-3-319-24574-4_28
11. Wang, Z.: Image quality assessment: from error visibility to structural similarity. IEEE Trans. Image Process. **13**(4), 600–612 (2004)
12. Wei, H., Cao, H., Cao, Y., Zhou, Y., Xue, W., Ni, D., Li, S.: Temporal-consistent segmentation of echocardiography with co-learning from appearance and shape. In: Martel, A.L., et al. (eds.) MICCAI 2020. LNCS, vol. 12262, pp. 623–632. Springer, Cham (2020). https://doi.org/10.1007/978-3-030-59713-9_60
13. Wei, H., Ma, J., Zhou, Y., Xue, W., Ni, D.: Co-learning of appearance and shape for precise ejection fraction estimation from echocardiographic sequences. Med. Image Anal. **84**, 102686 (2023)
14. Xue, W., Cao, H., Ma, J., Bai, T., Wang, T., Ni, D.: Improved segmentation of echocardiography with orientation-congruency of optical flow and motion-enhanced segmentation. IEEE J. Biomed. Health Inf. **26**(12), 6105–6115 (2022)
15. Xue, W., Islam, A., Bhaduri, M., Li, S.: Direct multitype cardiac indices estimation via joint representation and regression learning. IEEE Trans. Med. Imaging **36**(10), 2057–2067 (2017)
16. Xue, W., Nachum, I.B., Pandey, S., Warrington, J., Leung, S., Li, S.: Direct estimation of regional wall thicknesses via residual recurrent neural network. In: Niethammer, M., et al. (eds.) IPMI 2017. LNCS, vol. 10265, pp. 505–516. Springer, Cham (2017). https://doi.org/10.1007/978-3-319-59050-9_40
17. Zeng, Y., et al.: MAEF-Net: multi-attention efficient feature fusion network for left ventricular segmentation and quantitative analysis in two-dimensional echocardiography. Ultrasonics **127**, 106855 (2023)

Mitral Regurgitation Quantification from Multi-channel Ultrasound Images via Deep Learning

Keming Tang[1,2,3], Zhenyi Ge[4], Rongbo Ling[1,2,3], Jun Cheng[1,2,3], Wufeng Xue[1,2,3(✉)], Cuizhen Pan[4(✉)], Xianhong Shu[4], and Dong Ni[1,2,3]

[1] National-Regional Key Technology Engineering Laboratory for Medical Ultrasound, School of Biomedical Engineering, Shenzhen University Medical School, Shenzhen University, Shenzhen, China

[2] Medical Ultrasound Image Computing (MUSIC) Laboratory, Shenzhen University, Shenzhen, China

[3] Marshall Laboratory of Biomedical Engineering, Shenzhen University, Shenzhen, China

xuewf@szu.edu.cn

[4] Department of Echocardiography, Shanghai Institute of Cardiovascular disease, Zhongshan Hospital, Fudan University, Shanghai, China

pan.cuizhen@zs-hospital.sh.cn

Abstract. Mitral regurgitation (MR) is the most common heart valve disease. Prolonged regurgitation can cause changes in the heart size, lead to impaired systolic and diastolic capacity, and even threaten life. In clinical practice, MR is evaluated by the proximal isovelocity surface area (PISA) method, where manual measurements of the regurgitation velocity and the value of PISA radius from multiple ultrasound images are required to obtain the mitral regurgitant stroke volume (MRSV) and effective regurgitant orifice area (EROA). In this paper, we propose a fully automatic method for MR quantification, which follows the pipeline of ECG-based cycle detection, Doppler spectrum segmentation, PISA radius segmentation, and MR quantification. Specifically, for the Doppler spectrum segmentation, we proposed a novel adaptive-weighting multi-channel segmentation network, PISA-net, to accurately identify the upper and lower contours of the PISA radius from a pair of coupled M-mode PISA image and corresponding M-mode decolored image. Using the complementary information of the two coupled images and combing with the spatial attention module, the proposed PISA-net can well identify the contours of the PISA radius and therefore lead to accurate quantification of MR parameters. To the best of our knowledge, this is the first study of automatic MR quantification. Experimental results demonstrated the effectiveness of the whole pipeline, especially the PISA-net for PISA radius segmentation. The full method achieves a high Pearson correlation of 0.994 for both MRSV and EROA, implying its great potential in the clinical application of MR diagnosis.

Keywords: Mitral regurgitation · Multi-channel · Segmentation

H. Greenspan et al. (Eds.): MICCAI 2023, LNCS 14225, pp. 223–232, 2023.
https://doi.org/10.1007/978-3-031-43987-2_22

1 Introduction

Mitral regurgitation (MR) is the most common heart valve disease. The incidence increases significantly with age, with more than 13% prevalence in the population over 75 years old [12]. MR is a mitral valve lesion caused by organic or functional changes in the mitral leaflets, annulus, papillary muscles, or tendon cords. Prolonged regurgitation can cause changes in the heart size, and lead to impaired systolic and diastolic capacity, resulting in decreased cardiac function and even being life-threatening.

In the clinical diagnosis of MR, physicians assess the extent of MR by calculating the effective regurgitant orifice area (EROA) and mitral regurgitant stroke volume (MRSV) of MR patients from ultrasound images, including continuous wave Doppler images (CW) and color Doppler image (CD). The most commonly applied method for calculating EROA and MRSV uses measurements derived from the proximal isovelocity surface area (PISA) method [3,4,6,18,20]. It is a hemispherical isovelocity surface that points to the regurgitant blood flow at the valve orifice when accelerated, and this phenomenon is used for quantitative evaluation of regurgitant flow. Bargiggia et al. [1] used the single-point PISA method to estimate MRSV in the routine clinical diagnosis. However, the underlying assumption that the size of the regurgitant orifice (during systole) is constant can not be held, therefore usually leads to overestimation or underestimation of MRSV. To account for the dynamic variation, Chen et al. [2] proposed an M-mode PISA and Enriquez-Sarano et al. [5] proposed a Serial PISA. However, the average orifice area used in M-mode PISA and the temporal sampling in Serial PISA undermine the accuracy of the estimation. Militaru et al. [11] sought to evaluate the accuracy of a new postprocessing software to quantify MR that allows semi-automated computation of MR severity from 3D color Doppler transesophageal echocardiographic images. The method significantly underestimates MR and can only measure MRSV, ignoring other parameters like EROA, which may be a better predictor. Singh et al. [15] evaluated a semi-automated method using 3D color data sets of MR to quantify MRSV and transmitral dynamic flow curves. However, The EROA is subject to inaccuracy in the setting of altered tissue or color gain, which cannot extrapolate to the effectiveness of this method. Modified PISA [9,14,16] was proposed later to calculate more accurate MRSV and EROA with continued temporal curves of the blood velocity and the PISA radius obtained from multi-channel ultrasound images, including CW image, two-dimensional M-mode ultrasound image (M2D) and M-mode color Doppler image (MCD). However, it is still time-consuming and laborious to implement the method manually. In this paper, we aim to automatize this procedure, where automatic identification of the above-mentioned two curves during the regurgitation period is required.

Most of the existing methods for the automatic detection of Doppler image contours were based on noise reduction and boundary tracking algorithms. In [7,17], classic image processing techniques such as low-pass filtering, thresholding, and edge detection were used. However, robustness and generalization in the presence of severe image artifacts cannot be guaranteed. A probabilistic,

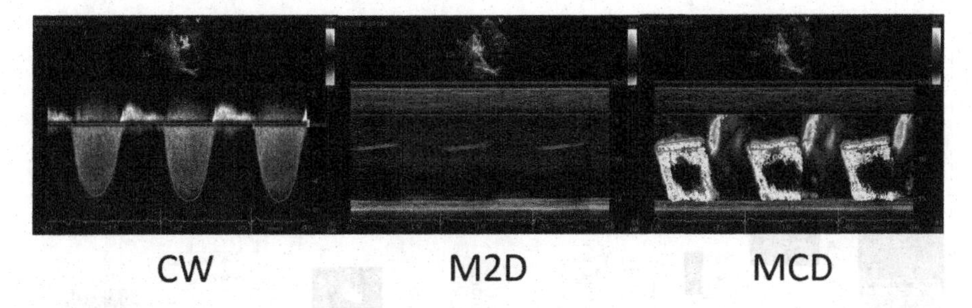

Fig. 1. Examples of the multi-channel ultrasound images, (1)continuous wave Doppler images (CW) are used to capture the blood velocity; (2)two-dimensional M-mode ultrasound images (M2D) provide the lower bound of the PISA radius; (3)M-mode color Doppler images (MCD) provide the upper bound of the PISA radius.

hierarchical, and discriminant(PHD) framework [21], was successfully applied to the automatic contour tracking of three kinds of Doppler blood flow images. Some related works have focused on model-based image segmentation algorithms. Indeed, knowing the expected shape can improve the tracking of velocity profiles. In the work of Wang et al. [19], a model-based feedback and adaptive weighted tracking algorithm was proposed. The algorithm combines a nonparametric statistical comparison of image intensities to estimate the edges of noisy impulse Doppler signals and a statistical shape model learned during manual tracking of the contours using. As for the M-mode ultrasound images, there is no existing automatic analysis method yet. In this method, we aim to estimate MRSV and EROA from multi-channel ultrasound images: CW, M2D, and MCD images (as illustrated in Fig. 1). While the CW image is used to estimate the blood velocity, the M2D and MCD images are used to estimate the contour of the PISA radius. Besides the presence of heavy noise in the images, a non-trivial challenge is that the M2D image is good at capturing the lower bound of the PISA radius, while the upper bound of the MCD image. To estimate the contour of the PISA radius, complementary information should be well extracted from the two images.

The contribution of the paper can be summarized as follows: First, we propose the first fully automatic pipeline for MR quantification from multi-channel ultrasound images based on the modified PISA method. The pipeline includes ECG-based cycle detection, Doppler spectrum segmentation, PISA radius segmentation, and MR quantification; Secondly, we propose a novel adaptive-weighting multi-channel segmentation network, PISA-net, to identify the lower and upper contours of the PISA radius from the complementary and coupled images, i.e., M2D and MCD. The network can adaptively select the related information of the corresponding input image and lead to an accurate estimation of the radius contour. Thirdly, after calculation based on the modified method, our method achieves accurate estimation of MRSV and EROA, with a Pearson correlation of 0.994 with the ground truths for both MR parameters.

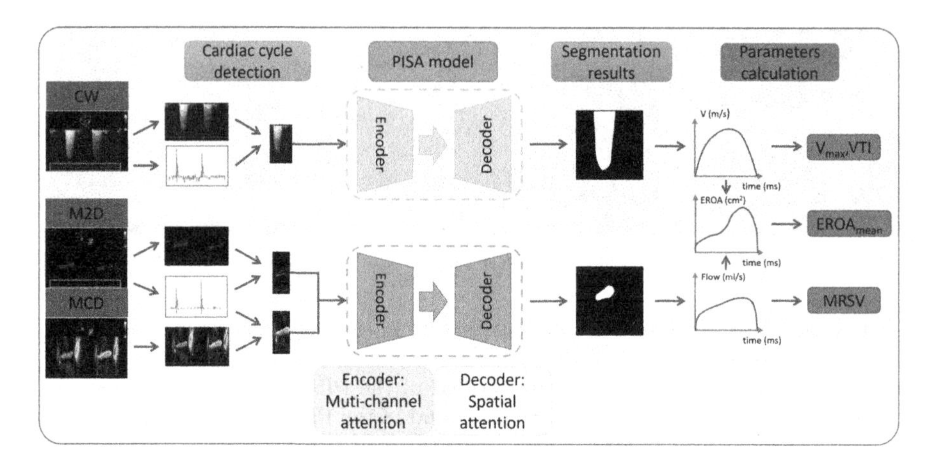

Fig. 2. Overview of the framework for MR quantification. This framework predicts four parameters from multi-channel ultrasound images: the peak flow velocity(Vmax), the flow velocity time integral(VTI), the mitral regurgitant stroke volume(MRSV), and the effective regurgitant orifice area(EROA).

2 Method

The overview of the proposed method illustrate in Fig. 2, and it contains three stages:(1) ECG-based cycle detection, (2) Doppler spectrum and PISA radius segmentation, and (3) MR quantification. The details of the proposed pipeline are as follows.

2.1 ECG-Based Cardiac Cycle Detection

Multiple cardiac cycles appear in the above-mentioned ultrasound images (Fig. 1). The first step in the pipeline is to segment the image content of each cycle, as shown in Fig. 2. The regions of interest in these ultrasound images, i.e., the blood spectrum in CW, and the texture content in the M2D and MCD are first cropped out according to the meta information of the DICOM file. The ECG signals as shown at the bottom of these images are first extracted and then used to segment the cropped images into multiple single-cycle images.

2.2 Doppler Spectrum and PISA Radius Segmentation

We proposed the PISA-net to obtain the Doppler spectrum and PISA radius segmentation from multi-channel images. The architecture of the PISA-net is shown in Fig. 3. We employ a classic U-net structure (Fig. 3(a)) as the baseline model for our task. Considering the above-mentioned challenges, we introduce the adaptive-weighting strategy for features from different images in the encoder so that PISA-net can learn which image should be used to extract the low-level

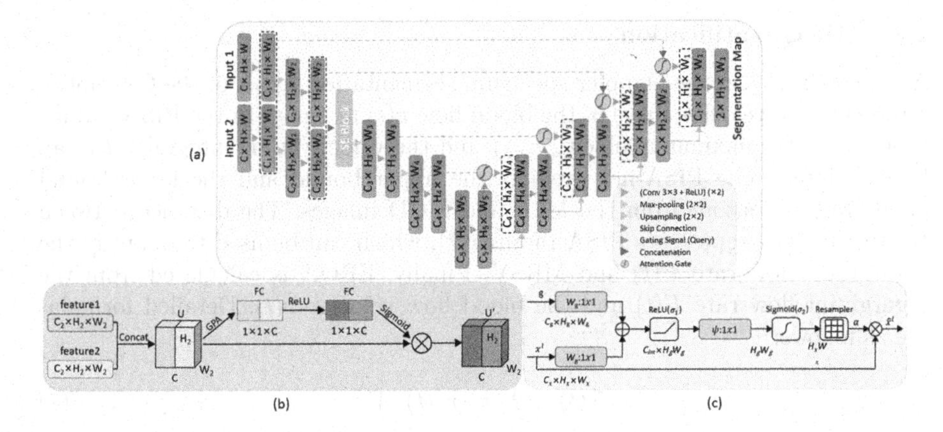

Fig. 3. Overview of the proposed PISA-net. (a) the architecture of PISA-net. (b) Squeeze-and-excitation(SE) block with attention mechanism on channel features. (c) Attention gate(AG) with attention mechanism on spatial features.

features for the lower and upper bounds of the PISA radius, respectively. To alleviate the effect of the heavy noise in these images, we introduce spatial attention mechanisms for features of the decoder layers, so that the global context can be used to suppress local noisy features.

Adaptive-Weighting Multi-channel Attention. PISA radius segmentation requires complementary information of both M2D and MCD images, and may also be affected by the image quality. We utilize the SE block [8] to adaptively weight features from different image channels. The structure of the SE block is shown in Fig. 3(b). Convolution features from M2D and MCD images are first contacted together. Through the global average pooling (GPA), each feature channel is compressed into a real number that can represent the global information of the channel. Then two fully connected layers and a Sigmoid layer are used to generate weights for each feature channel. Finally, the contacted features of M2D and MCD are weighted according to the weight vector to adjust the relative importance.

Spatial Attention. The features in the decoder are further enhanced with a spatial attention mechanism using features of the encoder. In this work, the attention gate block [13] (as shown in Fig. 3(c)) is used. Let x^l be the output feature map from layer l of the encoder, and g represents features from the previous block. Then, these features are fused by the addition operation and passed through a Sigmoid function to calculate the spatial attention map. The feature map x^l is then multiplied with the attention map to the enhanced features $\widehat{x}^l$, which makes the value of irrelevant regions smaller and the value of the target region larger, and therefore improves both the network prediction speed and the segmentation accuracy.

2.3 MR Quantification

As shown in Fig. 2, the Doppler spectrum segmentation result of the CW image represents the velocity curve of the blood flow $v(t)$ at time t in the PISA radius. From $v(t)$, the maximum velocity v_{max} and the velocity time integral VTI can be calculated. The PISA-net predicts the upper bound and the lower bound of the radius contour from the M2D and MCD images. The distance between the two bounds represents PISA radius $r(t)$, which can be used to quantify the regurgitant flow rate $F(t)$ and MRSV. Finally, EROA is calculated from the regurgitant flow rate $F(t)$ and the blood flow velocity $v(t)$. Detailed formulas are as follows:

$$F(t) = 2 \cdot \pi \cdot r^2(t) \cdot V_a \tag{1}$$

$$MRSV = \int_0^t F(t)\,\mathrm{d}t = 2 \cdot \pi \cdot V_a \cdot \int_0^t r^2(t)\,\mathrm{d}t \tag{2}$$

$$EROA(t) = \frac{F(t)}{v(t)} \tag{3}$$

$$EROA_{mean} = \frac{\int_0^T EROA(t)\,\mathrm{d}t}{T} \tag{4}$$

where V_a is a constant representing the aliasing velocity, and T denotes the duration length of the regurgitation.

3 Experiment and Results

3.1 Experimental Configuration

We obtained Doppler ultrasound images of 205 MR patients from a local hospital, and 157 of them were collected by GE VividE95 while the rest 48 were collected by PHILIPS CX50. For each patient, three images were included: CW, M2D, and MCD, as shown in Fig. 1. Data use declaration and acknowledgement: Our dataset was collected from Zhongshan Hospital, Fudan University. This study was approved by local institutional review boards. We divide these ultrasound images into a training dataset (159 patients) and a test dataset (46 patients). Among the test set, 45 patients had degenerative mitral regurgitation, and 1 had functional mitral regurgitation. All images were annotated by experienced doctors using the Pair annotation software package (https://www.aipair.com.cn/en/, Version 2.7, RayShape, Shenzhen, China) [10].

We used the Dice score to evaluate the segmentation accuracy, and the Pearson correlation coefficient (corr), mean absolute error (MAE), and mean relative error (MRE) to assess the performance of MR parameters estimation.

Our method was implemented using Pytorch 1.7.1 and trained on an NVIDIA A100 GPU. The size of the input image is $3 \times 256 \times 256$. The model was optimized by minimizing the binary cross-entropy loss function and using the Adam optimization algorithm. The learning rate was set as 0.001.

3.2 Results and Analysis

Effect of Adaptive Weighting and Multiple Channel. We test the effect of the adaptive weighting mechanism in PISA-net when placed in different positions of the encoder. Table 1 shows the segmentation performance for single inputs and multi-channel inputs, with the adaptive weighting in different layers. It can be drawn that 1)for both the spectrum segmentation and the PISA radius segmentation, the adaptive weighting performs best when placed after the second convolution block of the encoder; and 2) the multi-channel inputs for PISA radius segmentation do help improve the performance, implying that PISA-net can effectively make use of the complementary information in these images. Results in the third and fourth columns of Table 2 also validate this.

Table 1. Effect of the adaptive weighting and multiple channel inputs.

Image	layer1	layer2	layer3	layer4
CW	0.961	**0.962**	0.960	0.960
MCD	0.923	**0.925**	0.924	0.923
M2D+MCD	0.926	**0.937**	0.926	0.924

Table 2. Mean Dice of CW image input, MCD image input, and M2D and MCD inputs on the test set.

Method	CW	MCD	M2D+MCD
Unet	0.958	0.910	0.886
Unet+SE	0.958	0.911	0.918
Att-Unet	0.961	0.925	0.926
PISA-net	**0.962**	**0.925**	**0.937**

Segmentation Performance. We compared PISA-net with three different methods: Unet, Unet with SE block(Unet+SE), and Attention Unet(Att-Unet) with different images as the input, and the results are shown in Table 2. The proposed PISA-net achieves the best accuracy, while the Unet gets the lowest accuracy. When M2D and MCD are combined as the input, Unet+SE, Att-Unet, and PISA-net achieve better accuracy than that of single input MCD. These results demonstrate the effectiveness of multi-channel adaptive weighting and spatial attention. Figure 4 shows examples of the summation of the weights learned for features from M2D and MCD images, respectively. It can be observed that our method can learn the weights of the two images adaptively for different samples.

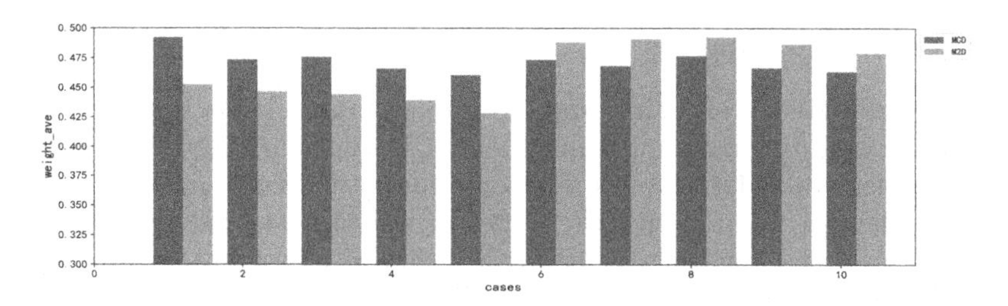

Fig. 4. Visualization of the average weights of the two inputs after using the SE block.

Table 3. Comparison of MR parameters quantification. The unit of MRE is %, For MAE, the unit of v_{max} is m/s, the unit of VTI is cm, the unit of $MRSV$ is ml, the unit of $EROA_{mean}$ is cm^2.

Method	Vmax			VTI			MRSV			EROA-mean		
	corr	MAE	MRE	corr	MAE	MRE	corr	MAE	MRE	corr	MAE	MRE
Unet	0.971	0.21	3.87	0.914	13.94	5.79	0.802	35.66	67.48	0.860	0.156	64.28
Unet+SE	0.894	0.20	3.70	0.930	19.96	8.63	0.960	14.63	18.12	0.981	0.054	15.05
Att-Unet	0.914	0.16	2.95	0.949	14.24	6.15	0.982	14.07	17.41	0.988	0.059	17.82
PISA-net	0.909	0.18	3.29	0.939	14.20	6.25	**0.994**	**8.49**	**9.95**	**0.994**	**0.040**	**11.27**

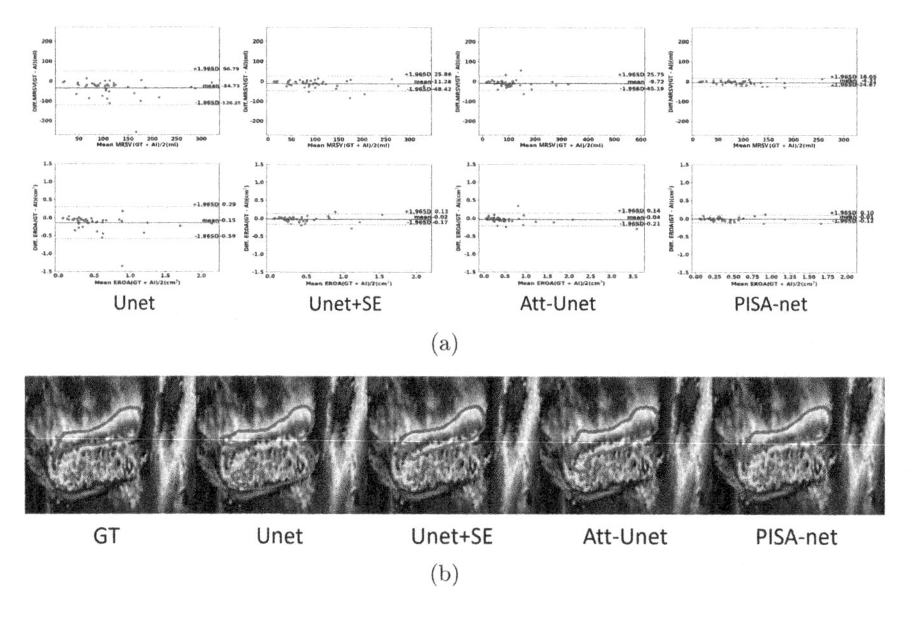

Fig. 5. (a) The MRSV and EROA calculated from the segmentation results of the four methods are compared with ground truth(GT). Bland-Altman plots show their bias. (b) Comparison of the segmentation results of a bad case in four methods.

Parameters Quantification. Table 3 is the comparison result for MR parameters quantification by different methods. PISA-net significantly outperforms the other methods for MRSV and EROA, with a Pearson correlation coefficient of 0.994 for both MR parameters. Figure 5(a) shows the Bland-Altman analysis of MRSV and EROA obtained. The PISA-net results in the least estimation bias. Figure 5(b) shows the segmentation results of a bad case for all four methods. PISA-net can still identify the lower and upper contours of the PISA radius from the complementary and coupled images accurately, showing the effectiveness of the adaptive weighting and the spatial attention.

4 Conclusion

In this work, we proposed the first fully automatic pipeline for MR quantification from multi-channel ultrasound images (CW, M2D, and MCD), based on the modified PISA method. An adaptive weighting mechanism and a spatial attention mechanism weighting were used to combine features of multi-channel inputs and enhance the local feature with a global context. Extensive experiments demonstrate that the proposed method is capable of delivering good segmentation results and excellent quantification of MR parameters, and has great potential in the clinical application of MR diagnosis.

Acknowledgement. The work is partially supported by the Natural Science Foundation of China (62171290), the Shenzhen Science and Technology Program (20220810145705001, JCYJ20190808115419619, SGDX20201103095613036), Medical Scientific Research Foundation of Guangdong Province (No. A2021370).

References

1. Bargiggia, G.S., et al.: A new method for quantitation of mitral regurgitation based on color flow doppler imaging of flow convergence proximal to regurgitant orifice. Circulation **84**, 1481–1489 (1991)
2. Chen, C., et al.: Noninvasive estimation of regurgitant flow rate and volume in patients with mitral regurgitation by doppler color mapping of accelerating flow field. J. Am. Coll. Cardiol. **21**(2), 374–83 (1993)
3. Dujardin, K.S., Enriquez-Sarano, M., Bailey, K.R., Nishimura, R.A., Seward, J.B., Tajik, A.J.: Grading of mitral regurgitation by quantitative doppler echocardiography: calibration by left ventricular angiography in routine clinical practice. Circulation **96**(10), 3409–15 (1997)
4. Enriquez-Sarano, M., Miller, F.A.J., Hayes, S.N., Bailey, K.R., Tajik, A.J., Seward, J.B.: Effective mitral regurgitant orifice area: clinical use and pitfalls of the proximal isovelocity surface area method. J. Am. Coll. Cardiol. **25**(3), 703–9 (1995)
5. Enriquez-Sarano, M., Sinak, L.J., Tajik, A.J., Bailey, K.R., Seward, J.B.: Changes in effective regurgitant orifice throughout systole in patients with mitral valve prolapse. a clinical study using the proximal isovelocity surface area method. Circulation **92**(10), 2951–2958 (1995)

6. Giesler, M., et al.: Color doppler echocardiographic determination of mitral regurgitant flow from the proximal velocity profile of the flow convergence region. Am. J. Cardiol. **71**(2), 217–24 (1993)
7. Greenspan, H., Shechner, O., Scheinowitz, M., Feinberg, M.: Doppler echocardiography flow-velocity image analysis for patients with atrial fibrillation. Ultrasound Med. Biol. **31**(8), 1031–40 (2005)
8. Hu, J., Shen, L., Albanie, S., Sun, G., Wu, E.: Squeeze-and-excitation networks. IEEE Trans. Pattern Anal. Mach. Intell. **42**, 2011–2023 (2017)
9. Hung, J.W., Otsuji, Y., Handschumacher, M.D., Schwammenthal, E., Levine, R.A.: Mechanism of dynamic regurgitant orifice area variation in functional mitral regurgitation: physiologic insights from the proximal flow convergence technique. J. Am. Coll. Cardiol. **33**(2), 538–45 (1999)
10. Liang, J., et al.: Sketch guided and progressive growing GAN for realistic and editable ultrasound image synthesis. Med. Image Anal. **79**, 102461 (2022)
11. Militaru, S., et al.: Validation of semiautomated quantification of mitral valve regurgitation by three-dimensional color doppler transesophageal echocardiography. J. Am. Soc. Echocardiogr. **33**(3), 342–354 (2020). https://doi.org/10.1016/j.echo.2019.10.013, https://www.sciencedirect.com/science/article/pii/S0894731719311150
12. Nkomo, V.T., Gardin, J.M., Skelton, T.N., Gottdiener, J.S., Scott, C.G., Enriquez-Sarano, M.: Burden of valvular heart diseases: a population-based study. Lancet **368**, 1005–1011 (2006)
13. Oktay, O., et al.: Attention U-Net: Learning where to look for the pancreas. ArXiv abs/1804.03999 (2018)
14. Schwammenthal, E., Chen, C., Benning, F., Block, M., Breithardt, G., Levine, R.A.: Dynamics of mitral regurgitant flow and orifice area: physiologic application of the proximal flow convergence method: Clinical data and experimental testing. Circulation **90**, 307–322 (1994)
15. Singh, A., et al.: A novel approach for semiautomated three-dimensional quantification of mitral regurgitant volume reflects a more physiologic approach to mitral regurgitation. J. Am. Soc. Echocardiogr. **35**(9), 940–946 (2022). https://doi.org/10.1016/j.echo.2022.05.005, https://www.sciencedirect.com/science/article/pii/S089473172200253X
16. Sun, H.L., Wu, T.J., Ng, C.C., Chien, C.C., Huang, C.C., Chie, W.C.: Efficacy of oropharyngeal lidocaine instillation on hemodynamic responses to orotracheal intubation. J. Clin. Anesth. **21**(2), 103–7 (2009)
17. Tschirren, J., Lauer, R.M., Sonka, M.: Automated analysis of doppler ultrasound velocity flow diagrams. IEEE Trans. Med. Imaging **20**, 1422–1425 (2001)
18. Vandervoort, P.M., et al.: Application of color doppler flow mapping to calculate effective regurgitant orifice area. an in vitro study and initial clinical observations. Circulation **88**(3), 1150–1156 (1993)
19. Wang, Z.W., Slabaugh, G.G., Zhou, M., Fang, T.: Automatic tracing of blood flow velocity in pulsed doppler images. In: 2008 IEEE International Conference on Automation Science and Engineering, pp. 218–222 (2008)
20. Yamachika, S., et al.: Usefulness of color doppler proximal isovelocity surface area method in quantitating valvular regurgitation. J. Am. Soc. Echocardiogr. **10**(2), 159–168 (1997). https://doi.org/10.1016/S0894-7317(97)70089-0, https://www.sciencedirect.com/science/article/pii/S0894731797700890
21. Zhou, S.K., et al.: A probabilistic, hierarchical, and discriminant framework for rapid and accurate detection of deformable anatomic structure. In: 2007 IEEE 11th International Conference on Computer Vision, pp. 1–8 (2007)

Progressive Attention Guidance for Whole Slide Vulvovaginal Candidiasis Screening

Jiangdong Cai[1], Honglin Xiong[1], Maosong Cao[1], Luyan Liu[1], Lichi Zhang[2], and Qian Wang[1](✉)

[1] School of Biomedical Engineering, ShanghaiTech University, Shanghai, China
qianwang@shanghaitech.edu.cn

[2] School of Biomedical Engineering, Shanghai Jiao Tong University, Shanghai, China

Abstract. Vulvovaginal candidiasis (VVC) is the most prevalent human candidal infection, estimated to afflict approximately 75% of all women at least once in their lifetime. It will lead to several symptoms including pruritus, vaginal soreness, and so on. Automatic whole slide image (WSI) classification is highly demanded, for the huge burden of disease control and prevention. However, the WSI-based computer-aided VCC screening method is still vacant due to the scarce labeled data and unique properties of candida. Candida in WSI is challenging to be captured by conventional classification models due to its distinctive elongated shape, the small proportion of their spatial distribution, and the style gap from WSIs. To make the model focus on the candida easier, we propose an attention-guided method, which can obtain a robust diagnosis classification model. Specifically, we first use a pre-trained detection model as prior instruction to initialize the classification model. Then we design a Skip Self-Attention module to refine the attention onto the fined-grained features of candida. Finally, we use a contrastive learning method to alleviate the overfitting caused by the style gap of WSIs and suppress the attention to false positive regions. Our experimental results demonstrate that our framework achieves state-of-the-art performance. Code and example data are available at https://github.com/caijd2000/MICCAI2023-VVC-Screening.

Keywords: Whole slide image · Vulvovaginal Candidiasis · Attention-Guided

1 Introduction

Vulvovaginal candidiasis (VVC) is a type of fungal infection caused by candida, which results in discomforting symptoms, including itching and burning in the genital area [4,18]. It is the most prevalent human candidal infection, estimated to afflict approximately 75% of all women at least once in their lifetime [1,20], resulting in huge consumption of medical resources. Currently, thin-layer cytology (TCT) [6] is one of the main tools for screening cervical abnormalities. Manual reading upon whole slide image (WSI) of TCT is time-consuming and

H. Greenspan et al. (Eds.): MICCAI 2023, LNCS 14225, pp. 233–242, 2023.
https://doi.org/10.1007/978-3-031-43987-2_23

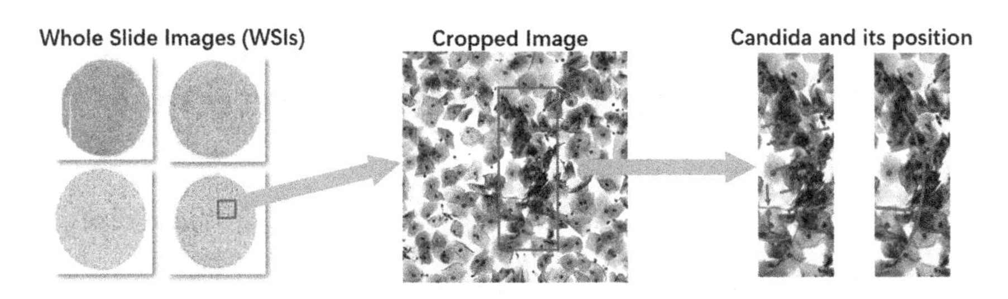

Fig. 1. Examples of WSIs (usually about 20000 × 20000 pixels), a cropped image of 1024 × 1024 pixels from the WSI, and zoom-in views of candida and its position (indicated by the red arrows and annotation). (Color figure online)

labor-intensive, which limits the efficiency and scale of disease screening. Therefore, automatic computer-aided screening for candida would be a valuable asset, which is low-cost and effective in the fight against infection.

Previous studies for computer-aided VVC diagnosis were mainly based on pap smears rather than WSIs. For example, Momenzadeh et al. [11] implemented automatic diagnosis based on machine learning. Peng et al. [13] compared different CNN models on VVC classification. Some works also applied deep learning to classify candida in other body parts [2,24]. In recent years, TCT has become mainstream in cervical disease screening compared to pap smear [8]. Many systems of automatic computer-aided WSI screening have been designed for cytopathology [22,23], and histopathology [17,21]. However, partially due to the limited data and annotation, screening for candidiasis is mostly understudied.

Computer-aided diagnosis for candidiasis through WSI is highly challenging (see examples in Fig. 1). (1) Candida is hard to localize in a large WSI, especially due to its long-stripe shape, low-contrast appearance, and often occlusion with respect to nearby cells. The representation of candida is easily dominated by other objects in deep layers of a network. (2) In addition to occupying only a small image space for each candida, the overall candida quantity in WSIs is also low compared to the number of other cells. The class imbalance makes it difficult to conduct discriminative learning and to find candida. (3) The staining of different samples leads to the huge style gap between WSIs. While collecting more candida data may contribute to a more robust network, such efforts are dwarfed by the inhomogeneity of WSIs, which adds to the risk of overfitting. All of the above issues make it difficult for diagnostic models to focus on candida, thus resulting in poor classification performance and generalization capability.

In this paper, we find that the attention for a deep network to focus on candida is the key to the high performance of the screening task. And we propose a series of strategies to make the model focus on candida progressively. Our contributions are summarized into three parts: (1) We use a detection task to pre-train the encoder of the classification model, moving the network's attention away from individual cells and onto candida-like objects; (2) We propose skip self-attention (SSA) to take into account multi-scale semantic and texture fea-

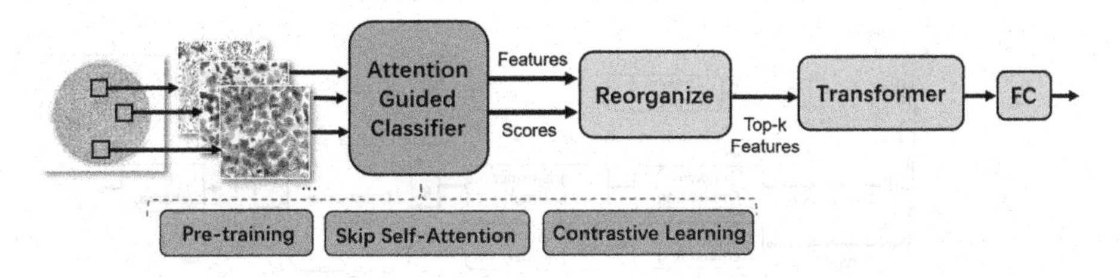

Fig. 2. The pipeline of proposed WSI-based VVC screening system.

tures, improving network attention to the candida that is severely occluded or with long hyphae; (3) Contrastive learning [3] is applied to alleviate the overfitting risk caused by the style gap and to improve the ability to discern candida.

2 Method

We use a hierarchical framework for cervical candida screening, concerning the huge size of WSI and the infeasibility of handling a WSI scan in one shot. The overall pipeline of our framework is presented in Fig. 2. Given a WSI, we first crop it into multiple images, each of which is sized 1024×1024. For each cropped image, we conduct image-level classification to find out whether it suffers from suspicious candida infection. The image-level classifier produces a score and feature representation of the image under consideration. Then scores and features from all cropped images are reorganized and aggregated by a transformer for final classification by a fully connected (FC) layer.

2.1 Detection Task for Pre-training

We use a pre-trained detection model as prior to initialize the classification model. In experimental exploration, we find that, if we train the detection network directly, the bounding-box annotation indicates the location of candida and can rapidly establish a rough understanding of the morphology of candida. However, the positioning task coming with detection lacks enough granularity, resulting in relatively low precision to discern cell edges or folding from candida. Meanwhile, directly training a classification model is usually easier to converge. However, in such a task, as candida occupies only a few pixels in an image, it is difficult for the classifier to focus on the target. That, the attention of the classifier may spread across the entire image, leading to overfitted training quickly.

Therefore, we argue that the detection and classification tasks are complementary to solve our problem. Particularly, we pre-train a detector and inherit its advantages in the classifier. We use Retinanet [10], which is composed of a backbone attached with FPN (Feature Pyramid Network, FPN, [9]) and a detection head, as shown in (Fig. 3). We chose the same encoder architecture (Resnet [5]) for the detection and classification networks. To train the encoder with the detection task (Fig. 3), we use bounding-box annotations to supervise

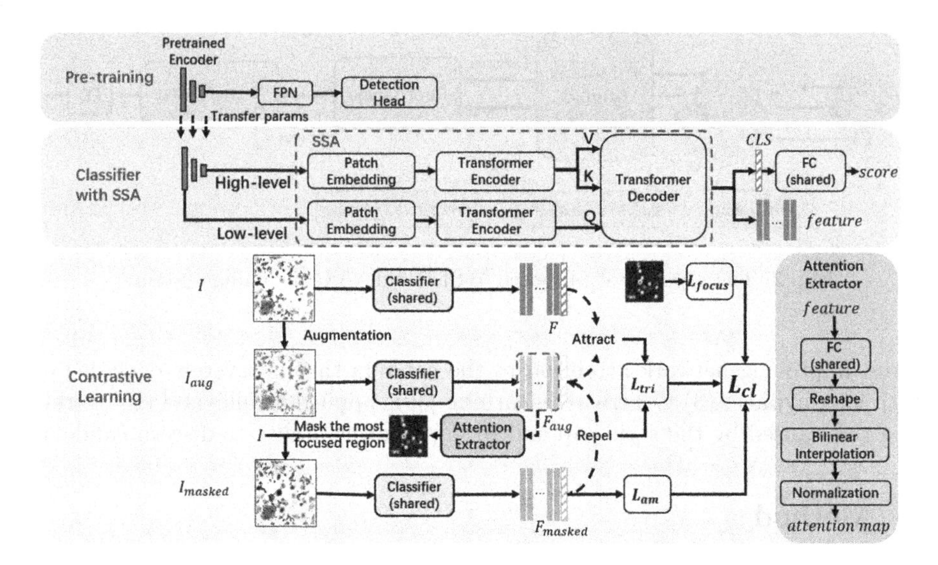

Fig. 3. Attention Guided Image-level Classification (corresponding to the classification model in Fig. 2). The same parameters are used between modules marked "shared".

Retinanet. We then initialize the classification network by directly loading the encoder parameters and freezing the first few layers during the training of the classification network. Note that pre-training not only discards the complex positioning task but also makes it easier for the classification network to converge especially in the early stage of training.

2.2 Transformer with Skip Self-Attention (SSA)

We design a novel skip self-attention (SSA) module to fuse discriminative features of candida from different scales. At a fine-grained level, the hyphae and spores of candida are usually the basis for judging. Yet we need to distinguish them from easily distorting factors such as contaminants in WSIs and edges of nearby cells. At a coarse-grained level, there is the phenomenon that a candida usually links multiple host cells and yields a string of them. Thus it is necessary to combine long-range visual cues that span several cells to derive the decision related to candida.

CNN-based methods have achieved excellent performance in computer-aided diagnosis including cervical cancer [22]. However, the unique shape and appearance of the candidate incur troubles for CNN-based classifiers, whose spatial field of view can be relatively limited. In recent years, vision transformer (ViT) has been widely used in visual tasks for its global attention mechanism [14], sensitivity to shape information in images [19], and robustness to occlusion [12]. Nevertheless, such a transformer can be hard to train for our task, due to the large image size, huge network parameters, and huge demand for training data. Therefore, to adapt to the shape and appearance of candida, we propose the SSA module and apply it to ViT for efficient learning.

Specifically, we use the pre-trained CNN-based encoder to extract features for each cropped image. The feature maps extracted after the first layer is considered *low-level*, which contains fine-grained texture information. On the contrary, the feature maps extracted from the last layer are *high-level*, which represents semantics regarding candida. To combine the low- and high-level features, we regard the low-level features as queries (Q), and the high-level features as keys (K) and values (V). For each patch in the ViT scheme, the transformer decoder computes the attention between low- and high-level features and combines them. The class token 'CLS' is used for the final classification. The combined feature maps can offer more representative information so that the classifier focuses more on different scales to long-range candida. Meanwhile, the extra SSA structure is simple, which causes a low computation burden.

2.3 Contrastive Learning

As mentioned in Sect. 1, the style gap is another problem, which makes overfitting more severe. In this part, we adopt the strategy of contrastive learning to alleviate such problems and further optimize the attention of the network. Our approach has two key goals: (1) to ensure that the features from the original image remain consistent after undergoing various image augmentations, and (2) to construct an image without the region of candida, resulting in highly dissimilar features compared to the original.

Inspired by a weakly supervised learning segmentation method [7], we construct a contrastive learning method, which will be described in detail in the following sections, as shown in Fig. 3.

To achieve this, we use augmentation and the attention map generated during the training process to construct three types of images and apply contrastive learning to the features extracted from them. For a given image I, we use image augmentation to generate I_{aug} and use the encoder attached with SSA to extract feature, F^c_{aug}. The attention map A is transformed from F^c_{aug} by an attention extractor. The attention extractor uses FC (the same params as that of the classifier) to reduce the channels of features (except the class token) to 2 (candida and others), then reshape the features representing candida to a feature map, and applies bilinear interpolation to upsample it to the same size of I. Equation 1 normalizes A to the interval [0,1], obtaining M to represent the likelihood of candida distribution. We get the masked image I_{masked} by subtracting M from I.

$$M = \frac{1}{1 + e^{-s(A-\sigma)}}, \tag{1}$$

where σ and s are used to adjust the range of values, set to 0.5 and 10.

As shown in Fig. 3, the features F, F_{aug} and F_{masked} from the three types of images I, I_{aug} and I_{masked} by the shared classifier. In our task, we hope that the style gap does not affect the feature extraction of the image, so the distance between F_{aug} and F should be attracted. At the same time, we hope that F_{masked} should not contain the characteristics of candida, which is repelled from F_{aug}. To achieve our goal, we introduce triplet loss [15] for contrastive learning as shown in the first part of Eq. 2.

In addition, we use two constraints, leading to more stable and robust training. If our network has effective attention, the masked image should not contain any candida, so the score of the Candida category after the mask $S(I_{masked})$ should be minimized. We use the attention mining loss to handle this, as shown in the second part of Eq. 2. Additionally, we need the attention to cover only the partial area around candida, without false positive regions. Otherwise, attention maps that cover the whole image can also result in low L_{tri}. We take the average grayscale of attention map $\bar{M}$ as a restriction, as shown in the last part of Eq. 2. L_{tri},L_{am}, and L_{focus} are combined as L_{cl} to constrain each other and take full advantage of contrastive learning, as shown in Eq. 2.

$$\begin{aligned} L_{cl} &= L_{tri} + L_{am} + L_{focus} \\ &= Triplet(F_{aug}, F_{orig}, F_{masked}) + S(I_{masked}) + \bar{M}. \end{aligned} \tag{2}$$

Finally, we use the cross-entropy loss L_{ce} to calculate the classification loss with $labels$. The total loss during training can be expressed as shown in Eq. 3. α is a hyper-parameter, set to 0.1.

$$L = L_{ce}(S(I_{aug}), labels) + \alpha L_{cl}. \tag{3}$$

2.4 Aggregated Classification for WSI

With the strategies above, we have built the classifier for all cropped images in a WSI. Then we can finish the pipeline of WSI-level classification, which is shown as part of Fig. 2. Specifically, for each cropped image, we conduct image-level classification to find out whether it suffers from suspicious candida infection. The image-level classification also produces a score, as well as the feature representation of the image under consideration. Then, we reorganize features from all images by ranking their scores and preserving that with top-k scores. We complete the aggregation of the top-k features by the transformer and make the WSI-level decision by an FC layer in the final.

3 Experimental Results

Datasets and Experimental Setup. Our samples were collected by a collaborating clinical institute in 2021. Each sample is scanned into a WSI following standard cytology protocol, which can be further cropped to 500 images sized 1024×1024.

For pre-training the detector, we prepare 1467 images with the size of 1024×1024 pixels, all of which have bounding-box annotations. The ratio of training and validation is 4:1.

For training of the image-level classification model, we use 1940 positive images (1467 of which are used in detector pre-training) and 2093 negative images. All images used to pre-train the detector are categorized as training data here. The rest 473 images are split in 5-fold cross-validation, from which we collect experimental results and report later. The ratio of training, validation, and testing is 3:1:1.

Table 1. Image-level classification results and ablation study on the three contributions of our method. (PT: pre-training; SSA: skip self-attention; CL: contrastive loss).

PT	SSA	CL	AUC	ACC	Sen	Spe	F1
			85.71 ± 2.14	79.79 ± 2.25	83.20 ± 2.35	76.55 ± 3.90	79.63 ± 1.66
		✓	87.24 ± 2.19	85.28 ± 1.45	90.88 ± 1.98	80.14 ± 3.07	85.43 ± 0.90
	✓		84.97 ± 3.78	81.55 ± 2.59	87.30 ± 3.05	76.27 ± 4.74	81.81 ± 2.63
	✓	✓	87.69 ± 3.77	86.54 ± 0.98	89.59 ± 0.33	83.54 ± 1.80	86.85 ± 0.89
✓			89.44 ± 3.27	89.32 ± 0.96	93.41 ± 1.40	85.56 ± 1.27	89.21 ± 1.32
✓		✓	94.31 ± 1.75	92.54 ± 2.51	$\mathbf{93.85 \pm 3.21}$	91.22 ± 3.36	92.27 ± 2.58
✓	✓		90.82 ± 2.57	89.84 ± 0.84	92.31 ± 1.41	87.54 ± 1.49	89.56 ± 1.27
✓	✓	✓	$\mathbf{97.20 \pm 1.46}$	$\mathbf{93.89 \pm 1.20}$	92.35 ± 1.75	$\mathbf{95.22 \pm 0.96}$	$\mathbf{93.47 \pm 1.18}$

At the WSI level, we use two datasets. Dataset-Small is balanced with 100 positive WSIs and 100 negative WSIs. We conduct a 5-fold cross-validation, and the ratio of training, validation, and testing is 3:1:1. We further validate upon an imbalanced Dataset-Large of 7654 WSIs. There are only 140 positive WSIs in this dataset, which is closer to real world. These two WSI-level datasets have no overlay with the data used to train the above detection and classification tasks.

For implementation details, the models are implemented by PyTorch and trained on 4 Nvidia Tesla V100S GPUs. All parameters are optimized by Adam for 100 epochs with the initial learning rate of 3×10^{-4}. The batch sizes of the detection task, image-level classification, and WSI-level classification are 8, 8, 16, respectively. To aggregate WSI classification, we use top-10 cropped images and their features. We report the performance using five common metrics: area under the receiver operating characteristic curve (AUC), accuracy (ACC), sensitivity (Sen), specificity (Spe), and F1-score.

Comparisons for Image-Level Classification. We conduct an ablation study to evaluate the contribution of pre-training (PT), skip self-attention (SSA), and contrastive learning (CL) for the image-level classification, as shown in Table 1. It is observed that with all our proposed components, the network reaches the highest AUC 97.20, which is 11.49% higher than the baseline. PT shows improvement in all situations, as a reasonable initial focus provides a solid foundation. SSA and CL can bring 2.89% and 6.38% improvement respectively compared to the method without each of them. It shows that SSA and CL can perform better when the model already has the basic ability to localize candida, i.e., after PT.

To verify whether our model focuses on important regions of the input image for accurate classification, we visualize the model's attention using Grad-CAM [16]. We present two examples in Fig. 4. We can see in Fig. 4(b) that the baseline's attention is very scattered spatially. After PT, the model can focus on the candida area, edges of cells, and folds that resemble candida, as shown in Fig. 4(c). After adding the SSA module, more texture information is used to

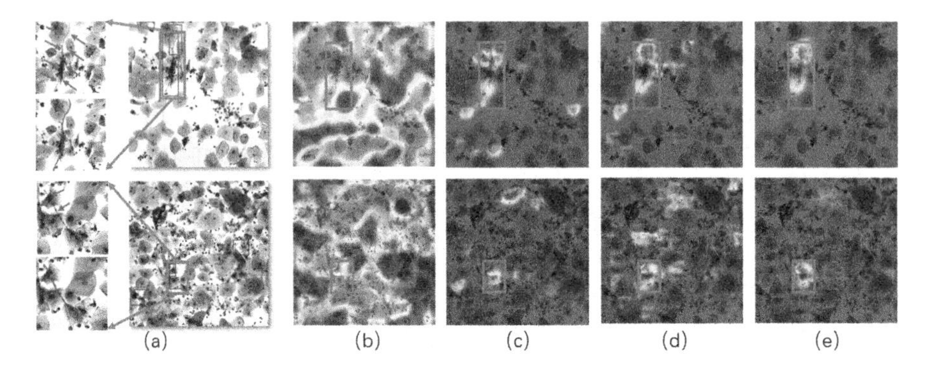

Fig. 4. (a) The original image, where the green box indicates candida (enlarged in the left) and the red box shows the prediction of the detection model. Other figures are Grad-CAM of (b) baseline, (c) baseline+PT, (d) baseline+PT+SSA, (e) baseline+PT+SSA+CL. (Color figure online)

distinguish with cells, as shown in Fig. 4(d). Finally, CL helps the model better narrow its attention, focusing on the most important part as shown in Fig. 4(e). These comparisons demonstrate that our proposed method effectively guides and corrects the model's attention.

Comparisons for WSI-Level Classification. We compare our proposed method to other methods in the whole slide of cervical disease screening. To save computation, we did not verify the performance of the methods that performed too poorly on Dataset-Small. The detection-based method [23] uses a detection network to get suspicious candida and classify WSIs with average confidence. Resnet trained without our method is the same as the baseline in Table 1. At the WSI level, we compare our method with traditional classifiers and a multi-instance learning method TransMIL [17]. We both considered the original TransMIL with pre-trained Resnet-50 and the modified version with our image-level encoder. Table 2 shows that our method reaches the highest AUC of 95.78%

Table 2. Comparision of different methods for WSI classification.

Method		Dataset-Small			Dataset-Large		
Image-level	WSI-level	AUC	ACC	Sen	AUC	ACC	Sen
Detection	Threshold	88.57 ± 9.56	80.00 ± 10.0	79.03 ± 14.4	\	\	\
Resnet	Threshold	88.75 ± 6.58	77.50 ± 9.35	82.17 ± 13.0	\	\	\
Resnet	TransMIL	93.85 ± 3.71	89.50 ± 3.67	$\mathbf{87.99 \pm 7.55}$	80.46	86.33	67.85
Ours	Threshold	92.50 ± 5.36	86.00 ± 6.44	83.04 ± 8.71	82.59	**88.14**	64.29
Ours	MLP	94.36 ± 3.50	91.00 ± 6.44	85.11 ± 13.16	83.40	87.32	67.86
Ours	TransMIL	95.35 ± 1.48	91.00 ± 3.21	85.74 ± 4.25	81.19	86.78	62.86
Ours	Transformer	$\mathbf{95.78 \pm 2.25}$	$\mathbf{91.64 \pm 3.17}$	85.55 ± 5.91	**84.18**	87.67	**68.57**

and is the most stable. Our attention-based method brings 6% improvement of accuracy on Data-Small compared to other methods with the same WSI-level method 'Threshold'. Transformer shows a better capacity of feature aggregation than other WSI-level classifiers, raising the AUC on Dataset-Large to 84.18%.

4 Conclusion

We introduced a novel attention-guided method for VVC screening, which can progressively correct the attention of the model. We pre-train a detection task for the initialization, then add SSA to fuse features from coarse and fine-grained, and finally narrower attention with contrastive learning. After obtaining accurate attention and good generalization for the image-level classifier, we reorganized and ensemble features from slices, and make a diagnosis. Both numerical metrics and visualization results show the effectiveness of our model. In the future, we would like to explore the method of weakly supervised learning to make use of a huge number of unlabeled images and jointly train the image-level and WSI-level models.

References

1. Benedict, K., Jackson, B.R., Chiller, T., Beer, K.D.: Estimation of direct healthcare costs of fungal diseases in the united states. Clin. Infect. Dis. **68**(11), 1791–1797 (2019)
2. Bettauer, V., et al.: A deep learning approach to capture the essence of candida albicans morphologies. Microbiol. Spectr. **10**(5), e01472–22 (2022)
3. Chen, T., Kornblith, S., Norouzi, M., Hinton, G.: A simple framework for contrastive learning of visual representations. In: International Conference on Machine Learning, pp. 1597–1607. PMLR (2020)
4. Gonçalves, B., Ferreira, C., Alves, C.T., Henriques, M., Azeredo, J., Silva, S.: Vulvovaginal candidiasis: epidemiology, microbiology and risk factors. Crit. Rev. Microbio. **42**(6), 905–927 (2016)
5. He, K., Zhang, X., Ren, S., Sun, J.: Deep residual learning for image recognition. In: Proceedings of the IEEE Conference on Computer Vision and Pattern Recognition, pp. 770–778 (2016)
6. Koss, L.G.: The papanicolaou test for cervical cancer detection: a triumph and a tragedy. JAMA **261**(5), 737–743 (1989)
7. Li, K., Wu, Z., Peng, K.C., Ernst, J., Fu, Y.: Tell me where to look: guided attention inference network. In: Proceedings of the IEEE Conference on Computer Vision and Pattern Recognition, pp. 9215–9223 (2018)
8. Li, T., Lai, Y., Yuan, J.: The diagnostic accuracy of TCT+ HPV-DNA for cervical cancer: systematic review and meta-analysis. Ann. Transl. Med. **10**(14), 761 (2022)
9. Lin, T.Y., Dollár, P., Girshick, R., He, K., Hariharan, B., Belongie, S.: Feature pyramid networks for object detection. In: Proceedings of the IEEE Conference on Computer Vision and Pattern Recognition, pp. 2117–2125 (2017)
10. Lin, T.-Y., Goyal, P., Girshick, R., He, K., Dollár, P.: Focal loss for dense object detection. In: Proceedings of the IEEE International Conference on Computer Vision, pp. 2980–2988 (2017)

11. Momenzadeh, M., Sehhati, M., Mehri Dehnavi, A., Talebi, A., Rabbani, H.: Automatic diagnosis of vulvovaginal candidiasis from pap smear images. J. Microsc. **267**(3), 299–308 (2017)
12. Naseer, M.M., Ranasinghe, K., Khan, S.H., Hayat, M., Shahbaz Khan, F., Yang, M.H.: Intriguing properties of vision transformers. Adv. Neural Inf. Process. Syst. **34**, 23296–23308 (2021)
13. Peng, S., Huang, H., Cheng, M., Yang, Y., Li, F.: Efficiently recognition of vaginal micro-ecological environment based on convolutional neural network. In: 2020 IEEE International Conference on E-health Networking, Application & Services (HEALTHCOM), pp. 1–6. IEEE (2021)
14. Raghu, M., Unterthiner, T., Kornblith, S., Zhang, C., Dosovitskiy, A.: Do vision transformers see like convolutional neural networks? Adv. Neural Inf. Process. Syst. **34**, 12116–12128 (2021)
15. Schroff, F., Kalenichenko, D., Philbin, J.: FaceNet: a unified embedding for face recognition and clustering. In: Proceedings of the IEEE Conference on Computer Vision and Pattern Recognition, pp. 815–823 (2015)
16. Selvaraju, R.R., Das, A., Vedantam, R., Cogswell, M., Parikh, D., Batra, D.: Grad-CAM: Why did you say that? arXiv preprint arXiv:1611.07450 (2016)
17. Shao, Z., Bian, H., Chen, Y., Wang, Y., Zhang, J., Ji, X., et al.: TransMIL: transformer based correlated multiple instance learning for whole slide image classification. Adv. Neural Inf. Process. Syst. **34**, 2136–2147 (2021)
18. Sobel, J.D.: Vulvovaginal candidosis. Lancet **369**(9577), 1961–1971 (2007)
19. Tuli, S., Dasgupta, I., Grant, E., Griffiths, T.L.: Are convolutional neural networks or transformers more like human vision? arXiv preprint arXiv:2105.07197 (2021)
20. Willems, H.M., Ahmed, S.S., Liu, J., Xu, Z., Peters, B.M.: Vulvovaginal candidiasis: a current understanding and burning questions. J. Fungi **6**(1), 27 (2020)
21. Zhang, H., et al.: DTFD-MIL: double-tier feature distillation multiple instance learning for histopathology whole slide image classification. In: Proceedings of the IEEE/CVF Conference on Computer Vision and Pattern Recognition, pp. 18802–18812 (2022)
22. Zhang, X., et al.: Whole slide cervical cancer screening using graph attention network and supervised contrastive learning. In: Wang, L., Dou, Q., Fletcher, P.T., Speidel, S., Li, S. (eds.) Medical Image Computing and Computer Assisted Intervention-MICCAI 2022. MICCAI 2022. Lecture Notes in Computer Science. vol. 13432. Springer, Cham (2022). https://doi.org/10.1007/978-3-031-16434-7_20
23. Zhou, M., et al.: Hierarchical pathology screening for cervical abnormality. Comput. Med. Imaging Graph. **89**, 101892 (2021)
24. Zieliński, B., Sroka-Oleksiak, A., Rymarczyk, D., Piekarczyk, A., Brzychczy-Włoch, M.: Deep learning approach to describe and classify fungi microscopic images. PloS One **15**(6), e0234806 (2020)

Detection-Free Pipeline for Cervical Cancer Screening of Whole Slide Images

Maosong Cao[1], Manman Fei[2], Jiangdong Cai[1], Luyan Liu[1], Lichi Zhang[2], and Qian Wang[1(✉)]

[1] School of Biomedical Engineering, ShanghaiTech University, Shanghai, China
qianwang@shanghaitech.edu.cn

[2] School of Biomedical Engineering, Shanghai Jiao Tong University, Shanghai, China

Abstract. Cervical cancer is a significant health burden worldwide, and computer-aided diagnosis (CAD) pipelines have the potential to improve diagnosis efficiency and treatment outcomes. However, traditional CAD pipelines have limitations due to the requirement of a detection model trained on a large annotated dataset, which can be expensive and time-consuming. They also have a clear performance limit and low data utilization efficiency. To address these issues, we introduce a two-stage detection-free pipeline, incorporating pooling transformer and MoCo pre-training strategies, that optimizes data utilization for whole slide images (WSIs) while relying solely on sample-level diagnosis labels for training. The experimental results demonstrate the effectiveness of our approach, with performance scaling up as the amount of data increases. Overall, our novel pipeline has the potential to fully utilize massive data in WSI classification and can significantly improve cancer diagnosis and treatment. By reducing the reliance on expensive data labeling and detection models, our approach could enable more widespread and cost-effective implementation of CAD pipelines in clinical settings. Our code and model is available at https://github.com/thebestannie/Detection-free-MICCAI2023.

Keywords: Detection-free · Contrastive Learning · Pathology Image Classification · Cervical Cancer

1 Introduction

Cervical cancer is a common and severe disease that affects millions of women globally, particularly in developing countries [9]. Early diagnosis is vital for successful treatment, which can significantly increase the cure rate [17]. In recent years, computer-aided diagnosis (CAD) methods have become an important tool in the fight against cervical cancer, as they aim to improve the accuracy and efficiency of diagnosis.

Several computer-aided cervical cancer screening methods have been proposed for whole slide images (WSIs) in the literature. Most of them are detection-based methods, which typically contain a detection model as well as some post-processing modules in their frameworks. For instance, Zhou et al. [29] proposed a

H. Greenspan et al. (Eds.): MICCAI 2023, LNCS 14225, pp. 243–252, 2023.
https://doi.org/10.1007/978-3-031-43987-2_24

three-step framework for cervical thin-prep cytologic test (TCT) [12]. The first step involves training a RetinaNet [13] as a cell detection network to localize suspiciously abnormal cervical cells from WSIs. In the second step, the patches centered on these detected cells are processed through a classification model, to refine the judgment of whether they are positive or negative. Finally, the positive patches refined by the patch-level classification are further combined to produce an overall positive/negative diagnosis for the WSI at the sample level.

Some methods improve the final classification performance by improving the detection model to identify positive cells more reliably. Cao et al. [1] improved the detection performance by incorporating clinical knowledge and attention mechanism into their cell detection model of AttFPN. Wei et al. [24] adopted the Yolo [20] architecture with a variety of convolution kernels of different sizes to accommodate diverse cell clusters. Other methods improve the classification performance by changing the post-processing modules behind the detection model. Cheng et al. [5] proposed a progressive identification method that leveraged multi-scale visual cues to identify abnormal cells and then an RNN [27] for sample-level classification. Zhang et al. [28] used GAT [23] to model the relation of the suspicious positive cells provided by detection, thus obtaining a global description of the WSI and performing sample-level classification.

These methods have achieved good results through continuous improvement on the detection-based pipeline, but there are some common drawbacks. First, they are not able to get rid of their reliance on detection models, which means they have a high need for expensive detection data labeling to train the detection model. Cervical cancer cell detection datasets involve labeling individual and small bounding boxes in a large number of cells. It often requires multiple experienced pathologists to annotate [15], which is very time-consuming and labor-intensive. Second, the widely used detection-based pipeline has not fully utilized the massive information in WSIs. A WSI is typically large (sized of about 20000×20000 pixels). A lot of data would be wasted if only a small part of annotated images (e.g., corresponding to positive cells and bounding boxes) was used as training data. Finally, many existing methods focus on detecting and classifying individual cells. The tendency to neglect effective integration of the overall information across the entire WSI results in poor performance in sample-level classification.

To address the aforementioned issues, we propose a detection-free pipeline in this paper, which does not rely on any detection model. Instead, our pipeline requires only sample-level diagnosis labels, which are naturally available in clinical scenarios and thus get rid of additional image labeling. To attain this goal, we have designed a two-stage pipeline as in Fig. 1. In the coarse-grained stage, we crop and downsample a WSI into multiple images, and conduct sample-level classification roughly based on all resized images. The coarse-grained classification yields attention scores, from which we perform attention guided selection to localize these key patches from the original WSI. Then, in the fine-grained stage, we use these key patches for fine prediction of the sample. The two stages in our pipeline adopt the same network design (i.e., encoder + pooling transformer), which makes our solution friendly to develop and to use. We also adopt contrastive learning to effectively utilize the massive information in WSIs when training the encoder for classification. As a summary, our pipeline surpasses previous detection-based methods and achieves state-of-the-art performance with large-scale training. Our experiments show that our method becomes more effective when increasing the data size for training. Moreover, while many pathological images are also based on WSIs, our pipeline has a high potential to extend to other pathological tasks.

2 Methodology

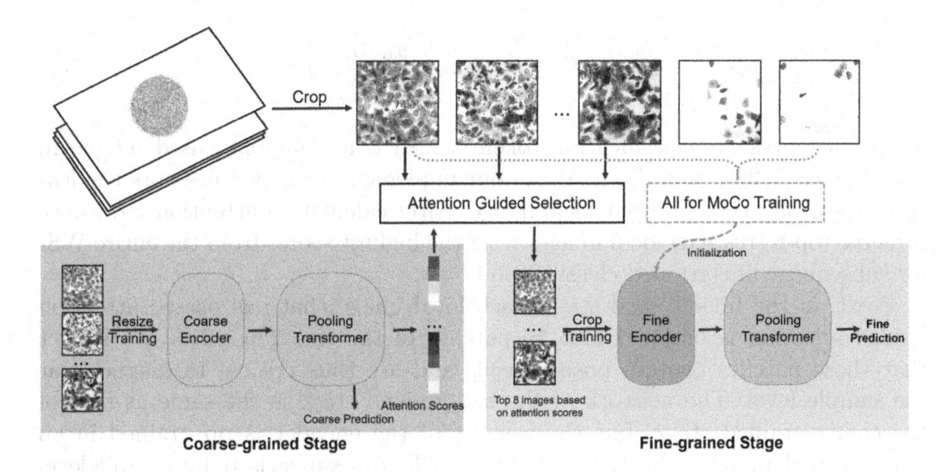

Fig. 1. The overview of our proposed method. For feasibility of computation, we crop a WSI into mutiple images. The cropped images are passed through the coarse-grained and fine-grained stages, where only sample-level diagnosis labels of WSIs, instead of any additional manual labeling, are required for training.

Two-Stage Pipeline with Attention Guided Selection. The overview of our two-stage pipeline is shown in Fig. 1. The input WSI is typically too big to be directly processed by a common deep learning pipeline [18], so we crop each WSI into local images sized 1024×1024 pixels. The images are then processed through the coarse-grained and fine-grained stages in order to obtain the WSI-level classification results, respectively. In general, the purpose of the coarse-grained stage is to replace the detection model and identify local images that may contain abnormal positive cells. The fine-grained stage then integrates these key regions, producing refined classification for the sample.

To complete sample-level classification, both stages share basically the same network architecture. The input images are first processed by a CNN encoder to extract features. Then, we propose the pooling transformer, which is modified from the basic transformer module in Sect. 2, to integrate these features for WSI classification. Additionally, the input images for both stages are 256×256. In the coarse-grained stage, in order to allow the model to examine as many local images as possible, we resize the cropped local images from 1024×1024 to 256×256. In the fine-grained stage, we enlarge suspicious local abnormality and thus crop input images to 256×256 from 1024×1024.

For the coarse-grained stage, after passing the resized local images through encoder and pooling transformer, we obtain a rough prediction result at the sample level. We then use the Cross-Entropy (CE) loss to minimize the difference between the predicted WSI label and the ground truth. In addition, we calculate the attention score to identify the local image inputs that are most likely to yield positive reading. We describe the attention score as

$$AS(x_0, f) = Softmax(\frac{x_0 \cdot f^T}{\sqrt{d_{x_0}}})f, \tag{1}$$

where x_0 represents classification token (which is a commonly used setting in transformer [7,22]), and d_{x_0} is 512 in our implementation, f represents the feature vector of a certain input local image. After calculating attention scores, we preserve top-8 (resized) local images with the highest scores from the entire WSI for subsequent fine-grained classification.

Next, in the fine-grained stage, each local image that has passed attention guided selection is cropped into 16 patches of the size 256×256. We expect that those patches contain positive cells and are thus critical to diagnosis at the sample level. The network of the fine-grained stage is the same as that of the coarse-grained stage, but the weights of the encoder is pre-trained in an unsupervised manner (Sect. 2). The same CE loss supervised by sample-level ground truth is used for the fine-grained stage here. For inference, the output of the fine-grained stage will be treated as the final result of the test WSI.

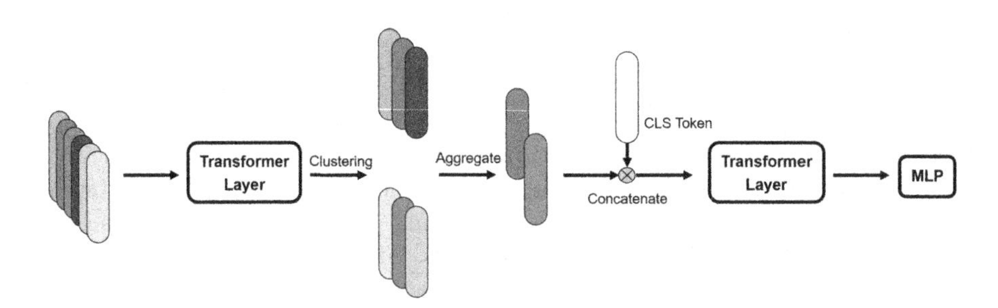

Fig. 2. Details of pooling transformer. It contains a token pooling layer in the middle of two transformer layers to cluster and aggregate redundant tokens.

Pooling Transformer. We use a transformer network to aggregate features of multiple inputs and to derive the sample-level outcome in both coarse-grained and fine-grained stages. We have observed that different local images of the same sample often have patterns of grouped similarity (such as the first two images in the upper-right of Fig. 1). For negative samples, most of the local images are similar with each other. For positive samples, the images of abnormal cells are inclined to be grouped into several clusters.

Therefore, inspired by [2,14], we propose pooling transformer that is effective to reduce the redundancy and distortion from the input images. The pooling transformer in Fig. 2 is designed to integrate all inputs toward the sample-level diagnosis. To remove redundant features, between two transformer layers, we use the affinity propagation algorithm [8] to cluster the inputs into several classes. Within each clustered class, we average the features and aggregate a single token. Finally, the classification (CLS) token is concatenated with all tokens after clustering-based pooling, and passed through the rest of the network to obtain the classification result. In this way, we find that the similar yet redundant input features can be fused, making the network more concise and efficient to calculate the attention between pooled features.

Contrastive Pre-training of Encoder. To make full use of WSI data and provide a better feature encoder, inspired by MoCo [4,11] and other contrastive learning methods [25,26] pre-training on ImageNet [6], we also perform pre-training for fine-grained encoder on a large scale of pathology images. Generally, large-scale pre-training usually requires a massive dataset and a suitable loss function. For data, WSI naturally has the advantage of having a large amount of training data. A WSI (20000×20000) can be cropped into about 5000–6000 patches (256×256). Therefore, we only need 2,000–3,000 WSI samples to obtain a dataset that can even be compared to ImageNet in quantity. For the loss function, there are typically two ways: one is like MAE [10] to model the loss function using masks, and the other is to use contrastive learning as in MoCo and CLIP [19]. In our task, since the structural features of cells are relatively weak compared to natural images, it is not suitable to model the loss function using masks. Therefore, we adopt a contrastive learning approach.

Specifically, in the same training batch, a patch (256×256, the same to the input size of the fine-grained stage) and its augmented patch are treated as a positive pair (note that here "positive/negative" is defined in the context of contrastive learning), and their features are required to be as similar as possible. Meanwhile, their features are required to be as dissimilar as possible from those of other patches. So the loss function can be described as

$$L = -\sum_{i=0}^{n}\left(\frac{f_i \cdot f_{i_a}}{|f_i||f_{i_a}|} - \sum_{j=1}^{n}\frac{f_i \cdot f_j}{|f_i||f_j|}\right) \tag{2}$$

f_i and f_{i_a} represent the positive pair, and f_j represents another patch negatively paired with f_i. Using this method, we can pre-train a feature encoder in an unsupervised manner and initialize it into our encoder for the fine-grained stage.

3 Experiment and Results

Dataset and Experimental Setup. In this study, we have collected 5384 cervical cytopathological WSI by 20x lens, each with 20000×20000 pixels, from our collaborating hospitals. Among them, there 2853 negative samples, and 2531 positive samples (962 ASCUS, and 1569 high-level positive samples). All WSIs only have diagnosis labels at the sample level, without annotation boxes at the cell level. And all sample labels are strictly diagnosed according to the TBS [16] criterion by a pathologist with 35 years of clinical experience. We conduct the experiment with initial learning rate of 1.0×10^{-4}, batch size of 4, and SGD optimizer [21] for 30 epochs each stage. For contrastive pre-training, we follow the settings of MoCov2 [3] and trained for 300 epochs.

Comparison to SOTA Methods. In this section, we experiment to compare our method with popular state-of-the-art (SOTA) methods, which are all fully supervised and detection-based. To the best of our knowledge, there are few good methods to train cervical cancer classification models in weakly supervised or unsupervised learning ways. No methods can achieve the detection-free goal either.

All the detection-based methods are evaluated in the following way. First, we label a dataset with cell-level bounding boxes to train a detection model. The detection dataset has 3761 images and 7623 cell-level annotations. After obtaining the suspicious cell patches provided by the detection model, we use the subsequent classification models used in these SOTA works to classify them and obtain the final classification results.

Table 1. Comparison with SOTA methods (%).

Method	Accuracy↑	Precision↑	Recall↑	F1-Score↑
AttFPN+Average [1]	78.33 ± 1.51	71.23 ± 2.10	79.39 ± 1.56	74.10 ± 2.33
RetinaNet+MLP [29]	74.31 ± 2.31	65.34 ± 3.57	78.12 ± 1.59	73.44 ± 2.78
RetinaNet+SVM [29]	72.37 ± 1.40	73.96 ± 0.79	77.38 ± 0.88	75.86 ± 0.95
LRModel+RNN [5]	80.55 ± 1.53	74.59 ± 1.66	82.31 ± 1.74	78.51 ± 1.32
RetinaNet+GAT [28]	83.74 ± 1.35	$\mathbf{80.38 \pm 1.43}$	84.58 ± 1.48	81.93 ± 1.02
Ours (Detection-Free)	$\mathbf{83.84 \pm 1.56}$	78.36 ± 1.23	$\mathbf{85.22 \pm 0.98}$	$\mathbf{82.12 \pm 0.93}$

As shown in Table 1 for fair five-fold cross-validation, our method outperforms all compared detection-based methods. While our method has a large margin with most methods in the table, the improvement against [28] (top-ranked in current detection-based methods) is relatively limited. On one hand, in [28], GAT aggregates local patches that are detected by Retinanet. And the attention mechanism of GAT is similar with the transformer used in our pipeline to certain extent. On the other hand, the result implies that our coarse-grained task has replaced the role of cell detection in early works. Thus, we conclude that a detection model trained with an expensive annotated dataset is not necessary to build a CAD pipeline for cervical abnormality.

Ablation Study. In this section, we experiment to demonstrate the effectiveness of all the proposed parts in our pipeline. We divide all 5384 samples into five independent parts for five-fold cross-validation, and the results are shown in Table 2. Here, CG means the classification passes only the coarse-grained stage. As can be seen, its performance is low, in that the resized images sacrifices the resolution and thus perform poorly for image-based classification. FG refers to classifying in the fine-grained stage. It is worth noting that without the attention scores provided by the coarse-grained stage, we have no way of knowing which local images might contain suspicious positive cells. Thus, we use random selection to experiment for FG only, as exhaustively checking all local images is computationally forbidden. As can be seen, the classification result is the lowest because it lacks enough access to the key image content in WSIs.

Table 2. Ablation study of our proposed methods. CG indicates coarse-grained classification, FG indicates fine-grained classification, PT indicates pooling transformer, and CL indicates contrastive learning.

Configuration				Metric (%)			
CG	FG	PT	CL	ACC	Precision	Recall	F1 Score
✔	–	–	–	74.96 ± 1.23	71.39 ± 1.21	82.49 ± 1.39	75.34 ± 1.66
–	✔	–	–	73.11 ± 2.10	70.48 ± 1.45	81.59 ± 1.45	74.09 ± 1.49
✔	✔	–	–	79.72 ± 1.30	74.64 ± 1.34	84.45 ± 1.95	78.48 ± 1.23
✔	✔	✔	–	81.34 ± 1.56	77.36 ± 1.23	84.79 ± 0.98	80.72 ± 0.93
✔	✔	✔	✔	83.84 ± 1.56	78.36 ± 1.23	85.22 ± 0.98	82.12 ± 0.93

By combining the two stages for attention guided selection, it is effective to improve the classification performance compared to the two previous experiments. Here, for the cases of CG, FG and CG+FG, an original transformer network without clustering-based pooling is used. In addition, as shown in the last two rows of the table, both pooling transformer (PT) and unsupervised pretraining (CL) contribute to our pipeline. Ultimately, we combine them together to achieve the best performance.

Sample Numbers and Inference Time. In order to further demonstrate the huge potential of our method, we also perform an ablation study on the number of samples used for training and compare the time consuming of the different methods. For the experiment of sample numbers, We compare the best fully supervised detection-based method (Retinanet+GAT [28]) with ours under the sample numbers of 500, 1000, 2000, and 5384. As shown by Table 3 and

Table 3. Alation study on sample number between Retinanet+GAT [28] and ours (%).

Number	Method	Accuracy↑	Precision↑	Recall↑	F1-Score↑
500	RetinaNet+GAT	77.59 ± 1.33	71.32 ± 1.39	75.23 ± 1.03	73.85 ± 1.45
	Ours	70.34 ± 2.58	64.49 ± 1.98	71.49 ± 2.49	68.88 ± 2.36
1000	RetinaNet+GAT	79.85 ± 1.70	75.60 ± 1.26	80.06 ± 0.96	77.96 ± 1.50
	Ours	74.47 ± 1.29	70.23 ± 1.21	76.45 ± 0.73	73.96 ± 0.78
2000	RetinaNet+GAT	81.59 ± 1.09	78.12 ± 1.73	82.37 ± 1.55	80.79 ± 1.38
	Ours	78.10 ± 1.32	73.88 ± 1.38	79.93 ± 1.06	77.39 ± 1.08
5384	RetinaNet+GAT	83.74 ± 1.35	80.38 ± 1.43	84.58 ± 1.48	81.93 ± 1.02
	Ours	83.84 ± 1.56	78.36 ± 1.23	85.22 ± 0.98	82.12 ± 0.93

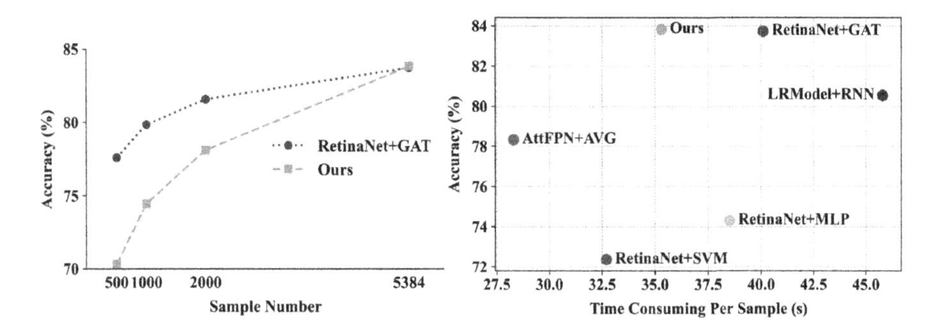

Fig. 3. Visualization of accuracy and inference time consuming for different methods.

left of Fig. 3, the traditional detection-based method has quickly encountered a saturation bottleneck as the amount of data increases. Although our method initially has poorer performance, it has shown an impressive growth trend. And at our current maximum data number (5384), the proposed pipeline has already exceeded the performance of the detection-based method. The above results also demonstrate that our new pipeline method has greater potential, even though it requires no cell-level image annotation. For inference time consuming, as shown in right of Fig. 3, our method has shorter inference time and a good balance between accuracy and inference time.

4 Conclusion and Discussion

In this paper, we propose a novel two-stage detection-free pipeline for WSI classification of cervical abnormality. Our method does not rely on detection models and eliminates the need for expensive cell-level data annotation. By leveraging just sample-level diagnosis labels, we achieve results that are competitive with fully supervised detection-based methods. Through the use of the proposed pooling transformer and unsupervised pre-training, our method makes full use of information within WSIs, resulting in improved efficiency in the use of pathological images. Importantly, our method offers even greater advantages with increasing amounts of data. And also, by utilizing attention weights, we can calculate attention scores to visually represent the importance of each image in the sample, making it easier for doctors to make judgments. Relevant visualization results can be found on our project homepage. Admittedly, our method has some limitations, such as slow training. Accelerating the training of massive data can be our next optimization direction.

References

1. Cao, L., et al.: A novel attention-guided convolutional network for the detection of abnormal cervical cells in cervical cancer screening. Med. Image Anal. **73**, 102197 (2021)
2. Chen, B., et al.: PSViT: better vision transformer via token pooling and attention sharing. arXiv preprint arXiv:2108.03428 (2021)
3. Chen, X., Fan, H., Girshick, R., He, K.: Improved baselines with momentum contrastive learning. arXiv preprint arXiv:2003.04297 (2020)
4. Chen, X., Xie, S., He, K.: An empirical study of training self-supervised vision transformers. In: Proceedings of the IEEE/CVF International Conference on Computer Vision, pp. 9640–9649 (2021)
5. Cheng, S., et al.: Robust whole slide image analysis for cervical cancer screening using deep learning. Nat. Commun. **12**(1), 1–10 (2021)
6. Deng, J., Dong, W., Socher, R., Li, L.J., Li, K., Fei-Fei, L.: Imagenet: a large-scale hierarchical image database. In: 2009 IEEE Conference on Computer Vision and Pattern Recognition, pp. 248–255. IEEE (2009)
7. Dosovitskiy, A., et al.: An image is worth 16x16 words: Transformers for image recognition at scale. arXiv preprint arXiv:2010.11929 (2020)
8. Frey, B.J., Dueck, D.: Clustering by passing messages between data points. Science **315**(5814), 972–976 (2007)
9. Gultekin, M., Ramirez, P.T., Broutet, N., Hutubessy, R.: World health organization call for action to eliminate cervical cancer globally. Int. J. Gynecol. Cancer **30**(4), 426–427 (2020)
10. He, K., Chen, X., Xie, S., Li, Y., Dollár, P., Girshick, R.: Masked autoencoders are scalable vision learners. In: Proceedings of the IEEE/CVF Conference on Computer Vision and Pattern Recognition, pp. 16000–16009 (2022)
11. He, K., Fan, H., Wu, Y., Xie, S., Girshick, R.: Momentum contrast for unsupervised visual representation learning. In: Proceedings of the IEEE/CVF Conference on Computer Vision and Pattern Recognition, pp. 9729–9738 (2020)
12. Koss, L.G.: The papanicolaou test for cervical cancer detection: a triumph and a tragedy. Jama **261**(5), 737–743 (1989)
13. Lin, T.-Y., Goyal, P., Girshick, R., He, K., Dollár, P.: Focal loss for dense object detection. In: Proceedings of the IEEE International Conference on Computer Vision, pp. 2980–2988 (2017)
14. Marin, D., Chang, J.H.R., Ranjan, A., Prabhu, A., Rastegari, M., Tuzel, O.: Token pooling in vision transformers. arXiv preprint arXiv:2110.03860 (2021)
15. Meng, Z., Zhao, Z., Li, B., Fei, S., Guo, L.: A cervical histopathology dataset for computer aided diagnosis of precancerous lesions. IEEE Trans. Med. Imaging **40**(6), 1531–1541 (2021)

16. Nayar, R., Wilbur, D.C.: The Bethesda System for Reporting Cervical Cytology: Definitions, Criteria, and Explanatory Notes. Springer, Cham (2015). https://doi.org/10.1007/978-3-319-11074-5
17. Patel, M.M., Pandya, A.N., Modi, J.: Cervical pap smear study and its utility in cancer screening, to specify the strategy for cervical cancer control. National J. Commun. Med. **2**(01), 49–51 (2011)
18. Qu, L., Luo, X., Liu, S., Wang, M., Song, Z.: DGMIL: distribution guided multiple instance learning for whole slide image classification. In: Wang, L., Dou, Q., Fletcher, P.T., Speidel, S., Li, S. (eds.) MICCAI 2022. LNCS, vol. 13432, pp. 24–34. Springer, Cham (2022). https://doi.org/10.1007/978-3-031-16434-7_3
19. Radford, A., et al.: Learning transferable visual models from natural language supervision. In: International Conference on Machine Learning, pp. 8748–8763. PMLR (2021)
20. Redmon, J., Farhadi, A.: Yolov3: an incremental improvement. arXiv preprint arXiv:1804.02767 (2018)
21. Robbins, H., Monro, S.: A stochastic approximation method. Ann. Math. Stat. 400–407 (1951)
22. Vaswani, A., et al.: Attention is all you need. In: Advances in Neural Information Processing Systems, vol. 30 (2017)
23. Veličković, P., Cucurull, G., Casanova, A., Romero, A., Lio, P., Bengio, Y.: Graph attention networks. arXiv preprint arXiv:1710.10903 (2017)
24. Wei, Z., Cheng, S., Liu, X., Zeng, S.: An efficient cervical whole slide image analysis framework based on multi-scale semantic and spatial deep features. arXiv preprint arXiv:2106.15113 (2021)
25. Wu, Z., Xiong, Y., Yu, S.X., Lin, D.: Unsupervised feature learning via non-parametric instance discrimination. In: Proceedings of the IEEE Conference on Computer Vision and Pattern Recognition, pp. 3733–3742 (2018)
26. Ye, M., Zhang, X., Yuen, P.C., Chang, S.-F.: Unsupervised embedding learning via invariant and spreading instance feature. In: Proceedings of the IEEE/CVF Conference on Computer Vision and Pattern Recognition, pp. 6210–6219 (2019)
27. Zaremba, W., Sutskever, I., Vinyals, O.: Recurrent neural network regularization. arXiv preprint arXiv:1409.2329 (2014)
28. Zhang, X., et al.: Whole slide cervical cancer screening using graph attention network and supervised contrastive learning. In: Wang, L., Dou, Q., Fletcher, P.T., Speidel, S., Li, S. (eds.) MICCAI 2022. LNCS, vol. 13432, pp. 202–211. Springer, Cham (2022). https://doi.org/10.1007/978-3-031-16434-7_20
29. Zhou, M., Zhang, L., Xiaping, D., Ouyang, X., Zhang, X., Shen, Q., Luo, D., Fan, X., Wang, Q.: Hierarchical pathology screening for cervical abnormality. Comput. Med. Imaging Graph. **89**, 101892 (2021)

Improving Pathology Localization: Multi-series Joint Attention Takes the Lead

Ashwin Raju(✉), Micha Kornreich, Colin Hansen, James Browning, Jayashri Pawar, Richard Herzog, Benjamin Odry, and Li Zhang

Covera Health, New York, NY, USA
ashwin.raju@coverahealth.com

Abstract. Automated magnetic resonance imaging (MRI) pathology localization can significantly reduce inter-reader variability and the time expert radiologists need to make a diagnosis. Many automated localization pipelines only operate on a single series at a time and are unable to capture inter-series relationships of pathology features. However, some pathologies require the joint consideration of multiple series to be accurately located in the face of highly anisotropic volumes and unique anatomies. To efficiently and accurately localize a pathology, we propose a **M**ulti-series j**O**int **AT**tention localization framework (MOAT) for MRI, which shares information among different MRI series to jointly predict the pathological location(s) in each MRI series. The framework allows different MRI series to share latent representations with each other allowing each series to get location guidance from the others and enforcing consistency between the predicted locations. Extensive experiments on three knee MRI pathology datasets, including medial compartment cartilage (MCC) high-grade defects, medial meniscus (MM) tear and displaced fragment/flap (DF) with 2729, 2355, and 4608 studies respectively, show that our proposed method outperforms the state of the art approaches by 3.4 to 8.0 mm on L1 distance, 6 to 27% on specificity and 5 to 14% on sensitivity across different pathologies.

Keywords: Pathology localization · Multi-series · Self-attention

1 Introduction

MRI is an essential diagnostic and investigative tool in clinical and research settings. Expert radiologists rely on multiple MRI series of varying acquisition parameters and orientations to capture different aspects of the underlying anatomy and diagnose any defect or pathology that may be present. For a knee

Supplementary Information The online version contains supplementary material available at https://doi.org/10.1007/978-3-031-43987-2_25.

H. Greenspan et al. (Eds.): MICCAI 2023, LNCS 14225, pp. 253–262, 2023.
https://doi.org/10.1007/978-3-031-43987-2_25

study, it is typical to acquire MRI series with coronal, sagittal, and axial orientations using proton density (PD), proton density fat suppressed (PDFS) or T2-weighted fat suppressed series (T2FS) for each study. When series are analyzed in concert, a radiologist can make a more effective diagnosis and mark down the location of any corresponding defect in each series. The defect location is typically represented as a single point [3] regardless of the defect size as a balance of effectiveness and efficiency.

In recent years, convolutional neural networks (CNNs) have achieved promising results in pathology localization. Many approaches rely on generating a multi-variate Gaussian heatmap, where the peak of the distribution represents the pathology localization. Hourglass [11,16], an encoder-decoder style architecture [9], is a mainstream model to generate a Gaussian heatmap. It uses a series of convolutional and pooling layers to extract features from the input image followed by upsampling and convolutional layers to generate the Gaussian heatmap. However, Hourglass-based methods can be overly resource-intensive when applied to 3D volumes [11]. To overcome this, regression-based models are becoming popular for detecting defects wherein a fully-connected layer is used on top of the encoder blocks to directly predict the location. These methods also alleviate the need for heatmap generation and post-processing methods to compute the location. Recently, transformer-based models have emerged as a promising trend in localization [4,6,14], and their performance has exceeded that of encoder-decoder based methods on single MRI volumes [4,7]. With the availability of multiple series, we propose a framework that imitates a clinical workflow, by simultaneously analyzing multiple series and paying attention to the location that corresponds to a pathology across multiple series.

To do this, we design a framework that utilizes self-attention across multiple series and we further add a mask to allow the model to focus on relevant areas, which we term as Masked Self-Attention (MSA). To predict the pathology location, we use a transformer decoder with an encoder-based initialization of the reference points. This approach provides a strong initial guess of the pathology location, improving the accuracy of the model's predictions. Overall, our framework leverages the strengths of both self-attention and encoder-decoder architectures to enhance the performance of pathology localization.

Specifically, our contributions are:

- We introduce a framework that enables the simultaneous use of multiple series from an MRI study, allowing for the sharing of pathology information across different series through Masked Self-Attention.
- We design a transformer-based decoder model to predict consistent locations across series in an MRI study, which reduces the network's parameters compared to standard heatmap-based approaches.
- Through extensive experiments on three knee pathologies, we demonstrate the effectiveness and efficiency of our framework, showing the benefits of Masked Self-Attention and a Pathology localization decoder to accurately predict pathology locations.

Overall, our framework represents a promising step towards more consistent and accurate localization, which could have important applications in medical diagnosis and treatment.

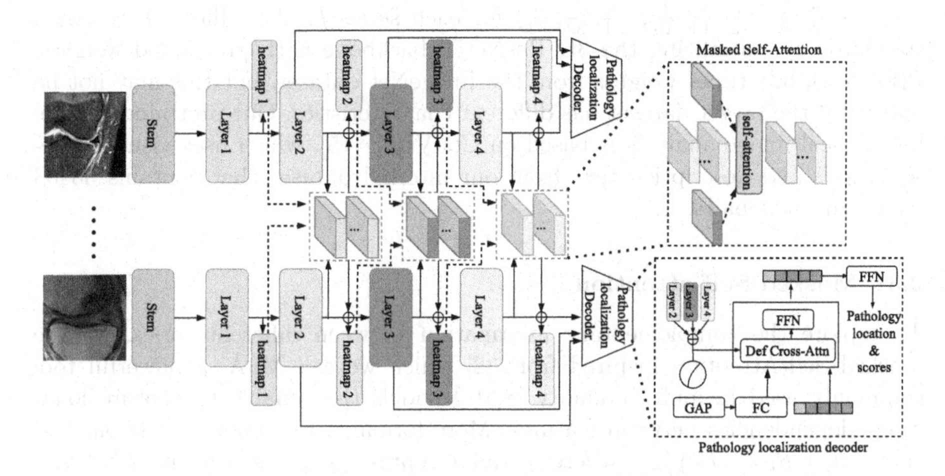

Fig. 1. Overview. More than 1 series are passed to encoders that have shared parameters. "Stem", "layer1", "layer2", "layer3" and "layer4" follows the ResNet [12] architecture convention. We perform Masked Self-Attention starting from layer 2. The Pathology localization decoder accepts feature maps from layer 2 to layer 4 and uses a query for each series to perform deformable cross attention to generate pathological landmarks.

2 Methods

2.1 Our Architecture

We aim to produce a reliable pathology location for each series in a given study if a location is available for that series. More formally, we assume that we are given a dataset, $\mathcal{D} = \{X_i, Y_i\}_{i=1}^{N}$, with N denoting the total number of studies in the dataset, X_i and Y_i denoting the set of series and corresponding location for each series. Due to different acquisition protocols, the number of series in each X_i can vary. Similarly, each Y_i can have a different number of location. Our goal is to predict a pathology location for each series and its corresponding confidence score. Figure 1 outlines our framework which can accept multiple series to generate a more accurate locations for each series.

2.2 Backbone

Our framework contains a backbone, which is responsible for generating multi-level feature maps. The multi-level feature maps are then fed into the pathology localization decoder. We use a 3D ResNet50 [12], which accepts the volume as the input and generates multiple feature maps. Each series has its own backbone with the weights been shared. Given an input $X_i^k \in \mathbb{R}^{d \times w \times h}$, denoting a series k from the study i, we extract multiple feature maps of resolutions $F^j \in (\frac{d}{1}, \frac{w}{8}, \frac{h}{8}), (\frac{d}{2}, \frac{w}{16}, \frac{h}{16}), (\frac{d}{4}, \frac{w}{32}, \frac{h}{32})$ for each series k. We adhere to common standards by initializing the 3D ResNet50 backbone with pretrained weights. Prior work fine-tunes weights from the ImageNet dataset, but this may not be optimal if the target dataset has different characteristics. Our pretrained model for medical image analysis is based on ConVIRT [15], which uses visual representations and descriptive text from our internal dataset that contains 35433 image and text pairs.

2.3 Masked Self-attention

To explore the complementary information between different series, we use Masked Self-Attention inspired from [2] which we call MSA, a powerful tool commonly used in multi-modality [8,10] models that enable to capture long-range dependencies between features. More formally, we denote the latent feature maps $R_l = \{F_l^j\}_{j=1}^J$, where j and l represents j^{th} series and l^{th} layer, $F_l^j \in (C_{in} \times d' \times w' \times h')$ with C_{in} representing the number of channels, d' representing the depth dimension, and w' and h' representing the width and height dimensions, respectively. We concatenate the features F_l^j along the depth dimension d' and add position embedding on the concatenated features. The transformer uses a linear projection for computing the set of queries, keys and values Q, K and V respectively. We adhere to the naming conventions used in [8].

$$Q = R_l.U^q, K = R_l.U^k, V = R_l.U^v \tag{1}$$

where $U^q \in \mathbb{R}^{C_{in} \times C_q}$, $U^k \in \mathbb{R}^{C_{in} \times C_k}$ and $U^v \in \mathbb{R}^{C_{in} \times C_v}$. The self-attention is calculated by taking the dot products between Q and K and then aggregating the values for each query,

$$A = \text{Softmax}\left(M_{l-1} + B + \frac{QK^T}{\sqrt{C_k}}\right)V \tag{2}$$

$$M_{l-1} = \begin{cases} 0 & \text{if } M_{l-1} = 1 \\ -\infty & \text{otherwise} \end{cases} \tag{3}$$

where, the attention mask $M_{l-1} \in \{0, 1\}$ is a binarized output (thresholded at δ_t) of the the resized mask prediction of the previous $(l-1)$-th layer. δ_t is empirically set to 0.15. The attention mask ignores the features that are not relevant to the pathology and attends to pathological features. B is a mask to handle missing series and it shares the same equation as 3.

Finally, the transformer uses a non-linear transformation to calculate the output features, R_{l+1}, which shares the same resolution as that of R_l.

$$R_{l+1} = \text{MLP}(A) + R_l \tag{4}$$

The transformer applies the attention mechanism 3 L times to generate a deep representation learning among the features. This approach allows the transformer model to effectively capture the relationships between different input positions.

2.4 Pathology Localization Decoder

The localization decoder follows the transformer decoder paradigm, using a query, reference points, and input feature maps to predict a location and corresponding score. The decoder has N identical layers, each consisting of cross-attention and feed forward networks (FFNs). The query $Q \in \mathbb{R}^{1\times 256}$ and reference points $R \in \mathbb{R}^3$ go through each layer, generating an updated Q as input for the next layer. Unlike Deformable DETR [17], the decoder initializes reference points by taking the last layer of the backbone feature map and applying Global Average Pooling, followed by a fully connected layer to generate the initial reference point. The localization refinement stage outputs location and scores for each layer N_i, similar to Deformable DETR, providing fast convergence.

2.5 Loss Functions

The model generates a single location $\hat{y_l} \in \mathbb{R}^3$, score y_s and auxiliary heatmap outputs H for each series in a given study. The goal of our framework is to generate one reasonable location and its corresponding score for each series. Since there may be multiple locations annotated for a series, we use the Hungarian Matching function [17] to find optimal matching with the prediction to one of many ground truth locations. This is similar to the approach used in DETR. The Masked Self-Attention in our framework uses heatmaps generated from the previous layers. To ensure accurate heatmap generation, we apply an auxiliary heatmap loss using Mean Square Error (MSE) between the generated heatmap and the ground truth Gaussian heatmap, where the loss is defined as,

$$L_{heatmap} = \sum_{i=1}^{K} (x - h_i)^2 \tag{5}$$

where K is the number of intermediate heatmaps generated, x and h_i are ground truth heatmap and predicted heatmap. To penalize the predicted location, we use the Huber loss defined as,

$$L_{point} = \sum_{i=1}^{N} \begin{cases} \frac{1}{2}(y_l^i - \hat{y_l^i})^2 & \text{if } \left|(y_l^i - \hat{y_l^i})\right| < \delta \\ \delta((y_l^i - \hat{y_l^i}) - \frac{1}{2}\delta) & \text{otherwise} \end{cases} \tag{6}$$

where δ is empirically set to 0.3. The distance of a pathology does not differ more than λ (which can be calculated from the dataset) across series. With this information, we enforce proximity between the world coordinates which can be converted from the predicted volume coordinates across different series. We employ a Margin L1 loss, which penalizes the distance between points if they exceed the margin. Formally,

$$L_{cons} = \sum_{i=1}^{N} \sum_{j=i+1}^{N} max(0, L1(wc_{\hat{y}_l^i}, wc_{\hat{y}_l^j}) - \lambda) \tag{7}$$

where N is the number of series in a given study, $wc_{\hat{y}_l}$ is the world coordinates converted from volume coordinates.

We then formulate the confidence score loss by considering the sum over the series of the binary cross entropy between the ground truth confidence score and predicted confidence score, formally defined as,

$$L_{score} = \sum_{i=1}^{N} -(y_s^i \log(p_i) + (1 - y_s^i) \log(1 - p_i)) \tag{8}$$

Overall, the entire loss for a given study is formulated as,

$$L = w_1 L_{point} + w_2 L_{score} + w_3 L_{cons} + w_4 L_{heatmap} \tag{9}$$

We set the hyper parameter w_1, w_2, w_3 and w_4 as 10, 1, 0.1, 1 respectively. These values are empirically set based on the validation loss.

3 Experiment

3.1 Implementation Details, Datasets and Evaluation Protocols

Implementation Details. Our model was implemented in Pytorch 1.13.1 on a NVIDIA A6000 GPU. We used an AdamW [5] optimizer with a weight decay of 10^{-4}. The initial learning rate for encoder was empirically set as 10^{-5} and 10^{-4} for all other modules. Before running the pathology detection, we perform a pre-processing step similar to [3] and resize the volume to $28 \times 128 \times 128$. Furthermore, we clip the intensity of the images at the 1st and 99th percentile, followed by an intensity normalization to ensure a mean of 0 and standard deviation of 1. Other hyper-parameters are mentioned in the supplementary paper.

Datasets. The study is limited to secondary use of existing HIPPA-based de-identified data. No IRB required. We primarily conduct our experiments using knee MRI datasets, with a specific focus on MM tear, MM displaced fragment flap (DF), and MCC defect. Studies were collected at over 25 different institutions, and differed in scanner manufacturers, magnetic field strengths, and imaging protocols. The pathological locations were annotated by American Board certified sub-specialists radiologists. The most common series types included fat-suppressed (FS) sagittal (Sag), coronal (Cor) and axial (Ax) orientations, using either T2-weighted (T2) or proton-density (PD) protocols. For pathology detection, we use CorFS, SagFS, and SagPD. The dataset statistics that we use for training, validation and test are shown in Table 1.

Table 1. Cor, Sag, SagPD refers to Coronal FS, Sagittal FS, Sagittal PD respectively. Values under series refer to number of series where defect locations are available. Negatives refer to number of studies that does not have a pathology.

Pathology	Train					Validation				Test			
	Negatives	Cor	Sag	SagPD	Union	Cor	Sag	SagPD	Union	Cor	Sag	SagPD	Union
MM Tear	1215	1466	1173	1146	**2679**	975	843	771	**975**	954	857	765	**954**
DF	673	387	306	243	**1862**	247	176	148	**277**	190	146	106	**216**
MCC	1000	759	797	364	**1926**	360	387	152	**437**	304	317	136	**366**

Evaluation Protocols. A useful pathology detection device should point the user to the correct location of a pathology. For model evaluation, we use the L1 distance between the predicted location to any annotation of the same pathology, labeled on the same series. To evaluate the pathology localization in a given study, we use the predicted pathology localization mask, which is obtained by thresholding the confidence score.

However, this alone does not provide a complete picture of the model's performance. To evaluate our confidence score's performance, we analyze the specificity and sensitivity of the confidence scores. We report the mean over all series in the test studies in Table 2

Table 2. Quantitative results. We show the L1 distance measured in (mm), Sensitivity (Sn), and Specificity (Sp) score for different models. "*" refers to the models that were trained with different hyper-parameters from their mentioned ones. The results are evaluated on the test dataset.

Methods	Param	FLOPs	MM Tear			MM DF			MCC defect		
	(M)	(G)	L1↓	Sn↑	Sp↑	L1↓	Sn↑	Sp↑	L1↓	Sn ↑	Sp↑
UNet* [3]	54.6	117.3	10.1	0.63	0.71	16.1	0.53	0.62	9.5	0.70	0.71
UNet w MSA	63.7	135.4	9.4	0.70	0.71	15.3	0.59	0.68	8.1	0.70	0.72
KNEEL* [11]	74.2	152.1	9.1	0.68	0.70	14.5	0.61	0.70	9.1	0.69	0.72
Regression	24.1	45.7	17.2	0.71	0.75	21.4	0.70	0.72	10.1	0.72	0.74
DETR* [1]	51.4	67.3	14.6	0.75	0.73	20.2	0.72	0.71	13.4	0.72	0.74
Def. DETR* [17]	37.1	71.4	15.8	0.77	0.80	21.3	0.73	0.75	12.9	0.75	0.79
Poseur* [6]	35.1	65.1	13.1	0.76	0.80	17.3	0.71	0.75	11.5	0.78	0.79
Poseur w MSA	44.2	71.3	10.3	0.80	0.81	14.9	0.73	0.76	7.2	0.80	**0.86**
MOAT	28.2	63.4	**4.7**	**0.85**	**0.86**	**8.1**	**0.80**	**0.81**	**3.9**	**0.88**	**0.86**

3.2 Comparison with SOTA Methods

Heatmap-Based Architectures. The proposed architecture was compared to two other models, the Gaussian Ball approach [3] which utilizes a UNet architecture to generate a heatmap and KNEEL [11], which uses an hourglass network architecture to predict the Gaussian heatmap. Two variants of UNet were compared, one with MSA and one without. The threshold was set for each model which balanced sensitivity and specificity on the validation data. The comparison revealed that the sensitivity and specificity of the proposed MOAT model were 14 to 27% and 15 to 17% higher, respectively, than those of the other models. Additionally, the L1 distance of the heatmap-based model was approximately 5.4 to 8.0 mm higher than that of MOAT for all true positives. Overall, the results suggest that MOAT outperforms the other models in terms of sensitivity, specificity, and L1 distance.

Regression-Based Architectures. We compared our proposed architecture with several other methods: 1) a simple regression method that removes the pathology localization decoder and uses a fully connected layer to predict the pathology locations, 2) DETR, 3) deformable DETR [17], and 4) Poseur [6], which uses Residual Log estimation. We adopt our ConVIRT pretrained encoder and add MSA to all the regression models to ensure a fair comparison. MOAT, which has 63.4G FLOPs, is highly efficient when compared to State-Of-The-Art (SOTA) regression models and has L1 distance lower than other models (4.7 mm) and the highest sensitivity and specificity among the models. We attach the standard deviation scores for each model in the supplementary section.

3.3 Ablation Study

We first analyze the importance of MSA to our framework by training models with and without MSA. As MSA is a variant of self-attention, we also experiment with self-attention and with an attention mechanism [13] that was popular prior to self-attention. Table 4 shows the L1 distance for Medial Meniscus Tear (MM Tear) pathology, where our MSA which is a variant of self-attention is able to achieve the lowest L1 distance. Similarly, we analyze the weight factor for consistency loss, as different weight factor yields different results. From Table 3, we can see that the lowest L1 distance was obtained when the weight factor was 0.1. All the ablation studies were performed on the MM Tear validation dataset.

Table 3. Ablation study on MM Tear dataset to analyze the need for Masked Self-Attention.

Methods	L1 distance ↓
No Masked Self-Attention	11.3
self-attention	8.8
Masked Self-Attention	**6.1**
CBAM [13]	16.2

Table 4. Ablation study on MM Tear to analyze the weight factor for the consistency loss.

Consistency(w_3)	L1 distance ↓
10	11.2
1	6.4
0.1	6.1
0.01	**5.1**

4 Conclusion

We propose MOAT, a framework for performing localization in multi-series MRI studies which benefits from the ability to share relevant information across series via a novel application of self-attention. We increase the efficiency of the MOAT model by using a pathology localization decoder which is a variant of deformable decoder and initializes the reference points from the backbone of the model. We evaluate the effectiveness of our proposed framework (MOAT) on three challenging pathologies from knee MRI and find that it represents a significant improvement over several SOTA localization techniques. Moving forward, we aim to apply our framework to pathologies from other body parts with multiple series.

References

1. Carion, N., Massa, F., Synnaeve, G., Usunier, N., Kirillov, A., Zagoruyko, S.: End-to-end object detection with transformers. In: Vedaldi, A., Bischof, H., Brox, T., Frahm, J.-M. (eds.) ECCV 2020. LNCS, vol. 12346, pp. 213–229. Springer, Cham (2020). https://doi.org/10.1007/978-3-030-58452-8_13
2. Cheng, B., Misra, I., Schwing, A.G., Kirillov, A., Girdhar, R.: Masked-attention mask transformer for universal image segmentation (2022)
3. Kornreich, M., et al.: Combining mixed-format labels for AI-based pathology detection pipeline in a large-scale knee MRI study. In: Wang, L., Dou, Q., Fletcher, P.T., Speidel, S., Li, S. (eds.) MICCAI 2022. LNCS, vol. 13438, pp. 183–192. Springer, Cham (2022). https://doi.org/10.1007/978-3-031-16452-1_18
4. Li, X., et al.: SDMT: spatial dependence multi-task transformer network for 3D knee MRI segmentation and landmark localization. IEEE Trans. Med. Imaging (2023)
5. Loshchilov, I., Hutter, F.: Decoupled weight decay regularization. arXiv preprint arXiv:1711.05101 (2017)
6. Mao, W., et al.: Poseur: direct human pose regression with transformers (2022)
7. Mathai, T.S., et al.: Lymph node detection in T2 MRI with transformers. In: Medical Imaging 2022: Computer-Aided Diagnosis, vol. 12033, pp. 855–859. SPIE (2022)
8. Prakash, A., Chitta, K., Geiger, A.: Multi-modal fusion transformer for end-to-end autonomous driving. In: Conference on Computer Vision and Pattern Recognition (CVPR) (2021)
9. Ronneberger, O., Fischer, P., Brox, T.: U-Net: convolutional networks for biomedical image segmentation. In: Navab, N., Hornegger, J., Wells, W.M., Frangi, A.F. (eds.) MICCAI 2015. LNCS, vol. 9351, pp. 234–241. Springer, Cham (2015). https://doi.org/10.1007/978-3-319-24574-4_28
10. Shvetsova, N., et al.: Everything at once - multi-modal fusion transformer for video retrieval. In: Proceedings of the IEEE/CVF Conference on Computer Vision and Pattern Recognition (CVPR), pp. 20020–20029 (2022)
11. Tiulpin, A., Melekhov, I., Saarakkala, S.: Kneel: knee anatomical landmark localization using hourglass networks. In: Proceedings of the IEEE/CVF International Conference on Computer Vision Workshops (2019)
12. Tran, D., Wang, H., Torresani, L., Ray, J., LeCun, Y., Paluri, M.: A closer look at spatiotemporal convolutions for action recognition. CoRR abs/1711.11248 (2017). http://arxiv.org/abs/1711.11248
13. Woo, S., Park, J., Lee, J.Y., Kweon, I.S.: CBAM: convolutional block attention module. In: Proceedings of the European Conference on Computer Vision (ECCV), pp. 3–19 (2018)

14. Xu, Y., Zhang, J., Zhang, Q., Tao, D.: Vitpose+: vision transformer foundation model for generic body pose estimation. arXiv preprint arXiv:2212.04246 (2022)
15. Zhang, Y., Jiang, H., Miura, Y., Manning, C.D., Langlotz, C.P.: Contrastive learning of medical visual representations from paired images and text. In: Machine Learning for Healthcare Conference, pp. 2–25. PMLR (2022)
16. Zhu, J., Zhao, Q., Zhu, J., Zhou, A., Shao, H.: A novel method for 3D knee anatomical landmark localization by combining global and local features. Mach. Vis. Appl. **33**(4), 52 (2022)
17. Zhu, X., Su, W., Lu, L., Li, B., Wang, X., Dai, J.: Deformable DETR: deformable transformers for end-to-end object detection. arXiv preprint arXiv:2010.04159 (2020)

Detection of Basal Cell Carcinoma in Whole Slide Images

Hongyan Xu[1,2], Dadong Wang[2(✉)], Arcot Sowmya[1], and Ian Katz[3]

[1] School of Computer Science and Engineering, University of New South Wales, Kensington, Australia
{hongyan.xu,a.sowmya}@unsw.edu.au

[2] Data61, The Commonwealth Scientific and Industrial Research Organisation, Canberra, Australia
Dadong.wang@csiro.au

[3] Southern Sun Pathology Pty Ltd, Thornleigh, Australia
ian.katz@southernsun.com.au

Abstract. Basal cell carcinoma (BCC) is a prevalent and increasingly diagnosed form of skin cancer that can benefit from automated whole slide image (WSI) analysis. However, traditional methods that utilize popular network structures designed for natural images, such as the ImageNet dataset, may result in reduced accuracy due to the significant differences between natural and pathology images. In this paper, we analyze skin cancer images using the optimal network obtained by neural architecture search (NAS) on the skin cancer dataset. Compared with traditional methods, our network is more applicable to the task of skin cancer detection. Furthermore, unlike traditional unilaterally augmented (UA) methods, the proposed supernet Skin-Cancer net (SC-net) considers the fairness of training and alleviates the effects of evaluation bias. We use the SC-net to fairly treat all the architectures in the search space and leveraged evolutionary search to obtain the optimal architecture for a skin cancer dataset. Our experiments involve 277,000 patches split from 194 slides. Under the same FLOPs budget (4.1G), our searched ResNet50 model achieves 96.2% accuracy and 96.5% area under the ROC curve (AUC), which are 4.8% and 4.7% higher than those with the baseline settings, respectively.

Keywords: Basal cell carcinoma (BCC) · whole-slide pathological images · deep learning · neural architecture search (NAS)

1 Introduction

Skin cancer, the most prevalent cancer globally, has seen increasing incidences over recent decades [1]. It constitutes a third of all cancer diagnoses, affecting one in five Americans [2]. Basal cell carcinoma (BCC), comprising 70% of cases, has surged by 20–80% in the last 30 years, exerting a significant healthcare strain.

H. Greenspan et al. (Eds.): MICCAI 2023, LNCS 14225, pp. 263–272, 2023.
https://doi.org/10.1007/978-3-031-43987-2_26

Timely BCC diagnosis is crucial to avoid complex treatments. Although histological evaluation remains the gold standard for detection [3], deep learning and computer vision advancements can streamline this process. Scanned traditional histology slides result in whole slide images (WSIs) that can be analyzed by deep learning models, significantly easing the histological evaluation burden. Recent advancements underscore the promise of this approach [4–6].

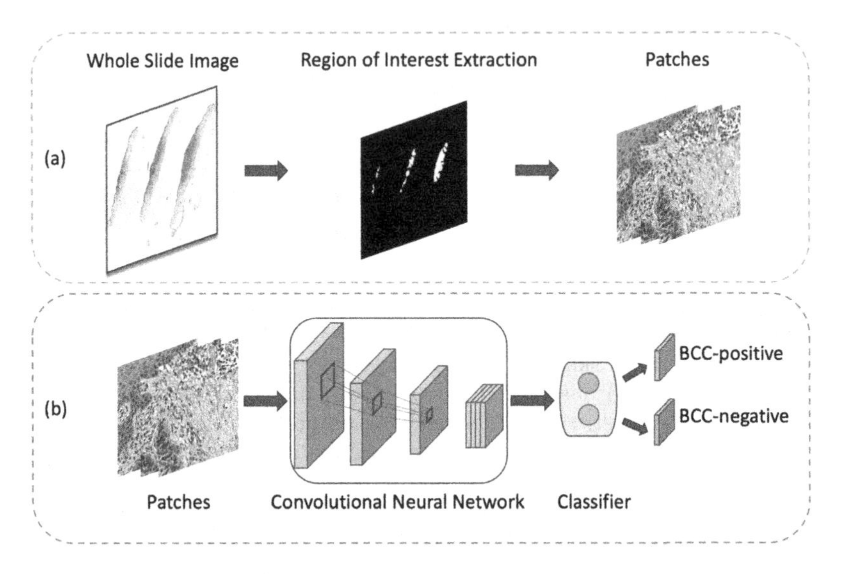

Fig. 1. Overall framework of the proposed model. (a) Region of interest (ROI) extraction and patch generation, and (b) patch detection and WSI classification.

Existing skin cancer detection methods [7–9] typically employs models like Inception Net and ResNet, designed for natural images like those in the ImageNet dataset. The significant variance in pathology and natural images can compromise these models' accuracy. Neural architecture search (NAS) addresses this issue by auto-designing superior models [10–13], exploring a vast architecture space. However, current NAS methods often overlook fairness in architecture ranking, impeding the discovery of top-performing models.

In this study, we utilized the NAS approach to identify the optimal network for skin cancer detection. To improve the efficiency and accuracy of the search, we developed a new framework named SC-net, which focuses on identifying highly valuable architectures. We observed that conventional NAS methods often overlook fairness ranking during the search, hindering the search for optimal solutions. Our SC-Net framework addresses this by ensuring fair training and precise ranking. The efficacy of SC-net was confirmed by our experimental results, with our ResNet50 achieving 96.2% top-1 accuracy and 96.5% AUC, outperforming baseline methods by 4.8% and 4.7% respectively.

Figure 1 shows the proposed framework, which integrates two modules. Module (a) extracts the region of interest (ROI) from WSI and generates patches, while Module (b) uses optimal model architecture from NAS to analyze features from patches and generate classifications.

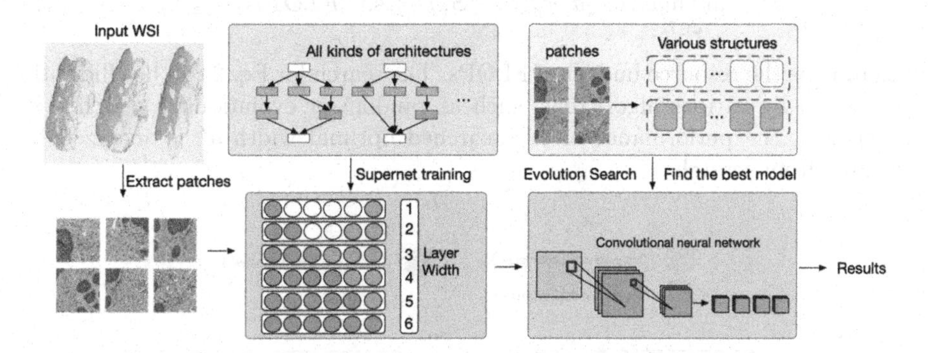

Fig. 2. The schematic diagram of the proposed method. Module (a) extracts ROI from WSI, creating 224×224 patches. Module (b) uses these patches to train and search optimal structure within a supernet via an evolutionary algorithm, yielding dataset predictions.

2 Methods

The proposed method (Fig. 2) involves dividing the input WSI into patches for training a supernet and the search for optimal architectures [14]. Section 2.2 provides further details about the supernet. A balanced evolutionary algorithm is then used to select the optimal structure from the search space, with the candidate structures' performance evaluated using mini-batch patch data. We evaluate the searched architectures on the skin cancer dataset.

2.1 One-Shot Channel Number Search

To extract an optimal architecture $\gamma \in \mathcal{G}$ from a vast search space $\mathcal{G}$, a weight-sharing strategy [15–17] is used to prevent training from scratch. The search leverages a supernet $\mathcal{S}$ with weights $\mathcal{W}$, with each path γ inheriting weights from $\mathcal{W}$. This makes one-shot NAS a two-step optimization process: supernet training and architecture search. The original dataset is typically split into training $\mathcal{D}_t$ and validation datasets $\mathcal{D}v$. The weights $\mathcal{W}$ of the supernet $\mathcal{S}$ are trained by uniformly sampling the network width d and optimizing the sub-network with weights $wd \subset \mathcal{W}$. The optimization function is defined as follows:

$$\mathcal{W}^* = \underset{w_d \in \mathcal{W}}{\arg\min} \underset{d \in U(\mathcal{D})}{\mathbb{E}} [\![\mathcal{L}_t(w_d; \mathcal{S}, d, \mathcal{D}_t]\!]) \quad (1)$$

where $U(\mathcal{D})$ is a uniform distribution of network widths, $\mathbb{E}$ is the expected value of random variables, and $\mathcal{L}_t$ is the training loss function. Then, the optimal network width d^* corresponds to the network width with the best performance (*e.g.* classification accuracy) on the validation dataset, *i.e.*,

$$d^* = \underset{d \in \mathcal{D}}{\arg\max} Acc(d, w_d^*; \mathcal{W}^*, \mathcal{S}, \mathcal{D}_v), \text{s.t. } FLOPs(d) \leq F_p, \tag{2}$$

where F_p is the resource budget of FLOPs. The search for Eq. 2 can be efficiently performed by various algorithms, such as random or evolutionary search [18]. Afterward, the performance of the searched optimal width d^* is analyzed by training from scratch.

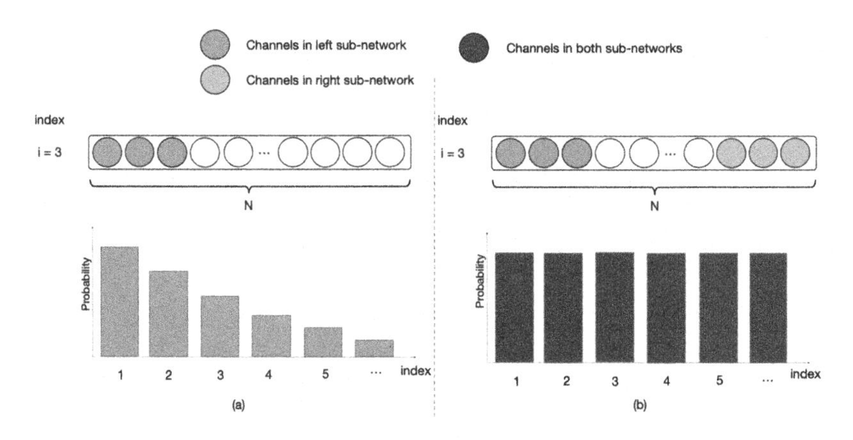

Fig. 3. Schematic diagram comparing (a) the UA method and (b) our method. In the UA principle, some channels are trained twice while others are trained only once or not at all, leading to channel training unfairness and evaluation bias. In contrast, our proposed method ensures that all channels are trained evenly (twice) by training both the width d and its complementary width.

2.2 SC-Net as a Balanced Supernet

Current approaches for neural architecture search [19–21] often employ a unilaterally augmented (UA) principle to evaluate each width, resulting in unfair training of channels in the supernet. As illustrated in Fig. 3(a), to search for a dimension d at a layer with a maximum of n channels, the UA principle assigns the left d channels in the supernet to indicate the corresponding architecture as

$$\gamma_A(d) = [1:d], d \leq n \tag{3}$$

where $\gamma_A(d)$ means the selected d channels from the left (smaller-index) side.

However, the UA principle leads to channel training imbalance in the supernet due to its constraints, as illustrated in Fig. 3(a). Channels with smaller indices are used for various sizes, resulting in over-training of the left channel kernels since widths are uniformly sampled. The unfairness can be quantified

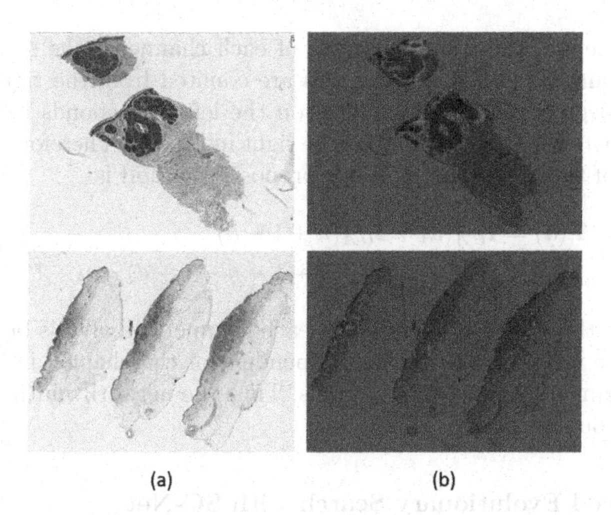

Fig. 4. Example of prediction probability heatmaps produced by the s_ResNet50 model. Column (a) shows the original WSIs that contain annotations represented by green circles. Column (b) displays the prediction probability heatmaps of our ResNet50 model. (Color figure online)

by $\mathcal{T}$, which represents how often a channel is utilized, reflecting its level of training. Given a layer has a maximum of n channels, the $\mathcal{T}$ for the i-th channel under the UA principle is

$$\mathcal{T}(i) = n - i + 1 \tag{4}$$

Correspondingly, the probability of i -th channel being trained can be expressed as $P_i = \frac{n-i+1}{n}$. Therefore, channels closer to the left will get more attempts during training, which leads the degree of training to vary widely between channels. This introduces evaluation bias and leads to sub-optimal results.

To mitigate evaluation bias on width, we propose a new SC-net that promotes the fairness of channels during training. As shown in Fig. 3(b), in the proposed supernet, each width is simultaneously evaluated by the sub-networks corresponding to the left and right channels. This can be seen as two identical networks S_l and S_r that are bilaterally coupled and evaluated using the UA principle, but counting channels in reverse order. Therefore, the number of all channels used for evaluating the width d can be expressed as:

$$\begin{aligned}\mathcal{H}(d) &= \mathcal{H}^l_{UA}(d) \uplus \mathcal{H}^r_{UA}(d) &(5)\\ &= [1 : d] \uplus [(n - d + 1) : (n - d)], &(6)\end{aligned}$$

where $\uplus$ represents the union of two lists with repeatable elements. In detail, the left channel in S_l follows the same UA principle setup as in Eq. (3), while for the right channel in S_r, we count channels from the right $\mathcal{H}^r(d) = [(n - d + 1) :$

$(n-d)]$. Therefore, the training degree of each channel is the sum of the two supernets S_l and S_r. Since the channels are counted from the right within S_r, the training degree of the d-th channel on the left corresponds to the training degree of the $(n-d+1)$-th channel on the right in Eq. (4). Therefore, the training degree $\mathcal{T}(d)$ of the d-th channel in our proposed method is

$$\mathcal{T}(d) = \mathcal{T}_{UA}(d) + \mathcal{T}_{UA}(n+1-d) \tag{7}$$

$$= (n-d+1) + (n+1-n-1+d) = n+1 \tag{8}$$

Therefore, the training degree $\mathcal{T}$ for each channel will always be equal to the same constant value of the width, independent of the channel index, ensuring fairness in terms of channel (filter) levels. Thus the network width can be fairly ranked using our network.

2.3 Balanced Evolutionary Search with SC-Net

Using a trained SC-net, the architecture can be evaluated. However, the search space involved in NAS is large, with more than 10^{20} possible architectures, requiring an evolutionary search using the multi-objective NSGA-II algorithm to improve the search performance. During the evolutionary search, the width d of each network is represented by the average precision of its corresponding left and right paths in the supernet $\mathcal{S}$, as shown in Eq. (9). The optimal width (not subnetwork) is determined as the one that achieves the best performance when trained from scratch. Here, $\mathcal{S}_l$ and $\mathcal{S}_r$ refer to the two paths of $\mathcal{S}$ that correspond to the width d during the training process.

$$Acc(\mathcal{W}, d, \mathcal{D}_v) = \frac{1}{2}\left(Acc(\mathcal{S}_l, d; \mathcal{D}_v) + Acc(\mathcal{S}_r, d; \mathcal{D}_v)\right) \tag{9}$$

3 Experiments

3.1 Experiment Settings

Table 1. Generated dataset split

	BCC-positive		BCC-negative	
Training	WSI	Patch	WSI	Patch
	118	132,981	37	90,291
Testing	WSI	Patch	WSI	Patch
	30	31,651	9	22,838

The dataset, comprised of 194 skin slides acquired from the Southern Sun Pathology laboratory, includes 148 BCC cases and 46 other types (common nevus,

SCC), all manually annotated by a dermatopathologist. BCC slides served as positive samples and the rest as negatives. These slides were scanned at ×20 magnification with a 0.44 μm pixel size using a Leica Aperio AT2 Scanner. The patient data were separated between training and testing to prevent overlap. Details are shown in Table 1. The experimental setup involved training models on two NVIDIA RTX A6000 GPUs using PyTorch. These models, initialized from a zero-mean Gaussian with standard deviation $\sigma = 0.001$, were trained for 200 epochs with a batch size of 256. Training used the Adam optimizer with a dynamic learning rate reduction strategy, starting with a learning rate of 5e-5 following a cosine schedule.

Table 2. Performance comparison on skin cancer dataset

Type	Model	FLOPs	Parameters	Acc	Se	Sp	F_1	AUC
WSI analysis-Related	Tian *et al.* [22]	4.1G	25.5M	92.5	92.3	92.1	92.2	92.7
	Hekler *et al.* [9]	4.1G	25.5M	93.4	92.9	92.2	92.8	93.5
	Jiang *et al.* [23]	2.9G	27.2M	92.1	91.2	91.7	91.4	92.4
NAS-Related	MetaPruning [18]	4.1G	25.5M	94.1	93.2	92.7	93.3	94.2
	AutoSlim [24]	4.1G	25.5M	93.8	92.9	92.2	93.0	93.0
Ours	ori_ResNet50	4.1G	25.5M	91.4	90.2	90.4	90.5	91.8
	s_ResNet50	4.1G	27.2M	**96.2**	**94.7**	**95.8**	**95.2**	**96.5**
	ori_MobileNetV2	300M	3.5M	86.7	86.2	85.8	86.2	87.3
	s_MobileNetV2	300M	4.0M	91.9	90.4	91.7	90.7	92.4

3.2 Performance Evaluation

We validated our algorithm using the curated skin cancer dataset and SC-net as a supernet, testing both heavy and light models. We performed a search on ResNet50 and MobileNetV2 models, compared against original ResNet50 (ori_ResNet50) and MobileNetV2 (ori_MobileNetV2) models as baselines. The resulting models are denoted as s_ResNet50 and s_MobileNetV2.

Comparison with Related Methods. To ensure a fair comparison on our dataset, we selected several papers in the field of pathological image analysis, such as [9,22,23], as well as others using the UA principle, such as [18,24].

Evaluation Metrics. Our model was evaluated on: (1) Accuracy (Acc): percentage of correct classifications. (2) Sensitivity (Se): proportion of true positives identified. (3) Specificity (Sp): proportion of true negatives identified. (4) F1 Score (F1): precision and recall's harmonic mean, indicating label alignment. (5) AUC: ROC curve area, reflecting the false/true positive rate trade-off.

As shown in Table 2, the s_ResNet50 model outperformed in all metrics, showing 4.8%, 4.5%, 5.4%, 4.7% and 4.7% improvements in accuracy, sensitivity, specificity, F_1 Score, and AUC, respectively, over ori_ResNet50, and surpassing

Table 3. Performance of searched models with different searching methods.

Evaluator	Searching		Models							
SC-net	Greedy search	Evolutionary search	ResNet50				MobileNetV2			
			Acc	Se	Sp	AUC	Acc	Se	Sp	AUC
	✓		93.8	92.9	92.2	93.0	89.7	90.3	89.9	90.2
✓	✓		94.9	93.7	93.1	95.3	90.7	89.8	90.0	91.2
		✓	94.0	93.5	92.7	94.2	89.8	90.2	90.5	90.5
✓		✓	**96.2**	**94.7**	**95.8**	**96.5**	91.9	90.4	91.7	92.4

the Hekler *et al.* [9] method by 2.8%, 1.8%, 3.6%, 2.4%, and 3.0%. It also compared favorably to the s_MobileNetV2, adding only 1.7M and 0.5M parameters, respectively. In terms of UA principle method, s_ResNet50 advanced by 2.1%, 1.5%, 3.1%, 1.9%, and 2.3% in the same metrics compared to MetaPruning [18], showcasing our method's efficacy while keeping model complexity reasonable.

Visualization of Probability Heatmaps. Figure 4 demonstrates an example of the prediction probability heatmaps produced by the s_ResNet50 model. A comparison of the labeled areas in Column (a) and the red areas in Column (b) indicates that the predicted areas are generally similar in scope to the corresponding labeled areas. This suggests that the model is accurately identifying regions of interest for BCC diagnosis.

3.3 Ablation Study

Effect of SC-net as a Supernet. Table 3 presents our experiments testing the SC-net on ResNet50 and MobileNetV2 using various supernets and search methods. The SC-net (second row) under greedy search improved accuracy by 1.1% (ResNet50) and 1.0% (MobileNetV2), while with evolutionary search, accuracy increased by 1.8% (ResNet50) and 2.1% (MobileNetV2). These results highlight SC-net's effectiveness as a supernet in bolstering evaluation and search performance.

Generalization Ability of SC-net. We tested our model's generalization on the ChestMNIST and DermaMNIST subsets of MedMNISTv2 [25], following established protocols. As shown in Table 4, our s_ResNet50 surpassed the original ResNet50 on all datasets, gaining 2.3% and 1.8% more AUC on ChestMNIST and DermaMNIST respectively, proving the model's robust generalization.

Table 4. Performance comparison on ChestMNIST and DermaMNIST datasets

Model	ChestMNIST		DermaMNIST	
	AUC	ACC	AUC	ACC
ResNet18	76.8	94.7	91.7	73.5
ResNet50	76.9	94.7	91.3	73.5
auto-sklearn	64.9	77.9	90.2	71.9
AutoKeras	74.2	93.7	91.5	74.9
Google AutoML	77.8	94.8	91.4	76.8
s_ResNet50	**79.2**	**95.5**	**93.1**	**77.8**

4 Conclusion and Future Work

In this paper, we introduce SC-net, a novel NAS framework for skin cancer detection in pathology images. By formulating SC-net as a balanced supernet, we ensure fair ranking and treatment of all potential architectures. With SC-net and evolutionary search, we obtained optimal architectures, achieving 96.2% Top-1 and 96.5% accuracy on a skin cancer dataset, improvements of 4.8% and 4.7% over baselines. Future work will apply our approach to larger datasets for wider-scale validation.

5 Compliance with Ethical Standards

This study was performed in line with the principles of the Declaration of Helsinki. Ethics approval was granted by CSIRO Health and Medical Human Research Ethics Committee (CHMHREC). The ethics approval number is 2021_030_LR, and the validity period is from 07 Apr 2021 to 31 Dec 2024.

References

1. Rogers, H.W., Weinstock, M.A., Feldman, S.R., Coldiron, B.M.: Incidence estimate of nonmelanoma skin cancer (keratinocyte carcinomas) in the us population, 2012. JAMA Dermatol. **151**(10), 1081–1086 (2015)
2. Stern, R.S.: Prevalence of a history of skin cancer in 2007: results of an incidence-based model. Arch. Dermatol. **146**(3), 279–282 (2010)
3. Xu, G., et al.: CAMEL: a weakly supervised learning framework for histopathology image segmentation. In: Proceedings of the IEEE/CVF International Conference on Computer Vision, pp. 10682–10691 (2019)
4. Lu, M.Y., Williamson, D.F., Chen, T.Y., Chen, R.J., Barbieri, M., Mahmood, F.: Data-efficient and weakly supervised computational pathology on whole-slide images. Nat. Biomed. Eng. **5**(6), 555–570 (2021)
5. Sharma, Y., Shrivastava, A., Ehsan, L., Moskaluk, C.A., Syed, S., Brown, D.: Cluster-to-conquer: a framework for end-to-end multi-instance learning for whole slide image classification. In: Medical Imaging with Deep Learning, pp. 682–698. PMLR (2021)

6. Xiang, T., et al.: DSNet: a dual-stream framework for weakly-supervised gigapixel pathology image analysis. IEEE Trans. Med. Imaging **41**(8), 2180–2190 (2022)
7. Lu, C., Mandal, M.: Automated analysis and diagnosis of skin melanoma on whole slide histopathological images. Pattern Recogn. **48**(8), 2738–2750 (2015)
8. Van Zon, M., et al.: Segmentation and classification of melanoma and nevus in whole slide images. In: 2020 IEEE 17th International Symposium on Biomedical Imaging (ISBI), pp. 263–266. IEEE (2020)
9. Hekler, A., et al.: Pathologist-level classification of histopathological melanoma images with deep neural networks. Eur. J. Cancer **115**, 79–83 (2019)
10. Su, X., You, S., Wang, F., Qian, C., Zhang, C., Xu, C.: BCNet: searching for network width with bilaterally coupled network. In: Proceedings of the IEEE/CVF Conference on Computer Vision and Pattern Recognition, pp. 2175–2184 (2021)
11. Liu, C., et al.: Progressive neural architecture search. In: Proceedings of the European Conference on Computer Vision (ECCV), pp. 19–34 (2018)
12. Xie, B., et al.: Multi-scale fusion with matching attention model: a novel decoding network cooperated with NAS for real-time semantic segmentation. IEEE Trans. Intell. Transp. Syst. **23**(8), 12622–12632 (2021)
13. Su, X., et al.: Vision transformer architecture search, arXiv preprint arXiv:2106.13700 (2021)
14. Cha, S., Kim, T., Lee, H., Yun, S.-Y.: Supernet in neural architecture search: a taxonomic survey, arXiv preprint arXiv:2204.03916 (2022)
15. Liu, H., Simonyan, K., Yang, Y.: DARTS: differentiable architecture search, arXiv preprint arXiv:1806.09055 (2018)
16. Xie, L., et al.: Weight-sharing neural architecture search: a battle to shrink the optimization gap. ACM Comput. Surv. (CSUR) **54**(9), 1–37 (2021)
17. Chen, X., Xie, L., Wu, J., Wei, L., Xu, Y., Tian, Q.: Fitting the search space of weight-sharing NAS with graph convolutional networks. In: Proceedings of the AAAI Conference on Artificial Intelligence, vol. 35, no. 8, pp. 7064–7072 (2021)
18. Liu, Z., et al.: Metapruning: meta learning for automatic neural network channel pruning. In: Proceedings of the IEEE/CVF International Conference on Computer Vision, pp. 3296–3305 (2019)
19. Chen, M., Peng, H., Fu, J., Ling, H.: Autoformer: searching transformers for visual recognition. In: Proceedings of the IEEE/CVF International Conference on Computer Vision, pp. 12270–12280 (2021)
20. Wan, A., et al.: Fbnetv2: differentiable neural architecture search for spatial and channel dimensions. In: Proceedings of the IEEE/CVF Conference on Computer Vision and Pattern Recognition, pp. 12965–12974 (2020)
21. Yan, Z., Dai, X., Zhang, P., Tian, Y., Wu, B., Feiszli, M.: FP-NAS: fast probabilistic neural architecture search. In: Proceedings of the IEEE/CVF Conference on Computer Vision and Pattern Recognition, pp. 15139–15148 (2021)
22. Tian, Y., et al.: Computer-aided detection of squamous carcinoma of the cervix in whole slide images, arXiv preprint arXiv:1905.10959 (2019)
23. Jiang, Y., et al.: Recognizing basal cell carcinoma on smartphone-captured digital histopathology images with a deep neural network. Br. J. Dermatol. **182**(3), 754–762 (2020)
24. Yu, J., Huang, T.: Autoslim: towards one-shot architecture search for channel numbers, arXiv preprint arXiv:1903.11728 (2019)
25. Yang, J., et al.: MedMNIST v2-a large-scale lightweight benchmark for 2D and 3D biomedical image classification. Sci. Data **10**(1), 41 (2023)

SCOL: Supervised Contrastive Ordinal Loss for Abdominal Aortic Calcification Scoring on Vertebral Fracture Assessment Scans

Afsah Saleem[1,2(✉)], Zaid Ilyas[1,2], David Suter[1,2], Ghulam Mubashar Hassan[3], Siobhan Reid[4], John T. Schousboe[5], Richard Prince[3], William D. Leslie[6], Joshua R. Lewis[1,2], and Syed Zulqarnain Gilani[1,2,3]

[1] Centre for AI & ML, School of Science, Edith Cowan University, Joondalup, Australia
afsah.saleem@ecu.edu.au

[2] Nutrition and Health Innovation Research Institute, Edith Cowan University, Joondalup, Australia

[3] Computer Science and Software Engineering, The University of Western Australia, Perth, Australia

[4] Department of Electrical and Computer Engineering, University of Manitoba, Winnipeg, Canada

[5] Park Nicollet Clinic and HealthPartners Institute, HealthPartners, Minneapolis, USA

[6] Departments of Medicine and Radiology, University of Manitoba, Winnipeg, Canada

Abstract. Abdominal Aortic Calcification (AAC) is a known marker of asymptomatic Atherosclerotic Cardiovascular Diseases (ASCVDs). AAC can be observed on Vertebral Fracture Assessment (VFA) scans acquired using Dual-Energy X-ray Absorptiometry (DXA) machines. Thus, the automatic quantification of AAC on VFA DXA scans may be used to screen for CVD risks, allowing early interventions. In this research, we formulate the quantification of AAC as an ordinal regression problem. We propose a novel Supervised Contrastive Ordinal Loss (SCOL) by incorporating a label-dependent distance metric with existing supervised contrastive loss to leverage the ordinal information inherent in discrete AAC regression labels. We develop a Dual-encoder Contrastive Ordinal Learning (DCOL) framework that learns the contrastive ordinal representation at global and local levels to improve the feature separability and class diversity in latent space among the AAC-24 genera. We evaluate the performance of the proposed framework using two clinical VFA DXA scan datasets and compare our work with state-of-the-art methods. Furthermore, for predicted AAC scores, we provide a clinical analysis to predict the future risk of a Major Acute Cardiovascular Event (MACE).

Supplementary Information The online version contains supplementary material available at https://doi.org/10.1007/978-3-031-43987-2_27.

H. Greenspan et al. (Eds.): MICCAI 2023, LNCS 14225, pp. 273–283, 2023.
https://doi.org/10.1007/978-3-031-43987-2_27

Our results demonstrate that this learning enhances inter-class separability and strengthens intra-class consistency, which results in predicting the high-risk AAC classes with high sensitivity and high accuracy.

Keywords: Supervised contrastive ordinal learning · Distance-metric learning · Cardiovascular Diseases · VFA DXA scans

1 Introduction

Abdominal Aortic Calcification (AAC) is an established marker of atherosclerotic cardiovascular disease (CVD) [19] and can help in identifying asymptomatic cases at risk for CVD-related hospitalizations and deaths [14]. CVDs are responsible for 32% of global deaths, with atherosclerotic CVD events being the leading cause [17]. Therefore, early detection and management of AAC can improve CVD prevention, and management [14]. AAC can be seen on lateral spine Vertebral Fracture Assessment (VFA) images acquired using Dual-energy X-ray Absorptiometry (DXA), X-rays or Computed Tomography (CT) [13,19]. However, DXA imaging is the most recommended and cost-effective approach for VF assessment, with the least radiation exposure [12,23]. In VFA DXA scans, AAC can be calculated manually using the Kauppila AAC-24 semi-quantitative scale [10]. However, the manual scoring of AAC in VFA images is arduous and subjective [18,19]. Thus, designing an automated system for detecting and quantifying AAC in VFA DXA scans may be the most feasible approach for obtaining valuable CVD risk information in asymptomatic individuals.

Very few attempts have been made for automated AAC scoring in VFA DXA scans [4,6,7,18]. Reid et al. [18] trained two CNN models using VFA DXA scans and reported results via ensembling on a limited test set. Gilani et al. [7] performed sequential scoring using a vision-to-language model, but achieved low sensitivity for a moderate-risk class, indicating difficulty in handling complex cases near class boundaries. These works [7,18] considered automated AAC scoring a regression task. However, simple regression losses depend on continuous regression labels and, thus, cannot make feature embeddings separable. Moreover, inter-class similarities, intra-class variations and image artifacts in low-resolution VFA DXA scan further complicate the task. Therefore, using these losses directly for AAC-24 score regression may not be optimal.

Contrastive representation learning has shown promising results in medical image classification [2,9,16] and segmentation tasks [3,8,24]. In the classification tasks, supervised contrastive learning (SupCon) [11] aims to bring feature embeddings with the same labels closer together in the latent space and move the dissimilar ones apart. However, SupCon cannot preserve the ordinal information of the regression labels in the latent space [5]. To address this, Dai et al. [5] propose supervised Adaptive Contrastive loss (AdaCon), which depends on an adaptive margin. For calculating the adaptive margin, they assumed that a regression label could be replaced with its Empirical Cumulative Distribution

Function (ECDF) [22]. Though this assumption might be valid for large datasets, it may not work with highly skewed, imbalanced and limited-size datasets [5].

To this end, we propose a novel Supervised Contrastive Ordinal Loss (SCOL), considering AAC scoring as an ordinal regression problem. We integrate a label-dependent distance metric with the supervised contrastive loss. Unlike AdaCon [5], this metric relies exclusively on discrete regression labels, making it possible to utilize the ordinal information inherent in these labels. Using SCOL, we design an effective Dual-encoder Contrastive Ordinal Learning (DCOL) framework. Unlike previous methods, which either use global [18] or local attention-based features [7], DCOL assimilates global and local feature embeddings to increase feature diversity, and class separability in latent space.

To the best of our knowledge, this is the first framework that explores contrastive learning for automated detection of AAC. Our contributions are summarized as follows: 1) We propose a novel supervised contrastive ordinal loss by incorporating distance metric learning with the supervised contrastive loss to improve inter-class separability and handle intra-class diversity among the AAC genera. 2) We design a Dual-Encoder Contrastive Ordinal Learning framework using the proposed loss to learn separable feature embeddings at global and local levels. 3) We achieve state-of-the-art results on two clinical datasets acquired using DXA machines from multiple manufacturers, demonstrating the generalizability and efficacy of our approach. 4) We compare the Major Adverse Cardiovascular Event (MACE) outcomes for machine-predicted AAC scores and the human-measured scores to explore the clinical relevance of our method. This work aims to contribute clinically in refining automated AAC prediction methods using low-energy VFA scans. Our code is available at [1].

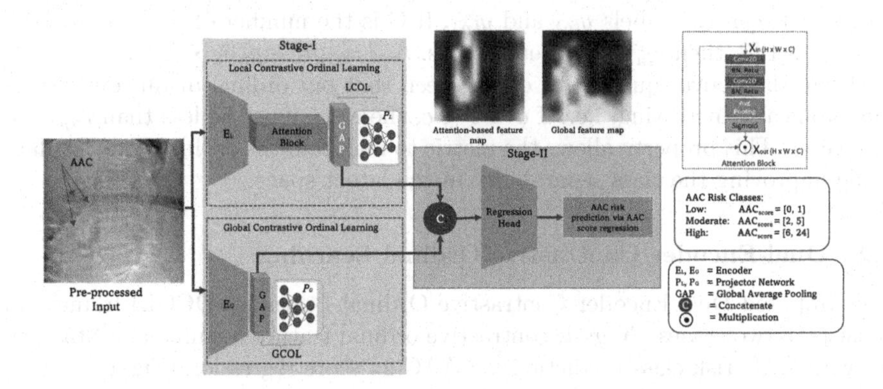

Fig. 1. Framework of our dual-encoder contrastive ordinal learning framework.

2 Methodology

2.1 Supervised Contrastive Ordinal Learning

Consider a training set T of M image-label pairs, such that $T = \{(x_i, y_i)\}_{i=1}^{M}$ where x_i is the i_{th} VFA DXA scan and y_i is the corresponding AAC score. Let a, p and n denote the indices of anchor, positive sample, and negative samples in a batch I. $P(i)$ is the set of indices of all the positive samples, i.e., having the same AAC score as y_a and $N(i)$ is the set of all other negative indices. Consider an encoder-projector network that maps the anchor image x_a in the embedding space such that $z_a = Proj(Enc(x_a))$, then the similarity between any two projections z_i and z_k in the latent space is: $sim(z_i, z_k) = z_i^T.z_k$.

Supervised contrastive loss [11] pulls images of the same class (positive samples) close together and pushes the negative ones apart. Following this strategy, we propose contrastive ordinal loss to move the negative sample x_n apart from the anchor x_a by a distance $r_{(a,n)}$: such that if $a < b < c$ then $r(a, b)$ must be less than $r(a, c)$ and vice versa. By incorporating this ordinal distance metric with the similarity function, we can maximize the benefit of ordinal information present in regression labels. Furthermore, in our AAC scoring task, this ordinal distance metric can help to increase inter-class separability and minimize the effect of intra-class variations. (For visualization see SM). Inspired by [5], we propose Supervised Contrastive Ordinal Loss (SCOL) for ordinal regression as:

$$L_{SCOL} = \sum_{i \in I} \frac{-1}{|P(i)|} \sum_{p \in P(i)} log \frac{exp(z_a.z_p/\tau)}{\sum_{n \in N(i)(n \neq a)} exp((z_a.z_n + r_{a,n})/\tau)} \quad (1)$$

where τ is a scaling hyper-parameter for contrastive loss and $r_{(a,n)}$ is the distance metric between two labels $y_{(a)}$ and $y_{(n)}$. If C is the number of ordinal labels in training set T, then $r_{(a,n)}$ is calculated as: $r_{a,n} = ||y_a - y_n||_2 \times 2/C$.

From the above equation, it can be seen that our ordinal distance metric is monotonically increasing, i.e., if $a < b < c$, then $r_{a,b}$ must be less than $r_{a,c}$ and vice versa. This property allows the metric to maintain the ordinality of the data while improving the class separability in the latent space.

2.2 Dual-Encoder Contrastive Ordinal Learning

The proposed Dual-Encoder Contrastive Ordinal Learning (DCOL) framework consists of two stages: Stage-I: contrastive ordinal feature learning and Stage-II: relevant AAC risk class prediction via AAC-24 score regression (Fig. 1).

Stage-I: It consists of two modules: Local Contrastive Ordinal Learning (LCOL) and Global Contrastive Ordinal Learning (GCOL). In these modules, we train the global and local encoder-projector networks individually in an end-to-end manner to extract contrastive feature embeddings.

Local Contrastive Ordinal Learning: In practice, to quantify AAC, clinicians focus on the aortic region adjacent to lumbar vertebrae L1-L4. Following this, in the Localized feature-based Contrastive Ordinal Learning module,

LCOL, we integrate a simple yet effective localized attention block with the baseline encoder E_l to roughly localize the aorta's position using only regression labels. This attention block is attached with E_l after extracting the deep feature map f_m from the last convolution layer. Our localized attention block consists of two 2D convolutional layers, followed by batch normalization and ReLu activation layers. This set of layers is then followed by an average pooling layer and sigmoid activation to create an activation map f_s for the most salient features in the given image. Multiplying this activation map f_s with the initial feature map f_m results in extracting the most significant features from f_m. These features are then projected into the latent space for processing by our SCOL loss. SCOL encourages the local contrastive embeddings with the same AAC score to move closer and the dissimilar ones apart based on the distance between their labels.

Global Contrastive Ordinal Learning: In the Global Contrastive Ordinal Learning module, we extract the global representation of a given VFA DXA scan. In encoder E_g, we replace the fully connected layers of the vanilla CNN model with a global average pooling (*GAP*) layer for feature extraction. These feature embeddings are then passed to the projection network P_g. SCOL operates on projected embeddings extracted from the whole lumbar region to maximize the feature separability while preserving the ordinal information in latent space. Both projector networks, P_l and P_g, consist of two Dense layers with 1280 and 128 neurons each, followed by ReLu activation.

Stage-II: In AAC-24 score regression, a small change in pixel-level information can move the patient from low to moderate or moderate to high-risk AAC class. Thus, to decrease the effect of intra-class variations and to increase the inter-class separability, we assimilate the features extracted from encoders E_l and E_g. The resultant feature vector is fed as input to a feed-forward network consisting of two Dense layers with 1280 and 128 neurons each, followed by ReLu activation. Finally, a linear layer predicts the final AAC regression score. This module is trained using root mean squared error loss L_{rmse} calculated as: $L_{rmse} = \sqrt{\sum_{i \in m}(y'_i - y_i)^2/m}$, where m is the number of samples, y_i are actual and y'_i are predicted AAC scores. The resulting AAC scores are then further classified into three AAC risk classes.

3 Experiments and Results

Dataset and Annotations: We conducted experiments on two de-identified clinical datasets acquired using the Hologic 4500A and GE iDXA scanners. Both datasets are manually annotated by clinicians using the AAC-24 scale [10] which divides the aortic walls into four segments based on the lumbar vertebrae bodies ($L1 - L4$). Each segment is assigned an AAC score 0: if no calcification (Cal), 1: Cal $\leq 1/3$ of aortic wall, 2: Cal $> 1/3$ but $< 2/3$ of aortic wall and 3: Cal $\geq 2/3$ of aortic wall. The total AAC-24 score can range from 0 to 24 and is further classified into three AAC risk classes using clinical thresholds [15]: Low-risk ($AAC < 2$), Moderate-risk ($2 \leq AAC \leq 5$), and High-risk ($AAC > 5$).

The **Hologic Dataset** [14] has 1,914 single-energy DXA scans acquired using a Hologic 4500A machine. Each scan has dimensions of at least 800 × 287 pixels. Among 1,914 scans, there are 764 scans belonging to the low-risk, 714 to moderate-risk and 436 scans to the high-risk AAC group. The **iDXA GE Dataset** [7,18] has 1,916 dual-energy VFA DXA scans. Among these, there are 829 belonging to low-risk, 445 to moderate-risk and 642 scans to the high-risk AAC group. These scans are acquired using an iDXA GE machine. These scans have dimensions of 1600 × 300 pixels.

Implementation Details: Each VFA scan in both datasets contains a full view of the thoracolumbar spine. To extract the region of interest (ROI), i.e., the abdominal aorta near the lumbar spine, we crop the upper half of the image, resize it to 300×300 pixels and rescale it between 0 and 1. We apply data augmentations including rotation, shear and translation. We implement all experiments in TensorFlow [21], using stratified 10-fold cross-validation on a workstation with NVIDIA RTX 3080 GPU. (For details, see SM.) In stage-I, we adopted an efficient, smaller and faster pre-trained model, EfficientNet-V2S [20], as the backbone for both encoders. We train stage-I for 200 epochs and stage II for 75 epochs using the RMSprops optimizer with the initial learning rate of 3×10^{-4} and a batch size of 16. τ in the proposed *SCOL* is 0.2. The inference time for each scan is less than 15 ms. To avoid overfitting, we used early stopping and reduce LR on the plateau while optimizing the loss.

Evaluation Metrics: The performance of the AAC regression score is evaluated in terms of Pearson's correlation, while for AAC risk classification task Accuracy, F1-Score, Sensitivity, Specificity, Negative Predictive Value (NPV) and Positive Predictive Value (PPV) are used in One Vs. Rest (OvR) setting. **Baseline:** EfficientNet-V2S model trained in regression mode using RMSE loss.

Table 1. Comparison of proposed loss SCOL with different losses using Hologic Dataset

Loss	Method		Pearson (%)	Accuracy (%)	F1-Score (%)	Sensitivity (%)	Specificity (%)
	GCL	LCL					
RMSE	✓	-	87.02	81.92	75.88	74.93	85.36
RMSE	-	✓	87.34	82.5	76.71	76.06	85.88
RMSE	✓	✓	88.12	83.52	77.91	77.00	86.64
SupCon [11]	✓	-	87.26	83.52	77.97	77.14	86.64
SupCon [11]	-	✓	86.39	83.38	77.74	77.02	86.60
SupCon [11]	✓	✓	88.24	83.94	78.69	77.82	87.00
AdaCon [5]	✓	-	88.14	82.51	76.67	75.82	85.86
AdaCon [5]	-	✓	88.60	83.69	78.16	77.28	86.79
AdaCon [5]	✓	✓	88.40	83.98	78.80	77.91	87.03
SCOL (This paper)	✓	-	88.74	84.08	78.82	78.13	87.10
SCOL (This paper)	-	✓	88.92	84.36	78.90	77.93	87.34
SCOL (This paper)	✓	✓	**89.04**	**85.27**	**80.25**	**79.46**	**88.05**

Ablation Study: Table 1 highlights the efficacy of our proposed loss *SCOL*. We train our dual-encoder contrastive learning framework with proposed SCOL,

Table 2. Class-wise performance comparison of the proposed framework with baseline. NPV: Negative Predicted Value, PPV: Positive Predicted Value

	AAC Class	Method	Accuracy (%)	F1-Score (%)	NPV (%)	PPV (%)	Sensitivity (%)	Specificity (%)
Hologic Dataset	Low	Baseline	75.96	69.00	78.89	71.11	67.01	81.91
	(n = 764)	DCOL	**80.56**	**75.52**	**83.59**	**75.93**	**75.13**	**84.17**
	Moderate	Baseline	72.94	65.92	80.78	62.16	70.17	74.58
	(n = 714)	DCOL	**77.90**	**71.36**	**83.75**	**69.06**	**73.80**	**80.33**
	High	Baseline	96.89	92.71	96.46	98.45	87.61	99.50
	(n = 436)	DCOL	**97.33**	**93.86**	**96.97**	**98.73**	**89.45**	**99.66**
	Average	Baseline	81.92	75.88	77.24	85.38	74.93	85.36
	(n = 1,914)	DCOL	**85.27**	**80.25**	**81.24**	**88.10**	**79.46**	**88.05**
iDXA GE Dataset	Low	Baseline	85.44	82.97	83.95	86.53	82.02	88.04
	(n = 829)	DCOL	**87.63**	**85.76**	**85.40**	**89.35**	**86.12**	88.37
	Moderate	Baseline	77.60	55.35	51.55	87.21	59.77	83.00
	(n = 445)	DCOL	**80.63**	**60.15**	**57.61**	**88.46**	**62.92**	**85.99**
	High	Baseline	90.18	84.74	88.47	90.95	81.30	94.66
	(n = 642)	DCOL	**90.92**	**85.99**	**89.39**	**91.60**	**82.71**	**95.05**
	Average	Baseline	84.41	74.35	74.65	88.23	74.37	88.56
	(n = 1,916)	DCOL	**86.39**	**77.28**	**77.47**	**89.80**	**77.25**	**89.94**

Table 3. Comparison of the proposed DCOL with State-of-the-Art methods [7, 18] on iDXA GE Dataset. NPV: Negative Predicted Value, PPV: Positive Predicted Value

Risk Class	Method	Pearson (%)	Accuracy (%)	Sens. (%)	Spec. (%)	NPV (%)	PPV (%)
Low	Reid et al. [18]	-	71.14	55.49	83.07	70.99	71.43
	Gilani et al. [7]	-	82.52	**86.37**	79.58	88.45	76.33
	DCOL (This paper)	-	**87.63**	86.12	**88.77**	**89.35**	**85.40**
Moderate	Reid et al. [18]	-	62.06	59.33	62.88	83.63	32.59
	Gilani et al. [7]	-	75.52	37.53	**87.02**	82.16	46.65
	DCOL (This paper)	-	**80.63**	**62.90**	85.99	**88.46**	**57.60**
High	Reid et al. [18]	-	79.12	54.83	91.37	80.06	76.19
	Gilani et al. [7]	-	87.89	80.22	91.76	90.20	83.06
	DCOL (This paper)	-	**90.90**	**82.70**	**95.05**	**91.60**	**89.39**
Average	Reid et al. [18]	65.00	70.77	56.55	79.11	78.23	60.07
	Gilani et al. [7]	84.00	81.98	68.04	86.12	86.93	68.68
	DCOL (This paper)	**91.03**	**86.39**	**77.25**	**89.94**	**89.80**	**77.47**

SupCon [11] and AdaCon [5], individually, on the Hologic dataset. We also evaluate the performance of the local and global contrastive modules (LCL and GCL) with each contrastive loss. Table 1 also shows the strength of integrating the localized attention block with the baseline model trained with RMSE loss.

Comparison with the Baseline: In Table 2, we compare the performance of our framework with the baseline on both datasets. For the Hologic dataset, our proposed method improved Pearson's correlation coefficient from a baseline of 0.87 to 0.89 and 3-class classification accuracy from 73%±3.82 to 78%±3.65 with ($p < 0.001$). While, for the iDXA GE dataset, the proposed method enhanced the Pearson's correlation from a baseline of 0.89 to 0.91 and averaged 3-class classification accuracy from 77% ± 3.9 to 80% ± 5.12 with ($p < 0.001$).

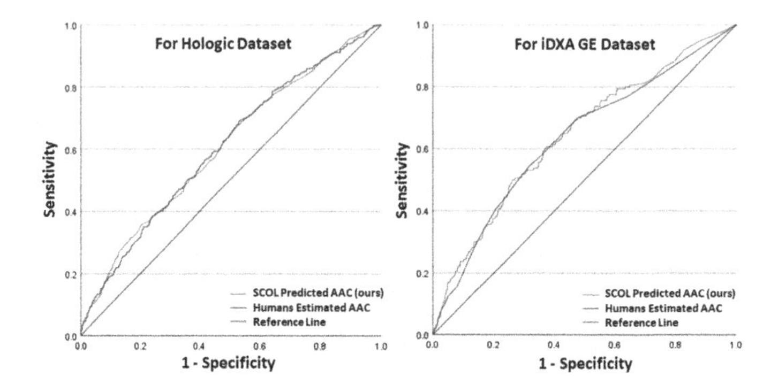

Fig. 2. Comparison of AUCs for Major Adverse Cardiovascular Events (MACE) associated with our predicted AAC-24 scores and human-measured AAC-24 scores.

Comparison with the State-of-the-Art (SOTA): Table 3 shows the comparison of our proposed framework with two SOTA methods [7,18] using the iDXA GE dataset. Our approach outperforms [18] by an average of 15.62% in accuracy and 20.7% in sensitivity with ($p < 0.001$), while in comparison to [7], accuracy is improved by 4.41% and sensitivity by 9.21%, with ($p < 0.001$).

Clinical Analysis and Discussion: To signify the clinical significance, we estimate the AUCs (Fig. 2) for Major Adverse Cardiovascular Events (MACE) associated with our predicted AAC-24 scores versus human-measured AAC-24 scores. For the Hologic dataset, in the cohort of 1083 women with clinical follow-up, 460 suffered a MACE event. Predicted AAC-24 scores provided similar AUCs to human AAC-24 scores (0.61 95%CI 0.57-0.64 vs. 0.61 95%CI 0.57-0.64). The three predicted AAC groups had 598 (55.2%), 394 (30.0%) and 159 (14.7%) of women in the low, moderate, and high AAC groups, respectively, with MACE events occurring in 37.8%, 46.2%, and 52.8% of these groups, respectively. The age-adjusted HRs for MACE events in the moderate and high AAC groups were 1.24 95%CI 1.01-1.53 and 1.45 95% CI 1.13-1.87, respectively, compared to the low predicted AAC group. These HRs were similar to the human AAC groups, i.e., for moderate and high AAC groups, HR 1.28 95%CI 1.04-1.57 and 1.47 95% CI 1.15-1.89, respectively, compared to the human low AAC group.

For the iDXA GE Dataset, in the cohort of 1877 patients with clinical follow-up, 160 experienced a MACE event. The AUCs of predicted AAC-24 scores were similar AUC to human AAC-24 (0.64 95%CI 0.60-0.69 vs. 0.63 95%CI 0.59-0.68). The predicted AAC groups had 877 (46.7%), 468 (24.9%), and 532 (28.3%) of people in the low, moderate, and high AAC groups, respectively, with MACE events occurring in 5.1%, 7.5%, and 15.0% of these groups, respectively. The age and sex-adjusted HR for MACE in the moderate AAC group was 1.21 95%CI 0.77-1.89, and 2.64 95% CI 1.80-3.86 for the high AAC group, compared to the low predicted AAC group, which were similar to the HRs of human AAC groups, i.e., for moderate and high AAC groups HR 1.15 95%CI 0.72-1.84 and 2.32 95% CI 1.59-3.38, respectively, compared to the human low AAC group.

4 Conclusion

We propose a novel Supervised Contrastive Ordinal Loss and developed a Dual-encoder Contrastive Ordinal Learning framework for AAC scoring and relevant AAC risk classification in low-energy VFA DXA scans. Our framework learns contrastive feature embeddings at the local and global levels. Our results demonstrate that the contrastive ordinal learning technique remarkably enhanced inter-class separability and strengthened intra-class consistency among the AAC-24 genera, which is particularly beneficial in handling challenging cases near the class boundaries. Our framework with SCOL loss demonstrates significant performance improvements, compared to state-of-the-art methods. Moreover, the ablation studies also establish the effectiveness of our dual-encoder strategy and localized attention block. These results suggest that our approach has great clinical potential for accurately predicting AAC scores and relevant risk classes.

Acknowledgement and Data Use Declaration. De-identified labelled images were sourced for the ML (Project no: 03349 LEWIS) from a number of existing studies collecting VFAs. For Hologic dataset, written informed consent was obtained from all participants. The Human Ethics Committee of the University of Western Australia approved the study protocol and consent form (approval no. 05/06/004/H50). The Human Research Ethics Committee of the Western Australian Department of Health also approved the data linkage study (approval no. 2009/24). For the GE images the study was approved by the Health Research Ethics Board for the University of Manitoba (HREB H2004:017L, HS20121). The Manitoba Health Information Privacy Committee approved access to the iDXA GE data and waived the requirement for signed consent (HIPC 2016/2017-29).

The study was supported by a National Health and Medical Research Council of Australia Ideas grant (APP1183570) and the Rady Innovation Fund, Rady Faculty of Health Sciences, University of Manitoba. The results and conclusions are those of the authors and no official endorsement by Manitoba Health and Seniors Care, or other data providers is intended or should be inferred. The salary of JRL is supported by a National Heart Foundation of Australia Future Leader Fellowship (ID: 102817). Also, SZG was partially funded by the Raine Priming Grant awarded by Raine Medical Research Foundation.

References

1. AS: Supervised-contrastive-ordinal-loss (2023). https://github.com/AfsahS/Supervised-Contrastive-Ordinal-Loss
2. Bhattacharya, D., et al.: Supervised contrastive learning to classify paranasal anomalies in the maxillary sinus. In: Wang, L., Dou, Q., Fletcher, P.T., Speidel, S., Li, S. (eds.) MICCAI 2022. LNCS, vol. 13433, pp. 42–438. Springer, Cham (2022). https://doi.org/10.1007/978-3-031-16437-8_41
3. Chaitanya, K., Erdil, E., Karani, N., Konukoglu, E.: Contrastive learning of global and local features for medical image segmentation with limited annotations. In: Advances in Neural Information Processing Systems (2020)
4. Chaplin, L., Cootes, T.: Automated scoring of aortic calcification in vertebral fracture assessment images. In: Medical Imaging 2019: Computer-Aided Diagnosis (2019)

5. Dai, W., Li, X., Chiu, W.H.K., Kuo, M.D., Cheng, K.T.: Adaptive contrast for image regression in computer-aided disease assessment. IEEE Trans. Med. Imaging **41**(5), 1255–1268 (2021)
6. Elmasri, K., Hicks, Y., Yang, X., Sun, X., Pettit, R., Evans, W.: Automatic detection and quantification of abdominal aortic calcification in dual energy x-ray absorptiometry. Procedia Comput. Sci. **96**, 1011–1021 (2016)
7. Gilani, S.Z., et al.: Show, attend and detect: towards fine-grained assessment of abdominal aortic calcification on vertebral fracture assessment scans. In: Wang, L., Dou, Q., Fletcher, P.T., Speidel, S., Li, S. (eds.) MICCAI 2022. LNCS, vol. 13433, pp. 439–450. Springer, Cham (2022). https://doi.org/10.1007/978-3-031-16437-8_42
8. Hua, Y., Shu, X., Wang, Z., Zhang, L.: Uncertainty-guided voxel-level supervised contrastive learning for semi-supervised medical image segmentation. Int. J. Neural Syst. **32**(04), 2250016 (2022)
9. Jaiswal, A., et al.: Scalp-supervised contrastive learning for cardiopulmonary disease classification and localization in chest X-rays using patient metadata. In: 2021 IEEE International Conference on Data Mining (ICDM). IEEE (2021)
10. Kauppila, L.I., Polak, J.F., Cupples, L.A., Hannan, M.T., Kiel, D.P., Wilson, P.W.: New indices to classify location, severity and progression of calcific lesions in the abdominal aorta: a 25-year follow-up study. Atherosclerosis **132**(2), 245–250 (1997)
11. Khosla, P., et al.: Supervised contrastive learning. In: Advances in Neural Information Processing Systems (2020)
12. Lems, W., et al.: Vertebral fracture: epidemiology, impact and use of DXA vertebral fracture assessment in fracture liaison services. Osteoporos. Int. **32**, 399–411 (2021)
13. Leow, K., et al.: Prognostic value of abdominal aortic calcification: a systematic review and meta-analysis of observational studies. J. Am. Heart Assoc. **10**(2), e017205 (2021)
14. Lewis, J.R., et al.: Long-term atherosclerotic vascular disease risk and prognosis in elderly women with abdominal aortic calcification on lateral spine images captured during bone density testing: a prospective study. J. Bone Mineral Res. **33**(6), 1001–1010 (2018)
15. Lewis, J.R., et al.: Abdominal aortic calcification identified on lateral spine images from bone densitometers are a marker of generalized atherosclerosis in elderly women. Arterioscler. Thromb. Vasc. Biol. **36**(1), 166–173 (2016)
16. Li, B., Li, Y., Eliceiri, K.W.: Dual-stream multiple instance learning network for whole slide image classification with self-supervised contrastive learning. In: Proceedings of the IEEE/CVF Conference on Computer Vision and Pattern Recognition (2021)
17. World Health Organization: Cardiovascular diseases (CVDS) (2023). https://www.who.int/en/news-room/fact-sheets/detail/cardiovascular-diseases-(cvds)
18. Reid, S., Schousboe, J.T., Kimelman, D., Monchka, B.A., Jozani, M.J., Leslie, W.D.: Machine learning for automated abdominal aortic calcification scoring of DXA vertebral fracture assessment images: a pilot study. Bone **148**, 115943 (2021)
19. Schousboe, J.T., Lewis, J.R., Kiel, D.P.: Abdominal aortic calcification on dual-energy X-ray absorptiometry: methods of assessment and clinical significance. Bone **104**, 91–100 (2017)
20. Tan, M., Le, Q.: Efficientnetv2: smaller models and faster training. In: International Conference on Machine Learning (2021)
21. Tensorflow.org: Tensorflow (2023). https://www.tensorflow.org/
22. Van der Vaart, A.W.: Asymptotic statistics, vol. 3 (2000)

23. Yang, J., Cosman, F., Stone, P., Li, M., Nieves, J.: Vertebral fracture assessment (VFA) for osteoporosis screening in us postmenopausal women: is it cost-effective? Osteoporos. Int. **31**, 2321–2335 (2020)
24. Zhao, X., Fang, C., Fan, D.J., Lin, X., Gao, F., Li, G.: Cross-level contrastive learning and consistency constraint for semi-supervised medical image segmentation. In: 2022 IEEE 19th International Symposium on Biomedical Imaging (ISBI). IEEE (2022)

STAR-Echo: A Novel Biomarker for Prognosis of MACE in Chronic Kidney Disease Patients Using Spatiotemporal Analysis and Transformer-Based Radiomics Models

Rohan Dhamdhere[1], Gourav Modanwal[1], Mohamed H. E. Makhlouf[2], Neda Shafiabadi Hassani[2], Satvika Bharadwaj[1], Pingfu Fu[3], Ioannis Milioglou[2], Mahboob Rahman[2,3], Sadeer Al-Kindi[2,3], and Anant Madabhushi[1,4](✉)

[1] Wallace H. Coulter Department of Biomedical Engineering, Georgia Institute of Technology and Emory University, Atlanta, GA 30322, USA
anantm@emory.edu
[2] University Hospitals, Cleveland, USA
[3] Case Western Reserve University, Cleveland, OH, USA
[4] Atlanta Veterans Affairs Medical Center, Atlanta, GA, USA

Abstract. Chronic Kidney Disease (CKD) patients are at higher risk of Major Adverse Cardiovascular Events (MACE). Echocardiography evaluates left ventricle (LV) function and heart abnormalities. LV Wall (LVW) pathophysiology and systolic/diastolic dysfunction are linked to MACE outcomes (O^- and O^+) in CKD patients. However, traditional LV volume-based measurements like ejection-fraction offer limited predictive value as they rely only on end-phase frames. We hypothesize that analyzing LVW morphology over time, through spatiotemporal analysis, can predict MACE risk in CKD patients. However, accurately delineating and analyzing LVW at every frame is challenging due to noise, poor resolution, and the need for manual intervention. Our contribution includes (a) developing an automated pipeline for identifying and standardizing heart-beat cycles and segmenting the LVW, (b) introducing a novel computational biomarker—STAR-Echo—which combines spatiotemporal risk from radiomic (M_R) and deep learning (M_T) models to predict MACE prognosis in CKD patients, and (c) demonstrating the superior prognostic performance of STAR-Echo compared to M_R, M_T, as well as clinical-biomarkers (EF, BNP, and NT-proBNP) for characterizing cardiac dysfunction. STAR-Echo captured the gray level intensity distribution, perimeter and sphericity of the LVW that changes differently over time in individuals who encounter MACE outcomes. STAR-

R. Dhamdhere and G. Modanwal—These authors contributed equally to this work.

Supplementary Information The online version contains supplementary material available at https://doi.org/10.1007/978-3-031-43987-2_28.

H. Greenspan et al. (Eds.): MICCAI 2023, LNCS 14225, pp. 284–294, 2023.
https://doi.org/10.1007/978-3-031-43987-2_28

Echo achieved an AUC of 0.71[0.53 − 0.89] for MACE outcome classification and also demonstrated prognostic ability in Kaplan-Meier survival analysis on a holdout cohort ($S_v = 44$) of CKD patients ($N = 150$). It achieved superior MACE prognostication (p-value = 0.037 (log-rank test)), compared to M_R (p-value = 0.042), M_T (p-value = 0.069), clinical biomarkers—EF, BNP, and NT-proBNP (p-value >0.05).

1 Introduction

Cardiovascular disease (CVD) is the most common cause of mortality among patients with Chronic Kidney disease (CKD), with CVD accounting for 40-50% deaths in patients with acute CKD [1]. Echocardiography (Echo), a non-invasive imaging modality, provides a quick critical assessment of cardiac structure and function for cardiovascular disease diagnosis. Despite its usefulness, due to its nature, echo data is noisy and have poor resolution, which presents challenges in effectively interpreting and analyzing them. Measurements like Left ventricle (LV) volume captured by ejection fraction (EF), LV mass and LV geometry have been standard biomarkers for diagnosing and predicting severity of CVD [2–7]. With the availability of large-scale public dataset like EchoNet [7], various works have even assimilated recent deep learning advances like graphs CNNs [8] and transformers [9] for EF estimation using automated LV segmentation from echo videos. However, recent studies on CKD patients demonstrated that standard echo measurements based on static LV volume and morphology, such as EF, may have limited prognostic value beyond baseline clinical characteristics, as adverse outcomes are to the common occurrence inspite of a preserved EF [10–12]. As a result, studies have utilized machine learning to provide more detailed links to cardiac structure and function using echo data (measurements, images and videos) [10–14]. LV Wall (LVW) alterations have been reported as non-traditional biomarker of CVD due to pathophysiological changes, in CKD patients [15,16]. Abnormalities of the LVW motion are known prognostic marker for MACE prediction in CVD patients. [17–19]. Longitudinal dysfunction common in CKD patients, is reflected in LVW with change in morphology [13]. This dysfunction has been associated with CKD progression [10,12] and abnormalities are evident even in early stages of CKD [13]. Thus, an investigation of LVW morphology and longitudinal changes is warranted [10].

Radiomic based interpretable features have been used to model cardiac morphology for predicting disease outcomes [2]. Deep learning based transformer architectures [20] have recently been popular for modelling of spatiotemporal changes in LV [9]. Though they lack interpretability, they are able to extract novel features to model the data using attention [20]. Thus, in CKD patients, a combination of radiomic and transformer based spatiotemporal models could associate LVW changes over time with progression of CVD and potentially provide some insight into the factors implicated for MACE outcomes.

2 Prior Work and Novel Contributions

The relationship between Chronic Kidney disease (CKD) and Cardiovascular disease (CVD) is well established [1,21]. Echo provides a noisy yet effective modality for CVD prediction. Echo data (image, video or clinical measurements) has been extensively analyzed using machine learning and deep learning based techniques, to help in diagnosing heart conditions, predicting their severity, and identifying cardiac disease states [2]. Several studies have employed machine learning to predict diagnostic measurements and global longitudinal strain [22, 23]. In recent years, deep learning-based approaches have been successfully used to tackle echocardiographic segmentation, view classification and phase detection tasks [24,25]. With advent of EchoNet [7], a plethora of studies estimating EF and predicting heart failure have come up [8,9] while few others have used deep learning for prognostic prediction using cardiac measurements [23] and to model wall motion abnormalities [18].

Echo coupled with machine learning, has been a potent tool for analysis of CVD in patients with CKD [13]. Machine learning based analysis has revealed poor prognostic value of EF and LV based biomarkers for CVD risk assessment in CKD patients. [10,12]. Thus, techniques developed using EchoNet [7] dataset are potentially less useful for modeling CVD outcomes in CKD patients. The need for a novel approach investigating the longitudinal changes in cardiac structure and function has been recommended for analyzing CVD in patients with CKD [10,13].

In this study, for the first time, we employ echo video based spatiotemporal analysis of LVW to prognosticate CVD outcomes in kidney disease patients. Our code is available here - https://github.com/rohand24/STAR_Echo
Key contributions of our work are as follows:

* STAR-Echo, a novel biomarker that combines spatiotemporal radiomics and transformer-based models to capture morphological differences in shape and texture of LVW and their longitudinal evolution over time.
* Unique interpretable features: spatiotemporal evolution of sphericity and perimeter (shape-based) and LongRunHighGrayLevelEmphasis (texture-based), prognostic for CVD risk in CKD patients
* Demonstrated the superiority of STAR-Echo in the prognosis of CVD in CKD patients, compared to individual spatiotemporal models and clinical biomarkers.
* An end-to-end automated pipeline for echo videos that can identify heartbeat cycles, segment the LVW, and predict a prognostic risk score for CVD in CKD patients.

3 STAR-ECHO: Methodological Description

3.1 Notation

We denote a patient's echo video as a series of two-dimensional frames, represented as $V = [I_1, I_2, ..., I_N]$, where $I_1, I_2, ..., I_N$ represent the N phases (frames)

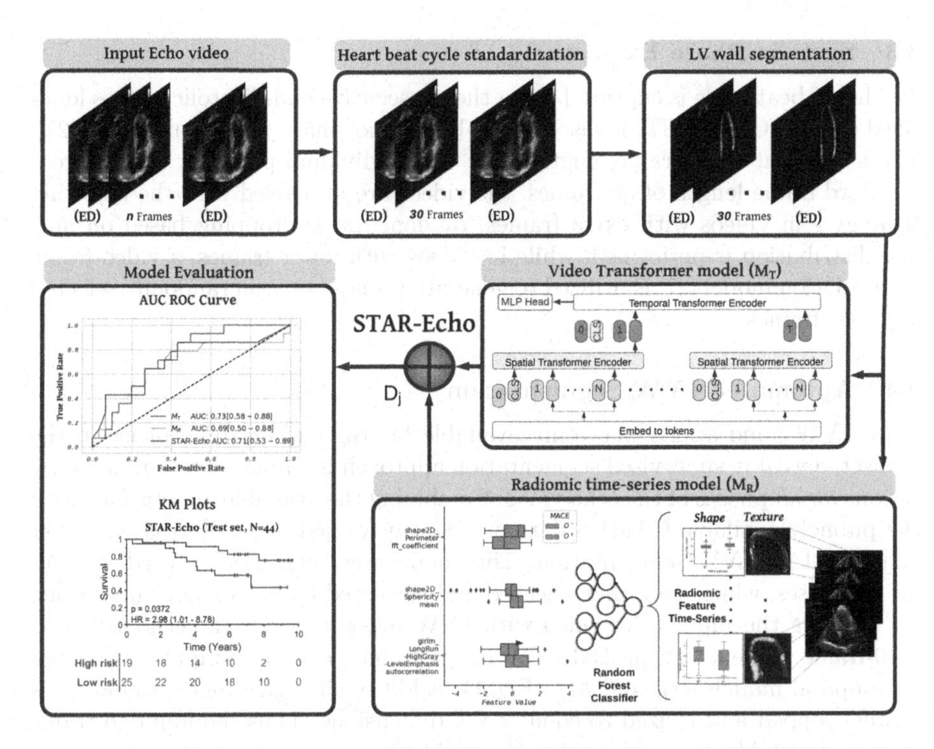

Fig. 1. Workflow of spatiotemporal analysis and STAR-Echo: STAR-Echo does fusion of complementary predictions from spatiotemporal models M_R and M_T [26]. Input echo videos are preprocessed to identify heart beat cycle between 2 end-diastolic(ED) frames. All echo videos are standardized to 30 frames. The segmented and masked LVW videos are input to the M_R and M_T pipelines.

showing the structure and function of the heart. These videos were masked with the region of interest i.e. LVW. This sequence V is preprocessed to represent one heartbeat cycle. The CVD outcomes for the patients are denoted as an event (O^+: occurrence of MACE) and non-event (O^-: no occurrence of MACE). The target outcome of a patient (O^+ or O^-) is represented by O^T.

3.2 Brief Overview

STAR-Echo comprises two spatiotemporal models (M_R, M_T). Model M_R grasps the spatiotemporal changes in the radiomic features extracted for each I_N in the echo video V for the course of one heartbeat cycle of the patient. Model M_T is a video transformer model extracting deep spatiotemporal features for echo video V of one beat cycle of the patient. M_R and M_T are separately trained and fused at Decision fusion junction D_j, to obtain STAR-Echo. The complete workflow is illustrated in Fig. 1.

3.3 Systolic Phase Extraction

The heart-beat cycle is captured using the consecutive end-diastolic phases identified by the CNN+LSTM based multi-beat echo phase detection model [25]. The heart-beat cycle frame lengths vary with individual patients. To achieve a standard frame length of 30 frames, the videos are processed with the following strategy - in videos with extra frames, random frame dropping based on normal distribution is performed, while in videos with fewer frames, a video-frame interpolation model [27] is utilized to generate frames between randomly selected adjacent frames.

3.4 Automated LVW Segmentation

Since LVW annotations were only available for the end-diastolic phase in the dataset, a weakly supervised segmentation approach is employed to obtain annotations for all phases of the echo video. Combining the available annotations with the publicly available CAMUS dataset [28], we trained a nnUNet-based U-Net [29] model for LVW segmentation. This model was then used to predict LVW for all phases, which were then checked and corrected by two expert annotators.

Masking the input echo video with LVW mask provides the echo video V, a 30-frame sequence of masked LVW image frames I_N, to be input to the spatiotemporal models M_R and M_T (Fig. 1). Additionally, each image frame I_N is center-cropped and resized to equal $k \times k$ dimensions. Thus the input to transformer model M_T is an echo video $V \epsilon \mathbb{R}^{30 \times k \times k}$.

3.5 Spatiotemporal Feature Representation

Radiomic Time-Series Model: To build this model, we employ a two-stage feature extraction process.

Radiomic Feature Extraction: In first stage, radiomic feature $R(I_N)$ is extracted on each I_N of the LVW for each phase (frame) of the echo video, V as given in (1). Radiomic features comprising of shape (10), texture (68) and first-order statistics (19) of the LVW in each phase of the echo video are extracted using pyradiomics [30] python package on the input echo phase. A total of 97 radiomic features are extracted per echo phase (frame) image I_N.

Time Series Feature Extraction: In the second stage, to model the temporal LVW motion, we consider the sequence of radiomic features from each phase of one heartbeat cycle as individual time-series. Thus, a radiomic feature time-series $t_R(V)$ is given by (2). Time-series feature T_R, as given by (3), is extracted for each radiomic feature time-series using the TSFresh library [31].

$$R(I_N) = f(I_N) \tag{1}$$

$$t_R(V) = [\, R(I_1),\, R(I_2),\, \ldots,\, R(I_N) \,] \tag{2}$$

$$T_R(V) = g(t_R(V)) \tag{3}$$

Thus, the radiomic time-series model M_R is trained on time-series features $T_R(V)$ obtained for each V of the patient to predict the outcome O^T.

Video Transformer Model: The Video Vision Transformer (ViViT) model [26] with factorised encoder is employed for this task. This architecture has two transformer encoders in series. The first encoder captures the spatial relationships between tokens from the same frame and generates a hidden representation for each frame. The second encoder captures the temporal relationships between frames, resulting in a "late fusion" of temporal and spatial information. The transformer model M_T is trained with input echo video V masked with LVW to predict the patient outcome O^T. Supplementary table S3 gives the important hyperparameters used for training the model M_T.

3.6 Model Fusion

The predictions from radiomic time-series model, M_R and video transformer model, M_T are fused at the junction D_j linear discriminant analysis (LDA) based fusion, with the output prediction probability P given by (4)

$$\log P(y = k|X_i) = \omega_k X_i + \omega_{k0} + C. \tag{4}$$

where ω_k, ω_{k0} are the learned parameters of the model and input X_i for patient i is the linear combination of the predictions of M_R and M_T given as in (5),

$$X_i = M_R(T(R(I_N))_i) + M_T(V_i) \tag{5}$$

4 Experimental Results and Discussion

4.1 Dataset Description

The dataset consisted of echo videos from patients with CKD ($N = 150$) and their composite CVD outcomes and clinical biomarker measurements for Ejection Fraction (EF), B-type natriuretic peptide (BNP) and N-terminal pro-B-type natriuretic peptide (NT-proBNP). It was stratified into 70% training ($S_t = 101$) and 30% holdout set ($S_v = 44$) The reported cardiovascular outcome, O^T, is composite of following cardiovascular events - chronic heart failure, myocardial infarction and stroke. The patients were participants of the Chronic Renal Insufficiency cohort (CRIC) [21] and their data was retrospectively made available

Table 1. Performance of M_R, M_T, STAR-Echo, clinical biomarkers in predicting O^T.

Model	Acc. (%)	AUC [95% CI]	SENS. (%)	SPEC. (%)	p-value	HR
STAR-Echo	70.45	0.71 [0.53-0.89]	64.29	73.33	0.037	2.98 [1.01-8.78]
M_R	70.45	0.69 [0.50-0.88]	64.29	73.33	0.043	3.40 [1.19-9.71]
M_T	68.18	0.73 [0.58-0.88]	68.18	35.63	0.069	2.56 [0.85-7.72]
EF(<50)	65.91	0.54 [0.36-0.72]	14.28	90.00	0.59	1.50 [0.26-8.6]
BNP(>35)	59.90	0.63 [0.44-0.81]	78.57	50.00	0.096	2.82 [0.98-8.11]
NT-proBNP(>125)	59.90	0.66 [0.47-0.84]	78.57	50.00	0.089	2.87 [1.0-8.26]

for this work. The CRIC study [21] is on individuals with mild to moderate CKD, not on dialysis. Diabetes is the main reported comorbidity for the study. The study followed the patients for CVD outcomes. Additional details along with inclusion-exclusion criteria of the dataset are included in the supplementary materials. In the dataset, a total of 46 patients experienced the composite event outcome. The dataset also included survival time information about the patients. The median survival time for patients, O^T was 6.5 years (median O^- = 6.7 years, median O^+ = 3.5 years).

4.2 Statistical and Survival Analysis

Statistical Feature Selection – Extracted times-series features (T_R) are normalized and multicollinear features are excluded. Analysis of Variance (ANOVA) based stable feature selection ($p < 0.002$) and Brouta [32] based all-relevant feature selection ($\alpha = 0.05, threshold = 0.98$) is applied sequentially. A Random forest classifier was trained on the training set using 5-fold cross-validation across 100 iterations, and the best model was selected based on AUC. The selected model was applied to the holdout set, and performance metrics were reported.

Survival Analysis – KM plots with log-rank test analyzed the prognostic ability of models. The curves showed survival time (in years) on the horizontal axis and probability of survival on the vertical axis, with each point representing survival probability of patients. Hazard ratios and p-values were reported for significance between low-risk and high-risk cohorts.

4.3 Evaluating Ability of STAR-ECHO to MACE Prognosis

MACE prognosis is a necessary task in CKD patients management due to high CVD events in CKD patients [33]. Figure 2 shows the KM curves for each indi-

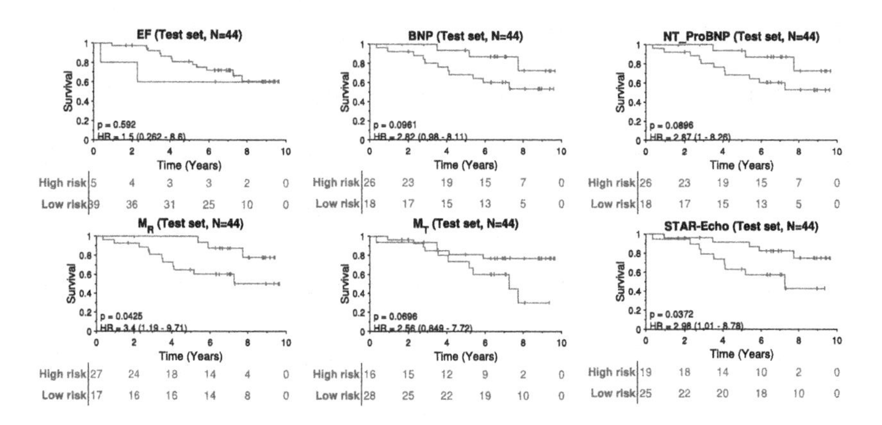

Fig. 2. Risk severity analysis using KM curves - Risk stratification using log-rank test shows significant separation only for M_R and *STAR-Echo* (bottom row). *STAR – Echo* model demonstrates improved significance (p = 0.037) and high hazard ratio (HR = 2.98[1.01 - 8.78]) indicative of prognosis for CVD risk prediction in CKD.

vidual models on the holdout set ($S_v = 44$). Significant separation($p < 0.05$) is observed for models M_R and STAR-Echo. None of the clinical biomarker models have significant separation. Thus, the models trained on EchoNet [7], relying on EF, would not provide a prognostic prediction. Thus, the radiomics features aid in prognosis of MACE risk in CKD patients. The combination model, STAR-Echo, outperforms all the individual models with better risk stratification ($p = 0.037$ and $HR = 2.98$) than individual spatiotemporal models($M_R : p = 0.042; M_T : p = 0.069$) and shows high accuracy in prediction compared to M_R and M_T models. Thus we are able to showcase that saptiotemporal changes in the shape and texture of LVW are prognostic markers for MACE outcomes in CKD patients.

Accuracy, area under the ROC curve (AUC), sensitivity, specificity were the prediction metrics and p-value and hazard ratios (HR) were the prognosis metrics observed (Table 1) for all the models, with the model STAR-Echo achieving the highest Accuracy of 70.45% with a significant p-value(= 0.0372) and high HR(= 2.98). No statistically significant difference is observed in AUC performances of the different spatiotemporal models, indicating that the video transformer and the radiomic time-series models are capturing similar spatiotemporal changes in the shape and intensities of the LVW. The individual clinical biomarker based models performed poorly consistent with observations in CKD patients [10,12]. Thus EchoNet [7] based models would perform poorly in predicting the MACE outcomes as well.

The top radiomic features (Fig. 3-A) include longitudinal changes in perimeter and sphericity of the LVW shape along with intensity changes in the LVW texture. The texture feature changes in systolic function and the shape differences over different frames can be observed in Fig. 3-B. Changes in the LVW shape can indicate left ventricular hypertrophy (LVH), a common complication of CKD linked to a higher risk of cardiovascular events. LVH can stem from

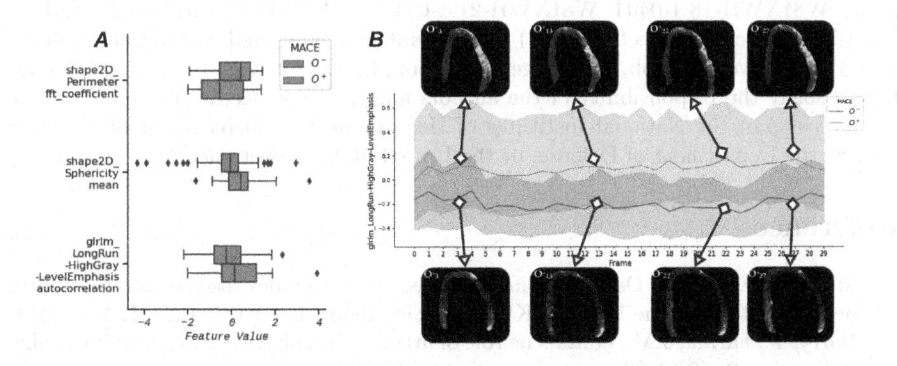

Fig. 3. (A) Top features with significant differences ($p < 0.05$) between MACE outcomes (O^- and O^+). (B) The mean feature value difference over time for selected texture feature, between O^- and O^+. The $3^{rd}, 13^{th}, 22^{th}, 27^{th}$ frames are shown for example O^- and O^+ cases. Clear differences in texture and shape are observed between O^- and O^+, evolving over time.

various factors that modify the ventricle's structure and geometry [15]. Texture changes in LV may also reflect alterations in collagen content or fibrosis which increase the risk of adverse events associated with cardiac remodeling [15].

5 Conclusion

In this study, we introduced STAR-Echo, a novel biomarker combining radiomics and video transformer-based descriptors to evaluate spatiotemporal changes in LVW morphology. STAR-Echo identifies novel features based on longitudinal changes in LVW shape (perimeter and sphericity) and texture (intensity variations) over a heartbeat cycle, similar to recent clinical pathophysiological findings of CVD in CKD [15]. Results show that STAR-Echo significantly improves CVD prognosis in CKD patients ($AUC = 0.71, p = 0.0372, HR = 2.98$) compared to clinical biomarkers, potentially outperforming EchoNet [7] based LV volume-based approaches. Future research will validate STAR-Echo in a larger patient population and incorporate clinical data for improved management of CVD and CKD.

Acknowledgement. Research reported in this publication was supported by the National Cancer Institute under award numbers R01CA268287A1, U01 CA269181, R01CA26820701A1, R01CA249992-01A1, R0CA202752-01A1, R01 CA208236-01A1, R01CA216579-01A1, R01CA220581-01A1, R01CA257612-01A1, 1U01CA239055-01, 1U01CA248226-01, 1U54CA254566-01, National Heart, Lung and Blood Institute 1R01HL15127701A1, R01HL15807101A1, National Institute of Biomedical Imaging and Bioengineering 1R43EB028736-01, VA Merit Review Award IBX004121A from the United States Department of Veterans Affairs Biomedical Laboratory Research and Development Service the Office of the Assistant Secretary of Defense for Health Affairs, through the Breast Cancer Research Program (W81XWH-19-1-0668), the Prostate Cancer Research Program (W81XWH-20-1-0851), the Lung Cancer Research Program (W81XWH-18-1-0440, W81XWH-20-1-0595), the Peer Reviewed Cancer Research Program (W81XWH-18-1-0404, W81XWH-21-1-0345, W81XWH-21-1-0160), the Kidney Precision Medicine Project (KPMP) Glue Grant and sponsored research agreements from Bristol Myers-Squibb, Boehringer-Ingelheim, Eli-Lilly and Astra-zeneca. The content is solely the responsibility of the authors and does not necessarily represent the official views of the National Institutes of Health, the U.S. Department of Veterans Affairs, the Department of Defense, or the United States Government.

References

1. Marx, N., Floege, J.: Dapagliflozin, advanced chronic kidney disease, and mortality: new insights from the DAPA-CKD trial. Eur. Heart J. **42**(13), 1228–1230 (2021)
2. Barry, T., Farina, J.M., et al.: The role of artificial intelligence in echocardiography. J. Imaging **9**, 50 (2023)
3. Zhang, J., Gajjala, S., et al.: Fully automated echocardiogram interpretation in clinical practice. Circulation **138**, 1623–1635 (2018)
4. Yang, F., Chen, X., et al.: Automated analysis of doppler echocardiographic videos as a screening tool for valvular heart diseases. JACC Cardiovasc. Imaging **15**, 551–563 (2022)

5. Hwang, I.-C., Choi, D., et al.: Differential diagnosis of common etiologies of left ventricular hypertrophy using a hybrid CNN-LSTM model. Sci. Rep. **12**, 20998 (2022)
6. Liu, B., Chang, H., et al.: A deep learning framework assisted echocardiography with diagnosis, lesion localization, phenogrouping heterogeneous disease, and anomaly detection. Sci. Rep. **13**, 3 (2023)
7. Ouyang, D., He, B., et al.: Video-based AI for beat-to-beat assessment of cardiac function. Nature **580**, 252–256 (2020)
8. Mokhtari, M., Tsang, T., et al.: EchoGNN: explainable ejection fraction estimation with graph neural networks. In: Wang, L., Dou, Q., Fletcher, P.T., Speidel, S., Li, S. (eds.) MICCAI 2022. LNCS, vol. 13434, pp. 360–369. Springer, Cham (2022). https://doi.org/10.1007/978-3-031-16440-8_35
9. Muhtaseb, R., Yaqub, M.: EchoCoTr: estimation of the left ventricular ejection fraction from spatiotemporal echocardiography. In: Wang, L., Dou, Q., Fletcher, P.T., Speidel, S., Li, S. (eds.) MICCAI 2022. LNCS, vol. 13434, pp. 370–379. Springer, Cham (2022). https://doi.org/10.1007/978-3-031-16440-8_36
10. Fitzpatrick, J.K., Ambrosy, A.P., et al.: Prognostic value of echocardiography for heart failure and death in adults with chronic kidney disease. Am. Heart J. **248**, 84–96 (2022)
11. Mark, P.B., Mangion, K., et al.: Left ventricular dysfunction with preserved ejection fraction: the most common left ventricular disorder in chronic kidney disease patients. Clin. Kidney J. **15**, 2186–2199 (2022)
12. Zelnick, L.R., Shlipak, M.G., et al.: Prediction of incident heart failure in CKD: the CRIC study. Kidney Int. Rep. **7**, 708–719 (2022)
13. Dohi, K.: Echocardiographic assessment of cardiac structure and function in chronic renal disease. J. Echocardiogr. **17**, 115–122 (2019)
14. Christensen, J., Landler, N.E., et al.: Left ventricular structure and function in patients with chronic kidney disease assessed by 3D echocardiography: the CPH-CKD ECHO study. Int. J. Cardiovasc. Imaging **38**, 1233–1244 (2022)
15. Jankowski, J., Floege, J., et al.: Cardiovascular disease in chronic kidney disease. Circulation **143**, 1157–1172 (2021)
16. Bongartz, L.G., Braam, B., et al.: Target organ cross talk in cardiorenal syndrome: animal models. Am. J. Physiol. Renal Physiol. **303**, F1253–F1263 (2012)
17. Kamran, S., Akhtar, N., et al.: Association of major adverse cardiovascular events in patients with stroke and cardiac wall motion abnormalities. J. Am. Heart Assoc. **10**, e020888 (2021)
18. Huang, M.-S., Wang, C.-S., et al.: Automated recognition of regional wall motion abnormalities through deep neural network interpretation of transthoracic echocardiography. Circulation **142**, 1510–1520 (2020)
19. Elhendy, A., Mahoney, D.W., et al.: Prognostic significance of the location of wall motion abnormalities during exercise echocardiography. J. Am. Coll. Cardiol. **40**, 1623–1629 (2002)
20. Vaswani, A., Shazeer, N., et al.: Attention is all you need. In: Advances in Neural Information Processing Systems, vol. 30, Curran Associates Inc. (2017)
21. Feldman, H., Dember, L.: Chronic renal insufficiency cohort study (2022). Artwork Size: 263268080 MB Pages: 263268080 MB Version Number: V11 Type: dataset
22. Salte, I.M., Østvik, A., et al.: Artificial intelligence for automatic measurement of left ventricular strain in echocardiography. JACC Cardiovasc. Imaging **14**, 1918–1928 (2021)

23. Pandey, A., Kagiyama, N., et al.: Deep-learning models for the echocardiographic assessment of diastolic dysfunction. JACC Cardiovasc. Imaging **14**, 1887–1900 (2021)
24. Zamzmi, G., Rajaraman, S., et al.: Real-time echocardiography image analysis and quantification of cardiac indices. Med. Image Anal. **80**, 102438 (2022)
25. Lane, E.S., Azarmehr, N., et al.: Multibeat echocardiographic phase detection using deep neural networks. Comput. Biol. Med. **133**, 104373 (2021)
26. Arnab, A., Dehghani, M., et al.: ViViT: a video vision transformer. In: 2021 IEEE/CVF International Conference on Computer Vision (ICCV), Montreal, QC, Canada, pp. 6816–6826. IEEE (2021)
27. Cheng, X., Chen, Z.: Multiple video frame interpolation via enhanced deformable separable convolution. IEEE Trans. Pattern Anal. Mach. Intell. **44**, 7029–7045 (2022)
28. Leclerc, S., Smistad, E., et al.: Deep learning for segmentation using an open large-scale dataset in 2D echocardiography. IEEE Trans. Med. Imaging **38**, 2198–2210 (2019)
29. Isensee, F., Jaeger, P.F., et al.: nnU-Net: a self-configuring method for deep learning-based biomedical image segmentation. Nat. Methods **18**, 203–211 (2021)
30. van Griethuysen, J.J., Fedorov, A., et al.: Computational radiomics system to decode the radiographic phenotype. Can. Res. **77**, e104–e107 (2017)
31. Christ, M., Braun, N., et al.: Time series feature extraction on basis of scalable hypothesis tests (tsfresh - a python package). Neurocomputing **307**, 72–77 (2018)
32. Kursa, M.B., Rudnicki, W.R.: Feature selection with the Boruta package. J. Stat. Softw. **36**, 1–13 (2010)
33. Jain, N., McAdams, M., et al.: Screening for cardiovascular disease in CKD: PRO. Kidney360 **3**, 1831 (2022)

Interpretable Deep Biomarker for Serial Monitoring of Carotid Atherosclerosis Based on Three-Dimensional Ultrasound Imaging

Xueli Chen[1], Xinqi Fan[1], and Bernard Chiu[1,2](✉)

[1] Department of Electrical Engineering, City University of Hong Kong, Kowloon, Hong Kong, China
{xuelichen3-c,xinqi.fan}@my.cityu.edu.hk

[2] Department of Physics and Computer Science, Wilfrid Laurier University, Waterloo, ON, Canada
bchiu@wlu.ca

Abstract. We developed an interpretable deep biomarker known as Siamese change biomarker generation network (SCBG-Net) to evaluate the effects of therapies on carotid atherosclerosis based on the vessel wall and plaque volume and texture features extracted from three-dimensional ultrasound (3DUS) images. To the best of our knowledge, SCBG-Net is the first deep network developed for serial monitoring of carotid plaque changes. SCBG-Net automatically integrates volume and textural features extracted from 3DUS to generate a change biomarker called *AutoVT* (standing for **Auto**matic integration of **V**olume and **T**extural features) that is sensitive to dietary treatments. The proposed *AutoVT* improves the cost-effectiveness of clinical trials required to establish the benefit of novel treatments, thereby decreasing the period that new anti-atherosclerotic treatments are withheld from patients needing them. To facilitate the interpretation of *AutoVT*, we developed an algorithm to generate change biomarker activation maps (CBAM) localizing regions having an important effect on *AutoVT*. The ability to visualize locations with prominent plaque progression/regression afforded by CBAM improves the interpretability of the proposed deep biomarker. Improvement in interpretability would allow the deep biomarker to gain sufficient trust from clinicians for them to incorporate the model into clinical workflow.

Keywords: 3D Ultrasound Imaging · Carotid Atherosclerosis · Deep Biomarker · Interpretable Machine Learning · Activation Map

1 Introduction

Although cardiovascular events are highly prevalent worldwide, it was estimated 75–80% of cardiovascular events in high-risk patients could be prevented through

---

Supplementary Information The online version contains supplementary material available at https://doi.org/10.1007/978-3-031-43987-2_29.

H. Greenspan et al. (Eds.): MICCAI 2023, LNCS 14225, pp. 295–305, 2023.
https://doi.org/10.1007/978-3-031-43987-2_29

lifestyle changes and medical/dietary interventions [22]. The opportunity to prevent cardiovascular events calls for the development of sensitive and cost-effective tools to identify high-risk patients and monitor serial changes in response to therapies. As carotid atherosclerosis is a major source of ischemic stroke and a major indicator of systematic atherosclerosis [9], the carotid artery has long served as a major imaging target for assessment of atherosclerotic diseases.

Carotid intima media thickness (IMT) is an early imaging biomarker measured from two-dimensional ultrasound (2DUS) images. The use of IMT in serial monitoring is limited by the small annual change (~0.015 mm) [4], which does not allow treatment effects to be measured in a clinically affordable timeframe. As plaques grow 2.4 times faster along the arteries than they thickens [3] and change circumferentially as well, volume measurements, such as total plaque volume (TPV) and vessel wall volume (VWV), afforded by 3DUS imaging techniques are more sensitive to treatment effects [1,12]. Biomarkers derived from plaque textural features extracted from 3DUS were also shown to be sensitive to medical [2] and dietary treatments [5,14]. However, few studies consider both volume and textural features, and the handcrafted textural features extracted in previous studies are independent of the subsequent biomarker generation. To address these issues, we propose an end-to-end Siamese change biomarker generation network (SCBG-Net) to extract features from the baseline and follow-up images for generating a biomarker, *AutoVT*, quantifying the degree of change in volume and texture automatically. Although deep networks have been proposed for carotid plaque composition characterization [13], plaque echogenicity classification [16,21], and plaque recognition [15,17], SCBG-Net is the first deep network developed for serial monitoring of carotid atherosclerosis.

A convolutional neural network (CNN) is typically represented as a black-box function that maps images to an output. However, a biomarker should be interpretable for it to be trusted by clinicians. One approach to promote the interpretability of the biomarker is to allow the visualization of regions that have a prominent effect on the biomarker. Class activation map (CAM) [27] and its variant [19,24] highlight regions having a strong contribution to classification results. Interpretability is not only desired in classification networks but also in networks focusing on quantifying the similarity of images, such as person re-identification [20,25]. The ranking activation map (RAM) [25] and its variant, such as gradient-weighted ranking activation map (CG-RAM) [20], highlight regions contributing to the similarity between a reference image and other images. For our application, there is a need to develop a technique to generate activation maps localizing regions with a prominent effect on the novel biomarker. Another contribution of this paper is the development of such an approach to generate change biomarker activation maps (CBAM).

2 Materials and Methods

2.1 3DUS Imaging and Preprocessing

In this work, we assessed the sensitivity of the proposed biomarker in evaluating the effect of pomegranate juice and tablets. Pomegranate is anti-oxidative, and

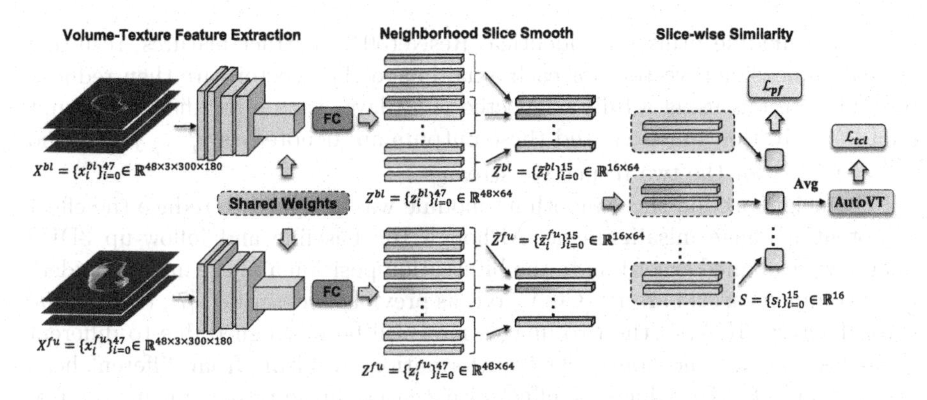

Fig. 1. Schematic of the Siamese change biomarker generation network (SCBG-Net).

previous studies have established that plaque texture features [5], the weighted average change of vessel-wall-plus-plaque thickness [26] and local vessel wall and plaque volume change [6] are able to detect the effect of pomegranate in the same cohort investigated in this study. Subjects were recruited by Stroke Prevention & Atherosclerosis Research Centre at Robarts Research Institute (London, Ontario, Canada) for a clinical trial (ISRCTN30768139). A total of 120 subjects involved in this study were randomized into two groups. 66 subjects received pomegranate extract and 54 subjects were given a placebo once daily for a year. There is no significant difference in the baseline characteristics between the two groups [26]. 3DUS images were acquired for participants at the baseline and a follow-up session, ranging from 283 to 428 days after the baseline scan. The reason for quantifying changes based on only two time points (i.e., baseline and follow-up) is that the rate of change of carotid plaque has been established as linear between the age of 50 to 75 in two studies involving over 6000 patients [11,23]. The 3DUS images were obtained by translating an ultrasound transducer (L12-5, Philips, Bothell, WA, USA) mounted on a mechanical assembly at a uniform speed of 3 mm/s along the neck for about 4 cm. Ultrasound frames acquired using an ultrasound machine (ATL HDI 5000, Philips) were digitized at the rate of 30 Hz and reconstructed into a 3D image. The input image to SCBG-Net was obtained by masking the ultrasound image with the manually segmented boundaries (Supplementary Fig. 1). Each 3DUS image was resliced into a stack of 2D axial images with a $1mm$ interslice distance (ISD) as described in [7].

2.2 Siamese Change Biomarker Generation Network (SCBG-Net)

Network Architecture. Figure 1 shows a schematic of the SCBG-Net. The baseline and follow-up image stacks consist of 48 axial images each, 24 from each of the left and right arteries. We denote the baseline (bl) and follow-up (fu) image stacks by $X^{bl} = \{x_i^{bl}\}_{i=0}^{47}$ and $X^{fu} = \{x_i^{fu}\}_{i=0}^{47}$, respectively, where x_i^{bl} and x_i^{fu} are axial slices with size 300×180. The baseline and follow-up images are processed by a Siamese architecture. The volume-texture feature extraction module utilizes an identical ResNet50 to extract features, resulting in 2048-dimensional vectors for each axial image. The vectors are then reduced to 64 dimensions using a fully connected (FC) layer with a rectified linear unit (ReLU) activation function, and these outputs are denoted as $z_i^{bl} = f(x_i^{bl})$ and $z_i^{fu} = f(x_i^{fu})$ for the two respective streams.

The neighborhood slice smoothing module was designed to reduce the effect of potential image misalignment. Although the baseline and follow-up 3DUS images were aligned according to the bifurcation position and the manually identified common carotid artery (CCA) axis as previously described [7], the internal carotid artery (ICA) of the two images may still be misaligned due to different bifurcation angles and non-linear transformations resulting from different head orientations [18]. To reduce the effect of potential image misalignment, the features of three neighboring slices in the baseline and follow-up images are averaged and denoted by $\bar{z}_i^{bl} = \frac{1}{3}\sum_{j=3i}^{3i+2} z_j^{bl}$ and $\bar{z}_i^{fu} = \frac{1}{3}\sum_{j=3i}^{3i+2} z_j^{fu}$, for $i = 0, 1, \ldots, 15$. The slice-wise cosine "dissimilarity" for a baseline-follow-up slice pair was defined by $d_c(\cdot,\cdot) = 1 - \cos(\cdot,\cdot)$. To represent disease progression and regression, the sign of the slice-wise vessel wall volume change ΔVol_i from baseline to follow-up was used to determine the sign of the slice-wise score. ΔVol_i of each smoothed slice pair was computed by averaging vessel wall volume change (i.e., area change × 1mm ISD) for a group of three neighbouring slices involved in the smoothing operation. The slice-wise score was obtained by:

$$s_i = s(\bar{z}_i^{bl}, \bar{z}_i^{fu}) = sgn(\Delta Vol_i) d_c(\bar{z}_i^{bl}, \bar{z}_i^{fu}), \tag{1}$$

where *sgn* represents the signed function. The use of ReLU in FC layers results in non-negative z_i, thereby limiting $d_c(\bar{z}_i{}^{bl}, \bar{z}_i^{fu})$ and $s(\bar{z}_i^{bl}, \bar{z}_i^{fu})$ to the range of $[0, 1]$ and $[-1, 1]$, respectively. Defined as such, s_i integrates vessel-wall-plus-plaque volume change with textural features extracted by the network. Finally, the *AutoVT* biomarker was obtained by averaging 16 slice-wise scores (i.e., $AutoVT = \frac{1}{16}\sum_{i=0}^{15} s_i$).

Loss Functions. We developed a treatment label contrastive loss (TCL) to promote discrimination between the pomegranate and placebo groups and a plaque-focus (PF) constraint that considers slice-based volume change.

(i) Treatment Label Contrastive Loss. The contrastive loss [8] maps similar pairs to nearby points and dissimilar pairs to distant points. In our biomarker learning problem, instead of separating similar and dissimilar pairs, we aim to discriminate baseline-follow-up image pairs of the pomegranate and placebo subjects. As changes occur in all patients, the baseline-follow-up image pairs are in general dissimilar for both groups. However, pomegranate subjects tend to experience a smaller plaque progression or even regression, whereas the placebo subjects have a larger progression [26]. As such, our focus is more on differentiating the two groups based on the signed difference between the baseline and follow-up images. We designed a treatment label contrastive loss (TCL) specifically for our biomarker learning problem:

$$\mathcal{L}_{tcl} = y \max(AutoVT, 0)^2 + (1 - y)\max(m - AutoVT, 0)^2, \tag{2}$$

where y is the group label of the input subject (pomegranate = 1, placebo = 0). For pomegranate subjects, instead of assigning a penalty based on the squared distance as in [8] (i.e., $AutoVT^2$ in our context), in which the penalty applies to both positive or negative $AutoVT$, we penalize only positive $AutoVT$ since it is expected that some pomegranate subjects would have a larger regression, which would be represented by a negative $AutoVT$. For placebo subjects, the penalty is applied only if $AutoVT$ is smaller than m.

(ii) Plaque-focus Constraint. We observe that slice pairs with high volume change are typically associated with a large plaque change (Supplementary Fig. 2). To incorporate such volume change information into $AutoVT$, we assigned a pseudo label ζ_i to each of the 16 smoothed slice pairs indexed by i based on the volume change ΔVol_i. $|\Delta Vol_i|$ was then ranked with the K_l slices with largest $|\Delta Vol_i|$ assigned $\zeta_i = 1$ and the K_s slices with smallest $|\Delta Vol_i|$ assigned $\zeta_i = 0$. The PF constraint was defined to promote the magnitude of s_i associated with slices with large $|\Delta Vol_i|$ and suppress that associated with slices with small $|\Delta Vol_i|$:

$$\mathcal{L}_{pf} = \frac{1}{K_s} \sum_{i \in \{i|\zeta_i=0\}} |s_i|^2 + \frac{1}{K_l} \sum_{j \in \{j|\zeta_j=1\}} (1.0 - |s_j|)^2. \tag{3}$$

The overall loss $\mathcal{L}$ is a weighted combination of $\mathcal{L}_{tcl}$ and $\mathcal{L}_{pf}$ (i.e., $\mathcal{L} = \mathcal{L}_{tcl} + w\mathcal{L}_{pf}$, where w is the weight).

2.3 Change Biomarker Activation Map (CBAM)

Figure 2 shows a schematic of the proposed CBAM developed for visualizing important regions contributing to the $AutoVT$ scoring. Like previous CAM methods, CBAM generates activation maps by linearly weighting feature maps at different levels of a network. However, the weights associated with the attention maps in CBAM are novel and tailored for the proposed SCBG-Net.

The importance of a feature map is determined by how much it affects the absolute value of slice-wise score s (Eq. 1). The reason for focusing on $|s|$ is that we would highlight regions that contribute to both positive and negative changes. We denote $A^{p,k}_{L,i}$ as the kth channel of feature maps from the inner convolution layer L of an image slice x^p_i, where $p \in \{bl, fu\}$. The importance of $A^{p,k}_{L,i}$ towards the slice-wise score is defined in a channel-wise pair-associated manner by:

$$R(A^{p,k}_{L,i}) = |s(f(x^p_i \circ M^{p,k}_{L,i}), f(x^q_i))| = d_c(f(x^p_i \circ M^{p,k}_{L,i}), f(x^q_i)), \tag{4}$$

with $M^{p,k}_{L,i} = \text{Norm}(\text{Up}(A^{p,k}_{L,i}))$, $\circ$ representing the Hadamard product and $(p,q) \in \{(bl, fu), (fu, bl)\}$. $\text{Up}(\cdot)$ upsamples $A^{p,k}_{L,i}$ into the size of x^p_i, and $\text{Norm}(\cdot)$ is a min-max normalization function mapping each element in the matrix into $[0, 1]$.

$A^{p,k}_{L,i}$ is first upsampled and normalized to $M^{p,k}_{L,i}$, which serves as an activation map to highlight regions in the input image. The importance of $A^{p,k}_{L,i}$ to the slice-wise score s is quantified by the cosine dissimilarity between the feature vectors

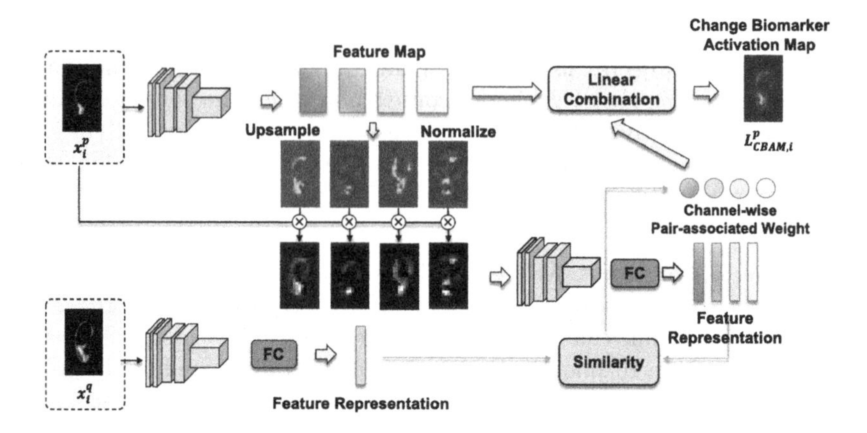

Fig. 2. Schematic of change biomarker activation map (CBAM).

generated by SCBG-Net for the highlighted input image and the corresponding image slice in the baseline-follow-up image pair. If the input image is a baseline image, the corresponding slice would be from the follow-up image, and vice versa.

For each slice x_i^p, the activation map from the convolutional layer L was generated as $L^p_{\text{CBAM},i} = \text{ReLU}(\sum_k \alpha^{p,k}_{L,i} A^{p,k}_{L,i})$, where the weight $\alpha^{p,k}_{L,i}$ is $R(A^{p,k}_{L,i})$ normalized by the softmax function: $\alpha^{p,k}_{L,i} = \frac{\exp(R(A^{p,k}_{L,i}))}{\sum_h \exp(R(A^{p,h}_{L,i}))}$.

3 Experiments and Results

Statistical Evaluation. The discriminative power of biomarkers was evaluated by p-values from two-sample t-tests for normally distributed measurements or Mann-Whitney U tests for non-normally distributed measurements. P-values quantify the ability of each biomarker to discriminate the change exhibited in the pomegranate and placebo groups.

Experimental Settings. Our model was developed using Keras on a computer with an Intel Core i7-6850K CPU and an NVIDIA RTX 1080Ti GPU. The ResNet50 was initialized by the ImageNet pretrained weights. The SGD optimizer was applied with an initial learning rate of 3×10^{-3}. An exponential decay learning rate scheduler was utilized to reduce the learning rate by 0.8 every 10 epochs. We set the number of slices with top/last $|\Delta Vol_i|$ in the definition of the PF constraint as $K_l = K_s = 3$. All models were evaluated by three-fold cross-validation with 80 labeled subjects and 40 test subjects. Labeled subjects are further partitioned into training and validation sets with 60 and 20 subjects, respectively. For the proposed SCBG-Net, the margin m and loss function weight w were tuned using the validation set. In all three trials, the optimized m and w were 0.8 and 0.15, respectively.

Comparison with Traditional Biomarkers. Table 1 shows p-values for *AutoVT* and traditional biomarkers. The proposed biomarker based on the

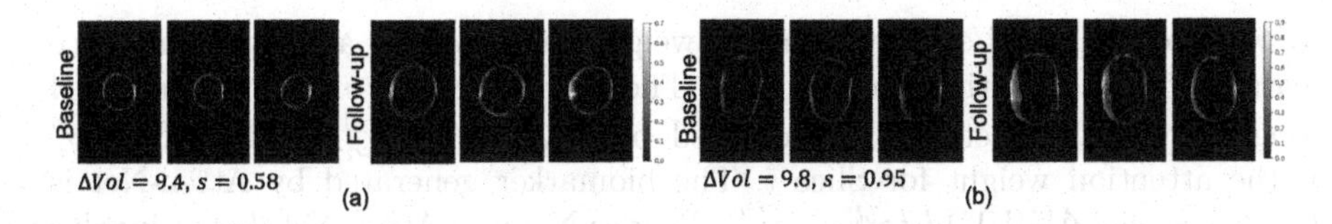

Fig. 3. Examples showing the sensitivity of *AutoVT* in highlighting texture-based plaque change.

overall loss function $\mathcal{L}$, *AutoVT*($\mathcal{L}$), was the most sensitive to the effect of pomegranate with the lowest p-value. This biomarker has learned the volumetric information from the input images, as demonstrated by the correlation coefficient of 0.84 between *AutoVT*($\mathcal{L}$) and ΔVWV. *AutoVT*($\mathcal{L}$) has also learned the texture information, as demonstrated in Fig. 3. While the slice pairs in Fig. 3(a) and (b) had similar ΔVol, the changes in slice pairs shown in (a) were more related to expansive remodeling of the vessel wall, whereas the changes in slice pairs shown in (b) were caused by plaque growth. This difference is characterized by the slice-wise score, which was higher for the example in Fig. 3(b) than in (a).

Comparison with Different Deep Models. We compared treatment effect sensitivity attained by SCBG-Net to those attained by two alternative deep models: the Aver-Net and Atten-Net. Given the smooth features $\{\bar{z}_i^{bl}\}_{i=0}^{15}$ and $\{\bar{z}_i^{fu}\}_{i=0}^{15}$, Aver-Net first computes a uniform average vector for each of the baseline and follow-up images by $z_{aver}^p = \frac{1}{16}\sum_{i=0}^{15} \bar{z}_i^p$ where $p \in \{bl, fu\}$ and then generates a biomarker by $bio_{aver} = sgn(\Delta VWV) d_c(z_{aver}^{bl}, z_{aver}^{fu})$, where VWV in the baseline/follow-up session was computed by summing the vessel wall areas at each axial image of the left and right arteries and multipled by the 1mm ISD. ΔVWV is the difference between VWV obtained from the baseline and follow-up images. In contrast, Atten-Net computes a weighted

Table 1. Comparison of *AutoVT* with other biomarkers.

	Measurements	Placebo (n = 54)	Pomegranate (n = 66)	P-value
Traditional	$\Delta TPV(mm^3)$	25.0 ± 76.5	16.1 ± 83.9	0.24
	$\Delta VWV\ (mm^3)$	112.4 ± 118.5	58.9 ± 104.9	0.01
Deep	bio_{aver}	0.151 ± 0.253	0.070 ± 0.138	0.028
	bio_{atten}	0.148 ± 0.227	0.065 ± 0.155	0.021
	$AutoVT(\mathcal{L}_{tcl})$	0.234 ± 0.262	0.119 ± 0.206	0.0087
Loss	$AutoVT(\mathcal{L}_{ce})$	0.216 ± 0.255	0.107 ± 0.243	0.018
	$AutoVT(\mathcal{L}_{bd})$	0.135 ± 0.165	0.060 ± 0.160	0.013
	$AutoVT(\mathcal{L}_{tcl})$	0.234 ± 0.262	0.119 ± 0.206	0.0087
	$AutoVT(\mathcal{L})$	0.241 ± 0.255	0.111 ± 0.209	0.0029

average vector based on the attention weight generated by a multiple instance learning-based attention module [10]. The baseline/follow-up attention-based weighted average vectors are computed by $z^p_{atten} = \frac{1}{16}\sum_{i=0}^{15}\gamma^p_i \bar{z}^p_i$, where γ^p_i is the attention weight for Slice i. The biomarker generated by Atten-Net is $bio_{atten} = sgn(\Delta VWV)d_c(z^{bl}_{atten}, z^{fu}_{atten})$. Aver-Net and Atten-Net do not involve slice-by-slice comparison, whereas slice-by-slice comparison was involved in two components of SCBG-Net: (i) the slice-wise score s_i (Eq. 1) and (ii) the PF constraint (Eq. 3). In this section, we focus on investigating the effect of Component (i) and that of Component (ii) will be studied in the next section focusing on loss functions. For this reason, the three models compared in this section were driven only by $\mathcal{L}_{tcl}$ (Eq. 2) for a fair comparison. Table 1 shows that SCBG-Net is the most sensitive to treatment effects among the three models.

Comparison with Different Losses. We compared our proposed loss with another two losses, including cross-entropy loss and bi-direction contrastive loss. Cross-entropy loss is expressed as $\mathcal{L}_{ce} = -y\log(\sigma(1 - AutoVT)) - (1 - y)\log(\sigma(AutoVT))$, where $\sigma(\cdot)$ is a sigmoid function. The bi-direction contrastive loss is a symmetric version of $\mathcal{L}_{tcl}$, expressed as $\mathcal{L}_{bd} = y\max(m + AutoVT, 0)^2 + (1-y)\max(m - AutoVT, 0)^2$. The margin m in $\mathcal{L}_{bd}$ was tuned in the same way as the proposed $\mathcal{L}$, with $m = 0.4$ being the optimized parameter in all three cross-validation trials. Table 1 shows p-values for different losses. Our proposed loss $\mathcal{L}_{tcl}$ has a higher sensitivity than $\mathcal{L}_{ce}$ and $\mathcal{L}_{bd}$, with further improvement attained by the incorporation of $\mathcal{L}_{pf}$. Pomegranate, as a dietary supplement, confers a weaker beneficial effect than intensive medical treatment [5,26]. $\mathcal{L}_{tcl}$ was designed to better model the weak benefit by not forcing the *AutoVT*s of pomegranate patients to get too negative; the *AutoVT*s of pomegranate patients would not be penalized as long as it is smaller than 0. In contrast, $\mathcal{L}_{ce}$ and $\mathcal{L}_{bd}$ promote more negative *AutoVT*s for pomegranate patients. $\mathcal{L}_{tcl}$ was designed to account for the weak beneficial effect of pomegranate, which may not lead to significant plaque regression in pomegranate patients compared to high-dose atorvastatin. Moreover, $\mathcal{L}_{pf}$ improves the discriminative power of *AutoVT* by using the ranking of $|\Delta VVol_i|$ among different axial images of the same patient.

Comparison with Other Activation Maps. Figure 4 compares the activation maps generated by CBAM and CG-RAM from features maps in the second convolutional layer. CBAM localizes regions with plaque changes accurately,

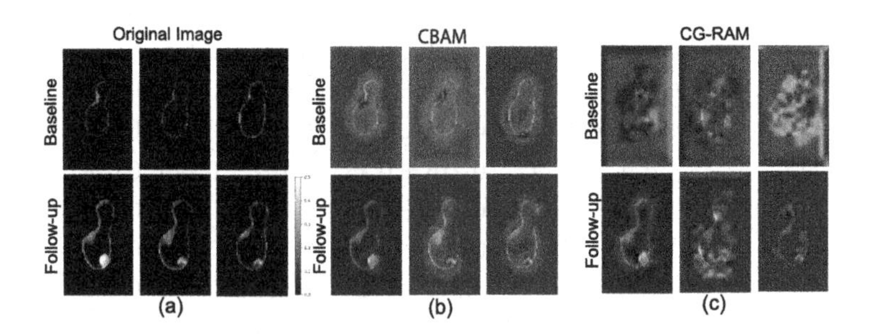

Fig. 4. Comparison between CBAM and CG-RAM.

whereas CG-RAM is less successful in highlighting regions with plaque changes. A possible reason for this observation is that CG-RAM is driven by gradient and may be adversely affected by gradient saturation issues, whereas CBAM is gradient-free and not affected by artifacts associated with gradient saturation.

4 Conclusion

We, for the first time, developed a deep biomarker to quantify the serial change of carotid atherosclerosis by integrating the vessel wall and plaque volume change and the change of textural features extracted by a CNN. We showed that the proposed biomarker, *AutoVT*, is more sensitive to treatment effect than vessel wall and plaque volume measurements. SCBG-Net involves slice-based comparison of textural features and vessel wall volume (Eq. 1) and we showed that this architecture results in a biomarker that is more sensitive than Aver-Net and Atten-Net that quantify global change for the left and right arteries. This result is expected as atherosclerosis is a focal disease with plaques predominantly occurring at the bifurcation. For the same reason, PF constraint that involves local slice-based assessment further improves the sensitivity of *AutoVT* in detecting treatment effects. We developed a technique to generate activation maps highlighting regions with a strong influence on *AutoVT*. The improvement in the interpretability of *AutoVT* afforded by the activation maps will help promote clinical acceptance of *AutoVT*.

Acknowledgement. Dr. Chiu is grateful for the funding support from the Research Grant Council of HKSAR, China (Project nos. CityU 11203218, CityU 11205822). The authors thank Dr. J. David Spence for providing the 3D ultrasound images investigated in this study.

References

1. Ainsworth, C.D., Blake, C.C., Tamayo, A., Beletsky, V., Fenster, A., Spence, J.D.: 3D ultrasound measurement of change in carotid plaque volume: a tool for rapid evaluation of new therapies. Stroke **36**(9), 1904–1909 (2005)
2. Awad, J., Krasinski, A., Parraga, G., Fenster, A.: Texture analysis of carotid artery atherosclerosis from three-dimensional ultrasound images. Med. Phys. **37**(4), 1382–1391 (2010)
3. Barnett, P.A., Spence, J.D., Manuck, S.B., Jennings, J.R.: Psychological stress and the progression of carotid artery disease. J. Hypertens. **15**(1), 49–55 (1997)
4. Bots, M.L., Evans, G.W., Riley, W.A., Grobbee, D.E.: Carotid intima-media thickness measurements in intervention studies: design options, progression rates, and sample size considerations: a point of view. Stroke **34**(12), 2985–2994 (2003)
5. Chen, X., et al.: Three-dimensional ultrasound evaluation of the effects of pomegranate therapy on carotid plaque texture using locality preserving projection. Comput. Methods Programs Biomed. **184**, 105276 (2020)
6. Chen, X., Zhao, Y., Spence, J.D., Chiu, B.: Quantification of local vessel wall and plaque volume change for assessment of effects of therapies on carotid atherosclerosis based on 3-D ultrasound imaging. Ultrasound Med. Biol. **49**(3), 773–786 (2023)

7. Egger, M., Spence, J.D., Fenster, A., Parraga, G.: Validation of 3D ultrasound vessel wall volume: an imaging phenotype of carotid atherosclerosis. Ultrasound Med. Biol. **33**(6), 905–914 (2007)
8. Hadsell, R., Chopra, S., LeCun, Y.: Dimensionality reduction by learning an invariant mapping. In: Proceedings of the IEEE Conference on Computer Vision and Pattern Recognition, vol. 2, pp. 1735–1742. IEEE (2006)
9. Hennerici, M., Hülsbömer, H.B., Hefter, H., Lammerts, D., Rautenberg, W.: Natural history of asymptomatic extracranial arterial disease: results of a long-term prospective study. Brain **110**(3), 777–791 (1987)
10. Ilse, M., Tomczak, J., Welling, M.: Attention-based deep multiple instance learning. In: International Conference on Machine Learning, pp. 2127–2136. PMLR (2018)
11. Johnsen, S.H., et al.: Carotid atherosclerosis is a stronger predictor of myocardial infarction in women than in men: a 6-year follow-up study of 6226 persons: the Tromsø study. Stroke **38**(11), 2873–2880 (2007)
12. Krasinski, A., Chiu, B., Spence, J.D., Fenster, A., Parraga, G.: Three-dimensional ultrasound quantification of intensive statin treatment of carotid atherosclerosis. Ultrasound Med. Biol. **35**(11), 1763–1772 (2009)
13. Lekadir, K., et al.: A convolutional neural network for automatic characterization of plaque composition in carotid ultrasound. IEEE J. Biomed. Health Inform. **21**(1), 48–55 (2016)
14. Lin, M., et al.: Longitudinal assessment of carotid plaque texture in three-dimensional ultrasound images based on semi-supervised graph-based dimensionality reduction and feature selection. Comput. Biol. Med. **116**, 103586 (2020)
15. Liu, J., et al.: Deep learning based on carotid transverse B-mode scan videos for the diagnosis of carotid plaque: a prospective multicenter study. Eur. Radiol. **33**, 1–10 (2022)
16. Ma, W., et al.: Multilevel strip pooling-based convolutional neural network for the classification of carotid plaque echogenicity. Comput. Math. Methods Med. **2021** (2021)
17. Ma, W., Zhou, R., Zhao, Y., Xia, Y., Fenster, A., Ding, M.: Plaque recognition of carotid ultrasound images based on deep residual network. In: IEEE 8th Joint International Information Technology and Artificial Intelligence Conference, pp. 931–934. IEEE (2019)
18. Nanayakkara, N.D., Chiu, B., Samani, A., Spence, J.D., Samarabandu, J., Fenster, A.: A "twisting and bending" model-based nonrigid image registration technique for 3-D ultrasound carotid images. IEEE Trans. Med. Imaging **27**(10), 1378–1388 (2008)
19. Selvaraju, R.R., Cogswell, M., Das, A., Vedantam, R., Parikh, D., Batra, D.: Grad-cam: visual explanations from deep networks via gradient-based localization. In: Proceedings of the IEEE International Conference on Computer Vision, pp. 618–626 (2017)
20. Shen, D., Zhao, S., Hu, J., Feng, H., Cai, D., He, X.: ES-Net: erasing salient parts to learn more in re-identification. IEEE Trans. Image Process. **30**, 1676–1686 (2020)
21. Skandha, S.S., et al.: 3-D optimized classification and characterization artificial intelligence paradigm for cardiovascular/stroke risk stratification using carotid ultrasound-based delineated plaque: atheromatic^{TM} 2.0. Comput. Biol. Med. **125**, 103958 (2020)
22. Spence, J.D.: Intensive management of risk factors for accelerated atherosclerosis: the role of multiple interventions. Curr. Neurol. Neurosci. Rep. **7**(1), 42–48 (2007)
23. Spence, J.D.: Determinants of carotid plaque burden. Atherosclerosis **255**, 122–123 (2016)

24. Wang, H., et al.: Score-CAM: score-weighted visual explanations for convolutional neural networks. In: Proceedings of the IEEE Conference on Computer Vision and Pattern Recognition Workshops, pp. 24–25 (2020)
25. Yang, W., Huang, H., Zhang, Z., Chen, X., Huang, K., Zhang, S.: Towards rich feature discovery with class activation maps augmentation for person re-identification. In: Proceedings of the IEEE Conference on Computer Vision and Pattern Recognition, pp. 1389–1398 (2019)
26. Zhao, Y., Spence, J.D., Chiu, B.: Three-dimensional ultrasound assessment of effects of therapies on carotid atherosclerosis using vessel wall thickness maps. Ultrasound Med. Biol. **47**(9), 2502–2513 (2021)
27. Zhou, B., Khosla, A., Lapedriza, A., Oliva, A., Torralba, A.: Learning deep features for discriminative localization. In: Proceedings of the IEEE Conference on Computer Vision and Pattern Recognition, pp. 2921–2929 (2016)

Learning Robust Classifier for Imbalanced Medical Image Dataset with Noisy Labels by Minimizing Invariant Risk

Jinpeng Li[1], Hanqun Cao[1], Jiaze Wang[1], Furui Liu[3], Qi Dou[1,2], Guangyong Chen[3(✉)], and Pheng-Ann Heng[1,2]

[1] Department of Computer Science and Engineering, The Chinese University of Hong Kong, Shatin, Hong Kong
[2] Institute of Medical Intelligence and XR, The Chinese University of Hong Kong, Shatin, Hong Kong
[3] Zhejiang Lab, Hangzhou, China
gychen@zhejianglab.com

Abstract. In medical image analysis, imbalanced noisy dataset classification poses a long-standing and critical problem since clinical large-scale datasets often attain noisy labels and imbalanced distributions through annotation and collection. Current approaches addressing noisy labels and long-tailed distributions separately may negatively impact real-world practices. Additionally, the factor of class hardness hindering label noise removal remains undiscovered, causing a critical necessity for an approach to enhance the classification performance of noisy imbalanced medical datasets with various class hardness. To address this paradox, we propose a robust classifier that trains on a multi-stage noise removal framework, which jointly rectifies the adverse effects of label noise, imbalanced distribution, and class hardness. The proposed noise removal framework consists of multiple phases. Multi-Environment Risk Minimization (MER) strategy captures data-to-label causal features for noise identification, and the Rescaling Class-aware Gaussian Mixture Modeling (RCGM) learns class-invariant detection mappings for noise removal. Extensive experiments on two imbalanced noisy clinical datasets demonstrate the capability and potential of our method for boosting the performance of medical image classification.

Keywords: Imbalanced Data · Noisy Labels · Medical Image Analysis

1 Introduction

Image classification is a significant challenge in medical image analysis. Although some classification methods achieve promising performance on balanced and clean medical datasets, balanced datasets with high-accuracy annotations are

---

Supplementary Information The online version contains supplementary material available at https://doi.org/10.1007/978-3-031-43987-2_30.

H. Greenspan et al. (Eds.): MICCAI 2023, LNCS 14225, pp. 306–316, 2023.
https://doi.org/10.1007/978-3-031-43987-2_30

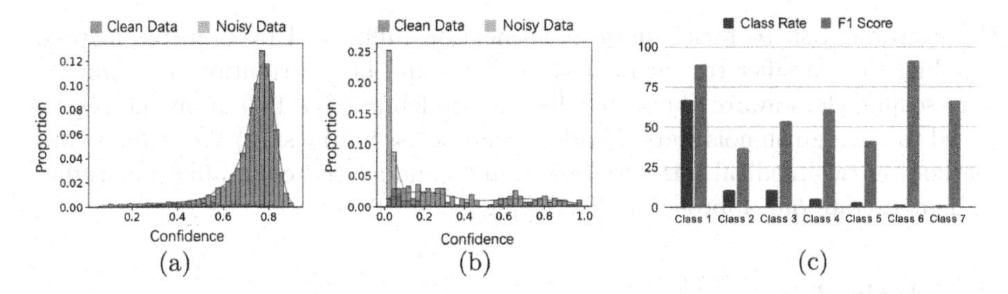

Fig. 1. Analysis of confidence distributions and class hardness on imbalanced noisy HAM10000 dataset [22]. (a) and (b) are confidence distributions of clean and noisy data on the majority class and the minority class, respectively (c) is the relationship between class rate and F1 score among different classes.

time-consuming and expensive. Besides, pruning clean and balanced datasets require a large amount of crucial clinical data, which is insufficient for large-scale deep learning. Therefore, we focus on a more practical yet unexplored setting for handling imbalanced medical data with noisy labels, utilizing all available low-cost data with possible noisy annotations. Noisy imbalanced datasets arise due to the lack of high-quality annotations [11] and skewed data distributions [18] where the number of instances largely varies across different classes. Besides, the class hardness problem where classification difficulties vary for different categories presents another challenge in removing label noise. Due to differences in disease epidemicity and collection difficulty, rare anomalies or anatomical features render diseases with low epidemicity easier to detect. However, existing techniques [12,23,24] fail to jointly address these scenarios, leading to inadequate classification outcomes. Therefore, noisy-labeled, imbalanced datasets with various class hardness remain a persistent challenge in medical classification.

Existing approaches for non-ideal medical image classification can be summarized into noisy classification, imbalanced recognition, and noisy imbalanced identification. Noisy classification approaches [3,7,23] conduct noise-invariant learning depending on the big-loss hypothesis, where classifiers trained with clean data with lower empirical loss aid with de-noising identification. However, imbalanced data creates different confidence distributions of clean and noisy data in the majority class and minority class as shown in Fig. 1, which invalidates the big-loss assumption [3,4]. Imbalanced recognition approaches [9,15,21] utilize augmented embeddings and imbalance-invariant training loss to re-balance the long-tailed medical data artificially, but the disturbance from noisy labels leads to uncasual feature learning, impeding the recognition of tail classes. Noisy long-tailed identification technique [25] has achieved promising results by addressing noise and imbalance concerns sequentially. However, the class hardness problem leads to vague decision boundaries that hinders accurate‘ noise identification.

In this work, we propose a multi-stage noise removal framework to address these concerns jointly. The main contributions of our work include: 1) We decompose the negative effects in practical medical image classification, 2) We minimize the invariant risk to tackle noise identification influenced by multiple factors, enabling the classifier to learn causal features and be distribution-invariant, 3) A re-scaling class-aware Gaussian Mixture Modeling (CGMM) approach is proposed to distinguish noise labels under various class hardness, 4) We evaluate our method on two medical image datasets, and conduct thorough ablation studies to demonstrate our approach's effectiveness.

2 Method

2.1 Problem Formulation

In the noisy imbalanced classification setting, we denote a medical dataset as $\{(x_i, y_i)\}_{i=1}^{N}$ where y_i is the corresponding label of data x_i and N is the total amount of instances. Here y_i may be noisy. Further, we split the dataset according to class categories. Then, we have $\{\mathcal{D}_j\}_{j=1}^{M}$, where M is the number of classes; $\mathcal{D}_j$ denotes the subset for class j. In each subset containing N_j samples, the data pairs are expressed as $\{(x_i^j, y_i^j)\}_{i=1}^{N_j}$. Without loss of generality, we order the classes as $N_1 > N_2 > ... > N_{M-1} > N_M$. Further, we denote the backbone as $\mathcal{H}(\cdot;\theta), \mathcal{X} \rightarrow \mathcal{Z}$ mapping data manifold to the latent manifold, the classifier head as $\mathcal{G}(\cdot;\gamma), \mathcal{Z} \rightarrow \mathcal{C}$ linking latent space to the category logit space, and the identifier as $\mathcal{F}(\cdot;\phi), \mathcal{Z} \rightarrow \mathcal{C}$. We aim to train a robust medical image classification model composed of a representation backbone and a classifier head on label noise and imbalance distribution, resulting in a minimized loss on the testing dataset:

$$\min \sum_{i \in D} L(\mathcal{G}\left[\mathcal{H}(x_i^{test};\theta);\gamma\right], y_i^{test}) \tag{1}$$

2.2 Mapping Correction Decomposition

We decompose the non-linear mapping $p(y = c|x)$ as a product of two space mappings $p_{\mathcal{G}}(y = c|z) \cdot p_{\mathcal{H}}(z|x)$. Given that backbone mapping is independent of noisy imbalanced effects, we conduct further disentanglement by defining e as the negative effects and $\mathcal{P}$ as constant for fixed probability mappings:

$$\begin{aligned}
p(y = c|x, e) &= p_{\mathcal{H}}(z|x) \cdot p_{\mathcal{G}}(y = c|z, e) \\
&= p_{\mathcal{H}}(z|x) \cdot \{p_{\mathcal{G}}(y = c|z) \cdot p_{\mathcal{G}}(y = c|e)\} \\
&= p_{\mathcal{H}}(z|x) \cdot p_{\mathcal{G}}(y = c|z) \cdot \{p_{\mathcal{G}}(y = c|[e_i, e_n, e_m])\} \\
&= \mathcal{P} \cdot \overbrace{p_{\mathcal{G}}(e_i|y = c)}^{\textbf{Imbalance}} \overbrace{p_{\mathcal{G}}(e_m|y = c, e_i)}^{\textbf{Hardness}} \overbrace{p_{\mathcal{G}}(e_n|y = c, e_i, e_m)}^{\textbf{Noise}}
\end{aligned} \tag{2}$$

The induction derives from the assumption that the incorrect mapping $p_{\mathcal{G}}(y = c|z, e)$ conditions on both pure latent to logits mapping $p_{\mathcal{G}}(y = c|z)$ and adverse effects $p_{\mathcal{G}}(y = c|e)$. By Bayes theorem, we decompose the effect into imbalance,

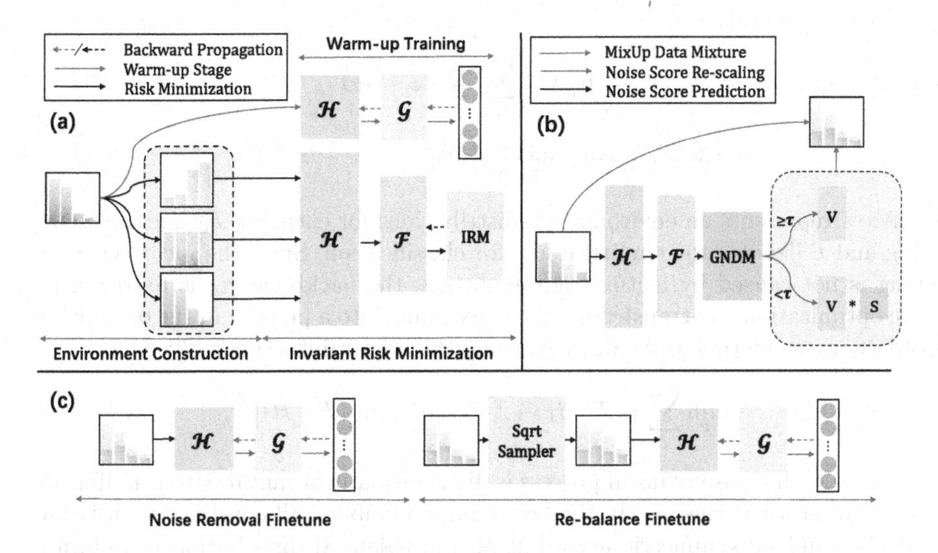

Fig. 2. Protocol for Noisy Long-tailed Recognition: (a) shows warm-up and MER schemes. Backbone $\mathcal{H}$ and classifier $\mathcal{G}$ are first trained in the warm-up phase. Noise identifier $\mathcal{F}$ is optimized across three constructed environments with $\mathcal{H}$ fixed during MER. (b) represents RCGM scheme for class-aware noise detection and score re-scaling. (c) displays final fine-tuning procedures including noise removal finetune and re-balanced finetune.

noise, and mode (hardness), where the noise effect depends on skew distribution and hardness effect; and the hardness effect is noise-invariant.

Currently, noise removal methods only address pure noise effects ($p_{\mathcal{G}}(e_n|y = c)$), while imbalance recognition methods can only resolve imbalanced distribution, which hinders the co-removal of adverse influences. Furthermore, the impact of hardness effects has not been considered in previous studies, which adds an extra dimension to noise removal. In essence, the fundamental idea of noisy classification involves utilizing clean data for classifier training, which determines the importance of noise identification and removal. To address these issues, we propose a mapping correction approach that combines independent noise detection and removal techniques to identify and remove noise effectively.

2.3 Minimizing Invariant Risk Across Multi-distributions

Traditional learning with noisy label methods mainly minimize empirical risk on training data. However, they fail to consider the influence of imbalanced distributions, which might cause a biased gradient direction on the optimization subspace. Following [25], we minimize the invariant risk [2] across multi-environment for independent detector learning. By assuming that the robust classifier performs well on every data distribution, we solve the optimizing object by finding the optima to reduce the averaged distance for gradient shift:

$$\begin{aligned}&\min_{\substack{\mathcal{H}_\theta:\mathcal{X}\rightarrow\mathcal{Z}\\ \mathcal{F}_\phi:\mathcal{Z}\rightarrow\mathcal{Y}}}\sum_{\varepsilon\in\mathcal{E}_{tr}}\mathcal{L}(\mathcal{F}_\phi\circ\mathcal{H}_\theta)\\ &\text{s.t. } \mathcal{F}_\phi\in\arg\min_{\bar{\mathcal{F}}_\phi:\mathcal{Z}\rightarrow\mathcal{Y}}\mathcal{L}(\bar{\mathcal{F}}_\phi\circ\mathcal{H}_\theta),\quad\forall\varepsilon\in\mathcal{E}_{tr},\end{aligned}\tag{3}$$

where ε represents an environment (distribution) for classifier $\mathcal{F}_\phi$ and backbone $\mathcal{H}_\theta$; and $\mathcal{L}$ denotes the empirical loss for classification. Since the incorrect mapping is not caused by feature representation, the backbone $\mathcal{H}_\theta$ is fixed during the optimization. By transferring the constraints into a penalty in the optimizing object, we solve this problem by learning the constraint scale ω [2]:

$$\min_{\mathcal{F}}\sum_{\epsilon}\mathcal{L}(\mathcal{F}\circ\mathcal{H})+\left\|\nabla_{w|w=1}\mathcal{L}(w\cdot\mathcal{F}\circ\mathcal{H})\right\|_2^2.\tag{4}$$

Ideally, the noise removal process is distribution-invariant if data is uniformly distributed w.r.t. classes. By the law of large numbers, all constructed distributions should be symmetric according to the balanced distribution to obtain a uniform expectation. To simplify this assumption, we construct three different data distributions [25] composed of one uniform distribution and two symmetric skewed distributions instead of theoretical settings. In practice, all environments are established from the training set with the same class categories.

2.4 Rescaling Class-Aware Gaussian Mixture

Existing noise labels learning methods [1,13] cluster all sample loss or confidence scores with Beta Mixture Model or Gaussian Mixture Model into noisy and clean distributions. From the perspective of clustering, definite and immense gaps between two congregate groups contribute to more accurate decisions. However, in medical image analysis, an overlooked mismatch exists between class hardness and difficulty in noise identification. This results in ineffectiveness of global cluster methods in detecting label noises across all categories. To resolve the challenge, we propose a novel method called rescaling class-aware Gaussian Mixture Modeling (RCGM) which clusters each category data independently by fitting confidence scores q_{ij} from ith class into two Gaussian distributions as $p_i^n(x^n|\mu^n,\Sigma^n)$ and $p_i^c(x^c|\mu^c,\Sigma^c)$. The mixed Gaussian $p_i^M(\cdot)$ is obtained by linear combinations α_{ik} for each distribution:

$$p_i^M(q_{ij}):=\sum_{k\in\{c,n\}}\alpha_{ik}p_i^k\left(q_{ij}\mid\mu_i^k,\Sigma_i^k\right),\tag{5}$$

which produces more accurate and independent measurements of label quality. Rather than relying on the assumption that confidence distributions of training samples depend solely on their label quality, RCGM solves the effect of class hardness in noisy detection by individually clustering the scores in each category. This overcomes the limitations of global clustering methods and significantly enhances the accuracy of noise identification even when class hardness varies.

Instead of assigning a hard label to the potential noisy data as [8] which also employs a class-specific GMM to cluster the uncertainty, we further re-scale the confidence score of class-wise noisy data. Let x_{ij} be the jth in class i, then its probability of having a clean label is:

$$\gamma_{ij} = \frac{\alpha_{ik} p_i^c (q_{ij} \mid \mu_i^c, \Sigma_i^c)}{p_i^M (q_{ij})}, \tag{6}$$

which is then multiplied by a hyperparameter s if the instance is predicted as noise to reduce its weight in the finetuning. With a pre-defined noise selection threshold as τ, we have the final clean score as:

$$v(x_{ij}) := \begin{cases} \gamma_{ij} & \text{if } \gamma_{ij} \geq \tau \\ s \cdot \gamma_{ij} & \text{if } \gamma_{ij} < \tau \end{cases} \tag{7}$$

2.5 Overall Learning Framework for Imbalanced and Noisy Data

In contrast to two-stage noise removal and imbalance classification techniques, our approach applies a multi-stage protocol: warm-up phases, noise removal phases, and fine-tuning phases as shown in Fig. 2. In the warm-up stage, we train backbone $\mathcal{H}$ and classifier $\mathcal{G}$ a few epochs by assuming that $\mathcal{G}$ only remembers clean images with less empirical loss. In the noise removal phases, we learn class-invariant probability distributions of noisy-label effect with MER and remove class hardness impact with RCGM. Finally, in the fine-tuning phases, we apply MixUp technique [13,25,26] to rebuild a hybrid distribution from noisy pairs and clean pairs by:

$$\begin{aligned} \hat{x}_{kl} &:= \alpha_{kl} x_k + (1 - \alpha_{kl}) x_l, \quad \forall x_k, x_l \in \mathcal{D} \\ \hat{y}_{kl} &:= \alpha_{kl} y_k + (1 - \alpha_{kl}) y_l, \quad \forall y_k, y_l \in \mathcal{D} \end{aligned} \tag{8}$$

where $\alpha_{kl} := \frac{v(x_k)}{v(x_l)}$ denotes the balanced scale; and $\{(\hat{x}_{kl}, \hat{y}_{kl})\}$ are the mixed clean data for classifier fine-tuning. Sqrt sampler is applied to re-balance the data, and cross-stage KL [12] and CE loss are the fine-tuning loss functions.

3 Experiment

3.1 Dataset and Evaluation Metric

We evaluated our approach on two medical image datasets with imbalanced class distributions and noisy labels. The first dataset, HAM10000 [22], is a dermatoscopic image dataset for skin-lesion classification with 10,015 images divided into seven categories. It contains a training set with 7,007 images, a validation set with 1,003 images, and a testing set with 2,005 images. Following the previous noisy label settings [25], we add 20% noise to its training set by randomly flipping labels. The second dataset, CHAOYANG [29], is a histopathology image dataset manually annotated into four cancer categories by three pathological

Table 1. Quantitative comparisons with state-of-the-art methods on HAM10000 and CHAOYANG datasets. The second-best performances are underlined.

Method	HAM10000			CHAOYANG		
	Macro-F1	B-ACC	MCC	Macro-F1	B-ACC	MCC
Focal Loss [14]	66.16	65.21	59.86	70.84	69.10	70.27
Sqrt-RS [17]	67.02	66.28	55.09	70.39	68.56	69.71
PG-RS [10]	62.91	63.29	51.14	71.03	69.35	70.18
CB-Focal [5]	63.41	72.63	52.21	<u>73.20</u>	<u>72.46</u>	71.37
EQL [21]	60.94	66.18	55.53	71.09	70.53	70.77
EQL V2 [20]	58.33	54.70	52.01	69.24	68.35	67.78
CECE [5]	40.92	56.75	37.46	47.12	47.56	43.50
CLAS [28]	69.61	70.85	63.67	71.91	71.46	70.71
FCD [12]	<u>71.08</u>	<u>72.85</u>	<u>66.58</u>	71.82	70.07	71.76
DivideMix [13]	69.72	70.84	65.33	50.44	49.34	50.29
NL [16]	44.42	42.52	55.81	71.75	69.99	71.63
NL+Sqrt-RS	62.46	61.42	52.44	71.77	70.88	71.46
GCE [27]	50.47	48.91	63.63	21.04	28.12	4.13
GCE+Sqrt-RS	70.81	70.76	65.86	70.83	69.77	70.21
GCE+Focal	66.19	68.71	61.82	72.91	71.25	<u>72.68</u>
Co-Learning [19]	58.05	51.02	57.73	60.73	59.78	64.13
H2E [25]	69.69	69.11	63.48	69.36	67.89	68.59
Ours	**76.43**	**75.60**	**70.19**	**74.50**	**72.75**	**73.08**

experts, with 40% of training samples having inconsistent annotations from the experts. To emulate imbalanced scenarios, we prune the class sizes of the training set into an imbalanced distribution as [5]. Consequently, CHAOYANG dataset consists of a training set with 2,181 images, a validation set with 713 images, and a testing set with 1,426 images, where the validation and testing sets have clean labels. The imbalanced ratios [12] of HAM10000 and CHAOYANG are 59 and 20, respectively. The evaluation metrics are Macro-F1, B-ACC, and MCC.

3.2 Implementation Details

We mainly follow the training settings of FCD [12]. ResNet-18 pretrained on the ImageNet is the backbone. The batch size is 48. Learning rates are 0.06, 0.001, 0.06 and 0.006 with the cosine schedule for four stages, respectively. We train our models by SGD optimizer with sharpness-aware term [6] for 90, 90, 90, and 20 epochs. The size of input image is 224×224. The scale and threshold in RCGM are 0.6 and 0.1, respectively.

3.3 Comparison with State-of-the-Art Methods

We compare our model with state-of-the-art methods which contain noisy methods (including DivideMix [13], NL [16], GCE [27], Co-Learning [19]), imbalance methods (including Focal Loss [14], Sqrt-RS [17], PG-RS [10], CB-Focal [5], EQL [21], EQL V2 [20], CECE [5], CLAS [28], FCD [12]), and noisy imbalanced classification methods (including H2E [25], NL+Sqrt-RS, GCE+Sqrt-RS, GCE+Focal). We train all approaches under the same data augmentations and network architecture. Table 1 exhibits the overall comparison of all approaches. We first conclude that noisy imbalanced setting does negatively affect learning with noise methods and imbalanced methods. In imbalanced methods, CECE only obtains 40.92 in Macro-F1 on HAM10000 and 47.12% Macro-F1 on CHAOYANG. In noisy methods, NL and GCE also suffer great performance declines. We mix these weakly-performed approaches with methods from the other category, observing the accuracy improvement. Compared to GCE, GCE+Sqrt-RS achieves +20.34% Macro-F1 on HAM10000 and +49.79% Macro-F1 on CHAOYANG. Similar increases happen in GCE & GCE+Focal and NL & NL+Sqrt-RS. Then, we compare our approach to state-of-the-art methods of the noisy (DivideMix), imbalanced (FCD), and noisy long-tailed (H2E) methods. Our framework achieves improvements in all metrics on both datasets, demonstrating the rationality of the assumption and the effectiveness of our framework.

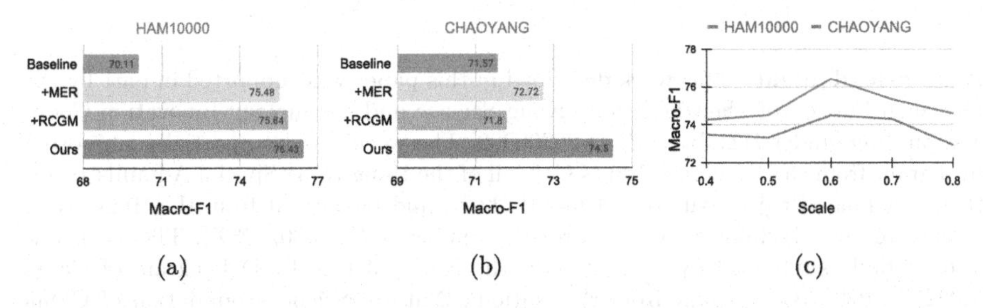

Fig. 3. Ablation analysis. (a) and (b) Quantitative performance comparison of different components of our method on HAM10000 and CHAOYANG datasets, respectively. (c) Comparative results of our approach with different s values.

3.4 Ablation Studies

As shown in Fig. 3, we evaluate the effectiveness of the components in our method by decomposing them on extensive experiments. We choose the first stage of FCD [12] as our baseline. Figure 3a and 3b show that only using MER or RCGM achieves better performance than our strong baseline on both datasets. For example, MER achieves 5.37% and 1.15% improvements on HAM10000 and CHAOYANG, respectively, demonstrating the effectiveness of our noise removal techniques. Further, our multi-stage noise removal technique outperforms single MER and RCGM, revealing that the decomposition for noise effect and hardness effect works on noisy imbalanced datasets. We find that the combination of MER and RCGM improves more on CHAOYANG dataset. This is because CHAOYANG has more possible label noise than HAM10000 caused by the high annotating procedure. From Fig. 3c, we observe the accuracy trends are as the scale increases and achieve the peak around 0.6. It indicates the re-scaling process for noise weight deduction contributes to balancing the feature learning and classification boundary disturbance from the mixture of noisy and clean data. Furthermore, similar performance trends reveal the robustness of scale s.

4 Conclusion and Discussion

We propose a multi-step framework for noisy long-imbalanced medical image classification. We address three practical adverse effects including data noise, imbalanced distribution, and class hardness. To solve these difficulties, we conduct Multi-Environment Risk Minimization (MER) and rescaling class-aware Gaussian Mixture Modeling (RCGM) together for robust feature learning. Extensive results on two public medical image datasets have verified that our framework works on the noisy imbalanced classification problem. The main limitation of our work is the manually designed multi-stage training protocol which lacks simplicity compared to end-to-end training and warrants future simplification.

Acknowlegdement. This work described in this paper was supported in part by the Shenzhen Portion of Shenzhen-Hong Kong Science and Technology Innovation Cooperation Zone under HZQB-KCZYB-20200089. The work was also partially supported by a grant from the Research Grants Council of the Hong Kong Special Administrative Region, China (Project Number: T45-401/22-N) and by a grant from the Hong Kong Innovation and Technology Fund (Project Number: GHP/080/20SZ). The work was also partially supported by a grant from the National Key R&D Program of China (2022YFE0200700), a grant from the National Natural Science Foundation of China (Project No. 62006219), and a grant from the Natural Science Foundation of Guangdong Province (2022A1515011579).

References

1. Arazo, E., Ortego, D., Albert, P., O'Connor, N.E., McGuinness, K.: Unsupervised label noise modeling and loss correction. In: Chaudhuri, K., Salakhutdinov, R. (eds.) ICML 2019 (2019)
2. Arjovsky, M., Bottou, L., Gulrajani, I., Lopez-Paz, D.: Invariant risk minimization. arXiv preprint arXiv:1907.02893 (2019)
3. Chen, P., Liao, B.B., Chen, G., Zhang, S.: Understanding and utilizing deep neural networks trained with noisy labels. In: ICML (2019)
4. Chen, X., Gupta, A.: Webly supervised learning of convolutional networks. In: ICCV (2015)

5. Cui, Y., Jia, M., Lin, T., Song, Y., Belongie, S.J.: Class-balanced loss based on effective number of samples. In: CVPR (2019)
6. Foret, P., Kleiner, A., Mobahi, H., Neyshabur, B.: Sharpness-aware minimization for efficiently improving generalization. In: 9th International Conference on Learning Representations, ICLR 2021, Virtual Event, Austria, 3–7 May 2021. OpenReview.net (2021). https://openreview.net/forum?id=6Tm1mposlrM
7. Frénay, B., Verleysen, M.: Classification in the presence of label noise: a survey. IEEE TNNLS **25**(5), 845–869 (2013)
8. Huang, Y., Bai, B., Zhao, S., Bai, K., Wang, F.: Uncertainty-aware learning against label noise on imbalanced datasets. In: Thirty-Sixth AAAI Conference on Artificial Intelligence, AAAI 2022, Thirty-Fourth Conference on Innovative Applications of Artificial Intelligence, IAAI 2022, The Twelveth Symposium on Educational Advances in Artificial Intelligence, EAAI 2022 Virtual Event, 22 February–1 March 2022, pp. 6960–6969. AAAI Press (2022). https://ojs.aaai.org/index.php/AAAI/article/view/20654
9. Kang, B., et al.: Decoupling representation and classifier for long-tailed recognition. arXiv preprint arXiv:1910.09217 (2019)
10. Kang, B., et al.: Decoupling representation and classifier for long-tailed recognition. In: ICLR (2020)
11. Karimi, D., Dou, H., Warfield, S.K., Gholipour, A.: Deep learning with noisy labels: exploring techniques and remedies in medical image analysis. Med. Image Anal. **65**, 101759 (2020)
12. Li, J., et al.: Flat-aware cross-stage distilled framework for imbalanced medical image classification. In: Wang, L., Dou, Q., Fletcher, P.T., Speidel, S., Li, S. (eds.) MICCAI 2022. LNCS, vol. 13433, pp. 217–226. Springer, Cham (2022). https://doi.org/10.1007/978-3-031-16437-8_21
13. Li, J., Socher, R., Hoi, S.C.H.: Dividemix: learning with noisy labels as semi-supervised learning. In: ICLR 2020 (2020)
14. Lin, T., Goyal, P., Girshick, R.B., He, K., Dollár, P.: Focal loss for dense object detection. In: ICCV (2017)
15. Liu, J., Sun, Y., Han, C., Dou, Z., Li, W.: Deep representation learning on long-tailed data: a learnable embedding augmentation perspective. In: CVPR (2020)
16. Ma, X., Huang, H., Wang, Y., Romano, S., Erfani, S.M., Bailey, J.: Normalized loss functions for deep learning with noisy labels. In: ICML 2020 (2020)
17. Mahajan, D., et al.: Exploring the limits of weakly supervised pretraining. In: ECCV (2018)
18. Song, H., Kim, M., Park, D., Shin, Y., Lee, J.G.: Learning from noisy labels with deep neural networks: a survey. IEEE TNNLS (2022)
19. Tan, C., Xia, J., Wu, L., Li, S.Z.: Co-learning: learning from noisy labels with self-supervision. In: Shen, H.T., et al. (eds.) ACM 2021 (2021)
20. Tan, J., Lu, X., Zhang, G., Yin, C., Li, Q.: Equalization loss V2: a new gradient balance approach for long-tailed object detection. In: CVPR (2021)
21. Tan, J., et al.: Equalization loss for long-tailed object recognition. In: CVPR 2020 (2020)
22. Tschandl, P., Rosendahl, C., Kittler, H.: The HAM10000 dataset, a large collection of multi-source dermatoscopic images of common pigmented skin lesions. Sci. Data **5**(1), 1–9 (2018)
23. Xue, C., Dou, Q., Shi, X., Chen, H., Heng, P.A.: Robust learning at noisy labeled medical images: applied to skin lesion classification. In: ISBI 2019 (2019)

24. Xue, C., Yu, L., Chen, P., Dou, Q., Heng, P.A.: Robust medical image classification from noisy labeled data with global and local representation guided co-training. IEEE TMI **41**(6), 1371–1382 (2022)
25. Yi, X., Tang, K., Hua, X.S., Lim, J.H., Zhang, H.: Identifying hard noise in long-tailed sample distribution. In: Avidan, S., Brostow, G., Cissé, M., Farinella, G.M., Hassner, T. (eds.) ECCV 2022. LNCS, vol. 13686, pp. 739–756. Springer, Cham (2022). https://doi.org/10.1007/978-3-031-19809-0_42
26. Zhang, H., Cisse, M., Dauphin, Y.N., Lopez-Paz, D.: mixup: beyond empirical risk minimization. arXiv preprint arXiv:1710.09412 (2017)
27. Zhang, Z., Sabuncu, M.R.: Generalized cross entropy loss for training deep neural networks with noisy labels. In: Bengio, S., Wallach, H.M., Larochelle, H., Grauman, K., Cesa-Bianchi, N., Garnett, R. (eds.) NIPS 2018 (2018)
28. Zhong, Z., Cui, J., Liu, S., Jia, J.: Improving calibration for long-tailed recognition. In: CVPR 2021 (2021)
29. Zhu, C., Chen, W., Peng, T., Wang, Y., Jin, M.: Hard sample aware noise robust learning for histopathology image classification. IEEE TMI **41**(4), 881–894 (2021)

TCEIP: Text Condition Embedded Regression Network for Dental Implant Position Prediction

Xinquan Yang[1,2,3], Jinheng Xie[1,2,3,5], Xuguang Li[4], Xuechen Li[1,2,3], Xin Li[4], Linlin Shen[1,2,3(✉)], and Yongqiang Deng[4]

[1] College of Computer Science and Software Engineering, Shenzhen University, Shenzhen, China
yangxinquan2021@email.szu.edu.cn, llshen@szu.edu.cn
[2] AI Research Center for Medical Image Analysis and Diagnosis, Shenzhen University, Shenzhen, China
[3] National Engineering Laboratory for Big Data System Computing Technology, Shenzhen University, Shenzhen, China
[4] Department of Stomatology, Shenzhen University General Hospital, Shenzhen, China
[5] National University of Singapore, Singapore , Singapore

Abstract. When deep neural network has been proposed to assist the dentist in designing the location of dental implant, most of them are targeting simple cases where only one missing tooth is available. As a result, literature works do not work well when there are multiple missing teeth and easily generate false predictions when the teeth are sparsely distributed. In this paper, we are trying to integrate a weak supervision text, the target region, to the implant position regression network, to address above issues. We propose a text condition embedded implant position regression network (TCEIP), to embed the text condition into the encoder-decoder framework for improvement of the regression performance. A cross-modal interaction that consists of cross-modal attention (CMA) and knowledge alignment module (KAM) is proposed to facilitate the interaction between features of images and texts. The CMA module performs a cross-attention between the image feature and the text condition, and the KAM mitigates the knowledge gap between the image feature and the image encoder of the CLIP. Extensive experiments on a dental implant dataset through five-fold cross-validation demonstrated that the proposed TCEIP achieves superior performance than existing methods.

Keywords: Dental Implant · Deep Learning · Text Guided Detection · Cross-Modal Interaction

---

Supplementary Information The online version contains supplementary material available at https://doi.org/10.1007/978-3-031-43987-2_31.

H. Greenspan et al. (Eds.): MICCAI 2023, LNCS 14225, pp. 317–326, 2023.
https://doi.org/10.1007/978-3-031-43987-2_31

1 Introduction

According to a systematic research study [2], periodontal disease is the world's 11th most prevalent oral condition, which potentially causes tooth loss in adults, especially the aged [8]. One of the most appropriate treatments for such a defect/dentition loss is prosthesis implanting, in which the surgical guide is usually used. However, dentists must load the Cone-beam computed tomography (CBCT) data into the surgical guide design software to estimate the implant position, which is tedious and inefficient. In contrast, deep learning-based methods show great potential to efficiently assist the dentist in locating the implant position [7].

Recently, deep learning-based methods have achieved great success in the task of implant position estimation. Kurt et al. [4] and Widiasri et al. [11] utilized the convolutional neural network (CNN) to locate the oral bone, e.g., the alveolar bone, maxillary sinus and jaw bone, which determines the implant position indirectly. Different from these implant depth measuring methods, Yang et al. [13] developed a transformer-based implant position regression network (Implant-Former), which directly predicts the implant position on the 2D axial view of tooth crown images and projects the prediction results back to the tooth root by the space transform algorithm. However, these methods generally consider simple situations, in which only one missing tooth is available. When confronting some special cases, such as multiple missing teeth and sparse teeth disturbance in Fig. 1(a), the above methods may fail to determine the correct implant position. In contrast, clinically, dentists have a subjective expertise about where the implant should be planted, which motivates us that, additional indications or conditions from dentists may help predict an accurate implant position.

In recent years, great success has been witnessed in Vision-Language Pretraining (VLP). For example, Radford [9] proposed Contrastive Language-Image Pretraining (CLIP) to learn diverse visual concepts from 400 million image-text pairs automatically, which can be used for vision tasks like object detection [17] and segmentation [12]. In this paper, we found that CLIP has the ability to learn the position relationship among instances. We showcase examples in Fig. 1(b) that the image-text pair with the word 'left' get a higher matching score than others, as the position of baby is on the left of the billboard.

Motivated by the above observation in dental implant and the property of CLIP, in this paper, we integrate a text condition from the CLIP to assist the implant position regression. According to the natural distribution, we divide teeth regions into three categories in Fig. 1(c), i.e., left, middle, and right. Specifically, during training, one of the text prompts, i.e., 'right', 'middle', and 'left' is paired with the crown image as input, in which the text prompt works as a guidance or condition. The crown image is processed by an encoder-decoder network for final location regression. In addition, to facilitate the interaction between features in two modalities, a cross-modal interaction that consists of cross-modal attention (CMA) and knowledge alignment module (KAM), is devised. The CMA module fuses conditional information, i.e., text prompt, to the encoder-decoder. This brings additional indications or conditions from the dentist to help

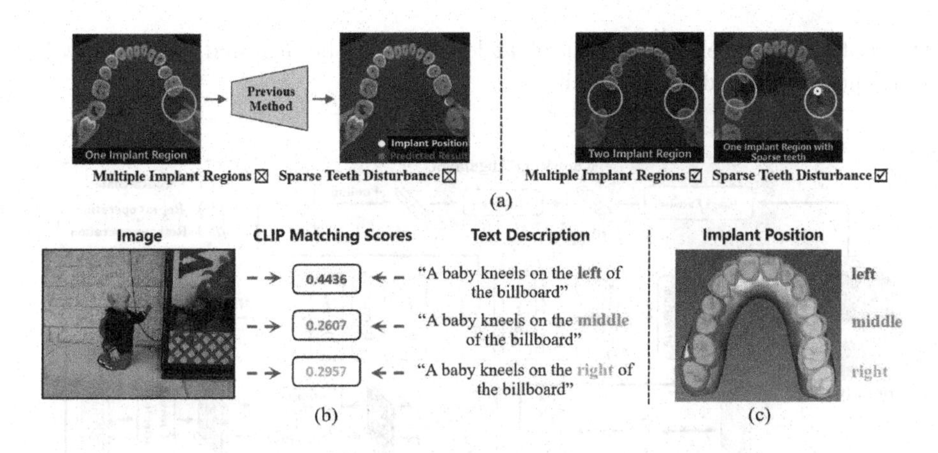

Fig. 1. (a) The 2D axial view of tooth crown images captured from different patients, where the pink and blue circles denote the implant and sparse teeth regions, respectively. (b) The matching score of the CLIP for a pair of image and text. (c) The division teeth region. (Color figure online)

the implant position regression. However, a knowledge gap may exist between our encoder-decoder and CLIP. To mitigate the problem, the KAM is proposed to distill the encoded-decoded features of crown images to the space of CLIP, which brings significant localization improvements. In inference, given an image, the dentist just simply gives a conditioning text like "let's implant a prosthesis on the left", the network will preferentially seek a suitable location on the left for implant prosthesis.

Main contributions of this paper can be summarized as follows: 1) To the best of our knowledge, the proposed TCEIP is the first text condition embedded implant position regression network that integrates a text embedding of CLIP to guide the prediction of implant position. (2) A cross-modal interaction that consists of a cross-modal attention (CMA) and knowledge alignment module (KAM) is devised to facilitate the interaction between features that representing image and text. (3) Extensive experiments on a dental implant dataset demonstrated the proposed TCEIP achieves superior performance than the existing methods, especially for patients with multiple missing teeth or sparse teeth.

2 Method

Given a tooth crown image with single or multiple implant regions, the proposed TCEIP aims to give a precise implant location conditioned by text indications from the dentist, i.e., a description of position like 'left', 'right', or 'middle'. An overview of TCEIP is presented in Fig. 2. It mainly consists of four parts: i) Encoder and Decoder, ii) Conditional Text Embedding, iii) Cross-Modal Interaction Module, and iv) Heatmap Regression Network. After obtaining the predicted coordinates of the implant at the tooth crown, we adopt the space transformation algorithm [13] to fit a centerline of implant to project the coordinates to the tooth root, where the real implant location can be acquired. Next, we will introduce these modules in detail.

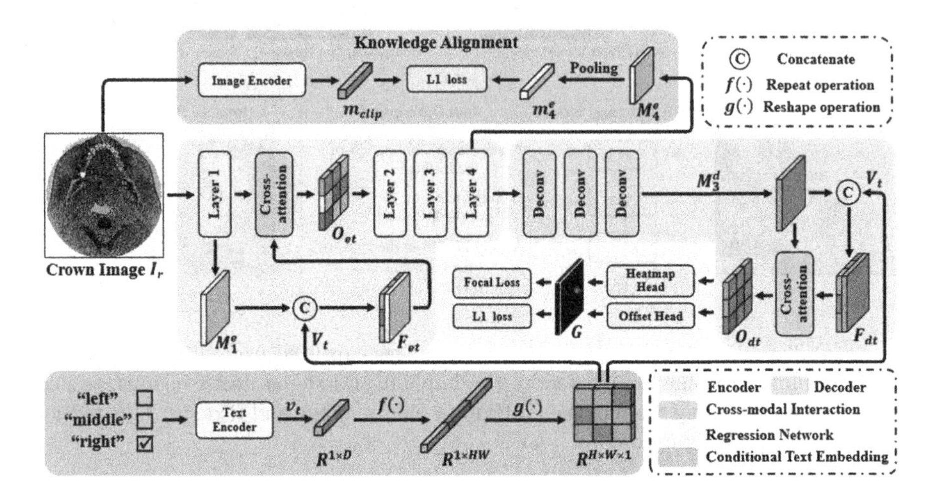

Fig. 2. The network architecture of the proposed prediction network.

2.1 Encoder and Decoder

We employ the widely used ResNet [3] as the encoder of TCEIP. It mainly consists of four layers and each layer contains multiple residual blocks. Given a tooth crown image $\mathbf{I}_r$, a set of feature maps, i.e., $\{\mathbf{M}_1^e, \mathbf{M}_2^e, \mathbf{M}_3^e, \mathbf{M}_4^e\}$, can be accordingly extracted by the ResNet layers. Each feature map has a spatial and channel dimension. To ensure fine-grained heatmap regression, three deconvolution layers are adopted as the Decoder to recover high-resolution features. It consecutively upsamples feature map $\mathbf{M}_4^e$ as high-resolution feature representations, in which a set of recovered features $\{\mathbf{M}_1^d, \mathbf{M}_2^d, \mathbf{M}_3^d\}$ can be extracted. Feature maps $\mathbf{M}_1^e$, $\mathbf{M}_4^e$ and $\mathbf{M}_3^d$ will be further employed in the proposed modules, where $\mathbf{M}_1^e$ and $\mathbf{M}_3^d$ have the same spatial dimension $\mathbb{R}^{128\times128\times C}$ and $\mathbf{M}_4^e \in \mathbb{R}^{16\times16\times\hat{C}}$.

2.2 Conditional Text Embedding

To integrate the text condition provided by a dentist, we utilize the CLIP to extract the text embedding. Specifically, additional input of text, e.g., 'left', 'middle', or 'right', is processed by the CLIP Text Encoder to obtain a conditional text embedding $\mathbf{v}_t \in \mathbb{R}^{1\times D}$. As shown in Fig. 2, to interact with the image features from ResNet layers, a series of transformation $f(\cdot)$ and $g(\cdot)$ over $\mathbf{v}_t$ are performed as follow:

$$\mathbf{V}_t = g(f(\mathbf{v}_t)) \in \mathbb{R}^{H\times W\times 1}, \tag{1}$$

where $f(\cdot)$ repeats text embedding $\mathbf{v}_t$ from $\mathbb{R}^{1\times D}$ to $\mathbb{R}^{1\times HW}$ and $g(\cdot)$ then reshapes it to $\mathbb{R}^{H\times W\times 1}$. This operation ensures better interaction between image and text in the same feature space.

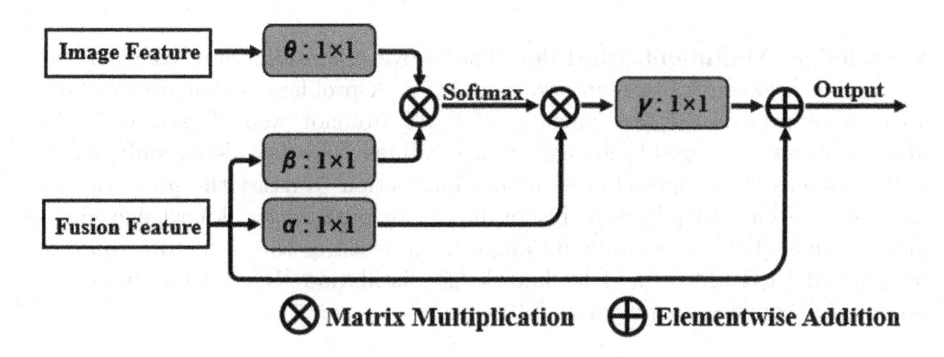

Fig. 3. The architecture of the proposed cross-modal attention module.

2.3 Cross-Modal Interaction

High-resolution features from the aforementioned decoder can be directly used to regress the implant position. However, it cannot work well in situations of multiple teeth loss or sparse teeth disturbance. In addition, although we have extracted the conditional text embedding from the CLIP to assist the network regression, there exists a big difference with the feature of encoder-decoder in the feature space. To tackle these issues, we propose cross-modal interaction, including i) Cross-Modal Attention and ii) Knowledge Alignment module, to integrate the text condition provided by the dentist.

Cross-Modal Attention Module. To enable the transformed text embedding $\mathbf{V}_t$ better interact with intermediate features of the encoder and decoder, we design and plug a cross-modal attention (CMA) module into the shallow layers of the encoder and the final deconvolution layer. The architecture of CMA is illustrated in Fig. 3. Specifically, the CMA module creates cross-attention between image features $\mathbf{M}_1^e$ and fusion feature $\mathbf{F}_{et} = [\mathbf{M}_1^e|\mathbf{V}_t]$ in the encoder, and image features $\mathbf{M}_3^d$ and fusion feature $\mathbf{F}_{dt} = [\mathbf{M}_3^d|\mathbf{V}_t]$ in the decoder, where $\mathbf{F}_{et}, \mathbf{F}_{dt} \in \mathbb{R}^{H\times W\times(C+1)}$. The CMA module can be formulated as follows:

$$\mathbf{O} = \gamma(\text{Softmax}(\theta(\mathbf{M})\beta(\mathbf{F}))\alpha(\mathbf{F})) + \mathbf{F}, \tag{2}$$

where four independent 1×1 convolutions α, β, θ, and γ are used to map image and fusion features to the space for cross-modal attention. At first, $\mathbf{M}$ and $\mathbf{F}$ are passed into $\theta(\cdot)$, $\beta(\cdot)$ and $\alpha(\cdot)$ for channel transformation, respectively. Following the transformed feature, $\mathbf{M}_\theta$ and $\mathbf{F}_\beta$ perform multiplication via a Softmax activation function to take a cross-attention with $\mathbf{F}_\alpha$. In the end, the output feature of the cross-attention via $\gamma(\cdot)$ for feature smoothing is added with $\mathbf{F}$. Given the above operations, the cross-modal features $\mathbf{O}_{et}$ and $\mathbf{O}_{dt}$ are obtained and passed to the next layer.

Knowledge Alignment Module. The above operations only consider the interaction between features in two modalities. A problem is that text embeddings from pre-trained text encoder of CLIP are not well aligned with the image features initialized by ImageNet pre-training. This knowledge shift potentially weakens the proposed cross-modal interaction to assist the prediction of implant position. To mitigate this problem, we propose the knowledge alignment module (KAM) to gradually align image features to the feature space of pre-trained CLIP. Motivated by knowledge distillation [10], we formulate the proposed knowledge alignment as follows:

$$\mathcal{L}_{align} = |\mathbf{m}_4^e - \mathbf{m}_{clip}|, \tag{3}$$

where $\mathbf{m}_4^e \in \mathbb{R}^{1\times D}$ is the transformed feature of $\mathbf{M}_4^e$ after attention pooling operation [1] and dimension reduction with convolution. $\mathbf{m}_{clip} \in \mathbb{R}^{1\times D}$ is the image embedding extracted by the CLIP Image Encoder. Using this criteria, the encoder of TCEIP approximates the CLIP image encoder and consequently aligns the image features of the encoder with the CLIP text embeddings.

2.4 Heatmap Regression Network

The heatmap regression network is used for locating the implant position, which consists of the heatmap and the offset head. The output of the heatmap head is the center localization of implant position, which is formed as a heatmap $\mathbf{G} \in [0,1]^{H\times W}$. Following [5], given coordinate of the ground truth implant location $(\tilde{t}_x, \tilde{t}_y)$, we apply a 2D Gaussian kernel to get the target heatmap:

$$\mathbf{G}_{xy} = \exp(-\frac{(x-\tilde{t}_x)^2 + (y-\tilde{t}_y)^2}{2\sigma^2}) \tag{4}$$

where σ is an object size-adaptive standard deviation. The predicted heatmap is optimized by the focal loss [6]:

$$\mathcal{L}_h = \frac{-1}{N}\sum_{xy}\begin{cases}(1-\hat{\mathbf{G}}_{xy})^\lambda \log(\hat{\mathbf{G}}_{xy}) & \text{if}\, \mathbf{G}_{xy}=1\\ (1-\hat{\mathbf{G}}_{xy})^\varphi \log(\hat{\mathbf{G}}_{xy})^\lambda \log(1-\hat{\mathbf{G}}_{xy}) & \text{otherwise}\end{cases} \tag{5}$$

where λ and φ are the hyper-parameters of the focal loss, $\hat{\mathbf{G}}$ is the predicted heatmap and N is the number of implant annotation in image. The offset head computes the discretization error caused by the downsampling operation, which is used to further refine the predicted location. The local offset loss $\mathcal{L}_o$ is optimized by the L1 loss. The overall training loss of network is:

$$\mathcal{L} = \mathcal{L}_h + \mathcal{L}_o + \mathcal{L}_{align} \tag{6}$$

2.5 Coordinate Projection

The output of TCEIP is the coordinate of implant at the tooth crown. To obtain the real implant location at the tooth root, we fit a centerline of implant using the predicted implant position of TCEIP and then extend the centerline to the root area, which is identical as [13]. By this means, the intersections of implant centerline with 2D slices of root image, i.e. the implant position at the tooth root area, can be obtained.

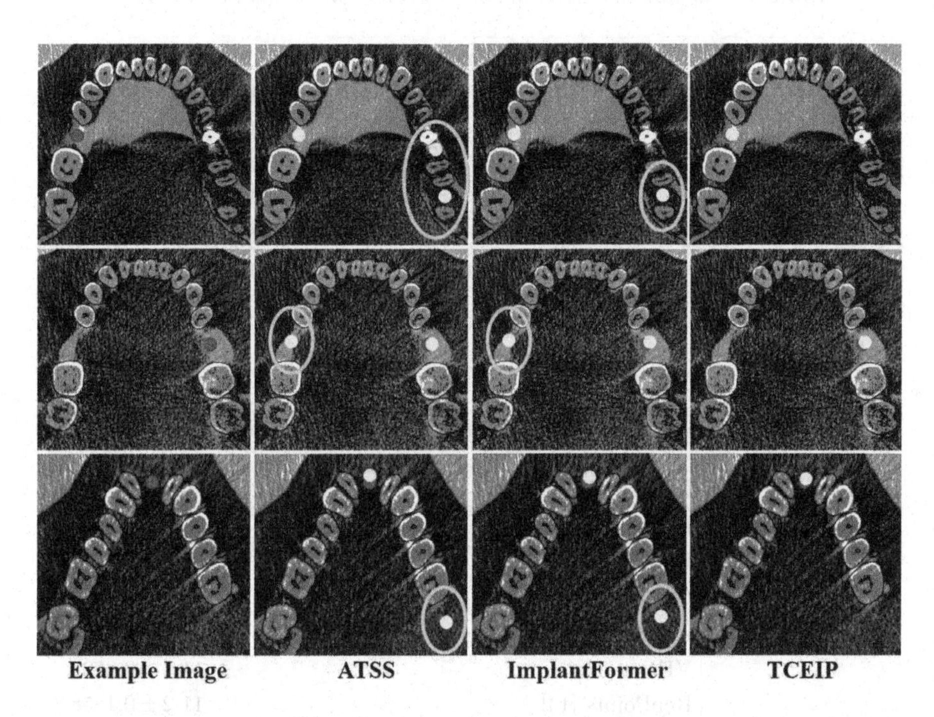

Fig. 4. Visual comparison of the predicted results with different detectors. The yellow and red circles represent the predicted implant position and ground-truth position, respectively. The blue ellipses denote false positive detections. (Color figure online)

3 Experiments and Results

3.1 Dataset and Implementation Details

The dental implant dataset was collected by [13], which contains 3045 2D slices of tooth crown images. The implant position annotations are annotated by three experienced dentists. The input image size of network is set as 512×512. We use a batch size of 8, Adam optimizer and a learning rate of 0.001 for the network training. Total training epochs is 80 and the learning rate is divided by 10 when epoch $= \{40, 60\}$. The same data augmentation methods in [13] was employed.

3.2 Performance Analysis

We use the same evaluation criteria in [13], i.e., average precision (AP) to evaluate the performance of our network. As high accurate position prediction is required in clinical practice, the IOU threshold is set as 0.75. Five-fold cross-validation was performed for all our experiments.

Table 1. The ablation experiments of each components in TCEIP.

Network	KAM	Text Condition	Feature Fusion	CMA	AP_{75}%
TCEIP					10.9 ± 0.2457
		✓			14.6 ± 0.4151
	✓	✓			15.7 ± 0.3524
	✓	✓		✓	16.5 ± 0.3891
	✓	✓	✓		17.1 ± 0.2958
	✓	✓	✓	✓	**17.8 ± 0.3956**

Table 2. Comparison of the proposed method with other mainstream detectors.

Methods	Network	Backbone	AP_{75}%
Transformer-based	ImplantFormer	ViT-Base-ResNet-50	13.7 ± 0.2045
	Deformable DETR [19]		12.8 ± 0.1417
CNN-based	CenterNet [18]	ResNet-50	10.9 ± 0.2457
	ATSS [16]		12.1 ± 0.2694
	VFNet [15]		11.8 ± 0.8734
	RepPoints [14]		11.2 ± 0.1858
	TCEIP		**17.8 ± 0.3956**

Ablation Studies. To evaluate the effectiveness of the proposed network, we conduct ablation experiments to investigate the effect of each component in Table 1. We can observe from the second row of the table that the introduction of text condition improves the performance by 3.7%, demonstrating the validity of using text condition to assist the implant position prediction. When combining both text condition and KAM, the improvement reaches 4.8%. As shown in the table's last three rows, both feature fusion operation and CAM improve AP value by 1.4% and 0.8%, respectively. When combining all these components, the improvement reaches 6.9%.

Comparison to the Mainstream Detectors. To demonstrate the superior performance of the proposed TCEIP, we compare the AP value with the mainstream detectors in Table 2. Only the anchor-free detector is used for comparison, due to the reason that no useful texture is available around the center of the implant. As the teeth are missing, the anchor-based detectors can not regress the implant position successfully. From the table we can observe that, the transformer-based methods perform better than the CNN-based networks (e.g., ImplantFormer achieved 13.7% AP, which is 1.6% higher than the best-performed anchor-free network - ATSS). The proposed TCEIP achieves the best AP value - 17.8%, among all benchmarks, which surpasses the ImplantFormer with a large gap. The experimental results proved the effectiveness of our method.

In Fig. 4, we choose two best-performed detectors from the CNN-based (e.g., ATSS) and transformer-based (e.g., ImplantFormer) methods for visual comparison, to further demonstrate the superiority of TCEIP in the implant position prediction. The first row of the figure is a patient with sparse teeth, and the second and third rows are a patient with two missing teeth. We can observe from the figure that both the ATSS and ImplantFormer generate false positive detection, except for the TCEIP. Moreover, the implant position predicted by the TCEIP is more accurate. These visual results demonstrated the effectiveness of using text condition to assist the implant position prediction.

4 Conclusions

In this paper, we introduce TCEIP, a text condition embedded implant position regression network, which integrate additional condition from the CLIP to guide the prediction of implant position. A cross-modal attention (CMA) and knowledge alignment module (KAM) is devised to facilitate the interaction between features in two modalities. Extensive experiments on a dental implant dataset through five-fold cross-validation demonstrated that the proposed TCEIP achieves superior performance than the existing methods.

Acknowledgments. This work was supported by the National Natural Science Foundation of China under Grant 82261138629; Guangdong Basic and Applied Basic Research Foundation under Grant 2023A1515010688 and 2021A1515220072; Shenzhen Municipal Science and Technology Innovation Council under Grant JCYJ2022053110141 2030 and JCYJ20220530155811025.

References

1. Dosovitskiy, A., et al.: An image is worth 16x16 words: transformers for image recognition at scale. arXiv preprint arXiv:2010.11929 (2020)
2. Elani, H., Starr, J., Da Silva, J., Gallucci, G.: Trends in dental implant use in the US, 1999–2016, and projections to 2026. J. Dent. Res. **97**(13), 1424–1430 (2018)

3. He, K., Zhang, X., Ren, S., Sun, J.: Deep residual learning for image recognition. In: Proceedings of the IEEE Conference on Computer Vision and Pattern Recognition, pp. 770–778 (2016)
4. Kurt Bayrakdar, S., et al.: A deep learning approach for dental implant planning in cone-beam computed tomography images. BMC Med. Imaging **21**(1), 86 (2021)
5. Law, H., Deng, J.: Cornernet: detecting objects as paired keypoints. In: Proceedings of the European Conference on Computer Vision (ECCV), pp. 734–750 (2018)
6. Lin, T.Y., Goyal, P., Girshick, R., He, K., Dollár, P.: Focal loss for dense object detection. In: Proceedings of the IEEE International Conference on Computer Vision, pp. 2980–2988 (2017)
7. Liu, Y., Chen, Z.C., Chu, C.H., Deng, F.L.: Transfer learning via artificial intelligence for guiding implant placement in the posterior mandible: an in vitro study (2021)
8. Nazir, M., Al-Ansari, A., Al-Khalifa, K., Alhareky, M., Gaffar, B., Almas, K.: Global prevalence of periodontal disease and lack of its surveillance. Sci. World J. 2020 (2020)
9. Radford, A., et al.: Learning transferable visual models from natural language supervision. In: International Conference on Machine Learning, pp. 8748–8763. PMLR (2021)
10. Rasheed, H., Maaz, M., Khattak, M.U., Khan, S., Khan, F.S.: Bridging the gap between object and image-level representations for open-vocabulary detection. arXiv preprint arXiv:2207.03482 (2022)
11. Widiasri, M., et al.: Dental-yolo: alveolar bone and mandibular canal detection on cone beam computed tomography images for dental implant planning. IEEE Access **10**, 101483–101494 (2022)
12. Xie, J., Hou, X., Ye, K., Shen, L.: Clims: cross language image matching for weakly supervised semantic segmentation. In: Proceedings of the IEEE/CVF Conference on Computer Vision and Pattern Recognition, pp. 4483–4492 (2022)
13. Yang, X., et al.: ImplantFormer: vision transformer based implant position regression using dental CBCT data. arXiv preprint arXiv:2210.16467 (2022)
14. Yang, Z., Liu, S., Hu, H., Wang, L., Lin, S.: RepPoints: point set representation for object detection. In: Proceedings of the IEEE/CVF International Conference on Computer Vision, pp. 9657–9666 (2019)
15. Zhang, H., Wang, Y., Dayoub, F., Sunderhauf, N.: VarifocalNet: an IoU-aware dense object detector. In: Proceedings of the IEEE/CVF Conference on Computer Vision and Pattern Recognition, pp. 8514–8523 (2021)
16. Zhang, S., Chi, C., Yao, Y., Lei, Z., Li, S.Z.: Bridging the gap between anchor-based and anchor-free detection via adaptive training sample selection. In: Proceedings of the IEEE/CVF Conference on Computer Vision and Pattern Recognition, pp. 9759–9768 (2020)
17. Zhou, X., Girdhar, R., Joulin, A., Krähenbühl, P., Misra, I.: Detecting twenty-thousand classes using image-level supervision. In: Avidan, S., Brostow, G., Cissé, M., Farinella, G.M., Hassner, T. (eds.) ECCV 2022. LNCS, vol. 1363, pp. 350–368. Springer, Cham (2022)
18. Zhou, X., Wang, D., Krähenbühl, P.: Objects as points. arXiv preprint arXiv:1904.07850 (2019)
19. Zhu, X., Su, W., Lu, L., Li, B., Wang, X., Dai, J.: Deformable DETR: deformable transformers for end-to-end object detection. arXiv preprint arXiv:2010.04159 (2020)

Vision Transformer Based Multi-class Lesion Detection in IVOCT

Zixuan Wang[1], Yifan Shao[2], Jingyi Sun[2], Zhili Huang[1], Su Wang[1(✉)], Qiyong Li[3], Jinsong Li[3], and Qian Yu[2(✉)]

[1] Sichuan University, Chengdu, China
hz1759156158@163.com
[2] Beihang University, Beijing, China
qianyu@buaa.edu.cn
[3] Sichuan Provincial People's Hospital, Chengdu, China

Abstract. Cardiovascular disease is a high-fatality illness. Intravascular Optical Coherence Tomography (IVOCT) technology can significantly assist in diagnosing and treating cardiovascular diseases. However, locating and classifying lesions from hundreds of IVOCT images is time-consuming and challenging, especially for junior physicians. An automatic lesion detection and classification model is desirable. To achieve this goal, in this work, we first collect an IVOCT dataset, including 2,988 images from 69 IVOCT data and 4,734 annotations of lesions spanning over three categories. Based on the newly-collected dataset, we propose a multi-class detection model based on Vision Transformer, called **G-Swin Transformer**. The essential part of our model is grid attention which is used to model relations among consecutive IVOCT images. Through extensive experiments, we show that the proposed G-Swin Transformer can effectively localize different types of lesions in IVOCT images, significantly outperforming baseline methods in all evaluation metrics. Our code is available via this link. https://github.com/Shao1Fan/G-Swin-Transformer

Keywords: IVOCT · Object Detection · Vision Transformer

1 Introduction

Despite the rapid development of new detection and treatment methods, the prevalence of cardiovascular disease continues to increase [1]. It is still reported to be the most prevalent and deadly disease worldwide, with more than 1 million people diagnosed with acute coronary syndrome (ACS) in the U.S. in 2016. The

Z. Wang and Y. Shao—Equal contribution.

Supplementary Information The online version contains supplementary material available at https://doi.org/10.1007/978-3-031-43987-2_32.

H. Greenspan et al. (Eds.): MICCAI 2023, LNCS 14225, pp. 327–336, 2023.
https://doi.org/10.1007/978-3-031-43987-2_32

average cost of hospital discharge for ACS patients is as high as $63,578 [2], which significantly increasing the financial burden on society and patients.

Optical coherence tomography (OCT) [3] is a new biomedical imaging technique born in the 19901990ss. Intravascular optical coherence tomography (IVOCT) [4] has a higher resolution compared with other imaging modalities in the vasculature and is considered to be the best imaging tool for plaque rupture, plaque erosion, and calcified nodules [5]. Therefore, most existing work on IVOCT images focuses on identifying vulnerable plaques in the vasculature [6–9], while neglecting other characteristic manifestations of atherosclerotic plaques in IVOCT images, such as macrophage infiltration and thrombus formation. These lesions are closely related to the development of plaque changes [10]. Studies have shown that atherosclerosis is an inflammatory disease dominated by macrophages and T lymphocytes, that a high density of macrophages usually represents a higher risk, and that thrombosis due to plaque rupture is a common cause of acute myocardial infarction [11,12]. In addition, some spontaneous coronary artery dissection (SCAD) can be detected in IVOCT images. The presence of the dissection predisposes to coronary occlusion, rupture, and even death [13,14]. These lesions are inextricably linked to ACS. All three types of features observed through IVOCT images are valuable for clinical treatment, as shown in Fig. 1. These lesions are inextricably linked to ACS and should be considered in clinical management.

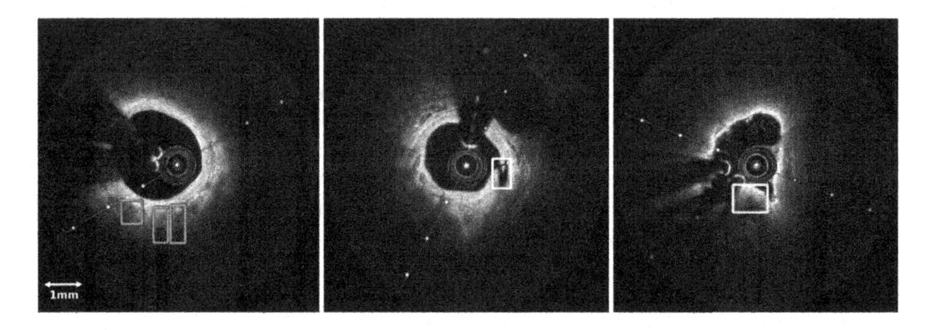

Fig. 1. Example images and annotations of our dataset. Each IVOCT data is converted to PNG images for annotation. The blue/green/red boxes represent bounding box of macrophages, cavities/dissections, thrombi, respectively. (Color figure online)

Achieving multi-class lesion detection in IVOCT images faces two challenges: 1) There is no public IVOCT dataset specifically designed for multi-class lesion detection. Most IVOCT datasets only focus on a single lesion, and research on the specific types of lesions in the cardiovascular system is still in its early stage. 2) It is difficult to distinguish between different lesions, even for senior radiologists. This is because these lesions vary in size and appearance within the same class, and some of them do not have regular form, as shown in Fig. 1. In clinical diagnosis, radiologists usually combine different pathological manifestations, lesion size, and the continuous range before and after in the IVOCT image to design accurate treatment strategies for patients. Unfortunately, most existing works ignore such information and do not consider the continuity of lesions in the 3D dimension. To address the above issues, we collaborated with the Cardiovascular Research Center of Sichuan Provincial People's Hospital to collect an IVOCT dataset and introduce a novel detection model that leverages the information from consecutive IVOCT images.

Overall, the contribution of this work can be summarized as follows: 1) We propose a new IVOCT dataset that is the first multi-class IVOCT dataset with bounding box annotations for macrophages, cavities/dissections, and thrombi. 2) We design a multi-class lesion detection model with a novel self-attention module that exploits the relationship between adjacent frames in IVOCT, resulting in improved performance. 3) We explore different data augmentation strategies for this task. 4) Through extensive experiments, we demonstrate the effectiveness of our proposed model.

2 Dataset

We collected and annotated a new IVOCT dataset consisting of 2,988 IVOCT images, including 2,811 macrophages, 812 cavities and dissections, and 1,111 thrombi. The collected data from 69 patients are divided into training/validation/test sets in a 55:7:7 ratio, respectively. Each split contains 2359/290/339 IVOCT frames. In this section, we will describe the data collection and annotation process in detail.

2.1 Data Collection

We collaborated with the Cardiovascular and Cerebrovascular Research Center of Sichuan Provincial People's Hospital, which provided us with IVOCT data collected between 2019 and 2022. The data include OCT examinations of primary patients and post-coronary stenting scenarios. Since DICOM is the most widely-used data format in medical image analysis, the collecting procedure was exported to DICOM, and the patient's name and other private information contained in DICOM were desensitized at the same time. Finally, the 69 DICOM format data were converted into PNG images with a size of 575 × 575 pixels. It is worth noting that the conversion from DICOM to PNG did not involve any downsampling operations to preserve as much information as possible.

2.2 Data Annotation

In order to label the lesions as accurately as possible, we designed a two-step annotation procedure. The first round was annotated by two expert physicians using the one-stop medical image labeling software *Pair*. Annotations of the two physicians may be different. Therefore, we asked them to discuss and reach agreement on each annotation. Next, the annotated data was sent to senior doctors to review. The review starts with one physician handling the labeling, including labeling error correction, labeling range modification, and adding missing labels. After that, another physician would continue to check and review the previous round's results to complete the final labeling. Through the above two steps, 2,988 IVOCT images with 4,734 valid annotations are collected.

3 Methodology

Recently, object detection models based on Vision Transformers have achieved state-of-the-art (SOTA) results on various object detection datasets, such as the MS-COCO dataset. Among them, the Swin Transformer [19] model is one of the best-performing models. Swin Transformer uses a self-attention mechanism within local windows to ensure computational efficiency. Moreover, its sliding window mechanism allows for global modeling by enabling self-attention computation between adjacent windows. Its hierarchical structure allows flexible modeling of information at different scales and is suitable for various downstream tasks, such as object detection.

3.1 G-Swin Transformer

In traditional object detection datasets such as the MS-COCO dataset, the images are typically isolated from each other without any correlation. However, in our proposed IVOCT dataset, each IVOCT scan contains around 370 frames with a strong inter-frame correlation. Specifically, for example, if a macrophage lesion is detected at the $[x, y, w, h]$ position in frame F_i of a certain IVOCT scan, it is highly likely that there is also a macrophage lesion near the $[x, y, w, h]$ position in frame F_{i-1} or F_{i+1}, due to the imaging and pathogenesis principles of IVOCT and ACS. Doctors also rely on the adjacent frames for diagnosis rather than a single frame when interpreting IVOCT scans. But, the design of the Swin-Transformer did not consider the utilization of inter-frame information. Though global modeling is enabled by using the sliding window mechanism. In the temporal dimension, it still has a locality because the model did not see adjacent frames.

Based on the Swin Transformer, we propose a backbone called G-Swin Transformer. Our proposed G-Swin Transformer is used as the basic module of the encoder in the full model, which is developed based on Faster R-CNN. The overall structure of the model is shown in Fig. 2. The model input consists of k 3-channel RGB images, and the input dimension is $[k * B, 3, H, W]$, where k indicates the number of frames that used in an iteration. After passing through Patch Partition and Linear Embedding layers, k feature maps belonging to frame F_0, F_1, ... F_{k-1}, respectively, are obtained, each with a size of H/4 * W/4 * C. These feature maps are then input to the G-Swin Transformer, where they go through 4 layers and a total of 12 Transformer blocks. Between each layer, a patch merging layer is used to reduce resolution, and model features of different dimensions. The output feature maps at different scales are then passed to a

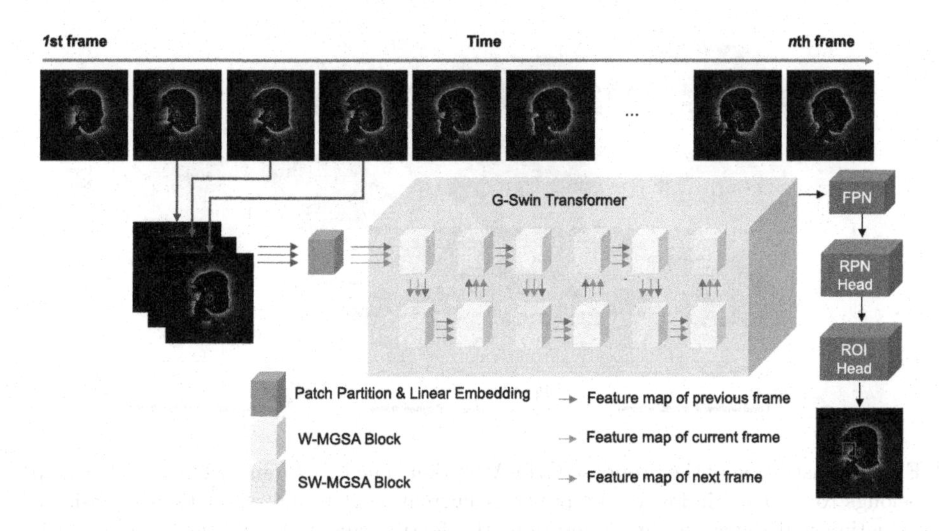

Fig. 2. The overall model structure. The proposed G-Swin Transformer is used as backbone network. The detection head follows Faster-RCNN's head. *W-MGSA* and *SW* refer to Window-Multihead Grid Self Attention and Shifted Window, respectively.

feature pyramid network (FPN) for fusion of features at different resolutions. The RPN Head is then applied to obtain candidate boxes, and finally, the ROI Head is used for classification and refinement of candidate boxes to obtain class and bbox (bounding box) predictions. The inter-frame feature fusing is happend in the attention block, introduced in the next subsection.

3.2 Grid Attention

To better utilize information from previous and future frames and perform feature fusion, we propose a self-attention calculation mode called "Grid Attention". The structure shown in Fig. 3 is an application of Grid Attention. The input of the block is 3 feature maps respectively from frames 0, 1, and 2. (Here we use $k = 3$.) Before entering the W-MSA module for multi-head self-attention calculation, the feature maps from different frames are fused together.

Based on the feature map of the key frame (orange color), the feature maps of the previous (blue) and next (green) frames first do a dimensional reduction from $[H, W, C]$ to $[H, W, C/2]$. Then they are down-sampled and a grid-like feature map are reserved. The grid-like feature map are then added to key-frame feature map, and the fusion progress finishes. In the W-MSA module, the self-attention within the local window and that between adjacent local windows are calculated, and the inter-frame information is fully used. The local window of key-frame has contained information from other frames, and self-attention calculation happens in inter-frames. The frame-level feature modeling can thus be achieved, simulating the way that doctors view IVOCT by combining information from previous and next frames.

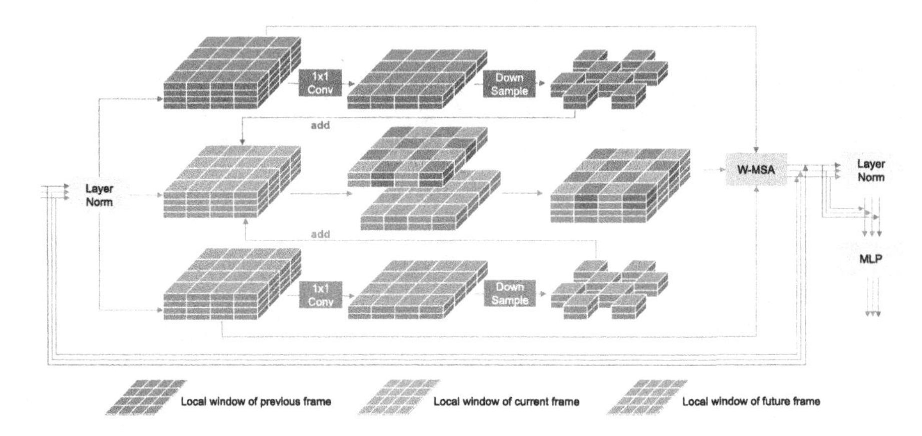

Fig. 3. Illustration of the proposed Grid Attention. The blue/orange/green feature map belongs to a local window of the previous/current/next frame. After the dimensional reduction and downsampling operation, the feature maps of previous/next frame is added to the current frame's feature map. (Color figure online)

During feature fusion with Grid Attention, the feature maps from different frames are fused together in a grid-like pattern (as shown in the figure). The purpose of this is to ensure that when dividing windows, half of the grid cells within a window come from the current frame, and the other half come from other frames. If the number of channels in the feature map is C, and the number of frames being fused is 3 (current frame + previous frame + next frame), then the first C/2 channels will be fused between the current frame and the previous frame, and the last C/2 channels will be fused between the current frame and the next frame. Therefore, the final feature map consists of 1/4 of the previous frame, 1/2 of the current frame, and 1/4 of the next frame. The impact of the current frame on the new feature map remains the largest, as the current frame is the most critical frame.

4 Experiments

Baseline Methods and Evaluation Metrics. The baseline is based on a PyTorch implementation of the open-source object detection toolbox MMDetection. We compare our proposed approach with Swin Transformer and four CNN-based network models including Faster-RCNN [15], YOLOv3 [16], YOLOv5 [17], Retinanet [18]. All the baseline model is pre-trained on the ImageNet dataset.

To ensure objective comparison, all experiments were conducted in the MMdetection framework. The metric we used is the AP/AR for each lesion and the mAP, based on the COCO metric and the COCO API (the default evaluation method in the MMdetection framework). We trained the model for 60 epochs with an AdamW optimizer following Swin Transformer. The learning

Table 1. Comparison of our proposed method and baseline methods.

	AP_{50}			mAP	$Recall_{50}$		
Method	Macrophage	cavities/ dissection	thrombus		Macrophage	cavities/ dissection	thrombus
Faster-RCNN	27.34	44.32	31.86	34.51	74.65	76.04	70.60
YOLOv3	20.25	35.42	33.17	29.61	67.37	75.52	61.82
YOLOv5	27.34	40.63	46.93	38.30	79.31	82.67	83.86
RetinaNet	25.86	38.93	30.17	31.65	84.48	88.54	73.03
Swin-Transformer	27.91	44.94	48.87	40.57	89.11	89.06	92.85
G-Swin-Transformer (Ours)	**30.55**	**52.25**	**51.92**	**44.91**	**91.49**	**89.58**	**95.45**

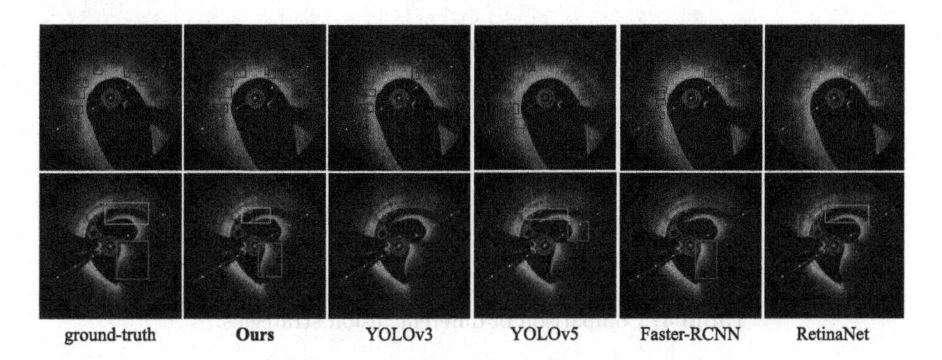

Fig. 4. Visualization results. From left to right are ground-truth, results of our model, Faster-RCNN, YOLOv3, YOLOv5 and RetinaNet. Our model achieves better results.

rate and weight decay is set to be 1e-4 and 1e-2, respectively. The batch size is set to be 2.

Quantitative and Qualitative Results. All experiments are conducted on our newly-collected dataset. Each model is trained on the training set, selected based on the performance of the validation set, and the reported results are obtained on the test set. Table 1 shows the comparison between the baseline methods and the G-Swin Transformer method. Methods based on Swin Transformer outperformed the four baseline methods in terms of precision and recall, and our proposed G-Swin Transformer outperforms the baseline method Swin Transformer by 2.15% in mAP.

Figure 4 compares some results of our method and baselines. The first row is the detection of the macrophage. Our method's prediction is the most closed to the ground truth. The second row is the detection of cavities/dissections and thrombi. Only our method gets the right prediction. The YOLOv3, YOLOv5, Faster-RCNN and RetinaNet model failed to detect all lesions, while RetinaNet model even produced some false positive lesions.

Table 2. Results of using different data augmentation strategies.

Augmentation Strategy					Metric	
Random Resize	Random Crop	Random Flip	Random Brightness	Random Contrast	mAP	AR
✗	✗	✓	✓	✓	33.13	79.12
✗	✓	✓	✓	✓	35.64	82.31
✓	✗	✓	✓	✓	38.52	85.81
✓	✓	✓	✗	✗	41.31	85.69
✓	✓	✓	✗	✓	41.92	85.31
✓	✓	✓	✓	✗	42.39	86.17
✓	✓	✗	✓	✓	43.34	91.41
✓	✓	✓	✓	✓	**44.91**	**92.18**

Table 3. Effect of different hyper-parameters in Grid Attention.

Fusion Layers	Fusion Strategy	mAP	AR
5	replace	41.69	89.28
5	add	40.31	90.52
3	replace	44.39	88.86
3	**add**	**44.91**	**92.18**

Table 4. Comparison of different fusion strategies.

Fusion methods	mAP	AR
No fusion	40.57	90.34
2.5D	38.25	78.44
Weighted sum	40.64	83.54
Ours	**44.91**	**92.18**

Effect of Different Data Augmentation Methods. We compared the impact of different data augmentation strategies in our task. As shown in Table 2, *Random Resize* and *Random Crop* had a significant impact on performance improvement. *Resize* had the greatest impact on the model's performance because different-sized OCT images were generated after data augmentation, and the lesions were also enlarged or reduced proportionally. Since the sizes of lesions in different images are usually different, different-sized lesions produced through data augmentation are advantageous for the model to utilize multi-scale features for learning.

Effect of Different Hyper-parameters. Table 3 shows the impact of hyper-parameters on the performance of the G-Swin Transformer model. The best mAP was achieved when using a 3-layer image input. Using the upper and lower 5 layers of image input not only increased the training/inference time, but also may not provide more valuable information since frame 0 and frame 4 are too far away from the key frame. The fusion strategy indicates how the feature map from other frames are combined with the key-frame feature map. We can find add them up gets better result then simply replacement. We think this is because by this way, the 1×1 convolutional layer can learn a residual weights, keeps more detail of the key-frame.

Effect of Fusion Methods. In addition to Grid Attention, there are other methods of feature fusion. The first method is like 2.5D convolution, in which multiple frames of images are mapped into 96-dimensional feature maps directly through convolution in the Linear Embedding layer. This method is the simplest, but since the features are fused only once at the initial stage of the network, the use of adjacent frame features is very limited. The second method is to weight and sum the feature maps of different frames before each Attention Block, giving higher weight to the current frame and lower weight to the reference frames. Table 4 shows the impact of other feature fusion methods on performance. Our method gets better mAP and AR.

5 Conclusion

In this work, we have presented the first multi-class lesion detection dataset of IVOCT scans. We have also proposed a Vision Transformer-based model, called G-Swin Transformer, which uses adjacent frames as input and leverages the temporary dimensional information inherent in IVOCT data. Our method outperforms traditional detection models in terms of accuracy. Clinical evaluation shows that our model's predictions provide significant value in assisting the diagnosis of acute coronary syndrome (ACS).

Acknowledgement. This work is supported by the National Key Research and Development Project of China (No. 2022ZD0117801).

References

1. Murphy, S., Xu, J., Kochanek, K., Arias, E., Tejada-Vera, B.: Deaths: final data for 2018 (2021)
2. Virani, S., et al.: Heart disease and stroke statistics-2021 update: a report from the American heart association. Circulation. **143**, e254–e743 (2021)
3. Huang, D., et al.: Optical coherence tomography. Science **254**, 1178–1181 (1991)
4. Bezerra, H., Costa, M., Guagliumi, G., Rollins, A., Simon, D.: Intracoronary optical coherence tomography: a comprehensive review: clinical and research applications. JACC: Cardiovas. Interv. **2**, 1035–1046 (2009)
5. Jia, H., et al.: In vivo diagnosis of plaque erosion and calcified nodule in patients with acute coronary syndrome by intravascular optical coherence tomography. J. Am. Coll. Cardiol. **62**, 1748–1758 (2013)
6. Li, C., et al.: Comprehensive assessment of coronary calcification in intravascular OCT using a spatial-temporal encoder-decoder network. IEEE Trans. Med. Imaging **41**, 857–868 (2021)

7. Liu, X., Du, J., Yang, J., Xiong, P., Liu, J., Lin, F.: Coronary artery fibrous plaque detection based on multi-scale convolutional neural networks. J. Signal Process. Syst. **92**, 325–333 (2020)
8. Gessert, N., et al.: Automatic plaque detection in IVOCT pullbacks using convolutional neural networks. IEEE Trans. Med. Imaging **38**, 426–434 (2018)
9. Cao, X., Zheng, J., Liu, Z., Jiang, P., Gao, D., Ma, R.: Improved U-net for plaque segmentation of intracoronary optical coherence tomography images. In: Farkaš, I., Masulli, P., Otte, S., Wermter, S. (eds.) ICANN 2021. LNCS, vol. 12893, pp. 598–609. Springer, Cham (2021). https://doi.org/10.1007/978-3-030-86365-4_48
10. Regar, E., Ligthart, J., Bruining, N., Soest, G.: The diagnostic value of intracoronary optical coherence tomography. Herz: Kardiovaskulaere Erkraenkungen **36**, 417–429 (2011)
11. Kubo, T., Xu, C., Wang, Z., Ditzhuijzen, N., Bezerra, H.: Plaque and thrombus evaluation by optical coherence tomography. Int. J. Cardiovasc. Imaging **27**, 289–298 (2011)
12. Falk, E., Nakano, M., Bentzon, J., Finn, A., Virmani, R.: Update on acute coronary syndromes: the pathologists' view. Eur. Heart J. **34**, 719–728 (2013)
13. Saw, J.: Spontaneous coronary artery dissection. Can. J. Cardiol. **29**, 1027–1033 (2013)
14. Pepe, A., et al.: Detection, segmentation, simulation and visualization of aortic dissections: a review. Med. Image Anal. **65**, 101773 (2020)
15. Ren, S., He, K., Girshick, R., Sun, J.: Faster R-CNN: towards real-time object detection with region proposal networks. In: Advances In Neural Information Processing Systems, vol. 28 (2015)
16. Redmon, J., Farhadi, A.: Yolov3: an incremental improvement. ArXiv Preprint ArXiv:1804.02767 (2018)
17. Jocher, G.: YOLOv5 by ultralytics (2020). https://github.com/ultralytics/yolov5
18. Lin, T., Goyal, P., Girshick, R., He, K., Dollár, P.: Focal loss for dense object detection. In: Proceedings of the IEEE International Conference On Computer Vision, pp. 2980–2988 (2017)
19. Liu, Z., Lin, Y., Cao, Y., Hu, H., Wei, Y., Zhang, Z., Lin, S., Guo, B.: Swin transformer: Hierarchical vision transformer using shifted windows. Proceedings of the IEEE/CVF International Conference On Computer Vision, pp. 10012–10022 (2021)

Accurate and Robust Patient Height and Weight Estimation in Clinical Imaging Using a Depth Camera

Birgi Tamersoy[1](✉), Felix Alexandru Pîrvan[2], Santosh Pai[3], and Ankur Kapoor[3]

[1] Digital Technology and Innovation, Siemens Healthineers, Erlangen, Germany
birgi.tamersoy@siemens-healthineers.com
[2] Siemens S.R.L. Romania, Bucuresti, Romania
felix.pirvan@siemens.com
[3] Digital Technology and Innovation, Siemens Healthineers, Princeton, NJ, USA
{santosh.pai,ankur.kapoor}@siemens-healthineers.com

Abstract. Accurate and robust estimation of the patient's height and weight is essential for many clinical imaging workflows. Patient's safety, as well as a number of scan optimizations, rely on this information. In this paper we present a deep-learning based method for estimating the patient's height and weight in unrestricted clinical environments using depth images from a 3-dimensional camera. We train and validate our method on a very large dataset of more than 1850 volunteers and/or patients captured in more than 7500 clinical workflows. Our method achieves a PH5 of 98.4% and a PH15 of 99.9% for height estimation, and a PW10 of 95.6% and a PW20 of 99.8% for weight estimation, making the proposed method state-of-the-art in clinical setting.

Keywords: Height Estimation · Weight Estimation · Clinical Workflow Optimization

1 Introduction

Many clinical imaging workflows require the patient's height and weight to be estimated in the beginning of the workflow. This information is essential for patient's safety and scan optimizations across modalities and workflows. It is used for accurate prediction of the Specific Absorption Rate (SAR) in Magnetic Resonance Imaging (MRI), contrast dose calculations in Computed Tomography (CT), and drug dose computations in Emergency Room (ER) workflows.

Contrary to its importance, there are no widely established methods for estimating the patient's height and weight. Measuring these values using an actual scale is not a common clinical practice since: 1) a measurement scale is not available in every scan room, 2) manual measurements add an overhead to the clinical workflow, and 3) manual measurements may not be feasibly for some

H. Greenspan et al. (Eds.): MICCAI 2023, LNCS 14225, pp. 337–346, 2023.
https://doi.org/10.1007/978-3-031-43987-2_33

patients with limited mobility. Alternative methods such as the Lorenz formulae [1] or the Crandall formulae [2] need additional body measurements (e.g. mid-arm circumference, waist circumference and/or hip circumference) and are neither very accurate nor simple. Consequently, clinical staff usually relies either on previously recorded patient information or their own experience in estimating the patient's height and weight, where the estimated values may significantly deviate from the actual values in both cases.

In this paper we present a deep-learning based method for accurately and robustly estimating the patient's height and weight in challenging *and* unrestricted clinical environments using depth images from a 3-dimensional (3D) camera. We aim to cover the patient demographics in common diagnostic imaging workflows. Our method is trained and validated on a very large dataset of more than 1850 volunteers and/or patients, captured in more than 7500 clinical scenarios, and consists of nearly 170k depth images. We achieve a PH5 (**p**ercentage of the **h**eight estimates within **5**% error) of 98.4% and a PH15 of 99.9% for height estimation, and a PW10 (**p**ercentage of the **w**eight estimates withing **10**% error) of 95.6% and a PW20 of 99.8% for weight estimation, making the proposed method state-of-the-art in clinical setting.

In addition to the clinical significance, our method has the following primary technical novelties: 1) we formulate the problem as an end-to-end single-value regression problem given only depth images as input (i.e. no error-prone intermediate stages such as volume computations), 2) we present a multi-stage training approach to ensure robustness in training (i.e. no need for hyper-parameter tunings at any stage), and 3) we evaluate our method on a very large dataset of both volunteers and patients, using 23-fold cross validation to ensure field generalization.

2 Related Work

A large number of previous methods have been proposed and independently evaluated for patient height and weight estimation. Since patient height estimation is considered to be an easier problem, the primary focus of the previous work has been on patient weight estimation.

Most of the existing work in patient weight estimation are formulae-based approaches where one or more anthropometric measurements are used for estimating the patient's weight. The Mercy method uses the humeral length and the mid-arm circumference (MAC) for estimating the paediatric body weight [4]. PAWPER XL-MAC is another height-based (in combination with MAC) method for estimating the body weight in paediatric patients [3]. Broca index [6] and Kokong formula [8] do not take into account a person's body habitus and estimates the "ideal weight" using only the body height information. Buckley method [7], Lorenz formulae [1], Crandall formulae [2] provide gender-specific weight estimation formulae given some anthropometric measurements such as the abdominal circumference, tight circumference, and MAC.

Formulae-based approaches are independently evaluated in numerous studies, both for paediatric patients [3] and adult patients [9,10]. A common conclusion of these studies is that the formulae-based methods usually perform poorly in the clinical setting with PW10 values below 70%.

More recently several methods have been proposed that leverage 3D camera input for estimating the patient's weight. In [11] an RGB-D camera and a thermal camera are used for precisely segmenting the patients and then extracting volume based features. These features are then fed into an artificial neural network (ANN) for patient weight estimation. In [12], first a 3D patient avatar is fitted to the acquired depth images, which is then used for part-volume based weight estimation. Both of these approaches require a number of additional algorithmic steps and the challenges of the clinical setup (such as heavy occlusions due to covers and/or additional devices like coils during an MRI examination) may affect the accuracy of the results.

Estimation of the patient weight by the clinical staff remains to be the most common approach in the clinical workflow. In [15], the performance of clinical staff is determined as PW10 of 78% for nurses and PW10 of 59% for physicians. In [16] the performance of the clinical staff is determined as PW10 of 66% for both nurses and physicians.

As a clinical acceptance criteria, Wells et al. [3] proposes a minimum accuracy for patient weight estimation as PW10 greater than 70% and PW20 greater than 95%.

3 Approach

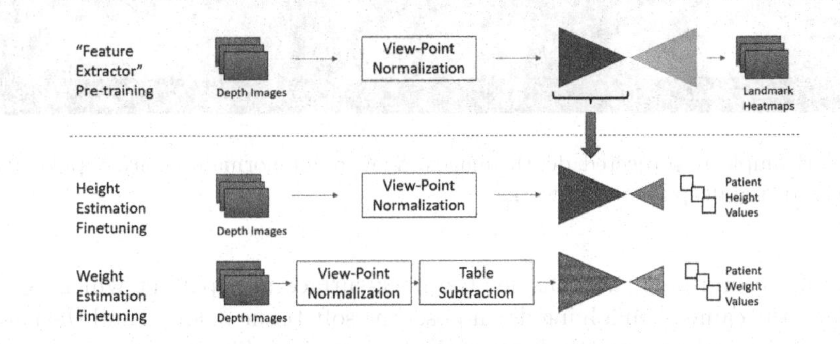

Fig. 1. Overview of the proposed method.

Overview of the proposed method is illustrated in Fig. 1. Our method takes in "normalized" depth images as input. This normalization covers two aspects: 1) normalization with respect to the view-point of the depth camera, and 2) normalization with respect to the variations in the patient tables. Input normalized depth images are then fed into a common feature extraction encoder network.

This encoder network is trained using landmark localization as an auxiliary task. In the last stage we train and utilize two separate single-value regression decoder networks for estimating the patient's height and weight. These steps are explained in-detail in the following sub-sections.

3.1 Obtaining "Normalized" Depth Images

When training deep neural networks it is more data efficient to eliminate as much of the foreseen variances in the problem as possible in the preprocessing steps. Conceptually this can be thought as reducing the "dimensionality" of the problem before the model training even starts.

Camera view-point is a good example when model inputs are images. With a known system calibration, a "virtual camera" may be placed in a consistent place in the scene (e.g. with respect to the patient table) and the "re-projected" depth images from this virtual camera may be used instead of the original depth images. This way the network training does not need to learn an invariance to the camera view-point, since this variance will be eliminated in the preprocessing. This process forms the first step in our depth image normalization. Figure 2 presents some examples.

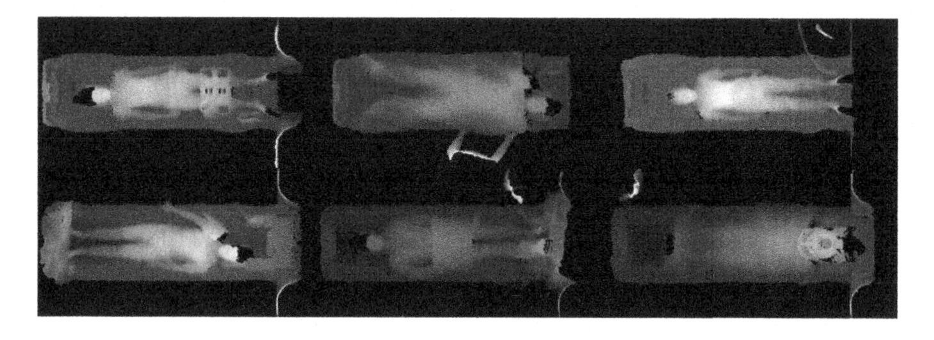

Fig. 2. Example re-projected depth images. View-point normalization simplifies the problem for the deep neural networks.

In the second step, we consider the back-surface of a patient which is not visible to the camera. In a lying down pose, the soft-tissue deforms and the back-surface of the patient takes the form of the table surface. Since there are a variety of patient tables (curved or flat) we eliminate this variance by performing a "table subtraction" from the view-point normalized depth images. This is especially important for accurate patient weight estimation across different systems.

For table subtraction, top surfaces extracted from 3D models of the corresponding patient tables are used. For a given input image, it is assumed that the corresponding patient table is known since this information is readily available in an integrated system. Even though we leveraged the actual 3D models of the patient tables, since the proposed approach only requires the top surface, this information may also be obtained during calibration by taking a depth image of the empty patient table. Figure 3 presents some examples of the table subtraction process.

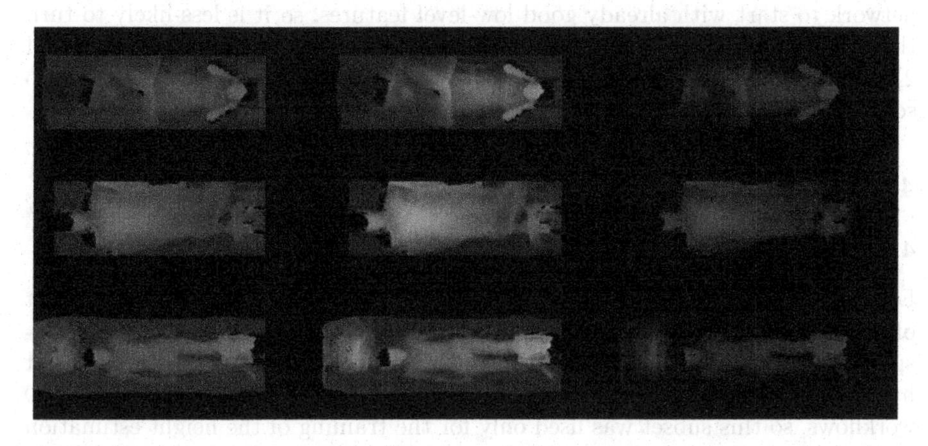

Fig. 3. "Table subtraction" for patient back-surface consistency. Left - original inputs. Center - original inputs overlayed with automatically aligned patient tables. Right - table subtracted depth images.

3.2 Learning Accurate Low-Level Features

Single-value regression problems require more attention during the training since the limited feedback provided through the loss function may result in the collapse of some of the features in the lower levels of the network, especially if a larger model is being trained.

In order to ensure that the learned low-level features are high-quality, we start with the training of a standard image-in image-out encoder-decoder network using the landmark localization as an auxiliary task. With this task, the network is expected to both capture local features (for getting precise landmark locations) *and* holistic features (for getting globally consistent landmark locations) better than the case where the network was to asked only to regress a single-value such as the height or the weight of the patient.

Once the encoder-decoder network is trained, we disregard the decoder part and use the encoder part as the pre-trained feature-extractor.

3.3 Task-Specific Decoders for Height and Weight Estimation

The final stage of our approach is the training of two separate single-value regression networks, one for the estimation of the patient's height and the other for the weight.

In this stage, we attach untrained decoder heads to the pre-trained encoder bases. During the training we allow *all* parameters of the model, including the pre-trained encoder, to be fine-tuned. The motivation for a complete fine-tuning comes from two primary reasons: 1) pre-training of the encoder part allows the network to start with already good low-level features, so it is less-likely to turn these good low-level features to degenerate features during the fine-tuning, and 2) by still allowing changes in the low-level features we have the potential to squeeze further performance from the networks.

4 Experiments

4.1 Dataset

For training and validation of our method we have collected a very large dataset of 1899 volunteers and/or patients captured in 7620 clinical workflows, corresponding to nearly 170k depth images. Within this large dataset, we did not have the patient table information for 909 patients and the corresponding 909 workflows, so this subset was used only for the training of the height estimation network. Some example depth snapshots of this extensive dataset is provided in Fig. 2.

This dataset is collected from multiple sites in multiple countries, over a span of several years. The target clinical workflows such as a variety of coils, a variety of positioning equipment, unrestricted covers (light and heavy blankets), and occlusions by technicians, are covered in this dataset. Due to volunteer and patient consents, this dataset cannot be made publicly available.

We consider the following inclusion criteria for the training and validation: patient weight between 45 kg to 120 kg, patient height between 140 cm to 200 cm, and patient body mass index (BMI) between 18.5 to 34.9. The distribution of the samples in our dataset, together with the above inclusion criteria, is illustrated in Fig. 4.

4.2 Training

Training of the feature extraction network is performed using a subset of nearly 2000 workflows. For these workflows, we also acquired the corresponding full-body 3D medical volumes (MRI acquisitions). A set of 10 anatomical landmarks corresponding to major joints (i.e. knees, elbows, shoulders, ankles, and wrists) are manually annotated by a group of experts in these 3D medical volumes. These 3D annotations are transferred to depth image coordinates for training using the known calibration information.

We use a modified version of ResNet [13] as our base feature extraction network. This is a smaller version compared to the originally proposed ResNet18, where the number of features in each block is kept constant at 32. We use only bottleneck blocks instead of the basic blocks used for the original ResNet18.

Training of the feature extraction network is done using the ADAM optimizer [14] with default parameters. Landmark locations are represented as 2D

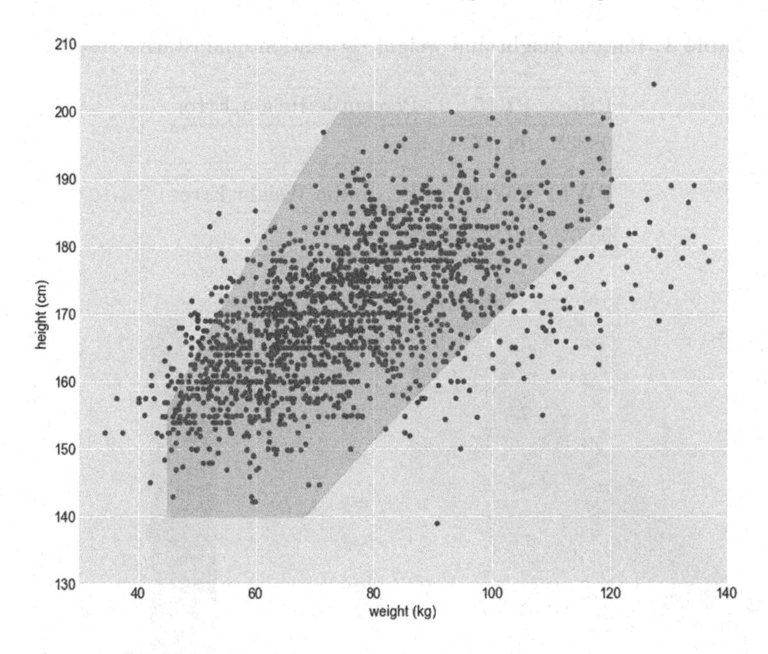

Fig. 4. Dataset distribution. Inclusion criteria is illustrated by the yellow region. (Color figure online)

heatmaps and mean squared error (MSE) loss is used. The feature extraction network is trained for 30 epochs with a patience of 10 epochs.

Once the feature extraction network is trained, we retain the encoder part and attach task-specific heads to form two separate networks, one for patient height estimation and the other one for patient weight estimation. Similar to the feature extraction network, we also train these networks using the ADAM optimizer with default parameters. As the loss function we use the symmetric mean absolute percentage error (SMAPE):

$$\text{SMAPE} = \frac{100}{n} \sum_{t=1}^{n} \frac{|P_t - A_t|}{(|P_t| + |P_t|)/2} \tag{1}$$

where A_t is the actual value and P_t is the predicted value. We train for 250 epochs with a relatively large patience of 50 epochs.

For height estimation we omitted the second step of depth normalization through table subtraction since the patient back-surface does not affect the height estimation significantly.

4.3 Results

We evaluate our model using 23-fold cross-validation. Since we have multiple workflows and depth images corresponding to the same volunteer or patient (e.g. same volunteer captured both in a "knee-scan" acquisition and a "hip-scan"

Table 1. Patient height and weight estimation quantitative results.

PH5	PH15	95-Percentile Height Error
98.4%	99.9%	3.4%

PW10	PW20	95-Percentile Weight Error
95.6%	99.8%	9.6%

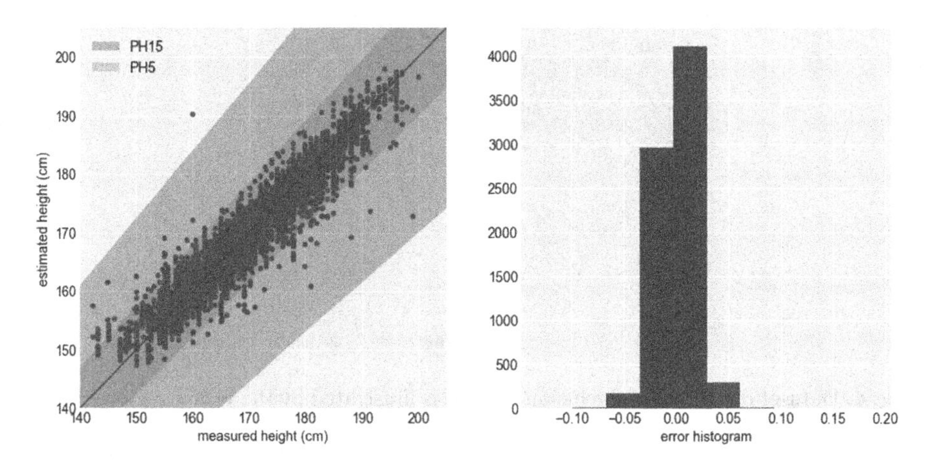

Fig. 5. Scatter plot and the error histogram for height estimation

acquisition), we ensure the splits are done based on the volunteer or patient identities. If a particular workflow contains more than one depth image, we apply a median filter on the frame-by-frame results to determine the final estimates for that workflow.

Table 1 provides our quantitative results. Our method achieved a PH5 of 98.4% and a PH15 of 99.9% for height estimation, and a PW10 of 95.6% and a PW20 of 99.8% for weight estimation, making the proposed method state-of-the-art in clinical setting. Figures 5 and 6 show the estimation scatter plots and the corresponding error histograms for height and weight estimation, respectively.

We also investigated the performance of our method for weight estimation for patient BMI groups outside our inclusion criteria. For patients with BMI $<$ 18.5, our method achieved a PW10 of 89.2% and a PW20 of 98.9%. For patients with BMI $>$ 34.9, our method achieved a PW10 of 96.2% and a PW20 of 99.8%. Even though the performance drops a little bit for underweight population, the main reason for keeping these populations outside the inclusion criteria is not the performance, but rather the limited support in the training dataset.

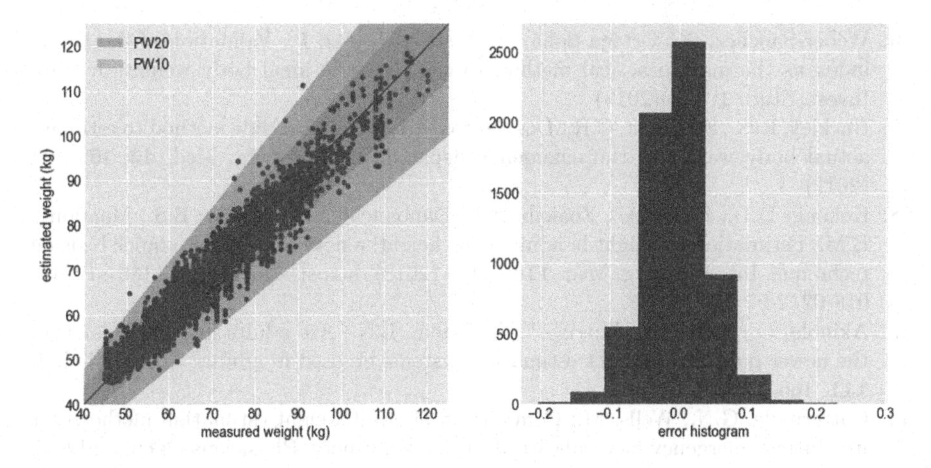

Fig. 6. Scatter plot and the error histogram for weight estimation

5 Conclusions

Accurate and rapid estimation of the patient's height and weight in clinical imaging workflows is essential for both the patient's safety and the possible patient-specific optimizations of the acquisition. In this paper, we present a deep-learning based method trained on a very large dataset of volunteer and patient depth images captured in unrestricted clinical workflows. Our method achieves a PH5 of 98.4% and a PH15 of 99.9% for height estimation, and a PW10 of 95.6% and a PW20 of 99.8% for weight estimation. These results out-perform all alternative methods to the best of our knowledge, including family estimates and clinical staff estimates.

Disclaimer. The concepts and information presented in this paper are based on research results that are not commercially available. Future availability cannot be guaranteed.

References

1. Lorenz, M.W., Graf, M., Henke, C., et al.: Anthropometric approximation of body weight in unresponsive stroke patients. J. Neurol. Neurosurg. Psychiatry **78**, 1331–1336 (2007)
2. Crandall, C.S., Gardner, S., Braude, D.A.: Estimation of total body weight in obese patients. Air Med. J. **28**, 139–145 (2007)
3. Wells, M., Goldstein, L.N., Bentley, A.: The accuracy of emergency weight estimation systems in children - a systematic review and meta-analysis. Int J. Emerg. Med. **10**, 1–43 (2017)
4. Abdel-Rahman, S.M., Ridge, A.L.: An improved pediatric weight estimation strategy. Open Medical Devices J. **4**, 87–97 (2012)
5. Wells, M.: A validation of the PAWPER XL-MAC tape for total body weight estimation in preschool children from low- and middle-income countries. PLoS One. **14**, e0210332 (2019)

6. Weber-Sanchez, A., Ortega Sofia, V., Weber-Alvarez, P.: Validation of the Broca index as the most practical method to calculate the ideal body weight. J. Clin. Invest. Stud. **1**, 1–4 (2018)
7. Buckley, R.G., Stehman, C.R., Dos Santos, F.L., et al.: Bedside method to estimate actual body weight in the emergency department. J. Emerg. Med. **42**, 100–104 (2012)
8. Kokong, D.D., Pam, I.C., Zoakah, A.I., Danbauchi, S.S., Mador, E.S., Mandong, B.M.: Estimation of weight in adults from height: a novel option for a quick bedside technique. Int. J. Emerg. Med. **11**(1), 1–9 (2018). https://doi.org/10.1186/s12245-018-0212-9
9. Akinola, O., Wells, M., Parris, P., Goldstein, L.N.: Are adults just big kids? Can the newer paediatric weight estimation systems be used in adults? S. Afr. Med. J. **111**, 166–170 (2021)
10. Cattermole, G.N., Wells, M.: Comparison of adult weight estimation methods for use during emergency medical care. J. Am. Coll. Emerg. Physicians Open **2**, e12515 (2021)
11. Pfitzner, C., May, S., Nüchter, A.: Body weight estimation for dose-finding and health monitoring of lying, standing and walking patients based on RGB-D data. Sensors **18**, 1311 (2018)
12. Dane, B., Singh, V., Nazarian, M., O'Donnell, T., Liu, S., Kapoor, A., Megibow, A.: Prediction of patient height and weight with a 3-dimensional camera. J. Comput. Assist. Tomogr. **45**, 427–430 (2021)
13. He, K., Zhang, X., Ren, S., Sun, J.: Deep residual learning for image recognition. In: IEEE Conference on Computer Vision and Pattern Recognition (CVPR), pp. 770–778 (2016)
14. Kingma, D.P., Ba, J.: Adam: a method for stochastic optimization. In: International Conference on Learning Representations (ICLR) (2015)
15. Menon, S., Kelly, A.M.: How accurate is weight estimation in the emergency department? Emerg. Med. Australas. **17**(2), 113–116 (2005)
16. Fernandes, C.M., Clark, S., Price, A., Innes, G.: How accurately do we estimate patients' weight in emergency departments? Can. Fam. Physician Medecin Famille Can. **45**, 2373 (1999)

AR^{2}T: Advanced Realistic Rendering Technique for Biomedical Volumes

Elena Denisova[1,3,4(✉)], Leonardo Manetti[3,4], Leonardo Bocchi[1,4], and Ernesto Iadanza[2]

[1] Department of Information Engineering, University of Florence, 50139 Florence, Italy
elena.denisova@unifi.it
[2] Department of Medical Biotechnologies, University of Siena, 53100 Siena, Italy
[3] Imaginalis S.r.l., 50019 Sesto Fiorentino, Italy
[4] Eidolab, Florence, Italy

Abstract. Three-dimensional (3D) rendering of biomedical volumes can be used to illustrate the diagnosis to patients, train inexperienced clinicians, or facilitate surgery planning for experts. The most realistic visualization can be achieved by the Monte-Carlo path tracing (MCPT) rendering technique which is based on the physical transport of light. However, this technique applied to biomedical volumes has received relatively little attention, because, naively implemented, it does not allow to interact with the data. In this paper, we present our application of MCPT to the biomedical volume rendering–Advanced Realistic Rendering Technique (AR^{2}T), in an attempt to achieve more realism and increase the level of detail in data representation. The main result of our research is a practical framework that includes different visualization techniques: iso-surface rendering, direct volume rendering (DVR) combined with local and global illumination, maximum intensity projection (MIP), and AR^{2}T. The framework allows interaction with the data in high quality for the deterministic algorithms, and in low quality for the stochastic AR^{2}T. A high-quality AR^{2}T image can be generated on user request; the quality improves in real-time, and the process is stopped automatically on the algorithm convergence, or by user, when the desired quality is achieved. The framework enables direct comparison of different rendering algorithms, i.e., utilizing the same view/light position and transfer functions. It therefore can be used by medical experts for immediate one-to-one visual comparison between different data representations in order to collect feedback about the usefulness of the realistic 3D visualization in clinical environment.

Keywords: Monte-Carlo Path Tracing · Realistic Rendering · Biomedical Volumes Visualization

Supplementary Information The online version contains supplementary material available at https://doi.org/10.1007/978-3-031-43987-2_34.

H. Greenspan et al. (Eds.): MICCAI 2023, LNCS 14225, pp. 347–357, 2023.
https://doi.org/10.1007/978-3-031-43987-2_34

1 Introduction

Now that the hardware performance has achieved a certain level, 3D rendering of biomedical volumes is becoming very popular, above all, with the younger generation of clinicians that uses to use the forefront technologies in their everyday life. 3D representation has proven useful for faster comprehension of traumas in areas of high anatomic complexity, for surgical planning, simulation, and training [2,9,15,25], [18,28]. Also, it improves communication with patients, which understand the diagnosis much better if illustrated in 3D [23].

The most popular techniques for volume data rendering are maximum-intensity projection (MIP), iso-surface rendering, and direct volume rendering (DVR). These techniques have their pros and cons, but essentially, they suffer from a lack of *photo realism*.

Our novel Advanced Realistic Rendering Technique (AR^2T) is based on Monte-Carlo path tracing (MCPT). Historically, MCPT is thought of as a technique suited to (iso)surfaces rendering [14,19]. Applied to biomedical volumes, this technique has received relatively little attention, probably because, if naively implemented, it does not allow interaction with the data in real-time [7]. However, due to continuous hardware improvement, the problems that can not be resolved in real-time today, will be resolved in real-time tomorrow. For this reason, we have audaciously decided to apply MCPT to the biomedical volumes in attempt to increases the realism and the level of detail in data representation.

In this paper, we present a practical framework that includes different visualization techniques, including AR^2T. Our framework allows the user to interact with the data in high quality for the deterministic algorithms (iso-surface, MIP, DVR), and in low quality for the stochastic AR^2T. Moreover, the framework supports a mixed modality that works as follows. By default, the data is rendered by DVR. It allows to interact with the data, adjust the transfer function, and apply clip planes. However, a high-quality AR^2T image can be generated at any moment by the user request without explicitly switching between rendering algorithms. The quality improves progressively, and the process can be stopped as soon as the desired quality is achieved. As an alternative, the improvement stops automatically, when the algorithm converged. The framework permits to compare different rendering techniques directly, i.e., using the same view/light position and transfer functions. It, therefore, promotes further research on the importance of realism in visualising biomedical volumes, providing medical experts with an immediate one-to-one visual comparison between different data representations.

Related Work. Various deterministic approaches were applied in an attempt to increase the realism of volume rendering. Above all, the direct volume rendering technique has been enriched with local and global illumination, combined with ambient occlusion and shadowing [13,21,24]. However, these approaches are not able to produce photo-realistic images, being based on a very simplified and far-fetched model.

One interesting technique for improved DVR, which includes realistic effects and physically based lighting, was proposed by Kroes et al. in 2012 [17]. They

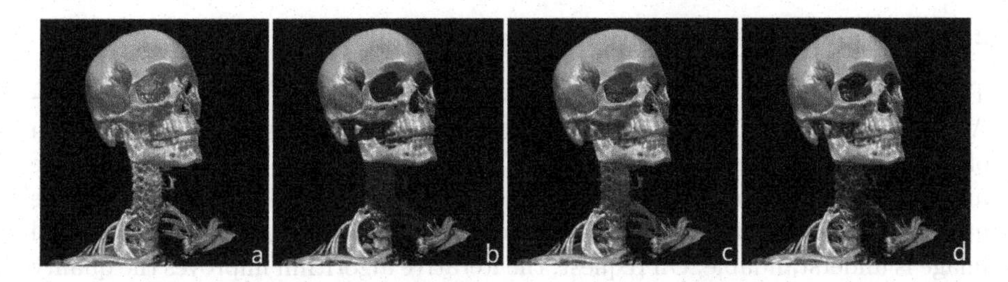

Fig. 1. Direct Volume Rendering. Human skull CBCT rendered with DVR: **a.** Local illumination; **b.** Local and global illumination, linear lighting; **c.** Local and global illumination, conical lighting; **d.** Local and global illumination with translucency, conical lighting. For all techniques, Phong local illumination was used.

were the first who demonstrated that, if properly optimized, ray tracing of volumetric data can be done interactively. However, their method is based on single scattering and consequently does not produce photo-realistic images. Despite that, this approach still arouses interest [5,30]. In fact, as far as we know, since then no one has presented any different reproducible technique on photo-realistic rendering of biomedical volumetric data.

Recently, the cinematic rendering (CR) 3D technique was introduced [4,10]. It is available as a part of commercial closed-source software, and the implementation details are not publicly available. Several studies compared DVR and CR images, produced by different software [6–8,29]. However, the one-to-one comparison is difficult, because it is not practically possible to align the data by means of visual settings and, above all, positioning.

2 Methods

The entire framework is written from scratch in C++ using Qt. The whole rendering runs on GPU and is implemented in OpenGL Shading Language (GLSL). In the following subsections, we describe the techniques we used within the framework, giving the details only for AR²T for the sake of brevity. The quality of the images produced with the different techniques is difficult to assess with a numerical index; following [4,6–8,29], we carried out a survey, based on the visual comparison of the proposed methods (see *Results*).

2.1 Deterministic Rendering Algorithms

First of all, we implemented the most popular rendering techniques for biomedical volumetric data: iso-surface, MIP, and DVR. They gave us the basis for the direct comparison of proposed methods. Then, we enriched our DVR model with local and global illumination, applying various approaches [11,16,20] in an attempt to improve realism and receive feedback from clinicians (see Fig. 1). It was immediately clear that despite these techniques can improve realism by introducing deep shadows, they are not suitable for the visualization of biomedical volumes because hide information in the shadowed areas without increasing anyhow the level of detail.

2.2 Advanced Realistic Rendering Technique

There are two modalities of AR^{2}T visualization: pure AR^{2}T and mixed DVR-AR^{2}T. When pure AR^{2}T is active, the interactivity is achieved by the execution of just one iteration of the algorithm. When the interaction is finished (e.g., the mouse button is released), 10 iterations of AR^{2}T are executed. To improve the quality, Gaussian blur filter [1] is applied during the interactions, so the overall image is understandable. On request, the iterative algorithm improves the quality until the user stops the process or the convergence is achieved (see subsection *Convergence*). In mixed modality, the interactions are in DVR, and the AR^{2}T runs on request. When the interaction restarts, the visualization automatically switches to DVR.

Our AR^{2}T is inspired by MCPT applied to analytically generated surfaces and isotropic volumes [12,26]. In our model, we provide advanced camera settings (aperture, focal distance, see Fig. 2) and support an unlimited number of light sources of any shape. For practical reasons, we limit the number of ray scatters to 10 (in our experiments, we did not see any improvement in realism for a larger number of scatters). In the following subsections, we step-by-step describe the implementation details of AR^{2}T, to allow the reproducibility of our results.

GPU Implementation. To be independent in the choice of hardware to run our framework, we implement AR^{2}T in GLSL. Unfortunately, there are two main issues to resolve for Monte-Carlo path tracing in OpenGL: 1. recursion, and 2. random number generation.

Recursion. As GLSL memory model does not allow for recursive function calls, which are essential for MCPT, we simulated the recursion by exploiting *multiple render targets* feature of modern GPUs. This feature allows the rendering pipeline to render images to multiple render target textures at once. Indeed, the information we need after every scatter of a ray is the resulting colour, the position where the scatter occurred, and the direction in which the ray scatters.

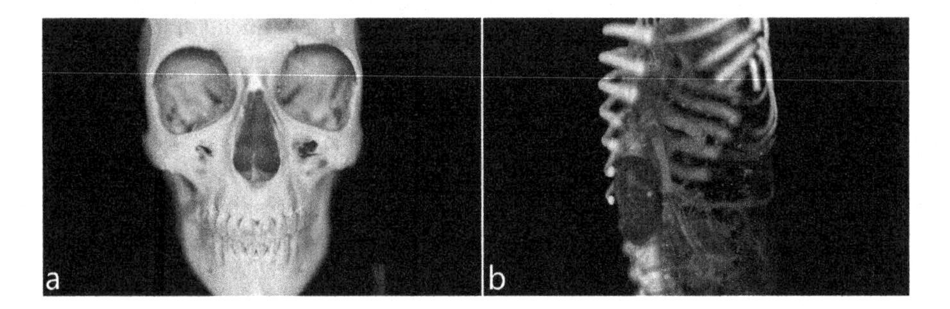

Fig. 2. AR^{2}T: Camera Aperture & Focal Distance. **a**. Human skull CBCT; in focus: nasal bone, infraorbital foramen, vomer. **b**. Dog abdomen CBCT with contrast; in focus: intrahepatic portocaval shunt, main portal vein, right kidney.

Therefore, three target textures are necessary for every rendering step. Moreover, two frame buffers are used in a *ping pong blending* manner (as described in [16]) to enable the reading of textures filled on the previous step. Thus, in the first step, the ray origin and direction are calculated according to the camera properties and position. In the subsequent steps, the ray origin and direction are read from the corresponding textures.

On any step, three situations are possible: (1) The ray does not hit the volume or the light source. In this case, a zero-length vector is saved to *direction* render target - it indicates that the ray scattering finishes here, and the resulting colour components are set to zeros (for ulterior speed up, and considering that the usual background for biomedical visualization is black, we do not model the Cornel box outside the volume). (2) The ray hits the light source. Then, the resulting colour, accumulated until this moment, is attenuated by the light's colour, and, again, the ray scattering finishes here. (3) The volume is hit. The scattering continues, until the ray encounters any of the stopping conditions, or the scatter number limit is achieved. In this case, the resulting colour components are set to zeros.

Random Number Generation. To provide the uniformly distributed random numbers on the fragment stage of the OpenGL pipeline, we generate a pool of 50 additional two-dimensional textures of *viewport* size and fill them with uniformly distributed random numbers generated with **std::uniform_int_distribution**. In each step, we randomly choose four textures from the pool to provide random numbers: two for advanced Woodcock tracking, one for scattering, and one for sampling direction. Every 100 iterations of the algorithm, we regenerate the pool of random numbers to avoid the quality improvement stuck. On Intel(R) Core(TM) i5-7600K CPU @ 3.80 GHz, the generation of the pool takes $\sim$ 1600 ms, for viewport size 1727 $\times$ 822. It occupies $\sim$270 Mb of RAM.

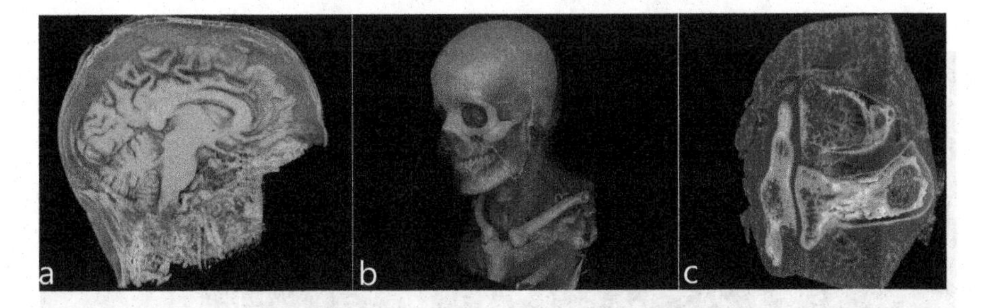

Fig. 3. AR^{2}T: Hard & Soft Tissues. **a**. Human brain MRI, https://openneuro.org/datasets/ds001780/versions/1.0.0: identified pre-central gyrus of frontal lobe, genu of the corpus callosum, lateral ventricle, optic nerve, pons, cerebellum, medulla oblongata, cerebellar tonsil; **b**. Spiral CT Manix, https://public.sethealth.app/manix.raw.gz; **c**. Human knee CBCT (post-mortem).

Advanced Woodcock Tracking. For volume sampling, we implemented the advanced Woodcock tracking technique, described in [3,27]. When the maximum volume density is much larger than the typical density in the volume, the Woodcock tracking can be improved by breaking the original volume into subvolumes, each having a small density variation. For this aim, every time the transfer function changes, we construct a voxel grid that has the local maxima of the density in its nodes. Then, the voxel grid is straightforwardly applied on Woodcock tracking, giving up to 5x speed-up respect to the basic implementation. On NVIDIA GeForce GTX 1060, the voxelization process, implemented in GLSL and executed on a fragment stage, takes up to 500 ms for a voxel grid node size $s = 4$, and depending on the volume size. The additional memory needed for the voxel grid storage is $1/s^3$ of the original volume size.

Phase Functions. In AR^{2}T, we use four well-known phase functions, described in [26]: Lambertian, metal, dielectric, and isotropic. Every time the ray hits a volume voxel, we choose which phase function to apply based on the density of the voxel: if the density is less than some threshold t_0, it causes dielectric scatter; when it is higher than some threshold t_1, it causes metal scatter; otherwise, as proposed in [26], we randomly decide if to sample toward the light sources or to pick direction according to the voxel reflection (*mixture probability density function*). When we decide to sample according to the hit voxel direction, we choose between surface and volumetric scattering. Following [17], we switch between Lambertian and isotropic phase functions, basing not only on the voxel density but also on the local gradient. Thus, Lambertian is chosen with the following probability:

$$p = 1 - \alpha(\overrightarrow{v}) \cdot (1 - e^{-s \cdot m(\overrightarrow{v})}), \tag{1}$$

where $\alpha(\overrightarrow{v}) \in [t_0, t_1]$ is the voxel density (or opacity), $m(\overrightarrow{v})$ is the normalized gradient magnitude, and s is the hybrid scattering factor.

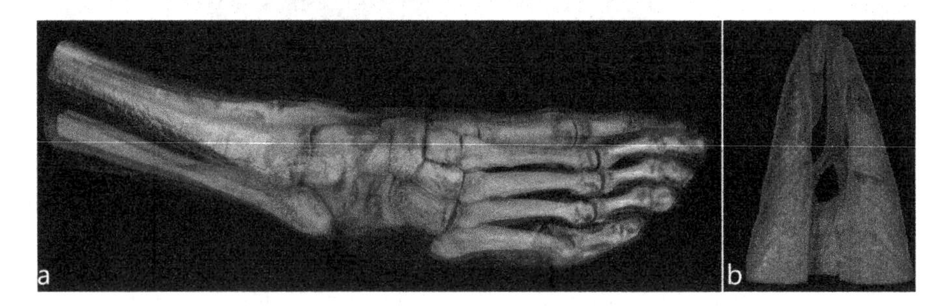

Fig. 4. AR^{2}T: Translucency. **a.** Human foot CBCT (post-mortem); well-distinguished distal tibiofibular joint (syndesmosis), talus, navicular, cuboid, middle cuneiform, fifth metatarsal, under the semitransparent skin. **b.** Cat thorax CBCT; semitransparent lungs and trachea.

Image Generation. When the first iteration of AR^{2}T is completed, the result contained in the colour texture (see *Recursion* for rehearse) is blit into the output rendering frame buffer to be immediately displayed. Moreover, it is saved locally to be summed with the results of the next iterations. On the next iterations, the accumulated result is saved into the local buffer, and then the medium (e.g. the sum divided by the iterations number) is blit into the output frame buffer and displayed.

Convergence. As a convergence criterion, we use mean square displacement (MSD) between the iterations [22]. After each iteration, the square displacement between the current and the previous pixel colour components is calculated directly on a fragment stage. When MSD becomes less than $\epsilon = 5 \cdot 10^{-7}$, the iterations stop, and the method is considered converged (see Fig. 5). In our experiments, the convergence was achieved within 800 iterations for all images, and it took up to 100 s.

3 Results

In our experiments, we have used spiral CT (Computed Tomography), MRI (Magnetic Resonance Imaging), and CBCT (Cone-Beam Computed Tomography) publicly available data sets. The CBCT data sets were acquired by SeeFactorCT3^{TM} (human) and Vimago^{TM}GT30 (vet) Multimodal Medical Imaging Platforms in our layout and are available on https://kaggle.com/datasets/imaginar2t/cbctdata.

To validate the superiority of the AR^{2}T over the other methods implemented in our platform, we ask a group of clinicians to participate into the survey. 22 participants (7 orthopedic surgeons, 1 trauma surgeon, 1 neurosurgeon, 7 interventional radiologists, 6 veterinaries) evaluated the data sets on Fig. 6, visualized

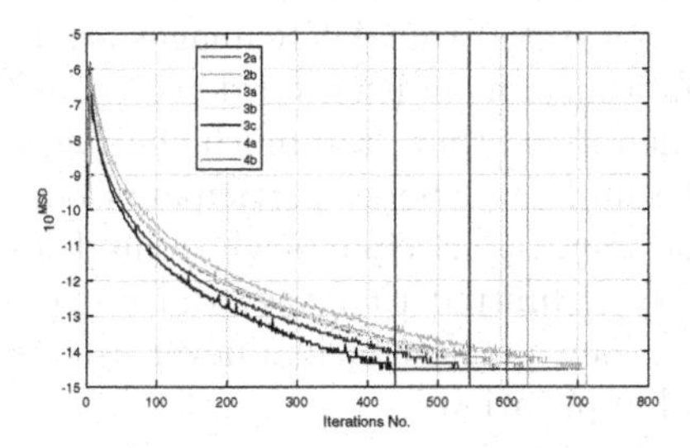

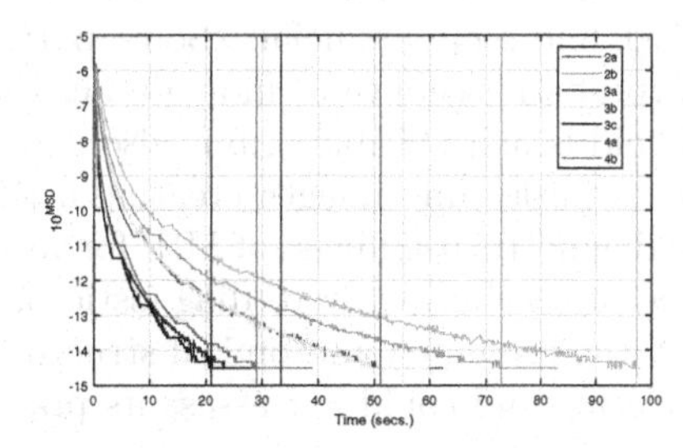

Fig. 5. Convergence Plots (Fig. 2a–4b). Vertical lines indicate the number of iterations needed for convergence (left) and the convergence time in seconds (right). All images were generated on NVIDIA GeForce GTX 1060 6 Gb, Intel(R) Core(TM) i5-7600K CPU @ 3.80 GHz RAM 16 Gb, viewport size 1727 × 822. All data sets are encoded in 16-bit format.

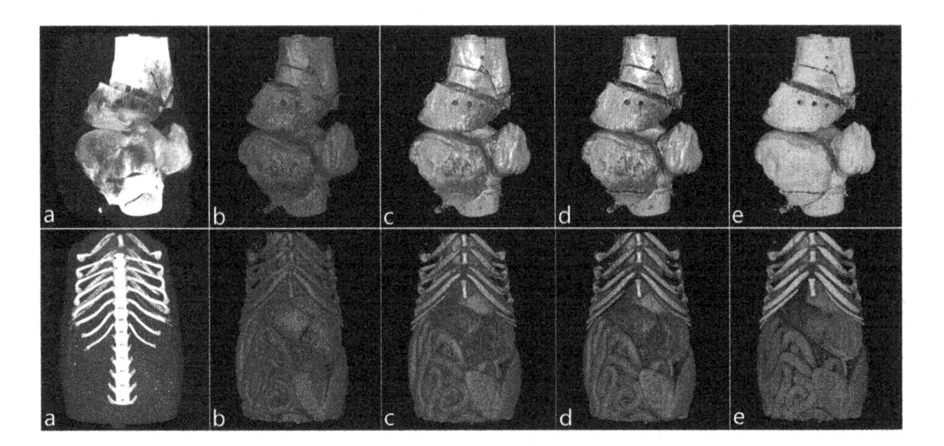

Fig. 6. One-To-One Rendering Comparison. Human knee CBCT post-mortem (top) and dog abdomen CBCT with contrast (bottom): MIP (**a**), Iso-surface (**b**), DVR with local (**c**) and global illumination, linear lighting (**d**), AR^2T (**e**). No light/view or transfer function changes across **a**–**e**.

using MIP, iso-surface, DVR with local illumination, DVR with local and global illumination (linear lighting), and AR^2T, voting the best overall image, the most realistic one, the more diagnostic (if any), and the more valuable in their practice.

According to Table 1, the AR^2T provides the best overall, the most realistic, diagnostic, and valuable images. Participants commented that the images produced with AR^2T "provide better resolution, amazing clarity with less artifacts and noise, better sharpness and contrast, are the closest in colour to real tissue and the most similar to a dissected body deprived of blood, help to understand the anatomy and the pathology of the district, have excellent qualities in general surgery for the definition of the splanchnic organs, for diagnosis and preoperative study". Meanwhile the others are "either glossy, or too colorized or not as sharp, and seem artificial". Some participants stated that DVR images are the sharpest and seem to be more detailed ("peritoneal meso is better detected, subxiphoid is more visible"), but also it was mentioned that "too much sharpening causes misleading images creating artificial findings". Some participants noted the diagnostic usefulness of MIP for vascular issues. Participants who stated that none of the images were diagnostic or useful admitted that they did not deal with the presented anatomical structures in their practice or had never used 3D rendering and could not assess its practical application.

Table 1. Survey Responses Indicating How Many of 22 Participants Voted the Method that Produces: Best Overall, Most Realistic, More Diagnostic (If Any), and More Valuable Images, by evaluating Fig. 6

Method	*Best Overall*		*Most Realistic*		*More Diagnostic*		*More Valuable*	
	Knee	*Abs*	*Knee*	*Abs*	*Knee*	*Abs*	*Knee*	*Abs*
(e) AR^2T	18	16	17	17	18	12	19	15
(d) $DVR^{\dagger}$	4	3	3	3	1	2	2	3
(c) DVR	—	3	1	2	—	2	—	2
(b) Iso-surf.	—	—	—	—	—	—	—	—
(a) MIP	—	—	1	—	1	—	—	—
None	—	—	—	—	2	6	1	2

Note: $DVR^{\dagger}$ – local and global illumination, linear lighting; DVR – local illumination.

4 Conclusions

The main result of our research is the novel advanced realistic rendering technique–AR^{2}T (see Fig. 2, 3, 4), implemented within a practical framework, that allows to compare different rendering techniques directly (Fig. 6).

Despite our model supports any number of light sources, all images (except Fig. 1), presented in this paper, were generated with a single spherical light source, placed right in front of the volume. We plan to dedicate extra time to find the best light configuration from the clinical point of view. Moreover, our future research will be focused on the ulterior improvement of AR^{2}T speed and quality, and the comparison metrics.

At the moment of writing this paper, we are evaluating free access to our framework's executable to share our results, facilitate the comparison with other approaches, and stimulate further research on the usefulness of photo-realistic 3D images in medicine.

References

1. Gaussian blur filter shader. https://web.archive.org/web/20150320024135/, http://www.gamerendering.com/2008/10/11/gaussian-blur-filter-shader/. Accessed 08 Mar 2023
2. Abou El-Seoud, S., Mady, A., Rashed, E.: An interactive mixed reality ray tracing rendering mobile application of medical data in minimally invasive surgeries (2019)
3. Behlouli, A., Visvikis, D., Bert, J.: Improved woodcock tracking on Monte Carlo simulations for medical applications. Phys. Med. Biol. **63**(22), 225005 (2018). https://doi.org/10.1088/1361-6560/aae937
4. Bueno, M.R., Estrela, C., Granjeiro, J.M., Estrela, M.R.D.A., Azevedo, B.C., Diogenes, A.: Cone-beam computed tomography cinematic rendering: clinical, teaching and research applications. Braz. Oral Research **35** (2021). https://doi.org/10.1590/1807-3107bor-2021.vol35.0024

5. Cheng, H., Xu, C., Wang, J., Chen, Z., Zhao, L.: Fast and accurate illumination estimation using LDR panoramic images for realistic rendering. IEEE Trans. Visual Comput. Graphics (2022). https://doi.org/10.1109/TVCG.2022.3205614
6. Dappa, E., Higashigaito, K., Fornaro, J., Leschka, S., Wildermuth, S., Alkadhi, H.: Cinematic rendering – an alternative to volume rendering for 3D computed tomography imaging. Insights Imaging **7**(6), 849–856 (2016). https://doi.org/10.1007/s13244-016-0518-1
7. Ebert, L.C., et al.: Forensic 3D visualization of CT data using cinematic volume rendering: a preliminary study. Am. J. Roentgenol. **208**(2), 233–240 (2017). https://doi.org/10.2214/AJR.16.16499
8. Eid, M., et al.: Cinematic rendering in CT: a novel, lifelike 3D visualization technique. Am. J. Roentgenol. **209**(2), 370–379 (2017). https://doi.org/10.2214/AJR.17.17850
9. Elshafei, M., et al.: Comparison of cinematic rendering and computed tomography for speed and comprehension of surgical anatomy. JAMA Surg. **154**(8), 738–744 (2019). https://doi.org/10.1001/jamasurg.2019.1168
10. Engel, K.: Real-time Monte-Carlo path tracing of medical volume data. In: GPU Technology Conference, 4–7 Apr 2016. San Jose Convention Center, CA, USA (2016)
11. Fernando, R., et al.: GPU Gems: Programming Techniques, Tips, and Tricks for Real-time Graphics, vol. 590. Addison-Wesley Reading (2004)
12. Fong, J., Wrenninge, M., Kulla, C., Habel, R.: Production volume rendering: Siggraph 2017 course. In: ACM SIGGRAPH 2017 Courses, pp. 1–79 (2017). https://doi.org/10.1145/3084873.3084907
13. Hernell, F., Ljung, P., Ynnerman, A.: Local ambient occlusion in direct volume rendering. IEEE Trans. Visual Comput. Graphics **16**(4), 548–559 (2009). https://doi.org/10.1109/TVCG.2009.45
14. Jensen, H.W., et al.: Monte Carlo ray tracing. In: ACM SIGGRAPH, vol. 5 (2003)
15. Johnson, P.T., Schneider, R., Lugo-Fagundo, C., Johnson, M.B., Fishman, E.K.: MDCT angiography with 3D rendering: a novel cinematic rendering algorithm for enhanced anatomic detail. Am. J. Roentgenol. **209**(2), 309–312 (2017). https://doi.org/10.2214/AJR.17.17903
16. Kniss, J., Premoze, S., Hansen, C., Shirley, P., McPherson, A.: A model for volume lighting and modeling. IEEE Trans. Visual Comput. Graphics **9**(2), 150–162 (2003). https://doi.org/10.1109/TVCG.2003.1196003
17. Kroes, T., Post, F.H., Botha, C.P.: Exposure render: an interactive photo-realistic volume rendering framework. PloS one **7**(7), e38596 (2012). https://doi.org/10.1371/journal.pone.0038586
18. Kutaish, H., Acker, A., Drittenbass, L., Stern, R., Assal, M.: Computer-assisted surgery and navigation in foot and ankle: state of the art and fields of application. EFORT Open Rev. **6**(7), 531–538 (2021). https://doi.org/10.1302/2058-5241.6.200024
19. Lafortune, E.P., Willems, Y.D.: Rendering participating media with bidirectional path tracing. In: Pueyo, X., Schröder, P. (eds.) EGSR 1996. E, pp. 91–100. Springer, Vienna (1996). https://doi.org/10.1007/978-3-7091-7484-5_10
20. Max, N., Chen, M.: Local and global illumination in the volume rendering integral. Technical report, Lawrence Livermore National Lab. (LLNL), Livermore, CA (United States) (2005)
21. McNamara, A.: Illumination in computer graphics. The University of Dublin (2003)

22. Michalet, X.: Mean square displacement analysis of single-particle trajectories with localization error: Brownian motion in an isotropic medium. Phys. Rev. E **82**(4), 041914 (2010). https://doi.org/10.1103/PhysRevE.82.041914
23. Pachowsky, M.L., et al.: Cinematic rendering in rheumatic diseases-photorealistic depiction of pathologies improves disease understanding for patients. Front. Med. **9**, 946106 (2022). https://doi.org/10.3389/fmed.2022.946106
24. Salama, C.R.: GPU-based Monte-Carlo volume raycasting. In: 15th Pacific Conference on Computer Graphics and Applications (PG 2007), pp. 411–414. IEEE (2007). https://doi.org/10.1109/PG.2007.27
25. Sariali, E., Mauprivez, R., Khiami, F., Pascal-Mousselard, H., Catonné, Y.: Accuracy of the preoperative planning for cementless total hip arthroplasty. a randomised comparison between three-dimensional computerised planning and conventional templating. Orthop. Traumatol. Surg. Res. **98**(2), 151–158 (2012). https://doi.org/10.1016/j.otsr.2011.09.023
26. Shirley, P., Morley, R.K.: Realistic Ray Tracing. AK Peters Ltd, Natick (2008)
27. Szirmay-Kalos, L., Tóth, B., Magdics, M.: Free path sampling in high resolution inhomogeneous participating media. In: Computer Graphics Forum, vol. 30, pp. 85–97. Wiley Online Library (2011). https://doi.org/10.1111/j.1467-8659.2010.01831.x
28. Wang, C., et al.: Patient-specific instrument-assisted minimally invasive internal fixation of calcaneal fracture for rapid and accurate execution of a preoperative plan: a retrospective study. BMC Musculoskelet. Disord. **21**, 1–11 (2020). https://doi.org/10.1186/s12891-020-03439-3
29. Xu, J., et al.: Interactive, in-browser cinematic volume rendering of medical images. Comput. Methods Biomech. Biomed. Eng. Imaging Visual. **11**, 1–8 (2022). https://doi.org/10.1080/21681163.2022.2145239
30. Zhou, S.: Woodcock tracking based fast Monte Carlo direct volume rendering method. J. Syst. Simul. **29**(5), 1125–1131 (2017). https://doi.org/10.16182/j.issn1004731x.joss.201705026

Transformer-Based End-to-End Classification of Variable-Length Volumetric Data

Marzieh Oghbaie[1,2](✉), Teresa Araújo[1,2], Taha Emre[2], Ursula Schmidt-Erfurth[1], and Hrvoje Bogunović[1,2]

[1] Christian Doppler Laboratory for Artificial Intelligence in Retina, Department of Ophthalmology and Optometry, Medical University of Vienna, Vienna, Austria
{marzieh.oghbaie,hrvoje.bogunovic}@meduniwien.ac.at
[2] Laboratory for Ophthalmic Image Analysis, Department of Ophthalmology and Optometry, Medical University of Vienna, Vienna, Austria

Abstract. The automatic classification of 3D medical data is memory-intensive. Also, variations in the number of slices between samples is common. Naïve solutions such as subsampling can solve these problems, but at the cost of potentially eliminating relevant diagnosis information. Transformers have shown promising performance for sequential data analysis. However, their application for long sequences is data, computationally, and memory demanding. In this paper, we propose an end-to-end Transformer-based framework that allows to classify volumetric data of variable length in an efficient fashion. Particularly, by randomizing the input volume-wise resolution(#slices) during training, we enhance the capacity of the learnable positional embedding assigned to each volume slice. Consequently, the accumulated positional information in each positional embedding can be generalized to the neighbouring slices, even for high-resolution volumes at the test time. By doing so, the model will be more robust to variable volume length and amenable to different computational budgets. We evaluated the proposed approach in retinal OCT volume classification and achieved 21.96% average improvement in balanced accuracy on a 9-class diagnostic task, compared to state-of-the-art video transformers. Our findings show that varying the volume-wise resolution of the input during training results in more informative volume representation as compared to training with fixed number of slices per volume.

Keywords: Optical coherence tomography · 3D volume classification · Transformers

1 Introduction

Volumetric medical scans allow for comprehensive diagnosis, but their manual interpretation is time consuming and error prone [19,21]. Deep learning methods

H. Greenspan et al. (Eds.): MICCAI 2023, LNCS 14225, pp. 358–367, 2023.
https://doi.org/10.1007/978-3-031-43987-2_35

have shown exceptional performance in automating this task [27], often at medical expert levels [6]. However, their application in the clinical practice is still limited, partially because they require rigid acquisition settings. In particular, variable volume length, i.e. number of slices, is common for imaging modalities such as computed tomography, magnetic resonance imaging or optical coherence tomography (OCT). Despite the advantages of having data diversity in terms of quality and size, automated classification of dense scans with variable input size is a challenge. Furthermore, the 3D nature of medical volumes results in a memory-intensive training procedure when processing the entire volume. To account for this constraint and make the input size uniform, volumes are usually subsampled, ignoring and potentially hiding relevant diagnostic information.

Among approaches for handling variable input size, Multiple Instance Learning (MIL) is commonly used. There, a model classifies each slice or subgroup of slices individually, and the final prediction is determined by aggregating sub-decisions via maximum or average pooling [16,17,23], or other more sophisticated fusion approaches [20,24]. However, they often do not take advantage of the 3D aspect of the data. The same problem occurs when stacking slice-wise embeddings [4,11,22], applying self-attention [5] for feature aggregation, or using principal component analysis (PCA) [9] to reduce the variable number of embeddings to a fixed size. As an alternative, recurrent neural networks (RNNs) [18] consider the volume as a sequence of arranged slices or the corresponding embeddings. However, their performance is overshadowed by arduous training and lack of parallelization.

Vision Transformers (ViTs) [8], on the other hand, allow parallel computation and effective analysis of longer sequences by benefiting from multi-head self-attention (MSA) and positional encoding. These components allow to model both local and global dependencies, playing a pivotal role for 3D medical tasks where the order of slices is important [13,14,26]. Moreover, ViTs are more flexible regarding input size. Ignoring slice positional information (bag-of-slices) or using sinusoidal positional encoding enables them to process sequences of arbitrary length with respect to computational resources. However, ViTs with learnable positional embeddings (PEs) have shown better performance [8]. In this case, the only restriction in processing variable length sequences is the number of PEs. Although interpolating the PE sequence helps overcome this restriction, the resultant sequence will not model the exact positional information of the corresponding slices in the input sequence, affecting ViTs performance [2]. Notably, Flexible ViT [2] (FlexiViT) handles patch sequences of variable sizes by randomizing the patch size during training and, accordingly, resizing the embedding weights and parameters corresponding to PEs.

Despite the merits of the aforementioned approaches, three fundamental challenges still remain. First, the model should be able to process inputs with variable volume resolutions, where throughout the paper we refer to the resolution in the dimension across slices (number of slices), and simultaneously capture the size-independent characteristics of the volume and similarities among the constituent slices. The second challenge is the scalability and the ability of the model to adapt to unseen volume-wise resolutions at inference time. Lastly, the training of deep learning models with high resolution volumes is both computationally expensive and memory-consuming.

In this paper, we propose a late fusion Transformer-based end-to-end framework for 3D volume classification whose local-similarity-aware PEs not only improve the model performance, but also make it more robust to interpolation of PEs sequence. We first embed each slice by a spatial feature extractor and then aggregate the corresponding sequence of slice-wise embeddings with a Feature Aggregator Transformer (FAT) module to capture 3D intrinsic characteristics of the volume and produce a volume-level representation. To enable the model to process volumes with variable resolutions, we propose a novel training strategy, Variable Length FAT (VLFAT), that enables FAT module to process volumes with different resolutions both at training and test times. VLFAT can be trained with a proportionally few #slices, an efficient trait in case of training time/memory constraints. Consequently, even with drastic slice subsampling during training, the model will be robust against extreme PEs interpolation for high-resolution volumes at the test time. The proposed approach is model-agnostic and can be deployed with Transformer-based backbones. VLFAT beats the state-of-the-art performance in retinal OCT volume classification on a private dataset with nine disease classes, and achieves competitive performance on a two-class public dataset.

2 Methods

Our end-to-end Transformer-based volume classification framework (Fig. 1) has three main components: 1) Slice feature extractor (SFE) to extract spatial biomarkers and create a representation of the corresponding slice; 2) Volume feature aggregator (VFA) to combine the slice-level representations into a volume-level representation, and 3) Volume classification. Trained with the proposed strategy, our approach is capable of processing and classifying volumes with varying volume-wise resolutions. Let's consider a full volume $v \in \mathbf{R}^{(N \times W \times H)}$, where (N, H, W) are the #slices, its width and height respectively. The input to the network is a subsampled volume by randomly selecting n slices.

Slice Feature Extractor (SFE). To obtain the slice representations, we use ViT as our SFE due to its recent success in medical interpretation tasks [10]. ViT mines crucial details from each slice and, using MSA and PE, accumulates the collected information in a learnable classification token, constituting the slice-wise embedding. For each slice token, we then add a learnable 1D PE [8] to retain the position of each slice in the volume.

Volume Feature Aggregator (VFA). The output of the previous step is a sequence of slice-wise embeddings, to which we append a learnable volume-level classification token [7]. The resulting sequence of embedding vectors is

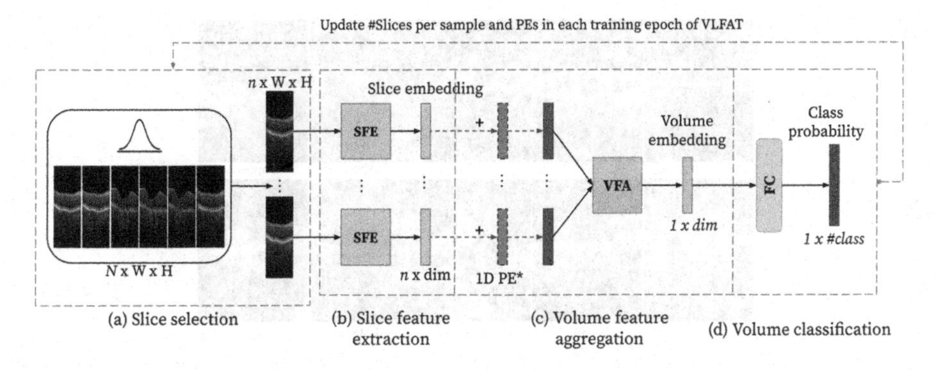

Fig. 1. The overview of the proposed Transformer-based approach for 3D volume classification. The shared SFE processes the input slices, and in line with VLFAT, both #slices and the PEs sequence are updated at each epoch. *1D PE is added to each slice embedding for FAT and VLFAT.

then processed by the FAT module to produce a volume-level embedding. In particular, we propose *VLFAT*, a FAT with enhanced learnable PEs, inspired on FlexiViT [2], where we modify #slices per input volume instead of patch sizes and correspondingly apply PEs interpolation. This allows handling arbitrary volume resolutions, which generally would not be possible except for an ensemble of models of different scales. Specifically, at initialization we set a fixed value, n, for #slices, resulting in PEs sequence with size $(n+1, dim)$, where an extra PE is assigned to the classification token and dim is the dimension of the slice representation. In each training step, we then randomly sample a new value for n from a predefined set and, accordingly, linearly interpolate the PEs sequence (Fig. 1), using the known adjacent PEs [3]. This allows to preserve the similarity between neighboring slices in the volume, sharing biomarkers in terms of locality, and propagating the corresponding positional information. The new PEs are then normalized according to a truncated normal distribution.

Volume Classification. Finally, the volume-level classification token is fed to a Fully Connected (FC) layer, which produces individual class scores. As a loss function, we employ the weighted cross-entropy.

3 Experiments

We tested our model for volume classification of macula-centered retinal OCT scans, where large variation in volume resolution (#B-scans) between samples is very common. For multiclass classification performance metrics, we relied on Balanced Accuracy (BAcc) and one-vs-all Area Under the Receiver Operating Curve (AUROC). The source code is available at: github.com/marziehoghbaie/VLFAT.

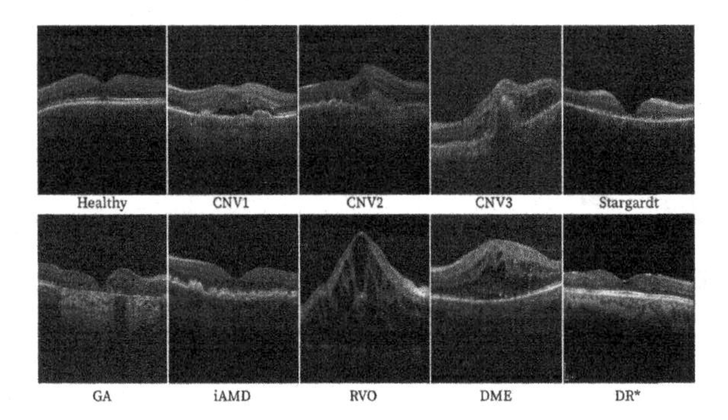

Fig. 2. Samples of central B-scan from all disease classes. *DR is only in OLIVES.

Datasets. We utilized three large retinal OCT volumetric datasets: *Duke* for pre-training all models, and *9C* and *OLIVES* for fine-tuning and testing purposes.

- *Duke:* Public dataset [22] with 269 intermediate age-related macular degeneration (iAMD) and 115 normal patients, acquired with a Bioptigen OCT device with a resolution of 100 B-scans per volume. Volumes were split patient-wise into 80% for training (384 samples) and 20% for validation (77 samples).
- *9C:* Private dataset with 4766 volumes (4711 patients) containing 9 disease classes: iAMD, three types of choroidal neovascularization (CNV1-3), geographic atrophy (GA), retinal vein occlusion (RVO), diabetic macular edema (DME), Stargardt disease, and healthy. Volumes were split patient-wise, for each class, into 70% training (3302 samples), 15% validation (742 samples), and 15% test (722 samples). The OCT volumes were acquired by four devices (Heidelberg Engineering, Zeiss, Topcon, Nidek), exhibiting large variation in #slices. Minimum, maximum, and average #slices per volume were 25, 261, and 81, respectively.
- *OLIVES:* Public dataset [15] with 3135 volumes (96 patients) labeled as diabetic retinopathy (DR) or DME. OCTs were acquired with Heidelberg Engineering device, and have resolutions of either 49 or 97, and were split patient-wise for each class into 80% training (1808 samples), 10% validation (222 samples), and 10% test (189 samples) (Fig. 2).

Comparison to State-of-the-Art Methods. We compared the performance of the proposed method with two state-of-the-art video ViTs (ViViT) [1], originally designed for natural video classification: 1) factorized encoder (FE) ViViT, that models the spatial and temporal dimensions separately; 2) factorised self-attention (FSA) ViViT, that simultaneously computes spatial and temporal interactions. We selected FE and FSA ViViTs as baselines to understand the importance of separate feature extractors and late fusion in our approach. FE ViViT, similar to ours, utilizes late fusion, while FSA ViViT is a slow-fusion model and processes spatiotemporal patches as tokens.

Ablation Studies. To investigate the contribution of SFE module, we deployed ViT and ResNet18 [26], a standard 2D convolutional neural network (CNN) in medical image analysis, with pooling methods as VFA where the quality of slice-wise features is more influential. For VFA, we explored average pooling (AP), max pooling (MP), and 1D convolution (1DConv). As MIL-based baselines, pooling methods can be viable alternatives to VLFAT for processing variable volume resolutions. In addition to learnable PE, we deployed sinusoidal PE (sinPE) and bag-of-slices (noPE) for FAT to examine the effect of positional information.

Robustness Analysis. We investigate the robustness of VFLAT and FAT to PEs sequence interpolation at inference time by changing the volume resolution. To process inputs with volume resolutions different from FAT's and VLFAT's input size, we linearly interpolate the sequence of PEs at the test time. For 9C dataset, we only assess samples with minimum #slices of 128 to better examine the PE's scalability to higher resolutions.

Implementation Details. The volume input size was $25 \times 224 \times 224$ for all experiments except for FSA ViViT where #slices was set to 24 based on the corresponding tublet size of $2 \times 16 \times 16$. During VLFAT training, the #slices varied between $\{5, 10, 15, 20, 25\}$, specified according to memory constraints. We randomly selected slices using a normal distribution with its mean at the central slice position, thus promoting the inclusion of the region near the fovea, essential for the diagnosis of macular diseases. Our ViT configuration is based on ViT-Base [8] with patch size 16×16, and 12 Transformer blocks and heads. For FAT and VLFAT, we set the number of Transformer blocks to 12 and heads to 3. The slice-wise and volume-wise embedding dimension were set to 768. The configuration of ViViT baselines was set according to the original papers [1]. Training was performed using AdamW [12] optimizer with learning rate of 6×10^{-6} with cosine annealing. All models were trained for 600 epochs with a batch size of 8. Data augmentation included random brightness enhancing, motion blur, salt/pepper noise, rotation, and random erasing [28]. The best model was selected based on the highest BAcc on the validation set. All experiments were performed using Pytorch 1.13.0+cu117 and timm library [25] on a server with 1 TB RAM, and NVIDIA RTX A6000 (48 GB VRAM).

4 Results and Discussion

In particular, on large 9C dataset our VLFAT achieved 21.4% and 22.51% BAcc improvement compared to FE ViViT and FSA ViViT, respectively. Incorporating our training strategy, VLFAT, improved FAT's performance by 16.12% on 9C, and 8.79% on OLIVES, which verifies the ability of VLFAT in learning more location-aware PEs, something that is also reflected in the increase of AUROCs (0.96$\rightarrow$ 0.98 on 9C dataset and 0.95$\rightarrow$ 0.97 on OLIVES). Per-class AUROCs are shown in Table 2. The results show that for most of the classes, our VLFAT has better diagnostic ability and collects more disease-specific clues from the volume.

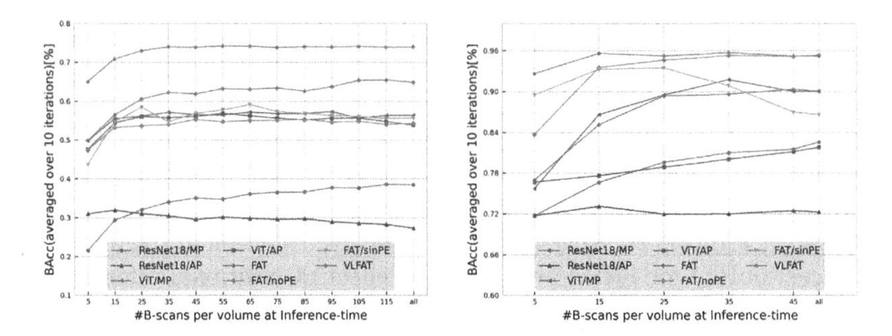

Fig. 3. Robustness analysis of VLFAT and vanilla FAT against PEs interpolation at the test time: (a) 9C dataset; (b) OLIVES

Table 1. Classification performance in terms of balanced accuracy (BAcc) and mean one-vs-all AUROC. FE: Factorised Encoder, FSA: Factorised Self-Attention, SFE: Slice Feature Extractor, VFA: Volume Feature Aggregator.

Method (SFE/VFA)	9C		OLIVES		#slices*
	BAcc	AUROC	BAcc	AUROC	
FE ViViT (baseline) [1]	0.64	0.96	0.93	0.98	25
FSA ViViT (baseline) [1]	0.63	0.95	0.92	0.98	24
ViT/1DConv	0.61	0.94	0.95	0.97	25
ResNet18/AP	0.34	0.73	0.72	0.83	all
ResNet18/MP	0.41	0.87	0.83	0.93	all
ViT/AP	0.59	0.95	0.82	0.97	all
ViT/MP	0.61	0.96	0.90	0.98	all
ViT/FAT (noPE)	0.58	0.93	**0.95**	**0.99**	all
ViT/FAT (sinPE)	0.62	0.93	0.87	0.96	all
ViT/FAT+	0.67	0.96	0.88	0.95	25
ViT/VLFAT (ours)	**0.78**	**0.98**	0.95	0.97	all

Legend: *#slices at the test time;
+the input length is fixed in both training and test time

The ablation study (Table 1) showed that each introduced component in the proposed model contributed to the performance improvement. In particular, ViT was shown as a better slice feature extractor compared to ResNet18, particularly on 9C dataset where the differences between disease-related biomarkers are more subtle. Additionally, the poor performance of the pooling methods as compared to FAT and 1DConv, emphasizes the importance of contextual volumetric information, the necessity of a learnable VFA, and the superiority of Transformers over 1DConv. Although, on OLIVES, less complicated VFAs (pooling/1DConv) and FAT (noPE) also achieved comparable results, which can be attributed primarily to DR vs. DME [15] being an easier classification task compared to the diverse disease severity in the 9C dataset. In addition, the competitive advantage

Table 2. Per-class classification performance (one-vs-all AUROC) on 9C dataset.

Method(SFE/VFA)	CNV1	CNV2	CNV3	DME	GA	Healthy	iAMD	RVO	Stargardt
FE ViViT [1]	0.93	0.91	0.95	0.94	0.99	0.95	0.92	0.95	0.99
FSA ViViT [1]	0.94	0.91	0.92	0.92	**1.0**	0.94	0.93	0.94	0.99
ViT/1DConv	0.88	0.92	0.94	0.91	0.98	0.92	0.92	0.92	**1.0**
ResNet18/AP	0.68	0.63	0.58	0.75	0.81	0.75	0.75	0.76	0.97
ResNet18/MP	0.78	0.77	0.79	0.87	0.91	0.84	0.84	0.84	0.91
ViT/AP	0.9	0.81	0.92	0.93	0.98	0.94	0.93	0.94	0.99
ViT/MP	0.95	0.85	0.95	0.94	0.98	0.94	0.93	0.96	0.99
ViT/FAT (noPE)	0.9	0.87	0.89	0.89	0.98	0.91	0.9	0.94	0.99
ViT/FAT (sinPE)	0.94	**0.93**	0.93	0.88	0.97	0.91	0.89	0.93	0.99
ViT/FAT	0.93	0.82	0.97	0.94	0.99	0.95	0.94	0.95	0.99
ViT/VLFAT (ours)	**0.98**	0.92	**0.98**	**0.98**	**1.0**	**0.99**	**0.98**	**0.98**	**1.0**

of VLFAT in handling different resolutions was not fully exploited in OLIVES since the large majority of cases had the same #slices. On 9C, however, the comparison of positional encoding strategies demonstrated that although ignoring PEs and sinusoidal approach provide deterministic predictions, the importance of learnable PEs in modeling the anatomical order of slices in the volume is crucial. The robustness analysis is shown in Fig. 3. VLFAT was observed to have more scalable and robust PEs when the volume-wise resolutions at the test time deviated from those used during training. This finding highlights the VLFAT's potential for resource-efficient training and inference.

5 Conclusions

In this paper, we propose an end-to-end framework for 3D volume classification of variable-length scans, benefiting from ViT to process volume slices and FAT to capture 3D information. Furthermore, we enhance the capacity of PE in FAT to capture sequential dependencies along volumes with variable resolutions. Our proposed approach, VLFAT, is more scalable and robust than vanilla FAT at classifying OCT volumes of different resolutions. On a large-scale retinal OCT datasets, our results indicate that this effective method performs in the majority of cases better than other common methods for volume classification.

Besides its applicability for volumetric medical data analysis, our VFLAT has potential to be applied on other medical tasks including video analysis (e.g. ultrasound videos) and high-resolution imaging, as is the case in histopathology. Future work would include adapting VLFAT to ViViT models to make them less computationally expensive. Furthermore, PEs in VLFAT could be leveraged for improving the visual interpretation of decision models by collecting positional information about the adjacent slices sharing anatomical similarities.

Acknowledgements. This work was supported in part by the Christian Doppler Research Association, Austrian Federal Ministry for Digital and Economic Affairs, the National Foundation for Research, Technology and Development, and Heidelberg Engineering.

References

1. Arnab, A., Dehghani, M., Heigold, G., Sun, C., Lučić, M., Schmid, C.: Vivit: a video vision transformer. In: Proceedings of the IEEE/CVF International Conference on Computer Vision, pp. 6836–6846 (2021)
2. Beyer, L., et al.: Flexivit: one model for all patch sizes. arXiv preprint arXiv:2212.08013 (2022)
3. Blu, T., Thévenaz, P., Unser, M.: Linear interpolation revitalized. IEEE Trans. Image Process. **13**(5), 710–719 (2004)
4. Chung, J.S., Zisserman, A.: Lip reading in the wild. In: Lai, S.-H., Lepetit, V., Nishino, K., Sato, Y. (eds.) ACCV 2016. LNCS, vol. 10112, pp. 87–103. Springer, Cham (2017). https://doi.org/10.1007/978-3-319-54184-6_6
5. Das, V., Prabhakararao, E., Dandapat, S., Bora, P.K.: B-scan attentive CNN for the classification of retinal optical coherence tomography volumes. IEEE Signal Process. Lett. **27**, 1025–1029 (2020)
6. De Fauw, J., et al.: Clinically applicable deep learning for diagnosis and referral in retinal disease. Nat. Med. **24**(9), 1342–1350 (2018)
7. Devlin, J., Chang, M.W., Lee, K., Toutanova, K.: Bert: pre-training of deep bidirectional transformers for language understanding. arXiv preprint arXiv:1810.04805 (2018)
8. Dosovitskiy, A., et al.: An image is worth 16x16 words: Transformers for image recognition at scale. arXiv preprint arXiv:2010.11929 (2020)
9. Fang, L., Wang, C., Li, S., Yan, J., Chen, X., Rabbani, H.: Automatic classification of retinal three-dimensional optical coherence tomography images using principal component analysis network with composite kernels. J. Biomed. Opt. **22**(11), 116011–116011 (2017)
10. He, K., et al.: Transformers in medical image analysis: a review. Intell. Med. (2022)
11. Howard, J.P., et al.: Improving ultrasound video classification: an evaluation of novel deep learning methods in echocardiography. J. Med. Artif. Intell. **3** (2020)
12. Loshchilov, I., Hutter, F.: Decoupled weight decay regularization. arXiv preprint arXiv:1711.05101 (2017)
13. Peiris, H., Hayat, M., Chen, Z., Egan, G., Harandi, M.: A robust volumetric transformer for accurate 3D tumor segmentation. In: Wang, L., Dou, Q., Fletcher, P.T., Speidel, S., Li, S. (eds.) MICCAI 2022. LNCS, vol. 13435, pp. 162–172. Springer, Cham (2022). https://doi.org/10.1007/978-3-031-16443-9_16
14. Playout, C., Duval, R., Boucher, M.C., Cheriet, F.: Focused attention in transformers for interpretable classification of retinal images. Med. Image Anal. **82**, 102608 (2022)
15. Prabhushankar, M., Kokilepersaud, K., Logan, Y.Y., Corona, S.T., AlRegib, G., Wykoff, C.: Olives dataset: Ophthalmic labels for investigating visual eye semantics. arXiv preprint arXiv:2209.11195 (2022)
16. Qiu, J., Sun, Y.: Self-supervised iterative refinement learning for macular oct volumetric data classification. Comput. Biol. Med. **111**, 103327 (2019)

17. Rasti, R., Rabbani, H., Mehridehnavi, A., Hajizadeh, F.: Macular OCT classification using a multi-scale convolutional neural network ensemble. IEEE Trans. Med. Imaging **37**(4), 1024–1034 (2017)
18. Romo-Bucheli, D., Erfurth, U.S., Bogunović, H.: End-to-end deep learning model for predicting treatment requirements in neovascular AMD from longitudinal retinal OCT imaging. IEEE J. Biomed. Health Inform. **24**(12), 3456–3465 (2020)
19. Semivariogram and semimadogram functions as descriptors for AMD diagnosis on SD-OCT topographic maps using support vector machine. Biomed. Eng. Online **17**(1), 1–20 (2018)
20. Simonyan, K., Zisserman, A.: Two-stream convolutional networks for action recognition in videos. In: Advances in Neural Information Processing Systems, vol. 27 (2014)
21. Singh, S.P., Wang, L., Gupta, S., Goli, H., Padmanabhan, P., Gulyás, B.: 3D deep learning on medical images: a review. Sensors **20**(18), 5097 (2020)
22. Sun, Y., Zhang, H., Yao, X.: Automatic diagnosis of macular diseases from OCT volume based on its two-dimensional feature map and convolutional neural network with attention mechanism. J. Biomed. Opt. **25**(9), 096004–096004 (2020)
23. de Vente, C., González-Gonzalo, C., Thee, E.F., van Grinsven, M., Klaver, C.C., Sánchez, C.I.: Making AI transferable across oct scanners from different vendors. Invest. Ophthalmol. Visual Sci. **62**(8), 2118–2118 (2021)
24. Wang, J., Cherian, A., Porikli, F., Gould, S.: Video representation learning using discriminative pooling. In: Proceedings of the IEEE Conference on Computer Vision and Pattern Recognition, pp. 1149–1158 (2018)
25. Wightman, R.: Pytorch image models (2019) https://doi.org/10.5281/zenodo.4414861. https://github.com/rwightman/pytorch-image-models
26. Windsor, R., Jamaludin, A., Kadir, T., Zisserman, A.: Context-aware transformers for spinal cancer detection and radiological grading. In: Wang, L., Dou, Q., Fletcher, P.T., Speidel, S., Li, S. (eds.) MICCAI 2022. LNCS, vol. 13433, pp. 271–281. Springer, Cham (2022). https://doi.org/10.1007/978-3-031-16437-8_26
27. Wulczyn, E., et al.: Deep learning-based survival prediction for multiple cancer types using histopathology images. PLoS ONE **15**(6), e0233678 (2020)
28. Zhong, Z., Zheng, L., Kang, G., Li, S., Yang, Y.: Random erasing data augmentation. In: Proceedings of the AAAI Conference on Artificial Intelligence, vol. 34, pp. 13001–13008 (2020)

ProtoASNet: Dynamic Prototypes for Inherently Interpretable and Uncertainty-Aware Aortic Stenosis Classification in Echocardiography

Hooman Vaseli[1], Ang Nan Gu[1], S. Neda Ahmadi Amiri[1], Michael Y. Tsang[2], Andrea Fung[1], Nima Kondori[1], Armin Saadat[1], Purang Abolmaesumi[1(✉)], and Teresa S. M. Tsang[2]

[1] Department of Electrical and Computer Engineering, The University of British Columbia, Vancouver, BC, Canada
{hoomanv,guangnan,purang}@ece.ubc.ca
[2] Vancouver General Hospital, Vancouver, BC, Canada

Abstract. Aortic stenosis (AS) is a common heart valve disease that requires accurate and timely diagnosis for appropriate treatment. Most current automatic AS severity detection methods rely on black-box models with a low level of trustworthiness, which hinders clinical adoption. To address this issue, we propose ProtoASNet, a prototypical network that directly detects AS from B-mode echocardiography videos, while making interpretable predictions based on the similarity between the input and learned spatio-temporal prototypes. This approach provides supporting evidence that is clinically relevant, as the prototypes typically highlight markers such as calcification and restricted movement of aortic valve leaflets. Moreover, ProtoASNet utilizes abstention loss to estimate aleatoric uncertainty by defining a set of prototypes that capture ambiguity and insufficient information in the observed data. This provides a reliable system that can detect and explain when it may fail. We evaluate ProtoASNet on a private dataset and the publicly available TMED-2 dataset, where it outperforms existing state-of-the-art methods with an accuracy of 80.0% and 79.7%, respectively. Furthermore, ProtoASNet provides interpretability and an uncertainty measure for each prediction, which can improve transparency and facilitate the interactive usage of deep networks to aid clinical decision-making. Our source code is available at: https://github.com/hooman007/ProtoASNet.

Keywords: Aleatoric Uncertainty · Aortic Stenosis · Echocardiography · Explainable AI · Prototypical Networks

T. S. M. Tsang and P. Abolmaesumi are joint senior authors.
H. Vaseli, A. Gu, and N. Ahmadi are joint first authors.

Supplementary Information The online version contains supplementary material available at https://doi.org/10.1007/978-3-031-43987-2_36.

H. Greenspan et al. (Eds.): MICCAI 2023, LNCS 14225, pp. 368–378, 2023.
https://doi.org/10.1007/978-3-031-43987-2_36

1 Introduction

Aortic stenosis (AS) is a common heart valve disease characterized by the calcification of the aortic valve (AV) and the restriction of its movement. It affects 5% of individuals aged 65 or older [2] and can progress rapidly from mild or moderate to severe, reducing life expectancy to 2 to 3 years [20]. Echocardiography (echo) is the primary diagnostic modality for AS. This technique measures Doppler-derived clinical markers [16] and captures valve motion from the parasternal long (PLAX) and short axis (PSAX) cross-section views. However, obtaining and interpreting Doppler measurements requires specialized training and is subject to significant inter-observer variability [14,15].

To alleviate this issue, deep neural network (DNN) models have been proposed for automatic assessment of AS directly from two-dimensional B-mode echo, a modality more commonly used in point-of-care settings. Huang et al. [9,10] proposed a multitask model to classify the severity of AS using echo images. Ginsberg et al. [6] proposed an ordinal regression-based method that predicts the severity of AS and provides an estimate of aleatoric uncertainty due to uncertainty in training labels. However, these works utilized black-box DNNs, which could not provide an explanation of their prediction process.

Explainable AI (XAI) methods can provide explanations of a DNN's decision making process and can generally be categorized into two classes. Post-hoc XAI methods explain the decisions of trained black-box DNNs. For example, gradient-based saliency maps [18,19] show where a model pays attention to, but these methods do not necessarily explain why one class is chosen over another [17], and at times result in misleading explanations [1]. Ante-hoc XAI methods are explicitly designed to be explainable. For instance, prototype-based models [4, 8,11,12,22,23], which the contributions of our paper fall under, analyze a given input based on its similarity to learned discriminative features (or "prototypes") for each class. Both the learned prototypes and salient image patches of the input can be visualized for users to validate the model's decision making.

There are two limitations to applying current prototype-based methods to the task of classifying AS severity from echo cine series. First, prototypes should be spatio-temporal instead of only spatial, since AS assessment requires attention to small anatomical regions in echo (such as the AV) at a particular phase of the heart rhythm (mid-systole). Second, user variability in cardiac view acquisition and poor image quality can complicate AV visualization in standard PLAX and PSAX views. The insufficient information in such cases can lead to more plausible diagnoses than one. Therefore, a robust solution should avoid direct prediction and notify the user. These issues have been largely unaddressed in previous work.

We propose ProtoASNet (Fig. 1), a prototype-based model for classifying AS severity from echo cine series. ProtoASNet discovers dynamic prototypes that describe shape- and movement-based phenomena relevant to AS severity, outperforming existing models that only utilize image-based prototypes. Additionally, our model can detect ambiguous decision-making scenarios based on similarity with less informative samples in the training set. This similarity is expressed as a measure of aleatoric uncertainty. To the best of our knowledge, the only prior

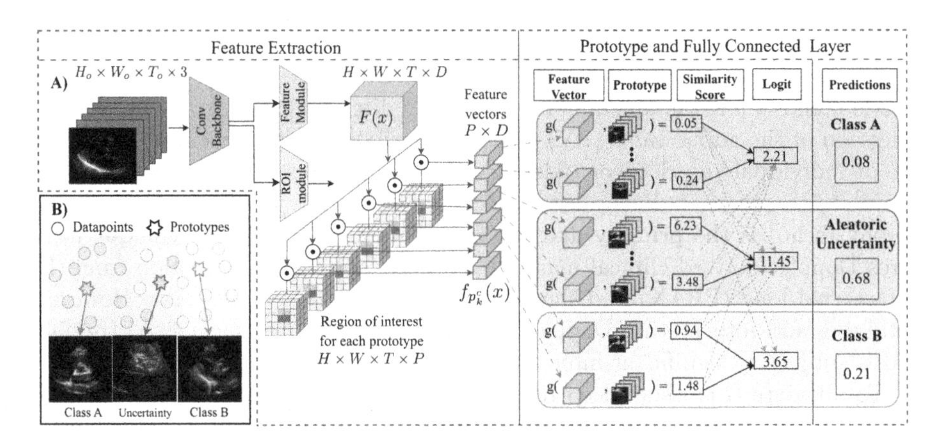

Fig. 1. **(A)** An overview of our proposed ProtoASNet architecture. ProtoASNet extracts spatio-temporal feature vectors $f_{p_k^c}(x)$ from the video, which are compared with learned prototypes. Similarity values between features and prototypes are aggregated to produce a score for class membership and aleatoric uncertainty. **(B)** Prototypes representing aleatoric uncertainty (blue) can capture regions of the data distribution with inherent ambiguity (intersection between green and yellow regions). In practice, this region consists of videos with poor visual quality. (Color figure online)

work for dynamic prototypes published to-date is [7]. ProtoASNet is the first work to use dynamic prototypes in medical imaging and the first to incorporate aleatoric uncertainty estimation with prototype-based networks.

2 Methods

2.1 Background: Prototype-Based Models

Prototype-based models explicitly make their decisions using similarities to cases in the training set. These models generally consist of three key components structured as $h(g(f(x)))$. Firstly, $f(.)$ is a feature encoder such as a ConvNet that maps images $x \in \mathbb{R}^{H_o \times W_o \times 3}$ to $f(x) \in \mathbb{R}^{H \times W \times D}$, where H, W, and D correspond to the height, width, and feature depth of the ConvNet's intermediate layer, respectively. Secondly, $g(.) \in \mathbb{R}^{H \times W \times D} \rightarrow \mathbb{R}^P$ is a prototype pooling function that computes the similarity of encoded features $f(x)$ to P prototype vectors. There are K learnable prototypes defined for each of C classes, denoted as p_k^c. Finally, $h(.) \in \mathbb{R}^P \rightarrow \mathbb{R}^C$ is a fully-connected layer that learns to weigh the input-prototype similarities against each other to produce a prediction score for each class. To ensure that the prototypes p_k^c reflect those of true examples in the training distribution, they are projected ("pushed") towards the embeddings of the closest training examples of class c.

$$p_k^c \leftarrow \underset{z \in \mathcal{Z}_c}{\arg\min} \|z - p_k^c\|_2, \text{where } \mathcal{Z}_c = \{z : z \in f_{p_k^c}(x_i) \ s.t. \ y_i \in c\} \tag{1}$$

Such models are inherently interpretable since they are enforced to first search for similar cases in the training set and then to compute how these similarities contribute to the classification. As a result, they offer a powerful approach for identifying and classifying similar patterns in data.

2.2 ProtoASNet

Feature Extraction. The overall structure of ProtoASNet is shown in Fig. 1. The feature extraction layer consists of a convolutional backbone, in our case the first three blocks of a pre-trained R(2+1)D-18 [21] model, followed by two branches of feature and region of interest (ROI) modules made up of two and three convolutional layers respectively. In both modules, the convolutional layers have ReLU activation function, except the last layers which have linear activations. Given an input video $x \in \mathbb{R}^{H_o \times W_o \times T_o \times 3}$ with T_o frames, the first branch learns a feature $F(x) \in \mathbb{R}^{H \times W \times T \times D}$, where each D-dimensional vector in $F(x)$ corresponds to a specific spatio-temporal region in the video. The second branch generates P regions of interest, $M_{p_k^c}(x) \in \mathbb{R}^{H \times W \times T}$, that specify which regions of $F(x)$ are relevant for comparing with each prototype p_k^c.

The features from different spatio-temporal regions must be pooled before being compared to prototypes. As in [12], we perform a weighted average pooling with the learned regions of interest as follows:

$$f_{p_k^c}(x) = \frac{1}{HWT} \sum_{H,W,T} |M_{p_k^c}(x)| \circ F(x), \tag{2}$$

where $|.|$ is the absolute value and $\circ$ is the Hadamard product.

Prototype Pooling. The similarity score of a feature vector $f_{p_k^c}$ and prototype p_k^c is calculated using cosine similarity, which is then shifted to $[0, 1]$:

$$g(x, p_k^c) = \frac{1}{2}(1 + \frac{< f_{p_k^c}(x), p_k^c >}{\|f_{p_k^c}(x)\|_2 \|p_k^c\|_2}). \tag{3}$$

Prototypes for Aleatoric Uncertainty Estimation. In Fig. 1, trainable uncertainty prototypes (denoted p_k^u) are added to capture regions in the data distribution that are inherently ambiguous (Fig. 1.B). We use similarity between $f_{p_k^u}(x)$ and p_k^u to quantify aleatoric uncertainty, denoted $\alpha \in [0, 1]$. We use an "abstention loss" (Eq. (6)) method inspired by [5] to learn α and thereby p_k^u. In this loss, α is used to interpolate between the ground truth and prediction, pushing the model to "abstain" from its own answer at a penalty.

$$\hat{y} = \sigma(h(g(x, p_k^c))), \quad \alpha = \sigma(h(g(x, p_k^u))); \tag{4}$$

$$\hat{y}' = (1 - \alpha)\hat{y} + \alpha y; \tag{5}$$

$$\mathcal{L}_{abs} = CrsEnt(\hat{y}', y) - \lambda_{abs} \log(1 - \alpha), \tag{6}$$

where σ denotes Softmax normalization in the output of $h(.)$, y and $\hat{y}$ are the ground truth and the predicted probabilities, respectively, and λ_{abs} is a regularization constant.

When projecting p_k^u to the nearest extracted feature from training examples, we relax the requirement in Eq. (1) allowing the uncertainty prototypes to be pushed to data with the ground truth of any AS severity class.

Class-Wise Similarity Score. The fully connected (FC) layer $h(.)$ is a dense mapping from prototype similarity scores to prediction logits. Its weights, w_h, are initialized to be 1 between class c and the corresponding prototypes and 0 otherwise to enforce the process to resemble positive reasoning. $h(.)$ produces a score for membership in each class and for α.

Loss Function. As in previous prototype-based methods [4,12], the following losses are introduced to improve performance: 1) Clustering and separation losses (Eq. (7)), which encourage clustering based on class, where $\mathcal{P}_y$ denotes the set of prototypes belonging to class y. Due to lack of ground truth uncertainties, these losses are only measured on p_k^c, not p_k^u; 2) Orthogonality loss (Eq. (8)), which encourages prototypes to be more diverse; 3) Transformation loss $\mathcal{L}_{trns}$ (described in [12]), which regularizes the consistency of the predicted occurrence regions under random affine transformations; 4) Finally, $\mathcal{L}_{norm}$ (described in [4]) regularizes w_h to be close to its initialization and penalizes relying on similarity to one class to influence the logits of other classes. Equation (9) describes the overall loss function where λ represent regularization coefficients for each loss term. The network is trained end-to-end. We conduct a "push" stage (see Eq. (1)) every 5 epochs to ensure that the learned prototypes are consistent with the embeddings from real examples.

$$\mathcal{L}_{clst} = -\max_{p_k^c \in \mathcal{P}_y} g(x, p_k^c), \quad \mathcal{L}_{sep} = \max_{p_k^c \notin \mathcal{P}_y} g(x, p_k^c); \tag{7}$$

$$\mathcal{L}_{orth} = \sum_{i>j} \frac{< p_i, p_j >}{\|p_i\|_2 \|p_j\|_2}; \tag{8}$$

$$\mathcal{L} = \mathcal{L}_{abs} + \lambda_{clst}\mathcal{L}_{clst} + \lambda_{sep}\mathcal{L}_{sep} + \lambda_{orth}\mathcal{L}_{orth} + \lambda_{trns}\mathcal{L}_{trns} + \lambda_{norm}\mathcal{L}_{norm}. \tag{9}$$

3 Experiments and Results

3.1 Datasets

We conducted experiments on a private AS dataset and the public TMED-2 dataset [10]. The private dataset was extracted from an echo study database of a tertiary care hospital with institutional review ethics board approval. Videos were acquired with Philips iE33, Vivid i, and Vivid E9 ultrasound machines. For each study, the AS severity was classified using clinically standard Doppler echo guidelines [3] by a level III echocardiographer, keeping only cases with concordant Doppler measurements. PLAX and PSAX view cines were extracted from each study using a view-detection algorithm [13], and subsequently screened by a level III echocardiographer to remove misclassified cines. For each cine, the echo beam area was isolated and image annotations were removed. The dataset consists of 5055 PLAX and 4062 PSAX view cines, with a total of 2572 studies. These studies were divided into training, validation, and test sets, ensuring patient exclusivity and following an 80-10-10 ratio. We performed randomized augmentations including resized cropping and rotation.

The TMED-2 dataset [10] consists of 599 fully labeled echo studies containing 17270 images in total. Each study consists of 2D echo images with clinician-annotated view labels (PLAX/PSAX/Other) and Doppler-derived study-level AS severity labels (no AS/early AS/significant AS). Though the dataset includes an unlabeled portion, we trained on the labeled set only. We performed data augmentation similar to the private dataset without time-domain operations.

3.2 Implementation Details

To better compare the results with TMED-2 dataset, we adopted their labeling scheme of no AS (normal), early AS (mild), and significant AS (moderate and severe) in our private dataset. We split longer cines into 32-frame clips which are approximately one heart cycle long. In both layers of the feature module, we used D convolutional filters, while the three layers in the ROI module had D, $\frac{D}{2}$, and P convolutional filters, preventing an abrupt reduction of channels to the relatively low value of P. In both modules, we used kernel size of $1{\times}1{\times}1$. We set $D = 256$ and $K = 10$ for AS class and aleatoric uncertainty prototypes. Derived from the hyperparameter selection of ProtoPNet [4], we assigned the values of 0.8, 0.08, and 10^{-4} to λ_{clst}, λ_{sep}, and λ_{norm} respectively. Through a search across five values of 0.1, 0.3, 0.5, 0.9, and 1.0, we found the optimal λ_{abs} to be 0.3 based on the mean F1 score of the validation set. Additionally, we found λ_{orth} and λ_{trns} to be empirically better as 10^{-2} and 10^{-3} respectively. We implemented our framework in PyTorch and trained the model end-to-end on one 16 GB NVIDIA Tesla V100 GPU.

3.3 Evaluations on Private Dataset

Quantitative Assessment. In Table 1, we report the performance of ProtoASNet in AS severity classification against the black-box baselines for image (Huang et al. [9]), video (Ginsberg et al. [6]), as well as other prototypical methods, i.e. ProtoPNet [4] and XProtoNet [12]. In particular, for ProtoASNet, ProtoPNet [4], and XProtoNet [12], we conduct both image-based and video-based experiments with ResNet-18 and R(2+1)D-18 backbones respectively. We apply softmax to normalize the ProtoASNet output scores, including α, to obtain class probabilities that account for the presence of aleatoric uncertainty. We aggregate model predictions by averaging their probabilities from the image- (or clip-) level to obtain cine- and study-level predictions. We believe the uncertainty probabilities reduce the effect of less informative datapoints on final aggregated results. Additionally, the video-based models perform better than the image-based ones because the learnt prototypes can also capture AV motion which is an indicator of AS severity. These two factors may explain why our proposed method, ProtoASNet, outperforms all other methods for study-level classification.

Table 1. Quantitative results on the test set of our private dataset in terms of balanced accuracy (bACC), mean F1 score, and balanced mean absolute error (bMAE). bMAE is the average of the MAE of each class, assuming labels of 0, 1, 2 for no AS, early AS and significant AS respectively. Study-level results were calculated by averaging the prediction probabilities over all cines of each study. Results are shown as "mean(std)" calculated across five repetitions for each experiment. Best results are in bold.

Method	Cine-level (N = 973)			Study-level (N=258)		
	bACC↑	F1 ↑	bMAE↓	bACC↑	F1 ↑	bMAE↓
Huang et al. [10]	70.2(1.5)	0.70(.02)	0.33(.02)	74.7(1.6)	0.75(.02)	0.28(.02)
ProtoPNet [4]	67.8(3.7)	0.66(.05)	0.36(.05)	70.9(4.7)	0.69(.07)	0.32(.05)
XProtoNet [12]	69.2(1.3)	0.69(.01)	0.34(.01)	73.8(0.8)	0.74(.01)	0.29(.01)
ProtoASNet (Image)*	70.1(1.6)	0.70(.02)	0.33(.02)	73.9(3.5)	0.74(.04)	0.29(.04)
Ginsberg et al. [6]	**76.0(1.4)**	**0.76(.01)**	**0.26(.01)**	78.3(1.6)	0.78(.01)	0.24(.02)
XProtoNet (Video)*	74.1(1.1)	0.74(.01)	0.29(.01)	77.2(1.4)	0.77(.01)	0.25(.02)
ProtoASNet	75.4(0.9)	0.75(.01)	0.27(.01)	**80.0(1.1)**	**0.80(.01)**	**0.22(.01)**

* Feature extraction modified to the corresponding input type.

Qualitative Assessment. The interpretable reasoning process of ProtoASNet for a video example is shown in Fig. 2. We observe that ProtoASNet places significant importance on prototypes corresponding to thickened AV leaflets due to calcification, which is a characteristic of both early and significant AS. Additionally, prototypes mostly capture the part of the heart cycle that aligns with the opening of the AV, providing a clinical indication of how well the valve opens up to be able to pump blood to the rest of the body. This makes ProtoASNet's reasoning process interpretable for the user. Note how the uncertainty prototypes focusing on AV regions where the valve leaflets are not visible, are contributing to the uncertainty measure, resulting in the case being flagged as uncertain.

Ablation Study. We assessed the effect of removing distinct components of our design: uncertainty prototypes ($\mathcal{L}_{abs}, p_k^u$), clustering and separation ($\mathcal{L}_{clst}, \mathcal{L}_{sep}$), and *push* mechanism. As shown in Table 2, keeping all the aforementioned components results in superior performance in terms of bACC and bMAE. We evaluated whether the model is capable of detecting its own misclassification using the value of α (or entropy of the class predictions in the case without $\mathcal{L}_{abs}, p_k^u$). This is measured by the AUROC of detecting ($y \neq \hat{y}$). Learning p_k^u may benefit accuracy by mitigating the overfitting of p_k^c to poor-quality videos. Furthermore, α seems to be a stronger indicator for misclassification than entropy. Moreover, we measured prototype quality using diversity and sparsity [8], normalized by the total number of prototypes. Ideally, each prediction can be explained by a low number of prototypes (low s_{spars}) but different predictions are explained with different prototypes (high Diversity). When $\mathcal{L}_{clst}$ and $\mathcal{L}_{sep}$ are removed, the protoypes are less constrained, which contributes to stronger misclassification detection and more diversity, but reduce accuracy and cause explanations to be less sparse. Finally, the *push* mechanism improves performance, countering the intuition of an interpretability-performance trade-off.

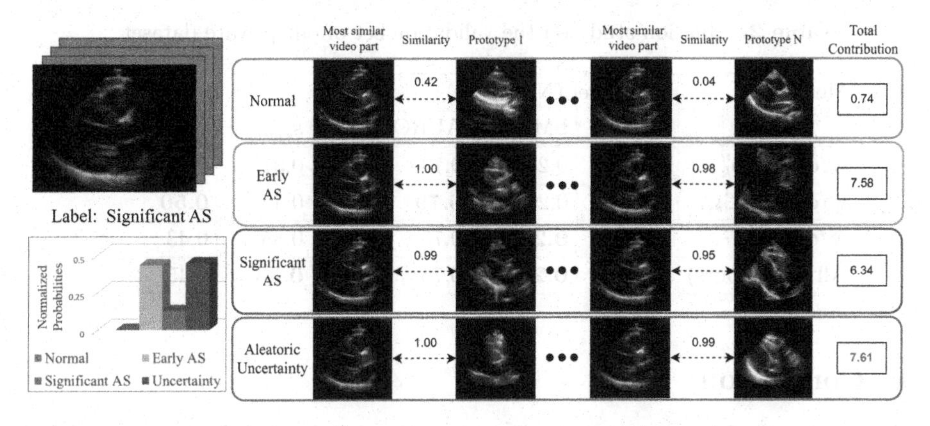

Fig. 2. Visualization of the ProtoASNet decision-making process for a test cine video showing significant AS but poor valve leaflet visualization. We visualize most similar video parts by overlaying the upsampled model-generated ROI, $M_{p_k^c}(x_{test})$, on the test cine video. Likewise, we visualize prototypes by finding the training clip each prototype is drawn from, x_p, and overlaying $M_{p_k^c}(x_p)$. ProtoASNet explains which spatio-temporal parts of the test echo are most similar to the prototypes and how accumulation of these supporting evidence results in the prediction probabilities. More visualizations of our model's performance are included in the supplementary material in video format.

3.4 Evaluation on TMED-2, a Public Dataset

We also applied our method to TMED-2, a public image-based dataset for AS diagnosis. Consistent with [10], images were fed to a WideResNet-based prototype model with two output branches. The view classifier branch used average-pooling of patches followed by a fully connected layer. However, the AS diagnosis branch used the prototype setup outlined in Methods. A diagram of the overall architecture is available in the supplementary material. We trained the model end-to-end with images from all views. During inference, images with high entropy in the predicted view and high aleatoric uncertainty for AS classification were discarded. Then, probabilities for PLAX and PSAX were used for weighted averaging to determine the study-level prediction. Addition of the prototypical layer and thresholding on predicted uncertainty achieves 79.7% accuracy for AS severity, outperforming existing black-box method [10] at 74.6%.

Table 2. Ablation study on the validation set of our private dataset.

Method	Clip-level (N = 1280)				
	bACC ↑	bMAE ↓	$\text{AUROC}_{y \neq \hat{y}}$ ↑	s_{spars} ↓	Diversity ↑
w/o $\mathcal{L}_{abs}, p_k^u$	76.1	0.25	0.73	0.37	**0.50**
w/o $\mathcal{L}_{clst}, \mathcal{L}_{sep}$	74.8	0.26	**0.79**	0.49	**0.50**
w/o *push*	77.9	**0.23**	0.75	0.35	0.43
All parts (ours)	**78.4**	**0.23**	0.75	**0.33**	0.45

4 Conclusion

We introduce ProtoASNet, an interpretable method for classifying AS severity using B-mode echo that outperforms existing black-box methods. ProtoASNet identifies clinically relevant spatio-temporal prototypes that can be visualized to improve algorithmic transparency. In addition, we introduce prototypes for estimating aleatoric uncertainty, which help flag difficult-to-diagnose scenarios, such as videos with poor visual quality. Future work will investigate methods to optimize the number of prototypes, or explore out-of-distribution detection using prototype-based methods.

Acknowledgements. This work was supported in part by the Canadian Institutes of Health Research (CIHR) and in part by the Natural Sciences and Engineering Research Council of Canada (NSERC).

References

1. Adebayo, J., Gilmer, J., Muelly, M., Goodfellow, I., Hardt, M., Kim, B.: Sanity checks for saliency maps. In: Advances in Neural Information Processing Systems, vol. 31. Curran Associates, Inc. (2018)
2. Ancona, R., Pinto, S.C.: Epidemiology of aortic valve stenosis (AS) and of aortic valve incompetence (AI): is the prevalence of AS/AI similar in different parts of the world. Eur. Soc. Cardiol. **18**(10) (2020)
3. Bonow, R.O., et al.: ACC/AHA 2006 guidelines for the management of patients with valvular heart disease: a report of the American college of cardiology/american heart association task force on practice guidelines (writing committee to revise the 1998 guidelines for the management of patients with valvular heart disease) developed in collaboration with the society of cardiovascular anesthesiologists endorsed by the society for cardiovascular angiography and interventions and the society of thoracic surgeons. J. Am. Coll. Cardiol. **48**(3), e1–e148 (2006)
4. Chen, C., Li, O., Tao, D., Barnett, A., Rudin, C., Su, J.K.: This looks like that: deep learning for interpretable image recognition. In: Advances in Neural Information Processing Systems, vol. 32 (2019)
5. DeVries, T., Taylor, G.W.: Learning confidence for out-of-distribution detection in neural networks. arXiv preprint arXiv:1802.04865 (2018)

6. Ginsberg, T., et al.: Deep video networks for automatic assessment of aortic stenosis in echocardiography. In: Noble, J.A., Aylward, S., Grimwood, A., Min, Z., Lee, S.-L., Hu, Y. (eds.) ASMUS 2021. LNCS, vol. 12967, pp. 202–210. Springer, Cham (2021). https://doi.org/10.1007/978-3-030-87583-1_20
7. Gulshad, S., Long, T., van Noord, N.: Hierarchical explanations for video action recognition. arXiv e-prints pp. arXiv-2301 (2023)
8. Hesse, L.S., Namburete, A.I.: InsightR-Net: interpretable neural network for regression using similarity-based comparisons to prototypical examples. In: Wang, L., Dou, Q., Fletcher, P.T., Speidel, S., Li, S. (eds.) MICCAI 2022. LNCS, vol. 13433, pp. 502–511. Springer, Cham (2022). https://doi.org/10.1007/978-3-031-16437-8_48
9. Huang, Z., Long, G., Wessler, B., Hughes, M.C.: A new semi-supervised learning benchmark for classifying view and diagnosing aortic stenosis from echocardiograms. In: Proceedings of the 6th Machine Learning for Healthcare Conference (2021)
10. Huang, Z., Long, G., Wessler, B., Hughes, M.C.: TMED 2: a dataset for semi-supervised classification of echocardiograms. In: DataPerf: Benchmarking Data for Data-Centric AI Workshop (2022)
11. Huang, Z., Li, Y.: Interpretable and accurate fine-grained recognition via region grouping. In: Proceedings of the IEEE/CVF Conference on Computer Vision and Pattern Recognition, pp. 8662–8672 (2020)
12. Kim, E., Kim, S., Seo, M., Yoon, S.: XProtoNet: diagnosis in chest radiography with global and local explanations. In: Proceedings of the IEEE/CVF Conference on Computer Vision and Pattern Recognition, pp. 15714–15723 (2021)
13. On modelling label uncertainty in deep neural networks: automatic estimation of intra-observer variability in 2D echocardiography quality assessment. IEEE Trans. Med. Imaging **39**(6), 1868–1883 (2019)
14. Minners, J., Allgeier, M., Gohlke-Baerwolf, C., Kienzle, R.P., Neumann, F.J., Jander, N.: Inconsistencies of echocardiographic criteria for the grading of aortic valve stenosis. Eur. Heart J. **29**(8), 1043–1048 (2008)
15. Minners, J., Allgeier, M., Gohlke-Baerwolf, C., Kienzle, R.P., Neumann, F.J., Jander, N.: Inconsistent grading of aortic valve stenosis by current guidelines: haemodynamic studies in patients with apparently normal left ventricular function. Heart **96**(18), 1463–1468 (2010)
16. Otto, C.M., et al.: 2020 ACC/AHA guideline for the management of patients with valvular heart disease: a report of the American college of cardiology/American heart association joint committee on clinical practice guidelines. Am. Coll. Cardiol. Found. Wash. DC **77**(4), e25–e197 (2021)
17. Rudin, C.: Stop explaining black box machine learning models for high stakes decisions and use interpretable models instead. Nat. Mach. Intell. **1**(5), 206–215 (2019)
18. Selvaraju, R.R., Cogswell, M., Das, A., et al.: Grad-CAM: visual explanations from deep networks via gradient-based localization. In: Proceedings of the IEEE International Conference on Computer Vision (2017)
19. Simonyan, K., Vedaldi, A., Zisserman, A.: Deep inside convolutional networks: visualising image classification models and saliency maps. In: 2nd International Conference on Learning Representations, Workshop Track Proceedings (2014)
20. Thoenes, M., et al.: Patient screening for early detection of aortic stenosis (AS)-review of current practice and future perspectives. J. Thorac. Dis. **10**(9), 5584 (2018)

21. Tran, D., Wang, H., Torresani, L., Ray, J., LeCun, Y., Paluri, M.: A closer look at spatiotemporal convolutions for action recognition. In: Proceedings of the IEEE Conference on Computer Vision and Pattern Recognition, pp. 6450–6459 (2018)
22. Trinh, L., Tsang, M., Rambhatla, S., Liu, Y.: Interpretable and trustworthy deepfake detection via dynamic prototypes. In: Proceedings of the IEEE/CVF Winter Conference on Applications of Computer Vision, pp. 1973–1983 (2021)
23. Wang, J., Liu, H., Wang, X., Jing, L.: Interpretable image recognition by constructing transparent embedding space. In: Proceedings of the IEEE/CVF International Conference on Computer Vision, pp. 895–904 (2021)

MPBD-LSTM: A Predictive Model for Colorectal Liver Metastases Using Time Series Multi-phase Contrast-Enhanced CT Scans

Xueyang Li[1], Han Xiao[2], Weixiang Weng[2], Xiaowei Xu[3], and Yiyu Shi[1](✉)

[1] University of Notre Dame, Notre Dame, USA
{xli34,yshi4}@nd.edu

[2] The First Affiliated Hospital of Sun Yat-sen University, Guangzhou, China
xiaoh69@mail.sysu.edu.cn, wengwx3@mail2.sysu.edu.cn

[3] Guangdong Provincial People's Hospital, Guangzhou, China

Abstract. Colorectal cancer is a prevalent form of cancer, and many patients develop colorectal cancer liver metastasis (CRLM) as a result. Early detection of CRLM is critical for improving survival rates. Radiologists usually rely on a series of multi-phase contrast-enhanced computed tomography (CECT) scans during follow-up visits to perform early detection of the potential CRLM. These scans form unique five-dimensional data (time, phase, and axial, sagittal, and coronal planes in 3D CT). Most of the existing deep learning models can readily handle four-dimensional data (e.g., time-series 3D CT images) and it is not clear how well they can be extended to handle the additional dimension of phase. In this paper, we build a dataset of time-series CECT scans to aid in the early diagnosis of CRLM, and build upon state-of-the-art deep learning techniques to evaluate how to best predict CRLM. Our experimental results show that a multi-plane architecture based on 3D bi-directional LSTM, which we call MPBD-LSTM, works best, achieving an area under curve (AUC) of 0.79. On the other hand, analysis of the results shows that there is still great room for further improvement. Our code is available at https://github.com/XueyangLiOSU/MPBD-LSTM.

Keywords: Colorectal cancer liver metastasis · Liver cancer prediction · Contrast-enhanced CT scan · Bi-directional LSTM

Supplementary Information The online version contains supplementary material available at https://doi.org/10.1007/978-3-031-43987-2_37.

H. Greenspan et al. (Eds.): MICCAI 2023, LNCS 14225, pp. 379–388, 2023.
https://doi.org/10.1007/978-3-031-43987-2_37

1 Introduction

Colorectal cancer is the third most common malignant tumor, and nearly half of all patients with colorectal cancer develop liver metastasis during the course of the disease [6,16]. Liver metastases after surgery of colorectal cancer is the major cause of disease-related death. Colorectal cancer liver metastases (CRLM) have therefore become one of the major focuses in the medical field. Patients with colorectal cancer typically undergo contrast-enhanced computed tomography (CECT) scans multiple times during follow-up visits after surgery for early detection of CRLM, generating a 5D dataset. In addition to the axial, sagittal, and coronal planes in 3D CT scans, the data comprises contrast-enhanced multiple phases as its 4th dimension, along with different timestamps as its 5th dimension. Radiologists heavily rely on this data to detect the CRLM in the very early stage [15].

Extensive existing works have demonstrated the power of deep learning on various spatial-temporal data, and can potentially be applied towards the problem of CRLM. For example, originally designed for natural data, several mainstream models such as E3D-LSTM [12], ConvLSTM [11] and PredRNN [13] use Convolutional Neural Networks (CNN) to capture spatial features and Long Short-Term Memory (LSTM) to process temporal features. Some other models, such as SimVP [4], replace LSTMs with CNNs but still have the capability of processing spatiotemporal information. These models can be adapted for classification tasks with the use of proper classification head.

However, all these methods have only demonstrated their effectiveness towards 3D/4D data (i.e., time-series 2D/3D images), and it is not clear how to best extend them to work with the 5D CECT data. Part of the reason is due to the lack of public availability of such data. When extending these models towards 5D CECT data, some decisions need to be made, for example: 1) What is the most effective way to incorporate the phase information? Simply concatenating different phases together may not be the optimal choice, because the positional information of the same CT slice in different phases would be lost. 2) Shall we use uni-directional LSTM or bi-direction LSTM? E3D-LSTM [12] shows uni-directional LSTM works well on natural videos while several other works show bi-directional LSTM is needed in certain medical image segmentation tasks [2,7].

In this paper, we investigate how state-of-art deep learning models can be applied to the CRLM prediction task using our 5D CECT dataset. We evaluate the effectiveness of bi-directional LSTM and explore the possible method of incorporating different phases in the CECT dataset. Specifically, we show that the best prediction accuracy can be achieved by enhancing E3D-LSTM [12] with a bi-directional LSTM and a multi-plane structure.

2 Dataset and Methodology

2.1 Dataset

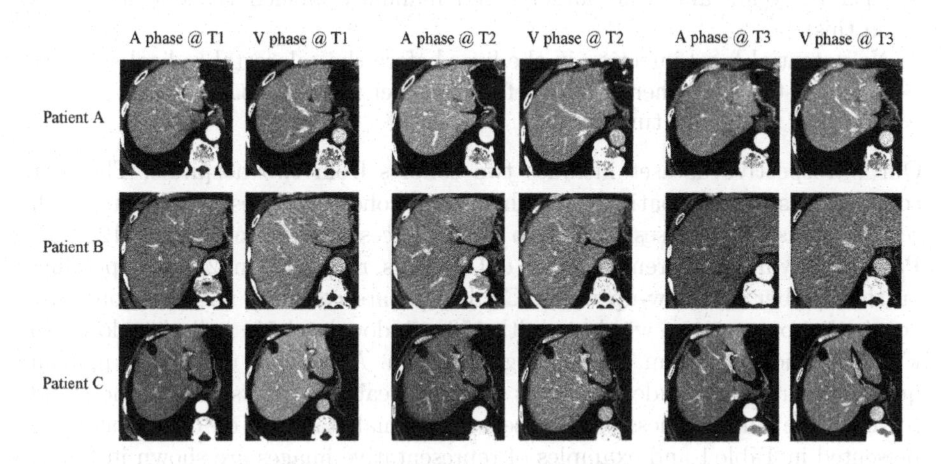

Fig. 1. Representative slices from 3D CT images of different patients in our dataset, at A/V phases and timestamps T0, T1, T2 (cropped to 256×256 for better view).

Table 1. Characreristics of our dataset

Cohort	# of positive cases	# of negative cases	total cases	positive rate
1st	60	141	201	0.299
2nd	9	59	68	0.132
Total	**69**	**200**	**269**	**0.257**

When patients undergo CECT scans to detect CRLM, typically three phases are captured: the unenhanced plain scan phase (P), the portal venous phase (V), and the arterial phase (A). The P phase provides the basic shape of the liver tissue, while the V and A phases provide additional information on the liver's normal and abnormal blood vessel patterns, respectively [10]. Professional radiologists often combine the A and V phases to determine the existence of metastases since blood in the liver is supplied by both portal venous and arterial routes.

Our dataset follows specific inclusion criteria:

- **No tumor** appears on the CT scans. That means patients **have not been diagnosed** as CRLM when they took the scans.

- Patients were previously diagnosed with colorectal cancer TNM stage I to stage III, and recovered from colorectal radical surgery.
- Patients have two or more times of CECT scans.
- We already determined whether or not the patients had liver metastases within 2 years after the surgery, and manually labeled the dataset based on this.
- No potential focal infection in the liver before the colorectal radical surgery.
- No metastases in other organs before the liver metastases.
- No other malignant tumors.

Our retrospective dataset includes two cohorts from two hospitals. The first cohort consists of 201 patients and the second cohort includes 68 patients. Each scan contains three phases and 100 to 200 CT slices with a resolution of 512×512. Patients may have different numbers of CT scans, ranging from 2 to 6, depending on the number of follow-up visits. CT images are collected with the following acquisition parameters: window width 150, window level 50, radiation dose 120 kV, slice thickness 1 mm, and slice gap 0.8 mm. All images underwent manual quality control to exclude any scans with noticeable artifacts or blurriness and to verify the completeness of all slices. Additional statistics on our dataset are presented in Table 1 and examples of representative images are shown in Fig. 1. The dataset is available upon request.

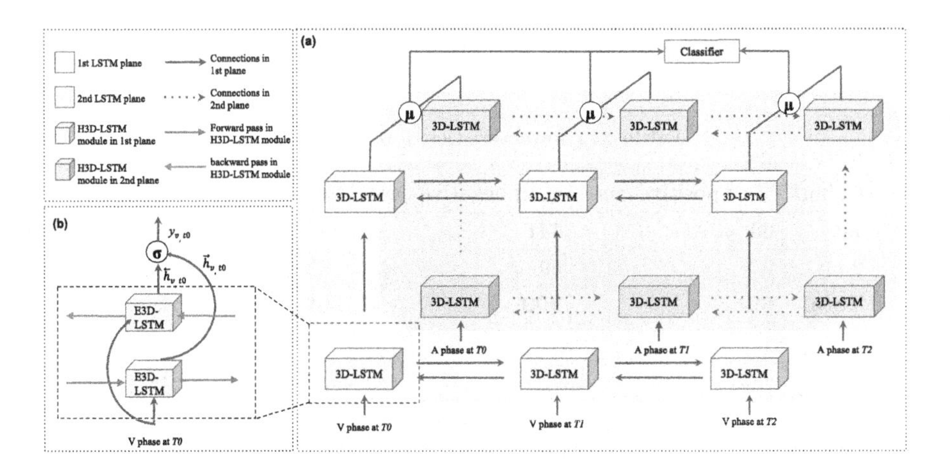

Fig. 2. (a) The general structure of MPBD-LSTM. The yellow plane is the 1st plane which is used to process the portal venous phase CT scans, and the gray plane is the second one used to process the arterial phase CT scans. μ is the average function. (b) The inner structure of a 3D-LSTM module. Blue arrow stands for the forward pass which generates the output of $\overrightarrow{h}_{v,t_0}$ and red arrow indicates the backward pass generating the output of $\overleftarrow{h}_{v,t_0}$. σ is the function used to combine two hidden-state outputs. y_{v,t_0} is the output of this 3D-LSTM module after processed by σ. (Color figure online)

2.2 Methods

Numerous state-of-the-art deep learning models are available to effectively process 4D data. In this paper, we will evaluate some of the most popular ones:

1) SaConvLSTM, introduced by Lin *et al.* [9], incorporates the self-attention mechanism into the ConvLSTM [11] structure, which improves the ability to capture spatiotemporal correlations compared to traditional LSTM.
2) E3D-LSTM, introduced by Wang *et al.* [12], integrates 3D CNNs into LSTM cells to capture both short- and long-term temporal relations. They used 3D-CNNs to handle the 3D data at each timestamp and LSTMs to compute information at different timestamps.
3) PredRNN-V2, introduced by Wang *et al.* [13,14], uses Spatiotemporal LSTM (ST-LSTM) by stacking multiple ConvLSTM units and connecting them in a zigzag pattern to handle spatiotemporal data of 4 dimensions.
4) SimVP [4], introduced by *Gao et al.*, uses CNN as the translator instead of LSTM.

All of these models need to be modified to handle 5D CECT datasets. A straightforward way to extend them is simply concatenating the A phase and V phase together, thus collapsing the 5D dataset to 4D. However, such an extension may not be the best way to incorporate the 5D spatiotemporal information, because the positional information of the same CT slice in different phases would be lost. Below we explore an alternative modification *multi-plane bi-directional LSTM* (MPBD-LSTM), based on E3D-LSTM, to handle the 5D data.

MPBD-LSTM. The most basic building block in MPBD-LSTM is the 3D-LSTM modules. Each 3D-LSTM module is composed of two E3D-LSTM cells [12]. Additionally, inspired by the bi-directional LSTM used in medical image segmentation task [2], we replace the uni-directional connections with bi-directional connections by using the backward pass in the 2nd E3D-LSTM cell in each 3D-LSTM module. This allows us to further jointly compute information from different timestamps and gives us more accurate modeling of temporal dynamics. The inner structure of one such module is shown in Fig. 2(b). Aside from the two E3D-LSTM cells, it also includes an output gate σ. Each 3D-LSTM module will generate an output $y_{v,t}$, which can be calculated as [3]:

$$y_{v,t} = \sigma(\overrightarrow{h}_{v,t}, \overleftarrow{h}_{v,t}) \tag{1}$$

where $\overrightarrow{h}_{v,t}$ and $\overleftarrow{h}_{v,t}$ are the output hidden state of the forward pass and backward pass of phase v at timestamp t, and σ is the function which is used to combine these two outputs, which we choose to use a summation function to get the summation product of these two hidden states. Therefore, the output of the bi-directional LSTM module presented in Fig. 2(b) can be represented as:

$$y_{v,t_0} = \overrightarrow{h}_{v,t_0} \oplus \overleftarrow{h}_{v,t_0} \tag{2}$$

in which $\oplus$ stands for summation. After this, the output y_{v,t_0} is passed into the bi-directional LSTM module in the next layer and viewed as input for this module.

Figure 2(a) illustrates how MPBD-LSTM uses these 3D-LSTM building blocks to handle the multiple phases in our CT scan dataset. We use two planes, one for the A phase and one for the V phase, each of which is based on a backbone of E3D-LSTM [12] with the same hyperparameters. We first use three 3D-CNN encoders (not displayed in Fig. 2(a)) as introduced in E3D-LSTM to extract the features. Each encoder is followed by a 3D-LSTM stack (the "columns") that processes the spatiotemporal data for each timestamp. The stacks are bidirectionally connected, as we described earlier, and consist of two layers of 3D-LSTM modules that are connected by their hidden states. When the spatiotemporal dataset enters the model, it is divided into smaller groups based on timestamps and phases. The 3D-LSTM stacks process these groups in parallel, ensuring that the CT slices from different phases are processed independently and in order, preserving the positional information. After the computation of the 3D-LSTM modules in each plane, we use an average function to combine the output hidden states from both planes.

An alternative approach is to additionally connect two planes by combining the hidden states of 3D-LSTM modules and taking their average if a module receives two inputs. However, we found that such design actually resulted in a worse performance. This issue will be demonstrated and discussed later in the ablation study.

In summary, the MPBD-LSTM model comprises two planes, each of which contains three 3D-LSTM stacks with two modules in each stack. It modifies E3D-LSTM by using bi-directional connected LSTMs to enhance communication between different timestamps, and a multi-plane structure to simultaneously process multiple phases.

3 Experiments

3.1 Data Augmentation and Selection

We selected 170 patients who underwent three or more CECT scans from our original dataset, and cropped the images to only include the liver area, as shown in Fig. 1. Among these cases, we identified 49 positive cases and 121 negative cases. To handle the imbalanced training dataset, we selected and duplicated 60% of positive cases and 20% of negative cases by applying Standard Scale Jittering (SSJ) [5]. For data augmentation, we randomly rotated the images from -30° to 30° and employed mixup [17]. We applied the same augmentation technique consistently to all phases and timestamps of each patient's data. We also used Spline Interpolated Zoom (SIZ) [18] to uniformly select 64 slices. For each slice, the dimension was 256×256 after cropping. We used the A and V phases of CECT for our CRLM prediction task since the P phase is only relevant when tumors are significantly present, which is not the case in our dataset. The dimension of our final input is $(3 \times 2 \times 64 \times 64 \times 64)$, representing

$(T \times P \times D \times H \times W)$, where T is the number of timestamps, P is the number of different phases, D is the slice depth, H is the height, and W is the width.

3.2 Experiment Setup

As the data size is limited, 10-fold cross-validation is adopted, and the ratio of training and testing dataset is 0.9 and 0.1, respectively. Adam optimizer [8] and Binary Cross Entropy loss function [1] are used for network training. For MPBD-LSTM, due to GPU memory constraints, we set the batch size to one and the number of hidden units in LSTM cells to 16, and trained the model till converge with a learning rate of 5e-4. Each training process required approximately 23 GB of memory and took about 20 h on an Nvidia Titan RTX GPU. We ran the 10 folds in parallel on five separate GPUs, which allowed us to complete the entire training process in approximately 40 h. We also evaluated E3D-LSTM [12], PredRNN-V2 [14], SaConvLSTM [9], and SimVP [4]. As this is a classification task, we evaluate all models' performance by their AUC scores.

4 Results and Discussion

Table 2. AUC scores of different models on our dataset

Model	AUC score
E3D-LSTM [12]	0.755
SaConvLSTM [9]	0.721
PredRNN-V2 [14]	0.765
SimVP [4]	0.662
MPBD-LSTM	**0.790**

Table 2 shows the AUC scores of all models tested on our dataset. Additional data on accuracy, sensitivity specificity, etc. can be found in the supplementary material. The MPBD-LSTM model outperforms all other models with an AUC score of 0.790. Notably, SimVP [4] is the only CNN-based model we tested, while all other models are LSTM-based. Our results suggest that LSTM networks are more effective in handling temporal features for our problem compared with CNN-based models. Furthermore, PredRNN-V2 [14], which passes memory flow in a zigzag manner of bi-directional hierarchies, outperforms the uni-directional LSTM-based SaConvLSTM [9]. Although the architecture of PredRNN-V2 is different from MPBD-LSTM, it potentially supports the efficacy of jointly computing spatiotemporal relations in different timestamps.

Table 3. Ablation study on bi-directional connection and multi-planes

Model	AUC score
MPBD-LSTM w/o multi-plane	0.774
MPBD-LSTM w/o bi-directional connection	0.768
MPBD-LSTM w/inter-plane connections	0.786
MPBD-LSTM	**0.790**

Ablation Study on Model Structures. As shown in Table 3, to evaluate the effectiveness of multi-plane and bi-directional connections, we performed ablation studies on both structures. First, we removed the multi-plane structure and concatenated the A and V phases as input. This produced a one-dimensional bi-directional LSTM (Fig. 2(a), without the gray plane) with an input dimension of $3 \times 128 \times 64 \times 64$, which is the same as we used on other models. The resulting AUC score of 0.774 is lower than the original model's score of 0.790, indicating that computing two phases in parallel is more effective than simply concatenating them. After this, we performed an ablation study to assess the effectiveness of the bi-directional connection. By replacing the bi-directional connection with a uni-directional connection, the MPBD-LSTM model's performance decreased to 0.768 on the original dataset. This result indicates that the bi-directional connection is crucial for computing temporal information effectively, and its inclusion is essential for achieving high performance in MPBD-LSTM.

Also, as mentioned previously, we initially connected the 3D-LSTM modules in two planes with their hidden states. However, as shown in Table 3, we observed that inter-plane connections actually decreased our AUC score to 0.786 compared to 0.790 without the connections. This may be due to the fact that when taking CT scans with contrast, different phases have a distinct focus, showing different blood vessels as seen in Fig. 1. Connecting them with hidden states in the early layers could disrupt feature extraction for the current phase. Therefore, we removed the inter-plane connections in the early stage, since their hidden states are still added together and averaged after they are processed by the LSTM layers.

Table 4. Ablation study on timestamps and phases

Model structure	AUC score
MPBD-LSTM @ T0	0.660
MPBD-LSTM @ T1	0.676
MPBD-LSTM @ T2	0.709
MPBD-LSTM @ all timestamps w/only A phase	0.653
MPBD-LSTM @ all timestamps w/only V phase	0.752
MPBD-LSTM @ All 3 timestamps	**0.790**

Ablation Study on Timestamps and Phases. We conducted ablation studies using CT images from different timestamps and phases to evaluate the effectiveness of time-series data and multi-phase data. The results, as shown in Table 4, indicate that MPBD-LSTM achieves AUC scores of 0.660, 0.676, and 0.709 if only images from timestamps T0, T1, and T2 are used, respectively. These scores suggest that predicting CRLM at earlier stages is more challenging since the features about potential metastases in CT images get more significant over time. However, all of these scores are significantly lower than the result using CT images from all timestamps. This confirms the effectiveness of using a time-series predictive model. Additionally, MPBD-LSTM obtains AUC scores of 0.653 and 0.752 on single A and V phases, respectively. These results suggest that the V phase is more effective when predicting CRLM, which is consistent with medical knowledge [15]. However, both of these scores are lower than the result of combining two phases, indicating that a multi-phase approach is more useful.

Error Analysis. In Fig. 1, Patients B and C are diagnosed with positive CRLM later. MPBD-LSTM correctly yields a positive prediction for Patient B with a confidence of 0.82, but incorrectly yields a negative prediction for Patient C with a confidence of 0.77. With similar confidence in the two cases, the error is likely due to the relatively smaller liver size of Patient C. Beyond this case, we find that small liver size is also present in most of the false negative cases. A possible explanation would be that smaller liver may provide less information for accurate prediction of CRLM. How to effectively address inter-patient variability in the dataset, perhaps by better fusing the 5D features, requires further research from the community in the future.

5 Conclusion

In this paper, we put forward a 5D CECT dataset for CRLM prediction. Based on the popular E3D-LSTM model, we established MPBD-LSTM model by replacing the uni-directional connection with the bi-directional connection to better capture the temporal information in the CECT dataset. Moreover, we used a multi-plane structure to incorporate the additional phase dimension. MPBD-LSTM achieves the highest AUC score of 0.790 among state-of-the-art approaches. Further research is still needed to improve the AUC.

References

1. Ba, J., Caruana, R.: Do deep nets really need to be deep? In: Advances in Neural Information Processing Systems, vol. 27 (2014)
2. Chen, J., Yang, L., Zhang, Y., Alber, M., Chen, D.Z.: Combining fully convolutional and recurrent neural networks for 3D biomedical image segmentation. In: Advances in Neural Information Processing Systems, vol. 29 (2016)

3. Cui, Z., Ke, R., Pu, Z., Wang, Y.: Deep bidirectional and unidirectional LSTM recurrent neural network for network-wide traffic speed prediction. arXiv preprint arXiv:1801.02143 (2018)
4. Gao, Z., Tan, C., Wu, L., Li, S.Z.: SimVP: simpler yet better video prediction. In: Proceedings of the IEEE/CVF Conference on Computer Vision and Pattern Recognition, pp. 3170–3180 (2022)
5. Ghiasi, G., et al.: Simple copy-paste is a strong data augmentation method for instance segmentation. In: Proceedings of the IEEE/CVF Conference on Computer Vision and Pattern Recognition, pp. 2918–2928 (2021)
6. Hao, M., Li, H., Wang, K., Liu, Y., Liang, X., Ding, L.: Predicting metachronous liver metastasis in patients with colorectal cancer: development and assessment of a new nomogram. World J. Surg. Oncol. **20**(1), 80 (2022)
7. Kim, S., An, S., Chikontwe, P., Park, S.H.: Bidirectional RNN-based few shot learning for 3D medical image segmentation. In: Proceedings of the AAAI Conference on Artificial Intelligence, vol. 35, pp. 1808–1816 (2021)
8. Kingma, D.P., Ba, J.: Adam: a method for stochastic optimization. arXiv preprint arXiv:1412.6980 (2014)
9. Lin, Z., Li, M., Zheng, Z., Cheng, Y., Yuan, C.: Self-attention ConvLSTM for spatiotemporal prediction. In: Proceedings of the AAAI Conference on Artificial Intelligence, vol. 34, pp. 11531–11538 (2020)
10. Patel, P.R., De Jesus, O.: Ct scan. In: StatPearls [Internet]. StatPearls Publishing (2022)
11. Shi, X., Chen, Z., Wang, H., Yeung, D.Y., Wong, W.K., Woo, W.C.: Convolutional lstm network: a machine learning approach for precipitation nowcasting. In: Advances in Neural Information Processing Systems, vol. 28 (2015)
12. Wang, Y., Jiang, L., Yang, M.H., Li, L.J., Long, M., Fei-Fei, L.: Eidetic 3D LSTM: a model for video prediction and beyond. In: International Conference on Learning Representations (2019)
13. Wang, Y., Long, M., Wang, J., Gao, Z., Yu, P.S.: PreDRNN: recurrent neural networks for predictive learning using spatiotemporal LSTMs. In: Advances in Neural Information Processing Systems, vol. 30 (2017)
14. Wang, Y., et al.: PredRNN: a recurrent neural network for spatiotemporal predictive learning (2021)
15. Xu, L.H., Cai, S.J., Cai, G.X., Peng, W.J.: Imaging diagnosis of colorectal liver metastases. World J. Gastroenterol: WJG **17**(42), 4654 (2011)
16. Yu, X., Zhu, L., Liu, J., Xie, M., Chen, J., Li, J.: Emerging role of immunotherapy for colorectal cancer with liver metastasis. Onco. Targets. Ther. **13**, 11645 (2020)
17. Zhang, H., Cisse, M., Dauphin, Y.N., Lopez-Paz, D.: mixup: beyond empirical risk minimization. arXiv preprint arXiv:1710.09412 (2017)
18. Zunair, H., Rahman, A., Mohammed, N., Cohen, J.P.: Uniformizing techniques to process CT scans with 3D CNNs for tuberculosis prediction. In: Rekik, I., Adeli, E., Park, S.H., Valdés Hernández, M.C. (eds.) PRIME 2020. LNCS, vol. 12329, pp. 156–168. Springer, Cham (2020). https://doi.org/10.1007/978-3-030-59354-4_15

Diffusion-Based Hierarchical Multi-label Object Detection to Analyze Panoramic Dental X-Rays

Ibrahim Ethem Hamamci[1(✉)], Sezgin Er[2], Enis Simsar[3], Anjany Sekuboyina[1], Mustafa Gundogar[4], Bernd Stadlinger[5], Albert Mehl[5], and Bjoern Menze[1]

[1] Department of Quantitative Biomedicine, University of Zurich, Zurich, Switzerland
ibrahim.hamamci@uzh.ch

[2] International School of Medicine, Istanbul Medipol University, Istanbul, Turkey

[3] Department of Computer Science, ETH Zurich, Zurich, Switzerland

[4] Department of Endodontics, Istanbul Medipol University, Istanbul, Turkey

[5] Center of Dental Medicine, University of Zurich, Zurich, Switzerland

Abstract. Due to the necessity for precise treatment planning, the use of panoramic X-rays to identify different dental diseases has tremendously increased. Although numerous ML models have been developed for the interpretation of panoramic X-rays, there has not been an end-to-end model developed that can identify problematic teeth with dental enumeration and associated diagnoses at the same time. To develop such a model, we structure the three distinct types of annotated data hierarchically following the FDI system, the first labeled with only quadrant, the second labeled with quadrant-enumeration, and the third fully labeled with quadrant-enumeration-diagnosis. To learn from all three hierarchies jointly, we introduce a novel diffusion-based hierarchical multi-label object detection framework by adapting a diffusion-based method that formulates object detection as a denoising diffusion process from noisy boxes to object boxes. Specifically, to take advantage of the hierarchically annotated data, our method utilizes a novel noisy box manipulation technique by adapting the denoising process in the diffusion network with the inference from the previously trained model in hierarchical order. We also utilize a multi-label object detection method to learn efficiently from partial annotations and to give all the needed information about each abnormal tooth for treatment planning. Experimental results show that our method significantly outperforms state-of-the-art object detection methods, including RetinaNet, Faster R-CNN, DETR, and DiffusionDet for the analysis of panoramic X-rays, demonstrating the great potential of our method for hierarchically and partially annotated datasets. The code and the datasets are available at https://github.com/ibrahimethemhamamci/HierarchicalDet.

Supplementary Information The online version contains supplementary material available at https://doi.org/10.1007/978-3-031-43987-2_38.

H. Greenspan et al. (Eds.): MICCAI 2023, LNCS 14225, pp. 389–399, 2023.
https://doi.org/10.1007/978-3-031-43987-2_38

Keywords: Diffusion Network · Hierarchical Learning · Multi-Label Object Detection · Panoramic Dental X-ray · Transformers

1 Introduction

The use of panoramic X-rays to diagnose numerous dental diseases has increased exponentially due to the demand for precise treatment planning [11]. However, visual interpretation of panoramic X-rays may consume a significant amount of essential clinical time [2] and interpreters may not always have dedicated training in reading scans as specialized radiologists have [13]. Thus, the diagnostic process can be automatized and enhanced by getting the help of Machine Learning (ML) models. For instance, an ML model that automatically detects abnormal teeth with dental enumeration and associated diagnoses would provide a tremendous advantage for dentists in making decisions quickly and saving their time.

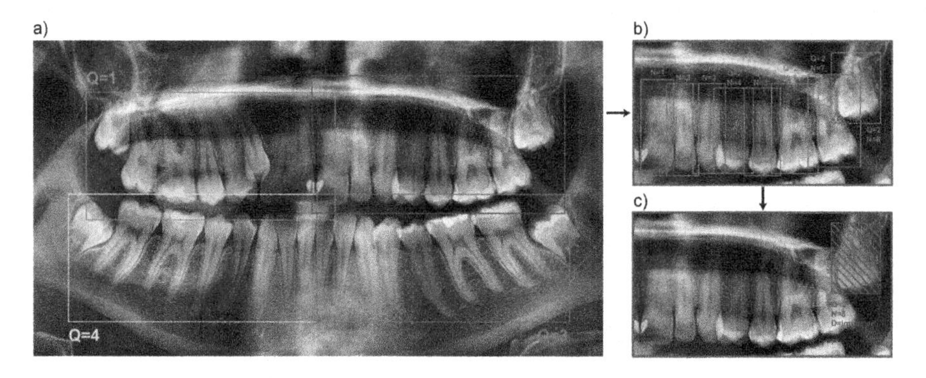

Fig. 1. The annotated datasets are organized hierarchically as (a) quadrant-only, (b) quadrant-enumeration, and (c) quadrant-enumeration-diagnosis respectively.

Many ML models to interpret panoramic X-rays have been developed specifically for individual tasks such as quadrant segmentation [19,29], tooth detection [6], dental enumeration [14,23], diagnosis of some abnormalities [12,30], as well as treatment planning [27]. Although many of these studies have achieved good results, three main issues still remain. *(1) Multi-label detection:* there has not been an end-to-end model developed that gives all the necessary information for treatment planning by detecting abnormal teeth with dental enumeration and multiple diagnoses simultaneously [1]. *(2) Data availability:* to train a model that performs this task with high accuracy, a large set of fully annotated data is needed [13]. Because labeling every tooth with all required classes may require expertise and take a long time, such kind of fully labeled large datasets do not always exist [24]. For instance, we structure three different available annotated data hierarchically shown in Fig. 1, using the Fédération Dentaire Internationale (FDI) system. The first data is partially labeled because it only included quadrant information. The second data is also partially labeled but contains additional enumeration information along with the quadrant. The third data is fully labeled because it includes all quadrant-enumeration-diagnosis information for each abnormal tooth. Thus, conventional object detection algorithms would not be well applicable to this kind of hierarchically and partially annotated data [21]. *(3) Model performance:* to the best of our knowledge, models designed to detect multiple diagnoses on panoramic X-rays have not achieved the same high level of accuracy as those specifically designed for individual tasks, such as tooth detection, dental enumeration, or detecting single abnormalities [18].

To circumvent the limitations of the existing methods, we propose a novel diffusion-based hierarchical multi-label object detection method to point out each abnormal tooth with dental enumeration and associated diagnosis concurrently on panoramic X-rays, see Fig. 2. Due to the partial annotated and hierarchical characteristics of our data, we adapt a diffusion-based method [5] that formulates object detection as a denoising diffusion process from noisy boxes to object boxes. Compared to the previous object detection methods that utilize conventional weight transfer [3] or cropping strategies [22] for hierarchical learning, the denoising process enables us to propose a novel hierarchical diffusion network by utilizing the inference from the previously trained model in hierarchical order to manipulate the noisy bounding boxes as in Fig. 2. Besides, instead of pseudo labeling techniques [28] for partially annotated data, we develop a multi-label object detection method to learn efficiently from partial annotations and to give all the needed information about each abnormal tooth for treatment planning. Finally, we demonstrate the effectiveness of our multi-label detection method on partially annotated data and the efficacy of our proposed bounding box manipulation technique in diffusion networks for hierarchical data.

The contributions of our work are three-fold. (1) We propose a multi-label detector to learn efficiently from partial annotations and to detect the abnormal tooth with all three necessary classes, as shown in Fig 3 for treatment planning. (2) We rely on the denoising process of diffusion models [5] and frame the detection problem as a hierarchical learning task by proposing a novel bounding box manipulation technique that outperforms conventional weight transfer as shown in Fig. 4. (3) Experimental results show that our model with bounding box manipulation and multi-label detection significantly outperforms state-of-the-art object detection methods on panoramic X-ray analysis, as shown in Table 1.

2 Methods

Figure 2 illustrates our proposed framework. We utilize the DiffusionDet [5] model, which formulates object detection as a denoising diffusion process from noisy boxes to object boxes. Unlike other state-of-the-art detection models, the denoising property of the model enables us to propose a novel manipulation technique to utilize a hierarchical learning architecture by using previously inferred boxes. Besides, to learn efficiently from partial annotations, we design a multi-label detector with adaptable classification layers based on available labels. In

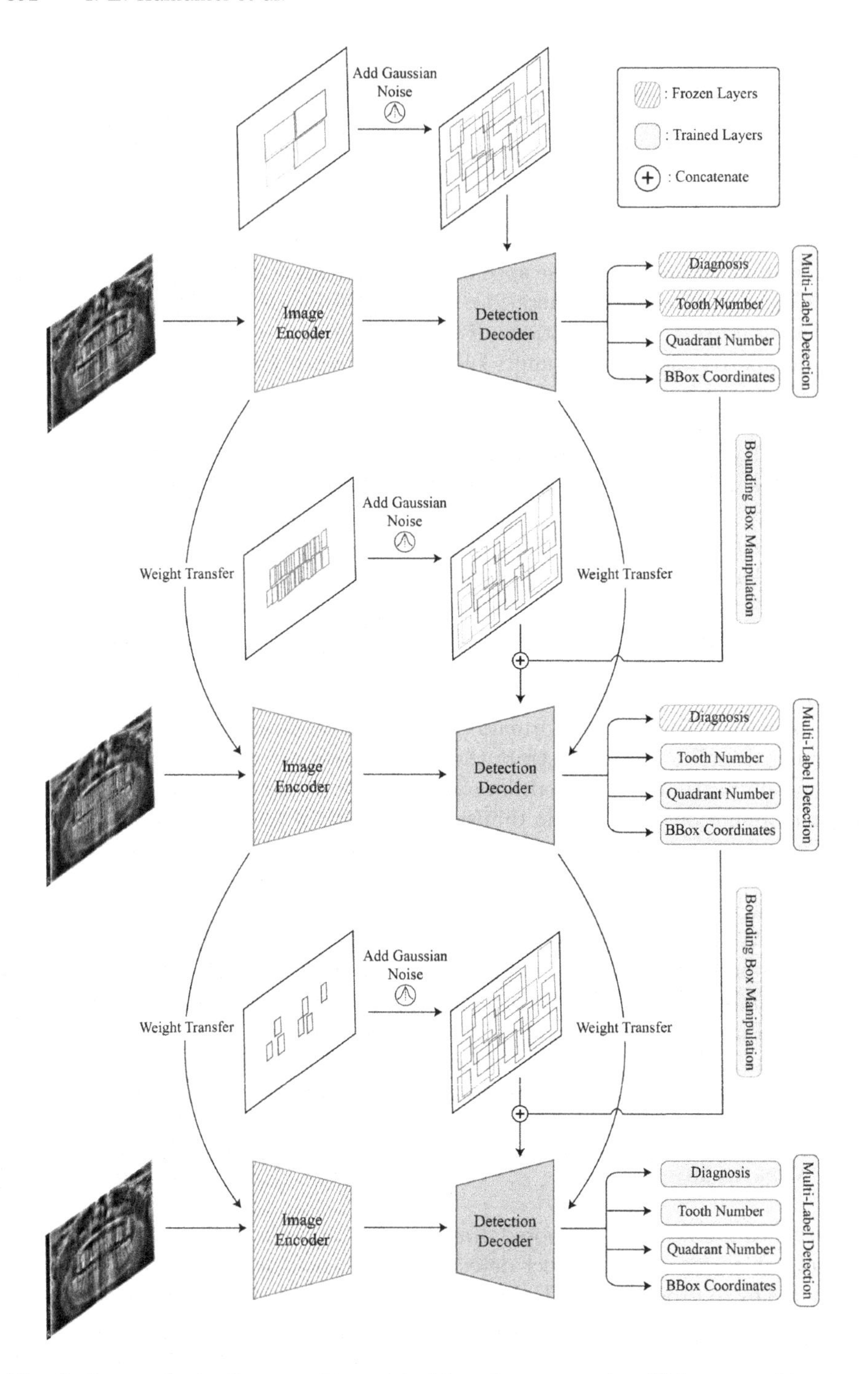

Fig. 2. Our method relies on a hierarchical learning approach utilizing a combination of multi-label detection, bounding box manipulation, and weight transfer.

addition, we designed our approach to serve as a foundational baseline for the Dental Enumeration and Diagnosis on Panoramic X-rays Challenge (DENTEX), set to take place at MICCAI 2023. Remarkably, the data and annotations we utilized for our method mirror exactly those employed for DENTEX [9].

2.1 Base Model

Our method employs the DiffusionDet [5] that comprises two essential components, an image encoder that extracts high-level features from the raw image and a detection decoder that refines the box predictions from the noisy boxes using those features. The set of initial noisy bounding boxes is defined as:

$$q(z_t|z_0) = \mathcal{N}(z_t|\sqrt{\bar{\alpha}_t}z_0, (1-\bar{\alpha}_t)I) \tag{1}$$

where z_0 represents the input bounding box b, and $b \in \mathbb{R}^{N\times 4}$ is a set of bounding boxes, z_t represents the latent noisy boxes, and $\bar{\alpha}_t$ represents the noise variance schedule. The DiffusionDet model [5] $f_\theta(z_t, t, x)$, is trained to predict the final bounding boxes defined as $b^i = (c_x^i, c_y^i, w^i, h^i)$ where (c_x^i, c_y^i) are the center coordinates of the bounding box and (w^i, h^i) are the width and height of the bounding boxes and category labels defined as y^i for objects.

2.2 Proposed Framework

To improve computational efficiency during the denoising process, DiffusionDet [5] is divided into two parts: an image encoder and a detection decoder. Iterative denoising is applied only for the detection decoder, using the outputs of the image encoder as a condition. Our method employs this approach with several adjustments, including multi-label detection and bounding box manipulation. Finally, we utilize conventional transfer learning for comparison.

Image Encoder. Our method utilizes a Swin-transformer [17] backbone pretrained on the ImageNet-22k [7] with a Feature Pyramid Network (FPN) architecture [15] as it was shown to outperform convolutional neural network-based models such as ResNet50 [10]. We also apply pre-training to the image encoder using our unlabeled data, as it is not trained during the training process. We utilize SimMIM [26] that uses masked image modeling to finetune the encoder.

Detection Decoder. Our method employs a detection decoder that inputs noisy initial boxes to extract Region of Interest (RoI) features from the encoder-generated feature map and predicts box coordinates and classifications using a detection head. However, our detection decoder has several differences from DiffusionDet [5]. Our proposed detection decoder (1) has three classification heads instead of one, which allows us to train the same model with partially annotated data by freezing the heads according to the unlabeled classes, (2) employs manipulated bounding boxes to extract RoI features, and (3) leverages transfer learning from previous training steps.

Multi-label Detection. We utilize three classification heads as quadrant-enumeration-diagnosis for each bounding box and freeze the heads for the unlabeled classes, shown in Fig. 2. Our model denoted by f_θ is trained to predict:

$$f_\theta(z_t, t, x, h_q, h_e, h_d) = \begin{cases} (y_q^i, b^i), & h_q = 1, h_e = 0, h_d = 0 \quad (a) \\ (y_q^i, y_e^i, b^i), & h_q = 1, h_e = 1, h_d = 0 \quad (b) \\ (y_q^i, y_e^i, y_d^i, b^i), & h_q = 1, h_e = 1, h_d = 1 \quad (c) \end{cases} \tag{2}$$

where y_q^i, y_e^i, and y_d^i represent the bounding box classifications for quadrant, enumeration, and diagnosis, respectively, and h_q, h_e, and h_d represent binary indicators of whether the labels are present in the training dataset. By adapting this approach, we leverage the full range of available information and improve our ability to handle partially labeled data. This stands in contrast to conventional object detection methods, which rely on a single classification head for each bounding box [25] and may not capture the full complexity of the underlying data. Besides, this approach enables the model to detect abnormal teeth with all three necessary classes for clinicians to plan the treatment, as seen in Fig. 3.

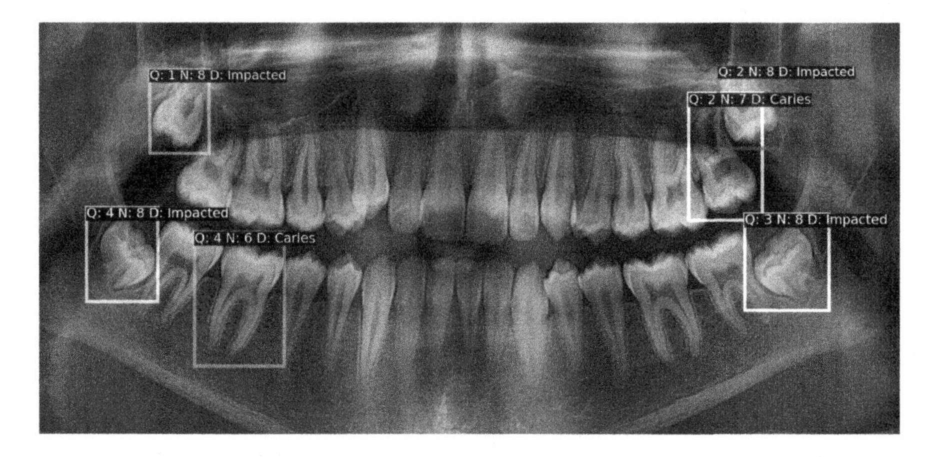

Fig. 3. Output from our final model showing well-defined boxes for diseased teeth with corresponding quadrant (Q), enumeration (N), and diagnosis (D) labels.

Bounding Box Manipulation. Instead of completely noisy boxes, we use manipulated bounding boxes to extract RoI features from the encoder-generated feature map and to learn efficiently from hierarchical annotations as shown in Fig. 2. Specifically, to train the model (b) in Eq. (2), we concatenate the noisy boxes described in Eq. (1) with the boxes inferred from the model (a) in Eq. (2) with a score greater than 0.5. Similarly, we manipulate the denoising process during the training of the model (c) in Eq. (2) by concatenating the noisy boxes with boxes inferred from the model (b) in Eq. (2) with a score greater than 0.5. The set of manipulated boxes b_m, and $b_m \in \mathbb{R}^{N\times4}$, can be defined as $b_m = [b_n[:-k], b_i]$, where b_n, and $b_n \in \mathbb{R}^{N\times4}$, represents the set of noisy boxes and, b_i, and $b_i \in \mathbb{R}^{k\times4}$, represents the set of inferred boxes from the previous training. Our framework utilizes completely noisy boxes during the inference.

3 Experiments and Results

We evaluate models' performances using a combination of Average Recall (AR) and Average Precision (AP) scores with various Intersection over Union (IoU) thresholds. This included $\text{AP}_{[0.5,0.95]}$, AP_{50}, AP_{75}, and separate AP scores for large objects (AP_l), and medium objects (AP_m).

Data. All panoramic X-rays were acquired from patients above 12 years of age using the VistaPano S X-ray unit (Durr Dental, Germany). To ensure patient privacy and confidentiality, panoramic X-rays were randomly selected from the hospital's database without considering any personal information.

To effectively utilize FDI system [8], three distinct types of data are organized hierarchically as in Fig. 1 (a) 693 X-rays labeled only for quadrant detection, (b) 634 X-rays labeled for tooth detection with both quadrant and tooth enumeration classifications, and (c) 1005 X-rays fully labeled for diseased tooth detection with quadrant, tooth enumeration, and diagnosis classifications. In the diagnosis, there are four specific classes corresponding to four different diagnoses: caries, deep caries, periapical lesions, and impacted teeth. The remaining 1571 unlabeled X-rays are used for pre-training. All necessary permissions were obtained from the ethics committee.

Experimental Design. To evaluate our proposed method, we conduct two experiments: (1) Comparison with state-of-the-art object detection models, including DETR [4], Faster R-CNN [20], RetinaNet [16], and DiffusionDet [5] in Table 1. (2) A comprehensive ablation study to assess the effect of our modifications to DiffusionDet in hierarchical detection performance in Fig. 4.

Evaluation. Fig. 3 presents the output prediction of the final trained model. As depicted in the figure, the model effectively assigns three distinct classes to each well-defined bounding box. Our approach that utilizes novel box manipulation and multi-label detection, significantly outperforms state-of-the-art methods. The box manipulation approach specifically leads to significantly higher AP and AR scores compared to other state-of-the-art methods, including RetinaNet, Faster-R-CNN, DETR, and DiffusionDet. Although the impact of conventional transfer learning on these scores can vary depending on the data, our bounding box manipulation outperforms it. Specifically, the bounding box manipulation approach is the sole factor that improves the accuracy of the model, while weight transfer does not improve the overall accuracy, as shown in Fig. 4.

Table 1. Our method outperforms state-of-the-art methods, and our bounding box manipulation approach outperforms the weight transfer. Results shown here indicate the different tasks in the test set which is multi-labeled (quadrant-enumeration-diagnosis) for abnormal tooth detection.

Method	AR	AP	AP_{50}	AP_{75}	AP_m	AP_l
Quadrant						
RetinaNet[16]	0.604	25.1	41.7	28.8	32.9	25.1
Faster R-CNN[20]	0.588	29.5	48.6	33.0	39.9	29.5
DETR[4]	0.659	39.1	60.5	47.6	55.0	39.1
Base (DiffusionDet)[5]	0.677	38.8	60.7	46.1	39.1	39.0
Ours w/o Transfer	0.699	42.7	64.7	**52.4**	50.5	42.8
Ours w/o Manipulation	**0.727**	40.0	60.7	48.2	59.3	40.0
Ours w/o Manipulation and Transfer	0.658	38.1	60.1	45.3	45.1	38.1
Ours (Manipulation+Transfer+Multilabel)	0.717	**43.2**	**65.1**	51.0	**68.3**	**43.1**
Enumeration						
RetinaNet[16]	0.560	25.4	41.5	28.5	55.1	25.2
Faster R-CNN[20]	0.496	25.6	43.7	27.0	53.3	25.2
DETR[4]	0.440	23.1	37.3	26.6	43.4	23.0
Base (DiffusionDet)[5]	0.617	29.9	47.4	34.2	48.6	29.7
Ours w/o Transfer	0.648	**32.8**	**49.4**	**39.4**	**60.1**	**32.9**
Ours w/o Manipulation	0.662	30.4	46.5	36.6	58.4	30.5
Ours w/o Manipulation and Transfer	0.557	26.8	42.4	29.5	51.4	26.5
Ours (Manipulation+Transfer+Multilabel)	**0.668**	30.5	47.6	37.1	51.8	30.4
Diagnosis						
RetinaNet[16]	0.587	32.5	54.2	35.6	41.7	32.5
Faster R-CNN[20]	0.533	33.2	54.3	38.0	24.2	33.3
DETR[4]	0.514	33.4	52.8	41.7	48.3	33.4
Base (DiffusionDet)[5]	0.644	37.0	58.1	42.6	31.8	37.2
Ours w/o Transfer	0.669	**39.4**	**61.3**	**47.9**	**49.7**	**39.5**
Ours w/o Manipulation	0.688	36.3	55.5	43.1	45.6	37.4
Ours w/o Manipulation and Transfer	0.648	37.3	59.5	42.8	33.6	36.4
Ours (Manipulation+Transfer+Multilabel)	**0.691**	37.6	60.2	44.0	36.0	37.7

Ablation Study. Our ablation study results, shown in Fig. 4 and Table 1, indicate that our approaches have a synergistic impact on the detection model's accuracy, with the highest increase seen through bounding box manipulation. We systematically remove every combination of bounding box manipulation and weight transfer, to demonstrate the efficacy of our methodology. Conventional transfer learning does not positively affect the models' performances compared to the bounding box manipulation, especially for enumeration and diagnosis.

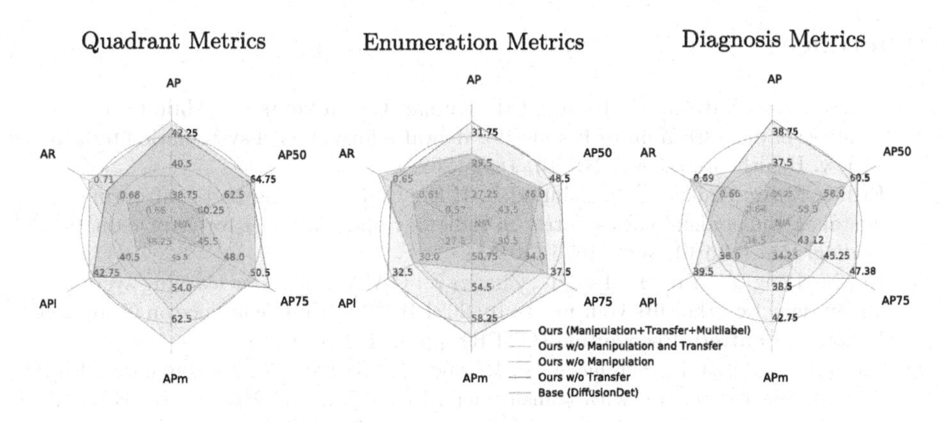

Fig. 4. The results of the ablation study reveals that our bounding box manipulation method outperforms conventional weight transfer.

4 Discussion and Conclusion

In this paper, we introduce a novel diffusion-based multi-label object detection framework to overcome one of the significant obstacles to the clinical application of ML models for medical and dental diagnosis, which is the difficulty in getting a large volume of fully labeled data. Specifically, we propose a novel bounding box manipulation technique during the denoising process of the diffusion networks with the inference from the previously trained model to take advantage of hierarchical data. Moreover, we utilize a multi-label detector to learn efficiently from partial annotations and to assign all necessary classes to each box for treatment planning. Our framework outperforms state-of-the-art object detection models for training with hierarchical and partially annotated panoramic X-ray data.

From the clinical perspective, we develop a novel framework that simultaneously points out abnormal teeth with dental enumeration and associated diagnosis on panoramic dental X-rays with the help of our novel diffusion-based hierarchical multi-label object detection method. With some limits due to partially annotated and limited amount of data, our model that provides three necessary classes for treatment planning has a wide range of applications in the real world, from being a clinical decision support system to being a guide for dentistry students.

Acknowledgements. We would like to thank the Helmut Horten Foundation for supporting our research.

References

1. AbuSalim, S., Zakaria, N., Islam, M.R., Kumar, G., Mokhtar, N., Abdulkadir, S.J.: Analysis of deep learning techniques for dental informatics: a systematic literature review. Healthcare (Basel) **10**(10), 1892 (2022)
2. Bruno, M.A., Walker, E.A., Abujudeh, H.H.: Understanding and confronting our mistakes: the epidemiology of error in radiology and strategies for error reduction. Radiographics **35**(6), 1668–1676 (2015)
3. Bu, X., Peng, J., Yan, J., Tan, T., Zhang, Z.: GAIA: a transfer learning system of object detection that fits your needs. In: 2021 IEEE/CVF Conference on Computer Vision and Pattern Recognition (CVPR), pp. 274–283 (2021)
4. Carion, N., Massa, F., Synnaeve, G., Usunier, N., Kirillov, A., Zagoruyko, S.: End-to-End object detection with transformers. In: Vedaldi, A., Bischof, H., Brox, T., Frahm, J.-M. (eds.) ECCV 2020. LNCS, vol. 12346, pp. 213–229. Springer, Cham (2020). https://doi.org/10.1007/978-3-030-58452-8_13
5. Chen, S., Sun, P., Song, Y., Luo, P.: DiffusionDet: diffusion model for object detection. arXiv preprint arXiv:2211.09788 (2022)
6. Chung, M., et al.: Individual tooth detection and identification from dental panoramic X-ray images via point-wise localization and distance regularization. Artif. Intell. Med. **111**, 101996 (2021)
7. Deng, J., Dong, W., Socher, R., Li, L.J., Li, K., Fei-Fei, L.: ImageNet: a large-scale hierarchical image database. In: 2009 IEEE Conference on Computer Vision and Pattern Recognition, Miami, FL, USA, pp. 248–255 (2009)
8. Glick, M., et al.: FDI vision 2020: shaping the future of oral health. Int. Dent. J. **62**(6), 278 (2012)
9. Hamamci, I.E., et al.: DENTEX: an abnormal tooth detection with dental enumeration and diagnosis benchmark for panoramic X-rays. arXiv preprint arXiv:2305.19112 (2023)
10. He, K., Zhang, X., Ren, S., Sun, J.: Deep residual learning for image recognition. arxiv 2015. arXiv preprint arXiv:1512.03385 (2015)
11. Hwang, J.J., Jung, Y.H., Cho, B.H., Heo, M.S.: An overview of deep learning in the field of dentistry. Imaging Sci. Dent. **49**(1), 1–7 (2019)
12. Krois, J.: Deep learning for the radiographic detection of periodontal bone loss. Sci. Rep. **9**(1), 8495 (2019)
13. Kumar, A., Bhadauria, H.S., Singh, A.: Descriptive analysis of dental X-ray images using various practical methods: a review. PeerJ Comput. Sci. **7**, e620 (2021)
14. Lin, S.Y., Chang, H.Y.: Tooth numbering and condition recognition on dental panoramic radiograph images using CNNs. IEEE Access **9**, 166008–166026 (2021)
15. Lin, T.Y., Dollár, P., Girshick, R., He, K., Hariharan, B., Belongie, S.: Feature pyramid networks for object detection. In: 2017 IEEE Conference on Computer Vision and Pattern Recognition (CVPR), Honolulu, HI, USA, pp. 2117–2125 (2017)
16. Lin, T.Y., Goyal, P., Girshick, R., He, K., Dollár, P.: Focal loss for dense object detection. In: 2017 IEEE International Conference on Computer Vision (ICCV), pp. 2980–2988 (2017)
17. Liu, Z., et al.: Swin transformer: hierarchical vision transformer using shifted windows. arXiv preprint arXiv:2103.14030 (2021)
18. Panetta, K., Rajendran, R., Ramesh, A., Rao, S.P., Agaian, S.: Tufts dental database: a multimodal panoramic X-ray dataset for benchmarking diagnostic systems. IEEE J. Biomed. Health Inform. **26**(4), 1650–1659 (2021)

19. Pati, S., et al.: GaNDLF: a generally nuanced deep learning framework for scalable end-to-end clinical workflows in medical imaging. arXiv preprint arXiv:2103.01006 (2021)
20. Ren, S., He, K., Girshick, R., Sun, J.: Faster R-CNN: towards real-time object detection with region proposal networks. In: Advances in neural information processing systems, vol. 28 (2015)
21. Shin, S.J., Kim, S., Kim, Y., Kim, S.: Hierarchical multi-label object detection framework for remote sensing images. Remote Sens. **12**(17), 2734 (2020)
22. Shin, S.J., Kim, S., Kim, Y., Kim, S.: Hierarchical multi-label object detection framework for remote sensing images. Remote Sens. **12**(17), 2734 (2020)
23. Tuzoff, D.V., et al.: Tooth detection and numbering in panoramic radiographs using convolutional neural networks. Dentomaxillofacial Radiol. **48**(4), 20180051 (2019)
24. Willemink, M.J., et al.: Preparing medical imaging data for machine learning. Radiology **295**(1), 4–15 (2020)
25. Wu, Y., Kirillov, A., Massa, F., Lo, W.Y., Girshick, R.: Detectron2 (2019)
26. Xie, Z., et al.: SimMIM: a simple framework for masked image modeling. arXiv preprint arXiv:2111.09886 (2021)
27. Yüksel, A.E., et al.: Dental enumeration and multiple treatment detection on panoramic X-rays using deep learning. Sci. Rep. **11**(1), 1–10 (2021)
28. Zhao, X., Schulter, S., Sharma, G., Tsai, Y.-H., Chandraker, M., Wu, Y.: Object detection with a unified label space from multiple datasets. In: Vedaldi, A., Bischof, H., Brox, T., Frahm, J.-M. (eds.) ECCV 2020. LNCS, vol. 12359, pp. 178–193. Springer, Cham (2020). https://doi.org/10.1007/978-3-030-58568-6_11
29. Zhao, Y., et al.: TsasNet: tooth segmentation on dental panoramic X-ray images by two-stage attention segmentation network. Knowl.-Based Syst. **206**, 106338 (2020)
30. Zhu, H., Cao, Z., Lian, L., Ye, G., Gao, H., Wu, J.: CariesNet: a deep learning approach for segmentation of multi-stage caries lesion from oral panoramic X-ray image. Neural Comput. Appl. **35**(22), 16051–16059 (2023). https://doi.org/10.1007/s00521-021-06684-2

Merging-Diverging Hybrid Transformer Networks for Survival Prediction in Head and Neck Cancer

Mingyuan Meng[1,2], Lei Bi[2](✉), Michael Fulham[1,3], Dagan Feng[1,4], and Jinman Kim[1]

[1] School of Computer Science, The University of Sydney, Sydney, Australia
[2] Institute of Translational Medicine, Shanghai Jiao Tong University, Shanghai, China
lei.bi@sjtu.edu.cn
[3] Department of Molecular Imaging, Royal Prince Alfred Hospital, Sydney, Australia
[4] Med-X Research Institute, Shanghai Jiao Tong University, Shanghai, China

Abstract. Survival prediction is crucial for cancer patients as it provides early prognostic information for treatment planning. Recently, deep survival models based on deep learning and medical images have shown promising performance for survival prediction. However, existing deep survival models are not well developed in utilizing multi-modality images (e.g., PET-CT) and in extracting region-specific information (e.g., the prognostic information in Primary Tumor (PT) and Metastatic Lymph Node (MLN) regions). In view of this, we propose a merging-diverging learning framework for survival prediction from multi-modality images. This framework has a merging encoder to fuse multi-modality information and a diverging decoder to extract region-specific information. In the merging encoder, we propose a Hybrid Parallel Cross-Attention (HPCA) block to effectively fuse multi-modality features via parallel convolutional layers and cross-attention transformers. In the diverging decoder, we propose a Region-specific Attention Gate (RAG) block to screen out the features related to lesion regions. Our framework is demonstrated on survival prediction from PET-CT images in Head and Neck (H&N) cancer, by designing an X-shape merging-diverging hybrid transformer network (named XSurv). Our XSurv combines the complementary information in PET and CT images and extracts the region-specific prognostic information in PT and MLN regions. Extensive experiments on the public dataset of HEad and neCK TumOR segmentation and outcome prediction challenge (HECKTOR 2022) demonstrate that our XSurv outperforms state-of-the-art survival prediction methods.

Keywords: Survival Prediction · Transformer · Head and Neck Cancer

Supplementary Information The online version contains supplementary material available at https://doi.org/10.1007/978-3-031-43987-2_39.

H. Greenspan et al. (Eds.): MICCAI 2023, LNCS 14225, pp. 400–410, 2023.
https://doi.org/10.1007/978-3-031-43987-2_39

1 Introduction

Head and Neck (H&N) cancer refers to malignant tumors in H&N regions, which is among the most common cancers worldwide [1]. Survival prediction, a regression task that models the survival outcomes of patients, is crucial for H&N cancer patients: it provides early prognostic information to guide treatment planning and potentially improves the overall survival outcomes of patients [2]. Multi-modality imaging of Positron Emission Tomography – Computed Tomography (PET-CT) has been shown to benefit survival prediction as it offers both anatomical (CT) and metabolic (PET) information about tumors [3, 4]. Therefore, survival prediction from PET-CT images in H&N cancer has attracted wide attention and serves as a key research area. For instance, HEad and neCK TumOR segmentation and outcome prediction challenges (HECKTOR) have been held for the last three years to facilitate the development of new algorithms for survival prediction from PET-CT images in H&N cancer [5–7].

Traditional survival prediction methods are usually based on radiomics [8], where handcrafted radiomics features are extracted from pre-segmented tumor regions and then are modeled by statistical survival models, such as the Cox Proportional Hazard (CoxPH) model [9]. In addition, deep survival models based on deep learning have been proposed to perform end-to-end survival prediction from medical images, where pre-segmented tumor masks are often unrequired [10]. Deep survival models usually adopt Convolutional Neural Networks (CNNs) to extract image features, and recently Visual Transformers (ViT) have been adopted for its capabilities to capture long-range dependency within images [11, 12]. These deep survival models have shown the potential to outperform traditional survival prediction methods [13]. For survival prediction in H&N cancer, deep survival models have achieved top performance in the HECKTOR 2021/2022 and are regarded as state-of-the-art [14–16]. Nevertheless, we identified that existing deep survival models still have two main limitations.

Firstly, existing deep survival models are underdeveloped in utilizing complementary multi-modality information, such as the metabolic and anatomical information in PET and CT images. For survival prediction in H&N cancer, existing methods usually use single imaging modality [17, 18] or rely on early fusion (i.e., concatenating multi-modality images as multi-channel inputs) to combine multi-modality information [11, 14–16, 19]. In addition, late fusion has been used for survival prediction in other diseases such as gliomas and tuberculosis [20, 21], where multi-modality features were extracted by multiple independent encoders with resultant features fused. However, early fusion has difficulties in extracting intra-modality information due to entangled (concatenated) images for feature extraction, while late fusion has difficulties in extracting inter-modality information due to fully independent feature extraction. Recently, Tang et al. [22] attempted to address this limitation by proposing a Multi-scale Non-local Attention Fusion (MNAF) block for survival prediction of glioma patients, in which multi-modality features were fused via non-local attention mechanism [23] at multiple scales. However, the performance of this method heavily relies on using tumor segmentation masks as inputs, which limits its generalizability.

Secondly, although deep survival models have advantages in performing end-to-end survival prediction without requiring tumor masks, this also incurs difficulties in extracting region-specific information, such as the prognostic information in Primary Tumor

(PT) and Metastatic Lymph Node (MLN) regions. To address this limitation, recent deep survival models adopted multi-task learning for joint tumor segmentation and survival prediction, to implicitly guide the model to extract features related to tumor regions [11, 16, 24–26]. However, most of them only considered PT segmentation and ignored the prognostic information in MLN regions [11, 24–26]. Meng et al. [16] performed survival prediction with joint PT-MLN segmentation and achieved one of the top performances in HECKTOR 2022. However, this method extracted entangled features related to both PT and MLN regions, which incurs difficulties in discovering the prognostic information in PT-/MLN-only regions.

In this study, we design an X-shape merging-diverging hybrid transformer network (named XSurv, Fig. 1) for survival prediction in H&N cancer. Our XSurv has a merging encoder to fuse complementary anatomical and metabolic information in PET and CT images and has a diverging decoder to extract region-specific prognostic information in PT and MLN regions. Our technical contributions in XSurv are three folds: (i) We propose a merging-diverging learning framework for survival prediction. This framework is specialized in leveraging multi-modality images and extracting region-specific information, which potentially could be applied to many survival prediction tasks with multi-modality imaging. (ii) We propose a Hybrid Parallel Cross-Attention (HPCA) block for multi-modality feature learning, where both local intra-modality and global inter-modality features are learned via parallel convolutional layers and cross-attention transformers. (iii) We propose a Region-specific Attention Gate (RAG) block for region-specific feature extraction, which screens out the features related to lesion regions. Extensive experiments on the public dataset of HECKTOR 2022 [7] demonstrate that our XSurv outperforms state-of-the-art survival prediction methods, including the top-performing methods in HECKTOR 2022.

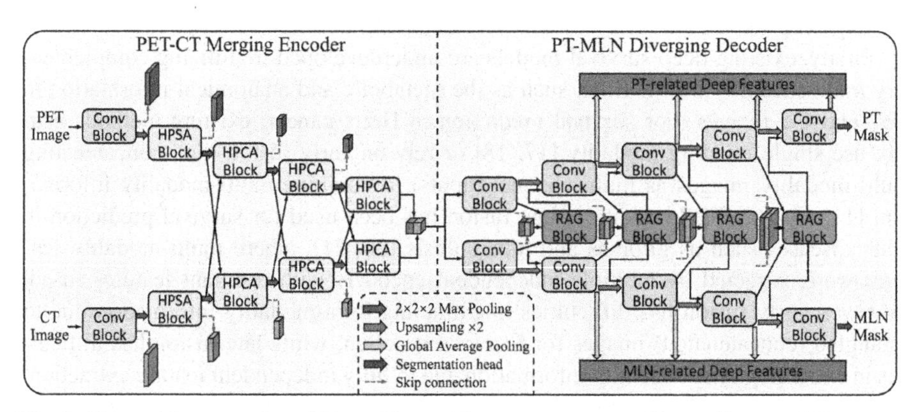

Fig. 1. The architecture of our XSurv. The architecture parameters N_{conv}, N_{self}, and N_{cross} are set as 1, 1, and 3 for illustration. Survival prediction head is omitted here for clarity.

2 Method

Figure 1 illustrates the overall architecture of our XSurv, which presents an X-shape architecture consisting of a merging encoder for multi-modality feature learning and a diverging decoder for region-specific feature extraction. The encoder includes two PET-/CT-specific feature learning branches with HPCA blocks (refer to Sect. 2.1), while the decoder includes two PT-/MLN-specific feature extraction branches with RAG blocks (refer to Sect. 2.2). Our XSurv performs joint survival prediction and segmentation, where the two decoder branches are trained to perform PT/MLN segmentation and provide PT-/MLN-related deep features for survival prediction (refer to Sect. 2.3). Our XSurv also can be enhanced by leveraging the radiomics features extracted from the XSurv-segmented PT/MLN regions (refer to Sect. 2.4). Our implementation is provided at https://github.com/MungoMeng/Survival-XSurv.

2.1 PET-CT Merging Encoder

Assuming N_{conv}, N_{self}, and N_{cross} are three architecture parameters, each encoder branch consists of N_{conv} Conv blocks, N_{self} Hybrid Parallel Self-Attention (HPSA) blocks, and N_{cross} HPCA blocks. Max pooling is applied between blocks and the features before max pooling are propagated to the decoder through skip connections. As shown in Fig. 2(a), HPCA blocks perform parallel convolution and cross-attention operations. The convolution operations are realized using successive convolutional layers with residual connections, while the cross-attention operations are realized using Swin Transformer [27] where the input x_{in} (from the same encoder branch) is projected as Q and the input x_{cross} (from the other encoder branch) is projected as K and V. In addition, Conv blocks perform the same convolution operations as HPCA blocks but discard cross-attention operations; HPSA blocks share the same overall architecture with HPCA blocks but perform self-attention within the input x_{in} (i.e., the x_{in} is projected as Q, K and V). Conv and HPSA blocks are used first and then followed by HPCA blocks, which enables the XSurv to learn both intra- and inter-modality information. In this study, we set N_{conv}, N_{self}, and N_{cross} as 1, 1, and 3, as this setting achieved the best validation results (refer to the supplementary materials). Other architecture details are also presented in the supplementary materials.

The idea of adopting convolutions and transformers in parallel has been explored for segmentation [28], which suggests that parallelly aggregating global and local information is beneficial for feature learning. In this study, we extend this idea to multi-modality feature learning, which parallelly aggregates global inter-modality and local intra-modality information via HPCA blocks, to discover inter-modality interactions while preserving intra-modality characteristics.

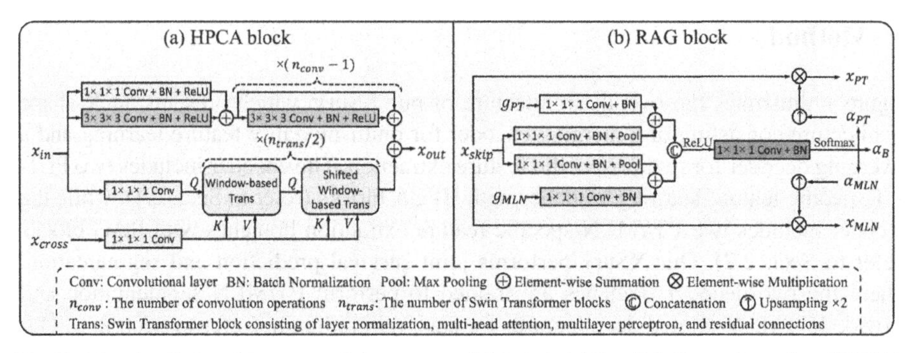

Fig. 2. The detailed architecture of the proposed (a) Hybrid Parallel Cross-Attention (HPCA) block and (b) Region-specific Attention Gate (RAG) block.

2.2 PT-MLN Diverging Decoder

As shown in Fig. 1, each decoder branch is symmetric to the encoder branch and thus includes a total of (N_{conv}+N_{self}+N_{cross}) Conv blocks. The features propagated from skip connections are fed into RAG blocks for feature diverging before entering the Conv blocks in two decoder branches, where the output of the former Conv block is upsampled and concatenated with the output of the RAG block. As shown in Fig. 2(b), RAG blocks generate three softmax-activated spatial attention maps α_{PT}, α_{MLN}, and α_B that correspond to PT, MLN, and background regions. These attention maps are computed based on the contextual information provided by the gating signals g_{PT} and g_{MLN} (which are the outputs of the former Conv blocks in the PT and MLN branches). The attention maps α_{PT} and α_{MLN} are multiplied with the features x_{skip} that are propagated from skip connections, which spatially diverge the features x_{skip} into PT- and MLN-related features x_{PT} and x_{MLN}. Different from the vanilla Attention Gate (AG) block [29], RAG blocks leverage the gating signals from two decoder branches and generate mutually exclusive (softmax-activated) attention maps.

The output of the last Conv block in the PT/MLN branch is fed into a segmentation head, which generates PT/MLN segmentation masks using a sigmoid-activated 1 × 1 × 1 convolutional layer. In addition, the outputs of all but not the first Conv blocks in the PT/MLN branches are fed into global averaging pooling layers to derive PT-/MLN-related deep features. Finally, all deep features are fed into a survival prediction head, which maps the deep features into a survival score using two fully-connected layers with dropout, L2 regularization, and sigmoid activation.

2.3 Multi-task Learning

Following existing multi-task deep survival models [11, 16, 24–26], our XSurv is end-to-end trained for survival prediction and PT-MLN segmentation using a combined loss: $\mathcal{L} = \mathcal{L}_{Surv} + \lambda(\mathcal{L}_{PT} + \mathcal{L}_{MLN})$, where the λ is a parameter to balance the survival prediction term $\mathcal{L}_{Surv}$ and the PT/MLN segmentation terms $\mathcal{L}_{PT/MLN}$. We follow [15] to adopt a negative log-likelihood loss [30] as the $\mathcal{L}_{Surv}$. For the $\mathcal{L}_{PT/MLN}$, we adopt the sum of Dice [31] and Focal [32] losses. The loss functions are detailed in the supplementary materials. The λ is set as 1 in the experiments as default.

2.4 Radiomics Enhancement

Our XSurv also can be enhanced by leveraging radiomics features (denoted as Radio-XSurv). Following [16], radiomics features are extracted from the XSurv-segmented PT/MLN regions via Pyradiomics [33] and selected by Least Absolute Shrinkage and Selection Operator (LASSO) regression. The process of radiomics feature extraction is provided in the supplementary materials. Then, a CoxPH model [9] is adopted to integrate the selected radiomics features and the XSurv-predicted survival score to make the final prediction. In addition, clinical indicators (e.g., age, gender) also can be integrated by the CoxPH model.

3 Experimental Setup

3.1 Dataset and Preprocessing

We adopted the training dataset of HECKTOR 2022 (refer to https://hecktor.grand-challenge.org/), including 488 H&N cancer patients acquired from seven medical centers [7], while the testing dataset was excluded as its ground-truth labels are not released. Each patient underwent pretreatment PET/CT and has clinical indicators. We present the distributions of all clinical indicators in the supplementary materials. Recurrence-Free Survival (RFS), including time-to-event in days and censored-or-not status, was provided as ground truth for survival prediction, while PT and MLN annotations were provided for segmentation. The patients from two centers (CHUM and CHUV) were used for testing and other patients for training, which split the data into 386/102 patients in training/testing sets. We trained and validated models using 5-fold cross-validation within the training set and evaluated them in the testing set.

We resampled PET-CT images into isotropic voxels where 1 voxel corresponds to 1 mm^3. Each image was cropped to $160 \times 160 \times 160$ voxels with the tumor located in the center. PET images were standardized using Z-score normalization, while CT images were clipped to $[-1024, 1024]$ and then mapped to $[-1, 1]$. In addition, we performed univariate and multivariate Cox analyses on the clinical indicators to screen out the prognostic indicators with significant relevance to RFS ($P < 0.05$).

3.2 Implementation Details

We implemented our XSurv using PyTorch on a 12 GB GeForce GTX Titan X GPU. Our XSurv was trained for 12,000 iterations using an Adam optimizer with a batch size of 2. Each training batch included the same number of censored and uncensored samples. The learning rate was set as 1e−4 initially and then reset to 5e−5 and 1e−5 at the 4,000th and 8,000th training iteration. Data augmentation was applied in real-time during training to minimize overfitting, including random affine transformations and random cropping to $112 \times 112 \times 112$ voxels. Validation was performed after every 200 training iterations and the model achieving the highest validation result was preserved. In our experiments, one training iteration (including data augmentation) took roughly 4.2 s, and one inference iteration took roughly 0.61 s.

3.3 Experimental Settings

We compared our XSurv to six state-of-the-art survival prediction methods, including two traditional radiomics-based methods and four deep survival models. The included traditional methods are CoxPH [9] and Individual Coefficient Approximation for Risk Estimation (ICARE) [34]. For traditional methods, radiomics features were extracted from the provided ground-truth tumor regions and selected by LASSO regression. The included deep survival models are Deep Multi-Task Logistic Regression and CoxPH ensemble (DeepMTLR-CoxPH) [14], Transformer-based Multimodal networks for Segmentation and Survival prediction (TMSS) [11], Deep Multi-task Survival model (DeepMTS) [24], and Radiomics-enhanced DeepMTS (Radio-DeepMTS) [16]. DeepMTLR-CoxPH, ICARE, and Radio-DeepMTS achieved top performance in HECKTOR 2021 and 2022. For a fair comparison, all methods took the same pre-processed images and clinical indicators as inputs. Survival prediction and segmentation were evaluated using Concordance index (C-index) and Dice Similarity Coefficient (DSC), which are the standard evaluation metrics in the challenges [6, 7, 35].

We also performed two ablation studies on the encoder and decoder separately: (i) We replaced HPCA/HPSA blocks with Conv blocks and compared different strategies to combine PET-CT images. (ii) We removed RAG blocks and compared different strategies to extract PT/MLN-related information.

Table 1. Comparison between XSurv and state-of-the-art survival prediction methods.

Methods		Survival prediction (C-index)	PT segmentation (DSC)	MLN segmentation (DSC)
CoxPH [9]	Radiomics	$0.745 \pm 0.024^{*}$	/	/
ICARE [34]	Radiomics	$0.765 \pm 0.019^{*}$	/	/
DeepMTLR-CoxPH [14]	CNN	$0.748 \pm 0.025^{*}$	/	/
TMSS [11]	ViT + CNN	$0.761 \pm 0.028^{*}$	$0.784 \pm 0.015^{*}$	$0.724 \pm 0.018^{*}$
DeepMTS [24]	CNN	$0.757 \pm 0.022^{*}$	$0.754 \pm 0.010^{*}$	$0.715 \pm 0.013^{*}$
XSurv (Ours)	Hybrid	0.782 ± 0.018	$\mathbf{0.800 \pm 0.006}$	$\mathbf{0.754 \pm 0.008}$
Radio-DeepMTS [16]	CNN + Radiomics	$0.776 \pm 0.018^{‡}$	$0.754 \pm 0.010^{‡}$	$0.715 \pm 0.013^{‡}$
Radio-XSurv (Ours)	Hybrid + Radiomics	$\mathbf{0.798 \pm 0.015}$	$\mathbf{0.800 \pm 0.006}$	$\mathbf{0.754 \pm 0.008}$

Bold: the best result in each column is in bold. ±: standard deviation.
*: P<0.05, in comparison to XSurv. ‡: P<0.05, in comparison to Radio-XSurv.

4 Results and Discussion

The comparison between our XSurv and the state-of-the-art methods is presented in Table 1. Our XSurv achieved a higher C-index than all compared methods, which demonstrates that our XSurv has achieved state-of-the-art performance in survival prediction of H&N cancer. When radiomics enhancement was adopted in XSurv and DeepMTS, our Radio-XSurv also outperformed the Radio-DeepMTS and achieved the highest C-index. Moreover, the segmentation results of multi-task deep survival models (TMSS, DeepMTS, and XSurv) are also reported in Table 1. Our XSurv achieved higher DSCs than TMSS and DeepMTS, which demonstrates that our XSurv can locate PT and MLN more precisely and this infers that our XSurv has better learning capability. We attribute these performance improvements to the use of our proposed merging-diverging learning framework, HPCA block, and RAG block, which can be evidenced by ablation studies.

The ablation study on the PET-CT merging encoder is shown in Table 2. We found that using PET alone resulted in a higher C-index than using both PET-CT with early or late fusion. This finding is consistent with Wang et al. [19]'s study, which suggests that early and late fusion cannot effectively leverage the complementary information in PET-CT images. As we have mentioned, early and late fusion have difficulties in extracting intra- and inter-modality information, respectively. Our encoder first adopts Conv/HPSA blocks to extract intra-modality information and then leverages HPCA blocks to discover their interactions, which achieved the highest C-index. For PT and MLN segmentation, our encoder also achieved the highest DSCs, which indicates that our encoder also can improve segmentation. In addition, MNAF blocks [22] were compared and showed poor performance. This is likely attributed to the fact that leveraging non-local attention at multiple scales has corrupted local spatial information, which degraded the segmentation performance and distracted the model from PT and MLN regions. To relieve this problem, in Tang et al.'s study [22], tumor segmentation masks were fed into the model as explicit guidance to tumor regions. However, it is intractable to have segmentation masks at the inference stage in clinical practice.

The ablation study on the PT-MLN diverging decoder is shown in Table 3. We found that, even without adopting AG, using a dual-branch decoder for PT and MLN segmentation resulted in a higher C-index than using a single-branch decoder, which demonstrates the effectiveness of our diverging decoder design. Adopting vanilla AG [29] or RAG in the dual-branch decoder further improved survival prediction. Compared to the vanilla AG, our RAG contributed to a larger improvement, and this enabled our decoder to achieve the highest C-index. In the supplementary materials, we visualized the attention maps produced by RAG blocks, where the attention maps can precisely locate PT/MLN regions and screen out PT-/MLN-related features. For PT and MLN segmentation, using a single-branch decoder for PT- or MLN-only segmentation achieved the highest DSCs. This is expected as the model can leverage all its capabilities to segment only one target. Nevertheless, our decoder still achieved the second-best DSCs in both PT and MLN segmentation with a small gap.

Table 2. Ablation study on the PET-CT merging encoder.

Methods		Survival prediction (C-index)	PT segmentation (DSC)	MLN segmentation (DSC)
SBE with C_e = [16, 32, 64, 128, 256]	Only PET	0.767	0.753	0.699
	Only CT	0.637	0.630	0.702
	Early fusion	0.755	0.783	0.722
DBE with C_e = [8, 16, 32, 64, 128]	Late fusion	0.762	0.796	0.744
	MNAF [22]	0.688	0.741	0.683
	Ours	**0.782**	**0.800**	**0.754**

Bold: the best result in each column is in bold. SBE: single-branch encoder. DBE: dual-branch encoder. C_e: the channel numbers or embedding dimensions used in the encoder.

Table 3. Ablation study on the PT-MLN diverging decoder.

Methods		Survival prediction (C-index)	PT segmentation (DSC)	MLN segmentation (DSC)
SBD with C_d = [256, 128, 64, 32, 16]	Only PT	0.751	**0.803**	/
	Only MLN	0.746	/	**0.758**
	PT and MLN	0.765	0.790	0.734
DBD with C_d = [128, 64, 32, 16, 8]	No AG	0.770	0.792	0.740
	Vanilla AG [29]	0.774	0.795	0.745
	Ours	**0.782**	0.800	0.754

Bold: the best result in each column is in bold. SBD: single-branch decoder. DBD: dual-branch decoder. C_d: the channel numbers used in the decoder.

5 Conclusion

We have outlined an X-shape merging-diverging hybrid transformer network (XSurv) for survival prediction from PET-CT images in H&N cancer. Within the XSurv, we propose a merging-diverging learning framework, a Hybrid Parallel Cross-Attention (HPCA) block, and a Region-specific Attention Gate (RAG) block, to learn complementary information from multi-modality images and extract region-specific prognostic information for survival prediction. Extensive experiments have shown that the proposed framework and blocks enable our XSurv to outperform state-of-the-art survival prediction methods on the well-benchmarked HECKTOR 2022 dataset.

Acknowledgement. This work was supported by Australian Research Council (ARC) under Grant DP200103748.

References

1. Parkin, D.M., Bray, F., Ferlay, J., Pisani, P.: Global cancer statistics, 2002. CA Cancer J. Clin. **55**(2), 74–108 (2005)
2. Wang, X., Li, B.B.: Deep learning in head and neck tumor multiomics diagnosis and analysis: review of the literature. Front. Genet. **12**, 624820 (2021)
3. Bogowicz, M., et al.: Comparison of PET and CT radiomics for prediction of local tumor control in head and neck squamous cell carcinoma. Acta Oncol. **56**(11), 1531–1536 (2017)
4. Gu, B., et al.: Prediction of 5-year progression-free survival in advanced nasopharyngeal carcinoma with pretreatment PET/CT using multi-modality deep learning-based radiomics. Front. Oncol. **12**, 899351 (2022)
5. Andrearczyk, V., et al.: Overview of the HECKTOR challenge at MICCAI 2020: automatic head and neck tumor segmentation in PET/CT. In: Andrearczyk, V., et al. (eds.) HECKTOR 2020. LNCS, vol. 12603, pp. 1–21. Springer, Cham (2021). https://doi.org/10.1007/978-3-030-67194-5_1
6. Andrearczyk, V., et al.: Overview of the HECKTOR challenge at MICCAI 2021: automatic head and neck tumor segmentation and outcome prediction in PET/CT images. In: Andrearczyk, V., et al. (eds.) HECKTOR 2021. LNCS, vol. 13209, pp. 1–37. Springer, Cham (2022). https://doi.org/10.1007/978-3-030-98253-9_1
7. Andrearczyk, V., et al.: Overview of the HECKTOR challenge at MICCAI 2022: automatic head and neck tumor segmentation and outcome prediction in PET/CT. In: Andrearczyk, V., et al. (eds.) HECKTOR 2022. LNCS, vol. 13626, pp. 1–30. Springer, Cham (2023). https://doi.org/10.1007/978-3-031-27420-6_1
8. Gillies, R.J., Kinahan, P.E., Hricak, H.: Radiomics: images are more than pictures, they are data. Radiology **278**(2), 563–577 (2016)
9. Cox, D.R.: Regression models and life-tables. J. Roy. Stat. Soc. Ser. B (Methodol.) **34**(2), 187–202 (1972)
10. Deepa, P., Gunavathi, C.: A systematic review on machine learning and deep learning techniques in cancer survival prediction. Prog. Biophys. Mol. Biol. **174**, 62–71 (2022)
11. Saeed, N., Sobirov, I., Al Majzoub, R., Yaqub, M.: TMSS: an end-to-end transformer-based multimodal network for segmentation and survival prediction. In: Wang, L., et al. (eds.) MICCAI 2022. LNCS, vol. 13437, pp. 319–329. Springer, Cham (2022). https://doi.org/10.1007/978-3-031-16449-1_31
12. Zheng, H., et al.: Multi-transSP: multimodal transformer for survival prediction of nasopharyngeal carcinoma patients. In: Wang, L., et al. (eds.) MICCAI 2022. LNCS, vol. 13437, pp. 234–243. Springer, Cham (2022). https://doi.org/10.1007/978-3-031-16449-1_23
13. Afshar, P., et al.: From handcrafted to deep-learning-based cancer radiomics: challenges and opportunities. IEEE Signal Process. Mag. **36**(4), 132–160 (2019)
14. Saeed, N., et al.: An ensemble approach for patient prognosis of head and neck tumor using multimodal data. In: Andrearczyk, V., et al. (eds.) HECKTOR 2021. LNCS, vol. 13209, pp. 278–286. Springer, Cham (2022). https://doi.org/10.1007/978-3-030-98253-9_26
15. Naser, M.A., et al.: Progression free survival prediction for head and neck cancer using deep learning based on clinical and PET/CT imaging data. In: Andrearczyk, V., et al. (eds.) HECKTOR 2021. LNCS, vol. 13209, pp. 287–299. Springer, Cham (2022). https://doi.org/10.1007/978-3-030-98253-9_27
16. Meng, M., Bi, L., Feng, D., Kim, J.: Radiomics-enhanced deep multi-task learning for outcome prediction in head and neck cancer. In: Andrearczyk, V., et al. (eds.) HECKTOR 2022. LNCS, vol. 13626, pp. 135–143. Springer, Cham (2023). https://doi.org/10.1007/978-3-031-27420-6_14

17. Diamant, A., et al.: Deep learning in head & neck cancer outcome prediction. Sci. Rep. **9**, 2764 (2019)
18. Fujima, N., et al.: Prediction of the local treatment outcome in patients with oropharyngeal squamous cell carcinoma using deep learning analysis of pretreatment FDG-PET images. BMC Cancer **21**, 900 (2021)
19. Wang, Y., et al.: Deep learning based time-to-event analysis with PET, CT and joint PET/CT for head and neck cancer prognosis. Comput. Methods Programs Biomed. **222**, 106948 (2022)
20. Zhou, T., et al.: M^2Net: multi-modal multi-channel network for overall survival time prediction of brain tumor patients. In: Martel, A.L., et al. (eds.) MICCAI 2020. LNCS, vol. 12263, pp. 221–231. Springer, Cham (2020). https://doi.org/10.1007/978-3-030-59713-9_22
21. D'Souza, N.S., et al.: Fusing modalities by multiplexed graph neural networks for outcome prediction in tuberculosis. In: Wang, L., et al. (eds.) MICCAI 2022. LNCS, vol. 13437, pp. 287–297. Springer, Cham (2022). https://doi.org/10.1007/978-3-031-16449-1_28
22. Tang, W., et al.: MMMNA-net for overall survival time prediction of brain tumor patients. In: Annual International Conference of the IEEE Engineering in Medicine & Biology Society, pp. 3805–3808 (2022)
23. Wang, X., Girshick, R., Gupta, A., He, K.: Non-local neural networks. In: IEEE Conference on Computer Vision and Pattern Recognition, pp. 7794–7803 (2018)
24. Meng, M., et al.: DeepMTS: deep multi-task learning for survival prediction in patients with advanced nasopharyngeal carcinoma using pretreatment PET/CT. IEEE J. Biomed. Health Inform. **26**(9), 4497–4507 (2022)
25. Meng, M., Peng, Y., Bi, L., Kim, J.: Multi-task deep learning for joint tumor segmentation and outcome prediction in head and neck cancer. In: Andrearczyk, V., et al. (eds.) HECKTOR 2021. LNCS, vol. 13209, pp. 160–167. Springer, Cham (2022). https://doi.org/10.1007/978-3-030-98253-9_15
26. Andrearczyk, V., et al.: Multi-task deep segmentation and radiomics for automatic prognosis in head and neck cancer. In: Rekik, I., et al. (eds.) PRIME 2021. LNCS, vol. 12928, pp. 147–156. Springer, Cham (2022). https://doi.org/10.1007/978-3-030-87602-9_14
27. Liu, Z., et al.: Swin transformer: hierarchical vision transformer using shifted windows. In: IEEE/CVF International Conference on Computer Vision, pp. 10012–10022 (2021)
28. Liu, W., et al.: PHTrans: parallelly aggregating global and local representations for medical image segmentation. In: Wang, L., et al. (eds.) MICCAI 2022. LNCS, vol. 13435, pp. 235–244. Springer, Cham (2022). https://doi.org/10.1007/978-3-031-16443-9_23
29. Schlemper, J., et al.: Attention gated networks: learning to leverage salient regions in medical images. Med. Image Anal. **53**, 197–207 (2019)
30. Gensheimer, M.F., Narasimhan, B.: A scalable discrete-time survival model for neural networks. PeerJ **7**, e6257 (2019)
31. Milletari, F., et al.: V-Net: fully convolutional neural networks for volumetric medical image segmentation. In: International Conference on 3D Vision, pp. 565–571 (2016)
32. Lin, T.Y., et al.: Focal loss for dense object detection. In: IEEE International Conference on Computer Vision, pp. 2980–2988 (2017)
33. Van Griethuysen, J.J., et al.: Computational radiomics system to decode the radiographic phenotype. Can. Res. **77**(21), e104–e107 (2017)
34. Rebaud, L., et al.: Simplicity is all you need: out-of-the-box nnUNet followed by binary-weighted radiomic model for segmentation and outcome prediction in head and neck PET/CT. In: Andrearczyk, V., et al. (eds.) HECKTOR 2022. LNCS, vol. 13626, pp. 121–134. Springer, Cham (2023). https://doi.org/10.1007/978-3-031-27420-6_13
35. Eisenmann, M., et al.: Biomedical image analysis competitions: the state of current participation practice. arXiv preprint, arXiv:2212.08568 (2022)

Coupling Bracket Segmentation and Tooth Surface Reconstruction on 3D Dental Models

Yuwen Tan[1], Xiang Xiang[1(✉)], Yifeng Chen[1], Hongyi Jing[1], Shiyang Ye[2], Chaoran Xue[2], and Hui Xu[2]

[1] Key Lab of Image Processing and Intelligent Control, Ministry of Education School of Artificial Intelligence and Automation, Huazhong University of Science and Technology, Wuhan, China
xex@hust.edu.cn

[2] State Key Lab of Oral Diseases; National Clinical Research Center for Oral Diseases; Department of Orthodontics, West China Hospital of Stomatology, Sichuan University, Chengdu, China

Abstract. Delineating and removing brackets on 3D dental models and then reconstructing the tooth surface can enable orthodontists to premake retainers for patients. It eliminates the waiting time and avoids the change of tooth position. However, it is time-consuming and labor-intensive to process 3D dental models manually. To automate the entire process, accurate bracket segmentation and tooth surface reconstruction algorithms are of high need. In this paper, we propose a graph-based network named BSegNet for bracket segmentation on 3D dental models. The dynamic dilated neighborhood construction and residual connection in the graph network promote the bracket segmentation performance. Then, we propose a simple yet effective projection-based method to reconstruct the tooth surface. We project the vertices of the hole boundary on the tooth surface onto a 2D plane and then triangulate the projected polygon. We evaluate the performance of BSegNet on the bracket segmentation dataset and the results show the superiority of our method. The framework integrating the segmentation and reconstruction achieves a low reconstruction error and can be used as an effective tool to assist orthodontists in orthodontic treatment.

Keywords: 3D dental surface · bracket segmentation · surface reconstruction · deep learning · orthodontic treatment

1 Introduction

With the rapid development of 3D scanning devices, the application of computer-aided diagnosis in orthodontics has gradually developed. In orthodontic treatment, an essential step for patients is to wear retainers. Orthodontists need to

---

Supplementary Information The online version contains supplementary material available at https://doi.org/10.1007/978-3-031-43987-2_40.

H. Greenspan et al. (Eds.): MICCAI 2023, LNCS 14225, pp. 411–420, 2023.
https://doi.org/10.1007/978-3-031-43987-2_40

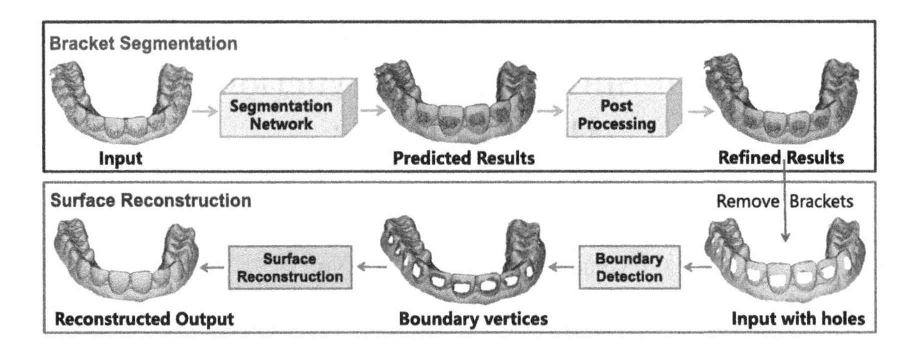

Fig. 1. The framework of bracket segmentation and tooth surface reconstruction. Integrating those two steps into a unified framework automates the whole process.

remove the brackets, utilize the intra-oral scanners to acquire the digital models, and then print models to fabricate retainers. However, there is a long waiting time for patients and the position of teeth would change, which affects the effectiveness of orthodontic treatment. As shown in Fig. 1, removing the brackets on 3D dental models and reconstructing the tooth surface can enable orthodontists to pre-make retainers. Patients can get retainers immediately after removing brackets which reduces the waiting time. When the retainer is lost or broken, a new one can be easily fabricated by using the archived digital model. The whole process helps to maintain the long-term stability of orthodontic treatment. However, it takes approximately 20 min for an orthodontist to precisely delineate brackets on a digital model. Besides that, the bracket segmentation requires a very high level of precision otherwise it will affect the reconstruction step. Therefore, an efficient and precise bracket segmentation method is crucial in orthodontic treatment. Orthodontists usually use CAD software (*i.e.*, Geomagic Studio) to reconstruct the tooth surface after removing the brackets due to the clinical applicability of its reconstructed results. However, it requires interactive manual operation and automatic segmentation algorithms cannot be deployed to the software. To automate the entire process, a 3D mesh reconstruction algorithm is also needed. Integrating segmentation and reconstruction into a unified framework can serve as an effective tool to assist orthodontic treatment.

Since 3D dental models can be transformed into point clouds, methods proposed for point cloud segmentation would provide guidance. Point cloud segmentation methods can be mainly summarized as MLP-based [10,12,13], graph-based [7,14], convolution-based [16], and attention-based [4]. Besides that, several methods have been proposed for computer-aided orthodontic processes such as tooth segmentation [3,9,17], landmark localization [5,6,15], and tooth completion [11,19] on 3D dental models. A two-stream graph-based network TSGCNet [17] is proposed for tooth segmentation. ToothCR [19] is proposed to recover the missing tooth which consists of point completion and surface reconstruction. The holes formed by the removal of the brackets are simple and regular polygons located on each tooth surface. An effective and fast method should be proposed to fill these characteristic holes and reconstruct the tooth surface. Different from the network [8] proposed for single-tooth bracket separation, our work focuses on the more challenging task of segmenting entire dental models. To the best of our knowledge, there exists no work proposed to integrate the segmentation and reconstruction of 3D dental models to automate the overall process.

As graph-based network [6,9,15,17,18] shows its superiority on various tasks on 3D dental models, we analyze the performance of different local operations and modules in the graph network. Based on these analyses, we propose a network named BSegNet for bracket segmentation. After segmenting the brackets, we adopt a simple yet effective projection-based method to reconstruct the tooth surface, which converts 3D holes into a 2D plane and triangulates the projected polygons on the 2D plane. We then estimate the z coordinate of the new vertices through the neighbor information and transform the vertices to the original space. The proposed method can better recover the surface of teeth.

The contributions of this paper are as follows: 1) We propose a graph-based network named BSegNet for bracket segmentation on 3D dental models which can reduce the burden on orthodontists; 2) An effective method is proposed to reconstruct the tooth surface where brackets are removed; 3) The low reconstruction error of the automatically processed models suggests that the framework integrating the segmentation and reconstruction can be used as a powerful tool to assist orthodontists.

2 Method

We propose a network named BSegNet for bracket segmentation on 3D dental models. BSegNet regards each mesh cell as a graph node and updates the node-wise feature via several local modules. For dental models with brackets removed, a simple yet effective reconstruction method is adopted to reconstruct the tooth surface. We will describe the network architecture of BSegNet (Fig. 2) and the process of tooth surface reconstruction in detail.

2.1 Bracket Segmentation Network

Compared to point clouds, more geometric spatial information can be obtained from 3D mesh data. Different input features will have different effects on the subsequent network training. In this paper, we use a 24-dimension vector as the initial input. The 24-dimensional input vector corresponds to the coordinates of three vertices, the normal vectors of three vertices, the normal vectors of the mesh cell, and the coordinates of the mesh cell centroid.

We use a MLP to map the input feature $F_0 \in \mathbb{R}^{24}$ into $F_1 \in \mathbb{R}^{64}$. Then a transform net is adopted to transform the feature F_1 into a canonical space to improve the robustness. The transformed feature is fed to several local modules which are designed to encode the local features of each mesh cell in semantic space. In each local module, the first step is to construct the graph $\mathcal{G}(V, E)$ where $V = \{c_1, c_2, ..., c_N\}$ denotes the set of mesh cells and E represents the

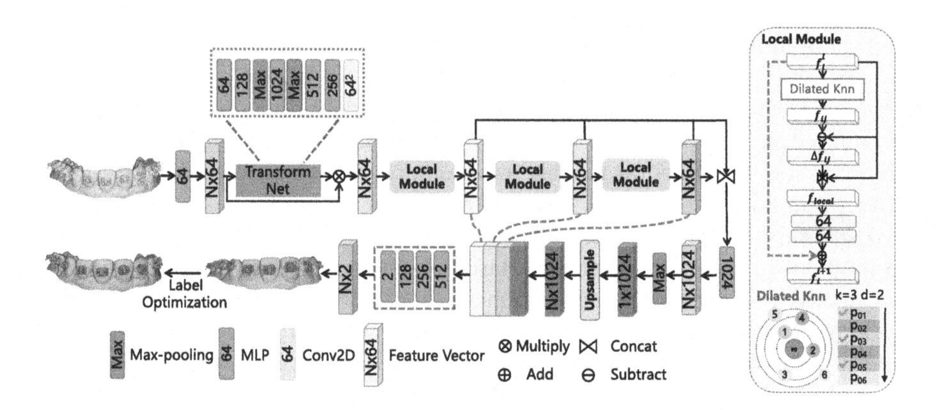

Fig. 2. The architecture of the bracket segmentation network (BSegNet).

connectivity relationship. To construct the $E \in |V| \times |V|$, we use the *dilated knn* algorithm to define the connectivity relationship where a mesh cell only connects to its nearest k neighbors. For a given mesh cell c_i, we calculate the feature distance between other mesh cells and select the k nearest neighbors within the range of the $k \times d$ neighbors, where d is the dilation factor to determine the dilation range. *Dilated knn* can get a large receptive field than *knn*. Let $\mathcal{N}_i$ denotes the set of neighbors, $D_s = \{c_1, c_2, ..., c_{k \times d}\}$ are the nearest $k \times d$ cells (except c_i). The dilated k neighbors of c_i is defined as

$$\mathcal{N}_i^k = \{c_1, c_{1+d}, c_{1+2\times d}, ...c_{1+(k-1)\times d}\} \tag{1}$$

After building the graph, we construct the local feature of each cell and update the feature through the graph convolution layers. Let the $f_i^l \in \mathbb{R}^{64}$ denote the feature vector of the mesh cell c_i in the l-th layer and $\triangle f_{ij} = \{f_l^i - f_l^j\}|_{j \in \mathcal{N}_i^k}$ denotes the edge feature which is used to capture the geometric relationship between each cell and its neighbors. Then the local feature $f_{local} = (\triangle f_{ij}^l \oplus f_i^l)$ goes through two Conv2D layers to further encode the feature. The update process of the mesh cell features (graph node) is defined as

$$f_i^{l+1} = R(h_\vartheta(\triangle f_{ij}^l \oplus f_i^l))|_{j \in \mathcal{N}_i^k} + f_i^l \tag{2}$$

where h_ϑ denotes the Conv2D layers and $R(.)$ stands for the feature aggregation function. To avoid the over-smoothing problem in the graph network, we use the residual connection in each local module the same as DeepGCN [7]. We concatenate the features of each local module and use an MLP layer to form a global high-dimension feature. Then the high-dimension global feature is concatenated with local features, forming the node-wise feature. Finally, several projection layers and one classifier layer are used to predict an N $\times 2$ probability matrix.

As the prediction results of the network may have some isolated labeled mesh cells, we use the graph-cut method to refine the results. The post-processing stage minimizes an energy function by combining the probability term and the smoothness term. The energy function to be optimized is defined as

$$E = \sum_{i=1}^{N} -log(max(p_i(l_i), \epsilon)) + \lambda \sum_{i}^{N} \sum_{j \in \mathcal{N}_i} S(p_i, p_j, l_i, l_j) \tag{3}$$

where $p_i(l_i)$ denotes the probability belongs to the l_i, ϵ is the minimal probability threshold, and λ denotes the smooth parameter. The local smoothness term is defined as

$$S(p_i, p_j, l_i, l_j) = \begin{cases} 0, & l_i = l_j \\ -log(\frac{\theta_{ij}}{\pi})d_{ij}, & l_i \neq l_j \end{cases} \tag{4}$$

where θ_{ij} denotes the dihedral angle of two adjacent facets and d_{ij} denotes the distance between the centroids of two adjacent facets.

2.2 Tooth Surface Reconstruction

First, we identify all the holes and extract their boundaries. Then we project the vertices of the boundary into the 2D plane. We triangulate the projected polygon without inserting new vertices. If the line segment inside the polygon exceeds the preset length, it should be n-equally divided. To get more uniform triangles, we use the optimal delaunay triangulation algorithm [1] to iteratively optimize the position of vertices. The optimization process of vertices is as follows

$$p^* = \frac{1}{\sum |T_j|} \sum_{T_j \in \Omega(p)} |T_j| c_j \tag{5}$$

where p is the vertex needs to be optimized, $\Omega(p)$ denotes the set of first-ring neighborhood mesh cells of p, $|T_j|$ denotes the area of mesh cell T_j, c_j is the circumcentre of T_j. After optimizing the positions of the vertices, the triangles have almost the same angles and the distribution is closer to the original models.

After the polygon triangulation, the x-y coordinates are determined and a layer-by-layer procedure is employed to estimate the corresponding z-values. To avoid the vertices at the gingiva, we only use the information of the first-ring neighborhoods of the boundary vertices. We first compute the slopes of the boundary vertices which are defined as

$$k_i^b = \frac{1}{|\mathcal{N}^1(i)|} \sum_{j \in \mathcal{N}^1(i)} \frac{\triangle z_{ij}}{\sqrt{\triangle x_{ij}^2 + \triangle y_{ij}^2}} \tag{6}$$

where $\mathcal{N}^1(i)$ denotes the set of first-ring neighborhood vertices of the boundary point b_i. Then the calculation of the z-coordinate value is denoted as

$$z_{new}^i = \frac{1}{\mathcal{N}^b(i)} \sum_{j \in \mathcal{N}^b(i)} (z_j^b + k_j^b \sqrt{\triangle x_{ij}^2 + \triangle y_{ij}^2}) \tag{7}$$

where $\mathcal{N}^b(i)$ denotes the set of adjacent boundary points and k_j^b denotes the slope of boundary vertex b_j. Then we regard the added points as new boundary vertices and repeat the above process until the z-values of all vertices are calculated.

Before transforming back to the original space, the extreme z values are removed by median filtering. Then we use a rotation matrix to obtain the coordinates in the original space. Finally, we employ Laplacian smoothing to enhance the smoothness of the reconstructed surface. To make the reconstructed surface blend better with the boundary, we need to reduce the effect of smoothing on the first-ring neighborhood of the boundary. The calculation equation is as follows

$$p_{new} = up_b + (1-u)p_s, \quad u = \frac{1}{1+exp(-\frac{d_h}{d_b})} \tag{8}$$

where p_b and p_s denote the vertex before and after smoothing, d_b is the average distance between the vertex and the adjacent boundary vertices, and d_h is the average length of the hole boundary.

3 Experiments and Results

3.1 Datasets and Implementation Details

We collect 80 dental mesh models in STL format from different patients. The number of mesh cells in each dental model is approximately 100,000 and all the dental models are down-sampled to nearly 24,000 mesh cells. The ground truth segmentations are annotated by professional orthodontists on the down-sampled dental models. We divide the dataset into a training, validation, and test set which consists of 45, 14, and 21 subjects, respectively. The performance of bracket segmentation is evaluated by mean Intersection-over-Union (mIoU) and Overall Accuracy (OA). The performance of reconstruction is evaluated by the Mean Distance (MD) and Standard Deviation (SD) of the distance between the models reconstructed by our method and by Geomagic Studio. We also evaluate the reconstruction error of models processed by the automatic framework and manually by orthodontists. We train all the networks by minimizing the cross-entropy loss for 400 epochs except for MeshSegNet which minimizes the dice loss. We use the Adam optimizer and set the mini-batch as 5. The initial learning rate is 0.001, and we anneal the learning rate using the cosine functions.

Table 1. The segmentation results of the test dataset in terms of both OA and mIoU. (w/p) denotes the segmentation results after label optimization. **Bold** is the best.

Method	Input	OA	mIoU	Bracket	Other	OA(w/p)	mIoU(w/p)
PointNet [12]	4p,4n	90.21	81.78	78.80	84.75	91.83	84.52
PointNet++ [13]	4p,4n	94.44	89.26	87.78	90.74	95.87	91.86
DGCNN [14]	4p,4n	96.14	92.42	91.42	93.41	97.47	94.96
PointConv [16]	4p,4n	95.30	90.83	89.65	92.00	96.81	93.66
PCT [4]	4p,4n	96.44	92.99	92.13	93.85	97.56	95.13
PointMLP [10]	4p,4n	96.51	93.15	92.29	94.00	97.60	95.23
MeshSegNet [9]	4p,1n	95.62	91.46	90.39	92.52	96.89	93.84
TSGCNet [17]	4p,4n	93.00	86.68	84.96	88.41	95.00	90.30
Ours	4p,4n	**97.28**	**94.60**	**93.95**	**95.26**	**98.13**	**96.25**

3.2 Experimental Evaluation

Comparison Results. We compare our method with SOTA point cloud segmentation methods and teeth segmentation methods. All the results are shown in Table 1 and BSegNet achieves the best performance. Since PointNet lacks the feature encoding of local regions, it performs worst among all the methods. The performance of PointNet++ and PointConv are close and outperform PointNet by a large margin. Although PointMLP has the best performance among all the compared methods, the mIoU of BSegNet is higher (94.60 vs. 93.15). We also compare our method with the attention-based method PCT and the mIoU of our method is higher than PCT (94.60 vs. 92.99). The tooth segmentation network on 3D dental models performs worse than several point segmentation methods in the bracket segmentation task, especially for TSGCNet which uses a two-stream network to encode the coordinates and normal vectors respectively. MeshSegNet performs better than TSGCNet but the proposed graph-constrained learning modules cause high computation complexity. Our method is based on the DGCNN but the mIoU of our method is much higher (94.60 vs. 92.42).

As shown in Table 2, the reconstruction error of the models predicted by BSegNet is significantly lower than PointMLP. When reconstructing the models processed by doctors, our reconstruction method achieves low values of SD and MD which reveals it can replace the interactive reconstruction operation to some extent. We also compare our method with another reconstruction method Meshfix and our method has a lower value of SD (0.032 vs. 0.049). Compared to the manual process by doctors, the reconstruction error of our automatic framework is clinically acceptable and it can assist in orthodontic treatment. Figure 3 displays the segmentation and tooth surface reconstruction results.

Ablation Study and Analysis. In this section, we analyze different operations and modules in the BSegNet. As shown in Table 3, the performance of the 24-dimension input is better than the 15-dimension input which suggests the

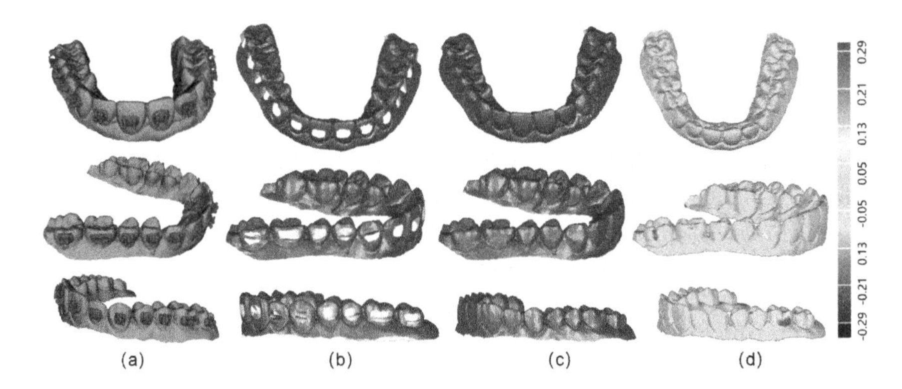

Fig. 3. (a) The bracket segmentation results after post-processing; (b) Dental models with brackets removed; (c) Models after tooth surface reconstruction; (d) Color-coded error map of automatically processed models compared with ground truth.

Table 2. The tooth surface reconstruction results compared to the ground truth in terms of MD(mm) and SD(mm).

Method	**Seg.**	**Recon.**	SD(mm)	MD+(mm)	MD-(mm)	Aver.(mm)
Semi.	PointMLP [10]	Geo	0.060	0.014	0.005	0.0095
Semi.	BSegNet	Geo	0.026	0.004	0.005	0.0045
Semi.	Doctors	Ours	0.015	0.001	0.004	0.0025
Auto.	BSegNet	Meshfix [2]	0.049	0.001	0.018	0.0095
Auto.	BSegNet	Ours	0.032	0.002	0.009	0.0055

normal vectors of the vertices help the subsequent network training. However, curvature information which is helpful for the teeth segmentation task does not improve and even hurts the bracket segmentation performance. An essential step in the graph-based network is to construct the local features of the center node. Many methods have been proposed to construct complex local features. However, our results show that using the edge feature with the central feature is sufficient for effectively representing local regions. We also analyze the results of different aggregation operations, and the results indicate that Max pooling is the most effective method for aggregation. As shown in Table 4, adopting the dynamic graph which updates the neighbor information in each local module, the mIoU has an improvement (+0.88) compared to the fixed graph. Under the same network structure, using the residual connection would further improve model performance by (+0.70). The use of *dilated knn* only brings slight performance improvement but does not require extra computation costs. With the post-processing, the predicted results can be further refined. Overall, using all the operations mentioned above acquires the best performance with 94.60 of mIoU and 96.25 after post-processing.

Table 3. Analysis experiments of different inputs, feature aggregation functions, and local feature constructions.

Input	OA	mIoU	R(.)	OA	mIoU	Feature	OA	mIoU
1p,1n	95.93	92.02	Att	95.40	91.03	$\{\triangle p_{ij}, f_j\}$	95.02	90.31
4p,1n	96.74	93.55	Max	**97.12**	**94.29**	$\{\triangle f_{ij}, f_i\}$	**97.12**	**94.29**
4p,4n	**97.12**	**94.29**	Sum	96.72	93.52	$\{\triangle f_{ij}, f_i, f_j\}$	96.56	93.20
4p,1n,3c	95.50	91.26	Mean	96.72	93.51	$\{\triangle p_{ij}, \triangle f_{ij}, f_i, f_j\}$	96.75	93.57

Table 4. Ablation study of different operators in the BSegNet.

Dynamic	Residual Con.	Dilated-knn	OA	mIoU	OA(w/p)	mIoU(w/p)
			96.28	92.71	97.55	95.14
✓			96.76	93.59	97.79	95.58
✓	✓		97.12	94.29	97.99	95.97
✓	✓	✓	**97.28**	**94.60**	**98.13**	**96.25**

4 Conclusion

In this paper, we propose a network named BSegNet for bracket segmentation on 3D dental models which can reduce the burden on orthodontists. BSegNet is a graph-based network that employs dynamic dilated neighborhood construction and residual connections to improve segmentation results. With label optimization, the segmentation results can be further refined. Experimental results on a clinical dataset demonstrate our method significantly outperforms related state-of-the-art methods. We also propose a simple yet effective method to reconstruct the tooth surface which can better recover the feature of the teeth. The whole framework achieves a low reconstruction error and can be used as a powerful tool to assist doctors in orthodontic diagnosis.

Acknowledgement. This research was supported by Sichuan Univ. Interdisciplinary Innovation Res. Fund (RD-03-202108), Natural Science Fund of Hubei Province (2022CFB823), HUST Independent Inno. Res. Fund (2021XXJS096), Alibaba Innovation Research (AIR) program (CRAQ7WHZ11220001-20978282), and grants from MoE Key Lab of Image Processing and Intelligent Control.

References

1. Alliez, P., Cohen-Steiner, D., Yvinec, M., Desbrun, M.: Variational tetrahedral meshing. In: ACM SIGGRAPH 2005 Papers, pp. 617–625 (2005)
2. Attene, M.: A lightweight approach to repairing digitized polygon meshes. Vis. Comput. **26**, 1393–1406 (2010)
3. Cui, Z., et al.: A fully automatic AI system for tooth and alveolar bone segmentation from cone-beam CT images. Nat. Commun. **13**(1), 2096 (2022)

4. Guo, M.H., Cai, J.X., Liu, Z.N., Mu, T.J., Martin, R.R., Hu, S.M.: PCT: point cloud transformer. Comput. Vis. Media **7**, 187–199 (2021)
5. Lang, Y., et al.: DentalPointNet: landmark localization on high-resolution 3D digital dental models. In: Wang, L., Dou, Q., Fletcher, P.T., Speidel, S., Li, S. (eds.) Medical Image Computing and Computer Assisted Intervention – MICCAI 2022. MICCAI 2022. Lecture Notes in Computer Science, vol. 13432, pp. 444–452. Springer, Cham (2022). https://doi.org/10.1007/978-3-031-16434-7_43
6. Lang, Y., et al.: DLLNet: an attention-based deep learning method for dental landmark localization on high-resolution 3D digital dental models. In: de Bruijne, M., et al. (eds.) MICCAI 2021. LNCS, vol. 12904, pp. 478–487. Springer, Cham (2021). https://doi.org/10.1007/978-3-030-87202-1_46
7. Li, G., Muller, M., Thabet, A., Ghanem, B.: DeepGCNs: can GCNs go as deep as CNNs? In: Proceedings of the IEEE/CVF International Conference on Computer Vision, pp. 9267–9276 (2019)
8. Li, R., et al.: Deep learning for separation and feature extraction of bonded teeth: tool establishment and application (2022)
9. Lian, C., et al.: Deep multi-scale mesh feature learning for automated labeling of raw dental surfaces from 3D intraoral scanners. IEEE Trans. Med. Imaging **39**(7), 2440–2450 (2020)
10. Ma, X., Qin, C., You, H., Ran, H., Fu, Y.: Rethinking network design and local geometry in point cloud: a simple residual MLP framework. arXiv preprint arXiv:2202.07123 (2022)
11. Ping, Y., Wei, G., Yang, L., Cui, Z., Wang, W.: Self-attention implicit function networks for 3D dental data completion. Comput. Aided Geom. Des. **90**, 102026 (2021)
12. Qi, C.R., Su, H., Mo, K., Guibas, L.J.: PointNet: deep learning on point sets for 3D classification and segmentation. In: Proceedings of the IEEE Conference on Computer Vision and Pattern Recognition, pp. 652–660 (2017)
13. Qi, C.R., Yi, L., Su, H., Guibas, L.J.: PointNet++: deep hierarchical feature learning on point sets in a metric space. In: Advances in Neural Information Processing Systems, vol. 30 (2017)
14. Wang, Y., Sun, Y., Liu, Z., Sarma, S.E., Bronstein, M.M., Solomon, J.M.: Dynamic graph CNN for learning on point clouds. ACM Trans. Graph. **38**(5), 1–12 (2019)
15. Wu, T.H., et al.: Two-stage mesh deep learning for automated tooth segmentation and landmark localization on 3D intraoral scans. IEEE Trans. Med. Imaging **41**(11), 3158–3166 (2022)
16. Wu, W., Qi, Z., Fuxin, L.: PointConv: deep convolutional networks on 3D point clouds. In: Proceedings of the IEEE/CVF Conference on Computer Vision and Pattern Recognition, pp. 9621–9630 (2019)
17. Zhang, L., et al.: TSGCNet: discriminative geometric feature learning with two-stream graph convolutional network for 3D dental model segmentation. In: Proceedings of the IEEE/CVF Conference on Computer Vision and Pattern Recognition, pp. 6699–6708 (2021)
18. Zheng, Y., Chen, B., Shen, Y., Shen, K.: TeethGNN: semantic 3D teeth segmentation with graph neural networks. IEEE Trans. Vis. Comput. Graph. **29**(7), 3158–3168 (2022)
19. Zhu, H., Jia, X., Zhang, C., Liu, T.: ToothCR: a two-stage completion and reconstruction approach on 3D dental model. In: Gama, J., Li, T., Yu, Y., Chen, E., Zheng, Y., Teng, F. (eds.) Advances in Knowledge Discovery and Data Mining. PAKDD 2022. Lecture Notes in Computer Science, vol. 13282, pp. 161–172. Springer, Cham (2022). https://doi.org/10.1007/978-3-031-05981-0_13

TSegFormer: 3D Tooth Segmentation in Intraoral Scans with Geometry Guided Transformer

Huimin Xiong[1,2], Kunle Li[1], Kaiyuan Tan[1], Yang Feng[4], Joey Tianyi Zhou[5,6], Jin Hao[7], Haochao Ying[3], Jian Wu[3], and Zuozhu Liu[1,2](✉)

[1] ZJU-UIUC Institute, Zhejiang University, Haining 314400, China
[2] Stomatology Hospital, School of Stomatology, Zhejiang University School of Medicine, Hangzhou 310058, China
zuozhuliu@intl.zju.edu.cn
[3] School of Public Health, Zhejiang University, Hangzhou 310058, China
[4] Angelalign Research Institute, Angel Align Inc., Shanghai 200011, China
[5] Centre for Frontier AI Research (CFAR), A*STAR, Singapore, Singapore
[6] Institute of High Performance Computing (IHPC), A*STAR, Singapore, Singapore
[7] ChohoTech Inc., Hangzhou, China

Abstract. Optical Intraoral Scanners (IOS) are widely used in digital dentistry to provide detailed 3D information of dental crowns and the gingiva. Accurate 3D tooth segmentation in IOSs is critical for various dental applications, while previous methods are error-prone at complicated boundaries and exhibit unsatisfactory results across patients. In this paper, we propose TSegFormer which captures both local and global dependencies among different teeth and the gingiva in the IOS point clouds with a multi-task 3D transformer architecture. Moreover, we design a geometry-guided loss based on a novel point curvature to refine boundaries in an end-to-end manner, avoiding time-consuming post-processing to reach clinically applicable segmentation. In addition, we create a dataset with 16,000 IOSs, the largest ever IOS dataset to the best of our knowledge. The experimental results demonstrate that our TSegFormer consistently surpasses existing state-of-the-art baselines. The superiority of TSegFormer is corroborated by extensive analysis, visualizations and real-world clinical applicability tests. Our code is available at https://github.com/huiminxiong/TSegFormer.

Keywords: 3D tooth segmentation · IOS mesh scans · Transformer

H. Xiong, K. Li, K. Tan—These authors contributed equally to this work.

Supplementary Information The online version contains supplementary material available at https://doi.org/10.1007/978-3-031-43987-2_41.

H. Greenspan et al. (Eds.): MICCAI 2023, LNCS 14225, pp. 421–432, 2023.
https://doi.org/10.1007/978-3-031-43987-2_41

1 Introduction

Deep learning is becoming increasingly popular in modern orthodontic treatments for tooth segmentation in intraoral scans (IOS), cone-beam CT (CBCT) and panoramic X-ray [9,15]. Accurate tooth segmentation in 3D IOS dental models is crucial for orthodontics treatment such as diagnosis, tooth crown-root analysis and treatment simulation [10,25]. Specifically, tooth segmentation classifies each triangular face of a 3D IOS tooth model with about 100,000 to 400,000 faces and a spatial resolution of 0.008-0.02mm into teeth and gingiva categories, following the Federation Dentaire Internationale (FDI) standard [8].

There are two main categories for tooth segmentation in IOS: conventional methods that handle 2D image projections [10,21,24] or directly operate on 3D IOS meshes [17,22,25,29], and deep learning methods that operate on meshes or point clouds [1,3,6,7,11,12,14,18,23,27,30]. However, many challenges persist. Complicated morphological topology or dental diseases (e.g. crowded or erupted teeth) can lead to unsatisfactory segmentation performance [6]. Additionally, current methods often fail to recognize mesh faces between adjacent teeth or the tooth and gingiva, requiring time-consuming post-processing to refine the noisy boundary segmentation [6,12,23]. Moreover, the state-of-the-art works such as MeshSegNet [12], TSGCNet [27] and DCNet [6] have only been evaluated with a limited amount of data samples and the clinical applicability need to be evaluated with large-scale dataset or in real-world scenarios.

Inspired by the success of transformers in various tasks [2,4,5,13,19,28], we propose a novel 3D transformer framework, named TSegFormer, to address the aforementioned challenges. In particular, the tooth segmentation task on 3D IOSs is formulated as a semantic segmentation task on point clouds sampled from raw IOS meshes. We design the 3D transformer with tailored self-attention layers to capture long-range dependencies among different teeth, learning expressive representations from inherently sophisticated structures across IOSs. In addition, we design a multi-task learning paradigm where another auxiliary segmentation head is introduced to assist in delimiting teeth and gingiva. Furthermore, in view of the confusing boundary segmentation, we devise a novel geometry guided loss based on a newly-defined point curvature to help learn accurate boundaries. The network is trained in an end-to-end manner and requires no complicated post-processing during inference, making it appealing to practical applications.

We collect a large-scale, high-resolution and heterogeneous 3D IOS dataset with 16,000 dental models where each contains over 100,000 triangular faces. To the best of our knowledge, it is the largest IOS dataset to date. Experimental results show that TSegFormer has reached 97.97% accuracy, 94.34% mean intersection over union (mIoU) and 96.01% dice similarity coefficient (DSC) on the large-scale dataset, outperforming previous works by a significant margin. To summarize, our main contributions are:

- We design a novel framework for 3D tooth segmentation with a tailed 3D transformer and a multi-task learning paradigm, aiming at distinguishing the permanent teeth with divergent anatomical structures and noisy boundaries.

- We design a geometry guided loss based on a novel point curvature for end-to-end boundary refinement, getting rid of the two-stage and time-consuming post-processing for boundary smoothing.
- We collect the largest ever 3D IOS dataset for compelling evaluation. Extensive experiments, ablation analysis and clinical applicability test demonstrate the superiority of our method, which is appealing in real-world applications.

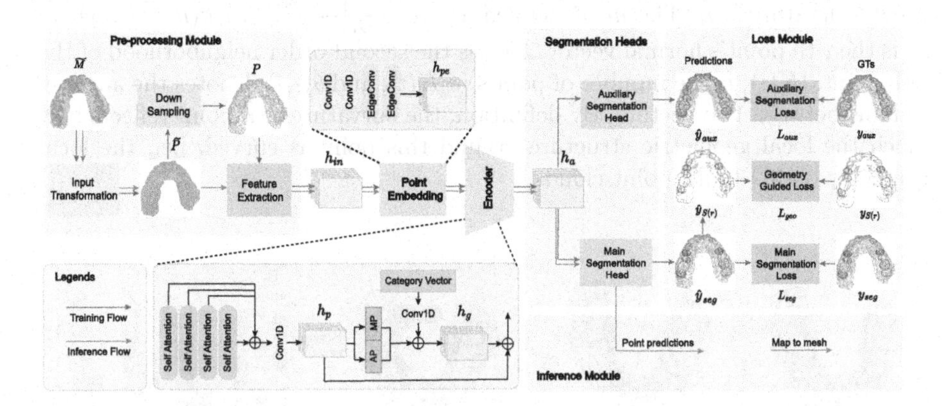

Fig. 1. The pipeline of our proposed TSegFormer for 3D tooth segmentation

2 Method

2.1 Overview

The overall pipeline is illustrated in Fig. 1. The original mesh $\widetilde{M}$ is converted to a point cloud $\widetilde{P}$ by taking the gravity center point of each mesh face. We down-sample a point cloud P with $N = 10,000$ points from $\widetilde{P}$, and extract the input feature matrix $h_{in} \in \mathbb{R}^{N\times 8}$ as defined below. The network first employs a point embedding module to capture abundant local structure information h_{pe} from h_{in}. Thereafter, we design the 3D transformer encoder with self-attention layers to capture high-level semantic representations h_a. With h_a, the main segmentation head produces prediction scores $\hat{y}_{seg} \in \mathbb{R}^{N\times 33}$ (32 permanant teeth and the gingiva), while the auxiliary head generates prediction scores $\hat{y}_{aux} \in \mathbb{R}^{N\times 2}$ to assist distinguishing the tooth-gingiva boundary. Furthermore, we devise a geometry guided loss L_{geo}, which is integrated with the main segmentation loss L_{seg} and the auxiliary loss L_{aux} to attain superior performance. During inference, we will extract the features $h_{in} \in \mathbb{R}^{\widetilde{N}\times 8}$ for all points in $\widetilde{P}$, process $\widetilde{P}$ into multiple sub-point clouds each with N points, then generate predictions for each point with $\lceil \frac{\widetilde{N}}{N} \rceil$ rounds of inference, and map them back to raw mesh $\widetilde{M}$.

2.2 TSegFormer Network Architecture

Feature Extraction. We first transform input meshes to point clouds as directly handling meshes with deep nets is computationally expensive, especially for high-resolution IOSs. To compensate for potential topology loss, we extract 8-dimensional feature vectors $h_{in} \in \mathbb{R}^{\widetilde{N} \times 8} / \mathbb{R}^{N \times 8}$ for each point to preserve sufficient geometric information, including the point's 3D Cartesian coordinates, 3-dimensional normal vector of mesh face, the Gaussian curvature and a novel point "curvature" m_i. The m_i is defined as $m_i = \frac{1}{|K(i)|} \sum_{j \in K(i)} \theta(n_i, n_j)$, where n_i is the i-th point's normal vector, $K(i)$ is the second-order neighborhood of the i-th point, $|K(i)|$ is the number of points in $K(i)$, and $\theta(\cdot, \cdot)$ denotes the angle in radians between two vectors. By definition, the curvature of a point reflects how much the local geometric structure around this point is curved, i.e., the local geometry on 3D tooth point clouds.

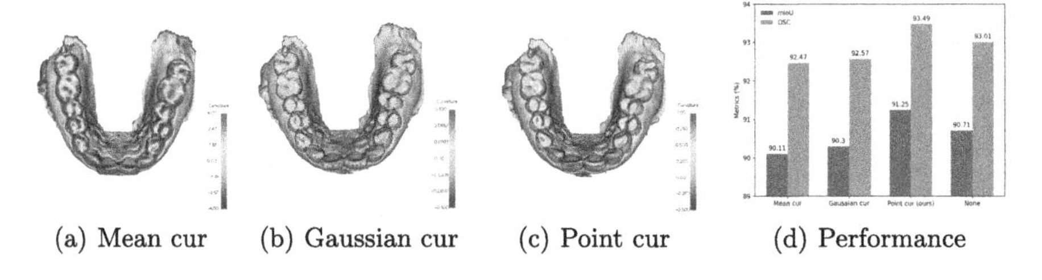

(a) Mean cur (b) Gaussian cur (c) Point cur (d) Performance

Fig. 2. Visualization and performance comparison of different curvatures(cur).

Backbone Network. Delineating complicated tooth-tooth or tooth-gingiva boundaries requires decent knowledge of local geometry in IOS. Hence, we first learn local dependencies from the input h_{in}. In particular, we design a point embedding module composed of two linear layers and two EdgeConv layers [20], which takes h_{in} as input and learn local features $h_{pe} \in \mathbb{R}^{N \times d_e}$. The point embedding module enriches point representations with local topological information, with ablation results in the Supplementary Material (*SM*) Table 1.

Meantime, in view of the inherently sophisticated and inconsistent shapes and structures of the teeth, and the ability of attention mechanism to capture long-range dependencies and suitability for handling unordered point cloud data [2,4,13,19,28], we build an encoder module based on it. The encoder module, composed of four successive self-attention layers and a linear transformation, further yields the high-level point feature maps h_p. To avoid misjudging jaw categories, an extra 2D category vector V is fed as input to help distinguish the maxillary and mandible and obtain the global feature maps h_g. Specifically, $h_g = \sigma(V) \oplus MP(h_p) \oplus AP(h_p)$, where "MP" and "AP" respectively denote the max and average pooling; $\oplus$ denotes concatenation and $\sigma(\cdot)$ is a linear layer. Finally, we obtain feature maps h_a for all points, where $h_a = h_p \oplus h_g$.

Segmentation Heads. To improve the network's ability to recognize different tooth and gingiva categories, we design two segmentation heads. The main segmentation head, an MLP (MLP_{seg}), generates point classification scores for 33 classes $\hat{y}_{seg} = MLP_{seg}(h_a) \in \mathbb{R}^{N\times 33}$ for tooth segmentation. Meanwhile, considering the prevalence of incorrect prediction of tooth-gingiva boundaries, we design an auxiliary segmentation head MLP_{aux} to provide binary classification scores for each point belonging to either tooth or gingiva, i.e., $\hat{y}_{aux} = MLP_{aux}(h_a) \in \mathbb{R}^{N\times 2}$. Experimental results indicate that the cooperation with MLP_{aux} can refine tooth-gingiva segmentation boundary.

Geometry Guided Loss. Previous methods are usually unsatisfactory to delineate the complicated tooth-tooth boundaries. Observing that points with high point curvatures often lie on the upper sharp ends of tooth crowns and the teeth boundaries (Fig. 2(c)), where mispredictions usually occur, we define the novel geometry guided loss L_{geo}. L_{geo} encourages TSegFormer to adaptively focus more on error-prone points with higher point curvatures with negligible extra computations. Specifically, we define it as

$$L_{geo} = -\sum_{i\in S(r)}\sum_{c=1}^{33}(1-\hat{p}_{ic}^{geo})^{\gamma}\cdot \Phi(y_{S(r)_i}, c)\cdot log(\hat{p}_{ic}^{geo}), \tag{1}$$

where γ is the modulating factor (empirically set to 2 in experiments); $y_{S(r)_i} \in \mathbb{R}^{33}$ represents the gold label of the i-th point in the point set $S(r)$; $\hat{p}_{ic}^{geo}$ denotes the predicted probability of the i-th point belonging to the c-th class, and $\Phi(y_i, c)$ is an indicator function which outputs 1 if $y_i = c$ and 0 otherwise. Concretely, $S(r)$ is a set of points whose point curvatures m_i are among the top $r \cdot 100\%$ $(0 < r \leq 1)$ of all N points, i.e., $S(r) := \{a_1,\ a_2,\ \cdots,\ \lceil a_{rN}\rceil\}$, where $m_{a_1} \geq m_{a_2} \geq \cdots > m_{a_{rN}} \geq m_{a_{rN+1}} \geq \cdots \geq m_{a_N}$. The experimental results (Fig. 2(d)) on a dataset of 2,000 cases indicate that L_{geo} is more effective with our point curvature over traditional mean and Gaussian curvatures, even they are worst than no curvature. This is because our point curvature provides more clear tooth-tooth and tooth-gingiva boundary indications (Fig. 2(a)-2(c)), thus avoiding misleading the model to focus too much on unimportant non-boundary points.

We employ the cross entropy loss as the loss of main segmentation head (L_{seg}) and the loss of auxiliary segmentation head (L_{aux}). The total loss L_{total} is computed by combining L_{seg}, L_{aux} for *all points* and L_{geo} for *hard points*: $L_{total} = L_{seg} + \omega_{geo}\cdot L_{geo} + \omega_{aux}\cdot L_{aux}$. We set the weights $\omega_{geo} = 0.001$, $\omega_{aux} = 1$ and the ratio $r = 0.4$, and detailed hyperparameter search results in *SM* Fig. 1 indicate that the performance is stable across different hyperparameter settings.

3 Experiments

3.1 Dataset and Experimental Setup

We construct a large-scale 3D IOS dataset consisting of 16,000 IOS meshes with full arches (each with 100,000 to 350,000 triangular faces) collected between 2018-

2021 in China, with evenly distributed maxillary and mandible scans labeled by human experts. Detailed data statistics are presented in *SM* Table 2, and 39.8% of the data have third-molars, 16.8% suffer from missing teeth, which all reveal the complexity of the dataset. The dataset is randomly split into training (12,000 IOSs), validation (2,000 IOSs) and test sets (2,000 IOSs). Furthermore, we collect an external dataset with 200 complex cases (disease statistics shown in *SM* Table 3) to evaluate the real-world clinical applicability of TSegFormer. Detailed training and architecture settings are in *SM* Tables 4 and 5.

3.2 Main Results on Tooth Segmentation

To our best knowledge, there has been no prior work on Transformer-based segmentation on non-Euclidean 3D tooth point clouds/meshes. Hence, we compare our TSegFormer to seven representative and state-of-the-art baselines from three categories: 1) neural networks for point clouds, including PointNet++ [16] and DGCNN [20]; 2) transformers for point clouds, including point transformer [28] and PVT [26]; 3) domain-specific architectures for 3D tooth segmentation, including MeshSegNet [12], TSGCNet [27] and DC-Net [6]. For fair comparison, baselines that cannot achieve raw-resolution mesh prediction followed the same inference protocols in 2.1, while the rest kept their original inference schemes.

Table 1. Main segmentation results (Tested on 1,000 patients).

Method	Mandible			Maxillary			All		
	mIoU ↑	DSC ↑	Acc ↑	mIoU ↑	DSC ↑	Acc ↑	mIoU ↑	DSC ↑	Acc ↑
PointNet++	81.11	85.33	94.96	83.89	87.12	96.28	82.57	86.27	95.65
DGCNN	92.41	94.49	97.68	93.82	95.61	98.01	93.15	95.08	97.85
point transformer	92.61	94.83	97.55	93.93	95.72	98.06	93.3	95.3	97.81
PVT	90.66	93.59	96.64	92.46	94.72	97.44	91.6	94.19	97.06
MeshSegNet	82.21	86.55	91.98	85.37	89.28	93.72	83.87	87.98	92.90
TSGCNet	80.71	85.23	92.78	80.97	85.28	93.86	80.85	85.25	93.34
DCNet	91.18	93.89	97.11	92.78	95.18	97.44	92.02	94.57	97.28
TSegFormer (Our)	**93.53**	**95.36**	**97.72**	**95.07**	**96.60**	**98.20**	**94.34**	**96.01**	**97.97**

Table 2. Ablation study on different main components.

Component		Mandible			Maxillary			All		
Geometry guided loss	Auxiliary branch	mIoU	DSC	Acc	mIoU	DSC	Acc	mIoU	DSC	Acc
		92.19	94.27	97.35	94.02	95.69	97.96	93.15	95.01	97.67
	✔	92.53	94.57	97.43	94.45	96.10	98.02	93.54	95.37	97.74
✔		92.46	94.47	97.37	94.45	96.12	98.02	93.51	95.33	97.71
✔	✔	**92.95**	**94.88**	**97.57**	**94.46**	**96.07**	**98.14**	**93.77**	**95.51**	**97.87**

We can firstly observe that TSegFormer outperforms existing best-performing point transformer model [28] by 0.16% in accuracy, 1.04% in mIoU and 0.71%

in DSC (Table 1). Such an improvement is surely significant considering the complicated real-world cases in our large-scale dataset and the relatively high performance of point transformer with an mIoU of 93.30%. Moreover, TSegFormer consistently surpassed all baselines on both mandible and maxillary in terms of all metrics, demonstrating its universal effectiveness.

It is important to integrate advanced architectures with domain-specific design for superior performance. We can notice that though MeshSegNet, TSGCNet, and DCNet are all domain-specific 3D tooth segmentation models, their performance, though on par with PVT and DGCNN, is worse than the point transformer. This is also consistent with the superior performance of transformer-based models on standard point cloud processing tasks, which could be mainly attributed to the larger dataset and powerful attention mechanism that better capture global dependencies. Hence, though models like MeshSegNet adopt some task-specific designs to achieve good performance, they still lag behind point transformer when a huge amount of data samples are available. In contrast, our TSegFormer employs the attention mechanism for point representation learning, and meanwhile, adopted task-specific architectures and geometry guided loss to further boost the performance. More statistical results are in *SM* Table 2.

Table 3. Segmentation performance of DCNet [6] with our geometry guided loss.

Model	Mandible			Maxillary			All		
	mIoU	DSC	Acc	mIoU	DSC	Acc	mIoU	DSC	Acc
DCNet	87.72	91.00	95.99	90.77	93.50	96.87	89.32	92.31	96.46
DCNet+L_{geo}	**89.66**	**92.58**	**96.46**	**91.75**	**94.28**	**97.14**	**90.75**	**93.47**	**96.82**

Table 4. Segmentation performance of TSegFormer under different training set scales.

Training set scale	Mandible			Maxillary			All		
	mIoU	DSC	Acc	mIoU	DSC	Acc	mIoU	DSC	Acc
500	86.36	89.74	95.45	89.22	91.69	96.57	87.86	90.76	96.04
1,000	90.60	93.13	96.72	92.84	94.77	97.50	91.78	93.99	97.13
2,000	92.15	94.28	97.31	94.05	95.79	97.92	93.15	95.08	97.63
4,000	93.10	95.05	97.69	94.68	96.28	98.12	93.93	95.70	97.92
8,000	93.27	95.15	97.67	94.83	96.40	98.16	94.09	95.81	97.92
12,000	**93.53**	**95.36**	**97.72**	**95.07**	**96.60**	**98.20**	**94.34**	**96.01**	**97.97**

3.3 Ablation Studies

Effectiveness of Geometry Guided Loss. Table 2 shows that introducing the geometry guided loss can improve the performance under all three metrics, e.g., around 0.4% improvement in mIoU. Besides, we show the universal effectiveness of the geometry guided loss by adding it to DCNet [6]. The performance of DCNet is also enhanced by 1.43% in mIoU (Table 3) with this additional loss.

Effectiveness of the Auxiliary Segmentation Head. The auxiliary segmentation head is designed to rectify the inaccuracy brought by mislabeling teeth and gingiva near their boundaries. Adding a loss for the auxiliary branch leads to about 0.4% mIoU performance improvement (Table 2).

Effectiveness on Training Data Efficiency. In real-world orthodontic applications, large-scale training data may not be directly accessible due to privacy concerns. Therefore, to show our model's data efficiency, we train our model on datasets with different sizes (Table 4). With only 500 training samples, TSegFormer is able to surpass PointNet++, MeshSegNet and TSGCNet trained on 12,000 samples. Furthermore, TSegFormer trained with only 2,000 samples can almost outperform all previous models trained on 12,000 samples. Overall, these results demonstrate the exceptional data efficiency of our TSegFormer.

Effectiveness of Local Point Embedding. *SM* Table 1 shows purely MLP-based structures perform worst due to the lack of local contexts, while EdgeConv layers can make up for this, and the collaboration of both performs best.

Table 5. Clinical applicability test on the external IOS dataset (200 cases). "#success"/"#fail": number of segmentation that meets/does not meet the clinical criteria. #param: number of parameters in the network. Inf-T: inference time for 200 cases.

Model	#success ↑	#fail ↓	clinical error rate (%) ↓	#param ↓	Inf-T(s) ↓
MeshSegNet	65	135	67.5	1.81M	128.56
TSGCNet	15	185	92.5	4.13M	31.40
Point Transformer	97	103	51.5	6.56M	437.21
DCNet	109	91	45.5	**1.70M**	**5.79**
TSegFormer (Our)	**152**	**48**	**24.0**	4.21M	23.15

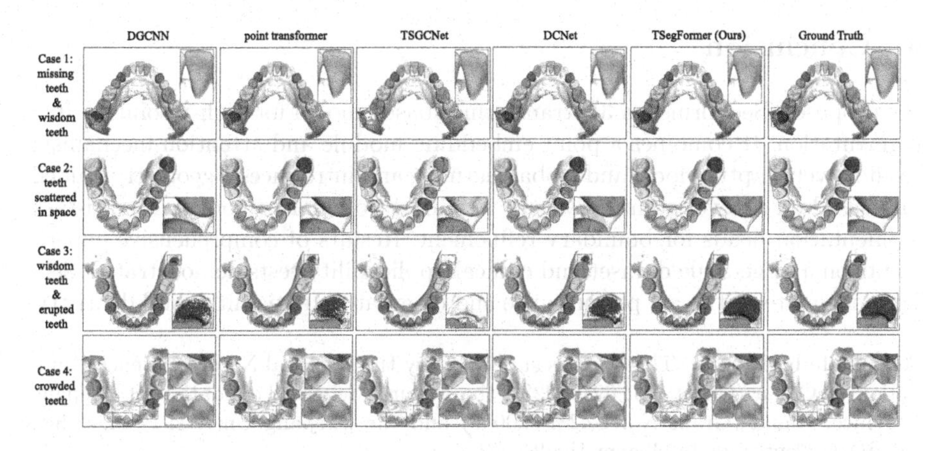

Fig. 3. Visualization of segmentation of different methods across different diseases.

3.4 Clinically Applicability Test and Visualization

To show the effectiveness of TSegFormer in real-world scenarios, we conducted a clinical applicability test (Table 5) on a dataset with 200 complex IOS scans, whose diseases statistics are in *SM* Table 3. The segmentation with five different models were evaluated by a committee of dentists with more than 5-year experience. We can notice that TSegFormer significantly outperforms the other models regarding the clinical error rate.

The feedback from dentists indicates that models such as TSGCNet cannot meet the requirement when dealing with complicated boundaries, while TSegFormer apparently handles them better. The point transformer and DCNet also showed promising performance, but they are yet far behind our TSegFormer. As for the number of parameters and inference time, though TSegFormer has the second most parameters among all methods we tested, it is the second fastest method that only takes around 23 s to complete inference for 200 cases, which is certainly acceptable in real-world clinical scenarios.

By visualization, we show the superiority of TSegFormer on various complicated dental diseases in Fig. 3. The baselines unavoidably produce false predictions or even fail to identify an entire third-molar, while TSegFormer can yield more accurate segmentation and smoother boundaries (see *SM* Fig. 2 for details), corroborating great potential for clinical applications. Specifically, *SM* Fig. 3 shows that with our geometry guided loss and auxiliary head, the isolated mispredictions and boundary errors are greatly reduced. However, TSegFormer fails in some complex samples, e.g. the missing tooth, the erupted wisdom tooth and sunken gingiva and alveolar bone regions, as illustrated in *SM* Fig. 4, which needs to be further studied.

4 Conclusion

We propose TSegFormer, a 3D transformer-based model for high-resolution IOS segmentation. It combines a point embedding module and attention mechanism to effectively capture local and global features, and introduces a geometry guided loss based on a novel point curvature to handle boundary errors and multi-task segmentation heads for boundary refinement. Results of comprehensive experiments on a large-scale dataset and clinical applicability tests demonstrate TSegFormer's state-of-the-art performance and its great potential in digital dentistry.

Acknowledgements. This work is supported by the National Natural Science Foundation of China (Grant No. 62106222), the Natural Science Foundation of Zhejiang Province, China (Grant No. LZ23F020008) and the Zhejiang University-Angelalign Inc. R&D Center for Intelligent Healthcare.

References

1. Cui, Z., et al.: TsegNet: an efficient and accurate tooth segmentation network on 3D dental model. Med. Image Anal. **69**, 101949 (2021)
2. Dosovitskiy, A., et al.: An image is worth 16×16 words: transformers for image recognition at scale. In: International Conference on Learning Representations (2021), https://openreview.net/forum?id=YicbFdNTTy
3. Ghazvinian Zanjani, F., et al.: Deep learning approach to semantic segmentation in 3D point cloud intra-oral scans of teeth. In: Cardoso, M.J., et al. (eds.) Proceedings of The 2nd International Conference on Medical Imaging with Deep Learning. Proceedings of Machine Learning Research, vol. 102, pp. 557–571. PMLR (2019)
4. Guo, M.H., Cai, J., Liu, Z.N., Mu, T.J., Martin, R.R., Hu, S.: PCT: point cloud transformer. Comput. Vis. Media **7**, 187–199 (2021)
5. Han, K., Xiao, A., Wu, E., Guo, J., Xu, C., Wang, Y.: Transformer in transformer. Adv. Neural. Inf. Process. Syst. **34**, 15908–15919 (2021)
6. Hao, J., et al.: Toward clinically applicable 3-Dimensional tooth segmentation via deep learning. J. Dent. Res. **101**(3), 304–311 (2022). https://doi.org/10.1177/00220345211040459. pMID: 34719980
7. He, X., et al.: Unsupervised pre-training improves tooth segmentation in 3-Dimensional intraoral mesh scans. In: International Conference on Medical Imaging with Deep Learning, pp. 493–507. PMLR (2022)
8. Herrmann, W.: On the completion of federation dentaire internationale specifications. Zahnarztliche Mitteilungen **57**(23), 1147–1149 (1967)
9. Jader, G., Fontineli, J., Ruiz, M., Abdalla, K., Pithon, M., Oliveira, L.: Deep instance segmentation of teeth in panoramic X-ray images. In: 2018 31st SIBGRAPI Conference on Graphics, Patterns and Images (SIBGRAPI), pp. 400–407 (2018)
10. Kondo, T., Ong, S.H., Foong, K.W.: Tooth segmentation of dental study models using range images. IEEE Trans. Med. Imaging **23**(3), 350–362 (2004)
11. Lian, C., et al.: MeshsNet: deep multi-scale mesh feature learning for end-to-end tooth labeling on 3D dental surfaces. In: Shen, D., et al. (eds.) Medical Image Computing and Computer Assisted Intervention – MICCAI 2019, pp. 837–845. Springer International Publishing, Cham (2019)

12. Lian, C., et al.: Deep multi-scale mesh feature learning for automated labeling of raw dental surfaces from 3d intraoral scanners. IEEE Trans. Med. Imaging **39**(7), 2440–2450 (2020)
13. Liu, Z., et al.: Swin transformer: hierarchical vision transformer using shifted windows. In: Proceedings of the IEEE/CVF International Conference on Computer Vision, pp. 10012–10022 (2021)
14. Liu, Z., et al.: Hierarchical self-supervised learning for 3D tooth segmentation in intra-oral mesh scans. IEEE Trans. Med. Imaging **42**(2), 467–480 (2023). https://doi.org/10.1109/TMI.2022.3222388
15. Miracle, A., Mukherji, S.: Conebeam CT of the head and neck, part 2: clinical applications. Am. J. Neuroradiol. **30**(7), 1285–1292 (2009). https://doi.org/10.3174/ajnr.A1654. http://www.ajnr.org/content/30/7/1285
16. Qi, C.R., Yi, L., Su, H., Guibas, L.J.: Pointnet++: deep hierarchical feature learning on point sets in a metric space. In: Proceedings of the 31st International Conference on Neural Information Processing Systems, pp. 5105–5114. NIPS 2017, Curran Associates Inc., Red Hook, NY, USA (2017)
17. Sinthanayothin, C., Tharanont, W.: Orthodontics treatment simulation by teeth segmentation and setup. In: 2008 5th International Conference on Electrical Engineering/Electronics, Computer, Telecommunications and Information Technology, vol. 1, pp. 81–84 (2008). https://doi.org/10.1109/ECTICON.2008.4600377
18. Sun, D., et al.: Tooth segmentation and labeling from digital dental casts. In: 2020 IEEE 17th International Symposium on Biomedical Imaging (ISBI), pp. 669–673 (2020). https://doi.org/10.1109/ISBI45749.2020.9098397
19. Vaswani, A., et al.: Attention is all you need. CoRR abs/1706.03762, http://arxiv.org/abs/1706.03762 (2017)
20. Wang, Y., Sun, Y., Liu, Z., Sarma, S.E., Bronstein, M.M., Solomon, J.M.: Dynamic graph CNN for learning on point clouds. ACM Trans. Graph. **38**(5) (2019)
21. Wongwaen, N., Sinthanayothin, C.: Computerized algorithm for 3D teeth segmentation. In: 2010 International Conference on Electronics and Information Engineering, vol. 1, pp. V1-277–V1-280 (2010)
22. Wu, K., Chen, L., Li, J., Zhou, Y.: Tooth segmentation on dental meshes using morphologic skeleton. Comput. Graph. **38**, 199–211 (2014)
23. Xu, X., Liu, C., Zheng, Y.: 3D tooth segmentation and labeling using deep convolutional neural networks. IEEE Trans. Visual Comput. Graphics **25**(7), 2336–2348 (2019)
24. Yamany, S.M., El-Bialy, A.M.: Efficient free-form surface representation with application in orthodontics. In: Nurre, J.H., Corner, B.D. (eds.) Three-Dimensional Image Capture and Applications II, vol. 3640, pp. 115–124. International Society for Optics and Photonics, SPIE (1999)
25. Yuan, T., Liao, W., Dai, N., Cheng, X., Yu, Q.: Single-tooth modeling for 3D dental model. Int. J. Biomed. Imaging **2010**, 535329 (2010)
26. Zhang, C., Wan, H., Shen, X., Wu, Z.: PVT: point-voxel transformer for point cloud learning. arXiv preprint arXiv:2108.06076 (2021)
27. Zhang, L., et al.: TSGCNet: discriminative geometric feature learning with two-stream graph convolutional network for 3D dental model segmentation. In: Proceedings of the IEEE/CVF Conference on Computer Vision and Pattern Recognition, pp. 6699–6708 (2021)
28. Zhao, H., Jiang, L., Jia, J., Torr, P.H., Koltun, V.: Point transformer. In: Proceedings of the IEEE/CVF International Conference on Computer Vision, pp. 16259–16268 (2021)

29. Zhao, M., Ma, L., Tan, W., Nie, D.: Interactive tooth segmentation of dental models. In: 2005 IEEE Engineering in Medicine and Biology 27th Annual Conference, pp. 654–657 (2005)
30. Zheng, Y., Chen, B., Shen, Y., Shen, K.: TeethGNN: semantic 3D teeth segmentation with graph neural networks. IEEE Trans. Vis. Comput. Graph. **29**, 3158–3168 (2022). https://doi.org/10.1109/TVCG.2022.3153501

Anatomical Landmark Detection Using a Multiresolution Learning Approach with a Hybrid Transformer-CNN Model

Thanaporn Viriyasaranon, Serie Ma, and Jang-Hwan Choi(✉)

Division of Mechanical and Biomedical Engineering, Graduate Program in System Health Science and Engineering, Ewha Womans University, Seoul, South Korea
choij@ewha.ac.kr

Abstract. Accurate localization of anatomical landmarks has a critical role in clinical diagnosis, treatment planning, and research. Most existing deep learning methods for anatomical landmark localization rely on heatmap regression-based learning, which generates label representations as 2D Gaussian distributions centered at the labeled coordinates of each of the landmarks and integrates them into a single spatial resolution heatmap. However, the accuracy of this method is limited by the resolution of the heatmap, which restricts its ability to capture finer details. In this study, we introduce a multiresolution heatmap learning strategy that enables the network to capture semantic feature representations precisely using multiresolution heatmaps generated from the feature representations at each resolution independently, resulting in improved localization accuracy. Moreover, we propose a novel network architecture called hybrid transformer-CNN (HTC), which combines the strengths of both CNN and vision transformer models to improve the network's ability to effectively extract both local and global representations. Extensive experiments demonstrated that our approach outperforms state-of-the-art deep learning-based anatomical landmark localization networks on the numerical XCAT 2D projection images and two public X-ray landmark detection benchmark datasets. Our code is available at https://github.com/seriee/Multiresolution-HTC.git.

Keywords: Anatomical landmark detection · Multiresolution learning · Hybrid transformer-CNN

1 Introduction

Anatomical landmark detection has been used successfully in parametric modeling [19], registration [22], and quantification of various anatomical abnormalities [6,21]. To detect landmarks automatically and accurately, advanced artificial intelligence technologies, including deep learning with convolutional neural net-

T. Viriyasaranon and S. Ma—Equally contributed.

H. Greenspan et al. (Eds.): MICCAI 2023, LNCS 14225, pp. 433–443, 2023.
https://doi.org/10.1007/978-3-031-43987-2_42

work (CNN)-based [13], transformer-based [7], and graph-convolution methods [8], have been developed and have attracted great interest from both academia and industry.

Generally, deep learning-based anatomical landmark detection is based on heatmap regression approaches [2,3], which decode the predicted landmark coordinates from the heatmap corresponding to the landmarks. In previous studies, the networks mostly generate only one resolution of the heatmap to decode the landmark coordinates. However, deriving the landmark coordinate from a high-resolution heatmap exhibits high bias and low variance, whereas the landmark coordinates obtained from a low-resolution heatmap demonstrate low bias and high variance. Typically, the heatmap regression-based detectors generate high-resolution heatmaps by utilizing the high-resolution coarse feature. However, this process results in the loss of specific landmark-related features, including crucial information regarding the geometric relationship between landmarks. Consequently, this impacts the network's ability to accurately localize landmarks.

In this study, we propose a multiresolution heatmap learning strategy that derives the predicted landmark coordinate from multiresolution heatmaps to balance the bias and variance of the predicted landmarks. Moreover, leveraging multiresolution feature representations to generate the heatmap can effectively increase the localization accuracy of the deep learning network.

Typically, the existing methods of anatomical landmark detection are formulated using CNN-based or transformer-based encoder-decoder architecture [16]. The convolution operation collects information by layer, which focus on the local feature information. Meanwhile, the vision transformer has the ability to encode global representations. To combine the advantages of CNNs and transformers, we introduce a novel hierarchical hybrid transformer and CNN architecture called the hybrid transformer-CNN (HTC). HTC introduces a stack of convolutional and transformer modules, which are applied to all stages of the encoder for extracting global information, and local information. Furthermore, we propose a lightweight positional-encoding-free transformer module. Instead of using multi-head attention, we introduce the bilinear pooling operation to capture second-order statistics of features and generate global representations. Moreover, general transformer encoders suffer from the fixed resolution of positional encoding, which results in decreased accuracy when interpolating the positional encoding during testing with resolutions different from the training data. To alleviate this problem, we remove the positional encoding from the transformer modules and employ a 3×3 convolutional operation as the patch embedding to capture location information and generate low-resolution fine features for the hierarchical encoder architecture design.

The main contributions of this paper are as follows:

- Introduction of a multiresolution heatmaps learning strategy, which increases the detection ability of the network by leveraging multiresolution information to derive the predicted landmark.

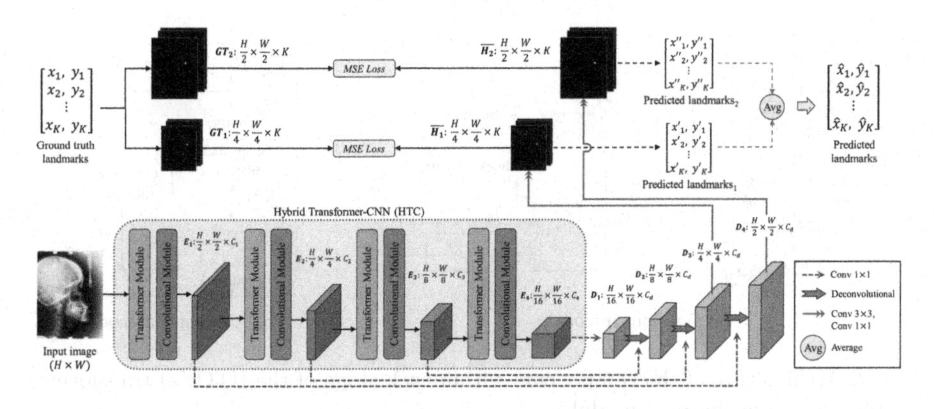

Fig. 1. Illustration of the overall architecture of the anatomical detector, including multiresolution heatmap learning and the hybrid transformer-CNN (HTC).

- Development of a hierarchical hybrid transformer and CNN architecture named the HTC, which sequentially combines transformer and convolutional modules.
- Our proposed HTC model trained with a multiresolution heatmap learning approach clearly outperforms previous state-of-the-art models on three datasets: XCAT 2D projections of head CBCT volumes, X-ray dataset from ISBI2023 Challenge [1], and hand X-ray dataset [13].

2 Methods

In this section, we introduce our anatomical landmark detector, which consists of the proposed multiresolution heatmap learning method and the HTC backbone network, as shown in Fig. 1.

2.1 Multiresolution Heatmap Learning

As depicted in Fig. 1, we generate the multiresolution prediction heatmap using multiresolution feature representations (D_i) from the decoder layers and extra convolutional layers. The output feature representations of the decoder layers are a combination of the feature representations from each of the stages of the encoder and upsampled feature representations from the previous stage. The number of channels of all stages of the decoder feature representations are set to be equal to C_d. In this work, we defined C_d as 256.

The lowest-resolution feature representation of the encoder E_4 of size $\frac{H}{16} \times \frac{W}{16} \times C_4$ is passed through 1×1 convolutional operation, and the output feature representation D_1 has size $\frac{H}{16} \times \frac{W}{16} \times C_d$. Then, D_1 is upsampled using 3×3 deconvolution operations and aggregated with the encoder feature representations E_3 to generate the next stage feature representations D_2 having size $\frac{H}{8} \times \frac{W}{8} \times C_d$.

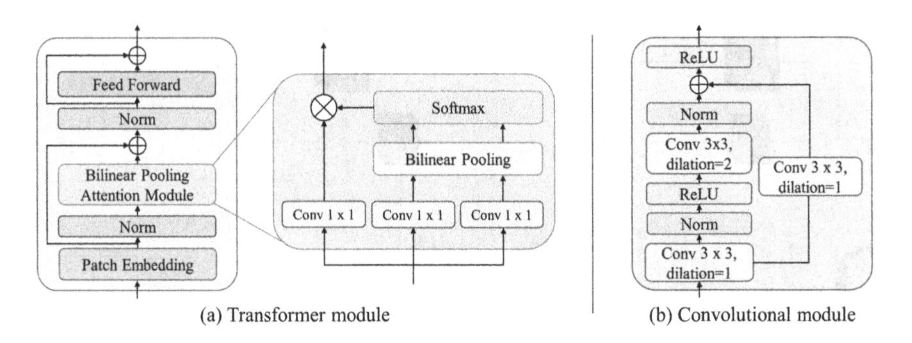

Fig. 2. Architecture of the components within each stage of the HTC (a) transformer modules; (b) convolutional modules.

Similarly, the high-resolution decoder feature representations D_3 and D_4 are the combinations of the previous decoder features D_2 and D_3, and encoder feature representation E_2 and E_1, respectively.

In this study, we generated predicted heatmaps H_1 and H_2 with two different resolution sizes $\frac{H}{4} \times \frac{W}{4} \times K$ and $\frac{H}{2} \times \frac{W}{2} \times K$ by utilizing 3×3 and 1×1 convolutional operations with decoder feature representations D_3 and D_4, respectively. Then, H_1 and H_2 are generated as follows:

$$\bar{H}_1 = Conv_1(Conv_3(D_3)),$$
$$\bar{H}_2 = Conv_1(Conv_3(D_4)),$$

where $Conv_1$ and $Conv_3$ are 1×1 and 3×3 convolutional operations, respectively. In addition, K is the number of landmarks.

During training, we use mean squared error (MSE) as the loss function between the predicted and ground-truth heatmaps for each resolution. Moreover, we enforce the detector to learn global and local information from the heatmap generated by the high-resolution coarse feature and the low-resolution fine-grained features, through the weighted summation of the heatmap loss from each resolution heatmap as follows:

$$\mathcal{L} = \mathcal{L}_{H_1} + \lambda \mathcal{L}_{H_2}, \tag{1}$$

where $\mathcal{L}_{H_1}$ and $\mathcal{L}_{H_2}$ are the respective losses calculated from the predicted heatmap H_1 and ground truth heatmap of size $\frac{H}{4} \times \frac{W}{4} \times K$ as well as the predicted heatmap H_2 and ground truth heatmap of size $\frac{H}{2} \times \frac{W}{2} \times K$. Additionally, λ is the loss weight, which is set as three in all the experiments. During inference, we calculate the output landmark prediction coordinates of the model by averaging the corresponding coordinates decoded from the heatmap at each resolution.

2.2 Hybrid Transformer-CNN (HTC)

We introduce a novel hierarchical encoder architecture named HTC, which consists of four stages of a stack of the convolutional $ConvU_i$ and transformer modules $TransU_i$ to generate multi-level feature representation. The overall architecture design of the HTC is shown in Fig. 1. At each stage, the transformer module captures global information such as the geometric relation between landmarks, while the convolutional module extracts local information. The architecture of the transformer and convolutional module are shown in Fig. 2 (a) and (b), respectively. In this study, we proposed a lightweight positional-encoding-free transformer module. Instead of using multi-head attention, we introduce the bilinear pooling operation to capture second-order statistics of features and generate global representations. Moreover, we eliminate the usage of positional encoding to address the challenge of reduced model performance when testing the model with resolutions that differ from the training data. This modification aims to mitigate the issue and improve the overall performance under varying resolution scenarios.

In the first stage, an input image of size $H \times W \times 3$ is fed to the patch embedding, which is a convolutional 3×3 operation to obtain a patch token of size $\frac{H}{2} \times \frac{W}{2} \times C_1$. Then, the patch token is passed through the bilinear pooling attention module, which requires three inputs: a query Q, a key K, and a value V, which are the patch token that passes through the 1×1 convolutional operations, separately. Then, Q, K, and V are flattened to size $\frac{HW}{2^2} \times C_1$. The key and query are then fed to bilinear pooling [9], which is an effective way of gathering the key features and capturing the global representations of the images as follows:

$$F = \sum_{n \in N} K_n Q_n^{\intercal}, \tag{2}$$

where N is the set of spatial locations (combinations of rows and columns). We further applied the softmax function to the output of bilinear pooling F to generate the attention weighting vector. Thereafter, matrix multiplication was performed between F and V, and the output of the bilinear pooling attention module was passed through to the feed-forward layer [23]. Then, the output of the feed-forward layer was reshaped as feature representations G_i of size $\frac{H}{2} \times \frac{W}{2} \times C_1$. In addition, the output of the transformer module $TransU_i$ was fed to the convolutional module $ConvU_i$, comprising 3×3 convolutional layers with dilated rates equal to one and two ($Conv_{3,d1}$ and $Conv_{3,d2}$), a batch normalization operation ($Norm$), and a rectified linear unit ($ReLU$) activation function as follows:

$$E_i = ReLU(Conv_{3,d1}(G_i) + Norm(Conv_{3,d2}(ReLU(Norm(Conv_{3,d1}(G_i)))))). \tag{3}$$

Similarly, using the feature representations from the previous stages as inputs, we obtained E_2, E_3, and E_4 with spatial reduction ratios of 4, 8, and 16 pixels, respectively, with respect to the input image.

3 Experiments

3.1 Dataset

To evaluate our method, we conducted experiments on a total of three datasets, including one 4D XCAT phantom CT dataset and two public X-ray datasets. Here, we generated head models from the 4D XCAT dataset [17] for 27 patients with varying anatomical sizes and genders. We manually labeled 13 cephalometric landmarks on CT phantom volumes. Moreover, we perform forward projection on both the 3D phantom CT volumes and landmarks at 360 angles per patient to obtain 2D images and landmark labels. We randomly selected 70% of the patients' CT scans as the training dataset (18 patients) and the remaining 30% as the test dataset (9 patients). The size of each image was 620×480, with a pixel spacing of 0.66 mm.

We also evaluated our method on the public X-ray dataset from the IEEE ISBI 2023 Challenge [1]. A total of 29 ground truth landmarks were labeled by two experts. The image sizes and pixel spacings vary over patients. We randomly selected 75% of the X-ray images of the provided training dataset as the network training dataset(525 images) and the remaining 25% as the test dataset.

Finally, we performed experiments on a public hand dataset containing X-ray images from 895 patients and having 37 landmarks [13]. The sizes of these images are not all the same, so we resized the images to 1024×1216. Owing to their lack of physical pixel resolution, we calculated the pixel spacing based on the assumption that the distance between two landmarks at both endpoints of the wrist is approximately 50 mm.

3.2 Implementation Details

In our experiments, we implemented our framework using MMPose [4], an open-source toolbox for pose estimation based on PyTorch. For XCAT CT landmark detection, our HTC with multiresolution heatmap learning was trained for 50 epochs using the AdamW optimizer with the initial learning rate set to 0.0003. Furthermore, we trained the ISBI2023 landmark detection method for 180 epochs using the AdamW optimizer, with an initial learning rate of 0.00045. In addition, for hand landmark detection, we trained for 300 epochs using the AdamW optimizer at an initial learning rate of 0.0004. For all the experiments, the evaluation metrics are the mean radial error (MRE, mm) and successful detection rate (SDR, %) under 2 mm, 2.5 mm, 3 mm, and 4 mm conditions.

3.3 Performance Evaluation

In this section, we compare the performances of the proposed and state-of-the-art methods as well as analyze ablation studies on the proposed method. In each of the tables below, the metrics showing the best and second-best performances are indicated by boldface and underlined, respectively.

Table 1. Performance comparison of the proposed method with previous state-of-the-art methods on the XCAT CT and ISBI2023 datasets.

Model	#Param(M)	XCAT CT dataset					ISBI2023 dataset				
		MRE(SD)	SDR(%)				MRE(SD)	SDR(%)			
			2 mm	2.5 mm	3 mm	4 mm		2 mm	2.5 mm	3 mm	4 mm
Hourglass [11]	94.85	5.35(10.21)	38.71	49.54	58.36	71.19	1.32(1.88)	82.90	88.41	92.12	95.53
HRNet-W48 [20]	65.33	3.91(6.22)	36.38	48.47	58.56	73.01	1.26(1.49)	83.94	88.85	92.26	96.00
HRFormer-S [26]	44.04	3.40(2.82)	36.50	48.58	58.71	72.67	1.34(1.45)	81.83	88.41	91.74	95.70
UNet [16]	35.35	4.08(9.49)	**47.28**	58.44	66.47	76.81	3.97(15.04)	82.13	86.58	89.28	92.22
PVT-Tiny [24]	16.91	5.93(7.72)	18.18	26.17	34.39	49.48	2.00(4.77)	72.99	80.43	85.34	90.54
Conformer-Ti [15]	22.32	3.88(4.17)	42.26	31.65	52.12	66.99	1.70(2.93)	75.69	83.23	87.35	93.10
GU2Net [27]	2.74	5.96(15.36)	38.66	48.41	56.39	66.67	1.78(5.10)	81.97	86.31	88.87	92.28
FARNet [2]	78.97	3.51(3.57)	35.62	46.54	56.21	71.33	1.10(**1.33**)	87.59	91.88	94.48	**96.87**
AFPF [3]	20.68	4.59(11.58)	41.57	52.98	61.74	72.51	4.13(16.35)	79.09	83.67	86.72	90.05
Multi-task UNet [25]	13.58	4.27(8.71)	40.80	51.46	59.74	71.87	5.91(15.63)	74.58	79.29	81.81	84.71
HTC+Multiresolution learning	16.20	**2.88(2.51)**	46.82	**59.02**	**67.97**	**79.68**	**1.08**(1.37)	**88.05**	**92.17**	**94.50**	96.69

Table 2. Performance comparison of the proposed method with other state-of-the-art models on the hand dataset. The symbol " * " indicates that we repeated the training with the same environment used to develop our method.

Model	Hand dataset			
	MRE(SD)	SDR(%)		
		2 mm	4 mm	10 mm
Payer et al. [12]	1.13(0.98)	87.60	98.66	99.96
GIRRF [5]	0.97(2.45)	91.60	97.84	99.31
DATR [28]	0.86(-)	94.04	99.20	99.97
Lindner et al. [10]	0.85(1.01)	93.68	98.95	99.94
Urschler et al. [21]	0.80(0.93)	92.19	98.46	98.46
Štern et al. [18]	0.80(0.91)	92.20	98.45	99.83
FARNet* [2]	0.67(0.64)	95.65	99.58	99.99
Payer et al. [14]	0.66(0.74)	94.99	99.27	99.99
HRNet-W48* [20]	0.65(0.62)	96.10	99.60	**100**
GU2Net* [27]	0.63(1.36)	96.01	99.39	99.98
HTC+Multiresolution learning	**0.56(0.58)**	**96.84**	**99.63**	**100**

Comparisons with State-of-the-art Methods: Performance comparisons on the XCAT CT and ISBI2023 dataset are shown in Table 1. We compared our proposed model with both natural and medical-domain landmark detectors. Our method remarkably achieved the best performance on the XCAT CT dataset with an MRE of 2.88 mm and on the ISBI2023 dataset with an MRE of 1.08 mm. Furthermore, our approach showed effects in lowering the standard deviation of the mean error on the XCAT CT dataset and improved the percentage of successful detection by 2.87%. Moreover, our method outperformed the highest detection rate among previous studies, reaching an SDR of 88.05% on the ISBI2023 dataset. Furthermore, the HTC was intentionally designed to have a smaller number of parameters compared to transformer-based architectures such as HRFormer and Conformer, despite achieving superior detection accuracy. Table 2 presents the performance comparison between the proposed and existing methods on the hand dataset. Our method significantly outperformed the best performance for all metrics on the hand dataset, with 0.56 mm MRE and 0.58 mm standard deviation of MRE. Additionally, qualitative comparisons of the images of the detection results are shown in Fig. 3.

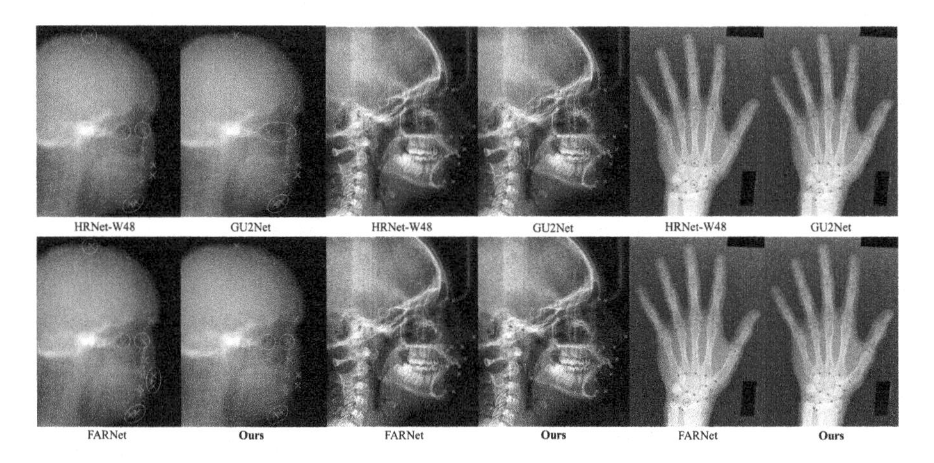

Fig. 3. Comparison of the proposed method and other models on the XCAT CT, ISBI2023, and hand datasets (from left to right). Ground-truth landmarks are marked in green, and the predictions are marked in red. We mark landmarks that do not overlap between predictions and ground truth with circles. (Color figure online)

Table 3. Comparison of the effects of the proposed HTC and multiresolution heatmap learning.

Model	Multi-resolution learning	XCAT CT dataset					ISBI2023 dataset				
		MRE(SD)	SDR(%)				MRE(SD)	SDR(%)			
			2 mm	2.5 mm	3 mm	4 mm		2 mm	2.5 mm	3 mm	4 mm
Conformer [15]	✗	3.88(4.16)	42.26	31.65	52.11	66.98	1.70(2.93)	75.69	83.23	87.35	93.10
HTC	✗	3.03(2.59)	43.64	55.72	64.73	76.99	1.11(1.37)	87.56	**91.80**	**94.26**	**96.65**
	✓	**2.88(2.51)**	**46.82**	**59.02**	**67.97**	**79.68**	**1.08(1.35)**	**87.80**	**91.80**	94.18	96.45

Ablation Study: We conducted an additional study to observe the effects of the proposed multiresolution heatmap learning and HTC. For this study, we compared the HTC with Conformer [15], which is a representative hybrid transformer and CNN architecture. As shown in Table 3, the HTC outperforms Conformer with MRE values of 3.03 and 1.11 mm for the XCAT CT and ISBI2023 datasets, respectively. Furthermore, the implementation of multiresolution heatmap learning can enhance the MRE of 0.15 and 0.03 mm for the XCAT CT and ISBI2023 datasets, respectively.

4 Conclusion

This study presents a new feature extraction architecture, referred to as the hybrid transformer-CNN (HTC), along with multiresolution heatmap learning for automatic anatomical landmark detection. The HTC architecture comprises of multiple stages of stacked transformer modules, which incorporate a bilinear pooling attention module to capture the global information of images and convolutional modules to extract local and specific feature representations relevant to landmarks. Additionally, we introduced multiresolution heatmap learning to improve the network's ability to capture global and local representations more accurately than learning from a single heatmap resolution, thereby enhancing network localization. Our experimental evaluations on three benchmark datasets demonstrate that the proposed method surpasses state-of-the-art approaches across various modalities and anatomical regions. These findings highlight the potential of our method for automatic anatomical landmark detection in various medical applications.

Acknowledgements. This research was partly supported by the BK21 FOUR (Fostering Outstanding Universities for Research) funded by the Ministry of Education (MOE, Korea) and National Research Foundation of Korea (NRF-5199990614253); by the Technology development Program of MSS [S3146559]; by the National Research Foundation of Korea (NRF-2022R1A2C1092072); and Institute of Information and communications Technology Planning and Evaluation (IITP) grant funded by the Korean government (MSIT) (No. RS-2022-00155966, Artificial Intelligence Convergence Innovation Human Resources Development (Ewha Womans University)).

References

1. Anwaar Khalid, M., et al.: CEPHA29: automatic cephalometric landmark detection challenge 2023. arXiv e-prints arxiv.org/abs/2212.04808 (2022)
2. Ao, Y., Wu, H.: Feature aggregation and refinement network for 2D anatomical landmark detection. J. Digit. Imaging **36**(2), 547–561 (2022). https://doi.org/10.1007/s10278-022-00718-4
3. Chen, R., Ma, Y., Chen, N., Lee, D., Wang, W.: Cephalometric landmark detection by attentive feature pyramid fusion and regression-voting. In: Shen, D., et al. (eds.) MICCAI 2019. LNCS, vol. 11766, pp. 873–881. Springer, Cham (2019). https://doi.org/10.1007/978-3-030-32248-9_97
4. Contributors, M.: OpenMMLab pose estimation toolbox and benchmark. https://github.com/open-mmlab/mmpose (2020)
5. Ebner, T., Stern, D., Donner, R., Bischof, H., Urschler, M.: Towards automatic bone age estimation from MRI: localization of 3D anatomical landmarks. In: Golland, P., Hata, N., Barillot, C., Hornegger, J., Howe, R. (eds.) MICCAI 2014. LNCS, vol. 8674, pp. 421–428. Springer, Cham (2014). https://doi.org/10.1007/978-3-319-10470-6_53
6. Ibragimov, B., Likar, B., Pernuš, F., Vrtovec, T.: Shape representation for efficient landmark-based segmentation in 3-D. IEEE Trans. Med. Imaging **33**(4), 861–874 (2014)

7. Jiang, Y., Li, Y., Wang, X., Tao, Y., Lin, J., Lin, H.: CephalFormer: incorporating global structure constraint into visual features for general cephalometric landmark detection. In: Wang, L., Dou, Q., Fletcher, P.T., Speidel, S., Li, S. (eds.) Medical Image Computing and Computer Assisted Intervention – MICCAI 2022. MICCAI 2022. Lecture Notes in Computer Science, vol. 13433, pp. 227–237. Springer, Cham (2022). https://doi.org/10.1007/978-3-031-16437-8_22
8. Lang, Y., et al.: Automatic localization of landmarks in craniomaxillofacial CBCT images using a local attention-based graph convolution network. In: Martel, A.L., et al. (eds.) MICCAI 2020. LNCS, vol. 12264, pp. 817–826. Springer, Cham (2020). https://doi.org/10.1007/978-3-030-59719-1_79
9. Lin, T.Y., RoyChowdhury, A., Maji, S.: Bilinear CNN models for fine-grained visual recognition. In: Proceedings of the IEEE International Conference on Computer Vision, pp. 1449–1457 (2015)
10. Lindner, C., Bromiley, P.A., Ionita, M.C., Cootes, T.F.: Robust and accurate shape model matching using random forest regression-voting. IEEE Trans. Pattern Anal. Mach. Intell. **37**(9), 1862–1874 (2014)
11. Newell, A., Yang, K., Deng, J.: Stacked hourglass networks for human pose estimation. In: Leibe, B., Matas, J., Sebe, N., Welling, M. (eds.) ECCV 2016. LNCS, vol. 9912, pp. 483–499. Springer, Cham (2016). https://doi.org/10.1007/978-3-319-46484-8_29
12. Payer, C., Štern, D., Bischof, H., Urschler, M.: Regressing heatmaps for multiple landmark localization using CNNs. In: Ourselin, S., Joskowicz, L., Sabuncu, M.R., Unal, G., Wells, W. (eds.) MICCAI 2016. LNCS, vol. 9901, pp. 230–238. Springer, Cham (2016). https://doi.org/10.1007/978-3-319-46723-8_27
13. Payer, C., Štern, D., Bischof, H., Urschler, M.: Integrating spatial configuration into heatmap regression based CNNs for landmark localization. Med. Image Anal. **54**, 207–219 (2019)
14. Payer, C., Štern, D., Bischof, H., Urschler, M.: Integrating spatial configuration into heatmap regression based CNNs for landmark localization. Med. Image Anal. **54**, 207–219 (2019)
15. Peng, Z., et al.: Conformer: local features coupling global representations for visual recognition. In: Proceedings of the IEEE/CVF International Conference on Computer Vision, pp. 367–376 (2021)
16. Ronneberger, O., Fischer, P., Brox, T.: U-Net: convolutional networks for biomedical image segmentation. In: Navab, N., Hornegger, J., Wells, W.M., Frangi, A.F. (eds.) MICCAI 2015. LNCS, vol. 9351, pp. 234–241. Springer, Cham (2015). https://doi.org/10.1007/978-3-319-24574-4_28
17. Segars, W.P., Sturgeon, G., Mendonca, S., Grimes, J., Tsui, B.M.: 4D XCAT phantom for multimodality imaging research. Med. Phys. **37**(9), 4902–4915 (2010)
18. Štern, D., Ebner, T., Urschler, M.: From local to global random regression forests: exploring anatomical landmark localization. In: Ourselin, S., Joskowicz, L., Sabuncu, M.R., Unal, G., Wells, W. (eds.) MICCAI 2016. LNCS, vol. 9901, pp. 221–229. Springer, Cham (2016). https://doi.org/10.1007/978-3-319-46723-8_26
19. Štern, D., Likar, B., Pernuš, F., Vrtovec, T.: Parametric modelling and segmentation of vertebral bodies in 3D CT and MR spine images. Phys. Med. Biol. **56**(23), 7505 (2011)
20. Sun, K., Xiao, B., Liu, D., Wang, J.: Deep high-resolution representation learning for human pose estimation. In: Proceedings of the IEEE/CVF Conference on Computer Vision and Pattern Recognition, pp. 5693–5703 (2019)

21. Urschler, M., Ebner, T., Štern, D.: Integrating geometric configuration and appearance information into a unified framework for anatomical landmark localization. Med. Image Anal. **43**, 23–36 (2018)
22. Urschler, M., Zach, C., Ditt, H., Bischof, H.: Automatic point landmark matching for regularizing nonlinear intensity registration: application to thoracic CT images. In: Larsen, R., Nielsen, M., Sporring, J. (eds.) MICCAI 2006. LNCS, vol. 4191, pp. 710–717. Springer, Heidelberg (2006). https://doi.org/10.1007/11866763_87
23. Vaswani, A., et al.: Attention is all you need. ArXiv abs/1706.03762 (2017)
24. Wang, W., et al.: Pyramid vision transformer: a versatile backbone for dense prediction without convolutions. In: Proceedings of the IEEE/CVF International Conference on Computer Vision, pp. 568–578 (2021)
25. Yao, Q., He, Z., Han, H., Zhou, S.K.: Miss the point: targeted adversarial attack on multiple landmark detection. In: Martel, A.L., et al. (eds.) MICCAI 2020. LNCS, vol. 12264, pp. 692–702. Springer, Cham (2020). https://doi.org/10.1007/978-3-030-59719-1_67
26. Yuan, Y., et al.: HRFormer: high-resolution vision transformer for dense predict. Adv. Neural. Inf. Process. Syst. **34**, 7281–7293 (2021)
27. Zhu, H., Yao, Q., Xiao, L., Zhou, S.K.: You only learn once: universal anatomical landmark detection. In: de Bruijne, M., et al. (eds.) MICCAI 2021. LNCS, vol. 12905, pp. 85–95. Springer, Cham (2021). https://doi.org/10.1007/978-3-030-87240-3_9
28. Zhu, H., Yao, Q., Zhou, S.K.: DATR: domain-adaptive transformer for multi-domain landmark detection. arXiv preprint arXiv:2203.06433 (2022)

Cross-View Deformable Transformer for Non-displaced Hip Fracture Classification from Frontal-Lateral X-Ray Pair

Zhonghang Zhu[1], Qichang Chen[1], Lequan Yu[2], Lianxin Wang[3], Defu Zhang[1], Baptiste Magnier[4], and Liansheng Wang[1](✉)

[1] Department of Computer Science at School of Informatics, Xiamen University, Xiamen, China
{zzhonghang,qcchen}@stu.xmu.edu.cn, dfzhang@xmu.edu.cn, dr_shepherd@sina.com

[2] Department of Statistics and Actuarial Science, The University of Hong Kong, Hong Kong SAR, China
lqyu@hku.hk

[3] Department of Orthopedics, The First Affiliated Hospital of Xiamen University, Xiamen, China
lswang@xmu.edu.cn

[4] Euromov Digital Health in Motion, University Montpellier, IMT Mines Ales, Ales, France
baptiste.magnier@mines-ales.fr

Abstract. Hip fractures are a common cause of morbidity and mortality and are usually diagnosed from the X-ray images in clinical routine. Deep learning has achieved promising progress for automatic hip fracture detection. However, for fractures where displacement appears not obvious (*i.e.*, non-displaced fracture), the single-view X-ray image can only provide limited diagnostic information and integrating features from cross-view X-ray images (*i.e.*, Frontal/Lateral-view) is needed for an accurate diagnosis. Nevertheless, it remains a technically challenging task to find reliable and discriminative cross-view representations for automatic diagnosis. First, it is difficult to locate discriminative task-related features in each X-ray view due to the weak supervision of image-level classification labels. Second, it is hard to extract reliable complementary information between different X-ray views as there is a displacement between them. To address the above challenges, this paper presents a novel cross-view deformable transformer framework to model relations of critical representations between different views for non-displaced hip fracture identification. Specifically, we adopt a deformable self-attention module to localize discriminative task-related features for each X-ray view only with the image-level label. Moreover, the located discriminative features are further adopted to explore correlated representations across views by taking advantage of the query of the dominated view as guidance. Furthermore, we build a dataset including 768 hip cases, in which each case has paired hip X-ray images (Frontal/Lateral-view), to evaluate our

H. Greenspan et al. (Eds.): MICCAI 2023, LNCS 14225, pp. 444–453, 2023.
https://doi.org/10.1007/978-3-031-43987-2_43

framework for the non-displaced fracture and normal hip classification task.

Keywords: Hip fracture diagnosis · X-ray image · Deformable transformer · Cross-view correspondence

1 Introduction

Hip fractures represent a life-changing event and carry a substantial risk of decreased functional status and death, especially in elderly patients [17]. Usually, they are diagnosed from X-ray images in clinical practice. Currently, proper X-ray fracture identification relies on the manual observation of board-certified radiologists, which leads to increased workload pressures to radiologists. However, accurate and timely diagnosis of hip fractures is critical, especially in emergency situations such as non-displaced hip fractures [13]. Therefore, automated X-ray image classification is of great significance to support the clinical assistant diagnosis.

Recently, Deep Learning (DL) methods for radiography analysis have gained popularity and shown promising results [4,7,14,18], which aims to distinguish normal radiography or prioritize urgent/critical cases with the goal of reducing the radiologist workload or improving the reporting time. For example, a triaging pipeline based on the urgency of exams has been proposed in [1] and Tang *et al.* [21] compared different DL models applied to several public chest radiography datasets for distinguishing abnormal cases. However, these works only focus on single-view radiography analysis. When the fracture displacement in the X-ray image is not apparent,*i.e.*, a non-displaced hip fracture as shown in Fig. 1, these methods may fail in extracting enough fracture features represented by a ridge [20] in the image and result in misdiagnosis. Therefore, it is necessary to develop cross-view learning approaches to diagnose fracture from paired views (Frontal/Lateral-images), which have been demonstrated to provide complementary features to promote the diagnostic performance [3,10]. Recent studies have been investigated for cross-view learning of X-ray images, which aims to exploit the value of paired X-ray images and fuse them to get a comprehensive anatomical representation for diagnosis [2,5,15,19,22]. However, these methods do not consider cross-view feature relations which is a quite important issue for accurate cross-view feature fusion.

Since the introduction of vision transformer models [9], more researches have been developed in the tokenization process and relation modeling among tokens in an image [8,16]. Recently, deformable self-attention has been proposed to refine visual tokens [6,23,25], which is powerful in focusing on relevant regions and capturing more informative features. Motivated by this, we propose a novel cross-view deformable transformer framework for hip fracture detection from cross-view X-ray images. Firstly, deformable self-attention modules are utilized to localize reliable task-related features of each view. Secondly, the dominated-view characteristics are used to explore informative representations in the other

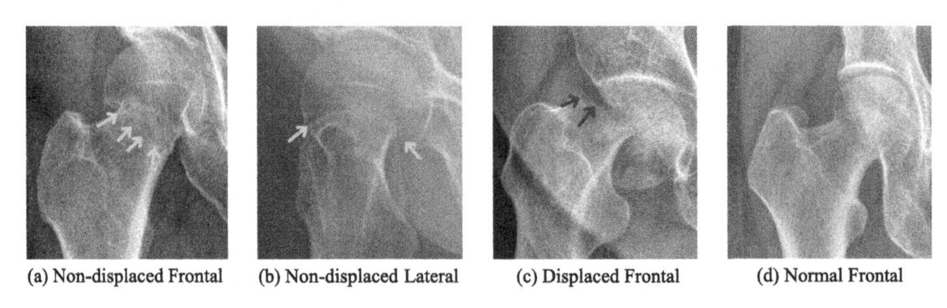

Fig. 1. Comparisons of non-displaced/displaced hip fracture and normal hip X-ray images. The fracture regions are marked by green arrows and red arrows for non-displaced/displaced fractures, respectively. (Color figure online)

view for effective feature fusion of cross-view X-ray images. Specifically, our contributions are three folds:

1. We propose a cross-view deformable transformer framework for non-displaced hip fracture classification, in which we take advantage of discriminative features of Frontal-view as a guidance to localize informative representations of Lateral-view for cross-view feature fusion.
2. For each view, we adopt the deformable self-attention module to select pivotal tokens in a data-dependent way.
3. We build a new non-displaced hip fracture X-ray dataset which includes both Frontal and Lateral views for each case to valid the proposed method. Our approach surpasses the state of the art in accuracy by over 1.5%.

2 Method

2.1 Overview

The detailed architecture of the proposed cross-view deformable attention framework is shown in Fig. 2. To model the relations among features across different views, the framework is designed as a joint of two view-specific deformable transformer branches with four stages. For each view-specific branch, the input image is firstly processed by shifted window attention modules presented in the left of Fig. 2 to aggregate information locally, followed by the last two stages to model the global relations among the locally augmented tokens with deformable self-attention modules. In the last two stages, the query features of the Frontal-view are adopted as the guidance to detect the relations among the Lateral-view tokens. The detailed design of each component is introduced below.

2.2 View-Specific Deformable Transformer Network

To discover the task-related regions of each view, the view-specific branch is designed as a deformable transformer network consisting of four stages. In each

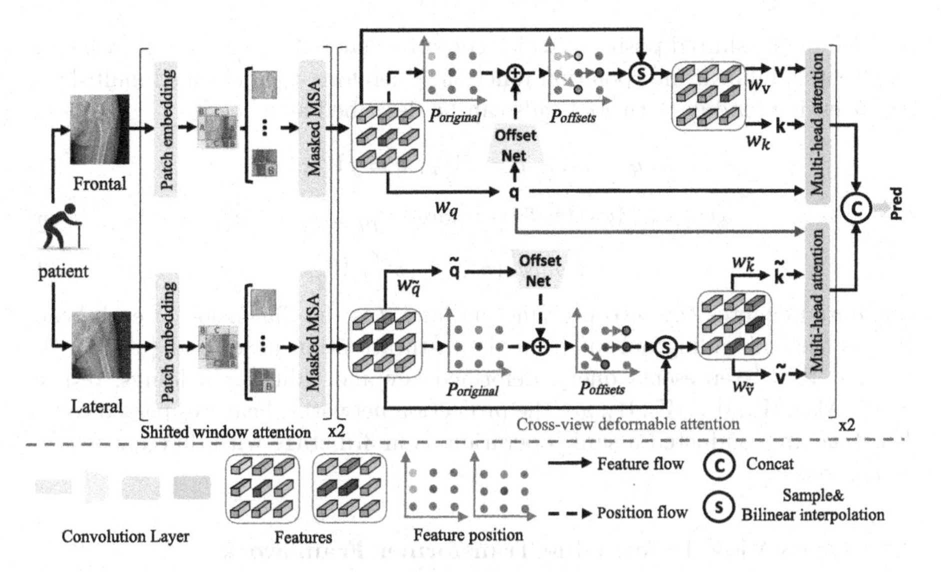

Fig. 2. Illustration of the proposed framework depicted in two view-specific branches with four stages. A pair of X-ray images, *i.e.*, Frontal and Lateral images are fed into two branches, respectively. The input images are processed with shifted window attention modules to aggregate discriminative local features (first two stages), while deformable self-attention modules are utilized to model the relations among tokens (last two stages). Moreover, Frontal queries are passed to model relations among Lateral features for cross-view deformable attention. The $p_{original}$ represents the original feature position of each view, while $p_{offsets}$ denotes the position offsets. The $W_{q/\tilde{q}}$, $W_{k/\tilde{k}}$ and $W_{v/\tilde{v}}$ are projection matrices for queries, keys and values, respectively.

branch, the first two stages explore the local representations of the input images with shift-window attention modules, followed by the last two stages exploit local tokens relation using deformable self-attention modules. Specifically, our framework takes an image of size $H \times W \times 3$ as input. After the first two stages, the input image will be embedded into feature maps $f_{layer3} \in H/4 \times W/4 \times C$, where the C denotes the channel number. The f_{layer3} will be passed to a query projection network W_q, which is a light network to obtain the query feature maps $f_{layer3;q}$. Moreover, a uniform grid $p_{original} \in \mathbb{R}^{H/4 \times W/4 \times 2}$ is generated as a position reference of points in f_{layer3}. The values of $p_{original}$ are linearly spaced and normalized to 2D coordinates range in $(-1,-1), \cdots, (1,1)$, in which the $(-1,\cdot), \cdots, (1,\cdot)$ and the $(\cdot,-1), \cdots, (\cdot,1)$ refers to the horizontal and vertical coordinates for reference points respectively. In the meanwhile, reference points offset $p_{offsets} \in \mathbb{R}^{H/4 \times W/4 \times 2}$ are generated from the $f_{layer3;q}$ by a light offset network consisted of two convolutional layers followed normalization layer, which are also normalized into $(-4/H, -4/W), \cdots, (4/H, 4/W)$. The shifted position of points in f_{layer3} are calculated as $p = \psi(p_{original} + a(p_{offsets}))$, where $a(\cdot)$ is a function (*i.e.*, $4\tanh(\cdot)$) to prevent the offset from becoming too large and $\psi(p_x, p_y) = (H * p_x, W * p_y)$. Then the deformed features of each point are sampled at the shifted position, which could be denoted as $\bar{f} = S(f, p)$, where S represents a bilinear interpolation function. Therefore, the deformed multi-head self-attention module with M heads can be described as:

$$q = fW_q,\ \bar{k} = \bar{f}W_k,\ \bar{v} = \bar{f}W_v, \tag{1}$$

$$z^m = \sigma(q^{(m)}\bar{k}^{(m)T}/\sqrt{d})\bar{v}^{(m)}, \quad m = 1, \cdots, M, \tag{2}$$

$$z = Concat\left(z^1, \cdots, z^M\right) W_o, \tag{3}$$

where $\sigma(\cdot)$ denotes the softmax function, and d is the dimension of each head. $z^{(m)}$ is the embedding output from the m-th attention head, and $\{q^{(m)}, \bar{k}^{(m)}, \bar{v}^{(m)}\} \in \mathbb{R}^{N \times d}$ represents query, deformed key and value embeddings, respectively. Also, W_q, W_k, W_v, W_o are the projection networks. Features passed to the 4th stage are conducted a same operation as in 3rd stage with different feature dimensions.

2.3 Cross-View Deformable Transformer Framework

The proposed cross-view framework is consisted of two joint view-specific branches, with a pair of X-ray images (Frontal-view and Lateral-view) from the same patient taken as the input of two individual view-specific branches, respectively. These input images will be embedded into primary representations in the first stage of view-specific network, then these primary features will be sent to the second stage to get representations with larger receptive field. To observe correlations between Frontal-view and Lateral-view, we opt for a simple solution to share queries from the Frontal-view to model token relations of Lateral-view in a self-attention manner as the Frontal-view contains dominated diagnosis features [24]. In this way, the focused regions of the Lateral-view are determined by the discriminative features of the Frontal-view. So for the Lateral-view branch, the multi-head self-attention can be denoted as:

$$q_{fr} = f_{fr}W_{q;fr},\ \bar{k_{la}} = \bar{f_{la}}W_{k;la},\ \bar{v_{la}} = \bar{f_{la}}W_{v;la}, \tag{4}$$

$$z_{la}^m = \sigma(q_{fr}^{(m)}\bar{k_{la}}^{(m)T}/\sqrt{d})\bar{v_{la}}^{(m)}, \quad m = 1, \cdots, M, \tag{5}$$

$$z_{la} = Concat\left(z_{la}^1, \cdots, z_{la}^M\right) W_{o;la}, \tag{6}$$

in which $(\cdot)_{fr}$ and $(\cdot)_{la}$ represent the features of Frontal-view and Lateral-view, respectively. While for the Frontal-view branch, the multi-head self-attention can be denoted as:

$$q_{fr} = f_{fr}W_{q;fr},\ \bar{k_{fr}} = \bar{f_{fr}}W_{k;fr},\ \bar{v_{fr}} = \bar{f_{fr}}W_{v;fr}, \tag{7}$$

$$z_{fr}^m = \sigma(q_{fr}^{(m)}\bar{k_{fr}}^{(m)T}/\sqrt{d})\bar{v_{fr}}^{(m)}, \quad m = 1, \cdots, M, \tag{8}$$

$$z_{fr} = Concat\left(z_{fr}^1, \cdots, z_{fr}^M\right) W_{o;fr}, \tag{9}$$

It is worth noting that view-specific reference points offset are derived from corresponding view-specific query feature maps which contain global view-specific position relations. By taking query feature maps from the Frontal-view as an informative clue, it makes sense to search relevant task-related features in the Lateral-view deformed values and keys embedding which are also discriminative features of Lateral-view. In this way, the cross-view transformer framework manages to localize task-related features in both views while exploring the cross-view related representations for feature aggregation. Then the final output can be denoted as $outputs = MLP(Concat(f_{fr}, f_{la}))$, where the f_{fr} and f_{la} represent the output features of the last layer of the Frontal-view and Lateral-view branches, respectively. The *Concat* is a concatenation operation and *MLP* is a projection head consisted of two fully connected layers to generate logit predictions.

2.4 Technical Details

The view-specific model shares a similar pyramid structure with DTA-T [23]. The first stage consists of one shift-window block whose head number is set as 3, followed the second stage with one shift-window block whose head number is set as 6. We adopt three deformable attention block with 12 heads in the 3rd stage and one deformable attention block with 24 heads in the 4th stage. To optimize the whole framework, we calculate the cross entropy loss between the label and final output of the cross-view deformable transformer for training.

3 Experiment

3.1 Experiment Setup

Dataset. The dataset used in this study includes 768 paired hip X-ray images (329 non-displaced fractures, 439 normal hips) from 4 different manufacturers of radiologic data sources: GE Healthcare, Philips Medical Systems, Kodak and Canon. All the hip radiographs are collected and labeled by experts with non-displaced fractures or normal for classification task.

Implementation Details. For experiments of our dataset, we manually locate the hip region and crop a 224 × 224 image that is centered on the original hip region whose size is 400 × 600. The learning rate is set as 3e-3 for the end-to-end training of the framework with a batch size of 32. We adopt a 10-fold cross-validation and report the average performance of 10 folds. For each fold, we further divide the data (the other 9 folds) into a training set (90%) and a validation set (10%) and take the best model on the validation part for testing.

Evaluation Metric. We evaluate our method with Accuracy (Acc), *Precision*, *Recall* and F1 score. The *Precision* and *Recall* are calculated with one-class-versus-all-other-classes and then calculate F1 score $\left(F1 = \frac{2 \cdot Precision \cdot Recall}{Precision + Recall}\right)$.

3.2 Experimental Results

Comparisons with the State of the Art. The proposed method is also compared with other cross-view fusion methods; results are reported in Table 1. *1) MVC-NET:* a network with back projection transposition branch to explicitly incorporate the spatial information from two views at the feature level. *2) DualNet:* an ensemble of two DenseNet-121 [12] networks followed a global average pooling operation of the final convolutional layer before a fully connected layer to simultaneously process multi-view images. *3) Auloss:* a DualNet regularized by auxiliary view-specific classification losses. *4) ResNet18-dual:* an ensemble of two ResNet18 [11] networks, and the predicted results are generated by concatenating logit outputs from each ResNet18 network. *5) Densenet-dual:* an ensemble of two DenseNet networks, and the predicted results are generated by concatenating logit outputs from each DenseNet network. *6) Swin-dual:* an ensemble of two swin-transformer networks [16], and the predicted results are generated by concatenating logit outputs from each swin-transformer network.

As shown in Table 1, we compare our method to different dual frameworks. It can be observed that the proposed method achieves better performance than others (compare *Ours* with *ResNet18-dual*, *Densenet-dual* and *Swin-dual*), which demonstrates that the accuracy boost is due to the deformable transformer network and the feature interaction design not the increased backbone size. In addition, the *MVC-NET* shares a similar feature-level interaction motivation with *Ours*, and the 19.1% accuracy improvement indicates that our cross-view deformable attention gains better performance. Otherwise, we demonstrate the effectiveness of the deformable transformer network by comparing the *Our w/o q* to *Swin-dual*, as the only difference between these two frameworks is that the *Our w/o q* change the last two stages of *Swin-dual* to deformable transformer modules.

Ablation Study. We also conduct ablation experiments to validate the design of our proposed different components. We compare the following different settings. *1) Frontal:* take the Frontal image as input of the view-specific deformable transformer network to generate the prediction. *2) Frontal_swin:* take the Frontal image as input of the swin-transformer network to generate the prediction. *3) Lateral:* take the Lateral image as input of the view-specific deformable transformer network to generate the prediction. *4) Lateral_swin:* take the Lateral image as input of the swin-transformer network to generate the prediction. *5) Ours w/o q:* the proposed framework without cross-view deformable attention.

Table 1 shows the ablation results. It is observed from *Ours w/o q* and *Ours* that the proposed method improves the performance of classification by adopting the proposed cross-view deformable attention, which demonstrates that the query of Frontal-view has a positive effect on mining the discrimination features of Lateral representations. Especially, cross-view learning contributes a minimum accuracy improvement of 3% compared *Ours* to *Frontal* and *Lateral* as discriminative features between different views can be complementary. Moreover, we

Table 1. Quantitative results (mean±standard deviation)% of different methods.

Method	Acc	*Precision*	*Recall*	F1 score
MVC-NET [26]	77.35±4.81	71.96±10.23	75.69±9.09	73.30±7.66
DualNet [19]	80.87±5.98	76.21±10.31	82.37±8.01	78.41±5.49
Auloss [10]	82.30±4.51	78.42±8.26	79.69±6.23	78.86±6.15
ResNet18-dual	88.41±2.86	95.87±3.27	76.14±8.29	84.58±4.61
Densenet-dual	90.36±3.42	94.37±3.11	82.20±7.95	87.66±4.60
Swin-dual	95.05±2.01	95.44±5.00	92.85±3.77	94.01±2.68
Lateral_swin	85.16±3.86	85.36±6.89	78.77±8.35	81.56±5.42
Lateral	84.37±4.43	84.51±7.78	78.01±9.47	80.68±6.41
Frontal_swin	91.67±3.53	94.12±5.22	85.55±8.64	89.33±5.34
Frontal	93.36±2.24	93.19±5.34	91.34±3.94	92.11±2.76
Ours w/o q	95.83±1.59	**96.10±4.82**	93.96±3.57	**96.36±1.31**
Ours	**96.48±1.83**	95.65±4.31	**96.22±3.69**	95.83±2.26

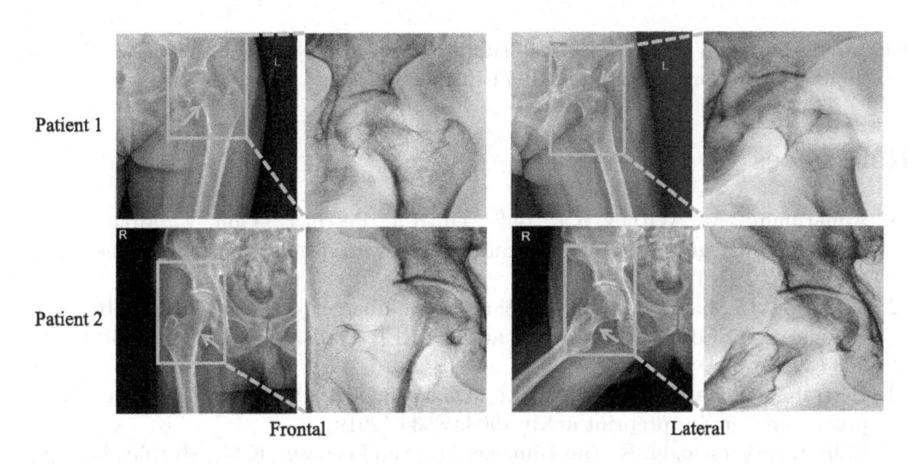

Fig. 3. Visualization results of different patients. The interest area is annotated by expert with green arrows and rectangles, whereas the highlighted areas show the interest regions of the model. (Color figure online)

present the performance of different view-specific networks by comparing *Frontal* to *Frontal_swin*, the results show that the deformable transformer network gains higher accuracy with about 1.7% increment. For the Lateral-view, the deformable transformer network also has comparable performance to Swin-transformer.

Visualization Results. To verify the effectiveness of the proposed framework, we visualize the interest regions of the model as shown in Fig. 3. It shows that the model could concentrate on the interested region of the diagnosis as labeled by expert. In addition, for the diagnosis of non-displaced hip fracture, the smoothness of the bone edge is a very important reference. As shown in Fig. 3, our model is also very good at focusing on bone smoothness in the same area from different perspectives in the same patient, indicating that the features from the Frontal view actually have a guidance to feature selection of the Lateral view.

4 Conclusion

This paper innovatively introduces a cross-view deformable transformer framework for non-displaced hip fracture classification from paired hip X-ray images. We adopt the deformable self-attention module to locate the interested regions of each view, while exploring feature relations among Lateral-view with the guidance of Frontal-view characteristics. In addition, the proposed deformable cross-view learning method is general and has great potential to boost the performance of detecting other complicated disease. Our future work will focus on more effective training strategies and extend our framework to other cross-view medical image analysis problems.

Acknowledgement. This work was supported by the National Key Research and Development Program of China (2019YFE0113900).

References

1. Annarumma, M., Withey, S.J., Bakewell, R.J., Pesce, E., Goh, V., Montana, G.: Automated triaging of adult chest radiographs with deep artificial neural networks. Radiology **291**(1), 196–202 (2019)
2. Bekker, A.J., Shalhon, M., Greenspan, H., Goldberger, J.: Multi-view probabilistic classification of breast microcalcifications. IEEE Trans. Med. Imaging **35**(2), 645–653 (2015)
3. Bertrand, H., Hashir, M., Cohen, J.P.: Do lateral views help automated chest X-ray predictions? arXiv preprint arXiv:1904.08534 (2019)
4. Çallı, E., Sogancioglu, E., van Ginneken, B., van Leeuwen, K.G., Murphy, K.: Deep learning for chest X-ray analysis: a survey. Med. Image Anal. **72**, 102125 (2021)
5. Carneiro, G., Nascimento, J., Bradley, A.P.: Deep learning models for classifying mammogram exams containing unregistered multi-view images and segmentation maps of lesions. In: Deep learning for medical image analysis, pp. 321–339 (2017)
6. Chen, Z., et al.: DPT: deformable patch-based transformer for visual recognition. In: Proceedings of the 29th ACM International Conference on Multimedia, pp. 2899–2907 (2021)
7. Cohen, J.P., et al.: Predicting COVID-19 pneumonia severity on chest X-ray with deep learning. Cureus **12**(7) (2020)
8. Dong, X., et al.: Cswin transformer: a general vision transformer backbone with cross-shaped windows. In: Proceedings of the IEEE/CVF Conference on Computer Vision and Pattern Recognition, pp. 12124–12134 (2022)
9. Dosovitskiy, A., et al.: An image is worth 16×16 words: transformers for image recognition at scale. arXiv preprint arXiv:2010.11929 (2020)

10. Hashir, M., Bertrand, H., Cohen, J.P.: Quantifying the value of lateral views in deep learning for chest X-rays. In: Medical Imaging with Deep Learning, pp. 288–303. PMLR (2020)
11. He, K., Zhang, X., Ren, S., Sun, J.: Deep residual learning for image recognition. In: 2016 IEEE Conference on Computer Vision and Pattern Recognition (CVPR) (2016)
12. Huang, G., Liu, Z., Laurens, V., Weinberger, K.Q.: Densely connected convolutional networks. IEEE Computer Society (2016)
13. Krogue, J.D., et al.: Automatic hip fracture identification and functional subclassification with deep learning. Radiol. Artif. Intell. **2**(2), e190023 (2020)
14. Li, M.D., et al.: Automated assessment of covid-19 pulmonary disease severity on chest radiographs using convolutional siamese neural networks. MedRxiv pp. 2020–05 (2020)
15. Liu, Y., Zhang, F., Zhang, Q., Wang, S., Wang, Y., Yu, Y.: Cross-view correspondence reasoning based on bipartite graph convolutional network for mammogram mass detection. In: Proceedings of the IEEE/CVF Conference on Computer Vision and Pattern Recognition, pp. 3812–3822 (2020)
16. Liu, Z., et al.: Swin transformer: hierarchical vision transformer using shifted windows. In: Proceedings of the IEEE/CVF International Conference on Computer Vision, pp. 10012–10022 (2021)
17. Mutasa, S., Varada, S., Goel, A., Wong, T.T., Rasiej, M.J.: Advanced deep learning techniques applied to automated femoral neck fracture detection and classification. J. Digit. Imaging **33**, 1209–1217 (2020)
18. Novikov, A.A., Lenis, D., Major, D., Hladvka, J., Wimmer, M., Bühler, K.: Fully convolutional architectures for multiclass segmentation in chest radiographs. IEEE Trans. Med. Imaging **37**(8), 1865–1876 (2018)
19. Rubin, J., Sanghavi, D., Zhao, C., Lee, K., Qadir, A., Xu-Wilson, M.: Large scale automated reading of frontal and lateral chest X-rays using dual convolutional neural networks. arXiv preprint arXiv:1804.07839 (2018)
20. Shokouh, G.S., Magnier, B., Xu, B., Montesinos, P.: Ridge detection by image filtering techniques: a review and an objective analysis. Pattern Recognit. Image Anal. **31**, 551–570 (2021)
21. Tang, Y.X.: Automated abnormality classification of chest radiographs using deep convolutional neural networks. NPJ Digit. Med. **3**(1), 70 (2020)
22. Van Tulder, G., Tong, Y., Marchiori, E.: Multi-view analysis of unregistered medical images using cross-view transformers. In: Medical Image Computing and Computer Assisted Intervention – MICCAI 2021, pp. 104–113 (2021)
23. Xia, Z., Pan, X., Song, S., Li, L.E., Huang, G.: Vision transformer with deformable attention. In: Proceedings of the IEEE/CVF Conference on Computer Vision and Pattern Recognition, pp. 4794–4803 (2022)
24. Yamada, Y., et al.: Automated classification of hip fractures using deep convolutional neural networks with orthopedic surgeon-level accuracy: ensemble decision-making with antero-posterior and lateral radiographs. Acta Orthop. **91**(6), 699–704 (2020)
25. Yue, X., et al.: Vision transformer with progressive sampling. In: Proceedings of the IEEE/CVF International Conference on Computer Vision, pp. 387–396 (2021)
26. Zhu, X., Feng, Q.: Mvc-Net: Multi-view chest radiograph classification network with deep fusion. In: 2021 IEEE 18th International Symposium on Biomedical Imaging (ISBI), pp. 554–558. IEEE (2021)

Computational Pathology

Multi-modal Pathological Pre-training via Masked Autoencoders for Breast Cancer Diagnosis

Mengkang Lu, Tianyi Wang, and Yong Xia(✉)

National Engineering Laboratory for Integrated Aero-Space-Ground-Ocean Big Data Application Technology, School of Computer Science and Engineering, Northwestern Polytechnical University, Xi'an 710072, China
yxia@nwpu.edu.cn

Abstract. Breast cancer (BC) is one of the most common cancers identified globally among women, which has become the leading cause of death. Multi-modal pathological images contain different information for BC diagnosis. Hematoxylin and eosin (H&E) staining images could reveal a considerable amount of microscopic anatomy. Immunohistochemical (IHC) staining images provide the evaluation of the expression of various biomarkers, such as the human epidermal growth factor receptor (HER2) hybridization. In this paper, we propose a multi-modal pre-training model via pathological images for BC diagnosis. The proposed pre-training model contains three modules: (1) the modal-fusion encoder, (2) the mixed attention, and (3) the modal-specific decoders. The pre-trained model could be performed on multiple relevant tasks (IHC Reconstruction and IHC classification). The experiments on two datasets (HEROHE Challenge and BCI Challenge) show state-of-the-art results.

Keywords: Breast cancer · Hematoxylin and eosin staining · Immunohistochemical staining · Multi-modal pre-training

1 Introduction

Breast cancer (BC) is one of the most common malignant tumors in women worldwide and it causes nearly 0.7 million deaths in 2020 [26]. The pathological process is usually the golden standard approach for BC diagnosis, which relies on leveraging diverse complementary information from multi-modal data. In addition to obtaining the histological characteristics of tumors from hematoxylin and eosin (H&E) staining images, immunohistochemical (IHC) staining images are also widely used for pathological diagnoses, such as the human epidermal growth factor receptor 2 (HER2), the estrogen receptor (ER), and the progesterone receptor (PR) [22]. With the development of deep learning, there are a lot of multi-modal fusion methods for cancer diagnosis [6,7,20,21].

Recently, with the development of Transformer, multi-modal pre-training has achieved great success in the fields of computer vision (CV) and natural language

H. Greenspan et al. (Eds.): MICCAI 2023, LNCS 14225, pp. 457–466, 2023.
https://doi.org/10.1007/978-3-031-43987-2_44

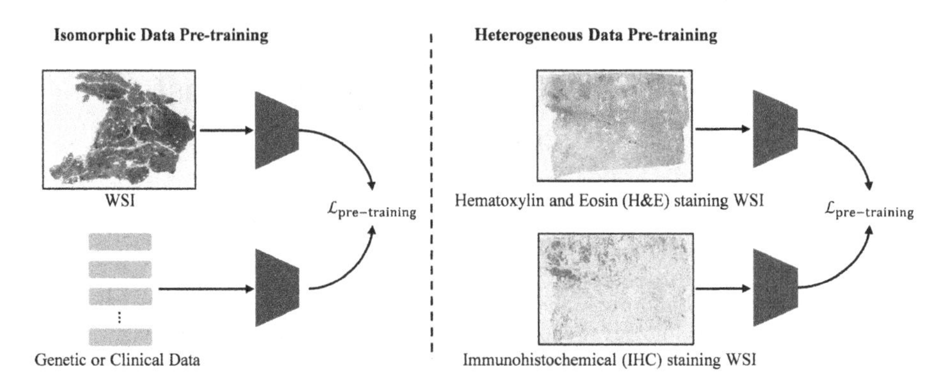

Fig. 1. Illustration of different multi-modal pre-training methods. The WSIs, genetic and clinical data from a patient could be used for isomorphic data pre-training. Pairs of H&E and IHC staining WSIs are used for heterogeneous data pre-training in our method.

processing (NLP). According to the data format, there are two main multi-modal pre-training approaches, as shown in Fig. 1. One is based on isomorphic data, such as vision-language pre-training [5] and vision-speech-text pre-training [3]. The other is based on heterogeneous data. Bachmann *et al.* [2] proposed Multi-MAE to pre-train models with intensity images, depth images, and segmentation maps. In the field of medical image analysis, it is widely recognized that using multi-modal data can produce more accurate diagnoses than using single-modal data. However, the development of multi-modal pre-training methods has been limited due to the scarcity of paired multi-modal data. Most methods focus on chest X-ray vision-language pre-training [8,11]. To our best knowledge, there is no work for multi-modal pre-training based on pathological heterogeneous data.

In this paper, we propose a multi-modal pre-training method based on masked autoencoders for BC downstream tasks. Our model consists of three parts, i.e., the modal-fusion encoder, the mixed attention, and the modal-specific decoder. We choose paired H&E and IHC (only HER2) staining images, which are cropped into non-overlapped patches as the input of our model. We randomly mask some patches by a ratio and feed the remaining patches into the modal-fusion encoder to get corresponding tokens. Then the mixed attention module is used to take the intra-modal and inter-modal correlation into account. Finally, we use modal-specific decoders to reconstruct the original H&E and IHC staining images respectively. Our contributions are summarized as follows:

- We propose a Multi-Modal Pre-training via Masked AutoEncoders MMP-MAE for BC diagnosis. To our best knowledge, this is the first pre-training work based on multi-modal pathological data.
- We evaluate the proposed method on two public datasets as HEROHE Challenge and BCI challenge, which shows that our method achieves state-of-the-art performance.

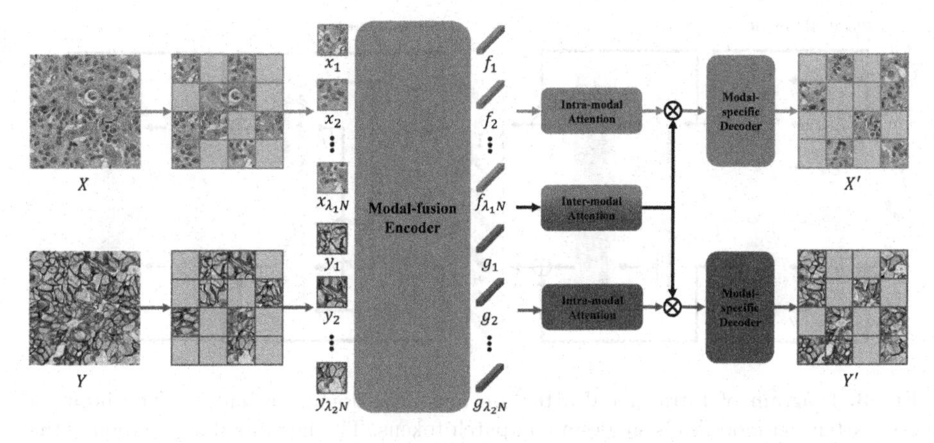

Fig. 2. Framework of our proposed MMP-MAE. A pair of images X and Y (H&E and IHC) are cropped into N non-overlapped patches, which are randomly masked by ratio λ_1 and λ_2. We feed the remaining patches $\{x_i\}_{i=1}^{\lambda_1 N}$ and $\{y_i\}_{i=1}^{\lambda_2 N}$ into the modal-fusion encoder to extract the patch tokens $\{f_i\}_{i=1}^{\lambda_1 N}$ and $\{g_i\}_{i=1}^{\lambda_2 N}$. Then we use intra-modal attention and inter-modal attention to take patch correlation into account. X' and Y' are reconstructed by modal-specific decoders respectively.

2 Method

2.1 Overview

The proposed MMP-MAE consists of three modules, i.e., the modal-fusion encoder, the mixed attention, and the modal-specific decoder, as shown in Fig. 2. A pair of H&E and HER2 images are cropped into regular non-overlapping patches. We mask some of the patches of two modalities with a ratio. The remained patches are fed into the modal-fusion encoder to get the corresponding tokens. Then we use the mixed attention module to extract intra-modal and inter-modal complementary information. Finally, the modal-specific tokens are fed into the modal-specific decoders to reconstruct the original H&E and HER2 images. The pre-trained modal-fusion encoder could be used for downstream tasks (e.g., HER2 status prediction and HER2 image generation based on H&E images).

2.2 MMP-MAE

Modal-Fusion Encoder. We use ViT-base [12] as the backbone of the modal-fusion encoder, which contains a linear projection, 12 transformer blocks, and a Multi-Layer Perceptron (MLP) head. We remove the MLP head and use the remained part to extract patch tokens. An image is cropped into several non-overlapping patches, and these patches are mapped to D dimension tokens with the linear projection and added position embeddings to retain positional information. Each transformer block consists of a multi-head self-attention layer (MHSA)

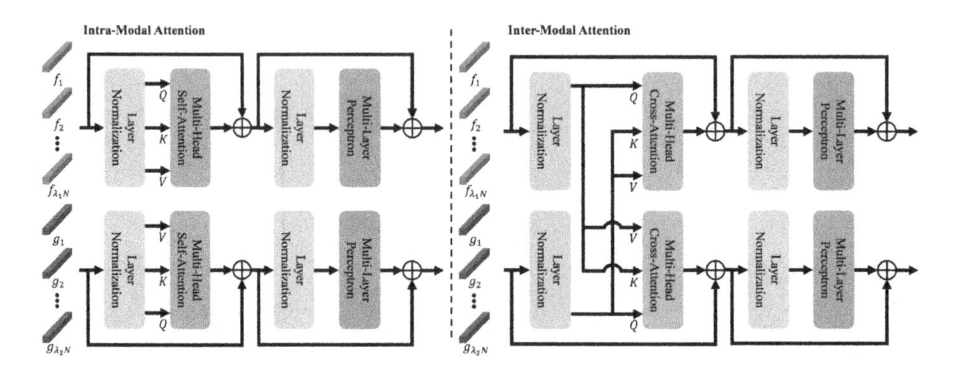

Fig. 3. Diagram of Intra-modal attention and inter-domain attention. The input of both attention modules is single-modal patch tokens. The intra-modal attention is the original transformer block, and there is no interaction between two modalities. We replace the MHSA with MHCA in the inter-modal attention to learn complementary information.

and a feedforward network (FFN). Layer normalization (LN) is applied before each layer, and the residual connection is used after each layer. The processing flow of ViT is shown in Alg. 1.

Mixed Attention. The mixed attention module contains intra-modal attention and inter-modal attention, as shown in Fig. 3. The intra-modal attention is the original transformer block, which consists of MHSA, MLP, LNs, and residual connections. It is defined as

$$A_x(F) = \mathrm{softmax}(\frac{Q_x K_x^\top}{\sqrt{d}})V_x,\ A_y(F) = \mathrm{softmax}(\frac{Q_y K_y^\top}{\sqrt{d}})V_y, \tag{1}$$

Algorithm 1. Transformer processing flow.

Input: A set of patches from one image $X = \{x_i\}_{i=1}^N$, $X \in \mathbb{R}^{N\times(R\times R\times C)}$
1: Transfer patches into linear embeddings
2: **for** $i = 1$ to N **do**
3: $f_i \leftarrow \mathrm{LP}(x_i)$, where $F = \{f_i\}_{i=1}^N$, $F \in \mathbb{R}^{N\times D}$
4: **end for**
5: Position encoding concatenation
6: $F_0 \leftarrow \mathrm{Concat}\,(F_p, F)$, where $F_p \in \mathbb{R}^{1\times D}$
7: **for** $l = 1$ to L **do**
8: $F_l^{'} \leftarrow \mathrm{MHSA}(\mathrm{LN}(F_{l-1})) + F_{l-1}$
9: $F_l \leftarrow \mathrm{FFN}(\mathrm{LN}(F_l^{'})) + F_l^{'}$
10: **end for**
Output: Class token and patch tokens $F_l \in \mathbb{R}^{(N+1)\times D}$

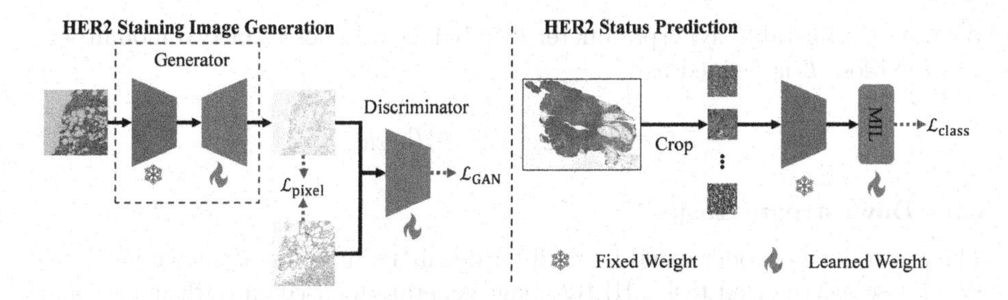

Fig. 4. Workflow of two downstream tasks. In the HER2 staining image generation task, we remain the structure of GAN and replace the generator with our pre-trained model. In the HER2 status prediction task, we replace the feature extractor with our pre-trained model to obtain representations with HER2 semantics.

In inter-modal attention, we replace MHSA with the multi-head cross-attention (MHCA) module. We use MHCA to leverage diverse complementary information between two modalities.

$$A_x(F) = \text{softmax}(\frac{Q_x K_y^\top}{\sqrt{d}})V_y,\ A_y(F) = \text{softmax}(\frac{Q_y K_x^\top}{\sqrt{d}})V_x, \tag{2}$$

Modal-Specific Decoder. Each modal-specific decoder is a shallow block with two transformer layers. Different from the transformer encoder, the target of the transformer decoder is used to reconstruct the original image.

Reconstruction Loss. Given a pair of H&E image X and HER2 image Y, which is cut into 16×16 non-overlapping patches $\{x_i\}_{i=1}^N$ and $\{y_i\}_{i=1}^N$. We mask some of the patches randomly with the ratio λ_1 and λ_2 $(\lambda_1 + \lambda_2 = 1)$. The remained patches are fed into the modal-fusion encoder and the output is corresponding patch tokens $\{f_i\}_{i=1}^{\lambda_1 N}$ and $\{g_i\}_{i=1}^{\lambda_2 N}$. We randomly generate masked patch tokens $\{e_j^x\}_{j=1}^{(1-\lambda_1)N}$ and $\{e_j^y\}_{j=1}^{(1-\lambda_2)N}$, which are learnable vectors for masked patch prediction. The input of the mixed attention module is the full set of tokens $\{f_i, e_j^x\}_{i=1,j=1}^{i=\lambda_1 N, j=(1-\lambda_1)N}$ and $\{g_i, e_j^y\}_{i=1}^{i=\lambda_2 N, j=(1-\lambda_2)N}$, which include both the remaining patch tokens and the masked patch tokens. After the process of the mixed attention module, H&E and HER2 patch tokens are fed into the modal-specific decoders respectively to reconstruct the original H&E image X' and HER2 image Y'. The reconstruction loss is computed by the mean squared error between the original images X, Y and the generative images X', Y', which is computed as

$$\mathcal{L}_{\text{H\&E}} = \frac{1}{T_1}\sum_{i=1}^{T_1} \mid p_i - p_i' \mid^2,\ \mathcal{L}_{\text{HER2}} = \frac{1}{T_2}\sum_{i=1}^{T_2} \mid q_i - q_i' \mid^2. \tag{3}$$

We use an adjustable hyperparameter θ to balance the losses of two modalities. The final loss $\mathcal{L}$ is defined as

$$\mathcal{L} = \theta\mathcal{L}_{\mathrm{H\&E}} + (1-\theta)\mathcal{L}_{\mathrm{HER2}} \tag{4}$$

2.3 Downstream Tasks

The pre-trained encoder could be used for downstream tasks, as shown in Fig. 4. We choose two relevant tasks: HER2 image generation based on H&E images and HER2 status prediction. In the HER2 generation task, we replace the generator of Pyramid Pix2pix, a generative adversarial network (GAN) in [16], with our pre-trained encoder and a light-weight decoder. The weights of the pre-trained encoder are fixed, and the light-weight decoder in the generator and the discriminator are learnable. We use pairs of H&E and IHC images for GAN training. In the HER2 status prediction task, we replace the universal extractor ResNet-50 [14] with our pre-trained encoder. We use CLAM-MIL [19] as the aggregator in our training process.

3 Experimental Results

3.1 Datasets

ACROBAT Challenge. The AutomatiC Registration Of Breast cAncer Tissue (ACROBAT) Challenge [27] provides H&E WSIs and matched IHC WSIs (ER, PR, HER2, and KI67), which consists of 750 training cases, 100 validation cases, and 300 testing cases. We choose paired H&E and HER2 WSIs for pre-training. We extract the key points and descriptors from paired WSIs using SIFT [18] and SuperPoint [10]. Then the extracted key points and descriptors are matched using RANSAC [13] and SuperGlue [25]. We repeat this procedure several times on the rotated, downsampled, or transformed moving WSI to fetch the best transformation based on mean squared error (MSE) loss between source and target WSIs' descriptors. After that, the selected transformation is optimized across different levels of WSIs by gradient descent with local normalized cross-correlation (NCC) as its cost function. In the final phase of nonrigid registration, we use the optimized transformation to get the initial displacement field, which is optimized across different levels of WSIs by gradient update. The loss function of which is the weighted sum of NCC and diffusive regularization. We resize the displacement field and apply it to the original moving WSI. After all the WSI pairs are well registered, we convert the padded H&E image to grayscale and apply median blur to it. Next, the Otsu threshold is applied to extract the foreground area, which is cropped into non-overlapping 256×256 images. Finally, all the chosen images (around 0.35 million) from WSI in the same pair are saved for MMP-MAE pre-training.

BCI Challenge. Breast Cancer Immunohistochemical Image Generation Challenge [16] consists of 3896 pairs of images for training and 977 pairs for testing, which are used to generate HER2 images based on H&E images.

Table 1. Performance comparison on BCI Challenge.

Method	PSNR(dB)	SSIM
cycleGAN [28]	16.20	0.373
Pix2pix [15]	18.65	0.419
Pyramid Pix2pix [16]	21.16	0.477
Proposed	**22.76**	**0.484**

HEROHE Challenge. HER2 On H&E (HEROHE) Challenge [9] is developed to predict the HER2 status in invasive BC cases via the analysis of HE slides. It contains 359 training samples and 150 test samples for WSI classification.

3.2 Experimental Setup

Experiments are implemented in PyTorch [24] and with 4 NVIDIA A100 Tensor Core GPUs. We pre-train our MMP-MAE on the ACROBAT dataset with AdamW [17] and the learning rate of $1e^{-4}$. The batch size of pre-training is 1024 and it takes about 30 h for 100 epochs. We use warmup for the first 10 epochs and the learning rate is set to $1e^{-6}$.

In the HER2 staining image generation task, we use 2 GPUs with a batch size of 4. The learning rate is $2e^{-4}$ and the optimizer is Adam. We use the learning rate decay strategy for stable training. Peak Signal to Noise Ratio (PSNR) and Structural Similarity (SSIM) are used as the evaluation indicators for the quality of the HER2 generated images.

In the HER2 status prediction task, we use 1 GPU with a batch size of 1 (WSI level). The learning rate is $1e^{-4}$ and the Adam optimizer is used. Four standard metrics are used to measure the HER2 status prediction results, including the area under the receiver operator characteristic curve (AUC), Precision, Recall, and F1-score.

3.3 Method Comparison

HER2 Staining Image Generation. Three methods on BCI datasets are compared in our experiments, as shown in Table 1. CycleGAN is a representative unsupervised method, which doesn't need paired images for training. So

Table 2. Performance comparison on HEROHE Challenge.

Method/Team	AUC	Precision	Recall	F1-Score
Macaroon	0.71	0.57	**0.83**	0.68
MITEL	0.74	0.58	0.78	0.67
Piaz	**0.84**	**0.77**	0.55	0.64
Dratur	0.75	0.57	0.70	0.63
IRISAI	0.67	0.58	0.67	0.62
Proposed	**0.84**	0.72	0.82	**0.74**

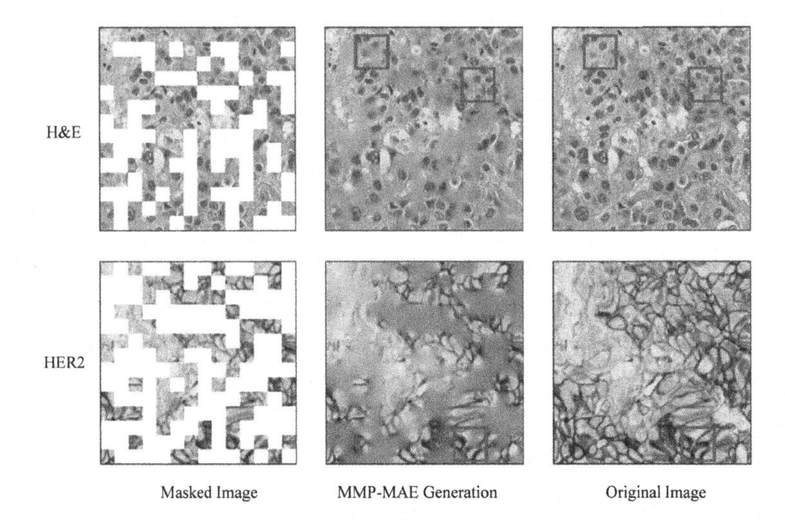

Fig. 5. Visualization of MMP-MAE generation results on the ACROBAT dataset. The region in the red box shows our MMP-MAE could learn the semantic information from the adjacent area. (Color figure online)

cycleGAN focuses more on style transformation, and it is difficult to match the cell-level information in detail. Pix2pix and Pyramid pix2pix use paired data, which obtain better results than cycleGAN. Pyramid pix2pix uses the multi-scale constraint, which performs better than pix2pix. Our method is based on the framework of Pyramid pix2pix and we replace the generator with our pre-trained encoder and a lightweight decoder. Our MMP-MAE further improves the performance, which achieves higher PSNR by 1.60, and SSIM by 0.007. The visualization on the ACROBAT dataset also shows our model could learn the modality-related information, as shown in Fig. 5.

HER2 Status Prediction. We compare our method with the top five methods reported in HEROHE challenge review [9]. Most of these methods use the multi-network ensemble strategy and extra datasets. Team Macaroon uses the CAMELYON dataset [4] for tumor classification. Team MITEL uses BACH dataset [1] for tumor classification. Team Piaz and Dratur both use a multi-network ensemble strategy to improve their performances. Team IRISAI first segment the tumor area and then predict the HER2 status. MMP-MAE still achieves competitive results by using a single pre-trained model, which is shown in Table 2. Our model improves and F1-Score by 6%. The results show our model pre-training has the ability to predict status from one modality.

4 Conclusion

In this paper, we propose a novel multi-modal pre-training framework, MMP-MAE for BC diagnosis. MMP-MAE use paired H&E and HER2 staining images for pre-training, which could be used for several downstream tasks such as HER2 staining image generation and HER2 status prediction only by H&E modality. Both the experiment results on BCI and HEROHE datasets show our pre-trained MMP-MAE demonstrates strong transfer ability. Our future work will expand our work to more modalities.

Acknowledgment. This work was supported in part by the Key Research and Development Program of Shaanxi Province, China, under Grant 2022GY-084, in part by the National Natural Science Foundation of China under Grant 62171377, and in part by the Key Technologies Research and Development Program under Grant 2022YFC2009903/2022YFC2009900.

References

1. Aresta, G., et al.: Bach: grand challenge on breast cancer histology images. Med. Image Anal. **56**, 122–139 (2019)
2. Bachmann, R., Mizrahi, D., Atanov, A., Zamir, A.: MultiMAE: multi-modal multi-task masked autoencoders. In: Avidan, S., Brostow, G., Cissé, M., Farinella, G.M., Hassner, T. (eds.) Computer Vision – ECCV 2022. ECCV 2022. Lecture Notes in Computer Science, vol. 13697, pp. 348–367. Springer, Cham (2022). https://doi.org/10.1007/978-3-031-19836-6_20
3. Baevski, A., Babu, A., Hsu, W.N., Auli, M.: Efficient self-supervised learning with contextualized target representations for vision, speech and language. arXiv preprint arXiv:2212.07525 (2022)
4. Bejnordi, B.E., et al.: Diagnostic assessment of deep learning algorithms for detection of lymph node metastases in women with breast cancer. JAMA **318**(22), 2199–2210 (2017)
5. Chen, F.L., et al.: VLP: a survey on vision-language pre-training. Mach. Intell. Res. **20**(1), 38–56 (2023)
6. Chen, R.J., et al.: Pathomic fusion: an integrated framework for fusing histopathology and genomic features for cancer diagnosis and prognosis. IEEE Trans. Med. Imaging **41**(4), 757–770 (2020)
7. Chen, R.J., et al.: Multimodal co-attention transformer for survival prediction in gigapixel whole slide images. In: Proceedings of the IEEE/CVF International Conference on Computer Vision, pp. 4015–4025 (2021)
8. Chen, Z., et al.: Multi-modal masked autoencoders for medical vision-and-language pre-training. In: Wang, L., Dou, Q., Fletcher, P.T., Speidel, S., Li, S. (eds.) Medical Image Computing and Computer Assisted Intervention – MICCAI 2022. MICCAI 2022. Lecture Notes in Computer Science, vol. 13435, pp. 679–689. Springer, Cham (2022). https://doi.org/10.1007/978-3-031-16443-9_65
9. Conde-Sousa, E., et al.: HEROHE challenge: predicting HER2 status in breast cancer from hematoxylin-eosin whole-slide imaging. J. Imaging **8**(8), 213 (2022)
10. DeTone, D., Malisiewicz, T., Rabinovich, A.: SuperPoint: self-supervised interest point detection and description. In: Proceedings of the IEEE Conference on Computer Vision and Pattern Recognition Workshops, pp. 224–236 (2018)
11. Do, T., Nguyen, B.X., Tjiputra, E., Tran, M., Tran, Q.D., Nguyen, A.: Multiple meta-model quantifying for medical visual question answering. In: de Bruijne, M., et al. (eds.) MICCAI 2021. LNCS, vol. 12905, pp. 64–74. Springer, Cham (2021). https://doi.org/10.1007/978-3-030-87240-3_7

12. Dosovitskiy, A., et al.: An image is worth 16×16 words: transformers for image recognition at scale. arXiv preprint arXiv:2010.11929 (2020)
13. Fischler, M.A., Bolles, R.C.: Random sample consensus: a paradigm for model fitting with applications to image analysis and automated cartography. Commun. ACM **24**(6), 381–395 (1981)
14. He, K., Zhang, X., Ren, S., Sun, J.: Deep residual learning for image recognition. In: Proceedings of the IEEE Conference on Computer Vision and Pattern Recognition, pp. 770–778 (2016)
15. Isola, P., Zhu, J.Y., Zhou, T., Efros, A.A.: Image-to-image translation with conditional adversarial networks. In: Proceedings of the IEEE Conference on Computer Vision and Pattern Recognition, pp. 1125–1134 (2017)
16. Liu, S., Zhu, C., Xu, F., Jia, X., Shi, Z., Jin, M.: BCI: breast cancer immunohistochemical image generation through pyramid pix2pix. In: Proceedings of the IEEE/CVF Conference on Computer Vision and Pattern Recognition, pp. 1815–1824 (2022)
17. Loshchilov, I., Hutter, F.: Decoupled weight decay regularization. arXiv preprint arXiv:1711.05101 (2017)
18. Lowe, D.G.: Object recognition from local scale-invariant features. In: Proceedings of the Seventh IEEE International Conference on Computer Vision, vol. 2, pp. 1150–1157. IEEE (1999)
19. Lu, M.Y., Williamson, D.F., Chen, T.Y., Chen, R.J., Barbieri, M., Mahmood, F.: Data-efficient and weakly supervised computational pathology on whole-slide images. Nature Biomed. Eng. **5**(6), 555–570 (2021)
20. Mobadersany, P., et al.: Predicting cancer outcomes from histology and genomics using convolutional networks. Proc. Natl. Acad. Sci. **115**(13), E2970–E2979 (2018)
21. Nakhli, R., et al.: Amigo: sparse multi-modal graph transformer with shared-context processing for representation learning of giga-pixel images. arXiv preprint arXiv:2303.00865 (2023)
22. Onitilo, A.A., Engel, J.M., Greenlee, R.T., Mukesh, B.N.: Breast cancer subtypes based on ER/PR and HER2 expression: comparison of clinicopathologic features and survival. Clin. Med. Res. **7**(1–2), 4–13 (2009)
23. Otsu, N.: A threshold selection method from gray-level histograms. IEEE Trans. Syst. Man Cybern. **9**(1), 62–66 (1979)
24. Paszke, A., Gross, S., Massa, F., Lerer, A., Bradbury, J., et al.: PyTorch: an imperative style, high-performance deep learning library. Adv. Neural. Inf. Process. Syst. **32**, 8026–8037 (2019)
25. Sarlin, P.E., DeTone, D., Malisiewicz, T., Rabinovich, A.: Superglue: learning feature matching with graph neural networks. In: Proceedings of the IEEE/CVF Conference on Computer Vision and Pattern Recognition, pp. 4938–4947 (2020)
26. Sung, H., et al.: Global cancer statistics 2020: globocan estimates of incidence and mortality worldwide for 36 cancers in 185 countries. CA Cancer J. Clin. **71**(3), 209–249 (2021)
27. Weitz, P., Valkonen, M., Solorzano, L., Hartman, J., Ruusuvuori, P., Rantalainen, M.: ACROBAT-automatic registration of breast cancer tissue. In: 10th Internatioal Workshop on Biomedical Image Registration (2022)
28. Zhu, J.Y., Park, T., Isola, P., Efros, A.A.: Unpaired image-to-image translation using cycle-consistent adversarial networks. In: Proceedings of the IEEE International Conference on Computer Vision, pp. 2223–2232 (2017)

Iteratively Coupled Multiple Instance Learning from Instance to Bag Classifier for Whole Slide Image Classification

Hongyi Wang[1], Luyang Luo[2], Fang Wang[3], Ruofeng Tong[1,4], Yen-Wei Chen[1,5], Hongjie Hu[3], Lanfen Lin[1(✉)], and Hao Chen[2,6(✉)]

[1] College of Computer Science and Technology, Zhejiang University, Hangzhou, China
llf@zju.edu.cn

[2] Department of Computer Science and Engineering, The Hong Kong University of Science and Technology, Hong Kong, China
jhc@cse.ust.hk

[3] Department of Radiology, Sir Run Run Shaw Hospital, Hangzhou, China

[4] Research Center for Healthcare Data Science, Zhejiang Lab, Hangzhou, China

[5] College of Information Science and Engineering, Ritsumeikan University, Kusatsu, Japan

[6] Department of Chemical and Biological Engineering, The Hong Kong University of Science and Technology, Hong Kong, China

Abstract. Whole Slide Image (WSI) classification remains a challenge due to their extremely high resolution and the absence of fine-grained labels. Presently, WSI classification is usually regarded as a Multiple Instance Learning (MIL) problem when only slide-level labels are available. MIL methods involve a patch embedding module and a bag-level classification module, but they are prohibitively expensive to be trained in an end-to-end manner. Therefore, existing methods usually train them separately, or directly skip the training of the embedder. Such schemes hinder the patch embedder's access to slide-level semantic labels, resulting in inconsistency within the entire MIL pipeline. To overcome this issue, we propose a novel framework called Iteratively Coupled MIL (ICMIL), which bridges the loss back-propagation process from the bag-level classifier to the patch embedder. In ICMIL, we use category information in the bag-level classifier to guide the patch-level fine-tuning of the patch feature extractor. The refined embedder then generates better instance representations for achieving a more accurate bag-level classifier. By coupling the patch embedder and bag classifier at a low cost, our proposed framework enables information exchange between the two modules, benefiting the entire MIL classification model. We tested our framework on two datasets using three different backbones, and our experimental results demonstrate consistent performance improvements over state-of-the-art MIL methods. The code is available at: https://github.com/Dootmaan/ICMIL.

Keywords: Multiple Instance Learning · Whole Slide Image · Deep Learning

---

Supplementary Information The online version contains supplementary material available at https://doi.org/10.1007/978-3-031-43987-2_45.

H. Greenspan et al. (Eds.): MICCAI 2023, LNCS 14225, pp. 467–476, 2023.
https://doi.org/10.1007/978-3-031-43987-2_45

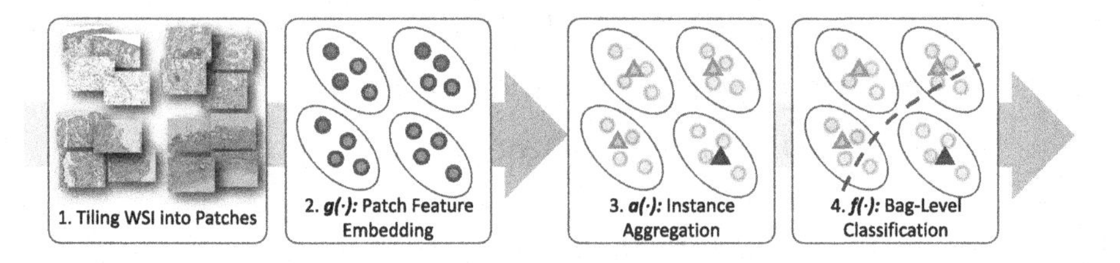

Fig. 1. The typical pipeline of traditional MIL methods on WSIs.

1 Introduction

Whole slide scanning is increasingly used in disease diagnosis and pathological research to visualize tissue samples. Compared to traditional microscope-based observation, whole slide scanning converts glass slides into gigapixel digital images that can be conveniently stored and analyzed. However, the high resolution of WSIs also makes their automated classification challenging [15]. Patch-based classification is a common solution to this problem [3,8,24]. It predicts the slide-level label by first predicting the labels of small, tiled patches in a WSI. This approach allows for the direct application of existing image classification models, but requires additional patch-level labeling. Unfortunately, patch-level labeling by histopathology experts is expensive and time-consuming. Therefore, many weakly-supervised [8,24] and semi-supervised [3,5] methods have been proposed to generate patch-level pseudo labels at a lower cost. However, the lack of reliable supervision directly hinders the performance of these methods, and serious class-imbalance problems could arise, as tumor patches may only account for a small portion of the entire WSI [12].

In contrast, MIL-based methods have become increasingly preferred due to their only demand for slide-level labels [18]. The typical pipeline of MIL methods is shown in Fig. 1, where WSIs are treated as bags, and tiled patches are considered as instances. The aim is to predict whether there are positive instances, such as tumor patches, in a bag, and if so, the bag is considered positive as well. In practice, a fixed ImageNet pre-trained feature extractor $g(\cdot)$ is usually used to convert the tiled patches in a WSI into feature maps due to limited GPU memory. These instance features are then aggregated by $a(\cdot)$ into a slide-level feature vector to be sent to the bag-level classifier $f(\cdot)$ for MIL training. Due to the high computational cost, end-to-end training of the feature extractor and bag classifier is prohibitive, especially for high-resolution WSIs. As a result, many methods focus solely on improving $a(\cdot)$ or $f(\cdot)$, leaving $g(\cdot)$ untrained on the WSI dataset (as shown in Fig. 2(b)). However, the domain shift between WSI and natural images may lead to sub-optimal representations, so recently there have been methods proposed to fine-tune $g(\cdot)$ using self-supervised techniques [4,12,21] or weakly-supervised techniques [10,13,23] (as shown in Fig. 2(c)). Nevertheless, since these two processes are still trained separately with different supervision signals, they lack joint optimization and may still leads to inconsistency within the entire MIL pipeline.

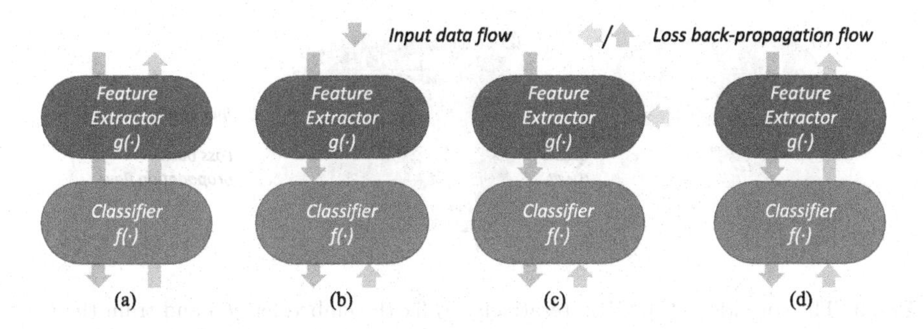

Fig. 2. Comparison between ICMIL and existing methods. (a) Ordinary end-to-end classification pipeline. (b) MIL methods that use fixed pre-trained ResNet50 as $g(\cdot)$. (c) MIL methods that introduce extra self-supervised fine-tuning of $g(\cdot)$. (d) Our proposed ICMIL which can bridge the loss back-propagation process from $f(\cdot)$ to $g(\cdot)$ by iteratively coupling them during training.

To address the challenges mentioned above, we propose a novel MIL framework called ICMIL, which can iteratively couple the patch feature embedding process with the bag-level classification process to enhance the effectiveness of MIL training (as illustrated in Fig. 2(d)). Unlike previous works that mainly focused on designing sophisticated instance aggregators $a(\cdot)$ [12,14,20] or bag classifiers $f(\cdot)$ [9,16,25], we aim to bridge the loss back-propagation process from $f(\cdot)$ to $g(\cdot)$ to improve $g(\cdot)$'s ability to perceive slide-level labels. Specifically, we propose to use the bag-level classifier $f(\cdot)$ to initialize an instance-level classifier $f'(\cdot)$, enabling $f(\cdot)$ to use the category knowledge learned from bag-level features to determine each instance's category. In this regard, we further propose a teacher-student [7] approach to effectively generate pseudo labels and simultaneously fine-tune $g(\cdot)$. After fine-tuning, the domain shift problem is alleviated in $g(\cdot)$, leading to better patch representations. The new representations can be used to train a better bag-level classifier in return for the next round of iteration.

In summary, our contributions are: (1) We propose ICMIL which bridges the loss propagation from the bag classifier to the patch embedder by iteratively coupling them during training. This framework fine-tunes the patch embedder based on the bag-level classifier, and the refined embeddings, in turn, help train a more accurate bag-level classifier. (2) We propose a teacher-student approach to achieve effective and robust knowledge transfer from the bag-level classifier $f(\cdot)$ to the instance-level representation embedder $g(\cdot)$. (3) We conduct extensive experiments on two datasets using three different backbones and demonstrate the effectiveness of our proposed framework.

2 Methodology

2.1 Iterative Coupling of Embedder and Bag Classifier in ICMIL

The general idea of ICMIL is shown in Fig. 3, which is inspired by the Expectation-Maximization (EM) algorithm. EM has been used with MIL in some

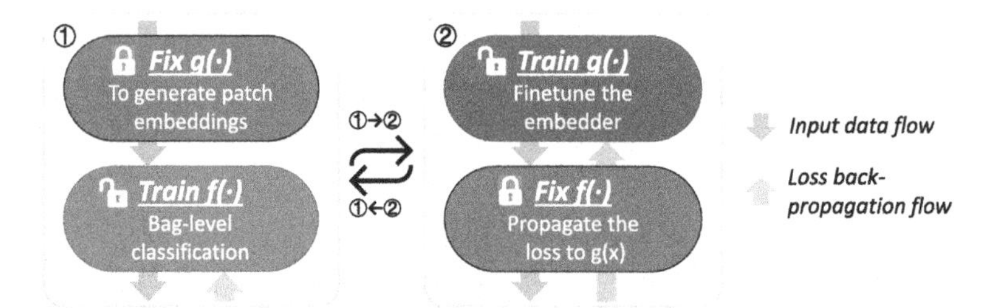

Fig. 3. The core idea of ICMIL: iteratively, ① fix the embedder $g(\cdot)$ and train the bag classifier $f(\cdot)$, ② fix the classifier $f(\cdot)$ and fine-tune the instance embedder $g(\cdot)$.

previous works [13,17,22], but it was only treated as an assisting tool for aiding the training of either $g(\cdot)$ or $f(\cdot)$ in the traditional MIL pipelines. In contrast, we are the first to consider the optimization of the entire MIL pipeline as an EM alike problem, utilizing EM for coupling $g(\cdot)$ and $f(\cdot)$ together iteratively. To begin with, we first employ a traditional approach to train a bag-level classifier $f(\cdot)$ on a given dataset, with patch embeddings generated by a fixed ResNet50 [6] pre-trained on ImageNet [19] (step ① in Fig. 3). Subsequently, this $f(\cdot)$ is considered as the initialization of a hidden instance classifer $f'(\cdot)$, generating pseudo-labels for each instance-level representation. This operation is feasible when the bag-level representations aggregated by $a(\cdot)$ are in the same hidden space as the instance representations, and most aggregation methods (e.g., max pooling, attention-based) satisfy this condition since they essentially make linear combinations of instance-level representations.

Next, we freeze the weights of $f(\cdot)$ and fine-tune $g(\cdot)$ with the generated pseudo-labels (step ② in Fig. 3), of which the detailed implementation is presented in Sect. 2.3. After this, $g(\cdot)$ is fine-tuned for the specific WSI dataset, which allows it to generate improved representations for each instance, thereby enhancing the performance of $f(\cdot)$. Moreover, with a better $f(\cdot)$, we can use the iterative coupling technique again, resulting in further performance gains and mitigation to the distribution inconsistencies between instance- and bag-level embeddings.

2.2 Instance Aggregation Method in ICMIL

Although most instance aggregators are compatible with ICMIL, they still have an impact on the efficiency and effectiveness of ICMIL. In addition to that $a(\cdot)$ has to project the bag representations to the same hidden space as the instance representations, it also should avoid being over-complicated. Otherwise, $a(\cdot)$ may lead to larger difference between the decision boundaries of bag-level classifer $f(\cdot)$ and instance-level classifier $f'(\cdot)$, which may cause ICMIL taking more time to converge.

Therefore, in our experiments, we choose to use the attention-based instance aggregation method [9] which has been widely used in many of the existing

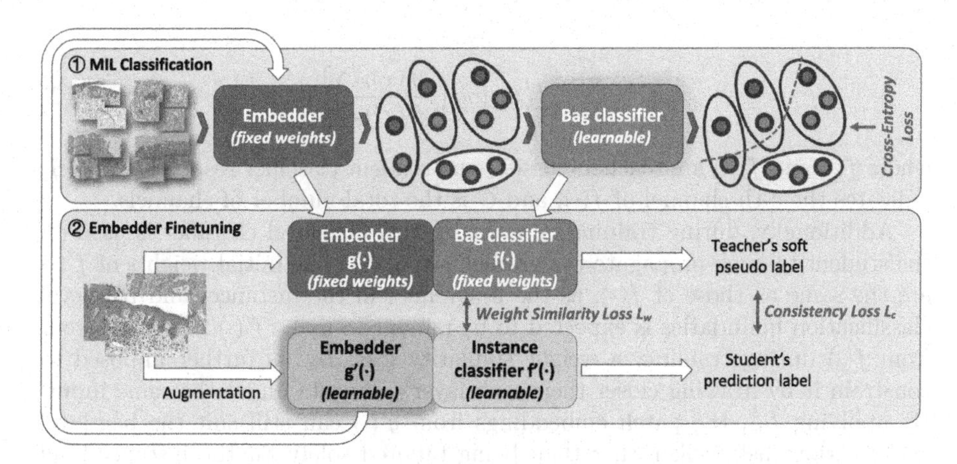

Fig. 4. A schematic view of the proposed teacher-student alike model for label propagation from $f(\cdot)$ to $g(\cdot)$ (mainly in step ②), and its position in ICMIL pipeline.

MIL frameworks [9,16,25]. For a bag that contains K instances, attention-based aggregation method firstly learns an attention score for each instance. Then, the aggregated bag-level representation H is defined as:

$$H = \sum_{k=1}^{K} a_k h_k, \tag{1}$$

where a_k is the attention score for the k-th instance h_k in the bag. Obviously, H and h_k remains in the same hidden space, satisfying the prerequisite of ICMIL.

2.3 Label Propagation from Bag Classifier to Embedder

We propose a novel teacher-student model for accurate and robust label propagation from $f(\cdot)$ to $g(\cdot)$. The model's architecture is depicted in Fig. 4. In contrast to the conventional approach of generating all pseudo labels and retraining $g(\cdot)$ from scratch, our proposed method can simultaneously process the pseudo label generation and $g(\cdot)$ fine-tuning tasks, making it more flexible. Moreover, incorporating augmented inputs in the training process allows for the better utilization of supervision signals, resulting in a more robust $g(\cdot)$. We also introduce a learnable $f'(\cdot)$ to self-adaptively modifying the instance-level decision boundary for more effective fine-tuning of the embedder.

Specifically, we freeze the weights of $g(\cdot)$ and $f(\cdot)$ and set them as the teacher. We then train a student patch embedding network, $g'(\cdot)$, to learn category knowledge from the teacher. For a given patch input x, the teacher generates the corresponding pseudo label, while the student receives an augmented image x' and attempts to generate a similar prediction to that of the teacher through a consistency loss L_c. This loss function is defined as:

$$L_c = \sum_{c=1}^{C} \left[f(x)_c log \left(\frac{f(x)_c}{f'(x')_c} \right) \right], \tag{2}$$

where $f(\cdot)$ and $f'(\cdot)$ are teacher classifer and student classifier respectively, $f(\cdot)_c$ indicates the c-th channel of $f(\cdot)$, and C is the total number of channels.

Additionally, during training, a learnable instance-level classifier is used on the student to back-propagate the gradients to $g'(\cdot)$. The initial weights of $f'(\cdot)$ are the same as those of $f(\cdot)$, as the differences in the instance- and bag-level classification boundaries is expected to be minor. To make $f'(\cdot)$ not so different from $f(\cdot)$ during training, a weight similarity loss, L_w, is further imposed to constrain it by drawing closer their each layer's outputs under the same input. By applying L_w, the patch embeddings from $g'(\cdot)$ can still suit the bag-level classification task well, rather than being tailored solely for the instance-level classifier $f'(\cdot)$. L_w is defined as:

$$L_w = \sum_{l=1}^{L} \sum_{c=1}^{C} \left[f(x)_c^l log \left(\frac{f(x)_c^l}{f'(x)_c^l} \right) \right], \tag{3}$$

where $f(\cdot)_c^l$ indicates the c-th channel of l-th layer's output in $f(\cdot)$. The overall loss function for this step is $L_c + \alpha L_w$, with α set to 0.5 in our experiments.

3 Experiments

3.1 Datasets

Our experiments utilized two datasets, with the first being the publicly available breast cancer dataset, Camelyon16 [1]. This dataset consists of a total of 399 WSIs, with 159 normal and 111 metastasis WSIs for the training set, and the remaining 129 for test. Although patch-level labels are officially provided in Camelyon16, they were not used in our experiments.

The second dataset is a private hepatocellular carcinoma (HCC) dataset collected from Sir Run Run Shaw Hospital, Hangzhou, China. This dataset comprises a total of 1140 valid tumor WSIs scanned at 40× magnification, and the objective is to identify the severity of each case based on the Edmondson-Steiner (ES) grading. The ground truth labels are binary classes of low risk and high risk, which were provided by experienced pathologists.

3.2 Implementation Details

For Camelyon16, we tiled the WSIs into 256×256 patches on 20× magnification using the official code of [25], while for the HCC dataset the patches are 384×384 on 40× magnification following the pathologists' advice. For both datasets, we used an ImageNet pre-trained ResNet50 to initialize $g(\cdot)$. The instance embedding process was the same of [16], which means for each patch, it would be

Table 1. Results of ablation studies on Camelyon16 with AB-MIL.

(a) Ablation study on the ICMIL iteration times

ICMIL Iterations	0	0.5	1	1.5	2	2.5	3
AUC	85.4	88.8	90.0	89.7	90.5	90.4	90.0
F1	78.0	79.4	80.5	80.1	82.0	80.7	81.7
Acc	84.5	85.0	86.6	86.0	85.8	86.9	86.6

(b) Loss Propagation

Method	Naïve	Ours
AUC	88.5	90.0
F1	78.8	80.5
Acc	83.9	86.6

Table 2. Comparison with other methods on Camelyon16 and HCC datasets, where † indicates the corresponding Camelyon16 results are cited from [25]. Best results are in bold, while the second best ones are underlined.

Method	Loss Propagation			Camelyon16			HCC		
	$g(\cdot)$	$f(\cdot)$	$f \rightarrow g$	AUC(%)	F1(%)	Acc(%)	AUC(%)	F1(%)	Acc(%)
Mean Pooling		✓		60.3	44.1	70.1	76.4	83.1	73.7
Max Pooling		✓		79.5	70.6	80.3	80.1	84.3	76.8
RNN-MIL† [2]		✓		87.5	79.8	84.4	79.4	84.1	75.5
AB-MIL† [9]		✓		85.4	78.0	84.5	81.2	86.0	78.1
DS-MIL† [12]	✓	✓		89.9	81.5	85.6	86.1	86.6	81.4
CLAM-SB† [16]		✓		87.1	77.5	83.7	82.1	84.3	77.1
CLAM-MB† [16]		✓		87.8	77.4	82.3	81.7	83.7	76.3
TransMIL† [20]		✓		90.6	79.7	85.8	81.2	84.4	76.7
DTFD-MIL [25]		✓		<u>93.2</u>	<u>84.9</u>	<u>89.0</u>	83.0	85.5	78.1
Ours (w/ Max Pooling)	✓	✓	✓	85.2 (+5.7)	74.7 (+4.1)	81.9 (+1.6)	86.6 (+6.5)	87.3 (+3.0)	82.0 (+5.2)
Ours (w/ AB-MIL)	✓	✓	✓	90.0 (+4.6)	80.5 (+2.5)	86.6 (+2.1)	<u>87.1 (+5.9)</u>	<u>88.3 (+2.3)</u>	<u>83.3 (+5.2)</u>
Ours (w/ DTFD-MIL)	✓	✓	✓	**93.7 (+0.5)**	**87.0 (+2.1)**	**90.6 (+1.6)**	**87.7 (+4.7)**	**89.1 (+3.6)**	**83.5 (+5.4)**

firstly embedded into a 1024-dimension vector, and then be projected to a 512-dimension hidden space for further bag-level training. For the training of bag classifier $f(\cdot)$, we used an initial learning rate of 2e-4 with Adam [11] optimizer for 200 epochs with batch size being 1. Camelyon16 results are reported on the official test split, while the HCC dataset used a 7:1:2 split for training, validation and test. For the training of patch embedder $g(\cdot)$, we used an initial learning rate of 1e-5 with Adam [11] optimizer with the batch size being 100. Three metrics were used for evaluation. Namely, area under curve (AUC), F1 score, and slide-level accuracy (Acc). Experiments were all conducted on a Nvidia Tesla M40 (12GB).

3.3 Experimental Results

Ablation Study. The results of ablation studies are presented in Table 1. From Table 1(a), we can learn that as the number of ICMIL iteration increases, the performance will also go up until reaching a stable point. Since the number of instances is very large in WSI datasets, we empirically recommend to choose to run ICMIL one iteration for fine-tuning $g(\cdot)$ to achieve the balance between performance gain and time consumption. From Table 1(b), it is shown that our

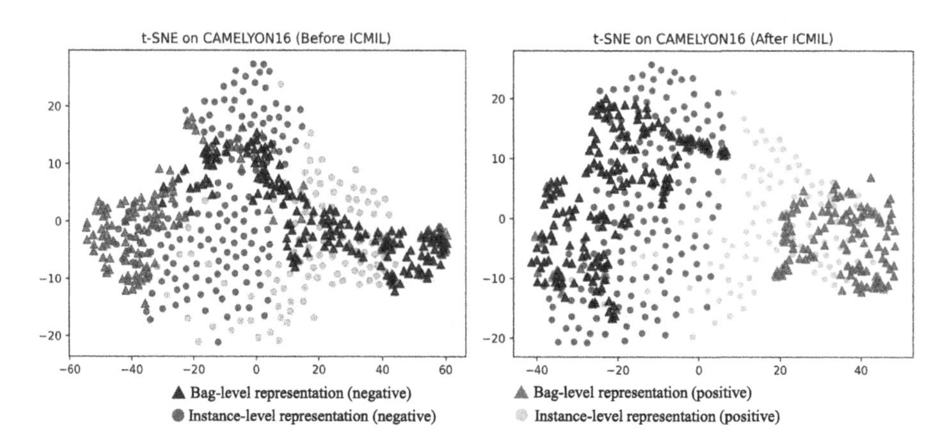

Fig. 5. Visualization of the instance- and bag-level representations before and after ICMIL training. We sample one instance from one bag w/ Max Pooling. Only one iteration of ICMIL is used to achieve the right figure.

teacher-student-based method outperforms the naïve "pseudo label generation" method for fine-tuning $g(\cdot)$, which demontrates the effectiveness of introducing the learnable instance-level classifier $f'(\cdot)$.

Comparison with Other Methods. Experimental results are presented in Table 2. As shown, our ICMIL framework consistently improves the performance of three different MIL baselines (i.e., Max Pool, AB-MIL, and DTFD-MIL), demonstrating the effectiveness of bridging the loss back-propagation from bag calssifier to embedder. It proves that a more suitable patch embedding can greatly enhance the overall MIL classification framework. When used with the state-of-the-art MIL method DTFD-MIL, ICMIL further increases its performance on Camelyon16 by 0.5% AUC, 2.1% F1, and 1.6% Acc.

Results on the HCC dataset also proves the effectiveness of ICMIL, despite the minor difference on the relative performance of baseline methods. Mean Pooling performs better on this dataset due to the large area of tumor in the WSIs (about 60% patches are tumor patches), which mitigates the impact of average pooling on instances. Also, the performance differences among different vanilla MIL methods tends to be smaller on this dataset since risk grading is a harder task than Camelyon16. In this situation, the quality of instance representations plays a crucial role in generating more separable bag-level representations. As a result, after applying ICMIL on the MIL baselines, these methods all gain great performance boost on the HCC dataset.

Furthermore, Fig. 5 displays the instance-level and bag-level representations of Camelyon16 dataset before and after applying ICMIL on AB-MIL backbone. The results indicate that one iteration of $g(\cdot)$ fine-tuning in ICMIL significantly improves the instance-level representations, leading to a better aggregated bag-level representation naturally. Besides, the bag-level representations are also more closely aligned with the instance representations, proving that ICMIL can reduce the inconsistencies between $g(\cdot)$ and $f(\cdot)$ by coupling them together for training, resulting in a better separability.

4 Conclusion

In this work, we propose ICMIL, a novel framework that iteratively couples the feature extraction and bag classification stages to improve the accuracy of MIL models. ICMIL leverages the category knowledge in the bag classifier as pseudo supervision for embedder fine-tuning, bridging the loss propagation from classifier to embedder. We also design a two-stream model to efficiently facilitate such knowledge transfer in ICMIL. The fine-tuned patch embedder can provide more accurate instance embeddings, in return benefiting the bag classifier. The experimental results show that our method brings consistent improvement to existing MIL backbones.

Acknowledgements. This work was supported by the National Key Research and Development Project (No. 2022YFC2504605), National Natural Science Foundation of China (No. 62202403) and Hong Kong Innovation and Technology Fund (No. PRP/034/22FX). It was also supported in part by the Grant in Aid for Scientific Research from the Japanese Ministry for Education, Science, Culture and Sports (MEXT) under the Grant No. 20KK0234, 21H03470.

References

1. Bejnordi, B.E., et al.: Diagnostic assessment of deep learning algorithms for detection of lymph node metastases in women with breast cancer. JAMA **318**(22), 2199–2210 (2017)
2. Campanella, G., et al.: Clinical-grade computational pathology using weakly supervised deep learning on whole slide images. Nature Med. **25**(8), 1301–1309 (2019)
3. Chen, Q., et al.: Deep learning for evaluation of microvascular invasion in hepatocellular carcinoma from tumor areas of histology images. Hepatol. Int. **16**(3), 590–602 (2022)
4. Chen, R.J., et al.: Scaling vision transformers to gigapixel images via hierarchical self-supervised learning. In: Proceedings of the IEEE/CVF Conference on Computer Vision and Pattern Recognition, pp. 16144–16155 (2022)
5. Cheng, N., et al.: Deep learning-based classification of hepatocellular nodular lesions on whole-slide histopathologic images. Gastroenterology **162**(7), 1948–1961 (2022)
6. He, K., Zhang, X., Ren, S., Sun, J.: Deep residual learning for image recognition. In: Proceedings of the IEEE Conference on Computer Vision and Pattern Recognition, pp. 770–778 (2016)
7. Hinton, G., Vinyals, O., Dean, J.: Distilling the knowledge in a neural network. arXiv preprint arXiv:1503.02531 (2015)
8. Hou, L., Samaras, D., Kurc, T.M., Gao, Y., Davis, J.E., Saltz, J.H.: Patch-based convolutional neural network for whole slide tissue image classification. In: Proceedings of the IEEE Conference on Computer Vision and Pattern Recognition, pp. 2424–2433 (2016)

9. Ilse, M., Tomczak, J., Welling, M.: Attention-based deep multiple instance learning. In: International Conference on Machine Learning, pp. 2127–2136. PMLR (2018)
10. Jin, C., Guo, Z., Lin, Y., Luo, L., Chen, H.: Label-efficient deep learning in medical image analysis: challenges and future directions. arXiv preprint arXiv:2303.12484 (2023)
11. Kingma, D.P., Ba, J.: Adam: a method for stochastic optimization. In: International Conference on Learning Representations (2015)
12. Li, B., Li, Y., Eliceiri, K.W.: Dual-stream multiple instance learning network for whole slide image classification with self-supervised contrastive learning. In: Proceedings of the IEEE/CVF Conference on Computer Vision and Pattern Recognition, pp. 14318–14328 (2021)
13. Liu, K., et al.: Multiple instance learning via iterative self-paced supervised contrastive learning. arXiv preprint arXiv:2210.09452 (2022)
14. Lu, M., et al.: Smile: sparse-attention based multiple instance contrastive learning for glioma sub-type classification using pathological images. In: MICCAI Workshop on Computational Pathology, pp. 159–169. PMLR (2021)
15. Lu, M.Y., et al.: AI-based pathology predicts origins for cancers of unknown primary. Nature **594**(7861), 106–110 (2021)
16. Lu, M.Y., Williamson, D.F., Chen, T.Y., Chen, R.J., Barbieri, M., Mahmood, F.: Data-efficient and weakly supervised computational pathology on whole-slide images. Nature Biomed. Eng. **5**(6), 555–570 (2021)
17. Luo, Z., et al.: Weakly-supervised action localization with expectation-maximization multi-instance learning. In: Vedaldi, A., Bischof, H., Brox, T., Frahm, J.-M. (eds.) ECCV 2020. LNCS, vol. 12374, pp. 729–745. Springer, Cham (2020). https://doi.org/10.1007/978-3-030-58526-6_43
18. Maron, O., Lozano-Pérez, T.: A framework for multiple-instance learning. In: Advances in Neural Information Processing Systems, vol. 10 (1997)
19. Russakovsky, O., et al.: ImageNet large scale visual recognition challenge. Int. J. Comput. Vision **115**(3), 211–252 (2015)
20. Shao, Z., Bian, H., Chen, Y., Wang, Y., Zhang, J., Ji, X., et al.: TransMIL: transformer based correlated multiple instance learning for whole slide image classification. Adv. Neural. Inf. Process. Syst. **34**, 2136–2147 (2021)
21. Srinidhi, C.L., Kim, S.W., Chen, F.D., Martel, A.L.: Self-supervised driven consistency training for annotation efficient histopathology image analysis. Med. Image Anal. **75**, 102256 (2022)
22. Wang, Q., Chechik, G., Sun, C., Shen, B.: Instance-level label propagation with multi-instance learning. In: Proceedings of the 26th International Joint Conference on Artificial Intelligence, pp. 2943–2949 (2017)
23. Wang, X., et al.: UD-MIL: uncertainty-driven deep multiple instance learning for oct image classification. IEEE J. Biomed. Health Inform. **24**(12), 3431–3442 (2020)
24. Zhang, C., Song, Y., Zhang, D., Liu, S., Chen, M., Cai, W.: Whole slide image classification via iterative patch labelling. In: 2018 25th IEEE International Conference on Image Processing (ICIP), pp. 1408–1412. IEEE (2018)
25. Zhang, H., et al.: DTFD-MIL: double-tier feature distillation multiple instance learning for histopathology whole slide image classification. In: Proceedings of the IEEE/CVF Conference on Computer Vision and Pattern Recognition, pp. 18802–18812 (2022)

MixUp-MIL: Novel Data Augmentation for Multiple Instance Learning and a Study on Thyroid Cancer Diagnosis

Michael Gadermayr[1(✉)], Lukas Koller[1], Maximilian Tschuchnig[1], Lea Maria Stangassinger[2], Christina Kreutzer[3], Sebastien Couillard-Despres[3], Gertie Janneke Oostingh[2], and Anton Hittmair[4]

[1] Department of Information Technology and Digitalization, Salzburg University of Applied Sciences, Salzburg, Austria
michael.gadermayr@fh-salzburg.ac.at
[2] Department of Biomedical Sciences, Salzburg University of Applied Sciences, Salzburg, Austria
[3] Spinal Cord Injury and Tissue Regeneration Center Salzburg, Research Institute of Experimental Neuroregeneration, Salzburg, Austria
[4] Department of Pathology and Microbiology, Kardinal Schwarzenberg Klinikum, Schwarzach, Austria

Abstract. Multiple instance learning is a powerful approach for whole slide image-based diagnosis in the absence of pixel- or patch-level annotations. In spite of the huge size of whole slide images, the number of individual slides is often rather small, leading to a small number of labeled samples. To improve training, we propose and investigate novel data augmentation strategies for multiple instance learning based on the idea of linear and multilinear interpolation of feature vectors within and between individual whole slide images. Based on state-of-the-art multiple instance learning architectures and two thyroid cancer data sets, an exhaustive study was conducted considering a range of common data augmentation strategies. Whereas a strategy based on to the original MixUp approach showed decreases in accuracy, a novel multilinear intra-slide interpolation method led to consistent increases in accuracy.

Keywords: Histopathology · Data augmentation · MixUp · Multiple Instance Learning

1 Motivation

Whole slide imaging is capable of effectively digitizing specimen slides, showing both the microscopic detail and the larger context, without any significant manual effort. Due to the enormous resolution of the whole slide images (WSIs), a classification based on straight-forward convolutional neural network architectures is not feasible. Multiple instance learning [8,10,13,18,20] (MIL) represents a methodology (with a high momentum indicated by a large number of recent publications) to deal with these huge images corresponding to single (global) labels. In the MIL setting, WSIs correspond

H. Greenspan et al. (Eds.): MICCAI 2023, LNCS 14225, pp. 477–486, 2023.
https://doi.org/10.1007/978-3-031-43987-2_46

to labeled bags, whereas extracted patches correspond to unlabeled bag instances. MIL approaches typically consist of a feature extraction stage, a MIL pooling stage and a following downstream classification. State-of-the-art approaches mainly rely on convolutional neural network architectures for feature extraction, often in combination with attention [10, 11] or self-attention [12]. For training the feature extraction stage, classical supervised and self-supervised learning is employed [10, 11]. While the majority of methods rely on separate learning stages, also end-to-end approaches have been proposed [3, 14]. In spite of the large amount of data, the number of labeled samples in MIL (represented by the number of individual, globally labelled WSIs) is often small and/or imbalanced [6]. General data augmentation strategies, such as rotations, flipping, stain augmentation and normalization and affine transformations, are applicable to increase the amount of data [15]. All of these methods are performed in the image domain. Here, we consider feature-level data augmentation directly applied to the representation extracted using a convolutional neural network. These methods can be easily combined with image-based augmentation and show the advantage of a high computational efficiency (since operations are efficient and pre-computed features can be used) [10, 11]. For example, Li et al. [11] proposed an augmentation strategy based on sampling the patch-descriptors to generate several bags for an individual WSI. In this paper, we focus on the interpolations of patch descriptors based on the idea of Zhang et al [21], which is referred to as MixUp. This method was originally proposed as data agnostic approach which also shows good results if applied to image data [2,4,16]. Variations were proposed, to be applied to latent representations [17] as well as to balance data sets [6]. Due to the structure of MIL training data, we identified several options to perform interpolation-based data augmentation.

The main contribution of this work is a set of novel data augmentation strategies for MIL, based on the interpolation of patch descriptors. Inspired by the (linear) MixUp approach [21], we investigated several ways to translate this idea to the MIL setting. Beyond linear interpolation, we also defined a more flexible and novel multilinear approach. For evaluation, a large experimental study was conducted, including 2 histological data sets, 5 deep learning configurations for MIL, 3 common data augmentation strategies and 4 MixUp settings. We investigated the classification of WSIs containing thyroid cancer tissues [1,5]. To obtain an improved understanding of reasons behind the experimental results, we also investigate the feature distributions.

2 Methods

In this paper, we consider MIL approaches relying on separately trained feature extraction and classification stages [9, 10, 12]. The proposed augmentation methods are applied to the patch descriptors obtained after the feature extraction stage. This strategy is highly efficient during training since the features are only computed once (per patch) and for augmentation only simple arithmetic operations are applied to the (smaller) feature vectors. Image-based data augmentation strategies (such as stain-augmentation, rotations or deformations) can be combined easily with the feature-based approaches but require individual feature extraction during training. However, to avoid the curse of meta-parameters and thereby experiments these methods are not considered here.

In the original MixUp formulation of Zhang et al. [21], synthetic samples $\boldsymbol{x}'$ are generated such that $\boldsymbol{x}' = \alpha \cdot \boldsymbol{x_i} + (1-\alpha) \cdot \boldsymbol{x_j}$, where $\boldsymbol{x_i}$ and $\boldsymbol{x_j}$ are randomly sampled raw input feature vectors. Corresponding labels $\boldsymbol{y}'$ are generated such that $\boldsymbol{y}' = \alpha \cdot \boldsymbol{y_i} + (1-\alpha) \cdot \boldsymbol{y_j}$, where $\boldsymbol{y_i}$ and $\boldsymbol{y_j}$ are the corresponding one-hot label encodings. The weight α is drawn from a uniform distribution between 0 and 1.

A single input (corresponding to a WSI) of a MIL approach with a separate feature extraction stage [10] can be expressed as a P-tupel $X = (\boldsymbol{x_1}, ..., \boldsymbol{x_P})$ with $\boldsymbol{x_i}$ being the feature vector of an individual patch and P being the number of patches per WSI. The method proposed by Zhang et al. cannot directly be applied to these tupels. However, there are several options to adapt the basic idea to the changed setting.

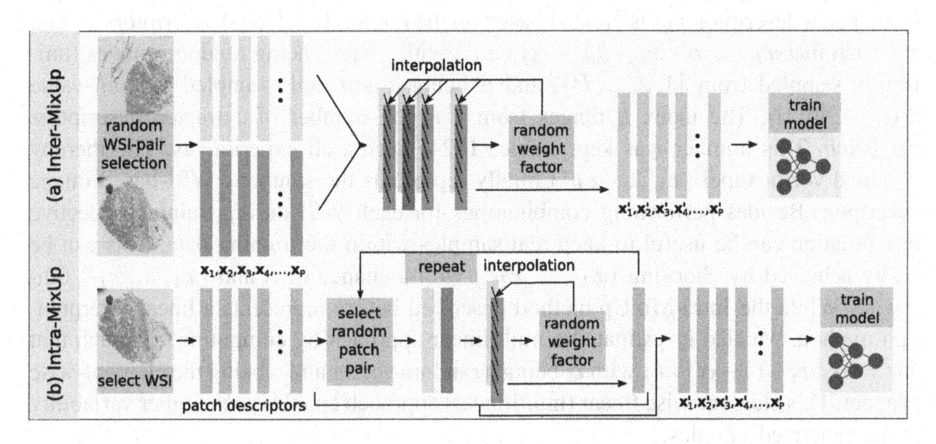

Fig. 1. Overview of the proposed feature-based data augmentation approaches. In the case of Inter-MixUp (a), a linear combination was applied on the pairs of WSI descriptors with a randomly selected weight factor. In the case of Intra-MixUp (b), patch-based descriptors from the same WSI were merged with individual random weights.

2.1 Inter-MixUp and Intra-MixUp

Inter-MixUp refers to the generation of synthetic feature vectors by linearly combining feature vectors of a pair of WSIs (see Fig. 1 (a)). All features of a WSI with index w can be represented by $X^{(w)}$, such that $X^{(w)} = (\boldsymbol{x_1}^{(w)}, \dots, \boldsymbol{x_P}^{(w)})$. To generate a new synthetic sample $X^{(u)'}$ based on two samples $X^{(w)}$ and $X^{(v)}$, we introduce the operation

$$X^{(u)'} = (\alpha \cdot \boldsymbol{x_1}^{(w)} + (1-\alpha) \cdot \boldsymbol{x_1}^{(v)},\ \alpha \cdot \boldsymbol{x_2}^{(w)} + (1-\alpha) \cdot \boldsymbol{x_2}^{(v)},\ \dots,\ \alpha \cdot \boldsymbol{x_P}^{(w)} + (1-\alpha) \cdot \boldsymbol{x_P}^{(v)})$$

with α being a uniformly sampled random weight ($\alpha \in [0, 1]$). The WSI indexes v and w are uniformly sampled from the set of indexes. The index u ranges from the 1 to the number of extracted WSI descriptors. Since the new synthetic descriptors are individually generated in each epoch, there is no benefit if the number of extracted WSI descriptors is increased. We fix this number to the number of WSIs in the training data set, in order to keep the number of training iterations per epoch consistent.

Two different configurations are considered. Firstly, we investigate the interpolation between WSIs of the same class (V1). Secondly, interpolation between all WSIs is performed, which also includes the interpolation between the labels (V2). In the case of V2, also the one-hot-encoded label vectors are linearly combined, such that $\boldsymbol{y}^{(u)\prime} = \alpha \cdot \boldsymbol{y}^{(w)} + (1 - \alpha) \cdot \boldsymbol{y}^{(v)}$ The random values, α, v and w are selected individually for each individual WSI and each epoch. Before applying the MixUp operation, the vector tupel is randomly shuffled (as performed in all experiments).

Intra-WSI combinations (*Intra-MixUp*) refers to the generation of synthetic descriptors by combining feature vectors within an individual WSI (see Fig. 1 (b)). A new synthetic patch descriptor $\boldsymbol{x_k}'$ is created based on the randomly selected descriptors $\boldsymbol{x_i}$ and $\boldsymbol{x_j}$, such that $\boldsymbol{x_k}' = \alpha \cdot \boldsymbol{x_i} + (1 - \alpha) \cdot \boldsymbol{x_j}$, with i and j being random indices (uniformly sampled from $\{1, 2, ..., P\}$) and α being a uniformly sampled random value a ($\alpha \in [0, 1]$). The index k ranges from 1 to the number of extracted descriptors per patch. This number was kept stable (1024) during all experiments. The thereby obtained vector tupel $(\boldsymbol{x_1}', ..., \boldsymbol{x_P}')$ finally represents the synthetic WSI-based image descriptor. Besides performing combinations for each WSI during training, selective interpolation can be useful to keep real samples within the training data. This can be easily achieved by choosing $(\boldsymbol{x_1}', ..., \boldsymbol{x_P}')$ with a chance of β and $(\boldsymbol{x_1}, ..., \boldsymbol{x_P})$ otherwise. While the Intra-MixUp method described before represents a linear interpolation method, we also investigated a multilinear approach by computing $\boldsymbol{x_k}'$ such that $\boldsymbol{x_k}' = \boldsymbol{\alpha} \circ \boldsymbol{x_i} + (1 - \boldsymbol{\alpha}) \circ \boldsymbol{x_j}$ with $\boldsymbol{\alpha}$ being a random vector and $\circ$ being the element-wise product. This element-wise linear (multilinear) approach enables even higher variability in the generated samples.

2.2 Experimental Setting

As experimental architecture, use the dual-stream MIL approach proposed by Li et al [10]. Since this model combines both, embedding-based and an instance-based encoding, the effect of both paths can be individually investigated without changing any other architectural details. Since the method represents a state-of-the-art approach, it further serves as well-performing baseline. In instance-based MIL, the information per patch is first condensed to a single scalar value, representing the classification per patch. Finally, all of these patch-based values are aggregated. In embedding-based MIL, the information per patch is translated into a feature vector. All feature vectors from a WSI are then aggregated followed by a classification. In the investigated model [10] an instance- and an embedding-based pathway are employed in parallel and are merged in the end by weighted addition. The embedding-based pathway contains an attention mechanism, to higher weight patches that are similar to the so-called critical instance. The model makes use of an individual feature extraction stage. Due to the limited number of WSIs, we did not train the feature extraction stage [7], but utilize a pre-trained network instead. Specifically, we applied a ResNet18 pre-trained on the image-net challenge data, due to the high performance in previous work on similar data [5]. ResNet18 was assessed as particularly appropriate due to the rather low dimensional output (512 dimensions). We actively decided not to use a self-supervised contrastive learning approach [10] as feature extraction stage since invariant features could interfere with the effect of data augmentation. We investigated various settings consisting of instance-based only (INST), embedding-based only (EMB) and the dual-stream approach with weightings 3/1, 2/2 (balanced) and 1/3 for the instance and the embedding-based pathways.

As comparison, several other augmentation methods on feature level are investigated including random sampling, selective random sampling and random noise. Random sampling corresponds to the random selection of patches (feature vectors) from each WSI. Thereby the amount of investigated data per WSI is reduced with the benefit of increasing the variability of the data. In the experiments, we adjust the sample ratio q between the patch-based features for training and testing. A q of 50 $\%$ indicates that 512 descriptors are used for training while for testing always a fixed number of 1024 is used. Selective random sampling corresponds to the random sampling strategy, with the difference that the ratio of features is not fixed but drawn from a uniform random distribution ($U(q, 100\ \%)$). Here, a q of 50 $\%$ indicates that for each WSI, between 512 and 1024 feature vectors are selected. In the case of the random noise setting, to each feature vector $\boldsymbol{x}_i$, a random noise vector $\boldsymbol{r}$ is added ($\boldsymbol{x}_i{}' = \boldsymbol{x}_i + \boldsymbol{r}$). The elements of r are randomly sampled (individually for each $\boldsymbol{x}_i$) from a normal distribution $N(0, \sigma')$. To incorporate for the fact that the feature dimensions show different magnitudes, σ' is computed as the product of the meta parameter σ and the standard deviation of the respective feature dimension.

In this work, we aimed at distinguishing different nodular lesions of the thyroid, focusing especially on benign follicular nodules (FN) and papillary carcinomas (PC). This differentiation is crucial, due to the different treatment options, in particular with respect to the extent of surgical resection of the thyroid gland [19]. The data set utilized in the experiments consists of 80 WSIs overall. One half (40) of the data set consists of frozen and the other half (40) of paraffin sections [5]), representing the different modalities. All images were acquired during clinical routine at the Kardinal Schwarzenberg Hospital. Procedures were approved by the ethics committee of the county of Salzburg (No. 1088/2021). The mean and median age of patients at the date of dissection was 47 and 50 years, respectively. The data set comprised 13 male and 27 female patients, corresponding to a slight gender imbalance. They were labeled by an expert pathologist with over 20 years experience. A total of 42 (21 per modality) slides were labeled as papillary carcinoma while 38 (19 per modality) were labeled as benign follicular nodule. For the frozen sections, fresh tissue was frozen at $-15°$ Celsius, slides were cut (thickness 5 μm) and stained immediately with hematoxylin and eosin. For the paraffin sections, tissue was fixed in 4 $\%$ phosphate-buffered formalin for 24 h. Subsequently formalin fixed paraffin embedded tissue was cut (thickness 2 μm) and stained with hematoxylin and eosin. The images were digitized with an Olympus VS120-LD100 slide loader system. Overviews at a 2x magnification were generated to manually define scan areas, focus points were automatically defined and adapted if needed. Scans were performed with a 20x objective (corresponding to a resolution of 344.57 nm/pixel). The image files were stored in the Oympus vsi format based on lossless compression.

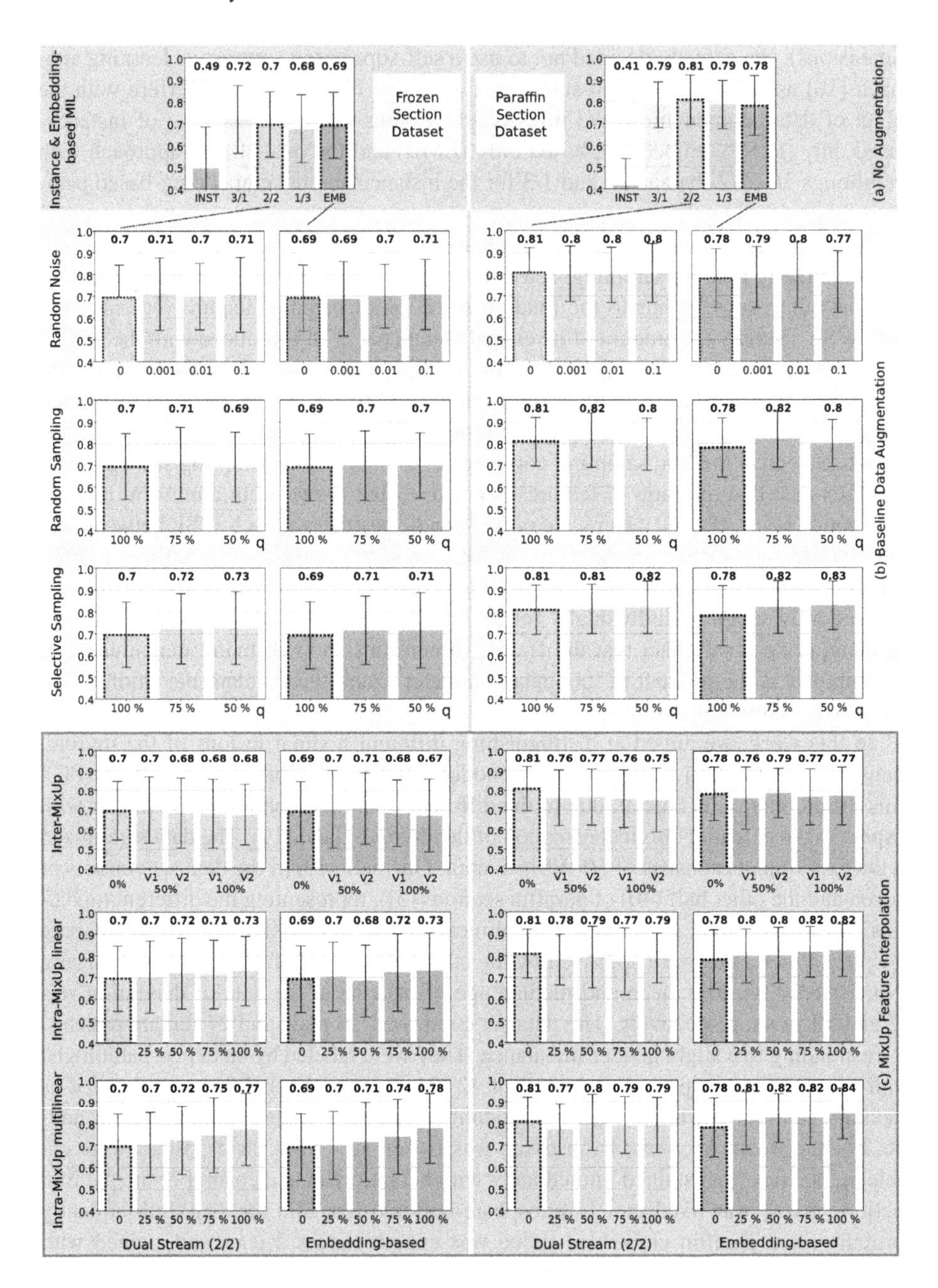

Fig. 2. Mean overall classification accuracy and standard deviation obtained with each individual combination. The columns represent the frozen (left) and paraffin data set (right). The top row (a) shows the baseline scores of embedding-based, instance-based and 3 combinations. Subfigure (b) shows the scores obtained with baseline data augmentation for embedding-based and dual-stream MIL. Subfigure (c) shows the scores obtained with interpolation between (Inter-MixUp) and within WSIs (Intra-MixUp).

The data set was randomly separated into training (80 %) and test data (20 %). The whole pipeline, including the separation, was repeated 32 times to achieve representative scores. Due to the almost balanced setting, the overall classification accuracy (mean and standard deviation) is finally reported. Adam was used as optimizer. The models were trained for 200 epochs with an initial learning rate of 0.0002. Random shuffling of the vector tupels (shuffling within the WSIs) was applied for all experiments.

The patches were randomly extracted from the WSI, based on uniform sampling. For each patch, we checked that at least 75 % of the area was covered with tissue (green color channel) in order to exclude empty areas [5]. To obtain a representation independent of the WSI size, we extracted 1024 patches with a size of 256×256 pixel per WSI, resulting in 1024 patch-descriptors per WSI [5]. For feature extraction, a ResNet18 network, pretrained on the image-net challenge was deployed [10]. Data and source code are publicly accessible via https://gitlab.com/mgadermayr/mixupmil. We use the reference implementation of the dual-stream MIL approach [10]. To obtain further insight into the feature distribution, we randomly selected patch descriptor pairs and computed the Euclidean distances. In detail, we selected 10,000 pairs (a) from different classes, (b) from different WSIs (similar and dissimilar classes), (c,d) from the same class and different WSIs, and (e) from the same WSI.

3 Results

Figure 2 shows the mean overall classification accuracy and standard deviations obtained with each individual combination. The columns represent the frozen (left) and paraffin data set (right). The top row (a) shows the baseline scores of embedding-based, instance-based and the 3 combinations. Subfigure (b) show the scores obtained with baseline data augmentation for embedding-based and dual-stream MIL. Subfigure (c) shows the scores obtained with interpolation between patches between (Inter-MixUp) and within WSIs (Intra-MixUp). Without data augmentation, scores between 0.49 and 0.72 were obtained for frozen and scores between 0.41 and 0.81 for the paraffin data set. To limit the number of figures and due to the fact that instance-based MIL showed weak scores only, in the following part the focus is on embedding-based and combined-MIL (2/2) only. With baseline data augmentation, scores between 0.69 and 0.73 were achieved for the frozen and between 0.78 and 0.83 for the paraffin data set. Inter-MixUp exhibited scores up to 0.71 for the frozen and up to 0.79 for the paraffin data set. Intra-MixUp showed average accuracy up to 0.78 for the frozen and up to 0.84 for the paraffin data set. The best scores were obtained with the multilinear setting. In Fig. 3, the distributions of the descriptor (Euclidean) distances between (a-d) patches from different different WSIs (inter-WSI) and (e) patches within a single WSI (intra-WSI) are provided. The mean distances range from 171.3 to 177.8 for the inter-WSI settings. In the intra-WSI setting, a mean distance of 134.8 was obtained. Based on the used common box plot variation (whiskers length is less than $1.5\times$ the interquartile range), a large number of data points was identified as outliers. However, these points are not considered as real outliers, but occur due to the asymmetrical data distribution (as indicated by the violin plot in the background).

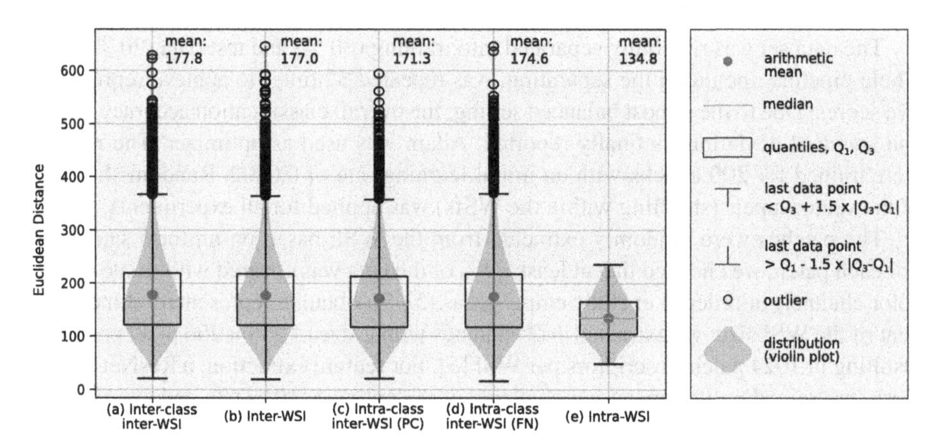

Fig. 3. Analysis of the distributions of the patch descriptor distances between (a) patches from different classes, (b) randomly selected patches from different WSIs, (c,d) patches from the same class and different WSIs (for both classes, PC and FN) and (e) patches within the WSIs.

4 Discussion

In this work, we proposed and examined novel data augmentation strategies based on the idea of interpolations of feature vectors in the MIL setting. Instance-based MIL did not show any competitive scores. Obviously the model reducing each patch to a single value is not adequate for the classification of frozen or paraffin sections from thyroid cancer tissues. The considered dual-stream approach, including an embedding and instance-based stream, exhibited slightly improved average scores, compared to embedding-based MIL only. In our analysis, we focused on the embedding-based configuration and on the balanced combined approach (referred to as 2/2). With the baseline data augmentation approaches, the maximum improvements were 0.03, and 0.02 for the frozen, and 0.01, and 0.05 for the paraffin data set. The Inter-MixUp approach did not show any systematic improvements. Independently of the chosen strategy (V1, V2), concerning the combination within or between classes, we did not notice any positive trend. The multilinear Intra-MixUp method, however, exhibited the best scores for 3 out of 4 combinations and the best overall mean accuracy for both, the frozen and the paraffin data set. Also a clear trend with increasing scores in the case of an increasing ratio of augmented data (β) is visible. The linear method showed a similar, but less pronounced trend. Obviously, the straightforward application of the MixUp scheme (as in case of the Inter-MixUp approach), is inappropriate for the considered setting. An inhibiting factor could be a high inter-WSI variability leading to incompatible feature vectors (which are too far away from realistic samples in the feature space). To particularly investigate this effect, we performed 2 different Inter-MixUp settings (V1 & V2), with the goal of identifying the effect of mixed (and thereby more dissimilar) or similar classes during interpolation. The analysis of the distance distributions between patch representations confirmed that, the variability between WSIs is clearly larger than the variability within WSIs. In addition, the results showed that the variability between classes is, on patch-level, not clearly larger than the variability within a class. Obviously variability due to the acquisition outweigh any disease specific variability. This could provide an explanation for the effectiveness of Intra-MixUp approach compared to the (similarly) poorly performing Inter-MixUp settings. We expect that stain normalization methods (but not stain augmentation) could be utilized to align the different WSIs to provide a more appropriate basis for inter-WSI interpolation. With regard to the different data sets, we noticed a stronger, positive effect in case of the frozen section data set. This is supposed to be due to the clearly higher variability of the frozen sections corresponding with a need for a higher variability in the training data. We also noticed a stronger effect of the solely embedding-based architecture (also showing the best overall scores). We suppose that this is due to the fact that the additional loss of the dual-stream architecture exhibits a valuable regularization tool to reduce the amount of needed training data. With the proposed Intra-MixUp augmentation strategy, this effect diminishes, since the amount and quality of training data is increased.

To conclude, we proposed novel data augmentation strategies based on the idea of interpolations of image descriptors in the MIL setting. Based on the experimental results, the multilinear Intra-MixUp setting proved to be highly effective, while the Inter-MixUp method showed inferior scores compared to a state-of-the-art baseline. We learned that there is a clear difference between combinations within and between WSIs with a noticeable effect on the final classification accuracy. This is supposedly due to the high variability between the WSIs compared to a rather low variability within the WSIs. In the future, additional experiments will be conducted including stain normalization methods and larger benchmark data sets to provide further insights.

Acknowledgement. This work was partially funded by the County of Salzburg (no. FHS2019-10-KIAMed)

References

1. Buddhavarapu, V.G., Jothi, A.A.: An experimental study on classification of thyroid histopathology images using transfer learning. Pattern Recognit. Lett. **140**, 1–9 (2020)
2. Chen, J.N., Sun, S., He, J., Torr, P.H., Yuille, A., Bai, S.: Transmix: attend to mix for vision transformers. In: Proceedings of the IEEE/CVF Conference on Computer Vision and Pattern Recognition (CVPR), pp. 12135–12144 (2022)
3. Chikontwe, P., Kim, M., Nam, S.J., Go, H., Park, S.H.: Multiple instance learning with center embeddings for histopathology classification. In: Proceedings of the International Conference on Medical Image Computing and Computer Assisted Intervention (MICCAI), pp. 519–528 (2020)
4. Dabouei, A., Soleymani, S., Taherkhani, F., Nasrabadi, N.M.: Supermix: supervising the mixing data augmentation. In: Proceedings of the IEEE/CVF Conference on Computer Vision and Pattern Recognition (CVPR), pp. 13794–13803 (2021)
5. Gadermayr, M., et al.: Frozen-to-paraffin: categorization of histological frozen sections by the aid of paraffin sections and generative adversarial networks. In: Proceedings of the MICCAI Workshop on Simulation and Synthesis in Medical Imaging (SASHIMI), pp. 99–109 (2021)
6. Galdran, A., Carneiro, G., Ballester, M.A.G.: Balanced-MixUp for highly imbalanced medical image classification. In: Proceedings of the Conference on Medical Image Computing and Computer Assisted Intervention (MICCAI), pp. 323–333 (2021)

7. Hou, L., Samaras, D., Kurc, T.M., Gao, Y., Davis, J.E., Saltz, J.H.: Patch-based convolutional neural network for whole slide tissue image classification. In: Proceedings of the Conference on Computer Vision and Pattern Recognition (CVPR), pp. 2424–2433 (2016)
8. Ilse, M., Tomczak, J., Welling, M.: Attention-based deep multiple instance learning. In: Proceedings of the International Conference on Machine Learning (ICML), pp. 2127–2136 (2018)
9. Lerousseau, M., et al.: Weakly supervised multiple instance learning histopathological tumor segmentation. In: Martel, A.L., et al. (eds.) MICCAI 2020. LNCS, vol. 12265, pp. 470–479. Springer, Cham (2020). https://doi.org/10.1007/978-3-030-59722-1_45
10. Li, B., Li, Y., Eliceiri, K.W.: Dual-stream multiple instance learning network for whole slide image classification with self-supervised contrastive learning. In: Proceedings of the Conference on Computer Vision and Pattern Recognition (CVPR), pp. 14318–14328 (2021). https://github.com/binli123/dsmil-wsi
11. Li, Z., et al.: A novel multiple instance learning framework for covid-19 severity assessment via data augmentation and self-supervised learning. Med. Image Anal. **69**, 101978 (2021)
12. Rymarczyk, D., Borowa, A., Tabor, J., Zielinski, B.: Kernel self-attention for weakly-supervised image classification using deep multiple instance learning. In: Proceedings of the IEEE/CVF Winter Conference on Applications of Computer Vision (WACV), pp. 1721–1730 (2021)
13. Shao, Z., et al.: Transmil: transformer based correlated multiple instance learning for whole slide image classification. In: Advances in Neural Information Processing Systems (NeurIPS), vol. 34, pp. 2136–2147 (2021)
14. Sharma, Y., Shrivastava, A., Ehsan, L., Moskaluk, C.A., Syed, S., Brown, D.: Cluster-to-conquer: a framework for end-to-end multi-instance learning for whole slide image classification. In: Proceedings of the Medical Imaging with Deep Learning Conference (MIDL), pp. 682–698 (2021)
15. Tellez, D., et al.: Quantifying the effects of data augmentation and stain color normalization in convolutional neural networks for computational pathology. Med. Image Anal. **58**, 101544 (2019)
16. Thulasidasan, S., Chennupati, G., Bilmes, J.A., Bhattacharya, T., Michalak, S.: On mixup training: improved calibration and predictive uncertainty for deep neural networks. In: Advances in Neural Information Processing Systems (NeurIPS), vol. 32 (2019)
17. Verma, V., et al.: Manifold mixup: better representations by interpolating hidden states. In: Proceedings of the International Conference on Machine Learning (ICML), vol. 97, pp. 6438–6447 (2019)
18. Wang, X., et al.: TransPath: transformer-based self-supervised learning for histopathological image classification. In: Proceedings of the Conference on Medical Image Computing and Computer Assisted Intervention (MICCAI), pp. 186–195 (2021)
19. Xi, N.M., Wang, L., Yang, C.: Improving the diagnosis of thyroid cancer by machine learning and clinical data. Sci. Rep. **12**(1), 11143 (2022)
20. Zhang, H., et al.: DTFD-MIL: double-tier feature distillation multiple instance learning for histopathology whole slide image classification. In: Proceedings of the IEEE/CVF Conference on Computer Vision and Pattern Recognition (CVPR), pp. 18802–18812 (2022)
21. Zhang, H., Cisse, M., Dauphin, Y.N., Lopez-Paz, D.: mixup: Beyond empirical risk minimization. In: Proceedings of the International Conference on Learning Representations (ICLR) (2018)

CellGAN: Conditional Cervical Cell Synthesis for Augmenting Cytopathological Image Classification

Zhenrong Shen[1], Maosong Cao[2], Sheng Wang[1,3], Lichi Zhang[1], and Qian Wang[2(✉)]

[1] School of Biomedical Engineering, Shanghai Jiao Tong University, Shanghai, China
[2] School of Biomedical Engineering, ShanghaiTech University, Shanghai, China
qianwang@shanghaitech.edu.cn
[3] Shanghai United Imaging Intelligence Co., Ltd., Shanghai, China

Abstract. Automatic examination of thin-prep cytologic test (TCT) slides can assist pathologists in finding cervical abnormality for accurate and efficient cancer screening. Current solutions mostly need to localize suspicious cells and classify abnormality based on local patches, concerning the fact that whole slide images of TCT are extremely large. It thus requires many annotations of normal and abnormal cervical cells, to supervise the training of the patch-level classifier for promising performance. In this paper, we propose CellGAN to synthesize cytopathological images of various cervical cell types for augmenting patch-level cell classification. Built upon a lightweight backbone, CellGAN is equipped with a non-linear class mapping network to effectively incorporate cell type information into image generation. We also propose the Skip-layer Global Context module to model the complex spatial relationship of the cells, and attain high fidelity of the synthesized images through adversarial learning. Our experiments demonstrate that CellGAN can produce visually plausible TCT cytopathological images for different cell types. We also validate the effectiveness of using CellGAN to greatly augment patch-level cell classification performance. Our code and model checkpoint are available at https://github.com/ZhenrongShen/CellGAN.

Keywords: Conditional Image Synthesis · Generative Adversarial Network · Cytopathological Image Classification · Data Augmentation

1 Introduction

Cervical cancer accounts for 6.6% of the total cancer deaths in females worldwide, making it a global threat to healthcare [6]. Early cytology screening is highly effective for the prevention and timely treatment of cervical cancer [23].

Supplementary Information The online version contains supplementary material available at https://doi.org/10.1007/978-3-031-43987-2_47.

H. Greenspan et al. (Eds.): MICCAI 2023, LNCS 14225, pp. 487–496, 2023.
https://doi.org/10.1007/978-3-031-43987-2_47

Nowadays, thin-prep cytologic test (TCT) [1] is widely used to screen cervical cancers according to the Bethesda system (TBS) rules [21]. Typically there are five types of cervical squamous cells under TCT examinations [5], including normal class or negative for intraepithelial malignancy (NILM), atypical squamous cells of undetermined significance (ASC-US), low-grade squamous intraepithelial lesion (LSIL), atypical squamous cells that cannot exclude HSIL (ASC-H), and high-grade squamous intraepithelial lesion (HSIL). The NILM cells have no cytological abnormalities while the others are manifestations of cervical abnormality to a different extent. By observing cellular features (*e.g.*, nucleus-cytoplasm ratio) and judging cell types, pathologists can provide a diagnosis that is critical to the clinical management of cervical abnormality.

After scanning whole-slide images (WSIs) from TCT samples, automatic TCT screening is highly desired due to the large population versus the limited number of pathologists. As the WSI data per sample has a huge size, the idea of identifying abnormal cells in a hierarchical manner has been proposed and investigated by several studies using deep learning [3,27,31]. In general, these solutions start with the extraction of suspicious cell patches and then conduct patch-level classification. The promising performance of cell classification at the patch level is critical, which contributes to sample-level diagnosis after integrating outcomes from many patches in a WSI. However, such a patch-level classification task requires a large number of annotated training data. And the efforts in collecting reliably annotated data can hardly be negligible, which requires high expertise due to the intrinsic difficulty of visually reading WSIs.

To alleviate the shortage of sufficient data to supervise classification, one may adopt traditional data augmentation techniques, which yet may bring little improvement due to scarcely expanded data diversity [26]. Thus, synthesizing cytopathological images for cervical cells is highly desired to effectively augment training data. Existing literature on pathological image synthesis has explored the generation of histopathological images [10,28]. In cytopathological images, on the contrary, cervical cells can be spatially isolated from each other, or are highly squeezed and even overlapped. The spatial relationship of individual cells is complex, adding diversity to the image appearance of color, morphology, texture, etc. In addition, the differences between cell types are mainly related to nuanced cellular attributes, thus requiring fine granularity in modulating synthesized images toward the expected cell types. Therefore, the task to synthesize realistic cytopathological images becomes very challenging.

Aiming at augmenting the performance of cervical abnormality screening, we develop a novel conditional generative adversarial network in this paper, namely CellGAN, to synthesize cytopathological images for various cell types. We leverage FastGAN [16] as the backbone for the sake of training stability and computational efficiency. To inject cell type for fine-grained conditioning, a non-linear mapping network embeds the class labels to perform layer-wise feature modulation in the generator. Meanwhile, we introduce the Skip-layer Global Context (SGC) module to capture the long-range dependency of cells for precisely modeling their spatial relationship. We adopt an adversarial learning scheme, where the discriminator is modified in a projection-based way [20] for matching condi-

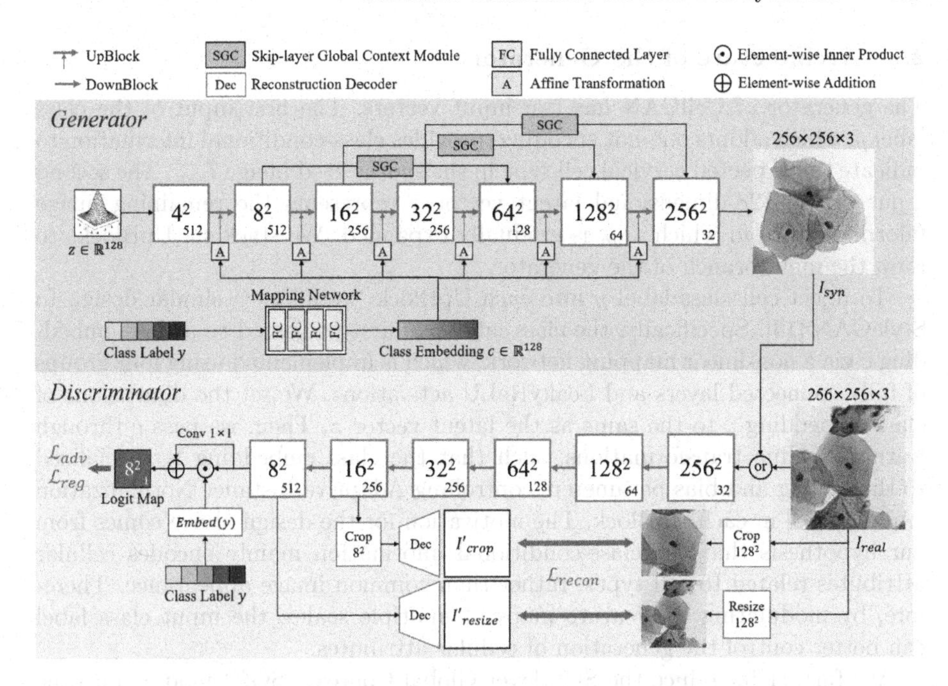

Fig. 1. Overall architecture of the proposed CellGAN. The numbers in the center and the bottom right corner of each square indicate the feature map size and the channel number, respectively.

tional data distribution. To the best of our knowledge, our proposed CellGAN is the first generative model with the capability to synthesize realistic cytopathological images for various cervical cell types. The experimental results validate the visual plausibility of CellGAN synthesized images, as well as demonstrate their data augmentation effectiveness on patch-level cell classification.

2 Method

The dilemma of medical image synthesis lies in the conflict between the limited availability of medical image data and the high demand for data amount to train reliable generative models. To ensure the synthesized image quality given relatively limited training samples, the proposed CellGAN is built upon FastGAN [16] towards stabilized and fast training for few-shot image synthesis. By working in a class-conditional manner, CellGAN can explicitly control the cervical squamous cell types in the synthesized cytopathological images, which is critical to augment the downstream classification task. The overall architecture of CellGAN is presented in Fig. 1, and more detailed structures of the key components are displayed in Supplementary Materials.

2.1 Architecture of the Generator

The generator of CellGAN has two input vectors. The first input of the class label y, which adopts one-hot encoding, provides class-conditional information to indicate the expected cervical cell type in the synthesized image I_{syn}. The second input of the 128-dimensional latent vector z represents the remaining image information, from which I_{syn} is gradually expanded. We stack six UpBlocks to form the main branch of the generator.

To inject cell class label y into each UpBlock, we follow a similar design to StyleGAN [13]. Specifically, the class label y is first projected to a class embedding c via a non-linear mapping network, which is implemented using four groups of fully connected layers and LeakyReLU activations. We set the dimensions of class embedding c to the same as the latent vector z. Then, we pass c through learnable affine transformations, such that the class embedding is specialized to the scaling and bias parameters controlling Adaptive Instance Normalization (AdaIN) [13] in each UpBlock. The motivation for the design above comes from our hypothesis that the class-conditional information mainly encodes cellular attributes related to cell types, rather than common image appearance. Therefore, by modulating the feature maps at multiple scales, the input class label can better control the generation of cellular attributes.

We further introduce the Skip-layer Global Context (SGC) module into the generator (see Fig. 2 in Supplementary Materials), to better handle the diversity of the spatial relationship of the cells. Our SGC module reformulates the idea of GCNet [4] with the design of SLE module from FastGAN [16]. It first performs global context modeling on the low-resolution feature maps, then transforms global context to capture channel-wise dependency, and finally merges the transformed features into high-resolution feature maps. In this way, the proposed SGC module learns a global understanding of the cell-to-cell spatial relationship and injects it into image generation via computationally efficient modeling of long-range dependency.

2.2 Discriminator and Adversarial Training

In an adversarial training setting, the discriminator forces the generator to faithfully match the conditional data distribution of real cervical cytopathological images, thus prompting the generator to produce visually and semantically realistic images. For training stability, the discriminator is trained as a feature encoder with two extra decoders. In particular, five ResNet-like [7] DownBlocks are employed to convert the input image into an $8 \times 8 \times 512$ feature map. Two simple decoders reconstruct downscaled and randomly cropped versions of input images I'_{crop} and I'_{resize} from 8^2 and 16^2 feature maps, respectively. These decoders are optimized together with the discriminator by using a reconstruction loss $\mathcal{L}_{recon}$ that is represented below:

$$\mathcal{L}_{recon} = \mathbb{E}_{f \sim Dis(x), x \sim I_{real}} \left[\| Dec(f) - \mathcal{T}(x) \|_{\ell_1} \right], \tag{1}$$

where $\mathcal{T}$ denotes the image processing (*i.e.*, $\frac{1}{2}$ downsampling and $\frac{1}{4}$ random cropping) on real image I_{real}, f is the processed intermediate feature map from the discriminator *Dis*, and *Dec* stands for the reconstruction decoder. This simple self-supervised technique provides a strong regularization in forcing the discriminator to extract a good image representation.

To provide more detailed feedback from the discriminator, PatchGAN [12] architecture is adopted to output an 8×8 logit map by using a 1×1 convolution on the last feature map. By penalizing image content at the scale of patches, the color fidelity of synthesized images is guaranteed as illustrated in our ablation study (see Fig. 3). To align the class-conditional fake and real data distributions in the adversarial setting, the discriminator directly incorporates class labels as additional inputs in the manner of projection discriminator [20]. The class label is projected to a learned 512-dimensional class embedding and takes inner-product at every spatial position of the $8 \times 8 \times 512$ feature map. The resulting 8×8 feature map is then added to the aforementioned 8×8 logit map, composing the final output of the discriminator.

For the objective function, we use the *hinge* version [15] of the standard adversarial loss $\mathcal{L}_{adv}$. We also employ R_1 regularization $\mathcal{L}_{reg}$ [17] as a slight gradient penalty for the discriminator. Combining all the loss functions above, the total objective $\mathcal{L}_{total}$ to train the proposed CellGAN in an adversarial manner can be expressed as:

$$\mathcal{L}_{total} = \mathcal{L}_{adv} + \mathcal{L}_{recon} + \lambda_{reg}\mathcal{L}_{reg}, \tag{2}$$

where λ_{reg} is empirically set to 0.01 in our experiments.

3 Experimental Results

3.1 Dataset and Experimental Setup

Dataset. In this study, we collect 14,477 images with 256×256 pixels from three collaborative clinical centers. All the images are manually inspected to contain different cervical squamous cell types. In total, there are 7,662 NILM, 2,275 ASC-US, 2,480 LSIL, 1,638 ASC-H, and 422 HSIL images. All the 256×256 images with their class labels are selected as the training data.

Implementation Details. We use the learning rate of 2.5×10^{-4}, batch size of 64, and Adam optimizer [14] to train both the generator and the discriminator for $100k$ iterations. Spectral normalization [19], differentiable augmentation [30] and exponential-moving-average optimization [29] are included in the training process. Fréchet Inception Distance (FID) [8] is used to measure the overall semantic realism of the synthesized images. All the experiments are conducted using an NVIDIA GeForce RTX 3090 GPU with PyTorch [22].

3.2 Evaluation of Image Synthesis Quality

We compare CellGAN with the state-of-the-art generative models for class-conditional image synthesis, *i.e.*, BigGAN [2] from cGANs [18] and Latent Diffusion Model (LDM) [25] from diffusion models [9]. As shown in Fig. 2, BigGAN

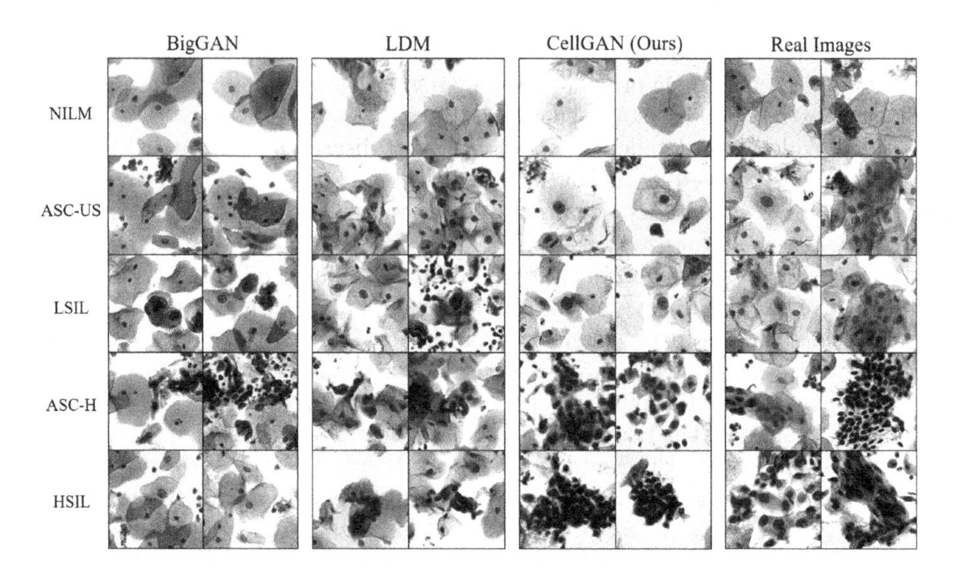

Fig. 2. Qualitative comparison between state-of-the-art generative models and the proposed CellGAN. Different rows stand for different cervical squamous cell types.

Table 1. Quantitative comparison between state-of-the-art generative models and the proposed CellGAN (↓: Lower is better).

Method	FID↓					
	NILM	ASC-US	LSIL	ASC-H	HSIL	Mean
BigGAN	29.5076	37.9543	35.5058	48.0228	85.6230	47.3227
LDM	53.4307	56.1689	49.0969	59.6406	84.9522	60.6579
CellGAN(Ours)	**26.0135**	**33.5718**	**33.3401**	**46.2965**	**68.3458**	**41.5136**

cannot generate individual cells with clearly defined cell boundaries. And it also fails to capture the morphological features of HSIL cells that are relatively limited in training data quantity. LDM only yields half-baked cell structures since the generated cells are mixed, and there exists negligible class separability among abnormal cell types. On the contrary, our proposed CellGAN is able to synthesize visually plausible cervical cells and accurately model distinguishable cellular features for each cell type. The quantitative comparison by FID in Table 1 also demonstrates the superiority of CellGAN in synthesized image quality.

To verify the effects of key components in the proposed CellGAN, we conduct an ablation study on four model settings in Table 2 and Fig. 3. We denote the models in Fig. 3 from left to right as *Model i*, *Model ii*, *Model iii*, and CellGAN. The visual results of *Model i* suffer from severe color distortions while the other models do not, indicating that the PatchGAN-based discriminator can guarantee color fidelity by patch-level image content penalty.

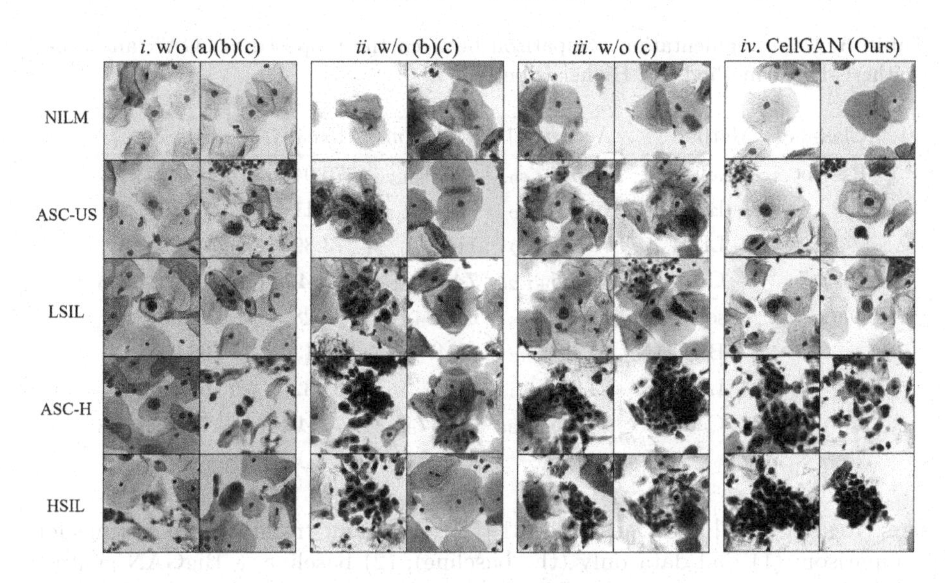

Fig. 3. Generated images from ablation study of the following key components: (a) PatchGAN architecture, (b) class mapping network, (c) SGC module.

Table 2. Quantitative ablation study of the following key component: (a) PatchGAN architecture, (b) class mapping network, (c) SGC module. (↓: Lower is better).

Model Setting			FID↓					
(a)	(b)	(c)	NILM	ASC-US	LSIL	ASC-H	HSIL	Mean
–	–	–	39.7048	45.9424	42.2336	58.3448	84.0265	54.0504
√	–	–	42.7974	36.1800	38.1507	52.0282	74.4304	48.7173
√	√	–	31.1720	38.1468	39.5540	47.2584	68.6068	44.9476
√	√	√	**26.0135**	**33.5718**	**33.3401**	**46.2965**	**68.3458**	**41.5136**

The abnormal cells generated by *Model i* and *Model ii* tend to have highly similar cellular features. In contrast, *Model iii* and CellGAN can accurately capture the morphological characteristics of different cell types. This phenomenon suggests that the implementation of the class mapping network facilitates more distinguishable feature representations for different cell types. By comparing the synthesized images from *Model iii* with CellGAN, it is observed that adopting SGC modules can yield more clear cell boundaries, which demonstrates the capability of SGC module in modeling complicated cell-to-cell relationships in image space. The quantitative results further state the effects of the components above.

3.3 Evaluation of Augmentation Effectiveness

To validate the data augmentation capacity of the proposed CellGAN, we conduct 5-fold cross-validations on the cell classification performances of two classi-

Table 3. Data augmentation comparison between the proposed CellGAN and other synthesis-based methods (↑: Higher is better).

Classifier	Method	Accuracy↑	Precision↑	Recall↑	F1-Score↑
ResNet	baseline	$74.30_{\pm 1.69}$	$68.00_{\pm 1.79}$	$70.94_{\pm 2.28}$	$68.88_{\pm 1.78}$
	+ BigGAN	$76.30_{\pm 2.80}$	$72.96_{\pm 2.58}$	$75.11_{\pm 2.20}$	$73.89_{\pm 2.44}$
	+ LDM	$75.80_{\pm 1.12}$	$71.14_{\pm 0.72}$	$73.89_{\pm 1.36}$	$72.29_{\pm 0.91}$
	+ CellGAN	$\mathbf{79.55}_{\pm \mathbf{1.20}}$	$\mathbf{74.88}_{\pm \mathbf{1.60}}$	$\mathbf{75.42}_{\pm \mathbf{1.74}}$	$\mathbf{74.70}_{\pm \mathbf{1.79}}$
DenseNet	baseline	$72.10_{\pm 0.66}$	$65.23_{\pm 1.17}$	$68.28_{\pm 1.26}$	$66.33_{\pm 1.25}$
	+ BigGAN	$75.40_{\pm 1.73}$	$68.47_{\pm 1.78}$	$70.13_{\pm 1.21}$	$68.94_{\pm 1.97}$
	+ LDM	$74.95_{\pm 1.94}$	$68.03_{\pm 2.11}$	$69.32_{\pm 1.65}$	$68.55_{\pm 2.37}$
	+ CellGAN	$\mathbf{76.15}_{\pm \mathbf{1.38}}$	$\mathbf{70.37}_{\pm \mathbf{1.52}}$	$\mathbf{72.42}_{\pm \mathbf{1.95}}$	$\mathbf{70.99}_{\pm \mathbf{1.75}}$

fiers (ResNet-34 [7] and DenseNet-121 [11]) using four training data settings for comparison: (1) real data only (the baseline); (2) baseline + BigGAN synthesized images; (3) baseline + LDM synthesized images; (4) baseline + CellGAN synthesized images. For each cell type, we randomly select 400 real images and divide them into 5 groups. In each fold, one group is selected as the testing data while the other four are used for training. For different data settings, we synthesize 2,000 images for each cell type using the corresponding generative method, and add them to the training data of each fold. We use the learning rate of 1.0×10^{-4}, batch size of 64, and SGD optimizer [24] to train all the classifiers for 30 epochs. Random flip is applied to all data settings since it is reasonable to use traditional data augmentation techniques simultaneously in practice.

The experimental accuracy, precision, recall, and F1 score are listed in Table 3. It is shown that both the classifiers achieve the best scores in all metrics using the additional synthesized data from CellGAN. Compared with the baselines, the accuracy values of ResNet-34 and DenseNet-121 are improved by 5.25% and 4.05%, respectively. Meanwhile, the scores of other metrics are all improved by more than 4%, indicating that our synthesized data can significantly enhance the overall classification performance. Thanks to the visually plausible and semantically realistic synthesized data, CellGAN is conducive to the improvement of cell classification, thus serving as an efficient tool for augmenting automatic abnormal cervical cell screening.

4 Conclusion and Discussion

In this paper, we propose CellGAN for class-conditional cytopathological image synthesis of different cervical cell types. Built upon FastGAN for training stability and computational efficiency, incorporating class-conditional information of cell types via non-linear mapping can better represent distinguishable cellular features. The proposed SGC module provides the global contexts of cell spatial relationships by capturing long-range dependencies. We have also found that the

PatchGAN-based discriminator can prevent potential color distortion. Qualitative and quantitative experiments validate the semantic realism as well as the data augmentation effectiveness of the synthesized images from CellGAN.

Meanwhile, our current CellGAN still has several limitations. First, we cannot explicitly control the detailed attributes of the synthesized cell type, *e.g.*, nucleus size, and nucleus-cytoplasm ratio. Second, in this paper, the synthesized image size is limited to 256×256. It is worth conducting more studies for expanding synthesized image size to contain much more cells, such that the potential applications can be extended to other clinical scenes (*e.g.*, interactively training pathologists) in the future.

Acknowledgement. This work was supported by the National Natural Science Foundation of China (No. 62001292).

References

1. Abulafia, O., Pezzullo, J.C., Sherer, D.M.: Performance of ThinPrep liquid-based cervical cytology in comparison with conventionally prepared Papanicolaou smears: a quantitative survey. Gynecol. Oncol. **90**(1), 137–144 (2003)
2. Brock, A., Donahue, J., Simonyan, K.: Large scale GAN training for high fidelity natural image synthesis. In: International Conference on Learning Representations (2018)
3. Cao, L., et al.: A novel attention-guided convolutional network for the detection of abnormal cervical cells in cervical cancer screening. Med. Image Anal. **73**, 102197 (2021)
4. Cao, Y., Xu, J., Lin, S., Wei, F., Hu, H.: GCNet: non-local networks meet squeeze-excitation networks and beyond. In: Proceedings of the IEEE/CVF International Conference on Computer Vision Workshops (2019)
5. Davey, D.D., Naryshkin, S., Nielsen, M.L., Kline, T.S.: Atypical squamous cells of undetermined significance: interlaboratory comparison and quality assurance monitors. Diagn. Cytopathol. **11**(4), 390–396 (1994)
6. Gultekin, M., Ramirez, P.T., Broutet, N., Hutubessy, R.: World health organization call for action to eliminate cervical cancer globally. Int. J. Gynecol. Cancer **30**(4), 426–427 (2020)
7. He, K., Zhang, X., Ren, S., Sun, J.: Deep residual learning for image recognition. In: Proceedings of the IEEE Conference on Computer Vision and Pattern Recognition, pp. 770–778 (2016)
8. Heusel, M., Ramsauer, H., Unterthiner, T., Nessler, B., Hochreiter, S.: GANs trained by a two time-scale update rule converge to a local nash equilibrium. In: Advances in Neural Information Processing Systems, vol. 30 (2017)
9. Ho, J., Jain, A., Abbeel, P.: Denoising diffusion probabilistic models. Adv. Neural. Inf. Process. Syst. **33**, 6840–6851 (2020)
10. Hou, L., Agarwal, A., Samaras, D., Kurc, T.M., Gupta, R.R., Saltz, J.H.: Robust histopathology image analysis: to label or to synthesize? In: Proceedings of the IEEE/CVF Conference on Computer Vision and Pattern Recognition, pp. 8533–8542 (2019)
11. Huang, G., Liu, Z., Van Der Maaten, L., Weinberger, K.Q.: Densely connected convolutional networks. In: Proceedings of the IEEE Conference on Computer Vision and Pattern Recognition, pp. 4700–4708 (2017)

12. Isola, P., Zhu, J.Y., Zhou, T., Efros, A.A.: Image-to-image translation with conditional adversarial networks. In: Proceedings of the IEEE Conference on Computer Vision and Pattern Recognition, pp. 1125–1134 (2017)
13. Karras, T., Laine, S., Aila, T.: A style-based generator architecture for generative adversarial networks. In: Proceedings of the IEEE/CVF Conference on Computer Vision and Pattern Recognition, pp. 4401–4410 (2019)
14. Kingma, D.P., Ba, J.: Adam: A method for stochastic optimization. arXiv preprint arXiv:1412.6980 (2014)
15. Lim, J.H., Ye, J.C.: Geometric GAN. arXiv preprint arXiv:1705.02894 (2017)
16. Liu, B., Zhu, Y., Song, K., Elgammal, A.: Towards faster and stabilized GAN training for high-fidelity few-shot image synthesis. In: International Conference on Learning Representations (2020)
17. Mescheder, L., Geiger, A., Nowozin, S.: Which training methods for GANs do actually converge? In: International Conference on Machine Learning, pp. 3481–3490. PMLR (2018)
18. Mirza, M., Osindero, S.: Conditional generative adversarial nets. arXiv preprint arXiv:1411.1784 (2014)
19. Miyato, T., Kataoka, T., Koyama, M., Yoshida, Y.: Spectral normalization for generative adversarial networks. In: International Conference on Learning Representations (2018)
20. Miyato, T., Koyama, M.: cGANs with projection discriminator. In: International Conference on Learning Representations (2018)
21. Nayar, R., Wilbur, D.C.: The Bethesda System for Reporting Cervical Cytology: Definitions, Criteria, and Explanatory Notes. Springer, Cham (2015). https://doi.org/10.1007/978-3-319-11074-5
22. Paszke, A., et al.: Pytorch: an imperative style, high-performance deep learning library. In: Advances in Neural Information Processing Systems, vol. 32 (2019)
23. Patel, M.M., Pandya, A.N., Modi, J.: Cervical pap smear study and its utility in cancer screening, to specify the strategy for cervical cancer control. Natl. J. Community Med. **2**(01), 49–51 (2011)
24. Robbins, H., Monro, S.: A stochastic approximation method. Ann. Math. Stat., 400–407 (1951)
25. Rombach, R., Blattmann, A., Lorenz, D., Esser, P., Ommer, B.: High-resolution image synthesis with latent diffusion models. In: Proceedings of the IEEE/CVF Conference on Computer Vision and Pattern Recognition, pp. 10684–10695 (2022)
26. Shorten, C., Khoshgoftaar, T.M.: A survey on image data augmentation for deep learning. J. Big Data **6**(1), 1–48 (2019)
27. Xiang, Y., Sun, W., Pan, C., Yan, M., Yin, Z., Liang, Y.: A novel automation-assisted cervical cancer reading method based on convolutional neural network. Biocybernetics Biomed. Eng. **40**(2), 611–623 (2020)
28. Xue, Y., et al.: Selective synthetic augmentation with histoGAN for improved histopathology image classification. Med. Image Anal. **67**, 101816 (2021)
29. Yazici, Y., Foo, C.S., Winkler, S., Yap, K.H., Piliouras, G., Chandrasekhar, V., et al.: The unusual effectiveness of averaging in GAN training. In: ICLR (Poster) (2019)
30. Zhao, S., Liu, Z., Lin, J., Zhu, J.Y., Han, S.: Differentiable augmentation for data-efficient GAN training. Adv. Neural. Inf. Process. Syst. **33**, 7559–7570 (2020)
31. Zhou, M., et al.: Hierarchical pathology screening for cervical abnormality. Comput. Med. Imaging Graph. **89**, 101892 (2021)

Democratizing Pathological Image Segmentation with Lay Annotators via Molecular-Empowered Learning

Ruining Deng[1], Yanwei Li[1], Peize Li[1], Jiacheng Wang[1], Lucas W. Remedios[1], Saydolimkhon Agzamkhodjaev[1], Zuhayr Asad[1], Quan Liu[1], Can Cui[1], Yaohong Wang[3], Yihan Wang[3], Yucheng Tang[2], Haichun Yang[3], and Yuankai Huo[1(✉)]

[1] Vanderbilt University, Nashville, TN 37215, USA
yuankai.huo@vanderbilt.edu
[2] NVIDIA Corporation, Santa Clara and Bethesda, USA
[3] Vanderbilt University Medical Center, Nashville, TN 37232, USA,

Abstract. Multi-class cell segmentation in high-resolution Giga-pixel whole slide images (WSI) is critical for various clinical applications. Training such an AI model typically requires labor-intensive pixel-wise manual annotation from experienced domain experts (e.g., pathologists). Moreover, such annotation is error-prone when differentiating fine-grained cell types (e.g., podocyte and mesangial cells) via the naked human eye. In this study, we assess the feasibility of democratizing pathological AI deployment by only using lay annotators (annotators without medical domain knowledge). The contribution of this paper is threefold: (1) We proposed a molecular-empowered learning scheme for multi-class cell segmentation using partial labels from lay annotators; (2) The proposed method integrated Giga-pixel level molecular-morphology cross-modality registration, molecular-informed annotation, and molecular-oriented segmentation model, so as to achieve significantly superior performance via 3 lay annotators as compared with 2 experienced pathologists; (3) A deep corrective learning (learning with imperfect labels) method is proposed to further improve the segmentation performance using partially annotated noisy data. From the experimental results, our learning method achieved F1 = 0.8496 using molecular-informed annotations from lay annotators, which is better than conventional morphology-based annotations (F1 = 0.7015) from experienced pathologists. Our method democratizes the development of a pathological segmentation deep model to the lay annotator level, which consequently scales up the learning process similar to a non-medical computer vision task. The official implementation and cell annotations are publicly available at https://github.com/hrlblab/MolecularEL.

Keywords: Image annotation · Registration · Noisy label learning · Pathology

1 Introduction

Multi-class cell segmentation is an essential technique for analyzing tissue samples in digital pathology. Accurate cell quantification assists pathologists in identifying and diagnosing diseases [5,29] as well as obtaining detailed information about the progression of the disease [23], its severity [28], and the effectiveness of treatment [15]. For

H. Greenspan et al. (Eds.): MICCAI 2023, LNCS 14225, pp. 497–507, 2023.
https://doi.org/10.1007/978-3-031-43987-2_48

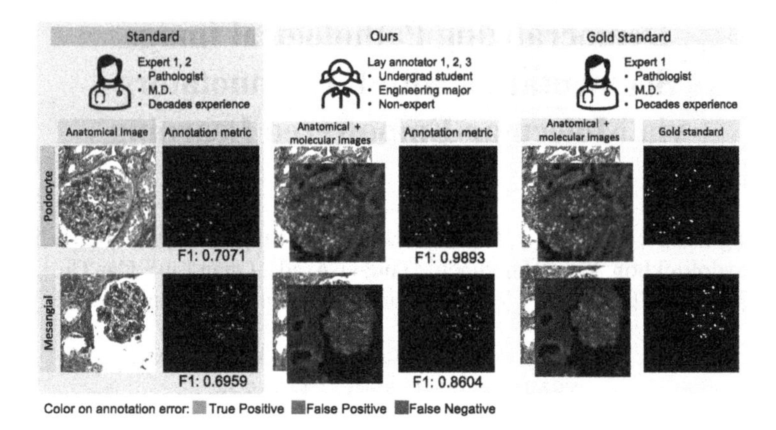

Fig. 1. The overall idea of this work. The left panel shows the standard annotation process (PAS only) for developing pathological segmentation models. The middle panel shows our molecular-informed annotation (with both PAS and IF images) that allows for better annotation quality from lay annotators as compared with the left panel. The right panel presents the gold standard annotation for this study, where the annotations are obtained by experienced pathologists upon both PAS and IF images.

example, the distribution and density of podocyte and mesangial cells in the glomerulus offer a faint signal of functional injury in renal pathology [14]. The cell-level characterization is challenging for experienced pathologists due to the decades of expensive medical training, long annotation time, large variability [30], and low accuracy, while it is impractical to hire massive experienced pathologists for cell annotation.

Previous works proposed several computer vision tools to perform automated or semi-automated cell segmentation on pathological images [17], including AnnotatorJ [12], NuClick [16], QuPath [2], etc. Such software is able to mark nuclei, cells, and multi-cellular structures by compiling pre-trained segmentation models [11], color deconvolution [25], or statistical analysis [22]. However, those automatic approaches still heavily rely on the morphology of cells from pathological Periodic acid-Schiff (PAS) images, thus demanding intensive human intervention for extra supervision and correction. Recently, immunofluorescence (IF) staining imaging has been widely used to visualize multiple biomolecules simultaneously in a single sample using fluorescently labeled antibodies [6,20]. Such technology can accurately serve as a guide to studying the heterogeneity of cellular populations, providing reliable information for cell annotation. Furthermore, crowd-sourcing technologies [1,13,19] were introduced generate better annotation for AI learning from multiple annotations.

In this paper, we proposed a holistic molecular-empowered learning scheme that democratizes AI pathological image segmentation by employing only lay annotators (Fig. 1). The learning pipeline consists of (1) morphology-molecular multi-modality image registration, (2) molecular-informed layman annotation, and (3) molecular-oriented corrective learning. The pipeline alleviates the difficulties at the R&D from the expert level (e.g., experienced pathologists) while relegating annotation to the lay annotator level (e.g., non-expert undergraduate students), all while enhancing both the accuracy and efficiency of the cell-level annotations. An efficient semi-supervised learning strategy is proposed to offset the impact of noisy label learning on lay annotations. The contribution of this paper is three-fold:

- We propose a molecular-empowered learning scheme for multi-class cell segmentation using partial labels from lay annotators;
- The molecular-empowered learning scheme integrates (1) Giga-pixel level molecular-morphology cross-modality registration, (2) molecular-informed annotation, and (3) molecular-oriented segmentation model to achieve statistically a significantly superior performance via lay annotators as compared with experienced pathologists;
- A deep corrective learning method is proposed to further maximize the cell segmentation accuracy using partially annotated noisy annotation from lay annotators.

2 Methods

The overall pipeline of the entire labeling and auto-quantification pipeline is presented in Fig. 2. Molecular images are aligned with anatomical images in order to provide accurate guidance for cell labeling by using multi-scale registration. After this registration, a functional unit segmentation model is implemented to localize the regions of glomeruli. Within those glomeruli, lay annotators label multiple cell types by using the pair-wise molecular images and anatomical images in ImageJ [12]. A partial-label learning model with a molecular-oriented corrective learning strategy is employed so as to diminish the gap between labels from lay annotators and gold standard labels.

2.1 Morphology-Molecular Multi-modality Registration

Multi-modality, multi-scale registration is deployed to ensure the pixel-to-pixel correspondence (alignment) between molecular IF and PAS images at both the WSI and regional levels. To maintain the morphological characteristics of the functional unit structure, a slide-wise multi-modality registration pipeline (Map3D) [8] is employed to register the molecular images to anatomical images. The first stage is global alignment. The Map3D approach was employed to achieve reliable translation on WSIs when encountering missing tissues and staining variations. The output of this stage is a pair-wise affine matrix $M_{Map3D}(t)$ from Eq. (1).

$$M_{Map3D} = \arg\min \sum_{i=1}^{N} ||A(x_i^{IF}, M) - x_i^{PAS}||_{Aff_{Map3D}} \quad (1)$$

To achieve a more precise pixel-level correspondence, Autograd Image Registration Laboratory (AIRLab) [27] was utilized to calibrate the registration performance at the second stage. The output of this step is $M_{AIRLab}(t)$ from Eq. (2).

$$M_{AIRLab} = \arg\min A_{M_{Map3D}} \sum_{i=1}^{N} ||A(x_i^{IF}, M) - x_i^{PAS}||_{Aff_{AIRLab}} \quad (2)$$

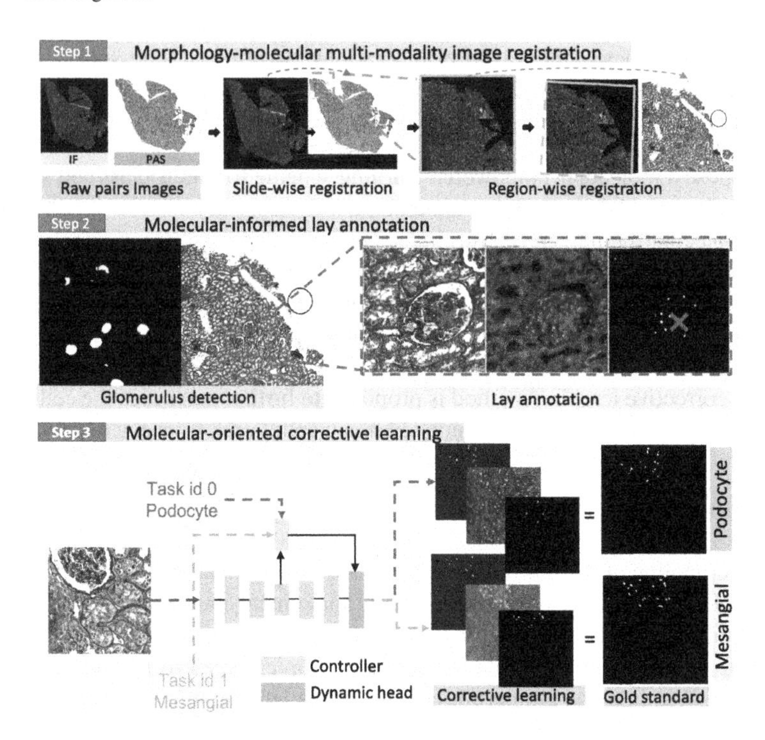

Fig. 2. The framework of the proposed molecular-empowered learning scheme. The molecular-empowered learning pipeline consists of (1) morphology-molecular multi-modality image registration, (2) molecular-informed layman annotation, and (3) molecular-oriented corrective learning. It democratizes AI pathological image segmentation by employing only lay annotators.

where i is the index of pixel x_i in the image I, with N pixels. The two-stage registration (Map3D + AIRLab) affine matrix for each pair is presented in Eq. (3).

$$M = (M_{Map3D}, M_{AIRLab}) \tag{3}$$

In Eq. (1) and (2), A indicates the affine registration. The affine matrix $M_{Map3D}(t)$ from Map3D is applied to obtain pair-wise image regions. The $||.||_{Aff_{Map3D}}$ and $||.||_{Aff_{AIRLab}}$ in Eq. (1) and (2) indicates the different similarity metrics for two affine registrations, respectively.

2.2 Molecular-Informed Annotation

After aligning molecular images with PAS images, an automatic multi-class functional units segmentation pipeline Omni-Seg [7] is deployed to locate the tuft unit on the images. With the tuft masks, the molecular images then manifest heterogeneous cells with different color signals on pathological images during the molecular-informed annotation. Each anatomical image attains a binary mask for each cell type, in the form

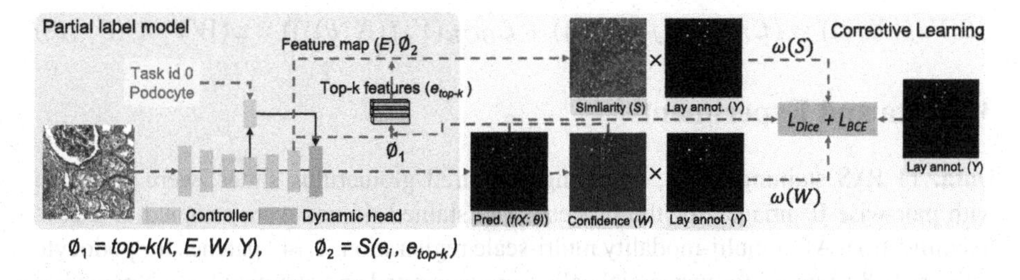

Fig. 3. The molecuar-oriented corrective learning in partial label model. A corrective learning are applied to highlight the regions where both the model and lay annotation agree on the current cell type, when calculating the loss function.

of a partial label. Following the same process, the pathologist examines both anatomical images and molecular images to generate a gold standard for this study (Fig. 1).

2.3 Molecular-Oriented Corrective Learning for Partial Label Segmentation

The lack of molecular expertise as well as the variability in the quality of staining in molecular images can cause annotations provided by non-specialists to be unreliable and error-prone. Therefore, we propose a corrective learning strategy (in Fig. 3) to efficiently train the model with noise labels, so as to achieve the comparable performance of training the same model using the gold standard annotations.

Inspired by confidence learning [21] and similarity attention [18], top-k pixel feature embeddings at the annotation regions with higher confidences from the prediction probability (W, defined as confidence score in Eq. (4)) are selected as critical representations for the current cell type from the decoder(in Eq. (5)).

$$W = f(X;\theta)[:,1] \tag{4}$$

$$top-k(k,E,W,Y) = (e_1,w_1),(e_2,w_2),...,(e_k,w_k) \cap Y \in (E,W) \tag{5}$$

where k denotes the number of selected embedding features. E is the embedding map from the last layer of the decoder, while Y is the lay annotation.

We then implement a cosine similarity score S between the embedding from an arbitrary pixel to those from critical embedding features as Eq. (6).

$$S(e_i, e_{top-k}) = \frac{\sum_{m=1}^{M}(e_i \times e_{top-k})}{\sqrt{\sum_{m=1}^{M}(e_i)^2} \times \sqrt{\sum_{m=1}^{M}(e_{top-k})^2}} \tag{6}$$

where m denotes the channel of the feature embeddings.

Since the labels from lay annotators might be noisy and erroneous, the W and S are applied in following Eq. (7) to highlight the regions where both the model and lay annotation agree on the current cell type, when calculating the loss function in Eq. (8).

$$\omega(W) = \exp(W) \times Y, \; \omega(S) = S \times Y \tag{7}$$

$$\mathcal{L}(Y, f(X;\theta)) = (\mathcal{L}_{Dice}(Y, f(X;\theta))) + \mathcal{L}_{BCE}(Y, f(X;\theta)))) \times \omega(W) \times \omega(S) \quad (8)$$

3 Data and Experiments

Data. 11 PAS staining WSIs, including 3 injured glomerulus slides, were collected with pair-wise IF images for the process. The stained tissues were scanned at a 20× magnification. After multi-modality multi-scale registration, 1,147 patches for podocyte cells, and 789 patches for mesangial cells were generated and annotated. Each patch has 512×512 pixels.

Morphology-Molecular Multi-modality Registration. The slide-level global translation from Map3D was deployed at a 5× magnification, which is 2 µm per pixel. The 4096×4096 pixels PAS image regions with 1024 pixels overlapping were tiled on anatomical WSIs at a 20× magnification, which is 0.5 µm per pixel.

Molecular-Empowered Annotation. The automatic tuft segmentation and molecular knowledge images assisted the lay annotators with identifying glomeruli and cells. ImageJ (version v1.53t) was used throughout the entire annotation process. "Synchronize Windows" was used to display cursors across the modalities with spatial correlations for annotation. "ROI Manager" was used to store all of the cell binary masks for each cell type.

Molecular-Oriented Corrective Learning. Patches were randomly split into training, validation, and testing sets - with a ratio of 6:1:3, respectively - at the WSI level. The distribution of injured glomeruli and normal glomeruli were balanced in the split.

Experimental Setting. 2 experienced pathologists and 3 lay annotators without any specialized knowledge were included in the experiment. All anatomical and molecular patches of glomerular structures are extracted from WSI on a workstation equipped with a 12-core Intel Xeon W-2265 Processor, and NVIDIA RTXA6000 GPU. An 8-core AMD Ryzen 7 5800X Processor workstation with XP-PEN Artist 15.6 Pro Wacom is used for drawing the contour of each cell. Annotating 1 cell type on 1 WSI requires 9 h, while staining and scanning 24 IF WSIs (as a batch) requires 3 h. The experimental setup for the 2 experts and the 3 lay annotators is kept strictly the same to ensure a fair comparison.

Evaluation Metrics. 100 patches from the testing set with a balanced number of injuries and normal glomeruli were captured by the pathologists for evaluating morphology-based annotation and molecuar-informed annotation. The annotation from one pathologist (over 20 years' experience) with both anatomical and molecular images as gold standard (Fig. 1). The balanced F-score (F1) was used as the major metric for this study. The Fleiss' kappa was used to compute the inter-rater variability between experts and lay annotators.

4 Results

Figure 4, Fig. 5 and Table 1 indicate the annotation performance from the naked human eye with expert knowledge and the lay annotator with molecular-informed learning. As

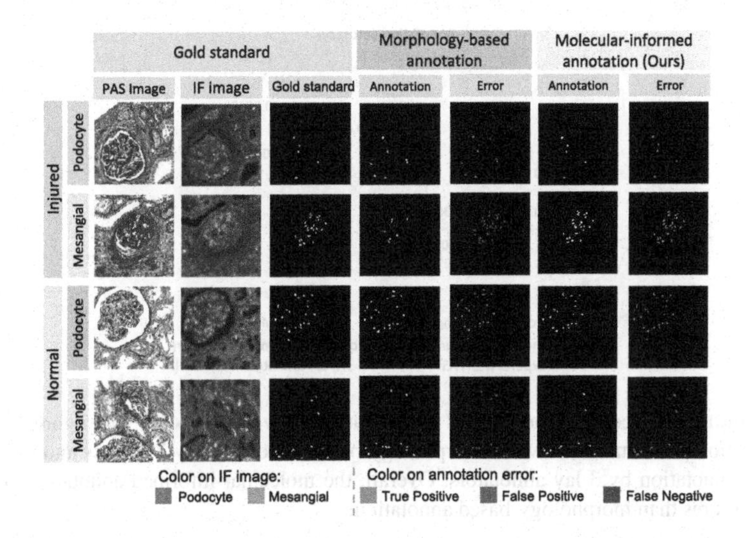

Fig. 4. Annotation accuracy using learning different strategies. This figure compares the annotation performance using different strategies. Note that the molecular-informed annotation only employed lay annotators, while the remaining results were from an experienced renal pathologist.

Table 1. Annotation accuracy from only anatomical morphology and molecular-informed annotation. Average F1 scores and Fleiss' kappa between 2 experts and 3 lay annotators are reported.

Method	Injured glomeruli		Normal glomeruli		Average		Fleiss' kappa	
	Podocyte	Mesangial	Podocyte	Mesangial	Podocyte	Mesangial	Podocyte	Mesangial
Morphology-based annotation (2 pathologists with PAS)	0.6964	0.6941	0.7067	0.6208	0.7015	0.6567	0.3973	0.4161
Molecular-informed annotation (3 lay annotators with PAS+IF)	**0.8374**	**0.8434**	**0.8619**	**0.8511**	**0.8496**	**0.8473**	**0.6406**	**0.5978**
p-value	p<0.001	p<0.001	p<0.001	p<0.001	p<0.001	p<0.001	N/A	N/A

shown, our learning method achieved better annotation with higher F1 scores with fewer false positive and false negative regions as compared with the pathologist's annotations. Statistically, the Fleiss' kappa test shows that the molecular-informed annotation by lay-annotators has higher annotation agreements than the morphology-based annotation by experts. This demonstrates the benefits of reducing the expertise requirement to a layman's level and improving accuracy in pathological cell annotation.

4.1 Performance on Multi-class Cell Segmentation

In Table 2, we compared the proposed partial label segmentation method to baseline models, including (1) multiple individual models (U-Nets [24], DeepLabv3s [4], and Residual U-Nets [26]), (2) multi-head models (Multi-class [10], Multi-Kidney [3]),

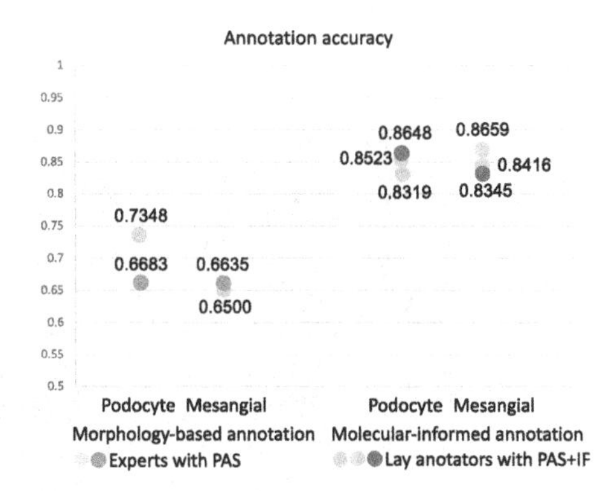

Fig. 5. Annotation accuracy between 2 experts and 3 lay annotators. This figure compares the annotation performance between morphology-based annotation by 2 experts and molecular-informed annotation by 3 lay annotators. Overall, the molecular-informed annotation achieved better F1 scores than morphology-based annotation.

and (3) single dynamic networks with noisy label learning (Omni-Seg [7]). Our results found that the partial label paradigm shows superior performance on multi-class cell segmentation. The proposed model particularly demonstrates better quantification in the normal glomeruli, which contain large amounts of cells.

To evaluate the performance of molecular-oriented corrective learning on imperfect lay annotation, we also implemented two noisy label learning strategies Confidence Learning (CL) [21] and Partial Label Loss (PLL) [18] with the proposed Molecular-oriented corrective learning (MOCL) on our proposed partial label model. As a result, the proposed molecular-oriented corrective learning alleviated the error between lay annotation and the gold standard in the learning stage, especially in the injured glomeruli that incorporate more blunders in the annotation due to the identification difficulty from morphology changing.

4.2 Ablation Study

The purpose of corrective learning is to alleviate the noise and distillate the correct information, so as to improve the model performance using lay annotation. Four designs of corrective learning with different utilization of similarity losses and confidence losses were evaluated with lay annotation in Table 3. Each score is used in either an exponential function or a linear function (Eq. (7)), when multiplying and calculating the loss function (Eq. (8)). The bold configuration was selected as the final design.

Table 2. Performance of deep learning based multi-class cell segmentation. F1 are reported.

Method	Data	Injured glomeruli		Normal glomeruli		Average	
		Podocyte	Mesangial	Podocyte	Mesangial	Podocyte	Mesangial
U-Nets [24]	G.S	0.6719	0.6867	0.7203	0.6229	0.6944	0.6617
DeepLabV3s [4]	G.S	**0.7127**	0.6680	0.7395	0.6163	0.7251	0.6476
Residual U-Nets [26]	G.S	0.6968	0.6913	0.7481	0.6601	0.7207	0.6790
Multi-class [10]	G.S	0.5201	0.4984	0.4992	0.4993	0.5214	0.4987
Multi-kidney [3]	G.S.	0.6735	0.6734	0.7542	0.6581	0.7108	0.6691
Omni-Seg [7]	G.S	0.7115	**0.6970**	**0.7746**	**0.6895**	**0.7407**	**0.6940**
Omni-Seg [7]	L.A	0.6941	0.7083	**0.7703**	0.6822	0.7295	0.6980
CL [21]	L.A	0.7047	0.6961	0.7536	0.6754	0.7274	0.6879
PLL [9]	L.A	0.6276	0.6853	0.6825	0.6268	0.6531	0.6622
MOCL(Ours)	L.A	**0.7198**	**0.7157**	0.7657	**0.6830**	**0.7411**	**0.7028**

*G.S. denotes gold standard dataset, *L.A. denotes lay annotation dataset

Table 3. Ablation study on different molecular-oriented corrective learning design.

Confidence score	Similarity score	Podocyte F1	Masengial F1	Average F1
Linear	Linear	0.7255	0.6843	0.7049
Linear	Exponent	0.7300	0.6987	0.7144
Exponent	Linear	**0.7411**	**0.7028**	**0.7219**
Exponent	Exponent	0.7304	0.6911	0.7108

5 Conclusion

In this work, we proposed a holistic, molecular-empowered learning solution to alleviate the difficulties of developing a multi-class cell segmentation deep learning model from the expert level to the lay annotator level, enhancing the accuracy and efficiency of cell-level annotation. An efficient corrective learning strategy is proposed to offset the impact of noisy label learning from lay annotation. The results demonstrate the feasibility of democratizing the deployment of a pathology AI model while only relying on lay annotators.

Acknowledgements. This work is supported in part by NIH R01DK135597(Huo), DoD HT9425-23-1-0003(HCY), and NIH NIDDK DK56942(ABF).

References

1. Amgad, M., et al.: NuCLS: a scalable crowdsourcing approach and dataset for nucleus classification and segmentation in breast cancer. GigaScience **11** (2022)
2. Bankhead, P., et al.: Qupath: open source software for digital pathology image analysis. Sci. Rep. **7**(1), 1–7 (2017)
3. Bouteldja, N., et al.: Deep learning-based segmentation and quantification in experimental kidney histopathology. J. Am. Soc. Nephrol. **32**(1), 52–68 (2021)

4. Chen, L.C., Papandreou, G., Schroff, F., Adam, H.: Rethinking atrous convolution for semantic image segmentation. arXiv preprint arXiv:1706.05587 (2017)
5. Comaniciu, D., Meer, P.: Cell image segmentation for diagnostic pathology. Advanced algorithmic approaches to medical image segmentation: State-of-the-art applications in cardiology, neurology, mammography and pathology, 541–558 (2002)
6. Day, K.E., Beck, L.N., Deep, N.L., Kovar, J., Zinn, K.R., Rosenthal, E.L.: Fluorescently labeled therapeutic antibodies for detection of microscopic melanoma. Laryngoscope **123**(11), 2681–2689 (2013)
7. Deng, R., Liu, Q., Cui, C., Asad, Z., Huo, Y., et al.: Single dynamic network for multi-label renal pathology image segmentation. In: International Conference on Medical Imaging with Deep Learning, pp. 304–314. PMLR (2022)
8. Deng, R., et al.: Map3d: registration-based multi-object tracking on 3d serial whole slide images. IEEE Trans. Med. Imaging **40**(7), 1924–1933 (2021)
9. Fan, J., Zhang, Z., Tan, T.: Pointly-supervised panoptic segmentation. In: Avidan, S., Brostow, G., Cisse, M., Farinella, G.M., Hassner, T. (eds.) ECCV 2022. LNCS, pp. 319–336. Springer, Cham (2022). https://doi.org/10.1007/978-3-031-20056-4_19
10. González, G., Washko, G.R., San José Estépar, R.: Multi-structure segmentation from partially labeled datasets. application to body composition measurements on CT scans. In: Stoyanov, D., et al. (eds.) RAMBO/BIA/TIA -2018. LNCS, vol. 11040, pp. 215–224. Springer, Cham (2018). https://doi.org/10.1007/978-3-030-00946-5_22
11. Graham, S., et al.: Hover-net: simultaneous segmentation and classification of nuclei in multi-tissue histology images. Med. Image Anal. **58**, 101563 (2019)
12. Hollandi, R., Diósdi, Á., Hollandi, G., Moshkov, N., Horváth, P.: AnnotatorJ: an imageJ plugin to ease hand annotation of cellular compartments. Mol. Biol. Cell **31**(20), 2179–2186 (2020)
13. Hsueh, P.Y., Melville, P., Sindhwani, V.: Data quality from crowdsourcing: a study of annotation selection criteria. In: Proceedings of the NAACL HLT 2009 Workshop on Active Learning for Natural Language Processing, pp. 27–35 (2009)
14. Imig, J.D., Zhao, X., Elmarakby, A.A., Pavlov, T.: Interactions between podocytes, mesangial cells, and glomerular endothelial cells in glomerular diseases. Front. Physiol. **13**, 488 (2022)
15. Jiménez-Heffernan, J., et al.: Mast cell quantification in normal peritoneum and during peritoneal dialysis treatment. Arch. Pathol. Lab. Med. **130**(8), 1188–1192 (2006)
16. Koohbanani, N.A., Jahanifar, M., Tajadin, N.Z., Rajpoot, N.: Nuclick: a deep learning framework for interactive segmentation of microscopic images. Med. Image Anal. **65**, 101771 (2020)
17. Korzynska, A., Roszkowiak, L., Zak, J., Siemion, K.: A review of current systems for annotation of cell and tissue images in digital pathology. Biocybernetics Biomed. Eng. **41**(4), 1436–1453 (2021)
18. Li, B., Li, Y., Eliceiri, K.W.: Dual-stream multiple instance learning network for whole slide image classification with self-supervised contrastive learning. In: Proceedings of the IEEE/CVF Conference on Computer Vision and Pattern Recognition, pp. 14318–14328 (2021)
19. Marzahl, C., et al.: Is crowd-algorithm collaboration an advanced alternative to crowdsourcing on cytology slides? In: Bildverarbeitung für die Medizin 2020. I, pp. 26–31. Springer, Wiesbaden (2020). https://doi.org/10.1007/978-3-658-29267-6_5
20. Moore, L.S., et al.: Effects of an unlabeled loading dose on tumor-specific uptake of a fluorescently labeled antibody for optical surgical navigation. Mol. Imaging Biol. **19**, 610–616 (2017)
21. Northcutt, C., Jiang, L., Chuang, I.: Confident learning: estimating uncertainty in dataset labels. J. Artif. Intell. Res. **70**, 1373–1411 (2021)

22. Oberg, A.L., Mahoney, D.W.: Statistical methods for quantitative mass spectrometry proteomic experiments with labeling. BMC Bioinf. **13**(16), 1–18 (2012)
23. Olindo, S.: Htlv-1 proviral load in peripheral blood mononuclear cells quantified in 100 ham/tsp patients: a marker of disease progression. J. Neurol. Sci. **237**(1–2), 53–59 (2005)
24. Ronneberger, O., Fischer, P., Brox, T.: U-Net: convolutional networks for biomedical image segmentation. In: Navab, N., Hornegger, J., Wells, W.M., Frangi, A.F. (eds.) MICCAI 2015. LNCS, vol. 9351, pp. 234–241. Springer, Cham (2015). https://doi.org/10.1007/978-3-319-24574-4_28
25. Ruifrok, A.C., Johnston, D.A., et al.: Quantification of histochemical staining by color deconvolution. Anal. Quant. Cytol. Histol. **23**(4), 291–299 (2001)
26. Salvi, M., et al.: Automated assessment of glomerulosclerosis and tubular atrophy using deep learning. Comput. Med. Imaging Graph. **90**, 101930 (2021)
27. Sandkühler, R., Jud, C., Andermatt, S., Cattin, P.C.: Airlab: autograd image registration laboratory. arXiv preprint arXiv:1806.09907 (2018)
28. Wijeratne, D.T., et al.: Quantification of dengue virus specific t cell responses and correlation with viral load and clinical disease severity in acute dengue infection. PLoS Neglected Trop. Dis. **12**(10), e0006540 (2018)
29. Xing, F., Yang, L.: Robust nucleus/cell detection and segmentation in digital pathology and microscopy images: a comprehensive review. IEEE Rev. Biomed. Eng. **9**, 234–263 (2016)
30. Zheng, Y., et al.: Deep-learning-driven quantification of interstitial fibrosis in digitized kidney biopsies. Am. J. Pathol. **191**(8), 1442–1453 (2021)

Gene-Induced Multimodal Pre-training for Image-Omic Classification

Ting Jin[1], Xingran Xie[1], Renjie Wan[2], Qingli Li[1], and Yan Wang[1](✉)

[1] Shanghai Key Laboratory of Multidimensional Information Processing, East China Normal University, Shanghai 200241, China
{51255904073,xrxie}@stu.ecnu.edu.cn, qlli@cs.ecnu.edu.cn, ywang@cee.ecnu.edu.cn

[2] Hong Kong Baptist University, Kowloon, Hong Kong

Abstract. Histology analysis of the tumor micro-environment integrated with genomic assays is the gold standard for most cancers in modern medicine. This paper proposes a Gene-induced Multimodal Pre-training (GiMP) framework, which jointly incorporates genomics and Whole Slide Images (WSIs) for classification tasks. Our work aims at dealing with the main challenges of multi-modality image-omic classification *w.r.t.* (1) the patient-level feature extraction difficulties from gigapixel WSIs and tens of thousands of genes, and (2) effective fusion considering high-order relevance modeling. Concretely, we first propose a group multi-head self-attention gene encoder to capture global structured features in gene expression cohorts. We design a masked patch modeling paradigm (MPM) to capture the latent pathological characteristics of different tissues. The mask strategy is randomly masking a fixed-length contiguous subsequence of patch embeddings of a WSI. Finally, we combine the classification tokens of paired modalities and propose a triplet learning module to learn high-order relevance and discriminative patient-level information. After pre-training, a simple fine-tuning can be adopted to obtain the classification results. Experimental results on the TCGA dataset show the superiority of our network architectures and our pre-training framework, achieving 99.47% in accuracy for image-omic classification. The code is publicly available at https://github.com/huangwudiduan/GIMP.

Keywords: Multimodal learning · Whole slide image classification

1 Introduction

Pathological image-omic analysis is the cornerstone of modern medicine and demonstrates promise in a variety of different tasks such as cancer diagnosis and prognosis [12]. With the recent advance of digital pathology and sequencing

---

Supplementary Information The online version contains supplementary material available at https://doi.org/10.1007/978-3-031-43987-2_49.

H. Greenspan et al. (Eds.): MICCAI 2023, LNCS 14225, pp. 508–517, 2023.
https://doi.org/10.1007/978-3-031-43987-2_49

technologies, modern cancer screening has jointly incorporated genomics and histology analysis of whole slide images (WSIs).

Though deep learning techniques have revolutionized medical imaging, designing a task-specific algorithm for image-omic multi-modality analysis is challenging. (1) The gigapixel WSIs, which generally yield 15,000 foreground patches during pre-processing, make attention-based backbones [6] hard to extract precise image (WSI)-level representations. (2) Learning features from genomics data which have tens of thousands of genes make models such as Transformer [16] impractical to use due to its quadratic computation complexity. (3) Image-omic feature fusion [2,3] may fail to model high-order relevance and the inherent structural characteristics of each modality, making the fusion less effective.

Specifically, to our knowledge, most multi-modality techniques have been designed for modalities such as chest X-ray and reports [1,17,23], CT and X-ray [18], CT and MRI [21], H&E cross-staining [22] via global feature, local feature or multi-granularity alignment. But, none of these works considers the challenges in WSIs and genes processing. Besides, vision-language models in the computer vision community stand out for their remarkable versatility [13,14]. Nevertheless, constrained by computing resources, the most commonly used multimodal representation learning strategy, contrastive learning, which relies on a large number of negative samples to avoid model collapse [8], is not affordable for gigapixel WSIs analysis. A big domain gap also hampers their usage in leveraging the structural characteristic of tumor micro-environment and genomic assay. Recently, the literature corpus has proposed some methods for accomplishing specific image-omic tasks via Kronecker Product fusion [2] or co-attention mapping between WSIs and genomics data [3]. But, the Kronecker product overly concerns feature interactions between modalities while ignoring high-order relevance, *w.r.t.* decision boundaries across multiple samples, which is critical to classification tasks. As for the co-attention module, it is unidirectional and cannot localize significant regions from genetic data with a large amount of information.

In this paper, we propose a task-specific framework dubbed Gene-induced Multimodal Pre-training (GiMP) for image-omic classification. Concretely, we first propose a transformer-based gene encoder, Group Multi-head Self Attention (GroupMSA), to capture global structured features in gene expression cohorts. Next, we design a pre-training paradigm for WSIs, Masked Patch Modeling (MPM), masking random patch embeddings from a fixed-length contiguous subsequence of a WSI. We assume that one patch-level feature embedding can be reconstructed by its adjacent patches, and this process enhances the learning ability for pathological characteristics of different tissues. Our MPM only needs to recover the masked patch embeddings in a fixed-length subsequence rather than processing all patches from WSIs. Furthermore, to model the high-order relevance of the two modalities, we combine CLS tokens of paired image and genomic data to form unified representations and propose a triplet learning module to differentiate patient-level positive and negative samples in a mini-batch. It is worth mentioning that although our unified representation fuses features from the whole gene expression cohort and partial WSIs in a mini-batch, we

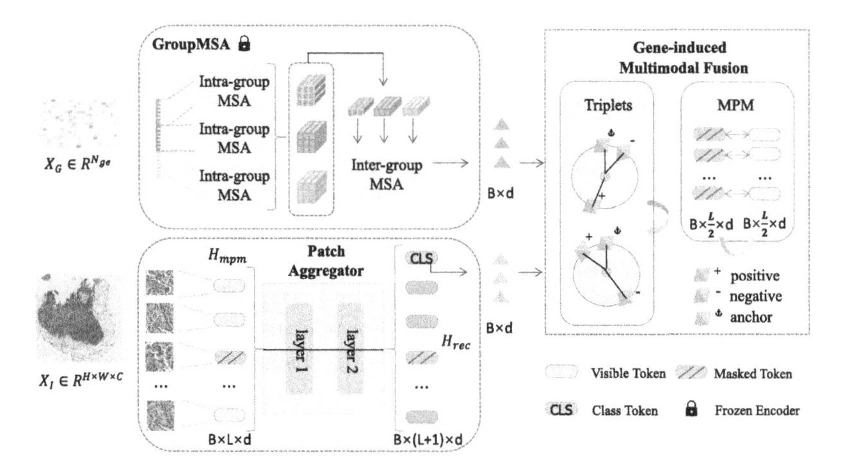

Fig. 1. Illustration of GiMP pre-training. Given a batch of image-omic pairs, we randomly select a fixed-length patch cohort and mask parts of the patch embeddings. Then we use two modality-specific encoders to capture unimodal features. Two pre-training objectives are considered: 1) building triplets by concatenated CLS tokens of each modality and enhancing the discriminability according to category relations, and 2) reconstructing the missing patch embeddings by its adjacent patches.

can still learn high-order relevance and discriminative patient-level information between these two modalities in pre-training thanks to the triplet learning module. In addition, note that our proposed method is different from self-supervised pre-training. Specifically, we focus not only on superior representation learning capability, but also category-related feature distributions, *w.r.t.* intra- and inter-class variation. With the training process going on, complete information from WSIs can be integrated and the fused multimodal representations with high discrimination will make it easier for the classifier to find the classification hyperplane. Experimental results demonstrate that our GiMP achieves significant improvement in accuracy than other image-omic competitors, and our multimodal framework shows competitive performance even without pre-training.

2 Method

Given a multimodal dataset $\mathcal{D}$ consisting of pairs of WSI pathological images and genomic data $(\mathbf{X}_I, \mathbf{X}_G)$, our GiMP learns feature representations via accomplishing masked patch modeling and triplets learning. As shown in Fig. 1, the overall framework consists of three parts: 1) group-based genetic encoder GroupMSA (Sect. 2.1), 2) efficient patch aggregator (Sect. 2.2) and 3) gene-induced multimodal fusion (Sect. 2.3). In the subsequent sections, we will introduce each part of our proposed framework in detail.

2.1 Group Multi-head Self Attention

In this section, we propose Group Multi-head Self Attention (GroupMSA), a specialized gene encoder to capture structured features in genomic data cohorts. Specifically, inspired by tokenisation techniques in natural language processing [16], the input expression cohort $\mathbf{X}_G \in \mathbb{R}^{N_{ge}}$ is partitioned into N_f non-overlapping fragments, and we then use a linear projection head to acquire fragment features $\mathbf{H}_f \in \mathbb{R}^{N_f \times d}$, where d is the hidden dimension. Next, we introduce an intra-and-inter attention module to capture local and global information in $\mathbf{H}_f$. Firstly, the fragment features are divided into groups and there are N_{gr} learnable group tokens linked to each group resulting in $(N_f/N_{gr}+1)$ tokens per group. Then the prepared tokens are fed to a vanilla multi-head self-attention (MSA) block to extract intra-group information. After that, we model cross-group interactions by another MSA layer on the global scale with the locally learned group tokens and a final classification token $\mathbf{CLS}_{ge} \in \mathbb{R}^d$. Finally, GroupMSA could learn dense semantics from the genomic data cohort.

2.2 Patch Aggregator with Efficient Attention Operation

Let's denote the whole slide pathological image with $H \times W$ spatial resolution and C channels by $\mathbf{X}_I \in \mathbb{R}^{H \times W \times C}$. We follow the preprocessing strategy of CLAM [11] to acquire patch-level embedding sequence, *i.e.*, each foreground patch with 256×256 pixels is fed into an ImageNet-pretrained ResNet50 and the background region is discarded. Let $\mathbf{H}_p = \{h_j \mid h_j \in \mathbb{R}^{1024}\}_{j=1}^{N_p}$ denote the sequence of patch embeddings corresponding to WSI $\mathbf{X}_I$ and note that the total patch number N_p is image-specific. Since the quadratic computational complexity of the standard self-attention mechanism is usually unaffordable in WSI analysis due to its long instances sequence, we employ Nystrom-based attention algorithm [20] to aggregate patch embeddings and yield image-level predictions. Specifically, the input sequence $\mathbf{H}_p$ is first embedded into a d-dimensional feature space and combined with a classification token $\mathbf{CLS}_{img}$, yielding $\mathbf{H}_p^0 \in \mathbb{R}^{(N_p+1)\times d}$. Then we perform different projection operations on $\mathbf{H}_p^0$:

$$\mathbf{Q}^l = \mathbf{H}_p^l \cdot W_Q^l, \mathbf{K}^l = \mathbf{H}_p^l \cdot W_K^l, \mathbf{V}^l = \mathbf{H}_p^l \cdot W_V^l, \tag{1}$$

$$\mathbf{H}_p^{l+1} = \text{softmax}(\frac{\mathbf{Q}^l \cdot \widetilde{\mathbf{K}}^{lT}}{\sqrt{d}}) \cdot \left(\text{softmax}(\frac{\widetilde{\mathbf{Q}}^l \cdot \widetilde{\mathbf{K}}^{lT}}{\sqrt{d}})\right)^{-1} \cdot \text{softmax}(\frac{\widetilde{\mathbf{Q}}^l \cdot \mathbf{K}^{lT}}{\sqrt{d}}) \cdot \mathbf{V}^l, \tag{2}$$

where $W_Q^l, W_K^l, W_V^l \in \mathbb{R}^{d \times d}$ are linear mapping matrices, $\widetilde{\mathbf{Q}}^l, \widetilde{\mathbf{K}}^l \in \mathbb{R}^{m \times d}$ ($m \ll N_p$) are downsampling matrices obtained from clustering tokens in $\mathbf{Q}^l$ and $\mathbf{K}^l$ for layer $l \in \{0, 1\}$.

2.3 Gene-Induced Multimodal Fusion

In this section, we first describe the formulation of masked patch modeling. Then we introduce the overall pipeline of our pre-training framework and illustrate how to apply it to downstream classification tasks.

Masked Patch Modeling. In WSIs, the foreground patches are spatially contiguous, which means the adjacent patches have similar feature embeddings. Thus, we propose a Masked Patch Modeling (MPM) pre-training strategy that masks random patch embeddings from a fixed-length contiguous subsequence $\mathbf{H}_{mpm} = \{h_j \mid h_j \in \mathbb{R}^{1024}\}_{j=i}^{L+i}$ in $\mathbf{H}_p$ and reconstruct the invisible information. The fixed subsequence length L is empirically set to 6,000 and the sequences shorter than L are duplicated to build mini batches. Besides, the masking ratio is set to 50% and the set of masked subscripts is denoted as $\mathcal{M} \in \mathbb{R}^{0.5L}$. Next, a two-layer Nystrom-based patch aggregator followed by a lightweight reconstruction decoder are adopted to process the masked sequence $\mathbf{H}_{mpm}$ and the reconstructed sequence is denoted as $\mathbf{H}_{rec} = \left\{\hat{h}_j \mid \hat{h}_j \in \mathbb{R}^{1024}\right\}_{j=1}^{L}$. Note that we reconstruct the missing feature embeddings rather than the raw pixels of the masked areas, which is different from traditional MIM methods like SimMIM [19] and MAE [5]. In this way, the model could consider latent pathological characteristics of different tissues, which makes the pretext task more challenging. The reconstruction L_1 loss is computed by:

$$\mathcal{L}_{rec} = \sum_{j=1}^{L} \mathbf{1}[j \in \mathcal{M}] \left\| h_j - \hat{h}_j \right\|_1, \tag{3}$$

where $\mathbf{1}[\cdot]$ is the indicator function.

Gene-Induced Triplet Learning. The transformer-based backbones in the classification task require the CLS token to be able to extract accurate global information, which is even more important yet difficult in WSIs due to the long sequence challenge. In addition, in order to construct the mini-batch, the subsequences we intercept in the MPM pre-training phase may not be sufficiently representative of the image-level characteristics. To overcome these issues, we further propose a gene-induced triplet learning module, which uses pathological images and genomic data as input and extracts high-order and discriminative features via CLS tokens. Firstly, we pre-train the GroupMSA module by patient-level annotations in advance and froze it in the following iterations. Next, a learnable CLS token $\mathbf{CLS}_{img}$ for WSIs is added to the input masked sequence $\mathbf{H}_{mpm}$. After extracting the input patch embeddings and gene sequence separately, we concatenate $\mathbf{CLS}_{img}$ and $\mathbf{CLS}_{ge}$ as $\mathbf{CLS}_{pat} \in \mathbb{R}^{2d}$ to represent patient-level characteristics.

Suppose we obtain a triplet list $\{x, x^+, x^-\}$ during current iteration, where x, x^+, x^- are concatenated tokens of anchor $\mathbf{CLS}_{pat}$, positive $\mathbf{CLS}_{pat}$, and negative $\mathbf{CLS}_{pat}$, respectively. To enhance the global modeling capability, *i.e.*, extracting more precise patient-level features, we expect that the distance between the anchor and the positive sample gets closer, while the negative sample is farther away. The loss function for optimizing triplet learning is computed by:

$$\mathcal{L}_{tri} = \max(\left\|x - x^+\right\|_2^2 + \delta - \left\|x - x^-\right\|_2^2, 0), \tag{4}$$

δ indicates a threshold, *e.g.*, $\delta = 0.8$. Finally, the loss function for GiMP pre-training is: $\mathcal{L}_{pre} = \mathcal{L}_{tri} + \mathcal{L}_{rec}$.

Multimodal Fine-Tuning. Applying the pre-trained backbone to image-omic classification task is straightforward, since GiMP pre-training allows it to learn representative patient-level features. We use a simple Multi-Layer Perceptron (MLP) head to map $\mathbf{CLS}_{pat}$ to the final class predictions $\hat{P}$, which can be written as $\hat{P} = \text{softmax}(\text{MLP}(\mathbf{CLS}_{pat}))$.

3 Experiments

3.1 Experimental Setup

Datasets. We verify the effectiveness of our method on The Caner Genome Atlas (TCGA) non-small cell lung cancer (NSCLC) dataset, which contains two cancer subtypes, *i.e.*, Lung Squamous Cell Carcinoma (LUSC) and Lung Adenocarcinoma (LUAD). After pre-processing [11], the patch number extracted from WSIs at 20× magnification varies from 485 to 148,569. We collect corresponding RNA-seq FPKM data for each patient and the length of the input genomic sequence is 60,480. Among 946 image-omic pairs, 470 of them belong to LUAD and 476 cases are LUSC. We randomly split the data into 567 for training, 189 for validation and 190 for testing.

Implementation Details. The pre-training process of all algorithms is conducted on the training set, without any extra data augmentation. Note that our genetic encoder, GroupMSA, is fully supervised pre-trained on unimodal genetic data to accelerate convergence and it is frozen during GiMP training process. The maximum pre-training epoch for all methods is set to 100 and we fine-tune the models at the last epoch. During fine-tuning, we evaluate the model on the validation set after every epoch, and save the parameters when it performs the best. AdamW [10] is used as our optimizer and the learning rate is 10^{-4} with cosine decline strategy. The maximum number of fine-tune epoch is 70. At last, we measure the performance on the test set. Training configurations are consistent throughout the fine-tuning process to ensure fair comparisons. All experiments are conducted on a single NVIDIA GeForce RTX 3090.

3.2 Comparison Between GiMP and Other Methods

We conduct comparisons between GiMP and three competitors under different settings. Firstly, we compare our proposed patch aggregator with the current state-of-the-art deep MIL models on *unimodal* TCGA-NSCLC dataset, *i.e.*, only pathological WSIs are included as input. As shown in Table 1, our proposed patch aggregator outperforms all the compared attention based multiple instance learning baselines in classification accuracy. In particular, 1.6% higher than the

Table 1. Accuracy comparison on the TCGA Lung Cancer dataset. The best results are marked in **bold**.

Modality	Pre-train	Method	Acc.
Pathology	w/o pre-train	ABMIL [6]	0.7737
		DSMIL [9]	0.7566
		CLAM-SB [11]	0.8519
		CLAM-MB [11]	0.8889
		TransMIL [15]	0.8836
		GiMP (w/o GroupMSA)	**0.8995**
Pathology & Genomic	w/o pre-train	PORPOISE [4]	0.9524
		Pathomic Fusion [2]	0.9684
		MCAT [3]	0.9632
		GiMP (ours)	**0.9737**
	w/ pre-train	MGCA [17]	0.9105
		BioViL [1]	0.9316
		REFERS [23]	0.9368
		GiMP (ours)	**0.9947**

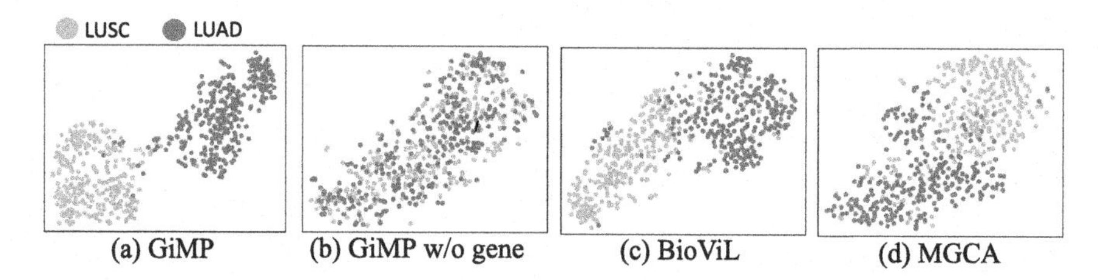

Fig. 2. t-SNE visualization of different methods with $\mathbf{CLS}_{pat}$ after pre-training. (a) image-omic GiMP pre-trained, (b) GiMP pre-trained without gene inducing, (c) BioViL [1] pre-trained, (d) MGCA [17] pre-trained.

second best compared method TransMIL [15]. We then explore the superiority of GiMP by comparing to state-of-the-art medical multi-modal approaches. We particularly compare our method to BioViL [1], MGCA [17] and REFERS [23], three popular multimodal pre-training algorithms in medical text-image classification task. We can observe in the table that, our GiMP raises ACC from 91.05% to 99.47% on TCGA-NSCLC dataset. Even without pre-training stage, GiMP shows competitive performance compared to PORPOISE [4], Pathomic Fusion [2], and MCAT [3], three influential image-omic classification architectures.

We further explore why GiMP works by insightful interpretation of the proposed method with t-SNE visualisation. Figure 2 shows the feature mixtureness of pre-trained $\mathbf{CLS}_{pat}$ extracting global information on training set. Compari-

Table 2. Ablation study on TCGA Lung Cancer dataset. "SNN" means replacing GroupMSA with SNN [7]. "Triplet" denotes our gene-induced triplet learning module.

Aggregator	GroupMSA	Triplet	MPM	Acc.
✓	SNN [7]			0.9684
✓	✓			0.9737
✓			✓	0.9579
✓		✓		0.9263
✓	✓	✓		0.9526
✓	✓	✓	✓	**0.9974**

son between Fig. 2 (a) and (b) indicates that the addition of the genomic data is indispensable in increasing the inter-class distance and reducing the intra-class distance, which confirms our motivation that gene-induced multimodal fusion could model high-order relevance and yield more discriminative representations. Moreover, compared to the mentioned self-supervised methods BioViL [1] and MGCA [17] in Fig. 2 (c) and (d), $\mathbf{CLS}_{pat}$ with GiMP pre-trained are well separated between LUAD and LUSC, *i.e.*, GiMP pays more attention to the category-related feature distribution and could extract more discriminative patient-level features during triplet learning.

3.3 Ablation Study

Table 2 summarizes the results of ablation study. We first evaluate the effectiveness of the proposed GroupMSA. In the first two rows, GroupMSA achieves 0.53% improvement compared to SNN [7], a popular genetic encoders used in PORPOISE [4] and Pathomic Fusion [2]. We then analyze the effect of adding genetic modality during pre-training. The evaluation protocol is first pre-training, and then fine-tuning on downstream multimodal classification task. "Aggregator + MPM" means GiMP only uses WSIs as input and reconstructs the missing patch embeddings during the pre-training phase. Since the fixed subsequence length $L = 6000$ is used in our setting, it is sometimes smaller than the original patch number, *e.g.*, the maximum size 148,569, the pre-trained model without genetic guidance may be not aware of sufficiently accurate patient-level characteristics, *i.e.*, ineffectively focused on normal tissues. "Aggregator + Triplet" indicates using unimodal image features to build triplets. We can likewise find that the lack of precise global representation leads to worse performance. Finally, we evaluate the necessity of the MPM module. "Aggregator + GroupMSA + Triplet" means GiMP only combines the CLS tokens of each modality and calculates triplet loss during pre-training. We can observe a performance drop without MPM module, *e.g.*, from 99.47% to 95.26%, which demonstrates that local pathological information is equally critical as high-order relevance.

4 Conclusion

In this paper, we propose a novel multimodal pre-training method to exploit the complementary relationship of genomic data and pathological images. Concretely, we introduce a genetic encoder with structured learning capabilities and an effective gene-induced multimodal fusion module which combines two pre-training objectives, triplet learning and masked patch modeling. Experimental results demonstrate the superior performance of the proposed GiMP compared to other state-of-the-art methods. The contribution of each proposed component of GiMP is also demonstrated in the experiments.

Acknowledgements. This work was supported by the National Natural Science Foundation of China (Grant No. 62101191), Shanghai Natural Science Foundation (Grant No. 21ZR1420800), and the Science and Technology Commission of Shanghai Municipality (Grant No. 22DZ2229004).

References

1. Boecking, B., et al.: Making the most of text semantics to improve biomedical vision-language processing. In: Avidan, S., Brostow, G., Cisse, M., Farinella, G.M., Hassner, T. (eds.) ECCV 2022. LNCS, vol. 13696, pp. 1–21. Springer, Cham (2022). https://doi.org/10.1007/978-3-031-20059-5_1
2. Chen, R.J., et al.: Pathomic fusion: an integrated framework for fusing histopathology and genomic features for cancer diagnosis and prognosis. IEEE Trans. Med. Imaging **41**(4), 757–770 (2020)
3. Chen, R.J., et al.: Multimodal co-attention transformer for survival prediction in gigapixel whole slide images. In: ICCV (2021)
4. Chen, R.J., et al.: Pan-cancer integrative histology-genomic analysis via multimodal deep learning. Cancer Cell **40**(8), 865–878 (2022)
5. He, K., Chen, X., Xie, S., Li, Y., Dollár, P., Girshick, R.: Masked autoencoders are scalable vision learners. In: CVPR (2022)
6. Ilse, M., Tomczak, J., Welling, M.: Attention-based deep multiple instance learning. In: ICML (2018)
7. Klambauer, G., Unterthiner, T., Mayr, A., Hochreiter, S.: Self-normalizing neural networks. In: Proceedings of the NeurIPS (2017)
8. Kong, L., de Masson d'Autume, C., Yu, L., Ling, W., Dai, Z., Yogatama, D.: A mutual information maximization perspective of language representation learning. In: ICLR (2020)
9. Li, B., Li, Y., Eliceiri, K.W.: Dual-stream multiple instance learning network for whole slide image classification with self-supervised contrastive learning. In: CVPR (2021)
10. Loshchilov, I., Hutter, F.: Fixing weight decay regularization in Adam. CoRR abs/1711.05101 (2017)
11. Lu, M.Y., Williamson, D.F., Chen, T.Y., Chen, R.J., Barbieri, M., Mahmood, F.: Data-efficient and weakly supervised computational pathology on whole-slide images. Nat. Biomed. Eng. **5**(6), 555–570 (2021)
12. Moch, H., et al.: The 2022 world health organization classification of tumours of the urinary system and male genital organs-part a: renal, penile, and testicular tumours. Eur. Urol. (2022)

13. Radford, A., et al.: Learning transferable visual models from natural language supervision. In: ICML (2021)
14. Ramesh, A., Dhariwal, P., Nichol, A., Chu, C., Chen, M.: Hierarchical text-conditional image generation with CLIP latents. CoRR abs/2204.06125 (2022)
15. Shao, Z., Bian, H., Chen, Y., Wang, Y., Zhang, J., Ji, X., et al.: TransMIL: transformer based correlated multiple instance learning for whole slide image classification. In: NeurIPS (2021)
16. Vaswani, A., et al.: Attention is all you need. In: NeurIPS (2017)
17. Wang, F., Zhou, Y., Wang, S., Vardhanabhuti, V., Yu, L.: Multi-granularity cross-modal alignment for generalized medical visual representation learning. CoRR abs/2210.06044 (2022)
18. Xie, Y., Zhang, J., Xia, Y., Wu, Q.: UniMiSS: universal medical self-supervised learning via breaking dimensionality barrier. In: Avidan, S., Brostow, G., Cisse, M., Farinella, G.M., Hassner, T. (eds.) ECCV 2022. LNCS, vol. 13681, pp. 558–575. Springer, Cham (2022). https://doi.org/10.1007/978-3-031-19803-8_33
19. Xie, Z., et al.: SimMIM: a simple framework for masked image modeling. In: CVPR (2022)
20. Xiong, Y., et al.: Nyströmformer: a nyström-based algorithm for approximating self-attention. In: AAAI (2021)
21. Yang, J., Zhang, R., Wang, C., Li, Z., Wan, X., Zhang, L.: Toward unpaired multi-modal medical image segmentation via learning structured semantic consistency. CoRR abs/2206.10571 (2022)
22. Yang, P., et al.: CS-CO: a hybrid self-supervised visual representation learning method for H&E-stained histopathological images. Med. Image Anal. **81**, 102539 (2022)
23. Zhou, H., Chen, X., Zhang, Y., Luo, R., Wang, L., Yu, Y.: Generalized radiograph representation learning via cross-supervision between images and free-text radiology reports. Nat. Mach. Intell. **4**(1), 32–40 (2022)

Artifact Restoration in Histology Images with Diffusion Probabilistic Models

Zhenqi He[1], Junjun He[2], Jin Ye[2], and Yiqing Shen[3](✉)

[1] The University of Hong Kong, Pokfulam, Hong Kong
[2] Shanghai AI Laboratory, Shanghai, China
[3] Johns Hopkins University, Baltimore, USA
yshen92@jhu.edu

Abstract. Histological whole slide images (WSIs) can be usually compromised by artifacts, such as tissue folding and bubbles, which will increase the examination difficulty for both pathologists and Computer-Aided Diagnosis (CAD) systems. Existing approaches to restoring artifact images are confined to Generative Adversarial Networks (GANs), where the restoration process is formulated as an image-to-image transfer. Those methods are prone to suffer from mode collapse and unexpected mistransfer in the stain style, leading to unsatisfied and unrealistic restored images. Innovatively, we make the first attempt at a denoising diffusion probabilistic model for histological artifact restoration, namely `ArtiFusion`. Specifically, `ArtiFusion` formulates the artifact region restoration as a gradual denoising process, and its training relies solely on artifact-free images to simplify the training complexity. Furthermore, to capture local-global correlations in the regional artifact restoration, a novel Swin-Transformer denoising architecture is designed, along with a time token scheme. Our extensive evaluations demonstrate the effectiveness of `ArtiFusion` as a pre-processing method for histology analysis, which can successfully preserve the tissue structures and stain style in artifact-free regions during the restoration. Code is available at https://github.com/zhenqi-he/ArtiFusion.

Keywords: Histological Artifact Restoration · Diffusion Probabilistic Model · Swin-Transformer Denoising Network

1 Introduction

Histology is critical for accurately diagnosing all cancers in modern medical imaging analysis. However, the complex scanning procedure for histological whole-slide images (WSIs) digitization may result in the alteration of tissue structures, due to improper removal, fixation, tissue processing, embedding, and storage [11]. Typically, these changes in tissue details can be caused by various extraneous factors such as bubbles, tissue folds, uneven illumination, pen marks, altered staining, and *etc* [13]. Formally, the changes in tissue structures are known as artifacts. The presence of artifacts not only makes the analysis more challenging

H. Greenspan et al. (Eds.): MICCAI 2023, LNCS 14225, pp. 518–527, 2023.
https://doi.org/10.1007/978-3-031-43987-2_50

for pathologists but also increases the risk of misdiagnosis for Computer-Aided Diagnosis (CAD) systems [14]. Particularly, deep learning models, which have become increasingly prevalent in histology analysis, have shown vulnerability to the artifact, resulting in a two-times increase in diagnosis errors [18].

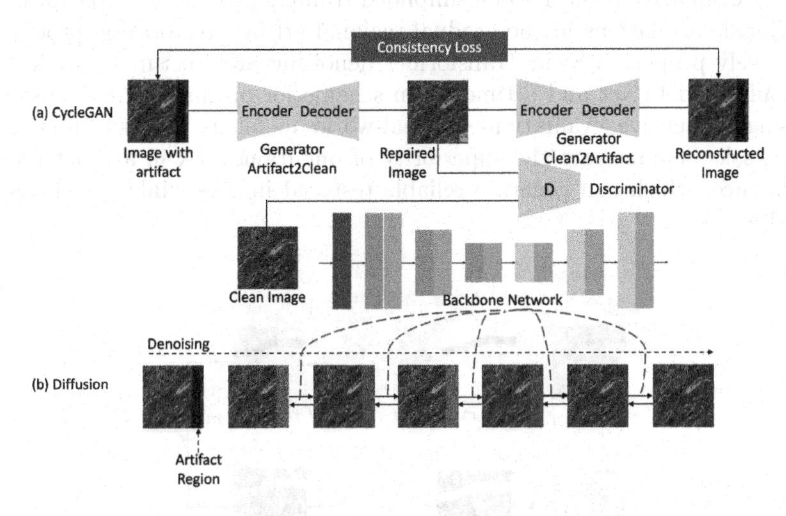

Fig. 1. Learning-based artifact restoration approaches. (a) CycleGAN [19] formulates the artifact restoration as an image-to-image transfer problem. It leverages two pairs of the generator and discriminator to learn the transfer between the artifact and artifact-free image domains. (b) Diffusion probabilistic model [5] (ours) formulates artifact restoration as a regional denoising process.

In real clinical practice, rescanning the WSIs that contain artifacts can partially address this issue. However, it may require multiple attempts before obtaining a satisfactory WSI, which can lead to a waste of time, medical resources, and deplete tissue samples. Discarding the local region with artifacts for deep learning models is another solution, but it may result in the loss of critical contextual information. Therefore, learning-based artifact restoration approaches have gained increasing attention. For example, CycleGAN [19] formulates the artifact restoration as an image-to-image transfer problem by learning the transfer between the artifact and artifact-free image domains from unpaired images, as depicted in Fig. 1(a). However, existing artifact restoration solutions are confined to Generative Adversarial Networks (GANs) [2], which are difficult to train due to the mode collapse and are prone to suffer from unexpected stain style mistransfer. To address these issues, we make the first attempt at a diffusion probabilistic model for artifact restoration approach [5], as shown in Fig. 1(b). Innovatively, our framework formulates the artifact restoration as a regional denoising process, which thus can to the most extent preserve the stain style and avoid the loss of contextual information in the non-artifact region. Furthermore, our approach is trained solely with artifact-free images, which reduces the difficulty in data collection.

The major contributions are two-fold. (1) We make the first attempt at a denoising diffusion probabilistic model for artifact removal, called `ArtiFusion`. This approach differs from GAN-based methods that require either paired or unpaired artifacts and artifact-free images, as our `ArtiFusion` relies solely on artifact-free images, resulting in a simplified training process. (2) To capture the local-global correlations in the gradual regional artifact restoration process, we innovatively propose a Swin-Transformer denoising architecture to replace the commonly-used U-Net and a time token scheme for optimal Swin-Transformer denoising. Extensive evaluations on real-world histology datasets and downstream tasks demonstrate the superiority of our framework in artifact removal performance, which can generate reliable restored images while preserving the stain style.

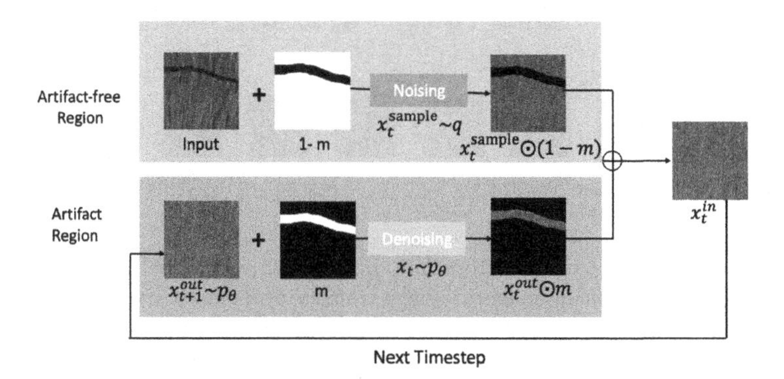

Fig. 2. The semantic illustration of inference stage in `ArtiFusion` for local regional artifact restoration.

2 Methodology

Overall Pipeline. The proposed histology artifact restoration diffusion model `ArtiFusion`, comprises two stages, namely the training, and inference. During the training stage, `ArtiFusion` learns to generate regional histology tissue structures based on the contextual information from artifact-free images. In the inference stage, `ArtiFusion` formulates the artifact restoration as a gradual denoising process. Specifically, it first replaces the artifact regions with Gaussian noise, and then gradually restores them to artifact-free images using the contextual information from nearby regions.

Diffusion Training Stage. The proposed `ArtiFusion` learns the capability of generating local tissue representation from contextual information during the training stage. To achieve this, we follow the formulations of DDPM [5], which involve a forward process that gradually injects Gaussian noise into an artifact-free image and a reverse process that aims to reconstruct images from noise. During the forward process, we can obtain a noisy version of $\mathbf{x}_t$ for arbitrary timestep $t \in \mathbb{N}[0, T]$ using a Gaussian transition kernel $q(\mathbf{x}t|\mathbf{x}t-1) = \mathcal{N}(x_t; \sqrt{1-\beta_t}\mathbf{x}t-1, \beta_t\mathbf{I})$, where $\beta_t \in (0, 1)$ are predefined hyper-parameters [5]. Simultaneously, the reverse process trains a denoising network $p_\theta(\mathbf{x}_{t-1}|\mathbf{x}_t^{in})$, which is parameterized by θ, to reverse the forward process $q(\mathbf{x}_t|\mathbf{x}_{t-1})$. The overall training objective L is defined as the variational lower bound of the negative log-likelihood, given by:

$$\mathbb{E}[-\log p_\theta(\mathbf{x}_0)] \leq \mathbb{E}_q[-\log p(\mathbf{x}_T) - \sum_{1 \leq t \leq T} \log \frac{p_\theta(\mathbf{x}_{t-1}|\mathbf{x}_t)}{q(\mathbf{x}_t|\mathbf{x}_{t-1})}] = L. \quad (1)$$

This formulation is extended in DDPM [5] to be further written as:

$$L = \mathbb{E}_q[\underbrace{D_{KL}(q(\mathbf{x}_T|x_0))||p(\mathbf{x}_T)}_{L_T} + \sum_{t>1} \underbrace{D_{KL}(q(\mathbf{x}_{t-1}|\mathbf{x}_t, \mathbf{x}_0))||p_\theta(\mathbf{x}_{t-1}|\mathbf{x}_t)}_{L_{t-1}} - \underbrace{\log p_\theta(\mathbf{x}_0|\mathbf{x}_1)}_{L_0}],$$

where $D_{KL}(\cdot||\cdot)$ is the KL divergence.

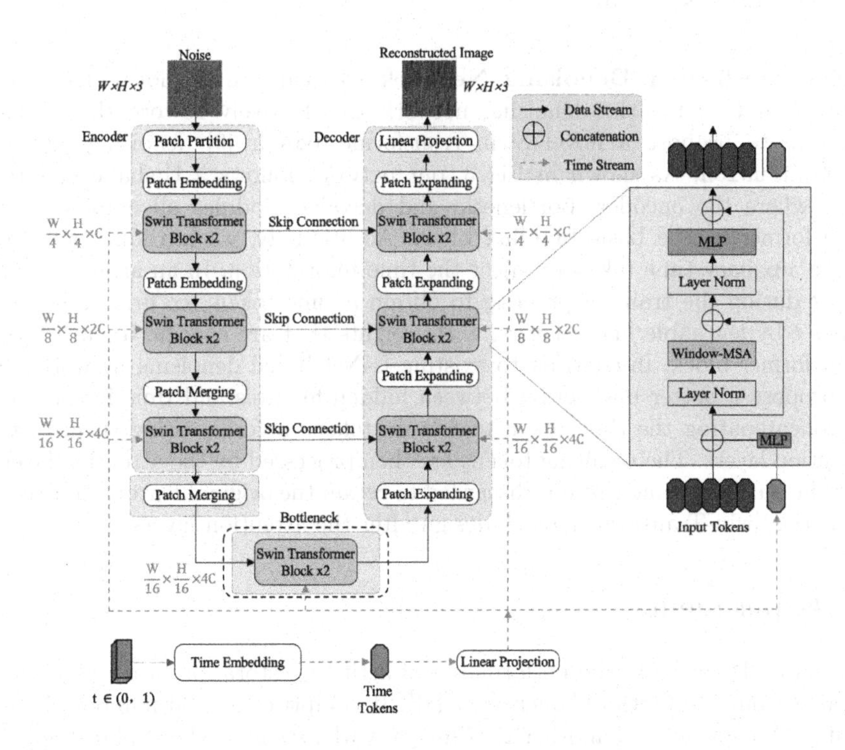

Fig. 3. The proposed Swin-Transformer denoising network.

Artifact Restoration in Inference Stage. During the inference stage, we first use a threshold method to detect the artifact region in the input image $\mathbf{x}_0$. Then, unlike the conventional diffusion models [5] that aim to generate the entire image, `ArtiFusion` selectively performs denoising resampling only in the artifact region to maximally preserve the original morphology and stain style in the artifact-free region, as shown in Fig. 2. Specifically, we represent the artifact-free region and the artifact region in the input image as $\mathbf{x}_0 \odot (1-\mathbf{m})$ and $\mathbf{x}_0 \odot \mathbf{m}$, respectively [10], where $\mathbf{m}$ is a Boolean mask indicating the artifact region and $\odot$ is the pixel-wise multiplication operator. To perform the denoising resampling, we write the input image $\mathbf{x}_t^{in}$ at each reverse step from t to $t-1$ as the sum of the diffused artifact-free region and the denoised artifact region, *i.e.*,

$$\mathbf{x}_t^{in} = \mathbf{x}_t^{sample} \odot (1 - \mathbf{m}) + \mathbf{x}_{t+1}^{out} \odot \mathbf{m}, \tag{2}$$

where $\mathbf{x}_t^{sample} o \odot (1-\mathbf{m})$ is artifact-free region diffused for t times using the Gaussian transition kernel *i.e.* $\mathbf{x}_t^{sample} \sim \mathcal{N}(\sqrt{\bar{\alpha}_t}\mathbf{x}_0, (1-\bar{\alpha}_t\mathbf{I}))$ with $\bar{\alpha}_t = \prod_{i=1}^{t}(1-\beta_i)$; and $\mathbf{x}_{t+1}^{out}$ is the output from the denoising network in the previous reverse step *i.e.*, $p_\theta(\mathbf{x}_{t+1}^{out}|\mathbf{x}_{t+1}^{in})$. Consequently, the final restored image is obtained as $\mathbf{x}_0 \odot (1 - \mathbf{m}) + \mathbf{x}_0^{out} \odot \mathbf{m}$.

Swin-Transformer Denoising Network. To capture the local-global correlation and enable the denoising network to effectively restore the artifact regions, we propose a novel Swin-Transformer-basedr [9] denoising network for `ArtiFusion`. As shown in Fig. 3, our network follows a U-shape architecture, where the encoder, bottleneck, and decoder modules all employ Swin-Transformer as the basic building block. Additionally, we introduce an innovative auxiliary time token to inject the time information. In an arbitrary time step t during the training process, to obtain a time token, we first embed the scalar t by learnable linear layers, with weights that are specific to each Swin-Transformer block. In contrast to existing U-Net based denoising networks [5], we propose a better interaction between hidden features and time information by concatenating the time token to feature tokens before passing them to the attention layers. The resulting tokens are then processed by the attention layers, and the auxiliary time token is discarded to retain the original feature dimension to fit the Swin-Transformer block design after the attention layers.

3 Experiments

Dataset. To evaluate the performance of artifact restoration, a training set is curated from a subset of Camelyon17 [8][1]. It comprises a total number of 2445 artifact-free images and another 2547 images with artifacts, where all histological images are scaled to the resolution of 256×256 pixels at the magnitude of $20\times$. The test set uses another public histology image dataset [6] with 462 artifact-free

[1] Available at https://camelyon17.grand-challenge.org.

images[2], where we obtain the paired artifact images by the manually-synthesized artifacts [18].

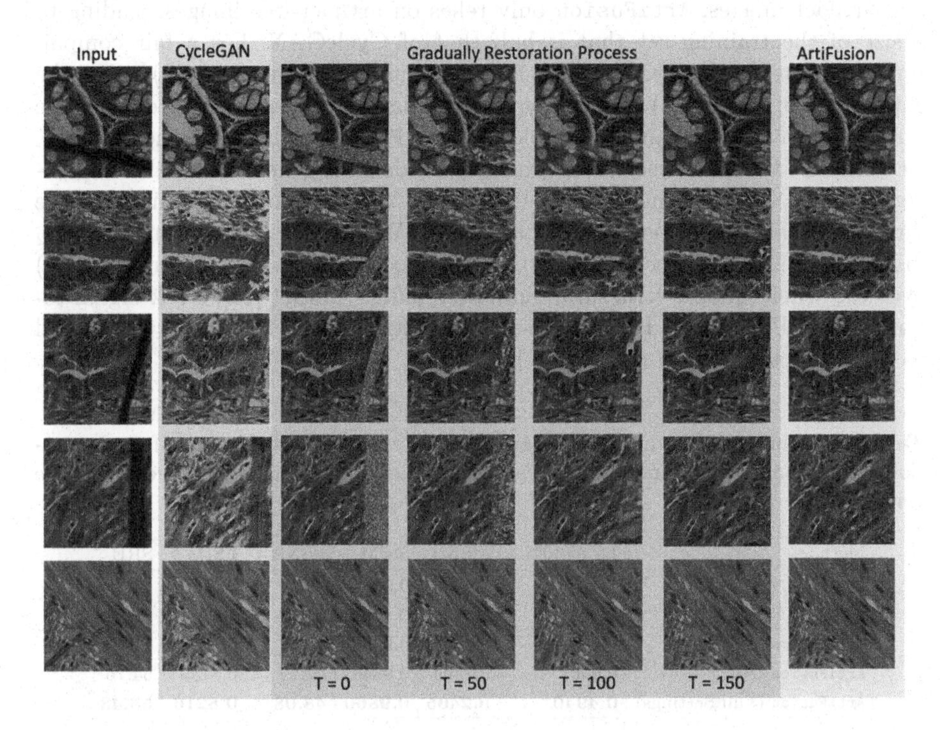

Fig. 4. Artifact restoration on five real-world artifact images. We observe that `ArtiFusion` can successfully overcome the drawback of stain style mistransfer in CycleGAN. We also illustrate the gradual denoising process in the artifact region by `ArtiFusion`, at time step $t = 0, 50, 100, 150$. It highlights the ability of `ArtiFusion` to progressively remove artifacts from the histology image, resulting in a final restored image that is both visually pleasing and scientifically accurate.

Implementations. We implement the proposed `ArtiFusion` and its counterpart in Python 3.8.10 and PyTorch 1.10.0. All experiments are carried out in parallel on two NVIDIA RTX A4000 GPU cards with 16 GiB memory. Hyper-parameters are as follows: a learning rate of 10^{-4} with Adam optimizer, the total timesteps is set to 250.

Compared Methods and Evaluation Metrics. As a proof-of-concept attempt at a generative-models-based artifact restoration framework in the histology domain, currently, there are limited available literature works and open-sourced codes for comparison. Consequently, we leverage the prevalent Cycle-GAN [19] as the baseline for comparison, because of its excellent performance

[2] Available at https://github.com/lu-yizhou/ClusterSeg.

in the image transfer, and also its nature that requires no paired data can fit our circumstance. Unlike CycleGAN which requires both artifact-free images and artifact images, `ArtiFusion` only relies on artifact-free images, leading to a size of the training set that is half that of CycleGAN. For a fair compaison, we train the CycleGAN with two configurations, namely (#1) using the entire dataset, and (#2) using only half the dataset, where the latter uses the same number of the training samples as `ArtiFusion`. Regarding the ablation, we compare the proposed Swin-Transformer denoising network with the conventional U-Net [5] (denoted as 'U-Net'), and the time token scheme with the direct summation scheme (denoted as 'Add'). We use the following metrics: L_2 distance (L2) with respect to the artifact region, the mean-squared error (MSE) over the whole image, structural similarity index (SSIM) [15], Peak signal-to-noise ratio (PSNR) [1], Feature-based similarity index (FSIM) [17] and Signal to reconstruction error ratio (SRE) [7].

Table 1. Quantitative comparison of `ArtiFusion` with CycleGAN on artifact restoration performance. The ↓ indicates the smaller value, the better performance; and vice versa.

Methods	L2 ($\times 10^4$) ↓	MSE ↓	SSIM ↑	PSNR ↑	FSIM ↑	SRE ↑
CycleGAN (#1) [19]	1.119	0.5583	0.9656	42.37	0.7188	51.42
CycleGAN (#2) [19]	1.893	0.5936	0.9622	42.12	0.7162	50.21
`ArtiFusion` (U-Net)	0.5027	0.2508	0.9850	47.61	0.8173	54.59
`ArtiFusion` (Add)	0.5007	0.2499	0.9850	47.79	0.8184	54.76
`ArtiFusion` (Full Settings)	**0.4940**	**0.2465**	**0.9860**	**48.08**	**0.8216**	**55.43**

Table 2. Comparison of the model complexity and efficiency in terms of the number of parameters, FLOPs, and averaged inference time.

Methods	#Params ($\times 10^6$)	FLOPs ($\times 10^9$)	Inference(s)
CycleGAN [19]	28.28	60.04	1.065
`ArtiFusion` (UNet)	108.41	247.01	112.37
`ArtiFusion` (Add)	27.74	7.69	30.14
`ArtiFusion` (Full Settings)	29.67	7.73	30.71

Evaluations on Artifact Restoration. The quantitative comparison with CycleGAN and `ArtiFusion` are shown in Table 1, where some exemplary images are illustrated in Fig. 4. Our results demonstrate the superiority of `ArtiFusion` over GAN in the context of artifact restoration, with a large margin observed in all evaluation metrics. For instance, `ArtiFusion` can reduce the L2 and MSE by more than 50%, namely from 1×10^4 to 0.5×10^4 and from 0.55 to 0.25 respectively. It implying that our method can to the large extent restore the artifact regions using the global information. In addition, `ArtiFusion` can improve other metrics, including SSIM, PSNR, FSIM and SRE by 0.0204, 5.72, 0.1028 and 4.02 respectively, indicating that it can preserve the stain style during the restoration process. Moreover, our ablation study shows that the Swin-Transformer denoising network can outperform the conventional U-Net, highlighting the significance of capturing global correlation for local artifact restoration. Finally, the concatenating time token with feature tokens can bring an improvement in terms of all evaluation matrices, making it a better fit for the transformer architecture than the direct summation scheme in U-Net [5]. In summary, our ablations confirm the effectiveness of all the components in our method.

Table 3. The effectiveness of the proposed artifact restoration framework in the downstream task-tissue classification task. We report the classification accuracy on the test set (%) with different network architectures including ResNet [4], RexNet [3] and EfficientNet [12].

Settings	ResNet18	ResNet34	ResNet50	RexNet100	EfficientNetB0
Clean	95.529	93.538	94.833	95.487	95.808
Artifacts	80.302	86.031	85.012	90.446	90.626
Restored w CycleGAN	86.326	88.273	87.994	90.776	91.811
Restored w `ArtiFusion`	92.376	91.252	90.408	92.310	94.232

Comparisons of Model Complexity. In Table 2, we compare the model complexity in terms of the number of parameters, Floating Point Operations Per second (FLOPs), and averaged inference time on one image. Our proposed model achieves a significant reduction in the number of parameters by 72.6%, namely from 108.41×10^6 to 29.67×10^6, compared with CycleGAN. This reduction in model size comes at the cost of longer inference time. However, a smaller model size can facilitate easier deployment in real clinical practice.

Evaluations by Downstream Classification Task. We further evaluate the proposed artifact restoration framework on a downstream tissue classification task. To this end, we use the public dataset NCT-CRC-HE-100K for training and CRC-VAL-HE-7K for testing, which together contains 100,000 training samples and 7,180 test samples. We consider the performance on the original unprocessed data, denoted as 'Clean', as the upper bound. Then, we manually synthesize the artifact (denoted as 'Artifact') and evaluate the classification performance with restoration approaches CycleGAN and `ArtiFusion`. In Table 3, comparisons show that the presence of artifacts can result in a significant performance decline of over 5% across all five network architectures. Importantly, the classification accuracy on images restored with `ArtiFusion` is consistently higher than those restored with CycleGAN, demonstrating the superiority of our model. These results highlight the effectiveness of `ArtiFusion` as a practical pre-processing method for histology analysis.

4 Conclusion

In this paper, we propose `ArtiFusion`, the first attempt at a diffusion-based artifact restoration framework for histology images. With a novel Swin-Transformer denoising backbone, `ArtiFusion` is able to restore regional artifacts using the context information, while preserving the tissue structures in artifact-free regions as well as the stain style. Experimental results on a public histological dataset demonstrate the superiority of our proposed method over the state-of-the-art GAN counterpart. Consequently, we believe that our proposed method has the potential to benefit the medical community by enabling more accurate diagnosis or treatment planning as a pre-processing method for histology analysis. Future work includes investigating the extension of `ArtiFusion` to more advanced diffusion models such as score-based or score-SDE models [16].

References

1. Osama, S., et al.: A comprehensive survey analysis for present solutions of medical image fusion and future directions. IEEE Access **9**, 11358–11371 (2021)
2. Ian, J., et al.: Generative adversarial networks (2014)
3. Han, D., Yun, S., Heo, B., Yoo, Y.: Rexnet: Diminishing representational bottleneck on convolutional neural network (2020)
4. He, K., Zhang, X., Ren, S., Sun, J.: Deep residual learning for image recognition. CoRR, abs/1512.03385 (2015)
5. Ho, J., Jain, A., Abbeel, P.: Denoising diffusion probabilistic models. CoRR, abs/2006.11239 (2020)
6. Ke, J., et al.: ClusterSeg: a crowd cluster pinpointed nucleus segmentation framework with cross-modality datasets. Med. Image Anal. **85**, 102758 (2023)
7. Lanaras, C., Bioucas-Dias, J., Galliani, S., Baltsavias, E., Schindler, K.: Super-resolution of sentinel-2 images: Learning a globally applicable deep neural network. ISPRS J. Photogrammetry Remote Sens. **146**, 305–319 (2018)
8. Litjens, G., et al.: 1399 H&E-stained sentinel lymph node sections of breast cancer patients: the CAMELYON dataset. GigaScience **7**(6), giy065 (2018)
9. Liu, Z., et al.: Swin transformer: hierarchical vision transformer using shifted windows. CoRR, abs/2103.14030 (2021)
10. Lugmayr, A., Danelljan, M., Romero, A., Yu, F., Timofte, R., Van Gool, L.: Repaint: inpainting using denoising diffusion probabilistic models. In: 2022 IEEE/CVF Conference on Computer Vision and Pattern Recognition (CVPR) (2022)
11. Seoane, J., Varela-Centelles, P.I., Ramírez, J.R., Cameselle-Teijeiro, J., Romero, M.A.: Artefacts in oral incisional biopsies in general dental practice: a pathology audit. Oral Dis. **10**(2), 113–117 (2004)
12. Tan, M., Le, Q.: EfficientNet: Rethinking model scaling for convolutional neural networks (2020)

13. Taqi, S.A., Sami, S.A., Sami, L.B., Zaki, S.A.: A review of artifacts in histopathology. J. Oral Maxillofacial Pathol.: JOMFP **22**(2), 279 (2018)
14. Wang, N.C., Kaplan, J., Lee, J., Hodgin, J., Udager, A., Rao, A.: Stress testing pathology models with generated artifacts. J. Pathol. Inf. **12**(1), 4 (2021)
15. Wang, Z., Bovik, A.C., Sheikh, H.R., Simoncelli, E.P.: Image quality assessment: from error visibility to structural similarity. IEEE Trans. Image Process. **13**(4), 600–612 (2004)
16. Yang, L., et al.: Diffusion models: A comprehensive survey of methods and applications. arXiv preprint arXiv:2209.00796 (2022)
17. Zhang, L., Zhang, L., Mou, X., Zhang, D.: FSIM: a feature similarity index for image quality assessment. IEEE Trans. Image Process. **20**(8), 2378–2386 (2011)
18. Zhang, Y., Sun, Y., Li, H., Zheng, S., Zhu, C., Yang, L.: Benchmarking the robustness of deep neural networks to common corruptions in digital pathology. In: Wang, L., Dou, Q., Fletcher, P.T., Speidel, S., Li, S. (eds.) MICCAI 2022. LNCS, vol. 13432, pp. 242–252. Springer, Cham (2022). https://doi.org/10.1007/978-3-031-16434-7_24
19. Zhu, J.Y., Park, T., Isola, P., Efros, A.A.:. Unpaired image-to-image translation using cycle-consistent adversarial networks. In: 2017 IEEE International Conference on Computer Vision (ICCV) (2017)

Forensic Histopathological Recognition via a Context-Aware MIL Network Powered by Self-supervised Contrastive Learning

Chen Shen[1], Jun Zhang[4], Xinggong Liang[1], Zeyi Hao[1], Kehan Li[3], Fan Wang[3(✉)], Zhenyuan Wang[1(✉)], and Chunfeng Lian[2(✉)]

[1] Key Laboratory of National Ministry of Health for Forensic Sciences, School of Medicine & Forensics, Health Science Center, Xi'an Jiaotong University, Xi'an, Shaanxi 710049, China
wzy218@xjtu.edu.cn

[2] School of Mathematics and Statistics, Xi'an Jiaotong University, Xi'an, Shaanxi 710149, China
chunfeng.lian@xjtu.edu.cn

[3] Key Laboratory of Biomedical Information Engineering of Ministry of Education, School of Life Science and Technology, Xi'an Jiaotong University, Xi'an, Shaanxi 710049, China
fan.wang@xjtu.edu.cn

[4] Tencent AI Lab, Shenzhen, China

Abstract. Forensic pathology is critical in analyzing death manner and time from the microscopic aspect to assist in the establishment of reliable factual bases for criminal investigation. In practice, even the manual differentiation between different postmortem organ tissues is challenging and relies on expertise, considering that changes like putrefaction and autolysis could significantly change typical histopathological appearance. Developing AI-based computational pathology techniques to assist forensic pathologists is practically meaningful, which requires reliable discriminative representation learning to capture tissues' fine-grained postmortem patterns. To this end, we propose a framework called FPath, in which a dedicated self-supervised contrastive learning strategy and a context-aware multiple-instance learning (MIL) block are designed to learn discriminative representations from postmortem histopathological images acquired at varying magnification scales. Our self-supervised learning step leverages multiple complementary contrastive losses and regularization terms to train a double-tier backbone for fine-grained and informative patch/instance embedding. Thereafter, the context-aware MIL adaptively distills from the local instances a holistic bag/image-level representation for the recognition task. On a large-scale database of 19,607 experimental rat postmortem images and 3,378 real-world human decedent images, our FPath led to state-of-the-art accuracy and promising cross-domain generalization in recognizing seven different postmortem tissues. The source code will be released on https://github.com/ladderlab-xjtu/forensic_pathology.

H. Greenspan et al. (Eds.): MICCAI 2023, LNCS 14225, pp. 528–538, 2023.
https://doi.org/10.1007/978-3-031-43987-2_51

Keywords: Forensic Pathology · Self-Supervised Learning · Multiple Instance Learning

1 Introduction

Computational pathology powered by artificial intelligence (AI) shows promising applications in various clinical studies [14,18], significantly easing the workload and promoting the development of clinical pathology. Inspired by such exciting progress, let's think step by step, so why not leverage advanced AI techniques to boost the research and applications in another important discipline, i.e., forensic pathology? Forensic pathology focuses on investigating the cause, manner, and time of (non-natural) deaths based on histopathological examinations of postmortem organ tissues [5]. As an indispensable part of the medicolegal autopsy, it provides critical evidence from the microscopic aspect to confirm, perfect, or refute macroscopic findings, establishing a reliable factual basis for future inferences [4]. Histopathological analysis in forensic pathology is challenging and time-consuming, since postmortem changes (e.g., putrefaction and autolysis) severely destroy tissues' typical image appearance, even making the manual differentiation between the tissues of different organs very difficult.

Although diverse deep-learning approaches have been proposed in clinical studies to process and analyze histopathological images [14,18], no similar work has yet in the forensic pathology community. The main reason could be threefold. **1)** Forensic and clinical pathology have distinct purposes. The former case analyzes the tissue images from multiple organs concurrently. In contrast, clinical diagnosis/prognosis usually focuses on one tissue type in one task [6]. **2)** Due to postmortem changes, histopathological images in forensic pathology have atypical appearances and more complex distributions than in clinical pathology, bringing additional challenges to deep representation learning [21,25]. **3)** Data in forensic pathology are more difficult to obtain and have relatively lower quality. Therefore, to deploy a reliable computational pathology system for forensic investigation, fine-grained discriminative representation learning from complex postmortem histopathological images is a very precondition.

In this paper, we introduce a deep computational pathology framework (*dubbed as* **FPath**) for forensic histopathological analysis. As shown in Fig. 1, FPath leverages the idea of self-supervised contrastive learning and multiple instance learning (MIL) to learn discriminative histopathological representations. Specifically, we propose a self-supervised contrastive learning strategy to learn a double-tier backbone network for fine-grained feature embedding of local image patches (i.e., instances in MIL). After that, a context-aware MIL block is designed, which adopts a self-attention mechanism to refine instance-level representations by aggregating contextual information, and then applies an adaptive-pooling operation to produce a holistic image-level representation for prediction. Our FPath performs efficient predictions without the need for tedious pre-processing (e.g., foreground extraction/segmentation). To the best of our knowledge, this paper is the first attempt that shows promising applications of advanced AI techniques (e.g., self-supervised contrastive learning) to forensic pathology.

The main technical contributions of our work are:

1) We design a double-tier backbone and a dedicated self-supervised learning strategy to capture discriminative instance-level histopathological patterns of postmortem organ tissues. The double-tier backbone combines CNN and transformer for local and non-local information fusion. To effectively train such a backbone to handle images acquired with varying microscopic magnifications, the dedicated self-supervised learning strategy leverages multiple complementary contrastive losses and regularization terms to concurrently maximize global and spatially fine-grained similarities between different views of the same instances/patches in an informative representation space.
2) We design a context-aware MIL branch to produce the bag-level discriminative representations for accurate and efficient postmortem histopathological recognition. Our MIL branch first refines instance embedding by leveraging a self-attention mechanism integrating positional embedding to model cross-patch associations for contextual information enhancement. Thereafter, an adaptive pooling operation is designed to learn deformable spatial attention to distill from contextually enhanced patch-level representations a holistic image-level representation for recognition.
3) Our FPath was applied to recognize postmortem organ tissues, a fundamental task in forensic pathology. To this end, we established a relatively large-scale multi-domain database consisting of an experimental rat postmortem dataset and a real-world human decedent dataset, each with 19,607 and 3,378 images acquired at a specific microscopic magnification (e.g., 5×, 10×, 20×, and 40×), respectively. On such a multi-domain database, our FPath led to promising cross-domain generalization and state-of-the-art accuracy in recognizing seven different postmortem organs.

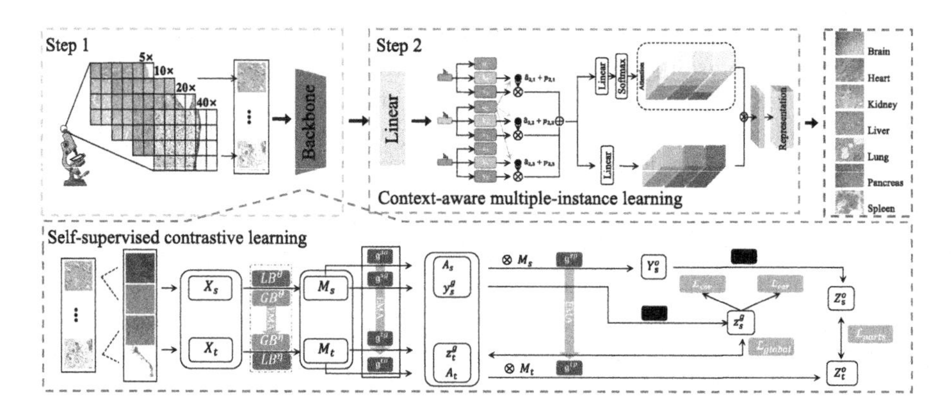

Fig. 1. Our FPath that consists of a self-supervised double-tier backbone (Step 1) and a context-aware MIL branch for postmortem recognition (Step 2).

2 Method

The schematic diagram of our FPath is shown in Fig. 1, which consists of two steps: **1)** Self-supervised contrastive learning of a double-tier backbone, and **2)** Context-aware multiple instance learning for postmortem tissue recognition.

2.1 Self-supervised Contrastive Patch Embedding

Double-Tier Backbone. Given patches from a postmortem histopathological image acquired at a specific magnification (i.e., 5×, 10×, 20×, or 40×), we adopt a backbone with a local branch (LB) and a global branch (GB) for instance/patch feature embedding. The LB is a ResNet50 [10] consisting of 16 successive bottlenecks, each with three convolutional layers with the kernel size of 1×1, 3×3, and 1×1, respectively. The GB is a Swin Transformer [17] that contains of a series of 12 window-based multi-head self-attention modules. Let an input patch be $\mathbf{X} \in \mathbb{R}^{H \times W \times 3}$. The corresponding feature embedding produced by the double-tier backbone will be $\mathbf{M} = \mathbf{M}_{\mathrm{LB}} \oplus \mathbf{M}_{\mathrm{GB}}$ ($\in \mathbb{R}^{h \times w \times C}$), where $\mathbf{M}_{\mathrm{LB}}$ and $\mathbf{M}_{\mathrm{GB}}$ denotes the representations from the LB and GB branch, respectively, and $\oplus$ stands for the channel-wise concatenation operation.

Self-supervised Contrastive Learning Strategy. We leverage the idea of self-supervised representation learning to establish the double-tier backbone. Referring to MoCo [9], our self-supervised learning is constructed by a teacher branch and a student branch. The student branch consists of six components, including a double-tier backbone (i.e., $\mathfrak{f}_\theta(\cdot)$), three projection layers (i.e., $\mathfrak{g}^{\mathrm{sg}}(\cdot)$, $\mathfrak{g}^{\mathrm{so}}(\cdot)$, $\mathfrak{g}^{\mathrm{sp}}(\cdot)$), and two prediction layers (i.e., $\mathfrak{p}^{\mathrm{sg}}(\cdot)$ and $\mathfrak{p}^{\mathrm{so}}(\cdot)$). The teacher branch contains four components, including a double-tier backbone $\mathfrak{f}_\eta(\cdot)$, and three projection layers (i.e., $\mathfrak{g}^{\mathrm{tg}}(\cdot)$, $\mathfrak{g}^{\mathrm{to}}(\cdot)$, and $\mathfrak{g}^{\mathrm{tp}}(\cdot)$). By feeding the two branches with different views of same patches, $\mathfrak{f}_\theta(\cdot)$ in the student branch (i.e., parameterized by θ) is trained via back-propagation to update $\mathfrak{f}_\eta(\cdot)$ in the teacher branch (i.e., parameterized by η) in a momentum-based moving average fashion, such as $\eta \leftarrow m \cdot \eta + (1 - m) \cdot \theta$, where $m = 0.99$ is the momentum parameter.

Another key issue that determines the quality of the embedding from such a self-supervised strategy is the formulation of respective contrastive loss functions and regularization terms. Accordingly, we design a thorough contrastive learning strategy to capture fine-grained discriminative patterns of postmortem tissues under varying microscopic magnifications. That is, let $\mathbf{X}_{\mathrm{s}}$ and $\mathbf{X}_{\mathrm{t}}$ be two different views of an image patch $\mathbf{X}$ generated by a random data augmentation process. Our contrastive learning strategy concurrently encourages the *global similarity* and *spatially fine-grained similarity* between the corresponding feature embedding $\mathbf{M}_{\mathrm{s}} = \mathfrak{f}_\theta(\mathbf{X}_{\mathrm{s}})$ and $\mathbf{M}_{\mathrm{t}} = \mathfrak{f}_\eta(\mathbf{X}_{\mathrm{t}})$ ($\in \mathbb{R}^{h \times w \times C}$). Also, *two regularization terms* are applied as auxiliary guidance to *protect the informativeness and avoid collapses* of the embedding learned by the backbone.

Specifically, the global similarity between $\mathbf{M}_s$ and $\mathbf{M}_t$ is encouraged by minimizing a general cosine contrastive loss, such as

$$\mathcal{L}_{\text{global}} = 2 - 2 \cdot \frac{< \mathbf{z}_s^g, \mathbf{z}_t^g >}{||\mathbf{z}_s^g||_2 \cdot ||\mathbf{z}_t^g||_2}, \tag{1}$$

where $\mathbf{z}_s^g = \mathfrak{p}^{sg}(\mathfrak{g}^{sg}(\text{GAP}(\mathbf{M}_s)))$ and $\mathbf{z}_t^g = \mathfrak{g}^{tg}(\text{GAP}(\mathbf{M}_t))$, with GAP$(\cdot)$ standing for the global average pooling that produces feature vectors.

In practice, forensic pathologists typically infer postmortem tissue type by evaluating the cellular compositions in multiple local regions. Accordingly, inspired by cross-view learning [11], we design a spatially fine-grained contrastive loss to explicitly encourage multi-parts similarity between $\mathbf{M}_s$ and $\mathbf{M}_t$. Assume $\mathbf{M}'_s$ and $\mathbf{M}'_t$ are two $(h \cdot w) \times C$ tensors flattened from $\mathbf{M}_s$ and $\mathbf{M}_t$ across the spatial dimension, respectively. They are further processed by $\mathfrak{g}^{so}(\cdot)$ and $\mathfrak{g}^{to}(\cdot)$ (followed by softmax normalization), respectively, to produce two $(h \cdot w) \times K$ attention matrices, i.e., $\mathbf{A}_s = \mathfrak{g}^{so}(\mathbf{M}'_s)$ and $\mathbf{A}_t = \mathfrak{g}^{to}(\mathbf{M}'_t)$, where K denotes the predefined number of parts. Thereafter, we aggregate the backbone representations in terms of the attention matrices to deduce multi-parts representations, i.e., $\mathbf{Z}_s^o = \mathfrak{p}^{so}(\mathfrak{g}^{sp}(\mathbf{A}_s^T \otimes \mathbf{M}'_s))$ and $\mathbf{Z}_t^o = \mathfrak{g}^{tp}(\mathbf{A}_t^T \otimes \mathbf{M}'_t)$, where $\otimes$ denotes tensor multiplication. Finally, the spatially fine-grained contrastive loss is quantified as

$$\mathcal{L}_{\text{parts}} = \sum_{k=1}^{K} \left(2 - 2 \cdot \frac{< \mathbf{Z}_s^o[k,:], \mathbf{Z}_t^o[k,:] >}{||\mathbf{Z}_s^o[k,:]||_2 \cdot ||\mathbf{Z}_t^o[k,:]||_2} \right) \tag{2}$$

where $\mathbf{Z}^o[k,:]$ denotes the kth part representation in $\mathbf{Z}^o \in \mathbb{R}^{K \times D}$.

Besides, two additional regularization terms are further included to stabilize contrastive representation learning. Following [1], we penalize small changes between the global representations of different image patches across each feature dimension. Also, we encourage the global representations to be diverse/orthogonal across different feature dimensions. Let $\mathcal{Z}_s^g$ be a set of feature representations for an input mini-batch of patches in the student branch, and $\widetilde{\mathcal{Z}_s^g}$ and $\overline{\mathcal{Z}_s^g}$ denote their channel-wise variation and mean. The regularization terms are defined as

$$\mathcal{L}_{\text{var}} = \frac{1}{D} \sum_{d=1}^{D} \max\left(0, 1 - \sqrt{\widetilde{\mathcal{Z}_s^g}[d] + \epsilon}\right) \tag{3}$$

$$\mathcal{L}_{\text{cov}} = \frac{1}{D^2 - D} \sum_{i \neq j} \left(\left\{ (\mathbf{z}_s^g - \overline{\mathcal{Z}_s^g})^T (\mathbf{z}_s^g - \overline{\mathcal{Z}_s^g}) \right\} [i,j] \right)^2 \tag{4}$$

where ϵ is a small scalar to stabilize numerical computation, $\widetilde{\mathcal{Z}_s^g}[d]$ denotes the dth dimension of $\widetilde{\mathcal{Z}_s^g}$, and $\{(\mathbf{z}_s^g - \overline{\mathcal{Z}_s^g})^T (\mathbf{z}_s^g - \overline{\mathcal{Z}_s^g})\}[i,j]$ is the $[i,j]$th element in such a covariance matrix. According to [1], Eqs. (3) and (4) jointly encourage the diversity across patches and feature dimensions, thus protecting the informativeness and avoid collapse of self-supervised contrastive learning.

Overall, we combine Eqs. (1) to (4) as the final loss function to train the double-tier backbone, such as $\mathcal{L}_{\text{all}} = \mathcal{L}_{\text{global}} + \mathcal{L}_{\text{parts}} + \gamma \mathcal{L}_{\text{var}} + \lambda \mathcal{L}_{\text{cov}}$, where γ and λ are two tuning parameters balancing different terms.

2.2 Context-Aware MIL

Given the patch/instance-level representations of a histopathological image from the double-tier backbone, we further design a context-aware MIL framework to aggregate their information for postmortem tissue recognition. Given patch embeddings of a Microscope image, our context-aware MIL part contains two main steps, i.e., a multi-head self-attention to refine each patch's feature and an adaptive pooling step to distill all patches' information.

In detail, we first adopt a multi-head self-attention (MSA) mechanism [19] integrating relative positional embedding to explicitly model cross-patch associations for contextual enhancement of the instance representations from the backbone. Let $\mathcal{Z} = \{\mathbf{z}_i\}_{i=1}^{I}$ be a set of the contextually enhanced instance embedding from an image. Thereafter, inspired by Deformable DETR [28], we further design an adaptive pooling operation, which is simple but effective to distill from $\mathcal{Z}$ a bag-level holistic representation for the classification purpose. Specifically, the bag-level holistic representation determined by the adaptive pooling is

$$\mathbf{z}_{\text{bag}} = \frac{1}{I}\sum_{i=1}^{I}(softmax(\mathfrak{h}_{\omega_1}(\mathbf{z}_i)) \circ \mathfrak{h}_{\omega_2}(\mathbf{z}_i)), \tag{5}$$

where $\mathfrak{h}_{\omega_1}(\cdot)$ and $\mathfrak{h}_{\omega_2}(\cdot)$ are two linear projections with the same number of output units, symbol $\circ$ denotes the Hadamard product between two tensors, and $softmax(\cdot)$ is performed across different instances to filter out uninformative patches and preserve discriminative patches in quantifying $\mathbf{z}_{\text{bag}}$ for classification.

3 Experiments

3.1 Data and Experimental Setup

Rat Postmortem Histopathology Dataset. Ninety Sprague-Dawley adult male rats were executed by the spinal cord dislocation and placed in a constant temperature and humidity environment for 6–8 h. The animal experiments were approved by the Laboratory Animal Care Committee of the anonymous institution. Seven organs, i.e., brain, heart, kidney, liver, lung, pancreas, and spleen, were removed and placed in the formalin solution. Briefly, paraffin sections of these organ tissues were stained with the H&E solution. The H&E-stained sections were then analyzed by three forensic pathologists, who used Lercai LAS EZ microscopes to record the areas according to their expertise. Overall, five to ten images were recorded from a section at each magnification (i.e., 5×, 10×, 20×, and 40×). Finally, we split the 90 rats as training, validation, and test sets of 60, 10, and 20 rats, respectively, each with 13,137, 2,235, and 4,325 images.

Human Forensic Histopathology Dataset. The real forensic images were provided by the Forensic Judicial Expertise Center of the anonymous institution, after getting the informed consent of relatives. All procedures followed the requirements of local laws and institutional guidelines, and were approved and supervised by the Ethics Committee. A total of 32 decedents participated in this study. Four to six images were recorded at each of three magnifications (5×, 10×, and 20×) per H&E stained section. Similar to the rat dataset, the human dataset was selected from the same seven organs. Finally, the training, validation and test sets contain 1,691 images, 628, and 1059 images, corresponding to 16, 6, and 10 different decedents, respectively.

Experimental Details. Notably, the double-tier backbone was self-supervised and learned on the rat training set for 100 epochs by setting the mini-batch size as 1024, with the parameters initialized by the ImageNet pre-trained models. The training data were augmented by a histopathology-oriented strategy by combining different kinds of staining jitters, random affine transformation, Gaussian blurring, resizing, etc. The image(patch) dimension in our implementation was 224*224. The tuning parameters γ and λ in $\mathcal{L}_{\text{all}}$ were set as 5 and 0.005, respectively. Thereafter, the MIL blocks on two different datasets were both trained by minimizing the cross-entropy loss for 20 epochs with the mini-batch size setting as 32. The experiments were conducted on three PCs with twenty NVIDIA GEFORCE RTX 3090 GPUs.

3.2 Results of Self-supervised Contrastive Learning

Our self-supervised double-tier backbone was compared with other state-of-the-art self-supervised learning approaches, including **balow twins** [27], **swin transformer (SSL)** [26], **TransPath** [23], **CTransPath** [24],**RetCCL** [22] and **MOCOV3** [3]. To evaluate the discriminative power of these competing methods, we adopted GAP to aggregate their instance representations from a whole image to train simple linear classifiers for the recognition of seven different organ tissues on both the rat and human datasets, with the test performance quantified in terms of four general classification metrics (i.e., **ACC**, **F1 score**, **MCC(Matthews Correlation Coefficient)**, and **Precision**). The corresponding results are summarized in Table 1, from which we can have two observations. *First*, our self-supervised double-tier backbone consistently outperformed all other competing methods in terms of all metrics on two datasets. *Second*, our method led to better generalization, as the backbone trained on the rat dataset shows promising performance on the challenging real-world human dataset (e.g., resulting in an ACC higher than 90%). These results suggest the effectiveness of our self-supervised learning strategy.

For a more detailed evaluation, we further conducted a series of ablation studies to evaluate the contributions of the contrastive losses (i.e., $\mathcal{L}_{\text{global}}$ and $\mathcal{L}_{\text{parts}}$) and regularization strategy (i.e., $\mathcal{L}_{\text{var}} + \mathcal{L}_{\text{cov}}$). The corresponding results are summarized in Table 2, from which we can see that, given the baseline of $\mathcal{L}_{\text{global}}$, both the inclusion of the spatially fine-grained contrastive loss (i.e., $\mathcal{L}_{\text{parts}}$) and informativeness regularization (i.e., $\mathcal{L}_{\text{var}}$ and $\mathcal{L}_{\text{cov}}$) led to respective performance gains. These results further justify our self-supervised design.

Table 1. Linear classification results obtained by different self-supervised learning approaches on the rat and human testing sets, respectively.

Competing methods	Rat dataset				Human dataset			
	ACC	F1	MCC	Precision	ACC	F1	MCC	Precision
balow twins [27]	0.9232	0.9123	0.9076	0.9070	0.7306	0.7311	0.6854	0.7345
swin transformer(SSL) [26]	0.9450	0.9369	0.9299	0.9330	0.8079	0.8088	0.7758	0.8125
Transpath [23]	0.7351	0.7397	0.6958	0.7481	0.5838	0.5657	0.5264	0.6050
CTransPath [24]	0.9635	0.9610	0.9535	0.9596	0.8794	0.8799	0.8591	0.8842
RetCCL [22]	0.9794	0.9801	0.9768	0.9810	0.7796	0.7789	0.7448	0.7961
MOCOV3 [3]	0.9732	0.9738	0.9681	0.9745	0.8103	0.8124	0.7790	0.8187
Ours	**0.9831**	**0.9831**	**0.9796**	**0.9831**	**0.9049**	**0.9044**	**0.8886**	**0.9056**

Table 2. Ablation studies to evaluate the contributions of different self-supervised contrastive losses and regularization terms.

Loss functions				Rat dataset				Human dataset			
$\mathcal{L}_{\text{global}}$	$\mathcal{L}_{\text{parts}}$	$\mathcal{L}_{\text{var}}$	$\mathcal{L}_{\text{cov}}$	ACC	F1	MCC	Precision	ACC	F1	MCC	Precision
✓				0.9732	0.9713	0.9660	0.9689	0.8918	0.8913	0.8734	0.8935
✓	✓			0.9817	0.9819	0.9778	0.9822	0.8953	0.8956	0.8779	0.8981
✓		✓	✓	0.9793	0.9799	0.9757	0.9806	0.8978	0.8976	0.8802	0.8983
✓	✓	✓	✓	**0.9831**	**0.9831**	**0.9796**	**0.9831**	**0.9049**	**0.9044**	**0.8886**	**0.9056**

3.3 Results of Multiple-Instance Learning

Based upon the double-tier backbone learned on the rat training set, we compared our context-aware MIL with other MIL methods, including the gated attention-based approach (i.e., **AB-MIL** [12], **DSMIL** [15], **Transmil** [19] and **MSA** [2,16]) approaches with/without different positional embedding strategies, i.e., relative position embedding (**MSA-RP** [17]), learnable position embedding (**MSA-LP** [7]), and 2D sine-cosine position embedding (**MSA-SP** [8]). Notably, our approach used MSA-RP as the baseline, based on which an adaptive pooling operation is designed to produce the final bag-level representation. To check the efficacy of **adaptive pool**, we further conducted a corresponding set of ablation studies by replacing it with other operations, including **max pool**, and **soft pool** [20]. These comparison and ablations results are shown in Table 3, from which we can observe that our method led to the best results on both datasets, with relatively more significant improvements on the challenging human dataset. Also, compared with other pooling operations, the adaptive pool design brought consistent performance gains. These results suggest the efficacy of our context-aware MIL for postmortem tissue recognition.

In addition, we conducted LayerCAM-based analysis [13] to check the explainability and reliability of our postmortem histopathological recognition results. From the representative examples shown in Fig. 2, we can have an interesting observation that our method tends to focus on tissue-specific postmortem patterns at different microscopic scales. For example, the spatial attention maps reliably highlighted the meningeal structures of the brain tissue, the glomeruli in the kidney cortex, and the central vein area between the liver lobules. On

Table 3. Multiple-instance learning results obtained by the competing methods and our Context-Aware MIL with different pooling strategies.

Competing methods	Rat dataset				Human dataset			
	ACC	F1	MCC	Precision	ACC	F1	MCC	Precision
AB-MIL [12]	0.9815	0.9828	0.9793	0.9844	0.9011	0.9005	0.8838	0.9050
DSMIL [15]	0.9951	0.9948	0.9937	0.9945	0.9176	0.9166	0.9030	0.9170
Transmil [19]	0.9899	0.9888	0.9875	0.9878	0.8824	0.8813	0.8622	0.8821
MSA [2,16]	0.9875	0.9883	0.9861	0.9892	0.9082	0.9082	0.8921	0.9100
MSA-LP [7]	0.9879	0.9875	0.9853	0.9873	0.9097	0.9087	0.8945	0.9109
MSA-SP [8]	0.9851	0.9839	0.981	0.9832	0.8915	0.8905	0.8748	0.8948
MSA-RP [17]	0.9915	0.9915	0.9896	0.9916	0.9218	0.9213	0.9085	0.9218
Ours + Max pool	0.9910	0.9909	0.9888	0.9909	0.9144	0.9147	0.9001	0.9191
Ours + Soft pool [20]	0.9935	0.9929	0.9915	0.9924	0.9047	0.9023	0.8883	0.9056
Ours + Adaptive pool	**0.9956**	**0.9952**	**0.9943**	**0.9949**	**0.9229**	**0.9218**	**0.9093**	**0.9263**

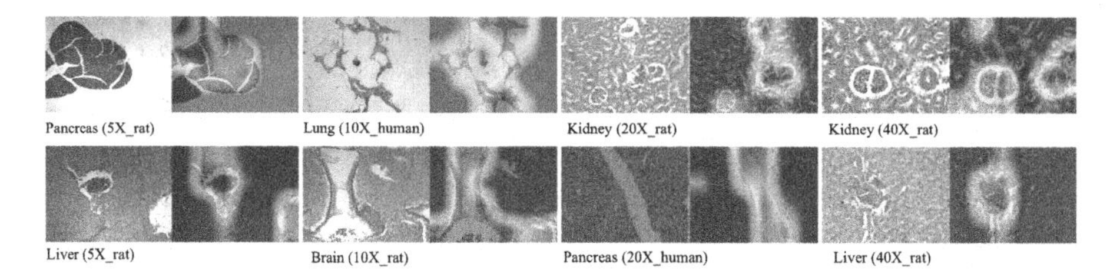

Fig. 2. Explainability analysis based on LayerCAM [13] for representative postmortem tissue images acquired at different microscopic scales.

the other hand, based on the pancreas example, we can see that our network can sensitively localize the pancreas glandular structure while filtering out the uninformative background in an end-to-end fashion, without the need for any pre-processing to segment first the foreground. These observations support our assumption that the proposed method is reliable and efficient in learning discriminative histopathological representations of postmortem organ tissues.

4 Conclusion

In this study, we have proposed a context-aware MIL framework powered by self-supervised contrastive learning to learn fine-grained discriminative representations for postmortem histopathological recognition. The dedicated self-supervised learning strategy concurrently maximizes multiple contrastive losses and regularization terms to deduce informative and discriminative instance embedding. Thereafter, the context-aware MIL framework adopts MSA followed by an adaptive pooling operation to distill from all instances a holistic bag/image-level representation. The experimental results on a relatively large-scale database suggest the state-of-the-art postmortem recognition performance of our method.

Funding Information. This work was supported in part by NSFC Grants (Nos. 62101430 & 62101431).

References

1. Bardes, A., Ponce, J., LeCun, Y.: VICReg: Variance-invariance-covariance regularization for self-supervised learning. arXiv preprint arXiv:2105.04906 (2021)
2. Chen, R.J., et al.: Scaling vision transformers to gigapixel images via hierarchical self-supervised learning. In: Proceedings of the IEEE/CVF Conference on Computer Vision and Pattern Recognition, pp. 16144–16155 (2022)
3. Chen, X., Xie, S., He, K.: An Empirical Study of Training Self-Supervised Vision Transformers. arXiv e-prints (2021)
4. De La Grandmaison, G.L., Charlier, P., Durigon, M.: Usefulness of systematic histological examination in routine forensic autopsy. J. Forensic Sci. **55**(1), 85–88 (2010)
5. DiMaio, D., DiMaio, V.J.: Forensic Pathology. CRC Press, Boca Raton (2001)
6. Dolinak, D., Matshes, E., Lew, E.O.: Forensic Pathology: Principles and Practice. Elsevier, Amsterdam (2005)
7. Dosovitskiy, A., et al.: An image is worth 16x16 words: Transformers for image recognition at scale. arXiv preprint arXiv:2010.11929 (2020)
8. He, K., Chen, X., Xie, S., Li, Y., Dollár, P., Girshick, R.: Masked autoencoders are scalable vision learners. In: Proceedings of the IEEE/CVF Conference on Computer Vision and Pattern Recognition, pp. 16000–16009 (2022)
9. He, K., Fan, H., Wu, Y., Xie, S., Girshick, R.: Momentum Contrast for Unsupervised Visual Representation Learning. arXiv e-prints (2019)
10. He, K., Zhang, X., Ren, S., Sun, J.: Deep residual learning for image recognition. In: 2016 IEEE Conference on Computer Vision and Pattern Recognition (CVPR), pp. 770–778 (2016)
11. Huang, L., You, S., Zheng, M., Wang, F., Qian, C., Yamasaki, T.: Learning where to learn in cross-view self-supervised learning. In: CVPR (2022)
12. Ilse, M., Tomczak, J.M., Welling, M.: Attention-based Deep Multiple Instance Learning. arXiv e-prints (2018)
13. Jiang, P.T., Zhang, C.B., Hou, Q., Cheng, M.M., Wei, Y.: LayerCAM: exploring hierarchical class activation maps for localization. IEEE Trans. Image Process. **30**, 5875–5888 (2021)
14. Lee, Y., et al.: Derivation of prognostic contextual histopathological features from whole-slide images of tumours via graph deep learning. Nat. Biomed. Eng. (2022)
15. Li, B., Li, Y., Eliceiri, K.W.: Dual-stream multiple instance learning network for whole slide image classification with self-supervised contrastive learning. In: Proceedings of the IEEE/CVF Conference on Computer Vision and Pattern Recognition, pp. 14318–14328 (2021)
16. Li, H., et al.: DT-MIL: deformable transformer for multi-instance learning on histopathological image. In: de Bruijne, M., et al. (eds.) MICCAI 2021. LNCS, vol. 12908, pp. 206–216. Springer, Cham (2021). https://doi.org/10.1007/978-3-030-87237-3_20
17. Liu, Z., et al.: Swin Transformer: Hierarchical Vision Transformer using Shifted Windows. arXiv e-prints (2021)
18. Lu, M.Y., Williamson, D.F.K., Chen, T.Y., Chen, R.J., Barbieri, M., Mahmood, F.: Data-efficient and weakly supervised computational pathology on whole-slide images. Nat. Biomed. Eng. **5**(6), 555–570 (2021)

19. Shao, Z., Bian, H., Chen, Y., Wang, Y., Zhang, J., Ji, X., et al.: TransMIL: transformer based correlated multiple instance learning for whole slide image classification. Adv. Neural. Inf. Process. Syst. **34**, 2136–2147 (2021)
20. Stergiou, A., Poppe, R., Kalliatakis, G.: Refining activation downsampling with softpool. In: Proceedings of the IEEE/CVF International Conference on Computer Vision, pp. 10357–10366 (2021)
21. Wang, G., et al.: An emerging strategy for muscle evanescent trauma discrimination by spectroscopy and chemometrics. Int. J. Mol. Sci. **23**(21), 13489 (2022)
22. Wang, X., et al.: RetCCL: clustering-guided contrastive learning for whole-slide image retrieval. Med. Image Anal. **83**, 102645 (2023)
23. Wang, X., et al.: TransPath: transformer-based self-supervised learning for histopathological image classification. In: de Bruijne, M., et al. (eds.) MICCAI 2021. LNCS, vol. 12908, pp. 186–195. Springer, Cham (2021). https://doi.org/10.1007/978-3-030-87237-3_18
24. Wang, X., et al.: Transformer-based unsupervised contrastive learning for histopathological image classification. Med. Image Anal. **81**, 102559 (2022)
25. Wu, H., et al.: Pathological and ATR-FTIR spectral changes of delayed splenic rupture and medical significance. Spectrochim. Acta. A Mol. Biomol. Spectrosc. **278**, 121286 (2022)
26. Xie, Z., et al.: Self-Supervised Learning with Swin Transformers. arXiv preprint arXiv:2105.04553 (2021)
27. Zbontar, J., Jing, L., Misra, I., LeCun, Y., Deny, S.: Barlow twins: Self-supervised learning via redundancy reduction. arXiv preprint arXiv:2103.03230 (2021)
28. Zhu, X., Su, W., Lu, L., Li, B., Wang, X., Dai, J.: Deformable DETR: Deformable Transformers for End-to-End Object Detection. arXiv e-prints (2020)

Segment Membranes and Nuclei from Histopathological Images via Nuclei Point-Level Supervision

Hansheng Li[1], Zhengyang Xu[1], Mo Zhou[1], Xiaoshuang Shi[2], Yuxin Kang[1], Qirong Bu[1], Hong Lv[3], Ming Li[3], Mingzhen Lin[4], Lei Cui[1](✉), Jun Feng[1], Wentao Yang[3], and Lin Yang[1]

[1] Northwest University, Xi'an, China
leicui@nwu.edu.cn
[2] University of Electronic Science and Technology, Chengdu, China
[3] Fudan University, Shanghai, China
[4] Shanghai Second Medical University, Shanghai, China

Abstract. Accurate segmentation and analysis of membranes from immunohistochemical (IHC) images are crucial for cancer diagnosis and prognosis. Although several fully-supervised deep learning methods for membrane segmentation from IHC images have been proposed recently, the high demand for pixel-level annotations makes this process time-consuming and labor-intensive. To overcome this issue, we propose a novel deep framework for membrane segmentation that utilizes nuclei point-level supervision. Our framework consists of two networks: a Seg-Net that generates segmentation results for membranes and nuclei, and a Tran-Net that transforms the segmentation into semantic points. In this way, the accuracy of the semantic points is closely related to the segmentation quality. Thus, the inconsistency between the semantic points and the point annotations can be used as effective supervision for cell segmentation. We evaluated the proposed method on two IHC membrane-stained datasets and achieved an 81.36% IoU and 85.51% F_1 score of the fully supervised method. *All source codes are available* here.

Keywords: Membrane segmentation · Point-based supervision · Immunohistochemical image

1 Introduction

Accurate quantification of immunohistochemistry (IHC) membrane staining images is a crucial aspect of disease assessment [14,15]. In clinical diagnosis, pathologists typically grade diseases by manually estimating the proportion of stained membrane area [14] or evaluating the completeness of stained membrane [23]. However, this manual

L. Cui, J. Feng, W. Yang and L. Yang–Equally contribution.
H. Li and Z. Xu—Equally first authors.

Supplementary Information The online version contains supplementary material available at https://doi.org/10.1007/978-3-031-43987-2_52.

H. Greenspan et al. (Eds.): MICCAI 2023, LNCS 14225, pp. 539–548, 2023.
https://doi.org/10.1007/978-3-031-43987-2_52

approach is tedious, time-consuming [22], and error-prone [13]. Therefore, there is an urgent need for precise automatic IHC membrane analysis methods to provide objective quantitative evidence and improve diagnostic efficiency.

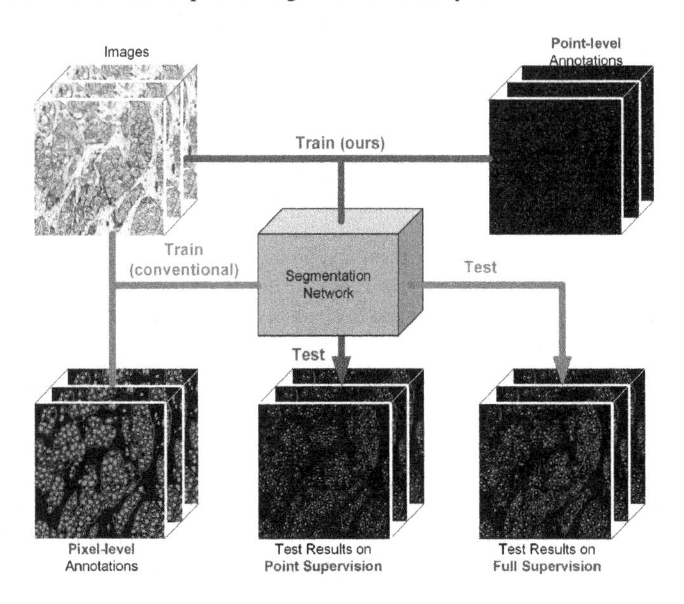

Fig. 1. Illustration of the full supervision (blue line) and point supervision (ours, red line) for membranes and nuclei segmentation. (Color figure online)

Despite numerous deep learning methods have been proposed for detecting cell nuclei [11,20] from hematoxylin-eosin (H&E) staining images, little attention has focused on analyzing cell membranes from IHC images. Currently, only a few fully supervised IHC membrane segmentation methods have been proposed [9,19], demonstrating the superiority of deep learning-based membrane segmentation. However, full supervision requires substantial time and effort for pixel-level annotations of cell membranes and nuclei. In contrast, as annotating the centers of nuclei requires much fewer efforts, weakly supervised learning has been studied for nuclei segmentation [16,21]. Nevertheless, how to utilize point annotations to supervise cell membrane segmentation is still under investigation.

This study proposes a novel point-based cell membrane segmentation method, which can significantly reduce the cost of pixel-level annotation required in conventional methods, as shown in Fig. 1. In this study, the category of the point annotation is used to describe the staining state of the cell membrane (e.g., complete-stained, incomplete-stained, and unstained). We employ a network named Seg-Net to segment the nuclei and membranes separately, followed by a Trans-Net to convert the segmentation results into semantic points. Since the accuracy of semantic points is directly related to the segmentation results, the segmentation quality can be implicitly supervised by the loss between the semantic points and the point annotations, as shown in Fig. 2

To the best of our knowledge, this is the first study that using point-level supervision for membrane segmentation from IHC images, which could significantly advance future

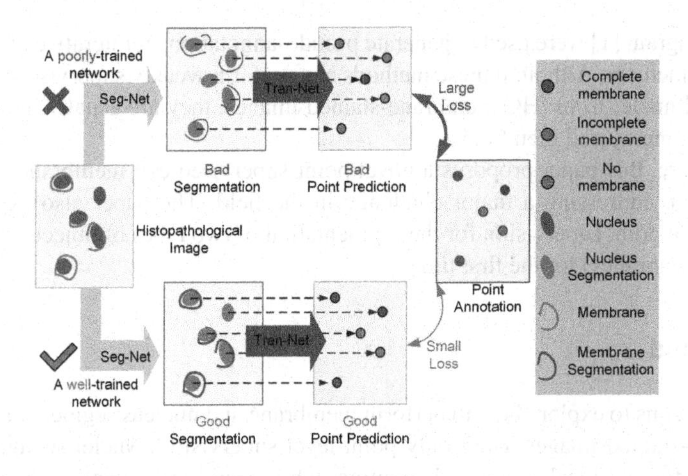

Fig. 2. The illustration of how point annotations are used to supervise the cell segmentation.

membrane analysis. Additionally, our method is the first to employ point annotations to simultaneously supervise the segmentation of two objects. Extensive experiments confirm the efficacy of the proposed method, attaining performance that is comparable to models trained with fully supervised data.

2 Related Works

Deep learning-based segmentation methods have been widely developed for cell nuclei segmentation from H&E images in recent years, ranging from convolutional neural networks [5,12,24] to Transformer-based architectures [8], resulting in continuously improved accuracy in nuclei segmentation.

For the task of analyzing IHC membrane-stained images, due to the challenge of pixel-level annotation, existing methods mostly adopt traditional unsupervised algorithms, such as watershed [17,26], active contour [3,25], and color deconvolution [2,4]. These traditional methods perform inadequately and are vulnerable to the effects of differences in staining intensity and abnormal staining impurities. In recent years, a few fully supervised cell membrane segmentation methods also have emerged [9,19], but the high cost of data annotation limits their applicability to various membrane staining image analysis tasks.

To reduce the annotation cost of nuclei segmentation in histopathological images, weakly supervised segmentation training methods have received attention, including: 1) Unsupervised cell nuclei segmentation methods represented by adversarial-based methods [6,7]. However, unsupervised methods are challenged by the difficulty of constraining the search space of model parameters, making it hard for the model to handle visually complex pathology images, such as H&E or IHC; 2) Weakly supervised cell nucleus segmentation algorithms with point annotation [16,21]. Because the cell nucleus shape in H&E images is almost elliptical, point annotation combined with

Voronoi diagram [1] were used to generate pseudo-annotations for iterative model training and refinement. Although these methods can perform weakly supervised segmentation of cell nuclei from IHC membrane-stained images, they are usually ineffective in segmenting messy cell membranes.

Therefore, this paper proposes a novel point-supervised cell membrane segmentation method, addressing a major challenge in the field. The paper also explores the feasibility of point supervision for the segmentation of two types of objects (cell nuclei and cell membranes) for the first time.

3 Method

This study aims to explore how to perform membrane and nucleus segmentation in IHC membrane-stained images using only point-level supervision. Nuclei segmentation is performed for cell localization and counting, while membrane quantification provides clinical evidence for diagnosis.

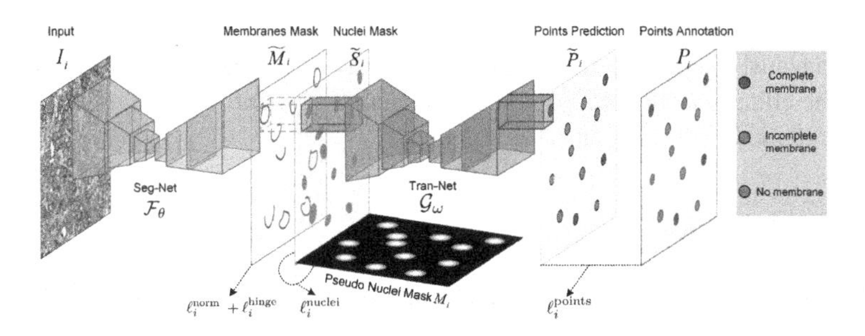

Fig. 3. The architecture of the proposed framework. The training stage employs two networks, namely a Segmentation network (Seg-Net) and a Transition network (Tran-Net). During inference, we only adopt the Seg-Net for segmentation.

3.1 Formulation of the Point-Level Supervised Segmentation Problem

Given an input cell image set $\{I_i\}_{i=1}^{N}$, where N is the number of images in this set, $I_i \in \mathbb{R}^{H\times W\times 3}$ with H, W representating the height and width of the image, respectively, and 3 being the number of channels of the image. Our goal is to obtain the mask of membranes ($\widetilde{M}_i \in \mathbb{R}^{H\times W\times 1}$) and nuclei ($\widetilde{S}_i \in \mathbb{R}^{H\times W\times 1}$), that is $\left\{\widetilde{\mathrm{M}}_i, \widetilde{\mathrm{S}}_i\right\} = \sigma(\mathcal{F}_\theta(I_i))$, where $\mathcal{F}_\theta$ is a segmentation network (Seg-Net) and with trainable parameters θ, and σ is the sigmoid activation function. We have point annotations $P_i \in \mathbb{R}^{H\times W\times(c+1)}$ for image I_i, in which c is the number of semantic categories used to describe the states of membrane staining.

In order to train $\mathcal{F}_\theta$ to segment membranes $\widetilde{M}_i$ and nuclei $\widetilde{S}_i$ using point annotations P_i in image I_i, we need to establish the relationship from input to segmentation, and then to point annotation, as shown in Eq. (1) and Fig. 3.

$$I_i \xrightarrow{\mathcal{F}_\theta} \left\{\widetilde{M}_i, \widetilde{S}_i\right\} \xrightarrow{\mathcal{G}_\omega} \widetilde{P}_i \sim= P_i, \tag{1}$$

where $\mathcal{G}_\omega$ is the transition network (Tran-Net) with parameters ω. $\mathcal{G}_\omega$ transforms $\left\{\widetilde{M}_i, \widetilde{S}_i\right\}$ to semantic points $\widetilde{P}_i \in \mathbb{R}^{H\times W\times(c+1)}$, it should be noted that $\widetilde{M}_i$ and $\widetilde{S}_i$ respectively provide the semantic and spatial information to $\mathcal{G}_\omega$ for semantic points prediction, so that the segmentation performance is crucial for $\mathcal{G}_\omega$. So that, by fitting $\widetilde{P}_i$ to P_i ($\widetilde{P}_i \sim= P_i$), the segmentation can be supervised.

3.2 Network Architecture

We adopt the U-Net [18] architecture for both $\mathcal{F}_\theta$ and $\mathcal{G}_\omega$ networks. The architecture of $\mathcal{F}_\theta$ is conventional and follows the original U-Net [18]. However, in the decoder of $\mathcal{G}_\omega$, we replace the size of the last convolution kernel from 1×1 to 9×9. This is because $\mathcal{G}_\omega$ is utilized to predict the category of semantic points, which are the center points of cells and related to the membrane. Therefore, a larger receptive field is required for the convolution to improve the accuracy of category prediction. During inference, only $\mathcal{F}_\theta$ is needed, and $\mathcal{G}_\omega$ is discarded.

3.3 Decouple the Membranes and Nuclei Segmentation

Our goal is to use Seg-Net to generate masks for both membranes ($\widetilde{M}_i \in \mathbb{R}^{H\times W\times 1}$) and nuclei ($\widetilde{S}_i \in \mathbb{R}^{H\times W\times 1}$),i.e., $\left\{\widetilde{\mathrm{M}}_\mathrm{i}, \widetilde{\mathrm{S}}_\mathrm{i}\right\} = \sigma(\mathcal{F}_\theta(I_i))$. However, the two Seg-Net channels are interdependent, which can result in nuclei and membranes being inseparably segmented. To overcome this issue, we enforce one channel to output the nuclei segmentation using a supervised mask $S_i \in \mathbb{R}^{H\times W\times 1}$. In this study, we create S_i by expanding each point annotation to a circle with a radius of 20. Thus, to provide semantic information to Tran-Net for predicting semantic points, the other channel must contain information describing the staining status of the membrane, which in turn decouples membrane segmentation.

Because both $\widetilde{S}_i$ and S_i are single-channel, we employ the naive L_1 loss to supervise the segmentation of the nuclei, as shown in Eq. (2)

$$\ell_i^{\text{nuclei}} = \left|S_i - \widetilde{S}_i\right|. \tag{2}$$

3.4 Constraints for Membranes Segmentation

As there are no annotations available for pixel-level membrane segmentation, the network could result in unwanted over-segmentation of membranes. This over-segmentation can take two forms: (1) segmentation of stained impurities, which can restrict the network's generalization performance by learning simple color features, and (2) segmentation of nuclei. To address these issues, we propose a normalization loss term, ℓ_i^{norm}, which is an L_1 normalization and is defined in Eq. (3). The purpose of this loss term is to encourage the network to learn a smoother membrane segmentation that does not capture small stained regions or nuclei.

$$\ell_i^{\text{norm}} = \left\| \widetilde{M}_i \right\|_1. \tag{3}$$

However, relying solely on the ℓ_i^{norm} normalization term might lead to a trivial solution, as it only minimizes the average value of the prediction result, potentially resulting in a minimal maximum confidence for the cell membrane segmentation (e.g., 0.03). To prevent this issue, we introduce a hinge-loss-like loss function, presented in Eq. (4), to constrain the distribution of membrane segmentation results. The hyper-parameter τ in the hinge-loss function determines the expected value of the result, where a larger τ corresponds to a smaller expected value. For instance, if τ is set to 1 or 2, the expected values of the result would be 1 or 0.5, respectively. By selecting an appropriate value for τ, we can mitigate the negative impact of ℓ_i^{norm}.

$$\ell_i^{\text{hinge}} = \max(0, 1 - \tau \cdot \widetilde{M}_i). \tag{4}$$

3.5 Point-Level Supervision

Using $\mathcal{G}_\omega$ to detect the central point of the cells is a typical semantic points detection task. The difference is that the input of $\mathcal{G}_\omega$ is the output of $\mathcal{F}_\theta$ rather than the image. Nevertheless, we can also employ the cross-entropy function for point-level supervision:

$$\ell_i^{\text{points}} = -\widetilde{P}_i \log\left(P_i\right). \tag{5}$$

To enhance the spatial supervision information of the point annotations P_i, it is worth noting that we extended each point annotation to a Gaussian circle with a radius of 5 pixels.

3.6 Total Loss

By leveraging the advantages of the above items, we can obtain the total loss as follows:

$$\mathcal{L}_i = \ell_i^{\text{nuclei}} + \ell_i^{\text{norm}} + \ell_i^{\text{hinge}} + \ell_i^{\text{points}}, \tag{6}$$

where ℓ_i^{norm} and ℓ_i^{hinge} are antagonistic, and their values are close to 0.5 in the ideal optimal state, which can be achieved at $\tau = 1$ in ℓ_i^{hinge}.

4 Experiments

In order to comprehensively verify the proposed method, we conduct extensive experiments on two IHC membrane-stained data sets, namely the PDL1 (Programmed cell depth 1 light 1) and HER2 (human epidermal growth factor receiver 2) datasets. The HER2 experiment is dedicated to validate segmentation performance, while the PDL1 experiment is utilized to verify the effectiveness of converting segmentation results into clinically relevant indicators.

4.1 Dataset

We collected 648 HER2 and 1076 PDL1 images at 40x magnification with a resolution of 1024×1024 from WSIs. The PDL1 data only has point-level annotations, where pathologists categorize cells as positive or negative based on membrane staining. The HER2 data has both pixel-level and point-level annotations, where pathologists delineate the nuclei and membranes for pixel-level annotations and categorize cells based on membrane staining for point-level annotations. Pixel-level annotations are used for testing and fully supervised experiments only. We split the data into training and test sets in a 1:1 ratio and do not use a validation set since our method is trained without pixel-level annotations.

4.2 Implementation Details and Evaluation Metric

We totally train the networks 50 epochs and employ Adam optimizer [10] with the initial learning rate of 5×10^{-4} and the momentum of 0.9. Images are randomly cropped to 512×512, and data augmentations such as random rotation and flip were employed during model training. The hyper-parameter τ in ℓ_i^{hinge} is set to 1.0.

We employ the Intersection over Union (IoU) segmentation metric and pixel-level F_1 score to validate the performance of the proposed method. However, only point-level annotations are equipped for the PDL1 dataset, we evaluate the segmentation performance at the point-level by converting the segmentation into key point predictions, and **the conversion process details are available in the supplementary materials**.

4.3 Result Analysis

HER2 Results. Table 1 shows the cell segmentation results of the proposed method and six comparison methods on the dataset HER2. Our proposed method can segment both cell nuclei and membranes simultaneously, outperforming both unsupervised methods and other point-level methods, with an IoU score of 0.5774 and an F_1 score of 0.6899 for membranes, and an IoU score of 0.5242 and an F_1 score of 0.6795 for nuclei. Furthermore, our ablation study shows that the hinge loss and normalization loss play important roles in improving the segmentation performance. Notably, other point-level methods not only fail to segment the cell membranes but also have limited performance in segmenting cell nuclei due to over-segmentation and under-segmentation errors, as shown in Fig. 4.

PDL1 Results. We chose to compare our method with unsupervised segmentation methods because existing point-supervised segmentation methods are unable to segment cell membranes. We present their results in Table 2, which illustrates that the proposed method outperforms other methods, achieving F_1 scores of 0.7131 and 0.7064 for negative and positive-stained cells, respectively. Among the unsupervised methods, color deconvolution [4] shows the best performance with F_1 scores of 0.5984 and 0.6136 for negative and positive cells, respectively. However, our proposed method significantly outperformed it. Besides, qualitative experimental results can be found in the supplementary materials.

Table 1. Comparison of the cell segmentation results on HER2 test data. w/o: without.

Supervised Settings	Methods	Membranes		Nuclei	
		IoU	F_1 score	IoU	F_1 score
Unspervised	USAR (our implementation) [6]	0.0865	0.1356	0.0832	0.1165
	Watershed [17]	0.3561	0.4427	0.0721	0.1285
	Active Contour [25]	0.3331	0.3938	0.0973	0.1571
	Color Deconvolution [4]	0.4242	0.5148	0.1418	0.2455
Point-level Supervised	C2FNet [21]	/	/	0.3007	0.4044
	Pseudoedgenet [27]	/	/	0.1548	0.2663
	Ours w/o hinge and norm loss	0.4877	0.5001	**0.5656**	**0.7155**
	Ours w/o hinge loss	0	0	0.5357	0.6908
	Ours	**0.5774**	**0.6899**	0.5242	0.6795
Fully supervised	Fully supervised	0.7096	0.8068	0.6873	0.7648

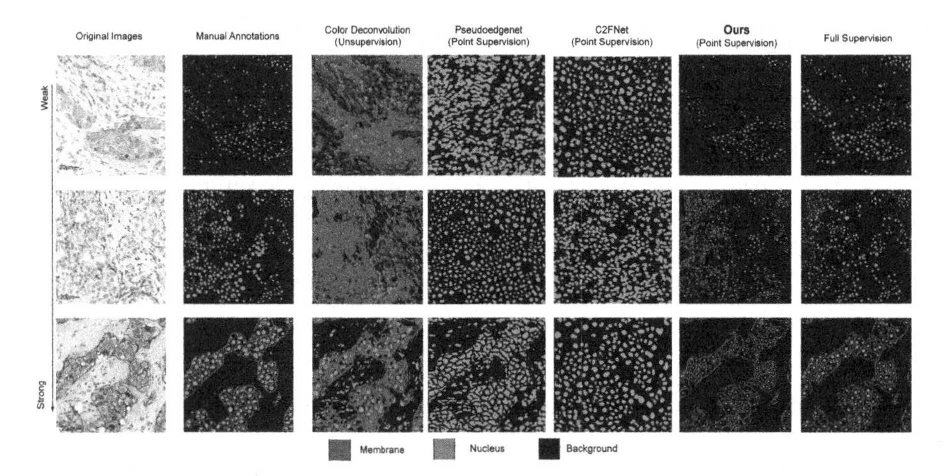

Fig. 4. Qualitative results on the HER2 test set.

Table 2. Comparison Point-level Results of the PDL1 Test Dataset.

Methods	Negative-stained Cells			Positive-stained Cells		
	Recall	Precision	F_1 score	Recall	Precision	F_1 score
Watershed [17]	0.6601	0.5597	0.5117	0.5650	0.5272	0.4853
Active Contour [25]	0.5876	0.5259	0.4956	0.5785	0.5266	0.4713
Color Deconvolution [4]	0.6583	0.5715	0.5984	0.7161	0.5758	0.6136
Ours	**0.7044**	**0.7791**	**0.7131**	**0.7308**	**0.6970**	**0.7064**

5 Conclusion

In this paper, we present a novel method for precise segmentation of cell membranes and nuclei in immunohistochemical (IHC) membrane staining images using only point-level supervision. Our method achieves comparable performance to fully supervised pixel-level annotation methods while significantly reducing annotation costs, only requiring one-tenth of the cost of pixel-level annotation. This approach effectively reduces the expenses involved in developing, deploying, and adapting IHC membrane-stained image analysis algorithms. In the future, we plan to further optimize the segmentation results of cell nuclei to further boost the performance of the proposed method, and extend it to the whole slide images (WSIs).

Acknowledgements. This work is supported by the National Natural Science Foundation of China (NSFC Grant No. 62073260, No.62106198 and No.62276052), and the Natural Science Foundation of Shaanxi Province of China (2021JQ-461), the General Project of Education Department of Shaanxi Provincial Government under Grant 21JK0927. Medical writing support is provided by AstraZeneca China. The technical and equipment support is provided by HangZhou DiYingJia Technology Co., Ltd (DeepInformatics++). The authors would like to thank the medical team at AstraZeneca China and techinical team at DeepInformatics++ for their scientific comments on this study.

References

1. Aurenhammer, F., Klein, R.: Voronoi diagrams. Handb. Comput. Geom. **5**(10), 201–290 (2000)
2. Di Cataldo, S., Ficarra, E., Macii, E.: Selection of tumor areas and segmentation of nuclear membranes in tissue confocal images: a fully automated approach. In: 2007 IEEE International Conference on Bioinformatics and Biomedicine (BIBM 2007), pp. 390–398. IEEE (2007)
3. Elmoataz, A., Schüpp, S., Clouard, R., Herlin, P., Bloyet, D.: Using active contours and mathematical morphology tools for quantification of immunohistochemical images. Signal Process. **71**(2), 215–226 (1998)
4. Ficarra, E., Di Cataldo, S., Acquaviva, A., Macii, E.: Automated segmentation of cells with ihc membrane staining. IEEE Trans. Biomed. Eng. **58**(5), 1421–1429 (2011)
5. Graham, S., et al.: Hover-net: simultaneous segmentation and classification of nuclei in multi-tissue histology images. Med. Image Anal. **58**, 101563 (2019)
6. Han, L., Yin, Z.: Unsupervised network learning for cell segmentation. In: de Bruijne, M., et al. (eds.) MICCAI 2021. LNCS, vol. 12901, pp. 282–292. Springer, Cham (2021). https://doi.org/10.1007/978-3-030-87193-2_27
7. Hou, L., Agarwal, A., Samaras, D., Kurc, T.M., Gupta, R.R., Saltz, J.H.: Robust histopathology image analysis: to label or to synthesize? In: Proceedings of the IEEE/CVF Conference on Computer Vision and Pattern Recognition, pp. 8533–8542 (2019)
8. Ji, Y., et al.: Multi-compound transformer for accurate biomedical image segmentation. In: de Bruijne, M., et al. (eds.) MICCAI 2021. LNCS, vol. 12901, pp. 326–336. Springer, Cham (2021). https://doi.org/10.1007/978-3-030-87193-2_31
9. Khameneh, F.D., Razavi, S., Kamasak, M.: Automated segmentation of cell membranes to evaluate her2 status in whole slide images using a modified deep learning network. Comput. Biol. Med. **110**, 164–174 (2019)
10. Kingma, D.P., Ba, J.: Adam: a method for stochastic optimization. arXiv preprint arXiv:1412.6980 (2014)
11. Lin, A., Chen, B., Xu, J., Zhang, Z., Lu, G., Zhang, D.: Ds-transunet: dual swin transformer u-net for medical image segmentation. IEEE Trans. Instrument. Meas. **71**, 1–15 (2022)

12. Luna, M., Kwon, M., Park, S.H.: Precise separation of adjacent nuclei using a siamese neural network. In: Shen, D., et al. (eds.) MICCAI 2019. LNCS, vol. 11764, pp. 577–585. Springer, Cham (2019). https://doi.org/10.1007/978-3-030-32239-7_64
13. Bueno-de Mesquita, J.M., Nuyten, D., Wesseling, J., van Tinteren, H., Linn, S., van De Vijver, M.: The impact of inter-observer variation in pathological assessment of node-negative breast cancer on clinical risk assessment and patient selection for adjuvant systemic treatment. Ann. Oncol. **21**(1), 40–47 (2010)
14. Mi, H., et al.: A quantitative analysis platform for pd-l1 immunohistochemistry based on point-level supervision model. In: IJCAI, pp. 6554–6556 (2019)
15. Qaiser, T., Rajpoot, N.M.: Learning where to see: a novel attention model for automated immunohistochemical scoring. IEEE Trans. Med. Imaging **38**(11), 2620–2631 (2019)
16. Qu, H., et al.: Weakly supervised deep nuclei segmentation using points annotation in histopathology images. In: International Conference on Medical Imaging with Deep Learning, pp. 390–400. PMLR (2019)
17. Roerdink, J.B., Meijster, A.: The watershed transform: definitions, algorithms and parallelization strategies. Fundamenta informaticae **41**(1–2), 187–228 (2000)
18. Ronneberger, O., Fischer, P., Brox, T.: U-Net: convolutional networks for biomedical image segmentation. In: Navab, N., Hornegger, J., Wells, W.M., Frangi, A.F. (eds.) MICCAI 2015. LNCS, vol. 9351, pp. 234–241. Springer, Cham (2015). https://doi.org/10.1007/978-3-319-24574-4_28
19. Saha, M., Chakraborty, C.: Her2net: a deep framework for semantic segmentation and classification of cell membranes and nuclei in breast cancer evaluation. IEEE Trans. Image Process. **27**(5), 2189–2200 (2018)
20. Swiderska-Chadaj, Z., et al.: Learning to detect lymphocytes in immunohistochemistry with deep learning. Med. Image Anal. **58**, 101547 (2019)
21. Tian, K., et al.: Weakly-supervised nucleus segmentation based on point annotations: a coarse-to-fine self-stimulated learning strategy. In: Martel, A.L., et al. (eds.) MICCAI 2020. LNCS, vol. 12265, pp. 299–308. Springer, Cham (2020). https://doi.org/10.1007/978-3-030-59722-1_29
22. Vogel, C., et al.: P1–07-02: discordance between central and local laboratory her2 testing from a large her2- negative population in virgo, a metastatic breast cancer registry (2011)
23. Wolff, A.C., et al.: Human epidermal growth factor receptor 2 testing in breast cancer: American society of clinical oncology/college of american pathologists clinical practice guideline focused update. Arch. Pathol. Lab. Med. **142**(11), 1364–1382 (2018)
24. Xu, J., et al.: Stacked sparse autoencoder (ssae) for nuclei detection on breast cancer histopathology images. IEEE Trans. Med. Imaging **35**(1), 119–130 (2015)
25. Yang, L., Meer, P., Foran, D.J.: Unsupervised segmentation based on robust estimation and color active contour models. IEEE Trans. Inf. Technol. Biomed. **9**(3), 475–486 (2005)
26. Yang, X., Li, H., Zhou, X.: Nuclei segmentation using marker-controlled watershed, tracking using mean-shift, and kalman filter in time-lapse microscopy. IEEE Trans. Circ. Syst. I: Regul. Papers **53**(11), 2405–2414 (2006)
27. Yoo, I., Yoo, D., Paeng, K.: PseudoEdgeNet: nuclei segmentation only with point annotations. In: Shen, D., et al. (eds.) MICCAI 2019. LNCS, vol. 11764, pp. 731–739. Springer, Cham (2019). https://doi.org/10.1007/978-3-030-32239-7_81

StainDiff: Transfer Stain Styles of Histology Images with Denoising Diffusion Probabilistic Models and Self-ensemble

Yiqing Shen[1] and Jing Ke[2,3](✉)

[1] Department of Computer Science, Johns Hopkins University, Baltimore, USA
yshen92@jhu.edu

[2] School of Electronic Information and Electrical Engineering, Shanghai Jiao Tong University, Shanghai, China
kejing@sjtu.edu.cn

[3] School of Computer Science and Engineering, University of New South Wales, Kensingt, Australia

Abstract. The commonly presented histology stain variation may moderately obstruct the diagnosis of human experts, but can considerably downgrade the reliability of deep learning models in various diagnostic tasks. Many stain style transfer methods have been proposed to eliminate the variance of stain styles across different medical institutions or even different batches. However, existing solutions are confined to Generative Adversarial Networks (GANs), AutoEncoders (AEs), or their variants, and often fell into the shortcomings of mode collapses or posterior mismatching issues. In this paper, we make the first attempt at a Diffusion Probabilistic Model to cope with the indispensable stain style transfer in histology image context, called `StainDiff`. Specifically, our diffusion framework enables learning from unpaired images by proposing a novel cycle-consistent constraint, whereas existing diffusion models are restricted to image generation or fully supervised pixel-to-pixel translation. Moreover, given the stochastic nature of `StainDiff` that multiple transferred results can be generated from one input histology image, we further boost and stabilize the performance by the proposal of a novel self-ensemble scheme. Our model can avoid the challenging issues in mainstream networks, such as the mode collapses in GANs or alignment between posterior distributions in AEs. In conclusion, `StainDiff` suffices to increase the stain style transfer quality, where the training is straightforward and the model is simplified for real-world clinical deployment.

Keywords: Diffusion Probabilistic Model · Histology Stain Transfer

H. Greenspan et al. (Eds.): MICCAI 2023, LNCS 14225, pp. 549–559, 2023.
https://doi.org/10.1007/978-3-031-43987-2_53

1 Introduction

Staining is a vital process in preparing tissue samples for histology studies. Specifically, with dyes such as Hematoxylin and Eosin, transparent tissue elements can be transformed into distinguishable features [1]. However, stain styles can vary significantly across different pathology labs or institutions. These variations can be due to the difference in staining materials, protocols, or processes among different pathologists or digital scanners [16,23]. Yet, the stain variations can cause inconsistencies between human domain experts [11]; and also hinder the performance of computer-aided diagnostic (CAD) systems [5,7]. Moreover, experiments have shown that stain variations can lead to a significant decrease in the accuracy and reproducibility of deep learning algorithms in histology analysis. Consequently, it is crucial to minimize staining variations to ensure reliable, consistent, and accurate CAD systems.

To address the issue of stain variations between different domains, stain style transfer has been proposed. While the conventional color matching [22] and stain separation methods [19] used to be popular; learning-based approaches have become increasingly dominant, because they eliminate the need for challenging manual selection of the template images. For example, Stain-to-Stain Translation (STST) [25] approaches stain style transfer within a fully supervised 'pix2pix' framework [12]. Another approach, called StainGAN [26], improves on STST by tailoring a CycleGAN [34] to get rid of the dependence on learning from paired histology images and enable an unsupervised learning manner. These methods have shown promising results in reducing staining variations.

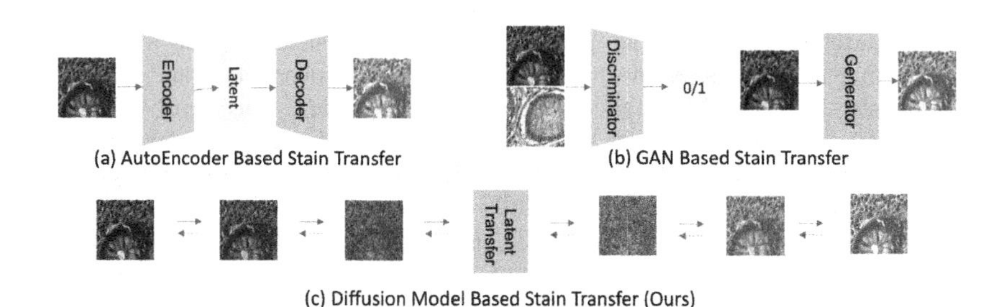

Fig. 1. Different generative model based stain style transfer solutions. (a) AutoEncoder (AE). (b) GAN. (c) Our proposed diffusion based `StainDiff`.

Existing learning-based methods for stain style transfer are primarily confined to Generative Adversarial Networks (GANs) [6] and AutoEncoder (AE) [2], as depicted in Fig. 1(a) and (b) respectively. However, GAN approaches and AE suffer from the training of extra discriminators and challenging alignment of the posterior distributions, respectively [27]. In contrast, diffusion models, such as the prevalent denoising diffusion probabilistic model (DDPM) [9], have emerged as an alternative approach that can achieve competitive performance in various image-related tasks, such as image generation, inpainting, super-resolution, and *etc* [3]. Importantly, diffusion models offer several advantages over GANs and AEs, including tractable probabilistic parameterization, stable training procedures, and theoretical guarantees [3]. Additionally, they can avoid some of the challenges encountered by GANs and AEs, such as the alignment of posterior distributions or training extra discriminators, leading to a simpler model and training process. However, the applicability of diffusion models to histology stain style transfer remains unexplored. While the current diffusion models focus on image synthesis [9] or supervised image-to-image transaction [24], they are not applicable to our circumstance, as obtaining paired histology slides with different stain styles is not feasible in real clinical practice [27]. Therefore, we design an innovative cycle-consistent diffusion model that allows the transfer of representations between latent spaces at different time steps with the same morphological structure preserved in an unsupervised manner, as shown in Fig. 1(c).

The major contributions are three-fold, summarized as follows. (1) We propose `StainDiff`, which is the first attempt at a pure denoising diffusion probabilistic model for stain transfer. More innovatively, unlike existing diffusion models, `StainDiff` is capable of learning from unpaired histology images, making it a more flexible and practical solution. The model is superior to GAN-based methods as the training of additional discriminators is free, and also spares for the difficulty in the alignment of posterior probabilities in AE-based approaches. (2) We also propose a self-ensemble scheme to further improve and stabilize the style transfer performance in `StainDiff`. This scheme utilizes the stochastic property of the diffusion model to generate multiple slightly different outputs from one input at the inference stage. (3) A broad range of histology tasks, such as stain normalization between multiple clients, can be conveniently achieved with minor adjustment to the loss in `StainDiff`.

2 Methods

Overview. The goal of this work is to design a diffusion model [9] to transfer the stain style between two domains, *i.e.*, $\mathcal{X}^A, \mathcal{X}^B$. However, the traditional training paradigm of conditional DDPMs with paired images $(\mathbf{x}_0^A, \mathbf{x}_0^B) \in \mathcal{X}^A \times \mathcal{X}^B$ is not feasible, as they are unavailable in the context of histology. To overcome this limitation, we design an innovative diffusion framework for stain style transfer, named `StainDiff`, which leverages the success of CycleGAN [34] and StyleGAN [26] and thus can be trained in an unsupervised manner with a novel cycle-consistency constraint. Specifically, `StainDiff` comprises two forward processes that perturb the histology image of two stain style domains to noise respectively, and two corresponding reverse processes that attempt to reconstruct noise back to original images from the perturbed ones. The overall training process is depicted in Fig. 2.

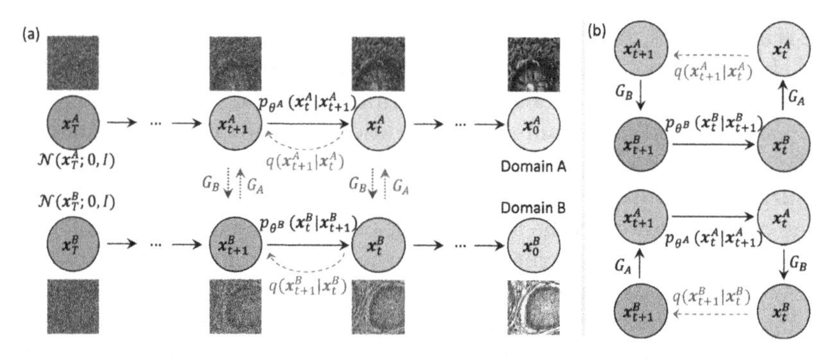

Fig. 2. (a) The directed graphical model for the training process of the proposed StainDiff. It comprises two diffusion paths, that each learns the histology image generation with respect to one stain style domain. The interplay between two domains is learned by a paired auxiliary transform network G_A and G_B through a cycle-consistency constraint. (b) We specify the consistency cycles to impose the regularization.

Forward Process. Parameterized by the Markov chain, the forward process in StainDiff follows the vanilla DDPM by perturbing the histology images gradually with Gaussian noise, until all structures and morphological context information are lost. Formally, given a histology image $\mathbf{x}_0^A$ with respect to the stain style domain A, a transition kernel q progressively generates a sequence of T latent variables $\mathbf{x}_1^A, \mathbf{x}_2^A, \cdots, \mathbf{x}_T^A$ thorough the following equation:

$$q(\mathbf{x}_t^A|\mathbf{x}_{t-1}^A) = \mathcal{N}(\mathbf{x}_t^A; \sqrt{1-\beta_t}\mathbf{x}_{t-1}^A, \beta_t\mathbf{I}), \tag{1}$$

where $\mathcal{N}(\cdot)$ denotes the Gaussian distribution, $\mathbf{I}$ is the identity matrix. The hyper-parameters β_ts follow a linear rule as defined in DDPM [9] to guarantee $q(\mathbf{x}_T^A) = \int q(\mathbf{x}_T^A|\mathbf{x}_0^A)q(\mathbf{x}_0^A)d\mathbf{x}_0^A \approx \mathcal{N}(\mathbf{x}_T^A; \mathbf{0}, \mathbf{I})$. Identically, we can progress the latent variables $\mathbf{x}_1^B, \mathbf{x}_2^B, \cdots, \mathbf{x}_T^B$ for the histology image $\mathbf{x}_0^B$ from the stain style domain B in the same fashion as Eq. (1).

Reverse Process and Cycle-Consistency Constraint. The reverse process in StainDiff optimizes two conditional diffusion models, namely $p(\mathbf{x}^A|\mathbf{x}^B)$ and $p(\mathbf{x}^B|\mathbf{x}^A)$, to transfer the stain style between two domains A and B. Concretely, two learnable transition kernels $p_{\theta^A}(\mathbf{x}_t^A|\mathbf{x}_{t+1}^A)$ and $p_{\theta^B}(\mathbf{x}_t^B|\mathbf{x}_{t+1}^B)$ learns to reverse the Eq. (1) and generate images characterized by stain style A and B respectively, by gradually removing the noise initialized from Gaussian prior. To ensure conservative outputs [24], L1-norm denoising objective $\mathcal{L}_d$ [4] is leveraged to train the denoising networks in the transition kernels. Due to the absence of pixel-to-pixel paired histology of both stain styles, it is infeasible to learn the interplay between them in a supervised manner as in most previous works. Consequently, a pair of auxiliary transform networks $G_A : \mathbf{x}_t^B \mapsto \mathbf{x}_t^A$ and $G_B : \mathbf{x}_t^A \mapsto \mathbf{x}_t^B$ are designed to learn the transfer between the latent variables across the two domains in an unsupervised fashion, using a novel cycle-consistency constraint. Formally, this constraint ensures that two cycles as depicted in Fig. 2(b), derive an identity mapping, *i.e.*,

$$\begin{cases} \mathbf{x}_{t+1}^A &= q \circ G_A \circ p_{\theta^B} \circ G_B(\mathbf{x}_{t+1}^A), \\ \mathbf{x}_{t+1}^B &= q \circ G_B \circ p_{\theta^A} \circ G_A(\mathbf{x}_{t+1}^B), \end{cases} \tag{2}$$

where $\circ$ denotes the composition of operations. It follows the cycle-consistency constraint formulated by

$$\mathcal{L}_c = \mathbb{E}_{t,\mathbf{x}_0^A} \|\tilde{\mathbf{x}}_{t+1}^A - \mathbf{x}_{t+1}^A\| + \mathbb{E}_{t,\mathbf{x}_0^B} \|\tilde{\mathbf{x}}_{t+1}^B - \mathbf{x}_{t+1}^B\|, \tag{3}$$

where $\tilde{\mathbf{x}}_{t+1}^A$ and $\tilde{\mathbf{x}}_{t+1}^B$ are defined by Eq. (2); $\mathbb{E}$ denotes the expectation; $\|\cdot\|$ is the L1-norm. Finally, the overall loss function is $\mathcal{L} = \mathcal{L}_d + \gamma\mathcal{L}_c$, balanced by the coefficient γ.

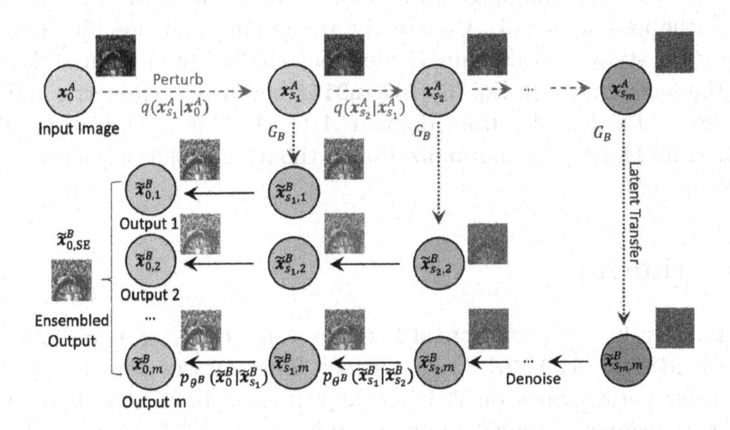

Fig. 3. The directed graphical model for the inference stage with self-ensemble scheme of the `StainDiff`.

Inference Process and Self-ensemble. We describe the inference stage of `StainDiff` by transferring the histology images from stain style A to B; while the inverse, namely from B to A, is similar. Given a histology image input $\mathbf{x}_0^A$ characterized by stain style A, we begin by perturbing it s steps with Eq. (1) to derive $\mathbf{x}_s^A$. Choosing the optimal value for s is important, as a large s (*e.g.*, $s = T$) leads to the loss of the contextual and structural information; while a small valued s (*e.g.*, $s = 1$) fails to inject sufficient noise for `StainDiff` to transfer style. An ideal range for s is a small subset from $[1, T]$ that is centered by $\frac{1}{2}T$. Consequently, in this work, we fix s in the range of $[S_1, S_2]$ with $S_1 = \frac{2}{5}T$ and $S_2 = \frac{3}{5}T$. Afterwards, the latent variable $\mathbf{x}_s^A$ is transferred into the corresponding latent space with respect to stain style B with auxiliary transform network G_B, which gives us $\tilde{\mathbf{x}}_s^B = G_B(\mathbf{x}_s^A)$. Next, we use the p-sample [9] iteratively to denoise the $\tilde{\mathbf{x}}_s^B$ and obtain the transferred image $\tilde{\mathbf{x}}_0^B$. As the sampling is a stochastic process, different values of s result in a slight difference in the transferred output. We exploit this property and propose a novel self-ensemble method that can implicitly generate an ensemble of transferred output without the need to train extra models. Specifically, we repeat the above process for m times with different s *i.e.*, $s_1, \cdots, s_m \in [S_1, S_2]$, and come to m outputs $\tilde{\mathbf{x}}_{0,i}^B$ with $i = 1, \cdots, m$. The self-ensemble transferred output is then given by $\tilde{\mathbf{x}}_{0,\mathrm{SE}}^B = \sum_{i=1}^m \tilde{\mathbf{x}}_{0,i}^B$. The graphical model for the inference process and the proposed self-ensemble scheme are summarized in Fig. 3.

Extension to Stain Normalization. The stain transfer primarily addresses the domain gap between two stain styles, which is mathematically formulated as a one-to-one mapping. Meanwhile, in some clinical settings, multiple institutions or hospitals are involved, where stain normalization is usually employed for multiple stain styles to one style alignment. The proposed symmetric `StainDiff` structure can be easily adapted to support stain normalization, with minimal change to the loss in Eq. (3). Concretely, we assume that domain A comprises multiple stain styles and domain B identifies the targeted stain style. By discarding the second term in Eq. (3), `StainDiff` becomes asymmetric and focuses specifically on the transfer from domain A to B. This modification allows us to use `StainDiff` for stain normalization without any other adjustments to the inference process.

3 Experiments

Datasets. Evaluations of `StainDiff` are conducted on two datasets. (1) Dataset-A: *MITOS-ATYPIA 14 Challenge*[1]. This dataset aims to measure the style transfer performance on 284 histology frames. Each slide is digitized by two different scanners, resulting in stain style variations. For a fair comparison, we follow the settings in previous work [26] by using 10,000 unpaired patches randomly cropped from the first 184 slides of both scanners as the training set. Meanwhile, 500 paired patches are generated from the remaining 100 slides as the test set, where we use Pearson correlation coefficient (PC), Structural Similarity index (SSIM) [31] and Feature Similarity Index for Image Quality Assessment (FSIM) [33] as the evaluation metrics. (2) Dataset-B: *The Cancer Genome Atlas (TCGA)*. This dataset evaluates the performance of stain normalization quantified by the downstream nine-category tissue structure classification accuracy [27]. Domain A contains histology of multiple stain styles, which are collected from 186 WSIs from TCGA-COAD and NCT-CRC-HE-100K [14]; and domain B is the target style, curated from 25 WSIs in CRC-VAL-HE-7K [14].

Implementations. All experiments are implemented in Python 3.8.13 with Pytorch 1.12.1 on two NVIDIA GeForce RTX 3090 GPU cards with 24GiB of

[1] https://mitos-atypia-14.grand-challenge.org.

Table 1. Comparison of stain style transfer performance on Dataset-A. To show the statistical significance, the p-values in terms of SSIM and FSIM are computed with respect to `StainDiff` (full setting). The 'w/o SE' denotes the exclusion of the self-ensemble scheme from the inference stage.

Method	PC (↑)	SSIM (↑)	FSIM (↑)	p_{SSIM}	p_{FSIM}
Reinhard [22]	$0.509_{\pm 0.091}$	$0.587_{\pm 0.041}$	$0.668_{\pm 0.032}$	1.0×10^{-5}	2.0×10^{-5}
Macenko [19]	$0.507_{\pm 0.108}$	$0.554_{\pm 0.084}$	$0.675_{\pm 0.024}$	3.2×10^{-4}	1.1×10^{-4}
Khan [17]	$0.563_{\pm 0.053}$	$0.643_{\pm 0.049}$	$0.702_{\pm 0.032}$	1.4×10^{-5}	5.7×10^{-5}
Vahadane [30]	$0.561_{\pm 0.058}$	$0.639_{\pm 0.063}$	$0.710_{\pm 0.031}$	8.8×10^{-5}	3.5×10^{-4}
StaNoSa [13]	$0.552_{\pm 0.081}$	$0.647_{\pm 0.070}$	$0.692_{\pm 0.044}$	4.1×10^{-4}	1.2×10^{-4}
StainGAN [26]	$0.572_{\pm 0.049}$	$0.692_{\pm 0.072}$	$0.723_{\pm 0.014}$	3.2×10^{-3}	1.4×10^{-4}
Harshal [21]	$0.552_{\pm 0.047}$	$0.685_{\pm 0.033}$	$0.705_{\pm 0.026}$	6.2×10^{-4}	2.7×10^{-4}
MWB [20]	$0.543_{\pm 0.107}$	$0.690_{\pm 0.052}$	$0.718_{\pm 0.032}$	1.2×10^{-3}	2.0×10^{-3}
CL-StainGAN [15]	$0.585_{\pm 0.039}$	$0.701_{\pm 0.034}$	$0.734_{\pm 0.022}$	2.4×10^{-3}	4.0×10^{-3}
`StainDiff` (w/o SE)	$\mathbf{0.590}_{\pm 0.019}$	$\mathbf{0.709}_{\pm 0.010}$	$\mathbf{0.742}_{\pm 0.007}$	2.3×10^{-3}	3.0×10^{-3}
`StainDiff` (Full)	$\mathbf{0.599}_{\pm 0.025}$	$\mathbf{0.721}_{\pm 0.017}$	$\mathbf{0.753}_{\pm 0.010}$	–	–

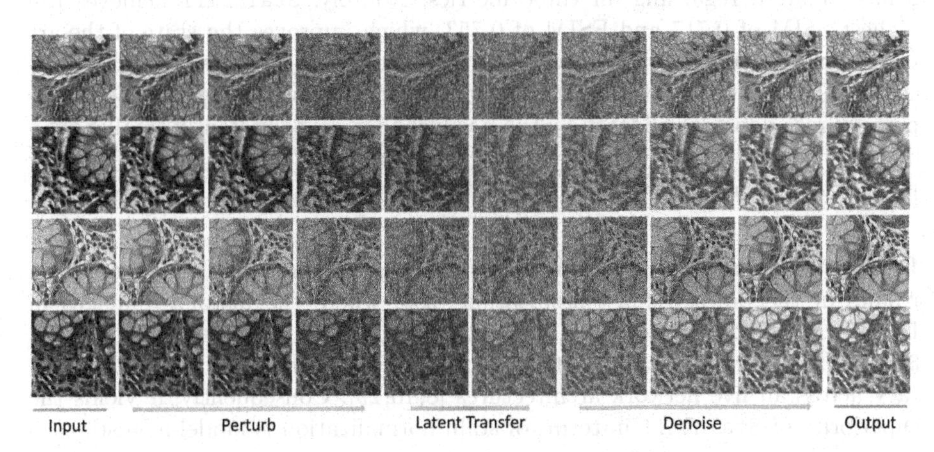

Fig. 4. Visualization of the progressive stain style transfer process in `StainDiff`.

memory each in parallel. We leverage the Adam optimizer with a learning rate of 2×10^{-4}, and a batch size of 4. The learning scheme follows previous work [18], where the training process continues for 100 epochs if the overall loss did not decrease to the average loss of the previous 20 epochs. For `StainDiff`, we set the diffusion time $T = 1000$, the balancing coefficient $\gamma = 1$, and ensemble number $m = 10$. All experiments are repeated for 7 runs with different fixed random seeds *i.e.*, $\{0, 1, 2, 3, 4, 5, 6\}$; and metrics are reported in the form of mean±standard deviation.

Table 2. Comparison of the downstream classification task w.r.t. accuracy (%). 'w/o SE' denotes excluding the self-ensemble scheme from the inference stage.

Method	ResNet-18	ResNet-50	DenseNet-121	DenseNet-169	EfficientNet
Reinhard [22]	92.10±0.58	93.11±0.39	92.04±0.74	91.17±0.33	92.48±0.99
Macenko [19]	90.35±1.20	90.55±1.43	90.47±1.53	89.99±1.04	91.24±0.85
Khan [17]	91.89±0.83	92.04±0.63	92.22±0.75	92.54±0.55	93.04±0.81
Vahadane [30]	92.87±0.35	93.01±0.42	92.94±0.50	93.31±0.52	93.51±0.39
StaNoSa [13]	93.25±0.78	93.88±0.21	94.41±0.78	94.60±0.99	94.48±0.53
StainGAN [26]	93.98±0.28	94.21±0.47	93.89±0.21	93.99±0.36	94.52±0.32
Harshal [21]	94.17±0.72	94.58±0.54	93.85±0.80	93.96±0.65	95.03±0.99
MWB [20]	94.56±0.93	95.15±0.79	94.14±0.68	94.32±0.39	95.25±0.55
CL-StainGAN [15]	96.22±0.73	96.89±0.38	95.98±0.58	96.04±0.29	96.49±0.57
StainDiff (w/o SE)	96.79±0.23	97.54±0.47	96.01±0.30	97.00±0.21	97.32±0.35
StainDiff (Full)	**97.48**±0.10	**98.12**±0.09	**96.98**±0.07	**97.93**±0.11	**98.42**±0.08

Evaluations on Style Transfer. The superiority of our diffusion model to GANs and AEs in histology stain style transfer is quantitatively reflected in Table 1. Specifically, on Dataset-A, `StainDiff` can surpass its counterparts with a large margin regarding all three metrics. Notably, `StainDiff` achieves the highest SSIM of 0.717 and FSIM of 0.753, which improves the state-of-the-art CL-StainGAN [15] by 0.016 and 0.019 respectively, without reliance on time-costly self-supervised pre-training. Moreover, the statistical significance of our performance boost is validated by the p-values that are consistently smaller than 0.005, as computed from the Wilcoxon signed-rank test. The progressive transfer process of `StainDiff` over time is visualized in Fig. 4.

Evaluations on Stain Normalization. Table 2 presents the comparison results of the downstream classification task, where the histology images in Dataset-B are normalized using different methods. The table clearly shows that `StainDiff` outperforms all the other methods and achieves the highest test accuracy across all five network architectures [8,10,29]. Consequently, it yields the superiority of `StainDiff` in terms of stain normalization is model-agnostic.

Ablation Study. Table 1 and 2 show that incorporating a self-ensemble scheme can both boost the performance of `StainDiff`, and bring down the variations, demonstrating its effectiveness in stabilizing the stain transfer and normalization. To further investigate the effect of ensemble number m, we conduct ablation on Dataset-A. Experimentally, the FSIM when $m = 1, 5, 10, 15, 20, 50$ are 0.742, 0.749, 0.753, 0.756, 0.759, 0.759 respectively. While a slight performance gain can be achieved with higher m values than 10, the ensemble becomes more time-consuming, as the cost time is linear to m. It implies an optimal m should be selected as a trade-off between the performance and computational time, such as 10 in this work.

4 Conclusion

In this paper, we propose `StainDiff`, a denoising diffusion model for histological stain style transfer, hence a model can get rid of the challenging issues in mainstream networks, such as the mode collapses in GANs or alignment between posterior distributions in AEs. Innovatively, by imposing a cycle-consistent constraint imposed on latent spaces, `StainDiff` enables learning from unpaired histology images, making it widely applicable to real clinical settings. One future work will explore efficient sampling diffusion models, *e.g.*, DDIM [28], to address the long sampling time issue as inherited from DDPM. Another direction is to investigate other formulations of the diffusion model in the context of stain transfer, such as score-based or score-SDE diffusion models [32]. These extensions will fully expand the scope of our work, hence further advancing towards a comprehensive solution of stain style transfer in histology images.

Acknowledgement. J. Ke was supported by National Natural Science Foundation of China (Grant No. 62102247) and Natural Science Foundation of Shanghai (No. 23ZR1430700).

References

1. Anderson, J.: An introduction to routine and special staining (2011). Accessed 18 Aug 2014
2. Bengio, Y., et al.: Greedy layer-wise training of deep networks. Adv. Neural Inf. Process. Syst. **19**, 1–8 (2006)
3. Cao, H., et al.: A survey on generative diffusion model. arXiv preprint arXiv:2209.02646 (2022)
4. Chen, N., et al.: Wavegrad: estimating gradients for waveform generation. In: International Conference on Learning Representations (2020)
5. Ciompi, F., et al.: The importance of stain normalization in colorectal tissue classification with convolutional networks. In: 2017 IEEE 14th International Symposium on Biomedical Imaging (ISBI 2017), pp. 160–163. IEEE (2017)
6. Goodfellow, I., et al.: Generative adversarial networks. Commun. ACM **63**(11), 139–144 (2020)
7. Gupta, V., Singh, A., Sharma, K., Bhavsar, A.: Automated classification for breast cancer histopathology images: is stain normalization important? In: Cardoso, M.J., et al. (eds.) CARE/CLIP -2017. LNCS, vol. 10550, pp. 160–169. Springer, Cham (2017). https://doi.org/10.1007/978-3-319-67543-5_16
8. He, K., et al.: Deep residual learning for image recognition. In: Proceedings of the IEEE Conference on Computer Vision and Pattern Recognition, pp. 770–778 (2016)
9. Ho, J., Jain, A., Abbeel, P.: Denoising diffusion probabilistic models. Adv. Neural Inf. Process. Syst. **33**, 6840–6851 (2020)
10. Huang, G., et al.: Densely connected convolutional networks. In: Proceedings of CVPR, pp. 4700–4708 (2017)
11. Ismail, S.M., et al.: Observer variation in histopathological diagnosis and grading of cervical intraepithelial neoplasia. Brit. Med. J. **298**(6675), 707–710 (1989)

12. Isola, P., et al.: Image-to-image translation with conditional adversarial networks. In: Proceedings of the IEEE Conference on Computer Vision and Pattern Recognition, pp. 1125–1134 (2017)
13. Janowczyk, A., et al.: Stain normalization using sparse autoencoders (stanosa): application to digital pathology. Comput. Med. Imaging Graph. **57**, 50–61 (2017)
14. Kather, J.N., Halama, N., Marx, A.: 100,000 histological images of human colorectal cancer and healthy tissue. Zenodo (2018)
15. Ke, J., Shen, Y., Liang, X., Shen, D.: Contrastive learning based stain normalization across multiple tumor in histopathology. In: de Bruijne, M., et al. (eds.) MICCAI 2021. LNCS, vol. 12908, pp. 571–580. Springer, Cham (2021). https://doi.org/10.1007/978-3-030-87237-3_55
16. Ke, J., et al.: Multiple-datasets and multiple-label based color normalization in histopathology with cgan. In: Medical Imaging 2021: Digital Pathology, vol. 11603, pp. 263–268. SPIE (2021)
17. Khan, A.M., et al.: A nonlinear mapping approach to stain normalization in digital histopathology images using image-specific color deconvolution. IEEE Trans. Biomed. Eng. **61**(6), 1729–1738 (2014)
18. Lyu, Q., Wang, G.: Conversion between ct and mri images using diffusion and score-matching models. arXiv preprint arXiv:2209.12104 (2022)
19. Macenko, M., et al.: A method for normalizing histology slides for quantitative analysis. In: 2009 IEEE International Symposium on Biomedical Imaging: From Nano to Macro, pp. 1107–1110. IEEE (2009)
20. Nadeem, S., Hollmann, T., Tannenbaum, A.: Multimarginal wasserstein barycenter for stain normalization and augmentation. In: Martel, A.L., et al. (eds.) MICCAI 2020. LNCS, vol. 12265, pp. 362–371. Springer, Cham (2020). https://doi.org/10.1007/978-3-030-59722-1_35
21. Nishar, H., Chavanke, N., Singhal, N.: Histopathological stain transfer using style transfer network with adversarial loss. In: Martel, A.L., et al. (eds.) MICCAI 2020. LNCS, vol. 12265, pp. 330–340. Springer, Cham (2020). https://doi.org/10.1007/978-3-030-59722-1_32
22. Reinhard, E., et al.: Color transfer between images. IEEE Comput. Graph. Appl. **21**(5), 34–41 (2001)
23. Rubin, R., Strayer, D.S., Rubin, E., et al.: Rubin's pathology: clinicopathologic foundations of medicine. Lippincott Williams & Wilkins (2008)
24. Saharia, C., et al.: Palette: image-to-image diffusion models. In: ACM SIGGRAPH 2022 Conference Proceedings, pp. 1–10 (2022)
25. Salehi, P., Chalechale, A.: Pix2pix-based stain-to-stain translation: a solution for robust stain normalization in histopathology images analysis. In: 2020 International Conference on Machine Vision and Image Processing (MVIP), pp. 1–7. IEEE (2020)
26. Shaban, M.T., et al.: Staingan: stain style transfer for digital histological images. In: 2019 IEEE 16th International Symposium on Biomedical Imaging (ISBI 2019), pp. 953–956. IEEE (2019)
27. Shen, Y., et al.: A federated learning system for histopathology image analysis with an orchestral stain-normalization gan. IEEE Trans. Med. Imaging **42**, 1969–1981 (2022)
28. Song, J., et al.: Denoising diffusion implicit models. arXiv preprint arXiv:2010.02502 (2020)
29. Tan, M., Le, Q.: Efficientnet: rethinking model scaling for convolutional neural networks. In: International Conference on Machine Learning, pp. 6105–6114. PMLR (2019)

30. Vahadane, A., et al.: Structure-preserving color normalization and sparse stain separation for histological images. IEEE Trans. Med. Imaging **35**(8), 1962–1971 (2016)
31. Wang, Z., et al.: Image quality assessment: from error visibility to structural similarity. IEEE Trans. Image Process. **13**(4), 600–612 (2004)
32. Yang, L., et al.: Diffusion models: a comprehensive survey of methods and applications. arXiv preprint arXiv:2209.00796 (2022)
33. Zhang, L., et al.: Fsim: a feature similarity index for image quality assessment. IEEE Trans. Image Process. **20**(8), 2378–2386 (2011)
34. Zhu, J.Y., et al.: Unpaired image-to-image translation using cycle-consistent adversarial networks. In: Proceedings of the IEEE International Conference on Computer Vision, pp. 2223–2232 (2017)

IIB-MIL: Integrated Instance-Level and Bag-Level Multiple Instances Learning with Label Disambiguation for Pathological Image Analysis

Qin Ren[1,2], Yu Zhao[1(✉)], Bing He[1(✉)], Bingzhe Wu[1], Sijie Mai[3], Fan Xu[1,4], Yueshan Huang[1,5], Yonghong He[2], Junzhou Huang[6], and Jianhua Yao[1(✉)]

[1] AI Lab, Tencent, Shenzhen 518000, China
jianhuayao@tencent.com
[2] Shenzhen International Graduate School, Tsinghua University, Shenzhen 518071, China
[3] School of Electronic and Information Technology, Sun Yat-sen University, Guangzhou 510006, China
[4] ShanghaiTech University, Shanghai 201210, China
[5] Shanghai Jiao Tong University, Shanghai 200240, China
[6] University of Texas at Arlington, Arlington, TX 76019, USA

Abstract. Digital pathology plays a pivotal role in the diagnosis and interpretation of diseases and has drawn increasing attention in modern healthcare. Due to the huge gigapixel-level size and diverse nature of whole-slide images (WSIs), analyzing them through multiple instance learning (MIL) has become a widely-used scheme, which, however, faces the challenges that come with the weakly supervised nature of MIL. Conventional MIL methods mostly either utilized instance-level or bag-level supervision to learn informative representations from WSIs for downstream tasks. In this work, we propose a novel MIL method for pathological image analysis with integrated instance-level and bag-level supervision (termed IIB-MIL). More importantly, to overcome the weakly supervised nature of MIL, we design a label-disambiguation-based instance-level supervision for MIL using Prototypes and Confidence Bank to reduce the impact of noisy labels. Extensive experiments demonstrate that IIB-MIL outperforms state-of-the-art approaches in both benchmarking datasets and addressing the challenging practical clinical task. The code is available at https://github.com/TencentAILabHealthcare/IIB-MIL.

Keywords: computational pathology · multi-instance learning · label disambiguation · prototype · confidence bank

Q. Ren and Y. Zhao—Equally-contributed authors.

Supplementary Information The online version contains supplementary material available at https://doi.org/10.1007/978-3-031-43987-2_54.

H. Greenspan et al. (Eds.): MICCAI 2023, LNCS 14225, pp. 560–569, 2023.
https://doi.org/10.1007/978-3-031-43987-2_54

1 Introduction

Pathology is widely recognized as the gold standard for disease diagnosis [15]. As the demand for intelligently pathological image analysis continues to grow, an increasing number of researchers have paid attention to this field [12,14, 25]. However, pathological image analysis remains a challenging task due to the complex and heterogeneous nature [19] of obtained whole slide images (WSIs), as well as their huge gigapixel-level size [20]. To address this issue, multiple instance learning (MIL) [1] is usually applied to formulate pathological image analysis tasks into weakly supervised learning problems. In the MIL setting, the entire WSI is regarded as a bag and tiled patches are instances. The primary challenge of MIL arises from its weakly supervised nature, i.e. only the bag-level label for the entire WSI is provided, while labels for individual patches are usually unavailable. Although MIL-based methods have shown impressive potential in solving a wide range of pathological image analysis tasks including cancer grading and subtype diagnosis [23], prognosis prediction [18], genotype-related tasks such as gene mutation prediction [4], etc., it is still an open question regarding learning an informative and effective representation of the entire WSI for down-streaming task based on MIL architecture.

Current MIL methods can be broadly categorized into two types: bag-level MIL and instance-level MIL. Bag-level MIL [9,17], also known as embedding-based MIL, involves converting patches (instances) into low-dimensional embeddings, which are then aggregated into WSI (bag)-level representations to conduct the analysis tasks [22]. The aggregator can take different architectures such as an attention module [7,13], convolutional neural network (CNN), Transformer [16], or graph neural network [10,28]. Instance-level MIL [2,8,24], on the other hand, focuses its learning process at the instance level, and then obtains the bag-level prediction by simply aggregating instance predictions. Bag-level MIL incorporates instance embeddings to create a bag representation, converting the MIL into a supervised learning problem. Furthermore, it can extract contextual information and correlations between instances. Nonetheless, Bag-level MIL needs to learn informative embeddings of instances and adjust the contributions of these instance embeddings to generate the bag representation simultaneously, which faces the risk of obtaining a suboptimal model given the limited training samples in practice. The instance-level MIL, however, faces the problem of noisy labels, which is caused by the common strategy of assigning the WSI labels to patches and the fact that there are lots of patches irrelevant to the WSI labels [3,6].

Considering these conventional MIL methods usually utilize either bag-level or instance-level supervision, leading to suboptimal performance. In this paper, we format the instance-level MIL as a noisy label learning task and propose to solve it by designing an instance-level supervision based on the label disambiguation [21]. Then we propose to combine bag-level and instance-level supervision to improve the performance of MIL. The bag-level and instance-level supervision can corporately optimize the instance embedding learning process and well-learned instance embeddings can facilitate the aggregation module to generate the bag representation. The co-supervision design also makes the MIL to be a multi-task learning framework, where the bag-level supervision channel works to globally summarise the WSI for prediction and the instance-level supervision channel can locally identify key relevant patches. The detailed contributions can be summarized as follows:

1) We propose a novel MIL method for pathological image analysis that leverages a specially-designed residual Transformer backbone and organically integrates both Transformer-based bag-level and label-disambiguation-based instance-level supervision for performance enhancement.
2) We develop a label-disambiguation module that leverages prototypes and confidence bank to tackle the weakly supervised nature of instance-level supervision and reduce the impact of assigned noisy labels.
3) The proposed framework outperforms state-of-the-art (SOTA) methods on public datasets and in a practical clinical task, demonstrating its superiors in WSI analysis. Besides, ablation studies illustrate the superiority of our co-supervision design compared to using only one type of supervision.

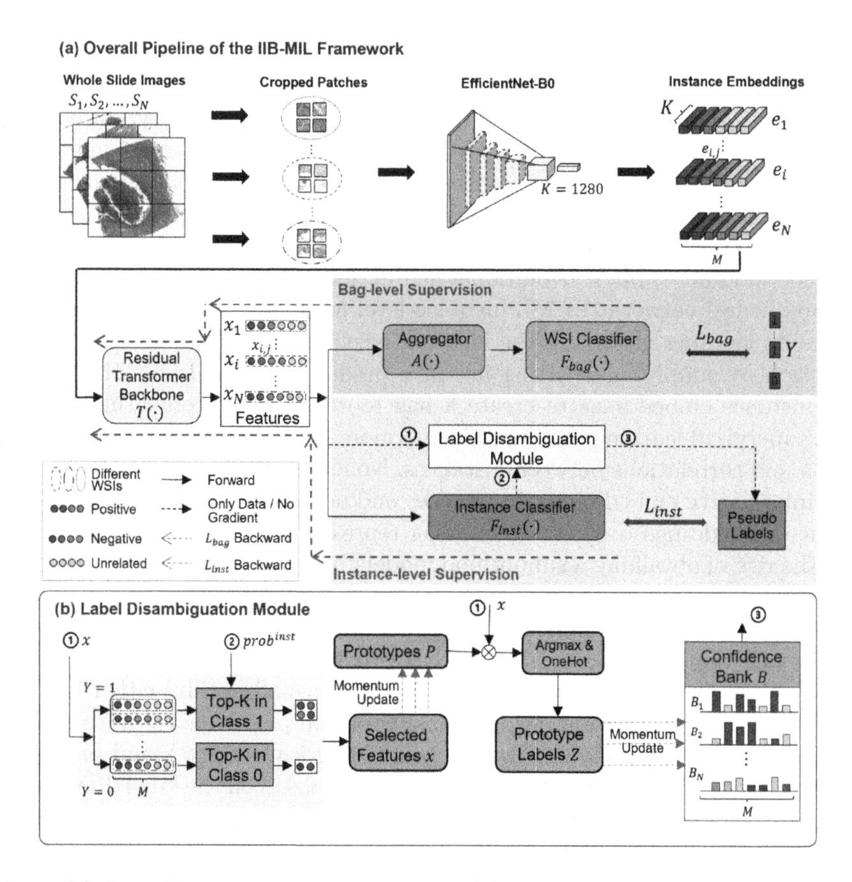

Fig. 1. (a) Overall framework of the IIB-MIL.(b) The detailed diagram of the label-disambiguation-based instance-level supervision. (Details are given in section: 2.1)

2 Method

2.1 Overview

The overall framework of the proposed IIB-MIL is shown in Fig. 1. Similar to previous works [27], IIB-MIL first transforms input huge-size WSI to a set of patch embeddings to simplify the following learning task using a pre-trained encoder, i.e. EfficientNet-B0. Then a specially-designed residual transformer backbone works to calibrate the obtained patch embeddings and encode the context information and correlation of patches. After that, IIB-MIL utilizes both a transformer-based bag-level and a label-disambiguation-based instance-level supervision to cooperatively optimize the model, where the bag-level loss is calculated referring to the WSI labels, while the instance loss is calculated referring to pseudo patch labels calibrated by the Label-Disambiguation module. Since bag-level supervision channel is trained to globally summarise information of all patches for prediction, the bag-level outputs are used as the final predictions during the test stage.

2.2 Problem Formulation

Assume there is a set of N WSIs denoted by $S = \{S_1, S_2, ..., S_N\}$. Each WSI S_i has a WSI-level label $Y_i \in \{1, ..., C\}$, where C represents category number. In each S_i, there exist M_i tiled patches without patch-level labels. To reduce the computational cost, we used a frozen pre-trained encoder to transform patches into K dimensional embeddings $\{e_{i,j} | e_{i,j} \in \mathbb{R}^K, i \in [1, N], j \in [1, M]\}$. Our proposed IIB-MIL comprehensively integrates obtained embeddings $\{e_{i,j}, ...\}$ to generate accurate WSI classification.

2.3 Backbone Network

Before bag-level and instance-level supervision channels, we design a residual transformer backbone $T(\cdot) : \mathbb{R}^K \to \mathbb{R}^D$ to calibrate the obtained patch embeddings and encode the context information and correlation of patches. $T(\cdot)$ maps patch embeddings $\{e_{i,j}, ...\}$ to a lower-dimensional feature space, denoted as $\{x_{i,j}, ...\}$, where $x_{i,j} = T(e_{i,j})$, $x_{i,j} \in \mathbb{R}^D$ is the calibrated embedding, $T(\cdot)$ is composed of transformer layers and skip connections (Details are given in the supplementary.).

2.4 Instance-Level Supervision

At the core of instance-level supervision is the label disambiguation module, which serves to rectify the imprecise labels that have been assigned to patches. It comprises prototypes and a confidence bank, takes instance features and instance classifier predictions as inputs, and generates soft labels as outputs (Fig. 1 (b)). The prototypes, denoted as $P \in \mathbb{R}^{C \times D}$, are initialized with all-zero vectors and employ momentum-based updates using selected instance features x with the highest probability $prob^{inst}$ of belonging to their corresponding categories. Prototype labels z are determined based on the proximity of patch features to the prototypes. Confidence $B \in \mathbb{R}^{N \times M \times C}$ is initialized with all WSI labels and uses momentum-based updates with z. Detailed steps are summarized as follows:

Step 1: Obtain the instance classifier output. The instance-level classifier, denoted as $F_{inst}(\cdot)$, takes $x_{i,j} \in \mathbb{R}^D$ as input and outputs the predicted instance probability $prob_{i,j}^{inst} \in \mathbb{R}^C$, as:

$$prob_{i,j}^{inst} = softmax(F_{inst}(x_{i,j})), \tag{1}$$

The probability that $x_{i,j}$ is predicted as class c is denoted as $prob_{i,j,c}^{inst} \in \mathbb{R}^1$.

Step 2: Obtain the prototype labels. At t time, the prototype vector for the category c is $P_{c,t} \in \mathbb{R}^D$. To update $P_{c,t}$, we select a set of instance features $Set_{c,t}$ that have the highest probabilities $prob_{i,j,c}^{inst}$ of belonging to category c. Specifically, we define $Set_{c,t}$ as:

$$Set_{c,t} = \{x_{i,j} | arg\ Top_K(prob_{i,j,c}^{inst}), j \in [1, M], i \in \{i | Y_i = c\}\}, \tag{2}$$

where K is the number of top instance features to select. Then, we use a momentum-based update rule to obtain $P_{c,t+1}$:

$$P_{c,t+1} = \alpha \cdot P_{c,t} + (1 - \alpha) \cdot x_{i,j}, \text{if } x_{i,j} \in Set_{c,t}, \tag{3}$$

where α is the momentum coefficient that automatically decreases from $\alpha = 0.95$ to $\alpha = 0.8$ across epochs. Then, we can obtain prototype labels $z_{i,j} \in \mathbb{R}^C$ using the following equation:

$$z_{i,j} = \text{OneHot}(\arg\max_c(P \cdot x_{i,j}^T)), \tag{4}$$

The resulting prototype label $z_{i,j} \in \mathbb{R}^C$ is a one-hot vector that indicates the category of the j-th instance in the i-th WSI.

Step 3: Obtain Soft Labels from the Confidence Bank Specifically, at time t, the pseudo-target $B_{i,j,t} \in \mathbb{R}^C$ of the instance embedding $e_{i,j}$ is updated by the following:

$$B_{i,j,t} = \beta \cdot B_{i,j,t-1} + (1 - \beta) \cdot z_{i,j}, \tag{5}$$

where β is the momentum update parameter with a default value of $\beta = 0.99$.

Step 4: Calculate Instance-Level Loss. We compute instance-Level Loss using the cross-entropy function:

$$L_{inst} = -\sum_{i=1}^{N}\sum_{j=1}^{M}\sum_{k=1}^{C} B_{i,j,c} \cdot log(prob_{i,j,c}^{inst}), \tag{6}$$

Here, $B_{i,j,c}$ and $prob_{i,j,c}^{inst}$ are the c-th component of the pseudo-target $B_{i,j}$ and predicted probability $prob_{i,j}^{inst}$, respectively.

2.5 Bag-Level Supervision

For bag-level supervision, instance features $x_i \in \mathbb{R}^{M\times D}$ go through a transformer-based aggregator $A(\cdot) : \mathbb{R}^{M\times D} \rightarrow \mathbb{R}^D$ and a WSI classifier $F_{bag}(\cdot) : \mathbb{R}^D \rightarrow \mathbb{R}^C$ in turn (Architecture details are given in the supplementary.). Then we obtain the predicted probability of WSI S_i as:

$$prob_i^{bag} = softmax(F_{bag}(A(x_i))). \tag{7}$$

The bag-level loss function is given by:

$$\mathcal{L}_{bag} = -\sum_{i=1}^{N} prob_i^{bag} \cdot \log(Y_i), \tag{8}$$

where $Y_i \in \mathbb{R}^C$ is the label of WSI S_i.

2.6 Training

In the training phase, We employ a warm-up strategy in which we update only the Prototypes and do not update the Confidence Bank during the first few epochs. Our approach is trained end-to-end, and the total loss function is :

$$\mathcal{L} = \mathcal{L}_{\text{bag}} + \lambda\mathcal{L}_{\text{inst}}, \tag{9}$$

where λ is the hyperparameter that controls the relative importance of the two losses.

3 Experiments

3.1 Dataset

We evaluate our model with three datasets. (1) LUAD-GM Dataset: The objective is to predict the epidermal growth factor receptor (EGFR) gene mutations in patients with lung adenocarcinoma (LUAD) using 723 Whole Slide Image (WSI) slices, where 47% of cases have EGFR mutations. (2) TCGA-NSCLC and TCGA-RCC Datasets: Cancer type classification is performed using The Cancer Genome Atlas (TCGA) dataset. The TCGA-NSCLC dataset comprised two subtypes, lung squamous cell carcinoma (LUSC) and lung adenocarcinoma (LUAD), while the TCGA-RCC dataset included three subtypes: renal chromophobe cell carcinoma (KICH), renal clear cell carcinoma (KIRC), and renal papillary cell carcinoma (KIRP).

3.2 Experiment Settings

The dataset was randomly split into three parts: training, validation, and testing, with 60%, 20%, and 20% of the samples, respectively. WSIs were preprocessed by cropping them into 1120 × 1120 patches, without overlap. The proposed model was implemented in Pytorch, trained on a 32GB TESLA V100 GPU, using AdamW [11] optimizer. The batch size was set to 4, with a learning rate of $1e^{-4}$ and a weight decay of $1e^{-5}$.

Table 1. The performance of IIB-MIL compared with other SOTA methods.

Models	LUAD-GM		TCGA-NSCLC		TCGA-RCC	
	AUC (%)	Accuracy	AUC (%)	Accuracy	AUC (%)	Accuracy
ABMIL [7,12]	52.44	54.55	86.56	77.19	97.02	89.34
CNN-MIL [20]	45.28	44.92	78.64	69.86	69.56	61.17
DSMIL [9]	78.53	71.53	89.25	80.58	98.4	92.94
CLAM-SB [13]	78.49	70.14	86.37	78.47	90.21	76.60
CLAM-MB [13]	82.33	75.70	88.18	81.80	97.23	88.16
ViT-MIL [5]	76.39	70.14	93.77	84.22	97.99	89.66
TransMIL [16]	77.29	74.45	96.03	88.35	98.82	94.66
SETMIL [27]	83.84	76.38	96.01	89.27	99.01	95.20
DTFD [26]	82.41	76.03	97.37	92.34	99.00	96.90
IIB-MIL(Ours)	**85.62 (+1.78)**	78.77	**98.11(+0.74)**	90.91	**99.57 (+0.56)**	95.24

Table 2. Ablation studies and model analysis of IIB-MIL.

Models	LUAD-GM		TCGA-NSCLC		TCGA-RCC	
	AUC (%)	Accuracy	AUC (%)	Accuracy	AUC (%)	Accuracy
w/o Instance	84.01 (−1.61)	76.20	95.89 (−2.22)	89.50	98.93 (−0.64)	93.12
w/o Label Disambiguation	84.23 (−1.39)	65.07	96.40 (−1.71)	89.47	98.97 (−0.6)	94.18
w/o Bag	84.67 (−0.95)	71.23	97.65 (−0.46)	91.39	99.01 (−0.56)	91.53
IIB-MIL	**85.62**	78.77	**98.11**	90.91	**99.57**	95.24
warmup = 1	84.81	75.34	96.14	89.47	99.06	92.59
warmup = 5	85.37	76.71	97.32	91.87	99.25	93.65
warmup = 10	**85.62**	78.77	**98.11**	90.91	**99.57**	95.24
warmup = 50	85.36	77.40	97.51	89.95	99.42	93.12
$\lambda = 0.1$	83.50	71.92	97.59	95.69	99.11	92.06
$\lambda = 1$	83.19	71.23	98.05	91.39	99.05	93.12
$\lambda = 5$	**85.62**	78.77	**98.11**	90.91	**99.57**	95.24
$\lambda = 10$	85.60	76.03	96.51	89.47	99.23	89.95

4 Results and Discussion

4.1 Comparison with State-of-the Art Methods

Table 1 presents a performance comparative analysis of IIB-MIL in relation to other SOTA methods, including ABMIL [7,12], CNN-MIL [20], DSMIL [9], CLAM [13], ViT-MIL [5], TransMIL [16], SETMIL [27], and DTFD [26]. All methods were evaluated in three tasks, namely gene mutation prediction (with or without EGFR mutation), TCGA-NSCLC subtype classification, and TCGA-RCC subtype classification. IIB-MIL achieved AUCs of 85.62%, 98.11%, and 99.57%. We can also find IIB-MIL outperformed other SOTA methods, in the three tasks with at least 1.78%, 0.74%, and 0.56% performance enhancement (AUC), respectively.

4.2 Ablation Studies

We conducted ablation studies to assess the efficacy of each component in IIB-MIL. The results, in Table 2, indicate that all of the designed components,

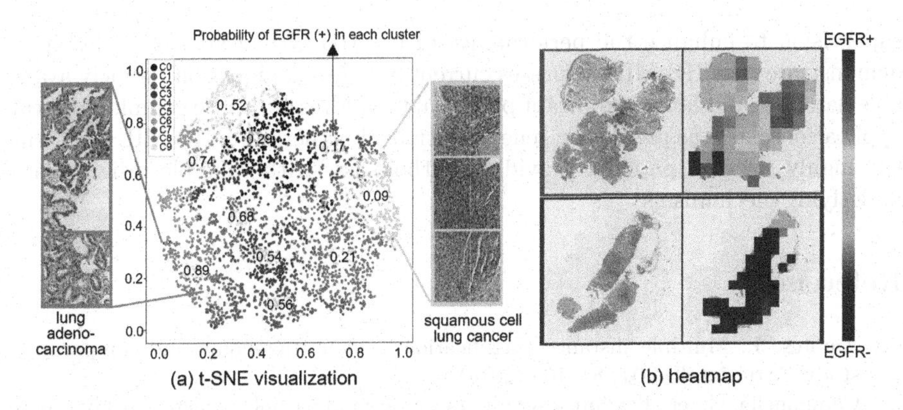

Fig. 2. (a)t-SNE plot of the patch features obtained from the backbone;(b) Example heatmaps of IIB-MIL on WSIs with known EFGR mutation labels.

including the label disambiguation module, instance-level supervision, and bag-level supervision, contribute to the success of IIB-MIL. We also investigated the impact of the warm-up epoch number and found that selecting an appropriate value, such as $warmup = 10$, can lead to better model performance. Furthermore, we examined the impact of the weighting factor λ, and the outcomes indicated that assigning greater importance to instance-level supervision ($\lambda = 5$) helps IIB-MIL enhance its performance, thus demonstrating the effectiveness of the designed label-disambiguation-based instance-level supervision.

4.3 Model Interpretation

Figure 2(a) shows the t-SNE plot of the obtained patch features from the backbone of the IIB-MIL. The patches are unsupervisedly clustered into groups based on their features, indicated by various colors. The numbers displayed within each group represent the average likelihood of the EGFR mutation predicted by the patches. With the help of the label-disambiguation-based instance-level supervision, IIB-MIL can identify highly positive and negative related patches to the WSI-label, i.e., the cyan-blue group and yellow group. Double-checked by pathologists, we find that the cyan-blue group consists of patches from lung adenocarcinoma and the yellow group consists of patches from the squamous cells. This finding aligns with the domain knowledge of pathologists. Figure 2(b) investigates the contribution of each patch in predicting EGFR mutation. The resulting heatmap shows the decision mechanism of IIB-MIL in the accurate distinguishment between EGFR mutation-positive and negative samples.

5 Conclusion

This paper presents IIB-MIL, a novel MIL approach for pathological image analysis. IIB-MIL utilizes a label disambiguation module to establish more precise instance-level supervision. It then combines the instance-level and bag-level supervision to enhance the performance of the IIB-MIL. Experimental results demonstrate that IIB-MIL surpasses current SOTA techniques on publicly available datasets, and holds significant potential for addressing more complex clinical applications, such as predicting gene mutations. Furthermore, IIB-MIL can identify highly relevant patches, providing pathologists with valuable insights into underlying mechanisms.

References

1. Amores, J.: Multiple instance classification: review, taxonomy and comparative study. Artif. intell. **201**, 81–105 (2013)
2. Campanella, G., et al.: Clinical-grade computational pathology using weakly supervised deep learning on whole slide images. Nat. Med. **25**(8), 1301–1309 (2019)
3. Chikontwe, P., Kim, M., Nam, S.J., Go, H., Park, S.H.: Multiple instance learning with center embeddings for histopathology classification. In: Martel, A.L., et al. (eds.) MICCAI 2020. LNCS, vol. 12265, pp. 519–528. Springer, Cham (2020). https://doi.org/10.1007/978-3-030-59722-1_50
4. Coudray, N., et al.: Classification and mutation prediction from non-small cell lung cancer histopathology images using deep learning. Nat. Med. **24**(10), 1559–1567 (2018)
5. Dosovitskiy, A., et al.: An image is worth 16×16 words: transformers for image recognition at scale. arXiv preprint arXiv:2010.11929 (2020)
6. Hou, L., Samaras, D., Kurc, T.M., Gao, Y., Davis, J.E., Saltz, J.H.: Patch-based convolutional neural network for whole slide tissue image classification. In: Proceedings of the IEEE Conference on Computer Vision and Pattern Recognition, pp. 2424–2433 (2016)
7. Ilse, M., Tomczak, J., Welling, M.: Attention-based deep multiple instance learning. In: International Conference on Machine Learning, pp. 2127–2136. PMLR (2018)
8. Lerousseau, M., Vakalopoulou, M., Classe, M., Adam, J., Battistella, E., Carré, A., Estienne, T., Henry, T., Deutsch, E., Paragios, N.: Weakly supervised multiple instance learning histopathological tumor segmentation. In: Martel, A.L., et al. (eds.) MICCAI 2020. LNCS, vol. 12265, pp. 470–479. Springer, Cham (2020). https://doi.org/10.1007/978-3-030-59722-1_45
9. Li, B., Li, Y., Eliceiri, K.W.: Dual-stream multiple instance learning network for whole slide image classification with self-supervised contrastive learning. In: Proceedings of the IEEE/CVF Conference on Computer Vision and Pattern Recognition, pp. 14318–14328 (2021)
10. Li, R., Yao, J., Zhu, X., Li, Y., Huang, J.: Graph CNN for survival analysis on whole slide pathological images. In: Frangi, A.F., Schnabel, J.A., Davatzikos, C., Alberola-López, C., Fichtinger, G. (eds.) MICCAI 2018. LNCS, vol. 11071, pp. 174–182. Springer, Cham (2018). https://doi.org/10.1007/978-3-030-00934-2_20
11. Loshchilov, I., Hutter, F.: Decoupled weight decay regularization. arXiv preprint arXiv:1711.05101 (2017)
12. Lu, M.Y., et al.: Ai-based pathology predicts origins for cancers of unknown primary. Nature **594**(7861), 106–110 (2021)
13. Lu, M.Y., Williamson, D.F., Chen, T.Y., Chen, R.J., Barbieri, M., Mahmood, F.: Data-efficient and weakly supervised computational pathology on whole-slide images. Nat. Biomed. Eng. **5**(6), 555–570 (2021)

14. Noorbakhsh, J., et al.: Deep learning-based cross-classifications reveal conserved spatial behaviors within tumor histological images. Nat. Commun. **11**(1), 6367 (2020)
15. Rubin, R., Strayer, D.S., Rubin, E., et al.: Rubin's Pathology: Clinicopathologic Foundations of Medicine. Lippincott Williams & Wilkins (2008)
16. Shao, Z., Bian, H., Chen, Y., Wang, Y., Zhang, J., Ji, X., et al.: Transmil: transformer based correlated multiple instance learning for whole slide image classification. Adv. Neural Inf. Process. Syst. **34**, 2136–2147 (2021)
17. Sharma, Y., Shrivastava, A., Ehsan, L., Moskaluk, C.A., Syed, S., Brown, D.: Cluster-to-conquer: a framework for end-to-end multi-instance learning for whole slide image classification. In: Medical Imaging with Deep Learning, pp. 682–698. PMLR (2021)
18. Skrede, O.J., et al.: Deep learning for prediction of colorectal cancer outcome: a discovery and validation study. The Lancet **395**(10221), 350–360 (2020)
19. Srinidhi, C.L., Ciga, O., Martel, A.L.: Deep neural network models for computational histopathology: a survey. Med. Image Anal. **67**, 101813 (2021)
20. Tellez, D., Litjens, G., van der Laak, J., Ciompi, F.: Neural image compression for gigapixel histopathology image analysis. IEEE Trans. Pattern Anal. Mach. Intell. **43**(2), 567–578 (2019)
21. Wang, H., et al.: Pico: contrastive label disambiguation for partial label learning. arXiv preprint arXiv:2201.08984 (2022)
22. Wang, X., Yan, Y., Tang, P., Bai, X., Liu, W.: Revisiting multiple instance neural networks. Pattern Recogn. **74**, 15–24 (2018)
23. Wetstein, S.C., et al.: Deep learning-based breast cancer grading and survival analysis on whole-slide histopathology images. Sci. Rep. **12**(1), 15102 (2022)
24. Xu, G., et al.: Camel: a weakly supervised learning framework for histopathology image segmentation. In: Proceedings of the IEEE/CVF International Conference on Computer Vision, pp. 10682–10691 (2019)
25. Zhang, H., Meng, Y., Qian, X., Yang, X., Coupland, S.E., Zheng, Y.: A regularization term for slide correlation reduction in whole slide image analysis with deep learning. In: Medical Imaging with Deep Learning, pp. 842–854. PMLR (2021)
26. Zhang, H., et al.: Dtfd-mil: double-tier feature distillation multiple instance learning for histopathology whole slide image classification. In: Proceedings of the IEEE/CVF Conference on Computer Vision and Pattern Recognition, pp. 18802–18812 (2022)
27. Zhao, Y., Lin, Z., Sun, K., Zhang, Y., Huang, J., Wang, L., Yao, J.: Setmil: spatial encoding transformer-based multiple instance learning for pathological image analysis. In: Medical Image Computing and Computer Assisted Intervention-MICCAI 2022: 25th International Conference, Singapore, 18–22 September 2022, Proceedings, Part II, pp. 66–76. Springer, Heidelberg (2022). https://doi.org/10.1007/978-3-031-16434-7_7
28. Zhao, Y., et al.: Predicting lymph node metastasis using histopathological images based on multiple instance learning with deep graph convolution. In: Proceedings of the IEEE/CVF Conference on Computer Vision and Pattern Recognition, pp. 4837–4846 (2020)

Semi-supervised Pathological Image Segmentation via Cross Distillation of Multiple Attentions

Lanfeng Zhong[1], Xin Liao[2], Shaoting Zhang[1,3], and Guotai Wang[1,3(✉)]

[1] University of Electronic Science and Technology of China, Chengdu, China
guotai.wang@uestc.edu.cn
[2] Department of Pathology, West China Second University Hospital, Sichuan University, Chengdu, China
[3] Shanghai Artificial Intelligence Laboratory, Shanghai, China

Abstract. Segmentation of pathological images is a crucial step for accurate cancer diagnosis. However, acquiring dense annotations of such images for training is labor-intensive and time-consuming. To address this issue, Semi-Supervised Learning (SSL) has the potential for reducing the annotation cost, but it is challenged by a large number of unlabeled training images. In this paper, we propose a novel SSL method based on Cross Distillation of Multiple Attentions (CDMA) to effectively leverage unlabeled images. Firstly, we propose a Multi-attention Tri-branch Network (MTNet) that consists of an encoder and a three-branch decoder, with each branch using a different attention mechanism that calibrates features in different aspects to generate diverse outputs. Secondly, we introduce Cross Decoder Knowledge Distillation (CDKD) between the three decoder branches, allowing them to learn from each other's soft labels to mitigate the negative impact of incorrect pseudo labels in training. Additionally, uncertainty minimization is applied to the average prediction of the three branches, which further regularizes predictions on unlabeled images and encourages inter-branch consistency. Our proposed CDMA was compared with eight state-of-the-art SSL methods on the public DigestPath dataset, and the experimental results showed that our method outperforms the other approaches under different annotation ratios. The code is available at https://github.com/HiLab-git/CDMA.

Keywords: Semi-supervised learning · Knowledge distillation · Attention · Uncertainty

1 Introduction

Automatic segmentation of tumor lesions from pathological images plays an important role in accurate diagnosis and quantitative evaluation of cancers. Recently, deep learning has achieved remarkable performance in pathological image segmentation when trained with a large and well-annotated dataset [6,13,

H. Greenspan et al. (Eds.): MICCAI 2023, LNCS 14225, pp. 570–579, 2023.
https://doi.org/10.1007/978-3-031-43987-2_55

20]. However, obtaining dense annotations for pathological images is challenging and time-consuming, due to the extremely large image size (e.g., 10000 $\times$ 10000 pixels), scattered spatial distribution, and complex shape of lesions.

Semi-Supervised Learning (SSL) is a potential technique to reduce the annotation cost via learning from a limited number of labeled data along with a large amount of unlabeled data. Existing SSL methods can be roughly divided into two categories: consistency-based [9,14,23] and pseudo label-based [2] methods. The consistency-based methods impose consistency constraints on the predictions of an unlabeled image under some perturbations. For example, Mean Teacher (MT)-based methods [14,23] encourage consistent predictions between a teacher and a student model with noises added to the input. Xie et al. [21] introduced a pairwise relation network to exploit semantic consistency between each pair of images in the feature space. Luo et al. [9] proposed an uncertainty rectified pyramid consistency between multi-scale predictions. Jin et al. [7] proposed to encourage the predictions of auxiliary decoders and a main decoder to be consistent under perturbed hierarchical features. Pseudo label-based methods typically generate pseudo labels for labeled images to supervise the network [4]. Since using a model's prediction to supervise itself may over-fit its bias, Chen et al. [2] proposed Cross Pseudo Supervision (CPS) where two networks learn from each other's pseudo labels generated by *argmax* of the output prediction. MC- Net+ [19] utilized multiple decoders with different upsampling strategies to obtain slightly different outputs, and each decoder's probability output was sharpened to serve as pseudo labels to supervise the others. However, the pseudo labels are not accurate and contain a lot of noise, using *argmax* or sharpening operation will lead to over-confidence of potentially wrong predictions, which limits the performance of the models. Additionally, some related works advocated the entropy-minimization methods. Typical entropy Minimization (EM) [15] that aims to reduce the uncertainty or entropy in a system. Wu et al. [17] directly applied entropy minimization to the segmentation results.

In this work, we propose a novel and efficient method based on Cross Distillation with Multiple Attentions (CDMA) for semi-supervised pathological image segmentation. Firstly, a Multi-attention Tri-branch Network (MTNet) is proposed to efficiently obtain diverse outputs for a given input. Unlike MC-Net+ [19] that is based on different upsampling strategies, our MTNet uses different attention mechanisms in three decoder branches that calibrate features in different aspects to obtain diverse and complementary outputs. Secondly, inspired by the observation that smoothed labels are more effective for noise-robust learning found in recent studies [10,22], we propose a Cross Decoder Knowledge Distillation (CDKD) strategy to better leverage the diverse predictions of unlabeled images. In CDKD, each branch serves as a teacher of the other two branches using soft label supervision, which reduces the effect of noise for more robust learning from inaccurate pseudo labels than *argmax* [2] and sharpening-based [19] pseudo supervision in existing methods. Differently from typical Knowledge Distillation (KD) methods [5,24] that require a pre-trained teacher to generate soft predictions, our method efficiently obtains the teacher and student's soft predictions simultaneously in a single forward pass. In addition, we apply an uncertainty minimization-based regularization to the average probability prediction across the decoders, which not only increases the network's confidence, but also improves the inter-decoder consistency for leveraging labeled images.

The contribution of this work is three-fold: 1) A novel framework named CDMA based on MTNet is introduced for semi-supervised pathological image segmentation, which leverages different attention mechanisms for generating diverse and complementary predictions for unlabeled images; 2) A Cross Decoder Knowledge Distillation method is proposed for robust and efficient learning from noisy pseudo labels, which is combined with an average prediction-based uncertainty minimization to improve the model's performance; 3) Experimental results show that the proposed CDMA outperforms eight state-of-the-art SSL methods on the public DigestPath dataset [3].

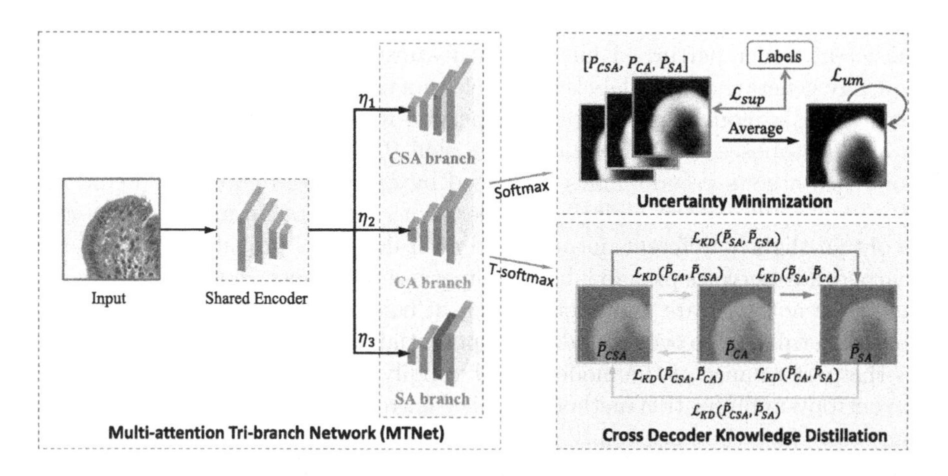

Fig. 1. Our CDMA for semi-supervised segmentation. Three decoder branches use different attentions to obtain diverse outputs. Cross Decoder Knowledge Distillation (CDKD) is proposed to better deal with noisy pseudo labels, and an uncertainty minimization is applied to the average probability prediction of the three branches. $\mathcal{L}_{sup}$ is only for labeled images.

2 Methods

As illustrated in Fig. 1, the proposed Cross Distillation of Multiple Attentions (CDMA) framework for semi-supervised pathological image segmentation consists of three core modules: 1) a tri-branch network MTNet that uses three different attention mechanisms to obtain diverse outputs, 2) a Cross Decoder Knowledge Distillation (CDKD) module to reduce the effect of noisy pseudo labels based on soft supervision, and 3) an average prediction-based uncertainty minimization loss to further regularize the predictions on unlabeled images.

2.1 Multi-attention Tri-Branch Network (MTNet)

Attention is an effective network structure design in fully supervised image segmentation [12,16]. It can calibrate the feature maps for better performance by paying more attention to the important spatial positions or channels with only a few extra parameters. However, it has been rarely investigated in semi-supervised segmentation tasks. To more effectively exploit attention mechanisms for semi-supervised pathological image segmentation, our proposed MTNet consists of a shared encoder and three decoder branches that are based on Channel Attention (CA), Spatial Attention (SA) and simultaneous Channel and Spatial Attention (CSA), respectively. The encoder consists of multiple convolutional blocks that are sequentially connected to a down-sampling layer, and each decoder has multiple convolutional blocks that are sequentially connected by an up-sampling layer. For a certain decoder, it uses CA, SA or SCA at the convolutional block at each resolution level to calibrate the features.

CA branch uses channel attention blocks to calibrate the features in the first decoder. A channel attention block highlights important channels in a feature map and it is formulated as:

$$F_c = F \cdot \sigma\Big(MLP\big(Pool^S_{avg}(F)\big) + MLP\big(Pool^S_{max}(F)\big)\Big) \tag{1}$$

where F represents an input feature map. $Pool^S_{avg}$ and $Pool^S_{max}$ represent average pooling and max-pooling across the spatial dimension, respectively. MLP and σ denote multi-layer perception and the sigmoid activation function respectively. F_c is the output feature map calibrated by channel attention.

SA branch leverages spatial attention to highlight the most relevant spatial positions and suppress the irrelevant regions in a feature map. An SA block is:

$$F_s = F \cdot \sigma\Big(Conv\big(Pool^C_{avg}(F) \oplus Pool^C_{max}(F)\big)\Big) \tag{2}$$

where $Conv$ denotes a convolutional layer. $Pool^C_{avg}$ and $Pool^C_{max}$ are average and max-pooling across the channel dimension, respectively. $\oplus$ means concatenation.

CSA branch calibrates the feature maps using a CSA block for each convolutional block. A CSA block consists of a CA block followed by an SA block, taking advantage of channel and spatial attention simultaneously.

Due to the different attention mechanisms, the three decoder branches pay attention to different aspects of feature maps and lead to different outputs. To further improve the diversity of the outputs and alleviate over-fitting, we add a dropout layer and a feature noise layer η [11] before each of the three decoders. For an input image, the logit predictions obtained by the three branches are denoted as Z_{CA}, Z_{SA} and Z_{CSA}, respectively. After using a standard Softmax operation, their corresponding probability prediction maps are denoted as P_{CA}, P_{SA} and P_{CSA}, respectively.

2.2 Cross Decoder Knowledge Distillation (CDKD)

Since the three branches have different decision boundaries, using the predictions from one branch as pseudo labels to supervise the others would avoid each branch over-fitting its bias. However, as the predictions for unlabeled training images are noisy and inaccurate, using hard or sharpened pseudo labels [2,19] would strengthen the confidence on incorrect predictions, leading the model to overfit the noise [10,22]. To address this problem, we introduce CDKD to enhance the ability of our MTNet to leverage unlabeled images and eliminate the negative impact of noisy pseudo labels. It forces each decoder to be supervised by the other two decoders' soft predictions. Following the practice of KD [5], a temperature calibrated Softmax (T-Softmax) is used to soften the probability maps:

$$\tilde{\mathbf{p}}_c = \frac{exp(\mathbf{z}_c/T)}{\sum_c exp(\mathbf{z}_c/T)} \tag{3}$$

where $\mathbf{z}_c$ represents the logit prediction for class c of a pixel, and $\tilde{\mathbf{p}}_c$ is the soft probability value for class c. Temperature T is a parameter to control the softness of the output probability. Note that $T = 1$ corresponds to a standard Softmax function, and a larger T value leads to a softer probability distribution with higher entropy. When $T < 1$, Eq. 3 is a sharpening function.

Let $\tilde{P}_{CA}$, $\tilde{P}_{SA}$ and $\tilde{P}_{CSA}$ represent the soft probability map obtained by T-Softmax for the three branches, respectively. With the other two branches being the teachers, the KD loss for the CSA branch is:

$$\mathcal{L}_{kd}^{CSA} = \mathbf{KL}(\tilde{P}_{CSA}, \tilde{P}_{CA}) + \mathbf{KL}(\tilde{P}_{CSA}, \tilde{P}_{SA}) \tag{4}$$

where $\mathbf{KL}()$ is the Kullback-Leibler divergence function. Note that the gradient of $\mathcal{L}_{kd}^{CSA}$ is only back-propagated to the CSA branch, so that the knowledge is distilled from the teachers to the student. Similarly, the KD losses for the CA and SA branches are denoted as $\mathcal{L}_{kd}^{CA}$ and $\mathcal{L}_{kd}^{SA}$, respectively. Then, the total distillation loss is defined as:

$$\mathcal{L}_{cdkd} = \frac{1}{3}(\mathcal{L}_{kd}^{CSA} + \mathcal{L}_{kd}^{CA} + \mathcal{L}_{kd}^{SA}) \tag{5}$$

2.3 Average Prediction-Based Uncertainty Minimization

Minimizing the uncertainty (e.g., entropy) [15] has been shown to be an effective regularization for predictions on unlabeled images, which increases the model's confidence on its predictions. However, applying uncertainty minimization to each branch independently may lead to inconsistent predictions between the decoders where each of them is very confident, e.g., two branches predict the foreground probability of a pixel as 0.0 and 1.0 respectively. To avoid this problem and further encourage inter-decoder consistency for regularization, we propose an average prediction-based uncertainty minimization:

$$\mathcal{L}_{um} = -\frac{1}{N}\sum_{i=0}^{N}\sum_{c=0}^{C}\bar{P}_i^c log(\bar{P}_i^c) \tag{6}$$

where $\bar{P} = (P_{CSA} + P_{CA} + P_{SA})/3$ is the average probability map. C and N are the class number and pixel number respectively. $\bar{P}_i^c$ is the average probability for class c at pixel i. Note that when $\mathcal{L}_{um}$ for a pixel is close to zero, the average probability for class c of that pixel is close to 0.0 (1.0), which drives all the decoders to predict it as 0.0 (1.0) and encourages inter-decoder consistency.

Finally, the overall loss function for our CDMA is:

$$\mathcal{L} = \mathcal{L}_{sup} + \lambda_1 \mathcal{L}_{cdkd} + \lambda_2 \mathcal{L}_{um} \tag{7}$$

where $\mathcal{L}_{sup} = (\mathcal{L}_{sup}^{CSA} + \mathcal{L}_{sup}^{CA} + \mathcal{L}_{sup}^{SA})/3$ is the average supervised learning loss for the three branches on the labeled training images, and the supervised loss for each branch calculates the Dice loss and cross entropy loss between the probability prediction (P_{CSA}, P_{CA} and P_{SA}) and the ground truth label. λ_1 and λ_2 are the weights of $\mathcal{L}_{cdkd}$ and $\mathcal{L}_{um}$ respectively. Note that $\mathcal{L}_{cdkd}$ and $\mathcal{L}_{um}$ are applied on both labeled and unlabeled training images.

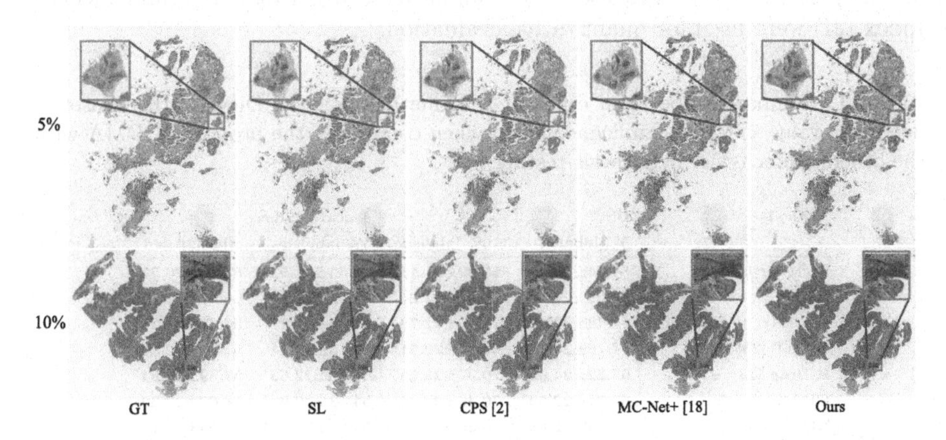

Fig. 2. Visual comparison between our proposed CDMA with state-of-the-art methods for semi-supervised semantic segmentation of WSIs. The green regions are lesions.

3 Experiments and Results

Dataset and Implementation Details. We used the public DigestPath dataset [3] for binary segmentation of colonoscopy tumor lesions from Whole Slide Images (WSI) in the experiment. The WSIs were collected from four medical institutions of ×20 magnification (0.475 μm/pixel) with an average size of 5000 × 5000. We randomly split 130 malignant WSIs into 100, 10, and 20 for training, validation and testing, respectively. For SSL, we investigated two annotation ratios: 5% and 10%, where only 5 and 10 WSIs in the training set were taken as annotated respectively. Labeled WSIs were randomly selected. For computational feasibility, we cropped the WSIs into patches with a size of 256 × 256.

At inference time for segmenting a WSI, we used a sliding window of size 256×256 with a stride of 192 × 192.

The CDMA framework was implemented in PyTorch, and all experiments were performed on one NVIDIA 2080Ti GPU. MTNet was implemented by extending DeepLabv3+ [1] into a tri-branch network, where the three decoders were equipped with CA, SA and CSA blocks respectively. The encoder used a backbone of ResNet50 pre-trained on ImageNet. The kernel size of *Conv* in the SA block is 7 × 7. SGD optimizer was used for training, with weight decay 5×10^{-4}, momentum 0.9 and epoch number 150. The learning rate was initialized to 10^{-3} and decayed by 0.1 every 50 epochs. The hyper-parameter setting was $\lambda_1 = \lambda_2 = 0.1$, $T = 10$ based on the best results on the validation set. The batch size was 16 (8 labeled and 8 unlabeled patches). For data augmentation, we adopted random flipping, random rotation, and random Gaussian noise. For inference, only the CSA branch was used due to the similar performance of the three branches after converge and the increased inference time of their ensemble, and no post-processing was used. Dice Similarity Coefficient (DSC) and Jaccard Index (JI) were used for quantitative evaluation.

Table 1. Comparison between different SSL methods on the DigestPath dataset. * denotes *p*-value < 0.05 (significance level) when comparing the proposed CDMA with the others under t-test hypothesis testing.

Methods	DSC		Jaccard Index	
	5% labeled	10% labeled	5% labeled	10% labeled
SL lower bound	64.74±23.24*	68.32±21.18*	52.35±21.53*	53.62±20.32*
EM [15]	67.09±24.28*	70.01±22.24*	54.55±22.40*	56.96±21.70*
MT [14]	67.46±23.10*	70.19±21.72*	54.68±21.27*	56.38±21.21*
UAMT [23]	67.76±23.44	69.64±22.41*	55.16±22.24	57.22±22.25*
R-Drop [18]	67.22±24.05*	70.37±23.58*	54.70±22.63*	57.39±22.94*
CPS [2]	67.71±22.50*	70.46±23.75	54.73±20.92*	58.67±23.30
HCE [7]	67.34±22.32*	70.29±22.62	54.58±20.37*	58.04±21.11
CNN&Transformer [8]	67.66±25.12	70.43±18.84*	55.74±23.38	57.89±19.48*
MC-Net+ [19]	67.81±24.22*	70.09±22.07*	55.40±22.54*	57.64±21.80*
Ours (CSA branch)	**69.72±22.06**	**72.24±21.21**	**57.09±21.23**	**60.17±21.98**
Full Supervision	77.47±12.49		64.97±14.09	

Comparison with State-of-the-Art Methods. Our CDMA was compared with eight existing SSL methods: 1) Entropy Minimization (EM) [15]; 2) Mean Teacher (MT) [14]; 3) Uncertaitny-Aware Mean Teacher (UAMT) [23]; 4) R-Drop [18] that introduces a dropout-based consistency regularization between two networks; 5) CPS [2]; 6) Hierarchical Consistency Enforcement (HCE) [7]; 7) CNN&Transformer [8] that introduces cross-supervision between CNN and Transformer; 8) MC-Net+ [19] that imposes mutual consistency between multiple slightly different decoders. They were also compared with the lower bound of Supervised Learning (SL) that only learns from the labeled images. All these methods used the same backbone of DeepLabv3+ [1] for a fair comparison.

Quantitative evaluation of these methods is shown in Table 1. In the existing methods, MC-Net+ [19] and CPS [2] showed the best performance for both of the two annotation ratios. Our proposed CDMA achieved a better performance than all the existing methods, with a DSC score of 69.72% and 72.24% when the annotation ratio was 5% and 10%, respectively. Figure 2 shows a qualitative comparison between different methods. It can be observed that our CDMA yields less mis-segmentation compared with CPS [2] and MC-Net+ [19].

Table 2. Ablative analysis of our proposed method.

Methods	Mean DSC		Mean JI	
	5% labeled	10% labeled	5% labeled	10% labeled
MTNet (Baseline)	65.02±23.94	68.61±22.10	52.59±22.54	55.47±21.81
MTNet + $\mathcal{L}_{cdkd}$ (argmax)	68.20±23.42	70.61±21.03	55.46±21.49	58.71±21.23
MTNet + $\mathcal{L}_{cdkd}$ (T=1)	68.22±23.55	70.32±21.67	55.48±21.57	58.45±21.32
MTNet + $\mathcal{L}_{cdkd}$	68.84±22.89	71.49±20.74	55.92±21.44	59.02±21.13
MTNet + $\mathcal{L}_{cdkd}$ + $\mathcal{L}'_{um}$	69.11±23.43	71.56±22.02	56.57±21.49	59.52±22.46
MTNet + $\mathcal{L}_{cdkd}$ + $\mathcal{L}_{um}$	**69.72±22.06**	72.24±21.21	**57.09±21.23**	60.17±21.98
MTNet(dual) +$\mathcal{L}_{cdkd}$+$\mathcal{L}_{um}$	69.49±22.42	71.65±20.48	56.96±21.85	59.13±21.10
MTNet(csa×3)+$\mathcal{L}_{cdkd}$+$\mathcal{L}_{um}$	69.24±23.57	71.50±20.54	56.93±22.34	59.04±21.25
MTNet(-atten)+$\mathcal{L}_{cdkd}$+$\mathcal{L}_{um}$	68.92±23.42	71.37±20.68	56.03±22.13	58.81±21.46
MTNet(ensb) +$\mathcal{L}_{cdkd}$+$\mathcal{L}_{um}$	69.66±22.08	**72.25±21.19**	57.01±21.25	**60.18±21.98**

Ablation Study. For ablation study, we set the baseline as using the proposed MTNet with three different decoders for supervised learning from labeled images only. It obtained an average DSC of 65.02% and 68.61% under the two annotation ratios respectively. The proposed $\mathcal{L}_{cdkd}$ was compared with two variants: $\mathcal{L}_{cdkd}$ (argmax) and $\mathcal{L}_{cdkd}$ (T=1) that represent using hard pseudo labels and standard probability output obtained by Softmax for CDKD respectively. Table 2 shows that our $\mathcal{L}_{cdkd}$ obtained an average DSC of 68.84% and 71.49% under the two annotation ratios respectively, and it outperformed $\mathcal{L}_{cdkd}$ (argmax) and $\mathcal{L}_{cdkd}$ (T=1), demonstrating that our CDKD based on softened probability prediction is more effective in dealing with noisy pseudo labels. By introducing our average prediction-based uncertainty minimization $\mathcal{L}_{um}$, the DSC was further improved to 69.72% and 72.24% under the two annotation ratios respectively. In addition, replacing our $\mathcal{L}_{um}$ by applying entropy minimization to each branch respectively ($\mathcal{L}'_{um}$) led to a DSC drop by around 0.65%.

Then, we compared different MTNet variants: 1) MTNet(dual) means a dual-branch structure (removing the CSA branch); 2) MTNet(csa×3) means all the three branches use CSA blocks; 3) MTNet(-atten) means no attention block is used in all the branches; and 4) MTNet(ensb) means using an ensemble of the three branches for inference. Note that all these variants were trained with $\mathcal{L}_{cdkd}$ and $\mathcal{L}_{um}$. The results in the second section of Table 2 show that using the same structures for different branches, i.e., MTNet(-atten) and MTNet(csa×3), had a lower performance than using different attention blocks, and using three attention branches outperformed just using two attention branches. It can also be found that using CSA branch for inference had a very close performance to MTNet(ensb), and it is more efficient than the later.

4 Conclusion

We have presented a novel semi-supervised framework based on Cross Distillation of Multiple Attentions (CDMA) for pathological image segmentation. It employs a Multi-attention Tri-branch network to generate diverse predictions based on channel attention, spatial attention, and simultaneous channel and spatial attention, respectively. Different attention-based decoder branches focus on various aspects of feature maps, resulting in disparate outputs, which is beneficial to semi-supervised learning. To eliminate the negative impact of incorrect pseudo labels in training, we employ a Cross Decoder Knowledge Distillation (CDKD) to enforce each branch to learn from soft labels generated by the other two branches. Experimental results on a colonoscopy tissue segmentation dataset demonstrated that our CDMA outperformed eight state-of-the-art SSL methods. In the future, it is of interest to apply our method to multi-class segmentation tasks and pathological images from different organs.

Acknowledgment. This work was supported by the National Natural Science Foundation of China (62271115).

References

1. Chen, L.C., Zhu, Y., Papandreou, G., Schroff, F., Adam, H.: Encoder-decoder with atrous separable convolution for semantic image segmentation. In: ECCV, pp. 801–818 (2018)
2. Chen, X., Yuan, Y., Zeng, G., Wang, J.: Semi-supervised semantic segmentation with cross pseudo supervision. In: CVPR, pp. 2613–2622 (2021)
3. Da, Q., et al.: Digestpath: a benchmark dataset with challenge review for the pathological detection and segmentation of digestive-system. Med. Image Anal. **80**, 102485 (2022)
4. Fan, D.P., et al.: Inf-Net: automatic covid-19 lung infection segmentation from CT images. IEEE Trans. Med. Imaging **39**(8), 2626–2637 (2020)
5. Hinton, G., Vinyals, O., Dean, J.: Distilling the knowledge in a neural network. In: NeurIPS, pp. 1–10 (2015)
6. Hou, X., et al.: Dual adaptive pyramid network for cross-stain histopathology image segmentation. In: Shen, D., et al. (eds.) MICCAI 2019. LNCS, vol. 11765, pp. 101–109. Springer, Cham (2019). https://doi.org/10.1007/978-3-030-32245-8_12
7. Jin, Q., et al.: Semi-supervised histological image segmentation via hierarchical consistency enforcement. In: Wang, L., Dou, Q., Fletcher, P.T., Speidel, S., Li, S. (eds.) MICCAI 2022. LNCS, vol. 13432, pp. 3–13. Springer, Heidelberg (2022). https://doi.org/10.1007/978-3-031-16434-7_1
8. Luo, X., Hu, M., Song, T., Wang, G., Zhang, S.: Semi-supervised medical image segmentation via cross teaching between CNN and transformer. In: MIDL, pp. 820–833. PMLR (2022)

9. Luo, X., et al.: Semi-supervised medical image segmentation via uncertainty rectified pyramid consistency. Med. Image Anal. **80**, 102517 (2022)
10. Müller, R., Kornblith, S., Hinton, G.E.: When does label smoothing help? In: NeurIPS, pp. 1–10 (2019)
11. Ouali, Y., Hudelot, C., Tami, M.: Semi-supervised semantic segmentation with cross-consistency training. In: CVPR, pp. 12674–12684 (2020)
12. Roy, A.G., Navab, N., Wachinger, C.: Recalibrating fully convolutional networks with spatial and channel "squeeze and excitation" blocks. IEEE Trans. Med. Imaging **38**(2), 540–549 (2019)
13. Shen, H., et al.: Deep active learning for breast cancer segmentation on immunohistochemistry images. In: Martel, A.L., et al. (eds.) MICCAI 2020. LNCS, vol. 12265, pp. 509–518. Springer, Cham (2020). https://doi.org/10.1007/978-3-030-59722-1_49
14. Tarvainen, A., Valpola, H.: Mean teachers are better role models: weight-averaged consistency targets improve semi-supervised deep learning results. In: NeurIPS, pp. 1–10 (2017)
15. Vu, T.H., Jain, H., Bucher, M., Cord, M., Pérez, P.: Advent: adversarial entropy minimization for domain adaptation in semantic segmentation. In: CVPR, pp. 2517–2526 (2019)
16. Woo, S., Park, J., Lee, J.Y., Kweon, I.S.: Cbam: convolutional block attention module. In: ECCV, pp. 3–19 (2018)
17. Wu, H., Wang, Z., Song, Y., Yang, L., Qin, J.: Cross-patch dense contrastive learning for semi-supervised segmentation of cellular nuclei in histopathologic images. In: CVPR, pp. 11666–11675 (2022)
18. , Wu, L., et al.: R-drop: regularized dropout for neural networks. In: NeurIPS, pp. 10890–10905 (2021)
19. Wu, Y., et al.: Mutual consistency learning for semi-supervised medical image segmentation. Med. Image Anal. **81**, 102530 (2022)
20. Xie, Y., Lu, H., Zhang, J., Shen, C., Xia, Y.: Deep segmentation-emendation model for gland instance segmentation. In: Shen, D., et al. (eds.) MICCAI 2019. LNCS, vol. 11764, pp. 469–477. Springer, Cham (2019). https://doi.org/10.1007/978-3-030-32239-7_52
21. Xie, Y., Zhang, J., Liao, Z., Verjans, J., Shen, C., Xia, Y.: Pairwise relation learning for semi-supervised gland segmentation. In: Martel, A.L., et al. (eds.) MICCAI 2020. LNCS, vol. 12265, pp. 417–427. Springer, Cham (2020). https://doi.org/10.1007/978-3-030-59722-1_40
22. Xu, K., Rui, L., Li, Y., Gu, L.: Feature normalized knowledge distillation for image classification. In: Vedaldi, A., Bischof, H., Brox, T., Frahm, J.-M. (eds.) ECCV 2020. LNCS, vol. 12370, pp. 664–680. Springer, Cham (2020). https://doi.org/10.1007/978-3-030-58595-2_40
23. Yu, L., Wang, S., Li, X., Fu, C.-W., Heng, P.-A.: Uncertainty-aware self-ensembling model for semi-supervised 3D left atrium segmentation. In: Shen, D., et al. (eds.) MICCAI 2019. LNCS, vol. 11765, pp. 605–613. Springer, Cham (2019). https://doi.org/10.1007/978-3-030-32245-8_67
24. Zhao, B., Cui, Q., Song, R., Qiu, Y., Liang, J.: Decoupled knowledge distillation. In: CVPR, pp. 11953–11962 (2022)

An Anti-biased TBSRTC-Category Aware Nuclei Segmentation Framework with a Multi-label Thyroid Cytology Benchmark

Junchao Zhu[1], Yiqing Shen[2], Haolin Zhang[3], and Jing Ke[4,5](✉)

[1] School of Life Sciences and Biotechnology, Shanghai Jiao Tong University, Shanghai, China
junchaozhu@sjtu.edu.cn

[2] Department of Computer Science, Johns Hopkins University, Baltimore, MD, USA
yshen92@jhu.edu

[3] School of Engineering, Shanghai Ocean University, Shanghai, China
m220851484@st.shou.edu.cn

[4] School of Electronic Information and Electrical Engineering, Shanghai Jiao Tong University, Shanghai, China
kejing@sjtu.edu.cn

[5] School of Computer Science and Engineering, University of New South Wales, Sydney, Australia

Abstract. The Bethesda System for Reporting Thyroid Cytopathology (TBSRTC) has been widely accepted as a reliable criterion for thyroid cytology diagnosis, where extensive diagnostic information can be deduced from the allocation and boundary of cell nuclei. However, two major challenges hinder accurate nuclei segmentation from thyroid cytology. Firstly, unbalanced distribution of nuclei morphology across different TBSRTC categories can lead to a biased model. Secondly, the insufficiency of densely annotated images results in a less generalized model. In contrast, image-wise TBSRTC labels, while containing lightweight information, can be deeply explored for segmentation guidance. To this end, we propose a TBSRTC-category aware nuclei segmentation framework (`TCSegNet`). To top up the small amount of pixel-wise annotations and eliminate the category preference, a larger amount of image-wise labels are taken in as the complementary supervision signal in `TCSegNet`. This integration of data can effectively guide the pixel-wise nuclei segmentation task with a latent global context. We also propose a semi-supervised extension of `TCSegNet` that leverages images with only TBSRTC-category labels. To evaluate the proposed framework and also for further cytology cell studies, we curated and elaborately annotated a multi-label thyroid cytology benchmark, collected clinically from 2019 to 2022, which will be made public upon acceptance. Our `TCSegNet` outperforms state-of-the-art segmentation approaches with an improvement of 2.0% Dice and 2.7% IoU; besides, the semi-supervised extension can further boost this margin. In conclusion, our study explores the weak annotations by constructing an image-

H. Greenspan et al. (Eds.): MICCAI 2023, LNCS 14225, pp. 580–590, 2023.
https://doi.org/10.1007/978-3-031-43987-2_56

wise-label-guided nuclei segmentation framework, which has the potential medical importance to assist thyroid abnormality examination. Code is available at https://github.com/Junchao-Zhu/TCSegNet.

Keywords: Unbalanced Nuclei Segmentation · Semi-Supervised Learning · Thyroid Cytology · TBSRTC Diagnostic Category

1 Introduction

Thyroid cancer is the most common cancer of the endocrine system, accounting for 2.1% of all malignant cancers [1]. Clinically, pathologists rely on the six-category of "The Bethesda System for Reporting Thyroid Cytopathology" (TBSRTC) [2,3] to distinguish the cell morphology in the stained cytopathologic sections. The emergence of computational pathology allows automatic diagnosis of thyroid cancer, and nuclei segmentation becomes one of the most critical diagnostic tasks [4,5], as the shapes of nuclei, whether round, oval, or elongated, can provide valuable information for further analysis [6]. For example, small and scattered thyroid cells with a light hue and relatively low cell density are usually low-grade and indicative of early-stage cancer; whereas large and dark cells with extreme-dense agglomeration are usually middle- or late-grade [3]. Correspondingly, accurate location of cell boundaries is essential for both pathologists and computer-aided diagnosis (CAD) systems to assist decision [7].

However, nuclei segmentation in thyroid cytopathology is still challenged by the varying cellularity of images from different TBSRTC categories [3,8]. For example, benign cells (I & II) present high sparsity and are difficult to be distinguished from background tissues, thus may account for a relatively small proportion when equal images are involved in a training set [3]. By contrast, high-grade cells (V & VI) are densely packed and severely clustered, thus much more are presented in a training set. In this way, an unbalanced distribution across different categories resulted, correspondingly, the training leads to biased models with lower accuracy [9,10]. Such distinct morphological differences can be characterized by the TBSRTC category, which thus inspires us to utilize the handy image-wise grading labels to guide the nuclei segmentation model learning from unbalanced datasets. We also noticed that another challenge for accurate nuclei identification is the heavy reliance on large-scale high-quality annotations [11]. Moreover, amongst multiple annotation paradigms [12], pixel-level labeling is the most time-consuming and laborious, whereas the image-wise diagnostic labels, *i.e.* TBSRTC categories, are comparatively simpler. Despite the labeling intensity, prevalent nuclei segmentation methods, e.g., CIA-Net [13], $CA^{2.5}$-Net [14], and ClusterSeg [15], are limited to pixel-wise annotations, where the potential benefits of integrating accessible image-wise labels are unaware.

To narrow the gap discussed, we propose a novel TBSRTC-category-aware nuclei segmentation framework. Our contributions are three-fold. (1) We propose a cytopathology nuclei segmentation network named `TCSegNet`, to provide supplementary guidance to facilitate the learning of nuclei boundaries.

Innovatively, our approach can help reduce bias in the learning process of the segmentation model with the routine unbalanced training set. (2) We expand `TCSegNet` to `Semi-TCSegNet` to leverage image-wise labels in a semi-supervised learning manner, which significantly reduces the reliance on annotation-intensive pixel-wise labels. Additionally, an HSV-intensity noise is designed specifically for cytopathology images to boost the generalization ability. (3) We establish a dataset of thyroid cytopathology image patches of 224×224, where 4,965 image labels are provided following TBSRTC, and 1,473 of them are densely annotated [3] (to be on GitHub upon acceptance). To the best of our knowledge, it is the first publicized thyroid cytopathology dataset of both image-wise and pixel-wise labels. The annotated dataset well alleviates the insufficiency of an open cytopathology dataset for computer-assisted analysis (Fig. 1).

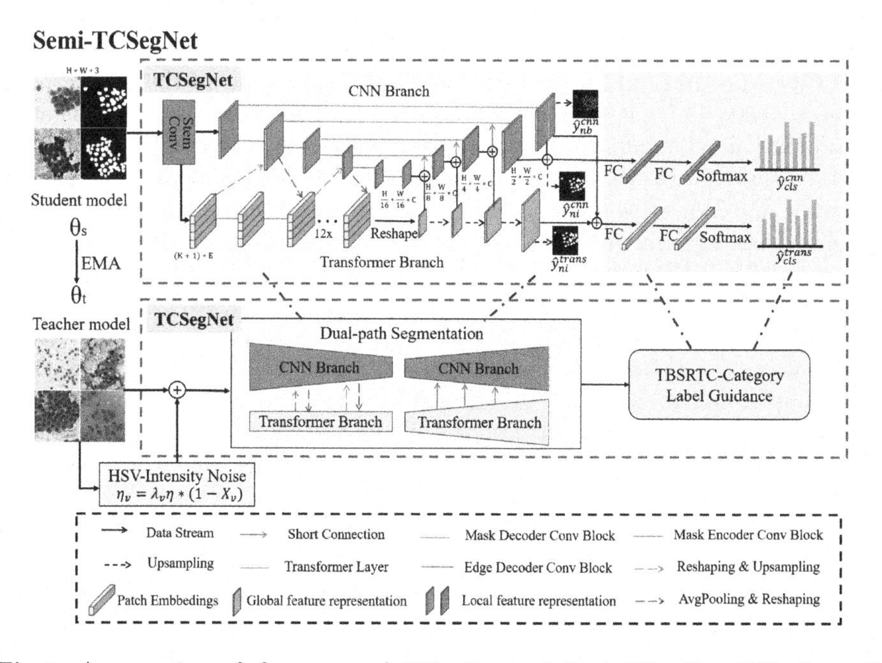

Fig. 1. An overview of the proposed `TCSegNet` and `Semi-TCSegNet`. `TCSegNet` utilizes annotation-lightweight image-wise TBSRTC-category labels to aid in the learning of unbalanced nuclei morphology in segmentation. `Semi-TCSegNet` reduces the heavy reliance on annotation-intensive pixel-wise labels through the use of a semi-supervised framework.

2 Methodology

Overview. We propose a novel TBSRTC-category Aware Segmentation Network (`TCSegNet`) to segment nuclei boundaries in cytopathology images, which is guided by TBSRTC-category label to learn from unbalanced data. Our model uses a CNN and Transformer dual-path U-shape architecture, where the CNN captures the local features, and the Transformer extracts the global features for a more comprehensive representation of nuclei allocation [16]. Considering the spatial distributions of thyroid cells in cytopathology images, our design provides extended global information for more accurate segmentation. Our approach employs short connections to allow effective communication of local and global representations [17]. Formally, the overall segmentation loss $\mathcal{L}_{seg}$ to train our model is a combination of the binary cross-entropy loss (BCE), *i.e.*

$$\mathcal{L}_{seg} = \gamma_{ni} \cdot \texttt{BCE}(\hat{y}_{ni}^{cnn}, y_{ni}) + \gamma_{ni} \cdot \texttt{BCE}(\hat{y}_{ni}^{trans}, y_{ni}) + \gamma_{nb} \cdot \texttt{BCE}(\hat{y}_{nb}^{cnn}, y_{nb}), \quad (1)$$

where $\hat{y}$ is the prediction from `TCSegNet` and y is and pixel-wise annotation, respectively. Subscript ni and nb denote the nuclei area and boundary, respectively. Superscripts cnn and $trans$ write for the CNN branch and Transformer branch, respectively. We set the balancing coefficient γ_{ni} to 1 and γ_{nb} to 10. Additionally, to ensure the consistency between the two branches, we impose a dice consistency loss ($\mathcal{L}_{cons}$) between the nuclei instance predictions from the CNN branch and the Transformer branch, namely $\mathcal{L}_{cons} = \texttt{Dice}(\hat{y}_{ni}^{cnn}, \hat{y}_{ni}^{trans})$.

TBSRTC-Category Label Guidance Block. In `TCSegNet`, we introduce a TBSRTC-category label guidance block to address the learning issue from unbalanced routine datasets. This block consists of two learnable fully connected layers that process the feature extracted by the CNN and Transformer branches separately, which obtains image-wise TBSRTC-category prediction denoted as $\hat{y}_{cls}^{cnn}$ and $\hat{y}_{cls}^{trans}$. Correspondingly, to train this block, we use a cross-entropy loss function (CE) that provides an extra supervision signal to help the network learn from unbalanced datasets, defined as follows:

$$\mathcal{L}_{cls} = \texttt{CE}(\hat{y}_{cls}^{cnn}, y_{cls}) + \gamma_{cls} \cdot \texttt{CE}(\hat{y}_{cls}^{trans}, y_{cls}), \quad (2)$$

where y_{cls} is the image-wise TBSRTC-category label, and the balancing coefficient γ_{cls} is set to 3, as the global feature captured by the Transformer branch is tightly correlated with the image-level classification tag. Finally, the overall loss for `TCSegNet` becomes

$$\mathcal{L}_s = \mathcal{L}_{seg} + \mathcal{L}_{cls} + \mathcal{L}_{cons}. \quad (3)$$

Extension to Semi-supervised Learning. To leverage images that only have image-wise labels, we extend to a semi-supervised mean teacher [18] framework called `Semi-TCSegNet`. In this framework, both the student and teacher share the same full-supervised nuclei segmentation architecture of `TCSegNet`. The weights of the teacher θ_t are updated with the exponential moving average (EMA) of the weights of student θ_s, and smoothing coefficient $\alpha = 0.99$, following the previous work [19]. Formally, the weights of the teacher at e-th epoch are updated by

$$\theta_t^e = \alpha\theta_t^{e-1} + (1 - \alpha)\theta_s^e. \quad (4)$$

During the training stage, the teacher model assigns pixel-wise soft labels to the images with exclusive image-wise labels, thus expanding the scale of labeled data to the student model. The overall loss function for training the student is a combination of the supervised loss $\mathcal{L}s$ in Eq. (3) computed on fully-annotated data, and the semi-supervised loss $\mathcal{L}ss$ computed on data with only image-wise labels. It follows that the overall training objective in `Semi-TCSegNet` is to minimize the loss function $\mathcal{L} = \mathcal{L}_s + \lambda_{ss} \cdot \mathcal{L}_{ss}$, where $\mathcal{L}_{ss}$ measures the segmentation consistency between the teacher and student models via L_2-norm. The balancing coefficient λ_{ss} changes for every epoch e, namely

$$\lambda_{ss}^{e} = \exp(-5\,(1 - e/e_{\max})^2), \tag{5}$$

where $e_{\max}$ is the maximum epoch number.

HSV-Intensity Noise. The traditional method of integrating Gaussian noise in the mean teacher [18] may be problematic when working with cytopathology images that have an imbalanced color distribution. To address this issue, we generate a novel intensity-based noise, which can adaptively behave stronger in the dark nuclei areas and weaker in bright cytoplasm or background regions. We first sample η from a Gaussian distribution $\mathcal{N}\left(0, \sigma^2\right)$, where σ is the standard deviation computed from the pixel values of the V channel in HSV space. The Gaussian noise η serves as the basis for generating the intensity-based noise, which is obtained by $\eta_v = \lambda_v \cdot \eta \cdot (1 - X_v)$. Specifically, X_v is the pixel value of the image's V channel in HSV space, and hyper-parameter λ_v is set as 0.5 to control the amplitude of the intensity-based noise. Finally, the value of the obtained noise is clamped to $[-0.2, 0.2]$ before being added to the images.

3 Experiments

Image Dataset. We construct a clinical thyroid cytopathology dataset with images of both image-wise and pixel-wise labels as a benchmark (appear in GitHub upon acceptance) Some representative images are presented in Fig. 2, together with the profile of the dataset. The dataset comprises 4,965 H&E stained image patches and labels of TBSRTC, where a subset of 1,473 images was densely annotated for nuclei boundaries by three experienced cytopathologists and reached a total number of 31,064 elaborately annotated nuclei. Patient-level images were partitioned first for training and test images, and patch-level curation was performed. We divided the dataset with image-wise labels into 80% training samples and the remaining 20% testing samples. Our collection of thyroid cytopathology images was granted with an Ethics Approval document.

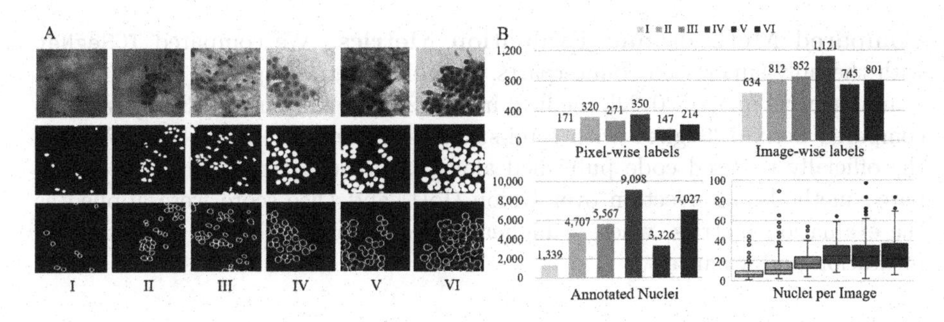

Fig. 2. The annotated thyroid cytopathology benchmark. A) Examples from I to VI of TBSRTC six diagnostic categories with pixel-wise nuclei mask and boundary annotations; B) The profile of the dataset.

Table 1. Quantitative comparisons in both fully-supervised and semi-supervised manners. The best performance is highlighted in **bold**, where we can observe that both `TCSegNet` and its semi-supervised extension outperform state-of-the-art.

Method		Dice	IoU
Fully-supervised	Mask R-CNN [20]	0.657	0.500
	Swin-Unet [21]	0.671	0.516
	SegNet [22]	0.676	0.587
	UNet++ [23]	0.784	0.691
	$Ca^{2.5}$-Net [14]	0.838	0.732
	CIA-Net [13]	0.854	0.775
	ClusterSeg [15]	0.857	0.761
	TCSegNet (Ours)	**0.877**	**0.788**
Semi-supervised	PseudoSeg [24]	0.734	0.612
	Cross Pseudo Seg [25]	0.737	0.618
	Cross Teaching [26]	0.795	0.704
	PS-ClusterSeg [15]	0.866	0.775
	MTMT-Net [27]	0.878	0.789
	Semi-TCSegNet (Ours)	**0.889**	**0.805**

Implementations. The proposed method and compared methods are implemented on a single NVIDIA GeForce RTX 3090 GPU card. We employ a conformer [16] with 12 Transformer layers and 5 CNN blocks as the encoder in `TCSegNet`. Both `TCSegNet` and `Semi-TCSegNet` use SGD optimizer with a momentum of 0.9 and a weight decay of 10^{-4}. The initial learning rate lr_0 is set to 5×10^{-3}, and the learning rate for e^{th} epoch is determined by the poly strategy [28], *i.e.*, $lr_e = lr_0 \times (1 - e/e_{max})^{0.9}$, where $e_{max} = 280$ is the total epoch number. We set the batch size for `TCSegNet` to 8, and for `Semi-TCSegNet` to 10, *i.e.* 8 fully-annotated images and 2 partially-annotated images per batch.

Compared Methods and Evaluation Metrics. We compared TCSegNet with the fully-supervised counterparts, including method specific for segmentation in general image [20,22], medical image [21,23], and nuclei [13–15]. We also compared Semi-TCSegNet with semi-supervised methods [15,24–27]. We used the officially released code published along with their papers for all the compared methods. Intersection over Union (IoU) and Dice score were applied as the evaluation metrics, where a higher value indicated a better semantic segmentation performance.

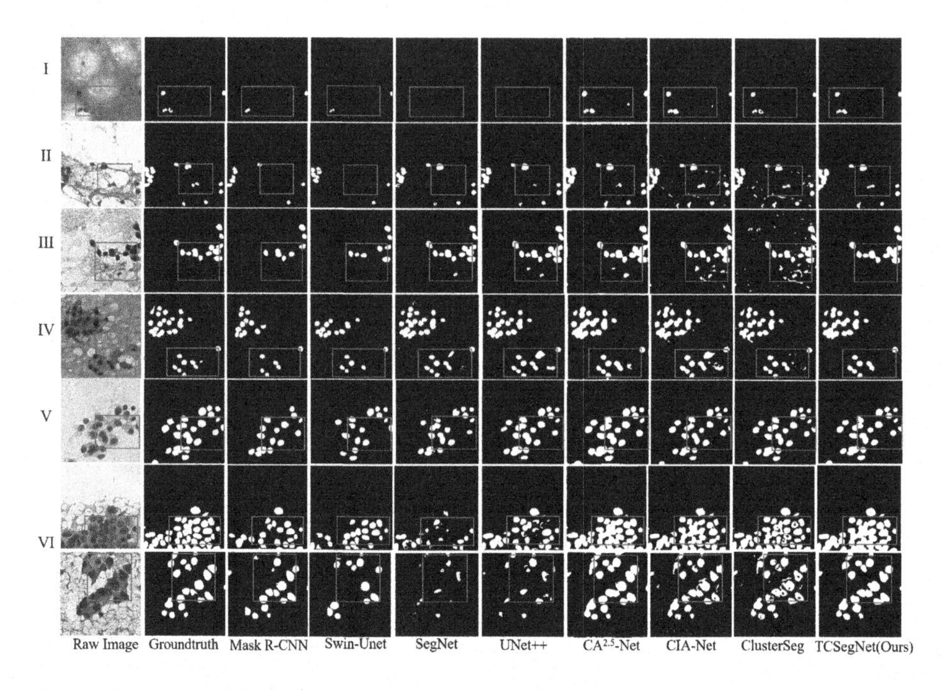

Fig. 3. Examples of segmented nuclei in thyroid cytopathology images by TCSegNet and prevent fully-supervised models methods. Observably, TCSegNet is more aware of scattered, small, or clustered nuclei.

Experimental Results. The results in Table 1 indicated that TCSegNet can achieve the highest performance by a Dice score of 87.7% and an IoU of 78.8%. The performance values in the challenging regions are highlighted with red boxes in Fig. 3, together with the line charts in Fig. 4 (A, B). Our approach is capable to address the current issue in the recognition and segmentation of small isolated cells graded in the I category, which is always ignored by the unbalanced pixel-wise cell morphology with other approaches. Also, it yields that the incorporation of TBSRTC-category can contribute to a partial alleviation of a biased model, resulting in more satisfying segmentation performance experimentally. Furthermore, the fact that the TBSRTC-category label is easy to obtain endows the applicability of our model to various circumstances that nuclei in various sizes, shapes, and dyeing styles can be accurately recognized and segmented. Consequently, it can serve as a guarantee for the validity and accuracy of the subsequent analysis in real clinical practice. Moreover, with the semi-supervised learning, `Semi-TCSegNet` can further boost the performance to an 88.9% Dice score, and 80.5% IoU, by leveraging additional data with image-wise TBSRTC-category labels solely. The performance improvement of 1.2% Dice, 1.7% IoU, together with the general improvement is shown in the boxplot in Fig. 4 (C, D), as a demonstration of the advantage using full data resources with `Semi-TCSegNet`.

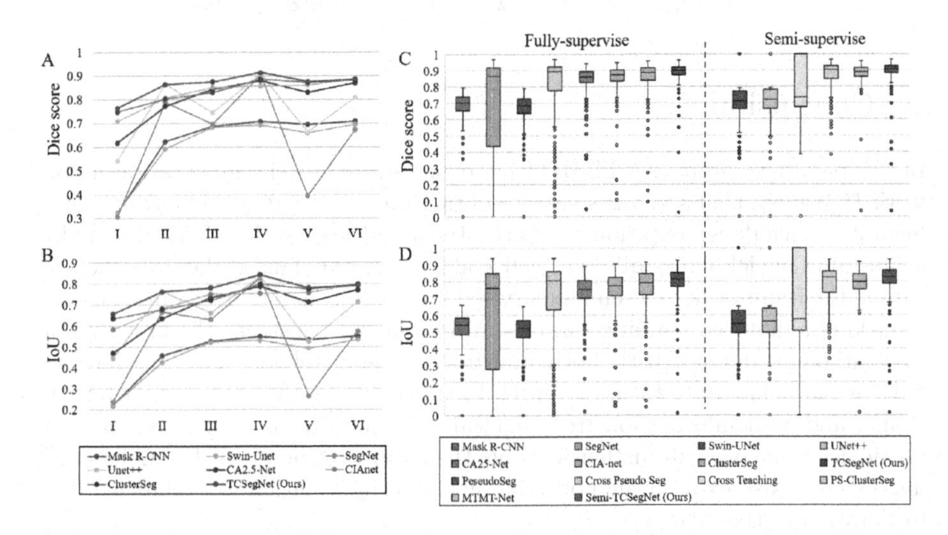

Fig. 4. Quantitative comparison presented by boxplot and line chart. A and B are the variation tendency of Dice score and IoU of six diagnostic categories in fully-supervised methods. C and D are the distribution of the Dice score and IoU of all mentioned models respectively. Our models presented a general improvement across the metrics.

Ablation Study. To evaluate the effectiveness of each functional block and demonstrate the functionality of semi-supervised learning, we illustrate the ablation study in Table 2. The results indicate that performance improvement is accumulated with increasing data size. Besides, training with a classification-learning block alone can increase the nuclei segmentation performance by 1.7% and 2.6% in the Dice score and IoU, respectively. Meanwhile, trained with specially designed HSV-Intensity noise can also increase the performance by 0.9% Dice and 1.4% IoU, showing its potential for generation ability improvement. Importantly, the benefits from the two blocks are orthonormal, where `Semi-TCSegNet` achieves the optimal performance with the utilization of both.

Table 2. Ablation study for our Semi-TCSegNet and functional blocks.

Classification Learning	HSV-Intensity Noise	Dice	IoU
		0.867	0.771
	✓	0.876	0.785
✓		0.884	0.797
✓	✓	**0.889**	**0.805**
W. image-wise data	+1k data	0.879	0.790
	+2k data	0.882	0.795

4 Conclusion

In this paper, we propose a TBSRTC-category aware nuclei segmentation framework `TCSegNet`, that leverages easy-to-obtain image-wise diagnostic category to facilitate nuclei segmentation. Importantly, it addresses the challenge of distinguishing nuclei across different cell scales in an unbalanced dataset. We also extend the framework to a semi-supervised learning fashion to overcome the issue of lacking annotated training samples. Moreover, we construct the first thyroid cytopathology dataset with both image-wise and pixel-wise labels, which we believe can it facilitate future research in this field. As the spatial distribution, shape, and area information from nuclear segmentation is supportive of diagnostic decisions, we will further leverage the segmentation result for malignancy analysis and also explore the potential of spatial information for unlabeled data exploration in the future.

Acknowledgement. This work was supported by National Natural Science Foundation of China (Grant No. 62102247) and Natural Science Foundation of Shanghai (No. 23ZR1430700).

References

1. Ferlay, J., et al.: Cancer incidence and mortality worldwide: sources, methods and major patterns in globocan 2012. Int. J. Cancer **136**(5), E359–E386 (2015)
2. Haugen, B.R., et al.: 2015 American thyroid association management guidelines for adult patients with thyroid nodules and differentiated thyroid cancer: the american thyroid association guidelines task force on thyroid nodules and differentiated thyroid cancer. Thyroid **26**(1), 1–133 (2016)
3. Cibas, E.S., Ali, S.Z.: The bethesda system for reporting thyroid cytopathology. Thyroid **19**(11), 1159–1165 (2009)
4. Kumar, N., et al.: A multi-organ nucleus segmentation challenge. IEEE Trans. Med. Imaging **39**(5), 1380–1391 (2019)
5. Veta, M., et al.: Prognostic value of automatically extracted nuclear morphometric features in whole slide images of male breast cancer. Mod. Pathol. **25**(12), 1559–1565 (2012)

6. Kakudo, K.: Thyroid FNA Cytology: Differential Diagnoses and Pitfalls. Springer, Heidelberg (2019). https://doi.org/10.1007/978-981-13-1897-9
7. Xing, F., Yang, L.: Robust nucleus/cell detection and segmentation in digital pathology and microscopy images: a comprehensive review. IEEE Rev. Biomed. Eng. **9**, 234–263 (2016)
8. Cibas, E.S., Ali, S.Z.: The 2017 bethesda system for reporting thyroid cytopathology. Thyroid **27**(11), 1341–1346 (2017)
9. Graham, S., et al.: Hover-net: Simultaneous segmentation and classification of nuclei in multi-tissue histology images. Med. Image Anal. **58**, 101563 (2019)
10. Gamper, J., Alemi Koohbanani, N., Benet, K., Khuram, A., Rajpoot, N.: PanNuke: an open pan-cancer histology dataset for nuclei instance segmentation and classification. In: Reyes-Aldasoro, C.C., Janowczyk, A., Veta, M., Bankhead, P., Sirinukunwattana, K. (eds.) ECDP 2019. LNCS, vol. 11435, pp. 11–19. Springer, Cham (2019). https://doi.org/10.1007/978-3-030-23937-4_2
11. Greenwald, N.F., et al.: Whole-cell segmentation of tissue images with human-level performance using large-scale data annotation and deep learning. Nat. Biotechnol. **40**(4), 555–565 (2022)
12. Everingham, M., Eslami, S.A., Van Gool, L., Williams, C.K., Winn, J., Zisserman, A.: The pascal visual object classes challenge: a retrospective. Int. J. Comput. Vision **111**, 98–136 (2015)
13. Zhou, Y., Onder, O.F., Dou, Q., Tsougenis, E., Chen, H., Heng, P.-A.: CIA-Net: robust nuclei instance segmentation with contour-aware information aggregation. In: Chung, A.C.S., Gee, J.C., Yushkevich, P.A., Bao, S. (eds.) IPMI 2019. LNCS, vol. 11492, pp. 682–693. Springer, Cham (2019). https://doi.org/10.1007/978-3-030-20351-1_53
14. Huang, J., Shen, Y., Shen, D., Ke, J.: CA$^{2.5}$-net nuclei segmentation framework with a microscopy cell benchmark collection. In: de Bruijne, M., et al. (eds.) MICCAI 2021. LNCS, vol. 12908, pp. 445–454. Springer, Cham (2021). https://doi.org/10.1007/978-3-030-87237-3_43
15. Ke, J., et al.: Clusterseg: a crowd cluster pinpointed nucleus segmentation framework with cross-modality datasets. Med. Image Anal. **85**, 102758 (2023)
16. Peng, Z., et al.: Conformer: local features coupling global representations for visual recognition. In: Proceedings of the IEEE/CVF International Conference on Computer Vision, pp. 367–376 (2021)
17. Hou, Q., Cheng, M.M., Hu, X., Borji, A., Tu, Z., Torr, P.H.: Deeply supervised salient object detection with short connections. In: Proceedings of the IEEE Conference on Computer Vision and Pattern Recognition, pp. 3203–3212 (2017)
18. Tarvainen, A., Valpola, H.: Mean teachers are better role models: weight-averaged consistency targets improve semi-supervised deep learning results. Adv. Neural Inf. Process. Syst. **30**, 1–10 (2017)
19. Laine, S., Aila, T.: Temporal ensembling for semi-supervised learning. arXiv preprint arXiv:1610.02242 (2016)
20. He, K., Gkioxari, G., Dollár, P., Girshick, R.: Mask r-cnn. In: Proceedings of the IEEE International Conference on Computer Vision, pp. 2961–2969 (2017)
21. Cao, H., et al.: Swin-unet: unet-like pure transformer for medical image segmentation. In: Computer Vision-ECCV 2022 Workshops: Tel Aviv, Israel, 23–27 October 2022, Proceedings, Part III, pp. 205–218. Springer, Heidelberg (2023). https://doi.org/10.1007/978-3-031-25066-8_9
22. Badrinarayanan, V., Kendall, A., Cipolla, R.: Segnet: a deep convolutional encoder-decoder architecture for image segmentation. IEEE Trans. Pattern Anal. Mach. Intell. **39**(12), 2481–2495 (2017)

23. Zhou, Z., Rahman Siddiquee, M.M., Tajbakhsh, N., Liang, J.: UNet++: a nested u-net architecture for medical image segmentation. In: Stoyanov, D., et al. (eds.) DLMIA/ML-CDS -2018. LNCS, vol. 11045, pp. 3–11. Springer, Cham (2018). https://doi.org/10.1007/978-3-030-00889-5_1
24. Zou, Y., et al.: Pseudoseg: designing pseudo labels for semantic segmentation. arXiv preprint arXiv:2010.09713 (2020)
25. Chen, X., Yuan, Y., Zeng, G., Wang, J.: Semi-supervised semantic segmentation with cross pseudo supervision. In: Proceedings of the IEEE/CVF Conference on Computer Vision and Pattern Recognition, pp. 2613–2622 (2021)
26. Luo, X., Hu, M., Song, T., Wang, G., Zhang, S.: Semi-supervised medical image segmentation via cross teaching between cnn and transformer. In: International Conference on Medical Imaging with Deep Learning, pp. 820–833. PMLR (2022)
27. Chen, Z., Zhu, L., Wan, L., Wang, S., Feng, W., Heng, P.A.: A multi-task mean teacher for semi-supervised shadow detection. In: Proceedings of the IEEE/CVF Conference on Computer Vision and Pattern Recognition, pp. 5611–5620 (2020)
28. Liu, W., Rabinovich, A., Berg, A.C.: Parsenet: looking wider to see better. arXiv preprint arXiv:1506.04579 (2015)

DARC: Distribution-Aware Re-Coloring Model for Generalizable Nucleus Segmentation

Shengcong Chen[1], Changxing Ding[1,2](✉), Dacheng Tao[3], and Hao Chen[4]

[1] School of Electronic and Information Engineering, South China University of Technology, Guangzhou 510000, China
c.shengcong@mail.scut.edu.cn, chxding@scut.edu.cn
[2] Pazhou Lab, Guangzhou 510330, China
[3] School of Computer Science, The Faculty of Engineering, The University of Sydney, Darlington, NSW 2008, Australia
[4] Department of Computer Science and Engineering and Department of Chemical and Biological Engineering, The Hong Kong University of Science and Technology, Hong Kong, China

Abstract. Nucleus segmentation is usually the first step in pathological image analysis tasks. Generalizable nucleus segmentation refers to the problem of training a segmentation model that is robust to domain gaps between the source and target domains. The domain gaps are usually believed to be caused by the varied image acquisition conditions, e.g., different scanners, tissues, or staining protocols. In this paper, we argue that domain gaps can also be caused by different foreground (nucleus)-background ratios, as this ratio significantly affects feature statistics that are critical to normalization layers. We propose a Distribution-Aware Re-Coloring (DARC) model that handles the above challenges from two perspectives. First, we introduce a re-coloring method that relieves dramatic image color variations between different domains. Second, we propose a new instance normalization method that is robust to the variation in foreground-background ratios. We evaluate the proposed methods on two H&E stained image datasets, named CoNSeP and CPM17, and two IHC stained image datasets, called DeepLIIF and BC-DeepLIIF. Extensive experimental results justify the effectiveness of our proposed DARC model. Codes are available at https://github.com/csccsccsccsc/DARC.

Keywords: Domain Generalization · Nucleus Segmentation · Instance Normalization

1 Introduction

Automatic nucleus segmentation has captured wide research interests in recent years due to its importance in pathological image analysis [1–4]. However, as

---

Supplementary Information The online version contains supplementary material available at https://doi.org/10.1007/978-3-031-43987-2_57.

H. Greenspan et al. (Eds.): MICCAI 2023, LNCS 14225, pp. 591–601, 2023.
https://doi.org/10.1007/978-3-031-43987-2_57

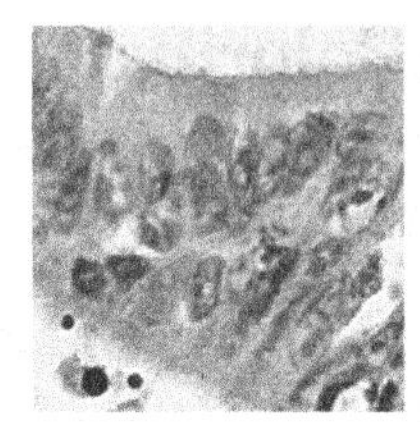 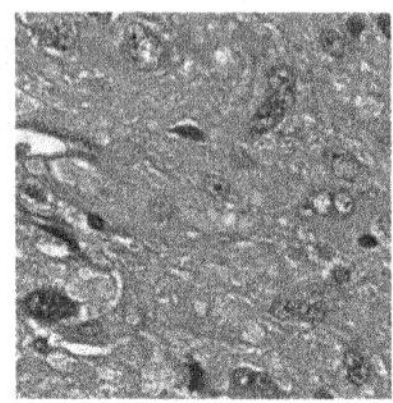 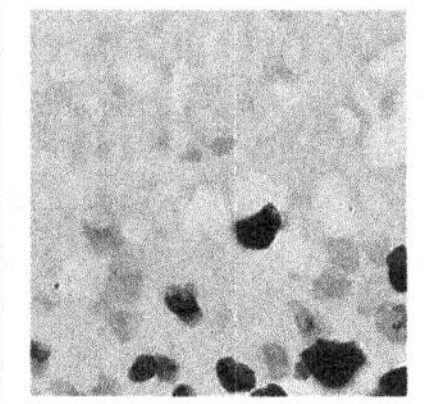 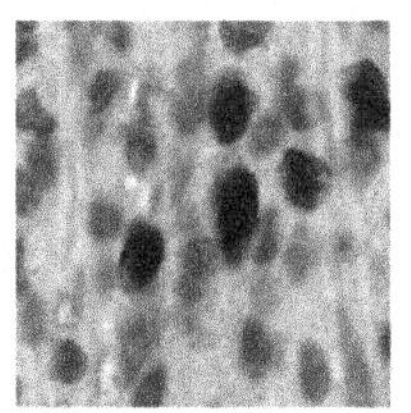

Fig. 1. Example image patches from different datasets. Their appearance differs significantly from each other due to variations in image modalities, staining protocols, scanner types, and tissues.

shown in Fig. 1, the variations in image modalities, staining protocols, scanner types, and tissues significantly affect the appearance of nucleus images, resulting in notable gap between source and target domains [5–7]. If a number of target domain samples are available before testing, one can adopt domain adaptation algorithms to transfer the knowledge learned from the source domain to the target domain [8–10]. Unfortunately, in real-world applications, it is usually expensive and time-consuming to collect new training sets for the ever changing target domains; moreover, extra computational cost is required, which is usually unrealistic for the end users. Therefore, it is highly desirable to train a robust nucleus segmentation model that is generalizable to different domains.

In recent years, the research on domain generalization (DG) has attracted wide attention. Most existing DG works are proposed for classification tasks [12,13] and they can be roughly grouped into data augmentation-, representation learning-, and optimization-based methods. The first category of methods [14–17] focus on the way to diversify training data styles and expect the enriched styles cover those appeared in target domains. The second category of methods aim to obtain domain-invariant features. This is usually achieved via improving model architectures [18–20] or introducing novel regularization terms [21,22]. The third category of methods [23–26] develop new model optimization strategies, e.g., meta-learning, that improve model robustness via artificially introducing domain shifts during training.

It is a consensus that a generalizable nucleus segmentation model should be robust to image appearance variation caused by the change in staining protocols, scanner types, and tissues, as illustrated in Fig. 1. In this paper, we argue that it is also desirable to be robust to the ratio between foreground (nucleus) and background pixel numbers. This ratio changes the statistics of each feature map channel, and affects the robustness of normalization layers, e.g., instance normalization (IN). We will empirically justify its impact in Sect. 2.3.

2 Method

2.1 Overview

In this paper, we adopt a U-Net-based model similar to that in [1] as the baseline. It performs both semantic segmentation and contour detection for nucleus

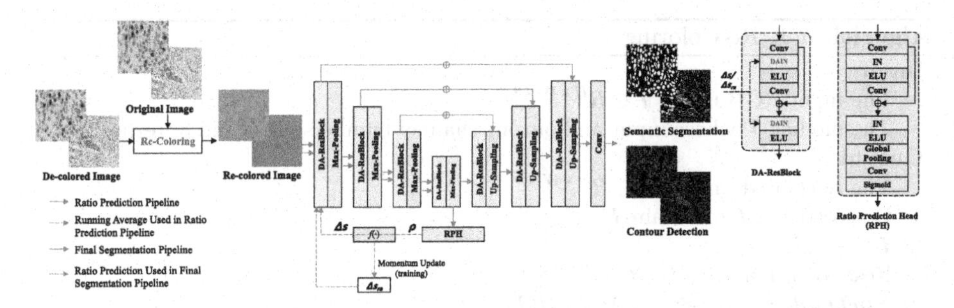

Fig. 2. Overview of the DARC model. The whole model is trained in an end-to-end manner. DARC first re-colors each image to relieve the impact caused by different image acquisition conditions. The re-colored image is then fed into the U-Net encoder and the ratio prediction head. This head predicts the foreground-background ratio ρ. Then, the re-colored image is fed into DARC again with ρ for final prediction. For simplicity, we only illustrate the data-flow of the first DA-ResBlock in details.

instances. The area of each nucleus instance is obtained via subtraction between the segmentation and contour prediction maps [1]. Details of the baseline model is provided in the supplementary material. To handle domain variations, we adopt IN rather than batch normalization (BN) in the U-Net model.

Our proposed Distribution-Aware Re-Coloring model (DARC) is illustrated in Fig. 2. Compared with the baseline, DARC replaces the IN layers with the proposed Distribution-Aware Instance Normalization (DAIN) layers. DARC first re-colors each image to relieve the influence caused by image acquisition conditions. The re-colored image is then fed into the U-Net encoder and the ratio prediction head. This head predicts the ratio between foreground and background pixel numbers. With the predicted ratio, the DAIN layers can estimate feature statistics more robustly and facilitate more accurate nucleus segmentation.

2.2 Nucleus Image Re-Coloring

We propose the Re-Coloring (RC) method to overcome the color change in different domains. Specifically, given a RGB image $\boldsymbol{I}$, e.g., an H&E or IHC stained image, we first obtain its grayscale image $\boldsymbol{I}_g$. We then feed $\boldsymbol{I}_g$ into a simple module T that consists of a single residual block and a 1×1 convolutional layer with output channel number of 3. In this way, we obtain an initial re-colored image $\boldsymbol{I}_r$.

However, de-colorization results in the loss of fine-grained textures and may harm the segmentation accuracy. To handle this problem, we compensate $\boldsymbol{I}_r$ with the original semantic information contained in $\boldsymbol{I}$. Recent works [40] show that semantic information can be reflected via the order of pixels according to their gray value. Therefore, we adopt the Sort-Matching algorithm [41] to combine the semantic information in $\boldsymbol{I}$ with the color values in $\boldsymbol{I}_r$. Details of RC is presented in Algorithm 1, in which *Sort* and *ArgSort* denote channel-wisely sorting the values and obtaining the sorted values and indices respectively, and

Algorithm 1. Re-Coloring

Input:

The input RGB image $\boldsymbol{I} \in R^{H \times W \times 3}$;

The module T whose input and output channel numbers are 1 and 3, respectively;

Output:

The re-colored image $\boldsymbol{I}_o \in R^{H \times W \times 3}$;

1: De-colorizing $\boldsymbol{I}$ to obtain $\boldsymbol{I}_g$;
2: $\boldsymbol{I}_r \leftarrow T(\boldsymbol{I}_g)$
3: Reshaping $\boldsymbol{I}$ and $\boldsymbol{I}_r$ to $R^{HW \times 3}$
4: $\boldsymbol{SortIndex} \leftarrow ArgSort(ArgSort(\boldsymbol{I}))$
5: $\boldsymbol{SortValue} \leftarrow Sort(\boldsymbol{I}_r)$
6: $\boldsymbol{I}_o \leftarrow AssignValue(\boldsymbol{SortIndex}, \boldsymbol{SortValue})$
7: **return** $\boldsymbol{I}_o$

Table 1. Evaluation on the impact of foreground-background ratio to model performance. Both training and testing samples are obtained from CPM17. B denotes the background expansion factor, which directly affects the foreground-background ratio.

B	1	2	4	6
AJI	65.11	59.11	53.41	54.13
Dice	86.14	84.05	80.97	79.75

$AssignValue$ denotes re-assembling the sorted values according to the provided indices. Details of the module T are included in the supplementary material.

Via RC, the original fine-grained structure information from $\boldsymbol{I}_g$ is recovered in $\boldsymbol{I}_r$. In this way, the re-colored image is advantageous in two aspects. First, the appearance difference between pathological images caused by the change in scanners and staining protocols is eliminated. Second, the re-colored image preserves fine-grained structure information, enabling precise instance segmentation to be possible.

2.3 Distribution-Aware Instance Normalization

Due to dramatic domain gaps, feature statistics may differ significantly between domains [5–7], which means that feature statistics obtained from the source domain may not apply to the target domain. Therefore, existing DG works usually replace BN with IN for feature normalization [12,19]. However, for dense-prediction tasks like semantic segmentation or contour detection, adopting IN alone cannot fully address the feature statistics variation problem. This is because feature statistics are also relevant to the ratio between foreground and background pixel numbers. Specifically, an image with more nucleus instances produces more responses in feature maps and thus higher feature statistic values, and vice versa. The difference in this ratio causes interference to nucleus segmentation.

Algorithm 2. Distribution-Aware Instance Normalization

Input:

Original feature maps $\boldsymbol{X} \in R^{H \times W \times C}$. The C-dimensional feature vector on its pixel (i, j) is denoted as $\boldsymbol{x}_{ij}$;

The modules E_μ and E_δ that re-estimate feature statistics;

$\boldsymbol{\Delta s}_{ra} \in R^{1 \times 1 \times C}$ that is obtained via running mean of $\boldsymbol{\Delta s}$ in the training stage;

The momentum factor α used to update $\boldsymbol{\Delta s}_{ra}$;

(Optional) $\boldsymbol{\Delta s} = f(\rho)$;

Output:

Normalized feature maps $\boldsymbol{Y} \in R^{H \times W \times C}$;

1: $\boldsymbol{\mu} \leftarrow \frac{1}{HW} \sum_{i=1}^{H} \sum_{j=1}^{W} \boldsymbol{x}_{ij}$

2: $\boldsymbol{\delta}^2 \leftarrow \frac{1}{HW} \sum_{i=1}^{H} \sum_{j=1}^{W} (\boldsymbol{x}_{ij} - \boldsymbol{\mu})^2$

3: **if** $\boldsymbol{\Delta s}$ is given **then**

4: // Using $\boldsymbol{\Delta s}$ to re-estimate feature statistics for final segmentation

5: $\boldsymbol{\mu}', \boldsymbol{\delta}' \leftarrow E_\mu(\boldsymbol{\mu}, \boldsymbol{\delta}, \boldsymbol{\Delta s}), E_\delta(\boldsymbol{\mu}, \boldsymbol{\delta}, \boldsymbol{\Delta s})$

6: **if** $Training$ **then**

7: // Updating $\boldsymbol{\Delta s}_{ra}$ during training

8: $\boldsymbol{\Delta s}_{ra} \leftarrow (1 - \alpha)\boldsymbol{\Delta s}_{ra} + \alpha \boldsymbol{\Delta s}$

9: **end if**

10: **else**

11: // Using $\boldsymbol{\Delta s}_{ra}$ to re-estimate feature statistics for ratio prediction

12: $\boldsymbol{\mu}', \boldsymbol{\delta}' \leftarrow E_\mu(\boldsymbol{\mu}, \boldsymbol{\delta}, \boldsymbol{\Delta s}_{ra}), E_\delta(\boldsymbol{\mu}, \boldsymbol{\delta}, \boldsymbol{\Delta s}_{ra})$

13: **end if**

14: $\boldsymbol{Y} \leftarrow (\boldsymbol{X} - \boldsymbol{\mu}')/\boldsymbol{\delta}'$

15: **return** $\boldsymbol{Y}$

To verify the above viewpoint, we evaluate the baseline model under different foreground-background ratios. Specifically, we first remove the foreground pixels via in-painting [27], and then pad the original testing images with the obtained background patches. We adopt B to denote the ratio between the size of the obtained new image and the original image size. Compared with the original images, the new images have the same foreground regions but more background pixels, and thus have different foreground-background ratios. Finally, we evaluate the performance of the baseline model with different B values. Experimental results are presented in Table 1. It is shown that the value of B affects the model performance significantly.

The above problem is common in nucleus segmentation because pathological images from different organs or tissues tend to have significantly different foreground-background ratios. However, this phenomenon is often ignored in existing research. To handle this problem, we propose the Distribution-Aware Instance Normalization (DAIN) method to re-estimate feature statistics that account for different ratios of foreground and background pixels. Details of DAIN is presented in Algorithm 2. The structures of E_μ and E_δ are included in the supplemental materials.

As shown in Fig. 2, to obtain the foreground-background ratio $\boldsymbol{\rho}$ of one input image, we first feed it to the model encoder with $\boldsymbol{\Delta s}_{ra}$ as the additional input. $\boldsymbol{\Delta s}_{ra}$ acts as pseudo residuals of feature statistics and is obtained in the training stage via averaging $\boldsymbol{\Delta s}$ in a momentum fashion. The output features by the encoder are used to predict the foreground-background ratio $\boldsymbol{\rho}$ with a Ratio-Prediction Head (RPH). $\boldsymbol{\rho}$ is then utilized to estimate the residuals of feature statistics: $\boldsymbol{\Delta s} = f(\boldsymbol{\rho})$. Here, f is a 1×1 convolutional layer that transforms $\boldsymbol{\rho}$ to a feature vector whose dimension is the same as the target layer's channel number. After that, the input image is fed into the model again with $\boldsymbol{\Delta s}$ as additional input and finally makes more accurate predictions.

The training of RPH requires an extra loss term L_{rph}, which is formulated as bellow:

$$L_{rph} = L_{BCE}(\boldsymbol{\rho}, \boldsymbol{\rho_g}) + L_{MSE}(f(\boldsymbol{\rho}), f(\boldsymbol{\rho_g})), \tag{1}$$

where $\boldsymbol{\rho_g}$ denotes the ground truth foreground-background ratio, and L_{BCE} and L_{MSE} denote the binary cross entropy loss and the mean squared error, respectively.

3 Experiments

3.1 Datasets

The proposed method is evaluated on four datasets, including two H&E stained image datasets CoNSeP [3] and CPM17 [28] and two IHC stained datasets DeepLIIF [29] and BC-DeepLIIF [29,32]. **CoNSeP** [3] contains 28 training and 14 validation images, whose sizes are 1000×1000 pixels. The images are extracted from 16 colorectal adenocarcinoma WSIs, each of which belongs to an individual patient, and scanned with an Omnyx VL120 scanner within the department of pathology at University Hospitals Coventry and Warwickshire, UK. **CPM17** [28] contains 32 training and 32 validation images, whose sizes are 500×500 pixels. The images are selected from a set of Glioblastoma Multiforme, Lower Grade Glioma, Head and Neck Squamous Cell Carcinoma, and non-small cell lung cancer whole slide tissue images. **DeepLIIF** [29] contains 575 training and 91 validation images, whose sizes are 512×512 pixels. The images are extracted from the slides of lung and bladder tissues. **BC-DeepLIIF** [29,32] contains 385 training and 66 validation Ki67 stained images of breast carcinoma, whose sizes are 512×512 pixels.

3.2 Implementation Details

In the training stage, patches of size 224×224 pixels are randomly cropped from the original samples. During training, the batch size is 4 and the total number of training iterations is 40,000. We use Adam algorithm for optimization, and the learning rate is initialized as $1e^{-3}$, which is gradually decreased to $1e^{-5}$ during training. We adopt the standard augmentation, like image color jittering and Gaussian blurring. In all experiments, the segmentation and contour detection predictions are penalized using the binary cross entropy loss.

Table 2. Comparisons in generalization performance on nucleus segmentation datasets. Results in each column are related to models trained on one domain and evaluated on the other three unseen domains. Methods marked by * are proposed in this paper. Results are in percentages.

Methods	CoNSeP		CPM17		DeepLIIF		BC-DeepLIIF		Average	
	AJI	Dice	AJI	Dice	AJI	Dice	AJI	Dice	AJI	Dice
Baseline (BN)	16.67	24.10	33.30	61.18	08.42	38.17	21.27	39.92	19.92	40.84
Baseline (IN)	32.13	48.67	33.94	65.83	41.48	67.17	21.52	37.49	32.27	54.79
BIN [19]	21.54	34.33	37.06	67.63	23.51	49.49	26.15	44.42	27.01	48.97
DSU [20]	21.42	34.66	39.12	66.55	27.21	55.10	25.09	41.83	28.21	49.53
SAN [35]	27.91	46.72	33.69	65.66	27.57	53.09	22.17	38.38	27.84	50.96
AmpNorm [36,37]	35.52	55.89	33.39	58.69	39.91	66.58	23.79	37.81	33.15	54.74
StainNorm [38]	41.06	60.81	32.75	64.68	38.55	63.95	25.41	43.81	34.44	58.11
StainMix [39]	34.22	51.07	35.05	65.49	38.48	64.92	26.88	45.62	33.66	56.78
TENT (BN) [34]	38.61	58.11	35.04	64.62	33.77	59.76	23.55	40.91	32.74	55.85
TENT (IN) [34]	32.34	48.87	33.24	65.73	42.08	66.87	22.38	38.04	32.51	54.88
EFDMix [40]	40.13	58.74	33.29	65.25	39.06	64.60	25.92	42.38	34.60	57.74
RC (IN)*	37.21	57.53	36.98	67.71	35.53	62.03	24.98	42.25	33.68	57.38
DAIN*	33.86	50.08	30.62	64.64	37.93	65.56	31.20	53.15	33.40	58.87
DAIN w/o Ratio*	27.37	40.35	33.25	65.05	40.21	66.82	29.16	48.30	32.50	55.13
$DARC_{all}$*	38.18	57.27	34.44	66.11	39.10	67.07	31.64	53.81	35.84	61.06
$DARC_{enc}$*	40.04	58.73	35.60	66.50	40.11	68.23	32.56	53.86	**37.08**	**61.83**

Table 3. Complexity comparison between the baseline model and DARC.

Models	#Parameters (M)	Inference Time (s/image)
Baseline (IN)	5.03	0.0164
$DARC_{enc}$	5.47	0.0253

3.3 Experimental Results and Analyses

In this paper, the models are compared using the AJI [33] and Dice scores. In the experiments, models trained on one of the datasets will be evaluated on the three unseen ones. To avoid the influence of the different sample numbers of the datasets, we calculate the average scores within each unseen domain respectively and then average them across domains.

In this paper, we re-implement some existing popular domain generalization algorithms for comparisons under the same training conditions. Specifically, we re-implement the TENT [34], BIN [19], DSU [20], Frequency Amplitude Normalization (AmpNorm) [36,37], SAN [35] and EFDMix [40]. We also evaluate the stain normalization [38] and stain mix-up [39] methods that are popular in pathological image analysis. Their performances are presented in Table 2. $DARC_{all}$ replaces all normalization layers with DAIN, while $DARC_{enc}$ replaces

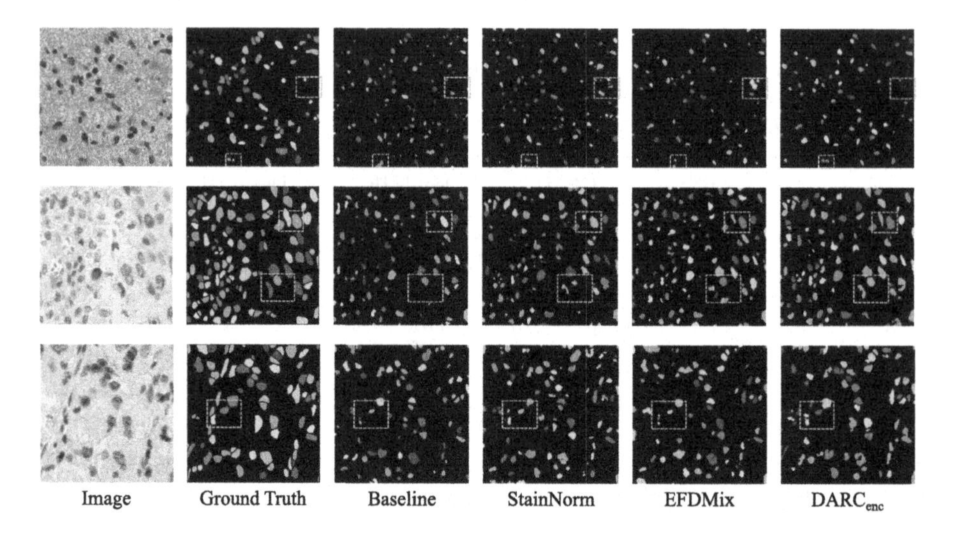

Fig. 3. Qualitative comparisons between Different Models.

the normalization layers in the encoder with DAIN and uses BN in its decoder. As shown in Table 2, DARC$_{enc}$ achieves the best average performance among all methods. Specifically, DARC$_{enc}$ improves the baseline model's average AJI and Dice scores by 4.81% and 7.04%. Compared with the other domain generalization methods, DAIN, DAIN w/o Ratio, DARC$_{all}$ and DARC$_{enc}$ achieve impressive performances on BC-DeepLIIF, which justify that re-estimating the instance-wise statistics is important for improving the domain generalization ability of models trained on BC-DeepLIIF. Qualitative comparisons are presented in Fig. 3. Moreover, the complexity analysis between the baseline model and DARC$_{enc}$ is presented in Table 3.

We separately evaluate the effectiveness of RC and DAIN, and present the results in Table 2. Also, we train a variant model without foreground-background ratio prediction, which is denoted as 'DAIN w/o Ratio' in Table 2. Compared with the baseline model, RC improves the average AJI and Dice scores by 1.41% and 2.59%, and DAIN improves the average AJI and Dice scores by 1.13% and 4.08%. Compared with the variant model without foreground-background ratio prediction, DAIN improves the average AJI and Dice scores by 0.90% and 3.74%. Finally, the combinations of RC and DAIN, i.e., DARC$_{all}$ and DARC$_{enc}$, achieve the best average scores. As shown in Table 2, DARC$_{enc}$ improves DARC$_{all}$ by 1.24% and 0.77% on AJI and Dice scores respectively. This is because after the operations by RC and DAIN in the encoder, the obtained feature maps are much more robust to the domain gaps, which enables the decoder to adopt the fixed statistics maintained during training. Moreover, using the fixed statistics is helpful to prevent the decoder from the influence of varied foreground-background ratios on feature statistics.

4 Conclusion

In this paper, we propose the DARC model for generalizable nucleus segmentation. To handle the domain gaps caused by varied image acquisition conditions, DARC first re-colors the input image while preserving its fine-grained structures as much as possible. Moreover, we find that the performance of instance normalization is sensitive to the varied ratios in foreground and background pixel numbers. This problem is well addressed by our proposed DAIN. Compared with existing works, DARC achieves significantly better performance on average across four benchmarks.

Acknowledgement. This work was supported in part by Guangdong Basic and Applied Basic Research Foundation under Grant 2023A1515010007, in part by the Guangdong Provincial Key Laboratory of Human Digital Twin under Grant 2022B1212010004, in part by CAAI-Huawei MindSpore Open Fund, and In part by National Natural Science Foundation of China (No. 62202403).

References

1. Chen, H., Qi, X., Yu, L., Dou, Q., Qin, J., Heng, P.-A.: DCAN: deep contour-aware networks for object instance segmentation from histology images. Med. Image Anal. **36**, 135–146 (2017)
2. Zhou, Y., Onder, O.F., Dou, Q., Tsougenis, E., Chen, H., Heng, P.A.: Cia-net: robust nuclei instance segmentation with contour-aware information aggregation. In: IPMI, pp. 682–693 (2019)
3. Graham, S., et al.: Hover-Net: simultaneous segmentation and classification of nuclei in multi-tissue histology images. Med. Image Anal. **58**, 101563 (2019)
4. Schmidt, U., Weigert, M., Broaddus, C., Myers, G.: Cell detection with star-convex polygons. In: Frangi, A.F., Schnabel, J.A., Davatzikos, C., Alberola-López, C., Fichtinger, G. (eds.) MICCAI 2018. LNCS, vol. 11071, pp. 265–273. Springer, Cham (2018). https://doi.org/10.1007/978-3-030-00934-2_30
5. Ben Taieb, A., Hamarneh, G.: Adversarial stain transfer for histopathology image analysis. IEEE Trans. Med. Imaging **37**(3), 792–802 (2018)
6. Stacke, K., Eilertsen, G., Unger, J., Lundström, C.: Measuring domain shift for deep learning in histopathology. IEEE J. Biomed. Health Inf. **25**(2), 325–336 (2021)
7. Aubreville, M., et al.: Mitosis domain generalization in histopathology images - the MIDOG challenge. Med. Image Anal. **84**, 102699 (2023)
8. Li, C., et al.: Domain adaptive nuclei instance segmentation and classification via category-aware feature alignment and pseudo-labelling. In: Wang, L., Dou, Q., Fletcher, P.T., Speidel, S., Li, S. (eds.) MICCAI 2022. LNCS, vol. 13437, pp. 715–724. Springer, Heidelberg (2022). https://doi.org/10.1007/978-3-031-16449-1_68
9. Liu, D., et al.: PDAM: a panoptic-level feature alignment framework for unsupervised domain adaptive instance segmentation in microscopy images. IEEE Trans. Med. Imaging **40**(1), 154–165 (2021)
10. Yang, S., Zhang, J., Huang, J., Lovell, B. C., Han, X.: Minimizing labeling cost for nuclei instance segmentation and classification with cross-domain images and weak labels. In: AAAI, vol. 35, no. 1, pp. 697–705 (2021)
11. Gulrajani I., Lopez-Paz D.: In search of lost domain generalization. In: ICLR, (2021)

12. Zhou, K., et al.: Domain generalization: a survey. IEEE Trans. Pattern Anal. Mach. Intell. **45**, 4396–4415 (2022)
13. Wang, J., et al.: Generalizing to unseen domains: a survey on domain generalization. IEEE Trans. Knowl. Data Eng. **35**, 8052–8072 (2022)
14. Huang, J., et al.: FSDR: frequency space domain randomization for domain generalization. In: CVPR, pp. 6891–6902 (2021)
15. Shu, Y., et al.: Open domain generalization with domain-augmented meta-learning. In: CVPR, pp. 9624–9633 (2021)
16. Zhou, Z., Qi, L., Shi, Y.: Generalizable medical image segmentation via random amplitude mixup and domain-specific image restoration. In: Avidan, S., Brostow, G., Cisse, M., Farinella, G.M., Hassner, T. (eds.) ECCV, vol. 13681, pp. 420–436. Springer, Heidelberg (2022). https://doi.org/10.1007/978-3-031-19803-8_25
17. Liu, Q., Chen, C., Qin, J., Dou, Q., Heng, P.-A.: FedDG: federated domain generalization on medical image segmentation via episodic learning in continuous frequency space. In: CVPR, pp. 1013–1023 (2021)
18. Jin, X., et al.: Style normalization and restitution for generalizable person re-identification. In: CVPR, pp. 3143–3152 (2020)
19. Nam, H., Kim, H.-E.: Batch-instance normalization for adaptively style-invariant neural networks. In: NeurIPS, pp. 2563–2572 (2018)
20. Li, X., Dai, Y., Ge, Y., Liu, J., Shan, Y., Duan, L.: Uncertainty modeling for out-of-distribution generalization. In ICLR (2022)
21. Li, H., Pan, S.J., Wang, S., Kot, A.C.: Domain generalization with adversarial feature learning. In: CVPR, pp. 5400–5409 (2018)
22. Tian, C.X., Li, H., Xie, X., Liu, Y., Wang, S.: Neuron coverage guided domain generalization. IEEE Trans. Pattern Anal. Mach. Intell. **45**(1), 1302–1311 (2023)
23. Robey, A., Pappas, G.J., Hassani, H.: Model-based domain generalization. In: NeurIPS, pp. 20210–20229 (2021)
24. Qiao, F., Zhao, L., Peng, X.: Learning to learn single domain generalization. In: CVPR, pp. 12556–12565 (2020)
25. Wang, Z., Luo, Y., Qiu, R., Huang, Z., Baktashmotlagh, M.: Learning to diversify for single domain generalization. In: ICCV, pp. 834–843 (2021)
26. Shi, Y., et al.: Gradient matching for domain generalization. In: ICLR (2022)
27. Telea, A.: An image inpainting technique based on the fast marching method. J. Graph. Tools **9**(1), 23–24 (2004)
28. Vu, Q.D., et al.: Methods for segmentation and classification of digital microscopy tissue images. Front. Bioeng. Biotechnol. **53** (2019)
29. Ghahremani, P., Marino, J., Dodds, R., Nadeem, S.: DeepLIIF: an online platform for quantification of clinical pathology slides. In CVPR, pp. 21399–21405 (2022)
30. Ronneberger, O., Fischer, P., Brox, T.: U-net: convolutional networks for biomedical image segmentation. In: Navab, N., Hornegger, J., Wells, W.M., Frangi, A.F. (eds.) MICCAI 2015. LNCS, vol. 9351, pp. 234–241. Springer, Cham (2015). https://doi.org/10.1007/978-3-319-24574-4_28
31. Llewellyn, B.D.: Nuclear staining with alum hematoxylin. Biotech. Histochem. **84**(4), 159–177 (2009)
32. Huang, Z., et al.: BCData: a large-scale dataset and benchmark for cell detection and counting. In: Martel, A.L., et al. (eds.) MICCAI 2020. LNCS, vol. 12265, pp. 289–298. Springer, Cham (2020). https://doi.org/10.1007/978-3-030-59722-1_28
33. Kumar, N., Verma, R., Sharma, S., Bhargava, S., Vahadane, A., Sethi, A.: A dataset and a technique for generalizable nuclear segmentation for computational pathology. IEEE Trans. Med. Imaging **36**(7), 1550–1560 (2017)

34. Wang, D., Shelhamer, E., Liu, S., Olshausen, B., Darrell, T.: Tent: fully test-time adaptation by entropy minimization. In: ICLR (2021)
35. Peng, D., Lei, Y., Hayat, M., Guo, Y., Li, W.: Semantic-aware domain generalizable segmentation. In CVPR, pp. 2594–2605 (2022)
36. Jiang, M., Wang, Z., Dou, Q.: Harmofl: harmonizing local and global drifts in federated learning on heterogeneous medical images. In: AAAI, vol. 36, no. 1, pp. 1087–1095 (2022)
37. Zhaol, X., Liu, C., Sicilia, A., Hwang, S. J., Fu, Y.: Test-time fourier style calibration for domain generalization. In: IJCAI (2022)
38. Macenko, M., et al.: A method for normalizing histology slides for quantitative analysis. In ISBI, pp. 1107–1110 (2009)
39. Chang, J.-R., et al.: Stain mix-up: unsupervised domain generalization for histopathology images. In: de Bruijne, M., de Bruijne, M., et al. (eds.) MICCAI 2021. LNCS, vol. 12903, pp. 117–126. Springer, Cham (2021). https://doi.org/10.1007/978-3-030-87199-4_11
40. Zhang, Y., Li, M., Li, R., Jia, K., Zhang L.: Exact feature distribution matching for arbitrary style transfer and domain generalization. In: CVPR, pp. 8035–8045 (2022)
41. Rolland, J.P., Vo, V., Bloss, B., Abbey, C.K.: Fast algorithms for histogram matching: application to texture synthesis. J. Electron. Imaging **9**(1), 39–45 (2000)

Multi-scale Prototypical Transformer for Whole Slide Image Classification

Saisai Ding, Jun Wang, Juncheng Li, and Jun Shi(✉)

School of Communication and Information Engineering, Shanghai University, Shanghai, China
junshi@shu.edu.cn

Abstract. Whole slide image (WSI) classification is an essential task in computational pathology. Despite the recent advances in multiple instance learning (MIL) for WSI classification, accurate classification of WSIs remains challenging due to the extreme imbalance between the positive and negative instances in bags, and the complicated pre-processing to fuse multi-scale information of WSI. To this end, we propose a novel multi-scale prototypical Transformer (MSPT) for WSI classification, which includes a prototypical Transformer (PT) module and a multi-scale feature fusion module (MFFM). The PT is developed to reduce redundant instances in bags by integrating prototypical learning into the Transformer architecture. It substitutes all instances with cluster prototypes, which are then re-calibrated through the self-attention mechanism of Transformer. Thereafter, an MFFM is proposed to fuse the clustered prototypes of different scales, which employs MLP-Mixer to enhance the information communication between prototypes. The experimental results on two public WSI datasets demonstrate that the pro-posed MSPT outperforms all the compared algorithms, suggesting its potential applications.

Keywords: Whole slide image · Multiple instance learning · Multi-scale feature · Prototypical Transformer

1 Introduction

Histopathological images are regarded as the 'gold standard' in the diagnosis of cancers. With the advent of the whole slide image (WSI) scanner, deep learning has gained its reputation in the field of computational pathology [1–3]. However, WSIs are extremely large in the size and lack of pixel-level annotations, making it difficult to adopt the traditional supervised learning methods for WSI classification [4].

To address this issue, multiple instance learning (MIL) has been successfully applied to the WSI classification task as a weakly supervised learning problem [5–7]. In this context, a WSI is considered as a bag, and the cropped patches within the slide are the

Supplementary Information The online version contains supplementary material available at https://doi.org/10.1007/978-3-031-43987-2_58.

H. Greenspan et al. (Eds.): MICCAI 2023, LNCS 14225, pp. 602–611, 2023.
https://doi.org/10.1007/978-3-031-43987-2_58

instances in this bag. However, the lesion regions usually only account for a small portion of the WSI, resulting in a large number of negative patches. When the positive and negative instances in the bag are highly imbalanced, the MIL models are prone to incorrectly discriminate these positive instances when using simple aggregation operations. To this end, several attention-based MIL models, such as ABMIL [8] and DSMIL [9], apply variants of the attention mechanism to re-weight instance features. Thereafter, the recent works develop the Transformer-based architectures to better model long-range instance correlations via self-attention [10–13]. However, since the average bag size of a WSI is more than 8000 at 20 $\times$ magnification, it is computationally infeasible to use the conventional Transformer and other stacked self-attention network architectures in MIL-related tasks.

Recently, prototypical learning is applied in WSI analysis to identify representative instances in the bag [14]. Some works adopt the K-means clustering on all instances in a bag to obtain K cluster centers i.e., instance prototypes, and then use these prototypes to represent the bags [15, 16]. These clustering-based MIL algorithms can significantly reduce the redundant instances, and thereby improving the training efficiency for WSI classification. However, it is different for K-means to specify the cluster number as well as the initial cluster centers, and different initial values may lead to different cluster results, thus affecting the performance of MIL. Besides, affected by the feature extractor, the clustering-based MIL algorithms may ignore the most important instances that contain critical diagnostic information. Therefore, it is necessary to develop a method that can fully exploit the potential complementary information between critical instances and prototypes to improve representation learning of prototypes.

On the other hand, when pathologists analysis the WSIs, they always observe the tissues at various resolutions [17]. Inspired by this diagnostic manner, some works use multi-scale information of WSIs to improve diagnostic accuracy. For example, Li et al. [9] adopted a pyramidal concatenation mechanism to fuse the multi-scale features of WSIs, in which the feature vectors of low-resolution patches are replicated and concatenated with the those of their corresponding high-resolution patches; Hou et al. [18] propose a heterogeneous graph neural network to learn the hierarchical representation of WSIs from a heterogeneous graph, which is constructed by the feature and spatial-scaling relationship of multi-resolution patches. However, since the number of patches at each resolution is quite different, it requires complex pre-processing to spatially align feature vectors of patches in different resolutions. Therefore, it is significant to develop an efficient and effective patch aggregation strategy to learn multi-scale information from WSIs.

In this work, we propose a Multi-Scale Prototypical Transformer (MSPT) for WSI classification. The MSPT includes two key components: a prototypical Transformer (PT) and a multi-scale feature fusion module (MFFM). The specifically developed PT uses a clustering algorithm to extract instance prototypes from the bags, and then re-calibrates these prototypes at each scale with the self-attention mechanism in Transformer [19]. MFFM is designed to effectively fuse multi-scale information of WSIs, which utilizes the MLP-Mixer [20] to learn effective representations by aggregating the multi-scale prototypes generated by the PT. The MLP-Mixer adopts two types of MLP layers to allow information communication in different dimensions of data.

The contributions of this work are summarized as follows:

1) A novel prototypical Transformer (PT) is proposed to learn superior prototype representation for WSI classification by integrating prototypical learning into the Transformer architecture. It can effectively re-calibrate the cluster prototypes as well as reduce the computational complexity of the Transformer.
2) A new multi-scale feature fusion module (MFFM) is developed based on the MLP-Mixer to enhance the information communication among phenotypes. It can effectively capture multi-scale information in WSI to improve the performance of WSI classification.

2 Method

2.1 MIL Problem Formulation

MIL is a typical weakly supervised learning method, where the training data consists of a set of bags, and each bag contains multiple instances. The goal of MIL is to learn a classifier that can predict the label of a bag based on the instances in it. In binary classification, a bag can be marked as negative if all in-stances in the bag are negative, otherwise, the bag is labeled as positive with at least one positive instance. In the MIL setting, a WSI is considered as a bag and the numerous cropped patches in WSI are regarded as instances in the bag. A WSI dataset T can be defined as:

$$T = \{x_i, y_i\}_{i=1}^{i=N}, \; x_i = \left\{I_i^j\right\}_{j=1}^{j=n}, \tag{1}$$

where x_i denotes a patient, y_i the label of x_i, I_i^j is the j-th instance of x_i, N is the number of patients and n is the number of instances.

2.2 Multi-scale Prototypical Transformer (MSPT)

The overall architecture of MSPT is shown in Fig. 1. A WSI is first divided into non-overlapping patches at different resolutions, and a pre-trained ResNet18 [21] is used to extract features from each patch. The learned multi-scale features are then fed into the proposed MSPT, which consists of a PT and an MFFM, to re-calibrate cluster prototypes at each scale and fuse multi-scale information of WSI. Finally, a WSI-level classifier is trained to predict the bag label.

Pre-training. It is a time consuming and tedious task for pathologists to annotate the patch-level labels in gigapixel WSIs, thus, a common practice is to use a pre-trained encoder network to extract instance-level features, such as an ImageNet pre-trained encoder or a self-supervised pre-trained encoder. In this work, we follow [9] to adopt SimCLR [22] to pre-training the patch encoder at different resolutions. SimCLR is a self-supervised learning algorithm to pre-trainng a network by maximizing the similarity between positive pairs and minimizing the similarity between negative pairs [22]. After pre-training, the extracted instances of different scales are fed into MSPT for prototype learning and multi-scale learning.

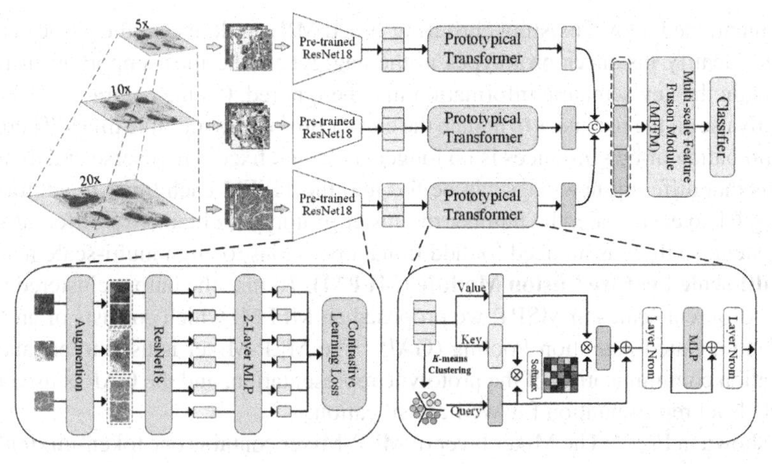

Fig. 1. Overview of the proposed MSPT.

Prototypical Transformer (PT). Most tissues in WSIs are redundancy, and therefore, we introduce the instance prototypes to reduce redundant instances. Specifically, for each instance bag $\boldsymbol{X}_{bag} \in \mathbb{R}^{n\times d_k}$, the K-means clustering algorithm is applied on all instances to get K centers (prototypes). These cluster prototypes can be used as instances to represent a new bag $\boldsymbol{P}_{bag} \in \mathbb{R}^{k\times d_k}$. However, the K-means clustering algorithm is sensitive to the initial selection of cluster centers, i.e. different initializations can lead to different results, and the final result may not be the global optimal solution. It is essential to try different initializations and choose the one with the lowest error. However, the WSI dataset generally has a long sequence of instances, which makes the clustering algorithms computationally expensive and slow down as the size of the bag increases.

To solve the issue above, we propose to apply the self-attention (SA) mechanism in Transformer to re-calibrate these cluster prototypes. As shown in Fig. 1, the optimization process can be divided into two steps: 1) the initial cluster prototype bag $\boldsymbol{P}_{bag}$ is obtained in the pre-processing stage by using the K-means clustering on $\boldsymbol{X}_{bag}$; ; 2) PT uses $\boldsymbol{X}_{bag}$ to optimize $\boldsymbol{P}_{bag}$ via the self-attention mechanism in Transformer. The detailed process is as follows:

$$\begin{aligned} \mathrm{SA}\left(\boldsymbol{P}_{bag}, \boldsymbol{X}_{bag}\right) &= softmax\left(\frac{QK^T}{\sqrt{d_k}}\right)\cdot V \\ &= softmax\left(\frac{\boldsymbol{W}_q\boldsymbol{P}_{bag}(\boldsymbol{X}_{bag}\boldsymbol{W}_k)^T}{\sqrt{d_k}}\right)\boldsymbol{W}_v\boldsymbol{X}_{bag} \rightarrow A_{map}\boldsymbol{W}_v\boldsymbol{X}_{bag} \rightarrow \hat{P} \end{aligned} \tag{2}$$

where $\boldsymbol{W}_q, \boldsymbol{W}_k, \boldsymbol{W}_v \in \mathbb{R}^{d_k\times d_k}$ are trainable matrices of query $\boldsymbol{P}_{bag}$ and the key-value pair $(\boldsymbol{X}_{bag}, \boldsymbol{X}_{bag})$, respectively, and $\boldsymbol{A}_{map} \in \mathbb{R}^{k\times n}$ is the attention matrix to compute the weight of $\boldsymbol{X}_{bag}$. Thus, the computational complexity of SA is $O(nm)$ instead of $O\left(n^2\right)$, and the k is much less than n. Specifically, for a single clustering prototype $p_k \in \boldsymbol{P}$, the SA layer scores the pairwise similarity between p_k and x_n for all $x_n \in \boldsymbol{X}$, which can be written as a row vector $[a_{k1}, a_{k2}, a_{k3}, \ldots, a_{kn}]$ in $\boldsymbol{A}_{map}$. These attention scores are then weighted to $\boldsymbol{X}_{bag}$ to update the $p_k \in \mathbb{R}^{1\times d_k}$ for completing the calibration of the clustering prototypes $\hat{\boldsymbol{P}} \in \mathbb{R}^{k\times d_k}$.

As mentioned above, existing clustering-based MIL methods use the K-means clustering to identify instances prototypes in the bag, where the most important instances that contain the key semantic information may be ignored. On the contrary, our PT can efficiently use all the instances to update the cluster prototypes multiple times. Therefore, the combination of bag instances is no longer static and fixed, but diverse and dynamic. It means that different new bags can be fed into the MFFM each time. In addition, by applying PT to each scale, the number of cluster prototypes obtained at different scales is consistent, so there is no need for additional operations to align multi-scale features.

Multi-scale Feature Fusion Module (MFFM). To fuse the output clustered prototypes at different scales in MSPT, we proposed an MFFM, which consists of an MLP-Mixer and a Gated Attention Pooling (GAP). The MLP-Mixer is used to enhance the information communication of the prototype representation, and the GAP is used to get the WSI-level representation for WSI classification.

As shown in Fig. 2, The Mixer layer of MLP-Mixer contains one token-mixing MLP and one channel-mixing MLP, each consisting of two fully-connected layers and a GELU activation function [23]. Token-mixing MLP is a cross-location operation to mix all prototypes, while channel-mixing MLP is a pre-location operation to mix features of each prototype. Thus, MLP-Mixer allows the information communication between different prototypes and prototype features to learn superior representation through information aggregation.

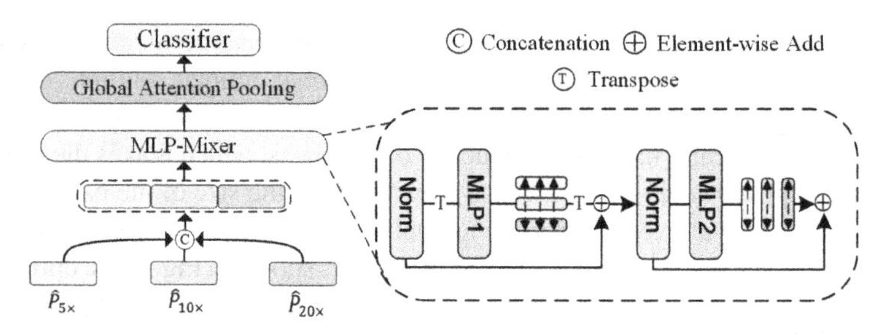

Fig. 2. The structure of MFFM.

Specifically, the procedure of MFFM is described as follows:

We first perform the feature concatenation operation on the multi-scale output clustering prototypes $\left[\hat{P}_{20\times}, \hat{P}_{10\times}, \hat{P}_{5\times}\right]$ to construct a feature pyramid $\breve{P}$:

$$concat\left[\hat{P}_{20\times}, \hat{P}_{10\times}, \hat{P}_{5\times}\right] \rightarrow \breve{P} \in \mathbb{R}^{k\times 3d_k} \tag{3}$$

where d_k is the feature vector dimension of the prototypes.

Then, the $\breve{P}$ is fed to the MLP-Mixer to obtain the corresponding hidden feature representation $H \in \mathbb{R}^{k \times 3d_k}$ as follows:

$$\begin{aligned} H_1 &= \breve{P}^T + W_2\sigma\left(W_1\text{LN}\left(\breve{P}^T\right)\right) \\ H &= H_1^T + W_4\sigma\left(W_3\text{LN}\left(H_1^T\right)\right) \end{aligned} \tag{4}$$

where LN denotes the layer normalization, σ denotes the activation function implemented by GELU, $W_1 \in \mathbb{R}^{k \times c}$, $W_2 \in \mathbb{R}^{c \times k}$, $W_3 \in \mathbb{R}^{3d_k \times d_s}$ *and* $W_4 \in \mathbb{R}^{d_s \times 3d_k}$ are the weight matrices of MLP layers.c and d_s are tunable hidden widths in the token-mixing and channel-mixing MLP, respectively.

Finally, the H is fed to the gated attention pooling (GAP) [8] to get the WSI-level representation $Z \in \mathbb{R}^{1 \times 3d_k}$ for WSI classification:

$$\begin{aligned} Z &= GAP(H) \\ \hat{Y} &= softmax(MLP(Z)) \end{aligned} \tag{5}$$

where $\widehat{Y} \in \mathbb{R}^{1 \times d_{out}}$ is the class label probability of the bag, and d_{out} is the number of classes.

3 Experiments and Results

3.1 Datasets

To evaluate the effectiveness of MSPT, we conducted experiments on two public dataset, namely Camelyon16 [24] and TCGA-NSCLC. Camelyon16 is a WSI dataset for the automated detection of metastases in lymph node tissue slides. It includes 270 training samples and 129 testing samples. After pre-processing, a total of 2.4 million patches at $\times 20$ magnification, 0.56 million patches at $\times 10$ magnification, and 0.16 million patches at $\times 5$ magnification, with an average of about 5900, 1400, and 400 patches per bag. The TCGA-NSCLC dataset includes two sub-types of lung cancer, i.e., Lung Squamous Cell Carcinoma (TGCA-LUSC) and Lung Adenocarcinoma (TCGA-LUAD). We collected a total of 854 diagnostic slides from the National Cancer Institute Data Portal (https://portal.gdc.cancer.gov). The dataset yields 4.3 million patches at $20\times$ magnification, 1.1 million patches at $10\times$ magnification, and 0.30 million patches at $5\times$ magnification with an average of about 5000, 1200, and 350 patches per bag.

3.2 Experiment Setup and Evaluation Metrics

In WSI pre-processing, each slide is cropped into non-overlapping 256×256 patches at different magnifications, and a threshold is set to filter out background ones. After patching, we use a pre-trained ResNet18 model to convert each 256×256 patch into a 512- dimensional feature vector. We selected accuracy (ACC) and area under curve (AUC) as evaluation metrics. For Camelyon16 dataset, we reported the results of the official testing set. For TCGA-NSCLC, we conducted five cross-validation on the 854 slides, and the results are reported in the format of mean $\pm$SD (standard deviation).

3.3 Implementation Details

For the feature extractor, we employed the SimCLR encoder trained by Lee et al. [9] for the Camelyon16 and TCGA datasets. But [9] only trained SimCLR encoders at 20× and 5× magnification, to align with that setting, we used the same settings to train the SimCLR encoder at 10× magnification on both datasets. For the proposed MSPT, the Adam optimizer was used to update the model weights, the initial learning rate of 1e-4 with a weight decay of 1e-5. The mini-batch size was set as 1. The MSPT models were trained for 150 epochs and they would early stop if the loss would not decrease in the past 30 epochs. All models were implemented by Python 3.8 with PyTorch toolkit 1.11.0 on a platform equipped with an NVIDIA GeForce RTX 3090 GPU.

3.4 Comparisons Experiment

Comparison Algorithms. The proposed MSPT was compared to state-of-the-art MIL-based algorithms: 1) The traditional pooling operators, such as mean-pooling and max-pooling; 2) the attention-based algorithms, including ABMIL [8] and DSMIL [9]; 3) the Transformer-based algorithm TransMIL [11]; 4) The clustering-based algorithm ReMix [16].

Table 1. Comparison results on the Camelyon16 and TCGA datasets.

Method	Camelyon16		TCGA-NSCLC	
	Accuracy	AUC	Accuracy	AUC
Mean-Pooling	0.8837	0.8916	0.8911 ± 0.011	0.9230 ± 0.010
Max-Pooling	0.9147	0.9666	0.9136 ± 0.014	0.9441 ± 0.016
ABMIL [8]	0.9302	0.9752	0.9123 ± 0.015	0.9457 ± 0.017
DSMIL [9]	0.9380	0.9762	0.9049 ± 0.010	0.9359 ± 0.011
TransMIL [11]	0.9225	0.9734	0.9095 ± 0.014	0.9432 ± 0.016
ReMix [16]	0.9458	0.9740	0.9167 ± 0.013	0.9509 ± 0.016
PT (Ours)	0.9458	0.9809	0.9257 ± 0.011	0.9567 ± 0.013
MSPT (Ours)	**0.9536**	**0.9869**	**0.9289 ± 0.011**	**0.9622 ± 0.015**

Experimental Results. Table 1 shows the comparison results on the Camelyon16 and TCGA-NSCLC datasets. In CAMELYON16, it can be found that the proposed MSPT outperforms all the compared algorithms with the best accuracy of 0.9536, and AUC of 0.9869. Compared to other algorithms, MSPT improves at least 0.78%, and 1.07% on classification ACC and AUC, indicating the effectiveness of MFFM to learn the multi-scale information of WSIs. In addition, PT achieves the best classification results in the single-resolution methods and outperforms ReMix on all indices, which proves PT can effectively re-calibrate the clustering prototypes.

In TCGA-NSCLC, the proposed MSPT algorithm again outperforms all the compared algorithms on all indices. It achieves the best classification performance of 0.9289

± 0.011 and 0.9622 ± 0.015 on the ACC and AUC. Moreover, MSPT improves at least 0.78% and 1.03%, respectively, on the corresponding indices compared with all other algorithms.

3.5 Ablation Study

To evaluate the contribution of PT and MFFM in the proposed MSPT, we further conducted a series of ablation studies.

Investigation of the Number of Prototypes in PT. To evaluate the effectiveness of the PT, we first changed the number of prototypes K in the range of $\{1, 2, 4, 8, 16, 32\}$ to get the optimal K for each dataset. Then, the following two variants were compared with PT: (1) Full-bag: the first variant was only trained on all the instances; (2) Prototype-bag: the second variant was only trained on the cluster prototypes.

As shown in Fig. 3, the horizontal axes denote the number of prototypes, and the vertical axes denote the classification accuracy. In the Camelyon16 dataset, the performance of both PT and Prototype-bag increases with the increase of K value, and achieves the best results with $K = 16$. In the TCGA-NSCLC dataset, PT always outperforms the Full-bag and Prototype-bag. These experimental results demonstrate that PT can effectively re-calibrate the clustering prototypes to achieve superior results.

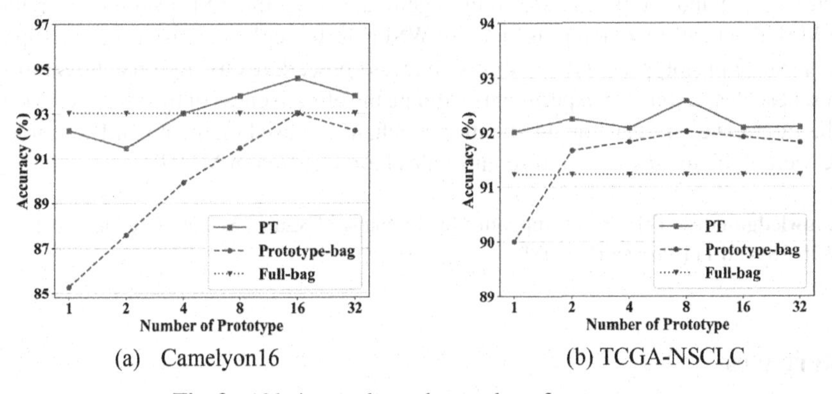

(a) Camelyon16 (b) TCGA-NSCLC

Fig. 3. Ablation study on the number of prototypes.

Investigation of Multi-scale Fusion. We further compared our MFFM with several other fusion strategies, including (1) Concatenation: this variant concatenated the cluster prototypes of each magnification before the classifier. (2) MS-Max: this variant used max-pooling on the cluster prototypes for each magnification, and then added them. (3) MS-Attention: this variant used attention-pooling [8] on the cluster prototypes for each magnification, and then added them.

Table 2 gives the results on the Camelyon16 and TCGA-NSCLC datasets. Compared with other multi-scale variants, the proposed MSPT improves ACC by at least 0.78% and 0.85% on Camelyon16 and TCGA-NSCLC, respectively, which proves that the MLP-Mixer in MFFM can effectively enhance the information communication among phenotypes and their features, thus improving the performance of feature aggregation.

Table 2. Classification results for evaluating different fusion strategies. All variants used the WSIs with three resolutions ($5\times + 10\times + 20\times$).

Method	Camelyon16		TCGA-NSCLC	
	ACC	AUC	ACC	AUC
Concatenation	0.9147	0.9598	0.9147 ± 0.018	0.9438 ± 0.016
MS-Max	0.9302	0.9729	0.9203 ± 0.014	0.9527 ± 0.019
MS-Attention	0.9458	0.9786	0.9204 ± 0.016	0.9571 ± 0.012
MFFM	**0.9536**	**0.9869**	**0.9289 ± 0.011**	**0.9722 ± 0.015**

More Studies. We provide more empirical studies, i.e., the effect of the multi-resolution scheme, the visualization results, and the training budgets, in Supplementary Materials to better understand MSPT.

4 Conclusion

In summary, we propose an MSPT for WSI classification that combine the prototype-based learning and multi-scale learning to generate powerful WSI-level representation. The MSPT reduces redundant instances in WSI bags by replacing instances with updatable instance prototypes, and avoids complicated procedures to align patch features at different scales. Extensive experiments validate the effectiveness of the proposed MSPT. In the future, we will develop an attention mechanism based on the magnification level to re-weight the features from different scales before fusion in MSPT.

Acknowledgments. This work is supported by the National Natural Science Foundation of China (81871428) and 111 Project (D20031).

References

1. Campanella, G., et al.: Clinical-grade computational pathology using weakly supervised deep learning on whole slide images. Nat. Med. **25**(8), 1301–1309 (2019)
2. Bera, K., Schalper, K.A., Rimm, D.L., Velcheti, V., Madabhushi, A.: Artificial intelligence in digital pathology—new tools for diagnosis and precision oncology. Nat. Rev. Clin. Oncol. **16**(11), 703–715 (2019)
3. Zarella, M.D., et al.: A practical guide to whole slide imaging: a white paper from the digital pathology association. Arch. Pathol. Lab. Med. **143**(2), 222–234 (2019)
4. Srinidhi, C.L., Ciga, O., Martel, A.L.: Deep neural network models for computational histopathology: a survey. Med. Image Anal. **67**, 101813 (2021)
5. Javed, S., et al.: Cellular community detection for tissue phenotyping in colorectal cancer histology images. Med. Image Anal. **63**, 101696 (2020)
6. Zheng, Y., et al.: A graph-transformer for whole slide image classification. IEEE Trans. Med. Imaging **41**(11), 3003–3015 (2022)

7. Lu, M.Y., Williamson, D.F., Chen, T.Y., Chen, R.J., Barbieri, M., Mahmood, F.: Data-efficient and weakly supervised computational pathology on whole-slide images. Nat. Biomed. Eng. **5**(6), 555–570 (2021)
8. Ilse, M., Tomczak, J., Welling, M.: Attention-based deep multiple instance learning. In: International Conference on Machine Learning, pp. 2127–2136. PMLR (2018)
9. Li, B., Li, Y., Eliceiri, K.W.: Dual-stream multiple instance learning network for whole slide image classification with self-supervised contrastive learning. In: Proceedings of the IEEE/CVF Conference on Computer Vision and Pattern Recognition, pp. 14318–14328 (2021)
10. Chen, R.J.: Scaling vision transformers to gigapixel images via hierarchical self-supervised learning. In: Proceedings of the IEEE/CVF Conference on Computer Vision and Pattern Recognition, pp. 16144–16155 (2022)
11. Shao, Z., Bian, H., Chen, Y., Wang, Y., Zhang, J., Ji, X.: TransMIL: transformer based correlated multiple instance learning for whole slide image classification. Adv. Neural. Inf. Process. Syst. **34**, 2136–2147 (2021)
12. Li, H., et al.: DT-MIL: deformable transformer for multi-instance learning on histopathological image. In: de Bruijne, M., et al. (eds.) MICCAI 2021. LNCS, vol. 12908, pp. 206–216. Springer, Cham (2021). https://doi.org/10.1007/978-3-030-87237-3_20
13. Huang, Z., Chai, H., Wang, R., Wang, H., Yang, Y., Wu, H.: Integration of patch features through self-supervised learning and transformer for survival analysis on whole slide images. In: de Bruijne, M., et al. (eds.) MICCAI 2021. LNCS, vol. 12908, pp. 561–570. Springer, Cham (2021). https://doi.org/10.1007/978-3-030-87237-3_54
14. Wang, Z., Yu, L., Ding, X., Liao, X., Wang, L.: Lymph node metastasis prediction from whole slide images with transformer-guided multiinstance learning and knowledge transfer. IEEE Trans. Med. Imaging **41**(10), 2777–2787 (2022)
15. Yao, J., Zhu, X., Jonnagaddala, J., Hawkins, N., Huang, J.: Whole slide images based cancer survival prediction using attention guided deep multiple instance learning networks. Med. Image Anal. **65**, 101789 (2020)
16. Yang, J., et al.: ReMix: a general and efficient framework for multiple instance learning based whole slide image classification. In: Wang, L., Dou, Q., Fletcher, P.T., Speidel, S., Li, S. (eds.) Medical Image Computing and Computer Assisted Intervention. pp. 35–45. Springer, Cham (2022). https://doi.org/10.1007/978-3-031-16434-7_4
17. Hashimoto, N., et al.: Multi-scale domain-adversarial multiple-instance CNN for cancer subtype classification with unannotated histopathological images. In: Proceedings of the IEEE/CVF Conference on Computer Vision and Pattern Recognition, pp. 3852–3861 (2020)
18. Hou, W., et al.: H^ 2-MIL: exploring hierarchical representation with heterogeneous multiple instance learning for whole slide image analysis. In: Proceedings of the AAAI Conference on Artificial Intelligence, pp. 933–941 (2022)
19. Dosovitskiy, A., et al.: An image is worth 16x16 words: transformers for image recognition at scale. In: ICLR (2021)
20. Tolstikhin, I.O., et al.: MLP-Mixer: an all-MLP architecture for vision. In: Advances in Neural Information Processing Systems, vol. 34, pp. 24261–24272 (2021)
21. Simonyan, K., Zisserman, A.: Very deep convolutional networks for large-scale image recognition. arXiv preprint arXiv:1409.1556 (2014)
22. Chen, T., Kornblith, S., Norouzi, M., Hinton, G.: A simple framework for contrastive learning of visual representations. In: International Conference on Machine Learning, pp. 1597–1607. PMLR (2020)
23. Hendrycks, D., Gimpel, K.: Gaussian error linear units (gelus). arXiv preprint arXiv:1606.08415 (2016)
24. Bejnordi, B.E., et al.: Diagnostic assessment of deep learning algorithms for detection of lymph node metastases in women with breast cancer. JAMA **318**(22), 2199–2210 (2017)

Transfer Learning-Assisted Survival Analysis of Breast Cancer Relying on the Spatial Interaction Between Tumor-Infiltrating Lymphocytes and Tumors

Yawen Wu[1], Yingli Zuo[1], Qi Zhu[1], Jianpeng Sheng[2], Daoqiang Zhang[1](✉), and Wei Shao[1](✉)

[1] College of Computer Science and Technology, Nanjing University of Aeronautics and Astronautics, MIIT Key Laboratory of Pattern Analysis and Machine Intelligence, Nanjing 211106, China
{dqzhang,shaowei20022005}@nuaa.edu.cn
[2] School of Medicine, Zhejiang University, Zhejiang 310058, China

Abstract. Whole-Slide Histopathology Image (WSI) is regarded as the gold standard for survival prediction of Breast Cancer (BC) across different subtypes. However, in cancer prognosis applications, the cost of acquiring patients' survival information is high and can be extremely difficult in practice. By considering that there exists a certain common mechanism for tumor progression among different subtypes of Breast Invasive Carcinoma(BRCA), it becomes critical to utilize data from a related subtype of BRCA to help predict the patients' survival in the target domain. To address this issue, we proposed a TILs-Tumor interactions guided unsupervised domain adaptation (T2UDA) algorithm to predict the patients' survival on the target BC subtype. Different from the existing feature-level or instance-level transfer learning strategy, our study considered the fact that the tumor-infiltrating lymphocytes (TILs) and its correlation with tumors reveal similar role in the prognosis of different BRCA subtypes. More specifically, T2UDA first employed the Graph Attention Network (GAT) to learn the node embeddings and the spatial interactions between tumor and TILs patches in WSI. Then, besides aligning the embeddings of different types of nodes across the source and target domains, we proposed a novel Tumor-TILs interaction alignment (TTIA) module to ensure that the distribution of interaction weights are similar in both domains. We evaluated the performance of our method on the BRCA cohort derived from the Cancer Genome Atlas (TCGA), and the experimental results indicated that T2UDA outperformed other domain adaption methods for predicting patients' clinical outcomes.

Y. Wu and Y. Zuo—Contribute equally to this work

H. Greenspan et al. (Eds.): MICCAI 2023, LNCS 14225, pp. 612–621, 2023.
https://doi.org/10.1007/978-3-031-43987-2_59

Keywords: Tumor-infiltrating Lymphocytes · Unsupervised Domain Adaption · Prognosis Prediction · Graph Attention Network · Breast Cancer

1 Introduction

Breast cancer (BC) is the most common cancer diagnosed among females and the second leading cause of cancer death among women after lung cancer [1]. The BC differs greatly in clinical behavior, ranging from carcinoma in site to aggressive metastatic disease [2,3]. Thus, effective and accurate prognosis of BC as well as stratifying cancer patients into different subgroups for personalized cancer management has attracted more attention than ever before.

Among different types of imaging biomarkers, histopathological images are generally considered the golden standard for BC prognosis since they can confer important cell-level information that can reflect the aggressiveness of BC [4]. Recently, with the availability of digitalized whole-slide pathological images (WSIs), many computational models have been employed for the prognosis prediction of various subtypes of BC. For instance, Lu et al [5] presented a novel approach for predicting the prognosis of ER-Positive BC patients by quantifying nuclear shape and orientation from histopathological images. Liu et al [6] developed a gradient boosting algorithm to predict the disease progression for various subtypes of BC. However, due to the high-cost of collecting survival information from the patients, it is still a challenge to build effective machine learning models for specific BC subtypes with limited annotation data.

To deal with the above challenges, several researchers began to design domain adaption algorithms, which utilize the labeled data from a related cancer subtype to help predict the patients' survival in the target domain. Specifically, Alirezazadeh et al [7] presented a new representation learning-based unsupervised domain adaption method to predict the clinical outcome of cancer patients on the target domain. Zhang et al [8] proposed a collaborative unsupervised domain adaptation algorithm, which conducts transferability-aware adaptation and conquers label noise in a collaborative way. Other studies include Xu et al [9] developed graph neural networks for unsupervised domain adaptation in histopathological image analysis, based on a backbone for embedding input images into a feature space, and a graph neural layer for propagating the supervision signals of images with labels.

Although much progress has been achieved, most of the existing studies applied the feature alignment strategy to reduce the distribution difference between source and target domains. However, such transfer learning methods neglected to take the interaction among different types of tissues into consideration. For example, it is widely recognized that tumor-infiltrating lymphocytes (TILs) and its correlation with tumors reveal a similar role in the prognosis of different BRCA subtypes. For instance, Kurozumi et al [10] revealed that high TILs expression was correlated with negative estrogen receptor (ER) expression and high histological grade ($P < 0.001$). Lu et al [11] utilized the TILs spatial pattern for survival analysis in different breast cancer subtypes including

ER-negative, ER-positive, and triple-negative. It can be expected that better prognosis performance can be achieved if we leveraged the TILs-Tumor interaction information to resolve the survival analysis task on the target domain.

Based on the above considerations, in this paper, we proposed a TILs-Tumor interactions guided unsupervised domain adaptation (T2UDA) algorithm to predict the patients' survival on the target BC subtype. Specifically, T2UDA first applied the graph attention network (GATs) to learn node embeddings and the spatial interactions between tumor and TILs patches in WSI. In order to preserve the node-level and interaction-level similarities across different domains, we not only aligned the embedding for different types of nodes but also designed a novel Tumor-TILs interaction alignment (TTIA) module to ensure that the distribution of the interaction weights are similar in both domains. We evaluated the performance of our method on the Breast Invasive Carcinoma (BRCA) cohort derived from the Cancer Genome Atlas (TCGA), and the experimental results indicated that T2UDA outperforms other domain adaption methods for predicting patients' clinical outcomes.

2 Method

We summarized the proposed T2UDA network in Fig. 1, which consists of three parts, *i.e.*, Graph Attention Network-based Framework, Feature Alignment(FA), and TILs-Tumor interaction alignment(TTIA). Next, we will introduce each part in detail.

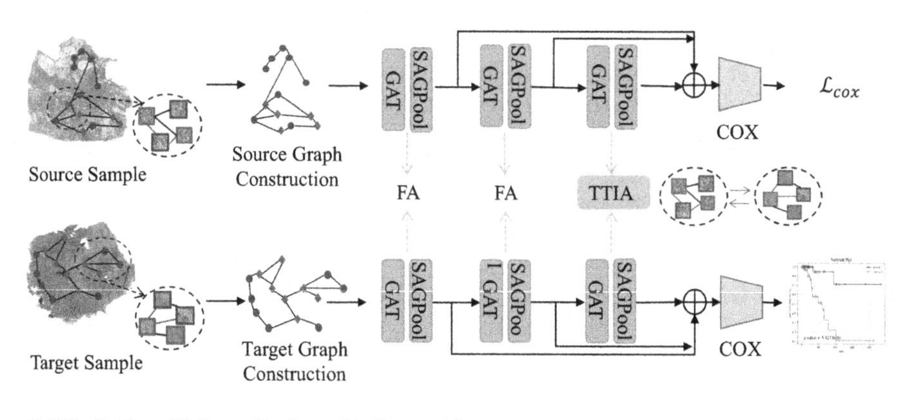

Fig. 1. The overall framework of the T2UDA network. T2UDA aligns TILs-Tumor edge interaction weights in the TTIA module and feature vectors in the FA module to reduce the discrepancy between different domains and achieving prognosis task on target BC patients.

Data Pre-processing. We obtained valid patches of 512×512 pixels from pathological images and segment the TILs and tumor tissues using a pre-trained U-Net++ model. Then we calculated the tumor and TILs area ratios in each patch and selected 300 patches with the largest ratios of each tissue type. Based on the selected patches, we constructed a graph $G = (V, E)$ for each WSI. Here, given patches as nodes V, we first calculated the pairwise distance among different nodes, and select the top 10 percent connections with the smallest distance values as edges E. For each node in V, we followed the study in [12], which applied the ResNet-101 model to extract node features. Then, the principal component analysis (PCA) is implemented to reduce dimensionalities of the node features to 128.

Calculating TILs-Tumor Interaction via Graph Attention Networks(GATs). To characterize the interaction between different TILs and tumor patches, we employed GAT [13], which has been proven to be useful in describing the spatial interaction between different tissues across WSIs. Our GAT-based framework consisted of 3 GAT layers interleaved with 3 graph pooling layers [14](shown in Fig. 1). The input of the GAT layer are $H = [H^L, H^T] \in R^{d\times 600}$, $H^w = [h_1^w, h_2^w, ..., h_{300}^w] \in R^{d\times 300}, w \in \{L, T\}$ represent the output features for TILs and tumor nodes after PCA. The GAT layer generate a new group of node features $H' = [(H^L)', (H^T)']$ via a weight matrix $W \in R^{d\times d'}$ and $(H^w)' \in R^{d'\times 300}, w \in \{L, T\}$. Next, with a shared attentional mechanism $\mathbb{R}^{d'} \times \mathbb{R}^{d'} \rightarrow \mathbb{R}$, we calculated the attention coefficients among different nodes, which can be formulated as:

$$e_{ij} = a\left([Wh_i \| Wh_j]\right), j \in \mathcal{N}_i. \tag{1}$$

Furthermore, a softmax function was then adopted to normalize the attention coefficients e_{ij}:

$$\alpha_{ij} = \text{softmax}_j\left(e_{ij}\right) = \frac{\exp\left(e_{ij}\right)}{\sum_{k\in\mathcal{N}_i}\exp\left(e_{ik}\right)}, \tag{2}$$

where N_i represents all neighbors of node i. The new feature vector v_i for node i was calculated via a weighted sum:

$$v_i = \sigma\left(\sum\nolimits_{v_j\in N(v_i)} \alpha_{ij} W h_j\right). \tag{3}$$

Finally, the output features of each GAT layer were aggregated in the readout layer. We fed the generated output features from each readout layer into the Cox hazard proportional regression model for the final prognosis predictions.

Feature Alignment. In the proposed GAT-based transfer learning framework, the feature alignment component was employed on its first two layers. Then, for the node embeddings with different types (TILs and Tumor) in both the source and target domain, we performed a mean pooling operation to obtain their aggregated features. Next, we aligned the aggregated tumor or TILs features from the two domains separately using Maximum Mean Discrepancy(MMD) [15]. Here, we adopted MMD for feature alignment due to its ability to measure the distance between two distributions without explicit assumptions on the data distribution, we showed the objective function of MMD in our method as follows:

$$L_{FA} = \sum_{r=1,2} \sum_{k \in L,T} \left\| \frac{1}{n} \sum_{i=1}^{n} (f_{i,k})^r - \frac{1}{m} \sum_{i=1}^{m} (f'_{i,k})^r \right\|_H^2 \tag{4}$$

where H is a Hilbert space, f represents the features from the source, f' represents the feature from the target, r represents the layer number, $k \in \{L, T\}$ referred to TILs or tumor node. In addition, n denotes the number of source samples, while m refers to the number of target samples.

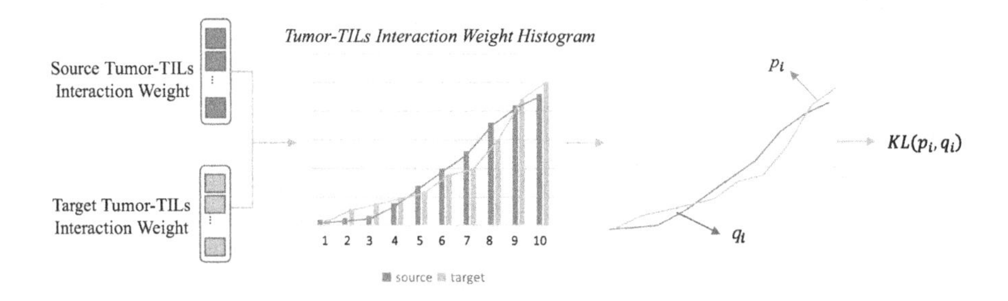

Fig. 2. The illustration of the proposed Interaction Weight Alignment module.

TILs-Tumor Interaction Alignment. To accurately characterize the interaction between TILs and tumors, we further analyzed the extracted interaction weights by dividing them into 10 intervals (i.e., bins). For each interval, we calculated the sum of all source domain interaction weights as i_k^s and the sum of all target domain interaction weights as i_k^t, where k represents the k-th interval. Consequently, we obtained two vectors and applied softmax on each of them for normalization that can be denoted as $p_i = [i_1^s, i_2^s, \cdots, i_{10}^s]$ and $q_i = [i_1^t, i_2^t, \cdots, i_{10}^t]$. In order to measure the dissimilarity between p_i and q_i, the Kullback-Leibler (KL) divergence is adapted on the third layer of GAT, which can be formulated as:

$$\boldsymbol{L_{TTIA}} = KL(\boldsymbol{p}_i, \boldsymbol{q}_i) = \sum p_i \log(\boldsymbol{p}_i / \boldsymbol{q}_i). \tag{5}$$

According to Eq.(5), we can ensure that the weight distributions for the TIL-Tumor interaction are consistent in the source and target domain, which will benefit the following survival analysis. It is beneficial for the target domain.

Prognosis Prediction by the Cox Proportional Hazard Model. The Cox proportional hazard model was applied to predict the patients' clinical outcome [16], and its negative log partial likelihood function can be formulated as:

$$l_{\text{prognosis}} = \sum_{i=1}^{N} \delta_i \left(\theta^T x_i - \log \sum_{j \in R(t_i)} \exp\left(\theta^T x_j\right) \right) \tag{6}$$

where x_i represents the output of the last layer for the prognosis task and $R(t_i)$ is the risk set at time t_i, which represents the set of patients that are still under risk before time t. In addition, δ_i is an indicator variable. Sample i refers to censored patient if $\delta_i = 0$, otherwise $\delta_i = 1$.

Overall Objective. To achieve domain-adaptive prognosis prediction, the final loss function included the Cox loss, FA loss, and TTIA loss as the following formula:

$$\boldsymbol{L_t} = \boldsymbol{L_{cox}} + \alpha \boldsymbol{L_{FA}} + \beta \boldsymbol{L_{TTIA}}, \tag{7}$$

where α and β represent the weights assigned to the importance of FA component and TTIA component respectively.

3 Experiments and Results

Datasets. We conducted our experiments on the breast invasive carcinoma (BRCA) dataset from The Cancer Genome Atlas (TCGA). Specifically, the BRCA dataset includes 661 patients with hematoxylin and eosin (HE)-stained pathological imaging and corresponding survival information. Among the collected BRCA patients in TCGA, the number of ER positive(ER+) and ER negative(ER−) patients are 515 and 146, respectively. We hope to investigate if the proposed T2UDA could be used to help improve the prognosis performance of (ER+) or (ER−) with the aid of the survival information on its counterpart.

3.1 Implementation Details and Evaluation Metrics

The dimension of intermediate layers in GAT was 256. The pooling ratio in SagPool was set to 0.7. α and β were tuned from $\{0.01, 0.1\}$. During training, the model was trained for 150 epochs for both the main experiment and all comparative experiments. We used the Adam optimizer with a learning rate tuned from $\{1e-5, 1e-4\}$. We evaluated the performance of our model using the Concordance Index (CI) and Area Under the Curve (AUC) as performance metrics. Both CI and AUC range from 0 to 1, with larger values indicating better prediction performance and vice versa [1].

Table 1. Quantitative Performance Comparison between Different Unsupervised Domain Adaptation Methods and Our Method.

	ER+→ER-		ER+→ER-	
	CI	AUC	CI	AUC
Source only	0.6187	0.6135	0.5771	0.6054
DDC	0.6118	0.6202	0.5854	0.6382
DANN	0.6414	0.6456	0.6274	0.6206
DeepJDOT	0.6535	0.6633	0.6357	0.6428
MDD	0.6498	0.6651	0.6261	0.6367
T2UDA-v1	0.6314	0.6450	0.6266	0.6427
T2UDA	**0.6933**	**0.7097**	**0.6803**	**0.6903**

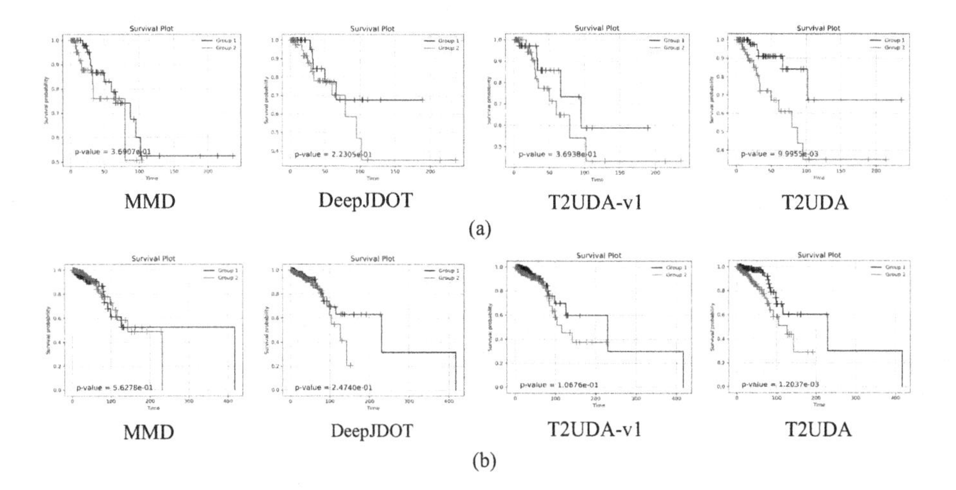

Fig. 3. The survival curves by applying different methods on two benchmark settings: (1) source domain is ER+ BC and Target domain is ER− BC in (a); (2) source domain is ER− BC and Target domain is ER+ BC in (b).

3.2 Result and Discussion

In this study, we compared the performance of our proposed model with several existing domain adaptation methods, including 1) DDC [17]: Utilize the Maximum Mean Discrepancy (MMD) to calculate the domain difference loss between source and target data and optimize both classification loss and disparity loss. 2) DANN [18]: An adversarial learning method that used gradient backpropagation to extract domain-independent features. 3) MDD [19]: An adversarial training method that combined metric learning and domain adaptation. 4) DeepJDOT [20]: An Unsupervised Domain Adaptation method based on optimal transport that simultaneously learns features and optimizes classifiers by measuring joint feature/label differences. 5) Source only: it was trained on

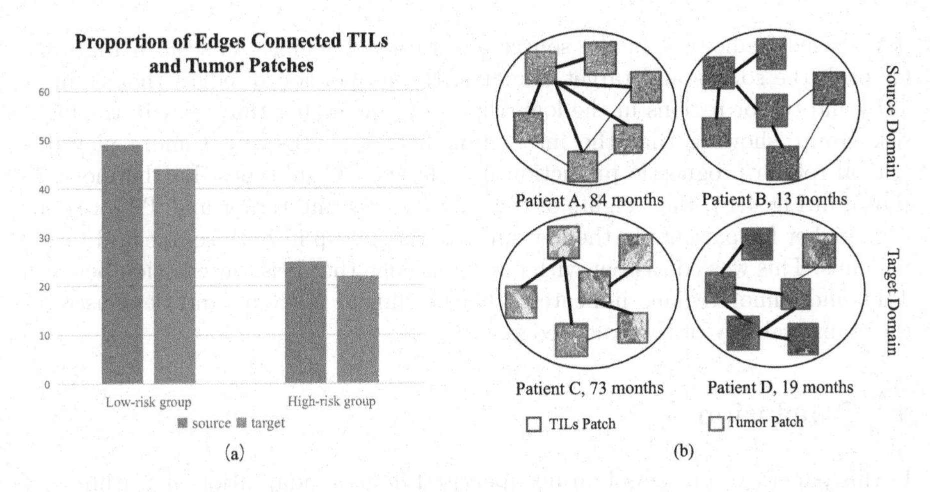

Fig. 4. (a): Compare proportion of Edges Connected TILs and Tumor patches of source and target domain (b): Compare interactions between TILs patches and tumor patches for long survival patients and short survival patients of source and target domain. The thick black line indicates the edges with higher weight.

the source domain and applied directly to the target domain. 6) T2UDA-v1: it was a variant of T2UDA which didn't use TTIA. The experimental results were presented in Table 1.

The results presented in Table 1 revealed several key observations. First, our proposed method outperformed feature alignment-based methods such as DDC and DeepJDOT in terms of both CI and AUC values. The reason lies in that these methods only transferred the knowledge at the feature level and neglected the inter-relationship between TILs and tumors. Second, our method outperformed adversarial-based methods such as DANN and MDD, as the high heterogeneity between the target and source domains results in negative transfer through adversarial training. Instead of directly aligning regions, our proposed method focused on similar TILs-Tumor interactions and aligning patches of the same tissue.

We also evaluated the contributions of the key components of our framework and found that T2UDA performed better than Source only and T2UDA-v1, which shows the advantage of minimizing differences in TILs-Tumor interaction weights.

In addition, we also evaluated the patient stratification performance of different methods. As shown in Fig. 3, our proposed T2UDA outperformed feature alignment-based methods (such as DDC and DeepJDOT), adversarial-based methods (such as DANN and MDD), and T2UDA-v1 in stratification performance, proving that considering the interaction between TILs and tumors as migration knowledge leads to better prognostic results.

We also examined the consistency of important edges in each group of stratified patients based on the TILs-Tumor interaction weights calculated by the

GAT-based framework in the source and target domains. As seen in Fig. 4(a), for both the source and target domains, the proportion of edges that connect TILs and tumor regions in the low-risk group was higher than that in the high-risk group, showing that the interaction between TILs and tumors played a critical role in prognostic prediction in different BC subtypes. Furthermore, as shown in Fig. 4(b), the weights of the edges connecting tumor and TILs regions were higher for patients in the low survival risk group in both source and target domains. This was consistent with our knowledge that brisk interaction between TILs and tumor regions indicates a better clinical outcome and demonstrates the transferability of this knowledge.

4 Conclusion

In this paper, we presented an unsupervised domain adaptation algorithm that leverages TILs-Tumor interactions to predict patients' survival in a target BC subtype(T2UDA). Our results demonstrated that the relationship between TILs and tumors is transferable and can be effectively used to improve the accuracy of survival prediction models. To the best of our knowledge, T2UDA was the first method to successfully achieve interrelationship transfer between TILs and tumors across different cancer subtypes for prognosis tasks.

Acknowledgements. This work was supported by the National Natural Science Foundation of China (Nos.62136004, 61902183, 61876082, 61732006, 62076129), the National Key R&D Program of China (Grant Nos.2018YFC2001600, 2018YFC2001602).

References

1. Shao, W., Wang, T., Huang, Z., Han, Z., Zhang, J., Huang, K.: Weakly supervised deep ordinal cox model for survival prediction from whole-slide pathological images. IEEE Trans. Med. Imaging **40**(12), 3739–3747 (2021)
2. Okabe, M., et al.: Predictive factors of the tumor immunological microenvironment for long-term follow-up in early stage breast cancer. Cancer Sci. **108**(1), 81–90 (2017)
3. Mizukami, Y., et al.: Detection of novel cancer-testis antigen-specific t-cell responses in til, regional lymph nodes, and pbl in patients with esophageal squamous cell carcinoma. Cancer Sci. **99**(7), 1448–1454 (2008)
4. Yawen, W., et al.: Recent advances of deep learning for computational histopathology: principles and applications. Cancers **14**(5), 1199 (2022)
5. Cheng, L., et al.: Nuclear shape and orientation features from h&e images predict survival in early-stage estrogen receptor-positive breast cancers. Lab. Invest. **98**(11), 1438–1448 (2018)
6. Liu, P., Fu, B., Yang, S.X., Deng, L., Zhong, X., Zheng, H.: Optimizing survival analysis of xgboost for ties to predict disease progression of breast cancer. IEEE Trans. Biomedical Eng. **68**(1), 148–160 (2020)

7. Alirezazadeh, P., Hejrati, B., Monsef-Esfahani, A., Fathi, A.: Representation learning-based unsupervised domain adaptation for classification of breast cancer histopathology images. Biocybern. Biomed. Eng. **38**(3), 671–683 (2018)
8. Zhang, Y., et al.: Collaborative unsupervised domain adaptation for medical image diagnosis. IEEE Trans. Image Process. **29**, 7834–7844 (2020)
9. Xu, D., Cai, C., Fang, C., Kong, B., Zhu, J., Li, Z.: Graph neural networks for unsupervised domain adaptation of histopathological imageanalytics. arXiv preprint arXiv:2008.09304 (2020)
10. Kurozumi, S., et al.: Prognostic significance of tumour-infiltrating lymphocytes for oestrogen receptor-negative breast cancer without lymph node metastasis. Oncol. Lett. **17**(3), 2647–2656 (2019)
11. Zixiao, L., et al.: Deep-learning-based characterization of tumor-infiltrating lymphocytes in breast cancers from histopathology images and multiomics data. JCO Clin. Cancer Informat. **4**, 480–490 (2020)
12. Zuo, Y.: Identify consistent imaging genomic biomarkers for characterizing the survival-associated interactions between tumor-infiltrating lymphocytes and tumors. In: Medical Image Computing and Computer Assisted Intervention-MICCAI 2022: 25th International Conference, Singapore, 18–22 September 2022, Proceedings, Part II, pp. 222–231. Springer (2022). https://doi.org/10.1007/978-3-031-16434-7_22
13. Veličković, P., Cucurull, G., Casanova, A., Romero, A., Lio, P., Bengio, Y.: Graph attention networks. arXiv preprint arXiv:1710.10903 (2017)
14. Lee, J., Lee, I., Kang, J.: Self-attention graph pooling. In: International Conference on Machine Learning, pp. 3734–3743. PMLR (2019)
15. Borgwardt, K.M., Gretton, A., Rasch, M.J., Kriegel, H.-P., Schölkopf, B., Smola, A.J.: Integrating structured biological data by kernel maximum mean discrepancy. Bioinformatics **22**(14), e49–e57 (2006)
16. Shao, W., et al.: Integrative analysis of pathological images and multi-dimensional genomic data for early-stage cancer prognosis. IEEE Trans. Med. Imaging **39**(1), 99–110 (2019)
17. Tzeng, E., Hoffman, J., Zhang, N., Saenko, K., Darrell, T.: Deep domain confusion: Maximizing for domain invariance. arXiv preprint arXiv:1412.3474 (2014)
18. Ganin, Y., Ustinova, E., Ajakan, H., Germain, P., Larochelle, H., Laviolette, F., Marchand, M., Lempitsky, V.: Domain-adversarial training of neural networks. J. Mach. Learn. Res. **17**(1), 2030–2096 (2016)
19. Zhang, Y., Liu, T., Long, M., Jordan, M.: Bridging theory and algorithm for domain adaptation. In: International Conference on Machine Learning, pp. 7404–7413. PMLR (2019)
20. Damodaran, B.B., Kellenberger, B., Flamary, R., Tuia, D., Courty, N.: DeepJDOT: deep joint distribution optimal transport for unsupervised domain adaptation. In: Ferrari, V., Hebert, M., Sminchisescu, C., Weiss, Y. (eds.) ECCV 2018. LNCS, vol. 11208, pp. 467–483. Springer, Cham (2018). https://doi.org/10.1007/978-3-030-01225-0_28

Pathology-and-Genomics Multimodal Transformer for Survival Outcome Prediction

Kexin Ding[1], Mu Zhou[2], Dimitris N. Metaxas[2], and Shaoting Zhang[3](✉)

[1] Department of Computer Science, UNC at Charlotte, Charlotte, NC, USA
[2] Department of Computer Science, Rutgers University, Piscataway, NJ, USA
[3] Shanghai Artificial Intelligence Laboratory, Shanghai, China
zhangshaoting@pjlab.org.cn

Abstract. Survival outcome assessment is challenging and inherently associated with multiple clinical factors (e.g., imaging and genomics biomarkers) in cancer. Enabling multimodal analytics promises to reveal novel predictive patterns of patient outcomes. In this study, we propose a multimodal transformer (**PathOmics**) integrating pathology and genomics insights into colon-related cancer survival prediction. We emphasize the unsupervised pretraining to capture the intrinsic interaction between tissue microenvironments in gigapixel whole slide images (WSIs) and a wide range of genomics data (e.g., mRNA-sequence, copy number variant, and methylation). After the multimodal knowledge aggregation in pretraining, our task-specific model finetuning could expand the scope of data utility applicable to both multi- and single-modal data (e.g., image- or genomics-only). We evaluate our approach on both TCGA colon and rectum cancer cohorts, showing that the proposed approach is competitive and outperforms state-of-the-art studies. Finally, our approach is desirable to utilize the limited number of finetuned samples towards data-efficient analytics for survival outcome prediction. The code is available at https://github.com/Cassie07/PathOmics.

Keywords: Histopathological image analysis · Multimodal learning · Cancer diagnosis · Survival prediction

1 Introduction

Cancers are a group of heterogeneous diseases reflecting deep interactions between pathological and genomics variants in tumor tissue environments [24]. Different cancer genotypes are translated into pathological phenotypes that could be assessed by pathologists [24]. High-resolution pathological images have proven their unique benefits for improving prognostic biomarkers prediction via

Supplementary Information The online version contains supplementary material available at https://doi.org/10.1007/978-3-031-43987-2_60.

H. Greenspan et al. (Eds.): MICCAI 2023, LNCS 14225, pp. 622–631, 2023.
https://doi.org/10.1007/978-3-031-43987-2_60

exploring the tissue microenvironmental features [1,10,12,13,18,25]. Meanwhile, genomics data (e.g., mRNA-sequence) display a high relevance to regulate cancer progression [3,29]. For instance, genome-wide molecular portraits are crucial for cancer prognostic stratification and targeted therapy [16]. Despite their importance, seldom efforts jointly exploit the multimodal value between cancer image morphology and molecular biomarkers. In a broader context, assessing cancer prognosis is essentially a multimodal task in association with pathological and genomics findings. Therefore, synergizing multimodal data could deepen a cross-scale understanding towards improved patient prognostication.

The major goal of multimodal data learning is to extract complementary contextual information across modalities [4]. Supervised studies [5–7] have allowed multimodal data fusion among image and non-image biomarkers. For instance, the Kronecker product is able to capture the interactions between WSIs and genomic features for survival outcome prediction [5,7]. Alternatively, the co-attention transformer [6] could capture the genotype-phenotype interactions for prognostic prediction. Yet these supervised approaches are limited by feature generalizability and have a high dependency on data labeling. To alleviate label requirement, unsupervised learning evaluates the intrinsic similarity among multimodal representations for data fusion. For example, integrating image, genomics, and clinical information can be achieved via a predefined unsupervised similarity evaluation [4]. To broaden the data utility, the study [28] leverages the pathology and genomic knowledge from the teacher model to guide the pathology-only student model for glioma grading. From these analyses, it is increasingly recognized that the lack of flexibility on model finetuning limits the data utility of multimodal learning. Meanwhile, the size of multimodal medical datasets is not as large as natural vision-language datasets, which necessitates the need for data-efficient analytics to address the training difficulty.

To tackle above challenges, we propose a pathology-and-genomics multimodal framework (i.e., **PathOmics**) for survival prediction (Fig. 1). We summarized our contributions as follows. **(1) Unsupervised multimodal data fusion.** Our unsupervised pretraining exploits the intrinsic interaction between morphological and molecular biomarkers (Fig. 1a). To overcome the gap of modality heterogeneity between images and genomics, we project the multimodal embeddings into the same latent space by evaluating the similarity among them. Particularly, the pretrained model offers a unique means by using similarity-guided modality fusion for extracting cross-modal patterns. **(2) Flexible modality finetuning.** A key contribution of our multimodal framework is that it combines benefits from both unsupervised pretraining and supervised finetuning data fusion (Fig. 1b). As a result, the task-specific finetuning broadens the dataset usage (Fig 1b and c), which is not limited by data modality (e.g., both single- and multi-modal data). **(3) Data efficiency with limited data size.** Our approach could achieve comparable performance even with fewer finetuned data (e.g., only use 50% of the finetuned data) when compared with using the entire finetuning dataset.

2 Methodology

Overview. Figure 1 illustrates our multimodal transformer framework. Our method includes an unsupervised multimodal data fusion pretraining and a supervised flexible-modal finetuning. From Fig. 1a, in the pretraining, our unsupervised data fusion aims to capture the interaction pattern of image and genomics features. Overall, we formulate the objective of multimodal feature learning by converting image patches and tabular genomics data into group-wise embeddings, and then extracting multimodal patient-wise embeddings. More specifically, we construct group-wise representations for both image and genomics modalities. For image feature representation, we randomly divide image patches into groups; Meanwhile, for each type of genomics data, we construct groups of genes depending on their clinical relevance [22]. Next, as seen in Fig. 1b and c, our approach enables three types of finetuning modal modes (i.e., multi-modal, image-only, and genomics-only) towards prognostic prediction, expanding the downstream data utility from the pretrained model.

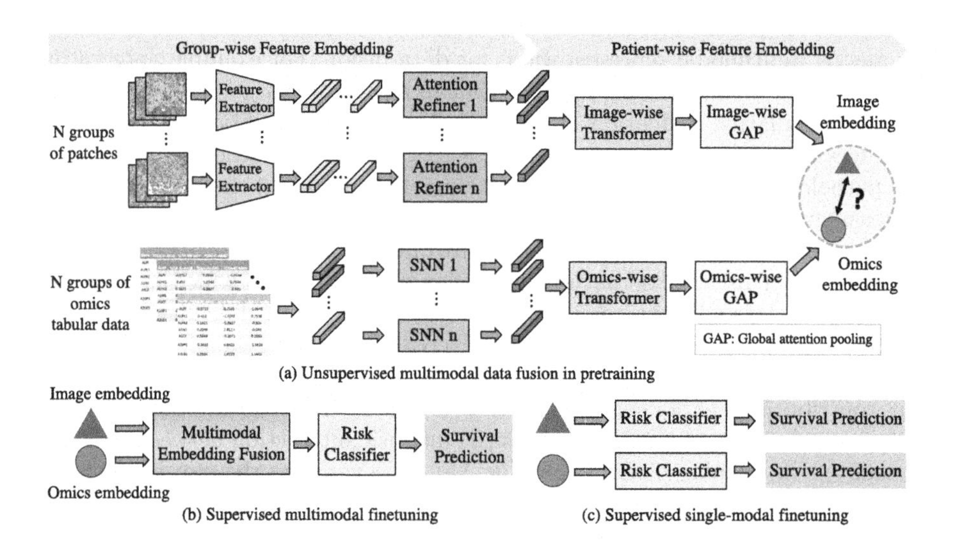

Fig. 1. Workflow overview of the pathology-and-genomics multimodal transformer (**PathOmics**) for survival prediction. In (a), we show the pipeline of extracting image and genomics feature embedding via an unsupervised pretraining towards multimodal data fusion. In (b) and (c), our supervised finetuning scheme could flexibly handle multiple types of data for prognostic prediction. With the multimodal pretrained model backbones, both multi- or single-modal data can be applicable for our model finetuning.

Group-Wise Image and Genomics Embedding. We define the group-wise genomics representation by referring to $N = 8$ major functional groups obtained from [22]. Each group contains a list of well-defined molecular features related to cancer biology, including transcription factors, tumor suppression, cytokines and growth factors, cell differentiation markers, homeodomain proteins, translocated cancer genes, and protein kinases. The group-wise genomics representation is defined as $G_n \in \mathbb{R}^{1\times d_g}$, where $n \in N$, d_g is the attribute dimension in each group which could be various. To better extract high-dimensional group-wise genomics representation, we use a Self-Normalizing Network (SNN) together with scaled exponential linear units (SeLU) and Alpha Dropout for feature extraction to generate the group-wise embedding $G_n \in \mathbb{R}^{1\times 256}$ for each group.

For group-wise WSIs representation, we first cropped all tissue-region image tiles from the entire WSI and extracted CNN-based (e.g., ResNet50) d_i-dimensional features for each image tile k as $h_k \in \mathbb{R}^{1\times d_i}$, where $d_i = 1,024$, $k \in K$ and K is the number of image patches. We construct the group-wise WSIs representation by randomly splitting image tile features into N groups (i.e., the same number as genomics categories). Therefore, group-wise image representation could be defined as $I_n \in \mathbb{R}^{k_n\times 1024}$, where $n \in N$ and k_n represents tile k in group n. Then we apply an attention-based refiner (ABR) [17], which is able to weight the feature embeddings in the group, together with a dimension deduction (e.g., fully-connected layers) to achieve the group-wise embedding. The ABR and the group-wise embedding $I_n \in \mathbb{R}^{1\times 256}$ are defined as:

$$a_k = \frac{epx\{w^T(tanh(V_1 h_k) \odot (sigm(V_2 h_k))\}}{\sum_{j=1}^{K} epx\{w^T(tanh(V_1 h_j) \odot (sigm(V_2 h_j))\}} \tag{1}$$

where w,V1 and V2 are the learnable parameters.

$$I_n = \sum_{k=1}^{K} a_k h_k \tag{2}$$

Patient-Wise Multimodal Feature Embedding. To aggregate patient-wise multimodal feature embedding from the group-wise representations, as shown in Fig. 1a, we propose a pathology-and-genomics multimodal model containing two model streams, including a pathological image and a genomics data stream. In each stream, we use the same architecture with different weights, which is updated separately in each modality stream. In the pathological image stream, the patient-wise image representation is aggregated by N group representations as $I_p \in \mathbb{R}^{N\times 256}$, where $p \in P$ and P is the number of patients. Similarly, the patient-wise genomics representation is aggregated as $G_p \in \mathbb{R}^{N\times 256}$. After generating patient-wise representation, we utilize two transformer layers [27] to extract feature embeddings for each modality as follows:

$$H_p^l = MSA(H_p) \tag{3}$$

where MSA denotes Multi-head Self-attention [27] (see Appendix 1), l denotes the layer index of the transformer, and H_p could either be I_p or G_p. Then, we construct global attention poolings [17] as Eq. 1 to adaptively compute a weighted sum of each modality feature embeddings to finally construct patient-wise embedding as $I_{embedding}^p \in \mathbb{R}^{1\times 256}$ and $G_{embedding}^p \in \mathbb{R}^{1\times 256}$ in each modality.

Multimodal Fusion in Pretraining and Finetuning. Due to the domain gap between image and molecular feature heterogeneity, a proper design of multimodal fusion is crucial to advance integrative analysis. In the pretraining stage, we develop an unsupervised data fusion strategy by decreasing the mean square error (MSE) loss to map images and genomics embeddings into the same space. Ideally, the image and genomics embeddings belonging to the same patient should have a higher relevance between each other. MSE measures the average squared difference between multimodal embeddings. In this way, the pretrained model is trained to map the paired image and genomics embeddings to be closer in the latent space, leading to strengthen the interaction between different modalities.

$$\mathcal{L}_{fusion} = argmin \frac{1}{P} \sum_{p=1}^{P} ((I^{p}_{embedding} - G^{p}_{embedding})^2) \tag{4}$$

In the single modality finetuning, even if we use image-only data, the model is able to produce genomic-related image feature embedding due to the multimodal knowledge aggregation already obtained from the model pretraining. As a result, our cross-modal information aggregation relaxes the modality requirement in the finetuning stage. As shown in Fig. 1b, for multimodal finetuning, we deploy a concatenation layer to obtain the fused multimodal feature representation and implement a risk classifier (FC layer) to achieve the final survival stratification (see Appendix 2). As for single-modality finetuning mode in Fig. 1c, we simply feed $I^{p}_{embedding}$ or $G^{p}_{embedding}$ into risk classifier for the final prognosis prediction. During the finetuning, we update the model parameters using a log-likelihood loss for the discrete-time survival model training [6](see Appendix 2).

3 Experiments and Results

Datasets. All image and genomics data are publicly available. We collected WSIs from The Cancer Genome Atlas Colon Adenocarcinoma (TCGA-COAD) dataset (CC-BY-3.0) [8,21] and Rectum Adenocarcinoma (TCGA-READ) dataset (CC-BY-3.0) [8,20], which contain 440 and 153 patients. We cropped each WSI into 512 × 512 non-overlapped patches. We also collected the corresponding tabular genomics data (e.g., mRNA sequence, copy number alteration, and methylation) with overall survival (OS) times and censorship statuses from Cbioportal [2,14]. We removed the samples without the corresponding genomics data or ground truth of survival outcomes. Finally, we included 426 patients of TCGA-COAD and 145 patients of TCGA-READ.

Experimental Settings and Implementations. We implement two types of settings that involve internal and external datasets for model pretraining and finetuning. As shown in Fig 2a, we pretrain and finetune the model on the same dataset (i.e., internal setting). We split TCGA-COAD into training (80%) and holdout testing set (20%). Then, we implement four-fold cross-validation on the

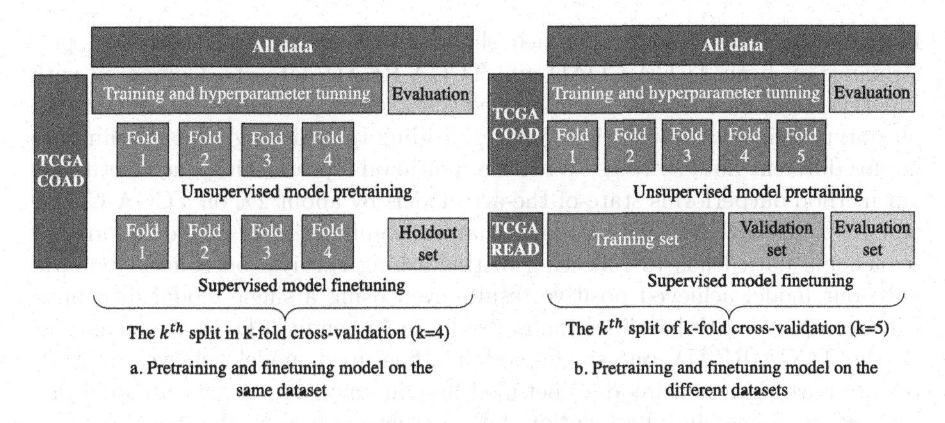

Fig. 2. Dataset usage. In a, we use TCGA-COAD dataset for model pretraining, finetuning, and evaluation. In b, we use TCGA-COAD dataset for model pretraining. Then, we use TCGA-READ dataset to finetune and evaluate the pretrained models.

training set for pretraining, finetuning, and hyperparameter-tuning. The test set is only used for evaluating the best finetuned models from each cross-validation split. For the external setting, we implement pretraining and finetuning on the different datasets, as shown in Fig 2b; we use TCGA-COAD for pretraining; Then, we only use TCGA-READ for finetuning and final evaluation. We implement a five-fold cross-validation for pretraining, and the best pretrained models are used for finetuning. We split TCGA-READ into finetuning (60%), validation (20%), and evaluation set (20%). For all experiments, we calculate the average performance on the evaluation set across the best models.

The number of epochs for pretraining and finetuning are 25, the batch size is 1, the optimizer is Adam [19], and the learning rate is 1e-4 for pretraining and 5e-5 for finetuning. We used one 32GB Tesla V100 SXM2 GPU and Pytorch. The concordance index (C-index) is used to measure the survival prediction performance. We followed the previous studies [5–7] to partition the overall survival (OS) months into four non-overlapping intervals by using the quartiles of event times of uncensored patients for discretized-survival C-index calculation (see Appendix 2). For each experiment, we reported the average C-index among three-times repeated experiments. Conceptionally, our method shares a similar idea to multiple instance learning (MIL) [9,23]. Therefore, we include two types of baseline models, including the MIL-based models (DeepSet [30], AB-MIL [17], and TransMIL [26]) and MIL multimodal-based models (MCAT [6], PORPOISE [7]). We follow the same data split and processing, as well as the identical training hyperparameters and supervised fusion as above. Notably, there is no need for supervised finetuning for the baselines when using TCGA-COAD (Table 1), because the supervised pretraining is already applied to the training set.

Results. In Table 1, our approach shows improved survival prediction performance on both TCGA-COAD and TCGA-READ datasets. Compared with supervised baselines, our unsupervised data fusion is able to extract the phenotype-genotype interaction features, leading to achieving a flexible finetuning for different data settings. With the multimodal pretraining and finetuning, our method outperforms state-of-the-art models by about 2% on TCGA-COAD and 4% TCGA-READ. We recognize that the combination of image and mRNA sequencing data leads to reflecting distinguishing survival outcomes. Remarkably, our model achieved positive results even using a single-modal finetuning when compared with baselines (more results in Appendix 3.1). In the meantime, on the TCGA-READ, our single-modality finetuned model achieves a better performance than multimodal finetuned baseline models (e.g., with model pretraining via image and methylation data, we have only used the image data for finetuning and achieved a C-index of 74.85%, which is about 4% higher than the best baseline models). We show that with a single-modal finetuning strategy, the model could generate meaningful embedding to combine image- and genomic-related patterns. In addition, our model reflects its efficiency on the limited finetuning data (e.g., 75 patients are used for finetuning on TCGA-READ, which are only 22% of TCGA-COAD finetuning data). In Table 1, our method could yield better performance compared with baselines on the small dataset across the combination of images and multiple types of genomics data.

Table 1. The comparison of C-index performance on TCGA-COAD and TCGA-READ dataset. "Methy" is used as the abbreviation of Methylation.

Model	Pretrain data modality	TCGA-COAD		TCGA-READ	
		Finetune data modality	C-index (%)	Finetune data modality	C-index (%)
DeepSets [30]	image+mRNA	-	58.70 ± 1.10	image+mRNA	70.19 ± 1.45
	image+CNA	–	51.50 ± 2.60	image+CNA	62.50 ± 2.52
	image+Methy	–	65.61 ± 1.86	image+Methy	55.78 ± 1.22
AB-MIL [17]	image+mRNA	–	54.12 ± 2.88	image+mRNA	68.79 ± 1.44
	image+CNA	–	54.68 ± 2.44	image+CNA	66.72 ± 0.81
	image+Methy	–	49.66 ± 1.58	image+Methy	55.78 ± 1.22
TransMIL [26]	image+mRNA	–	54.15 ± 1.02	image+mRNA	67.91 ± 2.35
	image+CNA	–	59.80 ± 0.98	image+CNA	62.75 ± 1.92
	image+Methy	–	53.35 ± 1.78	image+Methy	53.09 ± 1.46
MCAT [6]	image+mRNA	-	65.02 ± 3.10	image+mRNA	70.27 ± 2.75
	image+CNA	–	64.66 ± 2.31	image+CNA	60.50 ± 1.25
	image+Methy	–	60.98 ± 2.43	image+Methy	59.78 ± 1.20
PORPOI-SE [7]	image+mRNA	–	65.31 ± 1.26	image+mRNA	68.18 ± 1.62
	image+CNA	–	57.32 ± 1.78	image+CNA	60.19 ± 1.48
	image+Methy	–	61.84 ± 1.10	image+Methy	68.80 ± 0.92
Ours	image+mRNA	image+mRNA	$\mathbf{67.32 \pm 1.69}$	image+mRNA	74.35 ± 1.15
		image	63.78 ± 1.22	image	$\mathbf{74.85 \pm 0.37}$
		mRNA	60.76 ± 0.88	mRNA	59.61 ± 1.37
	image+CNA	image+CNA	61.19 ± 1.03	image+CNA	73.95 ± 1.05
		image	58.06 ± 1.54	image	71.18 ± 1.39
		CNA	56.43 ± 1.02	CNA	63.95 ± 0.55
	image+Methy	image+Methy	67.22 ± 1.67	image+Methy	71.80 ± 2.03
		image	60.43 ± 0.72	image	64.42 ± 0.72
		Methy	61.06 ± 1.34	Methy	65.42 ± 0.91

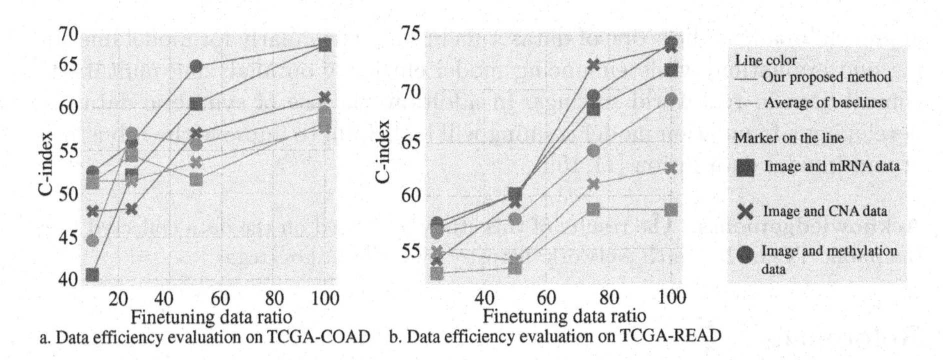

Fig. 3. Ablation study. In (a) and (b), we evaluate the model efficiency by using fewer data for model finetuning on TCGA-COAD and TCGA-READ. We show the average C-index of baselines, the detailed results are shown in the Appendix 3.2.

Ablation Analysis. We verify the model efficiency by using fewer amounts of finetuning data in finetuning. For TCGA-COAD dataset, we include 50%, 25%, and 10% of the finetuning data. For the TCGA-READ dataset, as the number of uncensored patients is limited, we use 75%, 50%, and 25% of the finetuning data to allow at least one uncensored patient to be included for finetuning. As shown in Fig. 3a, by using 50% of TCGA-COAD finetuning data, our approach achieves the C-index of 64.80%, which is higher than the average performance of baselines in several modalities. Similarly, in Fig. 3b, our model retains a good performance by using 50% or 75% of TCGA-READ finetuning data compared with the average of C-index across baselines (e.g., 72.32% versus 64.23%). For evaluating the effect of cross-modality information extraction in the pretraining, we kept supervised model training (i.e., the finetuning stage) while removing the unsupervised pretraining. The performance is lower 2%-10% than ours on multi- and single-modality data. For evaluating the genomics data usage, we designed two settings: (1) combining all types of genomics data and categorizing them by groups; (2) removing category information while keeping using different types of genomics data separately. Our approach outperforms the above ablation studies by 3%-7% on TCGA-READ and performs similarly on TCGA-COAD. In addition, we replaced our unsupervised loss with cosine similarity loss; our approach outperforms the setting of using cosine similarity loss by 3%-6%.

4 Conclusion

Developing data-efficient multimodal learning is crucial to advance the survival assessment of cancer patients in a variety of clinical data scenarios. We demonstrated that the proposed PathOmics framework is useful for improving the survival prediction of colon and rectum cancer patients. Importantly, our approach opens up perspectives for exploring the key insights of intrinsic genotype-phenotype interactions in complex cancer data across modalities. Our finetuning approach broadens the scope of dataset inclusion, particularly for model finetuning and evaluation, while enhancing model efficiency on analyzing multimodal clinical data in real-world settings. In addition, the use of synthetic data and developing a foundation model training will be helpful to improve the robustness of multimodal data fusion [11,15].

Acknowledgements. The results of this study are based on the data collected from the public TCGA Research Network: https://www.cancer.gov/tcga.

References

1. Bilal, M., et al.: Development and validation of a weakly supervised deep learning framework to predict the status of molecular pathways and key mutations in colorectal cancer from routine histology images: a retrospective study. Lancet Digital Health **3**(12), e763–e772 (2021)
2. Cerami, E., et al.: The cbio cancer genomics portal: an open platform for exploring multidimensional cancer genomics data. Cancer Discov. **2**(5), 401–404 (2012)
3. Chaudhary, K., Poirion, O.B., Lu, L., Garmire, L.X.: Deep learning-based multi-omics integration robustly predicts survival in liver cancer using deep learning to predict liver cancer prognosis. Clin. Cancer Res. **24**(6), 1248–1259 (2018)
4. Cheerla, A., Gevaert, O.: Deep learning with multimodal representation for pancancer prognosis prediction. Bioinformatics **35**(14), i446–i454 (2019)
5. Chen, R.J., et al.: Pathomic fusion: an integrated framework for fusing histopathology and genomic features for cancer diagnosis and prognosis. IEEE Trans. Med. Imaging **41**(4), 757–770 (2020)
6. Chen, R.J., et al.: Multimodal co-attention transformer for survival prediction in gigapixel whole slide images. In: Proceedings of the IEEE/CVF International Conference on Computer Vision, pp. 4015–4025 (2021)
7. Chen, R.J., et al.: Pan-cancer integrative histology-genomic analysis via multimodal deep learning. Cancer Cell **40**(8), 865–878 (2022)
8. Clark, K., et al.: The cancer imaging archive (tcia): maintaining and operating a public information repository. J. Digit. Imaging **26**, 1045–1057 (2013)
9. Dietterich, T.G., Lathrop, R.H., Lozano-Pérez, T.: Solving the multiple instance problem with axis-parallel rectangles. Artif. Intell. **89**(1–2), 31–71 (1997)
10. Ding, K., Liu, Q., Lee, E., Zhou, M., Lu, A., Zhang, S.: Feature-enhanced graph networks for genetic mutational prediction using histopathological images in colon cancer. In: Martel, A.L., et al. (eds.) MICCAI 2020. LNCS, vol. 12262, pp. 294–304. Springer, Cham (2020). https://doi.org/10.1007/978-3-030-59713-9_29
11. Ding, K., Zhou, M., Wang, H., Gevaert, O., Metaxas, D., Zhang, S.: A large-scale synthetic pathological dataset for deep learning-enabled segmentation of breast cancer. Sci. Data **10**(1), 231 (2023)
12. Ding, K., Zhou, M., Wang, H., Zhang, S., Metaxas, D.N.: Spatially aware graph neural networks and cross-level molecular profile prediction in colon cancer histopathology: a retrospective multi-cohort study. Lancet Digital Health **4**(11), e787–e795 (2022)
13. Ding, K., et al.: Graph convolutional networks for multi-modality medical imaging: Methods, architectures, and clinical applications. arXiv preprint arXiv:2202.08916 (2022)

14. Gao, J., et al.: Integrative analysis of complex cancer genomics and clinical profiles using the cbioportal. Sci. Signaling **6**(269), pl1-pl1 (2013)
15. Gao, Y., Li, Z., Liu, D., Zhou, M., Zhang, S., Meta, D.N.: Training like a medical resident: universal medical image segmentation via context prior learning. arXiv preprint arXiv:2306.02416 (2023)
16. Gentles, A.J., et al.: The prognostic landscape of genes and infiltrating immune cells across human cancers. Nat. Med. **21**(8), 938–945 (2015)
17. Ilse, M., Tomczak, J., Welling, M.: Attention-based deep multiple instance learning. In: International Conference on Machine Learning, pp. 2127–2136. PMLR (2018)
18. Kather, J.N., et al.: Deep learning can predict microsatellite instability directly from histology in gastrointestinal cancer. Nat. Med. **25**(7), 1054–1056 (2019)
19. Kingma, D.P., Ba, J.: Adam: a method for stochastic optimization. arXiv preprint arXiv:1412.6980 (2014)
20. Kirk, S., Lee, Y., Sadow, C., Levine: the cancer genome atlas rectum adenocarcinoma collection (tcga-read) (version 3) [data set]. The Cancer Imaging Archive (2016)
21. Kirk, S., et al.: Radiology data from the cancer genome atlas colon adenocarcinoma [tcga-coad] collection. The Cancer Imaging Archive (2016)
22. Liberzon, A., Birger, C., Thorvaldsdóttir, H., Ghandi, M., Mesirov, J.P., Tamayo, P.: The molecular signatures database hallmark gene set collection. Cell Syst. **1**(6), 417–425 (2015)
23. Maron, O., Lozano-Pérez, T.: A framework for multiple-instance learning. In: Advances in Neural Information Processing Systems 10 (1997)
24. Marusyk, A., Almendro, V., Polyak, K.: Intra-tumour heterogeneity: a looking glass for cancer? Nat. Rev. Cancer **12**(5), 323–334 (2012)
25. Qu, H., et al.: Genetic mutation and biological pathway prediction based on whole slide images in breast carcinoma using deep learning. NPJ Precision Oncol. **5**(1), 87 (2021)
26. Shao, Z., et al.: Transmil: transformer based correlated multiple instance learning for whole slide image classification. Adv. Neural. Inf. Process. Syst. **34**, 2136–2147 (2021)
27. Vaswani, A., et al.: Attention is all you need. In: Advances in Neural Information Processing Systems 30 (2017)
28. Xing, X., Chen, Z., Zhu, M., Hou, Y., Gao, Z., Yuan, Y.: Discrepancy and gradient-guided multi-modal knowledge distillation for pathological glioma grading. In: Medical Image Computing and Computer Assisted Intervention-MICCAI 2022: 25th International Conference, Singapore, 18–22 September 2022, Proceedings, Part V, pp. 636–646. Springer (2022). https://doi.org/10.1007/978-3-031-16443-9_61
29. Yang, M., et al.: A multi-omics machine learning framework in predicting the survival of colorectal cancer patients. Comput. Biol. Med. **146**, 105516 (2022)
30. Zaheer, M., Kottur, S., Ravanbakhsh, S., Poczos, B., Salakhutdinov, R.R., Smola, A.J.: Deep sets. In: Advances in Neural Information Processing Systems 30 (2017)

Adaptive Supervised PatchNCE Loss for Learning H&E-to-IHC Stain Translation with Inconsistent Groundtruth Image Pairs

Fangda Li[1(✉)], Zhiqiang Hu[2], Wen Chen[2], and Avinash Kak[1]

[1] Purdue University, West Lafayette, USA
{li1208,kak}@purdue.edu
[2] Sensetime Research, Beijing, China
{huzhiqiang,chenwen}@sensetime.com

Abstract. Immunohistochemical (IHC) staining highlights the molecular information critical to diagnostics in tissue samples. However, compared to H&E staining, IHC staining can be much more expensive in terms of both labor and the laboratory equipment required. This motivates recent research that demonstrates that the correlations between the morphological information present in the H&E-stained slides and the molecular information in the IHC-stained slides can be used for H&E-to-IHC stain translation. However, due to a lack of pixel-perfect H&E-IHC groundtruth pairs, most existing methods have resorted to relying on expert annotations. To remedy this situation, we present a new loss function, *Adaptive Supervised PatchNCE* (ASP), to directly deal with the input to target inconsistencies in a proposed H&E-to-IHC image-to-image translation framework. The ASP loss is built upon a patch-based contrastive learning criterion, named *Supervised PatchNCE* (SP), and augments it further with weight scheduling to mitigate the negative impact of noisy supervision. Lastly, we introduce the *Multi-IHC Stain Translation* (MIST) dataset, which contains aligned H&E-IHC patches for 4 different IHC stains critical to breast cancer diagnosis. In our experiment, we demonstrate that our proposed method outperforms existing image-to-image translation methods for stain translation to multiple IHC stains. All of our code and datasets are available at https://github.com/lifangda01/AdaptiveSupervisedPatchNCE.

Keywords: Generative Adversarial Network · Contrastive Learning · H&E-to-IHC Stain Translation

Supplementary Information The online version contains supplementary material available at https://doi.org/10.1007/978-3-031-43987-2_61.

H. Greenspan et al. (Eds.): MICCAI 2023, LNCS 14225, pp. 632–641, 2023.
https://doi.org/10.1007/978-3-031-43987-2_61

1 Introduction

Immunohistochemical (IHC) staining is a widely used technique in pathology for visualizing abnormal cells that are often found in tumors. IHC chromogens highlight the presence of certain antigens or proteins by staining their corresponding antibodies. For instance, the HER2 (human epidermal growth factor receptor 2) biomarker is associated with aggressive breast tumor development and is essential in forming a precise treatment plan. Despite its capability to provide highly valuable diagnostic information, the process of IHC staining is very labor-intensive, time-consuming and requires specialized histotechnologists and laboratory equipments [2]. Such factors hinder the general availability of IHC staining in histopathological applications.

At the other end of the spectrum, H&E (Hematoxylin and Eosin) staining, as the gold standard in histological staining, highlights the tissue structures and cell morphology. In routine diagnostics, on account of its much lower cost, an H&E-stained slide is prepared by pathologists in order to determine whether or not to also apply the IHC stains for a more precise assessment of the disease. Therefore, it is of great interest to have an algorithm that can automatically translate an H&E-stained slide into one that could be considered to have been stained with IHC while accurately predicting the target expression levels.

To that end, researchers have recently proposed to use GAN-based Image-to-Image Translation (I2IT) algorithms for transforming H&E-stained slides into IHC. Despite the progress, the outstanding challenge in training such I2IT frameworks is the lack of aligned H&E-IHC image pairs, or in other words, the inconsistencies in the H&E-IHC groundtruth pairs. To explain, since re-staining a slice is physically infeasible, a matching pair of H&E-IHC slices are taken from two depth-wise *consecutive cuts* of the same tissue then stained and scanned separately. This inevitably prevents pixel-perfect image correspondences due to the slice-to-slice changes in cell morphology, staining-induced degradation (*e.g.* tissue-tearing), imaging artifacts that may vary among slices (*e.g.* camera out-of-focus) and multi-slice registration errors. An example pair of patches is shown in Fig. 1 and another pair with significant inconsistencies is shown in Fig. 2(a)(c). In the latter, comparing the groundtruth IHC image to the input H&E image, one can clearly see the inconsistencies – nearly the entire left half of the tissue present in the H&E image is missing.

As a result, recent advances in H&E-to-IHC I2IT have mostly avoided using the inconsistent GT pairs and instead have imposed the cycle-consistency constraint [6,8,13]. Moreover, existing approaches have also exploited using expert annotations such as per-cell labels [9], semantic masks [8] and patch-level labels [8,13]. As for the prior works that directly utilize the H&E-IHC pairs for supervision, a variant of Pix2Pix [4] that uses a Gaussian Pyramid based reconstruction loss to accommodate the noisy GT is proposed in [7]. However, the robustness of such approaches that punish absolute errors in the generated image to dealing with GT inconsistencies remains unclear.

In this paper, we argue that the IHC slides, despite the disparities vis-a-vis their H&E counterparts, can still serve as useful targets for stain translation. The work we present in this paper is based on the important realization that even when pairs of consecutive tissue slices do not yield images that are pixel-perfect aligned, it is highly likely that the corresponding patches in the two stains share the same diagnostic label. For example, if the levels of expression in a region of the HER2 slide are high, the corresponding region in the H&E slide is highly likely to contain a high density of cancerous cells. Therefore, we set our goal to meaningfully leverage such correlations to benefit the H&E-to-IHC I2IT while being resilient to any inconsistencies.

Toward this goal, we propose a supervised patchwise contrastive loss named the *Adaptive Supervised PatchNCE* (ASP) loss. Our formulation of this loss was inspired by the recent research findings that contrastive loss benefits model robustness under label noise [3,12]. Furthermore, based on the observation that any dissimilarity between the patch embeddings at corresponding locations in the generated and groundtruth IHC images is indicative to the level of inconsistency of the GT at that location, we employ an adaptive weighting scheme in ASP. By down-weighting the contrastive loss at locations with low similarities, *i.e.* high inconsistencies, our proposed ASP loss helps the network learn more robustly.

Lastly, to support further research in virtual IHC-restaining, we present the *Multi-IHC Stain Translation* (MIST) as a new public dataset. The MIST dataset contains 4k+ training and 1k testing aligned H&E-IHC patches for each of the following IHC stains that are critical for *breast cancer* diagnostics: HER2, Ki67, ER (Estrogen Receptor) and PR (Progesterone Receptor). We evaluated existing I2IT methods and ours for multiple IHC stains and demonstrate the superior performance achieved by our method both qualitatively and quantitatively.

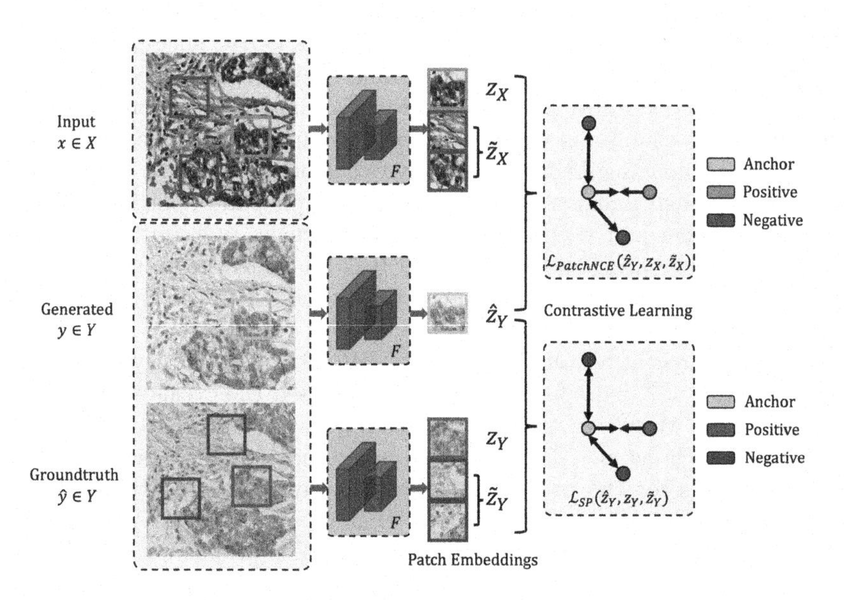

Fig. 1. Illustration of the PatchNCE loss from [11] and the Supervised PatchNCE (SP) loss. The patch embeddings $\boldsymbol{z}$ are extracted by a shared network F.

2 Method Description

2.1 The Supervised PatchNCE (SP) Loss

Before getting to our ASP loss, we need to first introduce the SP loss as a robust means to learning from inconsistent GT image pairs. The SP loss was inspired by the findings in recent literature that demonstrate the positive effect of contrastive learning on boosting model robustness against label noise [3,12,14]. It takes the same form as the PatchNCE loss as introduced in [11], except that it is applied on the generated-GT image pair (instead of the input-generated pair).

The goal of the PatchNCE loss is to ensure the content is consistent across translation by maximizing the mutual information between the input and the corresponding output. It does so by minimizing a patch-based InfoNCE loss [10], which encourages the network to associate the corresponding patches with each other in the learned embedding space, while disassociating them from the non-corresponding ones. Mathematically, the InfoNCE loss takes the form:

$$\mathcal{L}_{\text{InfoNCE}}(\boldsymbol{v}, \boldsymbol{v}^+, \boldsymbol{v}^-) = -\log \left[\frac{\exp(\boldsymbol{v} \cdot \boldsymbol{v}^+/\tau)}{\exp(\boldsymbol{v} \cdot \boldsymbol{v}^+/\tau) + \sum_{n=1}^{N} \exp(\boldsymbol{v} \cdot \boldsymbol{v}_n^-/\tau)} \right], \quad (1)$$

where $\boldsymbol{v}$, $\boldsymbol{v}^+$ and $\boldsymbol{v}^-$ are the embeddings of the anchor, positive and negative samples, respectively. With InfoNCE, the PatchNCE loss is set up as follows: given the anchor embedding $\hat{\boldsymbol{z}}_Y$ of a patch in the output image, the positive $\boldsymbol{z}_X$ is the embedding of the corresponding patch from the input image, while the negatives $\tilde{\boldsymbol{z}}_X$ are embeddings of the non-corresponding ones, *i.e.* $\mathcal{L}_{\text{PatchNCE}} = \mathcal{L}_{\text{InfoNCE}}(\hat{\boldsymbol{z}}_Y, \boldsymbol{z}_X, \tilde{\boldsymbol{z}}_X)$.

As for the SP loss, given the embedding of an output patch $\hat{\boldsymbol{z}}_Y$ as anchor, we now designate the embedding of the corresponding patch in the groundtruth image $\boldsymbol{z}_Y$ as the positive and the embeddings of the non-corresponding ones $\tilde{\boldsymbol{z}}_Y$ as the negatives. We then use the same InfoNCE-based contrastive learning objective, *i.e.* $\mathcal{L}_{\text{SP}} = \mathcal{L}_{\text{InfoNCE}}(\hat{\boldsymbol{z}}_Y, \boldsymbol{z}_Y, \tilde{\boldsymbol{z}}_Y)$. A depiction of both the PatchNCE loss and the SP loss is given in Fig. 1. It is worth noting that, despite the fact that a similar patchwise constrastive loss was proposed in [1] for supervised I2IT, it is one of our contributions in this paper to explicitly exploit the robustness of this contrastive loss in the context of H&E-to-IHC translation where the GT pairs can be highly inconsistent for reasons mentioned previously. We think that the key factor behind the robustness of $\mathcal{L}_{\text{SP}}$ to inconsistent GT compared to, say, the MSE loss, is its relativeness. Instead of using an absolute loss term that may not work well on inconsistent groundtruth pairs, $\mathcal{L}_{\text{SP}}$ punishes dissimilarities between the anchor and the positive in the learned latent space, relative to those between the anchor and the negatives.

2.2 The Adaptive Supervised PatchNCE (ASP) Loss

To learn selectively from more consistent groundtruth locations, we further propose to augment the Supervised PatchNCE loss in an adaptive manner. The key idea here is to automatically recognize patch locations that are inconsistent and adapt the SP loss so that the severely inconsistent patch locations will have lesser effects on training. To measure the consistency at a given patch location, we use the cosine similarity between the embeddings of the generated IHC patch and the corresponding GT patch. In Fig. 2, we show an example pair of generated vs GT IHC images that contain significant inconsistencies and their anchor-positive similarity heat map. For pairs of embeddings produced by a *trained* network, a high similarity value indicates good correspondence between the groundtruth patches while a low similarity value indicates inconsistencies.

Directly motivated by this observation, we first propose a weighting scheme for the SP loss. More specifically, we assign lower weights to patch locations that have low anchor-positive similarity values to alleviate the negative impacts the inconsistent targets may have on training. At training time t, the weight is a function of the anchor-positive cosine similarity. Examples of the weight function $h(\cdot)$ are shown in Fig. 3(b). The weight functions are monotonic increasing so that the more confident patch locations are always treated with more importance.

In order to make the weighting scheme work in practice, we must also account for the phase of training. The intuition is that, during the initial phase of training, the network is not going to be able to discriminate between consistent patch locations from those that are inconsistent. Additionally, as shown in Fig. 3(a), the histograms of the anchor-positive similarity evolve rather slowly over the training epochs. Therefore, it would not make sense to reinforce the weighting function in the beginning of the training as much as near the end of the training.

To that end, we further augment the weight so that it is also a function of the training iterations. Such scheduling of the weights is done so that in the beginning of the training, the weights are uniform in order not to wrongly bias the network when the embeddings are still indiscriminative. And as training progresses, the selective weighting scheme is gradually enforced so that the inconsistent patch locations are treated with reduced weights. We call this gradual process of shifting the learning focus *weight scheduling*. Let t denote the current iteration and T the total number of training iterations. Then weight scheduling is achieved by using a scheduling function $g(\frac{t}{T})$. Various options of $g(\cdot)$ are shown in Fig. 3(c). Subsequently, combining the weighting function with the scheduling function, we can write the following formula for the final weight:

$$w_t(\boldsymbol{v}, \boldsymbol{v}^+) = \left(1 - g\left(\frac{t}{T}\right)\right) \times 1.0 + g\left(\frac{t}{T}\right) \times h(\boldsymbol{v} \cdot \boldsymbol{v}^+). \tag{2}$$

We refer to the new augmented Supervised PatchNCE loss as the Adaptive Supervised PatchNCE (ASP) loss, which can be expressed as:

$$\mathcal{L}_{\text{ASP}}(G, H, X, Y, t) = \mathbb{E}_{(\boldsymbol{x},\boldsymbol{y})\sim(X,Y)} \sum_{l=1}^{L} \sum_{s=1}^{S_l} \frac{w_t(\hat{\boldsymbol{z}}_Y^{l,s}, \boldsymbol{z}_Y^{l,s})}{W_t^l} \cdot \mathcal{L}_{\text{InfoNCE}}(\hat{\boldsymbol{z}}_Y^{l,s}, \boldsymbol{z}_Y^{l,s}, \tilde{\boldsymbol{z}}_Y^{l,s}), \tag{3}$$

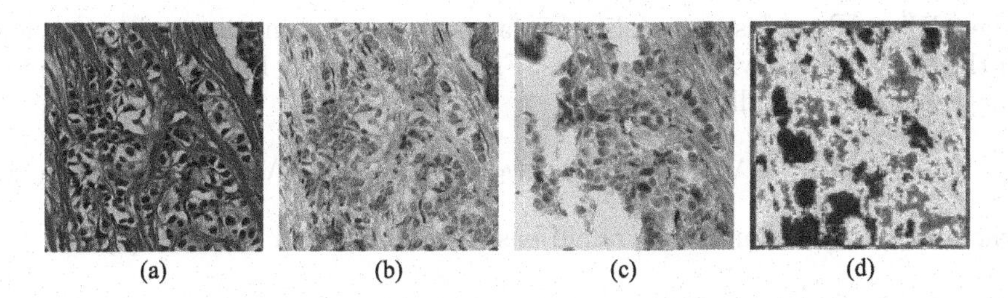

Fig. 2. (a) Input H&E image $\boldsymbol{x}$, (b) generated IHC image $\hat{\boldsymbol{y}}$, (c) groundtruth IHC image $\boldsymbol{y}$, and (d) heat map of the anchor-positive cosine similarities produced by a trained network at corresponding locations: $C_s = \boldsymbol{z}_{\hat{\boldsymbol{y}}}^s \cdot \boldsymbol{z}_{\boldsymbol{y}}^s$, where s is index of the spatial location.

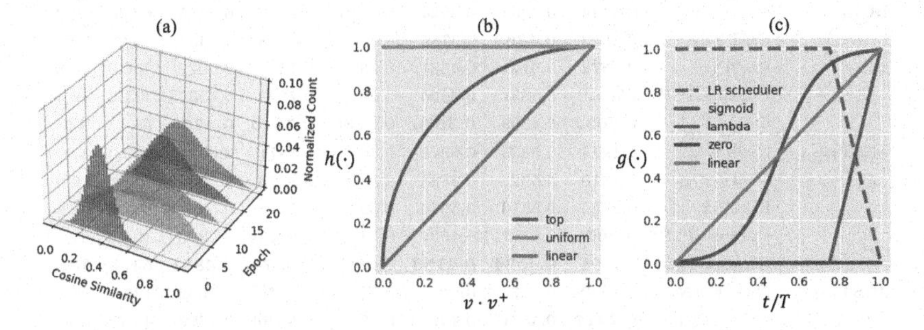

Fig. 3. (a) Histograms of the anchor-positive similarity values at different epochs; (b) The weight $h(\cdot)$ as a function of C_s; (c) The scheduling function $g(\cdot)$.

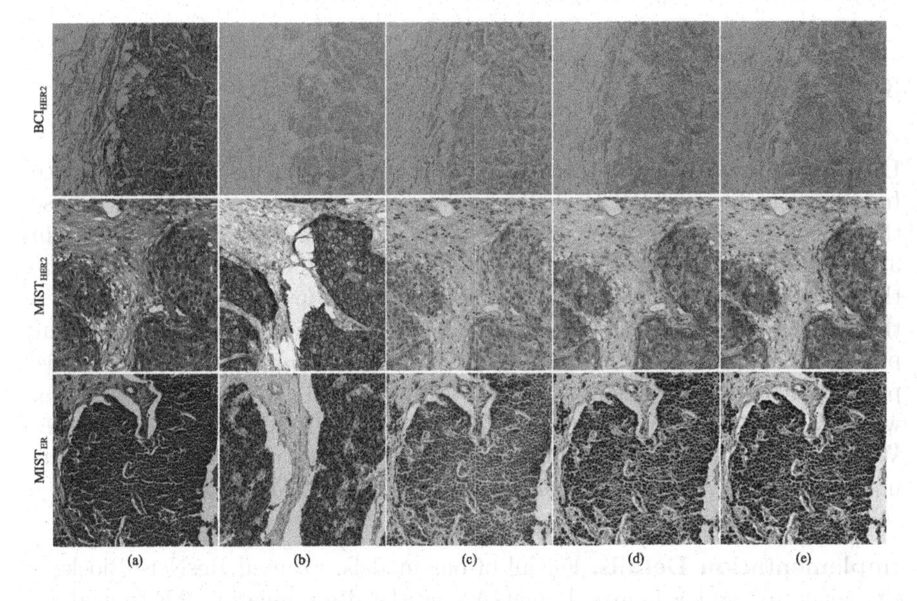

Fig. 4. Left to right: (a) Input H&E image; (b) Groundtruth IHC image; (c) Generated image without $\mathcal{L}_{\text{SP}}$; (d) With $\mathcal{L}_{\text{SP}}$; (e) With $\mathcal{L}_{\text{ASP}}^{(\text{lambda,linear})}$.

where $W_t^l = \sum_s w_t^{l,s}$ is a normalization factor that maintains the total magnitude of the loss after applying the weights. Finally, the overall learning objective for our generator is as follows:

$$\mathcal{L}_{\text{adv}} + \lambda_{\text{PatchNCE}}\mathcal{L}_{\text{PatchNCE}} + \lambda_{\text{ASP}}\mathcal{L}_{\text{ASP}} + \lambda_{\text{GP}}\mathcal{L}_{\text{GP}}, \tag{4}$$

where $\mathcal{L}_{\text{GP}}$ is the Gaussian Pyramid based reconstruction loss from [7].

Table 1. Quantitative evaluations on all three datasets using both paired and unpaired metrics. The **best** values are highlighted. Ours is with $\mathcal{L}_{\text{ASP}}^{(\text{lambda,linear})}$.

Dataset	Method	SSIM↑	$\text{PHV}_{T=0.01}$ ↓					FID↓	KID↓
			layer1	layer2	layer3	layer4	avg		
BCI_{HER2}	CycleGAN	0.4424	0.4264	0.3924	0.2610	0.7262	0.4515	**63.5**	10.7
	CUT+$\mathcal{L}_{\text{GP}}$	0.4802	**0.4263**	0.3784	0.2364	0.7328	0.4435	65.0	10.9
	Pix2Pix	0.4372	0.5121	0.4531	0.2953	0.7484	0.5022	100.0	44.6
	PyramidP2P	0.5001	0.4531	0.3826	0.2618	0.7293	0.4567	113.6	79.4
	Ours ($\mathcal{L}_{\text{ASP}}$)	**0.5032**	0.4308	**0.3670**	**0.2235**	**0.7210**	**0.4356**	65.1	**9.9**
$\text{MIST}_{\text{HER2}}$	CycleGAN	0.1914	0.5633	0.6346	0.4695	0.8871	0.6386	240.3	311.1
	CUT+$\mathcal{L}_{\text{GP}}$	0.1810	0.5321	0.4826	0.3060	0.8323	0.5383	66.8	19.0
	Pix2Pix	0.1559	0.5516	0.5070	0.3253	0.8511	0.5588	137.3	82.9
	PyramidP2P	**0.2078**	0.4787	0.4524	0.3313	0.8423	0.5262	104.0	61.8
	Ours ($\mathcal{L}_{\text{ASP}}$)	0.2004	**0.4534**	**0.4150**	**0.2665**	**0.8174**	**0.4881**	**51.4**	**12.4**
MIST_{ER}	CycleGAN	0.1982	0.5175	0.5092	0.3710	0.8672	0.5662	125.7	95.1
	CUT+$\mathcal{L}_{\text{GP}}$	**0.2217**	0.4531	0.4079	0.2725	**0.8194**	0.4882	43.7	8.7
	Pix2Pix	0.1500	0.5818	0.5282	0.3700	0.8620	0.5855	128.1	79.0
	PyramidP2P	0.2172	0.4767	0.4538	0.3757	0.8567	0.5407	107.4	84.2
	Ours ($\mathcal{L}_{\text{ASP}}$)	0.2061	**0.4336**	**0.4007**	**0.2649**	0.8205	**0.4799**	**41.4**	**5.8**

Note that KID values multiplied by 1000 are shown. CUT is from [11]

3 Experiments

Datasets. The following datasets are used in our experiments: the Breast Cancer Immunohistochemical (BCI) challenge dataset [7] and our own MIST dataset that is now in the public domain. The publicly available portion of BCI contains 3396 H H&E-HER2 patches for training and 500 of the same for testing. Note that we have additionally normalized the brightness levels of all BCI images to the same level. Due to the page limit, from the MIST dataset, here we only present detailed results on HER2 and ER. For $\text{MIST}_{\text{HER2}}$, we extracted 4642 paired patches for training and 1000 for testing from 64 WSIs. And for MIST_{ER}, we extracted 4153 patches for training, and 1000 for testing from 56 WSIs. All WSIs were taken at 20× magnification. All patches are of size 1024 × 1024 and non-overlapping. *Additional results on* $\text{MIST}_{\text{Ki67}}$ *and* MIST_{PR} *are provided in the Supplementary Materials.*

Implementation Details. For all of our models, we used ResNet-6Blocks as the generator and a 5-layer PatchGAN as the discriminator. We trained our networks with random 512 × 512 crops and a batch size of one. The Adam

Table 2. Ablation studies comparing different adaptive strategies for the SP loss. The **best** and second best values are highlighted. Note that $\mathcal{L}_{\text{SP}}$ is $\mathcal{L}_{\text{ASP}}^{(\text{zero,uniform})}$.

Dataset	Method	SSIM↑	$\text{PHV}_{T=0.01}$ ↓					FID↓	KID↓
			layer1	layer2	layer3	layer4	avg		
BCI_{HER2}	$\mathcal{L}_{\text{SP}}$	0.5094	0.4413	0.3771	0.2273	**0.7202**	0.4415	62.7	11.4
	$\mathcal{L}_{\text{ASP}}^{(\text{lambda,top})}$	0.5206	0.4411	0.3813	0.2282	0.7244	0.4438	62.1	12.8
	$\mathcal{L}_{\text{ASP}}^{(\text{lambda,lin.})}$	0.5032	**0.4308**	**0.3670**	**0.2235**	0.7210	**0.4356**	65.1	9.9
	$\mathcal{L}_{\text{ASP}}^{(\text{sigmoid,top})}$	**0.5236**	0.4503	0.3877	0.2331	0.7292	0.4501	65.9	12.3
	$\mathcal{L}_{\text{ASP}}^{(\text{linear,top})}$	0.4890	0.4327	0.3828	0.2305	0.7318	0.4445	**61.9**	**9.8**
$\text{MIST}_{\text{HER2}}$	$\mathcal{L}_{\text{SP}}$	**0.2159**	0.4712	0.4243	0.2611	0.8129	0.4924	55.6	20.8
	$\mathcal{L}_{\text{ASP}}^{(\text{lambda,top})}$	0.2035	**0.4451**	**0.4068**	**0.2554**	**0.8117**	**0.4798**	51.2	16.7
	$\mathcal{L}_{\text{ASP}}^{(\text{lambda,lin.})}$	0.2004	0.4534	0.4150	0.2665	0.8174	0.4881	51.5	12.4
	$\mathcal{L}_{\text{ASP}}^{(\text{sigmoid,top})}$	0.2086	0.4655	0.4191	0.2581	0.8138	0.4891	**45.2**	**11.5**
	$\mathcal{L}_{\text{ASP}}^{(\text{linear,top})}$	0.1809	0.4766	0.4262	0.2667	0.8178	0.4968	68.8	28.9
MIST_{ER}	$\mathcal{L}_{\text{SP}}$	**0.2236**	0.4517	0.4117	0.2714	0.8208	0.4889	46.4	12.5
	$\mathcal{L}_{\text{ASP}}^{(\text{lambda,top})}$	0.2096	0.4388	0.4052	0.2676	0.8215	0.4833	42.4	8.1
	$\mathcal{L}_{\text{ASP}}^{(\text{lambda,lin.})}$	0.2061	**0.4336**	0.4007	**0.2649**	**0.8205**	**0.4799**	**41.4**	**5.8**
	$\mathcal{L}_{\text{ASP}}^{(\text{sigmoid,top})}$	0.2192	0.4376	**0.3965**	0.2684	0.8215	0.4810	43.6	8.1
	$\mathcal{L}_{\text{ASP}}^{(\text{linear,top})}$	0.1981	0.4581	0.4072	0.2706	0.8217	0.4894	46.9	10.7

optimizer [5] was used with a linear decay scheduler (as shown in Fig. 3(c)) and an initial learning rate of 2×10^{-4}. The hyperparameters in Eq. (4) are set as: $\lambda_{\text{PatchNCE}} = 10.0$, $\lambda_{\text{ASP}} = 10.0$ and $\lambda_{\text{GP}} = 10.0$.

Evaluation Metrics. We compare the methods using both paired and unpaired evaluation metrics. To compare a pair of images, generated and groundtruth, we use the standard SSIM (Structural Similarity Index Measure) and PHV (Perceptual Hash Value) as described in [8]. As for the unpaired metrics, we use the FID (Fréchet Inception Distance) and the KID (Kernel Inception Distance).

Qualitative Evaluations. In Fig. 4, we compare visually the generated IHC images by our framework. It can be observed that by using either $\mathcal{L}_{\text{SP}}$ or $\mathcal{L}_{\text{ASP}}$, the pathological representations in the generated images are significantly more accurate. And by using $\mathcal{L}_{\text{ASP}}$, such representations appear to be more consistent.
Quantitative Evaluations. The full results comparing existing I2IT methods to ours are tabulated in Tab. 1. Overall, it can be observed that the proposed framework with the ASP loss consistently outperforms existing methods across all three datasets. For those methods, Fig. 5 visually illustrates the extent of hallucinations which we believe is the reason for their poor quantitative performance. Subsequently, in Tab. 2, we further provide results using different weighting and scheduling functions in our proposed $\mathcal{L}_{\text{ASP}}$. With $\mathcal{L}_{\text{SP}}$ already being a strong baseline, using different adaptive strategies can provide further gains in performance. It is also worth noting that if reinforced prematurely, adaptive weighting can lead to inferior convergence, *e.g.* $\mathcal{L}_{\text{ASP}}^{(\text{linear,top})}$.

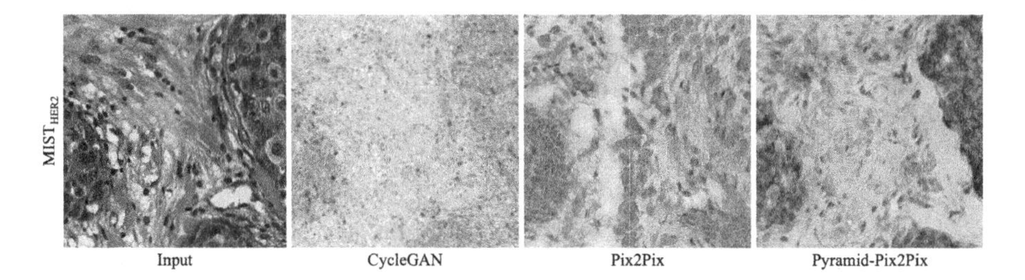

Fig. 5. Example H&E-to-HER2 translations by three baselines. False morphological alterations cause the tissue structures in the translated images to no longer match those in the input H&E image, especially the nuclei. Zoom in for sharper visualization.

4 Conclusion

In this paper, we have proposed the Adaptive Supervised PatchNCE (ASP) loss for learning H&E-to-IHC stain translation with inconsistent GT image pairs. The adaptive logic in ASP is based on the intuition that inconsistent patch locations should contribute less to learning. We demonstrated that our proposed framework is able to achieve significant improvements both qualitatively and quantitatively over the existing approaches for translations to multiple IHC stains. Finally, we have made public our Multi-IHC Stain Translation dataset with the hope to assist further research towards accurate H&E-to-IHC stain translation.

References

1. Andonian, A., Park, T., Russell, B., Isola, P., Zhu, J.Y., Zhang, R.: Contrastive feature loss for image prediction. In: Proceedings of the IEEE/CVF International Conference on Computer Vision, pp. 1934–1943 (2021)
2. Anglade, F., Milner, D.A., Jr., Brock, J.E.: Can pathology diagnostic services for cancer be stratified and serve global health? Cancer **126**, 2431–2438 (2020)
3. Ghosh, A., Lan, A.: Contrastive learning improves model robustness under label noise. In: Proceedings of the IEEE/CVF Conference on Computer Vision and Pattern Recognition, pp. 2703–2708 (2021)
4. Isola, P., Zhu, J.Y., Zhou, T., Efros, A.A.: Image-to-image translation with conditional adversarial networks. In: Proceedings of the IEEE Conference on Computer Vision and Pattern Recognition, pp. 1125–1134 (2017)
5. Kingma, D.P., Ba, J.: Adam: a method for stochastic optimization. arXiv preprint arXiv:1412.6980 (2014)
6. Lin, Y., et al.: Unpaired multi-domain stain transfer for kidney histopathological images. In: Proceedings of the AAAI Conference on Artificial Intelligence, vol. 36, pp. 1630–1637 (2022)
7. Liu, S., Zhu, C., Xu, F., Jia, X., Shi, Z., Jin, M.: Bci: breast cancer immunohistochemical image generation through pyramid pix2pix. In: Proceedings of the IEEE/CVF Conference on Computer Vision and Pattern Recognition, pp. 1815–1824 (2022)

8. Liu, S., et al.: Unpaired stain transfer using pathology-consistent constrained generative adversarial networks. IEEE Trans. Med. Imaging **40**(8), 1977–1989 (2021)
9. Liu, Y.: Predict ki-67 positive cells in h&e-stained images using deep learning independently from ihc-stained images. Front. Mol. Biosci. **7**, 183 (2020)
10. Oord, A.v.d., Li, Y., Vinyals, O.: Representation learning with contrastive predictive coding. arXiv preprint arXiv:1807.03748 (2018)
11. Park, T., Efros, A.A., Zhang, R., Zhu, J.-Y.: Contrastive learning for unpaired image-to-image translation. In: Vedaldi, A., Bischof, H., Brox, T., Frahm, J.-M. (eds.) ECCV 2020. LNCS, vol. 12354, pp. 319–345. Springer, Cham (2020). https://doi.org/10.1007/978-3-030-58545-7_19
12. Xue, Y., Whitecross, K., Mirzasoleiman, B.: Investigating why contrastive learning benefits robustness against label noise. In: International Conference on Machine Learning, pp. 24851–24871. PMLR (2022)
13. Zeng, B., et al.: Semi-supervised pr virtual staining for breast histopathological images. In: Medical Image Computing and Computer Assisted Intervention-MICCAI 2022: 25th International Conference, Singapore, 18–22 September 2022, Proceedings, Part II, pp. 232–241. Springer (2022). https://doi.org/10.1007/978-3-031-16434-7_23
14. Zheltonozhskii, E., Baskin, C., Mendelson, A., Bronstein, A.M., Litany, O.: Contrast to divide: self-supervised pre-training for learning with noisy labels. In: Proceedings of the IEEE/CVF Winter Conference on Applications of Computer Vision, pp. 1657–1667 (2022)

Deep Learning for Tumor-Associated Stroma Identification in Prostate Histopathology Slides

Zichen Wang[1,2], Mara Pleasure[1,3], Haoyue Zhang[1,2], Kimberly Flores[1,4], Anthony Sisk[1,4], William Speier[1,2,3,5], and Corey W. Arnold[1,2,3,4,5](✉)

[1] Computational Diagnostics Lab, UCLA, Los Angeles, USA
[2] The Department of Bioengineering, UCLA, Los Angeles, USA
[3] Medical Informatics Home Area, Graduate Programs in Bioscience, UCLA, Los Angeles, USA
[4] The Department of Pathology and Laboratory Medicine, UCLA, Los Angeles, USA
[5] The Department of Radiological Sciences, UCLA, Los Angeles, USA
cwarnold@ucla.edu

Abstract. The diagnosis of prostate cancer is driven by the histopathological appearance of epithelial cells and epithelial tissue architecture. Despite the fact that the appearance of the tumor-associated stroma contributes to diagnostic impressions, its assessment has not been standardized. Given the crucial role of the tumor microenvironment in tumor progression, it is hypothesized that the morphological analysis of stroma could have diagnostic and prognostic value. However, stromal alterations are often subtle and challenging to characterize through light microscopy alone. Emerging evidence suggests that computerized algorithms can be used to identify and characterize these changes. This paper presents a deep-learning approach to identify and characterize tumor-associated stroma in multi-modal prostate histopathology slides. The model achieved an average testing AUROC of 86.53% on a large curated dataset with over 1.1 million stroma patches. Our experimental results indicate that stromal alterations are detectable in the presence of prostate cancer and highlight the potential for tumor-associated stroma to serve as a diagnostic biomarker in prostate cancer. Furthermore, our research offers a promising computational framework for in-depth exploration of the field effect and tumor progression in prostate cancer.

Keywords: Field effect · Tumor-associated stroma · Prostate cancer

1 Introduction

Prostate cancer (PCa) diagnosis and grading rely on histopathology analysis of biopsy slides [1]. However, prostate biopsies are known to have sampling error as PCa is heterogenous and commonly multifocal, meaning cancer legions can be missed during the biopsy procedure [2]. If significant PCa is detected on

H. Greenspan et al. (Eds.): MICCAI 2023, LNCS 14225, pp. 642–651, 2023.
https://doi.org/10.1007/978-3-031-43987-2_62

biopsies and the patient has organ-confined cancer with no contraindications, radical prostatectomy (RP) is the standard of care [3,4]. Following RP, the prostate is processed and slices are mounted onto slides for analysis. Radical prostatectomy histopathology samples are essential for validating the biopsy-determined grade group [5,6]. Analysis of whole-mount slides, meaning slides that include slices of the entire prostate, provide more precise tumor boundary detection, identification of various tumor foci, and increased tissue for identifying morphological patterns not visible on biopsy due to a larger field of view.

Field effect refers to the spread of genetic and epigenetic alterations from a primary tumor site to surrounding normal tissues, leading to the formation of secondary tumors. Understanding field effect is essential for cancer research as it provides insights into the mechanisms underlying tumor development and progression. Tumor-associated stroma, which consists of various cell types, such as fibroblasts, smooth muscle cells, and nerve cells, is an integral component of the tumor microenvironment that plays a critical role in tumor development and progression. Reactive stroma, a distinct phenotype of stromal cells, arises in response to signaling pathways from cancerous cells and is characterized by altered stromal cells and increased extracellular matrix components [7,8]. Reactive stroma is often associated with tumor-associated stroma and is thought to be a result of field effects in prostate cancer. Altered stroma can create a pro-tumorigenic environment by producing a multitude of chemokines, growth factors, and releasing reactive oxygen species [9,10], which can lead to tumor development and aggressiveness [11]. Therefore, investigating the histological characterization of tumor-associated stroma is crucial in gaining insights into the field effect and tumor progression of prostate cancer.

Manual review for tumor-associated stroma is time-consuming and lacks quantitative metrics [12,13]. Several automated methods have been applied to analyze the tumor-stroma relationship; however, most of them focus on identifying a tumor-stroma ratio rather than finding reactive stroma tissue or require pathologist input. Machine learning algorithms have been used to quantify the percentage of tumor to stroma in bladder cancer patients, but required dichotomizing patients based on a threshold [14]. Software has been used to segment tumor and stroma tissue in breast cancer patient samples, but the method required constant supervision by a pathologist [15]. Similarly, Akoya Biosciences Inform software was used to quantify reactive stroma in PCa, but this method required substantial pathologist input to train the software [16]. Fully automated deep-learning methods have been developed to identify tumor-associated stroma in breast cancer biopsies, achieving an AUC of 0.962 in predicting invasive ductal cancer [13]. However, identifying tumor-associated stroma in prostate biopsies and whole-mount histopathology slides remains challenging.

Analyzing tumor-associated stroma in prostate cancer requires combining whole-mount and biopsy histopathology slides. Biopsy slides provide information on the presence of PCa, while whole-mount slides provide information on the extent and distribution of PCa, including more information on tumor-associated stroma. Combining the information from both modalities can provide a more accurate understanding of the tumor microenvironment. In this work, we explore the field effect in prostate cancer by analyzing tumor-associated stroma in multimodal histopathological images. Our main contributions can be summarized as follows:

- To the best of our knowledge, we present the first deep-learning approach to characterize prostate tumor-associated stroma by integrating histological image analysis from both whole-mount and biopsy slides. Our research offers a promising computational framework for in-depth exploration of the field effect and cancer progression in prostate cancer.
- We proposed a novel approach for stroma classification with spatial graphs modeling, which enable more accurate and efficient analysis of tumor microenvironment in prostate cancer pathology. Given the spatial nature of cancer field effect and tumor microenvironment, our graph-based method offers valuable insights into stroma region analysis.
- We developed a comprehensive pipeline for constructing tumor-associated stroma datasets across multiple data sources, and employed adversarial training and neighborhood consistency regularization techniques to learn robust multimodal-invariant image representations.

2 Method

2.1 Stroma Tissue Segmentation

Accurately analyzing tumor-associated stroma requires a critical pre-processing step of segmenting stromal tissue from the background, including epithelial tissue. This segmentation task is challenging due to the complex and heterogeneous appearance of the stroma. To address this, we propose utilizing the PointRend model [17], which can handle complex shapes and appearances and produce smooth and accurate segmentations through iterative object boundary refinement. Moreover, the model's efficiency and ability to process large images quickly make it suitable for analyzing whole-mount slides. By leveraging the PointRend model, we can generate stromal segmentation masks for more precise downstream analysis.

2.2 Stroma Classification with Spatial Patch Graphs

To capture the spatial nature of field effect and analyze tumor-associated stroma, modeling spatial relationships between stroma patches is essential. The spatial relationship can reveal valuable information about the tumor microenvironment, and neighboring stroma cells can undergo similar phenotypic changes in response to cancer. Therefore, we propose using a spatial patch graph to capture the high-order relationship among stroma tissue regions. We construct the stroma patch graph using a K-nearest neighbor (KNN) graph and neighbor sampling. The KNN graph connects each stroma patch to its K nearest neighboring patches. Given a central stroma patch, we iteratively add neighboring patches to construct

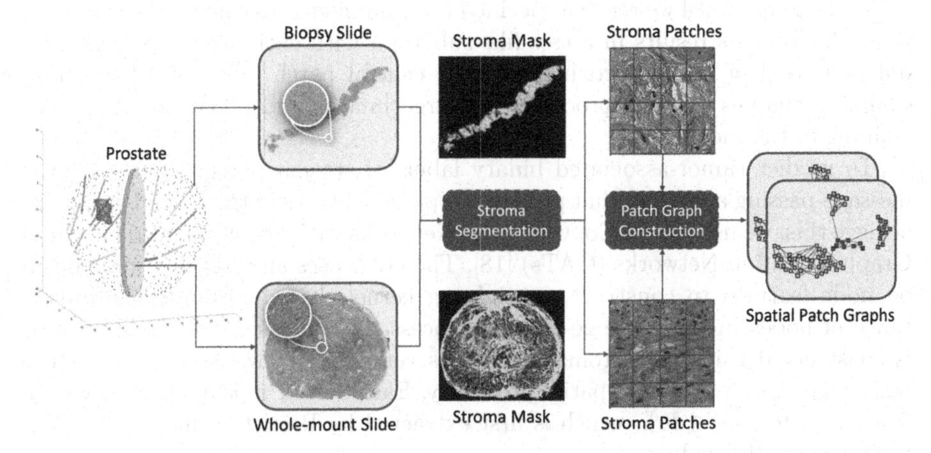

Fig. 1. Process of stroma patch graph construction: The left prostate model illustrates the locations of biopsy and whole-mount tissues in 3D space. Biopsy slides provide a targeted view while whole-mount slides offer a broader perspective of the tumor and surrounding tissue. The stroma segmentation module generates a stroma mask to isolate the stromal tissue, which is then used to construct spatial patch graphs for the proposed deep-learning model.

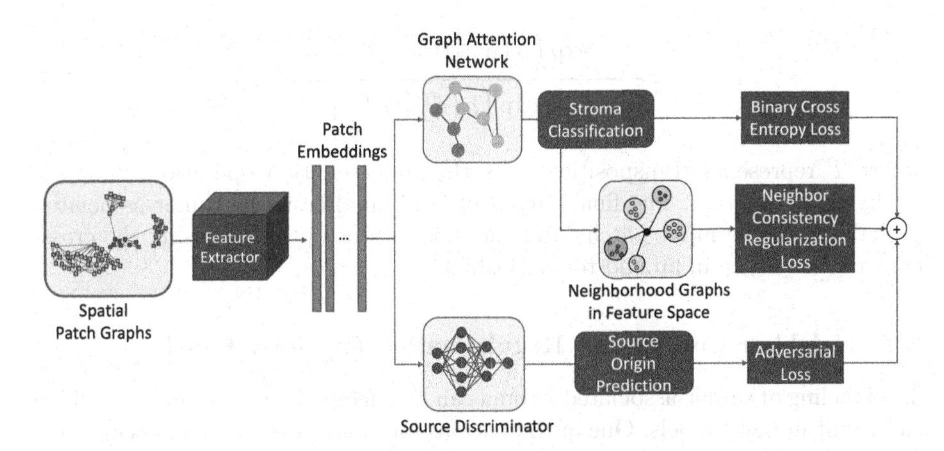

Fig. 2. Overview of the proposed model for identifying tumor-associated stroma in multi-modal prostate histopathology slides: The input patches are represented as spatial graphs and passed through a feature extractor. The patch embeddings are fed into a graph attention network (GAT) module to capture inter-patch relationships and refine the features with neighborhood consistency regularization (NCR) for handling noisy labels. The source discriminator serves as adversarial multi-modal learning (AML) module to predict data source (biopsy/whole-mount). The stroma classifier and source discriminator are trained simultaneously with the goal of successfully classifying tumor-associated stroma using multimodal-invariant features.

the patch graph until we reach a specified layer number L to control the subgraph size. This process results in a tree-like subgraph with each layer representing a different level of spatial proximity to the central patch. The use of neighbor sampling enables efficient processing of large images and allows for stochastic training of the model.

To predict tumor-associated binary labels of stroma patches, we employ a message-passing approach that propagates patch features in the spatial graph. To achieve this, we use Graph Convolutional Networks with attention, also known as Graph Attention Networks (GATs) [18]. The GAT uses an attention mechanism on node features to construct a weighting kernel that determines the importance of nodes in the message-passing process. In our case, the patch graph $\mathcal{G}$ is constructed using the stroma patches as vertices, and we connect the nodes with edges based on their spatial proximity. Each vertex v_i is associated with a feature vector $\vec{h}_{vi} \in \mathbb{R}^N$, which is first extracted by Resnet-50 model [19]. The GAT layer is defined as

$$g_{\mathcal{E}}(v_i) = \sum_{v_j \in \mathcal{N}_{v_i}^{\mathcal{E}} \cup \{v_i\}} \alpha_{v_i,v_j} W \vec{h}_{v_j} \tag{1}$$

where $W \in \mathbb{R}^{M \times N}$ is a learnable matrix transforming N-dimensional features to M-dimensional features. $\mathcal{N}_{v_i}^{\mathcal{E}}$ is the neighborhood of the node v_i connected by $\mathcal{E}$ in $\mathcal{G}$. GAT uses attention mechanism to construct the weighting coefficients as:

$$\alpha_{v_i,v_j} = \frac{\exp\left(\rho\left(\vec{a}^T\left[W\vec{h}_{v_i} \| W\vec{h}_{v_j}\right]\right)\right)}{\sum_{v_k \in \mathcal{N}_{v_i}^{\mathcal{E}}} \exp\left(\rho\left(\vec{a}^T\left[W\vec{h}_{v_i} \| W\vec{h}_{v_k}\right]\right)\right)} \tag{2}$$

where T represents transposition, $\|$ is the concatenation operation, and ρ is LeakyReLU function. The final output of GAT module is the tumor-associated probability of the input patch. And the module was optimized using the cross-entropy loss L_{GAT} in an end-to-end fashion.

2.3 Neighbor Consistency Regularization for Noisy Labels

The labeling of tumor-associated stroma can be affected by various factors, which can result in noisy labels. One of the reasons for noisy labels is the irregular distribution of the field effect, which makes it challenging to define a clear boundary between the tumor-associated and normal stroma regions. Additionally, the presence of tumor heterogeneity and the varied distribution of tumor foci can further complicate the labeling process.

To address this issue, we propose applying Neighbor Consistency Regularization (NCR) [20] to prevent the model from overfitting to incorrect labels. The assumption is that overfitting happens to a lesser degree before the final classifier, and this is supported by MOIT [21], which suggests that feature representations are capable of distinguishing between noisy and clean examples during model training. Based on this assumption, NCR introduces a neighbor consistency loss to encourage similar predictions of stroma patches that are similar in feature space. This loss penalizes the divergence of a patch prediction from a weighted combination of its neighbors' predictions in feature space, where the weights are determined by their feature similarity. Specifically, the loss function is designed as follows:

$$L_{\text{NCR}} = \frac{1}{m} \sum_{i=1}^{m} D_{\text{KL}} \left(\sigma\left(\mathbf{z}_i / T\right) \| \sum_{j \in \text{NN}_k(\mathbf{v}_i)} \frac{s_{i,j}}{\sum_k s_{i,k}} \cdot \sigma\left(\mathbf{z}_j / T\right) \right) \tag{3}$$

where D_{KL} is the KL-divergence loss to quantify the discrepancy between two probability distributions, T represents the temperature and $\text{NN}_k(\mathbf{v}_i)$ is the set of k nearest neighbors of v_i in the feature space.

2.4 Adversarial Multi-modal Learning

Biopsy and whole-mount slides provide complementary multi-modal information on the tumor microenvironment, and combining them can provide a more comprehensive understanding of tumor-associated stroma. However, using data from multiple modalities can introduce systematic shifts, which can impact the performance of a deep learning model. Specifically, whole-mount slides typically contain larger tissue sections and are processed using different protocols than biopsy slides, which can result in differences in image quality, brightness, and contrast. These technical differences can affect the pixel intensity distributions of the images, leading to systematic shifts in the features that the deep learning model learns to associate with tumor-associated stroma. For instance, a model trained on whole-mount slides only may not generalize well to biopsy slides due to systematic shifts, hindering model performance in the clinical application scenario.

To address the above issues, we propose an Adversarial Multi-modal Learning (AML) module to force the feature extractor to produce multimodal-invariant representations on multiple source images. Specifically, we incorporate a source discriminator adversarial neural network as auxiliary classifier. The module takes the stroma embedding as an input and predicts the source of the image (biopsy or whole-mount) using Multilayer Perceptron (MLP) with cross-entropy loss function L_{AML}. The overall loss function of the entire model is computed as:

$$L_{\text{Total}} = L_{\text{GAT}} + \alpha \cdot L_{\text{NCR}} - \beta \cdot L_{\text{AML}} \tag{4}$$

where hyper-parameters α and β control the impact of each loss term. All modules were concurrently optimized in an end-to-end manner. The stroma classifier and source discriminator are trained simultaneously, aiming to effectively classify tumor-associated stroma while impeding accurate source prediction by the discriminator. The optimization process aims to achieve a balance between these two goals, resulting in an embedding space that encodes as much information as possible about tumor-associated stroma identification while not encoding any information on the data source. By adopting the adversarial learning strategy, our model can maintain the correlated information and shared characteristics between two modalities, which will enhance the model's generalization and robustness.

3 Experiment

3.1 Dataset

In our study, we utilized three datasets for tumor-associated stroma analysis. (1) Dataset A comprises 513 tiles extracted from the whole mount slides of 40 patients, sourced from the archives of the Pathology Department at Cedars-Sinai Medical Center (IRB# Pro00029960). It combines two sets of tiles: 224 images from 20 patients featuring stroma, normal glands, low-grade and high-grade cancer [22], along with 289 images from 20 patients with dense high-grade cancer (Gleason grades 4 and 5) and cribriform/non-cribriform glands [23]. Each tile measures 1200×1200 pixels and is extracted from whole slide images captured at 20x magnification (0.5 microns per pixel). The tiles were annotated at the pixel-level by expert pathologists to generate stroma tissue segmentation masks and were cross-evaluated and normalized to account for stain variability. (2) Dataset B included 97 whole mount slides with an average size of over 174,000×142,000 pixels at 40x magnification. The prostate tissue within these slides had an average tumor area proportion of 9%, with an average tumor area of 77 square mm. An expert pathologist annotated the tumor region boundaries at the region-level, providing exhaustive annotations for all tumor foci. (3) Dataset C comprised 6134 negative biopsy slides obtained from 262 patients' biopsy procedures, where all samples were diagnosed as negative. These slides are presumed to contain predominantly normal stroma tissues without phenotypic alterations in response to cancer.

Dataset A was utilized for training the stroma segmentation model. Extensive data augmentation techniques, such as image scaling and staining perturbation, were employed during the training process. The model achieved an average test Dice score of 95.57 ± 0.29 through 5-fold cross-validation. This model was then applied to generate stroma masks for all slides in Datasets B and C. To precisely isolate stroma tissues and avoid data bleeding from epithelial tissues, we only extracted patches where over 99.5% of the regions were identified as stroma at 40X magnification to construct the stroma classification dataset.

For positive tumor-associated stroma patches, we sampled patches near tumor glands within annotated tumor region boundaries, as we presumed that tumor regions represent zones in which the greatest amount of damage has progressed. For negative stroma patches, we calculated the tumor distance for each patch by measuring the Euclidean distance from the patch center to the nearest edge of the labeled tumor regions. Negative stroma patches were then sampled from whole mount slides with a Gleason Group smaller than 3 and a tumor distance larger than 5 mm. This approach aims to minimize the risk of mislabeling tumor-associated stroma as normal tissue. Setting a 5mm threshold accounts for the typically minimal inflammatory responses induced by prostate cancers, particularly in lower-grade cases. To incorporate multi-modal information, we randomly sampled negative stroma patches from all biopsy slides in Dataset C. Overall, we selected over 1.1 million stroma patches of size 256×256 pixels at 40x magnification for experiments. During model training and testing, we performed stain normalization and standard image augmentation methods.

3.2 Model Training and Evaluation

For constructing KNN-based patch graphs, we limited the graph size by setting $K = 4$ and layer number $L = 3$. We controlled the strength of the NCR and AML terms by setting $\alpha = 0.25$ and $\beta = 0.5$, respectively. The Adam optimizer with a learning rate of 0.0005 was used for model training. All models were implemented using PyTorch on a single Tesla V100 GPU. To evaluate the model performance, we perform 5-fold cross-validation, where all slides are stratified by source origin and divided into 5 subsets. In each cross-validation trial, one subset was taken as the test set while the remaining subsets constituted the training set. We measure the prediction performance using the area under the receiver operating characteristic (AUROC), F1 score, precision, and recall.

4 Results and Discussions

Table 1. Performance comparison with model variants. Results are averaged over 5 folds and shown in terms of mean value ± standard deviation.

Methods	AUROC	F1	Precision	Recall
Base	76.49 ± 2.19	68.29 ± 1.04	69.73 ± 1.27	66.93 ± 1.39
Base+GAT	80.02 ± 1.96	72.56 ± 0.72	73.69 ± 0.63	71.47 ± 1.21
Base+GAT+AML	85.74 ± 1.18	76.73 ± 1.10	76.47 ± 1.55	77.02 ± 1.54
Base+GAT+AML+NCR	**86.53 ± 0.38**	**79.26 ± 0.36**	**78.45 ± 0.49**	**80.10 ± 0.32**

To evaluate the effectiveness of our proposed method, we conducted an ablation study by comparing the performance of different model variants presented in Table 1. Specifically, the base model is the ResNet-50 feature extractor for tumor-associated stroma classification. Each model variant included a different combination of modules presented in method sections. We systematically add one or more modules to the base model to evaluate their performance contribution. The results show that the full model outperforms the base model by a large margin with 10.04% in AUROC and 10.97% in F1 score, and each module contributes to the overall performance. Compared to the base model, the addition of the GAT module resulted in a significant improvement in all metrics, suggesting spatial information captured by the patch graph was valuable for stroma classification. The most notable performance improvement was achieved by the AML module, with a 5.72% increase in AUROC and 5.55% increase in

Recall. This improvement indicates that AML helps the model better capture the multimodal-invariant features that are associated with tumor-associated stroma while reducing the false negative prediction by eliminating the influence of systematic shift cross modalities. Finally, the addition of the NCR module further increased the average model performance and improved the model robustness across 5 folds. This suggests that NCR was effective in handling noisy labels and improving model's generalization ability.

In conclusion, our study introduced a deep learning approach to accurately characterize the tumor-associated stroma in multi-modal prostate histopathology slides. Our experimental results demonstrate the feasibility of using deep learning algorithms to identify and quantify subtle stromal alterations, offering a promising tool for discovering new diagnostic and prognostic biomarkers of prostate cancer. Through exploring field effect in prostate cancer, our work provides a computational system for further analysis of tumor development and progression. Future research can focus on validating our approach on larger and more diverse datasets and expanding the method to a patient-level prediction system, ultimately improving prostate cancer diagnosis and treatment.

References

1. Merriel, S.W.D., Funston, G., Hamilton, W.: Prostate cancer in primary care. Adv. Therapy **35**, 1285–1294 (2018)
2. Montironi, R., Beltran, A.L., Mazzucchelli, R., Cheng, L., Scarpelli, M.: Handling of radical prostatectomy specimens: total embedding with large-format histology. Inter. J. Breast Cancer, vol. 2012 (2012)
3. Han, W., et al.: Histologic tissue components provide major cues for machine learning-based prostate cancer detection and grading on prostatectomy specimens. Sci. Rep. **10**(1), 9911 (2020)
4. Kirby, R.S., Patel, M.I., Poon, D. M.C.: Fast facts: prostate cancer: If, when and how to intervene. Karger Medical and Scientific Publishers (2020)
5. Nayyar, R., et al.: Upgrading of gleason score on radical prostatectomy specimen compared to the pre-operative needle core biopsy: an indian experience. Indian J. Urology: IJU: J. Urological Soc. India **26**(1), 56 (2010)
6. Jang, W.S., et al.: The prognostic impact of downgrading and upgrading from biopsy to radical prostatectomy among men with gleason score 7 prostate cancer. Prostate **79**(16), 1805–1810 (2019)
7. Bonollo, F., Thalmann, G.N., Julio, M.K.-d, Karkampouna, S.: The role of cancer-associated fibroblasts in prostate cancer tumorigenesis. Cancers **12**(7):1887, 2020
8. Barron, D.A., Rowley, D.R.: The reactive stroma microenvironment and prostate cancer progression. Endocr. Relat. Cancer **19**(6), R187–R204 (2012)
9. Levesque, C., Nelson, P.S.: Cellular constituents of the prostate stroma: key contributors to prostate cancer progression and therapy resistance. Cold Spring Harbor Perspect. Med. **8**(8), a030510 (2018)
10. Liao, Z., Chua, D., Tan, N.S.: Reactive oxygen species: a volatile driver of field cancerization and metastasis. Mol. Cancer **18**, 1–10 (2019)
11. Hayward, S.W., et al.: Malignant transformation in a nontumorigenic human prostatic epithelial cell line. Cancer Res. **61**(22), 8135–8142 (2001)

12. Vivar, A.D.D., et al.: Histologic features of stromogenic carcinoma of the prostate (carcinomas with reactive stroma grade 3). Hum. Pathol. **63**, 202–211 (2017)
13. Bejnordi, B.E., et al.: Using deep convolutional neural networks to identify and classify tumor-associated stroma in diagnostic breast biopsies. Mod. Pathol. **31**(10), 1502–1512 (2018)
14. Zheng, Q., et al.: Machine learning quantified tumor-stroma ratio is an independent prognosticator in muscle-invasive bladder cancer. Int. J. Mol. Sci. **24**(3), 2746 (2023)
15. Millar, E.K.A., et al.: Tumour stroma ratio assessment using digital image analysis predicts survival in triple negative and luminal breast cancer. Cancers **12**(12), 3749 (2020)
16. Ruder, S., et al.: Development and validation of a quantitative reactive stroma biomarker (qrs) for prostate cancer prognosis. Hum. Pathol. **122**, 84–91 (2022)
17. Kirillov, A., Wu, Y., He, K., Girshick, R.: Pointrend: image segmentation as rendering. In: Proceedings of the IEEE/CVF Conference on Computer Vision and Pattern Recognition, pp. 9799–9808, 2020
18. Veličković, P., Cucurull, G., Casanova, A., Romero, A., Lio, P., Bengio, Y.: Graph attention networks. arXiv preprint arXiv:1710.10903 (2017)
19. He, K., Zhang, X., Ren, S., Sun, J.: Deep residual learning for image recognition. In: Proceedings of the IEEE Conference on Computer Vision and Pattern Recognition, pp. 770–778 (2016)
20. Iscen, A., Valmadre, J., Arnab, A., Schmid, C.: Learning with neighbor consistency for noisy labels. In: Proceedings of the IEEE/CVF Conference on Computer Vision and Pattern Recognition, pp. 4672–4681 (2022)
21. Ortego, D., Arazo, E., Albert, P., O'Connor, N.E., McGuinness, K.: Multi-objective interpolation training for robustness to label noise. In: Proceedings of the IEEE/CVF Conference on Computer Vision and Pattern Recognition, pp. 6606–6615 (2021)
22. Gertych, A., et al.: Machine learning approaches to analyze histological images of tissues from radical prostatectomies. Comput. Med. Imaging Graph. **46**, 197–208 (2015)
23. Ing, N., et al.: Semantic segmentation for prostate cancer grading by convolutional neural networks. In: Medical Imaging 2018: Digital Pathology, vol. 10581, pp. 343–355. SPIE (2018)

Robust Cervical Abnormal Cell Detection via Distillation from Local-Scale Consistency Refinement

Manman Fei[1], Xin Zhang[1], Maosong Cao[2], Zhenrong Shen[1], Xiangyu Zhao[1], Zhiyun Song[1], Qian Wang[2], and Lichi Zhang[1](✉)

[1] School of Biomedical Engineering, Shanghai Jiao Tong University, Shanghai, China
lichizhang@sjtu.edu.cn
[2] School of Biomedical Engineering, ShanghaiTech University, Shanghai, China

Abstract. Automated detection of cervical abnormal cells from Thinprep cytologic test (TCT) images is essential for efficient cervical abnormal screening by computer-aided diagnosis system. However, the detection performance is influenced by noise samples in the training dataset, mainly due to the subjective differences among cytologists in annotating the training samples. Besides, existing detection methods often neglect visual feature correlation information between cells, which can also be utilized to aid the detection model. In this paper, we propose a cervical abnormal cell detection method optimized by a novel distillation strategy based on local-scale consistency refinement. Firstly, we use a vanilla RetinaNet to detect top-K suspicious cells and extract region-of-interest (ROI) features. Then, a pre-trained Patch Correction Network (PCN) is leveraged to obtain local-scale features and conduct further refinement for these suspicious cell patches. We design a classification ranking loss to utilize refined scores for reducing the effects of the noisy label. Furthermore, the proposed ROI-correlation consistency loss is computed between extracted ROI features and local-scale features to exploit correlation information and optimize RetinaNet. Our experiments demonstrate that our distillation method can greatly optimize the performance of cervical abnormal cell detection without changing the detector's network structure in the inference. The code is publicly available at https://github.com/feimanman/Cervical-Abnormal-Cell-Detection.

Keywords: Cervical abnormal cell detection · Consistency learning · Cervical cytologic images

1 Introduction

Cervical cancer is the second most common cancer among adult women. If diagnosed early, it can be effectively treated and cured [19]. Nevertheless, delayed diagnosis of cervical cancer until an advanced stage will have a negative impact on patient prognosis and consume medical resources. Currently, early screening of cervical cancer is recommended worldwide as an effective method to prevent

H. Greenspan et al. (Eds.): MICCAI 2023, LNCS 14225, pp. 652–661, 2023.
https://doi.org/10.1007/978-3-031-43987-2_63

and treat cervical cancer. Thin-prep cytologic test (TCT) is the most common and effective screening method for detecting cervical abnormal and premalignant cervical lesions [5]. Conventionally it is performed by visually examining the stained cells collected through smearing on a glass slide, and generating a diagnosis report using the descriptive diagnosis method of the Bethesda system (TBS) [15]. Although TCT has been widely used in clinical applications and has significantly reduced the mortality rates caused by cervical cancer, it is still unavailable for population-wide screening [18]. This is partly due to its labor-intensive, time-consuming, and high cost [1]. Therefore, there is a high demand for automated cervical abnormality screening to facilitate efficient and accurate identification of cervical abnormalities.

With the development of deep learning [10], several attempts have been made to identify cervical abnormal cells using convolutional neural networks (CNNs). For example, Cao et al. [2] developed an attention feature pyramid network (AttFPN) for automatic abnormal cervical cell detection in cervical cytopathological images to assist pathologists in making more accurate diagnoses. Chen et al. [3] proposed a new framework that decomposes tasks and compares cells for cervical lesion cell detection. Liang et al. [11] proposed to explore contextual relationships to boost the performance of cervical abnormal cell detection. Lin et al. [22] presented an automatic cervical cell detection approach based on the Dense-Cascade R-CNN. It is worth mentioning that all of the aforementioned detection methods inevitably produce false positive results, which should be further refined by pathologists for manual checking or classification models established for automatic screening. To solve this problem, Zhou et al. [23] proposed a three-stage method including cell-level detection, image-level classification, and case-level diagnosis obtained by an SVM classifier. Zhu et al. [24] developed an artificial intelligence assistive diagnostic solution, which integrated YOLOv3 [16] for detection, Xception, and Patch-based models to boost classification.

Although the above-mentioned attempts can improve the screening performance significantly, there are several issues that need to be addressed: 1) Object detection methods often require accurate annotated data to guarantee performance with robustness and generalization. However, due to legal limitations, the scarcity of positive samples, and especially the subjectivity differences between cytopathologists for manual annotations [20], it is likely to generate noisy samples that affect the performance of the detection model. 2) Conventional object detection methods intend to directly extract the feature from the object area to locate and classify the object simultaneously. However, in clinical practice pathologists usually examine the target cells by comparing them to the surrounding cells to determine whether they are abnormal. Therefore, the visual feature correlations between the target cells and their surroundings can provide valuable information to aid the screening process, which also needs to be utilized when designing the cervical abnormal cell detection network.

To address these issues, we propose a novel method for cervical abnormal cell detection using distillation from local-scale consistency refinement. Inspired by knowledge distillation, we construct a pre-trained Patch Correction Network

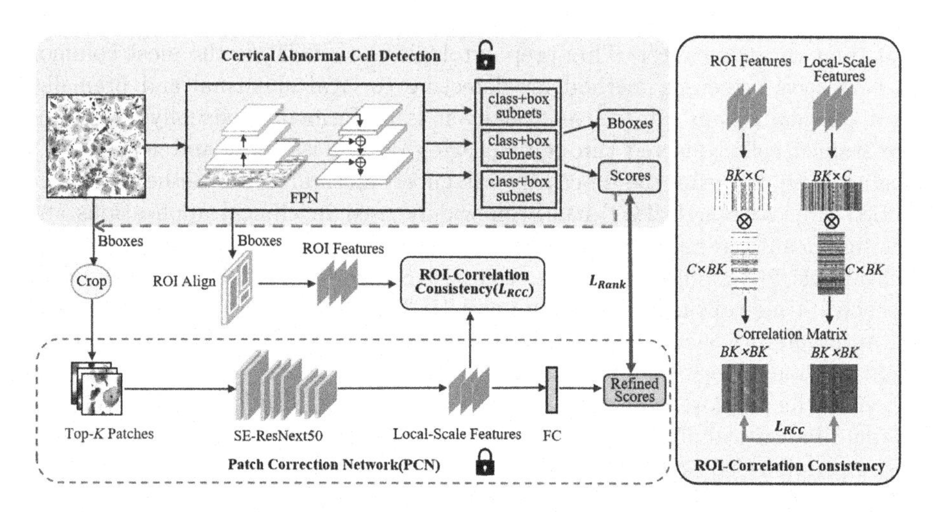

Fig. 1. The overview of our proposed framework, where PCN provides refined scores and local-scale features. Ranking Loss(L_{Rank}) is proposed to optimize the RetinaNet proposal classifier on detection scores and refined scores. And we incorporate consistency learning between ROI features and Local-scale features by RCC Loss(L_{RCC}). Note that PCN is frozen and the cervical abnormal cell detection is updated by L_{Rank} and L_{RCC} during training.

(PCN), which is designed to exploit the supervised information from the PCN to reduce the impact of noisy labels and utilize the contextual relationships between cells. In our approach, we begin by utilizing RetinaNet [12] to locate suspicious cells and crop the top-K suspicious cells into patches. Then we feed them into the PCN to obtain classification scores and propose a ranking loss to refine the classifier of the detection network by correcting the score of the detection model. In addition, we propose an ROI-Correlation Consistency (RCC) loss between ROI features and local-scale features from the PCN, which encourages the detector to explore the feature correlations of the suspicious cells. Our proposed method achieves improved performance during inference without changing the detector structure.

2 Method

The proposed framework is shown in Fig. 1, which includes cervical abnormal cell detection and the PCN. Concerning the huge size of the Whole Slide Image (WSI) and the infeasibility to handle a WSI scan for detection, we crop the WSI into images with the size of 1024 × 1024 as input to the detection. Firstly, We choose RetinaNet as our cervical abnormal cell detection, which uses a Feature Pyramid Network (FPN) backbone and attaches two subnetworks to obtain bounding boxes and classification scores. We implement the detection to locate the suspicious lesion cervical cells and extract the top K patches from the original image. Besides, we add the ROI Align layer [17] to the output of the FPN and generate ROI features. Then these patches are fed into the PCN to obtain refined scores and local-scale features. Subsequently, our ranking loss is employed to correct the score of the detection, followed by the RCC loss to capture the contextual relationships between the extracted cells for further optimizing the detection model. The distillation process involves leveraging the learned knowledge and expertise from the PCN to refine the detection results of RetinaNet.

2.1 Patch Correction Network(PCN)

In Fig. 1, the detection can automatically locate the suspicious cervical abnormal cells by providing their bounding boxes with the confidence scores. Due to the intrinsic architecture limitation of the detection and incomplete annotations, the confidence scores output by the RetinaNet may not be accurate, so we need another classification model to regrade the representative patches. Our framework leverages a local-scale classification refinement mechanism to guide the training of the detection model. We adopt SE-ResNext-50 [8] as the PCN, which has demonstrated its effectiveness in this field. The PCN is employed to refine and enhance the RetinaNet proposal classifier, which is trained from a large number of patches collected in advance with more excellent classification performance.

More specifically, the input image is processed by the base detector $F_d(\cdot)$ firstly to obtain the primary proposal information. The proposed PCN $F_c(\cdot)$ takes the top-K patches as inputs, which are cropped from original images according to the proposal location, denoted as $I_p = Cr(I, p)$, where $Cr(\cdot)$ denotes the crop function, I and p denote input image and proposal boxes predicted by $F_d(\cdot)$, respectively. Similar to the RetinaNet proposal classifier in $F_d(\cdot)$, the PCN $F_c(\cdot)$ outputs a classification distribution vector s_c. Therefore, the proposed PCN $F_c(\cdot)$ can be represented as:

$$s_c = F_c(I_p) = F_c(Cr(I, p)). \tag{1}$$

The key idea is to augment the base detector $F_d(\cdot)$ with the PCN $F_c(\cdot)$ in parallel to enhance the proposal classification capability.

2.2 Classification Ranking Loss

Due to the inaccurate confidence scores output by RetinaNet, false positive cells are inevitable after detection. Hence, a good correction network is required to generate more precise scores. In this work, the suspicious ranking of the detected patches is updated by applying PCN to them. The detector is optimized by inter-scale pairwise ranking loss. Specifically, the ranking loss is given by:

$$L_{Rank}(s_d, s_c) = max\left\{0, s_c - s_d + margin\right\}, \tag{2}$$

where s_c is the classification refinement score and s_d is the detection score, which enforces $s_d > s_c + margin$ in training. We set $margin = 0.05$. Such a design can enable RetinaNet to take the prediction score as references, and utilize refined scores from PCN to obtain more confident predictions. The ranking loss optimizes the detection to generate higher confidence scores than the previous prediction, thereby suppressing false positives and enabling the detection network to better distinguish between positive and negative cells.

2.3 ROI-Correlation Consistency (RCC) Learning

In order to solve the problem of mismatched inputs to the detection and classification models, we add the ROI Align layer to the output of the FPN. However, for cervical abnormal cell detection, normal and abnormal cells may have very similar appearances, which might not be sufficient for conducting effective differentiation. In clinical practice, to determine whether a cervical cell is normal or abnormal, cytopathologists usually compare it to the surrounding reference cells. Therefore, we studied the correlation between the top K ROIs to help more accurate classification of abnormal cells.

Based on the consistency strategy [14], which enhances the consistency of the intrinsic relation among different models, we propose ROI-correlation consistency, which regularizes the network to maintain the consistency of the semantic relation between patches under ROI features and local-scale features, and thereby encourage the detector to explore the feature interaction between cells from the extracted patches to improve the network performance.

We model the structured relation among different patches with a case-level Gram Matrix [6]. Given an input mini-batch with B samples, where B denotes the batch size. And each sample undergoes the ROI Align layer to obtain the top K ROIs, we denote the activation map of ROIs as $F^R \in \mathbb{R}^{B\times K\times H\times W\times C}$, where H and W are the spatial dimension of the feature map, and C is the channel number. We set $K = 10$, $H = 7$, $W = 7$, $C = 256$. We average pooling the feature map F^R along the spatial dimension and reshape it into $A^R \in \mathbb{R}^{BK\times C}$, and then the Case-wise Gram Matrix $G^R \in \mathbb{R}^{BK\times BK}$ is computed as:

$$G^R = A^R \cdot (A^R)^T, \tag{3}$$

where G_{ij} is the inner product between the vectorized activation map A_i^R and A_j^R, whose intuitive meaning is the similarity between the activations of i_{th} ROI and j_{th} ROI within the input mini-batch. The final ROI relation matrix R^R is obtained by conducting the L2 normalization for each row G_i^R of G^R, which is expressed as:

$$R^R = \left[\frac{G_1^R}{\|G_1^R\|_2}, \cdots, \frac{G_{BK}^R}{\|G_{BK}^R\|_2}\right]^T. \tag{4}$$

The proposed PCN $F_c(\cdot)$ takes the $B \times K$ proposals of box regressor as inputs, we denote the local-scale feature map by PCN as $F^C \in \mathbb{R}^{B\times K\times H'\times W'\times C}$, and set $H' = 56$, $W' = 56$. We perform average pooling on the feature map F^C across the spatial dimension and then reshape it into $A^C \in \mathbb{R}^{BK\times HWC}$, the Case-wise

Gram Matrix $G^C \in \mathbb{R}^{BK \times BK}$ and the final relation matrix R^C are computed as:

$$G^C = A^C \cdot (A^C)^T, \tag{5}$$

$$R^C = \left[\frac{G_1^C}{\|G_1^C\|_2}, \cdots, \frac{G_{BK}^C}{\|G_{BK}^C\|_2} \right]^T. \tag{6}$$

The RCC requires the correlation matrix to be stable under ROI features and local-scale features to preserve the semantic relation between patches. We then define the proposed RCC loss as:

$$L_{RCC} = \sum \frac{1}{BK} \|R^C(X) - R^R(X)\|_2^2, \tag{7}$$

where X is the proposals from the sampled mini-batch, $R^C(X)$ and $R^R(X)$ are the correlation matrices computed on X under different network. By minimizing L_{RCC} during the training process, the network could be enhanced to capture the intrinsic relation between patches, thus helping to extract additional semantic information from cells.

2.4 Optimization

To better optimize the Retinanet detector in a reinforced way, we take the following training strategy, which consists of three major stages. In the first stage, we collect images with doctors' labels for training and initialized the detection net. In the second stage, we train PCN with cross-entropy loss until convergence. In the last stage, we freeze the PCN and optimize the detector. The detector is optimized using the total objective function, which is written as follows:

$$L_{total} = L_{cls} + L_{reg} + \alpha L_{Rank} + \beta L_{RRC}, \tag{8}$$

where L_{cls} and L_{reg} are the ordinary detection loss for each detection head in RetinaNet. L_{cls} is a Cross-Entropy loss for classification and L_{reg} is a Smooth-L_1 loss for bounding box regression. L_{Rank} is the classification ranking loss,L_{RRC} is the RCC loss. α and β are hyper-parameters that denote the different weights of loss. During inference, only the optimized detector is used to output the final detection results without any additional modules.

3 Experimental Results

3.1 Dataset and Experimental Setup

Dataset. For cervical cell detection, our dataset includes 3761 images of 1024 × 1024 pixels cropped from WSIs. Our private dataset was collected and quality-controlled according to a standard protocol involving three pathologists: A, B, and C. Pathologist A had 33 years of experience in reading cervical cytology

Table 1. Performance comparison with state-of-the-art methods.

Method	AP	AP.5	AP.75	AR
Sparse R-CNN [21]	41.8	72.1	42.4	66.8
Deformable-DETR [25]	40.3	72.6	38.6	64.3
YoloV8 [4]	43.6	74.6	44.0	58.3
Faster R-CNN [17]	43.6	77.0	43.0	58.8
Cascade RRAM and GRAM [11]	44.6	77.5	47.7	60.0
RetinaNet [12]	45.7	81.3	46.2	58.8
Proposed method	**51.1**	**86.6**	**54.3**	**62.5**

images, while pathologists B and C had 10 years of experience each. Initially, the images were randomly assigned to pathologist B or C for initial labeling. Later, the assigned pathologist's annotations were reviewed and verified by the other pathologist. Any discrepancies found were checked and re-labeled by pathologist A. These images were divided into the training set and the testing set according to the ratio of 9:1. We also collect a new dataset of 5000 positive and negative 224×224 cell patches to train the PCN.

Implementation Details. The backbone of the suspicious cell detection network is RetinaNet with ResNet-50 [7]. The backbone of the pre-trained patch classification network is SE-ResNeXt-50. All parameters are optimized by Adam [9] with an initial learning rate of 4×10^{-5}. We set α to 0.25 and β to 1 during training. The model is implemented by PyTorch on 2 Nvidia Tesla P100 GPUs. We conduct a quantitative evaluation using two metrics: the COCO-style [13] average precision (AP) and average recall (AR). We calculate the average AP over multiple IoU thresholds from 0.5 to 0.95 with a step size of 0.05, and individually evaluated AP at the IoU thresholds of 0.5 and 0.75 (denoted as AP.5 and AP.75), respectively.

3.2 Evaluation of Cervical Abnormal Cell Detection

Comparison with SOTA Methods. We compare the performance of our proposed method against known methods for cervical lesion detection as well as representative methods for object detection. Table 1 presents the results, from which several observations can be drawn. (1) Among the models for object detection, Retinanet is generally superior to the other models. (2) Based on Retinanet, our method improves the detection performance significantly, especially AP.5 shows great performance improvement. This confirms the necessity and effectiveness of introducing the classification ranking and ROI-correlation consistency schemes for cervical lesion detection.

Ablation Study. We also perform an ablation study to further evaluate the contributions of each part in our method. Table 2 reports the detailed ablation results, from which several observations can be drawn. (1) Compared with the baseline model, Retinanet, our classification ranking loss achieves considerably

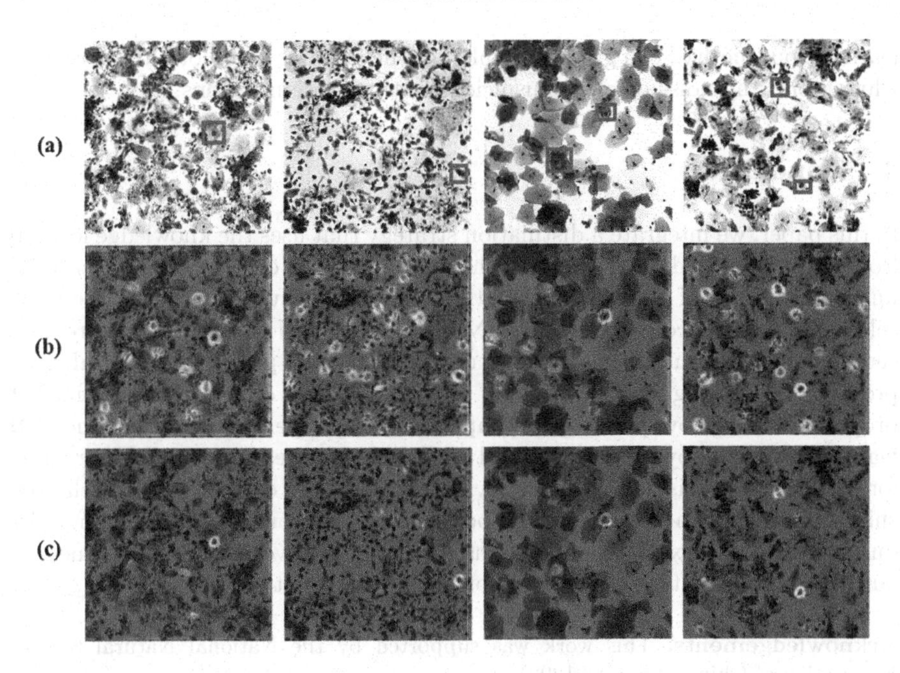

Fig. 2. Feature map visualization of RetinaNet and our method. (a) shows input images with ground-truth annotations. (b) shows feature maps from RetinaNet. (c) shows feature maps of our proposed method.

Table 2. Performance of ablation study for our local-scale consistency refinement.

Method	AP	AP.5	AP.75	AR
Baseline	45.7	81.3	46.2	58.8
+Ranking Loss	47.8	83.2	49.0	59.7
+RCC Loss	47.4	82.7	46.1	59.2
+Ranking Loss and RCC Loss	**51.1**	**86.6**	**54.3**	**62.5**

better performance, especially in AP.75, with an improvement of 2.8. (2) The RCC loss is also effective for learning better feature representations and distinguishing them well, and the AP is improved by 1.7. (3) With both ranking loss and RCC loss, our method has the best performance, which surpasses the baseline model by a large margin, validating the effectiveness of our method.

In addition, to further show the effectiveness of our method, we visualize the feature maps of Retinanet and the proposed method in Fig. 1. Those feature maps are from the Conv3 stages of the class-subnet backbone. Specifically, we sum and average the features in the channel dimension, and upsample them to the original image size. As shown in Fig. 2, our method can really learn better feature representations for abnormal cells, with the help of our proposed classification ranking refinement and ROI-correlation consistency learning. By model learning, our method can gradually enhance the features of abnormal cell regions while repressing noise or other suspicious but non-lesion regions.

4 Conclusion

In this paper, we integrate a distillation strategy that uses the knowledge learned from the pre-trained PCN to guide the training of the detection model to minimize the effects of noisy labels and explore the feature interaction between cells. Our method constructs RetinaNet with the PCN module which provides the refined scores and local-scale features of extracted patches. Specifically, we propose the ranking loss by utilizing refined scores to optimize the RetinaNet proposal classifier by reducing the impact of noisy labels. In addition, the ROI features generated by the detector and local-scale features from the PCN are used for correlation consistency learning, which explores the extracted cells' relationship. Our work can achieve better performance without adding new modules during inference. Experiments demonstrate the effectiveness and robustness of our method on the task of cervical abnormal cell detection.

Acknowledgements. This work was supported by the National Natural Science Foundation of China (No. 62001292).

References

1. Bengtsson, E., Malm, P.: Screening for cervical cancer using automated analysis of pap-smears. In: Computational and Mathematical Methods in Medicine 2014 (2014)
2. Cao, L., et al.: A novel attention-guided convolutional network for the detection of abnormal cervical cells in cervical cancer screening. Med. Image Anal. **73**, 102197 (2021)
3. Chen, T., et al.: A task decomposing and cell comparing method for cervical lesion cell detection. IEEE Trans. Med. Imaging **41**(9), 2432–2442 (2022)
4. Contributors, M.: Mmyolo: Openmmlab yolo series toolbox and benchmark (2022)
5. Davey, E., et al.: Effect of study design and quality on unsatisfactory rates, cytology classifications, and accuracy in liquid-based versus conventional cervical cytology: a systematic review. Lancet **367**(9505), 122–132 (2006)
6. Gatys, L.A., Ecker, A.S., Bethge, M.: A neural algorithm of artistic style. arXiv preprint arXiv:1508.06576 (2015)
7. He, K., Zhang, X., Ren, S., Sun, J.: Deep residual learning for image recognition. In: Proceedings of the IEEE Conference on Computer Vision and Pattern Recognition, pp. 770–778 (2016)
8. Hu, J., Shen, L., Sun, G.: Squeeze-and-excitation networks. In: Proceedings of the IEEE Conference on Computer Vision and Pattern Recognition, pp. 7132–7141 (2018)
9. Kingma, D.P., Ba, J.: Adam: a method for stochastic optimization. arXiv preprint arXiv:1412.6980 (2014)
10. LeCun, Y., Bengio, Y., Hinton, G.: Deep learning. Nature **521**(7553), 436–444 (2015)

11. Liang, Y., et al.: Exploring contextual relationships for cervical abnormal cell detection. arXiv preprint arXiv:2207.04693 (2022)
12. Lin, T.Y., Goyal, P., Girshick, R., He, K., Dollár, P.: Focal loss for dense object detection. In: Proceedings of the IEEE International Conference on Computer Vision, pp. 2980–2988 (2017)
13. Lin, T.-Y., et al.: Microsoft COCO: common objects in context. In: Fleet, D., Pajdla, T., Schiele, B., Tuytelaars, T. (eds.) ECCV 2014. LNCS, vol. 8693, pp. 740–755. Springer, Cham (2014). https://doi.org/10.1007/978-3-319-10602-1_48
14. Liu, Q., Yu, L., Luo, L., Dou, Q., Heng, P.A.: Semi-supervised medical image classification with relation-driven self-ensembling model. IEEE Trans. Med. Imaging **39**(11), 3429–3440 (2020)
15. Nayar, R., Wilbur, D.C. (eds.): The Bethesda System for Reporting Cervical Cytology. Springer, Cham (2015). https://doi.org/10.1007/978-3-319-11074-5
16. Redmon, J., Farhadi, A.: Yolov3: an incremental improvement. arXiv preprint arXiv:1804.02767 (2018)
17. Ren, S., He, K., Girshick, R., Sun, J.: Faster r-cnn: towards real-time object detection with region proposal networks. In: Advances in Neural Information Processing Systems 28 (2015)
18. Saslow, D., et al.: American cancer society, American society for colposcopy and cervical pathology, and American society for clinical pathology screening guidelines for the prevention and early detection of cervical cancer. Am. J. Clin. Pathol. **137**(4), 516–542 (2012)
19. Schiffman, M., Castle, P.E., Jeronimo, J., Rodriguez, A.C., Wacholder, S.: Human papillomavirus and cervical cancer. Lancet **370**(9590), 890–907 (2007)
20. Stoler, M.H., Schiffman, M., et al.: Interobserver reproducibility of cervical cytologic and histologic interpretations: realistic estimates from the ascus-lsil triage study. JAMA **285**(11), 1500–1505 (2001)
21. Sun, P., et al.: SparseR-CNN: end-to-end object detection with learnable proposals. arXiv preprint arXiv:2011.12450 (2020)
22. Yi, L., Lei, Y., Fan, Z., Zhou, Y., Chen, D., Liu, R.: Automatic detection of cervical cells using dense-cascade R-CNN. In: Peng, Y., et al. (eds.) PRCV 2020. LNCS, vol. 12306, pp. 602–613. Springer, Cham (2020). https://doi.org/10.1007/978-3-030-60639-8_50
23. Zhou, M., et al.: Hierarchical pathology screening for cervical abnormality. Comput. Med. Imaging Graph. **89**, 101892 (2021)
24. Zhu, X., et al.: Hybrid ai-assistive diagnostic model permits rapid tbs classification of cervical liquid-based thin-layer cell smears. Nat. Commun. **12**(1), 3541 (2021)
25. Zhu, X., Su, W., Lu, L., Li, B., Wang, X., Dai, J.: Deformable detr: deformable transformers for end-to-end object detection. arXiv preprint arXiv:2010.04159 (2020)

Instance-Aware Diffusion Model for Gland Segmentation in Colon Histology Images

Mengxue Sun, Wenhui Huang(✉), and Yuanjie Zheng

School of Information Science and Engineering, Shandong Normal University, Jinan, China
whhuang.sdu@gmail.com

Abstract. In pathological image analysis, determination of gland morphology in histology images of the colon is essential to determine the grade of colon cancer. However, manual segmentation of glands is extremely challenging and there is a need to develop automatic methods for segmenting gland instances. Recently, due to the powerful noise-to-image denoising pipeline, the diffusion model has become one of the hot spots in computer vision research and has been explored in the field of image segmentation. In this paper, we propose an instance segmentation method based on the diffusion model that can perform automatic gland instance segmentation. Firstly, we model the instance segmentation process for colon histology images as a denoising process based on the diffusion model. Secondly, to recover details lost during denoising, we use Instance Aware Filters and multi-scale Mask Branch to construct global mask instead of predicting only local masks. Thirdly, to improve the distinction between the object and the background, we apply Conditional Encoding to enhance the intermediate features with the original image encoding. To objectively validate the proposed method, we compared state-of-the-art deep learning model on the 2015 MICCAI Gland Segmentation challenge (GlaS) dataset and the Colorectal Adenocarcinoma Gland (CRAG) dataset. The experimental results show that our method improves the accuracy of segmentation and proves the efficacy of the method.

Keywords: Gland segmentation · Diffusion model · Colon histology images

1 Introduction

Colorectal cancer is a prevalent form of cancer characterized by colorectal adenocarcinoma, which develops in the colon or rectum's inner lining and exhibits glandular structures [5]. These glands play a critical role in protein and carbohydrate secretion across various organ systems. Histological examinations using Hematoxylin and Eosin staining are commonly conducted by pathologists to evaluate the differentiation of colorectal adenocarcinoma [15]. The extent of

H. Greenspan et al. (Eds.): MICCAI 2023, LNCS 14225, pp. 662–672, 2023.
https://doi.org/10.1007/978-3-031-43987-2_64

gland formation is a crucial factor in determining tumor grade and differentiation. Accurate segmentation of glandular instances on histological images is essential for evaluating glandular morphology and assessing colorectal adenocarcinoma malignancy. However, manual annotation of glandular instances is a time-consuming and expertise-demanding process. Hence, automated methods for glandular instance segmentation hold significant value in clinical practice.

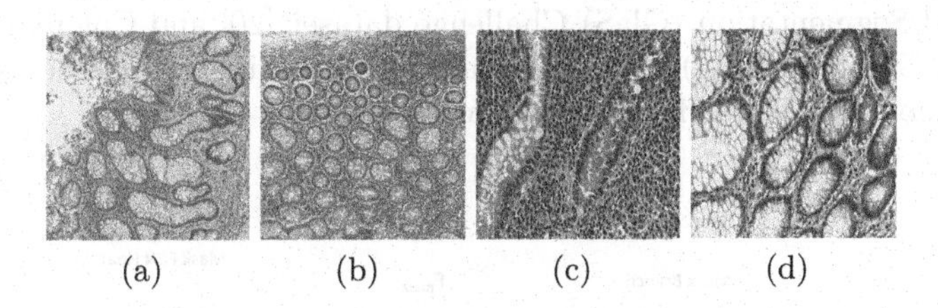

Fig. 1. (a–b) Example images from the CRAG dataset. (c–d) Example images from the GlaS dataset.

Automated segmentation has been explored using deep learning techniques [21,33], including U-Net [17], FCN [13], Siamese network [10,11] and their variations for semantic segmentation [31]. There are also methods that combine information bottleneck for detection and segmentation [23]. Additionally, two-stage instance segmentation methods like Mask R-CNN [7] and BlendMask [3] have been utilized, combining object detection and segmentation sub-networks. However, these methods may face difficulties in capturing different cell shapes and distinguishing tightly positioned gland boundaries. Limitations arise from image scaling and cropping, leading to information loss or distortion, resulting in ineffective boundary recognition and over-/under-segmentation. To overcome these limitations, we aim to perform gland instance segmentation to accurately identify the target location and prevent misclassification of background tissue.

Recently, diffusion model [9] has gained popularity as efficient generative models [16]. In the task of image synthesis, diffusion model has evolved to achieve state-of-the-art performance in terms of quality and mode coverage compared with GAN [32]. Furthermore, diffusion model has been applied to various other tasks [18]. DiffusionDet [4] treats the object detection task as a generative task on the bounding box space in images to handle projection detection. Several studies have explored the feasibility of using diffusion model in image segmentation [26]. These methods generate segmentation maps from noisy images and demonstrate better representation of segmentation details compared to previous deep learning methods.

In this paper, we propose a new method for gland instance segmentation based on the diffusion model. (1) Our method utilizes a diffusion model to perform denoising and tackle the task of gland instance segmentation in histology images. The noise boxes are generated from Gaussian noise, and the predicted ground truth (GT) boxes and segmentation masks are performed during the diffusion process. (2) To improve segmentation, we use instance-aware techniques to recover lost details during denoising. This includes employing a filter and a multi-scale Mask Branch to create a global mask and refine finer segmentation details. (3) To enhance object-background differentiation, we utilize Conditional Encoding to augment intermediate features with the original image encoding. This method effectively integrates the abundant information from the original image, thereby enhancing the distinction between the objects and the surrounding background. Our proposed method was trained and tested on the 2015 MICCAI Gland Segmentation (GlaS) Challenge dataset [20] and Colorectal Adenocarcinoma Gland (CRAG) dataset [6] (as shown in Fig. 1), and the experiment results demonstrate the efficacy of the method.

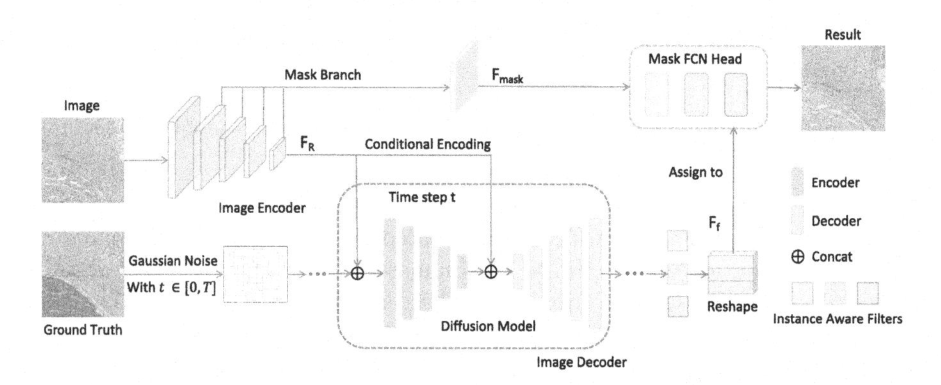

Fig. 2. The network architecture diagram of our model. The Image Encoder consists of a backbone that extracts multi-scale features from the input image. The Image Decoder based on a diffusion model incorporates the original image features as conditions to enhance the intermediate features. To preserve multi-scale information, we introduce a Mask Branch that operates on F_{mask}. By applying convolutions with weights assigned from filters to F_{mask}, we obtain instance masks.

2 Method

In this section, we present the architecture of our proposed method, which includes an Image Encoder, an Image Decoder, and a Mask Branch. The network structure is shown in Fig. 2.

2.1 Image Encoder

We propose to perform subsequent operations on the features of the original image, so we use an Image Encoder for advanced feature extraction. The Image Encoder takes the original image as input and we use a convolutional neural network such as ResNet [8] for feature extraction and a Feaure Pyramid Network

(FPN) [12] is used to generate a multi-scale feature map for ResNet backbone following.

The input image is $\mathbf{x}$ and the output is a high-level feature F_R.

$$F_R = \mathcal{F}(\mathbf{x}) \tag{1}$$

where $\mathcal{F}$ is the ResNet. The Image Encoder operates only once and uses the F_R as condition to progressively refine and generate predictions from the noisy boxes.

2.2 Image Decoder

We designed our model based on the diffusion model [30], which typically uses two Markov chains divided into two phases: a forward diffusion process and a reverse denoising process. The components of diffusion model are a learning reverse process called $p_\theta(\mathbf{z}_{t-1}|\mathbf{z}_t)$ that creates samples by converting noise into samples from $q(\mathbf{z}_0)$ and a forward diffusion process called $q(\mathbf{z}_t|\mathbf{z}_{t-1})$ that gradually corrupts data from some target distribution into a normal distribution. The forward diffusion process is defined as:

$$q(\mathbf{z}_t|\mathbf{z}_{t-1}) = \mathcal{N}(\mathbf{z}_t, \sqrt{1-\beta_t}\mathbf{z}_{t-1}, \beta_t\mathbf{I}) \tag{2}$$

A variance schedule $\beta_t \in (0, 1)$, $t \in \{1, ..., T\}$ determines the amount of noise that is introduced at each stage. Alternatively, we can obtain a sample of $\mathbf{z}_t$ from direct $\mathbf{z}_0$ as follows:

$$\mathbf{z}_t = \sqrt{\bar{\alpha_t}}\mathbf{z}_0 + (1 - \bar{\alpha_t})\epsilon \tag{3}$$

where $\bar{\alpha_t} = \prod_{s=0}^{t}(1 - \beta_s)$, $\epsilon \sim \mathcal{N}(0, \mathbf{I})$.

Our Image Decoder is based on diffusion model, which can be viewed as a noise-to-GT denoising process. In this setting, the data samples consist of a set of bounding boxes represented as $\mathbf{z}_0$, where $\mathbf{z}_0$ is a set of N boxes.

The neural network $f_\theta(\mathbf{z}_t, t)$ is trained to predict $\mathbf{z}_0$ from the $\mathbf{z}_t$ based on the corresponding image $\mathbf{x}$. In addition, to achieve complementary information by integrating the segmentation information from $\mathbf{z}_t$ into the original image encoding, we introduce Conditional Encoding, which uses the encoding features of the current step to enhance its intermediate features.

$$f_\theta(\mathbf{z}_t, F_R, t) = D((Concat(E(\mathbf{z}_t, F_R), F_R), t), t) \tag{4}$$

$$\mathbf{z}_0 = f(\cdots(f(\mathbf{z}_{T-m}, F_R, T - m)) \tag{5}$$

where D represent the decoder, E represent the encoder and $m \in \{1, ..., T\}$. We use Instance Aware Filters (IAF) during iterative sampling, which allows sharing parameters between steps.

$$F_f^t = IAF(f_\theta(\mathbf{z}_t, F_R, t), t) \tag{6}$$

where F_f^t is the output feature of the filter.

2.3 Mask Branch

We have also utilized dynamic mask head [22] to predict masks in our study. In this stage, we use the Mask Branch to fuse the different scale information of the FPN and output the mask feature F_{mask}. The diffusion process decodes RoI features into local masks, and multi-scale features can be supplemented with more detailed information for predicting global masks to compensate for the detail lost in the diffusion process, and we believe that instance masks require a larger perceptual domain because of the higher demands on instance edges. Specifically, the instance mask can be generated by convolving the feature map F_{mask} from the Mask Branch and F_f^t from the IAF, which is calculated as follows:

$$\mathbf{s} = MFH(F_{mask}, F_f^t) \tag{7}$$

where the predicted instance mask is denoted by $\mathbf{s} \in R^{H \times W}$. The Mask FCN Head, denoted by MFH, is comprised of three 1×1 convolutional layers.

We enhance our loss function by incorporating two components, L_d and L_s, and utilize the γ parameter to optimize the balance between these two losses.

$$L = L_d + \gamma L_s(\mathbf{s}, \mathbf{s}^{GT}) \tag{8}$$

where the L_s in our model represents the measure of overlap between the predicted instance mask and the ground truth $\mathbf{s}^{GT}$ [14], and the L_d is the loss of DiffusionDet. The optimal value for the parameter γ is usually determined based on achieving the best overall performance on the validation set. In this work, we chose $\gamma = 5$ to balance these two losses.

3 Experiments and Results

We presented the segmentation results of our model compared to the ground truth in Fig. 3, and provided both qualitative and quantitative evaluations that validate the effectiveness of our proposed network for gland instance segmentation.

Data and Evaluation Metrics: We evaluated the effectiveness of the proposed model on two datasets: the GlaS dataset and the CRAG dataset. The GlaS dataset comprises 85 training and 80 testing images, divided into 60 images in Test A and 20 images in Test B. The CRAG dataset consists of 173 training and 40 testing images. We have adopted Vahadane method for stain normalization [1]. Furthermore, to enhance the training dataset and mitigate the risk of overfitting, we employed random combinations of image flipping, translation, Gaussian blur, brightness variation, and other augmentation techniques.

We assessed the segmentation results using three metrics from the GlaS Challenge: (1) Object F1, which measures the accuracy of detecting individual glands,

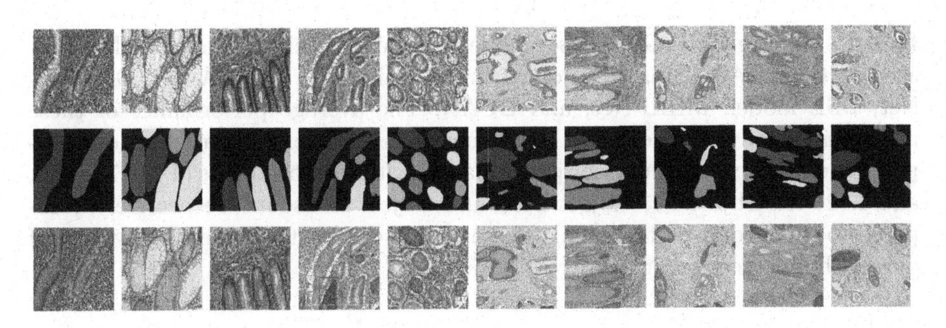

Fig. 3. The instance segmentation results on the GlaS dataset and CRAG dataset. From top to bottom: the original images, the ground truth, and the segmentation results produced by our method.

(2) Object Dice, which evaluates the volume-based accuracy of gland segmentation, and (3) Object Hausdorff, which assesses the shape similarity between the segmentation result and the ground truth. We assigned each method three ranking numbers based on these metrics and computed their sum to determine the final ranking for each method's overall performance.

Implementation Details: In our experiments, we choose the ResNet-50 with FPN as the backbone in the proposed method. The backbone is pretrained on ImageNet. Image decoder, Mask Branch and Mask FCN Head are trained end-to-end. We trained on the GlaS and CRAG datasets in a Python 3.8.3 environment on Ubuntu 18.04, using PyTorch 1.10 and CUDA 11.4. During training, we utilized an SGD optimizer with a learning rate of 2.5×10^{-5} and the weight decay as 10^{-4}. We set diffusion timesteps $T = 1000$ and chose a linear schedule from $\beta_1 = 10^{-4}$ to $\beta_T = 0.02$. Training was performed on A100 GPU with a batch size of 2.

Results on the GlaS Challenge Dataset: We conducted experiments to evaluate the performance of our proposed model by comparing it with the DSE model [27], the DMCN [28], the DCAN [2], the SPL-Net [29], the DoubleU-Net [24], the MILD-Net [6], the GCSBA-Net [25], and the MPCNN [19]. Table 1 provides an overview of the average performance of these models.

Our proposed model demonstrated a enhancement in performance, surpassing the second-best method on both Test A and Test B datasets. Specifically, on Test A, we observed an improvement of 0.006, 0.01, and 1.793 in Object F1, Object Dice, and Object Hausdorf. Similarly, on Test B, resulting in an improvement of 0.022, 0.014 and 3.694 in Object F1, Object Dice, and Object Hausdorf, respectively. Although Test B presented a more challenging task due to the presence of complex morphology in the images, our proposed model demonstrated accurate segmentation in all cases. The experimental results highlighted the effectiveness of our approach in improving the accuracy of gland instance segmentation.

Results on the CRAG Dataset: The proposed model was additionally evaluated on the CRAG dataset by comparing it against the GCSBA-Net, DoubleU-Net, DSE model, MILD-Net, and DCAN. The average performance of these models is shown in Table 2. Our experimental results demonstrate that our proposed method achieves superior performance, with improvements of 0.017, 0.012, and 4.026 for Object F1, Object Dice, and Object Hausdorff, respectively, compared to the second-best method. These results demonstrate the effectiveness of our method in segmenting different datasets.

Ablation Studies: Our network utilizes the Mask Branch and Conditional Encoding to enhance performance and segmentation quality. Ablation studies on the GlaS and CRAG datasets confirm the effectiveness of these modules (Table 3). The Mask Branch is responsible for multi-scale feature extraction and fusion with the backbone network, as well as refining the Image Decoder's output. Without the Mask Branch, direct usage of original image features lacks multi-scale information and results in less accurate segmentation. Conditional Encoding is employed to establish a connection between input image features

Table 1. The experimental results on GlaS Challenge dataset. The S represents the score, R represents the rank and Rank Sum refers to the sum of rank for each evaluation metric.

Models	Object F1				Object Dice				Object Hausdorff (pixels)				Rank Sum
	Test A		Test B		Test A		Test B		Test A		Test B		
	S	R	S	R	S	R	S	R	S	R	S	R	
Proposed	0.941	1	0.893	1	0.939	1	0.889	1	26.042	1	72.351	1	6
DoubleU-Net [24]	0.935	2	0.871	2	0.929	2	0.875	2	27.835	2	76.045	2	12
DSE model [27]	0.926	3	0.862	3	0.927	3	0.871	3	31.209	3	80.509	3	18
GCSBA-Net [25]	0.916	5	0.832	6	0.914	4	0.834	6	41.49	4	102.88	4	29
MILD-Net [6]	0.914	6	0.844	4	0.913	5	0.836	5	41.540	5	105.890	5	30
SPL-Net [29]	0.924	4	0.844	4	0.902	7	0.840	4	49.881	8	106.075	6	33
DMCN [28]	0.893	8	0.843	5	0.908	6	0.833	7	44.129	6	116.821	7	39
DCAN [2]	0.912	7	0.716	7	0.897	8	0.781	9	45.418	7	160.347	9	47
MPCNN [19]	0.891	9	0.703	8	0.882	9	0.786	8	57.413	9	145.575	8	51

Table 2. The experimental results on the CRAG dataset.

Models	Object F1		Object Dice		Object Hausdorff (pixels)		Rank Sum
	S	R	S	R	S	R	
Proposed	0.853	1	0.906	1	113.224	1	3
GCSBA-Net [25]	0.836	2	0.894	2	146.77	4	8
DoubleU-Net [24]	0.835	3	0.890	3	117.25	2	8
DSE model [27]	0.835	3	0.889	4	120.127	3	10
MILD-Net [6]	0.825	4	0.875	5	160.140	5	14
DCAN [2]	0.736	5	0.794	6	218.76	6	17

and the diffusion model. Performing reverse diffusion without any reference condition can introduce numerous errors and require multiple iterations to achieve the desired outcome. When employing Mask Branch, our approach resulted in an improvement of 0.082, 0.09, 0.07 in Object F1, and 0.07, 0.078, 0.07 in Object Dice, while Object Hausdorff decreased by 10.29, 11.11, 24.47 on GlaS Test A, GlaS Test B, and CRAG, respectively. Similarly, by utilizing Conditional Encoding, we observed an improvement of 0.048, 0.034, 0.052 in Object F1, and 0.026, 0.042, 0.057 in Object Dice, while Object Hausdorff decreased by 6.771, 8.115, 12.141 on GlaS Test A, GlaS Test B, and CRAG, respectively.

Table 3. The ablation study results on the CRAG and GlaS datasets demonstrate the impact of different modules on performance. The Mask Branch module contributes to multi-scale feature extraction, while the Conditional Encoding module establishes the connection between input image features and the diffusion model.

Dataset	Model		Object F1	Object Dice	Object Hausdorff (pixels)
	Mask Branch	Conditional Encoding			
GlaS Test A	✓	✓	0.941±0.041	0.939±0.057	26.042±1.688
	✓		0.893±0.105	0.913±0.092	32.813±2.731
		✓	0.859±0.098	0.869±0.107	36.332±1.460
GlaS Test B	✓	✓	0.893±0.046	0.889±0.072	72.351±1.271
	✓		0.859±0.061	0.847±0.072	80.466±4.604
		✓	0.803±0.077	0.811±0.082	83.461±3.704
CRAG	✓	✓	0.853±0.045	0.906±0.066	113.224±2.464
	✓		0.801±0.062	0.849±0.071	125.365±3.599
		✓	0.783±0.073	0.836±0.083	137.694±5.934

4 Conclusion and Discussion

In this paper, we propose a diffusion model based method for gland instance segmentation. By considering instance segmentation as a denoising process based on diffusion model. Our model contains three main parts: Image Encoder, Image Decoder, and Mask Branch. By utilizing a diffusion model with Conditional Encoding for denoising, we are able to improve the precision of instance localization while compensating for the missing details in the diffusion model. By incorporating multi-scale information fusion, our approach results in more accurate segmentation outcomes. Experimental results on the GlaS dataset and CRAG dataset show that our method surpasses state-of-the-art approach, demonstrating its effectiveness.

Although our method demonstrates excellent performance in gland instance segmentation, challenges arise in certain scenarios characterized by irregular shapes, flattening, and overlapping. In such cases, our network tends to classify multiple small targets with unclear boundaries as a single object, indicating limitations in segmentation accuracy when dealing with high aggregation or overlap. This limitation may stem from the difficulty in accurately distinguishing fine details between instances and the incorrect identification of boundaries.

To address these limitations, future work will focus on improving segmentation performance in challenging scenarios by specifically targeting three identified limitations: (1) Incorporate random noise during training to reduce reliance on bounding box information for denoising; (2) Explore more efficient methods for cross-step denoising in the diffusion model to improve processing time without compromising segmentation accuracy; and (3) Develop a more effective Conditional Encoding method to provide accurate instance context for noise filtering in discriminative tasks like nuclear segmentation.

Acknowledgments. This work is supported by funds from the National Natural Science Foundation of China (62003196, 62076249, 62072289, and 62073201) and the Provincial Natural Science Foundation of Shandong Province of China (ZR2020QF032).

References

1. Anand, D., Ramakrishnan, G., Sethi, A.: Fast gpu-enabled color normalization for digital pathology. In: 2019 International Conference on Systems, Signals and Image Processing (IWSSIP), pp. 219–224. IEEE (2019)
2. Chen, H., Qi, X., Yu, L., Heng, P.A.: Dcan: deep contour-aware networks for accurate gland segmentation. In: Proceedings of the IEEE Conference on Computer Vision and Pattern Recognition, pp. 2487–2496 (2016)
3. Chen, H., Sun, K., Tian, Z., Shen, C., Huang, Y., Yan, Y.: Blendmask: top-down meets bottom-up for instance segmentation. In: Proceedings of the IEEE/CVF Conference on Computer Vision and Pattern Recognition, pp. 8573–8581 (2020)
4. Chen, S., Sun, P., Song, Y., Luo, P.: Diffusiondet: diffusion model for object detection. arXiv preprint arXiv:2211.09788 (2022)
5. Fleming, M., Ravula, S., Tatishchev, S.F., Wang, H.L.: Colorectal carcinoma: pathologic aspects. J. Gastrointesti. Oncol. **3**(3), 153 (2012)
6. Graham, S., Chen, H., Gamper, J., Dou, Q., Heng, P.A., et al.: Mild-net: minimal information loss dilated network for gland instance segmentation in colon histology images. Med. Image Anal. **52**, 199–211 (2019)
7. He, K., Gkioxari, G., Dollár, P., Girshick, R.: Mask r-cnn. In: Proceedings of the IEEE International Conference on Computer Vision, pp. 2961–2969 (2017)
8. He, K., Zhang, X., Ren, S., Sun, J.: Deep residual learning for image recognition. In: Proceedings of the IEEE Conference on Computer Vision and Pattern Recognition, pp. 770–778 (2016)
9. Ho, J., Jain, A., Abbeel, P.: Denoising diffusion probabilistic models. Adv. Neural Inf. Process. Syst. **33**, 6840–6851 (2020)
10. Huang, W., Gu, J., Duan, P., Hou, S., Zheng, Y.: Exploiting probabilistic siamese visual tracking with a conditional variational autoencoder. In: 2021 IEEE International Conference on Robotics and Automation (ICRA), pp. 14213–14219. IEEE (2021)
11. Huang, W., Gu, J., Ma, X., Li, Y.: End-to-end multitask siamese network with residual hierarchical attention for real-time object tracking. Appl. Intell. **50**, 1908–1921 (2020)
12. Lin, T.Y., Dollár, P., Girshick, R., He, K., Hariharan, B., Belongie, S.: Feature pyramid networks for object detection. In: Proceedings of the IEEE Conference on Computer Vision and Pattern Recognition, pp. 2117–2125 (2017)

13. Long, J., Shelhamer, E., Darrell, T.: Fully convolutional networks for semantic segmentation. In: Proceedings of the IEEE Conference on Computer Vision and Pattern Recognition, pp. 3431–3440 (2015)
14. Milletari, F., Navab, N., Ahmadi, S.A.: V-net: fully convolutional neural networks for volumetric medical image segmentation. In: 2016 Fourth International Conference on 3D Vision (3DV), pp. 565–571. IEEE (2016)
15. Niazi, M.K.K., Parwani, A.V., Gurcan, M.N.: Digital pathology and artificial intelligence. Lancet Oncol. **20**(5), e253–e261 (2019)
16. Rombach, R., Blattmann, A., Lorenz, D., Esser, P., Ommer, B.: High-resolution image synthesis with latent diffusion models. In: Proceedings of the IEEE/CVF Conference on Computer Vision and Pattern Recognition, pp. 10684–10695 (2022)
17. Ronneberger, O., Fischer, P., Brox, T.: U-net: convolutional networks for biomedical image segmentation. In: Navab, N., Hornegger, J., Wells, W.M., Frangi, A.F. (eds.) MICCAI 2015. LNCS, vol. 9351, pp. 234–241. Springer, Cham (2015). https://doi.org/10.1007/978-3-319-24574-4_28
18. Sanchez, P., Kascenas, A., Liu, X., O'Neil, A.Q., Tsaftaris, S.A.: What is healthy? generative counterfactual diffusion for lesion localization. In: Deep Generative Models: Second MICCAI Workshop, DGM4MICCAI 2022, Held in Conjunction with MICCAI 2022, Singapore, 22 September 2022, Proceedings, pp. 34–44. Springer, Heidelberg (2022). https://doi.org/10.1007/978-3-031-18576-2_4
19. Sirinukunwattana, K., et al.: Gland segmentation in colon histology images: the glas challenge contest. Med. Image Anal. **35**, 489–502 (2017)
20. Sirinukunwattana, K., Snead, D.R., Rajpoot, N.M.: A stochastic polygons model for glandular structures in colon histology images. IEEE Trans. Med. Imaging **34**(11), 2366–2378 (2015)
21. Song, J., Zheng, Y., Xu, C., Zou, Z., Ding, G., Huang, W.: Improving the classification ability of network utilizing fusion technique in contrast-enhanced spectral mammography. Med. Phys. **49**(2), 966–977 (2022)
22. Tian, Z., Shen, C., Chen, H.: Conditional convolutions for instance segmentation. In: Vedaldi, A., Bischof, H., Brox, T., Frahm, J.-M. (eds.) ECCV 2020. LNCS, vol. 12346, pp. 282–298. Springer, Cham (2020). https://doi.org/10.1007/978-3-030-58452-8_17
23. Wang, J., et al.: Information bottleneck-based interpretable multitask network for breast cancer classification and segmentation. Med. Image Anal. **83**, 102687 (2023)
24. Wang, P., Chung, A.C.S.: DoubleU-net: colorectal cancer diagnosis and gland instance segmentation with text-guided feature control. In: Bartoli, A., Fusiello, A. (eds.) ECCV 2020. LNCS, vol. 12535, pp. 338–354. Springer, Cham (2020). https://doi.org/10.1007/978-3-030-66415-2_22
25. Wen, Z., Feng, R., Liu, J., Li, Y., Ying, S.: Gcsba-net: gabor-based and cascade squeeze bi-attention network for gland segmentation. IEEE J. Biomed. Health Inf. **25**(4), 1185–1196 (2020)
26. Wolleb, J., Sandkühler, R., Bieder, F., Valmaggia, P., Cattin, P.C.: Diffusion models for implicit image segmentation ensembles. In: International Conference on Medical Imaging with Deep Learning, pp. 1336–1348. PMLR (2022)
27. Xie, Y., Lu, H., Zhang, J., Shen, C., Xia, Y.: Deep segmentation-emendation model for gland instance segmentation. In: Shen, D., et al. (eds.) MICCAI 2019. LNCS, vol. 11764, pp. 469–477. Springer, Cham (2019). https://doi.org/10.1007/978-3-030-32239-7_52
28. Xu, Y., Li, Y., Wang, Y., Liu, M., Fan, Y., et al.: Gland instance segmentation using deep multichannel neural networks. IEEE Trans. Biomed. Eng. **64**(12), 2901–2912 (2017)

29. Yan, Z., Yang, X., Cheng, K.-T.T.: A deep model with shape-preserving loss for gland instance segmentation. In: Frangi, A.F., Schnabel, J.A., Davatzikos, C., Alberola-López, C., Fichtinger, G. (eds.) MICCAI 2018. LNCS, vol. 11071, pp. 138–146. Springer, Cham (2018). https://doi.org/10.1007/978-3-030-00934-2_16
30. Yang, L., Zhang, Z., Song, Y., Hong, S., Xu, R., et al.: Diffusion models: a comprehensive survey of methods and applications. arXiv preprint arXiv:2209.00796 (2022)
31. Zhang, Z., Tian, C., Bai, H.X., Jiao, Z., Tian, X.: Discriminative error prediction network for semi-supervised colon gland segmentation. Med. Image Anal. **79**, 102458 (2022)
32. Zheng, Y., et al.: Symreg-gan: symmetric image registration with generative adversarial networks. IEEE Trans. Pattern Anal. Mach. Intell. **44**(9), 5631–5646 (2021)
33. Zheng, Y., et al.: Image matting with deep gaussian process. IEEE Trans. Neural Netw. Learn. Syst. (2022)

Label-Free Nuclei Segmentation Using Intra-Image Self Similarity

Long Chen[1], Han Li[2,3], and S. Kevin Zhou[1,2,3](✉)

[1] Key Lab of Intelligent Information Processing of Chinese Academy of Sciences (CAS), Institute of Computing Technology, CAS, Beijing 100190, People's Republic of China

chenlong171@mails.ucas.ac.cn

[2] School of Biomedical Engineering, Division of Life Sciences and Medicine, University of Science and Technology of China, Hefei 230026, Anhui, People's Republic of China

hanli21@mail.ustc.edu.cn, skevinzhou@ustc.edu.cn

[3] Center for Medical Imaging, Robotics, Analytic Computing and Learning (MIRACLE), Suzhou Institute for Advanced Research, University of Science and Technology of China, Suzhou 215123, Jiangsu, People's Republic of China

Abstract. In computational pathology, nuclei segmentation from histology images is a fundamental task. While deep learning based nuclei segmentation methods yield excellent results, they rely on a large amount of annotated images; however, annotating nuclei from histology images is tedious and time-consuming. To get rid of labeling burden completely, we propose a **label-free** approach for nuclei segmentation, motivated from one pronounced yet omitted property that characterizes histology images and nuclei: **intra-image self similarity (IISS)**, that is, within an image, nuclei are similar in their shapes and appearances. First, we leverage traditional machine learning and image processing techniques to generate a pseudo segmentation map, whose connected components form candidate nuclei, both positive or negative. In particular, it is common that adjacent nuclei are merged into one candidate due to imperfect staining and imaging conditions, which violate the IISS property. Then, we filter the candidates based on a custom-designed index that roughly measures if a candidate contains multiple nuclei. The remaining candidates are used as pseudo labels, which we use to train a U-Net to discover the hierarchical features distinguish nuclei pixels from background. Finally, we apply the learned U-Net to produce final nuclei segmentation. We validate the proposed method on the public dataset MoNuSeg. Experimental results demonstrate the effectiveness of our design and, to the best of our knowledge, it achieves the **state-of-the-art performances of label-free segmentation** on the benchmark MoNuSeg dataset with a mean Dice score of 79.2%.

Keywords: Label-free · Nuclei segmentation · Pseudo Label

H. Greenspan et al. (Eds.): MICCAI 2023, LNCS 14225, pp. 673–682, 2023.
https://doi.org/10.1007/978-3-031-43987-2_65

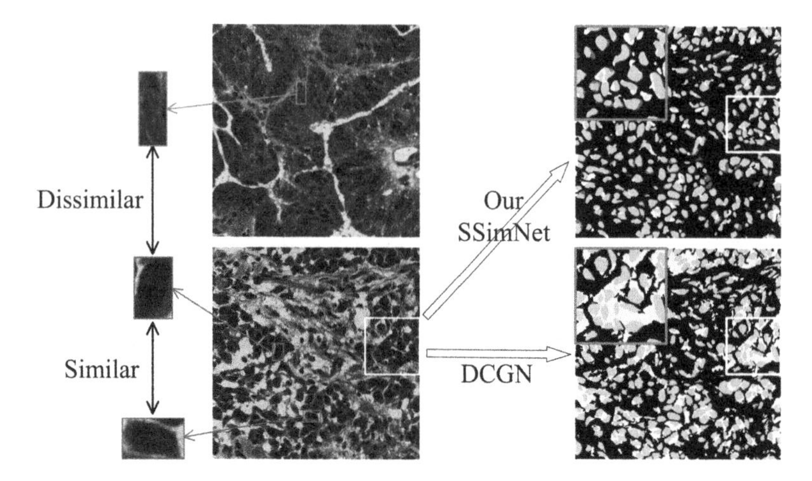

Fig. 1. The limitations of (deep) clustering model. Green, yellow and red colors refer to the true positive, the false positive and the false negative predictions. (Color figure online)

1 Introduction

Nuclei segmentation is a fundamental step in histology image analysis. In recent advances, with a large amount of labeled data, fully-supervised learning methods can easily achieve reasonable results [1–5]. However, accurate pixel-level annotation of nuclei is not always accessible for segmentation labeling is a labor-intensive and time-consuming procedure. Methods to relieve the high dependency on the accurate annotations of nuclei are highly needed.

Unsupervised learning (UL) methods achieved great success in the data dependency problem for nuclei segmentation, which learns from the structural properties in the data without any manual annotations. Based on the character of these methods, we can group them into two categories: the traditional UL methods and the deep learning UL methods. Traditional UL nuclei segmentation methods include watershed [6], contour detection [7], clustering [8,9] and random field [10]. These methods focus on either pixel value or shape information but fail to take advantage of both of them. Moreover, due to the heavily rely on preset parameters, these traditional methods also show weak robustness.

Therefore, some researchers [11–15] resort to deep UL segmentation models to better utilize both pixel value and shape information and develop a robust approach. The common and effective way is to employ image clustering by maximizing mutual information between image and predicted labels to distinguish foreground and background regions. Many image-clustering-based deep UL methods for natural tasks still achieve strong performances in nuclei segmentation. Kanezaki et al. [11] constrain a convolutional neural network (CNN) with superpixel level segmentation results. Ji et al. [12] propose the invariant information clustering. While reasonable results are obtained, these deep clustering-based methods still suffer difficulties: (i) Poor segmentation of the regions between adjacent nuclei. Deep clustering models succeed in transferring images to high-dimensional feature space and obtaining image segmentation results by means of clustering pixels' features. However, the regions between adjacent nuclei are similar to the nuclei regions in terms of color values and textures (as shown in Fig. 1). Deep clustering-based methods experience difficulties in dealing with these regions due to the lack of supervision. (ii) Underutilization of **intra-image self similarity (IISS)** information. As shown in Fig. 1, in terms of value, shape and texture, nuclei show a similar appearance within the same image but vary greatly among different images[1]. This phenomenon offers valuable information for networks to use but the current clustering models do not take this into account.

To address the above issues and motivated by the IISS property, we hereby propose a novel self-similarity-driven segmentation network (**SSimNet**) for unsupervised nuclei segmentation. As shown in Fig. 2, instead of designing complex discriminative network architectures, our framework derives knowledge from the IISS property to aid the segmentation. Specifically, we obtain candidate nuclei with some unsupervised image processing. For the obtained candidates, it is common that adjacent nuclei merged into one candidate due to imperfect staining and low image quality, which violate the IISS property. Hence, we filter the candidates based on a custom-designed index that roughly measures if a candidate contains multiple nuclei. The remaining candidates are used as pseudo labels, which we use to train a U-Net (aka SSimNet) to discover the hierarchical features that distinguish nuclei pixels from the background. Finally, we apply the learned SSimNet to produce the final nuclei segmentation.

To validate the effectiveness of our method, we conduct extensive experiments on the MoNuSeg dataset [16,17] based on ten existing unsupervised segmentation methods [9,11–15,18–20]. Our method outperforms all comparison methods with an average Dice score of 0.792 and aggregated Jaccard index of 0.498 on the MoNuSeg dataset which is close to the supervised method.

2 Method

As shown in Fig. 2, our SSimNet aims at unsupervised segmentation of nuclei from histology images. Specifically, by using a matrix factorization on hematoxylin and eosin (H&E) stained histology images, we get the hematoxylin channel image for clustering, active contour refining and softening to generate the final soft candidate label. Then according to the designed unsupervised evaluation metric driven from the IISS property, an SSimNet is trained with highly-rated soft pseudo labels and corresponding original patches. Last, while testing on the test image, to adapt the network to learn nucleus similarity within the same image, we fine tune the network with soft pseudo labels of some patches in current test images. In the following, we elaborate on each part in detail.

[1] Note that in our experiments, we use an image of size 1000^2 or 500^2.

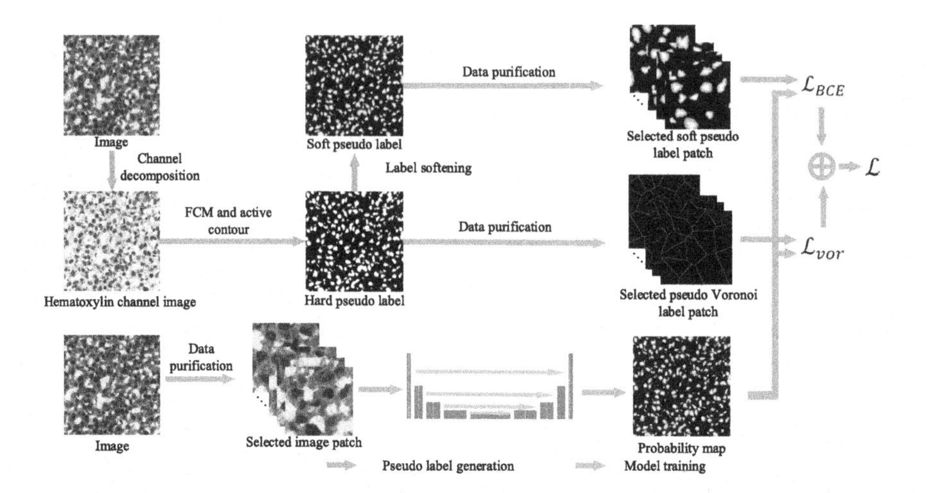

Fig. 2. The overview pipeline of the proposed method.

2.1 Candidate Nucleus Generation

Channel Decomposition. Suppose that we are given a training set $\mathcal{I}^S = \{I_i^S\}_{i=1}^N$ of histopathology images without any manual annotation. For each image, stained tissue colors are results from light attenuation, which depends on the type and amount of dyes that the tissues have absorbed. This property is prescribed by the Beer-Lambert law:

$$V = \log(I_0/I) = WH, \tag{1}$$

where $I \in \mathbb{R}^{3\times n}$ represents the histology image with three color channels and n pixels, I_0 is the illuminating light intensity of sample with $I_0 = 255$ for 8-bit images in our cases, $W \in \mathbb{R}^{3\times r}$ is the stain color matrix that encodes the color appearance of each stain with r representing the number of stains, and $H \in \mathbb{R}^{r\times n}$ is the stain density map. In this work, we follow the sparse non-negative matrix factorization in [21] to get the stain color matrices $\mathcal{W} = \{W_i\}_{i=1}^N$ and stain density maps $\mathcal{H} = \{H_i\}_{i=1}^N$ for $\mathcal{I}^S$. Note that usually histopathology images are stained with H&E and nuclei mainly absorb hematoxylin [22]; therefore, $r = 2$. We reconstruct the nuclei stain map with the first channel of $\mathcal{W}$ and $\mathcal{H}$:

$$\mathcal{I}^T = \{I_i^T\}_{i=1}^N = \{W_i[:,0] \cdot H_i[0,:]\}_{i=1}^N \tag{2}$$

Clustering and Active Contour. We transform $\mathcal{I}^T$ into CIELAB color space and invoke the Fuzzy C-Means method (FCM) with 2 clusters to obtain the candidate foreground pixels. To reduce the noise in clustering results, we use active contour method as a smoothing operation to get hard candidate labels:

$$\mathcal{P} = \{P_i\}_{i=1}^N = \{ActiveContour(FCM(I_i^T))\}_{i=1}^N \tag{3}$$

Label Smoothing. Since hard label is overconfident at the border of nuclei, which is detrimental to the training of the network, we soften the hard label one by one for each connected component in P_i using the following formulation:

$$B[k+1] = B[k] + \frac{P[k]}{2 \cdot A} \tag{4}$$

where k represents the kth epoch erosion of the connected component, $B[k]$ is the confidence score of pixels eroded in the kth epoch and $B[0] = 0.5$ as the initial condition, $P[k]$ means number of pixels eroded in the kth epoch, and A is the area of the connected component. As a terminal condition, we set the termination of erosion when $B[k] > 0.975$. Following Eq. (4), we obtain our soft candidate labels $\tilde{\mathcal{P}}$ from $\mathcal{P}$.

2.2 Data Purification and SSimNet Learning

So far, soft candidate labels $\tilde{P}_i$ have been acquired for each image I_i^S. However, it is common that adjacent nuclei are merged into one candidate due to imperfect staining and imaging conditions, which violate the IISS property. To this, we conduct data purification to build a reliable training set for subsequent learning.

Data Purification. We sample K patches with overlap from original image I_i^S. The sampled results are expressed as patch tissue $\mathcal{X} = \{x_i\}_{i=1}^{N \cdot K}$ and patch label $\mathcal{Y} = \{y_i\}_{i=1}^{N \cdot K}$. We design the Unsupervised Shape Measure Index (USMI) and calculate it using the algorithm in Fig. 3(left). Based on thresholding the USMI, we obtain pairs $(y_i, u_i)_{i=1}^{N \cdot K}$. Note that the smaller USMI is, the more the pseudo label conforms to prior knowledge. Sorting these pairs by USMI from the smallest to largest, only maintain the first $\alpha\%(0 < \alpha < 100)$ of data pairs as $(\tilde{\mathcal{X}} = \{x_{u(i)}\}_{i=1}^{\alpha \cdot N \cdot K}, \tilde{\mathcal{Y}} = \{y_{u(i)}\}_{i=1}^{\alpha \cdot N \cdot K})$. Figure 3(right) shows a separation of candidates into two groups (yellow and blue) with a typical yellow patch containing merger nuclei and a blue patch containing isolated nuclei.

To further separate possible adjacent nuclei in a blue patch, we follow [23] to construct the Voronoi label as in Fig. 2 by setting the center of connected component as 1, constructing Voronoi diagram, setting Voronoi edge as 0, and ignoring other pixels. Then, a Voronoi tri-label set $\tilde{\mathcal{Z}}$ can be acquired.

SSimNet Learning and Finetuning. By denoting our segmentation network as F, our final loss function to supervise the network training can be formulated as:

$$Loss = \sum_{\tilde{x} \in \tilde{\mathcal{X}}, \tilde{y} \in \tilde{\mathcal{Y}}, \tilde{z} \in \tilde{\mathcal{Z}}} \lambda L_{BCE}(F(\tilde{x}), \tilde{y}) + (1 - \lambda) L_{CE}(F(\tilde{x}), \tilde{z}), \tag{5}$$

where L_{BCE} is the binary cross-entropy loss and L_{CE} is the cross-entropy loss.

Also, we can obtain tissue patches and corresponding pseudo labels for each image in the test set termed as $SET_k = (\tilde{\mathcal{X}}_k^T, \tilde{\mathcal{Y}}_k^T, \tilde{\mathcal{Z}}_k^T)$. Before evaluation, we first fine tune our network F using SET_k for several epochs. As shown in the ablation study, this operation is simple but effective. And this fine tuning process can help the network capture the size and shape information in the current test slice.

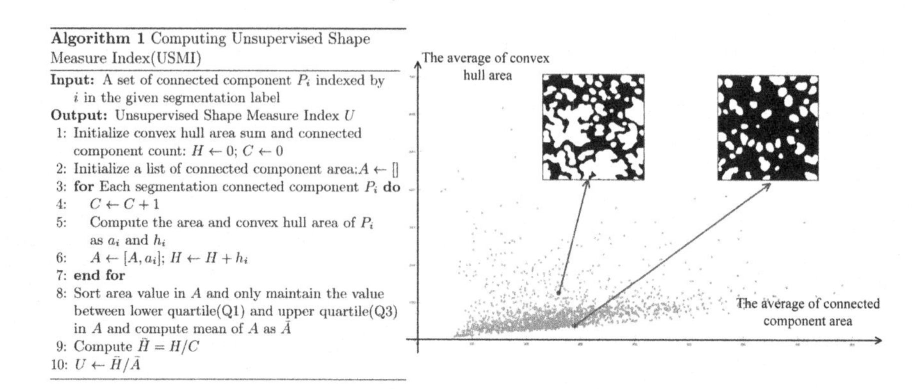

Algorithm 1 Computing Unsupervised Shape Measure Index(USMI)

Input: A set of connected component P_i indexed by i in the given segmentation label
Output: Unsupervised Shape Measure Index U
1: Initialize convex hull area sum and connected component count: $H \leftarrow 0$; $C \leftarrow 0$
2: Initialize a list of connected component area:$A \leftarrow []$
3: **for** Each segmentation connected component P_i **do**
4: $C \leftarrow C + 1$
5: Compute the area and convex hull area of P_i as a_i and h_i
6: $A \leftarrow [A, a_i]$; $H \leftarrow H + h_i$
7: **end for**
8: Sort area value in A and only maintain the value between lower quartile(Q1) and upper quartile(Q3) in A and compute mean of A as $\bar{A}$
9: Compute $\bar{H} = H/C$
10: $U \leftarrow \bar{H}/\bar{A}$

Fig. 3. Left: The algorithm that computes USMI. **Right:** Illustration of average convex hull area and the average of connected component area based on USMI. Yellow (or cyan) points denote the samples whose USMI is greater (or less) than the threshold. (Color figure online)

3 Experiments

3.1 Datasets and Settings

MoNuSeg. Multi-organ nuclei segmentation [16,17] (MoNuSeg) is used to evaluated our SSimNet. The MoNuSeg dataset consists of 44 H&E stained histopathology images with 28,846 manually annotated nuclei. With 1000×1000 pixel resolution, these images were extracted from whole slide images from the The Cancer Genome Atlas (TCGA) repository, representing 9 different organs from 44 individuals.

CPM17. The CPM17 dataset [24] is also derived from TCGA repository. The training and test set each consisted of 32 images tiles selected and extracted from a set of Non-Small Cell Lung Cancer (NSCLC), Head and Neck Squamous Cell Carcinoma (HNSCC), Glioblastoma Multiforme (GBM) and Lower Grade Glioma (LGG) tissue images. Moreover, each type cancer has 8 tiles and the size of patch is 500×500 or 600×600.

Settings. We compare our SSimNet with several current unsupervised segmentation methods. We follow the DCGN [15] to conduct comparison experiments. We crop the image indataset into patches of 256×256 pixels for training. All the methods were trained for 150 epochs on MoNuSeg and 200 epochs on CPM17 each time and experimented with an initial learning rate of $5e^{-5}$ and a decay of 0.98 per epoch. Our experiment repeated ten times on MoNuSeg dataset and only once on CPM17 dataset for an augmented convenience. Specially for our SSimNet training, we set $\alpha = 70\%$ for data purification and $\lambda = 0.9$ for loss in training. Moreover, we fine tune the network with only five epochs for each image on test set with optimizer parameter saved in checkpoint.

Table 1. Performance of the nuclei segmentation on MoNuSeg dataset. The best results are highlighted in **bold** and the second best underlined. 'ft' means fine tuning. The results are shown as "mean±standard deviation(upper-bound results)".

Methods	Precision%↑	Recall%↑	Dice%↑	AJI%↑
mKMeans	65.7±17.5(67.9)	79.2±17.4(77.3)	67.8± 9.4(68.2)	30.5±14.0(33.8)
GMM [9]	63.1±15.0(66.4)	82.2±10.9(81.9)	69.5± 8.5(71.7)	29.0±15.1(31.9)
IIC [12]	46.7± 9.2(51.6)	72.5±12.1(79.6)	56.0± 8.7(61.8)	5.6± 3.0(7.2)
Kim et al. [20]	57.5±24.9(69.8)	82.4±18.9(77.2)	60.6±17.1(69.4)	22.0±17.6(32.3)
Double DIP [18]	22.1± 5.1(22.1)	82.0±10.9(85.1)	34.4± 6.7(35.0)	1.3± 0.6(1.3)
Kanezaki et al. [11]	62.9±19.5(72.5)	82.2±16.2(78.3)	66.9±11.9(72.7)	26.0±16.6(35.1)
DCGMM [14]	69.3±13.5(69.8)	78.6±17.1(80.1)	70.7± 6.4(71.9)	31.4±12.4(34.5)
DIC [13]	51.1±24.9(59.5)	**84.8**±17.0(83.2)	57.1±16.5(64.4)	14.7±16.9(19.3)
DCAGMM [19]	61.9±13.7(69.1)	76.7±13.1(76.3)	66.4± 7.9(70.6)	30.0±12.6(36.5)
DCGN [15]	68.5±11.3(71.6)	83.4±11.5(80.8)	73.7± 4.3(74.3)	35.2±11.3(37.9)
Our SSimNet w/o ft	80.8± 2.1(79.7)	76.1± 3.3(79.3)	76.7± 1.4(78.5)	44.1± 1.5(45.3)
Our SSimNet	**82.0**± 1.7(82.0)	77.2± 2.5(78.6)	**79.2**± 0.6(80.0)	**49.8**± 0.9(51.2)
U-Net (supervised)	73.8± 1.3(75.5)	85.3± 0.4(85.8)	78.7± 1.0(80.0)	51.0± 0.9(52.4)

3.2 Experimental Results

To evaluate the effectiveness of SSimNet, we compare it with several deep learning based and conventional unsupervised segmentation methods on the mentioned datasets, including minibatch K-Means (termed as mKMeans), Gaussian Mixture Model [9] (termed as GMM), Invariant Information Clustering [12] (termed as IIC), Double DIP [18], Deep Clustering via Adaptive GMM model [19] (termed as DCAGMM), Deep Image Clustering [13] (termed as DIC), Kim's work [20], Kanezaki's work [11], Deep Conditional GMM [14] (termed as DCGMM), and Deep Constrained Gaussian Network [15] (termed as DCGN). For the methods without public codes, we report the results from the original publications for a fair comparison. The results are shown in Table 1.

As Table 1 shows, firstly, our SSimNet outperforms all other unsupervised model and performs even close to fully supervised U-Net under the metrics of Dice coefficient and Aggregated Jaccard Index (AJI). Secondly, while the recall of all comparison methods is higher than precision, our SSimNet's recall (0.772) is lower than precision (0.820) and also lower than the state-of-the-art method's recall (0.834). The reason lies in that our method considers mining as strong prior knowledge from tissue slice itself, which renders a tighter constraint on our model, leading the model to predict a lower confidence in the easily-confused region. Moreover, Figure 4 shows the visualization of two test slice. It also conforms the effectiveness of our method on eliminating the model confusion in the region between adjacent nuclei and the ability in capturing nuclei shape.

Besides, we conduct an additional comparison experiment based on CPM17 dataset to demonstrate the generalization of our method. As shown in Table 2, our method again achieves the top performances. Moreover, as the image size of CPM17 is smaller than that of MoNuSeg, the performance gain is not as big as on the MoNuSeg dataset.

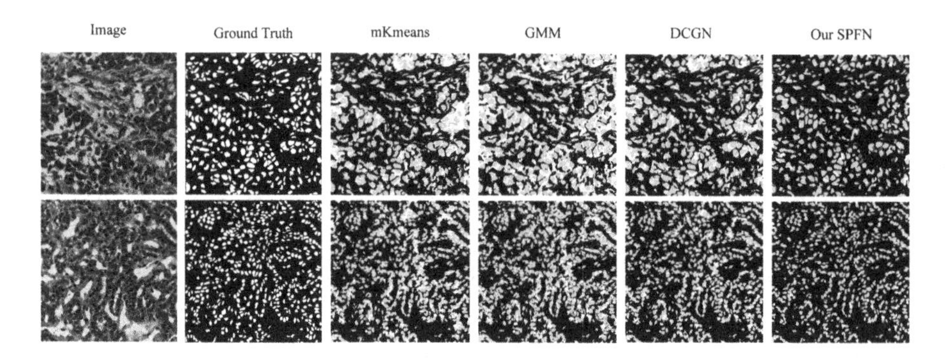

Fig. 4. Comparison of unsupervised nuclei segmentation results on MoNuSeg. Green, yellow and red colors refer to the true positive, the false positive and the false negative predictions. (Color figure online)

Table 2. Performance of the nuclei segmentation on CPM17 dataset. The best results are highlighted in **bold** and the second best underlined.

Methods	Precision%↑	Recall%↑	Dice%↑	AJI%↑
mKMeans	79.4	74.9	74.5	46.1
GMM [9]	79.0	75.5	72.7	43.0
Kanezaki et al. [11]	82.0	65.4	72.1	46.7
Our SSimNet w/o finetune	85.4	**80.7**	81.2	49.3
Our SSimNet	**85.8**	80.5	**81.6**	**49.8**

3.3 Ablation Study

We perform ablation studies by disabling each component to the SSimNet framework to evaluate their effectiveness. As shown in Table 3, each component in our SSimNet can bring different degrees of improvement, which shows that all of the label softening, data purification and finetuning process are significant parts of our SSimNet and play an indispensable role in achieving superior performance.

Table 3. Ablation study on SSimNet using MoNuSeg dataset.

LabelSoftening	DataPurification	Finetune	Precision%↑	Recall%↑	Dice%↑	AJI%↑
	✓	✓	78.8	78.7	78.3	45.6
✓		✓	79.6	77.3	77.9	47.6
✓	✓		80.8	76.1	76.7	44.1
✓	✓	✓	82.0	77.2	79.2	49.8

4 Conclusion

In this paper, we propose an SSimNet framework for label-free nuclei segmentation. Motivated by the intra-image self similarity (IISS) property, which characterize the histology images and nuclei, we design a series of operations to capture the prior knowledge and generate pseudo labels as supervision signal, which is used to learn the SSimNet for final nuclei segmentation. The IISS property renders us a tighter prior constraint for better model building compared to other unsupervised nuclei segmentation. Comprehensive experimental results demonstrate that SSimNet achieves the best performances on the benchmark MoNuSeg and CPM17 datasets, outperforming other unsupervised segmentation methods.

Acknowledgement. Supported by Natural Science Foundation of China under Grant 62271465 and Open Fund Project of Guangdong Academy of Medical Sciences, China (No. YKY-KF202206).

References

1. Graham, S., et al.: Hover-net: simultaneous segmentation and classification of nuclei in multi-tissue histology images. Med. Image Anal. **58**, 101563 (2019)
2. Graham, S., et al.: Mild-net: minimal information loss dilated network for gland instance segmentation in colon histology images. Med. Image Anal. **52**, 199–211 (2019)
3. Guo, Z., et al.: A fast and refined cancer regions segmentation framework in whole-slide breast pathological images. Sci. Rep. **9**(1), 882 (2019)
4. Mahbod, A., et al.: Cryonuseg: a dataset for nuclei instance segmentation of cryosectioned h&e-stained histological images. Comput. Biol. Med. **132**, 104349 (2021)
5. Xiang, T., Zhang, C., Liu, D., Song, Y., Huang, H., Cai, W.: BiO-Net: learning recurrent bi-directional connections for encoder-decoder architecture. In: Martel, A.L., et al. (eds.) MICCAI 2020. LNCS, vol. 12261, pp. 74–84. Springer, Cham (2020). https://doi.org/10.1007/978-3-030-59710-8_8
6. Umesh Adiga, P.S., Chaudhuri, B.B.: An efficient method based on watershed and rule-based merging for segmentation of 3-d histo-pathological images. Pattern Recogn. **34**(7), 1449–1458 (2001)
7. Wienert, S., et al.: Detection and segmentation of cell nuclei in virtual microscopy images: a minimum-model approach. Sci. Rep. **2**(1), 503 (2012)
8. Nogues, I., et al.: Automatic lymph node cluster segmentation using holistically-nested neural networks and structured optimization in CT images. In: Ourselin, S., Joskowicz, L., Sabuncu, M.R., Unal, G., Wells, W. (eds.) MICCAI 2016. LNCS, vol. 9901, pp. 388–397. Springer, Cham (2016). https://doi.org/10.1007/978-3-319-46723-8_45
9. Ragothaman, S., Narasimhan, S., Basavaraj, M.G., Dewar, R.: Unsupervised segmentation of cervical cell images using gaussian mixture model. In: Proceedings of the IEEE Conference on Computer Vision and Pattern Recognition Workshops, pp. 70–75 (2016)
10. Zhang, Y., Brady, M., Smith, S.: Segmentation of brain mr images through a hidden markov random field model and the expectation-maximization algorithm. IEEE Trans. Med. Imaging **20**(1), 45–57 (2001)

11. Kanezaki, A.: Unsupervised image segmentation by backpropagation. In 2018 IEEE International Conference on Acoustics, Speech and Signal Processing (ICASSP), pp. 1543–1547. IEEE (2018)
12. Ji, X., Henriques, J.F., Vedaldi, A.: Invariant information clustering for unsupervised image classification and segmentation. In: Proceedings of the IEEE/CVF International Conference on Computer Vision, pp. 9865–9874 (2019)
13. Zhou, L., Wei, W.: DIC: deep image clustering for unsupervised image segmentation. IEEE Access **8**, 34481–34491 (2020)
14. Zanjani, F.G., Zinger, S., Bejnordi, B.E., van der Laak, H.A.W.M., et al.: Histopathology stain-color normalization using deep generative models. In: Medical Imaging with Deep Learning (2018)
15. Nan, Y., et al.: Unsupervised tissue segmentation via deep constrained gaussian network. IEEE Trans. Med. Imaging **41**(12), 3799–3811 (2022)
16. Kumar, N., Verma, R., Sharma, S., Bhargava, S., Vahadane, A., Sethi, A.: A dataset and a technique for generalized nuclear segmentation for computational pathology. IEEE transactions on medical imaging **36**(7), 1550–1560 (2017)
17. Kumar, N., et al.: A multi-organ nucleus segmentation challenge. IEEE Trans. Med. Imaging **39**(5), 1380–1391 (2019)
18. Gandelsman, Y., Shocher, A., Irani, M.: "double-dip": unsupervised image decomposition via coupled deep-image-priors. In: Proceedings of the IEEE/CVF Conference on Computer Vision and Pattern Recognition, pp. 11026–11035 (2019)
19. Wang, J., Jiang, J.: Unsupervised deep clustering via adaptive gmm modeling and optimization. Neurocomputing **433**, 199–211 (2021)
20. Kim, W., Kanezaki, A., Tanaka, M.: Unsupervised learning of image segmentation based on differentiable feature clustering. IEEE Trans. Image Process. **29**, 8055–8068 (2020)
21. Vahadane, A., et al.: Structure-preserving color normalization and sparse stain separation for histological images. IEEE Trans. Med. Imaging **35**(8), 1962–1971 (2016)
22. Feldman, A.T., Wolfe, D.: Tissue processing and hematoxylin and eosin staining. In: Histopathology: Methods and Protocols, pp. 31–43 (2014)
23. Hui, Q., et al.: Weakly supervised deep nuclei segmentation using partial points annotation in histopathology images. IEEE Trans. Med. Imaging **39**(11), 3655–3666 (2020)
24. Vu, Q.D., et al.: Methods for segmentation and classification of digital microscopy tissue images. Front. Bioeng. Biotechnol. **53** (2019)

Histopathology Image Classification Using Deep Manifold Contrastive Learning

Jing Wei Tan and Won-Ki Jeong(✉)

Department of Computer Science and Engineering, Korea University, Seoul, South Korea
{jingwei_92,wkjeong}@korea.ac.kr

Abstract. Contrastive learning has gained popularity due to its robustness with good feature representation performance. However, cosine distance, the commonly used similarity metric in contrastive learning, is not well suited to represent the distance between two data points, especially on a nonlinear feature manifold. Inspired by manifold learning, we propose a novel extension of contrastive learning that leverages geodesic distance between features as a similarity metric for histopathology whole slide image classification. To reduce the computational overhead in manifold learning, we propose geodesic-distance-based feature clustering for efficient contrastive loss evaluation using prototypes without time-consuming pairwise feature similarity comparison. The efficacy of the proposed method is evaluated on two real-world histopathology image datasets. Results demonstrate that our method outperforms state-of-the-art cosine-distance-based contrastive learning methods.

Keywords: Contrastive learning · Manifold learning · Geodesic distance · Histopathology image classification · Multiple instance learning

1 Introduction

Whole slide image (WSI) classification is a crucial process to diagnose diseases in digital pathology. Owing to the huge size of a WSI, the conventional WSI classification process consists of patch decomposition and per-patch classification, followed by the aggregation of per-patch results using multiple instance learning (MIL) for the final per-slide decision [7]. MIL constructs *bag-of-features (BoF)* that effectively handles imperfect patch labels, allowing weakly supervised learning using per-slide labels for WSI classification. Although MIL does not require perfect per-patch label assignment, it is important to construct good feature vectors that are easily separated into different classes to make the classification more accurate. Therefore, extensive research has been conducted on metric and representation learning [11,13] aimed at developing improved feature representation.

H. Greenspan et al. (Eds.): MICCAI 2023, LNCS 14225, pp. 683–692, 2023.
https://doi.org/10.1007/978-3-031-43987-2_66

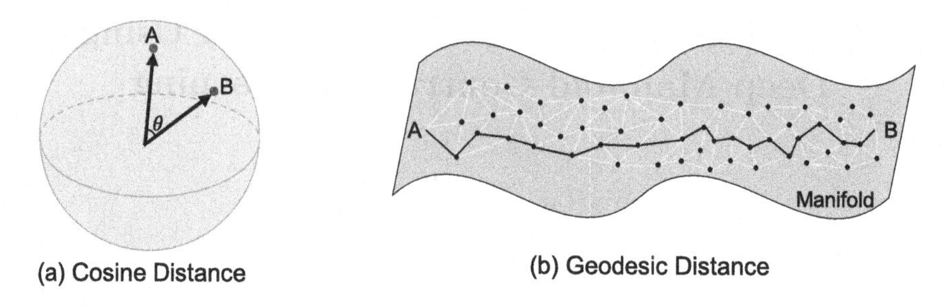

Fig. 1. Comparison of geodesic and cosine distance in n-dimensional space.

Recently, contrastive learning has demonstrated its robustness in the representational ability of the feature extractor, which employs self-supervised learning with a contrastive loss that forces samples from the same class to stay closer in the feature space (and vice versa). SimCLR [3] introduced the utilization of data augmentation and a learnable nonlinear transformation between the feature embedding and the contrastive loss to generally improve the quality of feature embedding. MoCo [6] employed a dynamic dictionary along with a momentum encoder in the contrastive learning model to serve as an alternative to the supervised pre-trained ImageNet model in various computer vision tasks. PCL [9] and HCSC [5] integrated the k-means clustering and contrastive learning model by introducing prototypes as latent variables and assigning each sample to multiple prototypes to learn the hierarchical semantic structure of the dataset. These prior works used cosine distance as their distance measurement, which computes the angle between two feature vectors as shown in Fig. 1(a). Although cosine distance is a commonly used distance metric in contrastive learning, we observed that the cosine distance approximates the difference between local neighbors and is insufficient to represent the distance between far-away points on a complicated, nonlinear manifold.

The main motivation of this work is to extend the current contrastive learning to represent the nonlinear feature manifold inspired by manifold learning. Owing to the manifold distribution hypothesis [8], the relative distance between high-dimensional data is preserved on a low-dimensional manifold. ISOMAP [12] is a well-known manifold learning approach that represents the manifold structure by using geodesic distance (i.e., the shortest path length between points on the manifold). There are several previous works that use manifold learning for image classification and reconstruction tasks, such as Lu *et al.* [10] and Zhu *et al.* [14]. However, the use of geodesic distance on the feature manifold for image classification is a recent development. Aziere *et al.* [2] applied the random walk algorithm on the nearest neighbor graph to compute the pairwise geodesic distance and proposed the N-pair loss to maximize the similarity between samples from the same class for image retrieval and clustering applications. Gong *et al.* [4] employed the geodesic distance computed using the Dijkstra algorithm on the k-nearest neighbor graph to measure the correlation between the original samples and then further divided each class into sub-classes to deal with the problems of high spectral dimension and channel redundancy in the hyperspectral images. However, this method captured the nonlinear data manifold structure on the original data (not on the feature vectors) only once at the beginning stage, which is not updated in the further training process.

In this study, we propose a hybrid method that combines manifold learning and contrastive learning to generate a good feature extractor (encoder) for histopathology image classification. Our method uses the sub-classes and prototypes as in conventional contrastive learning, but we propose the use of geodesic distance in generating the sub-classes to represent the non-linear feature manifold more accurately. By doing this, we achieve better separation between features with large margins, resulting in improved MIL classification performance. The main contributions of our work can be summarized as follows:

- We introduce a novel integration of manifold geodesic distance in contrastive learning, which results in better feature representation for the non-linear feature manifold. We demonstrate that the proposed method outperforms conventional cosine-distance-based contrastive learning methods.
- We propose a geodesic-distance-based feature clustering for efficient contrastive loss evaluation using prototypes without brute-force pairwise feature similarity comparison while approximating the overall manifold geometry well, which results in reduced computation.
- We demonstrate that the proposed method outperforms other state-of-the-art (SOTA) methods with a much smaller number of sub-classes without complicated prototype assignment (e.g., hierarchical clustering).

To the best of our knowledge, this work is the first attempt to leverage manifold geodesic distance in contrastive learning for histopathology WSI classification.

2 Method

The overview of our proposed model is illustrated in Fig. 2. It is composed of two stages: (1) train the feature extractor using deep manifold embedding learning and (2) train the WSI classifier using the deep manifold embedding extracted from the first stage. The input WSIs are pre-processed to extract $256 \times 256 \times 3$ dimensional patches from the tumor area at a $10\times$ magnification level. Patches with less than 50% tissue coverage are excluded from the experiment.

2.1 Deep Manifold Embedding Learning

As illustrated in Fig. 2(a), we first feed the patches into a feature extractor f, which is composed of an encoder, a pooling layer, and a multi-perceptron layer. The output is then passed through two different paths, namely, deep manifold and softmax paths.

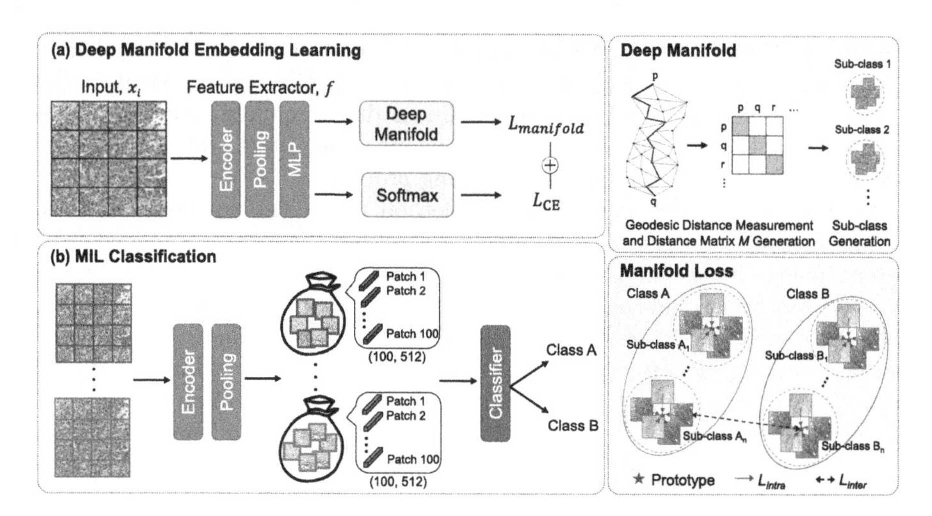

Fig. 2. Overview of our proposed method, which is composed of two stages: (a) deep manifold embedding learning and (b) MIL classification.

Deep Manifold. In this stage, the patches from each class are further grouped into sub-classes based on manifold geodesic distance. First, an undirected nearest neighbor graph $G_c = (V_c, E_c)$ is constructed, where V_c is a set of nodes made from the patch feature of the c-th class extracted by f, and E_c is the set of edges in the graph. Each node (patch feature) is connecting to its k-nearest neighbors (KNN) based on the weighted edges computed with Euclidean distance, given that the neighbor samples on the manifold should have a higher potential to be in the same sub-class. The geodesic distance matrix M on the manifold is then computed between each sample pair by using Dijkstra's algorithm based on the G_c. The samples of each class are then further clustered into several sub-classes with agglomerative clustering based on the geodesic distance. In agglomerative clustering, all the patch features are initially treated as individual clusters and the nearest two individual clusters are merged based on the geodesic distance matrix to form new clusters. The distance matrix M is then updated with the newly formed clusters. With the updated distance matrix M, the nearest two individual clusters are again merged to form new clusters. These steps are repeated until the desired number n of sub-classes is achieved.

Loss Functions. For the deep manifold training, we adopted two losses: (1) intra-subclass loss L_{intra} and (2) inter-subclass loss L_{inter}. The main idea in intra-subclass loss is to make the samples from the same sub-class stay near their respective sub-class prototype. L_{intra} is formulated as follows:

$$L_{intra} = \frac{1}{J \cdot I} \sum_{j=1}^{J} \sum_{i=1}^{I} (f(x_j^i) - p^+)^T (f(x_j^i) - p^+) \quad (1)$$

where x_j^i is the i-th patch in the j-th batch, J represents the total number of batches, I represents the total number of patches per batch, $f(\cdot)$ is the feature extractor, and p^+ indicates the positive prototype of the patch (i.e., the prototype of the subclass containing x_j^i). The prototype of each sub-class is computed by simply taking the mean of all the patch features that belong to each sub-class. Inter-subclass loss L_{inter} is proposed to make the sub-classes from a different class far apart from one another. The formulation of L_{inter} is as shown below:

$$L_{inter} = \frac{1}{J}\sum_{j=1}^{J}(\triangle - D(f(Q_j^A), P^B)) \tag{2}$$

$$D(Y, Z) = \max\{\sup_{y\in Y} d(y, Z), \sup_{z\in Z} d(z, Y)\} \tag{3}$$

where $f(Q_j^A)$ is a set of patch features in batch j from class A, P^B is a set of prototypes from the sub-classes of class B, and $\triangle$ is a positive margin between classes on data manifold. $D(\cdot)$ is the Hausdorff distance, where *sup* indicates supremum, *inf* indicates infimum, and $d(t, R) = \inf_{r\in R} ||t - r||$ which measures the distance from a data point $t \in Y$ to the subset $R \subseteq Y$. Then, the manifold loss is formulated as

$$L_{manifold} = L_{intra} + L_{inter} \tag{4}$$

Another path via softmax is simply trained on outputs from the feature extractor with the ground truth slide-level labels y by the cross-entropy loss L_{CE}, which is defined as follows:

$$L_{CE} = -\frac{1}{J\cdot I}\sum_{j=1}^{J}\sum_{i=1}^{I} y_j^i \cdot \log \hat{y}_j^i + (1 - y_j^i)\cdot \log(1 - \hat{y}_j^i) \tag{5}$$

where y is the ground truth slide-level label and $\hat{y}$ is predicted label.

Finally, the total loss for the first stage is defined as follows:

$$L_{total} = L_{manifold} + L_{CE} \tag{6}$$

2.2 MIL Classification

As illustrated in Fig. 2(b), in the second stage, the pre-trained feature extractor from the previous stage is then deployed to extract features for bag generation. A total of 50 bags are generated for each WSI, in which each bag is composed of the concatenation of the features from 100 patches in 512 dimensions. These bags are fed into a classifier with two layers of multiple perceptron layers (512 neurons) and a Softmax layer and then trained with a binary cross-entropy loss. After the classification, majority voting is applied to the predicted labels of the bags to derive the final predicted label for each WSI.

3 Result

3.1 Datasets

We tested our proposed method on two different tasks: (1) intrahepatic cholangiocarcinomas(IHCCs) subtype classification and (2) liver cancer type classification. The dataset for the former task was collected from 168 patients with 332 WSIs from Seoul National University hospital. IHCCs can be further categorized into small duct type (SDT) and large duct type (LDT). Using gene mutation information as prior knowledge, we collected WSIs with wild KRAS and mutated IDH genes for use as training samples in SDT, and WSIs with mutated KRAS and wild IDH genes for use in LDT. The rest of the WSIs were used as testing samples. The liver cancer dataset for the latter task was composed of 323 WSIs, in which the WSIs can be further classified into hepatocellular carcinomas (HCCs) (collected from Pathology AI Platform [1]) and IHCCs. We collected 121 WSIs for the training set, and the remaining WSIs were used as the testing set.

3.2 Implementation Detail

We used a pre-trained VGG16 with ImageNet as the initial encoder, which was further modified via deep manifold model training using the proposed manifold and cross-entropy loss functions. The number of nearest neighbors k and the number of sub-classes n were set to 5 and 10, respectively. In the deep manifold embedding learning model, the learning rates were set to 1e-4 with a decay rate of 1e-6 for the IHCCs subtype classification and to 1e-5 with a decay rate of 1e-8 for the liver cancer type classification. The k-nearest neighbors graph and the geodesic distance matrix are updated once every five training epochs, which is empirically chosen to balance running time and accuracy. To train the MIL classifier, we set the learning rate to 1e-3 and the decay rate to 1e-6. We used batch sizes 64 and 4 for training the deep manifold embedding learning model and the MIL classification model, respectively. The number of epochs for the deep manifold embedding learning model was 50, while 50 and 200 epochs for the IHCCs subtype classification and liver cancer type classification, respectively. As for the optimizer, we used stochastic gradient decay for both stages. The result shown in the tables is the average result from 10 iterations of the MIL classification model.

3.3 Experimental Results

The performance of different models from two different datasets is reported in this section. For the baseline model, we chose the pre-trained VGG16 feature extractor with an MIL classifier, which is the same as our proposed model except that the encoder is retrained using the proposed loss. Two SOTA methods using contrastive learning and clustering, PCL [9] and HCSC [5], are compared with our method in this study. The MIL classification result of the IHCCs subtype classification is shown in Table 1. Our proposed method outperformed the baseline CNN by about 4% increment in accuracy, precision, recall, and F1 score. Note that our method only used 20 sub-classes but outperformed PCL (using 2300 sub-classes) by 4% and HCSC (using 112 sub-classes) by 5% in accuracy.

Table 1. Classification performance on IHCCs subtype and liver cancer type dataset. (Acc.: Accuracy, Prec.: Precision, Rec.: Recall, F1: F1 Score, NA: Not Applicable)

Method	Prototype Number	IHCC Subtype				Liver Cancer Type			
		Acc.	Prec.	Rec.	F1	Acc.	Prec.	Rec.	F1
CNN	NA	0.7315	0.7372	0.7315	0.7270	0.7710	0.7781	0.7719	0.7657
PCL	500-800-1000	0.7386	0.7478	0.7394	0.7354	0.8146	0.7898	0.8146	0.7979
HCSC	2-10-100	0.7230	0.7265	0.7230	0.7231	0.7995	**0.8524**	0.7995	0.7825
Ours	20	**0.7703**	**0.7710**	**0.7678**	**0.7668**	**0.8239**	0.8351	**0.8239**	**0.8227**

Table 2. Ablation study of prototype assignment strategies.

Prototype	Prototype Number	Accuracy	Precision	Recall	F1 Score
Global	2	0.7365	0.7390	0.7365	0.7353
Local	20	**0.7703**	0.7710	0.7678	0.7668
Global + Local	22 (20 + 2)	0.7698	**0.7735**	**0.7698**	**0.7692**

The result of liver cancer type classification is also shown in Table 1. Our method achieved about 5% improvement in accuracy against the baseline and 1% to 2% improvement in accuracy against the SOTA methods. Moreover, it outperformed the SOTA methods with far fewer prototypes and without complicated hierarchical prototype assignments. To further evaluate the effect of prototypes, we conducted an ablation study for different prototype assignment strategies as shown in Table 2. Here, global prototypes imply assigning a single prototype per class while local prototypes imply assigning multiple prototypes per class (one per sub-class). When both are used together, it implies a hierarchical prototype assignment where local prototypes interact with the corresponding global prototype. As shown in this result, the model with local prototypes only performed about 4% higher than did the model with global prototypes only. Meanwhile, the combination of both prototypes achieved a similar performance to that of the model with local prototypes only. Since the hierarchical (global + local) assignment did not show a significant improvement but instead increased computation, we used only local sub-class prototypes in our final experiment setting.

Table 3. Classification performance of geodesic distance and cosine distance.

Method	Number of sub-classes	Accuracy	Precision	Recall	F1 Score
Cosine distance	2	0.7519	0.7552	0.7519	0.7503
Cosine distance	20	0.7576	0.7589	0.7576	0.7571
Ours	20	**0.7703**	**0.7710**	**0.7678**	**0.7668**

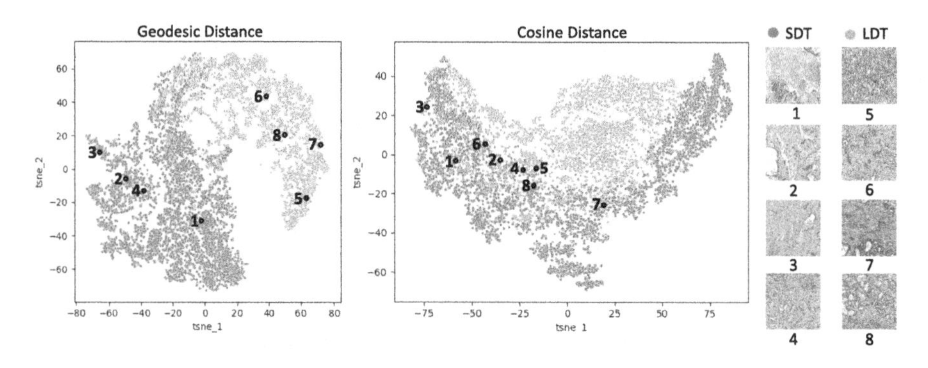

Fig. 3. Comparison of geodesic and cosine distance in feature space.(1)–(4) are the patches from SDT and (5)–(8) are the patches from LDT.

Since one of our contributions is the use of geodesic distance, we assessed the efficacy of the method by comparing it with the performance using cosine distance, as shown in Table 3. To measure the performance of the cosine-distance-based method, we simply replaced our proposed manifold loss with NT-Xent loss [3], which uses cosine distance in their feature similarity measurement. Two cosine distance experiments were conducted as follows: (1) use only their ground-truth class without further dividing the samples into sub-classes (i.e., global prototypes) and (2) divide the samples from each class into 10 sub-classes by using k-means clustering (i.e., local prototypes). As shown in Table 3, using multiple local prototypes shows slightly better performance compared to using global prototypes. By switching the NT-Xent loss with our geodesic-based manifold loss, the overall performance is increased by about 2%. Figure 3 visually compares the effect of the geodesic and cosine distance-based losses. Two scatter plots are t-SNE projections of feature vectors from the encoders trained using geodesic distance and cosine distance, respectively. Red dots represent SDT samples and blue dots represent LDT samples from the IHCCs dataset (corresponding histology thumbnail images are shown on the right). In this example, all eight cases are correctly classified by the method using geodesic distance while all cases are incorrectly classified by the method using cosine distance. It is clearly shown that geodesic distance can correctly measure the feature distance (similarity) on the manifold so that SDT and LDT groups are located far away in the t-SNE plot, whereas cosine distance failed to separate these groups and they are located nearby in the plot.

3.4 Conclusion and Future Work

In this paper, we proposed a novel geodesic-distance-based contrastive learning for histopathology image classification. Unlike conventional cosine-distance-based contrastive learning methods, our method can represent nonlinear feature manifold better and generate better discriminative features. One limitation of the proposed method is the extra computation time for graph generation and pairwise distance computation using the Dijkstra algorithm. In the future, we plan to optimize the algorithm and apply our method to other datasets and tasks, such as multi-class classification problems and natural image datasets.

Acknowledgements. This study was approved by the institutional review board of Seoul National University Hospital (IRB NO.H-1011-046-339). This work was partially supported by the National Research Foundation of Korea (NRF-2019M3E5D2A01063819, NRF-2021R1A6A1A13044830), the Institute for Information & Communications Technology Planning & Evaluation (IITP-2023-2020-0-01819), the Korea Health Industry Development Institute (HI18C0316), the Korea Institute of Science and Technology (KIST) Institutional Program (2E32210 and 2E32211) and a Korea University Grant.

References

1. Pathology AI platform (PAIP). http://wisepaip.org/paip
2. Aziere, N., Todorovic, S.: Ensemble deep manifold similarity learning using hard proxies. In: Proceedings of the IEEE/CVF Conference on Computer Vision and Pattern Recognition, pp. 7299–7307 (2019)
3. Chen, T., Kornblith, S., Norouzi, M., Hinton, G.: A simple framework for contrastive learning of visual representations. In: International Conference on Machine Learning, pp. 1597–1607. PMLR (2020)
4. Gong, Z., Hu, W., Du, X., Zhong, P., Hu, P.: Deep manifold embedding for hyperspectral image classification. IEEE Trans. Cybern. **52**(10), 10430–10443 (2021)
5. Guo, Y., et al.: HCSC: hierarchical contrastive selective coding. In: Proceedings of the IEEE/CVF Conference on Computer Vision and Pattern Recognition, pp. 9706–9715 (2022)
6. He, K., Fan, H., Wu, Y., Xie, S., Girshick, R.: Momentum contrast for unsupervised visual representation learning. In: Proceedings of the IEEE/CVF Conference on Computer Vision and Pattern Recognition, pp. 9729–9738 (2020)
7. Kather, J.N., et al.: Deep learning can predict microsatellite instability directly from histology in gastrointestinal cancer. Nat. Med. **25**(7), 1054–1056 (2019)
8. Lei, N.: A geometric understanding of deep learning. Engineering **6**(3), 361–374 (2020)
9. Li, J., Zhou, P., Xiong, C., Hoi, S.C.: Prototypical contrastive learning of unsupervised representations. arXiv preprint arXiv:2005.04966 (2020)
10. Lu, J., Wang, G., Deng, W., Moulin, P., Zhou, J.: Multi-manifold deep metric learning for image set classification. In: Proceedings of the IEEE conference on Computer Vision and Pattern Recognition, pp. 1137–1145 (2015)
11. Ommer, B.: Introduction to similarity and deep metric learning (2022). https://dvsml2022-tutorial.github.io/Talks/02_DML_tutorial_BjornOmmer_.pdf

12. Theodoridis, S., Koutroumbas, K.: Feature generation i: data transformation and dimensionality reduction. Pattern Recogn. **4** (2009)
13. Zheng, W., Chen, Z., Lu, J., Zhou, J.: Hardness-aware deep metric learning. In: Proceedings of the IEEE/CVF Conference on Computer Vision and Pattern Recognition, pp. 72–81 (2019)
14. Zhu, B., Liu, J.Z., Cauley, S.F., Rosen, B.R., Rosen, M.S.: Image reconstruction by domain-transform manifold learning. Nature **555**(7697), 487–492 (2018)

Zero-Shot Nuclei Detection via Visual-Language Pre-trained Models

Yongjian Wu[1], Yang Zhou[1], Jiya Saiyin[1], Bingzheng Wei[2], Maode Lai[3], Jianzhong Shou[4], Yubo Fan[1], and Yan Xu[1(✉)]

[1] School of Biological Science and Medical Engineering, State Key Laboratory of Software Development Environment, Key Laboratory of Biomechanics and Mechanobiology of Ministry of Education, Beijing Advanced Innovation Center for Biomedical Engineering, Beihang University, Beijing 100191, China
xuyan04@gmail.com

[2] Xiaomi Corporation, Beijing 100085, China

[3] Department of Pathology, School of Medicine, Zhejiang University, Zhejiang Provincial Key Laboratory of Disease Proteomics and Alibaba-Zhejiang University Joint Research Center of Future Digital Healthcare, Hangzhou 310053, China

[4] Chinese Academy of Medical Sciences and Peking Union Medical College, Beijing 100021, China

Abstract. Large-scale visual-language pre-trained models (VLPM) have proven their excellent performance in downstream object detection for natural scenes. However, zero-shot nuclei detection on H&E images via VLPMs remains underexplored. The large gap between medical images and the web-originated text-image pairs used for pre-training makes it a challenging task. In this paper, we attempt to explore the potential of the object-level VLPM, Grounded Language-Image Pre-training (GLIP) model, for zero-shot nuclei detection. Concretely, an automatic prompts design pipeline is devised based on the association binding trait of VLPM and the image-to-text VLPM BLIP, avoiding empirical manual prompts engineering. We further establish a self-training framework, using the automatically designed prompts to generate the preliminary results as pseudo labels from GLIP and refine the predicted boxes in an iterative manner. Our method achieves a remarkable performance for label-free nuclei detection, surpassing other comparison methods. Foremost, our work demonstrates that the VLPM pre-trained on natural image-text pairs exhibits astonishing potential for downstream tasks in the medical field as well. Code will be released at github.com/VLPMNuD.

Keywords: Nuclei Detection · Unsupervised Learning · Visual-Language Pre-trained Models · Prompt Designing · Zero-shot Learning

Y. Wu and Y. Zhou—Equal contribution.

Supplementary Information The online version contains supplementary material available at https://doi.org/10.1007/978-3-031-43987-2_67.

H. Greenspan et al. (Eds.): MICCAI 2023, LNCS 14225, pp. 693–703, 2023.
https://doi.org/10.1007/978-3-031-43987-2_67

1 Introduction

In the field of medical image processing, nuclei detection on Hematoxylin and Eosin (H&E)-stained images plays a crucial role in various areas of biomedical research and clinical applications [8]. While fully-supervised methods have been proposed for this task [9,20,28], the annotation remains labor-intensive and expensive. To address the aforementioned issue, several unsupervised methods have been proposed, including thresh-holding-based methods [12,22], self-supervised-based methods [26], and domain adaptation-based methods [14]. Among these, domain adaptation methods are mainstream and have demonstrated favorable performance by achieving adaptation through aligning the source and target domains [14]. However, current unsupervised methods exhibit strong empirical design and introduce subjective biases during the model design process, thus current unsupervised methods may lead to suboptimal results.

Yet, newly developed large-scale visual-language pre-trained models (VLPMs) have provided another possible unsupervised learning paradigm [11, 24]. VLPM learns aligned text and image features from massive text-image pairs acquired from the internet, making the learned visual features semantic-rich, general, and transferable. Zero-shot learning methods based on VLPM for downstream tasks such as text-driven image manipulation [23], image captioning [15], view synthesis [10], and object detection [16], have achieved excellent results.

Among VLPMs, Grounded Language-Image Pre-training (GLIP) model [16], pre-trained at the object level, can even rival fully-supervised counterparts in zero-shot object detection and phrase grounding tasks. Although VLPM has been utilized for object detection in natural scenes, zero-shot nuclei detection on H&E images via VLPM remains underexplored. The significant domain differences between medical H&E images and the natural images used for pre-training make this task challenging. It is wondered whether VLPM, with its rich semantic information, can facilitate direct prediction of nuclei detection through semantic-driven prompts, establishing an elegant, concise, clear but more efficient and transferable unsupervised system for label-free nuclei detection.

Building upon this concept, our goal is to establish a zero-shot nuclei detection framework based on VLPM. However, directly applying VLPM for this task poses two challenges. (1) Due to the gap between medical images and the web-originated text-image pairs used for pre-training, the text-encoder may lack prior knowledge of medical concept words, thus making the prompt design for zero-shot detection a challenging task. (2) Different from the objects in natural images, the high density and specialized morphology of nuclei in H&E stained images may lead to missed detection, false detection, and overlapping during zero-shot transfer solely with prompts.

To address the first challenge, Yamada *et al.* have analyzed and revealed that under the pre-training of vast text-image pairs, VLPM establishes a strong association binding between the object and its semantic attributes regardless of image domain, i.e., associated attribute text can fully describe the corresponding objects in an image through VLPM [27]. Therefore, it is feasible for VLPM to detect unseen medical objects in a label-free manner by constructing appropriate attribute texts. Manual prompting is a cumbersome and subjective process, which may lead to considerable bias. Yet, the VLPM network BLIP [15] has the capability to generate automatic descriptions for images. Therefore, we first use BLIP to automatically generate attribute words to describe the unseen nuclei object. This approach avoids the empirical manual prompt design and fully leverages the text-to-image aligning trait of VLPMs. We subsequently integrate these attribute words with medical nouns, i.e. "[shape][color][noun]", to create detection prompts. These prompts are then inputted into GLIP to realize zero-shot detection of nuclei. Our proposed automatic prompt designing method fully utilizes the text-to-image alignment of VLPM, and enables the automatic generation of the most suitable attribute text words describing the corresponding domain. Our approach offers excellent interpretability.

Through GLIP's strong object retrieval performance, we can obtain preliminary boxes. The precision of these preliminary boxes is relatively high, but there is still considerable room for improvement in recall. Therefore, we further establish a self-training framework. We use the preliminary boxes generated by GLIP as pseudo labels for further training YOLOX [7], to refine and polish the predicted boxes in an iterative manner. Together with the self-training strategy, the resulting model achieves a remarkable performance for label-free nuclei detection, surpassing other comparison methods. We demonstrate that VLPM, which is pre-trained on natural image-text pairs, also exhibits astonishing potential for downstream tasks in the medical field.

The contributions of this paper are threefold. (1) A novel zero-shot label-free nuclei detection framework is proposed based on VLPMs. Our method outperforms all existing unsupervised methods and demonstrates excellent transferability. (2) We leverage GLIP, which places more emphasis on object-level representation learning and generates more high-quality language-aware visual representations compared to Contrastive Language-Image Pre-training (CLIP) model, to achieve better nuclei retrieval. (3) An automatic prompt design process is established based on the association binding trait of VLPM to avoid non-trivial empirical manual prompt engineering.

2 Method

Our approach aims to establish a zero-shot nuclei detection framework based on VLPMs by directly using text prompts. We utilize GLIP for better object-level representation extraction. The overview of our framework is shown in Fig. 1.

2.1 Object-Level VLPM — GLIP

Recently, large VLPMs such as CLIP [24] and ALIGN [11] have made great progress in generic visual representation learning and demonstrated the enormous potential of utilizing prompts for zero-shot transfer.

A typical VLPM framework comprises two encoders: the text encoder, which encodes text prompts to semantic-rich text embeddings, and the image encoder,

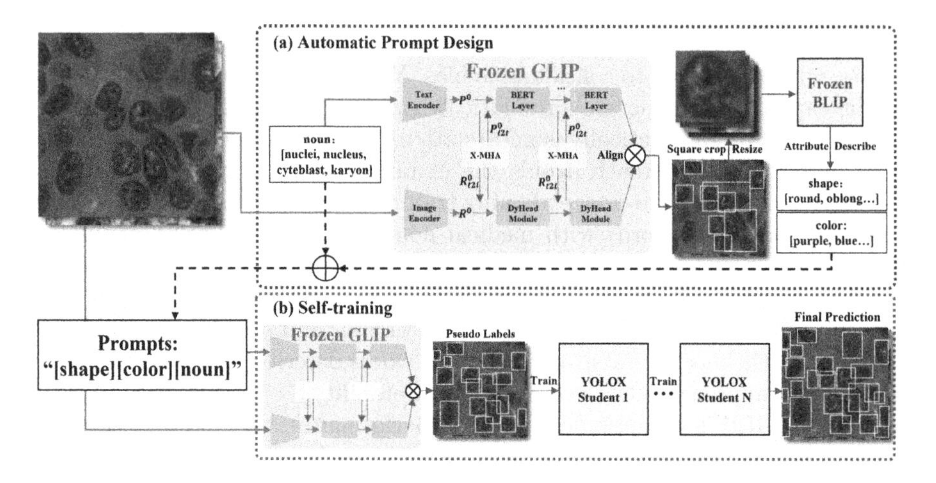

Fig. 1. An overview of our zero-shot nuclei detection framework based on the frozen object-level VLPM GLIP. (a) Given the original target nouns of the task, prompts are designed automatically to avoid non-trivial empirical manual prompt engineering, based on the association binding of VLPM and the image-to-text VLPM BLIP. (b) A self-training strategy is further adopted to refine and polish the predicted boxes in an iterative manner. The automatically designed prompts are used by GLIP to generate the preliminary results as pseudo labels.

which encodes images to visual embeddings. These VLPMs use a vast amount of web-originated text-image pairs $\{(X,T)\}_i$ to learn the text-to-image alignment through contrastive loss over text and visual embeddings. Denoting the image encoder as E_I, the text encoder as E_T, the cosine similarity function as $\cos(\cdot)$, and assuming that K text-image pairs are used in each training epoch, the objective of the contrastive learning can be formulated as:

$$\mathcal{L}_c = -\sum_{i=1}^{K} \log \frac{\exp\left(\cos\left(\cdot E_I\left(X_i\right) \cdot E_T\left(T_i\right)\right)/\tau\right)}{\sum_{j=1}^{K} \exp\left(\cos\left(E_I\left(X_i\right) \cdot E_T\left(T_j\right)\right)/\tau\right)}. \tag{1}$$

Through this aligning contrastive learning, VLPM aligns text and image in a common feature space, allowing one to directly transfer a trained VLPM to downstream tasks via manipulating text, i.e., prompt engineering. The visual representations of input images are semantic-rich and interpretability-friendly with the help of aligned text-image pairs.

However, the conventional VLPM pre-training process only aligns the image with the text from a whole perspective, which results in a lack of emphasis on object-level representations. Therefore, Li *et al.* proposed GLIP [16], whose image encoder generates visual embeddings for each object of different regions in the image to align with object words present in the text. Moreover, GLIP utilizes web-originated phrase grounding text-image pairs to extract novel object word entities, expanding the object concept in the text encoder. Additionally, unlike CLIP, which only aligns embeddings at the end of the model, GLIP builds a deep cross-modality fusion based on cross-attention [3] for multi-level alignment. As shown in Fig. 1(a), GLIP leverages DyheadModules [4] and BERTLayer [5] as the image and text encoding layers, respectively. With text embedding represented as R and visual embedding as P, the deep fusion process can be represented as:

$$R^i_{\mathrm{t2i}}, P^i_{\mathrm{i2t}} = \textit{X-MHA}\left(R^i, P^i\right), \cdots i \in \{0, 1, \ldots, L-1\}, \tag{2}$$

$$R^{i+1} = \textit{DyHeadModule}\left(R^i + R^i_{t2i}\right), \cdots R = R^L, \tag{3}$$

$$P^{i+1} = \textit{BERTLayer}\left(P^i + P^i_{t2i}\right), \cdots P = P^L, \tag{4}$$

where L is the total number of layers, R^0 denotes the visual features from swin-transformer-large [19], and P^0 denotes the token features from BERT [5]. *X-MHA* represents cross-attention. This architecture enables GLIP to attain superior object-level performance and semantic aggregation. Consequently, GLIP is better suited for object-level zero-shot transfer than conventional VLPMs. Thus, we adopt GLIP to extract better object-level representations for nuclei detection.

2.2 Automatic Prompt Design

The text input, known as the prompt, plays a crucial role in the zero-shot transfer of VLPM for downstream tasks. GLIP originally uses concatenated object nouns such as "object noun 1. Object noun 2..." as default text prompts for detection, and also allows for manual engineering to improve performance [16]. However, manual prompt engineering is a non-trivial challenge, demanding substantial effort and expertise. Furthermore, a notable disparity exists between the web-originated pre-training text-image pairs and medical images. Thus simple noun concatenation is insufficient for GLIP to retrieve nuclei.

We note that Yamada *et al.* prove that the pre-training on extensive text-image pairs has allowed VLPMs to establish a strong association binding between objects and their semantic attributes [27]. Thus, through VLPM, the associated attribute text can accurately depict the corresponding object. Based on this research, we propose that VLPM has the potential to detect unlabelled medical objects in a zero-shot manner by constructing relevant attribute text.

As shown in Fig. 1(a), we first use the image captioning VLPM BLIP [15] to automatically generate attribute words to describe the unseen nuclei object. BLIP allows us to avoid manual attribute design and generates attribute vocabulary that conforms to the text-to-image alignment of VLPM. This process involves three steps. (1) Directly input target medical nouns into GLIP for coarse box prediction. (2) Use the coarse boxes to squarely crop the image, the cropped objects are fed into a frozen BLIP to automatically generate attribute words that describe the object. (3) Word frequency statistics and classification are adopted to find the top M words that describe the object's shape and color, respectively, for that "shape" and "color" are the most relative two attributes that depict nuclei. For a thorough description, we augment the attribute words with synonyms retrieved by a pre-trained language model [25]. All these attribute words are combined with those medical nouns to automatically generate a triplet detection prompt of "[shape][color][noun]". Finally, all generated triplets were put into GLIP for refined detection. This method avoids the empirical manual prompt design and fully utilizes the text-to-image aligning trait of VLPMs.

Our automatic prompt design leverages the powerful text-to-image aligning capabilities of GLIP and BLIP. This approach also enables the automatic generation of the most appropriate attribute words for the specific domain, embodying excellent interpretability.

2.3 Self-training Boosting

Leveraging the strong object retrieval performance of GLIP, we obtain preliminary detection boxes with high precision but low recall. These boxes suffer from missed detection, false detection, and overlapping. To fully exploit the zero-shot potential of GLIP, a self-training framework is established. The automatic prompts are inputted into GLIP to generate the initial results which served as pseudo labels for training YOLOX [7]. Then, the converged YOLOX is used as a teacher to generate new pseudo labels, and iteratively trains students YOLOX. As self-training is based on the EM optimization algorithm [21], it propels our system to continuously refine the predicted boxes and achieve a better optimum.

3 Experiments

3.1 Dataset and Implementation

The dataset used in this study is the MoNuSeg dataset [13], which consists of 30 nuclei images of size 1000×1000, with an average of 658 nuclei per image. Following Kumar *et al.* [13], the dataset was split into training and testing sets with a ratio of 16:14. 16 training images served as inputs of GLIP to generate pseudo-labels for self-training, with 4 images randomly selected for validation. Annotations were solely employed for evaluation purposes on the test images. 16 overlapped image patches of size 256×256 were extracted from each image and randomly cropped into 224×224 as inputs.

In terms of experimental settings, four Nvidia RTX 3090 GPUs were utilized, each with 24 GB of memory. In the automatic prompt generating process, the VQA weights of BLIP finetuned on ViT-B and CapFilt-L [15] were used to generate [shape] and [color] attributes. These attribute words are augmented with synonyms by GPT [25], i.e. attribute augmentation. The target medical noun list was first set to ["nuclei"] straightforwardly, and was also augmented by GPT to ["nuclei", "nucleus", "cyteblast", "karyon"], i.e. noun augmentation. Attribute words were subsequently combined with the target medical nouns to "[shape][color][noun]" format as inputs of GLIP to generate bounding boxes as pseudo labels. The weights used for GLIP is GLIP-L. For self-training refinement, we used the default setting of YOLOX and followed the standard self-training methodology described in [6].

Table 1. Comparison results on MoNuSeg [13]. The best results of unsupervised methods are marked in **bold**.

supervision	method	mAP	AP50	AP75	AR
fully-supervised	**YOLOX (2021)** [7]	0.447	0.832	0.437	0.528
unsupervised	**SSNS (2020)** [26]	0.354	0.739	0.288	0.441
	SOP (2022) [14]	0.235	0.609	0.096	0.351
	VLDet (2023) [17]	0.173	0.407	0.112	0.263
	Ours	**0.416**	**0.808**	**0.382**	**0.502**

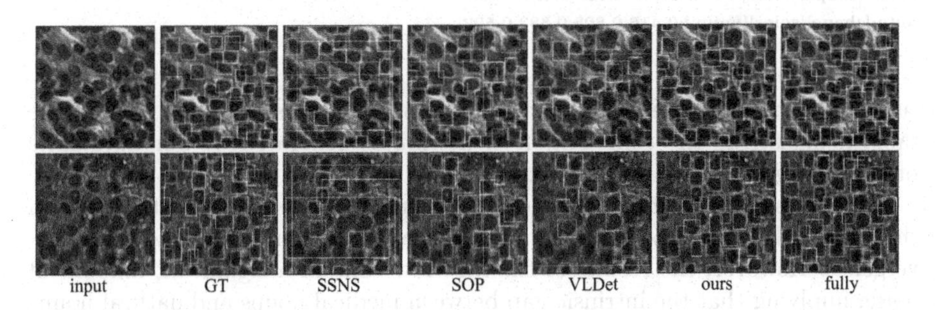

Fig. 2. Output visualizations of different models. The boxes are shown in white.

3.2 Comparison

Our proposed method was compared with the representative fully-supervised method YOLOX [7], as well as the current state-of-the-art (SOTA) methods in unsupervised object detection, including SSNS [26], SOP [14], and VLDet [17]. Among them, the fully-supervised method YOLOX represents the current SOTA on natural images, and VLDet is a newly proposed zero-shot object detection method based on CLIP. For evaluation, mAP, AP50, AP75, and AR are chosen as metrics, following COCO [18]. The final results are shown in Table 1.

Referring to the table, it is evident that our GLIP-based approach outperforms all unsupervised techniques, including domain adaptation-based and clip-based methods. Figure 2 depicts the visualization of the detection results.

3.3 Ablation Studies

Automatic Prompt Design. Firstly, we conducted an ablation study specifically targeting the prompt while ensuring that other conditions remained constant, the results are presented in Table 2.

The first row of the table displays the default noun-concatenation prompt GLIP originally adopted, i.e. "nuclei. nucleus. cyteblast. karyon". The second row represents the same set of nouns with some manual property descriptions added, like "Nuclei. Nucleus. cyteblast. karyon, which are round or oval, and purple or magenta". It is noteworthy that this manual approach is empirically

Table 2. Results of adopting different prompt design methods. The pompts of last 4 rows are automatically generated. The best results are marked in **bold**.

prompt design	mAP	AP50	AP75	AR
noun. [16]	0.064	0.152	0.036	0.150
manual design [16]	0.414	0.757	0.422	0.509
auto: [shape][noun.]	0.336	0.604	0.350	0.434
auto: [color][noun.]	0.213	0.447	0.176	0.325
auto: [shape][color]	0.413	0.726	0.438	0.498
auto:[shape][color][noun.]	**0.416**	**0.808**	**0.382**	**0.502**

Table 3. Ablation study on word augmentation. "A Aug." and "N Aug." refer to attribute augmentation and noun augmentation, respectively.

A Aug	N Aug	mAP	AP50	AP75	AR
		0.372	0.725	0.337	0.464
✓		**0.416**	**0.808**	**0.382**	**0.502**
	✓	0.316	0.754	0.182	0.419
✓	✓	0.332	0.659	0.296	0.434

subjective and therefore prone to significant biases. The subsequent rows in the table demonstrate the combination of attributes generated by our automatic prompt design method. In these experiments, M was set to 3.

It is worth noting that the predictions generated by the prompts shown in the first and second rows also employed the self-training strategy until convergence. However, the results of the first row contain a significant amount of noise, implying that the intrinsic gap between medical nouns and natural nouns impedes the directly zero-shot transfer of GLIP. The second row improves obviously, indicating the effectiveness of attribute description. But manual design is empirical and tedious. As for the automatically generated prompts, it is evident that as the description of attributes becomes comprehensive, from only including shape or color solely to encompassing both, GLIP's performance improves gradually. Furthermore, the second-to-last row indicates that even without nouns, attribute words alone can achieve good results, which also demonstrates the ability of BLIP-generated attribute words to effectively describe the target nuclei. Through the utilization of VLPM's text-to-image alignment capabilities, the proposed automatic prompt design method generates the most suitable attribute words for a given domain automatically, with a high degree of interpretability. Please refer to the supplement for a detailed list of [shape] and [color] attributes.

We further looked into the effect of word augmentation. The results are shown in Table 3. Without and with noun augmentation, the noun lists were ["nuclei"] and ["nuclei", "nucleus", "cyteblast", "karyon"], respectively. The first row of Table 3 uses non-augmented "[shape][color][noun]", while the first row of Table 2 uses noun-augmented concatenation. It is intriguing that applying noun augmentation may lead to suboptimum results compared with the counterparts using the straightforward ["nuclei"]. This is most likely because the augmented new synonym nouns are uncommon medical words and did not appear in the pre-training data that GLIP used. However, applying attribute word augmentation is generally effective because augmented attribute words are also common descriptions for natural scenes. These results suggest a general approach for leveraging the VLPM's potential in downstream medical tasks, that is identifying common attribute words that can be used for describing the target nouns.

Self-training and YOLOX. The box optimization process of the self-training stage is recorded, and the corresponding results are presented in the supplement. YOLOX and self-training are not the essential reasons for the superior performance of our method. The true key is the utilization of semantic-information-rich VLPMs. To illustrate this point, we employed another commonly used unsupervised detection method, superpixels [1], to generate pseudo labels in a zero-shot manner for a fair comparison. These pseudo labels were then fed into the self-training framework based on the YOLOX segmentation architecture, keeping the settings consistent with our approach except for the pseudo label generation. Additionally, we also used DETR [2] instead of YOLOX in our method. The results are shown in the supplement and demonstrate that the high performance of our method lies in the effective utilization of the knowledge from VLPMs rather than YOLOX or self-training.

4 Conclusion

In this work, we propose to use the object-level VLPM, GLIP, to realize zero-shot nuclei detection on H&E images. An automatic prompt design pipeline is proposed to avoid empirical manual prompt design. It fully utilizes the text-to-image alignment of BLIP and GLIP, and enables the automatic generation of the most suitable attribute describing words, offering excellent interpretability. Furthermore, we utilize the self-training strategy to polish the predicted boxes in an iterative manner. Our method achieves a remarkable performance for label-free nuclei detection, surpassing other comparison methods. We demonstrate that VLPMs pre-trained on natural image-text pairs still exhibit astonishing potential for downstream tasks in the medical field.

References

1. Achanta, R., Shaji, A., Smith, K., Lucchi, A., Fua, P., Süsstrunk, S.: Slic superpixels compared to state-of-the-art superpixel methods. IEEE Trans. Pattern Anal. Mach. Intell. **34**(11), 2274–2282 (2012)
2. Carion, N., Massa, F., Synnaeve, G., Usunier, N., Kirillov, A., Zagoruyko, S.: End-to-end object detection with transformers. In: Vedaldi, A., Bischof, H., Brox, T., Frahm, J.-M. (eds.) ECCV 2020. LNCS, vol. 12346, pp. 213–229. Springer, Cham (2020). https://doi.org/10.1007/978-3-030-58452-8_13
3. Chen, C.F.R., Fan, Q., Panda, R.: Crossvit: cross-attention multi-scale vision transformer for image classification. In: Proceedings of the IEEE/CVF International Conference on Computer Vision, pp. 357–366 (2021)
4. Dai, X., et al.: Dynamic head: unifying object detection heads with attentions. In: Proceedings of the IEEE/CVF Conference on Computer Vision and Pattern Recognition, pp. 7373–7382 (2021)
5. Devlin, J., Chang, M.W., Lee, K., Toutanova, K.: Bert: pre-training of deep bidirectional transformers for language understanding. arXiv preprint arXiv:1810.04805 (2018)

6. Dópido, I., Li, J., Marpu, P.R., Plaza, A., Dias, J.M.B., Benediktsson, J.A.: Semisupervised self-learning for hyperspectral image classification. IEEE Trans. Geosci. Remote Sens. **51**(7), 4032–4044 (2013)
7. Ge, Z., Liu, S., Wang, F., Li, Z., Sun, J.: Yolox: exceeding yolo series in 2021. arXiv preprint arXiv:2107.08430 (2021)
8. Gleason, D.F.: Histologic grading of prostate cancer: a perspective. Hum. Pathol. **23**(3), 273–279 (1992)
9. Graham, S., et al.: Hover-net: simultaneous segmentation and classification of nuclei in multi-tissue histology images. Med. Image Anal. **58**, 101563 (2019)
10. Jain, A., Tancik, M., Abbeel, P.: Putting nerf on a diet: semantically consistent few-shot view synthesis. In: Proceedings of the IEEE/CVF International Conference on Computer Vision, pp. 5885–5894 (2021)
11. Jia, C., et al.: Scaling up visual and vision-language representation learning with noisy text supervision. In: International Conference on Machine Learning, pp. 4904–4916. PMLR (2021)
12. Jiao, S., Li, X., Lu, X.: An improved OSTU method for image segmentation. In: 2006 8th International Conference on Signal Processing, vol. 2. IEEE (2006)
13. Kumar, N., Verma, R., Sharma, S., Bhargava, S., Vahadane, A., Sethi, A.: A dataset and a technique for generalized nuclear segmentation for computational pathology. IEEE Trans. Med. Imaging **36**(7), 1550–1560 (2017)
14. Le Bescond, L., et al.: Unsupervised nuclei segmentation using spatial organization priors. In: Wang, L., Dou, Q., Fletcher, P.T., Speidel, S., Li, S. (eds) Medical Image Computing and Computer Assisted Intervention – MICCAI 2022. MICCAI 2022. LNCS, vol. 13432, pp. 325–335. Springer, Cham (2022). https://doi.org/10.1007/978-3-031-16434-7_32
15. Li, J., Li, D., Xiong, C., Hoi, S.: Blip: bootstrapping language-image pre-training for unified vision-language understanding and generation. In: International Conference on Machine Learning, pp. 12888–12900. PMLR (2022)
16. Li, L.H., et al.: Grounded language-image pre-training. In: Proceedings of the IEEE/CVF Conference on Computer Vision and Pattern Recognition, pp. 10965–10975 (2022)
17. Lin, C., et al.: Learning object-language alignments for open-vocabulary object detection. arXiv preprint arXiv:2211.14843 (2022)
18. Lin, T.-Y., et al.: Microsoft COCO: common objects in context. In: Fleet, D., Pajdla, T., Schiele, B., Tuytelaars, T. (eds.) ECCV 2014. LNCS, vol. 8693, pp. 740–755. Springer, Cham (2014). https://doi.org/10.1007/978-3-319-10602-1_48
19. Liu, Z., et al.: Swin transformer: hierarchical vision transformer using shifted windows. In: Proceedings of the IEEE/CVF International Conference on Computer Vision, pp. 10012–10022 (2021)
20. Mahanta, L.B., Hussain, E., Das, N., Kakoti, L., Chowdhury, M.: IHC-net: a fully convolutional neural network for automated nuclear segmentation and ensemble classification for allred scoring in breast pathology. Appl. Soft Comput. **103**, 107136 (2021)
21. Moon, T.K.: The expectation-maximization algorithm. IEEE Signal Process. Magaz. **13**(6), 47–60 (1996)
22. Mouelhi, A., Rmili, H., Ali, J.B., Sayadi, M., Doghri, R., Mrad, K.: Fast unsupervised nuclear segmentation and classification scheme for automatic allred cancer scoring in immunohistochemical breast tissue images. Comput. Methods Prog. Biomed. **165**, 37–51 (2018)

23. Patashnik, O., Wu, Z., Shechtman, E., Cohen-Or, D., Lischinski, D.: Styleclip: text-driven manipulation of stylegan imagery. In: Proceedings of the IEEE/CVF International Conference on Computer Vision, pp. 2085–2094 (2021)
24. Radford, A., et al.: Learning transferable visual models from natural language supervision. In: International Conference on Machine Learning, pp. 8748–8763. PMLR (2021)
25. Radford, A., et al.: Language models are unsupervised multitask learners. OpenAI blog **1**(8), 9 (2019)
26. Sahasrabudhe, M., et al.: Self-supervised nuclei segmentation in histopathological images using attention. In: Martel, A.L., et al. (eds.) MICCAI 2020. LNCS, vol. 12265, pp. 393–402. Springer, Cham (2020). https://doi.org/10.1007/978-3-030-59722-1_38
27. Yamada, Y., Tang, Y., Yildirim, I.: When are lemons purple? The concept association bias of clip. arXiv preprint arXiv:2212.12043 (2022)
28. Yi, J., et al.: Multi-scale cell instance segmentation with keypoint graph based bounding boxes. In: Shen, D., et al. (eds.) MICCAI 2019. LNCS, vol. 11764, pp. 369–377. Springer, Cham (2019). https://doi.org/10.1007/978-3-030-32239-7_41

An AI-Ready Multiplex Staining Dataset for Reproducible and Accurate Characterization of Tumor Immune Microenvironment

Parmida Ghahremani[1], Joseph Marino[1], Juan Hernandez-Prera[2], Janis V. de la Iglesia[2], Robbert J. C. Slebos[2], Christine H. Chung[2], and Saad Nadeem[1(✉)]

[1] Memorial Sloan Kettering Cancer Center, New York, NY 10065, USA
nadeems@mskcc.org
[2] Moffitt Cancer Center, Tampa, FL 33612, USA
christine.chung@moffitt.org

Abstract. We introduce a new AI-ready computational pathology dataset containing restained and co-registered digitized images from eight head-and-neck squamous cell carcinoma patients. Specifically, the same tumor sections were stained with the expensive multiplex immunofluorescence (mIF) assay first and then restained with cheaper multiplex immunohistochemistry (mIHC). This is a first public dataset that demonstrates the equivalence of these two staining methods which in turn allows several use cases; due to the equivalence, our cheaper mIHC staining protocol can offset the need for expensive mIF staining/scanning which requires highly-skilled lab technicians. As opposed to subjective and error-prone immune cell annotations from individual pathologists (disagreement > 50%) to drive SOTA deep learning approaches, this dataset provides objective immune and tumor cell annotations via mIF/mIHC restaining for more reproducible and accurate characterization of tumor immune microenvironment (e.g. for immunotherapy). We demonstrate the effectiveness of this dataset in three use cases: (1) IHC quantification of CD3/CD8 tumor-infiltrating lymphocytes via style transfer, (2) virtual translation of cheap mIHC stains to more expensive mIF stains, and (3) virtual tumor/immune cellular phenotyping on standard hematoxylin images. The dataset is available at https://github.com/nadeemlab/DeepLIIF.

Keywords: multiplex immuofluorescence · multiplex immunohistochemistry · tumor microenvironment · virtual stain-to-stain translation

P. Ghahremani, J. Marino, C. H. Chung, and S. Nadeem—Equal contribution.

Supplementary Information The online version contains supplementary material available at https://doi.org/10.1007/978-3-031-43987-2_68.

H. Greenspan et al. (Eds.): MICCAI 2023, LNCS 14225, pp. 704–713, 2023.
https://doi.org/10.1007/978-3-031-43987-2_68

1 Introduction

Accurate spatial characterization of tumor immune microenvironment is critical for precise therapeutic stratification of cancer patients (e.g. via immunotherapy). Currently, this characterization is done manually by individual pathologists on standard hematoxylin-and-eosin (H&E) or singleplex immunohistochemistry (IHC) stained images. However, this results in high interobserver variability among pathologists, primarily due to the large ($> 50\%$) disagreement among pathologists for immune cell phenotyping [10]. This is also a big cause of concern for publicly available H&E/IHC cell segmentation datasets with immune cell annotations from single pathologists. Multiplex staining resolves this issue by allowing different tumor and immune cell markers to be stained on the same tissue section, avoiding any phenotyping guesswork from pathologists.

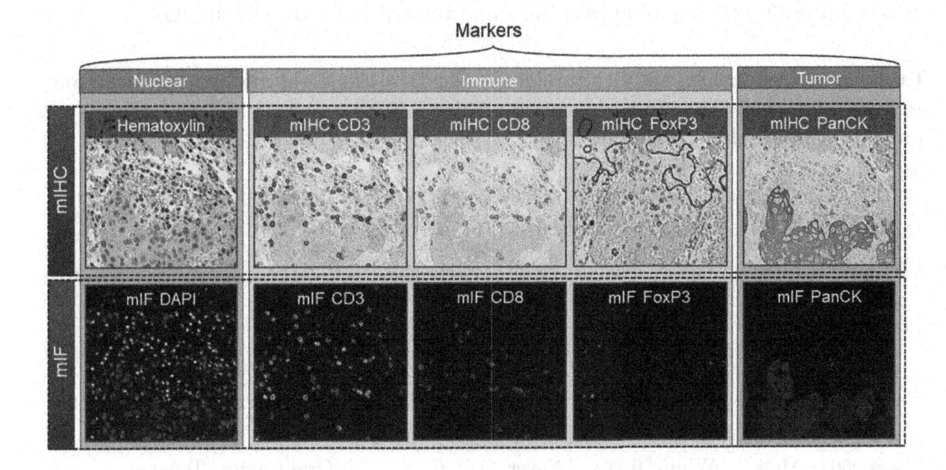

Fig. 1. Dataset overview. Restained and co-registered mIHC and mIF sample with nuclear (hematoxylin/DAPI), immune (CD3 - T-cell marker, CD8 - Cytotoxic T-cell, FoxP3 - regulatory T-cell), and tumor (PanCK) markers. CD3 = CD8 + FoxP3.

Multiplex staining can be performed using expensive multiplex immunofluorescence (mIF) or via cheaper multiplex immunohistochemistry (mIHC) assays. MIF staining (requiring expensive scanners and highly skilled lab technicians) allows multiple markers to be stained/expressed on the same tissue section (no co-registration needed) while also providing the utility to turn ON/OFF individual markers as needed. In contrast, current brightfield mIHC staining protocols relying on DAB (3,3'-Diaminobenzidine) alcohol-insoluble chromogen, even though easily implementable with current clinical staining protocols, suffer from occlusion of signal from sequential staining of additional markers. To this effect, we introduce a new brightfield mIHC staining protocol using alcohol-soluble aminoethyl carbazole (AEC) chromogen which allows repeated stripping, restaining, and scanning of the same tissue section with multiple markers. This requires only affine registration to align the digitized restained images to obtain non-occluded signal intensity profiles for all the markers, similar to mIF staining/scanning.

In this paper, we introduce a new dataset that can be readily used out-of-the-box with any artificial intelligence (AI)/deep learning algorithms for spatial characterization of tumor immune microenvironment and several other use cases. To date, only two denovo stained datasets have been released publicly: BCI H&E and singleplex IHC HER2 dataset [7] and DeepLIIF singleplex IHC Ki67 and mIF dataset [2], both without any immune or tumor markers. In contrast, we release the first denovo mIF/mIHC stained dataset with tumor and immune markers for more accurate characterization of tumor immune microenvironment. We also demonstrate several interesting use cases: (1) IHC quantification of CD3/CD8 tumor-infiltrating lymphocytes (TILs) via style transfer, (2) virtual translation of cheap mIHC stains to more expensive mIF stains, and (3) virtual tumor/immune cellular phenotyping on standard hematoxylin images.

Table 1. Demographics and other relevant details of the eight anonymized head-and-neck squamous cell carcinoma patients, including ECOG performance score, Pack-Year, and surgical pathology stage (AJCC8).

ID	Age	Gender	Race	ECOG	Smoking	PY	pStage	Cancer Site	Cancer Subsite
Case1	49	Male	White	3	Current	21	1	Oral Cavity	Ventral Tongue
Case2	64	Male	White	3	Former	20	4	Larynx	Vocal Cord
Case3	60	Male	Black	2	Current	45	4	Larynx	False Vocal Cord
Case4	53	Male	White	1	Current	68	4	Larynx	Supraglottic
Case5	38	Male	White	0	Never	0	4	Oral Cavity	Lateral Tongue
Case6	76	Female	White	1	Former	30	2	Oral Cavity	Lateral Tongue
Case7	73	Male	White	1	Former	100	3	Larynx	Glottis
Case8	56	Male	White	0	Never	0	2	Oral Cavity	Tongue

2 Dataset

The complete staining protocols for this dataset are given in the accompanying **supplementary material**. Images were acquired at 20× magnification at Moffitt Cancer Center. The demographics and other relevant information for all eight head-and-neck squamous cell carcinoma patients is given in Table 1.

2.1 Region-of-Interest Selection and Image Registration

After scanning the full images at low resolution, nine regions of interest (ROIs) from each slide were chosen by an experienced pathologist on both mIF and mIHC images: three in the tumor core (TC), three at the tumor margin (TM), and three outside in the adjacent stroma (S) area. The size of the ROIs was standardized at 1356×1012 pixels with a resolution of 0.5 μm/pixel for a total surface area of 0.343 mm^2. Hematoxylin-stained ROIs were first used to align all the mIHC marker images in the open source Fiji software using affine registration. After that, hematoxylin- and DAPI-stained ROIs were used as references to align mIHC and mIF ROIs again using Fiji and subdivided into 512×512 patches, resulting in total of 268 co-registered mIHC and mIF patches (∼33 co-registered mIF/mIHC images per patient).

2.2 Concordance Study

We compared mIF and mIHC assays for concordance in marker intensities. The results are shown in Fig. 2. This is the first direct comparison of mIF and mIHC

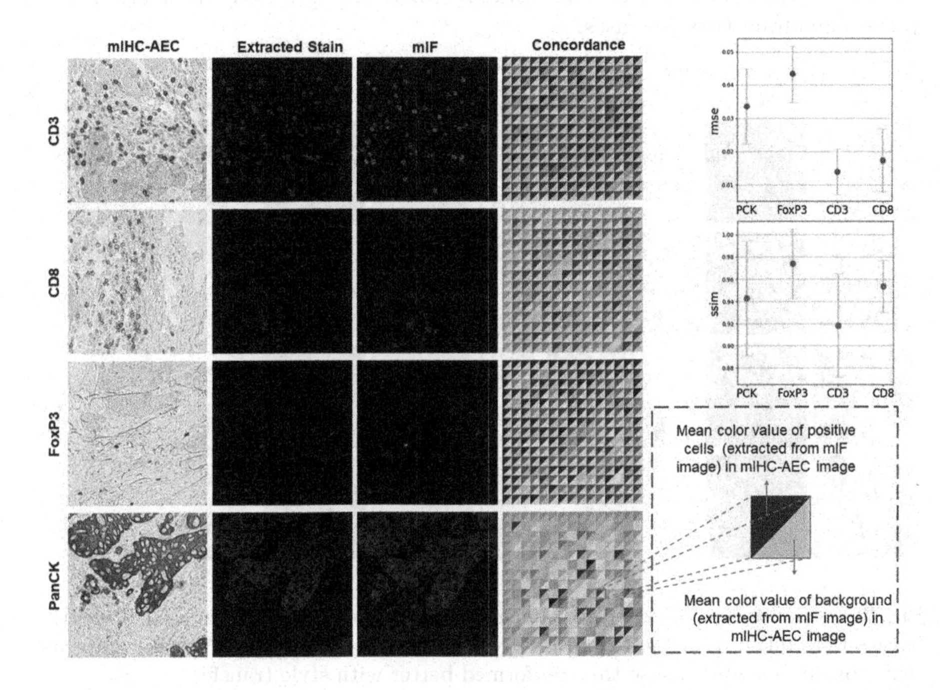

Fig. 2. Concordance Study. Second column shows stains extracted from first column mIHC-AEC images using Otsu thresholding. Third column shows the corresponding perfectly co-registered original mIF images. Using the mIF image, we separated foreground of the mIHC-AEC image from its background and calculated the mean value of the foreground pixels as well as the background pixels. The fourth column shows the results of the concordance study. Each square represents an image in the dataset and the top half of each square shows the mean color value of the positive cells, extracted from mIHC-AEC using its corresponding mIF image and the bottom half of it shows the mean color value of its background. *The high intensity of the top half of the squares represents positive cells and the low intensity of the bottom half represents non-positive cells (background), which is seen in almost all squares, demonstrating* ***high concordance among mIHC-AEC and mIF data***. The last column shows the RMSE and SSIM diagrams of all four stains calculated using the extracted stain from IHC-AEC images (second column) and the mIF images (third column). The low error rate of RMSE and high structural similarity seen in these diagrams show high concordance among mIHC-AEC and mIF images.

using identical slides. It provides a standardized dataset to demonstrate the equivalence of the two methods and a source that can be used to calibrate other methods.

3 Use Cases

In this section, we demonstrate some of the use cases enabled by this high-quality AI-ready dataset. We have used publicly available state-of-the-art tools such as Adaptive Attention Normalization (AdaAttN) [8] for style transfer in the IHC CD3/CD8 quantification use case and DeepLIIF virtual stain translation [2,3] in the remaining two use cases.

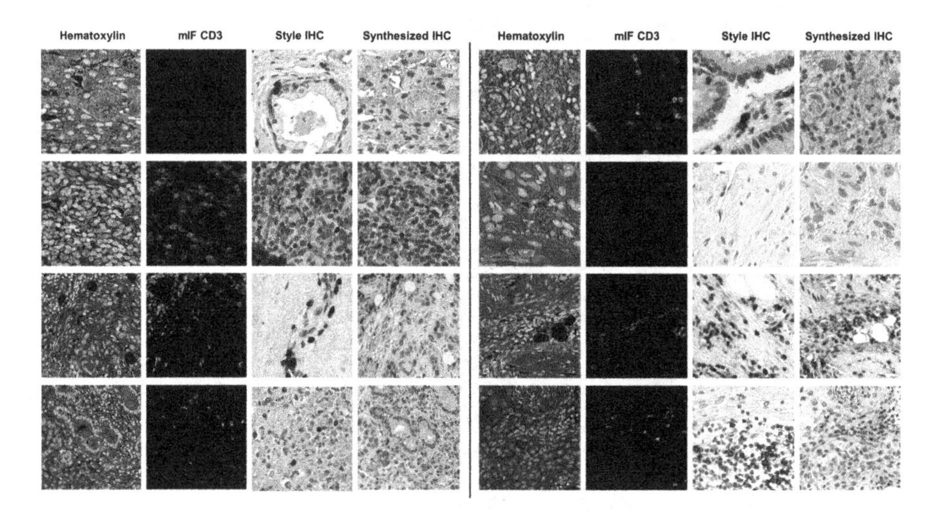

Fig. 3. Examples of synthesized IHC images and corresponding input images. Style IHC images were taken from the public LYON19 challenge dataset [14]. We used grayscale Hematoxylin images because they performed better with style transfer.

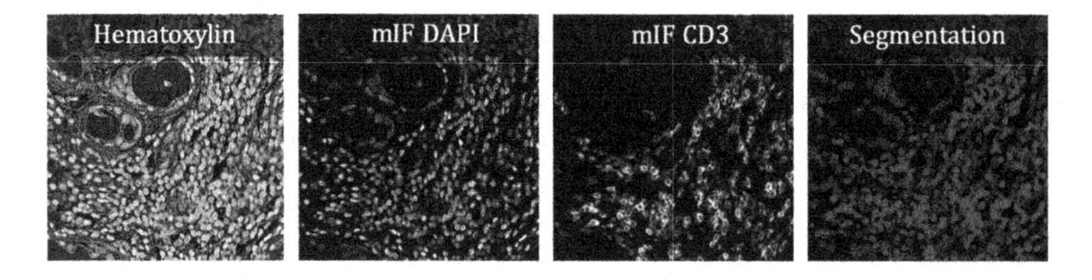

Fig. 4. Examples of Hematoxylin, mIF DAPI, mIF CD3 and classified segmentation mask for this marker. The DAPI images were segmented using Cellpose [13] and manually corrected by a trained technician and approved by a pathologist. The segmented masks were classified using the CD3 channel intensities.

Table 2. Quantitative metrics for NuClick and LYSTO testing sets. **F1** is the harmonic mean of recall and precision, **IOU** is intersection over union, and pixel accuracy (PixAcc) is $\frac{TP}{TP+FP+FN}$, where TP, FP, and FN represent the number of true positive, false positive, and false negative pixels, respectively.

Model	Dataset	NuClick			LYSTO
		F1↑	IOU↑	PixAcc↑	DiffCount↓
UNet [11]	NuClick	0.47 ± 0.30	0.36 ± 0.24	0.62 ± 0.37	10.06 ± 15.69
	Our Dataset	0.48 ± 0.29	0.36 ± 0.25	0.69 ± 0.37	2.91 ± 5.47
FPN [5]	NuClick	0.50 ± 0.31	0.39 ± 0.26	0.64 ± 0.38	2.82 ± 3.49
	Our Dataset	0.52 ± 0.31	0.40 ± 0.26	0.67 ± 0.36	1.90 ± 2.90
UNet++ [15]	NuClick	0.49 ± 0.30	0.37 ± 0.25	0.63 ± 0.37	2.75 ± 5.29
	Our Dataset	0.53 ± 0.30	0.41 ± 0.26	0.70 ± 0.36	2.19 ± 2.89

3.1 IHC CD3/CD8 Scoring Using mIF Style Transfer

We generate a stylized IHC image (Fig. 3) using three input images: (1) hematoxylin image (used for generating the underlying structure of cells in the stylized image), (2) its corresponding mIF CD3/CD8 marker image (used for staining positive cells as brown), and (3) sample IHC style image (used for transferring its style to the final image). The complete architecture diagram is given in the **supplementary material.** Specifically, the model consists of two sub-networks:

(a) Marker Generation: This sub-network is used for generating mIF marker data from the generated stylized image. We use a conditional generative adversarial network (cGAN) [4] for generating the marker images. The cGAN network consists of a generator, responsible for generating mIF marker images given an IHC image, and a discriminator, responsible for distinguishing the output of the generator from ground truth data. We first extract the brown (DAB channel) from the given style IHC image, using stain deconvolution. Then, we use pairs of the style images and their extracted brown DAB marker images to train this sub-network. This sub-network improves staining of the positive cells in the final stylized image by comparing the extracted DAB marker image from the stylized image and the input mIF marker image at each iteration.

(b) Style Transfer: This sub-network creates the stylized IHC image using an attention module, given (1) the input hematoxylin and the mIF marker images and (2) the style and its corresponding marker images. For synthetically generating stylized IHC images, we follow the approach outlined in AdaAttN [8]. We use a pre-trained VGG-19 network [12] as an encoder to extract multi-level feature maps and a decoder with a symmetric structure of VGG-19. We then use both shallow and deep level features by using AdaAttN modules on multiple layers of VGG. This sub-network is used to create a stylized image using the structure of the given hematoxylin image while transferring the overall color distribution of the style image to the final stylized image. The generated marker image from the first sub-network is used for a more accurate colorization of the positive cells against the blue hematoxylin counterstain/background; not defining loss functions based on the markers generated by the first sub-network leads to discrepancy in the final brown DAB channel synthesis.

For the stylized IHC images with ground truth CD3/CD8 marker images, we also segmented corresponding DAPI images using our interactive deep learning ImPartial [9] tool https://github.com/nadeemlab/ImPartial and then classified the segmented masks using the corresponding CD3/CD8 channel intensities, as shown in Fig. 4. We extracted 268 tiles of size 512×512 from this final segmented and co-registered dataset. For the purpose of training and testing all the models, we extract four images of size 256×256 from each tile due to the size of the external IHC images, resulting in a total of 1072 images. We randomly extracted tiles from the LYON19 challenge dataset [14] to use as style IHC images. Using these images, we created a dataset of synthetically generated IHC images from the hematoxylin and its marker image as shown in Fig. 3.

We evaluated the effectiveness of our synthetically generated dataset (stylized IHC images and corresponding segmented/classified masks) using our generated dataset with the NuClick training dataset (containing manually segmented CD3/CD8 cells) [6]. We randomly selected 840 and 230 patches of size 256×256 from the created dataset for training and validation, respectively. NuClick training and validation sets [6] comprise 671 and 200 patches, respectively, of size 256×256 extracted from LYON19 dataset [14]. LYON19 IHC CD3/CD8 images are taken from breast, colon, and prostate cancer patients. We split their training set into training and validation sets, containing 553 and 118 images, respectively, and use their validation set for testing our trained models. We trained three models including UNet [11], FPN [5], UNet++ [15] with the backbone of resnet50 for

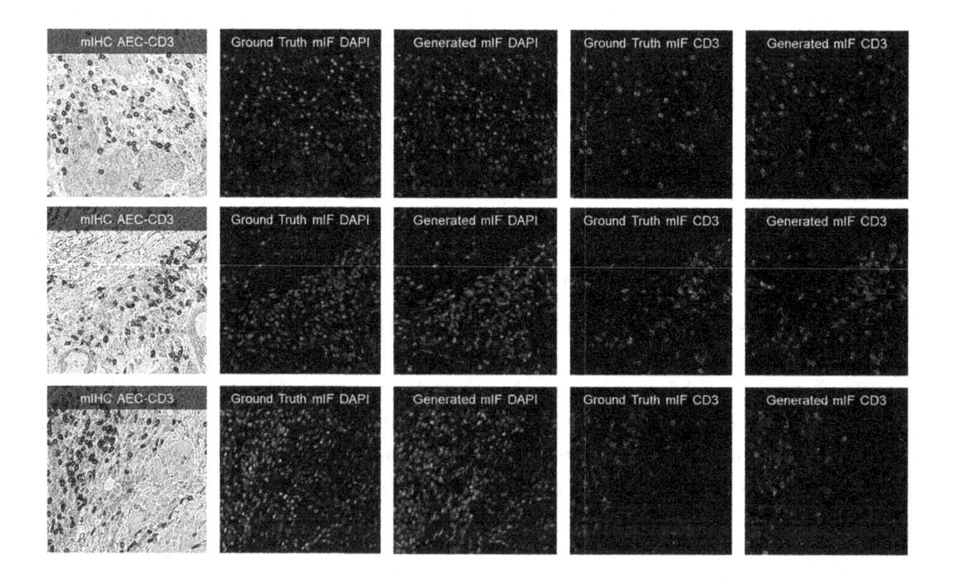

Fig. 5. Examples of ground-truth and generated mIF data from mIHC-AEC images.

200 epochs and early stopping on validation score with patience of 30 epochs, using binary cross entropy loss and Adam optimizer with learning rate of 0.0001. As shown in Table 2, models trained with our synthetic training set outperform those trained solely with NuClick data in all metrics.

We also tested the trained models on 1,500 randomly selected images from the training set of the Lymphocyte Assessment Hackathon (LYSTO) [1], containing image patches of size 299×299 obtained at a magnification of $40\times$ from breast, prostate, and colon cancer whole slide images stained with CD3 and CD8 markers. Only the total number of lymphocytes in each image patch are reported in this dataset. To evaluate the performance of trained models on this dataset, we counted the total number of marked lymphocytes in a predicted mask and calculated the difference between the reported number of lymphocytes in each image with the total number of lymphocytes in the predicted mask by the model. In Table 2, the average difference value (**DiffCount**) of lymphocyte number for the whole dataset is reported for each model. As seen, the trained models on our dataset outperform the models trained solely on NuClick data.

3.2 Virtual Translation of Cheap mIHC to Expensive mIF Stains

Unlike clinical DAB staining, as shown in style IHC images in Fig. 3, where brown marker channel has a blue hematoxylin nuclear counterstain to stain for all the cells, our mIHC AEC-stained marker images (Fig. 5) do not stain for all the cells including nuclei. In this use case, we show that mIHC marker images can be translated to higher quality mIF DAPI and marker images which

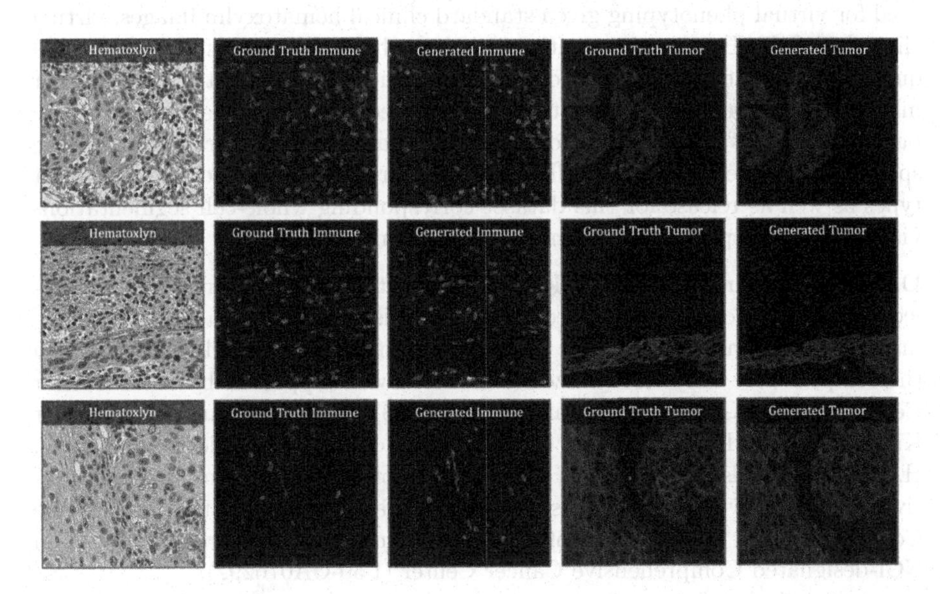

Fig. 6. Examples of ground-truth and generated mIF immune (CD3) and tumor (PanCK) markers from standard hematoxylin images.

stain effectively for all the cells. We used the publicly available DeepLIIF virtual translation module [2,3] for this task. We trained DeepLIIF on mIHC CD3 AEC-stained images to infer mIF DAPI and CD3 marker. Some examples of testing the trained model on CD3 images are shown in Fig. 5. We calculated the Mean Squared Error (MSE) and Structural Similarity Index (SSIM) to evaluate the quality of the inferred modalities by the trained model. The MSE and SSIM for mIF DAPI was 0.0070 and 0.9991 and for mIF CD3 was 0.0021 and 0.9997, indicating high accuracy of mIF inference.

3.3 Virtual Cellular Phenotyping on Standard Hematoxylin Images

There are several public H&E/IHC cell segmentation datasets with manual immune cell annotations from single pathologists. These are highly problematic given the large ($> 50\%$) disagreement among pathologists on immune cell phenotyping [10]. In this last use case, we infer immune and tumor markers from the standard hematoxylin images using again the public DeepLIIF virtual translation module [2,3]. We train the translation task of DeepLIIF model using the hematoxylin, immune (CD3) and tumor (PanCK) markers. Sample images/results taken from the testing dataset are shown in Fig. 6.

4 Conclusions and Future Work

We have released the first AI-ready restained and co-registered mIF and mIHC dataset for head-and-neck squamous cell carcinoma patients. This dataset can be used for virtual phenotyping given standard clinical hematoxylin images, virtual clinical IHC DAB generation with ground truth segmentations (to train high-quality segmentation models across multiple cancer types) created from cleaner mIF images, as well as for generating standardized clean mIF images from neighboring H&E and IHC sections for registration and 3D reconstruction of tissue specimens. In the future, we will release similar datasets for additional cancer types as well as release for this dataset corresponding whole-cell segmentations via ImPartial https://github.com/nadeemlab/ImPartial.

Data use Declaration and Acknowledgment: This study is not Human Subjects Research because it was a secondary analysis of results from biological specimens that were not collected for the purpose of the current study and for which the samples were fully anonymized. This work was supported by MSK Cancer Center Support Grant/Core Grant (P30 CA008748) and by James and Esther King Biomedical Research Grant (7JK02) and Moffitt Merit Society Award to C. H. Chung. It is also supported in part by the Moffitt's Total Cancer Care Initiative, Collaborative Data Services, Biostatistics and Bioinformatics, and Tissue Core Facilities at the H. Lee Moffitt Cancer Center and Research Institute, an NCI-designated Comprehensive Cancer Center (P30-CA076292).

References

1. Ciompi, F., Jiao, Y., Laak, J.: Lymphocyte assessment hackathon (LYSTO) (2019). https://zenodo.org/record/3513571
2. Ghahremani, P., et al.: Deep learning-inferred multiplex immunofluorescence for immunohistochemical image quantification. Nat. Mach. Intell. **4**, 401–412 (2022)
3. Ghahremani, P., Marino, J., Dodds, R., Nadeem, S.: Deepliif: an online platform for quantification of clinical pathology slides. In: Proceedings of the IEEE/CVF Conference on Computer Vision and Pattern Recognition, pp. 21399–21405 (2022)
4. Isola, P., Zhu, J.Y., Zhou, T., Efros, A.A.: Image-to-image translation with conditional adversarial networks. In: Proceedings of the IEEE Conference on Computer Vision and Pattern Recognition, pp. 1125–1134 (2017)
5. Kirillov, A., He, K., Girshick, R., Dollár, P.: A unified architecture for instance and semantic segmentation (2017). http://presentations.cocodataset.org/COCO17-Stuff-FAIR.pdf
6. Koohbanani, N.A., Jahanifar, M., Tajadin, N.Z., Rajpoot, N.: NuClick: a deep learning framework for interactive segmentation of microscopic images. Med. Image Anal. **65**, 101771 (2020)
7. Liu, S., Zhu, C., Xu, F., Jia, X., Shi, Z., Jin, M.: BCI: Breast cancer immunohistochemical image generation through pyramid pix2pix (Accepted CVPR Workshop). arXiv preprint arXiv:2204.11425 (2022)
8. Liu, S., et al.: AdaAttN: revisit attention mechanism in arbitrary neural style transfer. In: Proceedings of the IEEE/CVF International Conference on Computer Vision (ICCV), pp. 6649–6658 (2021)
9. Martinez, N., Sapiro, G., Tannenbaum, A., Hollmann, T.J., Nadeem, S.: Impartial: partial annotations for cell instance segmentation. bioRxiv, pp. 2021–01 (2021)
10. Reisenbichler, E.S., et al.: Prospective multi-institutional evaluation of pathologist assessment of pd-l1 assays for patient selection in triple negative breast cancer. Mod. Pathol. **33**(9), 1746–1752 (2020)
11. Ronneberger, O., Fischer, P., Brox, T.: U-net: convolutional networks for biomedical image segmentation. In: Navab, N., Hornegger, J., Wells, W.M., Frangi, A.F. (eds.) MICCAI 2015. LNCS, vol. 9351, pp. 234–241. Springer, Cham (2015). https://doi.org/10.1007/978-3-319-24574-4_28
12. Simonyan, K., Zisserman, A.: Very deep convolutional networks for large-scale image recognition. arXiv preprint arXiv:1409.1556 (2014)
13. Stringer, C., Wang, T., Michaelos, M., Pachitariu, M.: Cellpose: a generalist algorithm for cellular segmentation. Nat. Methods **18**(1), 100–106 (2021)
14. Swiderska-Chadaj, Z., et al.: Learning to detect lymphocytes in immunohistochemistry with deep learning. Med. Image Anal. **58**, 101547 (2019)
15. Zhou, Z., Rahman Siddiquee, M.M., Tajbakhsh, N., Liang, J.: UNet++: a nested U-net architecture for medical image segmentation. In: Stoyanov, D., et al. (eds.) DLMIA/ML-CDS -2018. LNCS, vol. 11045, pp. 3–11. Springer, Cham (2018). https://doi.org/10.1007/978-3-030-00889-5_1

Position-Aware Masked Autoencoder for Histopathology WSI Representation Learning

Kun Wu[1], Yushan Zheng[2(✉)], Jun Shi[3(✉)], Fengying Xie[1], and Zhiguo Jiang[1]

[1] Image Processing Center, School of Astronautics, Beihang University, Beijing 102206, China
[2] School of Engineering Medicine, Beijing Advanced Innovation Center on Biomedical Engineering, Beihang University, Beijing 100191, China
yszheng@buaa.edu.cn
[3] School of Software, Hefei University of Technology, Hefei 230601, China
juns@hfut.edu.cn

Abstract. Transformer-based multiple instance learning (MIL) framework has been proven advanced for whole slide image (WSI) analysis. However, existing spatial embedding strategies in Transformer can only represent fixed structural information, which are hard to tackle the scale-varying and isotropic characteristics of WSIs. Moreover, the current MIL cannot take advantage of a large number of unlabeled WSIs for training. In this paper, we propose a novel self-supervised whole slide image representation learning framework named position-aware masked autoencoder (PAMA), which can make full use of abundant unlabeled WSIs to improve the discrimination of slide features. Moreover, we propose a position-aware cross-attention (PACA) module with a kernel reorientation (KRO) strategy, which makes PAMA able to maintain spatial integrity and semantic enrichment during the training. We evaluated the proposed method on a public TCGA-Lung dataset with 3,064 WSIs and an in-house Endometrial dataset with 3,654 WSIs, and compared it with 6 state-of-the-art methods. The results of experiments show our PAMA is superior to SOTA MIL methods and SSL methods. The code will be available at https://github.com/WkEEn/PAMA.

Keywords: WSI representation learning · Self-supervised learning

1 Introduction

In the past few years, the development of histopathological whole slide image (WSI) analysis methods has dramatically contributed to the intelligent cancer diagnosis [4,10,15]. However, due to the limitation of hardware resources, it is

Supplementary Information The online version contains supplementary material available at https://doi.org/10.1007/978-3-031-43987-2_69.

H. Greenspan et al. (Eds.): MICCAI 2023, LNCS 14225, pp. 714–724, 2023.
https://doi.org/10.1007/978-3-031-43987-2_69

difficult to directly process gigapixel WSIs in an end-to-end framework. Recent studies usually divide the WSI analysis into multiple stages.

Generally, multiple instance learning (MIL) is one of the most popular solutions for WSI analysis [14,17,18]. MIL methods regard WSI recognition as a weakly supervised learning problem and focus on how to effectively and efficiently aggregate histopathological local features into a global representation. Several studies introduced attention mechanisms [9], recurrent neural networks [2] and graph neural network [8] to enhance the capacity of MIL in structural information mining. More recently, Transformer-based structures [13,19] are proposed to aggregate long-term relationships of tissue regions, especially for large-scale WSIs. These Transformer-based models achieved state-of-the-art performance in sub-type classification, survival prediction, gene mutant prediction, etc. However, these methods still rely on at least patient-level annotations. In the network-based consultation and communication platforms, there is a vast quantity of unlabeled WSIs not effectively utilized. These WSIs are usually without any annotations or definite diagnosis descriptions but are available for unsupervised learning. In this case, self-supervised learning (SSL) is gradually introduced into the MIL-based framework and is becoming a new paradigm for WSI analysis [1,11,16]. Typically, Chen et al. [5] explored and posed a new challenge referred to as slide-level self-learning and proposed HIPT, which leveraged the hierarchical structure inherent in WSIs and constructed multiple levels of the self-supervised learning framework to learn high-resolution image representations. This approach enables MIL-based frameworks to take advantage of abundant unlabeled WSIs, further improving the accuracy and robustness of tumor recognition.

However, HIPT is a hierarchical learning framework based on a greedy training strategy. The bias and error generated in each level of the representation model will accumulate in the final decision model. Moreover, the ViT [6] backbone used in HIPT is originally designed for nature sense images in fixed sizes whose positional information is consistent. However, histopathological WSIs are scale-varying and isotropic. The positional embedding strategy of ViT will bring ambiguity into the structural modeling. To relieve this problem, KAT [19] built hierarchical masks based on local anchors to maintain multi-scale relative distance information in the training. But these masks are manually defined which is not trainable and lacked orientation information. The current embedding strategy for WSI structural description is not complete.

In this paper, we propose a novel whole slide image representation learning framework named position-aware masked autoencoder (PAMA), which achieves slide-level representation learning by reconstructing the local representations of the WSI in the patch feature space. PAMA can be trained end-to-end from the local features to the WSI-level representation. Moreover, we designed a position-aware cross-attention mechanism to guarantee the correlation of local-to-global information in the WSIs while saving computational resources. The proposed approach was evaluated on a public TCGA-Lung dataset and an in-house Endometrial dataset and compared with 6 state-of-the-art methods. The results have demonstrated the effectiveness of the proposed method.

The contribution of this paper can be summarized into three aspects. (1) We propose a novel whole slide image representation learning framework named position-aware masked autoencoder (PAMA). PAMA can make full use of abundant unlabeled WSIs to learn discriminative WSI representations. (2) We propose a position-aware cross-attention (PACA) module with a kernel reorientation (KRO) strategy, which makes the framework able to maintain the spatial integrity and semantic enrichment of slide representation during the self-supervised training. (3) The experiments on two datasets show our PAMA can achieve competitive performance compared with SOTA MIL methods and SSL methods.

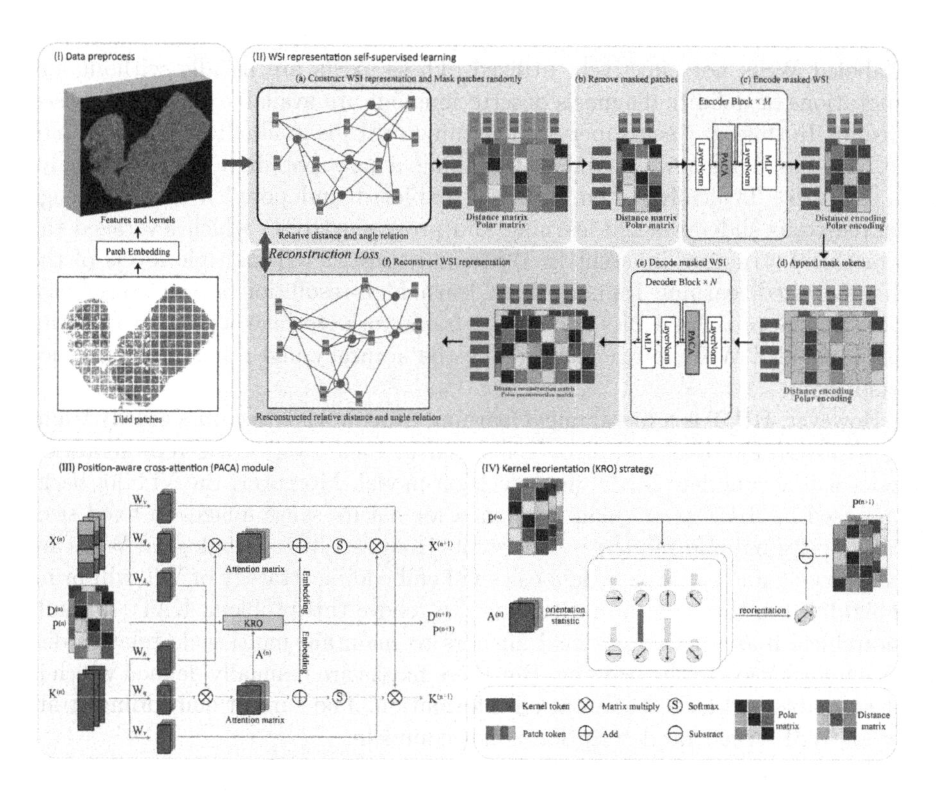

Fig. 1. The overview of the proposed whole slide image representation with position-aware masked autoencoder (PAMA), where (I) shows the data preprocessing including the patch embedding and anchors clustering, (II) describes the workflow of WSI representation self-supervised learning with PAMA, (III) is the structure of the position-aware cross-attention (PACA) module which is the core of the encoder and decoder, and (IV) shows the kernel reorientation (KRO) strategy and the detailed process is described in algorithm 1.

2 Methods

2.1 Problem Formulation and Data Preparation

MAE [7] is a successful SSL framework that learns image presentations by reconstructing the masked image in the original pixel space. We introduced this paradigm to WSI-level representation learning. The flowchart of the proposed work is illustrated in Fig. 1. First, we divided WSIs into non-overlapping image patches and meanwhile removed the background without tissue regions based on a threshold (as shown in Fig. 1(I)). Then, we applied the self-supervised learning framework DINO [3] for patch feature learning and extraction. Afterward, the features for a WSI are represented as $\mathbf{X} \in \mathbb{R}^{n_p \times d_f}$, where d_f is the dimension of the feature and n_p is the number of patches in the WSI. Inspired by KAT [19], we extracted multiple anchors by clustering the location coordinates of patches for the auxiliary description of the WSI structure. We assigned trainable representations for these anchors, which are formulated as $\mathbf{K} \in \mathbb{R}^{n_k \times d_f}$, where n_k is the number of anchors in the WSI. Here, we regard each anchor as an observation point of the tissue and assess the relative distance and orientation from the patch positions to the anchor positions. Specifically, a polar coordinate system is built on each anchor position, and the polar coordinates of all the patches on the system are recorded. Finally, a relative distance matrix $\mathbf{D} \in \mathbb{N}^{n_k \times n_p}$ and relative polar angle matrix $\mathbf{P} \in \mathbb{N}^{n_k \times n_p}$ are obtained, where $D_{ij} \in \mathbf{D}$ and $P_{ij} \in \mathbf{P}$ respectively represent the distance and polar angle of the i-th patch in the polar coordinate system that takes the position of the j-th anchor as the pole. Then, we can formulate a WSI as $S = \{\mathbf{X}, \mathbf{K}, \mathbf{D}, \mathbf{P}\}$.

2.2 Masked WSI Representation Autoencoder

Figure 1(II) illustrates the procedure of WSI representation learning. Referring to MAE [7], we random mask patch tokens with a high masking ratio (*i.e.* 75% in our experiments). The remaining tokens (as shown in Fig. 1(b)) are fed into the encoder. Each encoder block sequentially consists of LayerNorm, PACA module, LayerNorm, and multilayer perceptron (MLP), as shown in Fig. 1(c). Then, masked tokens are appended into encoded tokens to conduct the full set of tokens, which is shown in Fig. 1(d). Next, the decoder reconstructs the slide representation in feature space. Finally, mean squared error (MSE) loss is built between the reconstructed patch features and the original patch features. Referring to MAE [7], a trainable token is appended to the patch tokens to extract the global representation. After training, the pre-trained encoder will be employed as the backbone for various downstream tasks.

2.3 Position-Aware Cross-Attention

To preserve the structure information of the tissue, we propose the position-aware cross-attention (PACA) module, which is the core of the encoder and decoder blocks. The structure of PACA is shown in Fig. 1(III).

The message passing between the anchors and patches is achieved by a bi-directional cross-attention between the patches and anchors. First, the anchors collect the local information from the patches, which is formulated as

$$\mathbf{K}^{(n+1)} = \sigma(\frac{\mathbf{K}^{(n)}\mathbf{W}_q^{(n)} \cdot (\mathbf{X}^{(n)}\mathbf{W}_k^{(n)})^{\mathrm{T}}}{\sqrt{d_e}} + \varphi_d(\mathbf{D}^{(n)}) + \varphi_p(\mathbf{P}^{(n)})) \cdot (\mathbf{X}^{(n)}\mathbf{W}_v^{(n)}), \quad (1)$$

where $\mathbf{W}_l \in \mathbb{R}^{d_f \times d_e}, l = q, k, v$ are learnable parameters with d_e denoting the dimension of the head output, σ represents the softmax function, and φ_d and φ_p are the embedding functions that respectively take the distance and polar angle as input and output the corresponding trainable embedding values. Symmetrically, each patch token catches the information of all anchors into their own local representations by the equations

$$\mathbf{X}^{(n+1)} = \sigma(\frac{\mathbf{X}^{(n)}\mathbf{W}_q^{(n)} \cdot (\mathbf{K}^{(n)}\mathbf{W}_k^{(n)})^{\mathrm{T}}}{\sqrt{d_e}} + \varphi_d^{\mathrm{T}}(\mathbf{D}^{(n)}) + \varphi_p^{\mathrm{T}}(\mathbf{P}^{(n)})) \cdot (\mathbf{K}^{(n)}\mathbf{W}_v^{(n)}). \quad (2)$$

The two-way communication makes the patches and anchors timely transmit local information and perceive the dynamic change of global information. The embedding of relative distance and polar angle information helps the model maintain the semantic and structural integrity of the WSI and meanwhile prevents the WSI representation from collapsing to the local area throughout the training process.

In terms of efficiency, the computational complexity of self-attention is $O({n_p}^2)$ where n_p is the number of patch tokens. In contrast, our proposed PACA's complexity is $O(n_k \times n_p)$ where n_k is the number of anchors. Notice that $n_k << n_p$, the complexity is close to $O(n_p)$, i.e. linear correlation with the size of the WSI.

2.4 Kernel Reorientation

As for the polar angle matrix $\mathbf{P} \in \mathbb{N}^{n_k \times n_p}$, we specify the horizontal direction of all the anchors as the initial polar axis. In natural scene images, there is natural directional conspicuousness of semantics. For instance, in the case of a church, it is most likely to find a door below the windows rather than be located above them. But histopathology images have no absolute definition of direction. The semantics of WSI will not change with rotation and flip. Namely, it is isotropic. Embedding the orientation information with a fixed polar axis will lead to ambiguities in various slides.

To address this problem, we design a kernel reorientation (KRO) strategy to dynamically update the polar axis during the training. As shown in Fig. 1(IV), we equally divide the polar coordinate system into N bins and calculate the sum of the attention scores from each bin. Then, the orientation with the highest score is recognized as the new polar axis for the anchor. Based on the updated polar axis, we can then amend $\mathbf{P}^{(n)}$ to $\mathbf{P}^{(n+1)}$. The detailed algorithm is described in Algorithm 1.

3 Experiments and Results

3.1 Datasets

We evaluated the proposed method on two datasets, the public TCGA-Lung and the in-house Endometrial dataset, which are introduced as follows.

Algorithm 1: Kernel Reorientation algorithm.

Input:
$\mathbf{P}^{(n)} \in \mathbb{N}^{H \times n_k \times n_p}$: The relative polar angle matrix of n-th block, where H is the head number of multi-head attention, n_k is the number of anchors in the WSI, n_p is the number of patches in the WSI;
$\mathbf{A}^{(n)} \in \mathbb{R}^{H \times n_k \times n_p}$: The attention matrix from anchors to patches, defined as
$\mathbf{A}^{(n)} = \frac{\mathbf{K}^{(n)}\mathbf{W}_q^{(n)} \cdot (\mathbf{X}^{(n)}\mathbf{W}_k^{(n)})^{\mathrm{T}}}{\sqrt{d_e}}$;
D^{score}: A dictionary taking the angle as KEY for storing attention scores;
Output: $\mathbf{P}^{(n+1)} \in \mathbb{R}^{H \times n_k \times n_p}$: The updated polar angle matrix.
for h in H **do**
  for i in n_k **do**
    $D^{score} = 0$
    for j in n_p **do**
      $D^{score}[\mathbf{P}^{(n)}_{h,i,j}] \mathrel{+}= \mathbf{A}^{(n)}_{h,i,j}$;
    end
    $\mathbf{P}^{(n)}_{h,i,max} = \arg\max D^{score}$; `// Find the orientation that has the highest attention score.`
    for j in n_p **do**
      $\mathbf{P}^{(n+1)}_{h,i,j} = \mathbf{P}^{(n)}_{h,i,j} - \mathbf{P}^{(n)}_{h,i,max}$;`// Reorientation.`
    end
  end
end

TCGA-Lung dataset is collected from The Cancer Genome Atlas (TCGA) Data Portal. The dataset includes a total of 3,064 WSIs, which consist of three categories, namely Tumor-free (Normal), Lung Adenocarcinoma (LUAD), and Lung Squamous Cancer (LUSC),

Endometrial dataset includes 3,654 WSIs of endometrial pathology, which includes 8 categories, namely Well/Moderately/Low-differentiated endometrioid adenocarcinoma, Squamous differentiation carcinoma, Plasmacytoid carcinoma, Clear cell carcinoma, Mixed-cell adenocarcinoma, and benign tumor.

Each dataset was randomly divided into training, validation and test sets according to 6:1:3 while keeping each category of data proportionally. We conducted WSI multi-type classification experiments on the two datasets. The validation set was used to perform an early stop. The results of the test set were reported for comparison.

3.2 Implementation Details

The WSI representation pre-training stage uses all training data and does not involve any supervised information. During the downstream classification task,

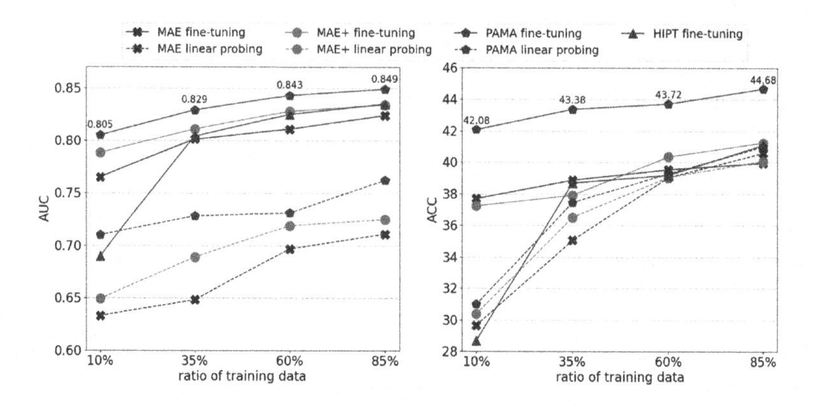

Fig. 2. Semi-supervised experiments with 10%, 35%, 60% and 85% of labelled data on the Endometrial dataset. Solid lines represent fine-tuning results and dotted lines represent liner probing results.

Table 1. Ablation study on 35% of labelled Endometrial dataset.

NO.	**Dis**	**Polar**	**KRO**	AUC	ACC
1	✓	✓	✓	0.829	43.38
2	✓	✓		0.809 (↓0.020)	39.52 (↓3.86)
3	✓			0.808 (↓0.021)	40.82 (↓2.56)
4		✓	✓	0.810 (↓0.019)	39.76 (↓3.62)
5				0.795 (↓0.034)	36.33 (↓7.05)

the pre-trained encoder is utilized as the slide representation extractor, and the $[CLS]$ token is fed into the following classifier consisting of a multilayer perceptron (MLP) and a fully connected layer. Following the protocol in self-supervised learning [7], we evaluated the quality of pre-training with the two approaches: 1) **Fine-tuning** is to train the whole network parameters, including WSI encoder and classifier; 2) **Linear probing** is to freeze the encoder and only train the classifier. The usage of $[CLS]$ token refers to the MAE [7] framework, which was concatenated with patch tokens. During pre-training, the $[CLS]$ token is not involved in loss computation, but it continuously interacts with kernels and receives global information. After pre-training, the pre-trained parameters of the $[CLS]$ token will be loaded for fine-tuning and linear probing.

To ensure the uniformity of patch features, we choose DINO [3] to extract patch features on the magnification under 20× lenses. Accuracy (ACC) and area under the ROC curve (AUC) are employed as evaluation metrics. We implemented all the models in Python 3.8 with PyTorch 1.7 and Cuda 10.2 and run the experiments on a computer with 4 GPUs of Nvidia Geforce 2080Ti.

3.3 Effectiveness of the WSI Representation Learning

We first conducted experiments on the Endometrial dataset to verify the effectiveness of self-supervised learning for WSI analysis under label-limited conditions. The results are shown in Fig. 2, where the performance obtained with different ratios of labeled training WSIs are compared. MAE [7] based on the patch features is implemented as the baseline. Furthermore, we applied the proposed distance and polar angle embedding to the self-attention module of MAE [7], which is referred to as MAE+ in Fig. 2.

Table 2. Comparison with weakly-supervised MIL and slide-level self-learning study on the two datasets for sub-type classification.

Methods	TCGA-Lung				Endometrial			
	35%		100%		35%		100%	
	AUC	ACC	AUC	ACC	AUC	ACC	AUC	ACC
DSMIL [12]	0.911	75.00	0.938	80.11	0.761	38.21	0.786	39.32
TransMIL [13]	0.932	79.62	0.959	84.35	0.783	38.43	0.798	40.01
SETMIL [18]	0.937	80.21	0.962	84.95	0.795	38.71	0.831	40.84
KAT [19]	0.951	83.37	0.965	85.81	0.799	38.89	0.835	41.93
HIPT [5]	0.967	84.23	0.977	87.83	0.804	38.69	0.842	40.63
MAE [7]	0.965	83.90	0.970	87.50	0.801	38.87	0.832	41.95
MAE+	0.969	85.07	0.981	88.25	0.811	37.91	0.845	42.85
PAMA	**0.982**	**90.84**	**0.988**	**92.48**	**0.829**	**43.38**	**0.851**	**43.64**

Overall, PAMA consistently achieves significantly better performance across all the label ratios than MAE [7] and HIPT [5]. These results have demonstrated the effectiveness of PAMA in WSI representation pre-training. Moreover, PAMA achieves the best stability in AUCs and ACCs when the label ratios are reduced from 85% to 10%. This is of practical importance as it reduces the dependence on a large number of labeled WSIs for training robust WSI analysis models. Meanwhile, it means that we can utilize the unlabeled WSIs to improve the capacity of the models with the help of PAMA. HIPT [5] is a two-stage self-learning framework, which first leverages DINO [3] to pre-train patches (256 × 256) divided from regions (4096 × 4096) and then utilizes DINO-4k [5] to pre-train regions of WSIs. The multi-stage framework accumulated the training bias and noise, which caused an AUC gap of HIPT [5] to MAE [7] and PAMA, especially trained with only 10% labeled WSIs. We also observed a significant improvement when comparing MAE+ with MAE [7]. It indicates the proposed distance and polar angle embedding strategy is more appreciated than the positional embedding of ViT [6] to describe the structure of histopathological WSIs. Please refer to the supplementary materials for more detailed results.

3.4 Ablation Study

Then, we conducted ablation experiments to verify the necessity of the proposed structural embedding strategy. The detailed results are shown in Table 1, where all the models were fine-tuned with 35% training WSIs. It shows that the AUC decreases by 0.019 and 0.021, respectively, when the distance or polar angle embedding is discarded. And, when removing both the distance and polar angle embedding, the AUC drops by 0.034. These results demonstrate that local and global spatial information is crucial for PAMA to learn WSI representations.

3.5 Comparison with SOTA Methods

Finally, we additionally compared the proposed PAMA with four weakly-supervised methods, DSMIL [12], TransMIL [13], SETMIL [18] and KAT [19]. The results are shown in Table 2. Overall, PAMA consistently achieves the best performance. In comparison with the second-best methods, PAMA achieves an increase of 0.015/0.011 and 0.025/0.009 in AUCs on TCGA and Endometrial datasets, respectively, by using 35%/100% labeled WSIs. Moreover, PAMA reveals the most robust capacity when reducing the training data from 100% to 35%, with AUC decreasing slightly from 0.988 to 0.982 and from 0.851 to 0.829 on the two datasets. TransMIL [13], SETMIL [18] and KAT [19] are state-of-the-art methods for histopathological image classification. They all considered the spatial adjacency of patches but neglected the orientation relationships of the patches. It is the main reason that the three methods cannot surpass our method even with 100% training WSIs.

4 Conclusion

In this paper, we proposed an effective self-supervised representation learning framework for WSI analysis. The experiments on two large-scale datasets have demonstrated the effectiveness of PAMA in the condition of limited-label. The results have shown superiority to the existing weakly-supervised and self-supervised MIL methods. Future work will focus on training the WSI representation model based on datasets across multiple organs, thus promoting the generalization ability of the model for different downstream tasks.

Acknowledgements. This work was partly supported by the National Natural Science Foundation of China (Grant No. 62171007, 61901018, and 61906058), and partly supported by the Fundamental Research Funds for the Central Universities of China (grant No. JZ2022HGTB0285).

References

1. Azizi, S., et al.: Robust and efficient medical imaging with self-supervision. arXiv preprint arXiv:2205.09723 (2022)

2. Campanella, G., et al.: Clinical-grade computational pathology using weakly supervised deep learning on whole slide images. Nat. Med. **25**(8), 1301–1309 (2019)
3. Caron, M., et al.: Emerging properties in self-supervised vision transformers. In: Proceedings of the IEEE/CVF International Conference on Computer Vision, pp. 9650–9660 (2021)
4. Chen, C., Lu, M.Y., Williamson, D.F., Chen, T.Y., Schaumberg, A.J., Mahmood, F.: Fast and scalable search of whole-slide images via self-supervised deep learning. Nat. Biomed. Eng. **6**(12), 1420–1434 (2022)
5. Chen, R.J., Chen, C., Li, Y., Chen, T.Y., Trister, A.D., Krishnan, R.G., Mahmood, F.: Scaling vision transformers to gigapixel images via hierarchical self-supervised learning. In: Proceedings of the IEEE/CVF Conference on Computer Vision and Pattern Recognition, pp. 16144–16155 (2022)
6. Dosovitskiy, A., et al.: An image is worth 16x16 words: transformers for image recognition at scale. arXiv preprint arXiv:2010.11929 (2020)
7. He, K., Chen, X., Xie, S., Li, Y., Dollár, P., Girshick, R.: Masked autoencoders are scalable vision learners. In: Proceedings of the IEEE/CVF Conference on Computer Vision and Pattern Recognition, pp. 16000–16009 (2022)
8. Huang, Z., Chai, H., Wang, R., Wang, H., Yang, Y., Wu, H.: Integration of patch features through self-supervised learning and transformer for survival analysis on whole slide images. In: de Bruijne, M., et al. (eds.) MICCAI 2021. LNCS, vol. 12908, pp. 561–570. Springer, Cham (2021). https://doi.org/10.1007/978-3-030-87237-3_54
9. Ilse, M., Tomczak, J., Welling, M.: Attention-based deep multiple instance learning. In: International Conference on Machine Learning, pp. 2127–2136. PMLR (2018)
10. Jaume, G., Song, A.H., Mahmood, F.: Integrating context for superior cancer prognosis. Nat. Biomed. Eng. 1–3 (2022)
11. Koohbanani, N.A., Unnikrishnan, B., Khurram, S.A., Krishnaswamy, P., Rajpoot, N.: Self-path: self-supervision for classification of pathology images with limited annotations. IEEE Trans. Med. Imaging **40**(10), 2845–2856 (2021)
12. Li, B., Li, Y., Eliceiri, K.W.: Dual-stream multiple instance learning network for whole slide image classification with self-supervised contrastive learning. In: Proceedings of the IEEE/CVF Conference on Computer Vision and Pattern Recognition, pp. 14318–14328 (2021)
13. Shao, Z., et al.: Transmil: transformer based correlated multiple instance learning for whole slide image classification. Adv. Neural Inf. Process. Syst. **34**, 2136–2147 (2021)
14. Su, Z., Tavolara, T.E., Carreno-Galeano, G., Lee, S.J., Gurcan, M.N., Niazi, M.: Attention2majority: weak multiple instance learning for regenerative kidney grading on whole slide images. Med. Image Anal. **79**, 102462 (2022)
15. Wu, Z., et al.: Graph deep learning for the characterization of tumour microenvironments from spatial protein profiles in tissue specimens. Nat. Biomed. Eng. 1–14 (2022)
16. Yang, P., Hong, Z., Yin, X., Zhu, C., Jiang, R.: Self-supervised visual representation learning for histopathological images. In: de Bruijne, M., et al. (eds.) MICCAI 2021. LNCS, vol. 12902, pp. 47–57. Springer, Cham (2021). https://doi.org/10.1007/978-3-030-87196-3_5
17. Yu, J.G., et al.: Prototypical multiple instance learning for predicting lymph node metastasis of breast cancer from whole-slide pathological images. Med. Image Anal. 102748 (2023)

18. Zhao, Y., et al.: Setmil: spatial encoding transformer-based multiple instance learning for pathological image analysis. In: Medical Image Computing and Computer Assisted Intervention-MICCAI 2022: 25th International Conference, Singapore, 18–22 September 2022, Proceedings, Part II, LNCS, pp. 66–76. Springer, Cham (2022). https://doi.org/10.1007/978-3-031-16434-7_7
19. Zheng, Y., Li, J., Shi, J., Xie, F., Jiang, Z.: Kernel attention transformer (KAT) for histopathology whole slide image classification. In: Medical Image Computing and Computer Assisted Intervention-MICCAI 2022: 25th International Conference, Singapore, September 18–22, 2022, Proceedings, Part II, LNCS. pp. 283–292. Springer, Cham (2022). https://doi.org/10.1007/978-3-031-16434-7_28

A One-Class Variational Autoencoder (OCVAE) Cascade for Classifying Atypical Bone Marrow Cell Sub-types

Jonathan Tarquino[1,2], Jhonathan Rodriguez[1,2], Charlems Alvarez-Jimenez[1,2], and Eduardo Romero[1,2(✉)]

[1] Universidad Nacional de Colombia, Bogotá, Colombia
[2] Computer Imaging and Medical Applications Laboratory-CIM@LAB, Bogotá, Colombia
edromero@unal.edu.co
http://cimalab.unal.edu.co/

Abstract. Atypical bone marrow (BM) cell-subtype characterization defines the diagnosis and follow up of different hematologic disorders. However, this process is basically a visual task, which is prone to inter- and intra-observer variability. The presented work introduces a new application of one-class variational autoencoders (OCVAE) for automatically classifying the 4 most common pathological atypical BM cell-subtypes, namely myelocytes, blasts, promyelocytes, and erythroblasts, regardless the disease they are associated with. The presented OCVAE-based representation is obtained by concatenating the bottleneck of 4 separated OCVAEs, specifically set to capture one-cell-sub-type pattern at a time. In addition, this strategy provides a complete validation scheme in a subset of an open access image dataset, demonstrating low requirements in terms of number of training images. Each particular OCVAE is trained to provide specific latent space parameters (64 means and 64 variances) for the corresponding atypical cell class. Afterwards, the obtained concatenated representation space feeds different classifiers which discriminate the proposed classes. Evaluation is done by using a subset ($n = 26,000$) of a public single-cell BM image database, including two independent partitions, one for setting the VAEs to extract features ($n = 20,800$), and one for training and testing a set classifiers ($n = 5,200$). Reported performance metrics show the concatenated-OCVAE characterization successfully differentiates the proposed atypical BM cell classes with accuracy = 0.938, precision = 0.935, recall = 0.935, f1-score = 0.932, outperforming previously published strategies for the same task (handcrafted features, ResNext, ResNet-50, XCeption, CoAtnet), while a more thorough experimental validation is included.

Keywords: One-class Variational autoencoder · Atypical bone marrow cell · artificial intelligence · cytology imaging · automatic classification

H. Greenspan et al. (Eds.): MICCAI 2023, LNCS 14225, pp. 725–734, 2023.
https://doi.org/10.1007/978-3-031-43987-2_70

1 Introduction

Bone marrow (BM) lineage classification and differential counting are at the very base of most diagnoses or clinical management of hematopoietic disorders [7]. Despite advances in cytogenetics, immunophenotyping and molecular technology, morphology of BM cellular lineages is still the gold standard in hematology. This cell morphometry estimation is performed by highly trained and experienced proffesionals and yet, this activity has been reported with high intra- and inter-operator variability [10,13], and even more variable when visually identifying atypical BM cells [11,32]. This examination not only defines the stage of proliferating disorders, but also it characterizes chronic diseases associated with particular distributions of leukocyte sub-types or atypical BM cells [7]. On the one hand cell morphological examination is fully dependent on the operator experience-skills and, on the other hand not objective quantitative measurements are available [1].

In order to provide a more quantitative BM cell characterization, different strategies have been proposed, either extracting hand-crafted features or applying deep learning based approaches. Importantly, most of them has been focused on leukemia detection [3,25], which narrows application of automated strategies to a single hematological disease. Regarding classic machine learning approaches using hand-crafted features, different representation spaces have been used in the multiclass task, including color [18], shape [19] and texture features [17]. These features are commonly combined with classification methods (support vector machines-SVM, random forest, linear discriminant analysis-LDA) [5]. Currently deep learning strategies are the best to discriminate only leukemia cells in peripheral blood images with the ResNext [22], the VGG-Net [4] or customized networks [4]. A much more complicated task relies on the discrimination of atypical cells which for peripheral blood samples has been tackled using a ResNext network (accuracy = 0.99) [22]. Likewise, they have been discriminated as part of a multiclass BM cell problem with more sub-types than the atypical ones, using the same ResNext architecture (accuracy = 0.90) [21], or the You Only Look Once (YOLO) [26]. Additionally, the emerging transformers have also being used to differentiate multiple cell sub-types in BM, particularly by using a CoAtNet [9] which integrates transformer-attention layers and ConvNet deep neural network architecture, and provides state-of-the-art performance (accuracy = 0.93) [27].

Unlike previous works, the presented approach is focused in quantifying changes in the most common pathological atypical BM cell subtypes, namely myelocytes (MYB), blasts (BLA), promyelocytes (PMO), and erythroblast (EBO), achieving a quantitative strategy to differentiate atypical BM cells regardless the disease they are associated with. Differentiating these classes is crucial since the proposed 4 cell subtypes are the ones which must be counted for a correct diagnosis and morphological characterization of acute myeloid leukemia and myelodysplasic disease [28], currently the most common and aggressive hematological disorders in adults. This classification task is so complex that even common immunohistochemical hematology stains like CD117 may also misclassify these cell sub-types [23]. The introduced methodology builds upon One-class variational autoencoders (OCVAE) and presents a new atypical BM cell representation by concatenating the latent spaces provided by 4 specialized OCVAE with the same architecture. In contrast to previous OCVAE applications focused on one-class training for binary classification [15], or anomaly detection based on the reconstruction quality [29], the introduced methodology presents a cascade of OCVAEs as feature extractor in a 4 class differentiation task. The methodology setting process splits data into two disjoint subsets, where the first one is used to set the OCVAEs parameters for capturing the particular patterns of each cell sub-type, separately. The second subset feeds all the trained OCVAEs for building a concatenated latent-space that serves to train different classifiers (support vector machine-SVM with linear and RBF kernel, and random forest), reporting accuracy, precision, recall and f1-score, as performance metrics. This approach was evaluated on a subset ($n = 26,000$) of a public image database [20], demonstrating to outperform previously published strategies, while requiring a lower number of training images.

2 Materials and Methods

2.1 The BM Smear Image Dataset

All experiments in the presented work were performed using a subset of the public image database "*An Expert-Annotated Dataset of Bone Marrow Cytology in Hematologic Malignancies*" [20]. This database is composed of 171,374 single-cells annotated by type, coming from the BM smears of a group of 945 patients covering a set of diseases, yet the individual hematological disease is not informed. All images were acquired using bright-field microscope with ×40 magnification and oil immersion, applied to May-Grünwald-Giemsa/Pappenheim stained samples. Each image was annotated by an expert morphologist at the Munich Leukemia Laboratory (MLL), assigning one out of 21 possible classes, including the atypical BM cells that the herein presented methodology is working with, i.e., myelocytes (MYB), blasts (BLA), promyelocytes (PMO), and erythroblast (EBO). Particularly, the subset herein used was composed of a balanced version of the aforementioned classes, making a new dataset of 26,000 images (6,500 for each of the four classes).

Data Use Declaration. The complete version of the public database used for validating the proposed strategy is provided by Matek et. al., under the TCIA Data Usage Policy and Restrictions, and it is publicly available at TCIA platform, https://doi.org/10.7937/TCIA.AXH3-T579. No ethical compliance statement is presented in this document since it is covered by the original dataset publication [20].

2.2 OCVAE Atypical BM Cell Differentiation Method

Single Cell Image Pre-processing. The first step of the presented approach is to apply a color-space transformation from RGB to Lab, for reducing possible

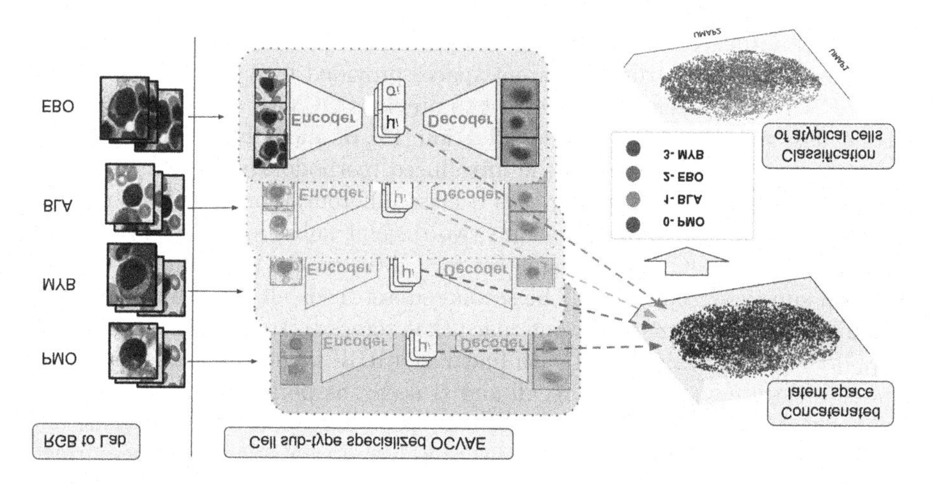

Fig. 1. Atypical BM cell characterization methodology: single cell images are transformed from RGB to Lab color space, b-channel image versions feed specialized OCVAES, and a cascade of latent space parameters (means-μ and variances-σ) are used to train classical classifiers (Random forest, SVM-linear and RBF kernel) that differentiates the 4 proposed classes (PMO, EBO, BLA and MYB).

illumination and stain variability effects. In the Lab color representation, the luminance component is isolated in the *L-channel*, and the remaining components (a and b) are more robust to the mentioned sources of variability [31]. Particularly, for the herein introduced work, only the *b-channel* is used to represent the whole set of images, taking into account that non-white blood cell image elements and background are homogenized in this particular Lab component.

Atypical BM Cell OCVAE Latent Space Representation. The main part of the introduced atypical BM cell characterization is based on a latent space representation of a variational autoencoder (VAE) bottleneck [16]. As commonly found in autoencoder strategies, the VAE encodes input data x by forcing dimensionality reduction to a bottleneck (latent space z) and decodes the compressed signal ($\hat{x} = f_{decoding}(z)$), aiming to minimize the output reconstruction error $\mathcal{L}_{rec} = |x - \hat{x}|_2^2$. The latent space is described by a set of parameterized distributions (mean-μ and variance-σ), and imposes these distributions to be as close as possible to unitary Gaussians. This guarantees a continuous approximation of the latent space by the Parzen theorem, i.e. $z \sim \mathcal{N}(0, 1)$, in terms of the Kullback-Leibler divergence (KLD) between the parametric posterior and the true posterior distributions.

Furthermore, to enhance VAE cell description, the presented approach takes advantage of the dedicated one-class characterization to force inter-class separability but increasing intra-class proximity. One-class classification strategies have been mainly used to find outlier samples in a given data space [6,24], with successful application in image related tasks [29], even using the reconstruction error of variational autoencoders (VAE) [14]. As shown in Fig. 1, unlike previously published strategies that are mainly set to identify anomalies or fake samples by using the autoencoder reconstruction error in a binary task [2], the presented strategy uses the regularized representation of a VAE bottleneck, without using the reconstruction quality, but the latent space ability to separate the proposed classes. The presented strategy trains a set of 4 specialized VAEs, each adjusted for estimating a representation of one atypical cell class (OCVAE). After that, all testing images pass through each specialized OCVAE (OCVAEs cascade), in a blind characterization process. Finally, all test encoded outputs, composed of means and variances, are concatenated in a single feature space which feeds different classifiers that discriminate image samples of the four atypical BM cell types. Additionally, the obtained feature matrix is normalized after concatenation, decreasing the bias possibility, given the separately trained OCVAEs.

In a more detailed description, the implemented VAE encoder is composed of 3 convolutional and 2 max-pooling, intercalated layers, followed by a flatten, a fully connected, and a lambda layer which is customized for sampling the latent space in terms of means μ and variances σ. Here the bottleneck is set to compress the input layer dimension ($256 \times 256 \times 1$) in terms of 64 Gaussian distributions, i.e., $\mu_{i:\{1-64\}}$ and $\sigma_{i:\{1-64\}}$. The decoder architecture follows similar encoder's layer organization (3 convolutional and 3 up-sampling layers), but returning the original dimensionality to the reconstructed images.

Evaluation. The introduced concatenated OCVAE representation is quantitatively evaluated by classifying the four proposed classes (PMO, BLA, MYB, EBO). The OCVAE cascade training is carried out by using 80% of the dataset ($n = 20,800$), i.e., parameters of each specialized OCVAE model are found by using 5,200 images coming from each of the proposed classes. The remaining 20% of the dataset ($n = 5,200$), composed of equal number of cells per class ($n = 1,300$), is used to obtain the feature space for evaluating the presented methodology while assuring independence to the parameterization image set. This independent data partition feeds the previously obtained specialized OCVAE encoder models, and the bottleneck values are concatenated to build an atypical BM cell representation. Afterward, the feature concatenation is used to train three different classifiers (SVM with linear and RBF kernels, and Random Forest) for differentiating cell types that compose the sample space. This experiment uses five iterations of a five-fold cross validation over the obtained feature space (20% of images), for reducing possible batch effect. Mean accuracy, precision, recall and f1-score, with their correspondent standard deviations, are presented as performance metrics. Finally, the best classifier, selected based on the performance metrics, is optimized by using different OCVAE parameter combinations as inputs, i.e., latent-space distribution means together with the corresponding variances, only means or only variances.

A second experiment is done to provide a baseline comparison in classifying the 4 atypical cell classes, by using different strategies reported in the literature for this task, including classical handcrafted image features and deep neural networks. For this evaluation procedure, classical image processing descriptors were included for providing a classification strategy that depends on both, the interpretability of the feature representation space and the classifier, like the one introduced in this work. This handcrafted representation comprises a set of 144 image features obtained from nucleus and cytoplasm as separated cell elements, and includes RGB-color space intensity statistics (mean, variance, kurtosis, skweness, entropy), Gray-level-Co-occurrence matrix statistics (contrast, dissimilarity, homogeneity, energy, correlation, angular second moment, Minkowski-Bouligand dimension), and shape descriptors (convexity, compactness, elongation, eccentricity, roundness, solidity, area, perimeter). All these features were used to train different classifiers from which an SVM classifier (linear kernel) provides the best performance with the proposed setup. Regarding deep learning approximations, different architectures were used, including two bench marking options Xception [8], ResNet50 [12], the network with the best published results on a related task based on attention-integrated network CoAtnet [9], and the ResNext [30] which was proposed by the database authors [21]. Furthermore, Xception and ResNet50 networks were trained by using imagenet weights along 30 epochs, while ResNext [21] and CoAtnet [27] evaluation uses the weights provided by the corresponding authors. Regarding the experimental setup, this comparison experiment follows the same data organization as presented in the first experiment, i.e., using 80% and 20% of the data, for training and testing respectively.

3 Results and Discussion

The results of the first experiment are shown in Table 1, where the overall classification performance of this multi-class problem demonstrates the best discrimination results were achieved with a SVM classifier (linear kernel), i.e., mean accuracy and recall of 0.881. Interestingly, regardless the implemented classifier, all the obtained performance values are higher than 0.86, indicating that differentiation stability relies on the concatenated OCVAE characterization space. In addition, as presented in Table 2, the OCVAE SVM-linear kernel results can be optimized by feeding the classifier only with the OCVAE latent variances, leading to an improvement of almost 0.06 in all metrics (accuracy = 0.938, precision = 0.935, recall = 0.935, f1-score = 0.932), when comparing the same classifier but using the whole latent space (means and variances). Finally, Fig. 2 presents the results per-class by evaluating the SVM classifier with a linear-kernel, using only variances. Such results evidenced model stability regardless the atypical cell class, with all metrics going above 0.93.

Finally, the baseline experiment results, presented in the table 2, demonstrates the performance achieved by ResNet50 (accuracy = 0.624, precision = 0.24, recall = 0.25 and f1-score = 0.248) and Xception network (accuracy = 0.708, precision = 0.697, recall = 0.899, f1-score = 0.72) are lower than the obtained with the proposed strategy. In contrast, handcrafted features (accuracy = 0.843, precision = 0.688, recall = 0.684, f1-score = 0.685) ResNext network (accuracy = 0.79, precision = 0.78, recall = 0.74, f1-score = 0.72), and

Table 1. Overall classification results in differentiating 4 atypical BM cell classes by using the whole OCVAE latent space representation with 3 classifiers: SVM (linear/RBF kernels) and Random forest.

	Mean Accuracy (std)	Mean Precision (std)	Mean Recall (std)	Mean f1-score (std)
SVM linear kernel	**0.881 (0.006)**	**0.882 (0.005)**	**0.881 (0.006)**	**0.882 (0.005)**
SVM RBF	0.866 (0.008)	0.863 (0.008)	0.866 (0.007)	0.863 (0.008)
Random forest	0.879 (0.011)	0.879 (0.01)	0.88 (0.013)	0.878 (0.013)

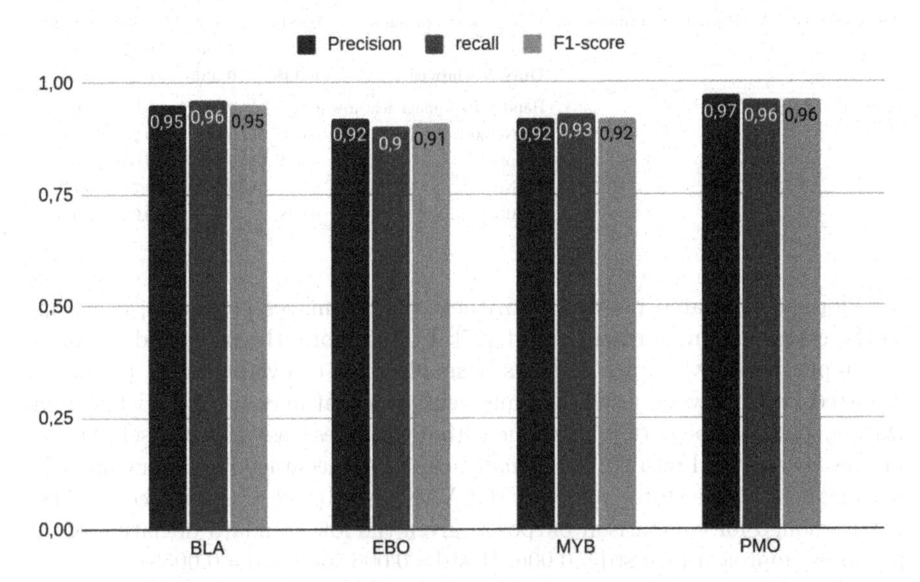

Fig. 2. Per class classification performance metrics (Precision, recall, F1-score), by using the concatenation of OCVAEs' variances as input of a SVM with a linear kernel.

CoAtnet (accuracy = 0.908, precision= 0.930, recall = 0.933, f1-score = 0.931) show better results among tested baseline methodologies, but still outperformed by optimized OCVAE cascade. Interestingly, the results of the bench marking networks are outperformed by the handcrafted-feature-based approach. These results are caused by the low number of training images, and the transference of learned weights from imagenet natural images, rather than cell related ones. Even where parameter optimization is applied to such networks, overfitting and 4 class complexity is affecting the predicting power of these architectures. Meanwhile, classical SVM take advantage of the separability given by pre-extracted features, which simplify the separation task and limits the overfitting risk. State-of-the-art networks, CoAtnet and ResNext, show better results by using images coming from the here in used database, instead of pre-trained network weights. Even when CoAtnet performance is closer to the introduced OCVAE cascade, it is important to highlight the low training data requirement of this approach (n = 20, 800 for setting OCVAE feature extractor and n = 5, 200 for training and testing the classifiers), compared to the amount of data used to train the

Table 2. Top panel: overall classification results by using different parameter combination for the introduced OCVAE (mean+variances, only means, only variances) and a SVM (linear kernel) to discriminate the 4 atypical BM cell types, showing the best performance results (blue) were obtained using just the variances of the OCVAE bottleneck. Bottom panel: the performance metrics of baseline methodologies including a classical image processing feature representation and 4 different deep learning architectures.

	Strategy	Accuracy	Precision	Recall	F1-score
Introduced OCVAE Parameter optimization	Variances and means	0.881	0.882	0.881	0.882
	Only Means	0.917	0.925	0.924	0.925
	Only Variances	**0.938**	**0.935**	**0.935**	**0.932**
Baseline comparison	Handcrafted image features	0.843	0.688	0.684	0.685
	ResNet50	0.624	0.24	0.25	0.248
	Xception	0.624	0.647	0.648	0.637
	ResNext	0.79	0.78	0.74	0.72
	CoAtnet	0.908	0.930	0.933	0.931

transformer-integrated network (CoAtnet- 20,000 images per class), according to the corresponding authors report [27]. Furthermore, the presented approach is computationally simpler and has a smaller risk of overfitting, a frequently reported problem when using data augmentation that may affect model generalization. This may be particularly true with myelocytes class since the set of 6,557 images is converted into 20,000. Finally, the mean accuracy improvement (3%), with respect to the state of the art (OCVAE_acc= 0.938, Coatnet_acc= 0.908), is fair enough for comparison purposes, given the low standard deviation of the proposed approach (acc_std= 0.006, f1_std= 0.006, prec_std= 0.0058).

Besides the classification performance, it is important to highlight that these results are obtained by the OCVAE bottleneck regularization, which reduces variance inflation and maximises the compactness of the feature space [15]. The previously mentioned OCVAE advantages, and the generative properties of this regularized latent-space prevent overfitting, but also as suggested by the results, provide a representation that keeps closer similar image concepts (cells that share similar patterns).

4 Conclusions

This work presents an atypical BM cell characterization strategy, which uses a concatenation of OCVAE's bottleneck parameters as a cell representation space. The combination of this space and an SVM classifier, demonstrates to successfully discriminate the 4 most common atypical cell types (PMO, BLA, MYO, EBO), while outperforms previously published strategies, with lower requirement of training images. This approach provides a tool for identifying these cell classes regardless the disease they are coming from, increasing the possibility of aiding differential blood counting in the presence of pathological conditions. Future work includes an independent evaluation by using other public databases and a more exhaustive baseline experimentation.

Acknowledgments. This work was supported in part by the project with code 110192092354 and entitle "Program for the Early Detection of Premalignant Lesions and Gastric Cancer in urban, rural and dispersed areas in the Department of Nariño" of call No. 920 of 2022 of MinCiencias.

References

1. Alférez, S., et al.: Automatic recognition of atypical lymphoid cells from peripheral blood by digital image analysis. Am. J. Clin. Pathol. **143**(2), 168–176 (2015)
2. An, J., Cho, S.: Variational autoencoder based anomaly detection using reconstruction probability. Spec. Lect. IE **2**(1), 1–18 (2015)
3. Anilkumar, K., Manoj, V., Sagi, T.: A survey on image segmentation of blood and bone marrow smear images with emphasis to automated detection of leukemia. Biocybernet. Biomed. Eng. **40**(4), 1406–1420 (2020)
4. Boldú, L., et al.: A deep learning model (ALNET) for the diagnosis of acute leukaemia lineage using peripheral blood cell images. Comput. Methods Prog. Biomed. **202**, 105999 (2021)
5. Boldú, L., et al.: Automatic recognition of different types of acute leukaemia in peripheral blood by image analysis. J. Clin. Pathol. **72**(11), 755–761 (2019)
6. Chalapathy, R., Menon, A.K., Chawla, S.: Anomaly detection using one-class neural networks. arXiv preprint arXiv:1802.06360 (2018)
7. Chen, W., et al.: The population characteristics of the main leukocyte subsets and their association with chronic diseases in a community-dwelling population: a cross-sectional study. Primary Health Care Res. Developm. **22** (2021)
8. Chollet, F.: Xception: deep learning with depthwise separable convolutions. In: Proceedings of the IEEE Conference on Computer Vision and Pattern Recognition, pp. 1251–1258 (2017)
9. Dai, Z., et al.: Coatnet: marrying convolution and attention for all data sizes. Adv. Neural Inf. Process. Syst. **34**, 3965–3977 (2021)
10. Fuentes-Arderiu, X., Dot-Bach, D.: Measurement uncertainty in manual differential leukocyte counting. Clin. Chem. Lab. Med. **47**(1), 112–115 (2009)
11. Gutiérrez, G., et al.: Eqas for peripheral blood morphology in Spain: a 6-year experience. Int. J. Lab. Hematol. **30**(6), 460–466 (2008)
12. He, K.,et al.: Deep residual learning for image recognition. In: Proceedings of the IEEE Conference on Computer Vision and Pattern Recognition, pp. 770–778 (2016)
13. Hodes, A., et al.: The challenging task of enumerating blasts in the bone marrow. In: Seminars in Hematology, vol. 56, pp. 58–64. Elsevier (2019)
14. Khalid, H., Woo, S.S.: Oc-fakedect: classifying deepfakes using one-class variational autoencoder. In: Proceedings of the IEEE/CVF Conference on Computer Vision and Pattern Recognition Workshops, pp. 656–657 (2020)
15. Kim, B., Ryu, K.H., Kim, J.H., Heo, S.: Feature variance regularization method for autoencoder-based one-class classification. Comput. Chem. Eng. **161**, 107776 (2022)
16. Kingma, D.P., Welling, M.: Auto-encoding variational bayes. arXiv preprint arXiv:1312.6114 (2013)
17. Krappe, S., et al.: Automated classification of bone marrow cells in microscopic images for diagnosis of leukemia: a comparison of two classification schemes with respect to the segmentation quality. In: SPIE Proceedings. SPIE (2015). https://doi.org/10.1117/12.2081946

18. Krappe, S., et al.: Automated morphological analysis of bone marrow cells in microscopic images for diagnosis of leukemia: nucleus-plasma separation and cell classification using a hierarchical tree model of hematopoesis. In: Medical Imaging 2016: Computer-Aided Diagnosis. SPIE (2016). https://doi.org/10.1117/12.2216037
19. Liu, H., Cao, H., Song, E.: Bone marrow cells detection: a technique for the microscopic image analysis. J. Med. Syst. **43**(4), 1–14 (2019)
20. Matek, C., et al.: An expert-annotated dataset of bone marrow cytology in hematologic malignancies (2021). https://doi.org/10.7937/TCIA.AXH3-T579, https://wiki.cancerimagingarchive.net/x/CoITBg https://wiki.cancerimagingarchive.net/x/CoITBg
21. Matek, C., et al.: Highly accurate differentiation of bone marrow cell morphologies using deep neural networks on a large image data set. Blood J. Am. Soc. Hematol. **138**(20), 1917–1927 (2021)
22. Matek, C., et al.: Human-level recognition of blast cells in acute myeloid leukemia with convolutional neural networks (2019). https://doi.org/10.1101/564039
23. Nedumannil, R., Sim, S., Westerman, D., Juneja, S.: Identification and quantitation of blasts in myeloid malignancies with marrow fibrosis or marrow hypoplasia and cd34 negativity. Pathology **53**(6), 795–798 (2021)
24. Ruff, L., et al.: Deep one-class classification. In: International Conference on Machine Learning, pp. 4393–4402. PMLR (2018)
25. Shah, A., et al.: Automated diagnosis of leukemia: a comprehensive review. IEEE Access **9**, 132097–132124 (2021)
26. Tayebi, R.M., et al.: Automated bone marrow cytology using deep learning to generate a histogram of cell types. Commun. Med. **2**(1), 45 (2022)
27. Tripathi, S., et al.: Hematonet: expert level classification of bone marrow cytology morphology in hematological malignancy with deep learning. Artif. Intell. Life Sci. **2**, 100043 (2022)
28. Vanna, R., et al.: Label-free imaging and identification of typical cells of acute myeloid leukaemia and myelodysplastic syndrome by Raman microspectroscopy. Analyst **140**(4), 1054–1064 (2015)
29. Wei, Q., et al.: Anomaly detection for medical images based on a one-class classification. In: Medical Imaging 2018: Computer-Aided Diagnosis, vol. 10575, pp. 375–380. SPIE (2018)
30. Xie, S., et al.: Aggregated residual transformations for deep neural networks. In: Proceedings of the IEEE Conference on Computer Vision and Pattern Recognition, pp. 1492–1500 (2017)
31. Zhang, C., et al.: White blood cell segmentation by color-space-based k-means clustering. Sensors **14**(9), 16128–16147 (2014)
32. Zini, G., Bain, B., Castoldi, G.: Others: European leukemianet (ELN) project diagnostic platform (wp10): final results of the first study of the european morphology consensus panel. Blood **112**(11), 1645 (2008)

MulHiST: Multiple Histological Staining for Thick Biological Samples via Unsupervised Image-to-Image Translation

Lulin Shi, Yan Zhang, Ivy H. M. Wong, Claudia T. K. Lo, and Terence T. W. Wong(✉)

Translational and Advanced Bioimaging Laboratory, Department of Chemical and Biological Engineering, Hong Kong University of Science and Technology, Hong Kong, China
ttwwong@ust.hk

Abstract. The conventional histopathology paradigm can provide the gold standard for clinical diagnosis, which, however, suffers from lengthy processing time and requires costly laboratory equipment. Recent advancements made in deep learning for computational histopathology have sparked lots of efforts in achieving a rapid chemical-free staining technique. Yet, existing approaches are limited to well-prepared thin sections, and invalid in handling more than one stain. In this paper, we present a multiple histological staining model for thick tissues (MulHiST), without any laborious sample preparation, sectioning, and staining process. We use the grey-scale light-sheet microscopy image of thick tissues as model input and transfer it into different histologically stained versions, including hematoxylin and eosin (H&E), Masson's trichrome (MT), and periodic acid-Schiff (PAS). This is the first work that enables the automatic and simultaneous generation of multiple histological staining for thick biological samples. Moreover, we empirically demonstrate that the AdaIN-based generator offers an advantage over other configurations to achieve higher-quality multi-style image generation. Extensive experiments also indicated that multi-domain data fusion is conducive to the model capturing shared pathological features. We believe that the proposed MulHiST can potentially be applied in clinical rapid pathology and will significantly improve the current histological workflow.

Keywords: Virtual staining · Multi-domain image translation · Thick tissues · Light-sheet microscopy

1 Introduction

Histological staining is regarded as the standard protocol in clinical pathological examination, which is used to label biological structures and morphological changes in tissues [1]. The most frequently used histological staining is H&E

H. Greenspan et al. (Eds.): MICCAI 2023, LNCS 14225, pp. 735–744, 2023.
https://doi.org/10.1007/978-3-031-43987-2_71

stain for the inspection of cell nuclei and the extracellular matrix, in addition to some special stains to complement specific biomarkers and particular structures, such as MT stain used for connective tissues and PAS stain used for mucopolysaccharides [2]. However, multiple tissue sections are required if special stains are desired since the same section cannot be stained several times in conventional pathology workflow. In general, pathologists need to check the H&E-stained images firstly for a basic examination, and then decide whether to prepare additional sections and perform special stains, which will increase the time for diagnosis. More importantly, the abovementioned traditional histochemical staining techniques can only be performed on thin sections of 2–10 μm. Therefore, sample preparation steps, including paraffin embedding, tissue slicing, and chemical dewaxing, will result in long turnaround times and high laboratory infrastructure demands.

The rapidly emerging field of digital virtual staining has shown great promise to revolutionize the decade-old staining workflow. Zhang et al. [3] have done pioneering works on multi-stain translation from unstained thin sections, and Yang et al. [4] also tried to achieve multiple stains generation from label-free tissue images. Both used an image registration to prepare pixel-level matched source-unstained and target-stained image pairs for supervised model training. However, obtaining such pixel-wise aligned data is not accessible for thick tissues as the traditional histological staining can only be performed on thin sections. Even though we can collect the surface cut from the thick specimen and then stain it with chemical reagents, there is still a huge morphological difference due to multiple-layer information captured by a slide-free microscope. Therefore, the virtual staining of thick tissues has to rely on unsupervised methods. There were some primary investigations on virtual staining of thick tissues that use slide-free imaging systems, such as MUSE [5], CHAMP [6], and UV-PAM [7]. However, those methods can only produce virtual H&E-stained images instead of multi-stained images. The emergency of starGAN opens new possibilities for multi-domain image translation [8], and they achieve flexible facial attribute transfer with the proposed domain label. [9,10] employ the idea of the domain label to represent different staining for multiple histological staining generations. However, those models use H&E staining as input and transfer H&E staining into other stains, which still require laborious tissue embedding and slide sectioning process. [11] focus on the unsupervised multiple virtual staining from autofluorescence images, yet the input should be images obtained from thin slides, which is not ideal for thick tissues.

In this paper, we propose MulHiST, a novel **Mul**tiple **Hi**stological **S**taining model for **T**hick biological tissues, which is not feasible in the traditional histochemical staining workflow. To our knowledge, this is the first attempt to achieve multiple histological staining generations for thick tissues.[1]. Our key contributions are: (1) we propose MulHiST: a generative adversarial network (GAN)-based multi-domain image translation model capable of mapping a given

[1] An implementation of MulHiST is available at https://github.com/TABLAB-HKUST/MulHiST.

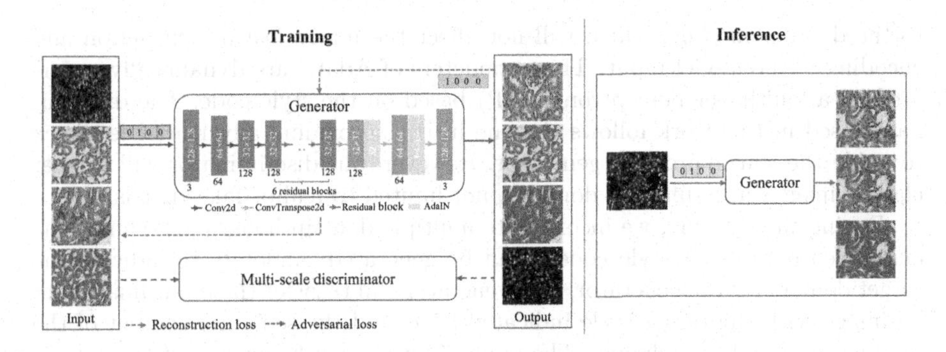

Fig. 1. Overview of the proposed **MulHiST** network.

light-sheet (LS) image of thick tissue into its histologically stained (HS) versions. We utilize unsupervised learning and do not require paired images, tailored for the virtual staining from slide-free imaging techniques; (2) we verify that the multi-domain translation can capture more reliable histopathological features for generating high-quality images, eliminating the ambiguity brought by multiple layers of information in thick tissue images; (3) both qualitative and quantitative results on H&E/PAS/MT staining generations show the superiority and efficiency of the proposed MulHiST over other baseline models.

2 Methodology

To represent different staining types, we follow the idea of the domain-specific attribute vector proposed in [8], aiming to use a one-hot vector c to indicate unstained and various stained domains. During the training, the generator aims to transfer the image style of input image x to the desired domain c while keeping the content of x: $G(x, c) \rightarrow y$, which means y has the same pathological context as x but with a different image style. When the training is finished, we use the well-trained generator to achieve multiple histological stain generations.

In general, the image of thick tissues contains several-layer information, and information from different layers interferes with each other, which will lead to ambiguity in the determination of pathological features, e.g., cell boundary and tissue content. We believe that all the image domains, i.e., LS images and H&E/PAS/MT-stained images, share some domain-invariant features that can facilitate the model training. Moreover, our generator learns the mapping between every two domains. Then, a reconstruction loss can be employed in the single-generator model (orange dashed arrows in Fig. 1). We only need to input the LS image for the inference to get its corresponding histological stained versions (right part in Fig. 1).

Unlike starGAN, we add the style code into the model with the Adaptive Instance Normalization (AdaIN) Layer [12] instead of concatenating the style code with the input image along the channel dimension. We only add the AdaIN to the decoder module, which will not affect the image feature extraction and encoding of the model input. The parameters of AdaIN are dynamically generated by a multilayer perceptron (MLP) based on the style code. The discriminator used in this work follows the one in [13], providing adversarial feedback to guide the learning of the generator. However, our discriminator will further classify images into different domains, not limited to real or fake signals.

During the training, we incorporate multiple domain images and shuffle the ensembled dataset. A style code c can be generated randomly to indicate the target domain for the generator. The generator will transfer the input image into an image with the target style indicated by c, and the semantic content of the original input will not change. The model forward can be expressed by:

$$G(x, c) = Dec(R(Enc(x)), MLP(c)) \tag{1}$$

where c is the style code that is randomly generated during the training. R is residual block. Enc and Dec are the encoder and decoder in the generator, respectively. The AdaIN layer in the decoder can be computed as:

$$AdaIN(x, s) = \sigma(s)\frac{x - \mu(x)}{\sigma(x)} + \mu(s) \tag{2}$$

where x here is the feature map results of the previous layer, and s is the output of MLP. AdaIN aims to align the mean and variance of the input to match those of the desired style. The overall loss formulation of the generator is:

$$L_G = \lambda_1 L_{adv}^G + \lambda_2 L_{rec} \tag{3}$$

$$L_{rec} = \|x - G(G(x, c_{trg}), c_{org})\| \tag{4}$$

where the L_{adv}^G is the adversarial loss and L_{rec} is the image reconstruction loss. The c_{trg} is the generated target style code and c_{org} is the style code of the original input images.

3 Experimental Results

3.1 Dataset and Implementation Details

In this work, we prepared six thick tissue slabs and obtained scanned images of ~15,000 × 15,000 pixels using an open-top LS microscope with an excitation wavelength of 266 nm. After imaging, the specimens were sectioned and histologically processed with the standard protocol to obtain HS images for the model training. We chose one set of scanned images as training data, and the others were used for the testing. For the training, we extracted small image patches with the size of 128 × 128 randomly. During testing, we divided the tested whole-slide image into 256 × 256 patches with 16 pixels overlap to avoid artifacts. There are 11,643 extracted image patches in the testing set.

Our model was implemented in PyTorch on a single NVIDIA GeForce RTX 3090 GPU. We trained our model with the Adam optimizer (with $\beta_1 = 0.5$ and $\beta_2 = 0.999$) [14]. The initial learning rate was set to 1×10^{-4} for both generator and discriminator with a linear decay scheduled after 50,000 iterations. The batch size was set to 16. The λ_1 in (3) was set to 5 and λ_2 was set to 10.

3.2 Evaluation Metrics

We quantitatively evaluate our model and results with Kernel Inception Distance (KID) [15] and Fréchet Inception Distance (FID) [16], which are prevalent for evaluating the quality of digitally synthesized images. Unlike natural image generation, biomedical image generation not only require these indicators that reflect the image quality but also require some more convincing metrics with clinical values. As shown in Fig. 3 (5th column, cycleGAN), the image style of generated results is similar to that of the real staining. However, the MT staining generated by cycleGAN is incorrect since the background and tissue content is reversed, which means that sometimes the model can produce target images with high fidelity, but it is difficult to keep correct semantic content or targeted biomarkers. It is not easy to identify pathological features from the LS image, therefore, we need a ground truth for a more reliable comparison. However, it is infeasible to obtain the well-matched ground truth of thick tissues.

In this work, we observed that LS images of thick tissues share a similar style as fluorescence images of thin sections i.e., the cell nuclei are highlighted with positive contrast for both thin and thick tissues with the help of fluorescent labels. In this case, we used the model trained on LS images of thick tissues to test the fluorescence images of thin sections. Then, we could prepare the ground truth of thin sections for comprehensive comparison, as well as quantitative indicators, such as mean square error (MSE), structural similarity (SSIM), and peak signal-to-noise ratio (PSNR). It is worth mentioning that when tested on the thin-section images, the model trained with thick-tissue images would underperform the model trained with thin-section images. Therefore, for virtual staining on thin slices, it is better to train on thin-section images as there are still some differences in details between data from thick tissues and thin sections. Here, we use the thin-section data only for model validation.

3.3 Quantitative and Qualitative Results

In this paper, we select cycleGAN [17], MUNIT [18], and starGAN [8] as baseline models. As shown in Fig. 2, our model surpasses all comparison methods on visual results. From the virtually stained PAS (top zoomed-in regions, yellow square), we can distinguish two different convoluted tubules according to the PAS-positive/negative patterns (yellow/green arrows) in our results. In general, the glomerular and tubular basement membrane, as well as the brush border of the proximal tubules, can be visualized by the PAS staining with pink color, whereas the interior of the distal tubule will not be stained [19]. There is something pink inside the tubules pointed by yellow arrows in our results and no pink area inside the tubules indicated by green arrows. However, from other model results, it is hard to recognize corresponding histopathological features. Here we train cycleGAN multiple times for every pair of source/target domains so that those three cycleGAN models are independent. For such single-domain image translation, it is hard to satisfy all domains with a correct transformation. Specifically, we can see that the H&E-stained images of cycleGAN are correct,

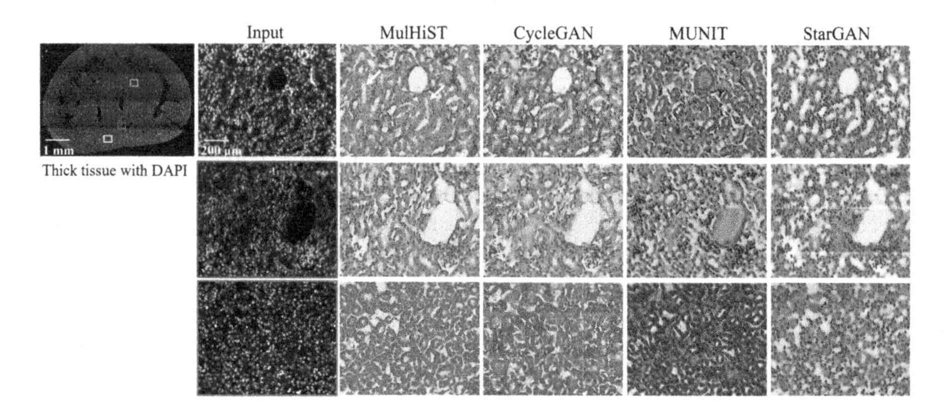

Fig. 2. Virtual generation of PAS-, H&E-, and MT-stained images of thick tissues (from the top to the bottom). Yellow arrows indicate proximal convoluted tubules (positive for PAS), and green ones mean distal convoluted tubules (negative for PAS). (Color figure online)

whereas the model reverses the background and cell nuclei in the MT domain. As shown in Fig. 2, neither MUNIT nor starGAN can perform well in this task.

Moreover, we quantitatively evaluate the model performance with FID and KID scores, as shown in Table 1 (top part). We can observe that our model outperforms the other baseline models significantly in three different image domains. Although the H&E staining of cycleGAN also achieves good FID and KID scores, the corresponding PAS and MT staining results are much worse than ours, which also agrees with the qualitative analysis shown in Fig. 2.

As no ground truth can be provided for the virtual staining of thick tissue. We collected 2 sets of thin mouse kidney sections for the model validation. We used the model trained with LS images of thick tissues to test the scanned images of prepared thin sections. In this situation, we can perform traditional histological staining to obtain ground truth for further comparison. Figure 3 confirms that our MulHiST can generate the correct pathological features, i.e., the PAS-positive proximal convoluted tubules indicated by yellow arrows and PAS-negative distal tubules pointed by green arrows, which are consistent with the ground truth. The same pathological representation can also be observed in the PAS result of cycleGAN, whereas the MT staining generated by cycleGAN presents an obvious error between the background and tissue content. Meanwhile, the quantitative evaluation in Table 1 (bottom part) also shows that our proposed model outperforms other baseline models in various evaluation metrics.

3.4 Ablation Analysis

In this paper, we have two main hypotheses, one is that the model can benefit from data fusion from multiple domains due to the domain-invariant features, and the other is that the AdaIN-based style transfer is better in source image feature extraction compared with channel-wise style code concatenation.

Table 1. Quantitative evaluation results on testing data. The top four rows are tested results of thick tissues, and the following ones are that of thin sections.

	H&E		PAS		MT		MSE	SSIM	PSNR
	FID	KID	FID	KID	FID	KID			
CycleGAN	68.06	**0.042**	111.29	0.062	150.77	0.096	–	–	–
MUNIT	150.20	0.114	131.93	0.086	128.80	0.098	–	–	–
StarGAN	165.20	0.114	207.67	0.158	208.89	0.180	–	–	–
MulHiST	**65.96**	0.049	**63.20**	**0.038**	**67.83**	**0.036**	–	–	–
CycleGAN	74.16	0.044	137.18	0.080	159.24	0.098	2257.73	0.5793	15.90
MUNIT	141.71	0.078	136.23	0.074	118.51	0.067	7749.29	0.4675	9.35
StarGAN	160.91	0.108	185.07	0.127	155.02	0.105	1266.57	**0.6811**	17.40
MulHiST	**69.90**	**0.039**	**82.49**	**0.048**	**84.44**	**0.055**	**1217.36**	0.6475	**17.50**

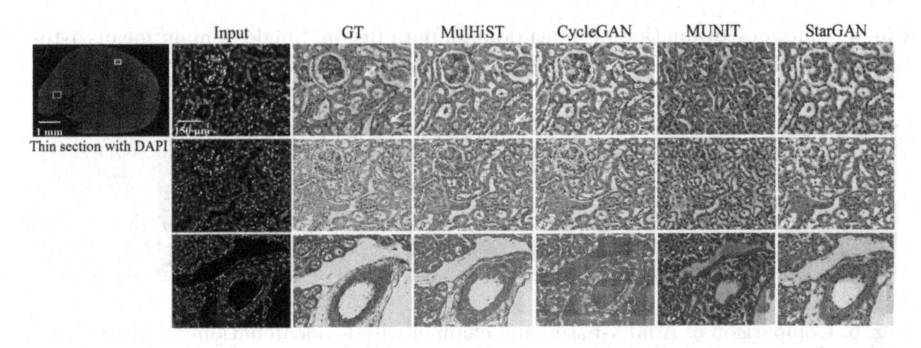

Fig. 3. Virtual staining output of thin section using different models. The 3rd column shows the ground truth (GT) obtained via the traditional histology staining protocol.

We first verify the importance of domain-invariant features shared by multiple domains. The starGAN [8] also reached a similar conclusion that different domains will share the same facial context, which is beneficial to facial expression synthesis. Similarly, there are also some shared features in our multi-domain dataset, such as cell nuclei and cytoplasm membranes. We claim that the model training can benefit from incorporating multiple data domains. From Fig. 4, the synthetic image quality improves with the increase of input data domains. For single-domain translation, the model is sensitive to different domains, where only H&E staining is correct and the other two cannot be determined. This also agrees with the performance of cycleGAN in Fig. 2, 3. The main reason is that the different models are built independently for single-domain image translation, resulting in the inability to share effective information among domains. When adding another domain, the model can correctly translate the background and tissue content. This can be attributed to the feature-sharing between different data domains. In addition, the triple-domain model is superior to the dual-domain ones, where all H&E, PAS, and MT show clear and natural structures.

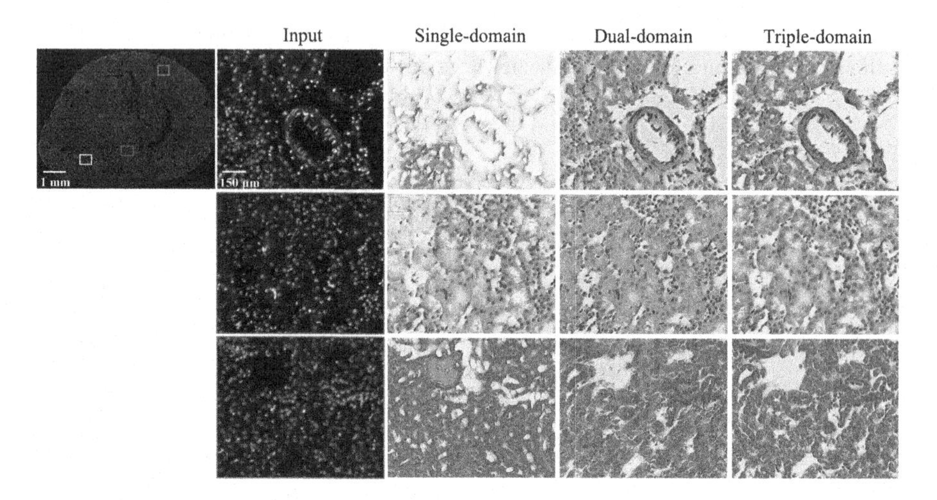

Fig. 4. Efficiency evaluation of multi-domain data fusion. Single-domain means 1-to-1 translation, dual-domain is 2-to-2 translation, and triple-domain refers to 3-to-3 one.

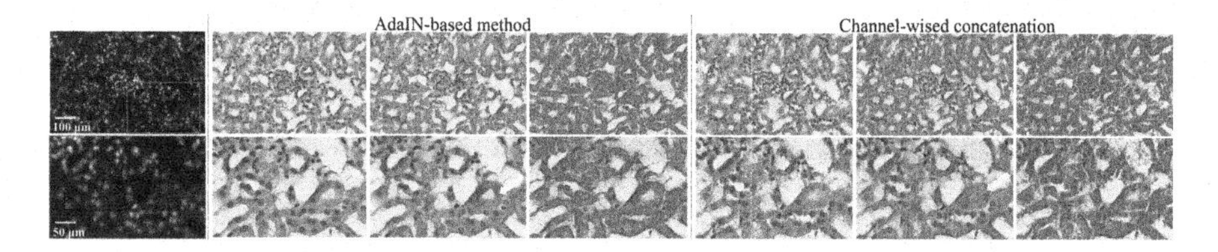

Fig. 5. Comparison of AdaIN-based and channel-wised concatenation-based generator.

Next, we compare the AdaIN-based method and channel-wise concatenation (Fig. 5). We can observe that both ways can achieve correct style transfer, but from the PAS staining, the AdaIN-based method can produce plausible pathological patterns that another model fails (yellow arrows). The PAS should be stained inside the tubules, and the results from the channel-wise concatenation method will cause ambiguity in the judgment by pathologists.

4 Conclusion

This paper proposes a multiple histological image generation model for thick biological samples. This is undesirable in the traditional histopathology workflow as the chemical histological staining should be performed on the thin sections, and the same section cannot be stained with various stains simultaneously. We use slide-free microscopy to capture the thick tissues and translate the scanned images into multiple stained-versions. The model is optimized in an unsupervised manner, fitting the issue of large morphological mismatches between the scanned thick tissue and histologically stained images. Experiment results demonstrated the superiority and the great promise of the proposed method in developing a slide-free, cost-effective, and chemical-free histopathology staining pipeline.

References

1. Gurcan, M.N., Boucheron, L.E., Can, A., Madabhushi, A., Rajpoot, N.M., Yener, B.: Histopathological image analysis: a review. IEEE Rev. Biomed. Eng. **2**, 147–171 (2009)
2. Alturkistani, H.A., Tashkandi, F.M., Mohammedsaleh, Z.M.: Histological stains: a literature review and case study. Glob. J. Health Sci. **8**(3), 72 (2016)
3. Zhang, Y., de Haan, K., Rivenson, Y., Li, J., Delis, A., Ozcan, A.: Digital synthesis of histological stains using micro-structured and multiplexed virtual staining of label-free tissue. Light: Sci. Appl. **9**(1), 78 (2020)
4. Yang, X., et al.: Virtual stain transfer in histology via cascaded deep neural networks. ACS Photonics **9**(9), 3134–3143 (2022)
5. Chen, Z., Yu, W., Wong, I.H., Wong, T.T.: Deep-learning-assisted microscopy with ultraviolet surface excitation for rapid slide-free histological imaging. Biomed. Opt. Exp. **12**(9), 5920–5938 (2021)
6. Zhang, Y., et al.: High-throughput, label-free and slide-free histological imaging by computational microscopy and unsupervised learning. Adv. Sci. **9**(2), 2102358 (2022)
7. Cao, R., et al.: Label-free intraoperative histology of bone tissue via deep-learning-assisted ultraviolet photoacoustic microscopy. Nat. Biomed. Eng. 1–11 (2022)
8. Choi, Y., Choi, M., Kim, M., Ha, J.W., Kim, S., Choo, J.: Stargan: unified generative adversarial networks for multi-domain image-to-image translation. In: Proceedings of the IEEE Conference on Computer Vision and Pattern Recognition, pp. 8789–8797 (2018)
9. Zhang, R., et al.: Mvfstain: multiple virtual functional stain histopathology images generation based on specific domain mapping. Med. Image Anal. **80**, 102520 (2022)
10. Lin, Y., et al.: Unpaired multi-domain stain transfer for kidney histopathological images. Proc. AAAI Conf. Artif. Intell. **36**, 1630–1637 (2022)
11. Shi, L., Wong, I.H., Lo, C.T., WK, T.L., Wong, T.T.: Unsupervised multiple virtual histological staining from label-free autofluorescence images. In: Proceedings of the IEEE International Symposium on Biomedical Imaging (2023)
12. Huang, X., Belongie, S.: Arbitrary style transfer in real-time with adaptive instance normalization. In: Proceedings of the IEEE International Conference on Computer Vision, pp. 1501–1510 (2017)
13. Wang, T.C., Liu, M.Y., Zhu, J.Y., Tao, A., Kautz, J., Catanzaro, B.: High-resolution image synthesis and semantic manipulation with conditional gans. In: Proceedings of the IEEE Conference on Computer Vision and Pattern Recognition, pp. 8798–8807 (2018)
14. Kingma, D.P., Ba, J.: Adam: a method for stochastic optimization. In: Proceedings of International Conference on Learning Representations (2015)
15. Bińkowski, M., Sutherland, D.J., Arbel, M., Gretton, A.: Demystifying mmd gans. arXiv preprint arXiv:1801.01401 (2018)
16. Heusel, M., Ramsauer, H., Unterthiner, T., Nessler, B., Hochreiter, S.: Gans trained by a two time-scale update rule converge to a local nash equilibrium. Adv. Neural Inf. Process. Syst. **30** (2017)

17. Zhu, J.Y., Park, T., Isola, P., Efros, A.A.: Unpaired image-to-image translation using cycle-consistent adversarial networks. In: Proceedings of the IEEE International Conference on Computer Vision, pp. 2223–2232 (2017)
18. Huang, X., Liu, M.Y., Belongie, S., Kautz, J.: Multimodal unsupervised image-to-image translation. In: Proceedings of the European Conference on Computer Vision (ECCV), pp. 172–189 (2018)
19. Kong, W., Haschler, T.N., Nürnberg, B., Krämer, S., Gollasch, M., Markó, L.: Renal fibrosis, immune cell infiltration and changes of trpc channel expression after unilateral ureteral obstruction in trpc6-/-mice. Cell. Physiol. Biochem. **52**, 1484–1502 (2019)

Multi-scope Analysis Driven Hierarchical Graph Transformer for Whole Slide Image Based Cancer Survival Prediction

Wentai Hou[1], Yan He[2], Bingjian Yao[2], Lequan Yu[3], Rongshan Yu[2], Feng Gao[4], and Liansheng Wang[2(✉)]

[1] Department of Information and Communication Engineering at School of Informatics, Xiamen University, Xiamen, China
houwt@stu.xmu.edu.cn

[2] Department of Computer Science at School of Informatics, Xiamen University, Xiamen, China
{yanhe56,yaobingjian}@stu.xmu.edu.cn, {rsyu,lswang}@xmu.edu.cn

[3] Department of Statistics and Actuarial Science, The University of Hong Kong, Hong Kong SAR, China
lqyu@hku.hk

[4] The Sixth Affiliated Hospital, Sun Yat-sen University, Guangzhou, China
gaof57@mail.sysu.edu.cn

Abstract. Cancer survival prediction requires considering not only the biological morphology but also the contextual interactions of tumor and surrounding tissues. The major limitation of previous learning frameworks for whole slide image (WSI) based survival prediction is that the contextual interactions of pathological components (*e.g.*, tumor, stroma, lymphocyte, *etc.*) lack sufficient representation and quantification. In this paper, we proposed a multi-scope analysis driven Hierarchical Graph Transformer (HGT) to overcome this limitation. Specifically, we first utilize a multi-scope analysis strategy, which leverages an in-slide superpixel and a cross-slide clustering, to mine the spatial and semantic priors of WSIs. Furthermore, based on the extracted spatial prior, a hierarchical graph convolutional network is proposed to progressively learn the topological features of the variant microenvironments ranging from patch-level to tissue-level. In addition, guided by the identified semantic prior, tissue-level features are further aggregated to represent the meaningful pathological components, whose contextual interactions are established and quantified by the designed Transformer-based prediction head. We evaluated the proposed framework on our collected Colorectal Cancer (CRC) cohort and two public cancer cohorts from the TCGA project, *i.e.*, Liver Hepatocellular Carcinoma (LIHC) and Kidney Clear Cell Carcinoma (KIRC). Experimental results demonstrate that our proposed method yields superior performance and richer interpretability compared to the state-of-the-art approaches.

Keywords: Whole slide image · Survival prediction · Contextual interaction · Graph neural network · Transformer

H. Greenspan et al. (Eds.): MICCAI 2023, LNCS 14225, pp. 745–754, 2023.
https://doi.org/10.1007/978-3-031-43987-2_72

1 Introduction

The ability to predict the future risk of patients with cancer can significantly assist clinical management decisions, such as treatment and monitoring [21]. Generally, pathologists need to manually assess the pathological images obtained by whole-slide scanning systems for clinical decision-making, *e.g.*, cancer diagnosis and prognosis [20]. However, due to the complex morphology and structure of human tissues and the continuum of histologic features phenotyped across the diagnostic spectrum, it is a tedious and time-consuming task to manually assess the whole slide image (WSI) [12]. Moreover, unlike cancer diagnosis and subtyping tasks, survival prediction is a future state prediction task with higher difficulty. Therefore, automated WSI analysis method for survival prediction task is highly demanded yet challenging in clinical practice.

Over the years, deep learning has greatly promoted the development of computational pathology, including WSI analysis [9,17,24]. Due to the huge size, WSIs are generally cropped to numerous patches with a fixed size and encoded to patch features by a CNN encoder (*e.g.*, Imagenet pretrained ResNet50 [11]) for further analysis. The attention-dominated learning frameworks (*e.g.*, ABMIL [13], CLAM [18], DSMIL [16], TransMIL [19], SCL-WC [25], HIPT [4], NAGCN [9]) mainly aim to find the key instances (*e.g.*, patches and tissues) for WSI representation and decision-making, which prefers the needle-in-a-haystack tasks, *e.g.*, cancer diagnosis, cancer subtyping, *etc.* To handle cancer survival prediction, some researchers integrated some attribute priors into the network design [5,26]. For example, Patch-GCN [5] treated the WSI as point cloud data, and the patch-level adjacent relationship of WSI is learned by a graph convolutional network (GCN). However, the fixed-size patches cropped from WSI mainly contain single-level biological entities (*e.g.*, cells), resulting in limited structural information. DeepAttnMISL [26] extracted the phenotype patterns of the patient via a clustering algorithm, which provides meaningful medical prior to guide the aggregation of patch features. However, this cluster analysis strategy only focuses on a single sample, which cannot describe the whole picture of the pathological components specific to the cancer type. Additionally, existing learning frameworks often ignore the capture of contextual interactions of pathological components (*e.g.*, tumor, stroma, lymphocyte, *etc.*), which is considered as important evidence for cancer survival prediction tasks [1,6]. Therefore, WSI-based cancer survival prediction still remains a challenging task.

In summary, to better capture the prognosis-related information in WSI, two technical key points should be fully investigated: (1) an analysis strategy to mine more comprehensive and in-depth prior of WSIs, and (2) a promising learning network to explore the contextual interactions of pathological components. To this end, this paper presents a novel multi-scope analysis driven learning framework, called Hierarchical Graph Transformer (HGT), to pertinently resolve the above technical key points for more reliable and interpretable

W. Hou and Y. He—Contributed equally to this work.

WSI-based survival prediction. First, to mine more comprehensive and in-depth attribute priors of WSI, we propose a multi-scope analysis strategy consisting of in-slide superpixels and cross-slide clustering, which can not only extract the spatial prior but also identify the semantic prior of WSIs. Second, to explore the contextual interactions of pathological components, we design a novel learning network, *i.e.*, HGT, which consists of a hierarchical graph convolution layer and a Transformer-based prediction head. Specifically, based on the extracted spatial topology, the hierarchical graph convolution layer in HGT progressively aggregate the patch-level features to the tissue-level features, so as to learn the topological features of variant microenvironments ranging from fine-grained (*e.g.*, cell) to coarse-grained (*e.g.*, necrosis, epithelium, *etc.*). Then, under the guidance of the identified semantic prior, the tissue-level features are further sorted and assigned to form the feature embedding of pathological components. Furthermore, the contextual interactions of pathological components are captured with the Transformer-based prediction head, leading to reliable survival prediction and richer interpretability. Extensive experiments on three cancer cohorts (*i.e.*, CRC, TCGA-LIHC and TCGA-KIRC) demonstrates the effectiveness and interpretability of our framework. Our codes are available at https://github.com/Baeksweety/superpixel_transformer.

2 Methodology

Figure 1 illustrates the pipeline of the proposed framework. Due to the huge size, WSIs are generally cropped to numerous patches with a fixed size (*i.e.*, 256×256) and encoded to patch features $V_{patch} \in \mathbb{R}^{n \times d}$ in the embedding space D by a CNN encoder (*i.e.*, ImageNet pretrained ResNet50 [11]) for further analysis, where n is the number of patches, $d = 1024$ is the feature dimension. The goal of WSI-based cancer survival prediction is to learn the feature embedding of V in a supervised manner and output the survival risk $O \in \mathbb{R}^1$.

However, conventional patch-level analysis cannot model complex pathological patterns (*e.g.*, tumor lymphocyte infiltration, immune cell composition, *etc.*), resulting in limited cancer survival prediction performance. To this end, we proposed a novel learning network, *i.e.*, HGT, which utilized the spatial and semantic priors mined by a multi-scope analysis strategy (*i.e.*, in-slide superpixel and cross-slide clustering) to represent and capture the contextual interaction of pathological components. Our framework consists two modules: a hierarchical graph convolutional network and a Transformer-based prediction head.

2.1 Hierarchical Graph Convolutional Network

Unlike cancer diagnosis and subtyping, cancer survival prediction is a quite more challenging task, as it is a future event prediction task which needs to consider complex pathological structures [20]. However, the conventional patch-level analysis is difficult to meet this requirement. Therefore, it is essential to extract and combine higher-level topology information for better WSI representation.

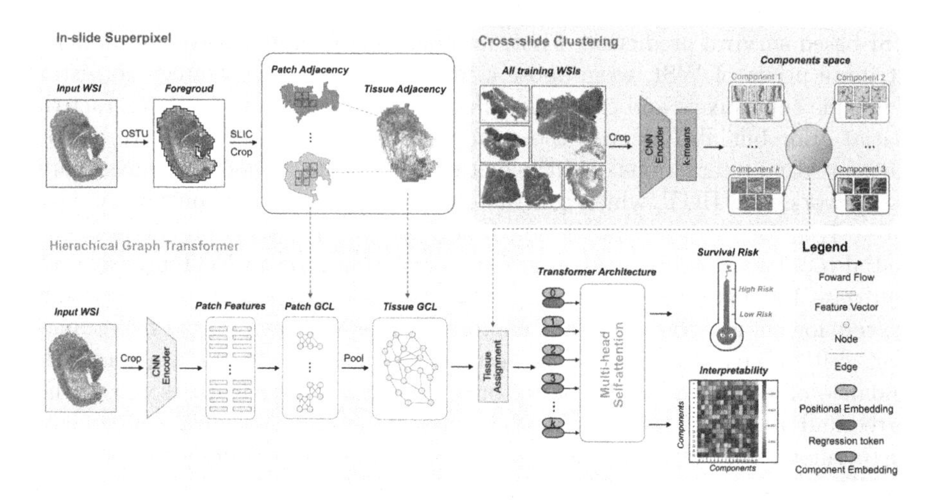

Fig. 1. Overview of the proposed multi-scope analysis (*i.e.*, in-slide superpixel and cross-slide clustering) driven Hierarchical Graph Transformer (HGT). Note that not all nodes and adjacent relationships are shown for visual clarity.

In-slide Superpixel. As shown in Fig .1, we first employ a Simple Linear Iterative Clustering (SLIC) [2] algorithm to detect non-overlapping homogeneous tissues of the foreground of WSI at a low magnification, which can be served as the spatial prior to mine the hierarchical topology of WSI. Intuitively, the cropped patches and segmented tissues in a WSI can be considered as hierarchical entities ranging from fine-grained level (*e.g.*, cell) to coarse-grained level (*e.g.*, necrosis, epithelium, *etc.*). Based on the in-slide superpixel, the tissue adjacency matrix $E_{tissue} \in \mathbb{R}^{m \times m}$ can be obtained, where m denote the number of superpixels. Then, the patches in each superpixel are further connected in an 8-adjacent manner, thus generating patch adjacency matrix $E_{patch} \in \mathbb{R}^{n \times n}$. The spatial assignment matrix between cropped patches and segmented tissues is denoted as $A_{spa} \in \mathbb{R}^{n \times m}$.

Patch Graph Convolutional Layer. Based on the spatial topology extracted by in-slide superpixel, the patch graph convolutional layer (Patch GCL) is designed to learn the feature of the fine-grained microenvironment (*e.g.*, cell) through the message passing between adjacent patches, which can be represented as:

$$V'_{patch} = \sigma(\mathrm{GraphConv}(V_{patch}, E_{patch})), \tag{1}$$

where $\sigma(\cdot)$ denotes the activation function, such as ReLU. GraphConv denotes the graph convolutional operation, *e.g.*, GCNConv [15], GraphSAGE [10], *etc.*

Tissue Graph Convolutional Layer. Third, based on the spatial assignment matrix A_{spa}, the learned patch-level features can be aggregated to the tissue-level features which contain the information of necrosis, epithelium, *etc.*

$$V_{tissue} = [A_{spa}]^{\mathrm{T}} V'_{patch}, \tag{2}$$

where $[\cdot]^{\mathrm{T}}$ denote the matrix transpose operation. The tissue graph convolutional layer (Tissue GCL) is further designed to learn the feature of this coarse-grained microenvironment, which can be represented as:

$$V'_{tissue} = \sigma(\mathrm{GraphConv}(V_{tissue}, E_{tissue})). \tag{3}$$

2.2 Transformer-Based Prediction Head

Clinical studies have shown that cancer survival prediction requires considering not only the biological morphology but also the contextual interactions of tumor and surrounding tissues [1]. However, existing analysis frameworks for WSI often ignore the capture of contextual interactions of pathological components (*e.g.*, tumor, stroma, lymphocyte, *etc.*), resulting in limited performance and interpretability. Therefore, it is necessary to determine the feature embedding of pathological components and investigate their contextual interactions for more reliable predictions.

Cross-Slide Clustering. As shown in Fig. 1, we perform the k-means algorithm on the encoded patch features of all training WSIs to generate k pathological components $P \in \mathbb{R}^{k \times d}$ in the embedding space D. P represents different pathological properties specific to the cancer type. Formally, the feature embedding of each tissue in space D is defined as the mean feature embeddings of the patches within the tissue. And then, the pathological component label of each tissue is determined as the component closest to the Euclidean distance of the tissue in space D. The semantic assignment matrix between segmented tissues and pathological components is denoted as $A_{sem} \in \mathbb{R}^{m \times k}$.

Transformer Architecture. Under the guidance of the semantic prior identified by cross-slide clustering, the learned tissue features V'_{tissue} can be further aggregated, forming a series meaningful component embeddings P' specific to the cancer type.

$$P' = [A_{sem}]^{\mathrm{T}} V'_{tissue}. \tag{4}$$

Then we employed a Transformer [22] architecture to mine the contextual interactions of P' and output the predicted survival risk. As shown in Fig. 1, P' is concatenated with an extra learnable regression token R and attached with positional embeddings E_{Pos}, which are processed by:

$$P'_{out} = \mathrm{MLP}(\mathrm{LN}(\mathrm{MHSA}(\mathrm{LN}([R; P'] + E_{Pos})))). \tag{5}$$

where P'_{out} is the output of Transformer, MHSA is the Multi-Headed Self-Attention [22], LN is Layer Normalization and MLP is Multilayer Perceptron. Finally, the representation of the regression token at the output layer of the Transformer, *i.e.*, $[P'_{out}]^0$, is served as the predicted survival risk O.

Loss Function and Training Strategy. For the network training, Cox loss [26] is adopted for the survival prediction task, which is defined as:

$$\mathcal{L}_{Cox} = \sum_{i=1}^{B} \delta_i \left(-O(i) + \log \sum_{j:t_j>=t_i} \exp\left(O(j)\right) \right), \tag{6}$$

where δ_i denote the censorship of i-th patient, $O(i)$ and $O(j)$ denote the survival output of i-th and j-th patient in a batch, respectively.

3 Experiments

3.1 Experimental Settings

Dataset. In this study, we used a **Colorectal Cancer (CRC)(385 cases)** cohort collected from co-operated hospital to evaluate the proposed method. Moreover, two public cancer cohorts from TCGA project, *i.e.*, **Liver Hepatocellular Carcinoma (LIHC)(371 cases)** and **Kidney Clear Cell Carcinoma (KIRC)(398 cases)** are also included as the censorship of these two data sets are relatively balanced. All WSIs are analyzed at $\times 10$ magnification and cropped into 256 $\times$ 256 patches. The average patch number of each WSI is 18727, 3680, 3742 for CRC, TCGA-LIHC, TCGA-KIRC, respectively. It should be noted that the largest WSI (from CRC) contains 117568 patches.

Implementation Details. All trials are conducted on a workstation with two Intel Xeon Silver 4210R CPUs and four NVIDIA GeForce RTX 3090 (24 GB) GPUs. Our graph convolutional model is implemented by Pytorch Geometric [7]. The initial number of superpixels of SLIC algorithm is set to {600, 700, 600}, and the number of clusters of k-means algorithm is set to {16, 16, 16} for CRC, TCGA-LIHC and TCGA-KICA cohorts. The non linearity of GCN is ReLU. The number of Transformer heads is 8, and the attention scores of all heads are averaged to produce the heatmap of contextual interactions. HGT is trained with a mini-batch size of 16, and a learning rate of $1e-5$ with Adam optimizer for 30 epochs.

Evaluation Metric. The concordance index (CI) [23] is used to measure the fraction of all pairs of patients whose survival risks are correctly ordered. CI ranges from 0 to 1, where a larger CI indicates better performance. Moreover, to evaluate the ability of patients stratification, the Kaplan-Meier (KM) analysis is used [23]. In this study, we conduct a 5-fold evaluation procedure with 5 runs to evaluate the survival prediction performance for each method. The result of $mean \pm std$ is reported.

3.2 Comparative Results

We compared seven state-of-the-art methods (SOTAs), *i.e.*, DeepSets [27], ABMIL [13], DeepAttnMISL [26], CLAM [18], DSMIL [16], PatchGCN [5], and TransMIL [19]. We also compared three baselines of our method, *i.e.*, w/o Patch GCL, w/o Tissue GCL and w/o Transformer. For fair comparison, same CNN extractor (*i.e.* ImageNet pretrained Resnet50 [11]), and survival prediction loss (*i.e.* Cox loss [26]) is adopt for all methods.

Table 1 and Fig. 2 show the results of CI and KM-analysis of each method, respectively. Generally, most MIL methods, *i.e.*, DeepSets, ABMIL, DSMIL, TransMIL mainly focus on a few key instances for prediction, but they do not have significant advantages in cancer prognosis. Furthermore, due to the large size of CRC dataset and relatively high model complexity, Patch-GCN and TransMIL encountered a memory overflow when processing the CRC dataset, which limits their clinical application. DeepAttnMISL has a certain semantic perception ability for patch, which achieves better performance in LIHC cohort. PatchGCN is capable to capture the local contextual interactions between patch, which also achieves satisfied performance in KIRC cohort. As our method has potential to explore the contextual interactions of pathological components, which more in line with the thinking of pathologists for cancer prognosis. Our method achieves higher CI and relatively low P-Value (< 0.05) of KM analysis on both three cancer cohorts, which consistently outperform the SOTAs and baselines. In addition, the feature aggregation of the lower levels (*i.e.*, patch and tissue) are guided by the priors, and the MHSA is only executed on pathological components, resulting in high efficiency even on the CRC dataset.

Table 1. Experimental results of CI. Results not significantly worse than the best (P-Value > 0.05, Two-sample t-test) are shown in bold. The second best results of SOTA methods are underlined. "-" denotes that the algorithm cannot be executed in this cohort due to memory overflow.

Type	Method	CRC	TCGA-LIHC	TCGA-KIRC
SOTAs	DeepSets [27]	0.504 ± 0.004	0.511 ± 0.011	0.483 ± 0.033
	ABMIL [13]	$\underline{0.580 \pm 0.005}$	0.634 ± 0.005	0.617 ± 0.094
	DeepAttnMISL [26]	0.570 ± 0.001	$\underline{0.644 \pm 0.009}$	0.584 ± 0.019
	CLAM [18]	0.575 ± 0.010	0.641 ± 0.002	0.635 ± 0.006
	DSMIL [16]	0.550 ± 0.016	0.626 ± 0.005	0.603 ± 0.022
	PatchGCN [5]	-	0.643 ± 0.003	$\underline{\mathbf{0.637 \pm 0.010}}$
	TransMIL [19]	-	0.641 ± 0.023	0.616 ± 0.014
Ablation study	w/o Patch GCL	0.597 ± 0.007	0.640 ± 0.002	0.626 ± 0.008
	w/o Tissue GCL	0.584 ± 0.010	0.644 ± 0.002	0.636 ± 0.008
	w/o Transformer	0.592 ± 0.010	0.647 ± 0.002	0.616 ± 0.005
Ours	HGT	$\mathbf{0.607 \pm 0.004}$	$\mathbf{0.657 \pm 0.003}$	$\mathbf{0.646 \pm 0.003}$

3.3 Interpretability of the Proposed Framework

We selected the CRC dataset for further interpretable analysis, as it is one of the leading causes of mortality in industrialized countries, and its prognosis-related factors have been widely studied [3,8]. We trained an encoded feature based classification model (*i.e.*, a MLP) on a open-source colorectal cancer dataset (*i.e.*, NCT-CRC-HE-100K [14]), which is annotated with 9 classes, including: adipose tissue (ADI); background (BACK); debris (DEB); lymphocytes (LYM); mucus (MUC); muscle (MUS); normal colon mucosa (NORM); stroma (STR); tumor (TUM). The trained classification model can be used to determine the biological semantics of the pathological components extracted by our model with a major voting rule. Figure 3 shows the original image, spatial topology, proportion and

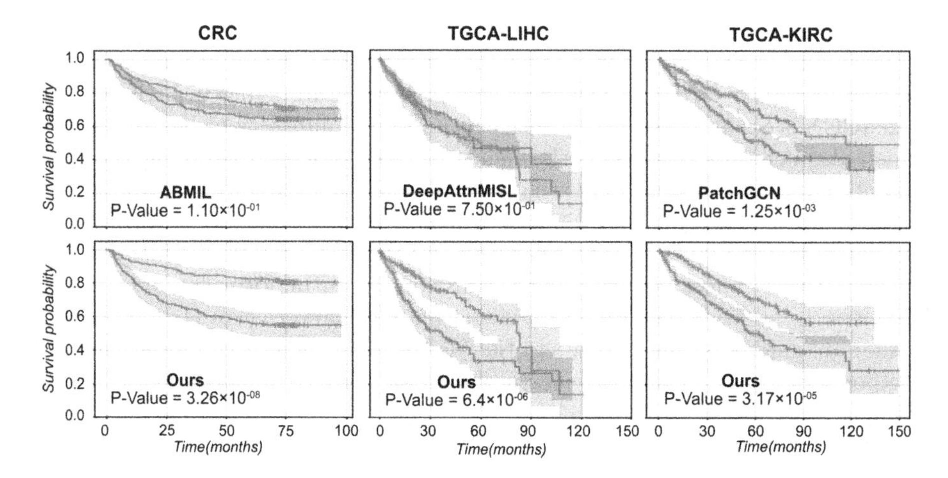

Fig. 2. KM analysis of second best SOTA method and our proposed framework for different datasets. All the patients across the five test folds are combined and analysis here. For each cohort, patients were stratified into high-risk (red curves) and low-risk (green curves) groups by the median score output by predictive models. (Color figure online)

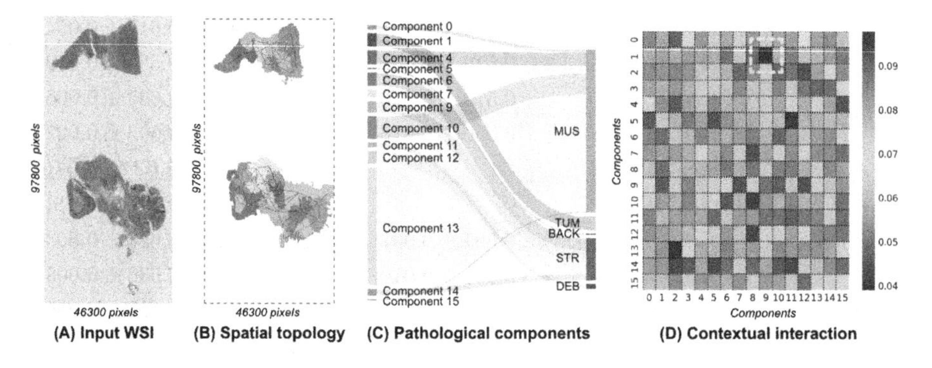

Fig. 3. Interpretability of the proposed method. A typical case in the test fold of CRC cohort is used for illustration. Best viewed by zoom in.

biological meaning of pathological components, and its contextual interactions of a typical case from CRC cohort. It can be seen that the interaction between component 1 (TUM) and component 9 (STR) has gained the highest attention of the network, which is consistent with the existing knowledge [3,8]. Moreover, there is also concentration of interaction in some other interactions, which may potentially imply some new biomarkers.

4 Conclusion

In this paper, we propose a novel learning framework, *i.e.*, multi-scope analysis driven HGT, to effectively represent and capture the contextual interaction of pathological components for improving the effectiveness and interpretability of WSI-based cancer survival prediction. Experimental results on three clinical cancer cohorts demonstrated our model achieves better performance and richer interpretability over the existing models. In the future, we will evaluate our framework on more tasks and further statistically analyze the interpretability of our model to find more pathological biomarkers related to cancer prognosis.

Acknowledgement. This work was supported by Ministry of Science and Technology of the People's Republic of China (2021ZD0201900)(2021ZD0201903).

References

1. AbdulJabbar, K., et al.: Geospatial immune variability illuminates differential evolution of lung adenocarcinoma. Nat. Med. **26**(7), 1054–1062 (2020)
2. Achanta, R., Shaji, A., Smith, K., Lucchi, A., Fua, P., Süsstrunk, S.: SLIC superpixels compared to state-of-the-art superpixel methods. IEEE Trans. Pattern Anal. Mach. Intell. **34**(11), 2274–2282 (2012)
3. Bilal, M., et al.: Development and validation of a weakly supervised deep learning framework to predict the status of molecular pathways and key mutations in colorectal cancer from routine histology images: a retrospective study. Lancet Digit. Health **3**(12), e763–e772 (2021)
4. Chen, R.J., et al.: Scaling vision transformers to gigapixel images via hierarchical self-supervised learning. In: Proceedings of the IEEE/CVF Conference on Computer Vision and Pattern Recognition, pp. 16144–16155 (2022)
5. Chen, R.J., et al.: Whole slide images are 2D point clouds: context-aware survival prediction using patch-based graph convolutional networks. In: de Bruijne, M., et al. (eds.) MICCAI 2021. LNCS, vol. 12908, pp. 339–349. Springer, Cham (2021). https://doi.org/10.1007/978-3-030-87237-3_33
6. Diao, J.A., et al.: Human-interpretable image features derived from densely mapped cancer pathology slides predict diverse molecular phenotypes. Nat. Commun. **12**(1), 1613 (2021)
7. Fey, M., Lenssen, J.E.: Fast graph representation learning with PyTorch geometric. In: ICLR Workshop on Representation Learning on Graphs and Manifolds (2019)
8. Foersch, S., et al.: Multistain deep learning for prediction of prognosis and therapy response in colorectal cancer. Nat. Med. **29**, 1–10 (2023)
9. Guan, Y., et al.: Node-aligned graph convolutional network for whole-slide image representation and classification. In: Proceedings of the IEEE/CVF Conference on Computer Vision and Pattern Recognition, pp. 18813–18823 (2022)

10. Hamilton, W., Ying, Z., Leskovec, J.: Inductive representation learning on large graphs. In: Advances in Neural Information Processing Systems, pp. 1024–1034 (2017)
11. He, K., Zhang, X., Ren, S., Sun, J.: Deep residual learning for image recognition. In: Proceedings of the IEEE Conference on Computer Vision and Pattern Recognition, pp. 770–778 (2016)
12. Hou, W., et al.: H^2-mil: exploring hierarchical representation with heterogeneous multiple instance learning for whole slide image analysis. In: Proceedings of the AAAI Conference on Artificial Intelligence, vol. 36, pp. 933–941 (2022)
13. Ilse, M., Tomczak, J., Welling, M.: Attention-based deep multiple instance learning. In: International Conference on Machine Learning, pp. 2127–2136. PMLR (2018)
14. Kather, J.N., et al.: Predicting survival from colorectal cancer histology slides using deep learning: a retrospective multicenter study. PLoS Med. **16**(1), e1002730 (2019)
15. Kipf, T.N., Welling, M.: Semi-supervised classification with graph convolutional networks. In: International Conference on Learning Representations (2017). https://openreview.net/forum?id=SJU4ayYgl
16. Li, B., Li, Y., Eliceiri, K.W.: Dual-stream multiple instance learning network for whole slide image classification with self-supervised contrastive learning. In: Proceedings of the IEEE/CVF Conference On Computer Vision and Pattern Recognition, pp. 14318–14328 (2021)
17. Liu, P., Fu, B., Ye, F., Yang, R., Ji, L.: DSCA: a dual-stream network with cross-attention on whole-slide image pyramids for cancer prognosis. Expert Syst. Appl. **227**, 120280 (2023)
18. Lu, M.Y., Williamson, D.F., Chen, T.Y., Chen, R.J., Barbieri, M., Mahmood, F.: Data-efficient and weakly supervised computational pathology on whole-slide images. Nature Biomed. Eng. **5**(6), 555–570 (2021)
19. Shao, Z., Bian, H., Chen, Y., Wang, Y., Zhang, J., Ji, X., et al.: TransMIL: Transformer based correlated multiple instance learning for whole slide image classification. Adv. Neural. Inf. Process. Syst. **34**, 2136–2147 (2021)
20. Sterlacci, W., Vieth, M.: Early colorectal cancer. In: Baatrup, G. (ed.) Multidisciplinary Treatment of Colorectal Cancer, pp. 263–277. Springer, Cham (2021). https://doi.org/10.1007/978-3-030-58846-5_28
21. Sung, H., et al.: Global cancer statistics 2020: GLOBOCAN estimates of incidence and mortality worldwide for 36 cancers in 185 countries. CA Cancer J. Clin. **71**(3), 209–249 (2021). https://doi.org/10.3322/caac.21660
22. Vaswani, A., et al.: Attention is all you need. In: Advances in Neural Information Processing Systems 30 (2017)
23. Wang, P., Li, Y., Reddy, C.K.: Machine learning for survival analysis: a survey. ACM Comput. Surv. **51**(6), 3214306 (2019). https://doi.org/10.1145/3214306
24. Wang, X., et al.: RetCCL: clustering-guided contrastive learning for whole-slide image retrieval. Med. Image Anal. **83**, 102645 (2023)
25. Wang, X., et al.: SCL-WC: cross-slide contrastive learning for weakly-supervised whole-slide image classification. In: Thirty-Sixth Conference on Neural Information Processing Systems (2022)
26. Yao, J., Zhu, X., Jonnagaddala, J., Hawkins, N., Huang, J.: Whole slide images based cancer survival prediction using attention guided deep multiple instance learning networks. Med. Image Anal. **65**, 101789 (2020)
27. Zaheer, M., Kottur, S., Ravanbakhsh, S., Poczos, B., Salakhutdinov, R.R., Smola, A.J.: Deep sets. In: Advances in Neural Information Processing Systems 30 (2017)

HIGT: Hierarchical Interaction Graph-Transformer for Whole Slide Image Analysis

Ziyu Guo[1], Weiqin Zhao[1], Shujun Wang[2], and Lequan Yu[1(✉)]

[1] The University of Hong Kong, Hong Kong SAR, China
{gzypro,wqzhao98}@connect.hku.hk, lqyu@hku.hk

[2] The Hong Kong Polytechnic University, Hong Kong SAR, China
shu-jun.wang@polyu.edu.hk

Abstract. In computation pathology, the pyramid structure of gigapixel Whole Slide Images (WSIs) has recently been studied for capturing various information from individual cell interactions to tissue microenvironments. This hierarchical structure is believed to be beneficial for cancer diagnosis and prognosis tasks. However, most previous hierarchical WSI analysis works (1) only characterize local or global correlations within the WSI pyramids and (2) use only unidirectional interaction between different resolutions, leading to an incomplete picture of WSI pyramids. To this end, this paper presents a novel Hierarchical Interaction Graph-Transformer (*i.e.*, HIGT) for WSI analysis. With Graph Neural Network and Transformer as the building commons, HIGT can learn both short-range local information and long-range global representation of the WSI pyramids. Considering that the information from different resolutions is complementary and can benefit each other during the learning process, we further design a novel Bidirectional Interaction block to establish communication between different levels within the WSI pyramids. Finally, we aggregate both coarse-grained and fine-grained features learned from different levels together for slide-level prediction. We evaluate our methods on two public WSI datasets from TCGA projects, *i.e.*, kidney carcinoma (KICA) and esophageal carcinoma (ESCA). Experimental results show that our HIGT outperforms both hierarchical and non-hierarchical state-of-the-art methods on both tumor subtyping and staging tasks.

Keywords: WSI analysis · Hierarchical representation · Interaction · Graph neural network · Vision transformer

Z. Guo and W. Zhao: Contributed equally to this work.

Supplementary Information The online version contains supplementary material available at https://doi.org/10.1007/978-3-031-43987-2_73.

H. Greenspan et al. (Eds.): MICCAI 2023, LNCS 14225, pp. 755–764, 2023.
https://doi.org/10.1007/978-3-031-43987-2_73

1 Introduction

Histopathology is considered the gold standard for diagnosing and treating many cancers [19]. The tissue slices are usually scanned into Whole Slide Images (WSIs) and serve as important references for pathologists. Unlike natural images, WSIs typically contain billions of pixels and also have a pyramid structure, as shown in Fig. 1. Such gigapixel resolution and expensive pixel-wise annotation efforts pose unique challenges to constructing effective and accurate models for WSI analysis. To overcome these challenges, Multiple Instance Learning (MIL) has become a popular paradigm for WSI analysis. Typically, MIL-based WSI analysis methods have three steps: (1) crop the huge WSI into numerous image patches; (2) extract instance features from the cropped patches; and (3) aggregate instance features together to obtain slide-level prediction results. Many advanced MIL models emerged in the past few years. For instance, ABMIL [9] and DeepAttnMIL [18] incorporated attention mechanisms into the aggregation step and achieved promising results. Recently, Graph-Transformer architecture [17] has been proposed to learn short-range local features through GNN and long-range global features through Transformer simultaneously. Such Graph-Transformer architecture has also been introduced into WSI analysis [15,20] to mine the thorough global and local correlations between different image patches. However, current Graph-Transformer-based WSI analysis models only consider the representation learning under one specific magnification, thus ignoring the rich multi-resolution information from the WSI pyramids.

Different resolution levels in the WSI pyramids contain different and complementary information [3]. The images at a high-resolution level contain cellular-level information, such as the nucleus and chromatin morphology features [10]. At a low-resolution level, tissue-related information like the extent of tumor-immune localization can be found [1], while the whole WSI describes the entire tissue microenvironment, such as intra-tumoral heterogeneity and tumor invasion [3]. Therefore, analyzing from only a single resolution would lead to an incomplete picture of WSIs. Some very recent works proposed to characterize and analyze WSIs in a pyramidal structure. H2-MIL [7] formulated WSI as a hierarchical heterogeneous graph and HIPT [3] proposed an inheritable ViT framework to model WSI at different resolutions. Whereas these methods only characterize local or global correlations within the WSI pyramids and use only unidirectional interaction between different resolutions, leading to insufficient capability to model the rich multi-resolution information of the WSI pyramids.

In this paper, we present a novel Hierarchical Interaction Graph-Transformer framework (*i.e.*, HIGT) to simultaneously capture both local and global information from WSI pyramids with a novel Bidirectional Interaction module. Specifically, we abstract the multi-resolution WSI pyramid as a heterogeneous hierarchical graph and devise a Hierarchical Interaction Graph-Transformer architecture to learn both short-range and long-range correlations among different image patches within different resolutions. Considering that the information from different resolutions is complementary and can benefit each other, we specially design a Bidirectional Interaction block in our Hierarchical Interaction ViT mod-

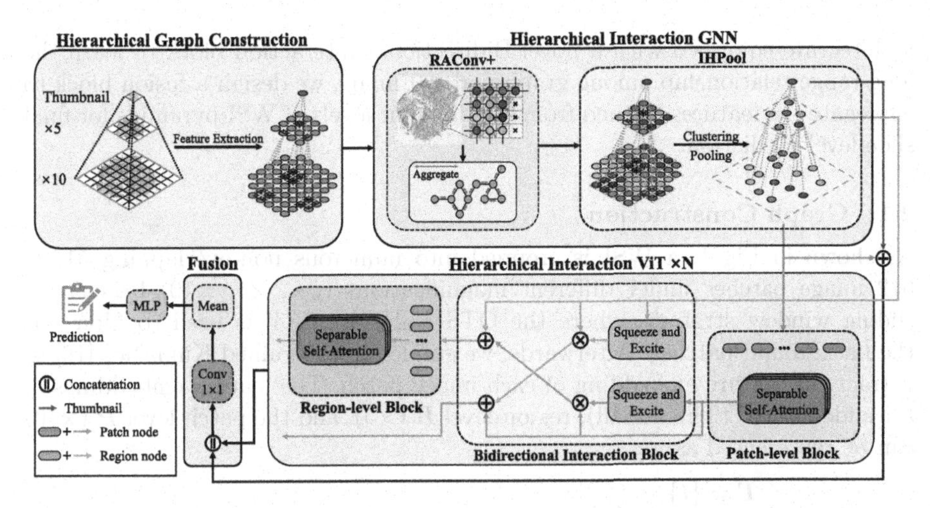

Fig. 1. Overview of the proposed HIGT framework. A WSI pyramid will be constructed as a hierarchical graph. Our proposed Hierarchical Interaction GNN and Hierarchical Interaction ViT block can capture the local and global features, and the Bidirectional Interaction module in the latter allows the nodes from different levels to interact. And finally, the Fusion block aggregates the coarse-grained and fine-grained features to generate the slide-level prediction.

ule to establish communication between different resolution levels. Moreover, a Fusion block is proposed to aggregate features learned from the different levels for slide-level prediction. To reduce the tremendous computation and memory cost, we further adopt the efficient pooling operation after the hierarchical GNN part to reduce the number of tokens and introduce the Separable Self-Attention Mechanism in Hierarchical Interaction ViT modules to reduce the computation burden. The extensive experiments with promising results on two public WSI datasets from TCGA projects, *i.e.*, kidney carcinoma (KICA) and esophageal carcinoma (ESCA), validate the effectiveness and efficiency of our framework on both tumor subtyping and staging tasks. The codes are available at https://github.com/HKU-MedAI/HIGT.

2 Methodology

Figure 1 depicts the pipeline of HIGT framework for better exploring the multi-scale information in hierarchical WSI pyramids. First, we abstract each WSI as a hierarchical graph, where the feature embeddings extracted from multi-resolution patches serve as nodes and the edge denotes the spatial and scaling relationships of patches within and across different resolution levels. Then, we feed the constructed graph into several hierarchical graph convolution blocks to learn the short-range relationship among graph nodes, following pooling operations to aggregate local context and reduce the number of nodes. We further devise a Separable Self-Attention-based Hierarchical Interaction Transformer architecture equipped with a novel Bidirectional Interaction block to learn the long-range relationship among graph nodes. Finally, we design a fusion block to aggregate the features learned from the different levels of WSI pyramids for final slide-level prediction.

2.1 Graph Construction

As shown in Fig. 1, a WSI is cropped into numerous non-overlapping 512×512 image patches under different magnifications (*i.e.*, $\times 5$, $\times 10$) by using a sliding window strategy, where the OTSU algorithm [4] is used to filter out the background patches. Afterwards, we employ a pre-trained KimiaNet [16] to extract the feature embedding of each image patch. The feature embeddings of the slide-level $\boldsymbol{T}$ (Thumbnail), region-level $\boldsymbol{R}$ ($\times 5$), and the patch-level $\boldsymbol{P}$ ($\times 10$) can be represented as,

$$
\begin{aligned}
&\boldsymbol{T} = \{\boldsymbol{t}\}, \\
&\boldsymbol{R} = \{\boldsymbol{r}_1, \boldsymbol{r}_2, \cdots, \boldsymbol{r}_N\}, \\
&\boldsymbol{P} = \{\boldsymbol{P}_1, \boldsymbol{P}_2, \cdots, \boldsymbol{P}_N\}, \boldsymbol{P_i} = \{\boldsymbol{p}_{i,1}, \boldsymbol{p}_{i,2}, \cdots, \boldsymbol{p}_{i,M}\},
\end{aligned} \tag{1}
$$

where $\boldsymbol{t}, \boldsymbol{r}_i, \boldsymbol{p}_{i,j} \in \mathbb{R}^{1 \times C}$ correspond to the feature embeddings of each patch in thumbnail, region, and patch levels, respectively. N is the total number of the region nodes and M is the number of patch nodes belonging to a certain region node, and C denotes the dimension of feature embedding (1,024 in our experiments). Based on the extracted feature embeddings, we construct a hierarchical graph to characterize the WSI, following previous H^2-MIL work [7]. Specifically, the cropped patches serve as the nodes of the graph and we employ the extracted feature embedding as the node embeddings. There are two kinds of edges in the graph: spatial edges to denote the 8-adjacent spatial relationships among different patches in the same levels, and scaling edges to denote the relationship between patches across different levels at the same location.

2.2 Hierarchical Graph Neural Network

To learn the short-range relationship among different patches within the WSI pyramid, we propose a new hierarchical graph message propagation operation, called RAConv+. Specifically, for any source node j in the hierarchical graph, we define the set of it all neighboring nodes at resolution k as $\mathcal{N}_k$ and $k \in K$. Here K means all resolutions. And the h_k is the mean embedding of the node j's neighboring nodes in resolution k. And $h_{j'}$ is the embedding of the neighboring nodes of node j in resolution k and $h_{j'} \in \mathcal{N}_k$. The formula for calculating the attention score of node j in resolution-level and node-level:

$$
\begin{aligned}
\alpha_k &= \frac{\exp\left(\boldsymbol{a}^\top \cdot \text{LeakyReLU}\left([\boldsymbol{U}\boldsymbol{h}_j \| \boldsymbol{U}\boldsymbol{h}_k]\right)\right)}{\sum_{k' \in K} \exp\left(\boldsymbol{a}^\top \cdot \text{LeakyReLU}\left([\boldsymbol{U}\boldsymbol{h}_j \| \boldsymbol{U}\boldsymbol{h}_{k'}]\right)\right)}, \\
\alpha_{j'} &= \frac{\exp\left(\boldsymbol{b}^\top \cdot \text{LeakyReLU}\left([\boldsymbol{V}\boldsymbol{h}_j \| \boldsymbol{V}\boldsymbol{h}_{j''}]\right)\right)}{\sum_{h_{j''} \in \mathcal{N}_k} \exp\left(\boldsymbol{b}^\top \cdot \text{LeakyReLU}\left([\boldsymbol{V}\boldsymbol{h}_j \| \boldsymbol{V}\boldsymbol{h}_{j''}]\right)\right)},
\end{aligned}
$$

$$\alpha_{j,j\prime} = \alpha_k + \alpha_{j\prime}, \tag{2}$$

where $\alpha_{j,j\prime}$ is the attention score of the node j to node $j\prime$ and h_j is the source node j embedding. And U, V, a and b are four learnable layers. The main difference from H2-MIL [6] is that we pose the non-linear *LeakyReLU* between a and U, b and V, to generate a more distinct attention score matrix which increases the feature differences between different types of nodes [2]. Therefore, the layer-wise graph message propagation can be represented as:

$$H^{(l+1)} = \sigma\left(\mathcal{A} \cdot H^{(l)} \cdot W^{(l)}\right), \tag{3}$$

where $\mathcal{A}$ represents the attention score matrix, and the attention score for the j-th row and j$\prime$-th column of the matrix is given by Eq. (2). At the end of the hierarchical GNN part, we use the IHPool [6] progressively aggregate the hierarchical graph.

2.3 Hierarchical Interaction ViT

We further propose a Hierarchical Interaction ViT (HIViT) to learn long-range correlation within the WSI pyramids, which includes three key components: Patch-level (PL) blocks, Bidirectional Interaction (BI) blocks, and Region-level (RL) blocks.

Patch-Level Block. Given the patch-level feature set $\boldsymbol{P} = \bigcup_{i=1}^{N} \boldsymbol{P}_i$, the PL block learns long-term relationships within the patch level:

$$\hat{\boldsymbol{P}}^{l+1} = PL(\boldsymbol{P}^l) \tag{4}$$

where $l = 1, 2, ..., L$ is the index of the HIViT block. $PL(\cdot)$ includes a Separable Self Attention (SSA) [13], 1×1 Convolution, and Layer Normalization in sequence. Note that here we introduced SSA into the PL block to reduce the computation complexity of attention calculation from quadratic to linear while maintaining the performance [13].

Bidirectional Interaction Block. We propose a Bidirectional Interaction (BI) block to establish communication between different levels within the WSI pyramids. The BI block performs bidirectional interaction, and the interaction progress from region nodes to patch nodes is:

$$\begin{aligned} &\boldsymbol{r}_i^{l'} \in \boldsymbol{R}^{l'}, \quad \boldsymbol{R}^{l'} = SE(\boldsymbol{R}^l) \cdot \boldsymbol{R}^l, \\ &\boldsymbol{P}_i^{l+1} = \{\boldsymbol{p}_{i,1}^{l+1}, \boldsymbol{p}_{i,2}^{l+1}, \cdots, \boldsymbol{p}_{i,k}^{l+1}\}, \quad \boldsymbol{p}_{i,k}^{l+1} = \hat{\boldsymbol{p}}_{i,k}^{l+1} + \boldsymbol{r}_i^{l'}, \end{aligned} \tag{5}$$

where the $SE(\cdot)$ means the Sequeeze-and-Excite layer [8] and the $\boldsymbol{r}_i^{l'}$ means the i-th region node in $\boldsymbol{R}^{l'}$, and $\hat{\boldsymbol{p}}_{i,k}^{l+1}$ is the k-th patch node linked to the i-th region node after the interaction. Besides, another direction of the interaction is,

$$\bar{\boldsymbol{P}} = \{\bar{\boldsymbol{P}}_1^{l+1}, \bar{\boldsymbol{P}}_2^{l+1}, \cdots, \bar{\boldsymbol{P}}_n^{l+1}\}, \quad \bar{\boldsymbol{P}}_i^{l+1} = MEAN(\hat{\boldsymbol{P}}_i^{l+1})$$
$$\hat{\boldsymbol{R}}^{l+1} = SE(\bar{\boldsymbol{P}}^{l+1}) \cdot \bar{\boldsymbol{P}}^{l+1} + \boldsymbol{R}^l, \tag{6}$$

where the $MEAN(\cdot)$ is the operation to get the mean value of patch nodes set $\hat{\boldsymbol{P}}_i^{l+1}$ associated with the i-th region node and $\bar{\boldsymbol{P}}_1^{l+1} \in \mathcal{R}^{1\times C}$ and the C is the feature channel of nodes, and $\hat{\boldsymbol{R}}^{l+1}$ is the region nodes set after interaction.

Region-Level Block. The final part of this module is to learn the long-range correlations of the interacted region-level nodes:

$$\boldsymbol{R}^{l+1} = RL(\hat{\boldsymbol{R}}^{l+1}) \tag{7}$$

where $l = 1, 2, ..., L$ is the index of the HIViT module, $\boldsymbol{R} = \{\boldsymbol{r}_1, \boldsymbol{r}_2, \cdots, \boldsymbol{r}_N\}$, and $RL(\cdot)$ has a similar structure to $PL(\cdot)$.

2.4 Slide-Level Prediction

In the final stage of our framework, we design a Fusion block to combine the coarse-grained and fine-grained features learned from the WSI pyramids. Specifically, we use an element-wise summation operation to fuse the coarse-grained thumbnail feature and patch-level features from the Hierarchical Interaction GNN part, and then further fuse the fine-grained patch-level features from the HIViT part with a concatenation operation. Finally, a 1×1 convolution and mean operation followed by a linear projection are employed to produce the slide-level prediction.

3 Experiments

Datasets and Evaluation Metrics. We assess the efficacy of the proposed HIGT framework by testing it on two publicly available datasets (KICA and ESCA) from The Cancer Genome Atlas (TCGA) repository. The datasets are described below in more detail:

- **KICA dataset.** The KICA dataset consists of 371 cases of kidney carcinoma, of which 279 are classified as early-stage and 92 as late-stage. For the tumor typing task, 259 cases are diagnosed as kidney renal papillary cell carcinoma, while 112 cases are diagnosed as kidney chromophobe.
- **ESCA dataset.** The ESCA dataset comprises 161 cases of esophageal carcinoma, with 96 cases classified as early-stage and 65 as late-stage. For the tumor typing task, there are 67 squamous cell carcinoma cases and 94 adenocarcinoma cases.

Experimental Setup. The proposed framework was implemented by PyTorch [14] and PyTorch Geometric [5]. All experiments were conducted on a workstation with eight NVIDIA GeForce RTX 3090 (24 GB) GPUs. The shape of all nodes' features extracted by KimiaNet is set to 1×1024. All methods are trained with a batch size of 8 for 50 epochs. The learning rate was set as 0.0005, with Adam optimizer. The accuracy (ACC) and area under the curve (AUC) are used as the evaluation metric. All approaches were evaluated with five-fold cross-validations (5-fold CVs) from five different initializations.

Table 1. Comparison with other methods on ESCA. Top results are shown in bold.

Method	Staging		Typing	
	AUC	ACC	AUC	ACC
ABMIL [9]	64.53 ± 4.80	64.39 ± 5.05	94.11 ± 2.69	93.07 ± 2.68
CLAM-SB [12]	67.45 ± 5.40	67.29 ± 5.18	93.79 ± 5.52	93.47 ± 5.77
DeepAttnMIL [18]	67.96 ± 5.52	67.53 ± 4.96	95.68 ± 1.94	94.43 ± 3.04
DS-MIL [11]	66.92 ± 5.28	66.83 ± 5.57	95.96 ± 3.07	94.77 ± 4.10
LA-MIL [15]	63.93 ± 6.19	63.45 ± 6.19	95.23 ± 3.75	94.69 ± 3.94
H2-MIL [7]	63.20 ± 8.36	62.72 ± 8.32	91.88 ± 4.17	91.31 ± 4.18
HIPT [3]	68.59 ± 5.62	68.45 ± 6.39	94.62 ± 2.34	93.01 ± 3.28
Ours	$\mathbf{71.11 \pm 6.04}$	$\mathbf{70.53 \pm 5.41}$	$\mathbf{96.81 \pm 2.49}$	$\mathbf{96.16 \pm 2.85}$

Table 2. Comparison with other methods on KICA. Top results are shown in bold.

Method	Staging		Typing	
	AUC	ACC	AUC	ACC
ABMIL [9]	77.40 ± 3.87	75.94 ± 5.06	97.76 ± 1.74	98.86 ± 0.69
CLAM-SB [12]	77.16 ± 3.64	76.61 ± 4.31	96.76 ± 3.42	97.13 ± 2.99
DeepAttnMIL [18]	76.77 ± 1.94	75.94 ± 2.41	97.44 ± 1.04	96.30 ± 2.63
DS-MIL [11]	77.33 ± 4.11	76.57 ± 5.14	98.03 ± 1.13	97.31 ± 1.85
LA-MIL [15]	69.37 ± 5.27	68.73 ± 5.09	98.34 ± 0.98	97.71 ± 1.76
H2-MIL [7]	65.59 ± 6.65	64.48 ± 6.20	98.06 ± 1.43	96.99 ± 3.01
HIPT [3]	75.93 ± 2.01	75.34 ± 2.31	98.71 ± 0.49	97.32 ± 2.24
Ours	$\mathbf{78.80 \pm 2.10}$	$\mathbf{76.80 \pm 2.30}$	$\mathbf{98.90 \pm 0.60}$	$\mathbf{97.90 \pm 1.40}$

Comparison with State-of-the-Art Methods. We first compared our proposed HIGT framework with two groups of state-of-the-art WSI analysis methods: (1) non-hierarchical methods including: ABMIL [9], CLAM-SB [12], DeepAttnMIL [18], DS-MIL [11], LA-MIL [15], and (2) hierarchical methods including: H2-MIL [7], HIPT [3]. For LA-MIL [15] method, it was introduced with a single-scale Graph-Transformer architecture. For H2-MIL [7] and HIPT [3], they were introduced with a hierarchical Graph Neural Network and hierarchical Transformer architecture, respectively. The results for ESCA and KICA datasets are summarized in Table 1 and Table 2, respectively. Overall, our model achieves a content result both in AUC and ACC of classifying the WSI, and especially in predicting the more complex task (i.e. Staging) compared with the SOTA approaches. Even for the non-hierarchical Graph-Transformer baseline LA-MIL and hierarchical transformer model HIPT, our model approaches at least around 3% and 2% improvement on AUC and ACC in the classification of the Staging of the KICA dataset. Therefore we believe that our model benefits a lot from its used modules and mechanisms.

Ablation Analysis. We further conduct an ablation study to demonstrate the effectiveness of the proposed components. The results are shown in Table 3. In its first row, we replace the RAConv+ with the original version of this operation. And in the second row, we replace the Separable Self Attention with a canonical transformer block. The third row changes the bidirectional interaction mechanism into just one direction from region-level to patch-level. And the last row, we remove the fusion block from our model. Finally, the ablation analysis results show that all of these modules we used actually improved the prediction effect of the model to a certain extent.

Table 3. Ablation analysis on KICA dataset.

Method	Staging		Typing	
	AUC	ACC	AUC	ACC
H2-MIL + HIViT	77.35 ± 3.41	**77.16 ± 3.29**	98.56 ± 1.01	95.00 ± 1.75
Ours w/o SSA	73.45 ± 8.48	71.47 ± 3.21	97.94 ± 2.51	97.42 ± 2.65
Ours w/o BI	72.42 ± 2.09	71.34 ± 7.23	98.04 ± 8.30	96.54 ± 2.80
Ours w/o Fusion	77.87 ± 2.09	76.80 ± 2.95	98.46 ± 0.88	97.35 ± 1.81
Ours	**78.80 ± 2.10**	76.80 ± 2.30	**98.90 ± 0.60**	**97.90 ± 1.40**

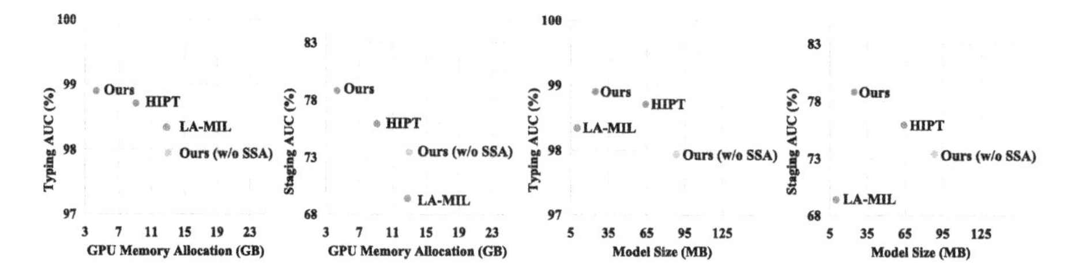

Fig. 2. Computational analysis of our framework and some selected SOTA methods. From left to right are scatter plots of Typing AUC v.s. GPU Memory Allocation, Staging AUC v.s. GPU Memory Allocation, Typing AUC v.s. Model Size, Staging AUC v.s. Model Size.

Computation Cost Analysis. We analyze the computation cost during the experiments to compare the efficiency between our methods and existing state-of-the-art approaches. Besides we visualized the model size (MB) and the training memory allocation of GPU (GB) v.s. performance in KICA's typing and staging task plots in Fig. 2. All results demonstrate that our model is able to maintain the promising prediction result while reducing the computational cost and model size effectively.

4 Conclusion

In this paper, we propose HIGT, a framework that simultaneously and effectively captures local and global information from the hierarchical WSI. Firstly, the constructed hierarchical data structure of the multi-resolution WSI is able to offer multi-scale information to the later model. Moreover, the redesigned H2-MIL and HIViT capture the short-range and long-range correlations among varying magnifications of WSI separately. And the bidirectional interaction mechanism and fusion block can facilitate communication between different levels in the Transformer part. We use IHPool and apply the Separable Self Attention to deal with the inherently high computational cost of the Graph-Transformer model. Extensive experimentation on two public WSI datasets demonstrates the effectiveness and efficiency of our designed framework, yielding promising results. In the future, we will evaluate on other complex tasks such as survival prediction and investigate other techniques to improve the efficiency of our framework.

Acknowledgement. The work described in this paper was partially supported by grants from the National Natural Science Fund (62201483), the Research Grants Council of the Hong Kong Special Administrative Region, China (T45-401/22-N), and The Hong Kong Polytechnic University (P0045999).

References

1. AbdulJabbar, K., et al.: Geospatial immune variability illuminates differential evolution of lung adenocarcinoma. Nature Med. **26**(7), 1054–1062 (2020)
2. Brody, S., Alon, U., Yahav, E.: How attentive are graph attention networks? arXiv preprint arXiv:2105.14491 (2021)
3. Chen, R.J., et al.: Scaling vision transformers to gigapixel images via hierarchical self-supervised learning. In: Proceedings of the IEEE/CVF Conference on Computer Vision and Pattern Recognition (CVPR), pp. 16144–16155 (June 2022)
4. Chen, R.J., et al.: Whole slide images are 2D Point Clouds: context-aware survival prediction using patch-based graph convolutional networks. In: de Bruijne, M., et al. (eds.) Medical Image Computing and Computer Assisted Intervention – MICCAI 2021: 24th International Conference, Strasbourg, France, September 27 – October 1, 2021, Proceedings, Part VIII, pp. 339–349. Springer International Publishing, Cham (2021). https://doi.org/10.1007/978-3-030-87237-3_33
5. Fey, M., Lenssen, J.E.: Fast graph representation learning with pytorch geometric. arXiv preprint arXiv:1903.02428 (2019)

6. Hou, W., Wang, L., Cai, S., Lin, Z., Yu, R., Qin, J.: Early neoplasia identification in Barrett's esophagus via attentive hierarchical aggregation and self-distillation. Medical Image Anal. **72**, 102092 (2021). https://doi.org/10.1016/j.media.2021.102092
7. Hou, W., et al.: H^2-mil: Exploring hierarchical representation with heterogeneous multiple instance learning for whole slide image analysis. In: Proceedings of the AAAI Conference on Artificial Intelligence, vol. 36, pp. 933–941 (2022)
8. Hu, J., Shen, L., Sun, G.: Squeeze-and-excitation networks. In: Proceedings of the IEEE Conference on Computer Vision and Pattern Recognition, pp. 7132–7141 (2018)
9. Ilse, M., Tomczak, J., Welling, M.: Attention-based deep multiple instance learning. In: International Conference on Machine Learning, pp. 2127–2136. PMLR (2018)
10. Kumar, N., Verma, R., Sharma, S., Bhargava, S., Vahadane, A., Sethi, A.: A dataset and a technique for generalized nuclear segmentation for computational pathology. IEEE Trans. Med. Imaging **36**(7), 1550–1560 (2017)
11. Li, B., Li, Y., Eliceiri, K.W.: Dual-stream multiple instance learning network for whole slide image classification with self-supervised contrastive learning. In: Proceedings of the IEEE/CVF Conference on Computer Vision and Pattern Recognition, pp. 14318–14328 (2021)
12. Lu, M.Y., Williamson, D.F., Chen, T.Y., Chen, R.J., Barbieri, M., Mahmood, F.: Data-efficient and weakly supervised computational pathology on whole-slide images. Nature Biomed. Eng. **5**(6), 555–570 (2021)
13. Mehta, S., Rastegari, M.: Separable self-attention for mobile vision transformers. arXiv preprint arXiv:2206.02680 (2022)
14. Paszke, A., et al.: Pytorch: an imperative style, high-performance deep learning library. Adv. Neural Inform. Process. Syst. **32** (2019)
15. Reisenbüchler, D., Wagner, S.J., Boxberg, M., Peng, T.: Local attention graph-based transformer for multi-target genetic alteration prediction. In: Wang, L., Dou, Q., Fletcher, P.T., Speidel, S., Li, S. (eds.) Medical Image Computing and Computer Assisted Intervention – MICCAI 2022: 25th International Conference, Singapore, September 18–22, 2022, Proceedings, Part II, pp. 377–386. Springer Nature Switzerland, Cham (2022). https://doi.org/10.1007/978-3-031-16434-7_37
16. Riasatian, A., et al.: Fine-tuning and training of densenet for histopathology image representation using TCGA diagnostic slides. Med. Image Anal. **70**, 102032 (2021)
17. Wu, Z., Jain, P., Wright, M., Mirhoseini, A., Gonzalez, J.E., Stoica, I.: Representing long-range context for graph neural networks with global attention. Adv. Neural. Inf. Process. Syst. **34**, 13266–13279 (2021)
18. Yao, J., Zhu, X., Jonnagaddala, J., Hawkins, N., Huang, J.: Whole slide images based cancer survival prediction using attention guided deep multiple instance learning networks. Med. Image Anal. **65**, 101789 (2020)
19. Yao, X.H., et al.: Pathological evidence for residual SARS-CoV-2 in pulmonary tissues of a ready-for-discharge patient. Cell Res. **30**(6), 541–543 (2020)
20. Zheng, Y., et al.: A graph-transformer for whole slide image classification. IEEE Trans. Med. Imaging **41**(11), 3003–3015 (2022)

ALL-IN: A Local GLobal Graph-Based DIstillatioN Model for Representation Learning of Gigapixel Histopathology Images With Application In Cancer Risk Assessment

Puria Azadi[1], Jonathan Suderman[1], Ramin Nakhli[1], Katherine Rich[1], Maryam Asadi[1], Sonia Kung[2], Htoo Oo[2], Mira Keyes[1], Hossein Farahani[1], Calum MacAulay[3], Larry Goldenberg[2], Peter Black[2], and Ali Bashashati[1](✉)

[1] University of British Columbia, Vancouver, BC, Canada
ali.bashashati@ubc.ca
[2] Vancouver Prostate Centre, Vancouver, BC, Canada
[3] BC Cancer Agency, Vancouver, BC, Canada

Abstract. The utility of machine learning models in histopathology image analysis for disease diagnosis has been extensively studied. However, efforts to stratify patient risk are relatively under-explored. While most current techniques utilize small fields of view (so-called local features) to link histopathology images to patient outcome, in this work we investigate the combination of global (i.e., contextual) and local features in a graph-based neural network for patient risk stratification. The proposed network not only combines both fine and coarse histological patterns but also utilizes their interactions for improved risk stratification. We compared the performance of our proposed model against the state-of-the-art (SOTA) techniques in histopathology risk stratification in two cancer datasets. Our results suggest that the proposed model is capable of stratifying patients into statistically significant risk groups ($p < 0.01$ across the two datasets) with clinical utility while competing models fail to achieve a statistical significance endpoint ($p = 0.148 - 0.494$).

Keywords: Histopathology · Risk Assessment · Graph Processing

1 Introduction

The examination of tissue and cells using microscope (referred to as histology) has been a key component of cancer diagnosis and prognostication since more than a hundred years ago. Histological features allow visual readout of cancer biology as they represent the overall impact of genetic changes on cells [20].

The great rise of deep learning in the past decade and our ability to digitize histopathology slides using high-throughput slide scanners have fueled interests in the applications of deep learning in histopathology image analysis. The

Supplementary Information The online version contains supplementary material available at https://doi.org/10.1007/978-3-031-43987-2_74.

H. Greenspan et al. (Eds.): MICCAI 2023, LNCS 14225, pp. 765–775, 2023.
https://doi.org/10.1007/978-3-031-43987-2_74

majority of the efforts, so far, focus on the deployment of these models for diagnosis and classification [27]. As such, there is a paucity of efforts that embark on utilizing machine learning models for patient prognostication and survival analysis (for example, predicting risk of cancer recurrence or expected patient survival). While prognostication and survival analysis offer invaluable insights for patient management, biological studies and drug development efforts, they require careful tracking of patients for a lengthy period of time; rendering this as a task that requires a significant amount of effort and funding.

In the machine learning domain, patient prognostication can be treated as a weakly supervised problem, which a model would predict the outcome (e.g., time to cancer recurrence) based on the histopathology images. Their majority have utilized Multiple Instance Learning (MIL) [8] that is a two-step learning method. First, representation maps for a set of patches (i.e., small fields of view), called a bag of instances, are extracted. Then, a second pooling model is applied to the feature maps for the final prediction. Different MIL variations have shown superior performances in grading or subtype classification in comparison to outcome prediction [10]. This is perhaps due to the fact that MIL-based technique do not incorporate patch locations and interactions as well as tissue heterogeneity which can potentially have a vital role in defining clinical outcomes [4,26].

To address this issue, graph neural networks (GNN) have recently received more attention in histology. They can model patch relations [17] by utilizing message passing mechanism via edges connecting the nodes (i.e., small patches in our case). However, most GNN-based models suffer from over smoothing [22] which limits nodes' receptive fields [3]. While local contexts mainly capture cell-cell interactions, global patterns such as immune cell infiltration patterns and tumor invasion in normal tissue structures (e.g., depth of invasion through myometrium in endometrial cancer [1]) could capture critical information about outcome [10]. Hence, locally focused methods are unable to benefit from the coarse properties of slides due to their high dimensions which may lead to poor performance.

This paper aims to investigate the potential of extracting fine and coarse features from histopathology slides and integrating them for risk stratification in cancer patients. Therefore, the contributions of this work can be summarized as: 1) a novel graph-based model for predicting survival that extracts both local and global properties by identifying morphological super-nodes; 2) introducing a fine-coarse feature distillation module with 3 various strategies to aggregate interactions at different scales; 3) outperforming SOTA approaches in both risk prediction and patient stratification scenarios on two datasets; 4) publishing two large and rare prostate cancer datasets containing more than 220 graphs for active surveillance and 240 graphs for brachytherapy cases. The code and graph embeddings are publicly available at https://github.com/pazadimo/ALL-IN

2 Related Works

2.1 Weakly Supervised Learning in Histopathology

Utilizing Weakly Supervised Learning for modeling histopathology problems has been getting popular due to the high resolution of slides and substantial time and financial costs associated with annotating them as well as the development of powerful deep discriminative models in the recent years [24].

Such models are used to perform nuclei segmentation [18], identify novel subtypes [12], or later descendants are even able to pinpoint sub-areas with a high diagnostic value [19].

2.2 Survival Analysis and GNNs in Histopathology

MIL-based models have been utilized for outcome prediction [29,32] which can also be integrated with attention-based variants [14]. GNNs due to their structural preserving capacity [28] have drawn attention in various histology domains by constructing the graph on cells or patches. However, current GNN-based risk assessment variants are only focused on short-range interactions [16,17] or consider local contexts [10]. We hypothesize that graph-based models' performance in survival prediction improves by leveraging both fine and coarse properties.

3 Method

Figure 1 summarizes our proposed end-to-end solution. Below, we have provided details of each module.

3.1 Problem Formulation

For P_n, which is the n-th patient, a set of patches $\{patch_j\}_{j=1}^{M}$ is extracted from the related whole slide images. In addition, a latent vector $z_j \in R^{1 \times d}$ is extracted from $patch_j$ using our encoder network (described in Sect. 3.2) that results in feature matrix $Z_n \in R^{M \times d}$ for P_n. Finally, a specific graph (G_n) for the n-th patient (P_n) can be constructed by assuming patches as nodes. Also, edges are connected based on the patches' k-nearest neighbour in the spatial domain resulting in an adjacency matrix A_n. Therefore, for each patient such as P_n, we have a graph defined by adjacency matrix A_n with size $M \times M$ and features matrix Z_n ($G_n = graph(Z_n, A_n)$). We estimate K super-nodes as matrix $S_n \in R^{K \times d}$ representing groups of local nodes with similar properties as coarse features for P_n's slides. The final model (ϵ_θ) with parameters θ utilizes G_n and S_n to predict the risk associated with this patient:

$$risk_n = \epsilon_\theta(G_n, S_n) = \epsilon_\theta(graph(X_n, A_n), S_n) \quad (1)$$

3.2 Self-supervised Encoder

Due to computational limits and large number of patches available for each patient, we utilize a self-supervised approach to train an encoder to reduce the inputs' feature space size. Therefore, We use DINO [9], a knowledge distillation model (KDM), with vision transformer (ViT) [13] as the backbone. It utilizes global and local augmentations of the input $patch_j$ and passes them to the student ($S_{\theta_1,ViT}$) and teacher ($T_{\theta_2,ViT}$) models to find their respective

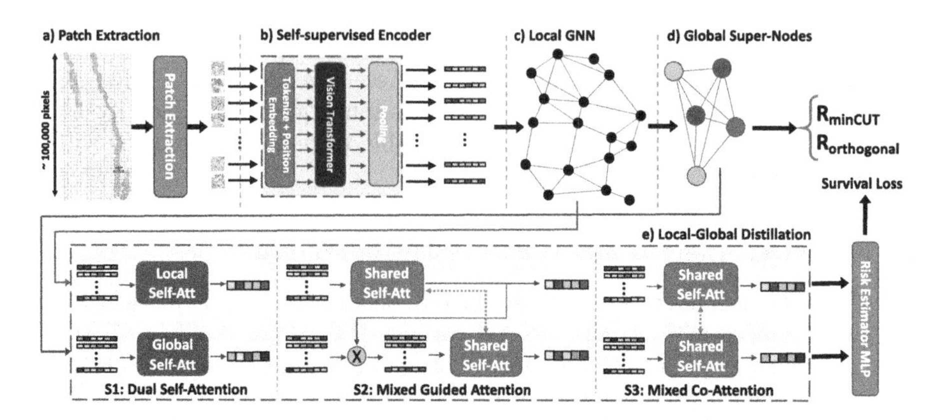

Fig. 1. The overview of our proposed method. a) The input slide is tiled into non-overlapping patches. b) The patches are fed into a self-supervised encoder to extract embeddings. c) A graph is constructed and the new local instance-level embeddings are obtained through the message-passing process. d) The global context representations in the form of super-nodes are extracted utilizing two unsupervised loss functions (R_{minCUT}, $R_{orthognal}$). e) The fine and coarse feature vectors are aggregated in the distillation module to obtain a representation that accounts for both local and global (contextual) histo-morphological features. Three different strategies (S1, S2, S3) are explored in this module. Finally, a Multilayer Perceptron (MLP) is deployed to estimate the risk using final resultant vectors.

representations without any labels. Then, by using distillation loss, it makes the representations' distribution similar to each other. Finally, the fixed weights of the teacher model are utilized in order to encode the input patches.

3.3 Local Graph Neural Network

GNN's objective is to find new nodes' embeddings via integrating local neighbors' interactions with individual properties of patches. By exploiting the message passing mechanism, this module iteratively aggregates features from neighbors of each vertex and generates the new node representations. We employ two graph convolution isomorphism operators (GINconv) [30] with the generalized form as:

$$X'_n = \phi\left(A_n + (1+\epsilon).I).X_n\right), \tag{2}$$

where ϵ is a small positive value and I is the identity matrix. Also, ϕ denotes the weights of two MLP layers. $X_n \in R^{M \times d}$ and $X'_n \in R^{M \times d}$ are GINconv's input and output feature matrices for P_n, which X_n equals Z_n for the first layer.

3.4 Super-Nodes Extractor

In order to find the coarse histo-morphological patterns disguised in the local graph, we propose extracting K Super-nodes, which each represents a weighted cluster of further processed local features. Intuitively, the number of super-nodes K should not be very large or small, as the former encourages them to only represent local clusters and the latter leads to larger clusters and loses subtle details. We exploit the minCUT [5] idea to extract super-nodes in a differentiable process after an auxiliary GINconv to focus more on large-scale interactions and to finally learn the most global correlated super-nodes. Inspired by the relaxation form of the known K-way minCUT problem, we create a continuous cluster matrix $C_n \in R^{M \times K}$ using MLP layers and can finally estimate the super-nodes features ($S_n \in R^{M \times d}$) as:

$$S_n = C_n^T.X'_n, \quad C_n = softmax\left(ReLU(X'_n.W_1).W_2\right), \tag{3}$$

where W_1, W_2 are MLPs' weights. Hence, the extracted nodes are directly dependent on the final survival-specific loss. In addition, two additional unsupervised weighted regularization terms are optimized to improve the process:

MinCut Regularizer. This term is motivated by the original minCUT problem and intends to solve it for the the patients' graph. It is defined as:

$$R_{minCUT} = -\frac{Tr(C_n^T.A_{n,norm}.C_n)}{Tr(C_n^T.D_n.C_n)}, \tag{4}$$

where D_n is the diagonal degree matrix for A_n. Also, $Tr(.)$ represents the trace of matrix and $A_{n,norm}$ is the normalized adjacency matrix. R_{minCUT}'s minimum value happens when $Tr(C_n^T.A_{n,norm}.C_n)$ equals $Tr(C_n^T.D_{g,n}.C_n)$. Therefore, minimizing R_{minCUT} causes assigning strongly similar nodes to a same super-node and prevent their association with others.

Orthogonality Regularizer. R_{minCUT} is non-convex and potent to local minima such as assigning all vertexes to a super-node or having multiple super-nodes with only a single vertex. $R_{orthogonal}$ penalizes such solutions and helps the model to distribute the graph's features between super-nodes. It can be formulated as:

$$R_{orthogonal} = \left\| \frac{C_n^T.C_n}{||C_n^T.C_n||_F} - \frac{I}{\sqrt{K}} \right\|_F, \tag{5}$$

where $||.||_F$ is the Frobenius norm, and I is the identity matrix. This term pushes the model's parameters to find coarse features that are orthogonal to each other resulting in having the most useful global features.

Overall, utilizing these two terms encourages the model to extract super-nodes by leaning more towards the strongly associated vertexes and keeping them against weakly connected ones [5], while the main survival loss still controls the global extraction process.

3.5 Fine-Coarse Distillation

We propose our fine-coarse morphological feature distillation module to leverage all-scale interactions in the final prediction by finding a local and a global patient-level representations ($\hat{h}_{l,n}$, $\hat{h}_{g,n}$). Assume that $X'_n \in R^{M \times d}$ and $S_n \in R^{K \times d}$ are the feature matrices taken from local GNN (Sect. 3.3) and super-nodes for P_n, respectively. We explore 3 different attention-based feature distillation strategies for this task, including:

- **Dual Attention (DA):** Two gated self-attention modules for local and global properties with separate weights $(W_{\phi,l}, W_{\phi,g}, W_{k,l}, W_{k,g}, W_{q,l}, W_{q,g})$ are utilized to find patches scores $\alpha_l \in R^{1\times M}$ and $\alpha_g \in R^{1\times K}$ and the final features $(\hat{h}_{l,n}, \hat{h}_{g,n})$ as:

$$\hat{h}_{l,n} = \sum_{i=1}^{M} W_{\phi,l}\alpha_{l,i}x'_{n,i}, \quad \alpha_l = softmax\left[W_{v,l}\left(tanh(W_{q,l}X_n^{'T}) \cdot sigm(W_{k,l}X_n^{'T})\right)\right], \tag{6}$$

$$\hat{h}_{g,n} = \sum_{i=1}^{K} W_{\phi,g}\alpha_{g,i}s_{n,i}, \quad \alpha_g = softmax\left[W_{v,g}\left(tanh(W_{q,g}S_n^T) \cdot sigm(W_{k,g}S_n^T)\right)\right], \tag{7}$$

  where $x'_{n,i}$ and $s_{n,i}$ are rows of X'_n and S_n, respectively, and the final representation $(\hat{h})$ is generated as $\hat{h} = cat(\hat{h}_l, \hat{h}_g)$.
- **Mixed Guided Attention (MGA):** In the first strategy, the information flows from local and global features to the final representations in parallel without mixing any knowledge. The purpose of this policy is the heavy fusion of fine and coarse knowledge by exploiting shared weights ($W_{\phi,shared}$, $W_{k,shared}$, $W_{q,shared}$, $W_{v,shared}$) in both routes and benefiting from the guidance of local representation on learning the global one by modifying Eq. (7) to:

$$\alpha_g = softmax\left[W_{\phi,g}\left(tanh(W_{q,g}S_n^T\hat{h}_{l,n}) \cdot sigm(W_{k,g}S_n^T\hat{h}_{l,n})\right)\right] \tag{8}$$

- **Mixed Co-Attention (MCA):** While the first strategy allows the extreme separation of two paths, the second one has the highest level of mixing information. Here, we take a balanced policy between the independence and knowledge mixture of the two routes by only sharing the weights without using any guidance.

4 Experiments and Results

4.1 Dataset

We utilize two prostate cancer (PCa) datasets to evaluate the performance of our proposed model. The first set (PCa-AS) includes 179 PCa patients who were managed with Active Surveillance (AS). Radical therapy is considered overtreatment in these patients, so they are instead monitored with regular serum prostate-specific antigen (PSA) measurements, physical examinations, sequential biopsies, and magnetic resonance imaging [23]. However, AS may be over- or under-utilized in low- and intermediate-risk PCa due to the uncertainty of current methods to distinguish indolent from aggressive cancers [11]. Although majority of patients in our cohort are classified as low-risk based on NCCN guidelines [21], a significant subset of them experienced disease upgrade that triggered definitive therapy (range: 6.2 to 224 months after diagnosis).

The second dataset (PCa-BT) includes 105 PCa patients with low to high risk disease who went through brachytherapy. This treatment involves placing a radioactive material inside the body to safely deliver larger dose of radiation at

Table 1. Comparison of our method against baselines and ablation study on policies.

Model	c-index ↑		p-value ↓		High ↓ - Low ↑ Median Time		Parameters
	PCa-AS	PCa-BT	PCa-AS	PCa-BT	PCa-AS	PCa-BT	PCa-AS
DeepSet	0.495 ± 0.017	0.50 ± 0.0	0.837	0.912	67.78–71.87	24.62–24.89	329K
AMIL	0.544 ± 0.06	0.533 ± 0.060	0.820	0.148	48.99–89.10	21.86–30.71	592K
DGC	0.522 ± 0.113	0.572 ± 0.150	0.494	0.223	47.61–96.66	23.44–24.85	626K
Patch-GCN	0.555 ± 0.059	0.541 ± 0.118	0.630	0.981	37.72–94.95	23.05–25.25	1,302K
ALL-IN + DA (ours)	0.631 ± 0.058	0.596 ± 0.062	< 0.01	< 0.01	37.72–115.91	21.86–35.77	850K
ALL-IN + MGA (ours)	0.632 ± 0.060	0.589 ± 0.074	< 0.01	< 0.01	47.61–101.39	21.86–35.77	653K
ALL-IN + MCA (ours)	$\mathbf{0.639 \pm 0.048}$	$\mathbf{0.600 \pm 0.077}$	$\mathbf{< 0.01}$	$\mathbf{< 0.01}$	**36.5–131.71**	**21.86–35.77**	653K

one time [25]. The recorded endpoint for this set is biochemical recurrence with time to recurrence ranging from 11.7 to 56.1 months.

We also utilized the Prostate cANcer graDe Assessment (PANDA) Challenge dataset [7] that includes more than 10,000 PCa needle biopsy slides (no outcome data) as an external dataset for training the encoder of our model.

4.2 Experiments

We evaluate the models' performance in two scenarios utilizing several objective metrics. Implementation details are available in supplementary material.

Hazard (Risk) Prediction. We utilize concordance-index (c-index) that measures the relative ordering of patients with observed events and un-censored cases relative to censored instances [2]. Using c-index, we compare the quality of hazard ranking against multiple methods including two MIL (DeepSet [31], AMIL [14]) and graph-based (DGC [17] and Patch-GCN [10]) models that were utilized recently for histopathology risk assessment. C-index values are available in Table 1. The proposed model with all strategies outperforms baselines across all sets and is able to achieve 0.639 and 0.600 on PCa-AS and PCa-BT, while the baselines, at best, obtain 0.555, and 0.572, respectively. Statistical tests (paired t-test) on c-indices also show that our model is statistically better than all baselines in PCa-AS and also superior to all models, except DGC, in PCa-BT. Superior performance of our MCA policy implies that balanced exploitation of fine and coarse features with shared weights may provide more robust contextual information compared to using mixed guided information or utilizing them independently.

Patient Stratification. The capacity of stratifying patients into risk groups (e.g., low and high risk) is another criterion that we employ to assess the utility of models in clinical practice. We evaluate model performances via Kaplan-Meier curve [15] (cut-off set as the ratio of patients with recurrence within 3

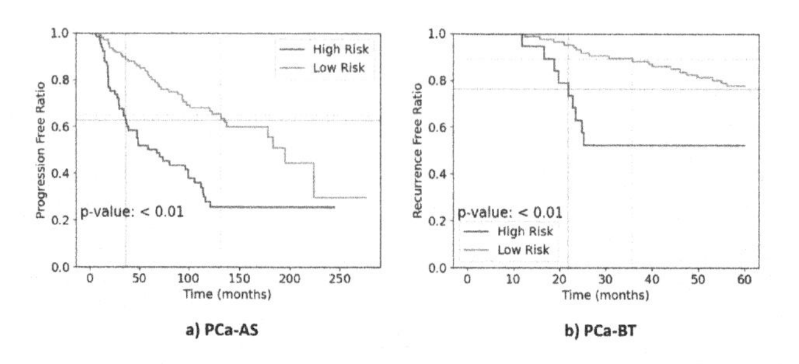

Fig. 2. Kaplan-Meier curves of mixed co-attention model for PCa-AS and PCa-BT.

years of therapy initiation for PCa-BT and the ratio of upgraded cases for PCa-AS), LogRank test [6] (with 0.05 as significance level), and median outcome associated with risk groups (Table 1 and Fig. 2). Our model stratified PCa-AS patients into high- and low-risk groups with median time to progression of 36.5 and 131.7 months, respectively. Moreover, PCa-BT cases assigned to high- and low-risk groups have median recurrence time of 21.86 and 35.7 months. While none of the baselines are capable of assigning patients into risk groups with statistical significance, our distillation policies achieve significant separation in both PCa-AS and PCa-BT datasets; suggesting that global histo-morphological properties improve patient stratification performance. Furthermore, our findings have significant clinical implications as they identify, for the first time, high-risk prostate cancer patients who are otherwise known to be low-risk based on clinico-pathological parameters. This group should be managed differently from the rest of the low-risk prostate cancer patients in the clinic. Therefore, providing evidence of the predictive (as opposed to prognostic) clinical information that our model provides. While a prognostic biomarker provides information about a patient's outcome (without specific recommendation on the next course of action), a predictive biomarker gives insights about the effect of a therapeutic intervention and potential actions that can be taken.

Ablation Study. We perform ablation study (Table 2) on various components of our framework including local nodes, self-supervised ViT-based encoder, and most importantly, super-nodes in addition to fine-coarse distillation module. Although our local-only model is still showing superior results compared to baselines, this analysis demonstrates that all modules are essential for learning the most effective representations. We also assess the impact of our ViT on the baselines (full-results in appendix), showing that it can, on average, improve their performance by an increase of ~ 0.03 in c-index for PCa-AS. However, the best baseline with ViT still has poorer performance compared to our model in both datasets, while the number of parameters (reported for ViT embeddings' size in Table 1) in our full-model is about half of this baseline. Achieving higher c-indices in our all model versions indicates the important role of coarse features and global context in patient risk estimation in addition to local patterns.

Table 2. Ablation study on different modules.

Modules				c-index ↑	
Model	Local-node	our KDM-ViT	Super-node + Distillation Model	PCa-AS	PCa-BT
Patch-GCN	✓	✓	✗	0.627 ± 0.046	0.588 ± 0.067
Ours	✓	✗	✗	0.584 ± 0.072	0.550 ± 0.109
	✓	✓	✗	0.622 ± 0.055	0.597 ± 0.045
	✓	✓	✓	**0.639 ± 0.048**	**0.600 ± 0.077**

5 Conclusion

While risk assessment is relatively under-explored, most existing methods are focused only on small fields of view. In this work, we introduce a novel graph-based model for integrating global and local features, which utilizes interactions at a larger scale for improved risk stratification. Using two cancer datasets, we evaluated the effectiveness of our model against the baseline methods for hazard prediction and patients stratification. Our results suggest that the proposed model outperforms them in risk assessment and is capable of separating patients into statistically significant risk groups with actionable clinical utility. The full capacity of this work can be revealed by extending it to other histology tasks.

Acknowledgment:. This work was supported by a Canadian Institutes of Health Research grant to AB, PB, and LG and Michael Smith Health Research BC Scholar grant to AB.

References

1. Abu-Rustum, N.R., et al.: The revised 2009 figo staging system for endometrial cancer: should the 1988 figo stages ia and ib be altered? Int. J. Gynecol. Cancer **21**(3) (2011)
2. Alabdallah, A., Ohlsson, M., Pashami, S., Rögnvaldsson, T.: The concordance index decomposition-a measure for a deeper understanding of survival prediction models. arXiv preprint arXiv:2203.00144 (2022)
3. Alon, U., Yahav, E.: On the bottleneck of graph neural networks and its practical implications. arXiv preprint arXiv:2006.05205 (2020)
4. Angell, H., Galon, J.: From the immune contexture to the immunoscore: the role of prognostic and predictive immune markers in cancer. Curr. Opin. Immunol. **25**(2), 261–267 (2013)
5. Bianchi, F.M., Grattarola, D., Alippi, C.: Spectral clustering with graph neural networks for graph pooling. In: International Conference on Machine Learning, pp. 874–883. PMLR (2020)
6. Bland, J.M., Altman, D.G.: The logrank test. BMJ **328**(7447), 1073 (2004)
7. Bulten, W., et al.: Artificial intelligence for diagnosis and gleason grading of prostate cancer: the panda challenge. Nat. Med. **28**(1), 154–163 (2022)

8. Carbonneau, M.A., Cheplygina, V., Granger, E., Gagnon, G.: Multiple instance learning: a survey of problem characteristics and applications. Pattern Recogn. **77**, 329–353 (2018)
9. Caron, M., Touvron, H., Misra, I., Jégou, H., Mairal, J., Bojanowski, P., Joulin, A.: Emerging properties in self-supervised vision transformers. In: Proceedings of the IEEE/CVF International Conference on Computer Vision, pp. 9650–9660 (2021)
10. Chen, R.J., et al.: Whole slide images are 2D point clouds: context-aware survival prediction using patch-based graph convolutional networks. In: de Bruijne, M., et al. (eds.) MICCAI 2021. LNCS, vol. 12908, pp. 339–349. Springer, Cham (2021). https://doi.org/10.1007/978-3-030-87237-3_33
11. Cooperberg, M.R., et al.: Outcomes of active surveillance for men with intermediate-risk prostate cancer. J. Clin. Oncol. Off. J. Am. Soc. Clin. Oncol. **29**(2), 228–234 (2011)
12. Darbandsari, A., et al.: Identification of a novel subtype of endometrial cancer with unfavorable outcome using artificial intelligence-based histopathology image analysis (2022)
13. Dosovitskiy, A., et al.: An image is worth 16×16 words: transformers for image recognition at scale. arXiv preprint arXiv:2010.11929 (2020)
14. Ilse, M., Tomczak, J., Welling, M.: Attention-based deep multiple instance learning. In: International Conference on Machine Learning, pp. 2127–2136. PMLR (2018)
15. Kaplan, E.L., Meier, P.: Nonparametric estimation from incomplete observations. J. Am. Stat. Assoc. **53**(282), 457–481 (1958)
16. Lee, Y., et al.: Derivation of prognostic contextual histopathological features from whole-slide images of tumours via graph deep learning. Nat. Biomed. Eng., 1–15 (2022)
17. Li, R., Yao, J., Zhu, X., Li, Y., Huang, J.: Graph CNN for survival analysis on whole slide pathological images. In: Frangi, A.F., Schnabel, J.A., Davatzikos, C., Alberola-López, C., Fichtinger, G. (eds.) MICCAI 2018. LNCS, vol. 11071, pp. 174–182. Springer, Cham (2018). https://doi.org/10.1007/978-3-030-00934-2_20
18. Liu, W., He, Q., He, X.: Weakly supervised nuclei segmentation via instance learning. In: 2022 IEEE 19th International Symposium on Biomedical Imaging (ISBI), pp. 1–5. IEEE (2022)
19. Lu, M.Y., Williamson, D.F., Chen, T.Y., Chen, R.J., Barbieri, M., Mahmood, F.: Data-efficient and weakly supervised computational pathology on whole-slide images. Nat. Biomed. Eng. **5**(6), 555–570 (2021)
20. Mobadersany, P., et al.: Predicting cancer outcomes from histology and genomics using convolutional networks. Proc. Natl. Acad. Sci. **115**(13), E2970–E2979 (2018)
21. Moses, K.A., et al.: Nccn guidelines® insights: prostate cancer early detection, version 1.2023: featured updates to the nccn guidelines. J. Natl. Comprehens. Cancer Netw. **21**(3), 236–246 (2023)
22. Oono, K., Suzuki, T.: Graph neural networks exponentially lose expressive power for node classification. arXiv preprint arXiv:1905.10947 (2019)
23. Ouzzane, A., et al.: Magnetic resonance imaging targeted biopsy improves selection of patients considered for active surveillance for clinically low risk prostate cancer based on systematic biopsies. J. Urol. **194**(2), 350–356 (2015)
24. Rony, J., Belharbi, S., Dolz, J., Ayed, I.B., McCaffrey, L., Granger, E.: Deep weakly-supervised learning methods for classification and localization in histology images: a survey. arXiv preprint arXiv:1909.03354 (2019)
25. Skowronek, J.: Current status of brachytherapy in cancer treatment-short overview. J. Contemp. Brachyther. **9**(6), 581–589 (2017)

26. Son, B., Lee, S., Youn, H., Kim, E., Kim, W., Youn, B.: The role of tumor microenvironment in therapeutic resistance. Oncotarget **8**(3), 3933 (2017)
27. Srinidhi, C.L., Ciga, O., Martel, A.L.: Deep neural network models for computational histopathology: a survey. Med. Image Anal. **67**, 101813 (2021)
28. Tang, S., Chen, D., Bai, L., Liu, K., Ge, Y., Ouyang, W.: Mutual crf-gnn for few-shot learning. In: Proceedings of the IEEE/CVF Conference on Computer Vision and Pattern Recognition, pp. 2329–2339 (2021)
29. Wetstein, S.C., et al.: Deep learning-based breast cancer grading and survival analysis on whole-slide histopathology images. Sci. Rep. **12**(1), 1–12 (2022)
30. Xu, K., Hu, W., Leskovec, J., Jegelka, S.: How powerful are graph neural networks? arXiv preprint arXiv:1810.00826 (2018)
31. Zaheer, M., Kottur, S., Ravanbakhsh, S., Poczos, B., Salakhutdinov, R.R., Smola, A.J.: Deep sets. Adv. Neural Inf. Process. Syst. **30**, 1–11 (2017)
32. Zhu, X., Yao, J., Zhu, F., Huang, J.: Wsisa: making survival prediction from whole slide histopathological images. In: Proceedings of the IEEE Conference on Computer Vision and Pattern Recognition, pp. 7234–7242 (2017)

Deep Cellular Embeddings: An Explainable Plug and Play Improvement for Feature Representation in Histopathology

Jacob Gildenblat[1,2], Anil Yüce[1(✉)], Samaneh Abbasi-Sureshjani[1], and Konstanty Korski[1]

[1] F. Hoffmann-La Roche AG, Basel, Switzerland
{jacob.gildenblat,anil.yuce,samaneh.abbasi, konstanty.korski}@roche.com
[2] DeePathology, Ra'anana, Israel

Abstract. Weakly supervised classification of whole slide images (WSIs) in digital pathology typically involves making slide-level predictions by aggregating predictions from embeddings extracted from multiple individual tiles. However, these embeddings can fail to capture valuable information contained within the individual cells in each tile. Here we describe an embedding extraction method that combines tile-level embeddings with a cell-level embedding summary. We validated the method using four hematoxylin and eosin stained WSI classification tasks: human epidermal growth factor receptor 2 status and estrogen receptor status in primary breast cancer, breast cancer metastasis in lymph node tissue, and cell of origin classification in diffuse large B-cell lymphoma. For all tasks, the new method outperformed embedding extraction methods that did not include cell-level representations. Using the publicly available HEROHE Challenge data set, the method achieved a state-of-the-art performance of 90% area under the receiver operating characteristic curve. Additionally, we present a novel model explainability method that could identify cells associated with different classification groups, thus providing supplementary validation of the classification model. This deep learning approach has the potential to provide morphological insights that may improve understanding of complex underlying tumor pathologies.

Keywords: Deep Learning · Whole Slide Images · Hematoxylin and Eosin

1 Introduction

Accurate diagnosis plays an important role in achieving the best treatment outcomes for people with cancer [1]. Identification of cancer biomarkers permits more granular classification of tumors, leading to better diagnosis, prognosis, and treatment decisions [2, 3]. For many cancers, clinically reliable genomic, molecular, or imaging biomarkers

Supplementary Information The online version contains supplementary material available at https://doi.org/10.1007/978-3-031-43987-2_75.

H. Greenspan et al. (Eds.): MICCAI 2023, LNCS 14225, pp. 776–785, 2023.
https://doi.org/10.1007/978-3-031-43987-2_75

have not been identified and biomarker identification techniques (e.g., fluorescence in situ hybridization) have limitations that can restrict their clinical use. On the other hand, histological analysis of hematoxylin and eosin (H&E)-stained pathology slides is widely used in cancer diagnosis and prognosis. However, visual examination of H&E-stained slides is insufficient for classification of some tumors because identifying morphological differences between molecularly defined subtypes is beyond the limit of human detection.

The introduction of digital pathology (DP) has enabled application of machine learning approaches to extract otherwise inaccessible diagnostic and prognostic information from H&E-stained whole slide images (WSIs) [4, 5]. Current deep learning approaches to WSI analysis typically operate at three different histopathological scales: whole slide-level, region-level, and cell-level [4]. Although cell-level analysis has the potential to produce more detailed and explainable data, it can be limited by the unavailability of sufficiently annotated training data. To overcome this problem, weakly supervised and multiple instance learning (MIL) based approaches have been applied to numerous WSI classification tasks [6–10]. However, many of these models use embeddings derived from tiles extracted using pretrained networks, and these often fail to capture useful information from individual cells. Here we describe a new embedding extraction method that combines tile-level embeddings with a cell-level embedding summary. Our new method achieved better performance on WSI classification tasks and had a greater level of explainability than models that used only tile-level embeddings.

2 Embedding Extraction Scheme

Transfer learning using backbones pretrained on natural images is a common method that addresses the challenge of using data sets that largely lack annotation. However, using backbones pretrained on natural images is not optimal for classification of clinical images [11]. Therefore, to enable the use of large unlabeled clinical imaging data sets, as the backbone of our neural network we used a ResNet50 model [12]. The backbone was trained with the bootstrap your own latent (BYOL) method [13] using four publicly available data sets from The Cancer Genome Atlas (TCGA) and three data sets from private vendors that included healthy and malignant tissue from a range of organs [14].

2.1 Tile-Level Embeddings

Following standard practice, we extracted tiles with dimensions of 256 × 256 pixels from WSIs (digitized at 40 × magnification) on a spatial grid without overlap. Extracted tiles that contained artifacts were discarded (e.g., tiles that had an overlap of >10% with background artifacts such blurred areas or pen markers). We normalized the tiles for stain color using a U-Net model for stain normalization [15] that was trained on a subset of data from one of the medical centers in the CAMELYON17 data set to ensure homogeneity of staining [16].

To create the tile-level embeddings, we used the method proposed by [17] to summarize the convolutional neural network (CNN) features with nonnegative matrix factorization (NMF) for K = 2 factors. We observed that the feature activations within the last layer of the network were not aligned with the cellular content. Although these features may still have been predictive, they were less interpretable, and it was more difficult to know what kind of information they captured. Conversely, we observed that the self-supervised network captured cellular content and highlighted cells within the tiles (Fig. 1). Therefore, the tile-level embeddings were extracted after dropping the last layer (i.e., dropping three bottleneck blocks in ResNet50) from the pretrained model.

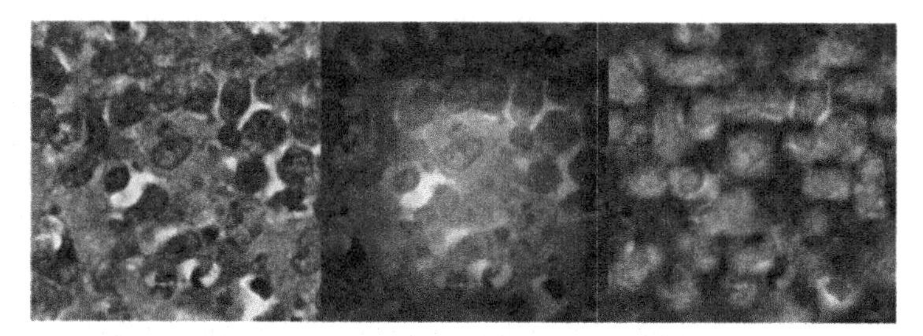

Fig. 1. A visualization of the output features of the backbone for a typical input tile (left), from the last layer (middle), and from the second to last layer (right) of the pretrained CNN summarized using NMF with $K = 2$ factors. Resolution: 0.25 μm/pixel.

2.2 Cell-Level Embeddings

Tiles extracted from WSIs may contain different types of cells, as well as noncellular tissue such as stroma and blood vessels and nonbiological features (e.g., glass). Cell-level embeddings may be able to extract useful information, based on the morphological appearance of individual cells, that is valuable for downstream classification tasks but would otherwise be masked by more dominant features within tile-level embeddings.

We extracted deep cell-level embeddings by first detecting individual cellular boundaries using StarDist [18] and extracting 32×32-pixel image crops centered around each segmented nucleus to create cell-patch images. We then used the pre-trained ResNet50 model to extract cell-level embeddings in a similar manner to the extraction of the tile-level embeddings. Since ResNet50 has a spatial reduction factor of 32 in the output of the CNN, the 32×32-pixel image had a 1:1 spatial resolution in the output. To ensure the cell-level embeddings contained features relevant to the cells, prior to the mean pooling in ResNet50 we increased the spatial image resolution to 16×16 pixels in the output from the CNN by enlarging the 32×32-pixel cell-patch images to 128×128 pixels and skipping the last 4-layers in the network.

Because of heterogeneity in the size of cells detected, each 32×32-pixel cell-patch image contained different proportions of cellular and noncellular features. Higher proportions of noncellular features in an image may cause the resultant embeddings to be dominated by noncellular tissue features or other background features. Therefore, to limit the information used to create the cell-level embeddings to only cellular features, we removed portions of the cell-patch images that were outside of the segmented nuclei by setting their pixel values to black (RGB 0, 0, 0). Finally, to prevent the size of individual nuclei or amount of background in each cell-patch image from dominating over the cell-level features, we modified the ResNet50 Global Average Pooling layer to only average the features inside the boundary of the segmented nuclei, rather than averaging across the whole output tensor from the CNN.

2.3 Combined Embeddings

To create a combined representation of the tile-level and cell-level embeddings, we first applied a nuclei segmentation network to each tile. Only tiles with ≥ 10 cells per tile, excluding any cells which overlapped the tile border, were included for embedding extraction. For the included tiles, we extracted the tile-level embeddings as described in Sect. 2.1 and for each detected cell we extracted the cell-level embeddings as described in Sect. 2.2. We then calculated the mean and standard deviation of the vectors of the cell-level embeddings for each tile and concatenated those to each tile-level embedding. This resulted in a combined embedding representation with a total size of 1536 pixels ($1024 + 256 + 256$).

In addition to the WSI classification results presented in the next sections, we also performed experiments to compare the ability of combined embeddings and tile-level embeddings to predict nuclei-related features that were manually extracted from the images and to identify tiles where nuclei had been ablated. The details and results of these experiments are available in supplementary materials and provide further evidence of the improved ability to capture cell-level information when using combined embeddings compared with tile-level embeddings alone.

3 WSI Classification Tasks

For each classification task we compared different combinations of tile-level and cell-level embeddings using a MIL framework. We also compared two different MIL architectures to aggregate the embeddings for WSI-level prediction.

The first architecture used an attention-MIL (A-MIL) network [19] (the code was adapted from a publicly available implementation [20]). We trained the network with a 0.001 learning rate and tuned the batch size (48 or 96) and bag sample size (512, 1024, or 2048) for each classification task separately. When comparing the combined embedding extraction method with the tile-level only embeddings, parameters were fixed to demonstrate differences in performance without additional parameter tuning.

Transformer (Xformer) was used as the second MIL architecture [21]. We used three Xformer layers, each with eight attention heads, 512 parameters per token, and 256 parameters in the multi-layer perceptron layers. The space complexity of the Xformer was quadratic with the number of tokens. While some WSIs had up to 100,000 tiles, we found, in practice, that we could not fit more than 6000 tokens in the memory. Consequently, we used the Nyströformer Xformer variant [22] since it consumes less memory (the code was adapted from a publicly available implementation [23]). This Xformer has two outputs, was trained with the Adam optimizer [24] with default parameters, and the loss was weighted with median frequency balancing [25] to assign a higher weight to the less frequent class. Like A-MIL, the batch and bag sample sizes were fixed for each classification task. During testing a maximum of 30,000 tiles per slide were used. The complete flow for WSI classification is shown in Fig. 2. The models were selected using a validation set, that was a random sample of 20% of the training data. All training was done using PyTorch version 1.12.1 (pytorch.org) on 8 NVIDIA Tesla V100 GPUs with Cuda version 10.2.

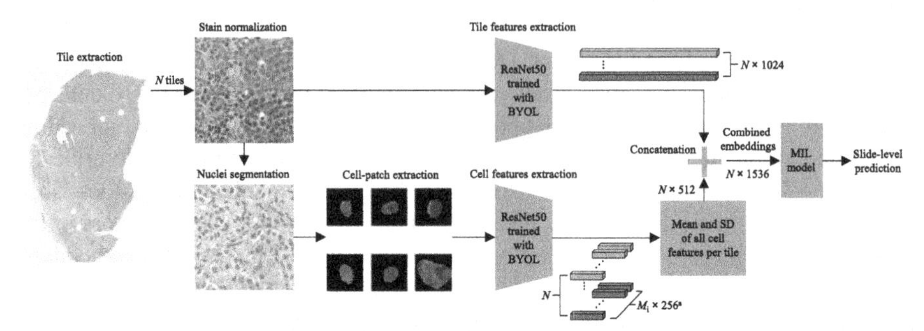

Fig. 2. Schematic visualization of the classification pipeline based on combined embeddings. Tile-level and cell-level embeddings are extracted in parallel and then concatenated embedding vectors are passed through the MIL model for the downstream task. $^{a}M_i$ equals the number of cells in tile i.

3.1 Data

We tested our feature representation method in several classification tasks involving WSIs of H&E-stained histopathology slides. The number of slides per class for each classification task are shown in Fig. 3.

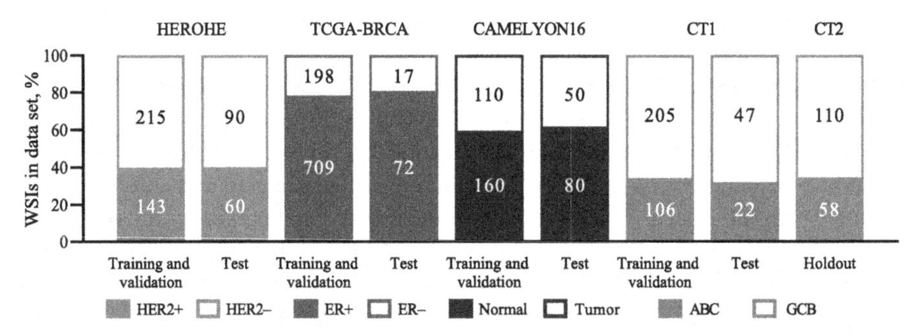

Fig. 3. Class distributions in the data used for WSI classification tasks. Numbers in the bars represent the number of WSIs by classification for each task.

For breast cancer human epidermal growth factor receptor 2 (HER2) prediction, we used data from the HEROHE Challenge data set [26]. To enable comparison with previous results we used the same test data set that was used in the challenge [27]. For prediction of estrogen receptor (ER) status, we used images from the TCGA-Breast Invasive Carcinoma (TCGA-BRCA) data set [28] for which the ER status was known.

For these two tasks we used artifact-free tiles from tumor regions detected with an in-house tumor detection model.

For breast cancer metastasis detection in lymph node tissue, we used WSIs of H&E-stained healthy lymph node tissue and lymph node tissue with breast cancer metastases from the publicly available CAMELYON16 challenge data set [16, 29]. All artifact-free tissue tiles were used.

For cell of origin (COO) prediction of activated B-cell like (ABC) or germinal center B-cell like (GCB) tumors in diffuse large B-cell lymphoma (DLBCL), we used data from the phase 3 GOYA (NCT01287741) and phase 2 CAVALLI (NCT02055820) clinical trials, hereafter referred to as CT1 and CT2, respectively. All slides were H&E-stained and scanned using Ventana DP200 scanners at 40× magnification. CT1 was used for training and testing the classifier and CT2 was used only as an independent holdout data set. For these data sets we used artifact-free tiles from regions annotated by expert pathologists to contain tumor tissue.

4 Model Classification Performance

For the HER2 prediction, ER prediction, and metastasis detection classification tasks, combined embeddings outperformed tile-level only embeddings irrespective of the downstream classifier architecture used (Fig. 4).

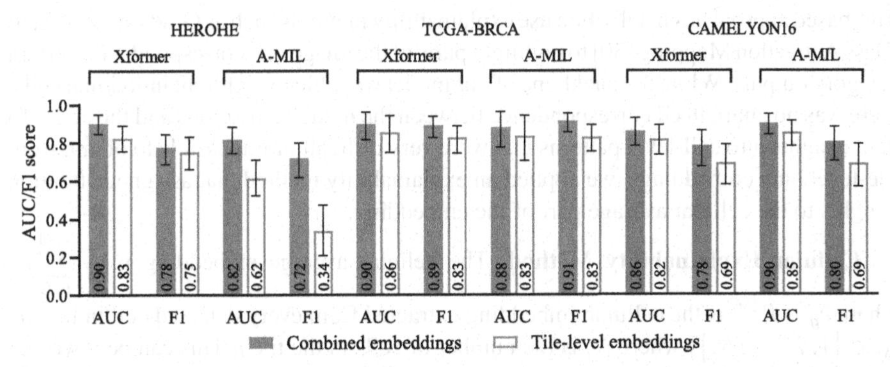

Fig. 4. Model performance using the Xformer and A-MIL architectures for the breast cancer HER2 status, breast cancer ER status, and breast cancer metastasis detection in lymph node tissue classification tasks. Error bars represent 95% confidence intervals computed by a 5000-sample bias-corrected and accelerated bootstrap.

In fact, for the HER2 classification task, combined embeddings obtained using the Xformer architecture achieved, to our knowledge, the best performance yet reported on the HEROHE Challenge data set (area under the receiver operating characteristic curve [AUC], 90%; F1 score, 82%).

For COO classification in DLBCL, not only did the combined embeddings achieve better performance than the tile-level only embeddings with both the Xformer and A-MIL architectures (Fig. 5) on the CT1 test set and CT2 holdout data set, but they also had a significant advantage versus tile-only level embeddings in respect of the additional insights they provided through cell-level model explainability (Sect. 4.1).

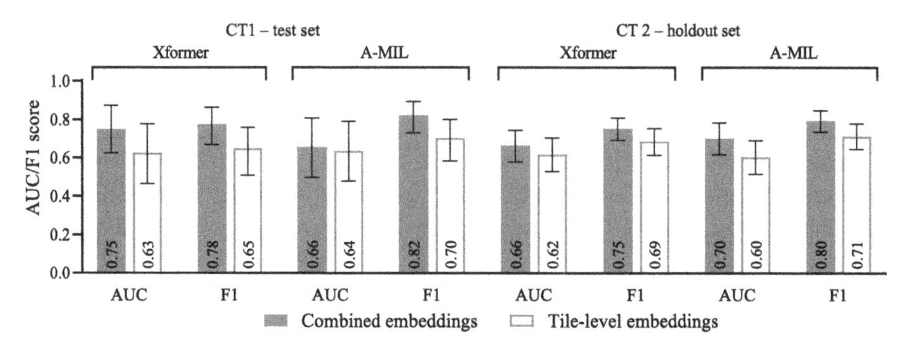

Fig. 5. Model performance using the Xformer and A-MIL architectures for the COO in DLBCL classification task. Error bars represent 95% confidence intervals computed by a 5000-sample bias-corrected and accelerated bootstrap.

4.1 Model Explainability

Tile-based approaches in DP often use explainability methods such as Gradient-weighted Class Activation Mapping [30] to highlight parts of the image that correspond with certain category outputs. While the backbone of our model was able to highlight individual cells, there was no guaranteed correspondence between the model activations and the cells. To gain insights into cell-level patterns that were very difficult or impossible to obtain from tile-level only embeddings, we applied an explainability method that assigned attention weights to the cellular average part of the embedding.

Cellular Explainability Method. The cellular average embedding is $\frac{1}{N}\sum_{i=0}^{N-1} e_{ij}$ where $e_{ij} \in R^{256}$ is the cellular embedding extracted from every detected cell in the tile $j(i \in \{1, 2, \ldots, N_j\})$ where N_j is the number of cells in the tile j. This can be rewritten as a weighted average of the cellular embeddings $\sum_{i=0}^{N-1} e_{ij} Sigmoid(w_i) / \sum_{i=0}^{N-1} Sigmoid(w_i)$ where $w_i \in R^{256}$ are the per cell attention weights that if initialized to 0 result in the original cellular average embedding. The re-formulation does not change the result of the forward pass since w_i are not all equal. Note that the weights are not learned through training but calculated per cell at inference time to get the per cell contribution. We computed the gradient of the output category (of the classification method applied on top of the computed embedding) with respect to the attention weights w_i: $grad_i = \partial Score_i / \partial w_i$ and visualized cells that received positive and negative gradients using different colors.

Visual Example Results. Examples of our cellular explainability method applied to weakly supervised tumor detection on WSIs from the CAMELYON16 data set using A-MIL are shown in Fig. 6. Cells with positive attention gradients shifted the output towards a classification of tumor and are labeled green. Cells with negative attention gradients are labeled red. When reviewed by a trained pathologist, cells with positive gradients had characteristics previously associated with breast cancer tumors (e.g., larger nuclei, more visible nucleoli, differences in size and shape). Conversely, negative cells had denser chromatin and resembled other cell types (e.g., lymphocytes). These repeatable findings demonstrate the benefit of using cell-level embeddings and our explainability method to gain a cell-level understanding of both correct and incorrect slide-level model predictions (Fig. 6). We also applied our explainability method to COO prediction in DLBCL.

In this case, cells with positive attention gradients that shifted the output towards a classification of GCB were labeled green and cells with negative attention gradients that shifted the classification towards ABC were labeled red. Cells with positive attention gradients were mostly smaller lymphoid cells with low grade morphology or were normal lymphocytes, whereas cells with negative attention gradients were more frequently larger lymphoid cells with high grade morphology (Fig. 6).

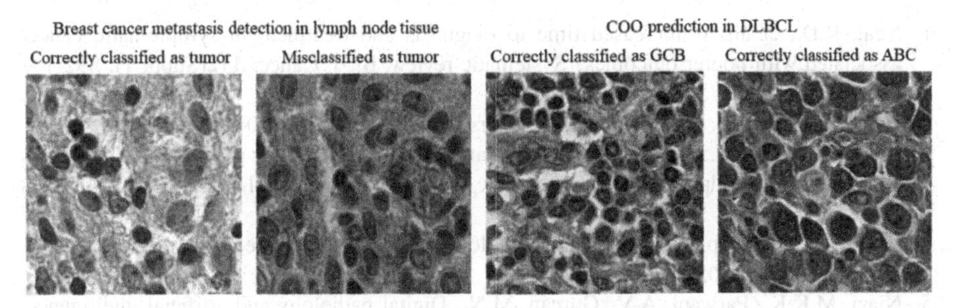

Fig. 6. Cellular explainability method applied to breast cancer metastasis detection in lymph nodes and COO prediction in DLBCL. Cells in the boundary margin were discarded.

5 Conclusions

We describe a method to capture both cellular and texture feature representations from WSIs that can be plugged into any MIL architecture (e.g., CNN or Xformer-based), as well as into fully supervised models (e.g., tile classification models). Our method is more flexible than other methods (e.g., Hierarchical Image Pyramid Transformer) that usually capture the hierarchical structure in WSIs by aggregating features at multiple levels in a complex set of steps to perform the final classification task. In addition, we describe a method to explain the output of the classification model that evaluates the contributions of histologically identifiable cells to the slide-level classification. Tile-level embeddings result in good performance for detection of tumor metastases in lymph nodes. However, introducing more cell-level information, using combined embeddings, resulted in improved classification performance. In HER2 and ER prediction tasks for breast cancer we demonstrate that addition of a cell-level embedding summary to tile-level embeddings can boost model performance by up to 8%. Finally, for COO prediction in DLBCL and breast cancer metastasis detection in lymph nodes, we demonstrated the potential of our explainability method to gain insights into previously unknown associations between cellular morphology and disease biology.

Acknowledgments. We thank the Roche Diagnostic Solutions and Genentech Research Pathology Core Laboratory staff for tissue procurement and immunohistochemistry verification. We thank the participants from the GOYA and CAVALLI trials. The results published here are in part based upon data generated by the TCGA Research Network: https://www.cancer.gov/tcga. We thank Maris Skujevskis, Uwe Schalles and Darta Busa for their help in curating the datasets and the annotations and Amal Lahiani for sharing the tumor segmentation model used for generating the results on HEROHE. The study was funded by F. Hoffmann-La Roche AG, Basel, Switzerland and writing support was provided by Adam Errington PhD of PharmaGenesis Cardiff, Cardiff, UK and was funded by F. Hoffmann-La Roche AG.

References

1. Neal, R.D., et al.: Is increased time to diagnosis and treatment in symptomatic cancer associated with poorer outcomes? Systematic review. Br. J. Cancer **112**(Suppl 1), S92-107 (2015)
2. Henry, N.L., Hayes, D.F.: Cancer biomarkers. Mol. Oncol. **6**(2), 140–146 (2012)
3. Park, J.E., Kim, H.S.: Radiomics as a quantitative imaging biomarker: practical considerations and the current standpoint in neuro-oncologic studies. Nucl. Med. Mol. Imaging **52**(2), 99–108 (2018)
4. Lee, K., et al.: Deep learning of histopathology images at the single cell level. Front. Artif. Intell. **4**, 754641 (2021)
5. Niazi, M.K.K., Parwani, A.V., Gurcan, M.N.: Digital pathology and artificial intelligence. Lancet Oncol. **20**(5), e253–e261 (2019)
6. van der Laak, J., Litjens, G., Ciompi, F.: Deep learning in histopathology: the path to the clinic. Nat. Med. **27**(5), 775–784 (2021)
7. Shao, Z., et al.: TransMIL: transformer based correlated multiple instance learning for whole slide image classification. In: Advances in Neural Information Processing Systems, vol. 34 pp. 2136–2147 (2021)
8. Wang, Y., et al.: CWC-transformer: a visual transformer approach for compressed whole slide image classification. Neural Comput. Appl. (2023)
9. Lu, M.Y., Williamson, D.F.K., Chen, T.Y., Chen, R.J., Barbieri, M., Mahmood, F.: Data-efficient and weakly supervised computational pathology on whole-slide images. Nat. Biomed. Eng. **5**(6), 555–570 (2021)
10. Chen, R.J., et al.: Scaling vision transformers to gigapixel images via hierarchical self-supervised learning. In: Proceedings of the IEEE/CVF Conference on Computer Vision and Pattern Recognition, pp. 16144–16155 (2022)
11. Kang, M., Song, H., Park, S., Yoo, D., Pereira, S.: Benchmarking self-supervised learning on diverse pathology datasets. In: Proceedings of the IEEE/CVF Conference on Computer Vision and Pattern Recognition (CVPR), pp. 3344–3354 (2023)
12. He, K., Zhang, X., Ren, S., Sun, J.: Deep residual learning for image recognition. In: Proceedings of the IEEE/CVF Conference on Computer Vision and Pattern Recognition, pp. 770–778 (2016)
13. Grill, J.-B., et al.: Bootstrap your own latent - a new approach to self-supervised learning. Adv. Neural. Inf. Process. Syst. **33**, 21271–21284 (2020)

14. Abbasi-Sureshjani, S., et al.: Molecular subtype prediction for breast cancer using H&E specialized backbone. In: MICCAI Workshop on Computational Pathology, pp. 1–9 (2021)
15. Tellez, D., et al.: Quantifying the effects of data augmentation and stain color normalization in convolutional neural networks for computational pathology. Med. Image Anal. **58**, 101544 (2019)
16. Litjens, G., et al.: H&E-stained sentinel lymph node sections of breast cancer patients: the CAMELYON dataset. GigaScience **7**(6), giy065 (2018)
17. Collins, E., Achanta, R., Süsstrunk, S.: Deep feature factorization for concept discovery. In: Ferrari, V., Hebert, M., Sminchisescu, C., Weiss, Y. (eds.) Computer Vision – ECCV 2018, pp. 352–368. Springer International Publishing, Cham (2018)
18. Schmidt, U., Weigert, M., Broaddus, C., Myers, G.: Cell detection with star-convex polygons. In: Frangi, A.F., Schnabel, J.A., Davatzikos, C., Alberola-López, C., Fichtinger, G. (eds.) MICCAI 2018. LNCS, vol. 11071, pp. 265–273. Springer, Cham (2018). https://doi.org/10.1007/978-3-030-00934-2_30
19. Ilse, M., Tomczak, J.M., Welling, M.: Attention-based deep multiple instance learning. ArXiv abs/1802.04712 (2018)
20. Attention-based Deep Multiple Instance Learning. https://github.com/AMLab-Amsterdam/AttentionDeepMIL. Accessed 24 Feb 2023
21. Vaswani, A., et al.: Attention is all you need. In: Guyon, I., et al. (eds.) Advances in Neural Information Processing Systems (NeurIPS 2017), vol. 31, pp. 5998–6008 (2017)
22. Xiong, Y., et al.: Nyströmformer: a Nystöm-based algorithm for approximating self-attention. Proc. Conf. AAAI Artif. Intell. **35**(16), 14138–14148 (2021)
23. Nyström Attention. https://github.com/lucidrains/nystrom-attention. Accessed 24 Feb 2023
24. Kingma, D.P., Ba, J.: Adam: a method for stochastic optimization. In: Proceedings of the 3rd International Conference on Learning Representations (ICLR 2015), abs/1412.6980 (2015)
25. Eigen, D., Fergus, R.: Predicting depth, surface normals and semantic labels with a common multi-scale convolutional architecture. In: Proceedings of the IEEE International Conference on Computer Vision, pp. 2650–2658 (2015)
26. HEROHE ECDP2020. https://ecdp2020.grand-challenge.org/. Accessed 24 Feb 2023
27. Conde-Sousa, E., et al.: HEROHE challenge: predicting HER2 status in breast cancer from hematoxylin-eosin whole-slide imaging. J. Imaging **8**(8) (2022)
28. National Cancer Institute GDC Data Portal. https://portal.gdc.cancer.gov/. Accessed 24 Feb 2023
29. CAMELYON17 Grand Challenge. https://camelyon17.grand-challenge.org/Data/. Accessed 24 Feb 2023
30. Selvaraju, R.R., Cogswell, M., Das, A., Vedantam, R., Parikh, D., Batra, D.: Grad-CAM: visual explanations from deep networks via gradient-based localization. In: 2017 IEEE International Conference on Computer Vision (ICCV), pp. 618–626 (2017)

NASDM: Nuclei-Aware Semantic Histopathology Image Generation Using Diffusion Models

Aman Shrivastava(✉) and P. Thomas Fletcher

University of Virginia, Charlottesville, VA, USA
{as3ek,ptf8v}@virginia.edu

Abstract. In recent years, computational pathology has seen tremendous progress driven by deep learning methods in segmentation and classification tasks aiding prognostic and diagnostic settings. Nuclei segmentation, for instance, is an important task for diagnosing different cancers. However, training deep learning models for nuclei segmentation requires large amounts of annotated data, which is expensive to collect and label. This necessitates explorations into generative modeling of histopathological images. In this work, we use recent advances in conditional diffusion modeling to formulate a first-of-its-kind nuclei-aware semantic tissue generation framework (NASDM) which can synthesize realistic tissue samples given a semantic instance mask of up to six different nuclei types, enabling pixel-perfect nuclei localization in generated samples. These synthetic images are useful in applications in pathology pedagogy, validation of models, and supplementation of existing nuclei segmentation datasets. We demonstrate that NASDM is able to synthesize high-quality histopathology images of the colon with superior quality and semantic controllability over existing generative methods. Implementation: https://github.com/4m4n5/NASDM.

Keywords: Generative Modeling · Histopathology · Diffusion Models

1 Introduction

Histopathology relies on hematoxylin and eosin (H&E) stained biopsies for microscopic inspection to identify visual evidence of diseases. Hematoxylin has a deep blue-purple color and stains acidic structures such as DNA in cell nuclei. Eosin, alternatively, is red-pink and stains nonspecific proteins in the cytoplasm and the stromal matrix. Pathologists then examine highlighted tissue characteristics to diagnose diseases, including different cancers. A correct diagnosis, therefore, is dependent on the pathologist's training and prior exposure to a wide variety of disease subtypes [30]. This presents a challenge, as some disease variants are extremely rare, making visual identification difficult. In recent years, deep learning methods have aimed to alleviate this problem by designing discriminative frameworks that aid diagnosis [15,28]. Segmentation models find applications in spatial identification of different nuclei types [6]. However, generative modeling in histopathology is relatively unexplored. Generative models can

H. Greenspan et al. (Eds.): MICCAI 2023, LNCS 14225, pp. 786–796, 2023.
https://doi.org/10.1007/978-3-031-43987-2_76

be used to generate histopathology images with specific characteristics, such as visual patterns identifying rare cancer subtypes [4]. As such, generative models can be sampled to emphasize each disease subtype equally and generate more balanced datasets, thus preventing dataset biases getting amplified by the models [7]. Generative models have the potential to improve the pedagogy, trustworthiness, generalization, and coverage of disease diagnosis in the field of histology by aiding both deep learning models and human pathologists. Synthetic datasets can also tackle privacy concerns surrounding medical data sharing. Additionally, conditional generation of annotated data adds even further value to the proposition as labeling medical images involves tremendous time, labor, and training costs. Recently, denoising diffusion probabilistic models (DDPMs) [8] have achieved tremendous success in conditional and unconditional generation of real-world images [3]. Further, the semantic diffusion model (SDM) demonstrated the use of DDPMs for generating images given semantic layout [27]. In this work, (1) we leverage recently discovered capabilities of DDPMs to design a first-of-its-kind nuclei-aware semantic diffusion model (NASDM) that can generate realistic tissue patches given a semantic mask comprising of multiple nuclei types, (2) we train our framework on the Lizard dataset [5] consisting of colon histology images and achieve state-of-the-art generation capabilities, and (3) we perform extensive ablative, qualitative, and quantitative analyses to establish the proficiency of our framework on this tissue generation task.

2 Related Work

Deep learning based generative models for histopathology images have seen tremendous progress in recent years due to advances in digital pathology, compute power, and neural network architectures. Several GAN-based generative models have been proposed to generate histology patches [16,31,33]. However, GANs suffer from problems of frequent mode collapse and overfitting their discriminator [29]. It is also challenging to capture long-tailed distributions and synthesize rare samples from imbalanced datasets using GANs. More recently, denoising diffusion models have been shown to generate highly compelling images by incrementally adding information to noise [8]. Success of diffusion models in generating realistic images led to various conditional [12,21,22] and unconditional [3,9,19] diffusion models that generate realistic samples with high fidelity. Following this, a morphology-focused diffusion model has been presented for generating tissue patches based on genotype [18]. Semantic image synthesis is a task involving generating diverse realistic images from semantic layouts. GAN-based semantic image synthesis works [20,24,25] generally struggled at generating high quality and enforcing semantic correspondence at the same time. To this end, a semantic diffusion model has been proposed that uses conditional denoising diffusion probabilistic model and achieves both better fidelity and diversity [27]. We use this progress in the field of conditional diffusion models and semantic image synthesis to formulate our NASDM framework.

3 Method

In this paper, we describe our framework for generating tissue patches conditioned on semantic layouts of nuclei. Given a nuclei segmentation mask, we intend to generate realistic synthetic patches. In this section, we (1) describe our data preparation, (2) detail our stain-normalization strategy, (3) review conditional denoising diffusion probabilistic models, (4) outline the network architecture used to condition on semantic label map, and (5) highlight the classifier-free guidance mechanism that we employ at sampling time.

3.1 Data Processing

We use the Lizard dataset [5] to demonstrate our framework. This dataset consists of histology image regions of colon tissue from six different data sources at 20× objective magnification. The images are accompanied by full segmentation annotation for different types of nuclei, namely, epithelial cells, connective tissue cells, lymphocytes, plasma cells, neutrophils, and eosinophils. A generative model trained on this dataset can be used to effectively synthesize the colonic tumor micro-environments. The dataset contains 238 image regions, with an average size of 1055×934 pixels. As there are substantial visual variations across images, we construct a representative test set by randomly sampling a 7.5% area from each image and its corresponding mask to be held-out for testing. The test and train image regions are further divided into smaller image patches of 128×128 pixels at two different objective magnifications: (1) at 20×, the images are directly split into 128×128 pixels patches, whereas (2) at 10×, we generate 256×256 patches and resize them to 128×128 for training. To use the data exhaustively, patching is performed with a 50% overlap in neighboring patches. As such, at (1) 20× we extract a total of 54,735 patches for training and 4,991 patches as a held-out set, while at (2) 20× magnification we generate 12,409 training patches and 655 patches are held out.

3.2 Stain Normalization

A common issue in deep learning with H&E stained histopathology slides is the visual bias introduced by variations in the staining protocol and the raw materials of chemicals leading to different colors across slides prepared at different labs [1]. As such, several stain-normalization methods have been proposed to tackle this issue by normalizing all the tissue samples to mimic the stain distribution of a given target slide [17,23,26]. In this work, we use the structure preserving color normalization scheme introduce by Vahadane et al. [26] to transform all the slides to match the stain distribution of an empirically chosen slide from the training dataset.

3.3 Conditional Denoising Diffusion Probabilistic Model

In this section, we describe the theory of conditional denoising diffusion probabilistic models, which serves as the backbone of our framework. A conditional

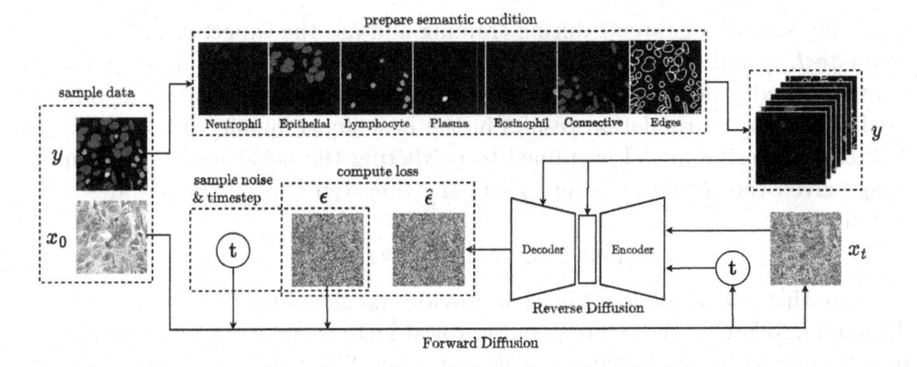

Fig. 1. NASDM training framework: Given a real image x_0 and semantic mask y, we construct the conditioning signal by expanding the mask and adding an instance edge map. We sample timestep t and noise ϵ to perform forward diffusion and generate the noised input x_t. The corrupted image x_t, timestep t, and semantic condition y are then fed into the denoising model which predicts $\hat{\epsilon}$ as the amount of noise added to the model. Original noise ϵ and prediction $\hat{\epsilon}$ are used to compute the loss in (4).

diffusion model aims to maximize the likelihood $p_\theta(x_0 \mid y)$, where data x_0 is sampled from the conditional data distribution, $x_0 \sim q(x_0 \mid y)$, and y represents the conditioning signal. A diffusion model consists of two intrinsic processes. The forward process is defined as a Markov chain, where Gaussian noise is gradually added to the data over T timesteps as

$$\begin{aligned} q(x_t \mid x_{t-1}) &= \mathcal{N}(x_t; \sqrt{1-\beta_t} x_{t-1}, \beta_t \mathbf{I}), \\ q(x_{1:T} \mid x_0) &= \prod_{t=1}^{T} q(x_t \mid x_{t-1}), \end{aligned} \tag{1}$$

where $\{\beta\}_{t=1:T}$ are constants defined based on the noise schedule. An interesting property of the Gaussian forward process is that we can sample x_t directly from x_0 in closed form. Now, the reverse process, $p_\theta(x_{0:T} \mid y)$, is defined as a Markov chain with learned Gaussian transitions starting from pure noise, $p(x_T) \sim \mathcal{N}(0, \mathbf{I})$, and is parameterized as a neural network with parameters θ as

$$p_\theta(x_{0:T} \mid y) = p(y_T) \prod_{t=1}^{T} p_\theta(x_{t-1} \mid x_t, y). \tag{2}$$

Hence, for each denoising step from t to $t-1$,

$$p_\theta(x_{t-1} \mid x_t, y) = \mathcal{N}(x_{t-1}; \mu_\theta(x_t, y, t), \Sigma_\theta(x_t, y, t)). \tag{3}$$

It has been shown that the combination of q and p here is a form of a variational auto-encoder [13], and hence the variational lower bound (VLB) can be described as a sum of independent terms, $L_{vlb} := L_0 + ... + L_{T-1} + L_T$, where each term corresponds to a noising step. As described in Ho et al. [8], we can randomly sample timestep t during training and use the expectation $E_{t,x_0,y,\epsilon}$ to estimate L_{vlb} and optimize parameters θ. The denoising neural network can be parameterized in several ways, however, it has been observed that using a noise-prediction based formulation results in the best image quality [8]. Overall, our NASDM denoising model is trained to predicting the noise added to the input image given the semantic layout y and the timestep t using the loss described as follows:

$$L_{\text{simple}} = E_{t,x,\epsilon}\left[\|\epsilon - \epsilon_\theta(x_t, y, t)\|_2\right]. \tag{4}$$

Note that the above loss function provides no signal for training $\Sigma_\theta(x_t, y, t)$. Therefore, following the strategy in improved DDPMs [8], we train a network to directly predict an interpolation coefficient v per dimension, which is turned into variances and optimized directly using the KL divergence between the estimated distribution $p_\theta(x_{t-1} \mid x_t, y)$ and the diffusion posterior $q(x_{t-1} \mid x_t, x_0)$ as $L_{\text{vlb}} = D_{KL}(p_\theta(x_{t-1} \mid x_t, y) \parallel q(x_{t-1} \mid x_t, x_0))$. This optimization is done while applying a stop gradient to $\epsilon(x_t, y, t)$ such that L_{vlb} can guide $\Sigma_\theta(x_t, y, t)$ and L_{simple} is the main guidance for $\epsilon(x_t, y, t)$. Overall, the loss is a weighted summation of the two objectives described above as follows:

$$L_{\text{hybrid}} = L_{\text{simple}} + \lambda L_{\text{vlb}}. \tag{5}$$

3.4 Conditioning on Semantic Mask

NASDM requires our neural network noise-predictor $\epsilon_\theta(x_t, y, t)$ to effectively process the information from the nuclei semantic map. For this purpose, we leverage a modified U-Net architecture described in Wang et al. [27], where semantic information is injected into the decoder of the denoising network using multi-layer, spatially-adaptive normalization operators. As denoted in Fig. 1, we construct the semantic mask such that each channel of the mask corresponds to a unique nuclei type. In addition, we also concatenate a mask comprising of the edges of all nuclei to further demarcate nuclei instances.

3.5 Classifier-Free Guidance

To improve the sample quality and agreement with the conditioning signal, we employ classifier-free guidance [10], which essentially amplifies the conditional distribution using unconditional outputs while sampling. During training, the conditioning signal, i.e., the semantic label map, is randomly replaced with a null mask for a certain percentage of samples. This leads to the diffusion model becoming stronger at generating samples both conditionally as well as unconditionally and can be used to implicitly infer the gradients of the log probability required for guidance as follows:

$$\begin{aligned} \epsilon_\theta(x_t \mid y) - \epsilon_\theta(x_t \mid \emptyset) &\propto \nabla_{x_t} \log p(x_t \mid y) - \nabla_{x_t} \log p(x_t), \\ &\propto \nabla_{x_t} \log p(y \mid x_t), \end{aligned} \tag{6}$$

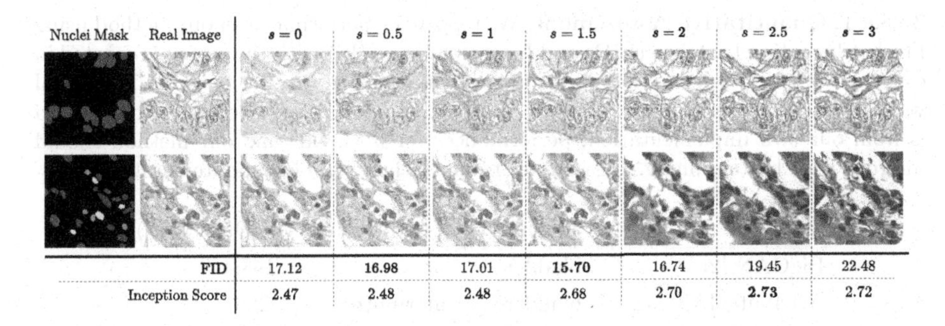

	s = 0	s = 0.5	s = 1	s = 1.5	s = 2	s = 2.5	s = 3
FID	17.12	**16.98**	17.01	**15.70**	16.74	19.45	22.48
Inception Score	2.47	2.48	2.48	2.68	2.70	**2.73**	2.72

Fig. 2. Guidance Scale Ablation: For a given mask, we generate images using different values of the guidance scale, s. The FID and IS metrics are computed by generating images for all masks in the test set at 20× magnification.

where $\emptyset$ denotes an empty semantic mask. During sampling, the conditional distribution is amplified using a guidance scale s as follows:

$$\hat{\epsilon}_\theta(x_t \mid y) = \epsilon_\theta(x_t \mid y) + s \cdot [\epsilon_\theta(x_t \mid y) - \epsilon_\theta(x_t \mid \emptyset)]. \tag{7}$$

4 Experiments

In this section, we first describe our implementation details and training procedure. Further, we establish the robustness of our model by performing an ablative study over objective magnification and classifier-guidance scale. We then perform quantitative and qualitative assessments to demonstrate the efficacy of our nuclei-aware semantic histopathology generation model. In all following experiments, we synthesize images using the semantic masks of the held-out dataset at the concerned objective magnification. We then compute Fréchet Inception Distance (FID) and Inception Score (IS) metrics between the synthetic and real images in the held-out set.

4.1 Implementation Details

Our diffusion model is implemented using a semantic UNet architecture (Sect. 3.4), trained using the objective in (5). Following previous works [19], we set the trade-off parameter λ as 0.001. We use the AdamW optimizer to train our model. Additionally, we adopt an exponential moving average (EMA) of the denoising network weights with 0.999 decay. Following DDPM [8], we set the total number of diffusion steps as 1000 and use a linear noising schedule with respect to timestep t for the forward process. After normal training with a learning rate of $1e-4$, we decay the learning rate to $2e-5$ to further finetune the model with a drop rate of 0.2 to enhance the classifier-free guidance capability during sampling. The whole framework is implemented using Pytorch and trained on 4 NVIDIA Tesla A100 GPUs with a batch-size of 40 per GPU. Code will be made public on publication or request.

Table 1. Quantitative Assessment: We report the performance of our method using Fréchet Inception Distance (FID) and Inception Score (IS) with the metrics reported in existing works. (-) denotes that corresponding information was not reported in original work. *Note that performance reported for best competing method on the colon data is from our own implementation, performances for both this and our method should improve with better tuning. Please refer to our github repo for updated statistics.

Method	Tissue type	Conditioning	FID(↓)	IS(↑)
BigGAN [2]	bladder	none	158.4	-
AttributeGAN [32]	bladder	attributes	53.6	-
ProGAN [11]	glioma	morphology	53.8	1.7
Morph-Diffusion [18]	glioma	morphology	20.1	2.1
Morph-Diffusion* [18]	colon	morphology	18.8	2.2
NASDM (Ours)	colon	semantic mask	**14.1**	**2.7**

4.2 Ablation over Guidance Scale (s)

In this study, we test the effectiveness of the classifier-free guidance strategy. We consider the variant without guidance as our baseline. As seen in Fig. 2, increase in guidance scale initially results in better image quality as more detail is added to visual structures of nuclei. However, with further increase, the image quality degrades as the model overemphasizes the nuclei and staining textures.

4.3 Ablation over Objective Magnification

Obj. Mag	FID(↓)	IS(↑)
10×	38.1	2.3
20×	**20.7**	**2.5**

As described in Sect. 3.1, we generate patches at two different objective magnifications of 10× and 20×. In this section, we contrast the generative performance of the models trained on these magnification levels respectively. From the table on right, we observe that the model trained at 20× objective magnification produces better generative metrics. Note that we only train on a subset on 20× mag. to keep the size of the training data constant.

4.4 Quantitative Analysis

To the best of our knowledge, ours is the only work that is able to synthesize histology images given a semantic mask, making a direct quantitative comparison tricky. However, the standard generative metric Fréchet Inception Distance (FID) measures the distance between distributions of generated and real images in the Inception-V3 [14] latent space, where a lower FID indicates that the model is able to generate images that are very similar to real data. Therefore, we compare FID and IS metrics with the values reported in existing works [18,32] (ref. Table 1) in their own settings. We can observe that our method outperforms all existing methods including both GANs-based methods as well as the recently proposed morphology-focused generative diffusion model.

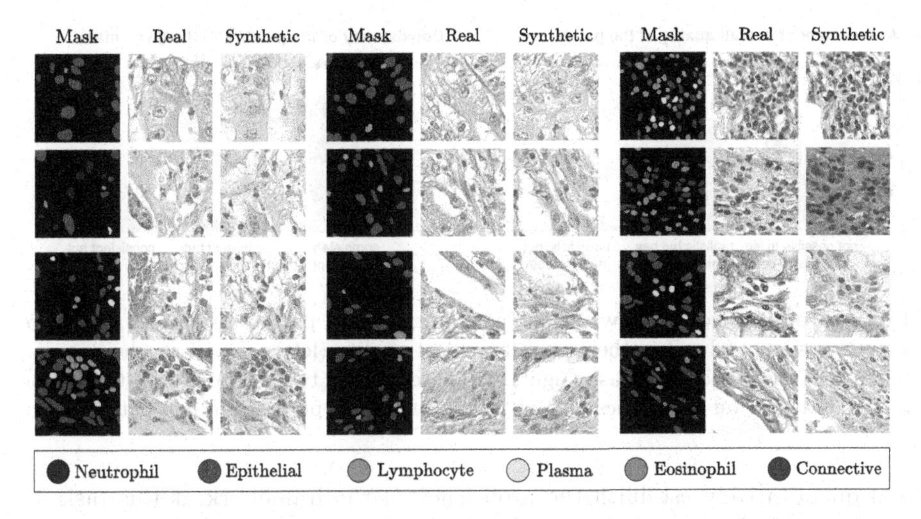

Fig. 3. Qualitative Results: We generate synthetic images given masks with each type of nuclei in different environments to demonstrate the proficiency of the model to generate realistic nuclei arrangements. Legend at bottom denotes the mask color for each type of nuclei.

4.5 Qualitative Analysis

We perform an expert pathologist review of the patches generated by the model. We use 30 patches, 17 synthetic and 13 real for this review. We have two experts assess the overall medical quality of the patches as well as their consistency with the associated nuclei masks on likert scale. The survey used for the review can be found on a public google survey[1]. It can be seen from this survey (Fig. 4) that the patches generated by the model are found to be more realistic than even the patches in our real set. We now qualitatively discuss the proficiency of our model in generating realistic visual patterns in synthetic histopathology images (refer Fig. 3). We can see that the model is able to capture convincing visual structure for each type of nuclei. In the synthetic images, we can see that the lymphocytes are accurately circular, while neutrophils and eosinophils have a more lobed structure. We also observe that the model is able to mimic correct nucleus-to-cytoplasm ratios for each type of nuclei. Epithelial cells are less dense, have a distinct chromatin structure, and are larger compared to other white blood cells. Epithelial cells are most difficult to generate in a convincing manner, however, we can see that model is able to capture the nuances well and generates accurate chromatin distributions.

5 Conclusion and Future Works

In this work, we present NASDM, a nuclei-aware semantic tissue generation framework. We demonstrate the model on a colon dataset and qualitatively

[1] https://forms.gle/1dLAdk9XKhp6FWMY6.

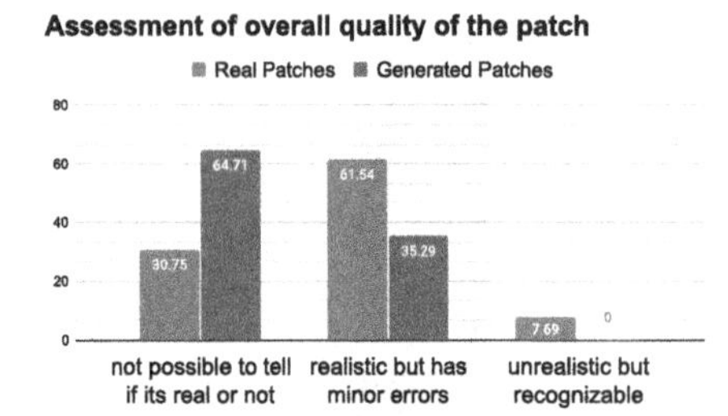

Fig. 4. Qualitative Review: Compiled results from a pathologist review. We have experts assess patches for, their overall medical quality (left), as well as, their consistency with the associated mask (right). We observe that the patches generated by the model do better on all metrics and majority are imperceptible from real patches.

and quantitatively establish the proficiency of the framework at this task. In future works, further conditioning on properties like stain-distribution, tissue-type, disease-type, etc. would enable patch generation in varied histopathological settings. Additionally, this framework can be extended to also generate semantic masks enabling an end-to-end tissue generation framework that first generates a mask and then synthesizes the corresponding patch. Further, future works can explore generation of patches conditioned on neighboring patches, as this enables generation of larger tissue areas by composing patches together.

Acknowledgements:. We would like to thank Dr. Shyam Raghavan, M.D., Fisher Rhoads, B.S., and Dr. Lubaina Ehsan, M.D. for their invaluable inputs for our qualitative analysis.

References

1. Bejnordi, B.E., Timofeeva, N., Otte-Höller, I., Karssemeijer, N., van der Laak, J.A.: Quantitative analysis of stain variability in histology slides and an algorithm for standardization. In: Medical Imaging 2014: Digital Pathology. vol. 9041, pp. 45–51. SPIE (2014)
2. Brock, A., Donahue, J., Simonyan, K.: Large scale GAN training for high fidelity natural image synthesis. arXiv preprint arXiv:1809.11096 (2018)
3. Dhariwal, P., Nichol, A.: Diffusion models beat GANs on image synthesis. Adv. Neural. Inf. Process. Syst. **34**, 8780–8794 (2021)
4. Fajardo, V.A., Findlay, D., Jaiswal, C., Yin, X., Houmanfar, R., Xie, H., Liang, J., She, X., Emerson, D.: On oversampling imbalanced data with deep conditional generative models. Expert Syst. Appl. **169**, 114463 (2021)
5. Graham, S., et al.: Lizard: a large-scale dataset for colonic nuclear instance segmentation and classification. In: Proceedings of the IEEE/CVF International Conference on Computer Vision, pp. 684–693 (2021)
6. Graham, S., et al.: Hover-net: simultaneous segmentation and classification of nuclei in multi-tissue histology images. Med. Image Anal. **58**, 101563 (2019)
7. Hall, M., van der Maaten, L., Gustafson, L., Adcock, A.: A systematic study of bias amplification. arXiv preprint arXiv:2201.11706 (2022)

8. Ho, J., Jain, A., Abbeel, P.: Denoising diffusion probabilistic models. Adv. Neural. Inf. Process. Syst. **33**, 6840–6851 (2020)
9. Ho, J., Saharia, C., Chan, W., Fleet, D.J., Norouzi, M., Salimans, T.: Cascaded diffusion models for high fidelity image generation. J. Mach. Learn. Res. **23**(47), 1–33 (2022)
10. Ho, J., Salimans, T.: Classifier-free diffusion guidance. arXiv preprint arXiv:2207.12598 (2022)
11. Karras, T., Aila, T., Laine, S., Lehtinen, J.: Progressive growing of gans for improved quality, stability, and variation. arXiv preprint arXiv:1710.10196 (2017)
12. Kawar, B., Elad, M., Ermon, S., Song, J.: Denoising diffusion restoration models. arXiv preprint arXiv:2201.11793 (2022)
13. Kingma, D.P., Welling, M.: Auto-encoding variational bayes. arXiv preprint arXiv:1312.6114 (2013)
14. Kynkäänniemi, T., Karras, T., Aittala, M., Aila, T., Lehtinen, J.: The role of imagenet classes in fr\'echet inception distance. arXiv preprint arXiv:2203.06026 (2022)
15. Van der Laak, J., Litjens, G., Ciompi, F.: Deep learning in histopathology: the path to the clinic. Nat. Med. **27**(5), 775–784 (2021)
16. Levine, A.B., et al.: Synthesis of diagnostic quality cancer pathology images by generative adversarial networks. J. Pathol. **252**(2), 178–188 (2020)
17. Macenko, M., et al.: A method for normalizing histology slides for quantitative analysis. In: 2009 IEEE International Symposium on Biomedical Imaging: from Nano to Macro, pp. 1107–1110. IEEE (2009)
18. Moghadam, P.A., et al.: A morphology focused diffusion probabilistic model for synthesis of histopathology images. In: Proceedings of the IEEE/CVF Winter Conference on Applications of Computer Vision, pp. 2000–2009 (2023)
19. Nichol, A.Q., Dhariwal, P.: Improved denoising diffusion probabilistic models. In: International Conference on Machine Learning, pp. 8162–8171. PMLR (2021)
20. Park, T., Liu, M.Y., Wang, T.C., Zhu, J.Y.: Semantic image synthesis with spatially-adaptive normalization. In: Proceedings of the IEEE/CVF Conference on Computer Vision and Pattern Recognition, pp. 2337–2346 (2019)
21. Saharia, C., et al.: Palette: Image-to-image diffusion models. In: ACM SIGGRAPH 2022 Conference Proceedings, pp. 1–10 (2022)
22. Saharia, C., Ho, J., Chan, W., Salimans, T., Fleet, D.J., Norouzi, M.: Image super-resolution via iterative refinement. IEEE Trans. Pattern Anal. Mach. Intell. **45**(4), 4713–4726 (2022)
23. Shrivastava, A., et al.: Self-attentive adversarial stain normalization. In: Del Bimbo, A., et al. (eds.) Pattern Recognition. ICPR International Workshops and Challenges: Virtual Event, January 10–15, 2021, Proceedings, Part I, pp. 120–140. Springer, Cham (2021). https://doi.org/10.1007/978-3-030-68763-2_10
24. Tan, Z., et al.: Diverse semantic image synthesis via probability distribution modeling. In: Proceedings of the IEEE/CVF Conference on Computer Vision and Pattern Recognition, pp. 7962–7971 (2021)
25. Tan, Z., et al.: Efficient semantic image synthesis via class-adaptive normalization. IEEE Trans. Pattern Anal. Mach. Intell. **44**(9), 4852–4866 (2021)
26. Vahadane, A., et al.: Structure-preserving color normalization and sparse stain separation for histological images. IEEE Trans. Med. Imaging **35**(8), 1962–1971 (2016)
27. Wang, W., et al.: Semantic image synthesis via diffusion models. arXiv preprint arXiv:2207.00050 (2022)

28. Wu, Y., et al.: Recent advances of deep learning for computational histopathology: principles and applications. Cancers **14**(5), 1199 (2022)
29. Xiao, Z., Kreis, K., Vahdat, A.: Tackling the generative learning trilemma with denoising diffusion gans. arXiv preprint arXiv:2112.07804 (2021)
30. Xie, L., Qi, J., Pan, L., Wali, S.: Integrating deep convolutional neural networks with marker-controlled watershed for overlapping nuclei segmentation in histopathology images. Neurocomputing **376**, 166–179 (2020)
31. Xue, Y., et al.: Selective synthetic augmentation with Histogan for improved histopathology image classification. Med. Image Anal. **67**, 101816 (2021)
32. Ye, J., Xue, Y., Liu, P., Zaino, R., Cheng, K.C., Huang, X.: A multi-attribute controllable generative model for histopathology image synthesis. In: de Bruijne, M., et al. (eds.) Medical Image Computing and Computer Assisted Intervention – MICCAI 2021: 24th International Conference, Strasbourg, France, September 27 – October 1, 2021, Proceedings, Part VIII, pp. 613–623. Springer, Cham (2021). https://doi.org/10.1007/978-3-030-87237-3_59
33. Zhou, Q., Yin, H.: A u-net based progressive gan for microscopic image augmentation. In: Medical Image Understanding and Analysis: 26th Annual Conference, MIUA 2022, Cambridge, UK, July 27–29, 2022, Proceedings, pp. 458–468. Springer (2022). https://doi.org/10.1007/978-3-031-12053-4_34

Triangular Analysis of Geographical Interplay of Lymphocytes (TriAnGIL): Predicting Immunotherapy Response in Lung Cancer

Sara Arabyarmohammadi[1,2], German Corredor[1], Yufei Zhou[2], Miguel López de Rodas[3], Kurt Schalper[3], and Anant Madabhushi[1,4](✉)

[1] Wallace H. Coulter Department of Biomedical Engineering, Georgia Institute of Technology and Emory University, Atlanta, USA
anantm@emory.edu

[2] Department of Computer and Data Sciences, Case Western Reserve University, Cleveland, USA

[3] Department of Pathology, School of Medicine, Yale University, New Haven, USA

[4] Atlanta Veterans Administration Medical Center, Atlanta, USA

Abstract. Quantitative immunofluorescence (QIF) enables identifying immune cell subtypes across histopathology images. There is substantial evidence to show that spatial architecture of immune cell populations (e.g. CD4+, CD8+, CD20+) is associated with therapy response in cancers, yet there is a paucity of approaches to quantify spatial statistics of interplay across immune subtypes. Previously, analyzing spatial cell interplay have been limited to either building subgraphs on individual cell types before feature extraction or capturing the interaction between two cell types. However, looking at the spatial interplay between more than two cell types reveals complex interactions and co-dependencies that might have implications in predicting response to therapies like immunotherapy. In this work we present, Triangular Analysis of Geographical Interplay of Lymphocytes (TriAnGIL), a novel approach involving building of heterogeneous subgraphs to precisely capture the spatial interplay between multiple cell families. Primarily, TriAnGIL focuses on triadic closures, and uses metrics to quantify triads instead of two-by-two relations and therefore considers both inter- and intra-family relationships between cells. The TriaAnGIL's efficacy for microenvironment characterization from QIF images is demonstrated in problems of predicting (1) response to immunotherapy (N = 122) and (2) overall survival (N = 135) in patients with lung cancer in comparison with four hand-crafted approaches namely DenTIL, GG, CCG, SpaTIL, and deep learning with GNN. For both tasks, TriaAnGIL outperformed hand-crafted approaches, and GNN with AUC = .70, C-index = .64. In terms of interpretability, TriAnGIL easily beats GNN, by pulling biological insights from immune cells interplay and shedding light on the triadic interaction of CD4+-Tumor-stromal cells.

Supplementary Information The online version contains supplementary material available at https://doi.org/10.1007/978-3-031-43987-2_77.

H. Greenspan et al. (Eds.): MICCAI 2023, LNCS 14225, pp. 797–807, 2023.
https://doi.org/10.1007/978-3-031-43987-2_77

1 Introduction

The tumor microenvironment (TME) is comprised of cancer, immune (e.g. B lymphocytes, and T lymphocytes), stromal, and other cells together with non-cellular tissue components [3,5,24,30]. It is well acknowledged that tumors evolve in close interaction with their microenvironment. Quantitatively characterizing TME has the potential to predict tumor aggressiveness and treatment response [3,23,24,30]. Different types of lymphocytes such as CD4+ (helper T cells), CD8+ (cytotoxic T cells), CD20+ (B cells), within the TME naturally interact with tumor and stromal cells. Studies [5,9] have shown that quantifying spatial interplay of these different cell families within the TME can provide more prognostic/predictive value compared to only measuring the density of a single biomarker such as tumor-infiltrating lymphocytes (TILs) [3,24]. Immunotherapy (IO) is the standard treatment for patients with advanced non-small cell lung cancer (NSCLC) [19] but only 27–45% of patients respond to this treatment [21]. Therefore, better algorithms and improved biomarkers are essential for identifying which cancer patients are most likely to respond to IO in advance of treatment. Quantitative features that relate to the complex spatial interplay between different types of B- and T-cells in the TME might unlock attributes that are associated with IO response. In this study, we introduce a novel approach called Triangular Analysis of Geographical Interplay of Lymphocytes (TriAnGIL), representing a unique and interpretable way to characterize the distribution, and higher-order interaction of various cell families (e.g., cancerous cells, stromal cells, lymphocyte subtypes) across digital histopathology slides. We demonstrate the efficacy of TriaAnGIL for characterizing TME in the context of predicting 1) response to IO with immune checkpoint inhibitors (ICI), 2) overall survival (OS), in patients with NSCLC, and 3) providing novel insights into the spatial interplay between different immune cell subtype. TriAnGIL source code is publicly available at http://github.com/sarayar/TriAnGIL.

2 Previous Related Work and Novel Contributions

Many studies have only looked at the density of a single biomarker (e.g. TILs), to show that a high density of TILs is associated with improved patient survival and treatment response in NSCLC [3,24]. Other works have attempted to characterize the spatial arrangement of cells in TME using computational graph-based approaches. These approaches include methods that connect cells regardless of their type (1) using global graphs (GG) such as Voronoi that connect all nuclei [2,14], or (2) using Cell cluster graphs (CCG) [16] to create multiple nuclear subgraphs based on cell-to-cell proximity to predict tumor aggressiveness and patient outcome [16]. Others have explored (3) the spatial interplay between two different cell types [5].One example approach is Spatial architecture of TIL (SpaTIL) [9] which attempted to characterize the interplay between immune and cancer cells and has proven to be helpful in predicting the recurrence in early-stage NSCLC. All of these approaches point to overwhelming evidence that spatial architecture of cells in TME is critical in predicting cancer outcome. However, these approaches have not been able to exploit higher-order interactions and dependencies between multiple cell types (> 2), relationships that might provide additional actionable insights. The contributions of this work include:

(1) TriAnGIL is a computational framework that characterizes the architecture and relationships of different cell types simultaneously. Instead of measuring only simple two-by-two relations between cells, it seeks to identify triadic spatial relations (hyperedges [18,20] instead of edges) between different cell types, thereby enabling the exploration of complex nuclear interactions of different cell types within TME. This in turn allows for development of machine classifiers to predict outcome and response in lung cancer patients treated with IO.
(2) TriAnGIL includes a set of quantitative metrics that capture the interplay within and between nuclei corresponding to different types/families. Previous works have focused primarily on intra-family relationships [2,5,9,14] while TriAnGIL measurements are able to consider inter- and intra-family relationships.
(3) Although deep learning (DL) models (e.g., graph neural networks(GNN)) have shown great capabilities in solving complex problems in the biomedical field, these tend to be black-box in nature. A key consideration in cancer immunology is the need for actionable insights into the spatial relationships between different types of immune cells. Not only does TriAnGIL provide predictions that are on par or superior compared to DL approaches, but also provides a way to glean insights into the spatial interplay of different immune cell types. These complex interactions enhance our understanding of the TME and will help pave the way for new therapeutic strategies that leverage these insights.

3 Description of TriAnGIL Methodology

3.1 Notation

Our approach consists of constructing heterogeneous graphs step by step and quantifying them by extracting features from them. The graphs are defined by G = (V, E), where V is the set of vertices (nodes) $V = \{v_1, ...v_N\}$ with τ_n vertex types, and E is the collection of pairs of vertices from V, $E = \{e_1, ...e_M\}$, which are called edges and ϕ_n is the mapping function that maps every vertex to one of n differential marker expressions in this dataset $\phi_n : V \rightarrow \tau_n$. G is represented by an adjacency matrix A that allows one to determine edges in constant time.

3.2 Node Linking and Computation of Graph-Interplay Features

The inputs of TriAnGIL are the coordinates of nuclear centroids and the corresponding cell types. In the TriAnGIL procedure, the centroid of each nucleus in a family is represented as a node of a graph. TriAnGIL is agnostic of the method used for identifying the coordinates and types. Once the different cell families are identified (Fig. 1-B), a list is generated for all possible sets comprising of membership from three [12,18] different cell families. By focusing on every set, TriAnGIL allows for capturing higher-order and balanced spatial triadic relationships [4] between cell families, while keeping the computational complexity relatively low. Therefore, we initiate the process with the first set on the list (α, β, γ), build heterogeneous graphs, extract features, and then select another set until we have explored all possible triads of cell families on the list. The three main steps of TriAnGIL is as follows:

1) **Quantifying in absence of one family:** First, we build a proximity graph (G_1) on nodes of α, β, γ based on the Euclidean distance of every two nodes. Two nodes will be connected if their distance is shorter than a given "interaction rate", θ regardless of their family and cell type (Fig. 1-C, 1-C1). The interaction rate is a hyper-parameter that controls how close we expect the distance to be so that we consider some interaction between the nodes.

$$\{G_1 = (V, E) \mid \forall e \in E : |e| <= \theta\} \quad (1)$$

We then exclude the nodes of α from all the interactions by removing its edges from G1 and characterize the relationship of β and γ (Fig. 1-C2). Next, we extract a series of features including clustering coefficient, average degree from the resulting subgraph. We repeat this process by removing all the edges of β (Fig. 1-C3) and then γ (Fig. 1-C4). In this manner, a total of 126 features (supplemental Table 1) are extracted (42 features for absence of one family $\times 3$).

2) **Triangulation-based connections:** A Delaunay triangulation is constructed by the nodes of α, β, γ (Fig. 1-D, and 1-D1). Delaunay triangulation is a planar graph formed by connecting the vertices in a way that ensures no point lies within the circumcircle of any triangle formed by the vertices [7]. We then extract 10 features relating to edge length and vertex count. Next, we prune long edges (D_1) where the Euclidean distance between connected nodes is more than the "interaction rate" (Fig. 1-D2). Next, a series of features were extracted from the remaining subgraph (e.g. number of edges between the nodes of α and β, β and γ, α and γ; complete list of features in supplemental Table 1).

$$\{D_1 = (V, E) \mid \forall e \in E : |e| <= \theta\} \quad (2)$$

3) **Triangular interactions:** As illustrated in Algorithm 1, from the unpruned Delaunay triangulation that includes the nodes of α, β, γ, we select those triangles (closed triads [8,25]) that link nodes from three distinct families (Fig. 1-E, 1-E1). In other words, we remove triangles with more than one vertex from a single family. Next, we call GetTriangleFeatures() function to quantify triangular relationships by extracting features from the resulting subgraphs (e.g. perimeter and area of triangles; complete list of features in supplemental Table 1).

4 Experimental Results and Discussion

4.1 Dataset

The cohort employed in this study was composed of pre-treatment tumor biopsy specimens from patients with NSCLC from five centers (two centers for training

Algorithm 1: Finding Triangles

INPUT: A jagged array Del : Delaunay graph with three vertices of every triangle in each row, A hashmap ϕ : maps nodes to their type
OUTPUT: Triangle features $TriFeatSet$
let $TriIndex \leftarrow \emptyset$ be the list for triangle indices
for $i = 1$ *to* $i = length(Del)$ **do**
  let $Marker \leftarrow \emptyset$ be a auxiliary list to keep the viewed markers
  for $j = 1$ *to* 3 **do**
    if $\phi(Del(i,j)) \notin Marker$ **then**
      $Marker \leftarrow Marker \bigcup \phi(Del(i,j))$
    end
  end
  if $length(Marker) == 3$ **then**
    $TriIndex \leftarrow TriIndex \bigcup i$
  end
end
$TriFeatSet \leftarrow$`GetTriangleFeatures`$(Del(TriIndex, :))$
return $TriFeatSet$

(S_t) and three centers for independent validation (S_v)). The entire analysis was carried out using 122 patients in Experiment 1 (73 in S_t, and 49 in S_v) and 135 patients in Experiment 2 (81 in S_t, and 54 in S_v). Specimens were analyzed with a multiplexed quantitative immunofluorescence (QIF) panel using the method described in [22]. From each whole slide image, 7 representative tiles were obtained and used to train the software InForm to define background, tumor and stromal compartments. Then, individual cells were segmented based on nuclear DAPI staining and the segmentation performance was controlled by direct visualization of samples by a trained observer. Next, the software was trained to identify cell subtypes based on marker expression (CD8, CD4, CD20, CK for tumor epithelial cells and absence of these markers for stromal cells).

4.2 Comparative Approaches

The efficacy of TriAnGIL was compared against five different approaches.

TIL density (DenTIL): For every patient, multiple density measures including the number of different cells types and their ratios are calculated [3,24] (supplemental Table 2).

GG: A Delaunay triangulation, a Voronoi diagram, and a Minimum Spanning Tree were constructed [2,14] on all nuclei regardless of their type. Architectural features (e.g., perimeter, triangle area, edge length) were then calculated on these global graphs for each patient.

CCG: For every patient, subgraphs are built on nuclei regardless of their type and only based on their Euclidean distance. Local graph metrics (e.g. clustering coefficient) [16] are then calculated from these subgraphs.

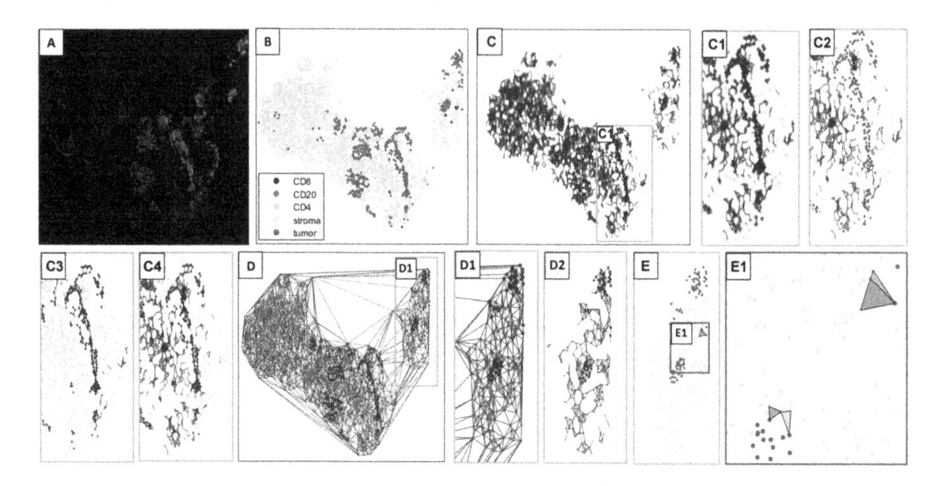

Fig. 1. Illustration of workflow for TriAnGIL. (A) Representative QIF image. (B) Nuclei subtypes based on differential marker expression. (C) Proximity graph on all CD4+-tumor-stroma nuclei. (C1) A zoomed-in region showing the edges based on Euclidean distances. (C2) The first family (tumor cells) is removed to enable characterization of other two nuclear types. (C3) Stroma cells are removed. (C4) CD4+ cells are removed. (D) A Delaunay triangulation on all CD4+, tumor, and stroma nuclei. (D1) Shows a zoomed-in region. (D2) Shows the same region in the pruned Delaunay subgraph. (E) From all Delaunay triangles, only those that connect CD4+-tumor-stroma are identified. (E1) A zoomed-in region showing the CD4+-tumor-stroma triangles.

SpaTIL: For each patient, first, subgraphs are built on individual cell types based on a distance parameter. The convex hulls are then constructed on these subgraphs. After selecting every two cell types, features are extracted from their convex hulls (e.g. the number of clusters of each cell type, area intersected between clusters [9]; complete list of combinations in supplemental Table 3).

GNN: A recent study [31] demonstrated that Transformer-based [29] GNNs are able to learn the arrangement of tiles across pathology images for survival analysis. Here, for each tile in the slide, a Delaunay graph was constructed regardless of cell subtypes, and tile-level feature representations (e.g.side length minimum, maximum, mean, and standard deviation, triangle area minimum, maximum, mean, and standard deviation) were aggregated by a Transformer according to their spatial arrangement [31]. Our approach utilized the Weisfeiler-Lehman (WL) test [15] for embedding graphs into Euclidean feature space. Well-known approaches, such as GraphSage [10], are considered as continuous approximations to the WL test. Therefore, our GNN is a valid baseline for heterogeneous graphs.

4.3 Experiment 1: Immunotherapy Response Prediction in Lung Cancer

Design: TriAnGIL was also trained to differentiate between patients who responded to IO and those who did not. For our study, the responders to IO were identified as those patients with complete response, partial response, and stable disease, and non-responders were patients with progressive disease. A linear discriminant analysis (LDA) classifier was trained on S_t to predict which patients would respond to IO. For creating the model, the minimum redundancy maximum relevance (mRMR) method [1] was used to select the top features. The same procedure using mRMR and LDA was performed for the comparative hand-crafted approaches. The ability to identify responders post-IO was assessed by the area under the receiver operating characteristic curve (AUC) in S_v.

Results: The two top predictive TriAnGIL features were found to be the number of edges between stroma and CD4+ cells, and the number of edges between stroma and tumor cells with more interactions between stromal cells and both CD4+ and tumor cells being associated with response to IO. This finding is concordant with other studies [13,17,22,27] that stromal TILs were significantly associated with improved OS. Therefore, TriAnGIL approach is not only predictive of treatment response but more critically it enables biological interpretations that a DL model might not be able to provide. In S_v, this LDA classifier was able to distinguish responders from non-responders to IO with AUC_{Tri}=0.70 that was higher than all other hand-crafted and DL approaches with $AUC_{Den} = 0.45$, $AUC_{GG} = 0.52$, $AUC_{Spa} = 0.53$, $AUC_{CCG} = 0.65$, $AUC_{GNN} = 0.65$.

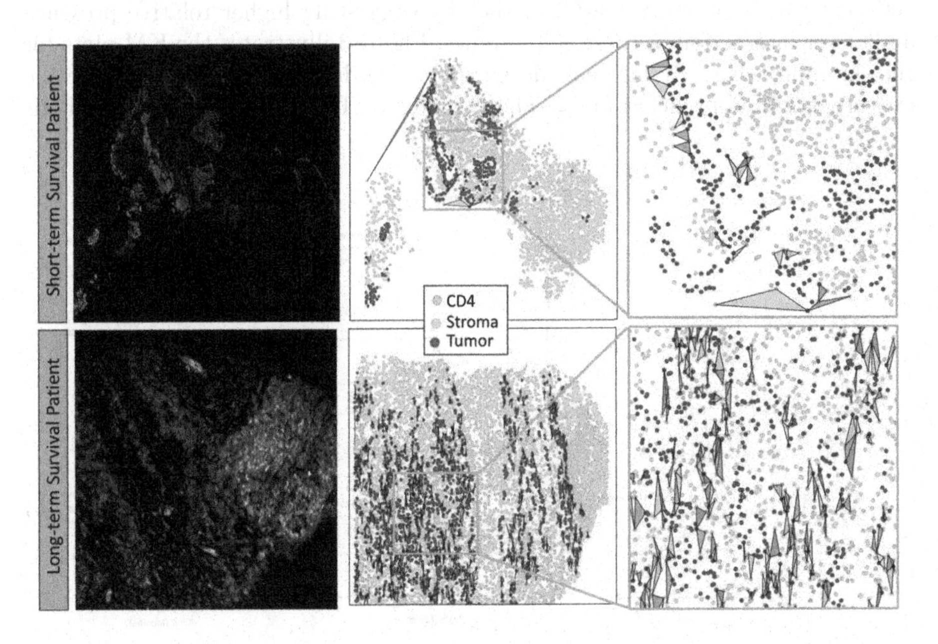

Fig. 2. Visual illustration of the qualitative difference in feature representations between NSCLC patients with short-term (top) and long-term survival (lower row). The leftmost column shows a part of a QIF image. The second column shows the triangular relationships formed by CD4, tumor, and stromal cell families. The third column shows a zoomed-in region. The number, size, and interplay of triangles formed on the three cell types are clearly different between the two cases. (Color figure online)

4.4 Experiment 2: Predicting Survival in Lung Cancer Patients Treated with Immunotherapy

Design: S_t was used to construct a least absolute shrinkage and selection operator (LASSO) [28] regularized Cox proportional hazards model [6] using the TriAnGIL features, to obtain risk score for each patient. LASSO features are listed in supplemental Table 4. The median risk score in S_t was used as a threshold in both S_t and S_v to dichotomize patients into low-risk/high-risk categories. Kaplan-Meier (KM) survival curves [26] were plotted and the model performance was summarized by hazard ratio (HR), with corresponding (95% confidence intervals (CI)) using the log-rank test, and Harrell's concordance index (C-index) on S_v. The C-index evaluates the correlation between risk predictions and survival times, aiming to maximize the discrimination between high-risk and low-risk patients [11]. OS is the time between the initiation of IO to the death of the patient. The patients were censored if the date of death was unknown.

Result: Figure 2 presents some TriAnGIL features in a field of view for a patient with long-term survival and another with short-term survival. More triangular relationships, shorter triangle edges, and smaller triangles with smaller perimeters are found in the long-term survival case when analyzing the triadic interactions within tumor-stroma-CD4, thereby suggesting higher relative presence and closer interaction of these cell families. Figure 3 illustrates the KM plots for the six approaches. We also calculated the concordance index (C-index) for the two prognostic approaches in S_v. The C-index for TriAnGIL and GNN methods

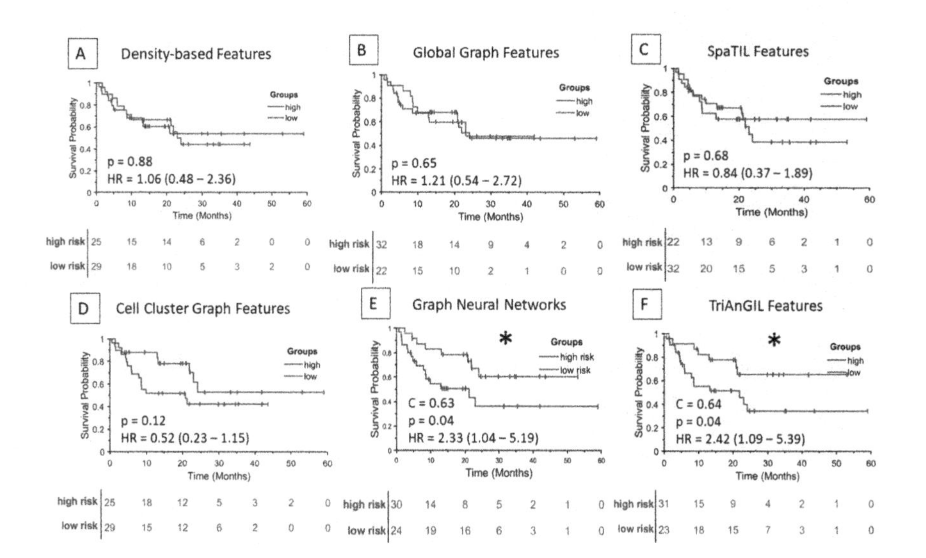

Fig. 3. The KM curves of the high-risk (red) and low-risk groups (blue) in S_v (N=54) using (A) DenTIL, (B) GG,(C) SpaTIL, (D) CCG,(E) GNN,(F) TriAnGIL features. Asterisks * indicate the models with $p < 0.05$ that statistically significantly prognosticate OS post-IO. (Color figure online)

were 0.64, and 0.63 respectively. Therefore, overall TriAnGIL worked marginally better than GNN, with much higher biological interpretability.

5 Concluding Remarks

We presented a new approach, Triangular Analysis of Geographical Interplay of Lymphocytes (TriAnGIL), to quantitatively chartacterize the spatial arrangement and relative geographical interplay of multiple cell families across pathological images. Compared to previous spatial graph-based methods, TriAnGIL quantifies the spatial interplay between multiple cell families, providing a more comprehensive portrait of the tumor microenvironment. TriAnGIL was predictive of response after IO (N = 122) and also demonstrated a strong correlation with OS in NSCLC patients treated with IO (N = 135). TriAnGIL outperformed other graph- and DL-based approaches, with the added benefit of provoding interpretability with regard to the spatial interplay between cell families. For instance, TriAnGIL yielded the insight that more interactions between stromal cells and both CD4+ and tumor cells appears to be associated with better response to IO. Although five cell families were studies in this work, TriAnGIL is flexible and could include other cell types (e.g., macrophages). Future work will entail larger validation studies and also evaluation on other use cases.

Acknowledgements. Research reported in this publication was supported by the National Cancer Institute under award numbers R01CA268287A1, U01 CA269181, R01 CA26820701A1, R01CA249992- 01A1, R01CA202752- 01A1, R01CA208236- 01A1, R01CA216579- 01A1, R01CA220581-01A1, R01CA257612- 01A1, 1U01CA239055- 01, 1U01CA248226- 01, 1U54CA254566- 01, National Heart, Lung and Blood Institute 1R01HL15127701A1, R01HL15807101A1, National Institute of Biomedical Imaging and Bioengineering 1R43EB028736- 01, VA Merit Review Award IBX004121A from the United States Department of Veterans Affairs Biomedical Laboratory Research and Development Service the Office of the Assistant Secretary of Defense for Health Affairs, through the Breast Cancer Research Program (W81XWH- 19- 1-0668), the Prostate Cancer Research Program (W81XWH- 20-1- 0851), the Lung Cancer Research Program (W81XWH-18-1-0440, W81XWH-20-1-0595), the Peer Reviewed Cancer Research Program (W81XWH- 18-1-0404, W81XWH- 21-1-0345, W81XWH- 21-1-0160), the Kidney Precision Medicine Project (KPMP) Glue Grant and sponsored research agreements from Bristol Myers-Squibb, Boehringer-Ingelheim, Eli-Lilly and Astrazeneca. The content is solely the responsibility of the authors and does not necessarily represent the official views of the National Institutes of Health, the U.S. Department of Veterans Affairs, the Department of Defense, or the United States Government.

References

1. Auffarth, B., López, M., Cerquides, J.: Comparison of redundancy and relevance measures for feature selection in tissue classification of CT images. In: Perner, P. (ed.) ICDM 2010. LNCS (LNAI), vol. 6171, pp. 248–262. Springer, Heidelberg (2010). https://doi.org/10.1007/978-3-642-14400-4_20

2. Basavanhally, A.N., et al.: Computerized image-based detection and grading of lymphocytic infiltration in HER2+ breast cancer histopathology. IEEE Trans. Biomed. Eng. **57**(3), 642–653 (2010)
3. Brambilla, E., et al.: Prognostic effect of tumor lymphocytic infiltration in resectable non-small-cell lung cancer. J. Clin. Oncol.: Official J. Am. Soc. Clin. Oncol. **34**, 1223–30 (2016)
4. Cartwright, D., Harary, F.: Structural balance: a generalization of Heider's theory. Psychol. Rev. **63**(5), 277 (1956)
5. Corredor, G., et al.: Spatial architecture and arrangement of tumor-infiltrating lymphocytes for predicting likelihood of recurrence in early-stage non-small cell lung cancer. Clin. Cancer Res. **25**(5), 1526–1534 (2019)
6. Cox, D.R.: Regression models and life-tables. J. Royal Stat. Soc.: Ser. B (Methodol.) **34**(2), 187–202 (1972)
7. Delaunay, B.: Sur la sphère vide. Izvestiya Akademii Nauk SSSR. Otdelenie Matematicheskikh i Estestvennykh Nauk **7**(4), 793–800 (1934)
8. Dimitrova, T., Petrovski, K., Kocarev, L.: Graphlets in multiplex networks. Sci. Rep. **10**(1), 1928 (2020)
9. Ding, R., et al.: Image analysis reveals molecularly distinct patterns of tils in NSCLC associated with treatment outcome. NPJ Precis. Oncol. **6**(1), 1–15 (2022)
10. Hamilton, W., Ying, Z., Leskovec, J.: Inductive representation learning on large graphs. In: Advances in Neural Information Processing Systems 30 (2017)
11. Harrell, F.E., Califf, R.M., Pryor, D.B., Lee, K.L., Rosati, R.A.: Evaluating the yield of medical tests. JAMA **247**(18), 2543–2546 (1982)
12. Kitts, J.A., Huang, J.: Triads. Encyclopedia Soc. Netw. **2**, 873–874 (2010)
13. Krishnamurti, U., Wetherilt, C.S., Yang, J., Peng, L., Li, X.: Tumor-infiltrating lymphocytes are significantly associated with better overall survival and disease-free survival in triple-negative but not estrogen receptor-positive breast cancers. Human Pathol. **64**, 7–12 (2017)
14. Lee, G., Veltri, R.W., Zhu, G., Ali, S., Epstein, J.I., Madabhushi, A.: Nuclear shape and architecture in benign fields predict biochemical recurrence in prostate cancer patients following radical prostatectomy: Preliminary findings. Eur. Urol. Focus **3**, 457–466 (2017)
15. Leman, A., Weisfeiler, B.: A reduction of a graph to a canonical form and an algebra arising during this reduction. Nauchno-Technicheskaya Informatsiya **2**(9), 12–16 (1968)
16. Lewis, J.S., Ali, S., Luo, J., Thorstad, W.L., Madabhushi, A.: A quantitative histomorphometric classifier (quhbic) identifies aggressive versus indolent p16-positive oropharyngeal squamous cell carcinoma. Am. J. Surg. Pathol. **38**(24145650), 128–137 (2014)
17. Luen, S., et al.: Prognostic implications of residual disease tumor-infiltrating lymphocytes and residual cancer burden in triple-negative breast cancer patients after neoadjuvant chemotherapy. Ann. Oncol. **30**(2), 236–242 (2019)
18. Ma, Y., Tang, J.: Deep Learning on Graphs. Cambridge University Press, Cambridge (2021)
19. Malhotra, J., Jabbour, S.K., Aisner, J.: Current state of immunotherapy for non-small cell lung cancer. Transl. Lung Cancer Res. **6**(2), 196 (2017)
20. Newman, M.: Networks. Oxford University Press, Oxford (2018)
21. Reck, M., et al.: Pembrolizumab versus chemotherapy for PD-L1-positive non-small-cell lung cancer. N. Engl. J. Med. **375**, 1823–1833 (2016)

22. de Rodas, M.L., et al.: Role of tumor infiltrating lymphocytes and spatial immune heterogeneity in sensitivity to PD-1 axis blockers in non-small cell lung cancer. J. ImmunoTherapy Cancer **10**(6), e004440 (2022)
23. Sato, J., et al.: CD20+ tumor-infiltrating immune cells and CD204+ M2 macrophages are associated with prognosis in thymic carcinoma. Cancer Sci. **111**(6), 1921–1932 (2020)
24. Schalper, K.A., et al.: Objective measurement and clinical significance of TILs in non-small cell lung cancer **107**(3). https://doi.org/10.1093/jnci/dju435
25. Sherwin, R.G.: Introduction to the graph theory and structural balance approaches to international relations. University of Southern California Los Angeles, Tech. Rep. (1971)
26. Simon, R.M., Subramanian, J., Li, M.C., Menezes, S.: Using cross-validation to evaluate predictive accuracy of survival risk classifiers based on high-dimensional data. Briefings Bioinform. **12**(3), 203–214 (2011)
27. Tavares, M.C., et al.: A high CD8 to FOXP3 ratio in the tumor stroma and expression of PTEN in tumor cells are associated with improved survival in non-metastatic triple-negative breast carcinoma. BMC Cancer **21**(1), 1–12 (2021)
28. Tibshirani, R.: The lasso method for variable selection in the cox model. Stat. Med. **16**(4), 385–395 (1997)
29. Vaswani, A., et al.: Attention is all you need (2017). https://doi.org/10.48550/arxiv.1706.03762
30. Whiteside, T.: The tumor microenvironment and its role in promoting tumor growth. Oncogene **27**(45), 5904–5912 (2008)
31. Zhou, Y., et al.: Transformer as a spatially aware multi-instance learning framework to predict the risk of death for early-stage non-small cell lung cancer. In: Digital and Computational Pathology. No. 12471–33, SPIE (TBD 2023), accepted for publication

Author Index

H. Greenspan et al. (Eds.): MICCAI 2023, LNCS 14225, pp. 809–814, 2023.
https://doi.org/10.1007/978-3-031-43987-2

Printed in the United States
by Baker & Taylor Publisher Services

Printed in the United States
by Baker & Taylor Publisher Services